The Potential of Dietary Antioxidants

The Potential of Dietary Antioxidants

Editors

Irene Dini

Domenico Montesano

MDPI • Basel • Beijing • Wuhan • Barcelona • Belgrade • Manchester • Tokyo • Cluj • Tianjin

Editors

Irene Dini
Pharmacy
Federico II
Naples
Italy

Domenico Montesano
Pharmacy
Federico II
Naples
Italy

Editorial Office
MDPI
St. Alban-Anlage 66
4052 Basel, Switzerland

This is a reprint of articles from the Special Issue published online in the open access journal *Antioxidants* (ISSN 2076-3921) (available at: www.mdpi.com/journal/antioxidants/special_issues/Potential_Dietary_Antioxidants).

For citation purposes, cite each article independently as indicated on the article page online and as indicated below:

LastName, A.A.; LastName, B.B.; LastName, C.C. Article Title. *Journal Name* **Year**, *Volume Number*, Page Range.

ISBN 978-3-0365-4952-1 (Hbk)
ISBN 978-3-0365-4951-4 (PDF)

Contents

About the Editors . . . ix

Irene Dini
The Potential of Dietary Antioxidants
Reprinted from: *Antioxidants* **2021**, *10*, 1752, doi:10.3390/antiox10111752 . . . 1

Nagwa I. El-Desoky, Nesrein M. Hashem, Antonio Gonzalez-Bulnes, Ahmed G. Elkomy and Zahraa R. Abo-Elezz
Effects of a Nanoencapsulated *Moringa* Leaf Ethanolic Extract on the Physiology, Metabolism and Reproductive Performance of Rabbit Does during Summer
Reprinted from: *Antioxidants* **2021**, *10*, 1326, doi:10.3390/antiox10081326 . . . 3

Sylvia Maina, Da Hye Ryu, Jwa Yeong Cho, Da Seul Jung, Jai-Eok Park and Chu Won Nho et al.
Exposure to Salinity and Light Spectra Regulates Glucosinolates, Phenolics, and Antioxidant Capacity of *Brassica carinata* L. Microgreens
Reprinted from: *Antioxidants* **2021**, *10*, 1183, doi:10.3390/antiox10081183 . . . 21

Yong Chool Boo
Arbutin as a Skin Depigmenting Agent with Antimelanogenic and Antioxidant Properties
Reprinted from: *Antioxidants* **2021**, *10*, 1129, doi:10.3390/antiox10071129 . . . 37

Sheng-Yao Peng, Li-Ching Lin, Shu-Rong Chen, Ammad A. Farooqi, Yuan-Bin Cheng and Jen-Yang Tang et al.
Pomegranate Extract (POMx) Induces Mitochondrial Dysfunction and Apoptosis of Oral Cancer Cells
Reprinted from: *Antioxidants* **2021**, *10*, 1117, doi:10.3390/antiox10071117 . . . 59

Mindaugas Liaudanskas, Vaidotas Žvikas and Vilma Petrikaitė
The Potential of Dietary Antioxidants from a Series of Plant Extracts as Anticancer Agents against Melanoma, Glioblastoma, and Breast Cancer
Reprinted from: *Antioxidants* **2021**, *10*, 1115, doi:10.3390/antiox10071115 . . . 79

Jaime Ortiz-Viedma, José M. Aguilera, Marcos Flores, Roberto Lemus-Mondaca, María José Larrazabal and José M. Miranda et al.
Protective Effect of Red Algae (*Rhodophyta*) Extracts on Essential Dietary Components of Heat-Treated Salmon
Reprinted from: *Antioxidants* **2021**, *10*, 1108, doi:10.3390/antiox10071108 . . . 93

Marko R. Antonijević, Dušica M. Simijonović, Edina H. Avdović, Andrija Ćirić, Zorica D. Petrović and Jasmina Dimitrić Marković et al.
Green One-Pot Synthesis of Coumarin-Hydroxybenzohydrazide Hybrids and Their Antioxidant Potency
Reprinted from: *Antioxidants* **2021**, *10*, 1106, doi:10.3390/antiox10071106 . . . 107

Éva Szabó, Tamás Marosvölgyi, Gábor Szilágyi, László Kőrösi, János Schmidt and Kristóf Csepregi et al.
Correlations between Total Antioxidant Capacity, Polyphenol and Fatty Acid Content of Native Grape Seed and Pomace of Four Different Grape Varieties in Hungary
Reprinted from: *Antioxidants* **2021**, *10*, 1101, doi:10.3390/antiox10071101 . . . 125

Irene Dini and Sonia Laneri
Spices, Condiments, Extra Virgin Olive Oil and Aromas as Not Only Flavorings, but Precious Allies for Our Wellbeing
Reprinted from: *Antioxidants* **2021**, *10*, 868, doi:10.3390/antiox10060868 . . . . **137**

Magdalena Keller, Elisa Manzocchi, Deborah Rentsch, Rosamaria Lugarà and Katrin Giller
Antioxidant and Inflammatory Gene Expression Profiles of Bovine Peripheral Blood Mononuclear Cells in Response to *Arthrospira platensis* before and after LPS Challenge
Reprinted from: *Antioxidants* **2021**, *10*, 814, doi:10.3390/antiox10050814 . . . . **169**

Salman Ul Islam, Muhammad Bilal Ahmed, Haseeb Ahsan and Young-Sup Lee
Recent Molecular Mechanisms and Beneficial Effects of Phytochemicals and Plant-Based Whole Foods in Reducing LDL-C and Preventing Cardiovascular Disease
Reprinted from: *Antioxidants* **2021**, *10*, 784, doi:10.3390/antiox10050784 . . . . **185**

Chiara Cavaliere, Angela Michela Immacolata Montone, Sara Elsa Aita, Rosanna Capparelli, Andrea Cerrato and Paola Cuomo et al.
Production and Characterization of Medium-Sized and Short Antioxidant Peptides from Soy Flour-Simulated Gastrointestinal Hydrolysate
Reprinted from: *Antioxidants* **2021**, *10*, 734, doi:10.3390/antiox10050734 . . . . **213**

Urszula Złotek, Sławomir Lewicki, Anna Markiewicz, Urszula Szymanowska and Anna Jakubczyk
Effects of Drying Methods on Antioxidant, Anti-Inflammatory, and Anticancer Potentials of Phenolic Acids in Lovage Elicited by Jasmonic Acid and Yeast Extract
Reprinted from: *Antioxidants* **2021**, *10*, 662, doi:10.3390/antiox10050662 . . . . **227**

Akanksha Tyagi, Su-Jung Yeon, Eric Banan-Mwine Daliri, Xiuqin Chen, Ramachandran Chelliah and Deog-Hwan Oh
Untargeted Metabolomics of Korean Fermented Brown Rice Using UHPLC Q-TOF MS/MS Reveal an Abundance of Potential Dietary Antioxidative and Stress-Reducing Compounds
Reprinted from: *Antioxidants* **2021**, *10*, 626, doi:10.3390/antiox10040626 . . . . **245**

Irene Dini, Danila Falanga, Ritamaria Di Lorenzo, Annalisa Tito, Gennaro Carotenuto and Claudia Zappelli et al.
An Extract from *Ficus carica* Cell Cultures Works as an Anti-Stress Ingredient for the Skin
Reprinted from: *Antioxidants* **2021**, *10*, 515, doi:10.3390/antiox10040515 . . . . **265**

Francisco Javier Martínez-Noguera, Cristian Marín-Pagán, Jorge Carlos-Vivas and Pedro E. Alcaraz
8-Week Supplementation of 2S-Hesperidin Modulates Antioxidant and Inflammatory Status after Exercise until Exhaustion in Amateur Cyclists
Reprinted from: *Antioxidants* **2021**, *10*, 432, doi:10.3390/antiox10030432 . . . . **287**

Katja Kramberger, Zala Jenko Pražnikar, Alenka Baruca Arbeiter, Ana Petelin, Dunja Bandelj and Saša Kenig
A Comparative Study of the Antioxidative Effects of *Helichrysum italicum* and *Helichrysum arenarium* Infusions
Reprinted from: *Antioxidants* **2021**, *10*, 380, doi:10.3390/antiox10030380 . . . . **309**

Rita Celano, Teresa Docimo, Anna Lisa Piccinelli, Patrizia Gazzerro, Marina Tucci and Rosa Di Sanzo et al.
Onion Peel: Turning a Food Waste into a Resource
Reprinted from: *Antioxidants* **2021**, *10*, 304, doi:10.3390/antiox10020304 . . . . **325**

Ionel Fizeșan, Marius Emil Rusu, Carmen Georgiu, Anca Pop, Maria-Georgia Ștefan and Dana-Maria Muntean et al.
Antitussive, Antioxidant, and Anti-Inflammatory Effects of a Walnut (*Juglans regia* L.) Septum Extract Rich in Bioactive Compounds
Reprinted from: *Antioxidants* **2021**, *10*, 119, doi:10.3390/antiox10010119 . . . . **343**

Evangelos Zoidis, Athanasios C. Pappas, Michael Goliomytis, Panagiotis E. Simitzis, Kyriaki Sotirakoglou and Savvina Tavrizelou et al.
Quercetin and Egg Metallome
Reprinted from: *Antioxidants* **2021**, *10*, 80, doi:10.3390/antiox10010080 . . . . **361**

Rozalia-Maria Anastasiadi, Federico Berti, Silvia Colomban, Claudio Tavagnacco, Luciano Navarini and Marina Resmini
Simultaneous Quantification of Antioxidants Paraxanthine and Caffeine in Human Saliva by Electrochemical Sensing for CYP1A2 Phenotyping
Reprinted from: *Antioxidants* **2020**, *10*, 10, doi:10.3390/antiox10010010 . . . . **375**

Klaudia Čobanová, Zora Váradyová, Ľubomíra Grešáková, Katarína Kucková, Dominika Mravčáková and Marián Várady
Does Herbal and/or Zinc Dietary Supplementation Improve the Antioxidant and Mineral Status of Lambs with Parasite Infection?
Reprinted from: *Antioxidants* **2020**, *9*, 1172, doi:10.3390/antiox9121172 . . . . **393**

Giulia Graziani, Teresa Docimo, Monica De Palma, Francesca Sparvoli, Luana Izzo and Marina Tucci et al.
Changes in Phenolics and Fatty Acids Composition and Related Gene Expression during the Development from Seed to Leaves of Three Cultivated Cardoon Genotypes
Reprinted from: *Antioxidants* **2020**, *9*, 1096, doi:10.3390/antiox9111096 . . . . **409**

Cinzia Ingallina, Alessandro Maccelli, Mattia Spano, Giacomo Di Matteo, Antonella Di Sotto and Anna Maria Giusti et al.
Chemico-Biological Characterization of Torpedino Di Fondi® Tomato Fruits: A Comparison with San Marzano Cultivar at Two Ripeness Stages
Reprinted from: *Antioxidants* **2020**, *9*, 1027, doi:10.3390/antiox9101027 . . . . **427**

Carmen Lammi, Nadia Mulinacci, Lorenzo Cecchi, Maria Bellumori, Carlotta Bollati and Martina Bartolomei et al.
Virgin Olive Oil Extracts Reduce Oxidative Stress and Modulate Cholesterol Metabolism: Comparison between Oils Obtained with Traditional and Innovative Processes
Reprinted from: *Antioxidants* **2020**, *9*, 798, doi:10.3390/antiox9090798 . . . . **457**

Stefania Garzoli, Francesco Cairone, Simone Carradori, Andrei Mocan, Luigi Menghini and Patrizia Paolicelli et al.
Effects of Processing on Polyphenolic and Volatile Composition and Fruit Quality of Clery Strawberries
Reprinted from: *Antioxidants* **2020**, *9*, 632, doi:10.3390/antiox9070632 . . . . **475**

Pierpaolo Cavallo, Irene Dini, Immacolata Sepe, Gennaro Galasso, Francesca Luisa Fedele and Andrea Sicari et al.
An Innovative Olive Pté with Nutraceutical Properties
Reprinted from: *Antioxidants* **2020**, *9*, 581, doi:10.3390/antiox9070581 . . . . **493**

Ivana Dimić, Nemanja Teslić, Predrag Putnik, Danijela Bursać Kovačević, Zoran Zeković and Branislav Šojić et al.
Innovative and Conventional Valorizations of Grape Seeds from Winery By-Products as Sustainable Source of Lipophilic Antioxidants
Reprinted from: *Antioxidants* **2020**, *9*, 568, doi:10.3390/antiox9070568 . . . . **507**

Irene Dini, Giulia Graziani, Francalisa Luisa Fedele, Andrea Sicari, Francesco Vinale and Luigi Castaldo et al.
An Environmentally Friendly Practice Used in Olive Cultivation Capable of Increasing Commercial Interest in Waste Products from Oil Processing
Reprinted from: *Antioxidants* **2020**, *9*, 466, doi:10.3390/antiox9060466 . . . . **527**

About the Editors

Irene Dini

Dr. Irene Dini is a researcher in Food Chemistry (CHIM 10) at the Department of Pharmacy of "Federico II" the University of Napoli and a specialist in "Food Science", with a Ph.D. in "Pharmacologically Active Natural Compounds".

She teaches "Food chemistry", "Spectroscopic methods in food analysis", and "Science of dietary supplements of plant origin" as a Research Professor at the Federico II University.

The scientific activity of Prof. Dini focuses on the study of natural substances and has mainly developed around four themes:

a) Chemical study of primary and secondary metabolites of interest for food;

b) Nutraceutical and nutricosmeceutical studies of phytochemicals from food plants;

c) Development of analytical methods for the dosage of metabolites of food and nutricosmeceutical interest;

d) Study of environmentally friendly agronomic techniques capable of improving the nutraceutical interest of food products.

Domenico Montesano

Domenico Montesano currently works at the Department of Pharmacy, University Federico II of Naples, Italy. Domenico carries out research in Food Chemistry, Analytical Chemistry, Environmental Chemistry and Biochemistry. Their current project is 'Food Safety and Food Quality'.

antioxidants MDPI

Editorial

The Potential of Dietary Antioxidants

Irene Dini

Department of Pharmacy, University of Naples Federico II, Via Domenico Montesano 49, 80131 Napoli, Italy; irdini@unina.it

Citation: Dini, I. The Potential of Dietary Antioxidants. *Antioxidants* **2021**, *10*, 1752. https://doi.org/10.3390/antiox10111752

Received: 28 October 2021
Accepted: 30 October 2021
Published: 1 November 2021

Oxidative stress happens when the levels of reactive species made from oxygen and nitrogen exceed the body's antioxidant capacity. Exogenous agents (i.e., UV rays and ionizing radiation) and endogenous processes (i.e., incomplete reduction of O_2 to H_2O in oxidative phosphorylation reactions, inflammation, and infections) [1] generate reactive nitrogen species (RNS), including nitric oxide ($^{\bullet}NO$), and reactive oxygen species (ROS), including superoxide ($O_2^{\bullet -}$), hydroxyl radical ($^{\bullet}OH$), hydrogen peroxide (H_2O_2), singlet oxygen, and ozone, which can produce oxidative stress. The regulation of ROS and RNS levels is indispensable to preserve cellular homeostasis. Low ROS and RNS levels stimulate immune responses, cell proliferation, apoptosis, differentiation, and stress-responsive pathways. High ROS and RNS levels damage lipids, proteins, and DNA (breaking single- or double-strands, modifying base, and DNA cross-links) [2]. Oxidative stress causes chronic diseases such as cardiovascular diseases, cancer, Alzheimer, chronic obstructive pulmonary, and neurodegenerative pathologies [3]. Antioxidant systems defend human cells from free radicals. They act by stopping free radicals, decreasing their development, and quenching the formed ROS and RNS [4]. Some enzymes are involved in the antioxidant defense system; among these, glutathione (GSHs) (reductase, peroxidases, and *S*-transferases) can neutralize ROS directly or with the help of metal cofactors (e.g., Mn, Cu, Se, and Zn) [5]. The antioxidant molecules are classified into primary and secondary defense molecules. The primary antioxidant molecules (i.e., vitamins C and E, ubiquinone, and glutathione) reduce oxidation effects by moving a proton to the free radical species or electron donors, or by terminating the chain reactions [6]. The secondary antioxidants (i.e., *N*-acetyl cysteine and lipoic acid) act as cofactors for some enzyme systems or neutralize the production of free radicals by transition metals [7]. In recent years, the consumption of foods rich in nutrients and phytochemicals (secondary plant metabolites) with antioxidant properties has been linked to positive health outcomes [2,8–10]. Food supplements containing natural antioxidant molecules have been formulated to satisfy the consumers' attention towards products containing natural molecules considered non-toxic, capable of preventing the pathologies related to oxidative stress, or improving skin and hair wellness [5]. Moreover, antioxidant molecules have been added to packaging materials to preserve food from oxidation and microorganism attack [11]. Unfortunately, in the plant kingdom, although the concentrations of antioxidants are widespread, they are low. Therefore, to increase the sources of antioxidant raw materials, it is necessary to improve the following:

1. Knowledge on the plants that express the greatest concentration of antioxidant molecules of interest,
2. Pedoclimatic conditions and cultivation stages that make most of their levels available in plants,
3. Extraction methods that can properly maximize their recovery,
4. The potential preventive and/or therapeutic effect of each chemical class of antioxidants,
5. Pharmaceutical delivery systems that allow for the full advantage of their actions in the body.

This work comprises original research papers and reviews on antioxidant molecules in food, the agricultural practices that maximize their levels in plants, the potential preventive effects of selected classes of antioxidant molecules, their potential use in functional foods, and the pharmaceutical delivery systems that maximize their potential activity when used as supplements. Finally, the recovery of food antioxidants from food waste is discussed, considering the need to reduce waste to protect the environment as an issue of primary importance [12].

Funding: This research received no external funding.

Conflicts of Interest: The author declare no conflict of interest.

References

1. Birben, E.; Sahiner, U.M.; Sackesen, C.; Erzurum, S.; Kalayci, O. Oxidative stress and antioxidant defense. *World Allergy Organ. J.* **2012**, *5*, 9–19.
2. Chikara, S.; Nagaprashantha, L.D.; Singhal, J.; Horne, D.; Awasthi, S.; Singhal, S.S. Oxidative stress and dietary phytochemicals: Role in cancer chemoprevention and treatment. *Cancer Lett.* **2018**, *413*, 122–134. [CrossRef] [PubMed]
3. Forman, H.J.; Zhang, H. Targeting oxidative stress in disease: Promise and limitations of antioxidant therapy. *Nat. Rev. Drug Discov.* **2021**, *20*, 689–709. [CrossRef] [PubMed]
4. Packer, L.; Valacchi, G. Antioxidants and the response of skin to oxidative stress: Vitamin E as a key indicator. *Skin Pharmacol. Appl. Skin Physiol.* **2002**, *15*, 282–290. [CrossRef] [PubMed]
5. Dini, I.; Laneri, S. Nutricosmetics: A brief overview. *Phytother. Res.* **2019**, *33*, 3054–3063. [CrossRef] [PubMed]
6. Pinnell, S.R. Cutaneous photodamage, oxidative stress, and topical antioxidant protection. *J. Am. Acad. Dermatol.* **2003**, *48*, 1–19. [CrossRef] [PubMed]
7. Dini, I.; Laneri, S. The New Challenge of Green Cosmetics: Natural Food Ingredients for Cosmetic Formulations. *Molecules* **2021**, *26*, 3921. [CrossRef] [PubMed]
8. Dini, I.; Laneri, S. Spices, Condiments, Extra Virgin Olive Oil and Aromas as Not Only Flavorings, but Precious Allies for Our Wellbeing. *Antioxidants* **2021**, *10*, 868. [CrossRef] [PubMed]
9. Cavallo, P.; Dini, I.; Sepe, I.; Galasso, G.; Fedele, F.L.; Sicari, A.; Bolletti Censi, S.; Gaspari, A.; Ritieni, A.; Lorito, M.; et al. An Innovative Olive Pâté with Nutraceutical Properties. *Antioxidants* **2020**, *9*, 581. [CrossRef] [PubMed]
10. Dini, I. Spices and herbs as therapeutic foods. In *Food Quality: Balancing Health and Disease*; Holban, A.M., Grumezescu, A.M., Eds.; Academic Press: Cambridge, MA, USA; Elsevier: London, UK, 2018; pp. 433–469.
11. Dini, I. Chapter 14—Use of Essential Oils in Food Packaging. In *Essential Oils in Food Preservation, Flavor and Safety*; Academic Press: Cambridge, MA, USA, 2016; pp. 139–147.
12. Dini, I.; Graziani, G.; Fedele, F.L.; Sicari, A.; Vinale, F.; Castaldo, L.; Ritieni, A. An Environmentally Friendly Practice Used in Olive Cultivation Capable of Increasing Commercial Interest in Waste Products from Oil Processing. *Antioxidants* **2020**, *9*, 466. [CrossRef] [PubMed]

antioxidants

MDPI

Article

Effects of a Nanoencapsulated *Moringa* Leaf Ethanolic Extract on the Physiology, Metabolism and Reproductive Performance of Rabbit Does during Summer

Nagwa I. El-Desoky [1], Nesrein M. Hashem [1,*], Antonio Gonzalez-Bulnes [2,*], Ahmed G. Elkomy [1] and Zahraa R. Abo-Elezz [1]

1 Department of Animal and Fish Production, Faculty of Agriculture (El-Shatby), Alexandria University, Alexandria 21545, Egypt; enagwa278@gmail.com (N.I.E.-D.); ahmed.elkoumi@alexu.edu.eg (A.G.E.); n.ebrahim10@yahoo.com (Z.R.A.-E.)

2 Departamento de Produccion y Sanidad Animal, Facultad de Veterinaria, Universidad Cardenal Herrera-CEU, C/Tirant lo Blanc, 7, 46115 Valencia, Spain

* Correspondence: nesreen.hashem@alexu.edu.eg (N.M.H.); antonio.gonzalezbulnes@uchceu.es (A.G.-B.)

Citation: El-Desoky, N.I.; Hashem, N.M.; Gonzalez-Bulnes, A.; Elkomy, A.G.; Abo-Elezz, Z.R. Effects of a Nanoencapsulated *Moringa* Leaf Ethanolic Extract on the Physiology, Metabolism and Reproductive Performance of Rabbit Does during Summer. *Antioxidants* **2021**, *10*, 1326. https://doi.org/10.3390/antiox10081326

Academic Editor: Maria Letizia Manca

Received: 29 June 2021
Accepted: 22 August 2021
Published: 23 August 2021

Abstract: This study investigated the effect of *Moringa* leaf ethanolic extract (MLEE) on heat-tolerance variables and the reproductive performance of rabbit does bred under hot climate conditions. Additionally, the effect of nanoencapsulation technology on the biological efficiency of MLEE was considered. A total of 56 rabbit does were randomly divided into four experimental groups and treated with 50 mg/kg body weight (BW) nonencapsulated MLEE, 25 or 10 mg/kg BW nanoencapsulated MLEE, or not treated (Control, C). The treatments continued for 50 days, including mating and pregnancy times. Physiological and hematochemical variables, hormonal profiles, and reproductive performance (kindling rate and litter characteristics) were determined. The active components of MLEE were identified. The results indicated that MLEE has 30 active components. All MLEE-based treatments reduced heat-stress-related indicators, such as rectal temperatures, respiratory rates and heart rate; improved hematochemical attributes, redox status, and hormones (progesterone and prolactin); and increased the total litter size, the kindling rate, litter size at birth and litter weight at birth. Adding MLEE can alleviate the negative impacts of heat stress by improving metabolism, redox status, and hormonal balance during pregnancy. These effects were seen whether MLLE was in free or encapsulated forms. However, the use of nanoencapsulated MLEE allowed 80% reduction (10 mg/kg BW) in the optimal dose (50 mg/kg BW) without affecting the efficiency of the treatment. These results support the importance of nanoencapsulation technology in improving the bioavailability of active components when they are orally administered.

Keywords: *Moringa* leaf extract; nanoencapsulation; rabbit; physiology; reproduction

1. Introduction

Rabbits are a good alternative to larger mammals for meat production and, thus, an adequate alternative source for the increasing population in developing countries [1]. Rabbits are characterized by high reproductive performance compared to that of other farm animals; unfortunately, they are sensitive to heat stress, which compromises productive rate and economic spin-off [2]. Increasing global warming represents a real challenge to the rabbit industry, particularly in arid and semi-arid regions; the breeding season could be restricted for several months due to the negative impacts of heat stress (high ambient temperatures) on fertility. Hence, it is necessary to investigate strategies for mitigating the consequences of heat stress on the production of rabbits.

Rabbit does are more susceptible to injury due to these effects as they exhibit critical physiological and hormonal changes during their reproductive cycles [3,4].The search for safe and effective naturally occurring feed supplements is considered to be a promising solution that can be applied to improve the harmful effects of heat stress. Phytochemicals are natural bioactive and non-nutritive plant chemicals that positively affect health [5–7].

One of the most effective sources of phytochemicals is the *Moringa* tree; it has several phytogenic bioactive compounds with antioxidant, antimicrobial and immunomodulatory activities that can improve the productive and reproductive performance and health of animals. Thus, this plant has gained considerable attention to be used as a feed additive or feedstuff in the livestock industry [8–13]. At the practical level, the use of a phytogenic substance that originated from active components is restricted by the stability of these molecules in the gastrointestinal tract (GIT), absorption, cellular uptake, and stability during handling and storage [14–16]. The nanoencapsulation of phytogenic active components is a current research need because it can be used to solve the previously mentioned problems related to the use of phytogenic components [17]. Accordingly, this study was designed to assess the effects of a non- or nanoencapsulated *Moringa* leaf ethanolic extract (MLEE) on reproductive performance, hematological variables, hormones, blood plasma metabolites, and the antioxidant indicators of rabbit does bred under environmental heat stress.

2. Materials and Methods

This study was conducted at the Laboratory of Rabbit Physiology Research, Agricultural Experimental Station, Faculty of Agriculture, Alexandria University, Egypt (31° 20′ N, 30° E); animals were cared for according to International Council for Laboratory Animal Science (ICLAS; http://iclas.org/members/member-list (accessed on 12 Aug 2021)).

2.1. Plant Extaction and Identification of Active Components

Moringa (*Moringa oleifera*) leaves were dried naturally until they reached approximately 90% dry matter and then were milled through a 1 mm screen. Moringa leaf powder (25 g/100 mL) was extracted using a 70% hydroethanolic solution at 40 °C for 72 h. The extract was filtered with Whatman No. 1 filter paper (Camlab, Cambridge, UK). The collected filtrate was evaporated to complete dryness at 45 °C. The residues were then stored at −20 °C pending use. The collected filtrate was evaporated to complete dryness at 45 °C [11]. The residues were then stored at −20 °C pending use.

Chemical constituents of MLEE were determined using gas chromatography/mass spectrometry (GC-MS) (Thermo Scientific TRACE-1300 series GC; Thermo Fisher Scientific Inc., Austin, TX, USA), fitted with a fused silica DB-5 capillary column (30 m × 0.32 mm id, 0.25 µm film thickness; Thermo Fisher Scientific Inc.) and coupled to a Triple Quadrupole Mass (TSQ 8000 Evo; Thermo Fisher Scientific Inc.). The column oven temperature was initially held at 50 °C and then increased by 5 °C/min to 250 °C, held for 2 min and then increased to the final temperature, 300 °C, by 30 °C/min and held for 2 min. Splitless injection mode (0.5 µL of a 1:1000 methanol solution) was used. The carrier gas was helium at a constant flow rate of 1 mL/min. The injector and detector temperatures were 250 °C and 290 °C, respectively. Mass spectra were scanned in the range 40 to 700 amu; the scan time was 5 scans/s. The constituents were identified by the combination of retention index data and mass spectra data using the NIST 14 mass spectral database.

2.2. Fabrication and Characterization of a Moringa Extract and Sodium Alginate Nanocomplex

The dried MLEE was used for the fabrication of a sodium alginate nanocomplex using calcium chloride ($CaCl_2$) as a cross-linking agent by adopting the ionic-gelation method [18]. Under continuous magnetic stirring, the MLEE (1.5 g) was first mixed with the sodium alginate solution (1%, *w/v*). Then, the mixture was added dropwise using a syringe into a $CaCl_2$ solution (2.2 mol/L) with a ratio of 2 sodium alginate and MLEE mixture:1 $CaCl_2$ solution. The synthesized nanoparticles were centrifuged at 8000 rpm for 20 min, and the resultant nanoparticles were collected and stored at −80 °C. The physicochemical characteristics, the size, polydispersity (PdI) and zeta potential of the sodium alginate-$CaCl_2$ nanoparticles conjugated or not with the extract was measured using a Scientific Nanoparticle Analyzer (Zetasizer Nano ZS, Malvern Instruments Ltd., Worcestershire, UK) at 25 °C.

The encapsulation efficiency (EE, %) of sodium alginate-$CaCl_2$ nanoparticles for the MLEE was estimated by determining the phenolic content of the raw MLEE (before encapsulation, C raw) and of the resultant supernatant following the collection of the nanocomplex particles (C supernatant), using the following equation: EE (%) = C raw − C supernatant/C raw × 100. The concentrations of total phenolic compounds in the raw MLEE and the supernatant were colorimetrically determined using the Folin–Ciocalteu assay at a 765 nm absorbance wavelength and with gallic acid (GA) as a standard [19]. The concentration of total phenolic compounds was 47.60 and 20.26 µg GA equivalent/gm for raw MLEE and the supernatant, respectively.

2.3. Animal Husbandry and Experimental Design

The rabbit breed used in this study was of V-line breeding, a maternal synthetic line selected based on pregnancy (Department of Animal Science, Universitat Politecnica de Valencia, Valencia, Spain; [20]). Fifty-six nulliparous (six-months-old) female rabbits, weighing 2.75 ± 0.18 kg, were housed in a naturally ventilated building under similar management and hygiene conditions. Rabbit does were kept in individual galvanized wire cages (60 L × 55 W × 40 H cm). Batteries were equipped with feeders for pelleted rations and automatic drinkers. Freshwater was available ad libitum. Rabbit does received their daily maintenance according to the National Research Council (NRC, [21]). The ingredients and chemical analysis of the experimental diet are shown in Table 1. Rabbit does were randomly divided into four experimental groups and treated with 50 mg/kg BW nonencapsulated MLEE (FM_{50}), 25 mg/kg BW nanoencapsulated MLEE (NM_{25}), 10 mg/kg BW nanoencapsulated MLEE (NM_{25}) or not treated (C). Rabbit does orally received different treatments for 50 d, starting from 20 d before insemination to the entire pregnancy period (30 d).

Table 1. Basal diet ingredients and chemical analysis (expressed as g/kg dry matter, DM).

Items	Basal Diet
Ingredients	
Alfalfa hay	280
Barley	180
Wheat bran	250
Yellow corn	60
Soybean meal	180
Barley grain	10
Molasses	30
Di-calcium phosphate	10
NaCl and premix †	10
Analyzed composition (on DM basis) ‡	
Crude protein	17.54
Ether extract	2.05
Crude fiber	12.53
Ash	9.43
Nitrogen-free extract	59.45

† For each 1 kg of premix (minerals and vitamins mixture) contains: vit. A, 20,000 IU; vit. D3, 15,000 IU; vit. B1, 0.33; vit. B2, 1.0 g; vit. B6, 0.33 g; vit. B5, 8.33 g; vit. B12, 1.7 mg; pantothenic acid, 3.33 g; biotin, 33 mg; folic acid, 0.83 g; choline chloride, 200 g; vit. E, 8.33 g; and vit. K, 0.33 g. ‡ Chemical analysis according to AOAC [22].

2.4. Metrological Data

Ambient temperature (°C) and relative humidity (RH, %) were recorded daily inside the rabbitry using an electronic digital thermo-hygrometer. The overall mean of the maximum and minimum temperatures, RH (%) and the temperature–humidity index (THI) during the experimental period is estimated. The temperature–humidity index (THI) was calculated according to the following equation: THI = db °C–[(0.31–0.31 × RH%) × (db °C–14.4)], where db °C = dry bulb temperature in Celsius, and RH% = percentage of relative humidity. The THI values were classified as the absence of heat stress (<27.8), moderate heat stress (27.8–28.8), severe heat stress (28.9–29.9) and very severe heat stress (>30.0) for rabbits, according to the classification of Marai et al. [23].

2.5. Physiological Parameters

Each rabbit doe was weighed weekly in the morning before feed was offered. The feed intake was calculated weekly by subtracting the unconsumed feed from the total amount offered and recorded as g/day. Rectal temperature was individually measured in the morning using a digital thermometer gently introduced into the rectum and attached to the rectal wall until a fixed reading was obtained. The respiration rate was measured by counting the number of flank movements during the complete inhalation–exhalation cycle (breath) per minute [24]. The heart rate was measured by counting the number of beats per minute using a *stethoscope* [23]. These data were used to estimate the mean of each variable during premating (day −10), mating (day 0) and days 10, 17 and 24 of pregnancy.

2.6. Hematochemical Variables and the Hormonal Profile

Blood samples were collected from the marginal ear vein of each rabbit using heparinized tubes during premating (day −10), mating (day 0) and days 10, 17 and 24 of pregnancy [25]. Each blood sample was divided into two subsamples: the first subsample was used to assess hematological variables, and the second subsample was centrifuged at 2000× *g* for 20 min at 4 °C to obtain plasma. The plasma samples were stored at −20 °C for further analyses. The counts of red blood cells (RBCs), packed corpuscular volume (PCV) and hemoglobin (Hb) were determined in the first subsamples (whole blood). The concentration of blood Hb was assessed by the colorimetric method using commercial kits (Reactivos GPL, Barcelona, Spain). Blood plasma metabolites, including total protein, albumin, glucose, cholesterol, triglycerides, activities of alanine aminotransferase (ALT) and aspartate aminotransferase (AST), were determined using the colorimetric method with commercial kits (BioSystem SA, Barcelona, Spain). Additionally, the total antioxidant capacity (TAC) and reduced glutathione enzyme (GSH-Px) activity were also determined as indicators of the antioxidant status of plasma (Biodiagnostic, Giza, Egypt). Progesterone and prolactin concentrations were analyzed in blood plasma samples collected during pregnancy (days 10, 17 and 24) using commercial solid-phase enzyme immuno-assay ELISA kits obtained from Pointe Scientific Inc., MI, USA and Cloud-Clone Corp, TX, USA, respectively. The methods' sensitivity was 0.0625 ng/mL and 0.063 ng/mL for progesterone and prolactin, respectively. The corresponding intra-assay and inter-assay coefficients of variation were 2.4–2.6% and <10%–<12%, respectively.

2.7. Productive and Reproductive Performance

After 20 days from starting of the treatment, does were mated with previously evaluated fertile bucks. The kindling rate ([number of delivered does/number of mated does] × 100), litter size at birth (total rabbits born, alive and dead) and litter weight at birth were recorded [24].

2.8. Statistical Analyses

All the statistical analyses were carried out using the Statistical Analysis Software package (SAS, Version 8. Cary, NC, USA: SAS Institute; 2001). The MIXED procedure for repeated measurement was used to assess the fixed effects of treatment (C, FM_{50}, NM_{25} and NM_{10}), status (physiological status at time of sampling and/or data collection) and the treatment by status interaction on physiological and hematochemical variables, redox status, hormonal profiles and reproductive performance. The rabbit dose effect was introduced as a random factor in this model. One-way ANOVA was used to assess the treatment effects on litter size, litter viability and litter weight, while a Chi-square test was used to assess the effects of treatments on the kindling rate. Differences among treatment means were tested by Duncan's multiple range tests. All the results are presented as the least square mean (±SEM), with significance accepted at $p < 0.05$.

3. Results

3.1. Chemical Compositions and Characterization of a Moringa Extract and Sodium Alginate Nanocomplex

The analysis of MLEE identified 30 chemical compounds belonging to different chemical families and with different biological activities. Among the detected chemical compounds, the abundant compounds were phytosphingosine (10.46%), N-acetylneuraminic acid,2,3-dehydro-2-deoxy- (6.50%), nialamide (5.45%), 1-Stearoyl-2-oleoyl-sn-glycero-3-phosphoethanolamine (4.83%), leukotriene E4 methyl ester (4.45%), D-erythro-sphingosine C-20 (4.31%) and exo-norbornyl alcohol (4.22%) (Table 2). Other 22 compounds were detected with a range of <4.0 to <1% (Table 2). The physicochemical characterization of alginate–$CaCL_2$ nanoparticles and alginate–$CaCL_2$ nanoencapsulated MLEE revealed that the mean size was 195.10 and 93.69 nm; the zeta potential was −3.41 and 8.95 mV; and PdI was 0.457 and 0.442, respectively. The encapsulation efficiency of alginate–$CaCL_2$ nanoparticles for MLEE was 57.43% (Table 3).

Table 2. Chemical composition of a *Moringa* leaf ethanolic extract.

Retention Time, min	Compounds	Area of Component, %	Chemical Formula	Molecular Weight
20.28	Phytosphingosine	10.46	$C_{18}H_{39}NO_3$	317
8.91	N-acetylneuraminic acid,2,3-dehydro-2-deoxy-	6.50	$C_{11}H_{17}NO_8$	291
7.43	Nialamide	5.45	$C_{16}H_{18}N4O_2$	298
25.75	1-Stearoyl-2-oleoyl-sn-glycero-3-phosphoethanolamine	4.83	$C_{41}H_{80}NO_8P$	745
9.16	Leukotriene E4 methyl ester	4.45	$C_{24}H_{39}NO_5S$	453
15.12	D-erythro-sphingosine C-20	4.31	$C_{20}H_{41}NO_2$	327
8.59	Exo-norbornyl alcohol	4.22	$C_7H_{12}O$	112
24.71	Ethidimuron	3.96	$C_7H_{12}N_4O_3S_2$	264
20.63	N-methylformamide	3.93	C_2H_5NO	59
3.37	Salinomycin	3.75	$C_{42}H_{70}O_{11}$	750
7.77	Folinic acid	3.69	$C_{20}H_{23}N_7O_7$	473
11.52	Lactacystin	3.66	$C_{15}H_{24}N_2O_7S$	376
25.59	Methyl alcohol	3.46	CH_4O	32
8.34	Pentadecanoic acid, 14-methyl-	3.40	$C_{16}H_{32}O_2$	256
17.15	L-arabinitol	3.37	$C_5H_{12}O_5$	152
17.83	1-Hexadecanoyl-2-(5Z,8Z,11Z,14Zeicosatetraenoyl)-sn-glycero-3-phosphoethanolamine	3.31	$C_{41}H_{74}NO_8P$	739
18.80	Deoxycholic acid	3.11	$C_{24}H_{40}O_4$	392
27.48	Glycan le-a trisaccharide	2.64	$C_{20}H_{35}NO_{15}$	529
9.56	2-furoic acid	2.52	$C_5H_4O_3$	112
18.07	Glycine	2.43	$C_2H_5NO_2$	75
7.53	Diapocynine	2.20	$C_{18}H_{18}O_6$	330
18.22	3-pentanol, 3-ethyl-	2.08	$C_7H_{16}O$	116
30.39	Rifabutin	1.93	$C_{46}H_{62}N_4O_{11}$	846
14.33	Palmatic acid	1.88	$C_4H_9NO_2$	103
20.48	Amino methyl propanol	1.73	$C_4H_{11}NO$	89
3.25	Adenosin' 5'-diphosphate	1.58	$C_{10}H_{15}N_5O_{10}P_2$	427
3.2	N-acetylleucyl-leucyl-methioninal	1.46	$C_{19}H_{35}N_3O_4S$	401
7.08	Thiocyanic acid	1.44	CHNS	59
9.23	9S,11,15S-trihydroxythrombox-13E-enoic acid	1.32	$C_{20}H_{36}O_6$	372
24.50	D-gluconic acid	0.94	$C_6H_{12}O_7$	196

Table 3. Physicochemical characteristics of the alginate–calcium chloride nanoparticles and prepared nanoencapsulated Moringa leaf ethanolic extract (MLEE).

Item	Alginate–Calcium Nanoparticles	Nanoencapsulated MLEE
Size, nm	195.10	93.69
Zeta potential, mV	−3.41	8.95
Poly disparity index	0.457	0.442
Entrapment efficiency, %	-	57.43
Total phenols, eq-mg gallic acid/kg DM	-	27.34

3.2. *Metrological Variables*

The means of ambient temperature, relative humidity, THI and daylight length (photoperiod) during the experimental period were 31.67 °C ± 1.21 °C, 82.67% ± 2.92%, 30.73 ± 1.23 and 13.66 h ± 0.14 h, respectively (Table 4).

Table 4. Changes in (mean ± SEM) ambient temperature (°C), relative humidity (%), temperature–humidity index (THI) and photoperiod (h) during the 10 d intervals of the experimental period (July–mid-August).

s	10	20	30	40	50	Overall	SEM	*p*-Value
Air temperature, °C	30.80 ^{cd}	30.20 ^{d}	31.64 ^{bc}	31.80 ^{b}	33.90 ^{a}	31.67	1.21	<0.001
Relative humidity, %	83.50 ^{a}	82.40 ^{ab}	83.36 ^{a}	80.90 ^{b}	83.60	82.67	2.92	0.163
THI	29.95 ^{bc}	29.33 ^{c}	30.69 ^{b}	30.77 ^{b}	32.90 ^{a}	30.73	1.23	<0.001
Day light length, h	14.6 ^{a}	13.75 ^{b}	13.47 ^{d}	13.59 ^{c}	13.43 ^{d}	13.66	0.14	<0.001

^{a,b,c,d} Means within a row with different superscript letters are significantly different ($p < 0.05$).

3.3. *Physiological Variables*

The effects of different treatments (C = 0 mg/kg BW MLEE, FM_{50} = 50 mg/kg BW MLEE, NM_{10} = 10 mg/kg BW nanoencapsulated MLEE and NM_{25} = 25 mg/kg BW nanoencapsulated MLEE) on BW, feed intake, rectal temperature, respiratory rate and heart rate are presented in Table 5 and Figure 1. Compared to C treatment, all other treatments significantly increased the BW and feed intake of does. Through the experimental period, FM_{50}, NM_{10} and NM_{25} significantly decreased rectal temperature, the respiratory rate and the heart rate compared to C treatment, and the lowest values were observed in NM_{10} (Table 5 and Figure 1).

Table 5. Effect of the nanoencapsulated *Moringa* leaf ethanolic extract (MLEE) on the physiological parameters of a female rabbit during the experimental period (mean ± SEM).

Variable	Treatment [1] (T) (n = 14/group)				SEM	*p*-Value		
	C	FM_{50}	NM_{25}	NM_{10}		T	Status (S)	T × S
Body weight, kg	2.60 ^{b}	2.69 ^{a}	2.80 ^{a}	2.74 ^{a}	0.04	0.009	<0.001	<0.001
Feed intake, g/day	116.52 ^{c}	120.35 ^{b}	120.27 ^{b}	126.45 ^{a}	7.47	<0.001	0.02	<0.001
Rectal temperature, °C	39.67 ^{a}	39.12 ^{b}	38.71 ^{c}	38.70 ^{c}	0.04	<0.001	<0.001	<0.001
Respiratory rate, breaths/min	90.02 ^{a}	78.84 ^{b}	78.65 ^{b}	75.71 ^{c}	1.34	<0.001	<0.001	<0.001
Heart rate, pulse/min	99.15 ^{a}	91.73 ^{b}	91.02 ^{b}	90.27 ^{b}	0.64	<0.001	<0.001	<0.001

[1] C = 0 mg/kg BW MLEE, FM_{50} = 50 mg/kg BW MLEE, NM_{25} = 25 mg/kg BW nanoencapsulated MLEE and NM_{10} = 10 mg/kg BW nanoencapsulated MLEE. Means within the same physiological status having different superscripts (a, b, c) differ significantly ($p < 0.05$).

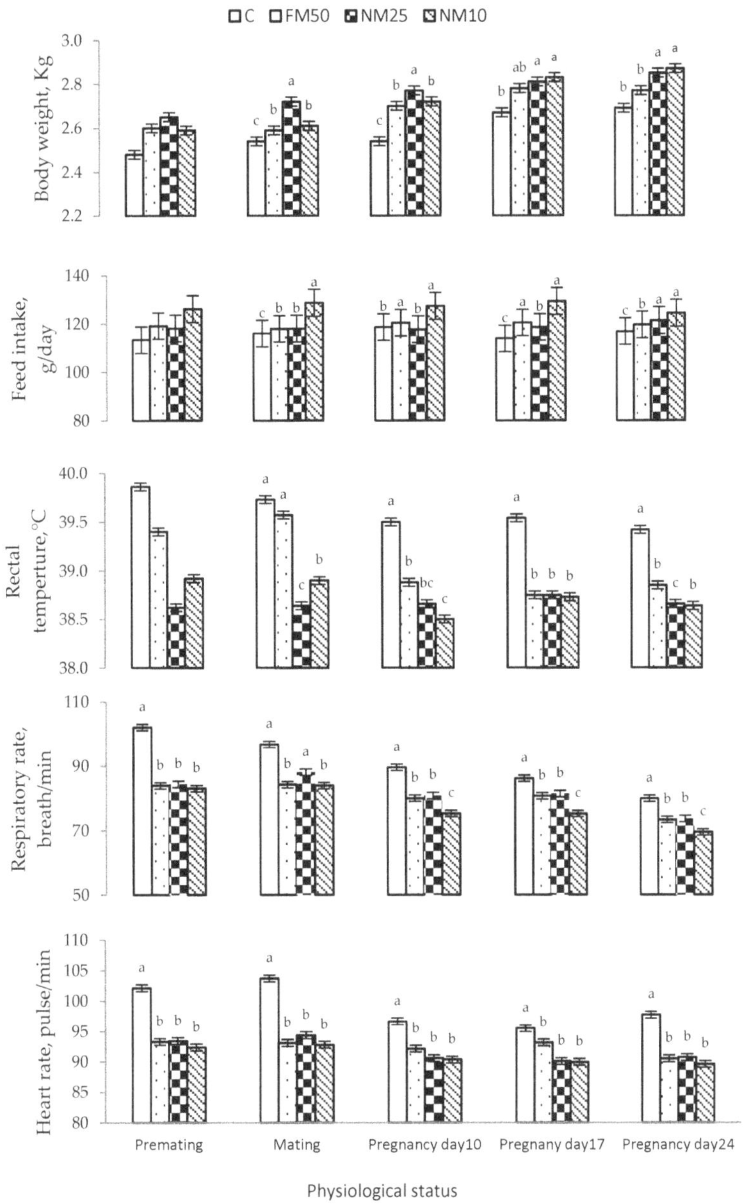

Figure 1. Mean (± SEM) treatment by physiological status effects on bodyweight, feed intake, rectal temperature, the respiratory rate and the heart rate of female rabbits. C = 0 mg/kg BW MLEE, FM_{50} = 50 mg/kg BW MLEE, NM_{25} = 25 mg/kg BW nanoencapsulated MLEE and NM_{10} = 10 mg/kg BW nanoencapsulated MLEE. Means within the same physiological status having different superscripts (a, b, c) differ significantly ($p < 0.05$).

3.4. Blood Plasma Metabolites

The effects of different treatments (C = 0 mg/kg BW MLEE, FM_{50} = 50 mg/kg BW MLEE, NM_{10} = 10 mg/kg BW nanoencapsulated MLEE and NM_{25} = 25 mg/kg BW nanoencapsulated MLEE) on the blood plasma of female rabbits during the experimental period are presented in Table 6 and Figure 2. Compared with the C treatment, all MLEE-based treatments significantly increased hematological variables, including RBC, Hb and PCV. The highest values were observed for a low concentration of encapsulated MLEE (NM_{10}). The same trend was observed for blood plasma metabolites (total protein, albumin and glucose), whereas all MLEE-based treatments significantly decreased the blood plasma concentrations of cholesterol, triglycerides, ALT and AST. All MLEE-based treatments significantly increased the TAC and GSH-Px compared to the C treatment, and the highest value was observed for the NM_{10} treatment. Both levels of nanoencapsulated MLEE treatments significantly increased blood plasma progesterone and prolactin compared to those due to the C treatment, whereas nonencapsulated MLEE resulted in intermediate values (Table 6 and Figure 3).

Table 6. Effect of nanoencapsulated *Moringa* leaf ethanolic extract (MLEE) on the blood plasma of female rabbits during the experimental period (mean ± SEM).

Variable	Treatment [1] (T) (*n*=14/group)					*p*-Value		
	C	FM_{50}	NM_{25}	NM_{10}		T	Status (S)	T × S
Hematology parameters								
Red blood cell count, $10^6/cm^3$	5.46 ^{d}	5.98 ^{c}	6.40 ^{b}	6.82 ^{a}	0.01	<0.001	<0.001	<0.001
Hemoglobin, g/dL	10.64 ^{d}	11.33 ^{c}	12.20 ^{b}	13.25 ^{a}	0.14	<0.001	<0.001	0.04
Packed corpuscular volume, %	32.27 ^{d}	34.2 ^{c}	36.22 ^{b}	38.96 ^{a}	0.35	<0.001	<0.001	<0.001
Plasma metabolites								
Total protein, g/dL	5.30 ^{d}	5.57 ^{c}	5.65 ^{b}	5.84 ^{a}	0.04	<0.001	0.05	0.71
Albumin, g/dL	3.31 ^{d}	3.52 ^{c}	3.75 ^{b}	3.86 ^{a}	0.04	<0.001	0.49	0.51
Glucose, mg/dL	73.54 ^{d}	80.20 ^{c}	83.40 ^{b}	87.93 ^{a}	0.87	<0.001	<0.001	0.0004
Cholesterol, mg/dL	220.24 ^{a}	210.50 ^{b}	205.01 ^{c}	196.70 ^{d}	1.19	<0.001	0.32	0.10
Triglycerides, mg/dL	188.79 ^{a}	179.82 ^{b}	174.26 ^{c}	167.91 ^{d}	1.48	<0.001	<0.001	0.43
Alanine aminotransferase, IU/L	68.09 ^{a}	66.11 ^{b}	65.35 ^{c}	63.25 ^{d}	0.27	<0.001	0.32	0.36
Aspartate aminotransferase, IU/L	60.58 ^{a}	58.38 ^{b}	57.81 ^{c}	53.51 ^{d}	0.60	<0.001	0.01	<0.001
Antioxidant								
Total antioxidant capacity, mmol/mL	416.86 ^{b}	421.89 ^{a}	421.86 ^{a}	422.13 ^{a}	0.47	<0.001	<0.001	<0.001
Glutathione peroxidase, mmol/mL	968.40 ^{c}	972.14 ^{b}	973.82 ^{ab}	975.76 ^{a}	0.65	<0.001	<0.001	<0.001
Hormone								
Progesterone, ng/mL	5.45 ^{c}	5.76 ^{b}	6.02 ^{a}	5.99 ^{a}	0.03	<0.001	<0.001	<0.001
Prolactin, ng/mL	2.80 ^{c}	3.11 ^{b}	3.33 ^{a}	3.39 ^{a}	0.07	<0.001	<0.001	<0.001

[1] C = 0 mg/kg BW MLEE, FM_{50} = 50 mg/kg BW MLEE, NM_{25} = 25 mg/kg BW nanoencapsulated MLEE and NM_{10} = 10 mg/kg BW nanoencapsulated MLEE. Means within the same physiological status having different superscripts (a, b, c, d) differ significantly ($p < 0.05$).

3.5. Reproductive Performance

The effects of different treatments (C = 0 mg/kg BW MLEE, FM_{50} = 50 mg/kg BW MLEE, NM_{10} = 10 mg/kg BW nanoencapsulated MLEE and NM_{25} = 25 mg/kg BW nanoencapsulated MLEE) on the reproductive performance of rabbit does are presented in Table 7. Rabbit does that received the NM_{10} treatment had the highest kidding rate, followed by those that received the FM_{50} and NM_{25}, compared with those that received the C treatment. All other treatments significantly increased total litter sizes and viable litters at birth (Table 7). Compared with all treatments, the NM_{10} treatment significantly increased the litter weight at birth.

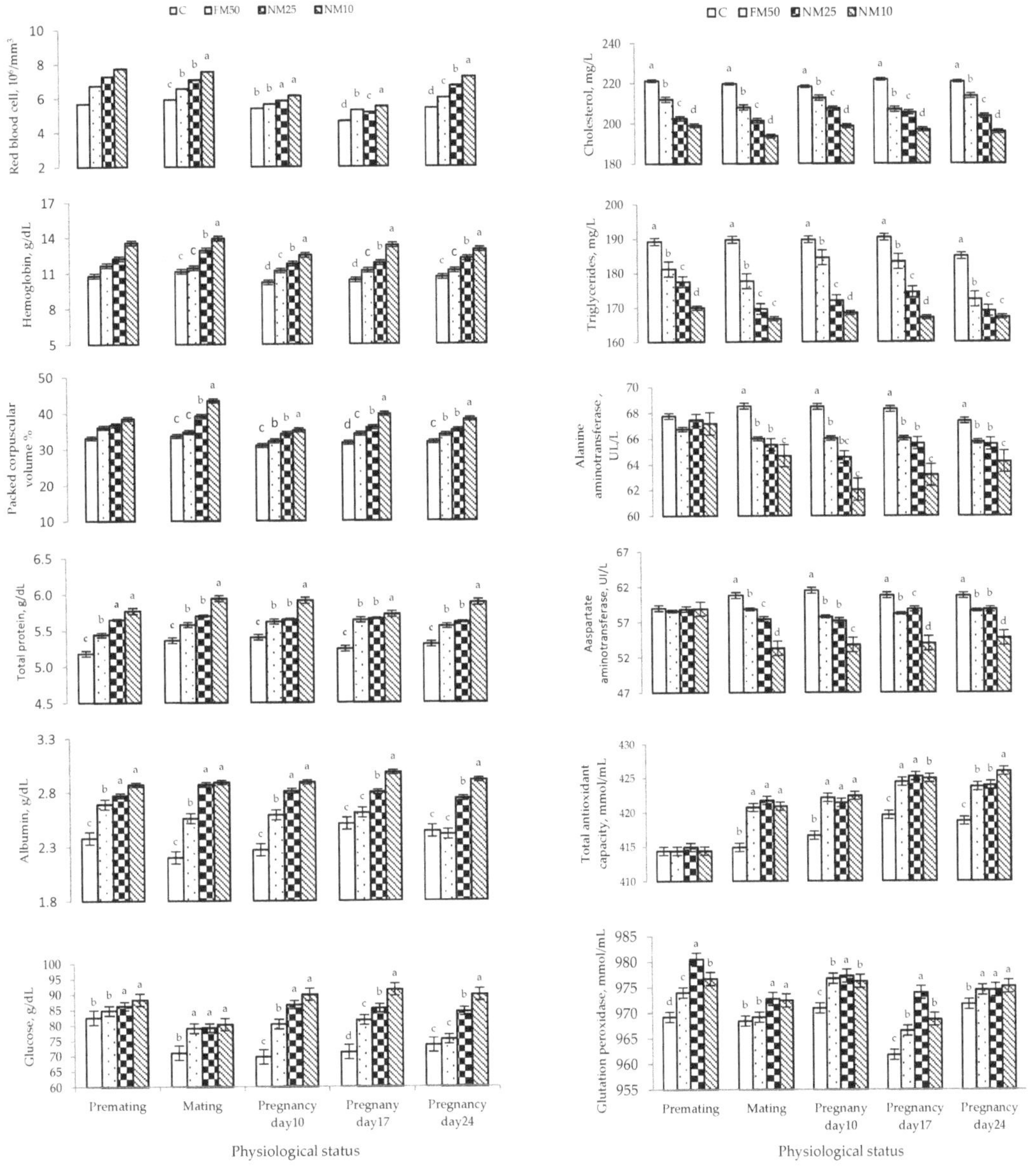

Figure 2. Changes (means ± SEM) in the blood plasma metabolites of rabbit does during the experimental period. C = 0 mg/kg BW MLEE, FM_{50} = 50 mg/kg BW MLEE, NM_{25} = 25 mg/kg BW nanoencapsulated MLEE and NM_{10} = 10 mg/kg BW nanoencapsulated MLEE. Means within the same physiological status having different superscripts (a, b, c, d) differ significantly ($p < 0.05$).

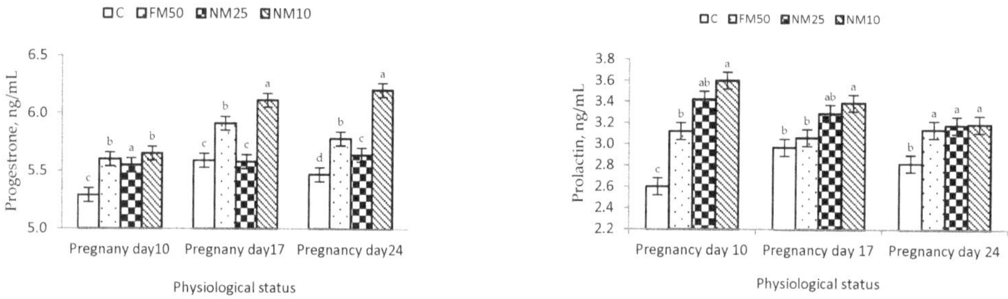

Figure 3. Changes (means ± SEM) in the blood plasma progesterone and prolactin of rabbit does during pregnancy. C = 0 mg/kg BW MLEE, FM_{50} = 50 mg/kg BW MLEE, NM_{25} = 25 mg/kg BW nanoencapsulated MLEE and NM_{10} = 10 mg/kg BW nanoencapsulated MLEE. Means within the same physiological status having different superscripts (a, b, c, d) differ significantly ($p < 0.05$).

Table 7. Effect of the nanoencapsulated *Moringa* leaf ethanolic extract (MLEE) on the reproductive performance of female rabbits during the experimental period (Mean ± SEM).

Variable	Treatment [1] (n = 14/Group)				SEM	p-Value
	C	FM_{50}	NM_{25}	NM_{10}		
Kindling rate, %	71.40 ^{c} (10/14)	92.85 ^{b} (13/14)	92.85 ^{b} (13/14)	100 ^{a} (14/14)	-	0.035
Litter size at birth	6.00 ^{b}	6.07 ^{b}	7.21 ^{a}	7.57 ^{a}	1.30	0.001
No. live litter sizes	5.64 ^{b}	6.00 ^{b}	7.07 ^{a}	7.36 ^{a}	1.13	<0.001
No. dead litter sizes	0.36	0.07	0.14	0.21	0.44	0.750
Litter weight at birth, g	258.18 ^{b}	293.77 ^{b}	316.43 ^{b}	393.50 ^{a}	74.93	<0.001

[1] C = 0 mg/kg BW MLEE, FM_{50} = 50 mg/kg BW MLEE, NM_{25} = 25 mg/kg BW nanoencapsulated MLEE and NM_{10} = 10 mg/kg BW nanoencapsulated MLEE. Means within the same physiological status having different superscripts (a, b, c) differ significantly ($p < 0.05$).

4. Discussion

This study is the first devoted to innovating a phytogenic feed additive using nanotechnology approaches in the livestock field. The development of encapsulation techniques facilitates the protection, as well as the controlled and targeted release, of bioactive molecules applied in the pharmaceutical, nutraceutical and food industries to improve their bioavailability (absorption and cellular intake) and to enhance the stability of bioactive compounds during processing and storage processes [14,15]. For this purpose, many natural polymers are used to encapsulate different bioactive components, including phytogenic crude extracts, used as natural-functioning nutritional supplements with health benefits [17,26]. This study used the nanoencapsulation ionic-gelation method for innovating a new feed additive that could be used in the rabbit industry during periods of heat stress, aiming at the phytochemicals of *Moringa* leaf extracts, specifically polyphenols. Based on several previous studies, a *Moringa* leaf extract has several phytogenic bioactive compounds with antioxidant, antimicrobial and immunomodulatory activities that can improve animals' reproductive performance and health [10,11,13].

In this study we used sodium alginate as a natural polymer for the nanoencapsulation process to ensure the safety of the final product to animal and human health. This natural polymer has been used as an efficient coating material with efficient protection ability due to its high encapsulation efficiency for active components of plant extracts, mainly phenolic compounds [27]. Alginate is a natural anionic polyelectrolyte polymer that encompasses unbranched binary copolymers of (1–4) linked D-mannuronic acid (M) and L-guluronic acid (G) residues with widely varying composition. Alginate is extracted from different species of brown algae (*Phaeophyceae*). Based on the unique physicochemical properties (biodegradability, biocompatibility and capability of forming three-dimensional gels in the presence of divalent cations such as CaCl2), low cost and simplicity of use, this polymer is considered one of the suitable choice materials for the encapsulation process [18].

The encapsulation technique used in this study was efficient to entrap approximately 57% of the bioactive compounds of the *Moringa* leaf extract, as indicated by the encapsulation efficiency of the *Moringa* extract phenolic compounds. The ionic-gelation method is an efficient and low-cost encapsulation technique that does not require specialized equipment, high temperature or organic solvents, making it suitable for encapsulating hydrophobic or hydrophilic compounds[15].

In this study, the conjugation of MLEE with alginate–$CaCl_2$ nanoparticles allowed an 80% reduction in dose with satisfactory positive effects on the reproductive performance of female rabbits. These results agree with those obtained in previous studies, aiming to encapsulate active phytogenic compounds, mainly polyphenols. This method was rapid and easily adapted to the industrial scale [28]. For example, the aqueous leaf extract of *Stevia rebaudiana Bertoni*, which is rich in phenolic compounds, was successfully encapsulated in $CaCl_2$ beads and showed high encapsulation efficiency (>60%) [18,28] values, as well as satisfactory antioxidant storage stability. Furthermore, Calvo et al. [29] found that using alginate–$CaCl_2$ as a nanocarrier for antioxidant compounds (betacyanin and polyphenols) derived from beetroot industrial wastes was efficient in encapsulating (between 20% and 40%) these active components, with good conservation of the antioxidant activity (up to 70%).

These findings support the relevance of the technique used to fabricate MLEE-conjugated alginate–$CaCl_2$ nanoparticles with physicochemical characteristics (<100 nm), allowing better bioavailability for target sites.

For oral delivery into the GIT (*oral pathway*), particle uptake in the GIT depends on diffusion and accessibility through mucus and contact with the cells of the GIT. The smaller particle diameter is fast diffused through GIT mucus to reach intestinal lining cells, followed by uptake through the GIT barrier to reach the blood [30]. Existing evidence indicates that particles smaller than 100 nm are absorbed in various tissues and organs. Smaller particles capable of being taken up by the villus epithelium may directly enter the bloodstream and then be predominantly scavenged by the liver and the spleen [31].

Heat stress induces various biological reactions and behavioral changes to cope with high ambient temperature and maintain thermal homeostasis [3,14]. Under heat stress, female rabbits express a high respiratory rate and water intake and low feed intake as adaptive mechanisms for high ambient temperature, which may negatively affect rabbit does' reproductive performance if maintained for a long time [32,33]. For example, exposure of New Zealand rabbits to 41 °C led to an 18% decrease in RBC count; 20%, hemoglobin content; 22%, blood platelet count; 11.2%, total protein; 24%, albumin, and 21%, globulin [4]. Under intensive production systems, as in most rabbit farms, these biological and behavioral responses could be more challenging when animals are housed in cages rather than in natural environments [34]. Furthermore, heat stress may be a threat to females, specifically during sensitive reproductive windows, such as mating and pregnancy periods. Female rabbits are sensitive to heat stress, which is considered an important factor influencing their reproduction, fertility and physiological traits [4,35]. Overall, heat stress and accompanying elevated oxidative stress increase the risk of spontaneous abortion and reduce milk production; litter size and litter performance; and the longevity, welfare, and health status of females.

This study was conducted during the summer, when THI was 29.20, which is classified as severe heat stress for rabbits. Notably, MLEE significantly reduced the rectal temperature of female rabbits in the treated groups. Treatment with MLEE in either free or encapsulated form around mating time and pregnancy reduced heat-stress-related indicators, such as rectal temperatures, respiratory rates and heart rates. It also improved hematochemical attributes (RBC, Hb, PCV, total protein, albumin and glucose), redox status (TAC and GSH-Px) and hormones (progesterone and prolactin) and decreased cholesterol, triglycerides, ALT and AST, suggesting an improvement in the heat tolerance of the animals.

These findings support the protective role of MLEE against the negative impacts of heat stress. Several mechanisms could mediate these effects, which could be due to several biologically active phytogenics in MLEE [11,13,36]. Notably, the enhancements in physiological events in MLEE-treated groups were not associated with low feed intake as one of the adaptive behavioral mechanisms for heat stress. This effect is essential for animals during their reproductive cycles, especially mating and lactation, to maintain adequate performance. Under heat stress, animals decrease feed intake to reduce metabolic heat production, leading to changes in energy balance and nutrient availability and affecting reproductive cyclicity, pregnancy and fetal development [37,38]. Given that active components of MLEE, N-acetylneuraminic acid 2,3-dehydro-2-deoxy- (6.50%) were identified in our extract, this component is a sialylated glycoprotein that has prebiotic properties, promotes neurodevelopment and boosts immune function and gut maturation [39,40]. Additionally, our MLEE has many active components with antimicrobial/anticoccidial activity, including salinomycin [41], 2- furoic acid [42], 9S,11,15 S-trihydroxythrombox-13E-enoic acid [43], rifabutin [44] and lactacystin [45]. Furthermore, active components can be used directly as valuable nutrients or to improve digestion (glycine, aminomethyl propanol, glycan le-a trisaccharide, N-acetyl leucyl-leucyl-methioninal and deoxycholic acid) or boost the energy status of animals (adenosine 5′-diphosphate and thiocyanic acid, L-arabinitol and D-gluconic acid) [46–49]. These components in MLEE explain the positive effects of the extract on digestion, the gut-intestinal microbiota ecosystem, nutrient availability and thus, the maintenance of the body weight of MLEE-treated does during pregnancy. These findings agree with that previously reported for *Moringa* plants having several nutrients that stimulate growth and increase the bioavailability of the nutrients and feed use, such as high-quality protein, vitamins, minerals, antioxidants and cytokine-type hormone antioxidant components, particularly vitamin C [12,13,50].

In this study, compared with the control, all MLEE-based treatments significantly increased hematological variables, including RBC and hemoglobin. The administration of MLEE improved redox status (higher TAC and GSH-Px), metabolism (higher energy-yielding nutrients: glucose; protein: total protein and albumin) and liver functions (ALT and AST) of does during mating and pregnancy.

Hemoglobin plays a vital role in carrying approximately 98% of oxygen throughout the animal body system, whereas PCV measures the proportion of blood made up of cells [51]. These findings might explain the improved heat-stress tolerance and maintain the rectal temperature of MLLE-treated does, without changes in respiration rates and heart rates by improving the oxygen delivery to different organs [4]. Furthermore, improved metabolism and the redox status of MLEE-treated does play vital roles in maintaining homeothermy during heat stress through different mechanisms. For example, albumin has osmoprotective properties in plasma, which regulates water balance and maintains protein and enzyme stability [52]. Moreover, glucose availability provides an easy energy source to all body cells without the need for sophisticated catabolic and/or anabolic processes that can be combined with the rise of heat, thus increasing heat stress [11]. The enhancements of the hematobiochemical attributes and redox status of MLEE-treated does could be attributed to the presence of some specific active components that stimulate the synthesis of some proteins/enzymes. For example, the increase in erythrocytes and hemoglobin concentration may be related to the effect of amino acids, vitamins [36,52] and minerals, particularly iron. Furthermore, a compound, such as ethidimuron, identified in our MLEE, acts as a precursor of intracellular glutathione. This is in line with some components' free radical scavenging activity, such as diapocynin [53]. These results explain the improved redox status of MLEE-treated does in our study.

Rabbit does are sensitive to heat stress, which is considered to be an important factor influencing their fertility and has negative effects on their reproductive and physiological traits [3,32,54,55]. In hot climates, the breeding of rabbits is stopped in most rabbit farms due to low reproductive performance and associated health problems. In this study, MLEE-treated does showed pronounced enhancements in sex hormones and reproductive performance compared with the control. These enhancements could be ascribed to the improved metabolism and health status of does around mating time and during pregnancy, as discussed earlier. In this context, some active components in MLEE positively affect specific reproductive events and functions. Sphingolipids, such as phytosphingosine (assembling 10.46% in our MLEE), are signaling molecules that regulate various physiological activities. In a study on pigs, exogenous phytosphingosine-1-phosphate (P1P) administration influenced animal reproduction by increasing porcine oocyte maturation and preimplantation embryo development through the regulation of oxidative stress and apoptosis signaling pathways [56]. P1P treatment upregulated the gene expression involved in cumulus expansion (Has2 and epidermal growth factor), antioxidant enzymes (superoxide dismutase and catalase) and developmental competence (octamer-binding transcription factor 4 (Oct4)) while activating extracellular signal-regulated kinase1/2 and (serine/threonine) protein kinase B signaling. P1P treatment also influenced oocyte survival by shifting the ratio of B-cell lymphoma 2 to Bax while inactivating (C-Jun N-terminal kinase) signaling. Other components, such as phosphatidylethanolamine (phospholipid depravities) and palmitic acid, play a vital role in cell membrane integrity. They act as structural components in a bilayer cell membrane [57,58]. Furthermore, nailamaid with an antidepressant activity[59]; folinic acid with antianemia, anticancer, anti-inflammatory, antitoxic activities[60]; and leukotriene E4 methyl ester [61] with anti-inflammatory and immunomodulatory effects were identified in MLEE. These components might attenuate the harmful effects of some inflammatory factors' production, such as histamine and prostaglandins, which might negatively affect reproductive performance [62–64]. These could be confirmed in our study, as kindling rates and litter sizes were significantly higher in MLEE-treated does. The enhancements in these two variables reflect the enhancement in oocyte quality, the fertilization rate and embryo survival through improved cell membrane integrity.

Finally, MLEE exerted positive effects on progesterone and prolactin levels during pregnancy. These results partially follow those obtained by Ajuogu et al. [65]. The addition of a 15 mg/kg leaf powder of *Moringa* significantly improved progesterone concentrations but did not affect prolactin concentrations during pregnancy. The authors suggested that *Moringa* leaves may contain active components that can directly act on the uterus and ovary, interfering with the release of prostaglandins [65]. The high level of progesterone during the first half of pregnancy plays a crucial role in preparing the uterus for embryo implantation by suppressing uterine motility/contractions during the first few days of pregnancy, thus increasing the opportunity for pregnancy maintenance [24]. In rabbits, the levels of prolactin hormone increase from 3–4 d of gestation and remain elevated for 15–21 d of gestation. Such an increase is essential to maintain adequate concentrations of progesterone during pregnancy [66].

Notably, the weights of litters born for encapsulated MLEE-treated does were higher than those recorded for control and nonencapsulated MLEE-treated does. This may refer to the higher bioavailability of MLEE components to fetuses through improved transfer via the placenta, as nanoencapsulation may facilitate the transfer of active components, specifically those with high molecular weight and low solubility, through the fetal–placental circulating system[67].

It is worth note that the low dose of nanoencapsulated MLEE (NM_{10}) resulted in better responses, as indicated by most determined variables, than the high dose of nanoencapsulated MLEE (NM_{25}). This could be due to the improved bioavailability of some active components, which, though it is an advantage of the nanoencapsulation process may interfere with the competence of some biological processes. For example, MLEE is rich with phthalate, which has endocrine-disrupting effects [11]. Moreover, increased antioxidant nutrients availability, mainly phenolic compounds and vitamins, can act as prooxidants by increasing oxidative stress [6,68].

5. Conclusions

This study confirms the positive role of MLEE as a supplement in heat-stress tolerance, metabolism and the reproductive performance of rabbit does bred under natural heat-stress conditions. The positive effects were due to the enrichment of MLEE with an impressive range of active components that can mend negative impacts of heat stress by improving digestion and feed use, energy and redox status, and hormonal balance during pregnancy. These effects have been seen whether MLLE was in free or in encapsulated form. However, the use of the nanoencapsulated form allowed an 80% reduction (10 mg/kg BW) in the optimal dose (50 mg/kg BW) without affecting the efficiency of the treatment. These results support the importance of nanoencapsulation technology in improving the bioavailability of active components when they are orally administered.

Author Contributions: Conceptualization, N.I.E.-D. and N.M.H.; methodology, N.I.E.-D. and N.M.H.; validation, N.I.E.-D. and N.M.H. formal analysis, N.M.H.; investigation, N.I.E.-D. and N.M.H.; data curation, N.I.E.-D. and N.M.H.; writing—original draft preparation, N.I.E.-D. and N.M.H.; writing—review and editing, N.M.H. and A.G.-B.; supervision, N.M.H., Z.R.A.-E. and A.G.E. All authors have read and agreed to the published version of the manuscript.

Funding: This research received no external funding.

Institutional Review Board Statement: The experimental procedures were previously assessed and approved by Alexandria University-Institutional Animal Care and Use Committee (AU-IACUC, AU 08 21 07 26 2 82).

Informed Consent Statement: Not applicable.

Data Availability Statement: The data presented in this study are available on request from the corresponding author. The data are not publicly available because of privacy.

Conflicts of Interest: The authors declare no conflict of interest.

References

1. Oseni, S.; Lukefahr, S. Rabbit production in low-input systems in Africa: Situation, knowledge and perspectives–A review. *World Rabbit Sci.* **2014**, *22*, 147–160. [CrossRef]
2. Hashem, N.; Abd El-Hady, A.; Hassan, O. Effect of vitamin E or propolis supplementation on semen quality, oxidative status and hemato-biochemical changes of rabbit bucks during hot season. *Livest. Sci.* **2013**, *157*, 520–526. [CrossRef]
3. Mady, E.; Karousa, M.; El-laithy, S.; Ahmed, S. Effect of Season on New Zealand White (NZW) Rabbits' Behaviorand Reproductive and productivePerformance. *Benha Vet. Med. J.* **2018**, *35*, 274–284. [CrossRef]
4. Mutwedu, V.; Nyongesa, A.; Oduma, J.; Kitaa, J.; Mbaria, J. Thermal stress causes oxidative stress and physiological changes in female rabbits. *J. Therm. Biol.* **2020**, *95*, 102780. [CrossRef] [PubMed]
5. Adesuyi, A.; Elumm, I.; Adaramola, F.; Nwokocha, A. Nutritional and phytochemical screening of Garcinia kola. *Adv. J. Food Sci. Technol.* **2012**, *4*, 9–14.
6. Hashem, N.M.; Gonzalez-Bulnes, A.; Simal-Gandara, J. Polyphenols in Farm Animals: Source of Reproductive Gain or Waste? *Antioxidants* **2020**, *9*, 1023. [CrossRef] [PubMed]
7. Hashem, N.M.; Hassanein, E.M.; Simal-Gandara, J. Improving reproductive performance and health of mammals using honeybee products. *Antioxidants* **2021**, *10*, 336. [CrossRef]
8. Amaglo, N.K.; Bennett, R.N.; Curto, R.B.L.; Rosa, E.A.; Turco, V.L.; Giuffrida, A.; Curto, A.L.; Crea, F.; Timpo, G.M. Profiling selected phytochemicals and nutrients in different tissues of the multipurpose tree *Moringa oleifera* L., grown in Ghana. *Food Chem.* **2010**, *122*, 1047–1054. [CrossRef]
9. Mbikay, M. Therapeutic potential of *Moringa oleifera* leaves in chronic hyperglycemia and dyslipidemia: A review. *Front. Pharmacol.* **2012**, *3*, 24. [CrossRef]
10. Abu, A.; Ahemen, T. Testicular Morphometry and Sperm Quality of Rabbit Bucks Fed Graded Levels of *Moringa oleifera* Leaf Meal (MOLM). *Agrosearch* **2013**, *13*, 49–56. [CrossRef]
11. El-Desoky, N.; Hashem, N.; Elkomy, A.; Abo-Elezz, Z. Physiological response and semen quality of rabbit bucks supplemented with Moringa leaves ethanolic extract during summer season. *Animal* **2017**, *11*, 1549–1557. [CrossRef] [PubMed]
12. Hashem, N.; Abd El-Hady, A.; Hassan, O. Inclusion of phytogenic feed additives comparable to vitamin E in diet of growing rabbits: Effects on metabolism and growth. *Ann. Agric. Sci.* **2017**, *62*, 161–167. [CrossRef]
13. Hashem, N.; Soltan, Y.; El-Desoky, N.; Morsy, A.; Sallam, S. Effects of *Moringa oleifera* extracts and monensin on performance of growing rabbits. *Livest. Sci.* **2019**, *228*, 136–143. [CrossRef]

14. McClements, D.J. Encapsulation, protection, and release of hydrophilic active components: Potential and limitations of colloidal delivery systems. *Adv. Colloid Interface Sci.* **2015**, *219*, 27–53. [CrossRef]
15. Nedovic, V.; Kalusevic, A.; Manojlovic, V., Levic, S.; Bugarski, B. An overview of encapsulation technologies for food applications. *Procedia Food Sci.* **2011**, *1*, 1806–1815. [CrossRef]
16. Embuscado, M.E. Spices and herbs: Natural sources of antioxidants–a mini review. *J. Funct. Foods* **2015**, *18*, 811–819. [CrossRef]
17. Kurozawa, L.E.; Hubinger, M.D. Hydrophilic food compounds encapsulation by ionic gelation. *Curr. Opin. Food Sci.* **2017**, *15*, 50–55. [CrossRef]
18. Arriola, N.D.A.; de Medeiros, P.M.; Prudencio, E.S.; Müller, C.M.O.; Amboni, R.D.d.M.C. Encapsulation of aqueous leaf extract of Stevia rebaudiana Bertoni with sodium alginate and its impact on phenolic content. *Food Biosci.* **2016**, *13*, 32–40. [CrossRef]
19. Sánchez-Rangel, J.C.; Benavides, J.; Heredia, J.B.; Cisneros-Zevallos, L.; Jacobo-Velázquez, D.A. The Folin–Ciocalteu assay revisited: Improvement of its specificity for total phenolic content determination. *Anal. Methods* **2013**, *5*, 5990–5999. [CrossRef]
20. Estany, J.; Baselga, M.; Blasco, A.; Camacho, J. Mixed model methodology for the estimation of genetic response to selection in litter size of rabbits. *Livest. Prod. Sci.* **1989**, *21*, 67–75. [CrossRef]
21. National Research Council (NRC). *Nutrient Requirements of Rabbits*, 2nd ed.; National Academy Press: Washington, DC, USA, 1977; p. 24.
22. Helrich, K. AOAC official methods of analysis. *Va. Assoc. Off. Anal. Chem.* **1990**, *2*, 1058–1059.
23. Marai, I.; Habeeb, A.; Gad, A. Rabbits' productive, reproductive and physiological performance traits as affected by heat stress: A review. *Livest. Prod. Sci.* **2002**, *78*, 71–90. [CrossRef]
24. Hashem, N.; Aboul-Ezz, Z. Effects of a single administration of different gonadotropins on day 7 post-insemination on pregnancy outcomes of rabbit does. *Theriogenology* **2018**, *105*, 1–6. [CrossRef]
25. Hosny, N.S.; Hashem, N.M.; Morsy, A.S.; Abo-Elezz, Z.R. Effects of Organic Selenium on the Physiological Response, Blood Metabolites, Redox Status, Semen Quality, and Fertility of Rabbit Bucks Kept Under Natural Heat Stress Conditions. *Front. Vet. Sci.* **2020**, *7*, 290. [CrossRef]
26. Hashem, N.M.; Gonzalez-Bulnes, A. Nanotechnology and Reproductive Management of Farm Animals: Challenges and Advances. *Animals* **2021**, *11*, 1932. [CrossRef]
27. Najafi-Soulari, S.; Shekarchizadeh, H.; Kadivar, M. Encapsulation optimization of lemon balm antioxidants in calcium alginate hydrogels. *J. Biomater. Sci. Polym. Ed.* **2016**, *27*, 1631–1644. [CrossRef] [PubMed]
28. Arriola, N.D.A.; Chater, P.I.; Wilcox, M.; Lucini, L.; Rocchetti, G.; Dalmina, M.; Pearson, J.P.; Amboni, R.D.d.M.C. Encapsulation of stevia rebaudiana Bertoni aqueous crude extracts by ionic gelation–Effects of alginate blends and gelling solutions on the polyphenolic profile. *Food Chem.* **2019**, *275*, 123–134. [CrossRef] [PubMed]
29. Calvo, T.R.A.; Perullini, M.; Santagapita, P.R. Encapsulation of betacyanins and polyphenols extracted from leaves and stems of beetroot in Ca (II)-alginate beads: A structural study. *J. Food Eng.* **2018**, *235*, 32–40. [CrossRef]
30. Abdelnour, S.A.; Alagawany, M.; Hashem, N.M.; Farag, M.R.; Alghamdi, E.S.; Hassan, F.U.; Bila, R.M.; Elnesr, S.S.; Dawood, M.A.; Nagadi, S.A. Nanominerals: Fabrication methods, benefits and hazards, and their applications in ruminants with special reference to selenium and zinc nanoparticles. *Animals* **2021**, *11*, 1916. [CrossRef]
31. Thulasi, A.; Rajendran, D.; Jash, S.; Selvaraju, S.; Jose, V.L.; Velusamy, S.; Mathivanan, S. Nanobiotechnology in animal nutrition. In *Animal Nutrition and Reproductive Physiology (Recent Concepts)*, 1st ed.; Sampath, K.T., Ghosh, J., Eds.; Satish Serial Publishing House: New Delhi, India, 2013; pp. 499–516.
32. Menchetti, L.; Brecchia, G.; Canali, C.; Cardinali, R.; Polisca, A.; Zerani, M.; Boiti, C. Food restriction during pregnancy in rabbits: Effects on hormones and metabolites involved in energy homeostasis and metabolic programming. *Res. Vet. Sci.* **2015**, *98*, 7–12. [CrossRef] [PubMed]
33. Menchetti, L.; Andoni, E.; Barbato, O.; Canali, C.; Quattrone, A.; Vigo, D.; Codini, M.; Curone, G.; Brecchia, G. Energy homeostasis in rabbit does during pregnancy and pseudopregnancy. *Anim. Reprod. Sci.* **2020**, *218*, 106505. [CrossRef] [PubMed]
34. Mousa-Balabel, T.M.; El-Sheikh, R.A.; Moustafa, E. New strategies for controlling heat stress in New Zealand White (NZW) rabbits in Egypt. *Life Sci. J.* **2017**, *14*, 64–70.
35. Piles, M.; Tusell, L.; Rafel, O.; Ramon, J.; Sánchez, J. Effect of heat intensity and persistency on prolificacy and preweaning kit growth at different stages of the rabbit production cycle. *J. Anim. Sci.* **2013**, *91*, 633–643. [CrossRef] [PubMed]
36. Makkar, H.; Francis, G.; Becker, K. Bioactivity of phytochemicals in some lesser-known plants and their effects and potential applications in livestock and aquaculture production systems. *Animal* **2007**, *1*, 1371–1391. [CrossRef]
37. Sharaf, A.; El-Darawany, A.; Nasr, A.; Habeeb, A. Alleviation the negative effects of summer heat stress by adding selenium with vitamin E or AD3E vitamins mixture in drinking water of female rabbits. *Biol. Rhythm Res.* **2021**, *52*, 535–548. [CrossRef]
38. Kumar, S.; Pandey, A.; Rao, M.M.; Razzaque, W. Role of ß carotene/vitamin A in animal reproduction. *Vet. World* **2010**, *3*, 236.
39. Jahan, M.; Wynn, P.; Wang, B. Molecular characterization of the level of sialic acids N-acetylneuraminic acid, N-glycolylneuraminic acid, and ketodeoxynonulosonic acid in porcine milk during lactation. *J. Dairy Sci.* **2016**, *99*, 8431–8442. [CrossRef]
40. Mohammed, A.; Iyeghe-Erakpotobor, G.; Zahraddeen, D.; Barje, P. Growth and Reproductive Performance of Rabbit Does Fed *Moringa oleifera* Leaf Meal Based Diets Supplemented with Garlic, Ginger or Black Pepper. *J. Anim. Prod. Res.* **2019**, *31*, 74–87.
41. Mao, J.; Fan, S.; Ma, W.; Fan, P.; Wang, B.; Zhang, J.; Wang, H.; Tang, B.; Zhang, Q.; Yu, X. Roles of Wnt/β-catenin signaling in the gastric cancer stem cells proliferation and salinomycin treatment. *Cell Death Dis.* **2014**, *5*, e1039. [CrossRef] [PubMed]

42. Perez, H.I.; Manjarrez, N.; Solis, A.; Luna, H.; Ramirez, M.A.; Cassani, J. Microbial biocatalytic preparation of 2-furoic acid by oxidation of 2-furfuryl alcohol and 2-furanaldehyde with Nocardia corallina. *Afr. J. Biotechnol.* **2009**, *8*, 2279–2282.
43. Shah, S.; Xue, Q.; Tang, L.; Carney, J.R.; Betlach, M.; McDaniel, R. Cloning, characterization and heterologous expression of a polyketide synthase and P-450 oxidase involved in the biosynthesis of the antibiotic oleandomycin. *J. Antibiot.* **2000**, *53*, 502–508. [CrossRef]
44. Rockwood, N.; Cerrone, M.; Barber, M.; Hill, A.M.; Pozniak, A.L. Global access of rifabutin for the treatment of tuberculosis–why should we prioritize this? *J. Int. AIDS Soc.* **2019**, *22*, e25333. [CrossRef]
45. van Tijn, P.; Verhage, M.C.; Hobo, B.; van Leeuwen, F.W.; Fischer, D.F. Low levels of mutant ubiquitin are degraded by the proteasome in vivo. *J. Neurosci. Res.* **2010**, *88*, 2325–2337. [CrossRef]
46. Meléndez-Hevia, E.; de Paz-Lugo, P.; Cornish-Bowden, A.; Cárdenas, M.L. A weak link in metabolism: The metabolic capacity for glycine biosynthesis does not satisfy the need for collagen synthesis. *J. Biosci.* **2009**, *34*, 853–872. [CrossRef]
47. Bougie, F.; Iliuta, M.C. Sterically hindered amine-based absorbents for the removal of CO_2 from gas streams. *J. Chem. Eng. Data* **2012**, *57*, 635–669. [CrossRef]
48. Dutta, S.; Aoki, K.; Doungkamchan, K.; Tiemeyer, M.; Bovin, N.; Miller, D.J. Sulfated Lewis A trisaccharide on oviduct membrane glycoproteins binds bovine sperm and lengthens sperm lifespan. *J. Biol. Chem.* **2019**, *294*, 13445–13463. [CrossRef]
49. Murugappa, S.; Kunapuli, S.P. The role of ADP receptors in platelet function. *Front. Biosci.* **2006**, *11*, 1977–1986. [CrossRef]
50. Abo El-Haded, R.; El-Rahim, A.; El-Kerdawy, D. Impact of Substituting Soya Bean Meal by Moringa (*Moringa oleifera*) Leaves Meal in the Diet on Growth Performance, Nutrients Digestibility, Blood Constituents and Carcass Traits of Growing Rabbits. *J. Product. Dev.* **2017**, *22*, 635–656.
51. Jensen, F.B. The dual roles of red blood cells in tissue oxygen delivery: Oxygen carriers and regulators of local blood flow. *J. Exp. Biol.* **2009**, *212*, 3387–3393. [CrossRef] [PubMed]
52. Abdelnour, S.A.; Al-Gabri, N.A.; Hashem, N.M.; Gonzalez-Bulnes, A. Supplementation with Proline Improves Haemato-Biochemical and Reproductive Indicators in Male Rabbits Affected by Environmental Heat-Stress. *Animals* **2021**, *11*, 373. [CrossRef]
53. Dranka, B.P.; Gifford, A.; Ghosh, A.; Zielonka, J.; Joseph, J.; Kanthasamy, A.G.; Kalyanaraman, B. Diapocynin prevents early Parkinson's disease symptoms in the leucine-rich repeat kinase 2 (LRRK2R1441G) transgenic mouse. *Neurosci. Lett.* **2013**, *549*, 57–62. [CrossRef]
54. Askar, A.; Ismail, E.I. Impact of heat stress exposure on some reproductive and physiological traits of rabbit does. *Egypt. J. Anim. Prod* **2012**, *49*, 151–159.
55. Habeeb, A.A.M.; Osman, S.F.; Gad, A.E. Signs of heat stress and some steps to reduce the negative effects on animals. *GSC Adv. Res. Rev.* **2020**, *4*, 046–058.
56. Park, K.M.; Wang, J.W.; Yoo, Y.M.; Choi, M.J.; Hwang, K.C.; Jeung, E.B.; Jeong, Y.W.; Hwang, W.S. Sphingosine-1-phosphate (S1P) analog phytosphingosine-1-phosphate (P1P) improves the in vitro maturation efficiency of porcine oocytes via regulation of oxidative stress and apoptosis. *Mol. Reprod. Dev.* **2019**, *86*, 1705–1719. [CrossRef]
57. Min, H.; Youn, E.; Kim, J.; Son, S.Y.; Lee, C.H.; Shim, Y.-H. Effects of Phosphoethanolamine Supplementation on Mitochondrial Activity and Lipogenesis in a Caffeine Ingestion Caenorhabditis elegans Model. *Nutrients* **2020**, *12*, 3348. [CrossRef] [PubMed]
58. Anneken, K.; Fischera, M.; Evers, S.; Kloska, S.; Husstedt, I.-W. Recurrent vacuolar myelopathy in HIV infection. *J. Infect.* **2006**, *52*, e181–e183. [CrossRef]
59. Wayment, H.K.; Schenk, J.O.; Sorg, B.A. Characterization of extracellular dopamine clearance in the medial prefrontal cortex: Role of monoamine uptake and monoamine oxidase inhibition. *J. Neurosci.* **2001**, *21*, 35–44. [CrossRef] [PubMed]
60. Shea, B.; Swinden, M.V.; Ghogomu, E.T.; Ortiz, Z.; Katchamart, W.; Rader, T.; Bombardier, C.; Wells, G.A.; Tugwell, P. Folic acid and folinic acid for reducing side effects in patients receiving methotrexate for rheumatoid arthritis. *J. Rheumatol* **2014**, *41*, 1049–1060. [CrossRef] [PubMed]
61. Scott, J.P.; Peters-Golden, M. Antileukotriene agents for the treatment of lung disease. *Am. J. Respir. Crit. Care Med.* **2013**, *188*, 538–544. [CrossRef]
62. Coppin, J.P.; Xu, Y.; Chen, H.; Pan, M.-H.; Ho, C.-T.; Juliani, R.; Simon, J.E.; Wu, Q. Determination of flavonoids by LC/MS and anti-inflammatory activity in *Moringa oleifera*. *J. Funct. Foods* **2013**, *5*, 1892–1899. [CrossRef]
63. Das, S.; Parida, U.K.; Bindhani, B.K. Green biosynthesis of silver nanoparticles using *Moringa oleifera* L. leaf. *Int. J. Nanotechnol. Appl.* **2013**, *3*, 51–62.
64. Leone, A.; Spada, A.; Battezzati, A.; Schiraldi, A.; Aristil, J.; Bertoli, S. Cultivation, genetic, ethnopharmacology, phytochemistry and pharmacology of *Moringa oleifera* leaves: An overview. *Int. J. Mol. Sci.* **2015**, *16*, 12791–12835. [CrossRef]
65. Ajuogu, P.K.; Mgbere, O.O.; Bila, D.S.; McFarlane, J.R. Hormonal changes, semen quality and variance in reproductive activity outcomes of post pubertal rabbits fed *Moringa oleifera* Lam. leaf powder. *J. Ethnopharmacol.* **2019**, *233*, 80–86. [CrossRef]
66. Nagwa, A.; Azoz, A.; Amina, F.; El-Shafie, M. Hormonal Changing during Pregnancy and Lactation of New Zealand White Female Rabbits. *Egypt. J. Anim. Prod.* **2004**, *41*, 501–513.

67. Hashem, N.M.; Gonzalez-Bulnes, A. State-of-the-art and prospective of nanotechnologies for smart reproductive management of farm animals. *Animals* **2020**, *10*, 840. [CrossRef]
68. Abo-Elsoud, M.A.; Hashem, N.M.; Nour El-Din, A.N.M.; Kamel, K.I.; Hassan, G.A. Soybean isoflavone affects in rabbits: Effects on metabolism, antioxidant capacity, hormonal balance and reproductive performance. *Anim. Reprod. Sci.* **2019**, *203*, 52–60. [CrossRef]

antioxidants

MDPI

Article

Exposure to Salinity and Light Spectra Regulates Glucosinolates, Phenolics, and Antioxidant Capacity of *Brassica carinata* L. Microgreens

Sylvia Maina [1,2,†], Da Hye Ryu [1,3,†], Jwa Yeong Cho [1,3], Da Seul Jung [1], Jai-Eok Park [1], Chu Won Nho [1,3], Gaymary Bakari [2], Gerald Misinzo [2], Je Hyeong Jung [1], Seung-Hoon Yang [4] and Ho-Youn Kim [1,*]

1 Smart Farm Research Center, Korea Institute of Science and Technology (KIST), Gangneung 25451, Korea; wairimusylvia@kist.re.kr (S.M.); dahye0507@kist.re.kr (D.H.R.); chocho7023@kist.re.kr (J.Y.C.); 118521@kist.re.kr (D.S.J.); j-park@kist.re.kr (J.-E.P.); cwnho@kist.re.kr (C.W.N.); jhjung@kist.re.kr (J.H.J.)
2 SACIDS Foundation for One Health, Sokoine University of Agriculture, Morogoro 25523, Tanzania; gaymary.bakari@sua.ac.tz (G.B.); gerald.misinzo@sacids.org (G.M.)
3 Division of Bio-Medical Science and Technology, KIST School, University of Science and Technology (UST), Daejeon 34113, Korea
4 Department of Medical Biotechnology, College of Life Science and Biotechnology, Dongguk University, Seoul 04620, Korea; shyang@dongguk.edu
* Correspondence: hykim@kist.re.kr; Tel.: +82-33-650-3580
† Authors contributed equally to this manuscript.

Citation: Maina, S.; Ryu, D.H.; Cho, J.Y.; Jung, D.S.; Park, J.-E.; Nho, C.W.; Bakari, G.; Misinzo, G.; Jung, J.H.; Yang, S.-H.; et al. Exposure to Salinity and Light Spectra Regulates Glucosinolates, Phenolics, and Antioxidant Capacity of *Brassica carinata* L. Microgreens. *Antioxidants* **2021**, *10*, 1183. https://doi.org/10.3390/antiox10081183

Academic Editors: Irene Dini and Domenico Montesano

Received: 30 June 2021
Accepted: 20 July 2021
Published: 26 July 2021

Abstract: The effect of salt treatment on *Brassica carinata* (BC) microgreens grown under different light wavelengths on glucosinolates (GLs) and phenolic compounds were evaluated. Quantifiable GLs were identified using ultra-high performance-quadrupole time of flight mass spectrometry. Extracts' ability to activate antioxidant enzymes (superoxide dismutase (SOD) and catalase (CAT)) was evaluated on human colorectal carcinoma cells (HCT116). Furthermore, BC compounds' ability to activate expression of nuclear transcription factor-erythroid 2 related factor (Nrf2) and heme-oxygenase-1 (HO-1) proteins was examined using specific antibodies on HCT116 cells. Sinigrin (SIN) was the abundant GLs of the six compounds identified and its content together with total aliphatic GLs increased in saline conditions. Fluorescent (FL) and blue plus red (B1R1) lights were identified as stable cultivation conditions for microgreens, promoting biomass and glucobrassicin contents, whereas other identified individual and total indole GLs behaved differently in saline and non-saline environments. Blue light-emitting diodes and FL light in saline treatments mostly enhanced SIN, phenolics and antioxidant activities. The increased SOD and CAT activities render the BC microgreens suitable for lowering oxidative stress. Additionally, activation of Nrf2, and HO-1 protein expression by the GLs rich extracts, demonstrate their potential to treat and prevent oxidative stress and inflammatory disorders. Therefore, effective salt treatments and light exposure to BC microgreens present an opportunity for targeted regulation of growth and accumulation of bioactive metabolites.

Keywords: *Brassicaceae*; light wavelength; bioactive compounds; reactive oxygen species (ROS); oxidative stress; antioxidant enzymes; antioxidant proteins

1. Introduction

Members of the family *Brassicaceae* have recently gained interest as nutraceutical foods and as a source of natural bioactive compounds, including phenolics and glucosinolates (GLs) [1]. Among these plants is Ethiopian mustard (*Brassica carinata* A. Braun), an orphan crop that originates from the highlands of Ethiopia, where it is known as "Gomenzer" (Yehabesha Gomen) and "Hamli Adri" in Amharic and Tigrigna languages [2]. *B. carinata* usefulness as an oilseed crop and as a vegetable has caused its cultivation to spread to other arid and semi areas of the world [3]. Also, the newly developed consumer-friendly varieties of Ethiopian mustard have caused increased acceptability of this nutritious vegetable [2].

The leaves and seeds of *B. carinata* are rich in nutrients, including proteins, carbohydrates, vitamins, and carotenoids [4,5]. Moreover, the vegetable has been reported to contain polyphenols and high contents of GLs, (such as sinigrin (SIN)) whose levels are approximated to be several folds higher compared to cabbage, broccoli, Chinese cabbage, and Korean leaf mustard [2,6]. These compounds owe most cruciferous vegetables their health-promoting potential with GLs hydrolysis products being associated with cancer prevention, antimicrobial, anti-inflammatory and polyphenols with antioxidant activities [7–9].

Polyphenols and GLs have been reported in several studies to be important natural antioxidants during oxidative stress to control the production of reactive oxidants such as reactive oxygen species (ROS) and reactive nitrogen species (RNS) [10]. The controlled production of the oxidants maintains their balanced concentration, which would otherwise cause them to nab cellular electrons resulting in chronic diseases such as diabetes, cancers and cardiovascular ailments [11,12]. Antioxidants also activate expression of phase II detoxifying proteins ((such as heme-oxygenase-1 (HO-1) through the activation of nuclear transcription factor-erythroid 2 related factor (Nrf2)); and antioxidant enzymes ((including superoxide dismutase (SOD), glutathione peroxidase (GPx) and catalase (CAT)). The antioxidant enzymes make it easier for the antioxidants to donate electrons and also, they assist in the recycling process by enabling reduction reactions of oxidized antioxidants [10]. Nrf2 pathway activation inhibits the progression of inflammation while expression of Nrf2 mediated antioxidant gene reduces the production of ROS [13].

In the majority of *Brassicaceae* species including kale, radish, and broccoli, higher contents of GLs and phenolic bioactive compounds are reported in juvenile stages of plants compared to mature plants [1,14,15], where the compounds decrease over time as a result of tissue expansion. The rapid changes in the compounds profiles during germination and early growth of vegetables make early harvesting a particularly relevant factor for maximizing the concentration of these desirable bioactive compounds [1,14]. This has contributed to the rising popularity of young vegetables (sprouts and microgreens), as a source of both health-promoting compounds and as functional foods [15–17].

During cultivation, elicitors possess the ability to induce changes that activate plant signaling pathways and enhance the production of specific phytochemicals [18,19]. Sodium chloride (NaCl) salt stress for instance causes physiological and biochemical perturbations in plants providing a strategy to increase GLs biosynthesis [20]. In particular, salt stress demonstrated to be effective in enhancing GLs in broccoli, kale, and radish [21] and thus has shown potential industrial applications [22,23].

Elicitation and production of healthier vegetables can be done stably in controlled environments, such as indoor vertical farm closed cultivation systems with regulated conditions of light irradiation, temperature, humidity, nutrients, and carbon dioxide [24–26]. Such farms are emerging as promising systems for the production of plant-derived medical ingredients and functional foods [27–29]. In these farming systems, light serves as an important factor due to its role in the plant's development, growth, morphogenesis, and biosynthesis of pigment [30]. In particular, light-emitting diodes (LEDs) are mainly used because of their high energy conservation efficiency, wavelength specificity, and adjustable light quality and intensity [31]. The responses exhibited by plants on exposure to monochromatic and combined lights are associated with distinct activation of phytochromes and photoreceptors that induce metabolic changes [32] therefore, some LED wavelengths are used to accumulate specific compounds and nutrients of interest in plants [33]. For instance, red (R) and blue (B) LEDs have been used in *Brassicaceae* species to increase anthocyanin, carotenoids, GLs, lutein, vitamin, soluble sugar, soluble protein, and polyphenolic contents [25,34–37].

Despite the perceived health importance of *B. carinata* vegetables from the reported available phytochemicals, and their applicability in a closed cultivation system, to date, this scientific information is lacking. It is, therefore, necessary to investigate the occurrence of the bioactive compounds, the health benefits, and conditions for enhancing the growth and production of high quality and phytochemical-rich microgreens. Thus, this study aimed to identify GLs profiles, determine the phenolic compound contents and assess the biological activities of *B. carinata* microgreens treated with salt stress and cultivated under different light wavelengths in an indoor closed system.

2. Materials and Methods

2.1. Plant Materials and Growth Conditions

Seeds of *B. carinata* obtained from the Kenya Resource Centre for Indigenous Knowledge, National Museums of Kenya, were cultivated in a vertical indoor farming system built at the SMART u-FARM at the Korea Institute of Science and Technology (Gangneung, South Korea). Before sowing, the seeds were washed and placed in a tray (30 cm × 20 cm × 5 cm) with a mesh plate. Plants were cultivated at 18–26 °C and relative humidity of 50–80% under closed and controlled cultivation conditions. Plants were exposed to 200 ± 11 μmol/m^{2}s of light intensity from fluorescent lamps (TL-D 18W/865; Philips, Amsterdam, The Netherlands) at 25 cm under a 14 h/10 h light/dark cycle for seven days after sowing (DAS) and then treated with elicitors. The plants were harvested at 14 DAS and their fresh biomass was weighed. The *B. carinata* microgreens were frozen in liquid nitrogen and stored at −80 °C before freeze-drying for five days to analyze chemical constituents and bioactivities of interest. The experiment was a randomized complete block design and was replicated thrice.

2.2. Elicitor Treatment and Sampling

After 7 DAS, *B. carinata* uniformly germinated plants were supplied with a nutrient solution and exposed to eight treatment of combinations of light wavelength and saline/non-saline solutions. The nutrient solution had an electrical conductivity of 1.2 ± 0.05 dS/m and contained 354 ppm N, 186 ppm P, 420 ppm K, 230 ppm Ca, 13 ppm Mg, and 83 ppm S (macronutrients), and 2.80 ppm Fe, 0.32 ppm B, 0.77 ppm Mn, 0.04 ppm Cu, 0.02 ppm Zn, and 0.02 ppm Mo (micronutrients). Fluorescent lamps (TL-D 18W/865; Philips Electronics, Seoul, South Korea) and LED lamps (KLB-40-2C; KAST Engineering, Gumi, South Korea) with a light intensity of 85 ± 11 μmol/m^{2}s were used as artificial light sources at 30 cm. The light conditions included fluorescent (FL)- the control, B LED (440 nm + warm white), R LED (660 nm), and B plus R LED at a 1:1 ratio (B1R1) for seven days. Additionally, a 100 mM NaCl solution (this concentration was selected based on a literature review [23,38]) was added to the nutrient solution and supplied as an elicitor for three days before the plants were harvested. At 14 DAS, 30 microgreens were randomly selected, collected, and combined to form each of the three replicates used in further processing and analysis.

2.3. Sample Extraction and Desulfation of Standards

Extracts of desulfated GLs were prepared following previously published protocols [21,39], with minor modifications. Freeze-dried sample (0.05 g) was immersed in 70% methanol (1 mL) and heated at 90 °C in a heating block for 30 min. After cooling, the samples were centrifuged at 2063× *g*, 4 °C for 15 min and the extraction process was repeated one more time after which 1.2 mL of the supernatant was transferred to new Falcon tubes containing 0.1 mM glucotropaeolin (GTR) (20 μL) as an internal standard and reacted with 0.15 mL of lead acetate and barium acetate mixture (1:1, *v/v*), then centrifuged at 13,475× *g*, 4 °C for 5 min. One milliliter of the supernatant was loaded onto a mini-column packed with diethyl-aminoethyl Sephadex A-25 anion exchange resin, which had been pre-activated with 0.1 M sodium acetate and rinsed with deionized water twice. The loaded sample was reacted with 200 μL of 0.1% purified arylsulphatase (*Helix pomatia* Type H-1; Sigma-Aldrich, St. Louis, MO, USA) at room temperature for 16 h for desulfation. Desulfo-GLs (DS-GLs) were eluted with 0.5 mL of distilled water and filtered through a 0.2 μm polyvinylidene difluoride filter for analysis.

The GLs were analyzed using desulfated standards of SIN (CAS 3952-98-5), GTR (CAS 499-26-3), glucobrassicin (GBS) (CAS 143231-38-3) (Extrasynthese, Genay, France), and neoglucobrassicin (GNBS) (CAS 5187-84-8) (Extrasynthese, Genay, France). One milliliter of each standard (2 mM) was loaded in pre-activated diethyl-aminoethyl resin and reacted with arylsulphatase for conversion into their desulfated forms.

For the phenolic contents and biological assays, 0.1 g of freeze-dried sample was extracted with 70% ethanol (2 mL) at 40 °C for 15 min. The extracts were filtered and concentrated to dryness before re-dissolving in dimethyl sulfoxide solvent.

2.4. GL Analysis Using Ultra-High-Performance Liquid Chromatography (UHPLC)-Diode Array Detection (DAD)-Quadrupole Time-of-Flight (QTOF)-Mass Spectrometry (MS)

DS-GLs were analyzed using an Agilent 1290 UHPLC system (Agilent, Waldbronn, Germany) combined with a QTOF instrument (Bruker Daltonics, Bremen, Germany). Samples (20 µL) were injected into a YMC-Triart C18 ExRS column (150 × 2.0 mm; particle size, 1.9 µm) and separated with mobile phases consisting of water (solvent A) and acetonitrile (solvent B), each acidified with 0.2% (*v/v*) formic acid. The gradient system composition was as follows: solvent B was kept at 0% for 1 min, and its ratio increased to 15%. This condition was continued for 4 min then; the B ratio increased to 100% for 7 min. 100% solvent A was applied for 1 min for column rinsing. The auto-sampler temperature was set at 4 °C, and the column temperature was maintained at 50 °C. Negative electrospray ionization mode was used for all compounds. The drying gas temperature was set at 300 °C with a flow rate of 8.0 L/min, while the nebulizer gas pressure and capillary voltage were maintained at 2.4 bar and 4000 V, respectively. Ionization quadrupole ion energy and collision energy were set at 2 eV and 20 eV, respectively and, for screening, the detection mass range was set at 100–700 *m/z*. All DS-GLs were detected at a wavelength of 229 nm and analyzed by comparing retention times with the available standards including the internal standard (GTR) and their specific fragmentation patterns. Their contents were calculated using the relative response factor (RRF) as previously reported [40,41] relative to the desulfoglucotropaeolin. After calculating the contents using Equation (1), described in the standard protocol [42], the contents were expressed in (µmoles/100 g).

$$\frac{\text{Relative peak area of desulfoglucosinolate}}{\text{Relative peak area of internal standard}} \times \frac{\mu\text{moles of the internal standard}}{\text{dry weight (g) of extracted plant material}} \times \text{RRF of desulfoglucosinolate} \quad (1)$$

Equation (1): Calculation of the content of each desulfoglucosinolate relative to the desulfoglucotropaeolin internal standard, expressed in µmoles/100 g of the sample dry weight.

2.5. Determination of the Total Phenolic Content (TPC)

Reagents and Chemicals

Folin-Ciocalteu reagent, gallic acid standard were obtained from Sigma-Aldrich Co. (St. Louis, MO, USA). Sodium carbonate was obtained from Fisher Scientific (Fair lawn, NJ, USA). Phenolic contents were determined by the Folin-Denis method, with some modification [43]. A 10-µL sample (4 mg/mL) was mixed with 2% Na_2CO_3 solvent (100 µL) in a 96-well plate that was then agitated for 3 min. Then, 10 µL of Folin–Ciocalteu reagent was added and the plate was incubated in the dark at room temperature for 30 min. The absorbance, at 750 nm was read using a multi detection microplate reader (Synergy HT; BioTek Instruments, Winooski, VT, USA). TPC was calculated based on a calibration curve determined using gallic acid as a standard and was expressed as mg gallic acid equivalents (GAE)/100 g of DW (dry weight).

2.6. Cell Culture, and Antioxidant Enzyme Superoxide Dismutase (SOD) and Catalase (CAT) Activity Assay

Human colon carcinoma cells (HCT116) were obtained from the ATCC (Manassas, VA, USA) and cultured in minimum essential media (Hyclone, Logan, UT, USA), supplemented with 10% fetal bovine serum (Thermo Fisher Scientific, Waltham, MA, USA) 1% penicillin-streptomycin mixture solution (Hyclone Logan, UT, USA) at 37 °C in a 5% CO_2 atmosphere. The cells were seeded (5×10^4 cells per well in a 24-well plate) and maintained for two days until their confluence reached 80%, the point at which media was removed. After removing the media, the cells were rinsed twice with phosphate-buffered saline (PBS; Hyclone Logan), and another media with or without 100 μg/mL extracts were added into the wells. The media was removed after 48 h, and the cells were washed twice with PBS. Cell SOD and CAT activity were evaluated using Abcam's commercial colorimetric assay kits (Abcam, Cambridge, UK). In brief, the cells were lysed with lysis buffer available in the assay kits, and SOD and CAT activity were measured according to the manufacturer's protocol.

2.7. Western Blotting of Nrf2 and HO-1

Cells treated with sample extracts were lysed using radioimmunoprecipitation assay buffer (RIPA buffer) (Thermo Fisher Scientific, Waltham, MA, USA) with a 1% protease inhibitor cocktail (Sigma-Aldrich, St. Louis, MO, USA). The protein concentrations in each lysate were quantified using the Bradford assay kit (Bio-Rad, Hercules, CA, USA). Proteins were denatured in sodium dodecyl sulfate (SDS) sample buffer separation with SDS polyacrylamide gel electrophoresis (SDS-PAGE). The proteins were then transferred to polyvinylidene fluoride (PVDF) membranes (Bio-Rad, Hercules, CA, USA) which were blocked with 3% bovine serum albumin in phosphate-buffered saline with tween solution. After 1 h, membranes were incubated with specific primary antibodies and horseradish peroxidase (HRP)-conjugated mouse and rabbit secondary antibodies. The antibodies against Nrf2 (Abcam, Cambridge, UK) and antibodies against β-actin and HO-1 (Santa Cruz Biotechnology, Dallas, TX, USA) were used. To visualize protein bands, ECL Western blotting detection kit (Thermo Fisher Scientific, Waltham, USA) and ImageQuant LAS-4000 (Fujifilm, Tokyo, Japan) were used. The quantification of protein bands was determined with ImageJ software (NIH, Bethesda, MD, USA).

2.8. Statistical Analysis

Data are expressed as the mean ± standard error. A comparison of means between two groups was done with the Student's t-test while among groups the means were compared by one-way analysis of variance (ANOVA) with Duncan's post-hoc tests in the SPSS software version 26. Principal component analysis (PCA) and orthogonal partial least squares discriminant analysis (OPLS-DA) was performed for UHPLC-QTOF-MS analysis results, total phenolic contents, antioxidant activity, and chemo-preventive ability, using the SIMCA program to understand the correlations between GLs a value of $p < 0.05$ was considered significant.

3. Results

3.1. Effects of LED Light and Salt Stress on Microgreens Biomass

The exposure of the microgreens to various LEDs and salinity had a significant impact on their fresh weight (Figure 1). Salinity treatment generally produced microgreens with lower biomass; although these plants did not show any visible signs of stress. The microgreens under B1R1 LED had the highest biomass, 1.10 folds higher compared to microgreens exposed to the control FL light (which is the commonly used light during cultivation). Lower biomass was generally observed in microgreens treated with salinity under all the various light irradiations with some specific lowest significant values in microgreens exposed to B and B1R1 LED.

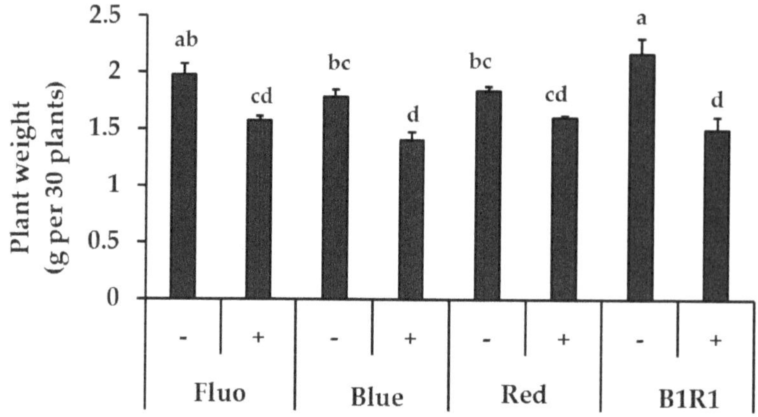

Figure 1. The effect of light, fluorescent (FL), blue, red, B1R1 (combination of blue plus red) on the biomass of microgreens cultivated in the absence (−) and presence (+) of salt, one week after treatments (n = 90). Data were expressed as mean ± standard error of three replicates. ANOVA analysis was done using Duncan's method and different letters were used to show statistical significance ($p < 0.05$).

3.2. Total Phenolic Content (TPC)

Figure 2 shows the TPC of microgreens grown under various conditions. The most evident result was that the tendency of change in TPC was different depending on the light condition. The light wavelength, in particular, influenced these contents, with FL light exhibiting higher levels while R and B1R1 LED exhibited lower levels of TPC. While salinity increased TPC for microgreens under B LED at 1.13-fold, p = 0.035 in comparison to their counterparts in non-saline conditions, only TPC of microgreens cultivated in saline conditions and exposed to FL light increased significantly (1.19-fold, p = 0.017) in comparison to the control. Microgreens grown under R and B1R1 LED, on the other hand, showed a significant negative effect of salt treatments, with a decrease of 0.56-fold and 0.75-fold compared to salt-free groups, p = 0.004 and 0.003, respectively.

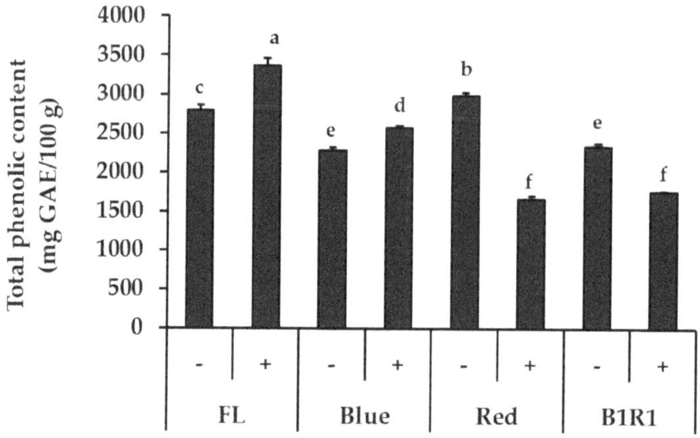

Figure 2. Total phenolic content of microgreens cultivated in the absence (−) and presence (+) of salt and exposed to fluorescent (FL), blue, red and combination of blue plus red (B1R1) light. Data were expressed as mean ± standard error (n = 3) in milligram(mg) gallic acid equivalent (GAE) per 100 g (g) of the extracts dry weight (DW). ANOVA analysis was done using Duncan's method and different letters (a–f) were used to show statistical significance ($p < 0.05$).

3.3. Identification of DS-GLs by UHPLC-QTOF-MS

The GLs identified in the *B. carinata* microgreen extracts are presented in Table 1 and Figure S1. Seven GLs, including the internal standard (GTR) and belonging to the aliphatic and indolic classes, were detected based on differences in their side-chain structures as illustrated in the cited papers. The aliphatic GLs (derived from methionine), included SIN; Rt = 5.51 min, $C_{10}H_{17}NO_6S$) and gluconapoleiferin (GNL; Rt = 6.72 min, $C_{12}H_{21}NO_{10}S_2$), whereas the indolic GLs (derived from tryptophan) identified were 4-hydroxy glucobrassicin (HGBS; Rt = 7.17 min, $C_{16}H_{20}N_2O_{10}S_2$), glucobrassicin (GBS; Rt = 9.33 min, $C_{16}H_{20}N_2O_9S_2$), 4-methoxy glucobrassicin (MGBS; Rt = 11.25 min, $C_{17}H_{22}N_2O_{10}S_2$), and neoglucobrassicin (GNBS; Rt = 12.44 min, $C_{17}H_{22}N_2O_{10}S_2$) [44–48].

Table 1. Desulfo glucosinolates determined through ultra-high performance-quadrupole time of flight mass spectrometry (UPLC QTOF-MS).

No.	Compound	Abbreviation	RT (min)	Formula	Precursor ion	Production	Relative Response Factor
1	Sinigrin	SIN	5.51	$C_{10}H_{17}NO_6S$	278.9960	195.0396 116.0270	1.05
2	Gluconapoleiferin	GNL	6.72	$C_{12}H_{21}NO_{10}S_2$	322.1013	195.0378 130.0408	1.05
3	4-Hydroxy glucobrassicin	HGBS	7.17	$C_{16}H_{20}N_2O_{10}S_2$	383.1020	221.0448 195.0394	0.29
4	Glucotropaeolin	GTR	8.31	$C_{14}H_{19}NO_9S_2$	328.0976	195.0421 166.0406	1.00
5	Glucobrassicin	GBS	9.33	$C_{16}H_{20}N_2O_9S_2$	367.1112	205.0524 195.0408	0.31
6	4-Methoxy glucobrassicin	MGBS	11.25	$C_{17}H_{22}N_2O_{10}S_2$	397.1167	235.0615 195.0389	0.26
7	Neoglucobrassicin	GNBS	12.44	$C_{17}H_{22}N_2O_{10}S_2$	397.1228	235.0633 195.0410	0.21

3.4. Quantification of Aliphatic GLs by UHPLC-DAD

In this study, GLs compounds were quantified by comparing the desulfated compounds' HPLC peak areas, retention times, and response factors relative to that of the internal standard GTR, which was included during the extraction process (Table 1 and Figure 3). The findings revealed distinct profiles of the identified GLs, as well as significant differences in their contents between treatments. When compared to indole group GLs, the total aliphatic GLs content was consistently high in all treatments, with specific values of SIN ranging from 16.15 μmol/100 g to 81.18 μmol/100 g and GNL ranging from 0.21 μmol/100 g to 0.73 μmol/100 g. These aliphatic GLs, both showed a similar pattern of increment in the salt treatments, regardless of the type of light irradiation and B LED specifically, significantly influenced their increase at 2.8 folds and 1.8-folds for SIN and GNL, respectively comparing to the contents of microgreen in the FL control group in absence of salt.

3.5. Quantification of Indole GLs by UHPLC-DAD

The four detected indole GLs: GBS, HGBS, MGBS and GNBS, responded differently to salt treatment and LED exposure (Figure 3). The content of GBS, the parent GL, was negatively correlated with the contents of the other compounds (HGBS, $r = -0.5421$; MGBS, $r = -0.2525$; GNBS, $r = -0.5627$); and, except for HGBS, salinity did not favor the accumulation of any of the other indole GLs. The highest amounts of GBS were detected in microgreens grown in control and non-saline conditions of B1R1 LED, whereas its byproducts accumulated differently in microgreens exposed to B and R LED, with HGBS and MGBS being conspicuously elevated in microgreens exposed to R LED and GNBS being elevated in microgreens exposed to both R and B LED.

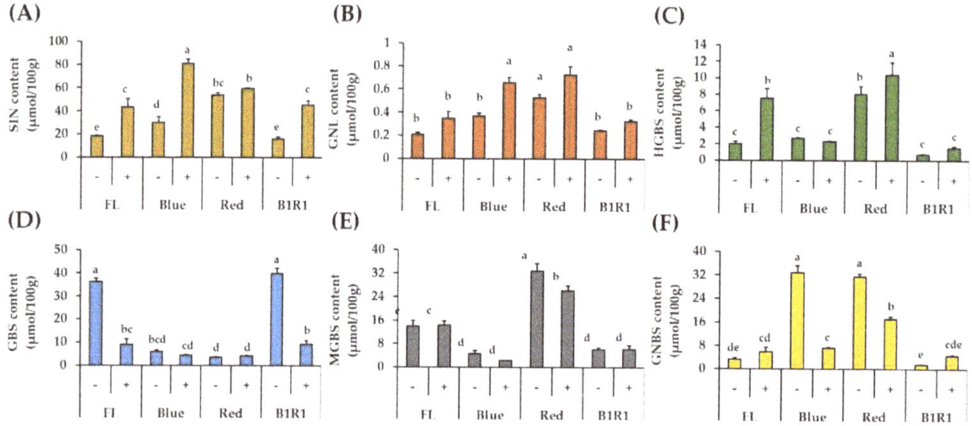

Figure 3. Quantification of DS-GLs (sinigrin (**A**), gluconapoleiferin (**B**), 4-hydroxy glucobrassicin (**C**), glucobrassicin (**D**), 4-methoxy glucobrassicin (**E**) and neoglucobrassicin (**F**)) in microgreens exposed to fluorescent (FL), blue, red and a combination of blue plus red (B1R1) lights and cultivated in the absence (−) and presence (+) of salt stress. Data were expressed as mean ± standard error ($n = 3$). ANOVA analysis was done using Duncan's method and different letters on columns of each graph were used to show statistical significance ($p < 0.05$).

3.6. Anti-Oxidant Enzyme Activity

The effect of the nontoxic microgreen extracts on the activities of SOD and CAT antioxidant enzymes was measured in HCT116 human colorectal carcinoma cells and compared against non-exposed cells. As shown in Figure 4A,B, stressed cells were treated with various microgreen extracts, which increased the activities of both enzymes, with microgreen extracts from saline treatments with higher overall activities than their counterparts. Specific high significantly different activities were observed in extracts of microgreens exposed to B and B1R1 LED, extracts with even higher activities than those from the control group. The microgreens from the saline treatment in R LED also showed significantly higher activity for CAT, as did the microgreens from the non-saline treatments in B LED for SOD, though their activity was not considerably higher than the activity of the control groups.

3.7. Effect of Microgreen Extracts on the Expression of Nrf2/HO-1 Pathway

We explored the antioxidative mechanisms of various extracts from cultivated microgreens, as well as their ability to activate the Nrf2 signaling pathway and up-regulate the HO-1 protein expression. Figure 5A–C shows that mainly *B carinata* microgreens rich in GLs activated the Nrf2 signaling pathway, causing an increase in the relative expression levels of HO-1 antioxidant protein. Extracts from salt-treated microgreens exposed to FL and B lights caused a significant increase in the expression of the Nrf2/HO-1 pathway, similar to extracts from microgreens exposed to R LED, which had a significant increase, particularly for HO-1 expression, compared to extracts from microgreens extracts from the control group. Salinity, on the other hand, decreased the relative expression levels of these antioxidants in the B1R1 LED exposed microgreens, with a significant decrease being observed for the Nrf2 (Figure 5A) relative expression but not for HO-1 (Figure 5B). Even though B1R1 LED exposed microgreens extracts had the least potential to activate the Nrf2/HO-1 system, we noticed a higher expression from the extracts from the microgreens in non-saline treatments of this LED, which had a higher total GLs content than their saline treatment counterparts.

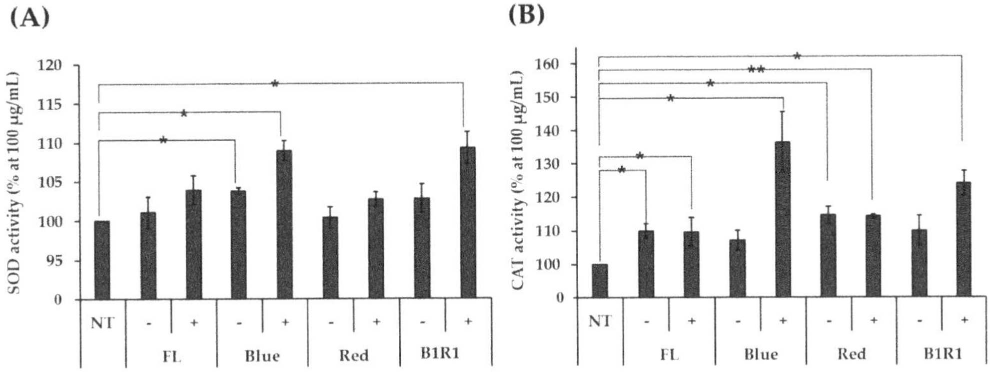

Figure 4. The activity of antioxidant enzymes superoxide dismutase (SOD (**A**)) and catalase (CAT (**B**)) was determined with various microgreen extracts at 100 µg/Ml using HCT116 human colorectal carcinoma cells. Extracts of microgreens exposed to fluorescent (FL), blue, red and blue plus red (B1R1) in the absence (−) or presence (+) of salt stress were treated to stressed cells while the control cells remained untreated. The experiment was conducted with three replications and data were expressed as mean ± standard error (n = 3). Activities were expressed as a percentage (%) compared to the non-treated (NT) group. Student t-test was conducted to see the difference between enzyme activity on untreated cells and cells treated with extracts from the various treatments. Extracts from microgreens exposed to B and B1R1 LED in the presence of salt exhibited higher activity than those from the microgreens cultivated in the control conditions. * = $p < 0.05$ and ** = $p < 0.01$ vs. the NT group.

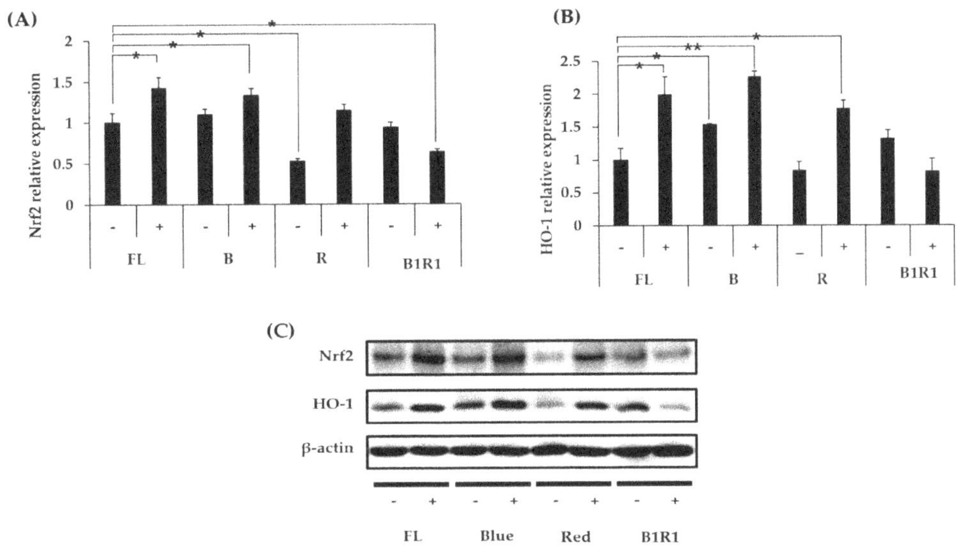

Figure 5. Effect of B. carinata microgreen extracts on the nuclear factor-erythroid 2-related factor (Nrf2) (**A**) and heme-oxygenase (HO-1) (**B**) protein expression in cells. Cells were treated with 100 µg/mL extracts for 48 h and the total expressed HO-1 protein and the Nrf2 were analyzed by Western blot (**C**) with β-Actin as the internal control. All data were presented as means ± standard error (n = 3. Student t-test was conducted to see the differences between relative expression of extracts from the control (fluorescence (FL) in absence of salt (−)) and extracts from treatment groups exposed to blue, red, B1R1 (blue plus red) in absence and presence (+) of salt, * = $p < 0.05$ and ** = $p < 0.01$ vs. the control.

3.8. Multivariate Analysis

A PLS-DA (Figure 6) explains 58.5% of the total variation and shows a strong link between salt stress in the B and B1R1 LEDs with the accumulation of GLs compounds such as SIN the major compound and the different biological analysis evaluated was developed. The Hotelling T2, which is the critical limit in PCA and which displays the normality region corresponding to 95% confidence, showed a normal area with no serious outliers (Figure S2B). In the PCA bi-plot analysis, the groups were not clearly discriminated because the factors were over-lapped and this explained 65% of the total variation (Figure S2A). Thus, the variance on the X matrix was split by PLS-DA model into predictive and orthogonal variances and a supervised multivariate analysis showed the clear discrimination between the sample groups in the scores space (Figure 6) than PCA. These were separated into two clusters one group with salt and the other without salt. The regression line generated on the permutation test (Figure S3B) implied that the OPLS-DA model was run well. These results show that the model fit (R^2) of X (R^2X) of OPLS-DA was 0.687 indicating that over 65% of the variation could be modeled by the selected components. The important features in the data were considered to be six variables: weight (1.5), SIN (1.29), total aliphatic GLs (1.29), total indolic GLs (1.22), GNL (1.05), and GBS (1.01) (Figure S3A). CAT, SOD, Nrf2, HO-1, and weight showed a positive correlation with the saline-treated groups more than the non-saline treated groups while indolic GL compounds showed a similar tendency for both groups. In particular, biomass and GBS were strongly correlated with B1R1 and FL light. On the other hand, except for GBS, the rest of the GL compounds indicated that they were strongly influenced by R light regardless of the presence of salt.

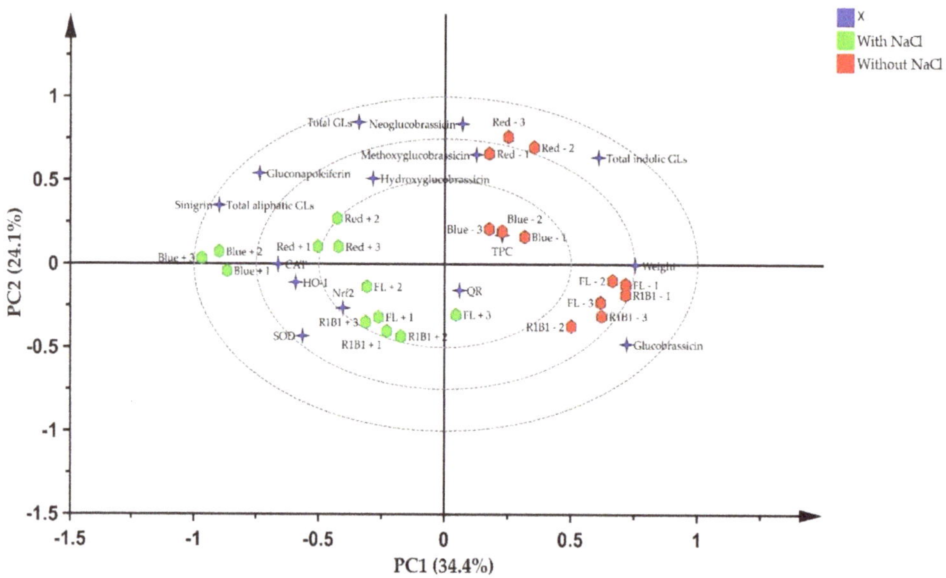

Figure 6. Partial least squares discriminant analysis (PLS-DA) score plots of variables and the treatment groups. The blue stars represent the variable and the green hexagons represent extracts from microgreens grown in the presence of salt stress while red hexagons are extracts from microgreens grown in absence of salt stress.

4. Discussion

Plants and plant extracts in the *Brassicaceae* family are rich in diverse health-benefiting compounds such as phenolics and GLs compounds. Specifically, phenolic compounds derived from tannins and sinapic acid, flavonoids, such as quercetin and kaempferol [24] and GLs such SIN, GBS, GNBS, progoitrin, gluconapin, and gluconasturtiin [44] have been found in *B. carinata*. The profiles of these compounds have been shown to differ among *Brassicaceae* species due to plant-specific factors such as salinity and light [49] light. As a result, such compounds, that reflect the biological potential of plant-derived natural products are relevant targets for improving plants functionality.

Elicitors, activate chemical defense and various biosynthetic pathways in plants and are used during the cultivation process as a strategy to increase specific plant phytochemicals [19]. Although the use of salt as an elicitor has been shown to disrupt the plants' osmotic balance and cause them to lose water to their surroundings [38], moderate salinity is especially effective in increasing compounds such as phenolic contents in brassica crops such as rapeseed [50], as seen in our results from the extracts of microgreens exposed to FL light and B LED. Comparable to our findings, salinity causes a decrease in weight, but the microgreens did not show any visible signs of stress, as previously observed in broccoli sprouts [38].

Light intensity and quality have a significant impact on plant growth and development, with the effects of these genotype-dependent factors causing different plant species to respond differently when exposed to specific wavelengths [29,51]. In this study, we found that microgreen biomass differed significantly after cultivating them under different light wavelengths, with the most noticeable differences on microgreens grown under B1R1, which had the highest biomass. However, our findings contradict those of a previous study in broccoli microgreens, in which B LED resulted in more shoots when compared to plants exposed to combined LEDs [37]. We speculate that R LED in our combination complemented the effects of B LED in increasing microgreens biomass, and given that we chose a different ratio from the previous study, it is also possible that this specific ratio had a positive influence on the biomass of microgreens plants.

Furthermore, light, which serves as an energy source for plants by influencing the photosynthetic light-dependent process, regulates the contents of some plant compounds such as phenolics and is used as an elicitor [29,52] as reported in broccoli microgreens [37], canola [29], lettuce [32] and Chinese mustard [52]. Several studies, however, have shown that light may also stress the plant during elicitation, resulting in photosynthesis suppression [53]. In our study, we found that B LED allowed for the accumulation of phenolic compounds while R LED did not. Previous studies indicate that B LED is a more powerful light source than R LED in terms of enhancing plant photosynthetic processes [37]. B LED influences plant leaf expansion, chlorophyll levels, light-induced stomatal opening, and photosynthesis [54], resulting in increased stomatal conductance, photo-synthetic electron transportability, and phenolic compounds [55,56].

Notably, previous studies have linked R light to increased starch levels in chloroplasts, which inhibits the translocation of photosynthates enzyme and ultimately reduces photosynthesis [57,58], possibly resulting in our observed lower phenolics in microgreens exposed to this LED. Moreover, the salt treatment of microgreen exposed R LED resulted in even lower phenolic contents due to the plants' reduced ability to cope with both stresses. Although B1R1 LED exposure increased biomass, it did not affect phenolic compounds. Previous studies have shown that rapid metabolite changes do not occur primarily when exposed to a mixture of LEDs because the changes are more sensitive to changing specific monochromatic lights; thus, proper light ratio selection is required to achieve changes [32].

The negative ion mode was used for identification based on previous reports of its suitability in detecting DS-GLs compounds due to their structural properties [59]. The GLs deprotonated ions $[M-H]^-$ (m/z), ion fragmentation patterns, and relative retention times of the compounds compared with that of available standards aids in the identification of the distinct compounds. We found out that the fragmentation of GLs revealed two groups of specific fragments. The MS^2 fragment corresponds to the aglycone portion and produces stable ions at *m/z* 195, 275, and 259 (by loss of R1-N=C=O, together with R1-N=C=S from the deprotonated molecule $[M-H]^-$), whereas *m/z* 241 ions are formed by the cleavage of water (H_2O, 18 Da). Fragmentation of the primary ions at *m/z* 259 resulted in another fragment that corresponds to the D-thioglucose group $[C_6H_{11}O_5S]^-$; which is shown to originate from a rearrangement reaction where the sulfate group is transferred to the thioglucosidase side chain in earlier studies [60].

Except for GNL, the other five compounds found in *B carinata* had previously been reported; however, we did not find progoitrin, gluconapin, or gluconasturtiin, which have also been reported in the plant [44]. The total aliphatic GLs content was consistently high in all treatments, owing to increasing levels of SIN, the most abundant compound in *B. carinata* [2]. Salinity stress significantly increased aliphatic GLs contents, which is consistent with previous research among *Brassicaceae* vegetables [22,61,62], and exposure to light wavelengths may have produced a synergistic reaction, leading to an even greater increase. In particular, B LED produced the highest SIN levels in microgreens, although in contrast to R LED, which increased SIN in canola [29] and broccoli microgreens [63]. These findings point to the possibility of a species-dependent metabolite increase as a result of light exposure. In addition, the decrease in GBS seen in the microgreens irradiated to B and R LEDs is in agreement with other studies involving elicitor treatments where GBS is converted to its byproducts [64]. The differences observed in the accumulation of GBS byproducts under various LED treatments, in saline and non-saline conditions are attributable to changes in the expression of several genes involved in their biosynthesis [64]. From our findings, exposure to R LED is likely to have caused the expression of more genes involved in indole GLs biosynthesis leading to the conversion of GBS to most of its products (HGBS, MGBS), whereas exposure to B LED was biased toward one arm of the biosynthetic pathway where GBS is converted to GNBS as seen in other studies involving plants cultivated in non-saline conditions [65].

The production of GLs in salt-treated plants is believed to protect plants from abiotic stress, and the salt concentration used in this study was within the salinity tolerance threshold [66] to activate genes involved in the biosynthesis of these aliphatic GLs in *B. carinata* microgreens, as expected. This is consistent with findings from other studies in which a salt concentration of 100 mM significantly increased the nutritional value of *Brassicaceae* plants such as radish sprouts [23], broccoli sprouts [38], and broccoli seedlings [61], demonstrating its utility in increasing the content of these health-promoting compounds [23,67].

Generally, cruciferous vegetables are consumed for their health-promoting benefits, including antioxidant, anti-proliferative, and chemo-preventive activities contributed mainly by GLs and phenolics [24,68–70]. Currently, there are growing pieces of evidence suggesting that oxidative stress from reactive oxidants plays a key role in the pathogenesis of several diseases [11]. In recent decades, multiple strategies were evaluated to reduce the impact of oxidative stresses including the use of naturally produced enzymes such as SOD and CAT as well as their overexpression by other compounds. Several pieces of literature have cited the importance of dietary polyphenols and the great roles they play in activating and up-regulating the expression of antioxidant enzymes including SOD, CAT, glutathione reductases, glutathione peroxidases and glutathione-*s*-transferases. An increase in the activities of these antioxidant enzymes in cells induced with ROS proves the enzymes' well adaptation in maintaining a balance of the reactive oxidants in the cells, therefore lowering oxidative stress [71].

The antioxidant enzymes are the first defense system of a cell against ROS [72]. The collapse of a balance in the production of reactive oxidants and scavenging ability causes an accumulation of ROS such as superoxides (O_2^-) and hydrogen peroxides (H_2O_2) which result in cell death or proliferation through DNA damage, protein oxidation, and lipid peroxidation [73]. In the cellular systems, antioxidant enzymes help to maintain this balance. For instance, SOD acts to convert the O_2^- radicals through dismutation to less toxic, H_2O_2 while CAT eliminates the H_2O_2 and changes it to water and oxygen [74].

Increased enzyme activity in in vitro assay is associated with reduced oxidative stress and reduced cell damage. Our observed increased antioxidant enzyme activity in cells treated with the microgreens extracts was facilitated by the available contents of natural antioxidants such as polyphenols and GLs as reported in other studies [71,75].

Our investigation on the mechanism of bioactive compounds in *B. carinata* microgreen also displayed an enhancement of Nrf2 and HO-1 expression. This enhancement is essential for cell protection due to the critical roles the Nrf2 pathway plays in reducing inflammation and oxidative stress. Plant compounds such as GSHPs activate the Nrf2/HO-1 antioxidant pathway and are therefore suitable as protective and treatment agents for oxidative stress, inflammatory diseases [76]. Several researchers are interested in identifying activators of this pathway as potential therapeutic strategies for various diseases including cancer [77,78]. Cruciferous vegetables have been identified as valuable sources of GSHPs, which activate Nrf2 and thus up-regulate phase II enzymes as well as antioxidant enzymes such as HO-1 [76]. In our study, microgreens cultivated in saline conditions under B, R, and FL light especially, accumulated the highest contents of GLs which possibly contributed to the pathway activation and resulted in adding the therapeutic value of microgreens.

5. Conclusions

In this study, we investigated whether *B carinata* is suited for the indoor farming system and whether we can achieve proper cultivation conditions through elicitation, for the production of microgreens with high biomass, rich in phytochemicals, and improved biological potential including antioxidant activities. Salinity and B LED were found to be important in promoting GLs and phenolics in *B. carinata* microgreens, as well as their overall biological activity, whereas R LED was a stressed condition that reduced the plants' ability to photosynthesize, resulting in low accumulation of beneficial phytochemicals, particularly in saline conditions. Furthermore, non-saline conditions with B1R1 LED were comparable to the control FL condition for producing microgreens with increased biomass. To increase and preserve the beneficial metabolite content and nutritional value of *B carinata*, B LED supplemental wavelength and proper blue plus red LED ratios may be beneficial in microgreens cultivation in the vertical farming system.

Supplementary Materials: The following are available online at https://www.mdpi.com/article/10.3390/antiox10081183/s1, Figure S1: UPLC-DAD chromatogram of available standards detected at 229 nm, Figure S2: PCA bi-plot of *B.carinata* microgreens (**A**) showing separation of groups on the space. Green hexagons represent the group of microgreen samples under different lights that were treated to salt stress. Red hexagons: represent the group of microgreen samples that were not treated with salt. Blue stars: X-variable. (**B**) Hotelling's T2 range plots of the PCA belong to the normal area, Figure S3: Multivariate statistical analysis done by SIMCA program of *B. carinata* microgreens. (**A**) The VIP score of data. (**B**) showing a validation plot for OPLS-DA. Green circle: R^2Y (Cum), Blue pentagons: Q^2 (Cum).

Author Contributions: Conceptualization, writing and methodology S.M. and D.H.R.; statistical analysis J.Y.C.; validation, resources, reviewing and editing J.-E.P. and C.W.N.; investigation D.S.J.; supervision, reviewing and editing, G.B. and G.M.; reviewing and editing J.H.J., S.-H.Y. and H.-Y.K.; funding acquisition and supervision H.-Y.K. All authors have read and agreed to the published version of the manuscript.

Funding: This research was supported both financially and with laboratory facilities from Korea Institute of Science and Technology in the Republic of Korea [Intramural grant number: 2Z06500] and Korea Institute for Advancement of Technology(KIAT) grant [2MR9730].

Institutional Review Board Statement: Not applicable.

Informed Consent Statement: Not applicable.

Data Availability Statement: The data presented in this study are available in article and supplementary material.

Acknowledgments: We acknowledge The Kenya Resource Centre for Indigenous Knowledge, National Museums of Kenya for providing us with the germplasm to perform this study. The first author also acknowledges the scholarship support from the Regional Scholarship and Innovation Fund of the Partnership for Skills in Applied Sciences, Engineering and Technology- (RSIF-PASET).

Conflicts of Interest: The authors declare no conflict of interest.

References

1. Vale, A.; Santos, J.; Brito, N.; Fernandes, D.; Rosa, E.; Oliveira, M.B.P. Evaluating the impact of sprouting conditions on the glucosinolate content of Brassica oleracea sprouts. *Phytochemistry* **2015**, *115*, 252–260. [CrossRef]
2. Hagos, R.; Shaibu, A.S.; Zhang, L.; Cai, X.; Liang, J.; Wu, J.; Lin, R.; Wang, X. Ethiopian Mustard (*Brassica carinata* A. Braun) as an Alternative Energy Source and Sustainable Crop. *Sustainability* **2020**, *12*, 7492. [CrossRef]
3. Marillia, E.-F.; Francis, T.; Falk, K.C.; Smith, M.; Taylor, D.C. Palliser's promise: Brassica carinata, an emerging western Canadian crop for delivery of new bio-industrial oil feedstocks. *Biocatal. Agric. Biotechnol.* **2014**, *3*, 65–74. [CrossRef]
4. Neugart, S.; Baldermann, S.; Ngwene, B.; Wesonga, J.; Schreiner, M. Indigenous leafy vegetables of Eastern Africa—A source of extraordinary secondary plant metabolites. *Food Res. Int.* **2017**, *100*, 411–422. [CrossRef] [PubMed]
5. Xin, H.; Falk, K.C.; Yu, P. Studies on Brassica carinata seed. 2. Carbohydrate molecular structure in relation to carbohydrate chemical profile, energy values, and biodegradation characteristics. *J. Agric. Food Chem.* **2013**, *61*, 10127–10134. [CrossRef]
6. Teklehaymanot, T.; Wang, H.; Liang, J.; Wu, J.; Lin, R.; Zhou, Z.; Cai, X.; Wang, X. Variation in Plant Morphology and Sinigrin Content in Ethiopian Mustard (*Brassica carinata* L.). *Hortic. Plant J.* **2019**, *5*, 205–212. [CrossRef]
7. Lozano-Baena, M.-D.; Tasset, I.; Obregón-Cano, S.; Haro-Bailon, D.; Muñoz-Serrano, A.; Alonso-Moraga, Á. Antigenotoxicity and tumor growing inhibition by leafy Brassica carinata and Sinigrin. *Molecules* **2015**, *20*, 15748–15765. [CrossRef] [PubMed]
8. Maina, S.; Misinzo, G.; Bakari, G.; Kim, H.-Y. Human, Animal and Plant Health Benefits of Glucosinolates and Strategies for Enhanced Bioactivity: A Systematic Review. *Molecules* **2020**, *25*, 3682. [CrossRef]
9. Mazumder, A.; Dwivedi, A.; Du Plessis, J. Sinigrin and its therapeutic benefits. *Molecules* **2016**, *21*, 416. [CrossRef]
10. Stephenie, S.; Chang, Y.P.; Gnanasekaran, A.; Esa, N.M.; Gnanaraj, C. An insight on superoxide dismutase (SOD) from plants for mammalian health enhancement. *J. Funct. Foods* **2020**, *68*, 103917. [CrossRef]
11. Vale, A.P.; Cidade, H.; Pinto, M.; Oliveira, M.B.P. Effect of sprouting and light cycle on antioxidant activity of Brassica oleracea varieties. *Food Chem.* **2014**, *165*, 379–387. [CrossRef]
12. Aires, A.; Fernandes, C.; Carvalho, R.; Bennett, R.N.; Saavedra, M.J.; Rosa, E.A. Seasonal effects on bioactive compounds and antioxidant capacity of six economically important Brassica vegetables. *Molecules* **2011**, *16*, 6816–6832. [CrossRef] [PubMed]
13. Kobayashi, E.H.; Suzuki, T.; Funayama, R.; Nagashima, T.; Hayashi, M.; Sekine, H.; Tanaka, N.; Moriguchi, T.; Motohashi, H.; Nakayama, K. Nrf2 suppresses macrophage inflammatory response by blocking proinflammatory cytokine transcription. *Nat. Commun.* **2016**, *7*, 1–14. [CrossRef] [PubMed]
14. Akhlaghi, M.; Bandy, B. Dietary broccoli sprouts protect against myocardial oxidative damage and cell death during ischemia-reperfusion. *Plant Foods Human Nutr.* **2010**, *65*, 193–199. [CrossRef] [PubMed]
15. Brazaitytė, A.; Viršilė, A.; Samuolienė, G.; Vaštakaitė-Kairienė, V.; Jankauskienė, J.; Miliauskienė, J.; Novičkovas, A.; Duchovskis, P. Response of Mustard Microgreens to Different Wavelengths and Durations of UV-A LEDs. *Front. Plant Sci.* **2019**, *10*, 1153. [CrossRef]
16. Kyriacou, M.C.; De Pascale, S.; Kyratzis, A.; Rouphael, Y. Microgreens as a Component of Space Life Support Systems: A Cornucopia of Functional Food. *Front. Plant Sci.* **2017**, *8*. [CrossRef]
17. Pinto, E.; Almeida, A.A.; Aguiar, A.A.; Ferreira, I. Comparison between the mineral profile and nitrate content of microgreens and mature lettuces. *J. Food Comp. Anal.* **2015**, *37*, 38–43. [CrossRef]
18. Baenas, N.; Piegholdt, S.; Schloesser, A.; Moreno, D.A.; Garcia-Viguera, C.; Rimbach, G.; Wagner, A.E. Metabolic Activity of Radish Sprouts Derived Isothiocyanates in Drosophila melanogaster. *Int. J. Mol. Sci.* **2016**, *17*, 251. [CrossRef]
19. Poulev, A.; O'Neal, J.M.; Logendra, S.; Pouleva, R.B.; Timeva, V.; Garvey, A.S.; Gleba, D.; Jenkins, I.S.; Halpern, B.T.; Kneer, R. Elicitation, a new window into plant chemodiversity and phytochemical drug discovery. *J. Med. Chem.* **2003**, *46*, 2542–2547. [CrossRef]
20. Rodriguez-Hernandez, M.d.C.; Moreno, D.A.; Carvajal, M.; Martinez-Ballesta, M.d.C. Genotype influences sulfur metabolism in broccoli (*Brassica oleracea* L.) under elevated CO2 and NaCl stress. *Plant Cell Physiol.* **2014**, *55*, 2047–2059. [CrossRef]
21. Kim, S.Y.; Park, J.-E.; Kim, E.O.; Lim, S.J.; Nam, E.J.; Yun, J.H.; Yoo, G.; Oh, S.-R.; Kim, H.S.; Nho, C.W. Exposure of kale root to NaCl and Na_2SeO_3 increases isothiocyanate levels and Nrf2 signalling without reducing plant root growth. *Sci. Rep.* **2018**, *8*, 1–11.
22. Del Carmen Martínez-Ballesta, M.; Moreno, D.A.; Carvajal, M. The physiological importance of glucosinolates on plant response to abiotic stress in Brassica. *Int. J. Mol. Sci.* **2013**, *14*, 11607–11625. [CrossRef]
23. Yuan, G.; Wang, X.; Guo, R.; Wang, Q. Effect of salt stress on phenolic compounds, glucosinolates, myrosinase and antioxidant activity in radish sprouts. *Food Chem.* **2010**, *121*, 1014–1019. [CrossRef]
24. Cartea, M.E.; Francisco, M.; Soengas, P.; Velasco, P. Phenolic compounds in Brassica vegetables. *Molecules* **2011**, *16*, 251–280. [CrossRef] [PubMed]
25. Kozai, T.; Fujiwara, K.; Runkle, E.S. *LED lighting for Urban Agriculture*; Springer: Singapore, 2016; pp. 1–454.

26. Kozai, T.; Niu, G. Chapter 1—Introduction. In *Plant Factory*; Academic Press: San Diego, CA, USA, 2016; pp. 3–5.
27. Goto, E. Chapter 15—Production of Pharmaceuticals in a Specially Designed Plant Factory A2—Kozai, Toyoki. In *Plant Factory*; Niu, G., Takagaki, M., Eds.; Academic Press: San Diego, CA, USA, 2016; pp. 193–200.
28. Azad, M.O.K.; Kim, W.W.; Park, C.H.; Cho, D.H. Effect of Artificial LED Light and Far Infrared Irradiation on Phenolic Compound, Isoflavones and Antioxidant Capacity in Soybean (*Glycine max* L.) Sprout. *Foods* **2018**, *7*, 174. [CrossRef] [PubMed]
29. Park, C.H.; Kim, N.S.; Park, J.S.; Lee, S.Y.; Lee, J.-W.; Park, S.U. Effects of Light-Emitting Diodes on the Accumulation of Glucosinolates and Phenolic Compounds in Sprouting Canola (*Brassica napus* L.). *Foods* **2019**, *8*, 76. [CrossRef]
30. Fukuda, N.; Kobayashi, M.; Ubukawa, M.; Takayanagi, K.; Sase, S. Effects of Light Quality, Intensity and Duration from Different Artificial Light Sources on the Growth of Petunia (Petunia × hybrida Vilm.). *J. Jpn. Soc. Hortic. Sci.* **2002**, *71*, 509–516. [CrossRef]
31. Yeh, N.; Chung, J.-P. High-brightness LEDs—Energy efficient lighting sources and their potential in indoor plant cultivation. *Renew. Sustain. Energy Rev.* **2009**, *13*, 2175–2180. [CrossRef]
32. Son, K.-H.; Lee, J.-H.; Oh, Y.; Kim, D.; Oh, M.-M.; In, B.-C. Growth and bioactive compound synthesis in cultivated lettuce subject to light-quality changes. *HortScience* **2017**, *52*, 584–591. [CrossRef]
33. Park, J.-E.; Kim, H.; Kim, J.; Choi, S.-J.; Ham, J.; Nho, C.W.; Yoo, G. A comparative study of ginseng berry production in a vertical farm and an open field. *Ind. Crops Prod.* **2019**, *140*, 111612. [CrossRef]
34. Brazaitytė, A.; Sakalauskienė, S.; Samuolienė, G.; Jankauskienė, J.; Viršilė, A.; Novičkovas, A.; Sirtautas, R.; Miliauskienė, J.; Vaštakaitė, V.; Dabašinskas, L.; et al. The effects of LED illumination spectra and intensity on carotenoid content in Brassicaceae microgreens. *Food Chem.* **2015**, *173*, 600–606. [CrossRef]
35. Dutta Gupta, S. *Light Emitting Diodes for Agriculture: Smart Lighting*; Springer: Singapore, 2017; pp. 1–334.
36. Hasan, M.M.; Bashir, T.; Ghosh, R.; Lee, S.K.; Bae, H. An Overview of LEDs' Effects on the Production of Bioactive Compounds and Crop Quality. *Molecules* **2017**, *22*, 1420. [CrossRef]
37. Kopsell, D.A.; Sams, C.E. Increases in shoot tissue pigments, glucosinolates, and mineral elements in sprouting broccoli after exposure to short-duration blue light from light emitting diodes. *J. Am. Soci. Hortic. Sci.* **2013**, *138*, 31–37. [CrossRef]
38. Guo, R.F.; Yuan, G.F.; Wang, Q.M. Effect of NaCl treatments on glucosinolate metabolism in broccoli sprouts. *J. Zhejiang Univ. Sci. B* **2013**, *14*, 124–131. [CrossRef]
39. Doheny-Adams, T.; Redeker, K.; Kittipol, V.; Bancroft, I.; Hartley, S.E. Development of an efficient glucosinolate extraction method. *Plant Methods* **2017**, *13*, 1–14. [CrossRef]
40. Clarke, D.B. Glucosinolates, structures and analysis in food. *Anal. Methods* **2010**, *2*, 310–325. [CrossRef]
41. Issa, R.A. Identification of Glucosinolate Profile in Brassica Oleracea for Quantitative Trait Locus Mapping. Ph.D. Thesis, University of Warwick, Coventry, UK, 2010.
42. European Community. Oilseeds-determination of glucosinolates-high performance liquid chromatography. *Off. J. Europ. Comm.* **1990**, *170*, 27–34.
43. Choi, Y.; Lee, S.; Chun, J.; Lee, H.; Lee, J. Influence of heat treatment on the antioxidant activities and polyphenolic compounds of Shiitake (*Lentinus edodes*) mushroom. *Food Chem.* **2006**, *99*, 381–387. [CrossRef]
44. Bellostas, N.; Sørensen, J.C.; Sørensen, H. Profiling glucosinolates in vegetative and reproductive tissues of four Brassica species of the U-triangle for their biofumigation potential. *J. Sci. Food Agric.* **2007**, *87*, 1586–1594. [CrossRef]
45. Bhandari, S.R.; Rhee, J.; Choi, C.S.; Jo, J.S.; Shin, Y.K.; Lee, J.G. Profiling of Individual Desulfo-Glucosinolate Content in Cabbage Head (*Brassica oleracea* var. *capitata*) Germplasm. *Molecules* **2020**, *25*, 1860. [CrossRef]
46. Hwang, E.S. Effect of Cooking Method on Antioxidant Compound Contents in Cauliflower. *Prev. Nutr. Food Sci.* **2019**, *24*, 210–216. [CrossRef] [PubMed]
47. Liang, X.; Lee, H.W.; Li, Z.; Lu, Y.; Zou, L.; Ong, C.N. Simultaneous Quantification of 22 Glucosinolates in 12 Brassicaceae Vegetables by Hydrophilic Interaction Chromatography–Tandem Mass Spectrometry. *ACS Omega* **2018**, *3*, 15546–15553. [CrossRef]
48. Thomas, M.; Badr, A.; Desjardins, Y.; Gosselin, A.; Angers, P. Characterization of industrial broccoli discards (*Brassica oleracea* var. *italica*) for their glucosinolate, polyphenol and flavonoid contents using UPLC MS/MS and spectrophotometric methods. *Food Chem.* **2018**, *245*, 1204–1211. [CrossRef]
49. Hassini, I.; Martinez-Ballesta, M.C.; Boughanmi, N.; Moreno, D.A.; Carvajal, M. Improvement of broccoli sprouts (*Brassica oleracea* L. var. italica) growth and quality by KCl seed priming and methyl jasmonate under salinity stress. *Sci. Hortic.* **2017**, *226*, 141–151. [CrossRef]
50. Falcinelli, B.; Sileoni, V.; Marconi, O.; Perretti, G.; Quinet, M.; Lutts, S.; Benincasa, P. Germination under moderate salinity increases phenolic content and antioxidant activity in rapeseed (*Brassica napus* var *oleifera* Del.) sprouts. *Molecules* **2017**, *22*, 1377. [CrossRef] [PubMed]
51. Nishimura, T.; Zobayed, S.M.; Kozai, T.; Goto, E. Medicinally important secondary metabolites and growth of Hypericum perforatum L. plants as affected by light quality and intensity. *Environ. Control Biol.* **2007**, *45*, 113–120. [CrossRef]
52. Park, C.H.; Park, Y.E.; Yeo, H.J.; Kim, J.K.; Park, S.U. Effects of Light-Emitting Diodes on the Accumulation of Phenolic Compounds and Glucosinolates in Brassica juncea Sprouts. *Horticulturae* **2020**, *6*, 77. [CrossRef]
53. Bayat, L.; Arab, M.; Aliniaeifard, S.; Seif, M.; Lastochkina, O.; Li, T. Effects of growth under different light spectra on the subsequent high light tolerance in rose plants. *AoB Plants* **2018**, *10*. [CrossRef]
54. Banaś, A.K.; Aggarwal, C.; Łabuz, J.; Sztatelman, O.; Gabryś, H. Blue light signalling in chloroplast movements. *J. Exp. Bot.* **2012**, *63*, 1559–1574. [CrossRef]

55. Muneer, S.; Kim, E.J.; Park, J.S.; Lee, J.H. Influence of green, red and blue light emitting diodes on multiprotein complex proteins and photosynthetic activity under different light intensities in lettuce leaves (*Lactuca sativa* L.). *Int. J. Mol. Sci.* **2014**, *15*, 4657–4670. [CrossRef]
56. Johkan, M.; Shoji, K.; Goto, F.; Hashida, S.-n.; Yoshihara, T. Blue light-emitting diode light irradiation of seedlings improves seedling quality and growth after transplanting in red leaf lettuce. *HortScience* **2010**, *45*, 1809–1814. [CrossRef]
57. Li, H.; Tang, C.; Xu, Z.; Liu, X.; Han, X. Effects of different light sources on the growth of non-heading Chinese cabbage (*Brassica campestris* L.). *J. Agric. Sci.* **2012**, *4*, 262. [CrossRef]
58. Trouwborst, G.; Hogewoning, S.W.; van Kooten, O.; Harbinson, J.; van Ieperen, W. Plasticity of photosynthesis after the 'red light syndrome'in cucumber. *Environ. Exp. Bot.* **2016**, *121*, 75–82. [CrossRef]
59. Bianco, G.; Pascale, R.; Lelario, F.; Bufo, S.; Cataldi, T. Investigation of glucosinolates by mass spectrometry. In *Glucosinolates. Reference Series in Phytochemistry*; Springer: Berlin/Heidelberg, Germany, 2017; pp. 431–461.
60. Velasco, P.; Francisco, M.; Moreno, D.A.; Ferreres, F.; García-Viguera, C.; Cartea, M.E. Phytochemical fingerprinting of vegetable Brassica oleracea and Brassica napus by simultaneous identification of glucosinolates and phenolics. *Phytochem. Anal.* **2011**, *22*, 144–152. [CrossRef] [PubMed]
61. SarikamiŞ, G. Influence of salinity on aliphatic and indole glucosinolates in broccoli (*Brassica oleracea* var. *italica*). *Appl. Ecol. Environ. Res.* **2017**, *15*, 1781–1788. [CrossRef]
62. Martínez-Ballesta, M.; Moreno-Fernández, D.A.; Castejón, D.; Ochando, C.; Morandini, P.A.; Carvajal, M.J.F. The impact of the absence of aliphatic glucosinolates on water transport under salt stress in Arabidopsis thaliana. *Front. Plant Sci.* **2015**, *6*, 524. [CrossRef]
63. Kopsell, D.A.; Sams, C.E.; Barickman, T.C.; Morrow, R.C. Sprouting broccoli accumulate higher concentrations of nutritionally important metabolites under narrow-band light-emitting diode lighting. *J. Am. Soc. Hortic.l Sci.* **2014**, *139*, 469–477. [CrossRef]
64. Cuong, D.M.; Park, S.U.; Park, C.H.; Kim, N.S.; Bong, S.J.; Lee, S.Y. Comparative analysis of glucosinolate production in hairy roots of green and red kale (*Brassica oleracea* var. *acephala*). *Prep. Biochem. Biotechnol.* **2019**, *49*, 775–782. [CrossRef]
65. Moon, J.; Jeong, M.J.; Lee, S.I.; Lee, J.G.; Hwang, H.; Yu, J.; Kim, Y.-R.; Park, S.W.; Kim, J.A. Effect of LED mixed light conditions on the glucosinolate pathway in brassica rapa. *J. Plant Biotechnol.* **2015**, *42*, 245–256. [CrossRef]
66. Petretto, G.L.; Urgeghe, P.P.; Massa, D.; Melito, S. Effect of salinity (NaCl) on plant growth, nutrient content, and glucosinolate hydrolysis products trends in rocket genotypes. *Plant Physiol. Biochem.* **2019**, *141*, 30–39. [CrossRef]
67. Lopez-Berenguer, C.; Martinez-Ballesta, M.C.; Garcia-Viguera, C.; Carvajal, M. Leaf water balance mediated by aquaporins under salt stress and associated glucosinolate synthesis in broccoli. *Plant Sci.* **2008**, *174*, 321–328. [CrossRef]
68. Boscaro, V.; Boffa, L.; Binello, A.; Amisano, G.; Fornasero, S.; Cravotto, G.; Gallicchio, M. Antiproliferative, Proapoptotic, Antioxidant and Antimicrobial Effects of *Sinapis nigra* L. and *Sinapis alba* L. Extracts. *Molecules* **2018**, *23*, 3004. [CrossRef]
69. Chaudhary, A.; Choudhary, S.; Sharma, U.; Vig, A.; Arora, S.J. In vitro Evaluation of Brassica sprouts for its Antioxidant and Antiproliferative Potential. *Ind. J. Pharm. Sci.* **2016**, *78*, 615–623. [CrossRef]
70. Le, T.N.; Chiu, C.-H.; Hsieh, P.-C. Bioactive Compounds and Bioactivities of *Brassica oleracea* L. var. *Italica* Sprouts and Microgreens: An Updated Overview from a Nutraceutical Perspective. *Plants* **2020**, *9*, 946. [CrossRef]
71. Hwang, J.H.; Ma, J.N.; Park, J.H.; Jung, H.W.; Park, Y.-K. Anti-inflammatory and antioxidant effects of MOK, a polyherbal extract, on lipopolysaccharide-stimulated RAW 264.7 macrophages. *Int. J. Mol. Med.* **2019**, *43*, 26–36. [CrossRef] [PubMed]
72. Mates, J. Effects of antioxidant enzymes in the molecular control of reactive oxygen species toxicology. *Toxicology* **2000**, *153*, 83–104. [CrossRef]
73. Huppi, P.S. The role of oxygen in health and disease–a series of reviews. *Pediatr. Res.* **2009**, *65*, 261–268.
74. NavaneethaKrishnan, S.; Rosales, J.L.; Lee, K.-Y. ROS mediated cancer cell killing through dietary phytochemicals. *Oxid. Med. Cell. Longev.* **2019**, *2019*. [CrossRef] [PubMed]
75. Kacem, M.; Simon, G.; Leschiera, R.; Misery, L.; ElFeki, A.; Lebonvallet, N. Antioxidant and anti-inflammatory effects of *Ruta chalepensis* L. extracts on LPS-stimulated RAW 264.7 cells. *In Vitro Cell. Dev. Biol. Animal* **2015**, *51*, 128–141. [CrossRef]
76. Sturm, C.; Wagner, A.E. Brassica-Derived Plant Bioactives as Modulators of Chemopreventive and Inflammatory Signaling Pathways. *Int. J. Mol. Sci.* **2017**, *18*, 1890. [CrossRef]
77. Catino, S.; Paciello, F.; Miceli, F.; Rolesi, R.; Troiani, D.; Calabrese, V.; Santangelo, R.; Mancuso, C. Ferulic acid regulates the Nrf2/heme oxygenase-1 system and counteracts trimethyltin-induced neuronal damage in the human neuroblastoma cell line SH-SY5Y. *Front. Pharmacol.* **2016**, *6*, 305. [CrossRef] [PubMed]
78. Luo, J.-F.; Shen, X.-Y.; Lio, C.K.; Dai, Y.; Cheng, C.-S.; Liu, J.-X.; Yao, Y.-D.; Yu, Y.; Xie, Y.; Luo, P.; et al. Activation of Nrf2/HO-1 Pathway by Nardochinoid C Inhibits Inflammation and Oxidative Stress in Lipopolysaccharide-Stimulated Macrophages. *Front. Pharmacol.* **2018**, *9*, 911. [CrossRef] [PubMed]

antioxidants

MDPI

Review

Arbutin as a Skin Depigmenting Agent with Antimelanogenic and Antioxidant Properties

Yong Chool Boo

Department of Molecular Medicine, Cell and Matrix Research Institute, BK21 Plus KNU Biomedical Convergence Program, School of Medicine, Kyungpook National University, Daegu 41944, Korea; ycboo@knu.ac.kr; Tel.: +82-53-420-4946

Abstract: Arbutin is a compound of hydroquinone and D-glucose, and it has been over 30 years since there have been serious studies on the skin lightening action of this substance. In the meantime, there have been debates and validation studies about the mechanism of action of this substance as well as its skin lightening efficacy and safety. Several analogs or derivatives of arbutin have been developed and studied for their melanin synthesis inhibitory action. Formulations have been developed to improve the stability, transdermal delivery, and release of arbutin, and device usage to promote skin absorption has been developed. Substances that inhibit melanin synthesis synergistically with arbutin have been explored. The skin lightening efficacy of arbutin alone or in combination with other active ingredients has been clinically evaluated. Combined therapy with arbutin and laser could give enhanced depigmenting efficacy. The use of arbutin causes dermatitis rarely, and caution is recommended for the use of arbutin-containing products, especially from the viewpoint that hydroquinone may be generated during product use. Studies on the antioxidant properties of arbutin are emerging, and these antioxidant properties are proposed to contribute to the skin depigmenting action of arbutin. It is hoped that this review will help to understand the pros and cons of arbutin as a cosmetic ingredient, and will lead to future research directions for developing advanced skin lightening and protecting cosmetic products.

Keywords: arbutin; melanin; pigment; melasma; skin lightening; cosmetic; hyperpigmentation; tyrosinase; antioxidant; nuclear factor erythroid 2-related factor 2 (Nrf2)

Citation: Boo, Y.C. Arbutin as a Skin Depigmenting Agent with Antimelanogenic and Antioxidant Properties. *Antioxidants* **2021**, *10*, 1129. https://doi.org/10.3390/antiox10071129

Academic Editors: Irene Dini and Domenico Montesano

Received: 11 June 2021
Accepted: 15 July 2021
Published: 15 July 2021

1. Introduction

Hydroquinone (Chemical structure 1, Figure 1) at concentrations varying from 2 to 5% has been prescribed as the primary therapy for hyperpigmentation disorders including melasma [1,2]. It is used alone or in combination with other active pharmaceutical ingredients, such as retinoids and steroids, for added benefits [3,4]. Hydroquinone not only inhibits tyrosinase (TYR) activity and destroys melanosomes, but also causes necrosis of melanocytes by modifying the membrane structure [5]. This is the potential mechanism of action of hydroquinone as a skin lightening agent as well as its toxicity mechanism. The use of this ingredient in cosmetics has been banned since 2001 because of the high risk of carcinogenesis in case of prolonged exposure to hydroquinone [6,7].

Arbutin (Chemical structure 2, Figure 1) is a compound with a structure in which one molecule of D-glucose is bound to hydroquinone. D-glucose exists in α, β, or γ-anomeric form in aqueous solution, with β-anomer being a dominant form. β-Arbutin (this stereoisomer is called arbutin) in which the β-anomer of D-glucose is bound to hydroquinone is mainly found in plants, such as wheat, pear, and bearberry [8]. α-arbutin (Chemical structure 3, Figure 1) is a compound of hydroquinone and the α-anomer of D-glucose [9]. It has been about 30 years since arbutin was studied in earnest to use as a hydroquinone alternative for skin lightening purposes [10]. It is thus timely to investigate the accumulated information on the efficacy and safety of arbutin and its mechanism of action.

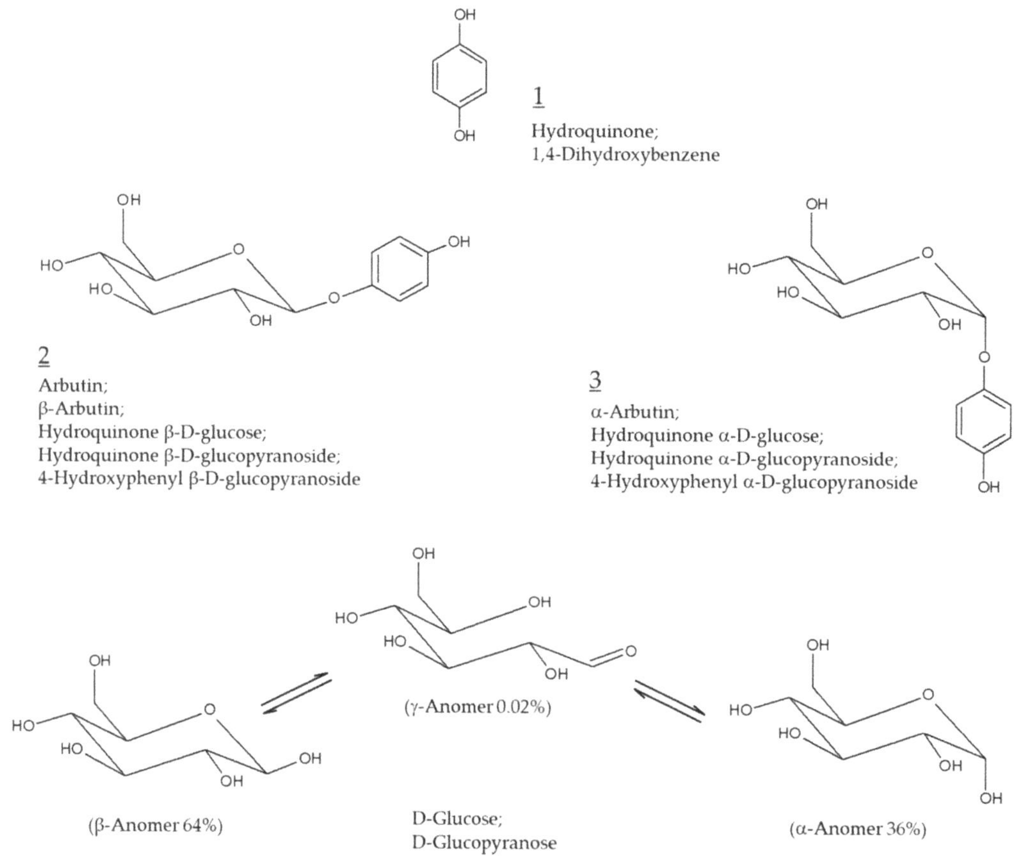

Figure 1. Chemical structures of hydroquinone, arbutin (β-arbutin), and α-arbutin. Anomeric structures of D-glucose are also shown for comparative purposes.

In this review, we will try to get answers to questions that have been controversial. The questions are as follows. Which arbutin or α-arbutin is more advantageous as a skin lightening agent? Do arbutin and α-arbutin have properties of inhibiting melanin synthesis without releasing hydroquinone? Is the toxicity of arbutin due to the release of hydroquinone?

There are other important questions to be answered. Is there a substance with a similar structure with more advantageous properties than arbutin? Could it be possible to increase the skin absorption efficiency of arbutin by a formulation or device-assisted method? Do clinical trial results support the skin lightening benefits of arbutin outweighs its potential harmful risks? What are the side effects of arbutin when applied to the skin? Do the antioxidant properties of arbutin and or α-arbutin contribute to their skin depigmenting action? Can arbutin be applied to skin disorders other than hyperpigmentation?

We hope that this review will help you find answers to these questions. It is also hoped that this review will help identify the pros and cons of arbutin as an active cosmetic ingredient, improve our understanding of its mechanism of action, and set the direction for future research on its cosmetic applications.

2. Modulation of Melanin Synthesis

2.1. Pigmentation and Melanin

Melanin is a polymeric, colored pigment distributed throughout the skin, hair, eye, and other tissues. It is synthesized in melanosomes, unique organelles located in epidermal melanocytes, and plays key roles in maintaining skin homeostasis [11,12], and photoprotection [13,14]. Dysregulated melanin metabolism results in skin pigmentary disorders, such as hyperpigmentation and hypopigmentation [15–17]. The melanocytes derived from individuals with different skin colors show different melanogenic activity although the distribution density of melanocytes in the skin is not much different [18,19]. There was a close association between the melanogenic activity of melanocytes and skin color in humans [20,21].

2.2. Regulation of Melanin Synthesis

Melanin synthesis is influenced by diverse factors including genetic background, epigenetic adaptation, hormonal changes, nutritional status, and environmental conditions [22,23]. For example, proopiomelanocortin-derived peptide hormones, such as α-melanocyte-stimulating hormone (MSH), β-MSH, and adrenocorticotrophic hormone, act as agonists of the melanocortin 1 receptor, a G protein-coupled receptor, and stimulate protein kinase A-mediated signaling pathway leading to the activation of cAMP response element-binding protein (CREB). [22,24]. In the nucleus, the active CREB binds to the cAMP response element on the promoter of microphthalmia-associated transcription factor (MITF), inducing transcription of the target gene [25,26]. MITF plays a key role in the regulation of melanin synthesis by regulating gene expression of melanogenic enzymes, including TYR, tyrosinase-related protein-1 (TYRP-1), and tyrosinase-related protein-2 (TYRP-2) [22,24]. MITF can also be activated by the stem cell factor/receptor tyrosine kinase protein c-Kit/mitogen-activated protein kinases *pathway*, and Wnt/frizzled/glycogen synthase kinase 3β/β-catenin pathway [27,28]. Other signaling pathways, such as phospholipase C/diacylglycerol/protein kinase Cβ cascade, and nitric oxide/cGMP/protein kinase G cascade also can regulate melanin synthesis [29,30]. For more details, please refer to other reviews on autocrine and paracrine regulation and cell signaling pathways associated with melanogenesis [30,31].

2.3. Melanin Synthesis Pathway

Melanin synthesis begins with the oxidation of L-tyrosine to DOPAquinone by monophenolase activity of TYR or the oxidation of L-3,4-dihydroxyphenylalanine (DOPA) to DOPAquinone by diphenolase activity of TYR [32–35]. DOPAquinone reacts with cysteine to produce 5-S-cysteinylDOPA or 2-S-cysteinylDOPA and enters the pheomelanin synthesis pathway. These two compounds are oxidized to quinones and through intramolecular cyclization, and benzothiazine and benzothiazole intermediates are produced. These intermediates are used as building blocks to synthesize the reddish-yellow polymer pheomelanin [36]. Alternatively, when DOPAquinone is oxidized to DOPAchrome via leukoDOPAchrome, it enters the eumelanin synthesis pathway. DOPAchrome is converted to 5,6-dihydroxyindole-2-carboxylic acid (DHICA) by DOPAchrome tautomerase activity of TYRP-2 or to 5,6-dihydroxyindole (DHI) by releasing CO_2. DHICA and DHI are oxidized to quinones by the DHICA oxidase activity of TYRP-1, and the DHI oxidase activity of TYR, respectively, and these are used as building blocks to synthesize the brownish-black polymer eumelanin [37].

2.4. Artificial Modulation of Melanogenesis

TYR-catalyzed enzyme reactions constitute key steps in the biosynthetic routes for melanin, and thus, the enzyme provides a useful target for the pharmacological control of skin hyper- and hypo-pigmentation in dermatology and cosmetology [38,39]. Gene expression of TYR could be enhanced or suppressed via pharmacological approaches [14,40]. Various synthetic and natural compounds are known to inhibit the catalytic activity of TYR in vitro [41–43]. For some active ingredients, the skin depigmenting efficacy was verified through clinical trials [44–48]. Various depigmenting active ingredients including arbutin are used in cosmetics [49].

3. Arbutin

3.1. Anti-Melanogenic Effect of Arbutin

Arbutin has the effect of reducing melanin content at a concentration that has little effect on the viability of cultured human melanocytes. Maeda et al. showed that arbutin dose-dependently reduced TYR activity in human melanocytes at concentrations between 0.1 and 1.0 mM without significantly decreasing cell viability, and its inhibitory effect against cellular melanin synthesis was more potent than that of kojic acid or L-ascorbic acid when compared at a fixed concentration (0.5 mM) [50]. Akiu et al. reported that the melanin content of cultured murine melanoma B16 cells was reduced by arbutin, and the effect was explained by the decrease in intracellular TYR activity [10]. Arbutin was shown to inhibit melanin production in B16 cells stimulated by α-MSH and abrogate the hyperpigmentation effects of α-MSH in brownish guinea pig and human skin explants in organ culture experiments [51].

The decrease in TYR activity in human melanocytes by arbutin does not appear to be due to the decrease in the expression level of this enzyme. Maeda et al. reported that arbutin at 0.5 mM reduced the activity of intracellular TYR by 50% but did not affect the mRNA expression level of TYR [52]. Chakraborty et al. showed that arbutin (0.37 mM) lowered cellular melanin content but did not reduce the protein levels of TYR, TYRP-1, and TYRP-2 [53]. If so, arbutin may inhibit post-translational modification or maturation of newly synthesized TYR or may induce irreversible inactivation of already synthesized mature TYR. In an experiment in which lysates obtained from human melanocytes treated with 1.0 mM arbutin or not were analyzed by zymography, the TYR activity was reduced to 87% in the former [52]. Therefore, it is reasonable to consider arbutin as an inactivator of cellular TYR, rather than a suppressor of TYR gene expression.

In an in vitro assay using crude protein extracts derived from murine melanoma B16 cells, arbutin was shown to be able to directly inhibit the catalytic activity of TYR [10]. In the experiment with mushroom TYR, when L-DOPA was used as a substrate, arbutin showed a lower inhibitory effect than kojic acid and L-ascorbic acid: Their 50% inhibitory (IC_{50}) values were 10 mM, 0.12 mM, and 0.2 mM, respectively [52]. In the experiment using human TYR (derived from melanocytes of Asian neonatal foreskins), when either L-tyrosine (for monophenolase activity) or L-DOPA (for diphenolase activity) were used as substrates, the IC_{50} values of arbutin were 5.7 mM and 18.9 mM, respectively: Arbutin appeared to be an inhibitor in a competitive relationship with L-tyrosine [52]. It is presumed that arbutin competes with a structurally similar substrate to bind to the active site of the TYR enzyme. More importantly, the concentration at which arbutin inhibits the catalytic activity of TYR in vitro is higher than the concentration that reduces cellular melanin, so there is no conviction as to whether this mechanism works in the cell.

It has been reported that arbutin can also act as a substrate for TYR. In the presence of a catalytic amount of L-DOPA as a cofactor, arbutin is oxidized by mushroom TYR to produce 3,4-dihydroxyphenyl-O-beta-D-glucopyranoside [54]. During the catalysis, the TYR enzyme exists as E_{met}, E_{deoxy}, and E_{oxy} forms that are mutually converted [34] (Figure 2). E_{deoxy} binds oxygen to form E_{oxy} that can capture a monophenol (M) or diphenol (D) substrate. When E_{oxy} encounters an M substrate, E_{oxy}-M is formed and it is converted to E_{met}-D that releases a quinone (Q) product and regenerates E_{deoxy}, completing the monophenolase cycle. If E_{oxy} encounters a D substrate, E_{oxy}-D is formed and then releases a Q product and E_{met}. E_{met} can bind to another D substrate, forming E_{met}-D that releases a Q product and E_{deoxy}, completing the monophenolase cycle. E_{deoxy} proceeds to the next cycles. However, in the absence of D substrates, E_{met} binds to a monophenol inhibitor (I), and this results in a dead-end pathway, causing the inactivation of the enzyme. Arbutin is considered to be able to inactivate the TYR enzyme by binding with E_{met} under L-DOPA-deficient conditions [55].

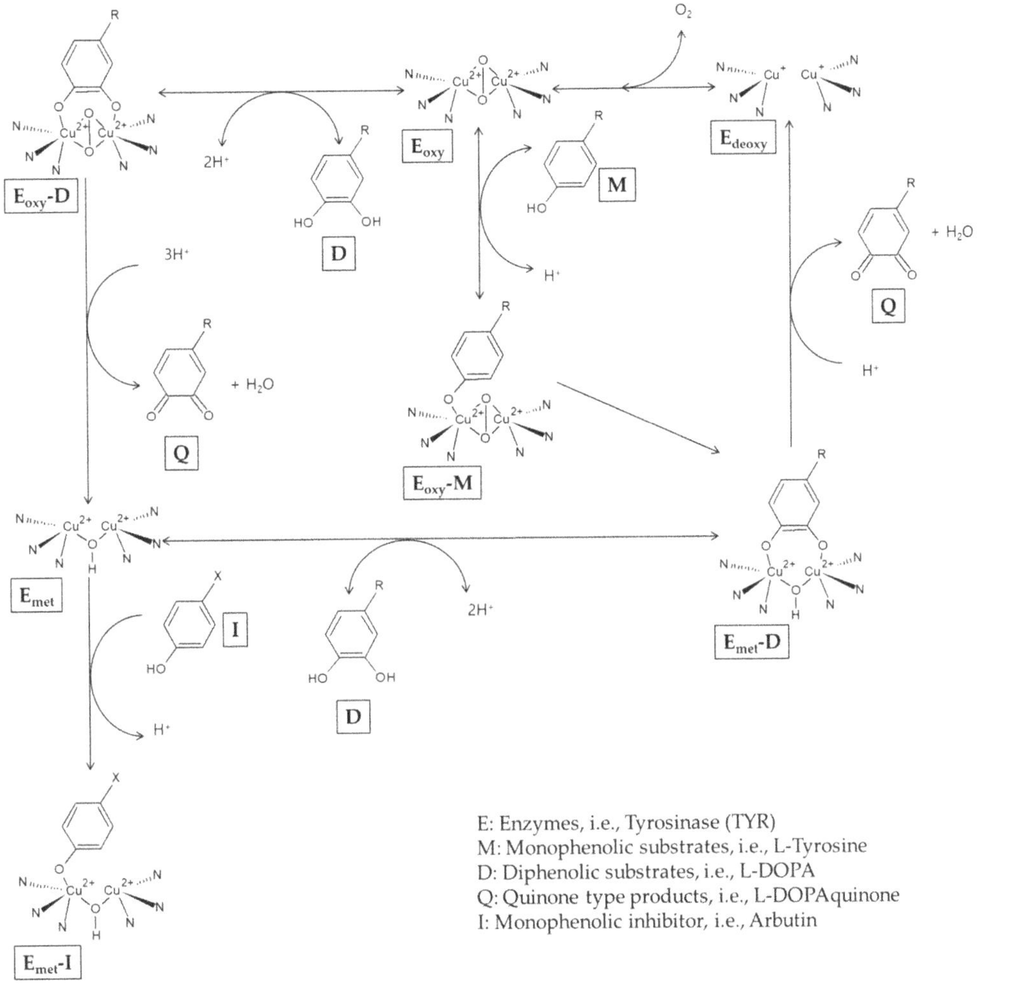

Figure 2. A potential mechanism for the inactivation of tyrosinase by arbutin.

Inoue et al. compared the effects of hydroquinone and arbutin on the differentiation of melanocytes [56]. The results showed that hydroquinone downregulated the early stage of differentiation of mouse embryonic stem cells to neural crest cells, and the late stage of differentiation to melanocytes with melanogenic capability. On the other hand, arbutin did not affect the early and late stages of differentiation of melanocytes and only suppressed elevations in TYR expression in the late stage of differentiation.

3.2. A Possible Production of Hydroquinone from Arbutin

There is a disagreement over whether arbutin works by being decomposed into hydroquinone and glucose or not. Akiu et al. stated that the decomposition of arbutin in the cell suspension could not be observed [10]. However, as shown in the experiment of Maeda et al., hydroquinone can reduce the activity of intracellular TYR by 50% at a concentration 100 times lower than arbutin [50,52]. When arbutin is added to cosmetic products, hydroquinone can be produced to a different level depending on storage conditions [57,58]. In addition, when arbutin is applied to the skin, hydroquinone can be produced by exposure to skin microorganisms [59] or ultraviolet radiation (UVR) [60]. Therefore, there remains a possibility that a small amount of hydroquinone, which may be produced as a decomposition product of arbutin, contributes to the inhibition of melanin synthesis or the inactivation of TYR in cells. Nevertheless, the majority of evidence supports that arbutin has intrinsic properties that inhibit cellular melanogenesis and reduce cellular TYR activity regardless of hydroquinone release.

3.3. Pro-Melanogenic Effect of Arbutin

There is another report that conflicts with many other studies that have reported the melanin-lowering action of arbutin. Nakajima et al. observed that in cultured normal human melanocytes (from neonatal Caucasian foreskins) treated with increasing concentrations of arbutin, the pigmentation became darker (effective concentrations, 2–8 mM), whereas the viability (2–8 mM) and the TYR activity (0.5–4 mM) of the cells decreased [61]. One possible explanation for the particular observation is that arbutin increases the synthesis of intracellular melanin by acting as a substrate of TYR as discussed above. Another possibility is that although arbutin inhibits the synthesis of new melanin, it blocks the release of synthesized melanin out of the cell, thereby causing the accumulation of melanin inside cells. Melanosome transfer is a unique biological process that delivers a package of organelles from melanocytes to keratinocytes [62], and various mechanisms, such as cytophagocytosis, membrane fusion, shedding-phagocytosis, and exocytosis-endocytosis, have been proposed to explain this process [63].

3.4. Preparation of Arbutin

Arbutin can be prepared by various methods, such as extraction from plants, bioconversion from hydroquinone, and chemical synthesis [64,65]. The content of arbutin in a plant varies depending on the species, parts of the plant, development stages, and harvest season [66,67]. The production efficiency for arbutin also varies depending on the extraction and purification methods [68–70].

Fermentation of soybeans with certain strains of Bacillus subtilis was shown to produce arbutin and other TYR inhibitory compounds [71]. Shang et al., engineered a yeast, Yarrowia lipolytica, to produce arbutin by expressing three exogenous genes, such as chorismate pyruvate lyase, 4-hydroxybenzoate 1-hydroxylase, and hydroquinone glucosyltransferase [72]. The engineered Y. lipolytica was capable of de novo biosynthesis of arbutin through the shikimate pathway.

4. α-Arbutin

4.1. Preparation of α-Arbutin

α-arbutin could be produced through transglycosylation from various glucose donors to hydroquinone catalyzed by purified enzymes. Sucrose phosphorylase from Leuconostoc mesenteroides [73], α-amylase from Bacillus subtilis X-23 [74], sucrose isomerase from Erwinia rhapontici [75], amylosucrase from Deinococcus geothermalis [76] and Cellulomonas carboniz T26 [77], and cyclodextrin glucanotransferase from *Thermoanaerobacter* sp. [78] were capable of α-arbutin synthesis.

Kurosu et al. synthesized α-arbutin through fermentation of Xanthomonas campestris WU-9701 [79], Xanthomonas CGMCC, and Xanthomonas BT-112 [80,81]. Wu et al. developed a fed-batch culture strategy for the production of α-arbutin using recombinant Escherichia coli cells anchoring surface-displayed transglucosidase as a whole-cell biocatalyst and lactose as an inducer of the recombinant protein [82]. α-arbutin can be chemically synthesized via an approach similar to that for arbutin [83,84].

4.2. Anti-Melanogenic Effect of α-Arbutin

Many studies have compared the inhibitory effects of arbutin (β-arbutin) and α-arbutin on TYR catalytic activity in an in vitro experiment, with inconsistent results (Table 1). Kiato et al. reported that α-arbutin inhibited monophenolase activity of mushroom TYR with potency slightly lower than arbutin or hydroquinone [73]. Funayama et al. reported that α-arbutin inhibited diphenolase activity of TYR derived from murine melanoma 10 times more potently than β-arbutin, and their IC_{50} values were 0.48 mM and 4.8 mM, respectively [85]. On the other hand, the inhibitory effect against mushroom TYR diphenolase was not observed in the case of α-arbutin, unlike β-arbutin (IC_{50}, 8.4 mM). Later, Qin et al. reported that α-arbutin inhibited monophenolase activity (IC_{50}, 4.5 mM), whereas it activated diphenolase activity of mushroom TYR [86].

Table 1. Tyrosinase inhibitory effects of arbutin and α-arbutin.

Literature	Compounds	Tyrosinase Inhibitory Effects		Enzymes and Substrates Used
		Monophenolase Activity	Diphenolase Activity	
[73]	Hydroquinone	97.2% inhibition at 3 mM		Mushroom tyrosinase; 0.3 mM L-tyrosine
	Arbutin	82.0% inhibition at 3 mM		
	α-arbutin	72.8% inhibition at 3 mM		
[85]	α-arbutin		IC_{50} = 0.48 mM	B16 mouse tyrosinase; 3.3 mM L-DOPA
	Arbutin		IC_{50} = 4.8 mM	
	α-arbutin		No inhibition	Mushroom tyrosinase; 0.83 mM L-DOPA
	Arbutin		IC_{50} = 8.4 mM	
[87]	α-arbutin	IC_{50} = 8 mM	IC_{50} = 8.87 mM	Mushroom tyrosinase; 0.25 mM L-tyrosine plus 0.01 mM L-DOPA for monophenolase activity; 0.5 mM L-DOPA for diphenolase activity
	Arbutin	IC_{50} = 0.9 mM	IC_{50} = 0.7 mM	

Garcia-Jimenez et al. demonstrated that both α and β-arbutin are apparent competitive inhibitors against both the monophenolase and diphenolase activities of TYR [87]. In their study, IC_{50} values of β-arbutin for TYR monophenolase and diphenolase activities were 0.9 and 0.7 mM, respectively, which were much lower than those for α-arbutin (IC_{50}, 8.0 and 8.87 mM for TYR monophenolase and diphenolase activities, respectively). They also kinetically characterized α and β-arbutin as substrates of TYR and obtained their Michaelis constant values of 6.5 mM and 3.0 mM, respectively, supporting that the TYR enzyme has a higher affinity for β-arbutin than α-arbutin.

As such, contradictory results have been reported regarding which arbutin or α-arbutin is a more potent TYR inhibitor. The causes for this extreme inconsistency between studies are not clear and it is only assumed that the inconsistent results may be due to differences in the origin and purity of the enzyme, the conformational state of the enzyme, the type and concentration of substrates, oxygen concentration, pH, temperature, the purity of arbutin and α-arbutin, and the possibility of hydroquinone contamination or production. A conclusion could be reached if several institutions conduct studies comparing the activity of the same substances under standardized experimental conditions.

The antimelanogenic effects of α-arbutin were reported by Sugimoto et al. [88]. They showed that α-arbutin decreased melanin content and TYR activity in cultured human melanoma cells, at a concentration below 1.0 mM, without significant effects on cell growth and TYR mRNA level. They further showed that treatment of the human skin model with α-arbutin (250 μg per tissue) reduced melanin content to a 40% level of the control, without causing cell death.

5. Other Related Compounds

5.1. Glycosidic Derivatives of Arbutin and α-Arbutin

Various glycosidic derivatives of arbutin and α-arbutin, which have additional molecules of glucose, were synthesized through a biological process (Table 2, Figure 3). It is interesting to note that their TYR inhibitory effects are different depending on the structure. However, they have a disadvantage in that the molecular weight is too large for industrial use.

Table 2. Tyrosinase inhibitory effects of arbutin and α-arbutin and their glucosides.

Literature	Compounds		Tyrosinase Inhibitory Effects		Enzymes and Substrates Used
	Name	Chemical Structure	Monophenolase Activity	Diphenolase Activity	
[89]	α-arbutin	3		IC_{50} = 2.1 mM	Human tyrosinase; 3.3 mM L-DOPA
	Arbutin	2		IC_{50} > 30 mM	
	4-Hydroxyphenyl β-maltoside	7		IC_{50} = 5.7 mM	
	4-Hydroxyphenyl β-maltotrioside	8		IC_{50} = 6.1 mM	
[90]	4-Hydroxyphenyl α-maltoside	11		IC_{50} = 4.9 mM	Human tyrosinase; 3.3 mM L-DOPA
	4-Hydroxyphenyl α-maltotrioside	12		IC_{50} = 13.9 mM	
[91]	Arbutin	2		K_i = 2.8 mM	Mushroom tyrosinase; 3.3 mM DOPA
	4-Hydroxyphenyl β-isomaltoside	9		K_i = 3.7 mM	
	4-Hydroxyphenyl β-isomaltotrioside	10		Not determined	
[92]	Arbutin	2	IC_{50} = 6 mM		Mushroom tyrosinase; 0.03% L-tyrosine
	β-D-Glucopyranosyl-(1→6)-arbutin	4	IC_{50} = 8 mM		
	β-D-Glucopyranosyl-(1→4)-arbutin	5	IC_{50} = 10 mM		
	β-D-Glucopyranosyl-(1→3)-arbutin	6	IC_{50} = 5 mM		
	α-D-Glucopyranosyl-(1→4)-arbutin	7	IC_{50} = 5 mM		
[93]	α-arbutin	3		IC_{50} = 2.1 mM	Human tyrosinase; 3 mM L-DOPA
	α-arbutin-α-glucoside	11		IC_{50} = 6.9 mM	
	α-arbutin-α-maltoside	12		IC_{50} = 15.6 mM	
	α-arbutin-α-maltotrioside	13		Not determined	

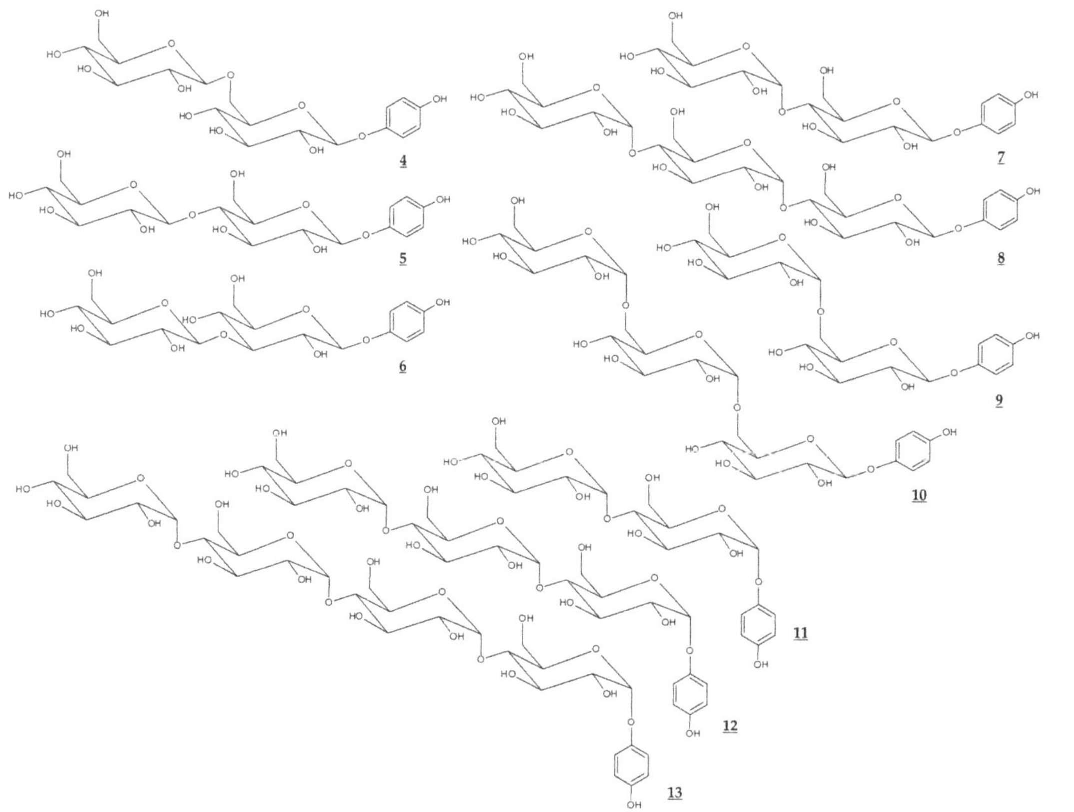

Figure 3. Chemical structures of glucosides of arbutin and α-arbutin.

5.2. Esters of Arbutin

Jiang et al. compared the effects of arbutin and acetylated arbutin (Chemical structure 14, Figure 4) on cell viability, apoptosis, and migration as well as melanin synthesis capacity [94]. After a 24 h treatment at 5.4 mM, both arbutin and acetylated arbutin reduced cell viability and melanogenic capacity, induced cell apoptosis, G1 cell cycle arrest, and mitochondrial disruption in B16 murine melanoma cells, with the latter compound being more effective.

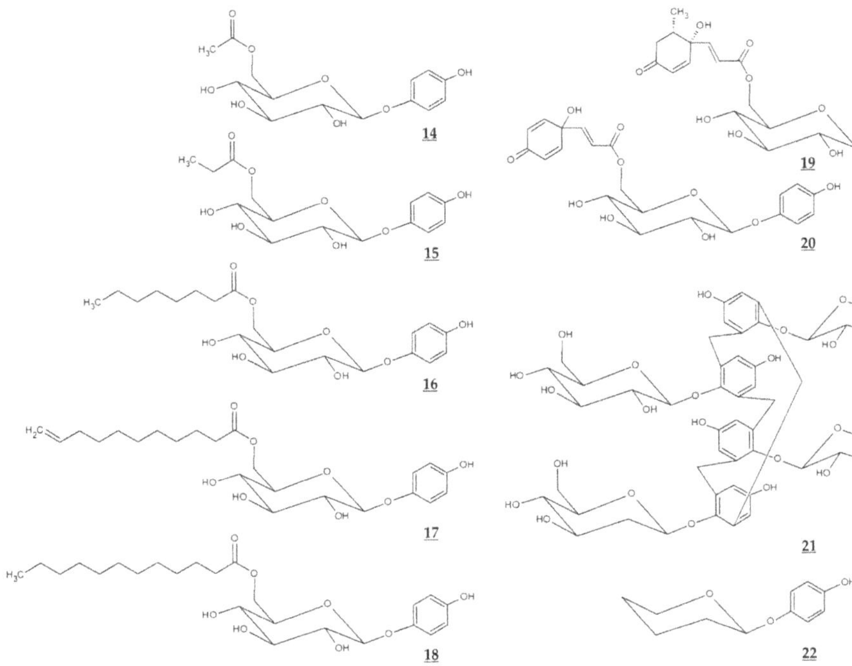

Figure 4. Chemical structures of various derivatives of arbutin and related compounds.

Various ester compounds of arbutin were also synthesized or extracted from plants, and some of them had TYR inhibitory effects or cell melanin synthesis inhibitory effects (Table 3, Figure 4). In particular, arbutin undecylenic acid ester (Chemical structure 17, Figure 4) was shown to be a more potent inhibitor of TYR than arbutin in several studies [95–97]. It is hoped to evaluate its skin lightening efficacy through in vivo studies or clinical trials.

Yamashita-Higuchi et al. isolated arbutin derivatives and related compounds from the leaves of Grevillea robusta [98]. Several compounds, such as grevilloside M and robustaside, showed potent antimelanogenic effects in B16 cells. These compounds were considered to inhibit melanin synthesis without involving TYR inhibition.

5.3. Calixarbutin

Ghaffarzadeh et al. synthesized calixarbutin (Chemical structure 21, Figure 4), a cyclic tetramer of arbutin that inhibits mushroom TYR activity and proliferation of A375 human malignant melanoma cell line, more potently than arbutin does [99]. The effects of calixarbutin on cellular melanogenesis and skin pigmentation are not known. Nonetheless, its cytotoxicity is very strong, and the molecular weight is too big to be used as an anti-melanogenic agent for topical application. Rather, its use for anticancer purposes is expected.

Table 3. Tyrosinase inhibitory effects of various esters of arbutin.

Literature	Compounds		Tyrosinase Inhibitory Effects		Enzymes and Substrates Used
	Name	Chemical Structure	Monophenolase Activity	Diphenolase Activity	
[95]	Arbutin	2	IC_{50} = 3 mM	IC_{50} = 40 mM	Mushroom tyrosinase; 1 mM catechol or 1 mM phenol
	Arbutin undecylenic acid ester	17	IC_{50} = 0.4 mM	IC_{50} = 0.4 mM	
[98]	Grevilloside M	19		No inhibition	Mushroom tyrosinase; 2.5 mM L-DOPA
	Robustaside D	20		No inhibition	
[96]	Arbutin	2	1.72% inhibition at 0.2 mM	Not detected	Mushroom tyrosinase; 2 mM L-tyrosine or 1 mM L-DOPA
	Arbutin propionate	15	0.86% inhibition at 0.2 mM	Not detected	
	Arbutin octylate	16	8.42% inhibition at 0.2 mM	Not detected	
	Arbutin undecenoate	17	15.64% inhibition at 0.2 mM	8.01% inhibition at 0.2 mM	
	Arbutin laurate	18	Not detected	Not detected	
[97]	Arbutin	2		IC_{50} = 29.4 mM	Silkworm hemolymph polyphenol oxidase; 14.4 mM L-DOPA
	Arbutin undecylenic acid ester	17		IC_{50} = 6.36 mM	

5.4. Deoxyarbutin

Deoxyarbutin (Chemical structure 22, Figure 4) was shown to be an effective inhibitor of mushroom TYR in vitro more potent than hydroquinone and arbutin [100,101]. In a hairless, pigmented guinea pig model, deoxyarbutin demonstrated rapid and sustained skin lightening whereas hydroquinone induced a transient skin lightening effect, and kojic acid and arbutin exhibited no significant skin lightening effect [100]. In a human clinical trial, topical treatment of 3% deoxyarbutin for 12 weeks improved solar lentigines in dark skin individuals, although it slightly reduced skin lightness in a population of light skin [100].

Another double-blind randomized controlled study was conducted treating a total of 59 women participants with 2% deoxyarbutin or 4% hydroquinone serum for 12 weeks [102]. During this period, the melanin index changed from 246.88 to 199.61 in the deoxyarbutin-treated group and from 244.22 to 203.98 in the hydroquinone-treated group, and skin lightness (L value) changed from 52.41 to 54.08 in the deoxyarbutin-treated group, and from 52.58 to 54.08 in the hydroquinone-treated group. Thus, it was concluded that that 2% deoxyarbutin and 4% hydroquinone sera showed comparable depigmenting efficacy. Despite these positive clinical results, there is a concern that deoxyarbutin may be degraded to produce hydroquinone, which may cause toxicity (Table 4).

Table 4. Opinion of the scientific committee on consumer safety (SCCS) on the safety of the use of arbutin, α-arbutin, and deoxyarbutin in cosmetic products.

Literature	Compounds		Statements
	Name	Chemical Structure	
[103]	Arbutin	2	"The SCCS considers the use of β-arbutin to be safe for consumers in cosmetic products in a concentration up to 7% in face creams provided that the contamination of hydroquinone in the cosmetic formulations remain below 1 ppm."
[104]	α-arbutin	3	"The SCCS considers the use of α-arbutin safe for consumers in cosmetic products in a concentration up to 2% in face creams and up to 0.5% in body lotions."
[105]	Deoxyarbutin	22	"Therefore, the overall conclusion of the SCCS is that the use of deoxyarbutin up to 3% in face creams is not safe."

6. Formulation and Devise

Arbutin was shown to be stabilized in lipid aggregates [106]. Monomyristoylphosphatidylcholine (14:0 lysoPC) and arbutin formed vesicles with interdigitated bilayers, stabilized by hydrogen bonds between glucose moiety of arbutin and the hydrated population of the carbonyl groups and the phosphates of lysoPC, and hydrophobic interactions between the phenyl group of arbutin and the acyl chain of lysoPC. The inclusion complex of arbutin in hydroxypropyl-β-cyclodextrin increased the water solubility and heat stability of arbutin [107].

Various approaches were developed to improve transdermal delivery and release of arbutin and α-arbutin. Encapsulation of arbutin in liposomes or micelles can enhance transdermal delivery and skin deposition [108,109]. Huang et al. designed nano-sized multi-phase emulsion using hydrocolloids for co-delivery of hydrophilic arbutin and hydrophobic p-coumaric acid (Chemical structure 23, Figure 5) [110]. The multi-phase emulsion showed a controlled release of both active ingredients. Chitosan nanoparticles and gold nanoparticles containing α-arbutin or β-arbutin have been prepared to enhance its transdermal delivery and release [111,112].

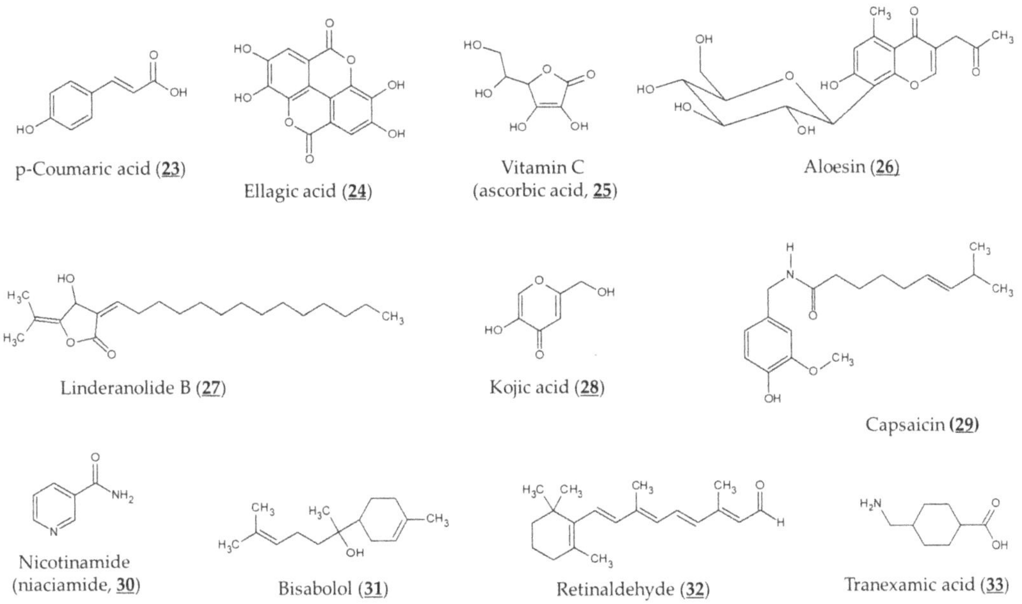

Figure 5. Chemical structures of various compounds that can affect the depigmenting activity of arbutin. Most compounds can additively or synergistically enhance the activity of arbutin, whereas capsaicin antagonizes the activity of arbutin.

Liao et al. reported that transdermal delivery of α-arbutin could be enhanced by ultrasound-assisted technology [113]. Ultrasound treatment using albumin-shelled microbubbles as a contrast agent improved in vitro penetration of α-arbutin through C57BL/6J mouse skin. The microbubble/ultrasound-assisted treatment of α-arbutin solution for 4 weeks enhanced skin lightness of mice, more effectively than conventional treatment of α-arbutin solution. Aung et al. introduced α-arbutin-loaded dissolving microneedles and hydrogel-forming microneedles [114,115]. The microneedles showed enhanced intradermal delivery and skin accumulations of α-arbutin in vivo in mice compared to commercial α-arbutin cream. Kim et al. constructed arbutin-imprinted biomaterials for use in facial mask products [116].

7. Clinical Evaluation of Skin Depigmenting Efficacy

7.1. Skin Depigmenting Efficacy of Arbutin

Choi et al. evaluated the depigmenting efficacy of aloesin and arbutin in a human study [117]. After irradiating the skin area of the forearm with UVR, a 10% solution of each substance was treated alone or together 4 times daily for 15 days. As a result, aleosin, arbutin, and their co-treatment reduced the UVR-induced hyperpigmentation by 34%, 43.5%, and 63.3%, respectively compared to the vehicle treatment.

A randomized, prospective, open-label study of female patients aged 26–50 years with epidermal or mixed melasma evaluated the skin-lightening effects of arbutin and ellagic acid (Chemical structure 24, Figure 5) [118]. A gel formulation containing arbutin (1%), ellagic acid (1%), or ellagic acid plus plant extract (each 1%) was applied to the face twice a day for 6 months, and the skin melanin index was measured before and after using the product. The above three gel formulations reduced the melanin index to 71% ($p = 0.05$), 79% ($p = 0.38$) and 76% ($p < 0.05$) levels of baseline, respectively. No evaluation of control products without an active ingredient was performed and this is a limitation of this study.

A randomized, placebo-controlled, double-blind trial involving 102 women, aged 26– 55, with melasma and solar lentigines, evaluated depigmenting efficacy of arbutin derived from Serratulae quinquefoliae [119]. The study group ($n = 54$) applied the cream containing the plant extract (final concentration of arbutin 2.51%) twice a day on the discolored side for 8 weeks. The results showed that the cream with the plant extract decreased melanin level in the skin pigmentation spot, compared to the control group ($n = 48$) applied with a placebo cream without the active ingredient. During 8 weeks of application, the melanin level of the test group decreased from 182.60 $\pm$ 39.41 to 168.76 $\pm$ 36.30 ($p < 0.000001$), and there was no significant change from 158.9 $\pm$ 34.41 to 166.84 $\pm$ 39.72 in the control group. Clinical improvement was observed in 75.86% of the female patients with melasma and 56.00% of the female patients with solar lentigines.

7.2. Combination with Other Active Ingredients

TYR inhibitory activity of arbutin was shown to be enhanced by combination with L-ascorbic acid (Chemical structure 25, Figure 5), particularly when oxygen is limited [55]. Co-treatment of aloesin (Chemical structure 26, Figure 5) and arbutin was shown to inhibit mushroom TYR activity synergistically [120]. These two compounds also reduced the TYR activity and melanogenesis of cultured human melanocytes synergistically [121]. Synergistic effects inhibiting TYR activity and reducing melanin content of B16F10 cells were also observed between linderanolide B (Chemical structure 27, Figure 5), a natural compound purified from Cinnamomum subavenium, and arbutin or kojic acid (Chemical structure 28, Figure 5) [122]. The inhibitory effect of arbutin on TYR expression and melanin synthesis was reversed by capsaicin (Chemical structure 29, Figure 5) in B16 mouse melanoma cells [123].

There have been several studies evaluating a combination formulation containing arbutin in clinical trials. In a prospective, single-arm, open-label study involving 33 participants with epidermal melasma [124], a cream formulation containing 4% nicotinamide (niacinamide, Chemical structure 30, Figure 5), 3% arbutin, 1% bisabolol (Chemical structure 31, Figure 5), and 0.05% retinaldehyde (Chemical structure 32, Figure 5) was applied to the entire face once daily for 60 days. There was a mean reduction in MASI (melasma area and severity index) score for 60 days (2.25 $\pm$ 1.87, $p < 0.0001$). The mean total melasma surface was significantly reduced from 1398.5 mm^2 at baseline to 923.4 mm^2 at day 60 ($p < 0.0001$).

In a three-arm randomized controlled trial involving a total of 44 subjects [125], treatment of a combination serum that contained 4% ferment filtrate, 2% niacinamide, 4% α-arbutin, and 2% or 3% tranexamic acid (Chemical structure 33, Figure 5) showed a significant improvement in skin brightness and pigmentation intensity after 4 weeks ($p < 0.001$). There were no differences in skin depigmenting efficacy between the combination serum treatment groups and the 4% hydroquinone treatment group.

A prospective, randomized, controlled split-face study [126] evaluated the hyperpigmentation improvement effect of a non-hydroquinone topical formulation (SKNB19) containing epidermal growth factor, tranexamic acid, vitamin C, arbutin, niacinamide, and other ingredients and a standard formulation containing 4% hydroquinone. Eighteen adult subjects with facial pigmentation were randomly assigned to apply SKNB19 twice daily to one side of their face and 4% hydroquinone at night to the other side. SKNB19-treated skin showed a statistically significant improvement in the overall appearance of hyperpigmentation and was rated better than hydroquinone-treated skin.

7.3. Combination Therapy with a Laser Treatment

A prospective study of 35 refractory melasma cases treated with 10 weekly laser sessions using a Q-switched Nd:YAG laser (MedLite C6), two monthly follow-up laser sessions, and topical 7% α-arbutin solution supported that the combination therapy is an effective and well-tolerated treatment for refractory melisma [127]. At 6 months, 66.7% of study subjects showed more than a 51% reduction of melisma even though mild and transient side effects were observed in several cases.

8. Antioxidant Properties of Arbutin and α-Arbutin

8.1. Reactive Oxygen Species in Melanin Synthesis

Melanin plays an important role in skin protection by absorbing UVR [128], but conversely, its synthetic process including reactions catalyzed by TYR generates reactive oxygen species (ROS) [129]. Independently of this, UVR, pollution, hormones, and drugs can stimulate the generation of ROS in melanocytes [130–133].

ROS of various origins can promote melanogenesis or cause melanocyte death, leading to hyperpigmentation or hypopigmentation [134]. Accumulation of oxidative damage leads to tumorigenesis [135]. Therefore, effective antioxidants are expected to reduce oxidative stress in cells, normalize the process of melanin production, and prevent melanocyte death and tumorigenesis. Oral administration of glutathione attenuated oxidative stress as well as hyperpigmentation [136,137].

8.2. Nuclear Factor Erythroid 2-Related Factor 2 (Nrf2)-Mediated Pathway

Cells have various forms of antioxidant defense, including the Nrf2-mediated pathway [40,138]. Nrf2 is a transcription factor that binds to antioxidant responsive elements (ARE) on the promoter regions of many phase II metabolism/antioxidant enzymes [139]. Activation of the Nrf2-ARE pathway leads to induction of phase II metabolic/antioxidant enzymes and increase of the low molecular antioxidants, such as glutathione [140].

Nrf2 was shown to negatively regulate melanin synthesis in cells through modulation of phosphoinositide 3-kinase/protein kinase B signaling pathway [141]. Dietary phenolics could suppress UVR-induced melanogenesis through stimulation of the Nrf2-ARE pathway in addition to their antioxidant and UVR absorption properties [142].

8.3. Antioxidant Properties of Arbutin and α-Arbutin

Takebayashi et al. examined the antioxidant activity of arbutin compared to that of hydroquinone [143]. Arbutin was a weaker scavenger against 1,1-diphenyl-2-picrylhydrazyl radical and a more potent scavenger against 2,2′-azinobis (3-ethylbenzothiazoline-6-sulphonic acid) cation radical compared to hydroquinone. Arbutin exerted slower but long-lasting scavenging activity against 2,2′-azobis (2-methylpropionamidine) dihydrochloride (AAPH)-derived peroxyl radical compared to hydroquinone. At 50 μM arbutin prevented AAPH-induced hemolysis of erythrocytes more effectively than hydroquinone. Arbutin (125–1000 μM) rescued skin fibroblasts exposed to AAPH, whereas hydroquinone (125 μM) exerted cytotoxicity.

Tada et al. detected hydroxyl radical generation from the TYR-catalyzed oxidations of L-tyrosine and L-DOPA, using an electron spin resonance-spin trapping technique [144]. They also observed that arbutin attenuated the hydroxyl radical generation in both reactions, suggesting that arbutin can reduce the levels of ROS derived from the melanogenic pathway. Arbutin at 500 μM was also shown to reduce intracellular hydroxyl radical production and prevent mitochondrial membrane potential loss and played an anti-apoptotic role in human lymphoma U937 cells irradiated with X-ray [145]. Arbutin attenuated the tert-butyl hydroperoxide-induced oxidative stress in human liver cancer HepG2 cell line (effective concentration, 150 μM) [146], human prostate cancer LNCaP cell line, and human fibroblasts (250 and 1000 μM) [147].

Polouliakh et al. conducted in silico comparative genomics analysis in human dermal fibroblasts treated with α-arbutin, and identified transcription factors with a potential role in tumor suppression, toxicity response, and wound healing [148]. α-arbutin upregulated Nrf2 transcription factor which consequently activates target genes involved in antioxidant defense. Arbutin attenuated lipopolysaccharide-induced acute kidney injury in rats, by inhibiting inflammation and apoptosis via the phosphoinositide 3-kinase/protein kinase B/Nrf2 pathway [149]. Arbutin also decreased the levels of pro-inflammatory cytokines and enhanced myocardial antioxidant status, attenuating isoproterenol-induced cardiac hypertrophy in mice [150].

Thus, arbutin and α-arbutin may reduce ROS levels by directly scavenging free radicals or indirectly enhancing the antioxidant capacity of cells through the activation of the Nrf2-ARE pathway. These antioxidant properties may contribute to the inhibitory action of arbutin and α-arbutin on melanin synthesis in cells.

9. Discussion

The generally agreed view on the mechanism by which arbutin inhibits the melanin synthesis in cells is that it inhibits the catalytic activity of already expressed TYR or irreversibly inactivates it rather than suppressing the new synthesis of TYR. Many studies reported that arbutin did not affect mRNA and protein expression of TYR in the concentration range where it inhibited cellular melanin synthesis. There are only a few reports that arbutin reduced the protein level of intracellular TYR [123].

Arbutin is structurally similar to L-tyrosine and can bind to the active site of TYR, thereby acting as a competitive inhibitor. In addition, arbutin can irreversibly inactivate TYR by binding to the enzyme in the form of E_{met} when diphenol substrates are deficient. Of course, there is a possibility that arbutin may act as a substrate for TYR in the form of E_{oxy} when diphenol substrates are present. The last mechanism may not be excluded, because arbutin has been shown to increase cellular melanin synthesis at certain conditions [61]. However, most studies support the mechanism by which arbutin inhibits the catalytic activity of TYR or causes its irreversible inactivation, thereby preventing the synthesis of melanin in cells.

Arbutin exhibits different levels of toxicity depending on the cell type and exposure time, but 1 mM is considered as a boundary concentration between cytotoxicity and safety (Table 5). Therefore, the alleviating action of arbutin on cellular melanin synthesis and oxidative stress observed at a concentration of 1 mM or lower is interpreted as physiologically significant. When arbutin is treated in vivo, its concentration in contact with cells must be maintained at 1 mM or lower, so that beneficial efficacy without the risk of serious side effects can be expected.

Studies on the antioxidant activity of arbutin are emerging [143,147]. Arbutin scavenges ROS, such as hydroxyl radicals [144], and activates the Nrf2-ARE pathway enhancing the antioxidant capacity of cells [149]. Arbutin can inhibit ROS-mediated signal transduction in melanocytes and prevent skin hyperpigmentation, like other dietary phenolic compounds [142]. In addition, the increased pool of intracellular thiol compounds may enhance pheomelanin synthesis and reduce eumelanin synthesis through the formation of DOPA-thiol conjugates [45,136,137]. The antioxidant action-based mechanism and the mechanism based on the TYR inhibitory action are not mutually exclusive and are assumed to work together for the inhibition of eumelanin synthesis (Figure 6).

Table 5. Effects of arbutin on the viability of different cells.

Literature	Cells	Effects of Arbutin on Cell Viability
[50,52]	Human melanocytes derived from neonatal Caucasian or Asian neonatal foreskins	Arbutin treatment at 0.01–1.0 mM for 3 d did not reduce cell viability whereas 5 mM treatment reduced cell viability by 26%.
[53]	Normal human melanocytes from foreskins of 18- to 40-year-old Japanese males	Cells grew well in the presence of 0.37 mM arbutin for 5 d, but 1.1 mM arbutin was cytotoxic and cells detached from the dish within 48 h.
[56]	BRUCE-4 embryonic stem cells of C57BL/6J mouse; Mouse bone marrow-derived stromal ST2 cells	Arbutin treatment for 24 h did not inhibit the proliferation of either cell at 1 mM.
[94]	Murine melanoma B16 cells	After 24 h of treatment, up to 3.6 mM arbutin had no significant effect on cell viability. After 48 h, up to 0.7 mM arbutin did not induce significant toxicity. After 72 h, 0.3–5.4 mM arbutin reduced cell viability by 24–45%. Arbutin at 5.4 mM induced apoptosis.
[143]	Normal human skin fibroblasts	Treatment with up to 1 mM arbutin for 24 h did not affect cell viability.
[151]	Human prostate carcinoma. The LNCaP cell line Human prostate carcinoma LNCaP cells	Treatment with 125–2000 µM for 24, 48, or 72 h did not significantly affect cell viability.Arbutin induced apoptosis at 1000 µM.
[147]	Fibroblast cell line from human newborn foreskins; LNCaP cells	Arbutin reduced the viability of these cells at doses above 1000 µM at 24 and 48 h post-exposure.

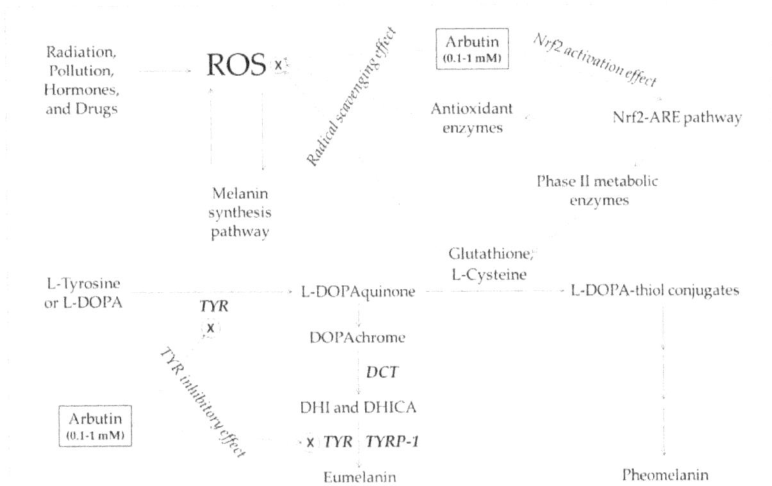

Figure 6. A hypothetical mechanism for the inhibition of eumelanin synthesis by arbutin involving its tyrosinase (TYR) inhibitory and antioxidant activities. Arbutin inhibits the catalytic activity of TYR. It also scavenges reactive oxygen species (ROS) from various sources that can induce melanin synthesis, apoptosis, or tumorigenesis. It can activate the erythroid 2-associated factor 2 (Nrf2)-antioxidant responsive elements (ARE) pathway to enhance the antioxidant capacity of cells. Increased thiol compounds, such as L-cysteine and glutathione, can react with L-DOPAquinon to form DOPA-thiol conjugates, which enter the pheomelanin synthesis pathway. As a result, eumelanin synthesis via DOPAchrome can be selectively downregulated by arbutin. Inhibitory targets of arbutin are indicated with ⊗.

There is still no consensus as to which arbutin or α-arbutin is better in terms of TYR inhibition, cell melanin synthesis inhibition, and skin lightening efficacy and safety. In addition, whether they inhibit cell melanin synthesis or show skin lightening action through the production of hydroquinone has not been completely resolved. Despite these controversies, the results of animal tests and clinical trials prove that arbutin and α-arbutin have the effect of alleviating skin pigmentation.

Arbutin and α-arbutin have the possibility to produce hydroquinone in the manufacturing process of raw materials, the manufacturing and storage process of cosmetic products, and the human use of the products. Whether they are chemically synthesized, extracted from plants, or produced by biological conversion processes, it is an important goal to achieve high purity in the final raw materials. The formulation is important to stabilize arbutin or α-arbutin in the final products. It should be remembered that exposure to microorganisms or UVR during storage and use of the product has the potential to generate hydroquinone [59,60]. Since dermatitis or allergic dermatitis has been observed in certain people using arbutin-containing cosmetic products [152,153], if any side effects are observed while using the product, one must discontinue use immediately and consult a doctor.

Substances and therapies that can synergize with arbutin are attracting attention [117,120,122]. Rather than using arbutin alone, a strategy for maximizing the skin lightening efficacy by combining it with other active ingredients with different mechanisms of action is preferred [124–126]. In future studies, it is hoped that more diverse active ingredients can be combined so that a highly synergistic increase in efficacy can be achieved. Techniques that aid in the transdermal absorption and efficient release of active ingredients are also needed. Nanoemulsions [110], nanoparticles [111,112], and microneedles [114] containing arbutin have been developed. Ultrasonic treatment with microbubbles is could promote percutaneous absorption [113]. Of course, a comprehensive treatment strategy that combines pharmacological treatment with laser treatment is expected to give added benefits [127].

10. Conclusions

In conclusion, the position of arbutin as a skin depigmenting agent is somewhere between the advantages of efficacy and the disadvantages of side effects. Therefore, knowledge of how cosmetic products containing arbutin are made, managed, and used is important. A hypothetical model for the depigmenting mechanism of arbutin involving antioxidant action as well as TYR inhibition is proposed. It is expected that more advanced depigmenting products will be developed based on accumulated information on arbutin and related substances. It is also expected that future research will examine whether arbutin can be applied to skin disorders other than hyperpigmentation.

Funding: This research was supported by a grant of the Korea Health Technology R&D Project through the Korea Health Industry Development Institute (KHIDI), funded by the Ministry of Health and Welfare, Republic of Korea (grant number: HP20C0004).

Conflicts of Interest: The authors declare no conflict of interest.

References

1. Saade, D.S.; Maymone, M.B.C.; De La Garza, H.; Secemsky, E.A.; Kennedy, K.F.; Vashi, N.A. Trends in Use of Prescription Skin Lightening Creams. *Int. J. Environ. Res. Public Health* **2021**, *18*, 5650. [CrossRef] [PubMed]
2. Neagu, N.; Conforti, C.; Agozzino, M.; Marangi, G.F.; Morariu, S.H.; Pellacani, G.; Persichetti, P.; Piccolo, D.; Segreto, F.; Zalaudek, I.; et al. Melasma treatment: A systematic review. *J. Dermatolog. Treat.* **2021**, 1–39. [CrossRef] [PubMed]
3. Jow, T.; Hantash, B.M. Hydroquinone-induced depigmentation: Case report and review of the literature. *Dermatitis* **2014**, *25*, e1–e5. [CrossRef]
4. Ozbey, R.; Okur, M.I. The use of 4% hydroquinone, 0.1% tretinoin, and 0.1% betamethasone creams to prevent hyperpigmentation of split-thickness skin grafts in Long-Evans rats. *J. Cosmet. Dermatol.* **2020**, *19*, 2663–2668. [CrossRef]
5. Jimbow, K.; Obata, H.; Pathak, M.A.; Fitzpatrick, T.B. Mechanism of depigmentation by hydroquinone. *J. Investig. Dermatol.* **1974**, *62*, 436–449. [CrossRef]
6. Kooyers, T.J.; Westerhof, W. Toxicology and health risks of hydroquinone in skin lightening formulations. *J. Eur. Acad. Dermatol. Venereol.* **2006**, *20*, 777–780. [CrossRef]

7. Draelos, Z.D. Skin lightening preparations and the hydroquinone controversy. *Dermatol. Ther.* **2007**, *20*, 308–313. [CrossRef]
8. Xu, W.H.; Liang, Q.; Zhang, Y.J.; Zhao, P. Naturally occurring arbutin derivatives and their bioactivities. *Chem. Biodivers.* **2015**, *12*, 54–81. [CrossRef]
9. Saeedi, M.; Khezri, K.; Seyed Zakaryaei, A.; Mohammadamini, H. A comprehensive review of the therapeutic potential of alpha-arbutin. *Phytother. Res.* **2021**. [CrossRef]
10. Akiu, S.; Suzuki, Y.; Asahara, T.; Fujinuma, Y.; Fukuda, M. Inhibitory effect of arbutin on melanogenesis-biochemical study using cultured B16 melanoma cells. *Nippon Hifuka Gakkai Zasshi* **1991**, *101*, 609–613. [PubMed]
11. Slominski, A.; Kim, T.K.; Brozyna, A.A.; Janjetovic, Z.; Brooks, D.L.; Schwab, L.P.; Skobowiat, C.; Jozwicki, W.; Seagroves, T.N. The role of melanogenesis in regulation of melanoma behavior: Melanogenesis leads to stimulation of HIF-1alpha expression and HIF-dependent attendant pathways. *Arch. Biochem. Biophys.* **2014**, *563*, 79–93. [CrossRef]
12. Slominski, R.M.; Zmijewski, M.A.; Slominski, A.T. The role of melanin pigment in melanoma. *Exp. Dermatol.* **2015**, *24*, 258–259. [CrossRef]
13. Yamaguchi, Y.; Beer, J.Z.; Hearing, V.J. Melanin mediated apoptosis of epidermal cells damaged by ultraviolet radiation: Factors influencing the incidence of skin cancer. *Arch. Dermatol. Res.* **2008**, *300* (Suppl. 1), S43–S50. [CrossRef]
14. Boo, Y.C. Emerging Strategies to Protect the Skin from Ultraviolet Rays Using Plant-Derived Materials. *Antioxidants* **2020**, *9*, 637. [CrossRef]
15. Rose, P.T. Pigmentary disorders. *Med. Clin. N. Am.* **2009**, *93*, 1225–1239. [CrossRef]
16. Ganju, P.; Nagpal, S.; Mohammed, M.H.; Nishal Kumar, P.; Pandey, R.; Natarajan, V.T.; Mande, S.S.; Gokhale, R.S. Microbial community profiling shows dysbiosis in the lesional skin of Vitiligo subjects. *Sci. Rep.* **2016**, *6*, 18761. [CrossRef]
17. Spritz, R.A.; Andersen, G.H. Genetics of Vitiligo. *Dermatol. Clin.* **2017**, *35*, 245–255. [CrossRef] [PubMed]
18. Costin, G.E.; Hearing, V.J. Human skin pigmentation: Melanocytes modulate skin color in response to stress. *FASEB J.* **2007**, *21*, 976–994. [CrossRef] [PubMed]
19. Slominski, A.T.; Zmijewski, M.A.; Skobowiat, C.; Zbytek, B.; Slominski, R.M.; Steketee, J.D. Sensing the environment: Regulation of local and global homeostasis by the skin's neuroendocrine system. *Adv. Anat. Embryol. Cell Biol.* **2012**, *212*, 1–115.
20. Iwata, M.; Corn, T.; Iwata, S.; Everett, M.A.; Fuller, B.B. The relationship between tyrosinase activity and skin color in human foreskins. *J. Investig. Dermatol.* **1990**, *95*, 9–15. [CrossRef] [PubMed]
21. Iozumi, K.; Hoganson, G.E.; Pennella, R.; Everett, M.A.; Fuller, B.B. Role of tyrosinase as the determinant of pigmentation in cultured human melanocytes. *J. Investig. Dermatol.* **1993**, *100*, 806–811. [CrossRef]
22. Slominski, A.; Tobin, D.J.; Shibahara, S.; Wortsman, J. Melanin pigmentation in mammalian skin and its hormonal regulation. *Physiol. Rev.* **2004**, *84*, 1155–1228. [CrossRef]
23. Videira, I.F.; Moura, D.F.; Magina, S. Mechanisms regulating melanogenesis. *An. Bras. Dermatol.* **2013**, *88*, 76–83. [CrossRef] [PubMed]
24. Steinhoff, M.; Stander, S.; Seeliger, S.; Ansel, J.C.; Schmelz, M.; Luger, T. Modern aspects of cutaneous neurogenic inflammation. *Arch. Dermatol.* **2003**, *139*, 1479–1488. [CrossRef] [PubMed]
25. Yasumoto, K.; Yokoyama, K.; Takahashi, K.; Tomita, Y.; Shibahara, S. Functional analysis of microphthalmia-associated transcription factor in pigment cell-specific transcription of the human tyrosinase family genes. *J. Biol. Chem.* **1997**, *272*, 503–509. [CrossRef] [PubMed]
26. Busca, R.; Ballotti, R. Cyclic AMP a key messenger in the regulation of skin pigmentation. *Pigment Cell Res.* **2000**, *13*, 60–69. [CrossRef] [PubMed]
27. Flaherty, K.T.; Hodi, F.S.; Fisher, D.E. From genes to drugs: Targeted strategies for melanoma. *Nat. Rev. Cancer* **2012**, *12*, 349–361. [CrossRef] [PubMed]
28. Serre, C.; Busuttil, V.; Botto, J.M. Intrinsic and extrinsic regulation of human skin melanogenesis and pigmentation. *Int. J. Cosmet. Sci.* **2018**, *40*, 328–347. [CrossRef] [PubMed]
29. Rzepka, Z.; Buszman, E.; Beberok, A.; Wrzesniok, D. From tyrosine to melanin: Signaling pathways and factors regulating melanogenesis. *Postepy Hig. Med. Dosw.* **2016**, *70*, 695–708. [CrossRef] [PubMed]
30. D'Mello, S.A.; Finlay, G.J.; Baguley, B.C.; Askarian-Amiri, M.E. Signaling Pathways in Melanogenesis. *Int. J. Mol. Sci.* **2016**, *17*, 1144. [CrossRef]
31. Yuan, X.H.; Jin, Z.H. Paracrine regulation of melanogenesis. *Br. J. Dermatol.* **2018**, *178*, 632–639. [CrossRef]
32. Cooksey, C.J.; Garratt, P.J.; Land, E.J.; Pavel, S.; Ramsden, C.A.; Riley, P.A.; Smit, N.P. Evidence of the indirect formation of the catecholic intermediate substrate responsible for the autoactivation kinetics of tyrosinase. *J. Biol. Chem.* **1997**, *272*, 26226–26235. [CrossRef] [PubMed]
33. Simon, J.D.; Peles, D.; Wakamatsu, K.; Ito, S. Current challenges in understanding melanogenesis: Bridging chemistry, biological control, morphology, and function. *Pigment Cell Melanoma Res.* **2009**, *22*, 563–579. [CrossRef] [PubMed]
34. Olivares, C.; Solano, F. New insights into the active site structure and catalytic mechanism of tyrosinase and its related proteins. *Pigment Cell Melanoma Res.* **2009**, *22*, 750–760. [CrossRef] [PubMed]
35. Longo, D.L.; Stefania, R.; Aime, S.; Oraevsky, A. Melanin-Based Contrast Agents for Biomedical Optoacoustic Imaging and Theranostic Applications. *Int. J. Mol. Sci.* **2017**, *18*, 1719. [CrossRef] [PubMed]
36. Davy, A.D.; Birch, D.J.S. Evidence for pheomelanin sheet structure. *Appl. Phys. Lett.* **2018**, *113*, 263701. [CrossRef]
37. Grieco, C.; Kohl, F.R.; Hanes, A.T.; Kohler, B. Probing the heterogeneous structure of eumelanin using ultrafast vibrational fingerprinting. *Nat. Commun.* **2020**, *11*, 4569. [CrossRef]

38. Ando, H.; Kondoh, H.; Ichihashi, M.; Hearing, V.J. Approaches to identify inhibitors of melanin biosynthesis via the quality control of tyrosinase. *J. Investig. Dermatol.* **2007**, *127*, 751–761. [CrossRef]
39. Pillaiyar, T.; Manickam, M.; Namasivayam, V. Skin whitening agents: Medicinal chemistry perspective of tyrosinase inhibitors. *J. Enzyme Inhib. Med. Chem.* **2017**, *32*, 403–425. [CrossRef]
40. Boo, Y.C. Natural Nrf2 Modulators for Skin Protection. *Antioxidants* **2020**, *9*, 812. [CrossRef]
41. Niu, C.; Aisa, H.A. Upregulation of Melanogenesis and Tyrosinase Activity: Potential Agents for Vitiligo. *Molecules* **2017**, *22*, 1303. [CrossRef]
42. Pillaiyar, T.; Namasivayam, V.; Manickam, M.; Jung, S.H. Inhibitors of Melanogenesis: An Updated Review. *J. Med. Chem.* **2018**, *61*, 7395–7418. [CrossRef]
43. Zolghadri, S.; Bahrami, A.; Hassan Khan, M.T.; Munoz-Munoz, J.; Garcia-Molina, F.; Garcia-Canovas, F.; Saboury, A.A. A comprehensive review on tyrosinase inhibitors. *J. Enzyme Inhib. Med. Chem.* **2019**, *34*, 279–309. [CrossRef]
44. Kolbe, L.; Mann, T.; Gerwat, W.; Batzer, J.; Ahlheit, S.; Scherner, C.; Wenck, H.; Stab, F. 4-n-butylresorcinol, a highly effective tyrosinase inhibitor for the topical treatment of hyperpigmentation. *J. Eur. Acad. Dermatol. Venereol.* **2013**, *27* (Suppl. S1), 19–23. [CrossRef]
45. Watanabe, F.; Hashizume, E.; Chan, G.P.; Kamimura, A. Skin-whitening and skin-condition-improving effects of topical oxidized glutathione: A double-blind and placebo-controlled clinical trial in healthy women. *Clin. Cosmet. Investig. Dermatol.* **2014**, *7*, 267–274. [CrossRef]
46. Boo, Y.C. p-Coumaric Acid as An Active Ingredient in Cosmetics: A Review Focusing on its Antimelanogenic Effects. *Antioxidants* **2019**, *8*, 275. [CrossRef] [PubMed]
47. Boo, Y.C. Human Skin Lightening Efficacy of Resveratrol and Its Analogs: From in Vitro Studies to Cosmetic Applications. *Antioxidants* **2019**, *8*, 332. [CrossRef] [PubMed]
48. Boo, Y.C.; Jo, D.J.; Oh, C.M.; Lee, S.Y.; Kim, Y.M. The First Human Clinical Trial on the Skin Depigmentation Efficacy of Glycinamide Hydrochloride. *Biomedicines* **2020**, *8*, 257. [CrossRef]
49. Hu, Z.M.; Zhou, Q.; Lei, T.C.; Ding, S.F.; Xu, S.Z. Effects of hydroquinone and its glucoside derivatives on melanogenesis and antioxidation: Biosafety as skin whitening agents. *J. Dermatol. Sci.* **2009**, *55*, 179–184. [CrossRef]
50. Maeda, K.; Fukuda, M. In vitro effectiveness of several whitening cosmetic components in human melanocytes. *J. Soc. Cosmet. Chem.* **1991**, *42*, 361–368.
51. Lim, Y.J.; Lee, E.H.; Kang, T.H.; Ha, S.K.; Oh, M.S.; Kim, S.M.; Yoon, T.J.; Kang, C.; Park, J.H.; Kim, S.Y. Inhibitory effects of arbutin on melanin biosynthesis of alpha-melanocyte stimulating hormone-induced hyperpigmentation in cultured brownish guinea pig skin tissues. *Arch. Pharm. Res.* **2009**, *32*, 367–373. [CrossRef]
52. Maeda, K.; Fukuda, M. Arbutin: Mechanism of its depigmenting action in human melanocyte culture. *J. Pharmacol. Exp. Ther.* **1996**, *276*, 765–769. [PubMed]
53. Chakraborty, A.K.; Funasaka, Y.; Komoto, M.; Ichihashi, M. Effect of arbutin on melanogenic proteins in human melanocytes. *Pigment Cell Res.* **1998**, *11*, 206–212. [CrossRef] [PubMed]
54. Nihei, K.; Kubo, I. Identification of oxidation product of arbutin in mushroom tyrosinase assay system. *Bioorg. Med. Chem. Lett.* **2003**, *13*, 2409–2412. [CrossRef]
55. Hori, I.; Nihei, K.; Kubo, I. Structural criteria for depigmenting mechanism of arbutin. *Phytother. Res.* **2004**, *18*, 475–479. [CrossRef] [PubMed]
56. Inoue, Y.; Hasegawa, S.; Yamada, T.; Date, Y.; Mizutani, H.; Nakata, S.; Matsunaga, K.; Akamatsu, H. Analysis of the effects of hydroquinone and arbutin on the differentiation of melanocytes. *Biol. Pharm. Bull.* **2013**, *36*, 1722–1730. [CrossRef]
57. Jeon, J.S.; Kim, B.H.; Lee, S.H.; Kwon, H.J.; Bae, H.J.; Kim, S.K.; Park, J.A.; Shim, J.H.; Abd El-Aty, A.M.; Shin, H.C. Simultaneous determination of arbutin and its decomposed product hydroquinone in whitening creams using high-performance liquid chromatography with photodiode array detection: Effect of temperature and pH on decomposition. *Int. J. Cosmet. Sci.* **2015**, *37*, 567–573. [CrossRef]
58. Avonto, C.; Wang, Y.H.; Avula, B.; Wang, M.; Rua, D.; Khan, I.A. Comparative studies on the chemical and enzymatic stability of alpha- and beta-arbutin. *Int. J. Cosmet. Sci.* **2016**, *38*, 187–193. [CrossRef]
59. Bang, S.H.; Han, S.J.; Kim, D.H. Hydrolysis of arbutin to hydroquinone by human skin bacteria and its effect on antioxidant activity. *J. Cosmet. Dermatol.* **2008**, *7*, 189–193. [CrossRef]
60. Chang, N.F.; Chen, Y.S.; Lin, Y.J.; Tai, T.H.; Chen, A.N.; Huang, C.H.; Lin, C.C. Study of Hydroquinone Mediated Cytotoxicity and Hypopigmentation Effects from UVB-Irradiated Arbutin and DeoxyArbutin. *Int. J. Mol. Sci.* **2017**, *18*, 969. [CrossRef]
61. Nakajima, M.; Shinoda, I.; Fukuwatari, Y.; Hayasawa, H. Arbutin increases the pigmentation of cultured human melanocytes through mechanisms other than the induction of tyrosinase activity. *Pigment Cell Res.* **1998**, *11*, 12–17. [CrossRef]
62. Rok, J.; Otręba, M.; Buszman, E.; Wrześniok, D. Melanin—From melanocyte to keratinocyte, that is how melanin is transported within the skin. *Ann. Acad. Med. Sil.* **2012**, *66*, 60–66.
63. Wu, X.F.; Hammer, J.A. Melanosome transfer: It is best to give and receive. *Curr. Opin. Cell Biol.* **2014**, *29*, 1–7. [CrossRef] [PubMed]
64. Zhu, X.; Tian, Y.; Zhang, W.; Zhang, T.; Guang, C.; Mu, W. Recent progress on biological production of alpha-arbutin. *Appl. Microbiol. Biotechnol.* **2018**, *102*, 8145–8152. [CrossRef] [PubMed]
65. Zhou, H.; Zhao, J.; Li, A.; Reetz, M.T. Chemical and Biocatalytic Routes to Arbutin (dagger). *Molecules* **2019**, *24*, 3303. [CrossRef] [PubMed]

66. Cui, T.; Nakamura, K.; Ma, L.; Li, J.Z.; Kayahara, H. Analyses of arbutin and chlorogenic acid, the major phenolic constituents in oriental pear. *J. Agric. Food Chem.* **2005**, *53*, 3882–3887. [CrossRef]
67. Tumova, L.; Doleckova, I.; Hendrychova, H.; Kasparova, M. Arbutin Content and Tyrosinase Activity of Bergenia Extracts. *Nat. Prod. Commun.* **2017**, *12*, 549–552.
68. Cho, J.-Y.; Park, K.Y.; Lee, K.H.; Lee, H.J.; Lee, S.-H.; Cho, J.A.; Kim, W.-S.; Shin, S.-C.; Park, K.-H.; Moon, J.-H. Recovery of arbutin in high purity from fruit peels of pear (Pyrus pyrifolia Nakai). *Food Sci. Biotechnol.* **2011**, *20*, 801–807. [CrossRef]
69. Sasaki, C.; Ichitani, M.; Kunimoto, K.K.; Asada, C.; Nakamura, Y. Extraction of arbutin and its comparative content in branches, leaves, stems, and fruits of Japanese pear Pyrus pyrifolia cv. Kousui. *Biosci. Biotechnol. Biochem.* **2014**, *78*, 874–877. [CrossRef]
70. Lee, B.-D.; Eun, J.-B. Optimum extraction conditions for arbutin from Asian pear peel by supercritical fluid extraction (SFE) using Box-Behnken design. *J. Med. Plants Res.* **2012**, *6*, 2348–2364.
71. Jin, Y.H.; Jeon, A.R.; Mah, J.H. Tyrosinase Inhibitory Activity of Soybeans Fermented with Bacillus subtilis Capable of Producing a Phenolic Glycoside, Arbutin. *Antioxidants* **2020**, *9*, 1301. [CrossRef]
72. Shang, Y.Z.; Wei, W.P.; Zhang, P.; Ye, B.C. Engineering Yarrowia lipolytica for Enhanced Production of Arbutin. *J. Agric. Food Chem.* **2020**, *68*, 1364–1372. [CrossRef]
73. Kitao, S.; Sekine, H. alpha-D-Glucosyl Transfer to Phenolic Compounds by Sucrose Phosphorylase from Leuconostoc mesenteroides and Production of alpha-Arbutin. *Biosci. Biotechnol. Biochem.* **1994**, *58*, 38–42. [CrossRef]
74. Nishimura, T.; Kometani, T.; Takii, H.; Terada, Y.; Okada, S. Purification and Some Properties of Alpha-Amylase from Bacillus-Subtilis X-23 That Glucosylates Phenolic-Compounds Such as Hydroquinone. *J. Ferment. Bioeng.* **1994**, *78*, 31–36. [CrossRef]
75. Zhou, X.; Zheng, Y.T.; Wei, X.M.; Yang, K.D.; Yang, X.K.; Wang, Y.T.; Xu, L.M.; Du, L.Q.; Huang, R.B. Sucrose Isomerase and Its Mutants from Erwinia rhapontici Can Synthesise alpha-Arbutin. *Protein Pept. Lett.* **2011**, *18*, 1028–1034. [CrossRef]
76. Seo, D.H.; Jung, J.H.; Ha, S.J.; Cho, H.K.; Jung, D.H.; Kim, T.J.; Baek, N.I.; Yoo, S.H.; Park, C.S. High-yield enzymatic bioconversion of hydroquinone to alpha-arbutin, a powerful skin lightening agent, by amylosucrase. *Appl. Microbiol. Biotechnol.* **2012**, *94*, 1189–1197. [CrossRef] [PubMed]
77. Yu, S.H.; Wang, Y.C.; Tian, Y.Q.; Xu, W.; Bai, Y.X.; Zhang, T.; Mu, W.M. Highly efficient biosynthesis of alpha-arbutin from hydroquinone by an amylosucrase from Cellulomonas carboniz. *Process Biochem.* **2018**, *68*, 93–99. [CrossRef]
78. Mathew, S.; Adlercreutz, P. Regioselective glycosylation of hydroquinone to alpha-arbutin by cyclodextrin glucanotransferase from *Thermoanaerobacter* sp. *Biochem. Eng. J.* **2013**, *79*, 187–193. [CrossRef]
79. Kurosu, J.; Sato, T.; Yoshida, K.; Tsugane, T.; Shimura, S.; Kirimura, K.; Kino, K.; Usami, S. Enzymatic synthesis of alpha-arbutin by alpha-anomer-selective-glucosylation of hydroquinone using lyophilized cells of Xanthomonas campestris WU. *J. Biosci. Bioeng.* **2002**, *93*, 328–330. [CrossRef]
80. Liu, C.; Zhang, P.; Liu, L.; Xu, T.; Tan, T.; Wang, F.; Deng, L. Isolation of alpha-arbutin from Xanthomonas CGMCC 1243 fermentation broth by macroporous resin adsorption chromatography. *J. Chromatogr. B Analyt. Technol. Biomed. Life Sci.* **2013**, *925*, 104–109. [CrossRef] [PubMed]
81. Wei, M.; Ren, Y.; Liu, C.; Liu, R.; Zhang, P.; Wei, Y.; Xu, T.; Wang, F.; Tan, T.; Liu, C. Fermentation scale up for alpha-arbutin production by Xanthomonas BT-112. *J. Biotechnol.* **2016**, *233*, 1–5. [CrossRef]
82. Wu, P.H.; Nair, G.R.; Chu, I.M.; Wu, W.T. High cell density cultivation of Escherichia coli with surface anchored transglucosidase for use as whole-cell biocatalyst for alpha-arbutin synthesis. *J. Ind. Microbiol. Biotechnol.* **2008**, *35*, 95–101. [CrossRef] [PubMed]
83. Wang, Z.X.; Shi, X.X.; Chen, G.R.; Ren, Z.H.; Luo, L.; Yan, J. A new synthesis of alpha-arbutin via Lewis acid catalyzed selective glycosylation of tetra-O-benzyl-alpha-D-glucopyranosyl trichloroacetimidate with hydroquinone. *Carbohydr. Res.* **2006**, *341*, 1945–1947. [CrossRef] [PubMed]
84. Cepanec, I.; Litvic, M. Simple and efficient synthesis of arbutin. *Arkivoc* **2008**, *2*, 19–24.
85. Funayama, M.; Arakawa, H.; Yamamoto, R.; Nishino, T.; Shin, T.; Murao, S. Effects of alpha- and beta-arbutin on activity of tyrosinases from mushroom and mouse melanoma. *Biosci. Biotechnol. Biochem.* **1995**, *59*, 143–144. [CrossRef] [PubMed]
86. Qin, L.; Wu, Y.; Liu, Y.; Chen, Y.; Zhang, P. Dual effects of alpha-arbutin on monophenolase and diphenolase activities of mushroom tyrosinase. *PLoS ONE* **2014**, *9*, e109398. [CrossRef]
87. Garcia-Jimenez, A.; Teruel-Puche, J.A.; Berna, J.; Rodriguez-Lopez, J.N.; Tudela, J.; Garcia-Canovas, F. Action of tyrosinase on alpha and beta-arbutin: A kinetic study. *PLoS ONE* **2017**, *12*, e0177330. [CrossRef] [PubMed]
88. Sugimoto, K.; Nishimura, T.; Nomura, K.; Sugimoto, K.; Kuriki, T. Inhibitory effects of alpha-arbutin on melanin synthesis in cultured human melanoma cells and a three-dimensional human skin model. *Biol. Pharm. Bull.* **2004**, *27*, 510–514. [CrossRef] [PubMed]
89. Sugimoto, K.; Nishimura, T.; Nomura, K.; Sugimoto, K.; Kuriki, T. Syntheses of arbutin-alpha-glycosides and a comparison of their inhibitory effects with those of alpha-arbutin and arbutin on human tyrosinase. *Chem. Pharm. Bull.* **2003**, *51*, 798–801. [CrossRef] [PubMed]
90. Sugimoto, K.; Nomura, K.; Nishimura, T.; Kiso, T.; Sugimoto, K.; Kuriki, T. Syntheses of alpha-arbutin-alpha-glycosides and their inhibitory effects on human tyrosinase. *J. Biosci. Bioeng.* **2005**, *99*, 272–276. [CrossRef] [PubMed]
91. Moon, Y.H.; Nam, S.H.; Kang, J.; Kim, Y.M.; Lee, J.H.; Kang, H.K.; Breton, V.; Jun, W.J.; Park, K.D.; Kimura, A.; et al. Enzymatic synthesis and characterization of arbutin glucosides using glucansucrase from Leuconostoc mesenteroides B-1299CB. *Appl. Microbiol. Biotechnol.* **2007**, *77*, 559–567. [CrossRef] [PubMed]

92. Jun, S.Y.; Park, K.M.; Choi, K.W.; Jang, M.K.; Kang, H.Y.; Lee, S.H.; Park, K.H.; Cha, J. Inhibitory effects of arbutin-beta-glycosides synthesized from enzymatic transglycosylation for melanogenesis. *Biotechnol. Lett.* **2008**, *30*, 743–748. [CrossRef] [PubMed]
93. Rudeekulthamrong, P.; Kaulpiboon, J. Optimization of amylomaltase for the synthesis of alpha-arbutin derivatives as tyrosinase inhibitors. *Carbohydr. Res.* **2020**, *494*, 108078. [CrossRef] [PubMed]
94. Jiang, L.; Wang, D.; Zhang, Y.; Li, J.; Wu, Z.; Wang, Z.; Wang, D. Investigation of the pro-apoptotic effects of arbutin and its acetylated derivative on murine melanoma cells. *Int. J. Mol. Med.* **2018**, *41*, 1048–1054. [CrossRef] [PubMed]
95. Tokiwa, Y.; Kitagawa, M.; Raku, T. Enzymatic synthesis of arbutin undecylenic acid ester and its inhibitory effect on mushroom tyrosinase. *Biotechnol. Lett.* **2007**, *29*, 481–486. [CrossRef]
96. Xu, H.; Li, X.; Xin, X.; Mo, L.; Zou, Y.; Zhao, G.; Yu, Y.; Chen, K. Antityrosinase Mechanism and Antimelanogenic Effect of Arbutin Esters Synthesis Catalyzed by Whole-Cell Biocatalyst. *J. Agric. Food Chem.* **2021**, *69*, 4243–4252. [CrossRef]
97. Masyita, A.; Salim, E.; Asri, R.M.; Nainu, F.; Hori, A.; Yulianty, R.; Hatta, M.; Rifai, Y.; Kuraishi, T. Molecular modeling and phenoloxidase inhibitory activity of arbutin and arbutin undecylenic acid ester. *Biochem. Biophys. Res. Commun.* **2021**, *547*, 75–81. [CrossRef]
98. Yamashita-Higuchi, Y.; Sugimoto, S.; Matsunami, K.; Otsuka, H.; Nakai, T. Grevillosides J-Q, arbutin derivatives from the leaves of Grevillea robusta and their melanogenesis inhibitory activity. *Chem. Pharm. Bull.* **2014**, *62*, 364–372. [CrossRef] [PubMed]
99. Ghaffarzadeh, J.; Nasuhi Pur, F. Calixarbutin: A Novel Calixarene-based Potential Water-soluble Anti-tyrosinase Agent with High Anti-melanoma Activity. *Iran. J. Pharm. Res.* **2020**, *19*, 236–241.
100. Boissy, R.E.; Visscher, M.; DeLong, M.A. DeoxyArbutin: A novel reversible tyrosinase inhibitor with effective in vivo skin lightening potency. *Exp. Dermatol.* **2005**, *14*, 601–608. [CrossRef]
101. Chawla, S.; de Long, M.A.; Visscher, M.O.; Wickett, R.R.; Manga, P.; Boissy, R.E. Mechanism of tyrosinase inhibition by deoxyArbutin and its second-generation derivatives. *Br. J. Dermatol.* **2008**, *159*, 1267–1274. [CrossRef]
102. Anwar, A.I.; Asmarani, Y.; Madjid, A.; Patellongi, I.; Adriani, A.; As'ad, S.; Kurniadi, I. Comparison of 2% deoxyarbutin and 4% hydroquinone as a depigmenting agent in healthy individuals: A double-blind randomized controlled clinical trial. *J. Cosmet. Dermatol.* **2021**. [CrossRef]
103. SCCS; Degen, G.H. Opinion of the Scientific Committee on Consumer Safety (SCCS)—Opinion on the safety of the use of beta-arbutin in cosmetic products. *Regul. Toxicol. Pharmacol.* **2015**, *73*, 866–867. [CrossRef]
104. Yang, Z.Q.; Wang, Z.H.; Zhang, T.L.; Tu, J.B.; Song, Y.; Hu, X.Y.; Li, G.G. The effect of aloesin on melanocytes in the pigmented skin equivalent model. *Zhonghua Zheng Xing Wai Ke Za Zhi* **2008**, *24*, 50–53. [PubMed]
105. SCCS; Degen, G.H. Opinion of the Scientific Committee on Consumer safety (SCCS)—Opinion on the safety of the use of deoxyarbutin in cosmetic products. *Regul. Toxicol. Pharmacol.* **2016**, *74*, 77–78.
106. Frias, M.A.; Winik, B.; Franzoni, M.B.; Levstein, P.R.; Nicastro, A.; Gennaro, A.M.; Diaz, S.B.; Disalvo, E.A. Lysophosphatidylcholine-arbutin complexes form bilayer-like structures. *Biochim. Biophys. Acta* **2008**, *1778*, 1259–1266. [CrossRef] [PubMed]
107. Li, Y.; Li, F.; Cai, H.Y.; Chen, X.; Sun, W.; Shen, W.Y. Structural characterization of inclusion complex of arbutin and hydroxypropyl-beta-cyclodextrin. *Trop. J. Pharm. Res.* **2016**, *15*, 2227–2233. [CrossRef]
108. Wen, A.H.; Choi, M.K.; Kim, D.D. Formulation of liposome for topical delivery of arbutin. *Arch. Pharm. Res.* **2006**, *29*, 1187–1192. [CrossRef] [PubMed]
109. Liang, K.; Xu, K.; Bessarab, D.; Obaje, J.; Xu, C. Arbutin encapsulated micelles improved transdermal delivery and suppression of cellular melanin production. *BMC Res. Notes* **2016**, *9*, 254. [CrossRef] [PubMed]
110. Huang, H.; Belwal, T.; Liu, S.B.; Duan, Z.H.; Luo, Z.S. Novel multi-phase nano-emulsion preparation for co-loading hydrophilic arbutin and hydrophobic coumaric acid using hydrocolloids. *Food Hydrocoll.* **2019**, *93*, 92–101. [CrossRef]
111. Ayumi, N.S.; Sahudin, S.; Hussain, Z.; Hussain, M.; Abu Samah, N.H. Polymeric nanoparticles for topical delivery of alpha and beta arbutin: Preparation and characterization. *Drug Deliv. Transl. Res.* **2019**, *9*, 482–496. [CrossRef]
112. Park, J.J.; Hwang, S.J.; Kang, Y.S.; Jung, J.; Park, S.; Hong, J.E.; Park, Y.; Lee, H.J. Synthesis of arbutin-gold nanoparticle complexes and their enhanced performance for whitening. *Arch. Pharm. Res.* **2019**, *42*, 977–989. [CrossRef]
113. Liao, A.H.; Ma, W.C.; Wang, C.H.; Yeh, M.K. Penetration depth, concentration and efficiency of transdermal alpha-arbutin delivery after ultrasound treatment with albumin-shelled microbubbles in mice. *Drug Deliv.* **2016**, *23*, 2173–2182. [CrossRef] [PubMed]
114. Aung, N.N.; Ngawhirunpat, T.; Rojanarata, T.; Patrojanasophon, P.; Opanasopit, P.; Pamornpathomkul, B. HPMC/PVP Dissolving Microneedles: A Promising Delivery Platform to Promote Trans-Epidermal Delivery of Alpha-Arbutin for Skin Lightening. *AAPS PharmSciTech* **2019**, *21*, 25. [CrossRef]
115. Aung, N.N.; Ngawhirunpat, T.; Rojanarata, T.; Patrojanasophon, P.; Pamornpathomkul, B.; Opanasopit, P. Fabrication, characterization and comparison of alpha-arbutin loaded dissolving and hydrogel forming microneedles. *Int. J. Pharm.* **2020**, *586*, 119508. [CrossRef] [PubMed]
116. Kim, H.S.; Kim, K.J.; Lee, M.W.; Lee, S.Y.; Yun, Y.H.; Shim, W.G.; Yoon, S.D. Preparation and release properties of arbutin imprinted inulin/polyvinyl alcohol biomaterials. *Int. J. Biol. Macromol.* **2020**, *161*, 763–770. [CrossRef]
117. Choi, S.; Park, Y.I.; Lee, S.K.; Kim, J.E.; Chung, M.H. Aloesin inhibits hyperpigmentation induced by UV radiation. *Clin. Exp. Dermatol.* **2002**, *27*, 513–515. [CrossRef] [PubMed]
118. Ertam, I.; Mutlu, B.; Unal, I.; Alper, S.; Kivcak, B.; Ozer, O. Efficiency of ellagic acid and arbutin in melasma: A randomized, prospective, open-label study. *J. Dermatol.* **2008**, *35*, 570–574. [CrossRef]

119. Morag, M.; Nawrot, J.; Siatkowski, I.; Adamski, Z.; Fedorowicz, T.; Dawid-Pac, R.; Urbanska, M.; Nowak, G. A double-blind, placebo-controlled randomized trial of Serratulae quinquefoliae folium, a new source of beta-arbutin, in selected skin hyperpigmentations. *J. Cosmet. Dermatol.* **2015**, *14*, 185–190. [CrossRef]
120. Jin, Y.H.; Lee, S.J.; Chung, M.H.; Park, J.H.; Park, Y.I.; Cho, T.H.; Lee, S.K. Aloesin and arbutin inhibit tyrosinase activity in a synergistic manner via a different action mechanism. *Arch. Pharm. Res.* **1999**, *22*, 232–236. [CrossRef]
121. Yang, Z.Q.; Wang, Z.H.; Tu, J.B.; Li, P.; Hu, X.Y. The effects of aloesin and arbutin on cultured melanocytes in a synergetic method. *Zhonghua Zheng Xing Wai Ke Za Zhi* **2004**, *20*, 369–371.
122. Hseu, Y.C.; Cheng, K.C.; Lin, Y.C.; Chen, C.Y.; Chou, H.Y.; Ma, D.L.; Leung, C.H.; Wen, Z.H.; Wang, H.M. Synergistic Effects of Linderanolide B Combined with Arbutin, PTU or Kojic Acid on Tyrosinase Inhibition. *Curr. Pharm. Biotechnol.* **2015**, *16*, 1120–1126. [CrossRef] [PubMed]
123. Hong, J.H.; Chen, H.J.; Xiang, S.J.; Cao, S.W.; An, B.C.; Ruan, S.F.; Zhang, B.; Weng, L.D.; Zhu, H.X.; Liu, Q. Capsaicin reverses the inhibitory effect of licochalcone A/beta-Arbutin on tyrosinase expression in b16 mouse melanoma cells. *Pharmacogn. Mag.* **2018**, *14*, 110–115. [PubMed]
124. Crocco, E.I.; Veasey, J.V.; Boin, M.F.F.D.; Lellis, R.F.; Alves, R.O. A Novel Cream Formulation Containing Nicotinamide 4%, Arbutin 3%, Bisabolol 1%, and Retinaldehyde 0.05% for Treatment of Epidermal Melasma. *Cutis* **2015**, *96*, 337–342.
125. Anwar, A.I.; Wahab, S.; Widita, W.; Nurdin, A.R.; Budhiani, S.; Seweng, A. Randomized control trial outcomes of tranexamic acid combination serum as a depigmenting agent for the use in healthy individuals. *Dermatol. Ther.* **2019**, *32*, e13146. [CrossRef] [PubMed]
126. Kalasho, B.D.; Minokadeh, A.; Zhang-Nunes, S.; Zoumalan, R.A.; Shemirani, N.L.; Waldman, A.R.; Pletzer, V.; Zoumalan, C.I. Evaluating the Safety and Efficacy of a Topical Formulation Containing Epidermal Growth Factor, Tranexamic Acid, Vitamin C, Arbutin, Niacinamide and Other Ingredients as Hydroquinone 4% Alternatives to Improve Hyperpigmentation: A Prospective, Randomize Controlled Split Face Study. *J. Cosmet. Sci.* **2020**, *71*, 263–290.
127. Polnikorn, N. Treatment of refractory melasma with the MedLite C6 Q-switched Nd:YAG laser and alpha arbutin: A prospective study. *J. Cosmet. Laser Ther.* **2010**, *12*, 126–131. [CrossRef]
128. Abbas, K.; Qadir, M.I.; Anwar, S. The Role of Melanin in Skin Cancer. *Crit. Rev. Eukaryot Gene Expr.* **2019**, *29*, 17–24. [CrossRef]
129. Smit, N.P.M.; Nieuwpoort, F.A.; Marrot, L.; Out, C.; Poorthuis, B.; van Pelt, H.; Meunier, J.R.; Pavel, S. Increased melanogenesis is a risk factor for oxidative DNA damage—Study on cultured melanocytes and atypical nevus cells. *Photochem. Photobiol.* **2008**, *84*, 550–555. [CrossRef]
130. Dumbuya, H.; Hafez, S.Y.; Oancea, E. Cross talk between calcium and ROS regulate the UVA-induced melanin response in human melanocytes. *FASEB J.* **2020**, *34*, 11605–11623. [CrossRef]
131. Suo, D.F.; Zeng, S.W.; Zhang, J.L.; Meng, L.H.; Weng, L.S. PM2.5 induces apoptosis, oxidative stress injury and melanin metabolic disorder in human melanocytes. *Exp. Ther. Med.* **2020**, *19*, 3227–3238. [CrossRef]
132. Perdomo, J.; Quintana, C.; Gonzalez, I.; Hernandez, I.; Rubio, S.; Loro, J.F.; Reiter, R.J.; Estevez, F.; Quintana, J. Melatonin Induces Melanogenesis in Human SK-MEL-1 Melanoma Cells Involving Glycogen Synthase Kinase-3 and Reactive Oxygen Species. *Int. J. Mol. Sci.* **2020**, *21*, 4970. [CrossRef]
133. Chang, S.P.; Huang, H.M.; Shen, S.C.; Lee, W.R.; Chen, Y.C. Nilotinib induction of melanogenesis via reactive oxygen species-dependent JNK activation in B16F0 mouse melanoma cells. *Exp. Dermatol.* **2018**, *27*, 1388–1394. [CrossRef]
134. Denat, L.; Kadekaro, A.L.; Marrot, L.; Leachman, S.A.; Abdel-Malek, Z.A. Melanocytes as instigators and victims of oxidative stress. *J. Investig. Dermatol.* **2014**, *134*, 1512–1518. [CrossRef] [PubMed]
135. Jenkins, N.C.; Grossman, D. Role of Melanin in Melanocyte Dysregulation of Reactive Oxygen Species. *Biomed. Res. Int.* **2013**, *2013*, 908797. [CrossRef] [PubMed]
136. Nagapan, T.S.; Lim, W.N.; Basri, D.; Ghazali, A.R. Oral supplementation of L-glutathione prevents ultraviolet B-induced melanogenesis and oxidative stress in BALB/c mice. *Exp. Anim.* **2019**, *68*, 541–548. [CrossRef] [PubMed]
137. Arjinpathana, N.; Asawanonda, P. Glutathione as an oral whitening agent: A randomized, double-blind, placebo-controlled study. *J. Dermatol. Treat.* **2012**, *23*, 97–102. [CrossRef] [PubMed]
138. Gegotek, A.; Skrzydlewska, E. The role of transcription factor Nrf2 in skin cells metabolism. *Arch. Dermatol. Res.* **2015**, *307*, 385–396. [CrossRef] [PubMed]
139. Raghunath, A.; Sundarraj, K.; Nagarajan, R.; Arfuso, F.; Bian, J.; Kumar, A.P.; Sethi, G.; Perumal, E. Antioxidant response elements: Discovery, classes, regulation and potential applications. *Redox Biol.* **2018**, *17*, 297–314. [CrossRef]
140. Schafer, M.; Dutsch, S.; Keller, U.A.D.; Navid, F.; Schwarz, A.; Johnson, D.A.; Johnson, J.A.; Werner, S. Nrf2 establishes a glutathione-mediated gradient of UVB cytoprotection in the epidermis. *Genes Dev.* **2010**, *24*, 1045–1058. [CrossRef]
141. Shin, J.M.; Kim, M.Y.; Sohn, K.C.; Jung, S.Y.; Lee, H.E.; Lim, J.W.; Kim, S.; Lee, Y.H.; Im, M.; Seo, Y.J.; et al. Nrf2 Negatively Regulates Melanogenesis by Modulating PI3K/Akt Signaling. *PLoS ONE* **2014**, *9*, e96035. [CrossRef] [PubMed]
142. Chaiprasongsuk, A.; Onkoksoong, T.; Pluemsamran, T.; Limsaengurai, S.; Panich, U. Photoprotection by dietary phenolics against melanogenesis induced by UVA through Nrf2-dependent antioxidant responses. *Redox Biol.* **2016**, *8*, 79–90. [CrossRef]
143. Takebayashi, J.; Ishii, R.; Chen, J.B.; Matsumoto, T.; Ishimi, Y.; Tai, A. Reassessment of antioxidant activity of arbutin: Multifaceted evaluation using five antioxidant assay systems. *Free Radic. Res.* **2010**, *44*, 473–478. [CrossRef] [PubMed]
144. Tada, M.; Kohno, M.; Niwano, Y. Alleviation effect of arbutin on oxidative stress generated through tyrosinase reaction with L-tyrosine and L-DOPA. *BMC Biochem.* **2014**, *15*, 23. [CrossRef]

145. Wu, L.H.; Li, P.; Zhao, Q.L.; Piao, J.L.; Jiao, Y.F.; Kadowaki, M.; Kondo, T. Arbutin, an intracellular hydroxyl radical scavenger, protects radiation-induced apoptosis in human lymphoma U937 cells. *Apoptosis* **2014**, *19*, 1654–1663. [CrossRef]
146. Seyfizadeh, N.; Tazehkand, M.Q.; Palideh, A.; Maroufi, N.F.; Hassanzadeh, D.; Rahmati-Yamchi, M.; Elahimanesh, F.; Borzoueisileh, S. Is arbutin an effective antioxidant for the discount of oxidative and nitrosative stress in Hep-G2 cells exposed to tert-butyl hydroperoxide? *Bratisl. Med. J. Bratisl. Lek. Listy* **2019**, *120*, 569–575. [CrossRef]
147. Ebadollahi, S.H.; Pouramir, M.; Zabihi, E.; Golpour, M.; Aghajanpour-Mir, M. The Effect of Arbutin on The Expression of Tumor Suppressor P53, BAX/BCL-2 Ratio and Oxidative Stress Induced by Tert-Butyl Hydroperoxide in Fibroblast and LNcap Cell Lines. *Cell J.* **2021**, *22*, 532–541.
148. Polouliakh, N.; Ludwig, V.; Meguro, A.; Kawagoe, T.; Heeb, O.; Mizuki, N. Alpha-Arbutin Promotes Wound Healing by Lowering ROS and Upregulating Insulin/IGF-1 Pathway in Human Dermal Fibroblast. *Front. Physiol.* **2020**, *11*, 586843. [CrossRef]
149. Zhang, B.; Zeng, M.; Li, B.; Kan, Y.; Wang, S.; Cao, B.; Huang, Y.; Zheng, X.; Feng, W. Arbutin attenuates LPS-induced acute kidney injury by inhibiting inflammation and apoptosis via the PI3K/Akt/Nrf2 pathway. *Phytomedicine* **2021**, *82*, 153466. [CrossRef]
150. Nalban, N.; Sangaraju, R.; Alavala, S.; Mir, S.M.; Jerald, M.K.; Sistla, R. Arbutin Attenuates Isoproterenol-Induced Cardiac Hypertrophy by Inhibiting TLR-4/NF-kappaB Pathway in Mice. *Cardiovasc. Toxicol.* **2020**, *20*, 235–248. [CrossRef] [PubMed]
151. Safari, H.; Zabihi, E.; Pouramir, M.; Morakabati, P.; Abedian, Z.; Karkhah, A.; Nouri, H.R. Decrease of intracellular ROS by arbutin is associated with apoptosis induction and downregulation of IL-1beta and TNF-alpha in LNCaP; prostate cancer. *J. Food Biochem.* **2020**, *44*, e13360. [CrossRef] [PubMed]
152. Matsuo, Y.; Ito, A.; Masui, Y.; Ito, M. A case of allergic contact dermatitis caused by arbutin. *Contact Dermat.* **2015**, *72*, 404–405. [CrossRef] [PubMed]
153. Numata, T.; Tobita, R.; Tsuboi, R.; Okubo, Y. Contact dermatitis caused by arbutin contained in skin-whitening cosmetics. *Contact Dermat.* **2016**, *75*, 187–188. [CrossRef] [PubMed]

antioxidants

MDPI

Article

Pomegranate Extract (POMx) Induces Mitochondrial Dysfunction and Apoptosis of Oral Cancer Cells

Sheng-Yao Peng [1], Li-Ching Lin [2,3,4], Shu-Rong Chen [5], Ammad A. Farooqi [6], Yuan-Bin Cheng [7], Jen-Yang Tang [8,9,*] and Hsueh-Wei Chang [1,10,11,*]

1 Department of Biomedical Science and Environmental Biology, Ph.D Program in Life Sciences, College of Life Sciences, Kaohsiung Medical University, Kaohsiung 80708, Taiwan; u109851101@kmu.edu.tw
2 Department of Radiation Oncology, Chi-Mei Foundation Medical Center, Tainan 71004, Taiwan; 8508a6@mail.chimei.org.tw
3 School of Medicine, Taipei Medical University, Taipei 11031, Taiwan
4 Chung Hwa University Medical Technology, Tainan 71703, Taiwan
5 Graduate Institute of Natural Products, Kaohsiung Medical University, Kaohsiung 80708, Taiwan; u106831002@kmu.edu.tw
6 Institute of Biomedical and Genetic Engineering (IBGE), Islamabad 54000, Pakistan; farooqiammadahmad@gmail.com
7 Department of Marine Biotechnology and Resources, National Sun Yat-sen University, Kaohsiung 80424, Taiwan; jmb@mail.nsysu.edu.tw
8 School of Post-Baccalaureate Medicine, Kaohsiung Medical University, Kaohsiung 80708, Taiwan
9 Department of Radiation Oncology, Kaohsiung Medical University Hospital, Kaohsiung 80708, Taiwan
10 Institute of Medical Science and Technology, National Sun Yat-sen University, Kaohsiung 80424, Taiwan
11 Center for Cancer Research, Kaohsiung Medical University, Kaohsiung 80708, Taiwan
* Correspondence: reyata@kmu.edu.tw (J.-Y.T.); changhw@kmu.edu.tw (H.-W.C.); Tel.: +886-7-312-1101 (ext. 7158) (J.-Y.T.); +886-7-312-1101 (ext. 2691) (H.-W.C.)

Citation: Peng, S.-Y.; Lin, L.-C.; Chen, S.-R.; Farooqi, A.A.; Cheng, Y.-B.; Tang, J.-Y.; Chang, H.-W. Pomegranate Extract (POMx) Induces Mitochondrial Dysfunction and Apoptosis of Oral Cancer Cells. *Antioxidants* **2021**, *10*, 1117. https://doi.org/10.3390/antiox10071117

Academic Editors: Irene Dini and Domenico Montesano

Received: 26 June 2021
Accepted: 10 July 2021
Published: 13 July 2021

Abstract: The anticancer effect of pomegranate polyphenolic extract POMx in oral cancer cells has rarely been explored, especially where its impact on mitochondrial functioning is concerned. Here, we attempt to evaluate the proliferation modulating function and mechanism of POMx against human oral cancer (Ca9-22, HSC-3, and OC-2) cells. POMx induced ATP depletion, subG1 accumulation, and annexin V/Western blotting-detected apoptosis in these three oral cancer cell lines but showed no toxicity to normal oral cell lines (HGF-1). POMx triggered mitochondrial membrane potential (MitoMP) disruption and mitochondrial superoxide (MitoSOX) generation associated with the differential downregulation of several antioxidant gene mRNA/protein expressions in oral cancer cells. POMx downregulated mitochondrial mass, mitochondrial DNA copy number, and mitochondrial biogenesis gene mRNA/protein expression in oral cancer cells. Moreover, POMx induced both PCR-based mitochondrial DNA damage and γH2AX-detected nuclear DNA damage in oral cancer cells. In conclusion, POMx provides antiproliferation and apoptosis of oral cancer cells through mechanisms of mitochondrial impairment.

Keywords: pomegranate; mitochondrial DNA; DNA damage; apoptosis; oral cancer

1. Introduction

Pomegranate has gained extraordinary appreciation because of its ability to inhibit/prevent a wide variety of cancers [1–3]. Pomegranate (*Punica granatum* L.) fruits contain abundant polyphenols [4,5]. The knowledge base in fields such as nutrigenetics and nutrigenomics continuously expands at a rapid rate. Emerging scientific evidence enables us to obtain a better understanding of the significant pharmacological properties of bioactive constituents derived from plants such as pomegranate [6]. Pomegranate, due to its bioactive compounds, belongs to a group of functional foods [7]. Its commercial dietary extract, POMx, is standardized with polyphenolic ellagitannin content, and its food safety is regarded as "generally recognized as being safe (GRAS)" by the U.S. Food and Drug

Administration (FDA) [8]. POMx has several cellular and clinically relevant functions on several types of cancer [9–13].

Since pomegranate is an antioxidant-rich natural product, POMx may have similar effects. Antioxidants commonly have a dual function for reducing or inducing cellular oxidative stress, coming along with low or high doses [14]. For example, low-dose POMx (2.5 to 40 μg/mL) suppresses UVB-induced oxidative stress in keratinocyte HaCaT cells. In contrast, high dose POMx (100 to 200 μg/mL) may trigger oxidative stress in several types of cancer cell lines such as lung cancer, leukemia, and fibrosarcoma [15–17].

The oxidative stress function in cancer cell lines after POMx incubation has rarely been investigated [15–17]. A detailed examination in oral cancer cells for the response and mechanism to POMx is warranted. The antiproliferation ability of POMx against oral cancer cells has rarely been investigated as well. Recently, we reported that low cytotoxic doses of POMx suppressed transwell migration ability and Matrigel invasion behavior of human oral cancer cells [18]. However, the antioral cancer function at high-dose POMx remains unclear.

Mitochondria serve as the powerhouse of the cells and are responsible for the central source of oxidative stress, which regulates cellular energy supplies, proliferation, and apoptosis [19,20]. Although POMx was reported to induce apoptosis in several types of cancer cell lines [10,11,21,22], only mitochondrial apoptosis signaling such as caspases were studied. Other examinations for evaluating mitochondrial function such as mitochondrial membrane potential (MitoMP), mitochondrial superoxide (MitoSOX), mitochondrial mass, mitochondrial DNA (mtDNA) copy number, mtDNA lesion, and mitochondrial biogenesis were rarely investigated.

We aimed to test the hypothesis that oxidative stress generated by POMx provides apoptosis resulting in antiproliferation against oral cancer cells via mitochondrial impairment. Therefore, we evaluated ATP content, apoptosis, mitochondrial function, and DNA damage in the example of human oral cancer cells following POMx incubation.

2. Materials and Methods

2.1. Cell Culture and Drug Source

One normal oral cell line, HGF-1 (human normal gingival fibroblast), and two oral cancer cell lines (HSC-3 and Ca9-22) were commercially available, and one oral cancer cell line (OC-2) was provided by Dr. Wan-Chi Tsai (Kaohsiung Medical University, Kaohsiung, Taiwan) [23]. These cell lines were kept in culture medium with DMEM/F-12 (Dulbecco's Modified Eagle Medium (DMEM)/Nutrient Mixture F-12) (Gibco, Grand Island, NY, USA) at a ratio of 3 vs. 2, containing penicillin, streptomycin, and 10% fetal bovine serum (Gibco).

POMx is a commercial pomegranate (*Punica granatum* L.)-derived polyphenols-rich aqueous extract powder (POM Wonderful, LLC, Los Angeles, CA, USA) [13,24]. The detailed characterization of chemicals in this POMx powder extract had been previously reported [9,24], such as ellagitannins (punicalagin and punicalin) and ellagic acid. POMx was immediately prepared in dimethyl sulfoxide (DMSO) before experiments.

2.2. Determination of Main Components of POMx by HPLC

Qualitative and quantitative analysis was performed on a Shimadzu HPLC (Kyoto, Japan) system, equipped with an LC-20AT prominence liquid chromatography, a SIL-40AD autosampler, and an SPD-M20A diode array detector. The determination was carried out with a reversed-phase column (Luna C_{18} column, 250 mm × 4.6 mm, 5 μm; Phenomenex; Torrance, CA, USA). The mobile phase consisted of (A) water with 0.1% (v/v) trifluoroacetic acid and (B) methanol. The gradient elution system was set as follows: 0−5 min, 1% B; 5−10 min 1−15% B; 10−15 min, 15% B; 15−35 min, 15−45% B; 35−40 min, 45−80% B; 40−45 min, 80−100% B; 45−50 min, 100% B. The flow rate was 1.0 mL/min. The injection was 10 μL, and the detection wavelength was 378 nm. The stock solutions were prepared by dissolving 1.0 mg of punicalin, punicalagin (Molnova; Ann Arbor, MI, USA), ellagic acid

(Sigma-Aldrich; St. Louis, MO, USA), and POMx in 1.0 mL methanol. Six concentrations of three standards were prepared by diluting with methanol.

2.3. Cell Viability and Morphology

Viability was analyzed by an intracellular ATP content assay (PerkinElmer Life Sciences, Boston, MA, USA) [25] and trypan blue assay [26]. Cell morphology was observed at 100x magnification.

2.4. Cell Cycle Analysis

Cells were incubated with Biotium 7-aminoactinomycin D (7AAD) (Hayward, CA, USA) at the requirement of 1 μg/mL, 30 min, 37 °C, and DNA content was analyzed by Accuri C6 flow cytometer using FL3 channel (Becton-Dickinson, Mansfield, MA, USA) [27].

2.5. Annexin V/7AAD for Apoptosis Analysis

Cells were stained with annexin V mixed 7AAD kit (Strong Biotech Corp., Taipei, Taiwan) according to the user's instructions, and signals analyzed by Accuri C6 flow cytometer (Becton-Dickinson) using FL1/FL3 channels as previously described [28].

2.6. Acridine Orange (AO) Staining for Autophagy Analysis

AO staining was used to detect acidic vesicular organelles as the fast screening for autophagy [29]. Cells were stained by 10 ng/mL AO (Sigma, St Louis, MO, USA) at the requirement (37 °C, 30 min) and analyzed by Accuri C6 flow cytometer (Becton-Dickinson) using FL3 channel as previously described [30].

2.7. Mitochondrial Membrane Potential (MitoMP)

Cells were stained by 5 nM MitoProbe™ $DiOC_2(3)$ (Thermo Fisher Scientific, Carlsbad, CA, USA) at the requirement (37 °C, 30 min) and analyzed by Accuri C6 flow cytometer (Becton-Dickinson) using FL1 channel as previously described [31].

2.8. Mitochondrial Superoxide (MitoSOX) Generation

Cells were stained by 50 nM MitoSOX™ Red (Thermo Fisher Scientific) at the requirement (37 °C, 30 min) and analyzed by Accuri C6 flow cytometer (Becton-Dickinson) using FL3 channel as previously described [32].

2.9. Quantitative RT-PCR (qRT-PCR) Analysis: Antioxidant- and Mitochondrial Biogenesis-Related Genes

RNA was extracted and reverse-transcribed [33] to cDNA for qRT-PCR as described [34]. In addition, for antioxidant-related genes [35,36], nuclear factor erythroid 2-like 2 (*NFE2L2*), glutamate-cysteine ligase catalytic subunit (*GCLC*), thioredoxin (*TXN*), catalase (*CAT*), superoxide dismutase 1 (*SOD1*) [37], heme oxygenase 1 (*HMOX1*), and quinone dehydrogenase 1 (*NQO1*) were selected.

For mitochondrial biogenesis-related genes [38], transcription factor B2, mitochondrial (*TFB2M*), transcription factor A, mitochondrial (*TFAM*), RNA polymerase mitochondrial (*POLRMT*), and Tu translation elongation factor, mitochondrial (*TUFM*) [39] were selected. Touch-down PCR program (running for 50 cycles) [34] was applied to qRT-PCR reactions.

The qRT-PCR primer information for antioxidant- and mitochondrial biogenesis-related genes, as well as GAPDH, are provided in the top and bottom of Table 1, respectively. In reference to housekeeping gene *GAPDH*, the relative mRNA expression ($\log_2$) of these genes was calculated according to the $2^{-\Delta\Delta Ct}$ method [40]. In brief, ΔCt is calculated as (Ct value of a target gene—Ct value of *GAPDH* gene), where the target genes are antioxidant- and mitochondrial biogenesis-related genes. When the Ct value of a target gene is undetectable (>50 cycles; qRT-PCR is performed for 50 cycles), it was assigned 50 cycles for further calculation. The ΔΔCt is the difference in ΔCt between the drug treatment and untreated control, which is ΔΔCt = ΔCt (drug treatment) − ΔCt (control).

Table 1. Primer sequences and amplicon lengths for antioxidant- and mitochondrial biogenesis-related genes.

Genes	Forward Primers (5′→3′)	Reverse Primers (5′→3′)	Length
NFE2L2	GATCTGCCAACTACTCCCAGGTT [36]	CTGTAACTCAGGAATGGATAATAGCTCC [36]	302 bp
GCLC	ACAAGCACCCTCGCTTCAGTACC [36]	CTGCAGGCTTGGAATGTCACCT [36]	232 bp
TXN	GAAGCAGATCGAGAGCAAGACTG [36]	GCTCCAGAAAATTCACCCACCT [36]	270 bp
CAT	ATGCAGGACAATCAGGGTGGT [36]	CCTCAGTGAAGTTCTTGACCGCT [36]	274 bp
SOD1	AGGGCATCATCAATTTCGAGC [37]	CCCAAGTCTCCAACATGCCTC [36]	211 bp
HMOX1	CCTTCTTCACCTTCCCCAACAT [36]	GGCAGAATCTTGCACTTTGTTGC [36]	251 bp
NQO1	GAAGGACCCTGCGAACTTTCAGTA [36]	GAAAGCACTGCCTTCTTACTCCG [36]	258 bp
TFB2M	CTGCTGGAGTGCAATCCAGGTC	TCCAACTACTTTTAAAGGGATGTCTGC	285 bp
TFAM	TTAAAGCTCAGAACCCAGATGCA [39]	TTACAGTCTTCAGCTTTTCCTGCG	354 bp
POLRMT	CTGAGCGACTTTCCCCAGGAGT	CTTACGTGTGTTGGGCTTTCGG	294 bp
TUFM	TGCTCTCTGTGCCCTTGAGGGT	CTTGTGGAACATCTCAATGCCTGTC	277 bp
GAPDH	CCTCAACTACATGGTTTACATGTTCC [41]	CAAATGAGCCCCAGCCTTCT [42]	220 bp

2.10. Mitochondrial Mass

For mitochondrial mass measurement, cells were stained by 300 nM MitoTracker™ Green FM (Thermo Fisher Scientific) at the requirement (37 °C, 30 min) and analyzed by Accuri C6 flow cytometer (Becton-Dickinson) using FL1 channel as described [43].

2.11. Quantitative PCR (qPCR): mtDNA Copy Number

Total genomic DNA from cells incubated with POMx for 24 and 72 h was prepared according to the OMEGA Bio-Tek user manual of the E.Z.N.A.® Tissue DNA kit (Norcross, GA, USA) [44]. Using the nuclear DNA (nDNA) gene *GAPDH* as a reference, the relative copy numbers of mtDNA such as NADH-ubiquinone oxidoreductase chain 1 (*ND1*) and *ND5* genes [45] were analyzed using the $2^{-\Delta\Delta Ct}$ method [40] after qPCR reaction in a touch-down program [34]. The PCR information for the mtDNA copy number is listed in Table 2.

Table 2. Primer sequences and amplicon lengths for mitochondrial DNA copy number and DNA damage related genes.

Genes	Forward Primers (5′→3′)	Reverse Primers (5′→3′)	Length
ND1	CCTCCTACTCCTCATTGTACCCATTC	TGAAGAGTTTTATGGCGTCAGCG	155 bp
ND1-L	CCTCCTACTCCTCATTGTACCCATTC	GAGTGTGCCTGCAAAGATGGTAGAG	1203 bp
ND5	GTTTCATCCTCGCCTTAGCATGA	AGTCAGGGGTGGAGACCTAATTGG	157 bp
ND5-L	GTTTCATCCTCGCCTTAGCATGA	GGTGATGATGGAGGTGGAGATTTG	1190 bp
GAPDH	GAAGCTGAGTCATGGGTAGTTGG [44]	GATCTGGTTTCCGGAAGACG [44]	220 bp

L indicates the long-run PCR for target genes such as *ND1* and *ND5*.

2.12. Semi-Long Run Quantitative PCR (SLR-qPCR): mtDNA Damage

SLR-qPCR was applied to assess mtDNA damages [46]. Using SLR-qPCR, the copy numbers of two DNA fragments with different lengths, i.e., small (*ND1*/*ND5*) and long (*ND1*-L/*ND5*-L) fragments, were measured for calculating mtDNA damage (lesions per 10 kb DNA between *ND1* and *ND5*) by the formula: $(1–2^{-\Delta\,(\text{long Ct-short Ct})}) \times 10{,}000$ (bp)/length of the long fragment (bp) [46]. The primer and PCR amplicon information for mitochondrial DNA damage is provided in Table 2.

2.13. DNA Damage: γH2AX

The level of double-strand break marker for DNA damage (γH2AX) was analyzed [47]. p-Histone H2A.X (Ser 139) at 500X dilution was chosen as the primary antibody purchased from Santa Cruz Biotechnology (Santa Cruz, CA, USA) to detect γH2AX at 4 °C for 1 h. Subsequently, a secondary antibody conjugated by Alexa 488 was used in its flow cytometry application (BD Accuri C6; FL1 channel).

2.14. Western Blotting Analysis for Apoptosis, Antioxidant Signaling, Mitochondrial Resident Proteins, and Mitochondrial Biogenesis

All Western blotting routine steps were mentioned previously [48]. Apoptosis antibodies included cleaved poly (ADP-ribose) polymerase (c-PARP), Bcl-xL, Bcl-2, and Bax (Cell signaling; Danvers, MA, USA). Antioxidant signaling antibodies included nuclear factor erythroid 2-related factor 2 (NRF2) (Fine Test; Wuhan, China), catalase (Merck; Darmstadt, Germany), peroxiredoxin 1 (PRX1) (GeneTex; Irvine, CA, USA), and superoxide dismutase 1 (SOD1) (Abcam; Cambridge, UK). Mitochondrial resident protein antibodies included translocase of the inner membrane (TIMM22) (Proteintech; Rosemont, IL, USA) and translocase of outer mitochondrial membrane 20 (TOMM20) (Cell signaling). Mitochondrial biogenesis antibodies (Biorbyt; Cambridge, UK) included RNA polymerase mitochondrial (POLRMT), Tu translation elongation factor, mitochondrial (TUFM), transcription factor B2, mitochondrial (TFB2M), transcription factor A, and mitochondrial (TFAM). Except for antibodies against TFB2M (1:5000) and β-actin (Sigma-Aldrich; St. Louis, MO, USA) (1:10,000), all antibodies were used in 1:1000 dilution.

2.15. Statistics

One-way ANOVA processed all the statistics after Tukey's HSD post hoc tests using JMP® 12 software to compare different groups [49]. Treatments without overlapping low cases are regarded as significant differences.

3. Results

3.1. HPLC profile of POMx and Three Main Bioactive Components

The contents for punicalin, punicalagin, and ellagic acid of POMx-capsules were analyzed by HPLC using authentic reference compounds. The linear equations of three main compounds were $y = 10^7x - 69{,}990$ ($R^2 = 0.9997$), $y = 5 \times 10^6x - 47{,}955$ ($R^2 = 0.9998$), and $y = 7 \times 10^6x - 21{,}969$ ($R^2 = 0.9999$), respectively. The results show that POMx contains punicalagin 26.582 mg/g, ellagic acid 47.857 mg/g, and punicalin 8.375 mg/g (Figure 1).

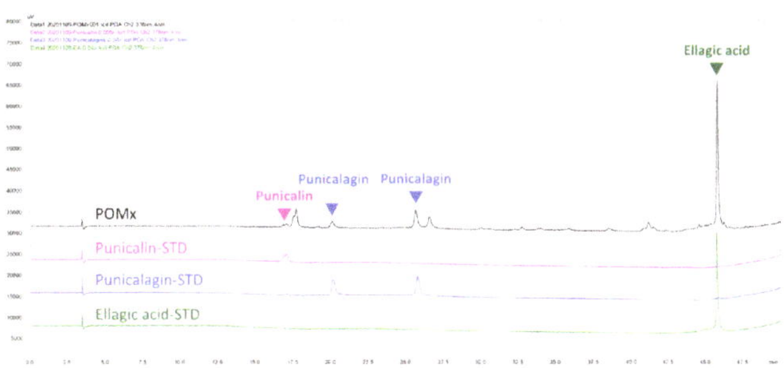

	Punicalin	Punicalagin	Ellagic acid
POMx (mg/g)	8.375 ± 0.003	26.582 ± 0.564	47.857 ± 3.951

Figure 1. HPLC profile of POMx and the contents of its three main bioactive components. The HPLC profile for punicalagin, ellagic acid, and punicalin was provided as well as their contents within POMx (mg/g). STD means standard.

3.2. Antiproliferation of Oral Cancer Cells Following POMx Incubation

Cell viability detected by ATP assay in oral cancer cells after POMx (0, 50, 75, 100, and 125 μg/mL) treatment for 24 h is dose-responsively decreased (Figure 2A). IC_{50} value at 24 h ATP assay for POMx in oral cancer cells (Ca9-22, HSC-3, and OC-2) are 80.53, 100.34, and 108.12 μg/mL, respectively. Moreover, longer exposure to POMx for 72 h decreases more viability to oral cancer cells than that of the 24 h treatment. In contrast, normal oral cells (HGF-1) show only a mild decrease after 72 h exposure to POMx.

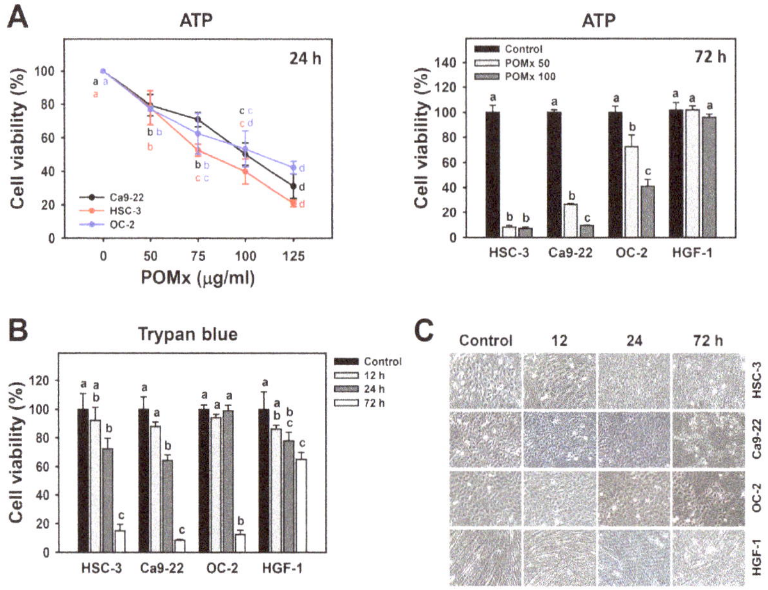

Figure 2. Cell viability and morphology of oral cancer cells after POMx incubation. Cells were incubated with 0, 50, and 100 μg/mL of POMx (control, POMx 50, and POMx 100) or other indicated concentrations for 24 or 72 h. (**A**) Cell viability for three oral cancer cell lines (HSC-3, Ca9-22, and OC-2) at 24 h and 72 h ATP assays. (**B**) Cell viability for three oral cancer cell lines (HSC-3, Ca9-22, and OC-2), and a normal oral cell line (HGF-1) at 0, 12, 24, and 72 h trypan blue assays. (**C**) Morphology for oral cancer cells and normal oral cells at 0, 12, 24, and 72 h for 100 μg/mL POMx incubation. Treatments without overlapping low cases (a to d) are significant differences for the same cell lines. $p < 0.05$. Data, mean ± SD ($n = 3$). The morphology image was photographed at 100× magnification.

Similarly, cell viability detected by trypan blue assay in oral cancer and normal oral cells after POMx (0 and 100 μg/mL) treatment for 0, 12, 24, and 72 h are time-dependently decreased (Figure 2B). In addition, it was noted that cell viabilities for oral cancer cells (Ca9-22, HSC-3, and OC-2) are lower than that of normal oral cells.

Figure 2C shows that 24 and 72 h POMx incubations of oral cancer cells induce abnormal cell morphology while normal oral cells (HGF-1) retain normal morphology. Accordingly, POMx has a selective killing effect on oral cancer cells but less harmful to normal oral cells.

3.3. Cell Cycle Change of Oral Cancer Cells Following POMx Incubation

After POMx incubations (0, 50, and 100 μg/mL) for 24 and 72 h, the patterns for cell cycles in three oral cancer cell lines are shown (Figure 3A). For 24 h POMx incubation, HSC-3 and OC-2 cells show slightly sub-G1 accumulations but not for Ca9-22 (Figure 3B). For 100 μg/mL POMx incubation, all these three cell lines show a decrease in the G1 phase. HSC-3 and Ca9-22 cells show an increase to G2/M, while OC-2 cells show an increase to S phase and G2/M decrease.

For 72 h POMx incubation, HSC-3 cells show dramatic sub-G1 and S phase accumulations but show a decrease in G1 and G2/M phases (Figure 3B). Ca9-22 and OC-2 cells show moderate subG1 and G2/M accumulation but show decreased G1 phase compared with the control.

Accordingly, POMx differentially disturbs cell cycle distribution of oral cancer cells between 24 and 72 h, and POMx at 72 h induces more subG1 accumulation (apoptosis-like) than at 24 h.

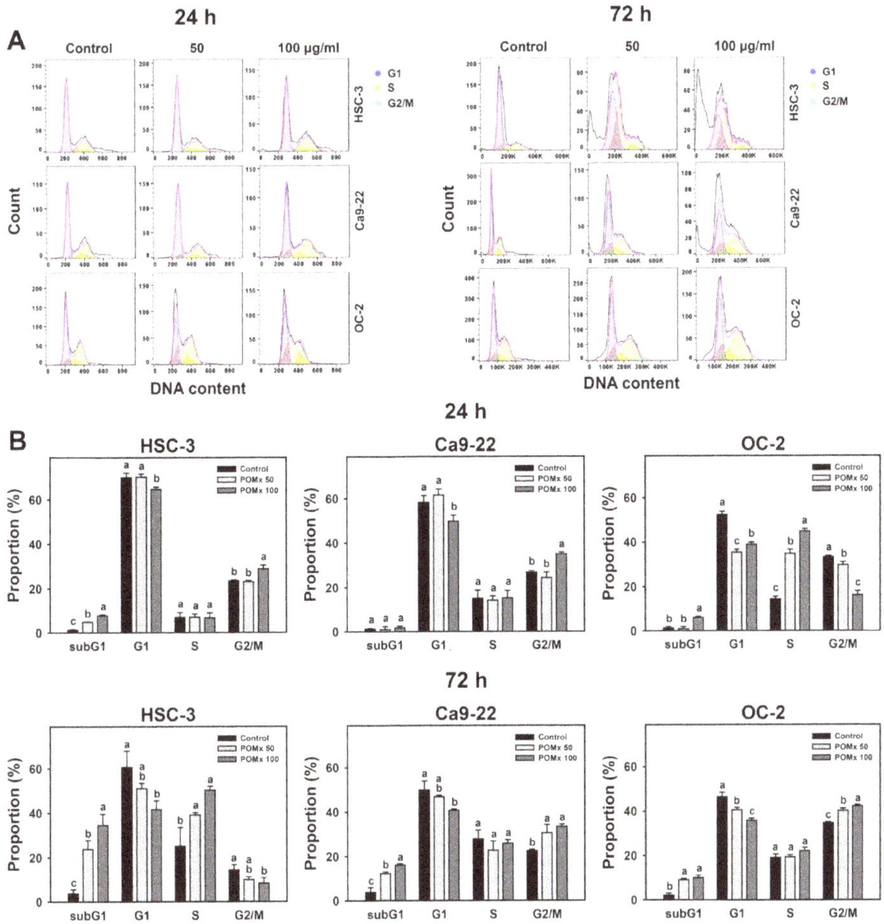

Figure 3. Cell cycle phase of oral cancer cells after POMx incubation. (**A**) Flow cytometry patterns. Cells (HSC-3, Ca9-22, and OC-2) were incubated with control, 50, and 100 µg/mL of POMx (control, POMx 50, and POMx 100) for 24 and 72 h. (**B**) Statistics for (**A**). Treatments without overlapping low cases (a to c) represent significant differences for the same cell lines. $p < 0.05$. Data, mean ± SD ($n = 3$).

3.4. Apoptosis and Autophagy Changes of Oral Cancer Cells Following POMx Incubation

After POMx incubations (0, 50, and 100 µg/mL) for 24 and 72 h, the dual staining patterns for annexin V/7AAD in oral cancer and normal cell lines are shown (Figure 4A). For 24 h POMx incubation, apoptosis (%) counting for annexin V (+)/7AAD (+ or −) population in oral cancer (HSC-3, Ca9-22, and OC-2) and normal oral cells (HGF-1) are weakly changed (Figure 4B).

In addition, POMx induces relatively more apoptosis in oral cancer cells than in normal oral cells. Moreover, apoptosis proteins such as cleaved PARP and BAX are increased, and the anti-apoptosis proteins such as Bcl-2 and Bcl-xL are decreased after 72 h POMx incubation (Figure 4C).

Moreover, the AO-detected autophagy of three oral cancer cell lines is decreased by POMx during 12, 24, and 72 h incubations compared with the control, suggesting that POMx may inter-regulate apoptosis and autophagy. Accordingly, 72 h POMx incubation induces more apoptosis than 24 h for oral cancer cells. Moreover, POMx induces more apoptosis in oral cancer cells than in normal oral cells, especially for 100 µg/mL at 72 h.

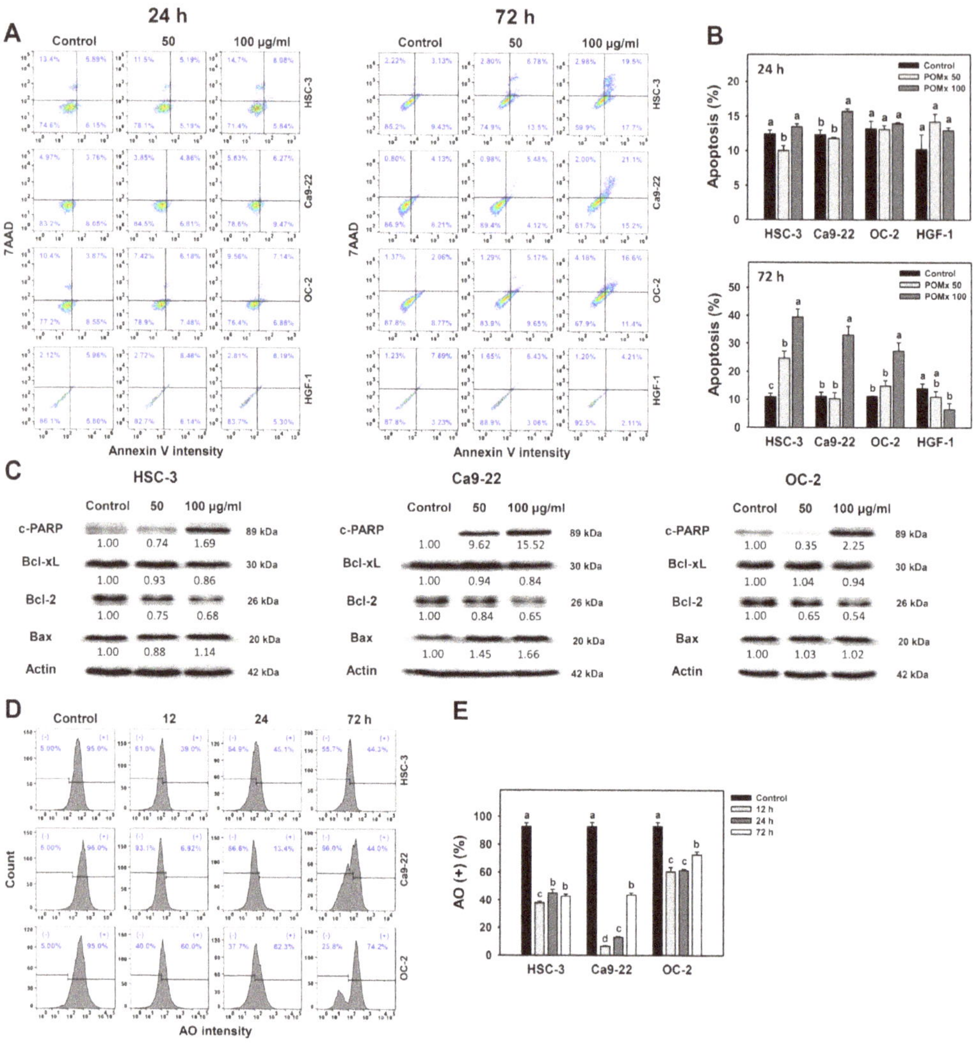

Figure 4. Apoptosis and autophagy changes of oral cancer cells after POMx incubation. Cells (HSC-3, Ca9-22, and OC-2) were incubated with control, 50, and 100 μg/mL of POMx (control, POMx 50, and POMx 100) for 24 and 72 h. (**A**) Annexin V/7AAD flow cytometry patterns. (**B**) Statistics for (**A**). (**C**) Western blotting for apoptosis marker expressions after 72 h POMx (0, 50, and 100 μg/mL) incubation. (**D**) Acridine orange (AO) flow cytometry patterns at 0, 12, 24, and 72 h for 100 μg/mL POMx incubation (**E**) Statistics for (**D**). Treatments without overlapping low cases (a to d) are significant differences for the same cell type. $p < 0.001$. Data, mean ± SD ($n = 3$).

3.5. MitoMP of Oral Cancer Cells Following POMx Incubation

After POMx incubation (0, 50, and 100 μg/mL) for 24 h, the patterns for MitoMP in oral cancer and normal (HGF-1) cell lines are shown (Figure 5A). The MitoMP (−) (%) of these three oral cancer cells dose-responsively increase after POMx incubation (Figure 5B). Moreover, POMx induces more MitoMP (−) (%) in three oral cancer cells than normal oral cells.

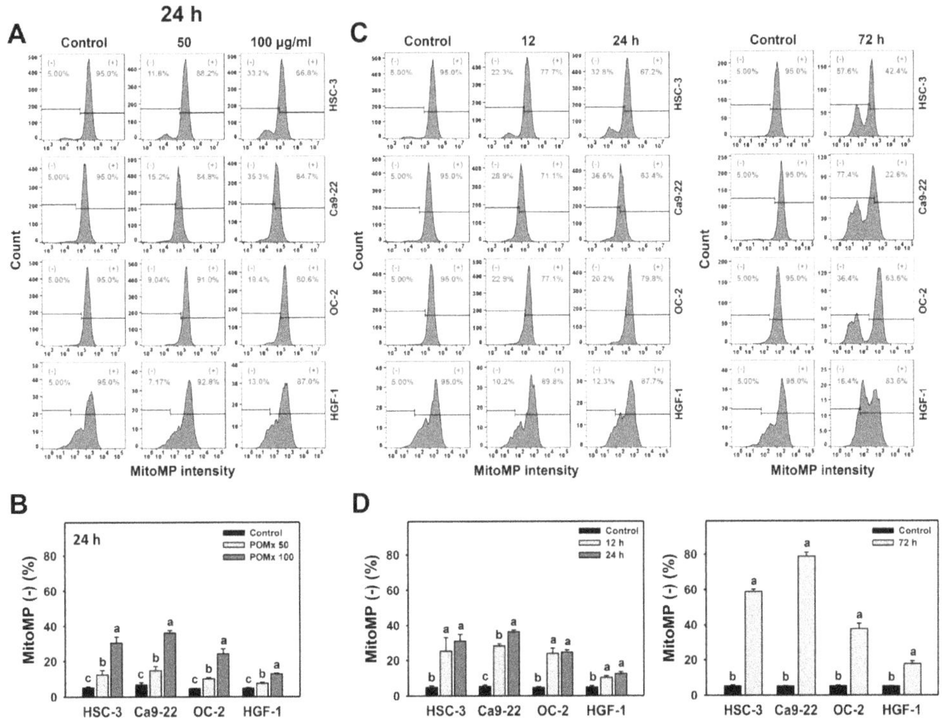

Figure 5. MitoMP change of oral cancer cells after POMx incubation. (**A**) Dose-response to MitoMP flow cytometry patterns. Cells (HSC-3, Ca9-22, and OC-2) were incubated with control, 50, and 100 µg/mL of POMx (control, POMx 50, and POMx 100) for 24 h. The MitoMP-negative (−) population was defined on the left part. (**B**) Statistics for (**A**). (**C**) Time course to MitoMP flow cytometry patterns. Cells were incubated with control and 100 µg/mL of POMx for 0, 12, 24, and 72 h. (**D**) Statistics for (**C**). Treatments without overlapping low cases (a to c) represent significant differences for the same cell lines. $p < 0.01$. Data, mean ± SD ($n = 3$).

After time course treatments of POMx, the dynamics of flow cytometry patterns for MitoMP in these oral cancer cells are shown (Figure 5C). The MitoMP (−) (%) of these three oral cancer cells is increased over time (12, 24, and 72 h) after POMx incubation compared with the control (Figure 5D). Moreover, POMx induces more MitoMP (−) (%) in three oral cancer cells than in normal oral cells throughout the time course. Accordingly, POMx causes higher MitoMP destruction in oral cancer cells than in normal oral cells.

3.6. MitoSOX Generation of Oral Cancer Cells Following POMx Incubation

After POMx incubations (0, 50, and 100 µg/mL) for 24 h, the patterns for MitoSOX in oral cancer (Ca9-22, HSC-3, and OC-2) and normal oral (HGF-1) cells are shown (Figure 6A). The MitoSOX (+) (%) of these three oral cancer cells were dose-responsively increased after POMx incubation while remaining unchanged in normal cells (Figure 6B).

After time course treatments of POMx, the flow cytometry patterns for MitoSOX in oral cancer and normal oral cells are shown (Figure 6C). The MitoSOX (+) (%) of these three oral cancer cells was increased over time (12, 24, and 72 h) after POMx incubation compared with the control, while it was unchanged in normal cells at 12 and 24 h and decreased at 72 h (Figure 6D). Accordingly, POMx induced higher MitoSOX generation in oral cancer cells than normal oral cells.

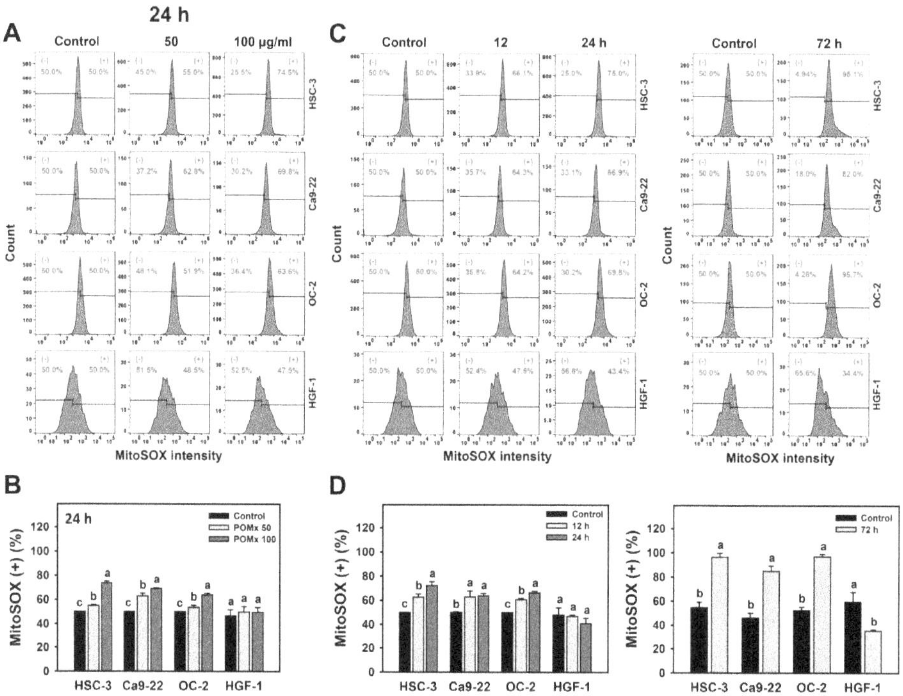

Figure 6. MitoSOX generation of oral cancer cells after POMx incubation. (**A**) Dose-response to MitoSOX flow cytometry patterns. Cells (HSC-3, Ca9-22, and OC-2) were incubated with 0, 50, and 100 μg/mL of POMx (control, POMx 50, and POMx 100) for 24 h. The MitoSOX (+) population was defined in the right part. (**B**) Statistics for (**A**). (**C**) Time course to MitoSOX flow cytometry patterns. Cells were incubated with control and 100 μg/mL of POMx for 0, 12, 24, and 72 h. (**D**) Statistics for (**C**). Treatments without overlapping low cases (a to c) show significant differences for the same cell lines. $p < 0.05$. Data, mean ± SD ($n = 3$).

3.7. Antioxidant Gene Expression of Oral Cancer Cells Following POMx Incubation

Inhibition of antioxidant pathways may induce increased oxidative stress [50]. Accordingly, the antioxidant signaling gene expression for mRNA [35], including *NFE2L2*, *GCLC*, *TXN*, *CAT*, *SOD1*, *HMOX1*, and *NQO1*, was examined for POMx-incubated oral cancer cells. After 24 h POMx incubations (0, 50, and 100 μg/mL), the relative mRNA gene expressions of these antioxidant genes in oral cancer (Ca9-22, HSC-3, and OC-2) cells were downregulated compared with the control (Figure 7A, top). However, for 24 h POMx incubations, the protein expressions of these antioxidant genes in oral cancer (Ca9-22, HSC-3, and OC-2) cells were almost unchanged (Figure 7B, top). Accordingly, mRNA and protein expressions were regulated differentially at 24 h POMx incubation.

For 72 h POMx incubations, the relative gene expressions of these antioxidant genes show differential expressions in these three oral cancer cell lines (Figure 7A, bottom). For example, 72 h POMx suppressed antioxidant mRNA gene expressions in HSC-3 and OC-2 cells, but these were induced in Ca9-22 cells. For 72 h POMx incubations, the protein expressions of these antioxidant genes were suppressed in HSC-3 and OC-2 cells but induced in Ca9-22 cells (Figure 7B, bottom). Accordingly, mRNA and protein expressions were consistently expressed at 72 h POMx incubation.

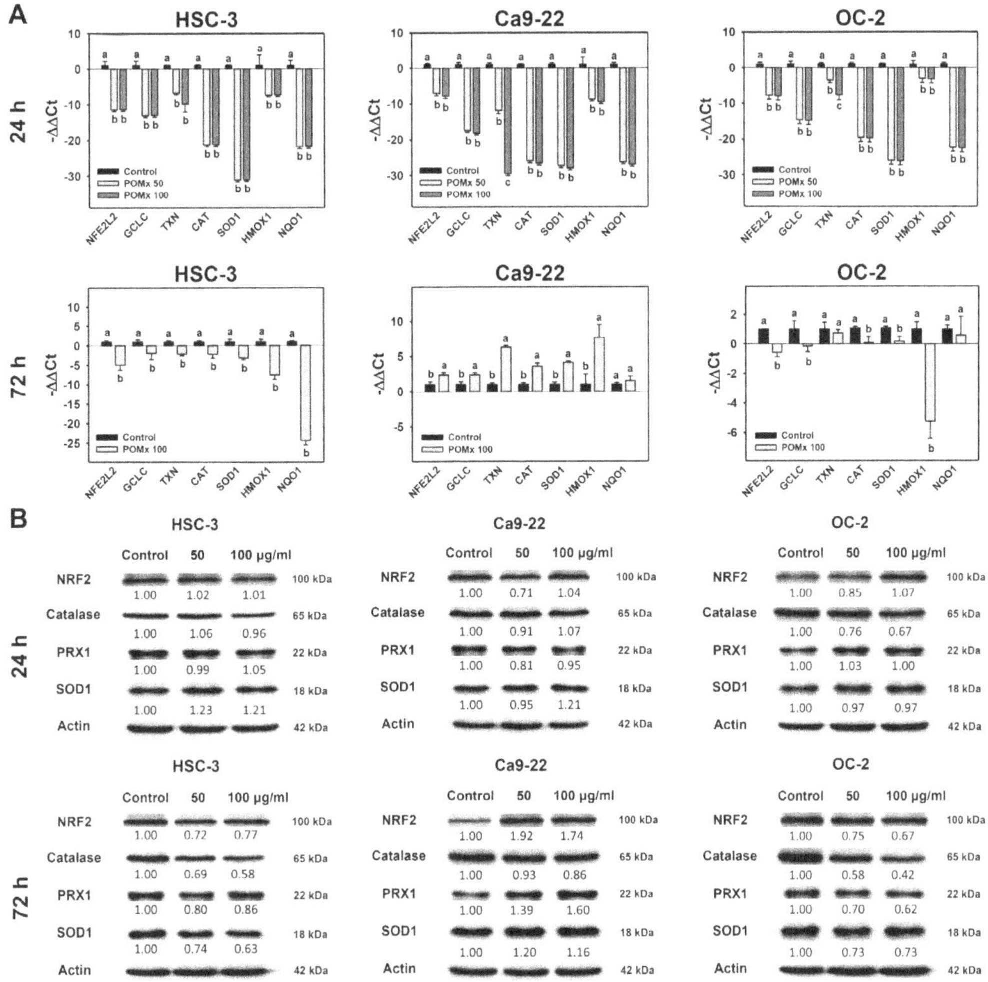

Figure 7. mRNA and protein expressions of antioxidant genes of oral cancer cells after POMx incubation. (**A**) Relative mRNA expressions (log_2) of antioxidant genes of oral cancer cells after POMx incubation for 24 and 72 h. (**B**) Western blotting for antioxidant signaling proteins for 24 and 72 h. Cells were incubated with 0, 50, and 100 μg/mL of POMx (control, POMx 50, and POMx 100) for 24 and 72 h. Treatments without overlapping low cases (a to c) indicate significant differences for the same cell lines. $p < 0.05$. Data, mean ± SD ($n = 3$).

3.8. Mitochondrial Mass of Oral Cancer Cells Following POMx Incubation

After 24 h POMx incubation (0, 50, and 100 μg/mL), the patterns for Mitotracker in three oral cancer cell lines are shown (Figure 8A). The Mitotracker (+) (%) of these three oral cancer cells were decreased after POMx incubation compared with the control (Figure 8B).

After time course treatments of POMx, the flow cytometry patterns for Mitotracker in these oral cancer cells are shown (Figure 8C). The Mitotracker (+) (%) of these three oral cancer cells are decreased at 12 and 24 h POMx incubation compared with the control, although it was slightly increased at 72 h (Figure 8D).

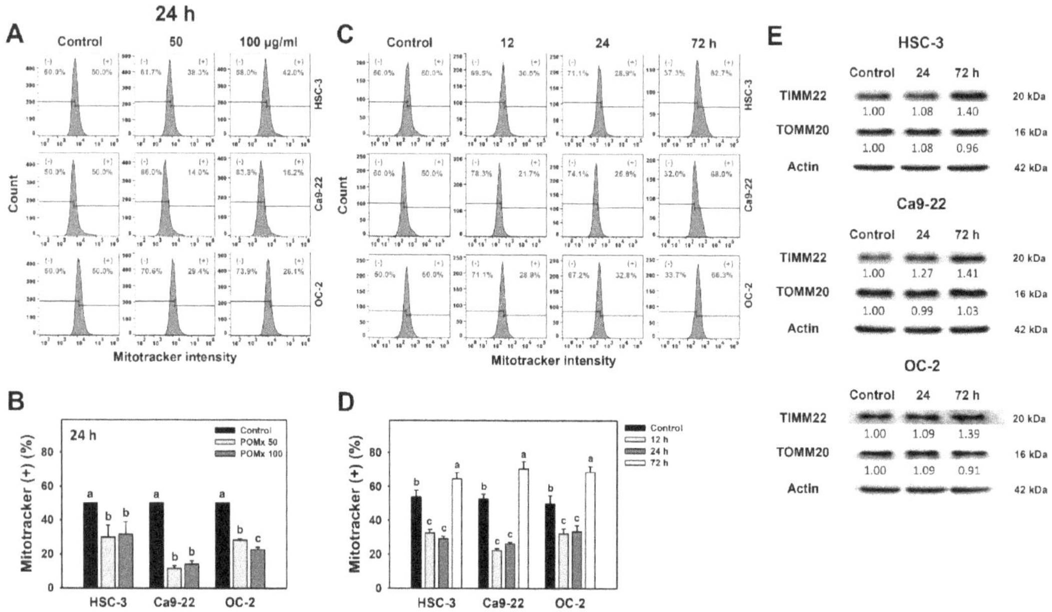

Figure 8. Mitochondrial mass of oral cancer cells after POMx incubation. (**A**) Dose-response to Mitotracker flow cytometry patterns. Cells (HSC-3, Ca9-22, and OC-2) were incubated with control, 50, and 100 μg/mL of POMx (control, POMx 50, and POMx 100) for 24 h. The Mitotracker (+) population was defined on the right side of each panel. (**B**) Statistics for (**A**). (**C**) Time course to Mitotracker flow cytometry patterns. Cells were incubated with control and 100 μg/mL of POMx for 0, 12, 24, and 72 h. (**D**) Statistics for (**C**). Treatments without overlapping low cases (a to c) are significantly different for the same cell lines. $p < 0.05$. Data, mean ± SD ($n = 3$). (**E**) Western blotting for mitochondrial resident proteins for 24 and 72 h. Cells were incubated with control and 100 μg/mL of POMx for 24 and 72 h.

After time course treatments of POMx, the mitochondrial resident protein (TIMM22 and TOMM20) expressions were detected in oral cancer cells (Figure 8E). For 24 h POMx incubations, the TIMM22 and TOMM20 are almost unchanged in oral cancer cells. However, although TOMM20 remains unchanged, TIMM22 is upregulated at 72 h POMx incubations, consistent with Mitotracker detection. Accordingly, Mitotracker detections and protein expressions are differentially regulated at 24 h POMx incubation but in a consistently regulated manner at 72 h.

3.9. Mitochondrial DNA Copy Number, Lesion and Biogenesis of Oral Cancer Cells Following POMx Incubation

In addition to MitoMP, MitoSOX, and mitochondrial mass as described above (Figures 5–8), other mitochondrial functions such as mitochondrial DNA copy number, lesion, and biogenesis were further examined in POMx-incubated oral cancer cells (Figure 9). After POMx incubations (0, 50, and 100 μg/mL) for 24 and 72 h, the relative mtDNA copy numbers of three oral cancer cell lines were dose-responsively decreased (Figure 9A).

mtDNA damages between *ND1* and *ND5* genes were higher in oral cancer cells following 24 and 72 h POMx incubation than those of the control (Figure 9B). Moreover, the mRNA expressions of all tested mitochondrial biogenesis genes (*TFB2M, TFAM, POLRMT,* and *TUFM*) were downregulated by 24 h POMx compared with the control (Figure 9C, left). Among these biogenesis genes, TUFM was dramatically downregulated by 72 h POMx (Figure 9C, right). The protein expressions of these mitochondrial biogenesis genes were consistently downregulated at 24 and 72 h POMx (Figure 9D). Therefore, POMx downregulates gene expressions for mitochondrial biogenesis in oral cancer cells.

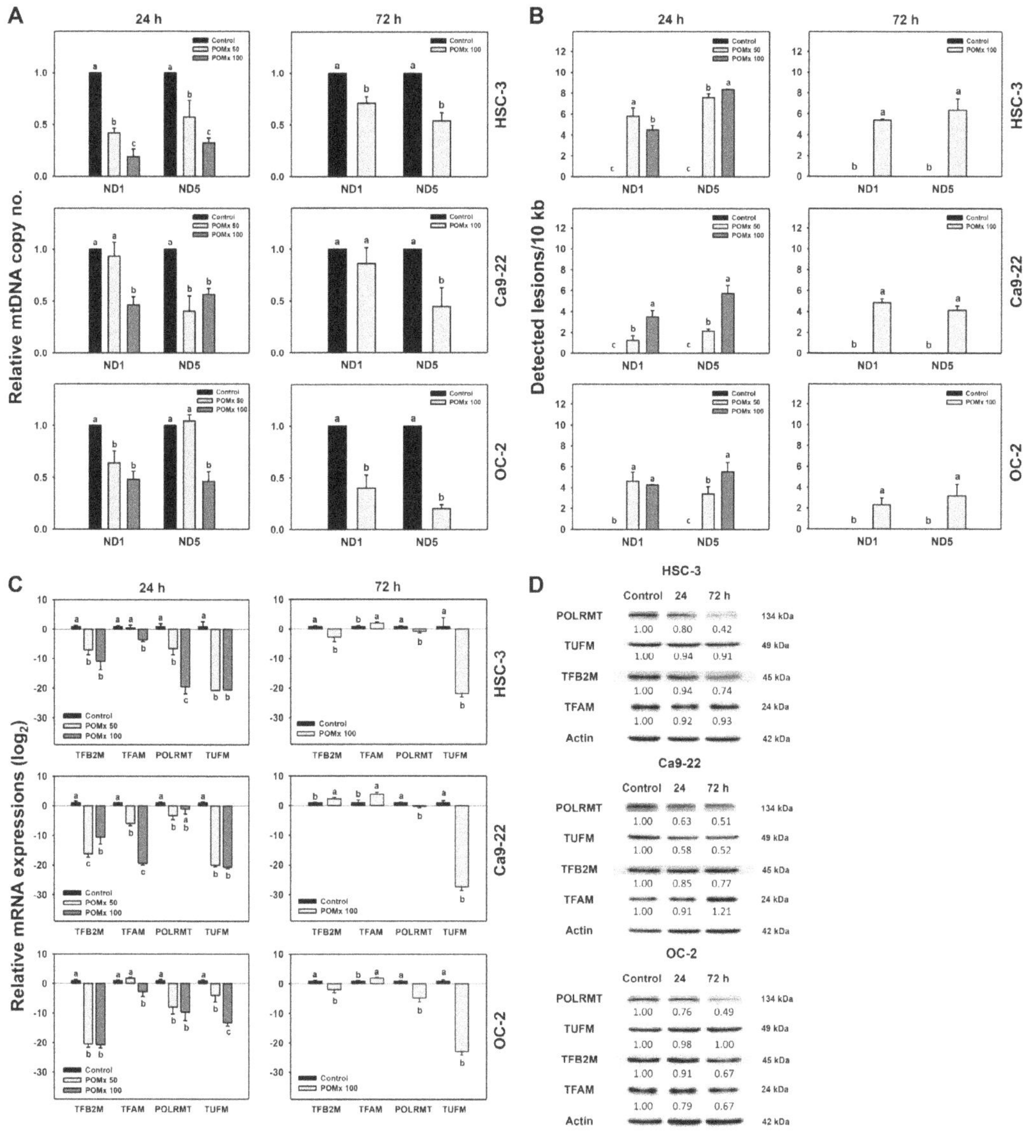

Figure 9. mtDNA copy number, mtDNA lesion, and mitochondrial biogenesis of oral cancer cells after POMx incubation. Cells (HSC-3, Ca9-22, and OC-2) were incubated with 0, 50, and 100 μg/mL of POMx (control, POMx 50, and POMx 100) for 24 and 72 h. (**A**) Relative mtDNA copy number in *ND1* and *ND5* genes. (**B**) mtDNA lesion frequency per 10 kb DNA between *ND1* to *ND5* genes. (**C**) Relative mRNA expressions (log_2) for mitochondrial biogenesis genes. Treatments without overlapping low cases (a to c) are significantly different for the same cell lines. $p < 0.05$. Data, mean ± SD ($n = 3$). (**D**) Western blotting for mitochondrial biogenesis genes at 24 and 72 h POMx incubations. Cells were incubated with control and 100 μg/mL of POMx for 24 and 72 h.

3.10. γH2AX-Detected DNA Damage of Oral Cancer Cells Following POMx Incubation

After POMx incubations (0, 50, and 100 μg/mL) for 0, 24, and 72 h, the patterns for γH2AX in three oral cancer cell lines were shown (Figure 10A). The γH2AX (+) (%) of these three oral cancer cells was slightly increased at 24 h POMx incubation and dramatically increased at 72 h POMx incubation compared with the control (Figure 10B).

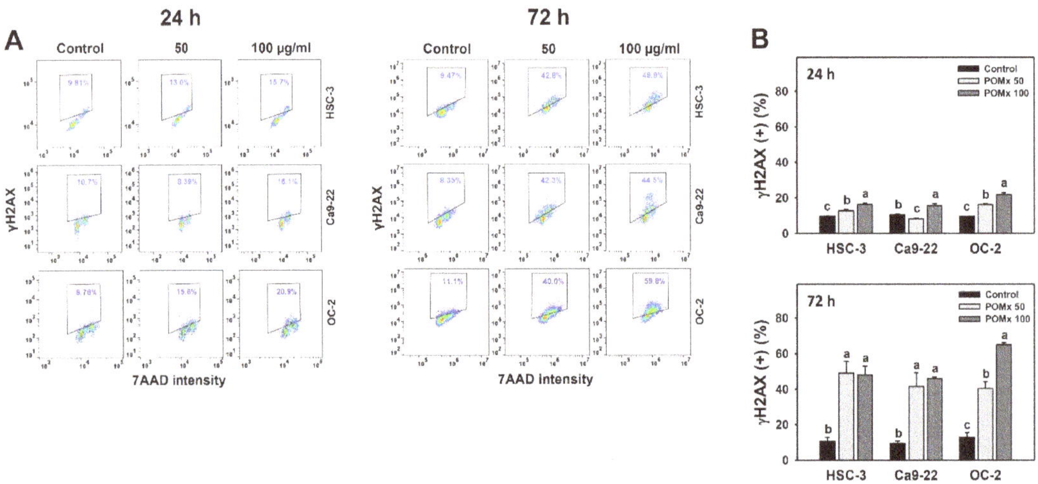

Figure 10. γH2AX change of oral cancer cells after POMx incubation. (**A**) Dose-response to γH2AX flow cytometry patterns. Cells (HSC-3, Ca9-22, and OC-2) were incubated with control, 50, and 100 μg/mL of POMx (control, POMx 50, and POMx 100) for 0, 24, and 72 h. The γH2AX (+) population (%) is marked within a dashed line. (**B**) Statistics of γH2AX (+) (%) in (**A**). Treatments without overlapping low cases (a to c) are significantly different for the same cell lines. $p < 0.001$. Data, mean ± SD (n = 3).

Accordingly, POMx triggers γH2AX-detected DNA damage in oral cancer cells.

4. Discussion

We found that POMx showed antiproliferation, apoptosis, oxidative stress, mitochondrial impairment, and DNA damage to several kinds of oral cancer cells. The detailed mechanisms for the POMx-induced antiproliferation are discussed in the following.

4.1. POMx Has a Selective Antiproliferation Function towards Cancer Cells with Safety to Normal Cells

POMx provides antioxidant-rich natural products [51] and shows anticancer effects on several cancer cells [11,16,52]. This is partly explained by antioxidants having dual functions to reduce or induce oxidative stress at physiological or high concentrations [14]. The present study shows that the IC_{50} values at the 24 h ATP assay for POMx incubated three oral cancer cell lines (Ca9-22, HSC-3, and OC-2) were 80.53, 100.34, and 108.12 μg/mL, respectively (Figure 2). Similarly, IC_{50} values at 72 h MTS assay for POMx for prostate (C4-2, PC3, and ARCaPM) [11] were 42, 78, and 161 μg/mL, respectively. Moreover, trypan blue assay in addition to ATP assay confirms viability results of POMx in oral cancer and normal cells.

The safety of POMx is well documented. For example, normal human prostatic epithelial PrEC cells showed no cytotoxicity (95% viability) to POMx [11]. At the 72 h MTT assay, pomegranate fruit extract (PFE) (50–150 μg/mL) showed antiproliferation against lung cancer cells with 53% viability but no cytotoxic effects on normal bronchial epithelial cells with 90% viability [15]. The pomegranate juice and oil showed antiproliferation and apoptosis in prostate cancer cells but no cytotoxicity in normal prostate epithelial cells [53]. Similarly, the normal oral cells (HGF-1) show higher viability in both ATP assay and trypan blue assay (Figure 2A,B) than the three oral cancer cell lines of this study. Punicalagin and ellagic acid, two main components of POMx, induce apoptosis of colon cancer cells without affecting normal colon cells [54]. Therefore, POMx and other pomegranate-derived natural products provided selective killing against several cancer cells and did not show side effects on normal oral cells.

4.2. POMx Inhibits Antioxidant Signaling to Generate Oxidative Stress

When the pro-oxidant level is higher than the antioxidant level, cellular oxidative stress is generated. In addition, mitochondrial impairment may change antioxidant gene expressions [55]. For example, advanced glycation end products were reported to inhibit the cellular antioxidant system and trigger oxidative stress [50].

Similar to the present study, the mRNA expressions for several antioxidant genes (*NFE2L2, GCLC, TXN, CAT, SOD1, HMOX1*, and *NQO1*) were downregulated at 24 h POMx for three oral cancer cell lines (Figure 7A). Moreover, protein expressions for these antioxidant genes are downregulated at 72 h POMx treatment for oral cancer cells (HSC-3 and OC-2) but slightly upregulated for Ca9-22 cells (Figure 7B). These results suggest that mRNA and protein expressions for antioxidant signaling may be differentially regulated between different oral cancer cell lines. Under these differential regulation modes, oxidative stress such as MitoMP depletion and MitoSOX generation were upregulated at 12, 24, and 72 h POMx in three oral cancer cells (Figures 5 and 6). Therefore, antioxidant pathways play a vital function in POMx induced oxidative stress in the present study.

4.3. POMx Induces Mitochondrial Impairment in Oral Cancer Cells

In addition to MitoMP and MitoSOX, mitochondrial mass, mtDNA copy number, mtDNA lesion, and mitochondrial biogenesis were also changed after POMx incubation in the present study. Similarly, Resveratrol may mitigate neurotoxicity following Rotenone treatment through promoting mitochondrial mass and DNA copy number [56]. Thus, there is a complex interaction between these mitochondrial functions.

Modulating mitochondrial function is associated with apoptosis. In view of mitochondrial mass change, several treatments may subsequently induce apoptosis. For example, TNFα decreases mitochondrial mass and induces apoptosis in human dermal microvascular endothelial cells (HMEC-1) [57]. In the present study, POMx shows similar results for oral cancer cells. The Mitotracker-detected mitochondrial mass is downregulated at 12 and 24 h POMx treatment but slightly upregulated at 72 h (Figure 8C,D). Similarly, Western blotting shows that mitochondrial resident protein TIM22 is upregulated at 72 h POMx. Accordingly, the mitochondrial mass is dynamically changed over time after POMx treatment of oral cancer cells. The role of POMx-induced mitochondrial mass change warrants a detailed investigation in the future.

mtDNA copy number change may regulate apoptosis. Increasing mtDNA copy number may inhibit apoptosis. In contrast, reducing mtDNA copy number was shown to induce ROS generation and apoptosis in tumor cells [58]. Similarly, 24 and 72 h POMx treatment increased oxidative stress and decreased mtDNA copy number (Figure 9A) in oral cancer cells, leading to apoptosis.

A change of mtDNA damage regulates apoptosis. Single [59] or double [60] strand breaks in mtDNA may induce apoptosis. Moreover, mtDNA damage induces MitoSOX generation and subsequent apoptosis [61]. Similarly, 24 and 72 h POMx treatment causes mtDNA damage (Figure 9B), MitoSOX (Figure 6), and apoptosis (Figure 4). Moreover, oxidative stress also induces oxidative DNA damage of nuclear DNA [62]. Our finding supported this because POMx caused DNA double-strand breaks (γH2AX) in oral cancer cells (Figure 10).

Change of mitochondrial biogenesis change regulates apoptosis. Biogenesis may increase mitochondrial mass and DNA copy number [56] and is associated with apoptosis [63,64]. Similarly, 24 and 72 h POMx treatment inhibits mRNA and protein expressions for mitochondrial biogenesis of gene expression (*TFB2M, TFAM, POLRMT*, and *TUFM*) in oral cancer cells (Figure 9C,D). This finding supports the notion that a decrease in mitochondrial biogenesis reduces the mitochondrial mass (Figure 8). Moreover, mitochondrial fission factor (MFF) overexpression in breast cancer cells decreases both mitochondrial mass and activity [65]. Since POMx downregulates mitochondrial biogenesis (Figure 9C,D) and mass (Figure 8), it is possible that POMx treatment causes mitochondrial fission and leads to apoptosis of oral cancer cells. It warrants a detailed investigation of the role of mitochondrial fission in POMx treatment for oral cancer cells in the future.

4.4. POMx Induces Apoptosis but Inhibits Autophagy in Oral Cancer Cells

POMx and pomegranate leaf extract (PLE) respectively induce apoptosis in human prostate [10] and lung [21] cancer cells. However, no caspase experiments were performed before the present study. Ethanol extracts of pomegranate fruit (PEE) induced apoptosis by cleaving Cas-3 and raising Bax/Bcl-2 ratio in urinary bladder cancer T24 cells [22]. Consistently, 72 h POMx induced apoptosis for oral cancer cells by the results of annexin V expression (Figure 4A) and Western blotting (Figure 4C).

The autophagy pathway is activated to guarantee the elimination of damaged mitochondria to maintain cell survival. In the case where autophagy is reduced, this may lead to cell death without the elimination of damaged mitochondria. This rationale is partly supported by our finding that the increase of apoptosis is accompanied by a decrease in AO-detected autophagy ranging from 12 to 72 h POMx treatment (Figure 4E). These results warrant a detailed investigation of the impact of mitophagy or autophagy upon POMx treatment of oral cancer cells.

5. Conclusions

In the present study, the antiproliferation of POMx was evaluated using several types of oral cancer cells, and its detailed mechanisms related to mitochondrial function were explored. POMx treatment shows antiproliferation and apoptosis associated with downregulating antioxidant gene expression and triggering mitochondrial impairment, causing ATP depletion, MitoMP disruption, and MitoSOX generation as well as decreases in mitochondrial mass, mtDNA copy number, and mitochondrial biogenesis. Moreover, both nuclear and mitochondrial DNA damages were induced by POMx incubation in oral cancer cells. In conclusion, POMx provides antiproliferation and apoptosis effects on oral cancer cells through impaired mitochondrial functioning.

Author Contributions: Conceptualization—H.-W.C. and J.-Y.T.; data curation—S.-Y.P.; formal analysis—S.-Y.P. and S.-R.C.; Methodology—L.-C.L. and Y.-B.C.; Supervision—H.-W.C. and J.-Y.T.; Writing—original draft—S.-Y.P., A.A.F. and H.-W.C.; Writing—review & editing—H.-W.C. and J.-Y.T. All authors have read and agreed to the published version of the manuscript.

Funding: This work was partly supported by funds from the Ministry of Science and Technology (MOST 108-2320-B-037-015-MY3), the National Sun Yat-sen University-KMU Joint Research Project (#NSYSUKMU 110-P016), the Kaohsiung Medical University Hospital (KMUH109-9M56), the Kaohsiung Medical University Research Center (KMU-TC108A04), and the NPUST-KMU joint research project (#NPUST-KMU-110-P003).

Institutional Review Board Statement: Not applicable.

Informed Consent Statement: Not applicable.

Data Availability Statement: Data is contained within the article.

Acknowledgments: The authors thank our colleague Dr. Hans-Uwe Dahms (https://biology.kmu.edu.tw/index.php/en-GB/faculty-members/92-hans-uwe-dahms) for editing the manuscript.

Conflicts of Interest: The authors declare that there are no conflict of interest among them.

Gene Accession Numbers

	Gene	Accession number
RNA	*NFE2L2*	NM_006164.5
	GCLC	NM_001498.4
	TXN	NM_003329.4
	CAT	NM_001752.4
	SOD1	NM_000454.4
	HMOX1	NM_002133.3
	NQO1	NM_000903.3
	TFB2M	NM_022366.3
	TFAM	NM_003201.3
	POLRMT	NM_005035.4
	TUFM	NM_003321.5
	GAPDH	NM_002046.7
	Gene	**Accession number**
DNA	*ND1*	NC_012920.1
	ND1-L	
	ND5	
	ND5-L	
	GAPDH	NG_007073.2

References

1. Ahmadiankia, N. Molecular targets of pomegranate (*Punica granatum*) in preventing cancer metastasis. *Iran J. Basic Med. Sci.* **2019**, *22*, 977–988.
2. Sharma, P.; McClees, S.F.; Afaq, F. Pomegranate for prevention and treatment of cancer: An update. *Molecules* **2017**, *22*, 177. [CrossRef] [PubMed]
3. Mandal, A.; Bhatia, D.; Bishayee, A. Anti-inflammatory mechanism involved in pomegranate-mediated prevention of breast cancer: The role of NF-kappaB and Nrf2 signaling pathways. *Nutrients* **2017**, *9*, 436. [CrossRef]
4. Singh, B.; Singh, J.P.; Kaur, A.; Singh, N. Phenolic compounds as beneficial phytochemicals in pomegranate (*Punica granatum* L.) peel: A review. *Food Chem.* **2018**, *261*, 75–86. [CrossRef] [PubMed]
5. Russo, M.; Cacciola, F.; Arena, K.; Mangraviti, D.; de Gara, L.; Dugo, P.; Mondello, L. Characterization of the polyphenolic fraction of pomegranate samples by comprehensive two-dimensional liquid chromatography coupled to mass spectrometry detection. *Nat. Prod. Res.* **2020**, *34*, 39–45. [CrossRef] [PubMed]
6. Wong, T.L.; Strandberg, K.R.; Croley, C.R.; Fraser, S.E.; Venkata, K.C.N.; Fimognari, C.; Sethi, G.; Bishayee, A. Pomegranate bioactive constituents target multiple oncogenic and oncosuppressive signaling for cancer prevention and intervention. *Semin Cancer Biol.* **2021**. [CrossRef] [PubMed]
7. Kandylis, P.; Kokkinomagoulos, E. Food applications and potential health benefits of pomegranate and its derivatives. *Foods* **2020**, *9*, 122. [CrossRef]
8. Wang, Y.; Zhang, H.; Liang, H.; Yuan, Q. Purification, antioxidant activity and protein-precipitating capacity of punicalin from pomegranate husk. *Food Chem.* **2013**, *138*, 437–443. [CrossRef]
9. Rasheed, Z.; Akhtar, N.; Anbazhagan, A.N.; Ramamurthy, S.; Shukla, M.; Haqqi, T.M. Polyphenol-rich pomegranate fruit extract (POMx) suppresses PMACI-induced expression of pro-inflammatory cytokines by inhibiting the activation of MAP Kinases and NF-kappaB in human KU812 cells. *J. Inflamm.* **2009**, *6*, 1. [CrossRef]
10. Koyama, S.; Cobb, L.J.; Mehta, H.H.; Seeram, N.P.; Heber, D.; Pantuck, A.J.; Cohen, P. Pomegranate extract induces apoptosis in human prostate cancer cells by modulation of the IGF-IGFBP axis. *Growth Horm. IGF Res.* **2010**, *20*, 55–62. [CrossRef]
11. Wang, Y.; Zhang, S.; Iqbal, S.; Chen, Z.; Wang, X.; Wang, Y.A.; Liu, D.; Bai, K.; Ritenour, C.; Kucuk, O.; et al. Pomegranate extract inhibits the bone metastatic growth of human prostate cancer cells and enhances the in vivo efficacy of docetaxel chemotherapy. *Prostate* **2014**, *74*, 497–508. [CrossRef]
12. Hamoud, S.; Hayek, T.; Volkova, N.; Attias, J.; Moscoviz, D.; Rosenblat, M.; Aviram, M. Pomegranate extract (POMx) decreases the atherogenicity of serum and of human monocyte-derived macrophages (HMDM) in simvastatin-treated hypercholesterolemic patients: A double-blinded, placebo-controlled, randomized, prospective pilot study. *Atherosclerosis* **2014**, *232*, 204–210. [CrossRef] [PubMed]
13. Nallanthighal, S.; Elmaliki, K.M.; Reliene, R. Pomegranate extract alters breast cancer stem cell properties in association with inhibition of epithelial-to-mesenchymal transition. *Nutr. Cancer* **2017**, *69*, 1088–1098. [CrossRef] [PubMed]
14. Bouayed, J.; Bohn, T. Exogenous antioxidants—Double-edged swords in cellular redox state: Health beneficial effects at physiologic doses versus deleterious effects at high doses. *Oxid. Med. Cell Longev.* **2010**, *3*, 228–237. [CrossRef] [PubMed]

15. Khan, N.; Hadi, N.; Afaq, F.; Syed, D.N.; Kweon, M.H.; Mukhtar, H. Pomegranate fruit extract inhibits prosurvival pathways in human A549 lung carcinoma cells and tumor growth in athymic nude mice. *Carcinogenesis* **2007**, *28*, 163–173. [CrossRef] [PubMed]
16. Asmaa, M.J.; Ali, A.J.; Farid, J.M.; Azman, S. Growth inhibitory effects of crude pomegranate peel extract on chronic myeloid leukemia, K562 cells. *Int. J. Appl. Basic Med. Res.* **2015**, *5*, 100–105.
17. Sineh Sepehr, K.; Baradaran, B.; Mazandarani, M.; Yousefi, B.; Abdollahpour Alitappeh, M.; Khori, V. Growth-inhibitory and apoptosis-inducing effects of *Punica granatum* L. var. spinosa (apple punice) on fibrosarcoma cell lines. *Adv. Pharm. Bull.* **2014**, *4*, 583–590. [PubMed]
18. Peng, S.Y.; Hsiao, C.C.; Lan, T.H.; Yen, C.Y.; Farooqi, A.A.; Cheng, C.M.; Tang, J.Y.; Yu, T.J.; Yeh, Y.C.; Chuang, Y.T.; et al. Pomegranate extract inhibits migration and invasion of oral cancer cells by downregulating matrix metalloproteinase-2/9 and epithelial-mesenchymal transition. *Environ. Toxicol.* **2020**, *35*, 673–682. [CrossRef]
19. Jezek, J.; Cooper, K.F.; Strich, R. Reactive oxygen species and mitochondrial dynamics: The yin and yang of mitochondrial dysfunction and cancer progression. *Antioxidants* **2018**, *7*, 13. [CrossRef]
20. Sung, Y.J.; Kao, T.Y.; Kuo, C.L.; Fan, C.C.; Cheng, A.N.; Fang, W.C.; Chou, H.Y.; Lo, Y.K.; Chen, C.H.; Jiang, S.S.; et al. Mitochondrial Lon sequesters and stabilizes p53 in the matrix to restrain apoptosis under oxidative stress via its chaperone activity. *Cell Death Dis.* **2018**, *9*, 697. [CrossRef]
21. Li, Y.; Yang, F.; Zheng, W.; Hu, M.; Wang, J.; Ma, S.; Deng, Y.; Luo, Y.; Ye, T.; Yin, W. *Punica granatum* (pomegranate) leaves extract induces apoptosis through mitochondrial intrinsic pathway and inhibits migration and invasion in non-small cell lung cancer in vitro. *Biomed. Pharm.* **2016**, *80*, 227–235. [CrossRef]
22. Lee, S.T.; Lu, M.H.; Chien, L.H.; Wu, T.F.; Huang, L.C.; Liao, G.I. Suppression of urinary bladder urothelial carcinoma cell by the ethanol extract of pomegranate fruit through cell cycle arrest and apoptosis. *BMC Complement. Altern. Med.* **2013**, *13*, 364. [CrossRef]
23. Wong, D.Y.; Chang, K.W.; Chen, C.F.; Chang, R.C. Characterization of two new cell lines derived from oral cavity human squamous cell carcinomas—OC1 and OC2. *J. Oral Maxillofac. Surg.* **1990**, *48*, 385–390. [CrossRef]
24. Vlachojannis, C.; Zimmermann, B.F.; Chrubasik-Hausmann, S. Efficacy and safety of pomegranate medicinal products for cancer. *Evid. Based Complement. Altern. Med.* **2015**, *2015*, 258598. [CrossRef] [PubMed]
25. Chen, C.Y.; Yen, C.Y.; Wang, H.R.; Yang, H.P.; Tang, J.Y.; Huang, H.W.; Hsu, S.H.; Chang, H.W. Tenuifolide B from *Cinnamomum tenuifolium* stem selectively inhibits proliferation of oral cancer cells via apoptosis, ROS generation, mitochondrial depolarization, and DNA damage. *Toxins* **2016**, *8*, 319. [CrossRef] [PubMed]
26. Yen, C.Y.; Chiu, C.C.; Chang, F.R.; Chen, J.Y.; Hwang, C.C.; Hseu, Y.C.; Yang, H.L.; Lee, A.Y.; Tsai, M.T.; Guo, Z.L.; et al. 4beta-Hydroxywithanolide E from *Physalis peruviana* (golden berry) inhibits growth of human lung cancer cells through DNA damage, apoptosis and G2/M arrest. *BMC Cancer* **2010**, *10*, 46. [CrossRef]
27. Chang, H.W.; Li, R.N.; Wang, H.R.; Liu, J.R.; Tang, J.Y.; Huang, H.W.; Chan, Y.H.; Yen, C.Y. Withaferin A induces oxidative stress-mediated apoptosis and DNA damage in oral cancer cells. *Front. Physiol.* **2017**, *8*, 634. [CrossRef]
28. Wang, H.R.; Tang, J.Y.; Wang, Y.Y.; Farooqi, A.A.; Yen, C.Y.; Yuan, S.F.; Huang, H.W.; Chang, H.W. Manoalide preferentially provides antiproliferation of oral cancer cells by oxidative stress-mediated apoptosis and DNA damage. *Cancers* **2019**, *11*, 1303. [CrossRef]
29. Wang, S.H.; Shih, Y.L.; Kuo, T.C.; Ko, W.C.; Shih, C.M. Cadmium toxicity toward autophagy through ROS-activated GSK-3beta in mesangial cells. *Toxicol. Sci.* **2009**, *108*, 124–131. [CrossRef] [PubMed]
30. Chou, H.L.; Lin, Y.H.; Liu, W.; Wu, C.Y.; Li, R.N.; Huang, H.W.; Chou, C.H.; Chiou, S.J.; Chiu, C.C. Combination therapy of chloroquine and C(2)-ceramide enhances cytotoxicity in lung cancer H460 and H1299 cells. *Cancers* **2019**, *11*, 370. [CrossRef] [PubMed]
31. Chiu, C.C.; Huang, J.W.; Chang, F.R.; Huang, K.J.; Huang, H.M.; Huang, H.W.; Chou, C.K.; Wu, Y.C.; Chang, H.W. Golden berry-derived 4beta-hydroxywithanolide E for selectively killing oral cancer cells by generating ROS, DNA damage, and apoptotic pathways. *PLoS ONE* **2013**, *8*, e64739. [CrossRef]
32. Chang, Y.T.; Huang, C.Y.; Tang, J.Y.; Liaw, C.C.; Li, R.N.; Liu, J.R.; Sheu, J.H.; Chang, H.W. Reactive oxygen species mediate soft corals-derived sinuleptolide-induced antiproliferation and DNA damage in oral cancer cells. *Oncol. Targets Ther.* **2017**, *10*, 3289–3297. [CrossRef]
33. Chang, H.W.; Yen, C.Y.; Chen, C.H.; Tsai, J.H.; Tang, J.Y.; Chang, Y.T.; Kao, Y.H.; Wang, Y.Y.; Yuan, S.F.; Lee, S.Y. Evaluation of the mRNA expression levels of integrins alpha3, alpha5, beta1 and beta6 as tumor biomarkers of oral squamous cell carcinoma. *Oncol. Lett.* **2018**, *16*, 4773–4781.
34. Yen, C.Y.; Huang, C.Y.; Hou, M.F.; Yang, Y.H.; Chang, C.H.; Huang, H.W.; Chen, C.H.; Chang, H.W. Evaluating the performance of fibronectin 1 (FN1), integrin alpha4beta1 (ITGA4), syndecan-2 (SDC2), and glycoprotein CD44 as the potential biomarkers of oral squamous cell carcinoma (OSCC). *Biomarkers* **2013**, *18*, 63–72. [CrossRef]
35. Stagos, D.; Balabanos, D.; Savva, S.; Skaperda, Z.; Priftis, A.; Kerasioti, E.; Mikropoulou, E.V.; Vougogiannopoulou, K.; Mitakou, S.; Halabalaki, M.; et al. Extracts from the mediterranean food plants *Carthamus lanatus*, *Cichorium intybus*, and *Cichorium spinosum* enhanced GSH levels and increased Nrf2 expression in human endothelial cells. *Oxid. Med. Cell Longev.* **2018**, *2018*, 6594101. [CrossRef]

36. Yu, T.J.; Tang, J.Y.; Ou-Yang, F.; Wang, Y.Y.; Yuan, S.F.; Tseng, K.; Lin, L.C.; Chang, H.W. Low concentration of withaferin A inhibits oxidative stress-mediated migration and invasion in oral cancer cells. *Biomolecules* **2020**, *10*, 777. [CrossRef] [PubMed]
37. Nishio, S.; Teshima, Y.; Takahashi, N.; Thuc, L.C.; Saito, S.; Fukui, A.; Kume, O.; Fukunaga, N.; Hara, M.; Nakagawa, M.; et al. Activation of CaMKII as a key regulator of reactive oxygen species production in diabetic rat heart. *J. Mol. Cell Cardiol.* **2012**, *52*, 1103–1111. [CrossRef]
38. Jeong, S.H.; Kim, H.K.; Song, I.S.; Noh, S.J.; Marquez, J.; Ko, K.S.; Rhee, B.D.; Kim, N.; Mishchenko, N.P.; Fedoreyev, S.A.; et al. Echinochrome a increases mitochondrial mass and function by modulating mitochondrial biogenesis regulatory genes. *Mar. Drugs* **2014**, *12*, 4602–4615. [CrossRef] [PubMed]
39. Shin, J.W.; Wu, Y.; Kang, Y.G.; Kim, J.K.; Choi, H.J.; Shin, J.W. The effects of epigallocatechin-3-gallate and mechanical stimulation on osteogenic differentiation of human mesenchymal stem cells: Individual or synergistic effects. *Tissue Eng. Regen. Med.* **2017**, *14*, 307–315. [CrossRef] [PubMed]
40. Livak, K.J.; Schmittgen, T.D. Analysis of relative gene expression data using real-time quantitative PCR and the 2(-Delta Delta C(T)) Method. *Methods* **2001**, *25*, 402–408. [CrossRef] [PubMed]
41. Fujii, Y.; Yoshihashi, K.; Suzuki, H.; Tsutsumi, S.; Mutoh, H.; Maeda, S.; Yamagata, Y.; Seto, Y.; Aburatani, H.; Hatakeyama, M. CDX1 confers intestinal phenotype on gastric epithelial cells via induction of stemness-associated reprogramming factors SALL4 and KLF5. *Proc. Natl. Acad. Sci. USA* **2012**, *109*, 20584–20589. [CrossRef] [PubMed]
42. Laddha, N.C.; Dwivedi, M.; Mansuri, M.S.; Singh, M.; Patel, H.H.; Agarwal, N.; Shah, A.M.; Begum, R. Association of neuropeptide Y (NPY), interleukin-1B (IL1B) genetic variants and correlation of IL1B transcript levels with vitiligo susceptibility. *PLoS ONE* **2014**, *9*, e107020. [CrossRef] [PubMed]
43. Wang, S.C.; Wang, Y.Y.; Lin, L.C.; Chang, M.Y.; Yuan, S.F.; Tang, J.Y.; Chang, H.W. Combined treatment of sulfonyl chromen-4-ones (CHW09) and ultraviolet-C (UVC) enhances proliferation inhibition, apoptosis, oxidative stress, and DNA damage against oral cancer cells. *Int. J. Mol. Sci.* **2020**, *21*, 6443. [CrossRef] [PubMed]
44. Lin, S.D.; Huang, S.H.; Lin, Y.N.; Wu, S.H.; Chang, H.W.; Lin, T.M.; Chai, C.Y.; Lai, C.S. Engineering adipose tissue from uncultured human adipose stromal vascular fraction on collagen matrix and gelatin sponge scaffolds. *Tissue Eng. Part A* **2011**, *17*, 1489–1498. [CrossRef] [PubMed]
45. Bijak, M.; Synowiec, E.; Sitarek, P.; Sliwinski, T.; Saluk-Bijak, J. Evaluation of the cytotoxicity and genotoxicity of flavonolignans in different cellular models. *Nutrients* **2017**, *9*, 1356. [CrossRef] [PubMed]
46. Rothfuss, O.; Gasser, T.; Patenge, N. Analysis of differential DNA damage in the mitochondrial genome employing a semi-long run real-time PCR approach. *Nucleic Acids Res.* **2010**, *38*, e24. [CrossRef] [PubMed]
47. Tang, J.Y.; Peng, S.Y.; Cheng, Y.B.; Wang, C.L.; Farooqi, A.A.; Yu, T.J.; Hou, M.F.; Wang, S.C.; Yen, C.H.; Chan, L.P.; et al. Ethyl acetate extract of *Nepenthes adrianii* x *clipeata* induces antiproliferation, apoptosis, and DNA damage against oral cancer cells through oxidative stress. *Environ. Toxicol.* **2019**, *34*, 891–901. [CrossRef]
48. Tang, J.Y.; Shu, C.W.; Wang, C.L.; Wang, S.C.; Chang, M.Y.; Lin, L.C.; Chang, H.W. Sulfonyl chromen-4-ones (CHW09) shows an additive effect to inhibit cell growth of X-ray irradiated oral cancer cells, involving apoptosis and ROS generation. *Int. J. Radiat. Biol.* **2019**, *95*, 1226–1235. [CrossRef]
49. Chang, H.S.; Tang, J.Y.; Yen, C.Y.; Huang, H.W.; Wu, C.Y.; Chung, Y.A.; Wang, H.R.; Chen, I.S.; Huang, M.Y.; Chang, H.W. Antiproliferation of *Cryptocarya concinna*-derived cryptocaryone against oral cancer cells involving apoptosis, oxidative stress, and DNA damage. *BMC Complement. Altern. Med.* **2016**, *16*, 94. [CrossRef]
50. Vasdev, S.; Gill, V.D.; Singal, P.K. Modulation of oxidative stress-induced changes in hypertension and atherosclerosis by antioxidants. *Exp. Clin. Cardiol.* **2006**, *11*, 206–216. [PubMed]
51. Hajleh, M.A.; Al-Dujaili, A. Anti-cancer activity of pomegranate and its biophenols; general review. *EC Nutr.* **2016**, *6*, 28–52.
52. Shirode, A.B.; Kovvuru, P.; Chittur, S.V.; Henning, S.M.; Heber, D.; Reliene, R. Antiproliferative effects of pomegranate extract in MCF-7 breast cancer cells are associated with reduced DNA repair gene expression and induction of double strand breaks. *Mol. Carcinog.* **2014**, *53*, 458–470. [CrossRef] [PubMed]
53. Hong, M.Y.; Seeram, N.P.; Heber, D. Pomegranate polyphenols down-regulate expression of androgen-synthesizing genes in human prostate cancer cells overexpressing the androgen receptor. *J. Nutr. Biochem.* **2008**, *19*, 848–855. [CrossRef] [PubMed]
54. Larrosa, M.; Tomas-Barberan, F.A.; Espin, J.C. The dietary hydrolysable tannin punicalagin releases ellagic acid that induces apoptosis in human colon adenocarcinoma Caco-2 cells by using the mitochondrial pathway. *J. Nutr. Biochem.* **2006**, *17*, 611–625. [CrossRef]
55. Singh, K.K. Mitochondria damage checkpoint in apoptosis and genome stability. *FEMS Yeast Res.* **2004**, *5*, 127–132. [CrossRef]
56. Peng, K.; Tao, Y.; Zhang, J.; Wang, J.; Ye, F.; Dan, G.; Zhao, Y.; Cai, Y.; Zhao, J.; Wu, Q.; et al. Resveratrol regulates mitochondrial biogenesis and fission/fusion to attenuate rotenone-induced neurotoxicity. *Oxid. Med. Cell Longev.* **2016**, *2016*, 6705621. [CrossRef]
57. Kalogeris, T.J.; Baines, C.; Korthuis, R.J. Adenosine prevents TNFalpha-induced decrease in endothelial mitochondrial mass via activation of eNOS-PGC-1alpha regulatory axis. *PLoS ONE* **2014**, *9*, e98459. [CrossRef]
58. Mei, H.; Sun, S.; Bai, Y.; Chen, Y.; Chai, R.; Li, H. Reduced mtDNA copy number increases the sensitivity of tumor cells to chemotherapeutic drugs. *Cell Death Dis.* **2015**, *6*, e1710. [CrossRef]
59. Tann, A.W.; Boldogh, I.; Meiss, G.; Qian, W.; Van Houten, B.; Mitra, S.; Szczesny, B. Apoptosis induced by persistent single-strand breaks in mitochondrial genome: Critical role of EXOG (5′-EXO/endonuclease) in their repair. *J. Biol. Chem.* **2011**, *286*, 31975–31983. [CrossRef]

60. Van Houten, B.; Hunter, S.E.; Meyer, J.N. Mitochondrial DNA damage induced autophagy, cell death, and disease. *Front. Biosci.* **2016**, *21*, 42–54. [CrossRef] [PubMed]
61. Ricci, C.; Pastukh, V.; Leonard, J.; Turrens, J.; Wilson, G.; Schaffer, D.; Schaffer, S.W. Mitochondrial DNA damage triggers mitochondrial-superoxide generation and apoptosis. *Am. J. Physiol. Cell Physiol.* **2008**, *294*, C413–C422. [CrossRef] [PubMed]
62. Bohr, V.A.; Dianov, G.L. Oxidative DNA damage processing in nuclear and mitochondrial DNA. *Biochimie* **1999**, *81*, 155–160. [CrossRef]
63. Vayssiere, J.L.; Petit, P.X.; Risler, Y.; Mignotte, B. Commitment to apoptosis is associated with changes in mitochondrial biogenesis and activity in cell lines conditionally immortalized with simian virus 40. *Proc. Natl. Acad. Sci. USA* **1994**, *91*, 11752–11756. [CrossRef] [PubMed]
64. Meira Martins, L.A.; Vieira, M.Q.; Ilha, M.; de Vasconcelos, M.; Biehl, H.B.; Lima, D.B.; Schein, V.; Barbe-Tuana, F.; Borojevic, R.; Guma, F.C. The interplay between apoptosis, mitophagy and mitochondrial biogenesis induced by resveratrol can determine activated hepatic stellate cells death or survival. *Cell Biochem. Biophys.* **2015**, *71*, 657–672. [CrossRef] [PubMed]
65. Sanchez-Alvarez, R.; De Francesco, E.M.; Fiorillo, M.; Sotgia, F.; Lisanti, M.P. Mitochondrial fission factor (MFF) inhibits mitochondrial metabolism and reduces breast cancer stem cell (CSC) activity. *Front. Oncol.* **2020**, *10*, 1776. [CrossRef] [PubMed]

antioxidants

MDPI

Article

The Potential of Dietary Antioxidants from a Series of Plant Extracts as Anticancer Agents against Melanoma, Glioblastoma, and Breast Cancer

Mindaugas Liaudanskas [1], Vaidotas Žvikas [1] and Vilma Petrikaitė [2,3,*]

[1] Faculty of Pharmacy, Lithuanian University of Health Sciences, 44307 Kaunas, Lithuania; Mindaugas.Liaudanskas@lsmuni.lt (M.L.); Vaidotas.zvikas@lsmuni.lt (V.Ž.)

[2] Institute of Cardiology, Lithuanian University of Health Sciences, 44307 Kaunas, Lithuania

[3] Faculty of Medicine, Lithuanian University of Health Sciences, 44307 Kaunas, Lithuania

* Correspondence: Vilma.Petrikaite@lsmuni.lt; Tel.: +370-6862-9383

Abstract: In modern society, cancer is one of the most relevant medical problems. It is important to search for promising plant raw materials whose extracts have strong antioxidant and anticancer effects. The aim of this study was to determine the composition of phenolic compounds in plant extracts, to evaluate their antioxidant and anticancer activity, and to find the correlations between those activities. Extracts of calendula, sage, bearberry, eucalyptus, yarrow, and apple were selected for the study. The phenolic compounds of these extracts were determined by the UPLC-ESI-MS/MS method and the antioxidant activity was evaluated in vitro by four different UV-VIS spectrophotometric methods (ABTS, DPPH, CUPRAC, FRAP). The anticancer activity of extracts was tested against melanoma IGR39, glioblastoma U-87, and triple-negative breast cancer MDA-MB-231 cell lines in vitro by MTT assay. The highest content of identified and quantified phenolic compounds was found in sage leaf extract and the lowest in ethanol eucalyptus leaf extract. The highest antioxidant activity was determined by all applied methods for the acetone eucalyptus leaf extract. The majority of extracts were mostly active against the melanoma IGR39 cell line, and possessed the lowest activity against the glioblastoma U-87 cell line. Acetone extract of eucalyptus leaf samples exhibited the highest anticancer activity against all tested cell lines. Strong and reliable correlation has been found between antioxidant and anticancer activity in breast cancer and glioblastoma cell lines, especially when evaluating antioxidant activity by the FRAP method.

Keywords: antioxidant; anticancer; marigold; sage; bearberry; eucalyptus; yarrow; apples

Citation: Liaudanskas, M.; Žvikas, V.; Petrikaitė, V. The Potential of Dietary Antioxidants from a Series of Plant Extracts as Anticancer Agents against Melanoma, Glioblastoma, and Breast Cancer. *Antioxidants* **2021**, *10*, 1115. https://doi.org/10.3390/antiox10071115

Academic Editors: Irene Dini and Domenico Montesano

Received: 16 June 2021
Accepted: 9 July 2021
Published: 12 July 2021

Publisher's Note: MDPI stays neutral with regard to jurisdictional claims in published maps and institutional affiliations.

1. Introduction

Numerous scientific data confirm the benefits of natural antioxidants for human health [1]. A relationship between the consumption of plant-rich foods and the incidence of oncological [2], cardiovascular [3–5], and neurodegenerative diseases [6] has been established.

Evaluation of the antioxidant activity of plant extracts is extremely important. It provides a scientific basis for the use of herbal preparations for the prevention and treatment of oxidative stress disorders. Herbal extracts have a variety of compounds that act as antioxidants through different reaction mechanisms. Therefore, the scientific literature notes that performing a single test of antioxidant activity is not adequate, thus it is recommended to use at least two different methods [7]. Hence, we evaluated the antiradical and reductive activity of the plant raw material extracts by four different methods in vitro (ABTS, CUPRAC, DPPH, and FRAP). ABTS and DPPH assays are based on the ability of antioxidants to scavenge free radicals [8,9], whereas FRAP and CUPRAC assays measure the antioxidant reducing activity [10,11].

ABTS and CUPRAC methods allow to assess the antioxidant activity of both hydrophilic and lipophilic antioxidants, and their pH is close to that of body fluids [12,13].

An acidic medium (pH = 3.6) is used with the FRAP method [11], which can strongly influence the antioxidant properties of some natural antioxidants [14]. This method is not applicable to antioxidants whose mechanism of action is based on hydrogen transfer reactions [15]. Free DPPH radicals are soluble exclusively in organic solvents, which limits the evaluation of the antiradical activity of hydrophilic antioxidants by this method. One of the major limitations of the ABTS method is that any compound, even without being an antioxidant (e.g., various sugars and citric acid), that has a redox potential less than ABTS+ radical-cation may be involved in the reaction, and thus can distort the results.

Herbal extracts are multicomponent matrices that can contain thousands of different biologically active compounds that have a wide variety of biological effects on the human body. The search for promising plant-based raw materials and analysis of their chemical composition and biological activity is particularly important when developing new drugs for the prevention and treatment of various diseases. For this reason, it is important to assess de novo the chemical composition of herbal extracts used in traditional and folk medicine using modern research methods, in order to determine which raw materials have the highest antioxidant, anticancer, and other biological activity in vitro and in vivo. Extracts of plant raw materials for this study (eucalyptus, sage and bearberry leaves, yarrow grass, marigold flowers, and apples) were selected owing to their strong antioxidant and anticancer effects. González-Burgos et al. indicate that eucalyptus leaf extracts have strong antioxidant and neuroprotective effects [16]. Sage leaves are especially rich in phenolic compounds—natural antioxidants. Many studies have proven the anticancer effects of their extracts [17]. Scientific literature provides data on the anticancer effects [18–21] and antioxidant activity [22–25] of yarrow grass, calendula flowers, bearberry leaves, and apple fruit extracts.

One of the most important bioactive compounds accumulated in plants is phenolic compounds, which are classified as natural antioxidants. Oxidative stress causes changes in cellular metabolism associated with DNA, protein damage, and lipid peroxidation [26]. These changes can lead to inflammatory processes as well as cardiovascular, cancerous, and other diseases [27]. Phenolic compounds effectively neutralize reactive forms of oxygen and nitrogen [28], and are thus valuable in the prevention and treatment of many diseases. Studies on the qualitative and quantitative composition and biological activity of plant raw materials that accumulate phenolic compounds are important and relevant.

Cancer is one of the most relevant societal, scientific, and medical issues in modern society, as morbidity and mortality are constantly increasing throughout the world [29]. Although various cancer prevention measures and modern malignancy research methods are being implemented, it is still not possible to claim that cancerous diseases are defeated. Therefore, efforts have been made to find more effective treatments for cancer. Herbal extracts provide promising preparations for cancer prevention and treatment, the potential of which has not yet been fully evaluated and exploited. Researchers have shown that plant extracts, their isolated fractions, or individual components can inhibit cancer cell-stimulating enzymes, stimulate the production of anticancer enzymes, enhance the immune system response, and protect DNA and other cellular structural molecules through their antioxidant effects [30]. In this context, it is important to conduct anticancer activity studies of plant extracts by selecting extracts with the highest activity.

The relationship between the antioxidant effect of plant extracts and their anticancer activity is controversial [31,32]. There is information in the scientific literature regarding the association of antioxidant activity of plant extracts with their anticancer activity [33], but data denying this direct relationship have also been published [34]. For this reason, it is extremely important to evaluate the relationship between the chemical composition of specific plant extracts under investigation and their antioxidant and anti-cancer activity. To accomplish this task, a correlation analysis was performed, the results of which allow statistically reliable estimation of the existence of the above-mentioned relationships.

The main aim of this study is to determine the composition of phenolic compounds in selected plant extracts in order to evaluate their antioxidant and anticancer activity and correlation between the estimates of these parameters. The obtained research results will provide new knowledge about the qualitative and quantitative composition as well as antioxidant and anticancer activity of selected plant extracts widely used in traditional and folk medicine and their correlation. This up-to-date information will be valuable from a theoretical and practical point of view to identify plant extracts with the strongest antioxidant and anticancer activity that could be promising for further in vivo studies, the manufacture of pharmaceuticals for the prophylaxis and treatment of cancer, and their use in clinical practice.

2. Materials and Methods

2.1. Plant Material

Eucalyptus globulus Labill. and *Salvia officinalis* L. leaves, *Achillea millefolium* L. herb, and *Calendula officinalis* L. flowers (manufacturer JSC Acorus Calamus, Švenčionys, Lithuania) were bought in the local pharmacy. Dried material was ground to a powder using a mill (IKA® A11 basic, Staufen im Breisgau, Germany). The apples samples (cultivar 'Ligol') were supplied by the Institute of Horticulture, Lithuanian Research Centre for Agriculture and Forestry, Babtai, Lithuania (55°60′ N, 23°48′ E). Each apple was cut into slices of equal size (up to 1 cm in thickness), and the stalks and seeds were removed. The slices were immediately frozen in a freezer (at −35 °C) with air circulation and then lyophilized with a Zirbus sublimator (ZIRBUS technology, Bad Grund, Germany) at the pressure of 0.01 mbar (condenser temperature, −85 °C). The lyophilized slices were ground to fine powder using a Retsch 200 mill (Haan, Germany). Loss on drying before analysis was determined by drying about 1 g of milled leaves in a moisture analyzer (Precisa HA 300, "Precisa Instruments AG, Dietikon, Switzerland) to complete evaporation of water and volatile compounds (drying temperature: 105 °C).

2.2. Cell Culture

Human melanoma cell line IGR39, human breast adenocarcinoma cell line MDA-MB-231, and human glioblastoma cell line U-87 (a kind gift from Dr. Manel Esteller, Bellvitge Biomedical Research Institute (IDIBELL)) were grown in DMEM Glutamax medium supplemented with 10% FBS and 1% antibiotics at 37 °C in a humidified atmosphere containing 5% CO_2. All cell cultures routinely were grown to 70% confluence and trypsinized with 0.125% TrypLE™ Express solution before passage. They were used until passage 20.

2.3. Chemicals and Materials

All chemical solvents, reagents, and standards used were of analytical grade. Acetonitrile, acetone, and formic acid were obtained from Sigma-Aldrich GmbH (Buchs, Switzerland) and ethanol from JSC Vilniaus degtinė (Vilnius, Lithuania). 1,1-Diphenyl-2-picrylhydrazyl (DPPH•) radical, 2,2′-azino-bis(3-ethylbenzothiazoline-6-sulphonic acid) (ABTS), potassium persulfate, ammonium acetate, sodium acetate trihydrate, iron (III) chloride hexahydrate, and 2,4,6-tripyridyl-s-triazine (TPTZ) were obtained from Sigma-Aldrich (Steinheim, Germany). Neocuproine was purchased from Alfa Aesar (Kandel, Germany) and copper (II) chloride—from Carl Roth Carl Roth GmbH + Co. KG (Karlsruhe, Germany). Folin–Ciocalteu reagent, gallic acid monohydrate, sodium carbonate, aluminum chloride hexahydrate, and hexamethylenetetramine were purchased from Sigma-Aldrich GmbH (Buchs, Switzerland). All analytical standards for UPLC-ESI-MS/MS analysis were HPLC grade and were also purchased from Sigma–Aldrich (St. Louis, MO, USA), except for and isorhamnetin 3-O-glucoside, which was from Extrasynthese (Genay, France). Deionized water, produced by the Milli–Q® (Millipore, Bedford, MA, USA) high-performance liquid chromatography water purification system, was used.

DMSO (≥99%, Ph. Eur. grade) was obtained from Sigma-Aldrich (St. Louis, MO, USA). 3-(4,5-dimethylthiazol-2-yl)-2,5-diphenyltetrazolium bromide (MTT, ≥97%) was purchased from Sigma-Aldrich (St. Louis, MO, USA).

All cell culture plastic ware was purchased from Thermo Fisher Scientific, Corning and Techno Plastic Products. TrypLE^{TM} Express, Dulbecco's modified Eagle high glucose medium (DMEM Glutamax), fetal bovine serum (FBS), penicillin/streptomycin solution (100×), and phosphate buffered saline (PBS) were obtained from Gibco.

2.4. Extraction

The powder from *E. globulus* and *S. officinalis* leaves, *A. millefolium* herb, and *C. officinalis* flowers was extracted as described by González-Burgos et al. [16]. Raw material was soaked for 3 h in 40% or 70% (*v*/*v*) ethanol or acetone. Here, 40 and 70% ethanol was selected as an extractant based on literature data, providing the best extraction yield of phenolic compounds [35,36]. Further, 70% acetone was used for comparative studies of the extraction efficiency of eucalyptus leaf samples, as this solvent is proposed by some scientists as an effective extractant for the extraction of phenolic compounds [37,38].

Soaked raw material was transferred to a percolator, covered with extractant, and left to macerate (48 h). Then, it was percolated (rate 0.3 mL min^{-1}) and high concentration extract (85% of total extract amount) was obtained. Low concentration extract was decanted and it was evaporated using a rotary evaporator (IKA® HB 10, Staufen im Breisgau, Germany) up to 15% of the total liquid extract amount. The remaining part of the low concentration extract was transferred to a single container with high concentration extract. The organic phase of liquid extract was evaporated using a rotary evaporator and the remaining aqueous phase was lyophilised using lyophilisator Zirbus (Zirbus technology GmbH, Bad Grund, Germany) at 0.01 mbar pressure and condenser temperature of −85 °C.

The powder from lyophilized apples was extracted as described by Liaudanskas et al. 2014 [39]. For the extraction of apple samples, 70% ethanol was selected based on the results of apple flavonoid extraction efficiency studies published in this article. An amount of 2.5 g of lyophilized apple powder (exact weight) was weighed, added to 30 mL of ethanol (70%, *v*/*v*), and extracted in a Sonorex Digital 10 P ultrasonic bath (Bandelin Electronic GmbH & Co. KG, Berlin, Germany) for 20 min at 40 °C. The extract obtained was filtered through a paper filter; the apple lyophilizate on the filter was washed twice with 10 mL of ethanol (70%, *v*/*v*) in a 50 mL flask.

Extraction conditions (the solvent and the ratio are shown in Table 1.

Table 1. Extracts prepared for chemical composition and biological activity analysis.

No.	Raw Material	Extraction Solvent	Concentration of Extraction Solvent, % (*v*/*v*)	The Ratio of Extraction Solvent to Raw Material
1	Dried calendula flowers	Ethanol	40	1:5
2	Dried sage leaves	Ethanol	40	1:5
3	Dried yarrow herb	Ethanol	40	1:5
4	Dried bearberry leaves	Ethanol	70	1:20
5	Dried eucalyptus leaves	Acetone	70	1:2
6	Lyophilized apples	Ethanol	70	1:20
7	Dried eucalyptus leaves	Ethanol	40	1:2

All extracts were filtered through a membrane filter with a pore size of 0.22 μm (Carl Roth GmbH, Karlsruhe, Germany).

2.5. UPLC–ESI–MS/MS Conditions

Qualitative and quantitative analysis of phenolic compounds was performed according to the previously validated and described UPLC–ESI–MS/MS method [16]. Separation of phenolic compounds was performed with Acquity H-class UPLC system (Waters, Milford, MA, USA) equipped with a triple quadrupole tandem mass spectrometer (Xevo, Waters, Milford, MA, USA) with an electrospray ionization source (ESI) to obtain MS/MS data. YMC Triart C18 (100 × 2.0 mm; 1.9 μm) column (YMC Europe GmbH, Dislanken, Germany) was used for analysis. The column temperature was maintained at 40 °C. Gradient elution was performed with mobile phase consisting of 0.1% formic acid water solution (solvent A) and acetonitrile (solvent B) with a flow rate set to 0.5 mL min^{-1}. Linear gradient profile was applied as follows for solvent A: initially 95% for 1 min; to 70% over 4 min; 50% over 7 min; and 95% over 2 min. Negative electrospray ionization was applied for analysis: capillary voltage −2 kV, source temperature 150 °C, desolvation temperature 400 °C, desolvation gas flow 700 L h^{-1}, and cone gas flow 20 L h^{-1}. Collision energy and cone voltage were optimized for each compound separately.

2.6. Determination of Antioxidant Activity

ABTS$^{\bullet+}$ radical cation decolorization assay. An ABTS$^{\bullet+}$ radical cation decolorization assay was applied according to the methodology described by Re et al. (9)]. A volume of 3 mL of ABTS$^{\bullet+}$ solution (absorbance 0.800 ± 0.02) was mixed with 10 µL of the ethanol extract of apple leaves. A decrease in absorbance was measured at a wavelength of 734 nm after keeping the samples for 30 min in the dark. The regression equation for this assay was established to be $y = 0.000084x - 0.002068$; $R^2 = 0.998777$.

DPPH$^{\bullet}$ free radical scavenging assay. The DPPH$^{\bullet}$ free radical scavenging activity was determined using the method proposed by Brand-Williams et al. [8]. DPPH$^{\bullet}$ solution in 96.3% v/v ethanol (3 mL, 6×10^{-5} M) was mixed with 10 µL of the ethanol extract of apple leaves. A decrease in absorbance was determined at a wavelength of 517 nm after keeping the samples for 30 min in the dark. The regression equation for this assay was established to be $y = 0.000111x - 0.000709$; $R^2 = 0.998024$.

CUPRAC assay. The CUPRAC solution included copper (II) chloride (0.01 M in water), ammonium acetate buffer solution (0.001 M, pH = 7), and neocuproine (0.0075 M in ethanol) (ratio 1:1:1). Here, 3 mL of CUPRAC reagent was mixed with 10 µL of extracts. An increase in absorbance was recorded at $\lambda = 450$ nm. The regression equation for this assay was established to be $y = 0.000048x - 0.001273$; $R^2 = 0.998580$ [10].

FRAP assay. The ferric reducing antioxidant power (FRAP) assay was carried out as described by Benzie and Strain (11).The working FRAP solution included TPTZ (0.01 M dissolved in 0.04 M HCl), $FeCl_3 \times 6H_2O$ (0.02 M in water), and acetate buffer (0.3 M, pH 3.6) at the ratio of 1:1:10. A volume of 3 mL of a freshly prepared FRAP reagent was mixed with 10 µL of the apple leaf extract. An increase in absorbance was recorded after 30 min at a wavelength of 593 nm. The regression equation for this assay was established to be $y = 0.000128x - 0.034745$; $R^2 = 0.998682$.

Calculation of antioxidant activity of the extract. The antioxidant activity of extracts was calculated from the Trolox calibration curve and expressed as µmol Trolox equivalent (*TE*) per gram of absolutely dry weight (DW). *TE* was calculated according to the following formula:

$$TE = \frac{c \times V}{m}, \quad (1)$$

c—the concentration of Trolox established from the calibration curve (in µM); V—the volume of leaf extract (in L); m—the weight (precise) of lyophilized leaf powder (in g).

2.7. Cell Viability Assay

Cell viability was studied using the method of MTT. Here, 100 µL of cells was seeded in 96-well plates in triplicate (5000 cells/well) and incubated at 37 °C for 24 h. Then, serial double dilutions of tested extracts (from 10 mg/mL to 0.156 mg/mL) were made in microplates. Cells treated only with medium containing 1.0% of ethanol served as a negative control. Free medium without cells was used as a positive control. After 72 h incubating at 37 °C, the cell growth medium in all the wells was replaced with the new one containing 0.5 mg/mL of MTT. After 4 h, the liquid was aspirated from the wells and discarded. Formazan crystals were dissolved in 100 µL of DMSO, and absorbance was measured at a test wavelength of 490 nm and a reference wavelength of 630 nm using a multi-detection microplate reader. The experiments were repeated three times independently and the results were given as means ± SD.

Applying Hill fit to compound dose–cell metabolic activity (absorbance) curves, the effective concentration (EC_{50}) values, reducing cell viability by 50%, were calculated.

2.8. Statistical Analysis

All the data obtained from this study were analysed statistically using SPSS™ software for Windows, Version 20.0 (SPSS Inc., Chicago, IL, USA) and the Microsoft Excel software package (Microsoft Corp, Redmond, WA, USA). Data are presented as mean ± standard error (S.E.) of at least three independent experiments. The correlation between antioxidant and anticancer activity was assessed by calculating Pearson's coefficient r and its statistical reliability. Correlation was considered as very strong when r = 0.90–0.99 (positive) or −0.99–(−0.90) (negative); strong when r = 0.70–0.89 (positive) or −0.89–(−0.70) (negative); and moderate when r = 0.40–0.69 (positive) or −0.69–(−0.40) (negative).

3. Results and Discussion

3.1. Extract Composition

Qualitative and quantitative phenolic composition of tested extracts was investigated using the UPLC–ESI–MS/MS method (Table 2).

Table 2. Chemical composition of extracts. Grey colour—tested substance was not detected, blue colour—very high amount of tested substance was detected.

Active Substances	Extracts						
mg/g DW	1	2	3	4	5	6	7
Apigenin	12.99 ± 0.47	20.32 ± 0.77	92.88 ± 4.32				
Aviculiarin				276.9 ± 10.23	0.75 ± 0.03	118.61 ± 5.42	2.58 ± 0.09
Caffeic acid	50.65 ± 2.03	30.74 ± 1.22	112.38 ± 4.93				0.06 ± 0.01
(+)-Catechin			0.03 ± 0.01	220.2 ± 10.41	0.32 ± 0.01	70.09 ± 2.98	0.03 ± 0.01
Chlorogenic acid	206.16 ± 8.35	11.84 ± 0.48	538.5 ± 24.11	184.05 ± 7.36	94.97 ± 4.01	677.6 ± 30.23	80.24 ± 2.47
p-Coumaric acid		16.24 ± 0.05					1.56 ± 0.05
Galangin							0.90 ± 0.04
Hyperoside	93.96 ± 4.50		7.35 ± 0.31	273.6 ± 13.26	2.65 ± 0.07	114.24 ± 5.63	1.59 ± 0.06
Isorhamnetin	241.65 ± 11.0	0.40 ± 0.02	10.60 ± 0.44	1.81 ± 0.07			0.66 ± 0.02
Isorhamnetin 3-*O*-glucoside	171.15 ± 8.03		0.78 ± 0.03	17.92 ± 0.82			
Isorhamnetin 3-*O*-rutinoside	320.41 ± 15.3	2.26 ± 0.09		82.54 ± 4.03		1.66 ± 0.06	
Kaempherol	1.35 ± 0.06	0.15 ± 0.01	0.47 ± 0.02	0.49 ± 0.02	0.22 ± 0.01		1.49 ± 0.05
Kaempherol 3-*O*-glucoside	8.79 ± 0.37	1.17 ± 0.05	0.12 ± 0.01	11.17 ± 0.47	0.70 ± 0.02	ND	0.32 ± 0.01
Luteolin 7-*O*-glucoside	7.54 ± 0.27			9.74 ± 0.41			
Orientin	0.48 ± 0.02	0.82 ± 0.03	14.60 ± 0.67	2.99 ± 0.14			3.75 ± 0.17
Phloridzin			3.83 ± 0.14	3.30 ± 0.12	35.51 ± 1.42	64.90 ± 3.01	20.93 ± 0.88
Quercetin	32.14 ± 1.47		5.76 ± 0.20	38.20 ± 1.70	2.59 ± 0.10	1.61 ± 0.06	28.40 ± 1.33
Quinic acid	381.56 ± 16.3	4.09 ± 0.18	2342.1 ± 101.4	7824.8 ± 350.6	703.6 ± 32.03	5398.6 ± 213.36	627.94 ± 28.9
Rosmarinic acid	0.39 ± 0.01	1799.2 ± 80.5	7.17 ± 0.29	9.80 ± 0.42	0.34 ± 0.02	106.35 ± 4.69	0.07 ± 0.01
Rutin	153.32 ± 7.12	6.87 ± 0.29	29.21 ± 1.20	71.97 ± 2.98	12.53 ± 0.46	22.92 ± 1.03	7.30 ± 0.28
Syringic acid	48.59 ± 1.97	40.45 ± 1.76					
Tiliroside		1.21 ± 0.04		3.01 ± 0.11	ND	0.65 ± 0.02	0.15 ± 0.01
Vitexin		0.32 ± 0.02	1.46 ± 0.05		1.90 ± 0.08	ND	1.76 ± 0.06

The total amount of identified and quantified phenolic compounds in the tested extracts ranged from 151.79 ± 6.21 mg/g (ethanol extract of eucalyptus leaves) up to 1931.95 ± 80.37 mg/g (sage leaf extract). The total amount of phenolic compounds in eucalyptus leaf samples extracted with different extraction solvents (acetone and ethanol) did not differ statistically significantly and was significantly lower than that of other plant raw material extracts (Table 2); the extraction solvent had no statistically significant effect on the extraction yield.

The UPLC–ESI–MS/MS analysis of selected plant extracts showed that flavonoid compounds were the major component among all identified phenolic compounds, and 17 different flavonoids were identified and quantified. The total amount of flavonoids detected in the extracts of the analyzed samples varied significantly. The highest total amount of the identified and quantified compounds of the flavonoid group was found in marigold blossom extracts (1043.78 ± 47.5 mg/g), and the lowest in sage leaf extracts (33.52 ± 1.42 mg/g).

The most abundant group of flavonoids in the investigated extracts is quercetin derivatives. The scientific literature indicates that quercetin and its glycosides have a broad biological activity. They might reduce the risk of cardiovascular diseases [40], metabolic disorders [3], and certain types of cancer [41]. Aglycone quercetin and its glycosides—avicularin, hyperoside, and rutin—were identified after analysis of selected plant extracts. Aglicones and glycosides of kaempherol and isorhamnetine luteolin glycosides, apigenin and its glycoside vitexin, and other flavonoids were detected in the extracts of the studied plant raw materials. The amount of flavonoids identified and quantified in the extracts of different plant raw materials varied widely. The acetone extract of eucalyptus leaf samples was dominated by phloridzin and the ethanol extract by quercetin. Marigold leaf blossom extracts were dominated by isorhamnetin 3-O-rutinoside, whereas all of the eucalyptus leaf sample extracts and yarrow grass extract extracts did not even contain this compound. Apigenin predominated in the extracts of sage leaves and yarrow grass. Quercetin group glycosides avicularin and hyperoside dominated in the extracts of bearberry berries and apples, and apigenin was not identified in these extracts. The obtained results confirm the claims of other scientists that the plant species has a great influence on the diversity of the composition of biologically active compounds [42].

Another group of phenolic compounds found in tested plant extracts was phenolic acids. The vast majority of the identified phenolic acids were hydroxycinnamic acid derivatives. Only one derivative of hydroxybenzoic acid, syringic acid, was identified. The highest total amount of identified and quantified phenolic acids (1898.43 ± 85.21 mg/g) was found in sage leaf extracts, and the lowest (81.93 ± 3.12 mg/g) in ethanol eucalyptus leaf extracts. Chlorogenic acid was predominant in most of the plant material sample extracts studied for a wide range of potential health benefits, including its strong antioxidant [43], anti-diabetic [44], anti-carcinogenic [45], anti-inflammatory [46], and anti-obesity impacts [47]. An exception was the sage leaf extracts, which were large enriched with rosmarinic acid and contained a relatively low amount of chlorogenic acid (Table 2).

Quinic acid (cyclitol), which is not classified as phenolic compound, was identified and quantified in the tested extracts. This acid was predominantly found in most of the tested plant raw materials, except for marigold and sage leaf extracts (Table 2).

3.2. *Antioxidant Activity*

Acetone extract of eucalyptus leaf samples exhibited the strongest antiradical activity assessed by the ABTS and DPPH methods (1.56 ± 0.03 mmol TE/g DW and 5.20 ± 0.40 mmol TE/g DW, respectively) and reducing activity assessed by the CUPRAC and FRAP methods (1.98 ± 0.11 mmol TE/g DW and 16.26 ± 0.67 mmol TE/g DW, respectively). The lowest antiradical and reducing activity in vitro, as assessed by all methods applied, was determined by examination of apple sample extracts; Figure 1.

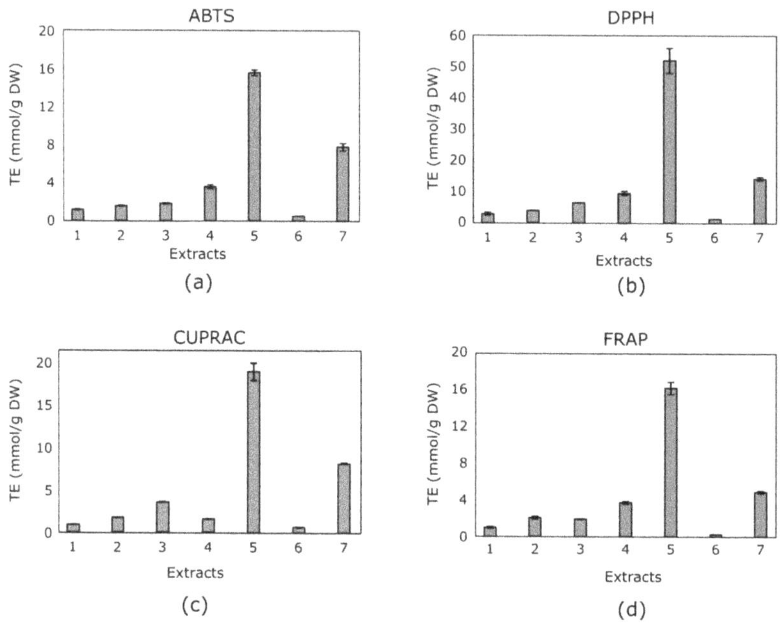

Figure 1. The antioxidant activity of all tested extracts by the (**a**) ABTS, (**b**) DPPH, (**c**) CUPRAC, and (**d**) FRAP methods.

The differences in the results obtained may have been influenced by a variety of factors—different media or pH [48], different lipophilic-hydrophilic properties of different antioxidants [49], and differences in qualitative and quantitative composition between the extracts and mechanisms of applied methods.

The antioxidant activity of plant extracts is determined by a complex of biologically active compounds. The scientific literature indicates that one of the strongest antioxidants that can determine the anti-radical and reducing properties of extracts is phenolic compounds [50]. However, in our study, there was no clear relationship between the quantitative composition of phenolic compounds and the in vitro antioxidant activity of the plant extracts tested. Contrary to expectations, extracts of sage leaves and calendula flowers, which contained the highest levels of identified and quantified compounds, did not exhibit strong antioxidant activity. In contrast, eucalyptus leaf extracts, which contained small amounts of phenolic compounds, exhibited particularly strong antiradical and reductive activity in vitro. These results can be explained by the fact that only part of the phenolic compounds responsible for such potent antioxidant activity of eucalyptus leaf extracts have been identified in vitro. The antioxidant activity of these extracts may have been due to phenolic compounds other than the antioxidants that we did not detect.

In order to evaluate the correlation strength between the in vitro antioxidant activities of tested plant extracts by different methods, the Pearson correlation coefficient was calculated (Table 3). A strong correlation between the antiradical and reducing activity was established. A strong correlation was found between the in vitro antioxidant activity assessed by ABTS and DPPH. Such results can be explained by the fact that the mechanism of action of both methods is based on the binding (inactivation) of simulated synthetic free radicals—$ABTS^{•+}$ radical cation or $DPPH^{•}$ free radical [8,9]. The weakest correlation was found between antioxidant activity assessed by FRAP and CUPRAC methods. Such data are quite unexpected, as both methods are based on the same mechanism of action—they determine the reducing activity of extracts or other solutions. Such a result could be explained by the differences between these methods. The pH of the reaction medium is neutral (pH = 7) with the CUPRAC method and acidic (pH = 3.6) with the

FRAP method [10,11]. The pH of the medium is a critical factor in assessing the in vitro antioxidant activity of plant antioxidants, especially phenolic compounds [36]. Various scientific studies have shown that the antioxidant activity of phenolic compounds is highly dependent on the pH of the medium [14]. The antioxidant activity of phenolic compounds is mediated by functional hydroxyl groups [39]. As the pH of the medium decreases, the dissociation process of the hydroxyl group weakens, leading to a lower antioxidant activity of the phenolic compounds [10]. The antioxidant activity of plant antioxidants based on hydrogen atom transfer reactions (thiols, carotenoids) cannot be determined by the FRAP method [40,41]. Perhaps, the above mentioned differences between the FRAP and CUPRAC methods may have led to a slightly weaker correlation between their in vitro antioxidant activity.

Table 3. Correlation between the different methods applied for testing antioxidant activity in this research.

		ABTS	DPPH	CUPRAC	FRAP
ABTS	Correlation coefficient	1	0.972 **	0.981 **	0.976 **
	p value (two-tailed)		0.000	0.000	0.000
DPPH	Correlation coefficient	0.972 **	1	0.976 **	0.997 **
	p value (two-tailed)	0.000		0.000	0.000
CUPRAC	Correlation coefficient	0.981 **	0.976 **	1	0.971 **
	p value (two-tailed)	0.000	0.000		0.005
FRAP	Correlation coefficient	0.976 **	0.997 **	0.971 **	1
	p value (two-tailed)	0.000	0.000	0.005	

** Correlation is significant at the 0.01 level (two-tailed).

3.3. Anticancer Effect

The effect of the tested extracts on cancer cell viability was variable (Figure 2). Six out of eight extracts possessed a cytotoxic effect against all cancer cell lines. Extract of apples and extract of calendula flowers did not possess anticancer activity at the tested concentration (up to 15 mg/mL). Sak et al. in their experiments in vitro against human melanoma SK-MEL-2 cells also observed relatively low activity of marigold extract [51]. A weak antiproliferative effect on several breast cancer cell lines of extracts made from apple peels was established by other scientists too [52].

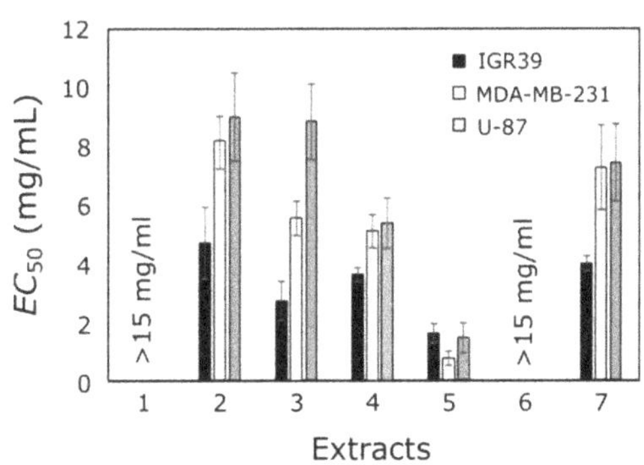

Figure 2. The EC_{50} values of tested extracts against IGR39, MDA-MB-231, and U-87 cancer cell lines.

The majority of extracts were mostly active against the melanoma IGR39 cell line, and possessed the lowest activity against the glioblastoma U-87 cell line. However, this cell line is constitutively resistant to many chemotherapeutic agents owing to over-expression of P-glycoprotein and its stemness properties [53]. Only one extract showed a cell viability reducing effect against glioblastoma cells at lower than 2 mg/mL concentration. Triple-negative breast cancer cell line MDA-MB-231 was also quite resistant to the tested extracts. Only one extract showed an anticancer effect on these cells at lower than 2 mg/mL concentration (dried eucalyptus leaf acetone extract). It could be explained that this cell line does not have estrogen, progesterone, and HER-2 receptors, and is characterized by a more aggressive nature than other types of breast cancer cells [54].

Acetone extract of eucalyptus leaf samples exhibited the highest anticancer activity against melanoma, triple-negative breast cancer, and glioblastoma cell lines (1.6 ± 0.3 mg/mL, 0.8 ± 0.3 mg/mL, and 1.5 ± 0.5 mg/mL, respectively). These results are consistent with the antioxidant activity, as this extract showed the highest antioxidant effect by all the methods used (Figure 1). The lowest anticancer effect was determined for the same extracts that showed the lowest antioxidant activity (extract of apples and extract of calendula flowers).

The most active extract, made from eucalyptus leaves with acetone, possessed different activity on tested cell lines (Figure 3). The highest anticancer effect was observed on the triple-negative breast cancer cell line MDA-MB-231.

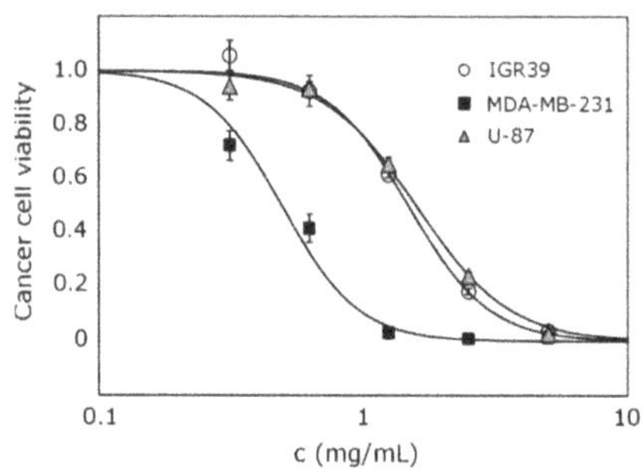

Figure 3. Comparison of extract 6 activity against IGR39, MDA-MB-231, and U-87 cell lines.

However, contrary to expectations from antioxidant activity assays, ethanol extract of eucalyptus leaves did not exhibit high anticancer activity. In contrast to other eucalyptus leaf extracts made with acetone, this extract was from 2.4 to 9.5 times less active against different cell lines. These results can be explained in that the anticancer effect is not necessarily related to antioxidant properties of active substances, and such a phenomenon was observed previously with different substances with antioxidant activity in vitro [55]. In our study, there was no clear relationship between the quantitative composition of phenolic compounds and the in vitro anticancer activity of the tested plant extracts.

Apigenin, caffeic acid, catechin, chlorogenic acid, quercetin, kaempherol, and other compounds that are widely known to possess activity against cancer cells [56–59] were not found in the most active extract or were detected only at quite low amounts (Table 2). This could mean that this extract contains other substances that could be responsible for its anticancer effect, or the complex of all substances could possess strong synergistic activity. In contrary, some moderately active extracts contained the above-mentioned compounds in high levels, but still did not show high activity against tested cancer cell lines, and their effect was different between them. It was noticed previously by other scientists, that some compounds, such as apigenin, may show different activity depending on cell prototype [60].

Our results showed that anticancer activity could be related to antioxidant activity of extracts (Table 4).

Table 4. Correlation between antioxidant and anticancer activity of tested plant extracts.

		Antioxidant Activity, mmol TE/g DW			
Anticancer Activity		**ABTS**	**DPPH**	**CUPRAC**	**FRAP**
IGR39	Correlation coefficient	−0.583	−0.558	−0.517	−0.600
	p value (two-tailed)	0.129	0.151	0.190	0.116
MDA-MB-231	Correlation coefficient	−0.791 *	−0.793 *	−0.744 *	−0.803 **
	p value (two-tailed)	0.019	0.019	0.034	0.016
U-87	Correlation coefficient	−0.827 *	−0.817 *	−0.738 *	−0.850 **
	p value (two-tailed)	0.011	0.013	0.036	0.008

* Correlation is significant at the 0.05 level (two-tailed). ** Correlation is significant at the 0.01 level (two-tailed).

A strong and reliable correlation was found between antioxidant and anticancer activity in breast cancer and glioblastoma cell lines, especially when evaluating antioxidant activity by the FRAP method. The same correlation between antioxidant and anticancer activity in melanoma cells was established as moderate; however, it was not found to be reliable. This could mean that the mechanism of anticancer activity in melanoma cells could be different and is less related to the scavenging free radicals or the antioxidant reducing activities.

4. Conclusions

The extracts of calendula, sage, bearberry, eucalyptus, yarrow, and apple widely used in traditional and fold medicine contain 23 individual phenolic compounds, including 17 flavonoids. The highest amount of identified and quantified compounds was found in sage leaf extract. Rosmarinic acid was predominant in these extracts, while chlorogenic acid was predominant in the extracts of yarrow grass and apple tree fruits. Quinic acid, a non-phenolic compound, was identified and quantified and was found in the highest amount in the extracts of bearberry leaves.

The highest antioxidant activity in vitro was determined using all applied methods in acetone extracts of eucalyptus leaves. The weakest antiradical and reductive activity in vitro by all methods was detected in apple extracts.

Six out of eight extracts possessed cytotoxic effect against triple-negative breast cancer (MDA-MB-231), melanoma (IGR39), and glioblastoma (U-87) cell lines. Acetone extract of eucalyptus leaf samples exhibited the highest anticancer activity against all tested cell lines. A strong and reliable correlation was found between antioxidant and anticancer activity in breast cancer and glioblastoma cell lines, especially when evaluating antioxidant activity by the FRAP method.

Author Contributions: M.L. and V.P. designed the study; M.L. performed extraction and analysed antioxidant activity; V.Ž. performed composition analysis of tested extracts; V.P. performed anticancer activity experiments; M.L. and V.P. analysed the data, wrote the manuscript, and revised it. All authors have read and agreed to the published version of the manuscript.

Funding: This research received no external funding.

Institutional Review Board Statement: Not applicable.

Informed Consent Statement: Not applicable.

Data Availability Statement: All datasets generated for this study are included in the article.

Conflicts of Interest: The authors declare no conflict of interest.

References

1. Wilson, D.W.; Nash, P.; Buttar, H.S.; Griffiths, K.; Singh, R.; De Meester, F.; Horiuchi, R.; Takahashi, T. The Role of Food Antioxidants, Benefits of Functional Foods, and Influence of Feeding Habits on the Health of the Older Person: An Overview. *Antioxidants* **2017**, *6*, 81. [CrossRef]
2. Aghajanpour, M.; Nazer, M.R.; Obeidavi, Z.; Akbari, M.; Ezati, P.; Kor, N.M. Functional foods and their role in cancer prevention and health promotion: A comprehensive review. *Am. J. Cancer Res.* **2017**, *7*, 740–769.

3. Anand David, A.V.; Arulmoli, R.; Parasuraman, S. Overviews of Biological Importance of Quercetin: A Bioactive Flavonoid. *Pharmacogn. Rev.* **2016**, *10*, 84–89.
4. Kim, H.; Caulfield, L.E.; Garcia-Larsen, V.; Steffen, L.M.; Coresh, J.; Rebholz, C.M. Plant-Based Diets Are Associated with a Lower Risk of Incident Cardiovascular Disease, Cardiovascular Disease Mortality, and All-Cause Mortality in a General Population of Middle-Aged Adults. *J. Am. Heart Assoc.* **2019**, *8*, e012865. [CrossRef]
5. Kerley, C.P. A Review of Plant-based Diets to Prevent and Treat Heart Failure. *Card. Fail. Rev.* **2018**, *4*, 54–61. [CrossRef]
6. Hu, N.; Yu, J.-T.; Tan, L.; Wang, Y.-L.; Sun, L.; Tan, L. Nutrition and the risk of Alzheimer's disease. *Biomed. Res. Int.* **2013**, *2013*, 524820. [CrossRef] [PubMed]
7. Prior, R.L. Fruits and vegetables in the prevention of cellular oxidative damage. *Am. J. Clin. Nutr.* **2003**, *78*, 570S–578S. [CrossRef]
8. Brand-Williams, W.; Cuvelier, M.E.; Berset, C. Use of a free radical method to evaluate antioxidant activity. *LWT Food Sci. Technol.* **1995**, *28*, 25–30. [CrossRef]
9. Re, R.; Pellegrini, N.; Proteggente, A.; Pannala, A.; Yang, M.; Rice-Evans, C. Antioxidant activity applying an improved ABTS radical cation decolorization assay. *Free Radic. Biol. Med.* **1999**, *26*, 1231–1237. [CrossRef]
10. Apak, R.; Güçlü, K.; Demirata, B.; Ozyürek, M.; Celik, S.E.; Bektaşoğlu, B.; Berker, K.I.; Özyurt, D. Comparative evaluation of various total antioxidant capacity assays applied to phenolic compounds with the CUPRAC assay. *Molecules* **2007**, *12*, 1496–1547. [CrossRef]
11. Benzie, I.F.F.; Strain, J.J. The ferric reducing ability of plasma (FRAP) as a measure of "antioxidant power": The FRAP assay. *Anal. Biochem.* **1996**, *239*, 70–76. [CrossRef]
12. Shalaby, E.A.; Shanab, S.M.M. Comparison of DPPH and ABTS assays for determining antioxidant potential of water and methanol extracts of *Spirulina platensis*. *Indian J. Mar. Sci.* **2013**, *42*, 9.
13. Çekiç, S.D.; Başkan, K.S.; Tütem, E.; Apak, R. Modified cupric reducing antioxidant capacity (CUPRAC) assay for measuring the antioxidant capacities of thiol-containing proteins in admixture with polyphenols. *Talanta* **2009**, *79*, 344–351. [CrossRef]
14. Amorati, R.; Pedulli, G.F.; Cabrini, L.; Zambonin, L.; Landi, L. Solvent and pH Effects on the Antioxidant Activity of Caffeic and Other Phenolic Acids. *J. Agric. Food Chem.* **2006**, *54*, 2932–2937. [CrossRef]
15. Staško, A.; Brezová, V.; Biskupič, S.; Mišík, V. The potential pitfalls of using 1,1-diphenyl-2-picrylhydrazyl to characterize antioxidants in mixed water solvents. *Free. Radic. Res.* **2007**, *41*, 379–390. [CrossRef] [PubMed]
16. González-Burgos, E.; Liaudanskas, M.; Viškelis, J.; Žvikas, V.; Janulis, V.; Gómez-Serranillos, M.P. Antioxidant activity, neuroprotective properties and bioactive constituents analysis of varying polarity extracts from *Eucalyptus globulus* leaves. *J. Food Drug Anal.* **2018**, *26*, 1293–1302. [CrossRef] [PubMed]
17. Jiang, Y.; Zhang, L.; Rupasinghe, H.V. The anticancer properties of phytochemical extracts from *Salvia* plants. *Bot. Targets Ther.* **2016**, *6*, 25–44.
18. Mouhid, L.; De Cedrón, M.G.; García-Carrascosa, E.; Reglero, G.; Fornari, T.; De Molina, A.R. Yarrow supercritical extract exerts antitumoral properties by targeting lipid metabolism in pancreatic cancer. *PLoS ONE* **2019**, *14*, e0214294. [CrossRef]
19. Jiménez-Medina, E.; Garcia-Lora, A.; Paco, L.; Algarra, I.; Collado, A.; Garrido, F. A new extract of the plant *Calendula officinalis* produces a dual in vitro effect: Cytotoxic anti-tumor activity and lymphocyte activation. *BMC Cancer* **2006**, *6*, 119. [CrossRef] [PubMed]
20. Martino, E.; Vuoso, D.C.; D'Angelo, S.; Mele, L.; D'Onofrio, N.; Porcelli, M.; Cacciapuoti, G. Annurca apple polyphenol extract selectively kills MDA-MB-231 cells through ROS generation, sustained JNK activation and cell growth and survival inhibition. *Sci. Rep.* **2019**, *9*, 1–15. [CrossRef]
21. Amarowicz, R.; Pegg, R.B. Inhibition of proliferation of human carcinoma cell lines by phenolic compounds from a bear-berry-leaf crude extract and its fractions. *J. Funct. Foods* **2013**, *5*, 660–667. [CrossRef]
22. Trumbeckaite, S.; Benetis, R.; Bumblauskiene, L.; Burdulis, D.; Janulis, V.; Toleikis, A.; Viškelis, P.; Jakštas, V. *Achillea millefolium* L. s.l. herb extract: Antioxidant activity and effect on the rat heart mitochondrial functions. *Food Chem.* **2011**, *127*, 1540–1548. [CrossRef]
23. Preethi, K.C.; Kuttan, G.; Kuttan, R. Antioxidant Potential of an Extract of *Calendula officinalis*. Flowers in Vitro and in Vivo. *Pharm. Biol.* **2006**, *44*, 691–697. [CrossRef]
24. Raudone, L.; Raudonis, R.; Liaudanskas, M.; Janulis, V.; Viskelis, P. Phenolic antioxidant profiles in the whole fruit, flesh and peel of apple cultivars grown in Lithuania. *Sci. Hortic.* **2017**, *216*, 186–192. [CrossRef]
25. Mohd Azman, N.A.; Gallego, M.G.; Segovia, F.; Abdullah, S.; Shaarani, S.M.; Almajano Pablos, M.P. Study of the Properties of Bearberry Leaf Extract as a Natural Antioxidant in Model Foods. *Antioxidants* **2016**, *5*, 11. [CrossRef]
26. Pizzimenti, S.; Toaldo, C.; Pettazzoni, P.; Dianzani, M.U.; Barrera, G. The "Two-Faced" Effects of Reactive Oxygen Species and the Lipid Peroxidation Product 4-Hydroxynonenal in the Hallmarks of Cancer. *Cancers* **2010**, *2*, 338–363. [CrossRef]
27. Reuter, S.; Gupta, S.C.; Chaturvedi, M.M.; Aggarwal, B.B. Oxidative stress, inflammation, and cancer: How are they linked? *Free Radic. Biol. Med.* **2010**, *49*, 1603–1616. [CrossRef]
28. Pandey, K.B.; Rizvi, S.I. Plant Polyphenols as Dietary Antioxidants in Human Health and Disease. *Oxid. Med. Cell. Longev.* **2009**, *2*, 270–278. [CrossRef]
29. Kooti, W.; Servatyari, K.; Behzadifar, M.; Asadi-Samani, M.; Sadeghi, F.; Nouri, B.; Marzouni, H.Z. Effective Medicinal Plant in Cancer Treatment, Part 2: Review Study. *J. Evid. Based Integr. Med.* **2017**, *22*, 982–995. [CrossRef]

30. Deshmukh, V.N.; Sakarkar, D.M. Ethnopharmacological review of traditional medicinal plants for anticancer activity. *Int. J. PharmTech Res.* **2011**, *3*, 298–308.
31. Roleira, F.M.F.; Tavares-da-Silva, E.J.; Varela, C.L.; Costa, S.C.; Silva, T.; Garrido, J.; Borges, F. Plant derived and dietary phenolic antioxidants: Anticancer properties. *Food Chem.* **2015**, *183*, 235–258. [CrossRef] [PubMed]
32. Hu, M.-L. Dietary polyphenols as antioxidants and anticancer agents: More questions than answers. *Chang. Gung Med. J.* **2011**, *34*, 1–12.
33. Cai, Y.; Luo, Q.; Sun, M.; Corke, H. Antioxidant activity and phenolic compounds of 112 traditional Chinese medicinal plants associated with anticancer. *Life Sci.* **2004**, *74*, 2157–2184. [CrossRef]
34. Wang, S.; Meckling, K.A.; Marcone, M.F.; Kakuda, Y.; Tsao, R. Can phytochemical antioxidant rich foods act as anti-cancer agents? *Food Res. Int.* **2011**, *44*, 2545–2554. [CrossRef]
35. Nepote, V.; Grosso, N.R.; Guzmán, C.A. Optimization of extraction of phenolic antioxidants from peanut skins. *J. Sci. Food Agric.* **2004**, *85*, 33–38. [CrossRef]
36. Vatai, T.; Škerget, M.; Knez, Ž. Extraction of phenolic compounds from elder berry and different grape marc varieties using organic solvents and/or supercritical carbon dioxide. *J. Food Eng.* **2009**, *90*, 246–254. [CrossRef]
37. Boulekbache-Makhlouf, L.; Medouni, L.; Medouni-Adrar, S.; Arkoub, L.; Madani, K. Effect of solvents extraction on phenolic content and antioxidant activity of the byproduct of eggplant. *Ind. Crop. Prod.* **2013**, *49*, 668–674. [CrossRef]
38. Nasr, A.; Zhou, X.; Liu, T.; Yang, J.; Zhu, G.-P. Acetone-water mixture is a competent solvent to extract phenolics and antioxidants from four organs of *Eucalyptus camaldulensis*. *Turk. J. Biochem.* **2019**, *44*, 231–239. [CrossRef]
39. Liaudanskas, M.; Viškelis, P.; Jakštas, V.; Raudonis, R.; Kviklys, D.; Milašius, A.; Janulis, V. Application of an Optimized HPLC Method for the Detection of Various Phenolic Compounds in Apples from Lithuanian Cultivars. *J. Chem.* **2014**, *2014*. Available online: https://www.hindawi.com/journals/jchem/2014/542121/ (accessed on 10 January 2020). [CrossRef]
40. Peterson, J.J.; Dwyer, J.T.; Jacques, P.F.; McCullough, M.L. Associations between flavonoids and cardiovascular disease incidence or mortality in European and US populations. *Nutr. Rev.* **2012**, *70*, 491–508. [CrossRef]
41. Su, C.; Haskins, A.H.; Omata, C.; Aizawa, Y.; Kato, T.A. PARP Inhibition by Flavonoids Induced Selective Cell Killing to BRCA2-Deficient Cells. *Pharmaceuticals* **2017**, *10*, 80. [CrossRef]
42. Konieczynski, P.; Viapiana, A.; Lysiuk, R.; Wesolowski, M. Chemical Composition of Selected Commercial Herbal Remedies in Relation to Geographical Origin and Inter-Species Diversity. *Biol. Trace Elem. Res.* **2017**, *182*, 169–177. [CrossRef] [PubMed]
43. Liang, N.; Kitts, D.D. Role of Chlorogenic Acids in Controlling Oxidative and Inflammatory Stress Conditions. *Nutrients* **2015**, *8*, 16. [CrossRef]
44. Ong, K.W.; Hsu, A.; Tan, B.K.H. Anti-diabetic and anti-lipidemic effects of chlorogenic acid are mediated by ampk activation. *Biochem. Pharmacol.* **2013**, *85*, 1341–1351. [CrossRef]
45. Sadeghi Ekbatan, S.; Li, X.-Q.; Ghorbani, M.; Azadi, B.; Kubow, S. Chlorogenic Acid and Its Microbial Metabolites Exert Anti-Proliferative Effects, S-Phase Cell-Cycle Arrest and Apoptosis in Human Colon Cancer Caco-2 Cells. *Int. J. Mol. Sci.* **2018**, *19*, 723. [CrossRef] [PubMed]
46. Shin, H.S.; Satsu, H.; Bae, M.-J.; Zhao, Z.; Ogiwara, H.; Totsuka, M.; Shimizu, M. Anti-inflammatory effect of chlorogenic acid on the IL-8 production in Caco-2 cells and the dextran sulphate sodium-induced colitis symptoms in C57BL/6 mice. *Food Chem.* **2015**, *168*, 167–175. [CrossRef] [PubMed]
47. Naveed, M.; Hejazi, V.; Abbas, M.; Kamboh, A.A.; Khan, G.J.; Shumzaid, M.; Ahmad, F.; Babazadeh, D.; FangFang, X.; Modarresi-Ghazani, F.; et al. Chlorogenic acid (CGA): A pharmacological review and call for further research. *Biomed. Pharmacother.* **2018**, *97*, 67–74. [CrossRef] [PubMed]
48. Sun, H.-N.; Mu, T.-H.; Xi, L.-S. Effect of pH, heat, and light treatments on the antioxidant activity of sweet potato leaf polyphenols. *Int. J. Food Prop.* **2017**, *20*, 318–332. [CrossRef]
49. Cano, A.; Alcaraz, O.; Acosta, M.; Arnao, M.B. On-line antioxidant activity determination: Comparison of hydrophilic and lipophilic antioxidant activity using the $ABTS^{\bullet+}$ assay. *Redox Rep.* **2002**, *7*, 103–109. [CrossRef]
50. Kaurinovic, B.; Vastag, D. Flavonoids and Phenolic Acids as Potential Natural Antioxidants. In *Antioxidants*; IntechOpen: London, UK, 2019. Available online: https://www.intechopen.com/books/antioxidants/flavonoids-and-phenolic-acids-as-potential-natural-antioxidants (accessed on 3 January 2020).
51. Sak, K.; Nguyen, T.H.; Ho, V.D.; Do, T.T.; Raal, A. Cytotoxic effect of chamomile (*Matricaria recutita*) and marigold (*Calendula officinalis*) extracts on human melanoma SK-MEL-2 and epidermoid carcinoma KB cells. *Cogent Med.* **2017**, *4*. [CrossRef]
52. Sun, J.; Liu, R.H. Apple phytochemical extracts inhibit proliferation of estrogen-dependent and estrogen-independent human breast cancer cells through cell cycle modulation. *J. Agric. Food Chem.* **2008**, *56*, 11661–11667. [CrossRef] [PubMed]
53. Balça-Silva, J.; Matias, D.; Carmo, A.; Dubois, L.G.F.; Gonçalves, A.C.; Girao, H.; Canedo, N.H.S.; Correia, A.H.; De Souza, J.M.; Ribeiro, A.B.S.; et al. Glioblastoma entities express subtle differences in molecular composition and response to treatment. *Oncol. Rep.* **2017**, *38*, 1341–1352. [CrossRef] [PubMed]
54. Lee, K.-L.; Kuo, Y.-C.; Ho, Y.-S.; Huang, Y.-H. Triple-Negative Breast Cancer: Current Understanding and Future Therapeutic Breakthrough Targeting Cancer Stemness. *Cancers* **2019**, *11*, 1334. [CrossRef]
55. Grigalius, I.; Petrikaite, V. Relationship between Antioxidant and Anticancer Activity of Trihydroxyflavones. *Molecules* **2017**, *22*, 2169. Available online: https://www.ncbi.nlm.nih.gov/pmc/articles/PMC6149854/ (accessed on 5 January 2020). [CrossRef]

56. Shahidi, F.; Yeo, J. Bioactivities of Phenolics by Focusing on Suppression of Chronic Diseases: A Review. *Int. J. Mol. Sci.* **2018**, *19*, 1573. Available online: https://www.ncbi.nlm.nih.gov/pmc/articles/PMC6032343/ (accessed on 5 January 2020). [CrossRef]
57. Rauf, A.; Imran, M.; Ali Khan, I.; Ur-Rehman, M.; Gilani, S.A.; Mehmood, Z.; Mubarak, M.S. Anticancer potential of quercetin: A comprehensive review. *Phytother. Res.* **2018**, *32*, 2109–2130. [CrossRef] [PubMed]
58. Imran, M.; Salehi, B.; Sharifi-Rad, J.; Aslam Gondal, T.; Saeed, F.; Imran, A.; Shahbaz, M.; Tsouh Fokou, P.V.; Arshad, M.U.; Khan, H.; et al. Kaempferol: A Key Emphasis to Its Anticancer Potential. *Molecules* **2019**, *24*, 2277. Available online: https://www.ncbi.nlm.nih.gov/pmc/articles/PMC6631472/ (accessed on 10 January 2020). [CrossRef]
59. Yan, X.; Qi, M.; Li, P.; Zhan, Y.; Shao, H. Apigenin in cancer therapy: Anti-cancer effects and mechanisms of action. *Cell Biosci.* **2017**, *7*, 1–16. Available online: https://www.ncbi.nlm.nih.gov/pmc/articles/PMC5629766/ (accessed on 6 January 2020). [CrossRef]
60. Shukla, S.; Gupta, S. Apigenin: A Promising Molecule for Cancer Prevention. *Pharm. Res.* **2010**, *27*, 962–978. [CrossRef]

antioxidants

MDPI

Article

Protective Effect of Red Algae (*Rhodophyta*) Extracts on Essential Dietary Components of Heat-Treated Salmon

Jaime Ortiz-Viedma [1,*], José M. Aguilera [2], Marcos Flores [3,*], Roberto Lemus-Mondaca [1], María José Larrazabal [4], José M. Miranda [5] and Santiago P. Aubourg [6]

1 Departamento de Ciencia de los Alimentos y Tecnología Química, Facultad de Ciencias Químicas y Farmacéuticas, Universidad de Chile, Santos Dumont 964, Santiago 8320000, Chile; rlemus@uchile.cl
2 Departamento de Ingeniería Química y Bioprocesos, Facultad de Ingeniería, Pontificia Universidad Católica de Chile, Vicuña Mackenna 4860, Santiago 8940000, Chile; jmaguile@ing.puc.cl
3 Departamento de Ciencias Básicas, Facultad de Ciencias, Universidad Santo Tomás, Talca 3460000, Chile
4 Departamento de Ciencia de los Alimentos y Nutrición, Facultad de Ciencias de la Salud, Universidad de Antofagasta, Avenida Angamos 601, Antofagasta 1240000, Chile; maria.larrazabal@uantof.cl
5 Departamento de Química Analítica, Nutrición y Bromatología, Facultad de Veterinaria, Universidad de Santiago de Compostela, 27002 Lugo, Spain; josemanuel.miranda@usc.es
6 Department of Food Science and Technology, Marine Research Institute (CSIC), Calle Eduardo Cabello, 6, 36208 Vigo, Spain; saubourg@iim.csic.es
* Correspondence: jaortiz@uchile.cl (J.O.-V.); marcosflores@santotomas.cl (M.F.); Tel.: +56-22-978-1663 (J.O.-V.)

Citation: Ortiz-Viedma, J.; Aguilera, J.M.; Flores, M.; Lemus-Mondaca, R.; Larrazabal, M.J.; Miranda, J.M.; Aubourg, S.P. Protective Effect of Red Algae (*Rhodophyta*) Extracts on Essential Dietary Components of Heat-Treated Salmon. *Antioxidants* **2021**, *10*, 1108. https://doi.org/10.3390/antiox10071108

Academic Editors: Irene Dini and Domenico Montesano

Received: 25 May 2021
Accepted: 6 July 2021
Published: 11 July 2021

Abstract: Salmon paste contains nutritious components such as essential fatty acids (EPA, DHA), vitamin E and astaxanthin, which can be protected with the addition of red algae extracts. Phenolic extracts were prepared with an ethanol: water mixture (1:1) from the red seaweeds *Gracilaria chilensis*, *Gelidium chilense*, *Iridaea larga*, *Gigartina chamissoi*, *Gigartina skottsbergii* and *Gigartina radula*, obtained from the Pacific Ocean. Most algae had a high content of protein (>7.2%), fiber (>55%) and β-glucans (>4.9%), all expressed on a dry weight basis. Total polyphenols (TP), total flavonoids (TF), antioxidant (DPPH, FRAP) and antibacterial power of the extracts were measured. In addition, the nutritional components of the algae were determined. Results showed that the content of TP in the six algae varied between 2.6 and 11.3 mg EAG/g dw and between 2.2 and 9.6 for TF. Also, the extracts of *G. skottsbergii*, *G. chamissoi*, *G. radula* and *G. chilensis* showed the highest antiradical activity (DPPH, FRAP). All samples exhibited a low production of primary oxidation products, and protection of the essential components and the endogenous antioxidants tocopherols and astaxanthin, particularly in the case of *G. skottsbergii*, *G. chamissoi*, *G. radula* and *G. chilensis*. Furthermore, all algae had inhibitory activity against the tested microorganisms, coincident with their antioxidant capacity. Results show that the extracts may have future applications in the development and preservation of essential dietary components of healthy foods.

Keywords: red algae; antioxidant and antimicrobial ability; lipoperoxidation; salmon

1. Introduction

For centuries, seaweeds have been an important source of entrenched food in Asian countries such as Japan, China, and Korea, and other coastal geographic areas in other regions of the world.

Although there is extensive harvesting of seaweeds at the artisanal level and the development of cultivation technologies has led to algae being a safe food, their use as staple foods and a source of bioactive components has not transcended to the western diet [1]. Many experts suggest that algae (seaweeds) can be an excellent alternative to the current trend of consumers to have natural and healthy plant-based foods and ingredients [1–5]. In recent times, there has been a wide interest in the functional properties of seaweeds and their application as bioactive components in the preparation of ingredients and additives that improve the conservation of different types of foods. Among the typical functional ingredients that are being investigated for their application in the formulation of healthy foods are antioxidants such as polyphenols, flavonoids, florotannins, carotenoids, etc., as well as immune protectants such as b-glucans, oligosaccharides, dietary fiber, lignans, peptides, etc. Moreover, seaweeds provide essential nutrients such as polyunsaturated fatty acids (PUFAS) and vitamins (folic acid and ascorbic acid, retinol), [2,3,6–9] and components related to sensory and gastronomic aspects that provide the umami flavor [1,9]. Therefore, the application of ingredients derived from algae may be a safe alternative to replace artificial food additives [10]. Recently, research regarding the use of red, green and brown algae extracts in the preservation of food has shown that the different bioactive antioxidant components derived from seaweed are effective in preserving fish and shellfish [11–14]. In addition, it has recently been observed that phenolic extracts from different red algae show the ability to inhibit the development of different pathogenic bacterial strains and the toxic products generated by their proteolytic activity, among which biogenic amines stand out due to their risk to human health [12,13,15]. Salmon is a fish of high nutritional richness due to its content in PUFAS, carotenoids, minerals, vitamins, etc., and is highly demanded by consumers. However, the high PUFAS content makes it very susceptible to deterioration by oxidation under refrigerated, frozen, cooked and dried conditions [16]. The presence of oxygen, metals, temperature, microbial contamination, manipulation, among others, are direct factors that catalyze the oxidative process of polyunsaturated lipids of marine species [14,16,17]. Therefore, it is necessary to search for natural antioxidants capable of inhibiting oxidation and the consequent loss of essential components (e.g., PUFAS, vitamins, amino acids, etc.) of salmon during thermal processing either as a processed product or as a culinary preparation.

The red seaweeds *Gracilaria chilensis, Gelidium chilense, Iridaea long, Gigartina chamissoi, Gigartina skottsbergii* and *Gigartina radula* occur in great abundance along the central and southern coast of Chile. They are normally exported at a low price to Asian countries without exploiting their use as food, functional ingredients and nutraceuticals at the local level. Given the many unknowns about the potential of chemical, bacteriostatic and food preservation properties of these algae species, the objective of this work was to develop a preliminary chemical study of the inhibitory power of phenolic extracts of these algae against lipid oxidation of salmon paste subjected to a cooking heat treatment. In addition, the ability of phenolic extracts to inhibit the development of different pathogenic bacterial strains was determined.

2. Materials and Methods

2.1. Red Algae Samples

Six red algae species were collected from coastal areas in the central and southern regions of Chile (Figure 1) and dried at 60 °C for 3 h. Raw algae samples were provided by Algamar S.A. (Santiago, Chile). The dried algae were ground and sieved to a particle size of 0.8 ± 0.2 mm. Samples were stored at room temperature (25 °C) in sealed polyethylene bags.

2.2. Nutritional Content

The standard methods Association of Official Analytical Chemists (AOAC) and American Oil Chemists Society (AOCS) were used to quantify the moisture, ash, lipid and protein content in seaweed meals [18,19].. Considering that the sum of percentages of lipids, proteins, minerals and carbohydrates are equivalent to 100% of the dried algae. Once the percentages of lipids, proteins and minerals were determined, the carbohydrates were determined by subtraction (carbohydrates = 100 -% lipids -% minerals -% proteins). Total dietary fiber content was determined using half the amount of the algae sample and by applying AOAC procedures. The content of β-glucans was determined using an enzymatic assay (K-YBGL, Megazyme, Ireland) based on official methods [20]. Results were expressed as g/100 g dry basis.

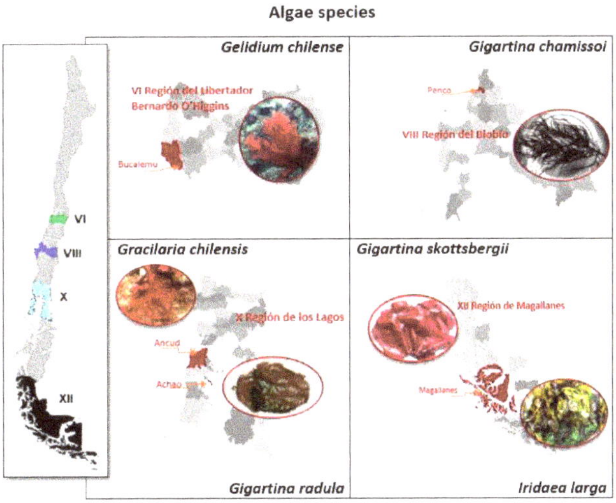

Figure 1. Red algae studied and region of Chile from which they were collected.

2.3. Polyphenol Content

Hydroalcoholic extraction was used to obtain the phenolic extracts of algae. Dried algae (50 g) were soaked in 100 mL of extractant (ethanol/water: 50/50 *v*/*v*) and then stirred in a shaker (Wrist Action, Burell, Pittsburgh, PA, USA) for 8 h in a closed flask. Samples were then sonicated in a bath for 15 min at 25 °C, stirred for an additional 15 min, and filtered with a Whatman No. 1 paper. The entire procedure was repeated two more times but using 50 mL of extractant and only 30 min of shaking. The resulting extracts were finally concentrated at 40 °C in a vacuum rotary evaporator Büchi (Stuttgart, Germany) and then re-diluted with 200 mL of water (d) before storing at 5 ± 0.5 °C in glass bottles lined with aluminum foil. The total phenol (TP) of the algae was determined using the methodology proposed by Yildiz et al. (2011), which applies the Folin-Ciocalteu reagent and the colored solution was measured at 765 nm [21]. The TP of the extracts is expressed as micrograms of gallic acid equivalents per gram of dried algae (mg GAE/g dw).

2.4. Determination of Flavonoids

The flavonoids in the extract were measured by mixing 600 µL of the algae extract with 2.58 mL of solution A (1.8 mL of 5% NaNO2 and 24 mL of water). After standing for 5 min, 180 µL of 10% $AlCl_3$ was added and the mixture allowed to stand for 1 min. Finally, 2.52 mL of solution B (12 mL of 1M NaOH plus 14.4 mL of water) was added and the contents were immediately read at 415 nm. Total flavonoids (TF) were obtained using a standard calibration curve of catechin (Sigma-Aldrich Co., Madrid, Spain) in a concentration range of 5 at 50 µg/mL and expressed as mg of catechin equivalents per gram of dried algae (mg CE /100 g dw) [22].

2.5. Antiradical Activity

Antiradical activity was evaluated by the DPPH test (2,2-diphenyl-1-picrylhydrazyl) radical assay according to Brand-Williams et al. (1995) [23]. Loss of color (purple) in the radical solution when exposed to each extract was measured at 517 nm. Results were expressed as % inhibition of DPPH radical according to the following formula:

$$DPPH\ (\%) = [(Ac - As)/(Ac)] \times 100 \quad (1)$$

where, Ac = control absorbance, As = Sample absorbance

2.6. Antioxidant Capacity by FRAP Method

The algae extracts were evaluated by the Iron Reduction/Antioxidant Power (FRAP) method [24]. The FRAP reagent was mixed in acetate buffer with the TPTZ solution and $FeCl_3 \times 6\ H_2O$. The mixture (1500 µL) was reacted at 37 °C with the extracts of algae 50 µL and 150 µL of water (d) After 4 min, the absorbance was measured at 593 nm. To determine the reduced Fe concentration, a $FeSO_4$ calibration curve was used. Results were expressed in Mmol Fe^{2+}/100 g.

2.7. Antibacterial Activity

To determine the antibacterial activity plating tests were carried out in agar wells according to the method proposed by González del Val [25]. The following bacterial strains were used: *Bacillus cereus* (ATCC 6633), *Escherichia coli* (ATCC 25922), *Staphylococcus aureus* (ATCC 25923), *Pseudomonas aeruginosa* (ATCC 27853), *Proteus mirabilis*, and a clinical strain of *Salmonella enteritidis*. These strains were grown in Mueller-Hinton broth for 24 h under agitation at 35 °C ± 2 °C. Isolated colonies were obtained for each strain. Bacteria were then streaked onto Mueller-Hinton agar plates (in triplicate) and grown for 24 h at 35 °C ± 2 °C. An isolated colony was selected from each inoculated dish and seeded into a tube containing a nutrient broth; this was followed by incubation at 35 °C ± 2 °C. Turbidity in the tubes was then adjusted to 0.5 units (108 CFU/mL) according to the McFarland standard. A sterile sachet (6 mm diameter) was used to make wells on the surface of agar plates. Prior to deposition into the wells, bacterial smears were prepared on Mueller-Hinton agar plates with the grass-planting technique and strain-dependent adjustments of bacterial concentration. Then, extract samples (25 µL) were deposited on each well (in triplicate), allowed to absorb for 30 min, and incubated for 24 h at 37 °C. Control tests were performed using pure (98% v/v) and diluted (50% v/v) methanol. The results were measured qualitatively by either the presence or absence of an inhibition halo.

2.8. Antioxidant Effect of Extracts in Cooked Salmon Paste

Cooked salmon pastes were prepared as follows. Salmon fillets obtained from the company Fiordo Austral (Puerto Montt, Chile) were kept in refrigeration after eliminating the inedible parts. Then the fillets were ground (Moulinex Mincer, AD6011, Shanghai, China) to obtain a homogeneous paste and kept refrigerated at 5 °C. Subsequently, extracts of each alga were prepared, dissolving 40 mg of dry extract in 50 mL of water (d). Each solution (50 mL) of extract was added individually in different samples of salmon paste (200 g each) to obtain a final concentration of 160 ppm of extract in the paste. This procedure was performed in triplicate. A paste control was prepared to which 50 mL of distilled water were added. A paste control was prepared to which 50 mL of distilled water was added. The pastes (10 g) with the addition of the algae extracts were placed in glass tubes and cooked for 30 min at 90 ± 5 °C by immersing them in a hot water bath. Subsequently, to determine the degree of oxidation of the lipid and loss of essential components (EPA, DHA, alpha-tocopherol and astaxanthin) from the heat-treated salmon muscle. Lipids were extracted from the muscle by the method of Bligh and Dyer [26].

2.9. Analysis of Polyunsaturated Fatty Acids ω-3

To determine the protective effect of algae on polyunsaturated fatty acids w-3; EPA and DHA (eicosapentaenoic acid and docosahexaenoic acid) in the cooked salmon pasta, the identification and quantification of the fatty acids was carried out by gas chromatography (GLC), after extraction of the lipids from the fish paste by the method of Bligh and Dyer [26] and their derivatization to methyl esters.

The derivatization consisted of an alkaline methylation of 100 mg of oil added to 10 mL of 0.2 N sodium methylate under reflux conditions for 10 min. Followed by acid methylation (H_2SO_4/methanol) in boiling condition for 20 min, cooling and final extraction of the methyl esters with hexane followed by dissolution with 10% NaCL (Spanish Standard UNE 55-037-73).

A gas chromatograph HP 5890 (Hewlett-Packard, Palo Alto, CA, USA) and a fused silica capillary column BPX70 (50 m, 0.25 μm film; SGE, Incorporated, Austin, TX, USA) were used. Conditions were: FID detector, injector and detector temperature 240 °C, initial temperature 160 °C, gradient 2 °C/min up to 230 °C. Carrier gas H2. FAME were identified based on sample standards (QualmixFish, Larodan, Malmo, Sweden; FAME Mix, Supelco, Inc., Merck, Darmstadt, Germany). In addition to determining the percentage content of the essential fatty acids EPA and DHA of the salmon paste, the polyunsaturation of the fat was expressed as a polyene index (PI) according to: PI = EPA + DHA/C16: 0 (palmitic acid).

2.10. Tocopherols

Measurement of tocopherols and provitamin E (α-Tocopherol) in the fat of salmon paste was carried out by HPLC analysis, with a fluorescence detector according to AOCS Official Method Ce 8-89 (1993) [19]. Detection was carried out at an excitation wavelength of 290 nm and emission 330 nm. The mobile phase was 2-propanol (0.05%) in hexane, with a flow of 1 mL/min. External standards α, β, γ and δ tocopherol (Merck, Darmstadt, Germany) were used and results expressed as μg tocol/g lipid.

2.11. Lipid Oxidation

The peroxide index (PV) was measured according to the AOCS Official Method, Cd 8-53 (1993) [19]. Five grams of extracted salmon fat were introduced into a 250 mL flask and mixed with 30 mL of acetic acid-chloroform solution (3:2), stirring the mixture with 3 g of KI and 0.5 mL of distilled water. The stoppered flask was shaken in the dark for 1 minute stopping the reaction with 30 mL of distilled water. The mixture was titrated with 0.01 N sodium thiosulfate and starch as indicator. The results were expressed in meq of active oxygen/kg of fat.

2.12. Statistical Analysis

Data from the different chemical assays were statistically analyzed using Statgraphics Centurion XV (StatPoint Technologies Inc., Warrenton, VA). Results were evaluated using analysis of variance (ANOVA). Mean values (n = 3) were compared through the multiple range test, using the procedure of honestly significant difference (HSD) of Tukey. In all cases, differences were considered significant at a confidence level of 95% ($p < 0.05$).

3. Results and Discussion

3.1. Nutritional Composition

Table 1 presents the nutritional composition of each algae species. Most algae presented low amounts of total lipids (0.2–1.4%), similar to percentages reported in some terrestrial edible plants, such as spinach (0.8%) and chard (0.4%) [27]. The obtained values also coincide with low lipid contents found in other macroalgae, where the reported maximum value is 4% of dry weight (dw) [28–30].

Table 1. Nutritional composition of assessed algae species (g/100 g dw).

Algae	Ash	Lipids	Proteins	Carbohydrates	Fiber	β-Glucans
I. larga (a)	21.29 ± 1.69c	0.73 ± 0.19b	1.23 ± 1.45b	66.74 ± 4.49b	59.76 ± 2.09c	7.0 ± 0.4 b,d,f
G. chilensis (b)	12.28 ± 2.45b	1.36 ± 0.25c	19.94 ± 1.13d	66.42 ± 0.16b	59.99 ± 5.24c	4.9 ± 0.1 a,c,d,f
G. chilense (c)	8.77 ± 1.44a	1.40 ± 0.26c	20.26 ± 1.48d	69.57 ± 4.30a	55.45 ± 2.10b	6.0 ± 0.6 b,d,f
G. chamissoi(d)	13.66 ± 1.21d	3.73 ± 0.05d	14.08 ± 1.50c	68.53 ± 3.45a	55.16 ± 1.32b	3.2 ± 0.8 a,b,ce,f
G. rádula (e)	19.58 ± 1.52c	0.95 ± 0.14b	11.18 ± 1.18b	68.29 ± 2.20a	48.91 ± 1.71a	6.4 ± 0.2 b,d,f
G.skottsbergii (f)	25.72 ± 1.24b	0.20 ± 0.03a	7.57 ± 0.91a	66.51 ± 5.32d	59.18 ± 5.11c	5.6 ± 0.3 a,b,d,e

Values correspond to the average of triplicates ± standard deviation. Letters indicate significant differences ($p < 0.05$). Proteins were estimated by converting nitrogen content using a factor of 6.25. Total carbohydrates were estimated as the difference of subtracting 100 other components (i.e., lipids, minerals, proteins, moisture).

Algae presented notable differences in lipid contents, with the lowest and highest contents found in *G. skottsbergii* (0.20%) and *G. chamissoi* (3.73%), respectively. Variability in lipid contents may be due to differences between species and phenomena associated with growth conditions in each geographic zone. Likewise, protein contents varied widely. The highest protein contents were found in *G. chilense* (20.26%), and the lowest protein contents were found in *G. skottsbergii*. Nevertheless, all of the studied algae presented values that are comparable to some common foods in the human diet, such as eggs (13%) [27]. Previous reports have estimated up to 34% (dry weight) in protein contents of some seaweeds [30,31], which would be comparable to foods high in proteins, such as soy (36%). Protein contents are, however, dependent on species, collection time, and season, among other factors [28,32]. The percentage of total minerals in the six red algae varied between 8% and 25% with *G. skottsbergii* having the highest content and *G. chilense* the lowest. These results are lower than previous reports on mineral contents in algae. The mineral content in algae normally represents close to 35% dw [30,33,34], which is higher than found in most terrestrial plants (excepting spinach, which has a value similar to algae). Mineral content found in the studied algae suggest they would be an important source of microelements (e.g., Ca, Fe, Zn). Mineral accumulation is due to the ability of algae to selectively absorb inorganic substances from the ocean through surface polysaccharides [29,33].

The amount of total dietary fiber found in the analyzed samples was close to 60%, excepting *G. radula* (see Table 1). Fiber content in studied algae surpasses foods traditionally considered to be high in dietary fiber, such as chard (47.7%), spinach (47.3%), and raw carrots (48.74%) [32,33]. Results are in line with the range (25-75%) reported for different varieties of brown and red algae [2]. Similar values in red algae were reported by Jiménez-Escrig and Goñi (1999) [29], indicating that the primary contents of fiber would be soluble dietary fiber comprised of sulfated galactans, agar, or carrageenans. The amount of dietary fiber in the assessed algae surpasses daily recommended values for humans (i.e., 30 g/portion), serving as a determinant of the principal nutritional and physiological effects of consuming this marine plant.

The six assessed algae were noteworthy concerning protein, mineral, and fiber contents. Lipid contents, in turn, were lower. This would translate into fewer calories, meaning that the assessed algae could be candidates in the development of new food products for the management of weight loss. All of the red algae studied had significant levels of β-glucans (Table 1), which represented 3-7% dw. β-glucans have been shown to have beneficial effects for the immune system, as they are essential for the intestinal microbiota, favoring human health [35,36]. Bobadilla et al. (2013) determined similar contents of β-glucans in Chilean algae, finding the highest levels of β-glucans in the fronds of brown algae [6]. It should be noted that the β-glucan values obtained for *G. chilensis* in the present study even exceeded the highest values reported by Bobadilla et al. (2013) [6]. Since it has been shown that the β-glucans of *G. chilensis* can activate the cellular immune system of lymphocytes, it has been proposed that this alga could reduce the mortality of fingerlings in aquaculture, particularly for species of high commercial value, such as salmon or trout [6,36].

3.2. Antioxidant Activity of Phenolic Extracts

Results obtained for total polyphenols expressed in gallic acid equivalents (Table 2) indicate that polyphenols content in the six assessed algae species vary between 2.6 and 11.3 mg GAE/100 g algae. The highest polyphenolic contents were found in *G. skottsbergii* and *G. chilense*. On the other hand, *I. larga* and *G. radula* presents an intermediate value, whereas the lowest contents were found in *G. chilensis* and *G. chamissoi*. The total polyphenol contents were lower than in other types of red algae, where values between 37 and 178 mg/g dw have been reported [15]. The flavonoid content for the six algae varied between 2.2 and 9.6. Both polyphenol and flavonoid contents were similar to those determined in the algae *Hypnea musciformis* and *Acanthophora muscoides* [15]. While the samples of *G. radula*, *G. skottsbergii* and *G. Chilense*, presented a greater antiradical capacity of DPPH according to their higher content of TP and TF (Table 2), they did not present a good reducing power of Fe^{3+}. These results coincide with those obtained in other red algae studied by Arulkumar et al (2020) [15]. It should be noted that in most of the studies carried out on free radical or in lipid oxidation model systems, the antioxidant capacity is proportional to the polyphenol content [37–40]. Although *G. chamissoi* and *G. chilensis* did not have high levels of total polyphenols and flavonoids, they did present a good reducing value (FRAP).

Table 2. Total polyphenols (TP), Total flavonoids (TF) contents and antiradical capacity (IC50) of algae extracts. Values correspond to the mean of three independent analyses ± standard deviation.

Algae	TP	TF	DPPH	FRAP
	mgGAE/g dw	mgCE/g dw	(%)	g Fe^{2+}/100 g
I. larga (a)	6.9 ± 1.2 b,c,d,f	5.8 ± 1.2 b,c,d,f	26.6 ± 3.5 b,c,e,f	0.36 ± 0.01 b,c,d
G. chilensis (b)	2.6 ± 0.6 a,c,e,f	2.2 ± 1.2 a,c,e,f	34.2 ± 6.0 a,c,e,d,e	0.57 ± 0.11 a,c,d,e,f
G. chilense (c)	9.9 ± 1.3 a,b,c,d,e	8.4 ± 1.2 a,b,d,e,f	51.2 ± 9.1 a,b,c,d,e,d	0.47 ± 0.04 a,b,d,e,f
G. chamissoi (d)	3.4 ± 0.4 a,b,d,e,f	3.1 ± 1.2 a,b,c,d,e	21.6 ± 4.1 b,c,d,e	0.62 ± 0.08 a,c,d,f
G. rádula (e)	6.1 ± 0.9 a,b,c,d	5.1 ± 1.2 b,c,d,e	72.2 ± 10.8 a,b,c,d	0.31 ± 0.01 b,c,d
G. skottsbergii (f)	11.3 ± 2.1 a,b,c,d,e	9.6 ± 1.2 a,b,c,d,e	66.9 ± 16.4 a,b,c,d	0.34 ± 0.01 b,c,d

Values correspond to the average of triplicates ± standard deviation. Letters indicate significant differences ($p < 0.05$).

Exceptionally, *I. larga* despite having a good level of polyphenols and flavonoids, did not present good anti-radical and reducing capacity. Some authors have reported that the higher the polarity of the solvents used, the higher the content of polyphenols extracted [40]. Thus, the polarity of the solvents used can help to produce a selective extraction of different bioactive compounds which have a different antiradical response capacity for the same type of sample. The previous treatment of the raw material could also cause the loss of the antioxidant power of the polyphenols present in these commercial red algae. In particular, the effect of drying could decrease the total content of polyphenols due to oxidation phenomena and structural changes, giving lower values than those present in fresh algae [29,37,40]. The most abundant polyphenols in algae known as florotannins, which can be in the form of floroglucinol polymers in various types of red, brown and green algae [37,41], could be affected by the conditions of extraction, drying, storage, among others. In addition, other bioactive components responsible for the antiradical properties such as catechins, flavonols and flavonol glucosides [42], which have been identified from methanolic extracts of red and brown algae could have been sensitive to the handling conditions of the samples.

3.3. Antibacterial Assays

Table 3 presents the results obtained from the bacterial inhibition tests. The six hydroalcoholic algae extracts presented antimicrobial properties. A halo of inhibition was formed for most of the strains tested and over 90% of the plaques presented a halo of inhibition. The positive results of antimicrobial activity obtained for *G. radula, G. skottsbergii, G. chilense* and *G. chilensis* coincide with the high antioxidant capacities of these species. However, this result could be more related to the structure of the bioactive compounds than to the total content of polyphenols or antioxidant capacity. A similar inhibitory capacity has been observed in methanolic and acetone extracts from brown algae. More specifically, the methanolic extracts of *Sargassum latifolium* B and *Sargassum platycarpum* A were more active against Gram (+) and Gram (−) bacteria, the acetone extract of *S. latifolium* B being more inhibitory against *Salmonella* sp. [43]. Other studies with red algae extracts using methanol did not demonstrate bacterial inhibition but agglutinated Gram (+) bacteria from *S. aureus* and Gram (−) cells from E. *coli*, multi-resistant *Salmonella* and *Vibrio harveyi* [44].

The effectiveness of the red algae studied here may be due to the inhibitory action of other components present in the extracts such as pigments, organic acids and mineral salts, etc. Other studies in extracts of the red algae *Spergofusiforme* and *Sargassum vulgare* have used mass spectrometry identifying phenols, terpenoids, acetogenins, indoles, fatty acids and volatile halogenated hydrocarbons. Said extracts registered antimicrobial activity against the microorganisms *S. aureus* 2 and *Klebsiella pneumonia* [13] Among the studied algae, *I. larga* only presented antimicrobial activity against the stains *S. enteritidis*, *E. coli*, and *B. cereus* (Table 3). While this alga did not have the lowest total polyphenols contents, nor did it present the lowest antioxidant capacity, it is probable that the polyphenols present did not affect the bacterial strains due to the complexity of the bacterial cell membrane, which can be more selective and less permeable than the strains in which positive results were achieved.

Table 3. Detected antimicrobial reactions of assessed algae extracts.

Algae Extract	Bacterial Strain					
	S. enteritidis	*E. coli*	*B. cereus*	*P. aeruginosa*	*St. aureus*	*P. mirabilis*
Iridaea larga	+	+	+	−	−	−
Gracilaria chilensis	+	+	+	+	+	+
Gelidium chilense	+	+	+	+	+	+
Gigartina chamissoi	+	+	+	+	+	+
Gigartina radula	+	+	+	+	+	+
Gigartina skottsbergii	+	+	+	+	+	+

+: Positive antibacterial activity. −: Negative antibacterial activity.

The antimicrobial effect of algae extracts could be corroborated with transmission electron microscopy, looking at the morphological changes of microorganisms. For example, El Shafay et al. (2015) [13] observed that treatment of the bacteria *S. aureus* 2 and *K. pneumonia* with extracts from *S. fusiforme* and *S. vulgare* resulted in perforation of the cell wall, content-escape to the cytoplasm, and external distortion of the cell shape, among other damages to the cell structure [13]. Presumably, the bioactive components present in the six currently assessed red algae extracts would have permeated the interior of the microbial cell to produce cell-level damages, resulting in the inhibition of the bacterial strains used for assays.

3.4. Protective Effect of Lipid Oxidation in Cooked Salmon Paste

Figure 2 shows the inhibitory effect of red algae extracts on the oxidation of the fat of cooked salmon pasta, expressed by the peroxide number. Results indicated that all the algae extracts presented a protective effect against oxidation of salmon paste, given by an inhibition of the formation of peroxides that are primary oxidation products of lipids. The most efficient algae were *G. chamissoi*, *G. radula* and *G. chilensis*.

This protective effect was also observed in canned salmon prepared with liquid packaging containing extracts of the brown algae *Durvilleae antarctica*, *Ulva lactuca* and the red algae *Porphyra columbina* [14]. The protective effect can be attributed to the action of the bioactive components present in the seaweed extracts, such as, polyphenols, florotannins, diterpenes, carotenoids, phytosterols and possibly, low molecular weight hydrocolloids. On the other hand, extracts of other types of red and brown algae, added to chilled fish, have also shown effects against the oxidation and evolution of biogenic amines, trimethylamine, etc. [11,15,45,46].

Other plant-based foods that contain similar bioactive components have also shown protection in thermally processed fish [47,48]. The protective effect against thermal oxidation has been attributed to the solubilisation of phenols and other hydrophilic components at the water-muscle tissue interface, which could be maximized in a homogenized system such as the salmon paste used in this study.

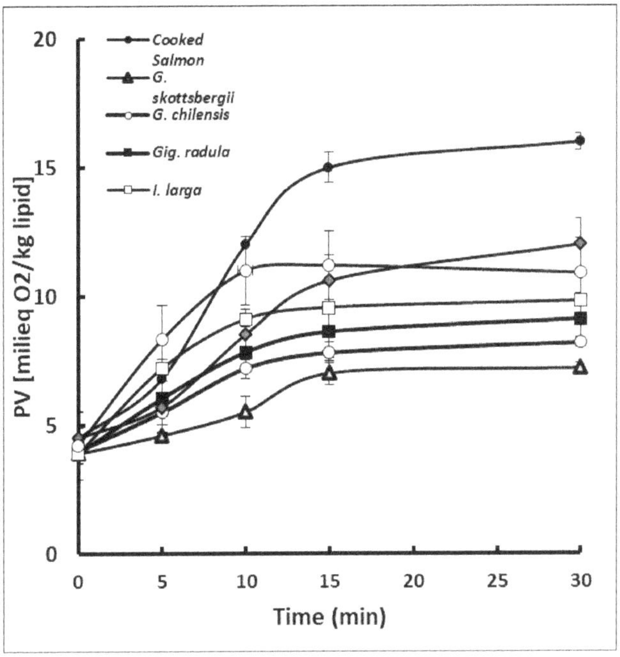

Figure 2. Inhibition of PV formation during cooking of salmon paste samples added with different extracts of red algae. Values are expressed as means ($n = 3$) ± standard deviation.

3.5. Protective Effect of EPA and DHA (PUFAS) in Cooked Salmon Paste

Figure 3 shows the variation of the essential fatty acids EPA and DHA and the polyene index of the cooked salmon pasta samples versus the same pasta added with each of the red algae extracts. The results indicated that cooking the salmon paste without extracts generated a high decrease of DHA close to 50% and of 25% of EPA.

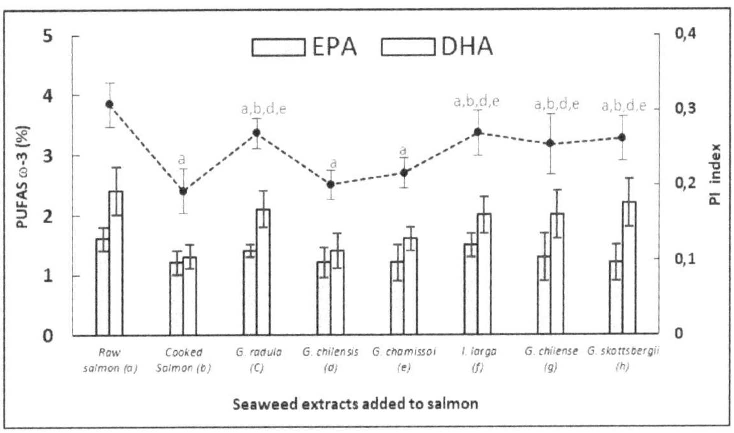

Figure 3. Polyene index of cooked and raw salmon pastes added with different extracts of red algae: (a) Raw salmon paste, (b) Cooked salmon paste, (c) *G. chamissoi*, (d) *G. skottsbergii*, (e) *G. radula*, (f) *I. larga*, (g) *G. chilense*, (h) *G. chilensis*. Values are expressed as means ($n = 3$) ± standard deviation. Different letters correspond to mean values with significant difference ($p < 0.05$).

On the other hand, most of the fish pastes treated with seaweed extracts presented a significant protection ($p < 0.05$) given by a high content of both essential fatty acids in the cooked product. Losses were reduced to less than 25 %, probably due to the inhibition of the fatty acid oxidation process in triglycerides, which was reflected in the maintenance of the polyene index. On the other hand, the percentage contents of EPA and DHA in the fat of salmon paste were as follows: (a) 1.6 ± 0.4; 2.4 ± 1.3; (b) 1.2 ± 0.2, 1.3 ± 0.1; (c) 1.4 ± 0.2, 2.1 ± 0.4; (d) 1.2 ± 0.3, 1.4 ± 0.3; (e) 1.2 ± 0.3, 1.6 ± 0.5; (f) 1.5 ± 0.3, 2.1 ± 0.1; (g) 1.3 ± 0.4, 2.0 ± 0.5; (h) 1.2 ± 0.3, 2.2 ± 0.6, respectively. C16: 0 remained constant at an average 10 ± 0.6. Similar results using plant extracts, algae and oils have been evidenced in other fish conservation studies at low and high temperatures [14,47,48].

3.6. *Protective Effect of Pro Vitamin E in Cooked Salmon Paste*

Figure 4 shows the level of tocopherols in control raw and cooked salmon pasta versus pastas added with red algae extracts and cooked at 90 °C for 30 min. Results indicated that in general the content of α, γ, δ-tocopherols had a significant loss ($p < 0.05$) due to cooking, presented in salmon paste cooked with seaweed extracts. On the other hand, all samples of salmon paste added with extracts of red algae significantly decreased the loss of tocopherols from the paste ($p < 0.05$) compared to the cooked control pasta. The algae *G. skottsbergii* maintained a higher content of tocopherols in salmon than the other red algae. The *G. chilensis* extract was the one that maintained a higher content of α, γ and δ-tocopherol. The other salmon pastes presented a similar content of tocopherols. Similar results have been observed in extracts of other types of vegetables such as stevia and grape skins applied to fish, shellfish and crustaceans [14,15,47–50].

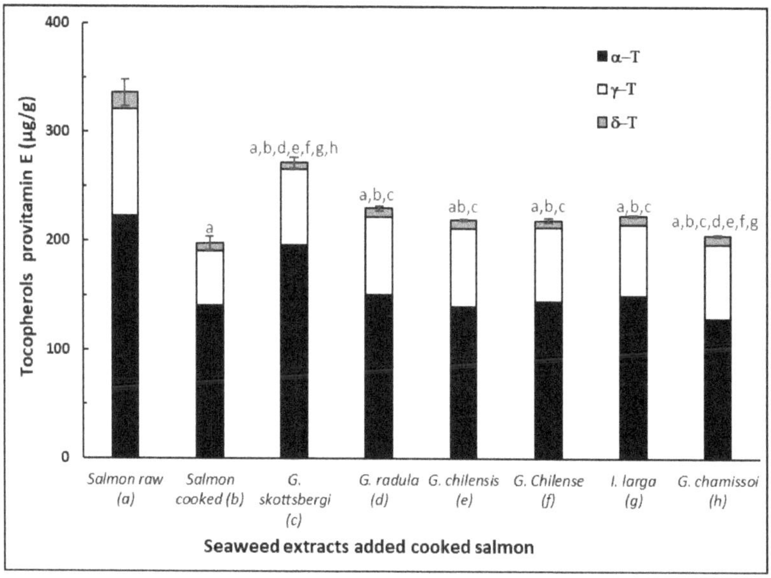

Figure 4. Tocopherols of cooked and raw salmon pastes added with different extracts of red algae: (a) Raw salmon paste, (b) Cooked salmon paste, (c) *G. chamissoi*, (d) *G. skottsbergii*, (e) *G. radula*, (f) *I. larga*, (g) *G. chilense*, (h) *G. chilensis*. Values are expressed as means ($n = 3$) ± standard deviation. Different letters correspond to mean values with significant difference ($p < 0.05$).

Algae extracts may have contributed with tocopherols [29] in addition to flavonoids and polyphenols from the salmon pastes. Losses of tocopherols during cooking are due to their protective action against the thermo-oxidation of polyunsaturated lipids of the adipose tissue of salmon, which can be associated with the maintenance presented in the content of EPA and DHA and in the polyene index (PI) and (Figure 3). At the same time, these results are correlated with the inhibition of the primary products given by the attenuation of the peroxide curves (Figure 2). Similar results have been also obtained in canned salmon heat treated in the presence of red and brown algae [14].

3.7. Protective Effect of Astaxanthin in Cooked Salmon Paste

Figure 5 shows the level of astaxanthin in raw and cooked salmon pasta versus the same pasta added with red algae extracts and cooked at 90 °C for 30 min. Results indicated a loss of more than 50% of astaxanthin in the control salmon paste. A moderate inhibitory effect of astaxanthin degradation was observed only in the samples with *G. skottsbergii*, *G. radula* and *G. chilense*. The relative protection of seaweed extracts against astaxanthin and tocopherols in cooked salmon pastes could be due to a synergistic effect of seaweed phenols and other endogenous antioxidant compounds from salmon muscle (such as ascorbic acid, uric acid, thiols, biliru-bin, etc.) capable of reversing the degradation of astaxanthin during lipid oxidation and the transition from tocopherol to tocopherylquinone. Therefore, part of the astaxanthin and tocopherol would be recovered and remain active, thus more effectively inhibiting the formation of peroxides and at the same time protecting the content of polyunsaturated fatty acids. [48,50,51].

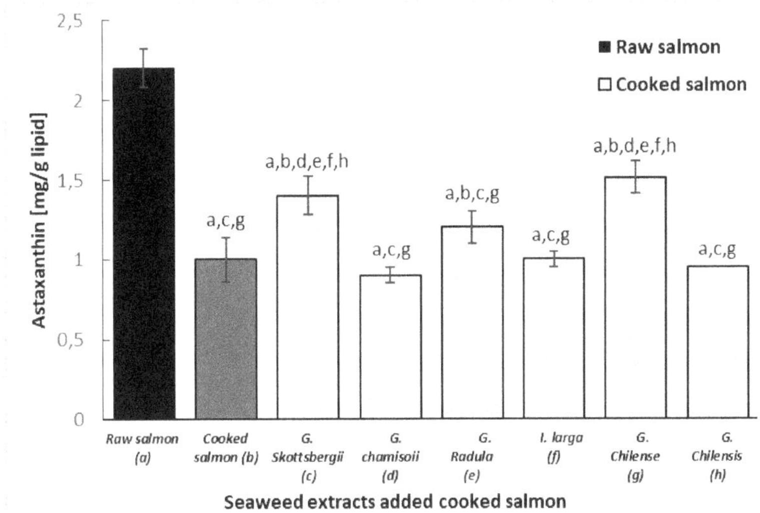

Figure 5. Astaxanthin of cooked and raw salmon pastes added with different extracts of red algae: (a) Raw salmon paste, (b) Cooked salmon paste, (c) *G. chamissoi*, (d) *G. skottsbergii*, (e) *G. radula*, (f) *I. larga*, (g), *G. chilense*, (h) *G. chilensis*. Values are expressed as means ($n = 3$) ± standard deviation. Different letters correspond to mean values with significant difference ($p < 0.05$).

4. Conclusions

This study suggests that beneficial dietary components present in red algae have some bioactive properties, especially a protective activity against the loss of essential dietary components present in salmon paste under thermo-oxidation conditions. The antioxidant activity and the protection of the essential fatty acids EPA and DHA, tocopherols (vitamin E) and astaxanthin present in salmon pastes, attributable to polyphenols, flavonoids and other bioactive components of the red algae, opens new possibilities of applications these extracts in fresh and processed foods. However, the evidence obtained requires of new studies to increase the available scientific knowledge on the functional and health-promoting properties of algae. In addition, it is necessary to deepen in sensory evaluation studies for marine products fortified with edible algae to determine the acceptability by consumers.

Author Contributions: Conceptualization, M.J.L., J.O.-V. and M.F.; methodology, J.O.-V., J.M.M., and S.P.A.; software, R.L.-M. and M.J.L; validation, J.O.-V., J.M.A. and M.F.; formal analysis, J.M.M., and S.P.A.; investigation, J.O.-V., R.L.-M. and M.F. resources, J.O.-V. and J.M.A.; data curation, J.O.-V., J.M.A., R.L.-M., M.J.L., and M.F.; writ-ing—original draft preparation, M.F. and S.P.A.; writing—review and editing, J.O.-V., J.M.M., S.P.A.; writing—review and editing, J.O.-V., J.M.A., and M.J.L.; visualization, J.O.-V., and J.M.A.; final revision of manuscript, J.M.A.; supervision: J.O.-V.; project administration, J.O.-V.; funding acquisition, J.O.-V. All authors have read and agreed to the published version of the manuscript.

Funding: This research was funded by the Fondo Nacional de Desarrollo Científico y Tecnológico grant number Fondecyt 1180082. The authors thank the company Algamar S.A. (Santiago, Chile) for the collection, processing and donation of the six algae used in this study.

Institutional Review Board Statement: Not applicable.

Informed Consent Statement: Not applicable.

Data Availability Statement: Data is contained within the article.

Acknowledgments: The author Marcos Flores thanks to Dirección General de Investigación Aplicada e Innovación of the Universidad Santo Tomás, Chile for the support in the translation of the manuscript.

Conflicts of Interest: The authors declare no conflict of interest.

References

1. Figueroa, V.; Farfán, M.; Aguilera, J.M. Seaweeds as Novel Foods and Source of Culinary Flavors. *Food Rev. Int.* **2021**. [CrossRef]
2. Ortiz, J.; Uquiche, E.; Robert, P.; Romero, N.; Quitral, V.; Llanten, C. Functional and nutritional value of the Chilean seaweeds *Codium fragile, Gracilaria chilensis* and *Macrocystis pyrifera*. *Eur. J. Lipid Sci. Technol.* **2009**, *111*, 320–327. [CrossRef]
3. MacArtain, P.; Gill, C.I.; Brooks, M.; Campbell, R.; Rowland, I.R. Nutritional value of edible seaweeds. *Nutr. Rev.* **2007**, *65*, 535–543. [CrossRef] [PubMed]
4. Rebours, C.; Marinho-Soriano, E.; Zertuche-González, J.A.; Hayashi, L.; Vásquez, J.; Kradolfer, P.; Soriano, G.; Ugarte, R.; Abreu, M.H.; Bay-Larsen, I.; et al. Seaweeds: An opportunity for wealth and sustainable livelihood for coastal communities. *J Appl. Phycol.* **2014**, *26*, 1939–1951. [CrossRef] [PubMed]
5. Lange, K.W.; Hauser, J.; Nakamura, Y.; Kanaya, S. Dietary seaweeds and obesity. *Food Sci. Hum. Wellness* **2015**, *4*, 87–96. [CrossRef]
6. Bobadilla, F.; Rodríguez-Tirado, C.; Imarai, M.; Galotto, M.G.; Andersson, R. Soluble -1,3/1,6-glucan in seaweed from the southern hemisphere and it immunomodulatory effect. *Carbohyd. Polym.* **2013**, *92*, 241–248. [CrossRef]
7. Dovale-Rosabal, G.; Rodríguez, A.; Contreras, E.; Ortiz-Viedma, J.; Muñoz, M.; Trigo, M.; Aubourg, S.P.; Espinosa, A. Concentration of EPA and DHA from Refined Salmon Oil by Optimizing the Urea–Fatty Acid Adduction Reaction Conditions Using Response Surface Methodology. *Molecules* **2019**, *24*, 1642. [CrossRef] [PubMed]
8. Pando, M.E.; Rodríguez, A.; Galdames, A.; Berríos, M.; Rivera, M.; Romero, N.; Valenzuela, M.A.; Ortiz, J.; Aubourg, S. Maximization of the docosahexaenoic and eicosapentaenoic acids content in concentrates obtained from a by-product of rainbow trout (*Oncorhynchus mykiss*) processing. *Eur. Food Res. Technol.* **2020**, *144*, 536–543. [CrossRef]
9. Mateluna, C.; Figueroa, V.; Ortiz, J.; Aguilera, J.M. Effect of processing on the texture and microstructure of the algae *Durvillaea antarctica*. *J. Appl. Phycology* **2020**, *32*, 4211–4219. [CrossRef]
10. Alcicek, Z. The effects of thyme (*Thymus vulgaris* L.) oil concentration on liquid-smoked vacuum-packed rainbow trout (*Oncorhynchus mykiss* Walbaum) fillets during chilled storage. *Food Chem.* **2011**, *128*, 683–688. [CrossRef]
11. Trigo, M.; López, M.; Dovale, G.; Ortiz, J.; Rodríguez, A.; Aubourg, S.P. Enhancement of sensory acceptance of frozen mackerel by alga-extract glazing. *Bulg. Chem. Commun.* **2019**, *51*, 216–223.
12. Miranda, J.M.; Trigo, M.; Barros-Velazquez, J.; Aubourg, S.P. Effect of icing medium containing the alga focus spiralis on the microbiological activity and lipid oxidation in chilled megrim (*Lepidorhombus whiffiagonis*). *Food Cont.* **2016**, *59*, 290–297. [CrossRef]
13. El Shafay, S.M.; Ali, S.S.; El-Sheekh, M.M. Antimicrobial activity of some seaweeds species from Red sea, against multidrug resistant bacteria. *Egypt. J. Aquat. Res.* **2016**, *42*, 65–74. [CrossRef]
14. Ortiz, J.; Vivanco, J.P.; Aubourg, S.P. Lipid and Sensory quality of canned Atlantic salmon (*Salmo salar*): Effect of the use of different seaweed extracts as covering liquids. *European J. Lipid Sci. and Technol.* **2014**, *116*, 596–605. [CrossRef]
15. Arulkumar, A.; Satheeshkumar, K.; Paramasivam, S.; Rameshthangam, P.; Miranda, J.M. Chemical Biopreservative Effects of Red Seaweed on the Shelf Life of Black Tiger Shrimp (*Penaeus monodon*). *Foods* **2020**, *9*, 634. [CrossRef]

16. Ortiz, J.; Lemus-Mondaca, R.; Vega-Gálvez, A.; Ah-Hen, K.; Puente-Diaz, L.; Zura-Bravo, L.; Aubourg, S.P. Influence of air-drying temperature on drying kinetics, colour, firmness and biochemical characteristics of Atlantic salmon (*Salmo salar* L.) fillets. *Food Chem.* **2013**, *4*, 162–169. [CrossRef] [PubMed]
17. Ortiz, J.; Vivanco, J.P.; Quitral, V.; Larraín, M.A.; Concha, G.; Aubourg, S.P. Effect of the antioxidant profile in the diet of farmed coho salmon (*Oncorhynchus kisutch*) on the nutritional value retention during frozen storage. *Grasas Aceites* **2013**, *64*, 3. [CrossRef]
18. AOAC. *Official Methods of Analysis of Association of Official Analytical Chemists International*, 16th ed.; AOAC: Gaithersburg, MD, USA, 1995; Volume I, II, p. 870.
19. AOCS. *Official Methods and Recommended Practices of the American Oil Chemists Society*, 4th ed.; AOCS Press: Champaign, IL, USA, 1993; pp. 54–56.
20. AOAC. *Official Methods of Analysis of Association of Official Analytical Chemists International*; Official Method 995.16; AOAC Press: Champaign, IL, USA, 2005.
21. Yildiz, G.; Vatan, O.; Çelikler, S.; Dere, Ş. Determination of the Phenolic Compounds and Antioxidative Capacity in Red Algae *Gracilaria bursa-pastoris*. *Int. J. Food Prop.* **2011**, *14*, 496–502. [CrossRef]
22. Zhishen, J.; Mengcheng, T.; Jianming, W. The determination of flavonoid contents in mulberry and their scavenging effects on superoxide radicals. *Food Chem.* **1999**, *64*, 555–559. [CrossRef]
23. Brand-Williams, W.; Cuvelier, M.-E.; Berset, C. Use of a free radical method to evaluate antioxidant activity. *LWT Food Sci. Technol.* **1995**, *28*, 25–30. [CrossRef]
24. Benzie, I.F.; y Strain, J.J. The Ferric Reducing Ability of Plasma (FRAP) as a Measure of Antioxidant Power China. *Anal. Biochem.* **1996**, *239*, 70–76. [CrossRef]
25. González del Val, A.; Platas, G.; Basilio, A.; Cabello, A.; Gorrochategui, J.; Suay, I.; Vicente, F.; Portillo, E.; Jiménez del Río, M.; García Reina, G.; et al. Screening of antimicrobial activities in red, green and Brown macroalga from Gran Canaria (Canary Islands, Spain). *Int. J. Microbiol.* **2001**, *4*, 35–40. [CrossRef]
26. Bligh, E.G.; Dyer, W.J. A rapid method of extraction and purification of total lipids. *Can. J. Biochem Physiol.* **1959**, *37*, 911–917. [CrossRef]
27. Ortiz, J.; Romero, N.; Robert, P.; Araya, J.; Lopez-Hernández, J.; Bozzo, C.; Navarrete, E.; Osorio, A.; Rios, A. Dietary fiber, amino acid, fatty acid and tocopherol contents of the edible seaweeds *Ulva lactuca* and *Durvillaea Antarctica*. *Food Chem.* **2006**, *99*, 98–104. [CrossRef]
28. Sanchez-Machado, D.; Lopez Hernandez, J.; Paseiro Lozada, P. High performance liquid chromatographic determination of alfa-tocopherol in macroalgae. *J. Chromatogr. A* **2002**, *976*, 277–284. [CrossRef]
29. Jiménez-Escrig, A.; Goñi, I. Nutritional evaluation and physiological effects of edible marine macroalgae. *Arch. Latinoam Nutr.* **1999**, *49*, 114–120.
30. Schmidt-Hebbel, H.; Pennacchiotti, I.; Masson, L.; Mella, M.A. *Tabla de Composición Química de los Alimentos*, 8th ed.; Universidad de Chile: Santiago, Chile, 1992.
31. Chandini, S.K.; Ganesan, P.; Suresh, P.V.; Bhaskar, N. Seaweeds as a source of nutritionally beneficial compounds—A review. *J Food Sci Tech.* **2008**, *45*, 1–13.
32. Cherry, P.; O'Hara, C.; Magee, P.J.; McSorley, E.M.; Allsopp, P.J. Risks and benefits of consuming edible seaweeds. *Nutr. Rev.* **2019**, *77*, 307–329. [CrossRef]
33. Rohani-Ghadikolaei, K.; Abdulalian, E.; Ng, W. Evaluation of the proximate, fatty acid and mineral composition of representative green, brown and red seaweeds from the Persian Gulf of Iran as potential food and feed resources. *J Food Sci Tech.* **2012**, *49*, 774–780. [CrossRef]
34. Dhargalkar, V.K.; Neelam, P. Seaweed: Promising plant of the millennium. *Sci. Cult.* **2005**, *71*, 60–66.
35. Lechat, H.; Amat, M.; Mazoyer, J.; Buleón, A.; Lahaye, M. Structure and distribution of glucomannan and sulfated glucan in the cell walls of the red alga *Kappaphycus alvarezii* (Gigartinales, Rhodophyta). *J. Phycol.* **2001**, *36*, 891–902. [CrossRef]
36. Yang, Y.; Zhao, X.; Li, J.; Jiang, H.; Shan, X.; Wang, Y.; Ma, W.; Hao, J.; Yu, G. A β-glucan from Durvillaea Antarctica has immunomodulatory effects on RAW264.7 macrophages via Toll-like receptor 4. *Carbohyd. Polym.* **2018**, *191*, 255–265. [CrossRef]
37. Sathya, R.; Kanaga, N.; Sankar, P.; Jeeva, S. Antioxidant properties of phlorotannins from brown seaweed *Cystoseira trinodis* (Forsskål) C. Agardh. *Arab. J. Chem.* **2017**, *10*, S2608–S2614. [CrossRef]
38. Devi, G.K.; Manivannan, K.; Thirumaran, G.; Rajathi, G.; Anantharaman, P. In vitro antioxidant activities of selected seaweeds from Southeast coast of India. *Asian Pac. J. Trop. Med.* **2011**, *4*, 205–211. [CrossRef]
39. Kajal, C.; Deepu, J.; Nammunayathuputhenkotta, K.P. Antioxidant activities and phenolic contents of three red seaweeds (Division: Rhodophyta) harvested from the Gulf of Mannar of Peninsular India. *J Food Sci Tech.* **2015**, *52*, 1924–1935. [CrossRef]
40. Chakraborty, K.; Krishnankartha, N.; Praveen, K.; KizekadathVijayan, K.; Syda Rao, G. Evaluation of phenolic contents and antioxidant activities of brown seaweeds belonging to *Turbinaria* spp. (Phaeophyta, Sargassaceae) collected from Gulf of Mannar. Asian Pac. *J. Trop. Biomed.* **2013**, *3*, 8–16. [CrossRef]
41. Wang, T.; Jónsdóttir, R.; Olafsdóttir, G. Total phenolic compounds, radical scavenging and metal chelation of extracts from Icelandic seaweeds. *Food Chem.* **2009**, *116*, 240–248. [CrossRef]

42. Yoshie-Stark, Y.; Hsieh, Y.P.; Suzuki, T. Distribution of flavonoids and related compounds from seaweeds in Japan. *J. Tokyo Univ. Fish.* **2003**, *89*, 1–6.
43. Moubayed, N.M.S.; Al Houri, H.J.M.; Khulaifi, A.; Al Farraj, D.A. Antimicrobial, antioxidant properties and chemical composition of seaweeds collected from Saudi Arabia (Red Sea and Arabian Gulf). *Saudi J. Biol. Sci.* **2017**, *24*, 162–169. [CrossRef]
44. De Alencar, D.B.; De Carvalho, F.C.T.; Rebouças, R.H.; Dos Santos, D.R.; Dos Santos Pires-Cavalcante, K.M.; De Lima, R.L.; Baracho, B.M.; Bezerra, R.M.; Viana, F.A.; Dos Fernandes Vieira, R.H.S.; et al. Bioactive extracts of red seaweeds *Pterocladiella capillacea* and *Osmundaria obtusiloba* (Floridophyceae: Rhodophyta) with antioxidant and bacterial agglutination potential. *Asian Pac. J. Trop. Med.* **2016**, *9*, 372–379. [CrossRef] [PubMed]
45. Oucif, H.; Miranda, J.M.; Trigo, M.; Iglesias, R.; Toro, J.; Al-Mehidi, S.; Barros-Velázquez, J.; Aubourg, S.P. Antimicrobial and antioxidant effects of alga *Cystoseira compressa* extract during the chilled storage of horse mackerel (*Trachurus trachurus*). In Proceedings of the 47th West European Fish Technologists Association Conference, Dublin, Ireland, 9–12 October 2017.
46. Miranda, J.M.; Ortiz, J.; Barros-Velázquez, J.; Aubourg, S.P. Quality Enhancement of Chilled Fish by Including Alga *Bifurcaria bifurcata* Extract in the Icing Medium. *Food Bioprocess Technol.* **2016**, *9*, 387–395. [CrossRef]
47. Medina, I.; Sacchi, R.; Aubourg, S.P. A 13C-NMR study of lipid alterations during fish canning: Effect of filling medium. *J. Sci. Food Agric.* **1995**, *69*, 445–450. [CrossRef]
48. Rodríguez, A.; Cruz, J.M.; Paseiro-Losada, P.; Aubourg, S.P. Effect of a polyphenol-vacuum packaging on lipid deterioration during an 18-month frozen storage of coho salmon (*Oncorhynchus kisutch*). *Food Bioprocess Technol.* **2012**, *5*, 2602–2611. [CrossRef]
49. Ortiz-Viedma, J.; Romero, N.; Puente, L.; Burgos, K.; Toro, M.; Ramirez, L.; Rodriguez, A.; J Barros-Velazquez, J.; Aubourg, S.P. Antioxidant and antimicrobial effects of stevia (*Stevia rebaudiana* Bert.) extracts during preservation of refrigerated salmon paste. *Eur. J. Lipid Sci. Technol.* **2017**, *119*, 1600467. [CrossRef]
50. Iglesias, J.; Pazos, M.; Andersen, M.L.; Skibsted, L.H.; Medina, I. Caffeic Acid as Antioxidant in Fish Muscle: Mechanism of Synergism with Endogenous Ascorbic Acid and α-Tocopherol. *Food Chem.* **2006**, *99*, 98–104. [CrossRef] [PubMed]
51. Iglesias-Neira, J.; Pazos, M.; Maestre, R.; Torres, J.L.; Medina, I. Galloylated Polyphenols as Inhibitors of Hemoglobin-Catalyzed Lipid Oxidation in Fish Muscle. *J. Agric. Food Chem.* **2011**, *59*, 5684–5691. [CrossRef] [PubMed]

antioxidants

Article

Green One-Pot Synthesis of Coumarin-Hydroxybenzohydrazide Hybrids and Their Antioxidant Potency

Marko R. Antonijević [1,2], Dušica M. Simijonović [1], Edina H. Avdović [1,*], Andrija Ćirić [2], Zorica D. Petrović [2], Jasmina Dimitrić Marković [3], Višnja Stepanić [4] and Zoran S. Marković [1,*]

1 Department of Science, Institute for Information Technologies, University of Kragujevac, Jovana Cvijića bb, 34000 Kragujevac, Serbia; mantonijevic@uni.kg.ac.rs (M.R.A.); dusicachem@kg.ac.rs (D.M.S.)
2 Department of Chemistry, Faculty of Science, University of Kragujevac, Radoja Domanovića 12, 34000 Kragujevac, Serbia; andrija.ciric@pmf.kg.ac.rs (A.Ć.); zorica.petrovic@pmf.kg.ac.rs (Z.D.P.)
3 Faculty of Physical Chemistry, University of Belgrade, Studentski trg 12-16, 11000 Belgrade, Serbia; markovich@ffh.bg.ac.rs
4 Ruđer Bošković Institute, Bijenička Cesta 54, 10000 Zagreb, Croatia; visnja.stepanic@irb.hr
* Correspondence: edina.avdovic@pmf.kg.ac.rs (E.H.A.); zmarkovic@uni.kg.ac.rs (Z.S.M.); Tel.: +381-34-610-01-95 (Z.S.M.)

Abstract: Compounds from the plant world that possess antioxidant abilities are of special importance for the food and pharmaceutical industry. Coumarins are a large, widely distributed group of natural compounds, usually found in plants, often with good antioxidant capacity. The coumarin-hydroxybenzohydrazide derivatives were synthesized using a green, one-pot protocol. This procedure includes the use of an environmentally benign mixture (vinegar and ethanol) as a catalyst and solvent, as well as very easy isolation of the desired products. The obtained compounds were structurally characterized by IR and NMR spectroscopy. The purity of all compounds was determined by HPLC and by elemental microanalysis. In addition, these compounds were evaluated for their in vitro antioxidant activity. Mechanisms of antioxidative activity were theoretically investigated by the density functional theory approach and the calculated values of various thermodynamic parameters, such as bond dissociation enthalpy, proton affinity, frontier molecular orbitals, and ionization potential. In silico calculations indicated that hydrogen atom transfer and sequential proton loss–electron transfer reaction mechanisms are probable, in non-polar and polar solvents respectively. Additionally, it was found that the single-electron transfer followed by proton transfer was not an operative mechanism in either solvent. The conducted tests indicate the excellent antioxidant activity, as well as the low potential toxicity, of the investigated compounds, which makes them good candidates for potential use in food chemistry.

Keywords: coumarins; green synthesis; antioxidants; DFT

Citation: Antonijević, M.R.; Simijonović, D.M.; Avdović, E.H.; Ćirić, A.; Petrović, Z.D.; Marković, J.D.; Stepanić, V.; Marković, Z.S. Green One-Pot Synthesis of Coumarin-Hydroxybenzohydrazide Hybrids and Their Antioxidant Potency. *Antioxidants* **2021**, *10*, 1106. https://doi.org/10.3390/antiox10071106

Academic Editors: Irene Dini and Domenico Montesano

Received: 7 June 2021
Accepted: 7 July 2021
Published: 10 July 2021

Publisher's Note: MDPI stays neutral with regard to jurisdictional claims in published maps and institutional affiliations.

1. Introduction

The oxygen molecule is an important electron acceptor in metabolic processes in the cells of living organisms. It plays an important role in the cell's respiratory processes, especially in a process of oxidative phosphorylation. This molecule is included in almost all electron transfer processes in organisms, and is one of the key components of the electron-transport chain, which has a crucial role in energy production [1]. However, besides its many positive effects, as a result of its bi-radical properties, it also enables the formation of partially reduced chemical species known as reactive oxygen species (ROS), which can be involved in starting a chain reaction that could be potentially dangerous for the cell [2]. The production of ROS in the organism is a natural process that allows the immune system to remove foreign bodies from blood, manages cell-signaling, leads to acceleration of the aging process, etc. ROS are constantly forming within the cells upon exposure to drugs, air pollutants, ultraviolet rays, ionizing radiation, smoke, and some endogenous metabolites of the redox and respiratory chain during the transfer of electrons. Under normal conditions, the organism is capable of regulating the production of ROS in such a manner that damage caused by influences of chemical agents on healthy cells is reduced to a minimum. The human body has developed a complex defense system against oxidative stress that includes preventive, restorative, antioxidant mechanisms, and physiological defense [3]. The natural system of antioxidant protection includes enzymatic and non-enzymatic antioxidants. Enzymatic antioxidants consist of a limited number of proteins such as catalase, glutathione peroxidase, superoxide dismutase, along with some auxiliary enzymes. They work by converting reactive species into hydrogen peroxide and later into water, in the presence of cofactors such as copper, zinc, manganese, and iron. Another group of antioxidants can be divided into direct and indirect, depending on the mechanism of action. Direct-acting antioxidants are extremely important in the body's defense against oxidative stress. Most of them are taken into the body through diet, and only a small number are synthesized in the body itself. These include vitamins, polyphenols, carotenoids, etc. Most of the naturally occurring antioxidants are found in the plant world, often in plants used in traditional medicine. Extraction, identification, and evaluation of the biological activity of these compounds has been a hot topic in the last couple of decades [4–6].

Despite the great efforts of many researchers over many years, creating new and potent antioxidant agents, which will also satisfy other criteria, like solubility and selective biological activity/toxicity, is quite challenging. Coumarins are relatively simple, naturally occurring phenolic compounds consisting of fused γ-pyrone and benzene rings, along with the carbonyl group on the pyrone ring at position C2. They are found in some essential oils like lavender oil, cinnamon bark oil, and cassia leaf oil. They mostly occur in higher plants, in the fruits, followed by the roots, stems, and leaves. The most frequent source for human exposure to coumarins is certain types of cinnamon. Coumarins have important biological functions, such as growth regulation, respiratory control, protection against herbivores and microorganisms. Simple coumarins act as hormones and signaling molecules. Over the last decade, antioxidant, anti-inflammatory, anticoagulant, enzyme inhibitory, antimicrobial, and anticancer activities of coumarins have been widely investigated. Some of them, like novobiocin and armillarisin A (antibiotics), warfarin, phenprocoumon, and acenocoumarol (anticoagulants), and hymecromone (antispasmodic and choleretic) have been in clinical use for many years [7–15].

Synthetic coumarins have great potential for developing new compounds with the aforementioned desired activities. It is worth pointing out that the reactions regarding 3-acetyl-4-hydroxycoumarin and various benzoyl hydrazides have been described in only a few articles [16–18]. The synthesis of coumarin-hydrazide derivatives was achieved via condensation of 3-acetyl-4-hydroxycoumarin with appropriate hydrazides under reflux in *n*-propanol or by heating in ethanol and in the presence of acetic acid as catalyst. Additionally, in these studies, the isolated products were evaluated for their antioxidant, anti-LOX, and anticancer activity. The investigated compounds exhibited good to moderate cytotoxicity and anti-inflammatory activity, while the radical scavenging activity was bad.

The aim of this research is the synthesis of new coumarin hydroxybenzohydrazide derivatives with expected good antioxidant activity. For this synthesis green, simple and low-cost protocol was used. The synthesized compounds are structurally characterized, and their in vitro antiradical activity is investigated. Since they can be potent antioxidants, a theoretical investigation of the structure and reaction mechanisms involved in free radical scavenging is performed. DFT approach allowed the computation of various thermodynamic properties, such as bond dissociation enthalpy (BDE), proton affinity (PA), frontier molecular orbitals and HOMO-LUMO gap, and ionization potential (IP), to predict the mechanisms of antioxidative action.

2. Materials and Methods

2.1. Chemical Reagents and Instruments

Chemicals (purity > 98%) used for the synthesis were acquired from Merck, and vinegar (acetic acid 90 g/L) was purchased from the local market. Compounds 3-acetyl-4-hydroxycoumarin (**1**) and hydrazides (**2**) were prepared according to the previously reported methods [19,20]. All chemicals used in HPLC analysis were HPLC grade.

IR spectroscopy was performed on a Perkin-Elmer Spectrum One FT-IR spectrometer using the KBr disc. The NMR spectra were recorded on a Varian Gemini spectrometer (200 MHz for ^{1}H and 50 MHz for ^{13}C) in DMSO-d_6. The UV-Vis measurements were recorded on Agilent Technologies, Cary 300 Series UV/Vis Spectrophotometer. Melting points were determined on a Mel-Temp capillary melting points apparatus, model 1001. Elemental microanalysis for carbon, hydrogen, and nitrogen was done in the Centre for Instrumental Analysis, at the Faculty of Chemistry, Belgrade. Elemental (C, H, N) analysis of the samples was carried out on an Elemental analysis system VARIO EL III CHNOS, model—Elementar Analysensysteme GmbH, 2003. To confirm the purity of isolated products, HPLC analyses were performed using an HPLC system (Shimadzu Prominence, Kyoto, Japan) with a PDA detector (SPD-M20A). The separation was carried out using a Hypersil Gold aQ C18 column (150 × 4.6 mm, 5 μm) at a flow rate of 1 mL/min. Mobile phases: (A) 0.1% formic acid in water, (B) 0.1% formic acid in acetonitrile. Gradient profile 0–2 min 2% B, 45 min 95% B.

2.2. Synthesis of Coumarin-Hydroxybenzohydrazide Derivatives

The starting compounds 3-acetyl-4-hydroxycoumarin (**1**) (1 mmoL) and benzoyl hydrazides (**2**) (1 mmoL), were dissolved in a mixture of vinegar and ethanol (1:1) (20 mL) and stirred at reflux for 5 h. Reaction progress was monitored using thin-layer chromatography (TLC). When the reaction was completed, the resulting mixture was cooled to room temperature and the precipitate was collected by filtration. All of the coumarin-hydroxybenzohydrazide products (**3a–e**) were characterized with melting points, ^{1}H NMR, ^{13}C NMR, and IR spectra, as well as with elemental and HPLC analysis.

(E)-N'-(1-(2,4-dioxochroman-3-ylidene)ethyl)-2-hydroxybenzohydrazide (**3a**). Beige solid, m.p. 248–249 °C; Isolated yield: 0.221 g (65.31%), HPLC purity: 91.44%; ^{1}H NMR (200 MHz, DMSO-d_6) δ: 15.95 (s, 1H), 8.00 (dd, *J* = 7.8, 1.6 Hz, 1H), 7.87 (dd, *J* = 7.8, 1.7 Hz, 1H), 7.67 (m, 1H), 7.53–7.40 (m, 1H), 7.40–7.25 (m, 2H), 7.07–6.90 (m, 2H), 2.74 (s, 3H); ^{13}C NMR (50 MHz, DMSO-d_6) δ: 178.8, 170.5, 164.3, 161.5, 157.5, 153.1, 134.3, 134.1, 129.9, 125.7, 123.9, 119.6, 119.5, 117.1, 116.4, 116.3, 95.4, 17.6; IR (KBr): 3521 (OH), 3435, 3267 (NH), 3049 (=CH), 2930, 2862 (CH), 1692, 1629, 1608 (C=O), 1547, 1455 (C=C), 1233 (C–O) cm^{-1}; $C_{18}H_{14}N_2O_5$ (FW = 338.32): C, 63.90; N, 8.28; H, 4.17%; found: C, 63.74; N, 8.45; H, 4.28%.

(E)-N'-(1-(2,4-dioxochroman-3-ylidene)ethyl)-4-hydroxybenzohydrazide (**3b**).White solid, m.p. 264–265 °C; Isolated yield: 0.215 g (63.51%), HPLC purity: 97.79%; ^{1}H NMR (200 MHz, DMSO-d_6) δ: 15.72 (s, 1H), 11.56 (s, 1H), 10.34 (s, 1H), 7.99 (dd, *J* = 7.8, 1.6 Hz, 1H), 7.90–7.78 (m, 2H), 7.73–7.57 (m, 1H), 7.39–7.23 (m, 2H), 6.96–6.85 (m, 2H), 2.74 (s, 3H); ^{13}C NMR (50 MHz, DMSO-d_6) δ: 179.2, 171.8, 164.6, 161.6, 161.5, 153.2, 134.3, 130.1, 125.7, 123.9, 121.6, 119.8, 116.4, 115.4, 95.3, 17.8; IR (KBr): 3565 (OH), 3435, 3183 (NH), 3057 (=CH), 2964, 2807 (CH), 1666, 1639, 1605 (C=O), 1566, 1490, 1467(C=C), 1232 (C–O) cm^{-1}; $C_{18}H_{14}N_2O_5$ (FW = 338.32): C, 63.90; N, 8.28; H, 4.17%; found: C, 63.68; N, 8.40; H, 4.37%.

(E)-N'-(1-(2,4-dioxochroman-3-ylidene)ethyl)-4-hydroxy-3-methoxybenzohydrazide (**3c**). Light yellow solid, m.p. 223–224 °C; Isolated yield: 0.246 g (66.83%), HPLC purity: 98.50%; ^{1}H NMR (200 MHz, DMSO-d_6) δ: 15.67 (s, 1H), 11.55 (s, 1H), 9.96 (s, 1H), 8.00 (dd, *J* = 7.8, 1.3 Hz, 1H), 7.95 (m, 1H), 7.67 (ddd, *J* = 8.1, 7.4, 1.7 Hz, 1H), 7.54–7.43 (m, 2H), 7.32 (ddd, *J* = 8.0, 5.3, 1.6 Hz, 2H), 6.92 (d, *J* = 11.3 Hz, 1H), 3.86 (s, 3H), 2.75 (s, 3H); ^{13}C NMR (50 MHz, DMSO-d_6) δ: 179.3, 175.9, 172.2, 164.6, 161.6, 153.2, 151.1, 147.5, 134.4, 125.8, 123.9, 122.0, 121.8, 119.8, 116.4, 115.3, 111.9, 95.4, 55.9, 17.9; IR (KBr): 3524 (OH), 3388, 3235 (NH), 3090 (=CH), 2944, 2782 (CH), 1675, 1605, 1591 (C=O), 1527, 1484, 1410 (C=C), 1221 (C–O) cm^{-1}; $C_{19}H_{16}N_2O_6$ (FW = 368.10): C, 61.96; N, 7.61; H, 4.38%; found: C, 61.62; N, 7.43; H, 4.17%.

(E)-N′-(1-(2,4-dioxochroman-3-ylidene)ethyl)-2,3-dihydroxybenzohydrazide (**3d**). Beige solid, m.p. 245–247 °C; Isolated yield: 0.223 g (62.85%), HPLC purity: 99.10%; ^{1}H NMR (200 MHz, DMSO-d_6) δ: 15.88 (s, 1H), 8.00 (dd, J = 7.8, 1.4 Hz, 1H), 7.76–7.58 (m, 1H), 7.47–7.20 (m, 3H), 7.02 (dd, J = 7.8, 1.6 Hz, 1H), 6.80 (t, J = 7.9 Hz, 1H), 2.73 (s, 3H); ^{13}C NMR (50 MHz, DMSO-d_6) δ: 178.9, 171.0, 165.5, 161.5, 153.2, 147.2, 146.2, 134.4, 125.7, 124.0, 119.6, 119.4, 119.1, 119.1, 116.4, 116.2, 95.5, 17.6; IR (KBr): 3498 (OH), 3386, 3269 (NH), 3050 (=CH), 2964, 2863 (CH), 1703, 1659, 1609 (C=O), 1583, 1524, 1484, 1466 (C=C), 1222 (C–O) cm^{-1}; $C_{18}H_{14}N_2O_6$ (FW = 354.32): C, 61.02; N, 7.91; H, 3.98%; found: C, 61.22; N, 7.79; H, 3.67%.

(E)-N′-(1-(2,4-dioxochroman-3-ylidene)ethyl)-3,4-dihydroxybenzohydrazide (**3e**). Beige solid, m.p. 255–257 °C; Isolated yield: 0.233 g (65.71%), HPLC purity: 97.46%; ^{1}H NMR (200 MHz, DMSO-d_6) δ: 15.71 (s, 1H), 11.51 (s, 1H), 9.86 (s, 1H), 9.45 (s, 1H), 7.99 (dd, J = 7.9, 1.5 Hz, 1H), 7.66 (ddd, J = 8.1, 7.4, 1.7 Hz, 1H), 7.44–7.14 (m, 4H), 6.87 (d, J = 8.1 Hz, 1H), 2.73 (s, 3H); ^{13}C NMR (50 MHz, DMSO-d_6) δ: 179.2, 171.6, 164.7, 161.6, 153.2, 150.1, 145.3, 134.3, 125.7, 123.9, 121.9, 120.3, 119.8, 116.4, 115.4, 115.4, 95.3, 17.8; IR (KBr): 3478 (OH), 3419, 3156 (NH), 3066 (=CH), 2973 (CH), 1668, 1641, 1602 (C=O), 1568, 1489, 1457 (C=C), 1222 (C–O) cm^{-1}. $C_{18}H_{14}N_2O_6$ (FW = 354.32): C, 61.02; N, 7.91; H, 3.98%; found: C, 61.37; N, 7.73; H, 3.69%.

2.3. DPPH Radical Scavenging Assay

The free radical scavenging activities of obtained products were determined by 2,2-diphenyl-1-picrylhydrazyl (DPPH) assay [21]. In detail, the tested compounds (20 μL of different concentrations dissolved in DMSO and 980 mL of methanol) were mixed with an equal volume of the solution of DPPH in methanol (0.05 mM). The prepared samples were shaken well and left at room temperature in the dark for 20 and 60 min. After the incubation period absorbance was determined at 517 nm by using the methanol as a blank control. All tests were run in triplicate and averaged. Nordihydroguaiaretic acid (NDGA) and quercetin were used as positive controls. For products that exert good activity, IC_{50} values were determined. Detailed calculations of IC_{50} value for the most active compound is given in the Supplementary Material. The results are presented as mean ± SD. For calculation of the stoichiometric factor [22,23], the following equation was used:

$$stoichiometric\ factor = \frac{[DPPH]_0}{(2 \times IC_{50})}$$

2.4. Computational Methods

The calculations were carried out by employing the *Gaussian09* software package [24]. For optimization of neutral molecules, the corresponding radicals, anions, and radical cations quantum chemical calculations based on density functional theory (DFT) were used. Since it has been shown that the M062-X method well describes short-range and medium-range intramolecular and intermolecular interactions (<500 pm) [25], the M06-2X method is applied in combination with 6–311++G(d,p) basis set. This theoretical model is suitable for thermodynamic and kinetic investigation of reaction mechanisms of examined compounds with free radicals [22,26]. All investigated structures were reoptimized in methanol (ε = 32.61), and benzene (ε = 2.27). The solvent effect was included by the Conductor-like Polarizable Continuum Model (CPCM) [27], without any geometrical constraints. Solvents were selected to simulate the polar and non-polar environments, as well as conditions of experimental measurements.

Three antioxidant mechanisms were selected for the evaluation of the antioxidant activity: Hydrogen Atom Transfer (HAT), Single-Electron Transfer followed by Proton Transfer (SET-PT), and Sequential Proton Loss-Electron Transfer (SPLET) [28–31].

The abstraction of hydrogen atoms can be described by two mechanisms, HAT and Proton-Coupled Electron Transfer (PCET) [29]. The HAT mechanism is defined as the simultaneous transfer of an electron and a proton between the donor and acceptor. On the other hand, the PCET mechanism involves a significant redistribution of molecular charge, during the transfer of hydrogen atoms. In this study, only the HAT mechanism was considered because only thermodynamic parameters were used to test antioxidant capacity. It should be noted that both reaction pathways are necessary for the kinetic examination of the mechanisms of antioxidant activity. In the HAT mechanism, the homolytic cleavage of the O-H bond occurs leading to the separation between radical species (Ar-O$^\bullet$) and a hydrogen atom (Equation (1)) [30,31].

$$\text{Ar-OH} \rightarrow \text{Ar-O}^\bullet + \text{H}^\bullet \quad (1)$$

The SET-PT mechanism is a two-step process that includes electron loss from a molecule and generation of the radical cation (Ar–OH$^{+\bullet}$), which in the second step loses protons (Equation (2a,b)).

$$\text{Ar-OH} \rightarrow \text{Ar-OH}^{\bullet+} + e^- \quad (2a)$$

$$\text{Ar-OH}^{\bullet+} \rightarrow \text{Ar-O}^\bullet + \text{H}^+ \quad (2b)$$

The SPLET mechanism is also a two-step mechanism. In the first step, an anion (Ar–O$^-$) is formed from the antioxidant molecule, while the corresponding radical is formed after the electron transfer in the second step of this mechanism (Equation (3a,b)).

$$\text{Ar-OH} \rightarrow \text{Ar-O}^- + \text{H}^+ \quad (3a)$$

$$\text{Ar-O}^- \rightarrow \text{Ar-O}^\bullet + e^- \quad (3b)$$

The thermodynamic parameters governing the mentioned antioxidative mechanisms are BDE (Bond Dissociation Enthalpy) describing HAT, PA (Proton Affinity), and ETE (Electron Transfer Enthalpy) describing SPLET, and IP (Ionization Potential) and PDE (Proton Dissociation Enthalpy) describing SET-PT mechanism. The contribution of these parameters was calculated from the enthalpies of the optimized chemical species using the following equations [32,33]:

$$\text{BDE} = H(\text{Ar-O}^\bullet) + H(\text{H}^\bullet) - H(\text{Ar-OH}) \quad (4)$$

$$\text{IP} = H(\text{Ar-OH}^{\bullet+}) + H(e^-) - H(\text{Ar-OH}) \quad (5a)$$

$$\text{PDE} = H(\text{Ar-O}^\bullet) + H(\text{H}^+) - H(\text{Ar-OH}^{\bullet+}) \quad (5b)$$

$$\text{PA} = H(\text{Ar-O}^-) + H(\text{H}^+) - H(\text{Ar-OH}) \quad (6a)$$

$$\text{ETE} = H(\text{Ar-O}^\bullet) + H(e^-) - H(\text{Ar-O}^-) \quad (6b)$$

All reaction enthalpies used in the equations were calculated at 298 K. The calculated enthalpy values of the solvated proton, $H(\text{H}^+)$ and electron, $H(e^-)$, for the M062-X method in benzene (−856.9 and −11.9 kJ moL^{-1}) and methanol (−1010.5 and −93.5 kJ moL^{-1}) were taken from the literature [28,34].

The free radical scavenging potency of the synthesized coumarin derivatives toward the DPPH radical was investigated in two different solvents: benzene and methanol. Mechanisms of radical scavenging activity are described by Equations (1s)–(5s), given in the Supplementary Material. Thermodynamic parameters of the investigated mechanisms were calculated according to Equations (6s)–(10s) from Supplementary material.

3. Results and Discussion

3.1. Synthesis

The synthesis of coumarin-hydroxybenzohydrazide derivatives (**3**) was achieved by treatment of 3-acetyl-4-hydroxycoumarin (**1**) with corresponding benzoyl hydrazides (**2a–e**), Scheme 1. The reaction medium for these reactions was a mixture of vinegar and ethanol (1:1). This medium provided good solubility of the reactants and acted as a catalyst. The coumarin-hydroxybenzohydrazide derivatives **3a–e** were obtained in moderate yields under reflux for 5 h. All products **3a–e** were precipitated during the reaction and isolated by filtration. Three of five obtained products (**3c–e**) are reported in this study for the first time.

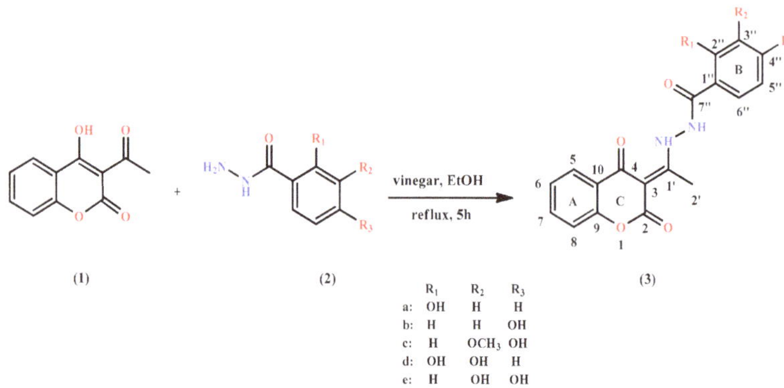

Scheme 1. Synthesis of coumarin-hydroxybenzohydrazide derivatives.

The synthesized coumarin-hydroxybenzohydrazide derivatives (**3**) were characterized using IR, ^{1}H and ^{13}C NMR spectroscopy, as well as with melting points. The purity of all isolated compounds was determined by HPLC and elemental analysis. The coumarin-hydroxybenzohydrazide derivatives (**3**) were obtained in yields 62.85–66.83% and with HPLC purity in the range 91.44–99.10%, Table 1. HPLC data and chromatograms can be found in Supplementary Material, Tables S1–S5 and Figures S11–S15, respectively.

Table 1. Isolated yield and HPLC purity of synthesized coumarin-hydroxybenzohydrazide derivatives **3**.

Entry	Compound	Isolated Yield (%)	HPLC Purity (%)
1	3a	65.31	91.44
2	3b	63.51	97.79
3	3c	66.83	98.50
4	3d	62.85	99.10
5	3e	65.71	97.46

The 2,4-dione tautomeric form of these derivatives was assumed on the basis of the obtained spectroscopy data and X-ray structure which was determined for a similar type of compounds in our previous studies [35–37].

All synthesized compounds were characterized by IR spectroscopy. The obtained spectra are very similar to each other. In all spectra, bands of about 3500 cm^{-1} belonging to phenolic OH groups are observed, while bands of NH groups appear at lower wavenumbers, in the range of 3200–3400 cm^{-1}. Bands in the region 1700–1600 cm^{-1} are assigned to different carbonyl groups. The stretching vibrations corresponding to the C-O group were identified between 1220 and 1230 cm^{-1}.

The NMR spectra of coumarin-hydroxybenzohydrazide derivatives presented in the Supplementary Material (Figures S1–S10) are in good agreement with the proposed structures. In the ^{1}H NMR spectra of all obtained compounds (**3a–e**) characteristic singlets positioned at 2.73–2.75 ppm are assigned to the protons of the methyl group in position C2′. In addition, the broad singlet appearing in the 9.45–10.34 ppm range is assigned to phenolic protons, and the singlets positioned at a higher chemical shift, between 11.32–11.56 and 15.67–15.95 ppm, are assigned to two NH groups. The NMR spectra clearly show multiple resonance maxima of aromatic protons of the 2,4-dioxochroman and benzoyl moieties in the range of 6.80–8.00 ppm. In addition, the singlet at 3.86 ppm in the spectrum of compound **3c** is assigned to the proton of the methoxy group attached to the aromatic ring of the hydrazide moiety.

In the ^{13}C NMR spectra of compound **3a–e**, the peak at about 18 ppm was assigned to the carbon atom of the methyl group in position C2′. The peak at about 95 ppm is attributed to the C3 atom. The aromatic carbons of both parts have resonance maxima in the 112–153 ppm region. The resonance maxima with the highest chemical shifts, at about 179 ppm, originate from carbon atoms of the C4-carbonyl group, while the signals of the amide and lactone carbonyl groups occur at lower chemical shifts, at about 161–165 ppm. The C1 atom showed a resonance in the range of 171–176 ppm.

The predicted mechanism of synthesis of new coumarin-hydrazide hybrids is presented in Scheme 2. The first step of the reaction is the nucleophilic attack of the hydrazide on the carbon of the carbonyl group of coumarin (**I**), whereby intermediate (**II**) is formed. In the next step, the proton from the NH_2 group is transferred to a partially negative oxygen atom of the C=O group, which is followed by the formation of a hemiaminal (**III**). In the further course of the reaction, the protonation of the -OH group takes place in the presence of acetic acid, which is followed by releasing of the water molecules (**IV**). Deprotonation of the iminium ion (**V**) leads to the formation of the enol form of the coumarin-hydroxybenzohydrazide hybrids (**VI**). In the last step, by acid-catalyzed keto-enol tautomerization, a more stable keto form (**VII**) is obtained. The confirmation of this reaction path is supported by the fact that ^{1}H NMR spectra for all obtained compounds showed signals at about 15 ppm, which originate from the proton-NH group [35–37].

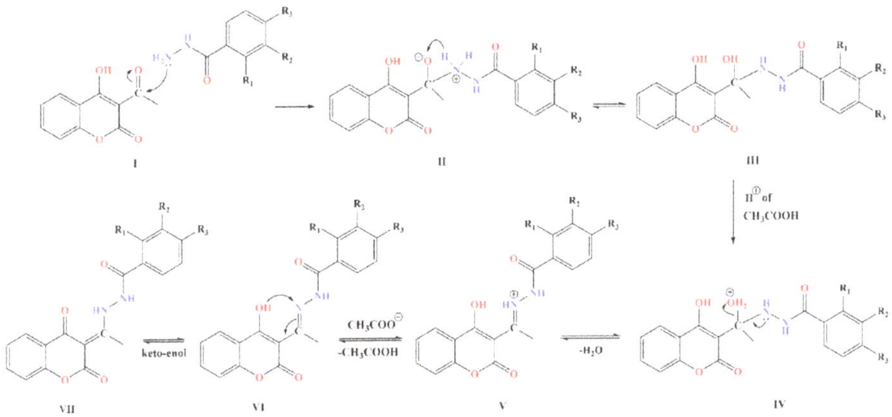

Scheme 2. The suggested mechanism of the formation of coumarin-hydroxybenzohydrazide hybrids by the reaction between 3-acetyl-4-hydroxycoumarin and benzoyl hydrazides.

3.2. *In Vitro Antioxidative Activity of Tested Compounds*

The results of the in vitro DPPH test are presented in Table 2. The obtained results show that two out of five investigated compounds, **3d**, and **3e**, expressed high antioxidant activity, with IC_{50} values of 2.9 and 12.9 μM, respectively. It should be mentioned that the compound **3d** reduced DPPH well and exhibited high activity, slightly lower than the reference compounds NDGA and quercetin. The compound **3e** also showed high antioxidative activity, a little bit lower than compound **3d**. Namely, based on the obtained results, the new compounds **3d** and **3e** can be considered good antioxidants. Additionally, the stoichiometric factor (SF) of **3d**, which is equal to 4.4, indicates very good antioxidative ability, taking into account the fact that significant radical scavengers have SF bigger than 2 [22,23,38,39].

Table 2. In vitro DPPH antioxidant activity of products **3** and reference compounds (%, IC_{50} and SF values).

Compound	DPPH Scavenging Ability (%)										
	100 µM		50 µM		25 µM		* 150 µM	* 10 µM	* 2.5 µM	IC_{50} (µM)	SF
	20 Min	60 Min	20 Min	60 Min	20 Min	60 Min	20 Min	20 Min	20 Min		
3a	2.9 ± 0.7	8.7 ± 0.7	0.6 ± 0.8	1.2 ± 0.7	0.6 ± 0.8	1.2 ± 0.7	nd	nd	nd	nd	nd
3b	4.5 ± 1.8	15.8 ± 0.2	0.1 ± 1.0	4.7 ± 1.7	0.1 ± 1.0	4.7 ± 1.7	nd	nd	nd	nd	nd
3c	31.5 ± 0.4	41.5 ± 0.6	20.4 ± 1.2	26.7 ± 0.9	20.4 ± 1.2	26.7 ± 0.9	56.4 ± 1.0	nd	nd	142.2 ± 2.7	nd
3d	94.9 ± 0.6	94.3 ± 0.9	92.9 ± 0.1	92.5 ± 0.4	92.9 ± 0.1	92.5 ± 0.4	nd	nd	42.9 ± 1.3	2.9 ± 0.1	4.4
3e	66.2 ± 1.8	75.3 ± 0.4	63.1 ± 1.1	73.6 ± 0.9	63.1 ± 1.1	73.6 ± 0.9	nd	47.7 ± 0.3	32.5 ± 0.7	12.9 ± 0.4	1.0
NDGA	94.5 ± 0.2	94.1 ± 0.7	94.6 ± 0.7	94.6 ± 0.6	94.6 ± 0.7	94.6 ± 0.6	nd	nd	59.9 ± 0.1	1.7 ± 0.1	7.4
Quercetin	95.1 ± 0.9	95.4 ± 0.8	95.3 ± 0.8	95.1 ± 0.9	95.3 ± 0.8	95.1 ± 0.9	nd	nd	60.9 ± 7.2	1.9 ± 0.1	6.6

* Relevant concentrations used in DPPH assay for determination of IC_{50} values; nd—Not determined.

The obtained results for compounds **3a–c** revealed that the presence of only one phenolic OH group had no significant effect on the DPPH radical scavenging activity. However, for compounds **3d** and **3e**, with additional phenolic OH groups in neighboring positions, a significant improvement of antioxidative activity was observed. This is in agreement with the fact that the pronounced antioxidative capacity of compounds with catechol moiety stems from resonance and electron-donating effects of these groups, which increase the stability of the formed phenoxy radical. In these compounds, the hydroxy group supports the homolytic cleavage of the neighboring O–H bond and enables the formation of a hydrogen bond with formed phenoxy radical [40–42].

In addition, appropriate in silico methods were used to determine the most likely mechanism of radical scavenging activity of coumarin-hydroxybenzohydrazide derivatives (**3**).

3.3. *Thermodynamic Parameters of Antioxidative Activity*

Geometry optimization of the examined compounds was performed using the M062X/6-311++G(d,p) model. The most stable conformations of **3a–e** are shown in Figure S16. The obtained values for bond lengths, bond angles, and dihedral angles are shown in Tables S8–S10. To confirm that the obtained geometries were the minimums on the potential energy surface, the vibration frequencies were calculated at the same theoretical level. This means that the corresponding stationary points were without negative eigenvalues in the force constant matrix. From the optimized molecular structure of the five coumarin derivatives, it can be seen that the N-N bond is a partial double bond that exists in all compounds and provides delocalization of the π-electron between the coumarin base and the aromatic ring (B), which contributes to the stabilization of the formed radicals after H-abstraction. Compared to all phenol-hydroxy and amino groups on all considered molecules **3a–e**, the bond length C1′N-H is noticeably longer than C7″N-H. This may be associated with a ketone group at the C4 position that forms an HB between C1′NH and O4, resulting in an extension of the bond length of the C1′N-H bond. Alongside this hydrogen bond, compounds **3c–e** are capable of establishing additional hydrogen bonds, due to the fact that they possess additional OH groups in different positions. The additional hydrogen bond in **3c** is located between the C4″-OH and C3″-OCH_3 groups, while there are three HB in **3d**, from which two are formed between C2″-OH, the adjacent carbonyl group and C3″-OH group. There are only two HBs in compound **3e**. Besides C1′NH-O4, an HB is formed in the catechol moiety with adjacent C3″OH-O4″. The number and position of the hydrogen bonds play a major role in the stabilization of the molecule and consequently have an impact on its antioxidative capacity [43]. Careful analysis of the structures of the considered compounds showed that benzopyrone and phenyl rings were not in the planar conformation. The calculated dihedral angles C-N-N-C were found in the interval between 74.71°–76.51° for all scrutinized compounds. These values clearly show that the examined compounds lose coplanarity to varying degrees. These results indicate the weakening of the electronic distribution within the compounds, which significantly affects their antioxidant activity.

3.3.1. Bond Dissociation Energy

The BDE values for OH and NH groups are known to be of particular importance for understanding the mechanism of free radical scavenging activity [44,45]. This means that a weak O-H or N-H bond has a rapid response and thus potential for antioxidative activity. The capacity of coumarin-hydroxybenzohydrazide derivatives to remove free radicals is generally related to the existence of OH and NH groups in a particular position on the core of coumarin-hydroxybenzohydrazide derivatives. Optimization of the geometry of radical species formed after homolytic cleavage of either OH or NH bonds for all coumarin-hydroxybenzohydrazide derivatives is performed starting from the optimized geometry of neutral molecules. Removal of H-atoms from the C2″-OH, C3″-OH, C4″-OH, and C7″-NH positions in coumarin-hydroxybenzohydrazide derivatives yields various radical species.

For example, the radical species created by removing the H-atom from the C2″-OH group of **3a** is called **3a**C2″O•. The remaining radical species are generated and named in the same way. It should be noted that the geometry optimization of free radical species was performed using the same theoretical model and in the same solvents as for the parent molecules.

The calculated BDE values for all coumarin-hydroxybenzohydrazide derivatives are shown in Table 3. From the calculated values, it can be seen that the smallest BDE values are obtained for the C7″-NH position, in both solvents, for all investigated molecules, except in the case of compound **3e**, where slightly lower BDE values are obtained for the OH group in the para position. Such BDE values for **3e** are a consequence of catechol structure and the formation of intramolecular hydrogen bonds between both, adjacent OH groups and the resulting radicals with the adjacent OH group. In addition, the BDE values of the neighboring OH groups in **3d** are somewhat higher, despite the fact that the groups are in *o*- and *m*-positions.

Table 3. Thermodynamic parameters of antioxidant mechanism for investigated coumarin-hydroxybenzohydrazide derivatives in kJ moL^{-1}.

Position	BENZENE *HAT* BDE	*SET-PT* IP	PDE	*SPLET* PA	ETE	METHANOL *HAT* BDE	*SET-PT* IP	PDE	*SPLET* PA	ETE
					3a					
C2″-OH	404	714	123	446	391	373	552	19	191	380
C7″-NH	383		102	413	402	362		7	161	398
					3b					
C4″-OH	382	688	127	407	408	381	551	28	185	394
C7″-NH	348		94	372	409	355		3	156	398
					3c					
C4″-OH	377	693	118	422	388	377	528	47	200	375
C7″-NH	348		88	376	405	355		25	158	395
					3d					
C2″-OH	372		99	420	385	362		41	191	369
C3″-OH	374	706	101	430	376	365	519	44	200	363
C7″-NH	348		75	343	438	355		34	133	420
					3e					
C3″-OH	342		72	393	382	346		9	178	367
C4″-OH	344	703	74	380	397	348	535	11	168	378
C7″-NH	346		77	370	410	355		19	154	399

This is a consequence of better delocalization of the unpaired electron in the **3e**C4″O• than in the **3d**C2″O• radical (Figure 1). However, it should be mentioned that BDE values for the position C7″-NH, calculated in benzene, are a little bit lower than those calculated in methanol are.

3.3.2. Proton Affinity

The heterolytic cleavage of N-H and O-H bonds leads to the formation of the corresponding anions. This process is considered to be the first step of the SPLET mechanistic pathway and is described by the thermodynamic parameter PA. The high calculated PA values indicate that the SPLET mechanism is not operative in non-polar conditions. On the other hand, PA values obtained in methanol were significantly lower, due to increased anion stabilization in polar solvents. This is a consequence of the fact that polar protic solvents stabilize anions via hydrogen bonding. The obtained results are presented in Table 3. By analyzing the PA values, it is clear that the abstraction of the proton from nitrogen is much easier than from an oxygen atom. The lowest PA value, found for **3d**, implies that this compound is the most reactive when operating following the SPLET mechanism. Such results are in good agreement with the results obtained in vitro.

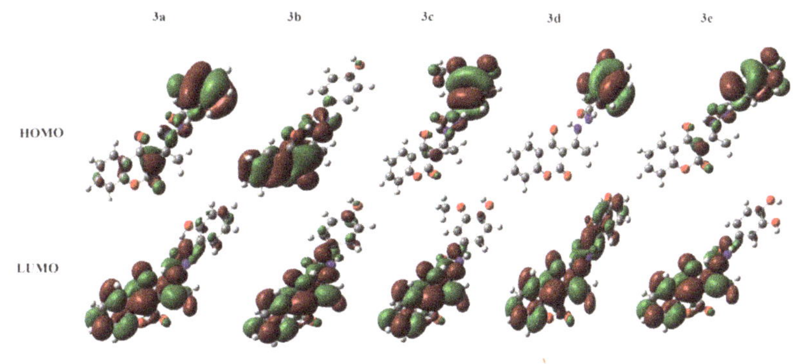

Figure 1. Frontier orbitals of the investigated compounds.

3.3.3. Frontier Molecular Orbitals and HOMO–LUMO Gap

Investigating the energies of the frontier molecular orbitals is very important for understanding the chemical behavior of tested compounds, because these orbitals usually take part in reactions. The ability of molecules with respect to electron donation or electron affinity can be determined using HOMO or LUMO energy values. The highest occupied molecular orbital (HOMO) usually indicates that the molecule is a good electron donor, and characterizes the susceptibility of the molecule to attack by electrophiles. The energy of the lowest unoccupied molecular orbital (LUMO) is related to electron affinity and describes the sensitivity of molecules toward nucleophile attack. The HOMO–LUMO gap (H-L), calculated as the energy difference between the HOMO and LUMO orbitals, is an important index of stability and chemical reactivity. A high value of H-L implies high stability of molecules and lower reactivity in chemical reactions, while a low value of H-L implies low stability and higher molecular reactivity [22,33].

The frontier orbitals for the investigated compounds in methanol are shown in Figure 1, and the frontier orbital energies, as well as H-L, are indicated in Table 4. The active site can be visually demonstrated by the distribution of the frontier orbital. Based on the results presented in Figure 1, it is obvious that the HOMO of almost all investigated compounds is mainly distributed in the aromatic ring of benzoyl-hydrazide, while LUMO is mainly assigned in the coumarin part of the molecule. The electron donation ability of a molecule can be determined by HOMO values; HOMO with high energy corresponds to a strong electron donation ability. From Table 4, it can be seen that **3c** possesses the highest HOMO among all compounds, while **3a** possesses the lowest one. The H-L of **3d**, in methanol, is 6.39 eV, while for the least reactive **3b** it is 6.74 eV. Values for the remaining compounds lie in the range of 0.35 eV. It should be noted that almost identical values are obtained for the H-L energy in the nonpolar solvent. The lower the value of H-L, the more pronounced the antioxidant abilities of the molecules. The relatively low energy of H-L indicates that the scrutinized compounds could be a highly reactive system. The obtained values for the H-L are in accordance with the in vitro results obtained by the DPPH test (Table 2).

Table 4. The frontier orbital energy and energy gap (eV) of the investigated compounds.

Ligand	METHANOL			BENZENE		
	HOMO	LUMO	HL Gap (eV)	HOMO	LUMO	HL Gap (eV)
3a	−8.09	−1.41	6.68	−8.10	−1.37	6.72
3b	−8.09	−1.36	6.74	−8.01	−1.26	6.75
3c	−7.80	−1.35	6.45	−7.85	−1.26	6.59
3d	−7.81	−1.42	6.39	−7.87	−1.41	6.46
3e	−7.86	−1.36	6.50	−7.94	−1.28	6.66

When the HOMO orbital is localized on the B ring (phenol/catechol part of the molecule), as is the case for the compounds **3e**, **3d**, and **3c**, values of the H-L, as well as the E_{HOMO}, are higher than in the case of the compounds **3a** and **3b**. Therefore, compounds **3e**, **3d**, and **3c** are more reactive than compounds **3a** and **3b**. The difference in HOMO distribution between **3a** and **3b** is a consequence of HB forming between OH group in position C2″-OH and carbonyl group in position C7″. Besides the position of the OH group and its ability to form HB, several electron-rich substituents on the B ring play an important part in HOMO and LUMO distribution, as well as the values of the H-L. As seen from Figure 1, HOMO tends to be localized on the part of the molecule with electron-rich substituents, because of the increased electron delocalization. This is why **3c**, **3d**, and **3e** have HOMO distributed only over the B ring, which makes the HOMO orbital less stable, thus making the H-L smaller. Expectedly, compounds with a smaller HL-gap are more reactive towards free radical species.

The data presented in Table 4 suggest that compound **3d** has the highest antioxidative potential since it has the lowest value of H-L.

3.3.4. Ionization Potential

Removal of free radicals can also be achieved by donating one electron from the parent compound to the free radical, which is followed by the formation of a radical cation. The ability of the investigated compound to donate an electron is associated with prolonged electron delocalization across the entire molecule. It is known that many natural products such as flavonoids, which have a high degree of π-delocalization, are more active as radical scavengers. The value that measures this ability is IP. The calculated IP values for the investigated compounds in methanol and benzene are given in Table 3.

Generally speaking, molecules with lower IP values are more easily engaged in radical-scavenging reactions [46,47]. All investigated compounds show mutually similar and high values for this thermodynamic parameter (Table 3). Based on the presented values for IP, it can be concluded that this mechanistic pathway is unlikely, and should be excluded from further discussion.

3.4. *Theoretical Assessment of Mechanisms of Antioxidative Action*

Based on the values of BDE, IP, PDE, PA, and ETE, the dominant mechanism of antioxidant action of coumarin-hydroxybenzohydrazide derivatives **3a–e** can be assumed. The thermodynamically most preferable mechanism is the one with the lowest value for the parameter describing the first step. All thermodynamic parameters for the examined coumarin-hydroxybenzohydrazide derivatives **3a–e** were calculated by applying the same level of theory in benzene and methanol as solvents (Table 3). The obtained values of thermodynamic parameters indicate that the good antioxidative activity of these compounds can be expected. It was found that PA values in methanol were considerably lower than the corresponding BDEs. This implies that SPLET is the predominant mechanism of antioxidant action in a polar environment. The obtained PA values suggest that compound **3d** shows the highest antioxidative capacity, which is in good agreement with IC_{50} values (Table 2). On the other hand, the values of BDE and PA (Table 3) indicate the competition of HAT and SPLET mechanisms in a non-polar medium.

3.5. *Radicals and Anions of Investigated Compounds*

The potential antioxidative action of the investigated compounds requires homolytic and heterolytic bond cleavage between the hydrogen and heteroatom. This leads to the formation of corresponding radicals and anions. The relative stability of these chemical species depends on their ability for the delocalization of unpaired electrons or charges. Depending on their relative stability, it can be determined which functional group or position is the most likely to contribute to the antioxidant activity, i.e., to the ROS inactivation. According to the thermodynamic parameters presented in Table 3, radical species derived from NH groups are generally more stable than those formed by homolytic cleavage of the OH bond. This behavior can be explained by examining the radical structures of the investigated compounds presented in Figure 2. Radicals formed by removing hydrogen atoms from the nitrogen of the hydrazide group are more stable. The main reason for this lies in the fact that nitrogen atom rehybridizes from sp^3 to sp^2 hybrid, which causes the investigated compounds to become planar.

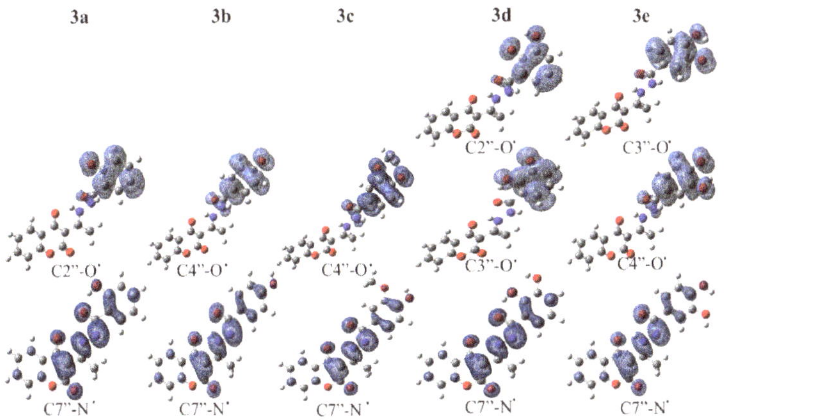

Figure 2. Spin density maps of the radicals obtained in the radical scavenging reactions of investigated compounds.

The planar conformation enables the delocalization of the spin over the whole molecule (Figure 2), and better delocalization leads to the higher radical stability of all investigated compounds. On the other hand, in cases where the antioxidant action is achieved through the OH group, the spin is delocalized only on the B ring of the molecule.

The maps of electrostatic potentials (ESP) of the investigated anions show that in the planar conformation, negative charge is better delocalized than in the non-planar one (Figure S18). As the atomic charges of ESP are important for understanding the second step of the SPLET mechanism, ESP maps of all coumarin hybrids were calculated to verify high electron density regions that can indicate the atoms responsible for the next step and, consequently, the electron transfer. Figure S17 shows the ESP maps of the tested compounds, while Figure S18 shows ESP maps of the corresponding anions.

It can be seen, from Figure S18, that a localized homogeneous negative surface (represented in red) is on the atom from which the single proton was abstracted. The best delocalization effect was found for the nitrogen atom anion in position C7″-N$^-$, which delocalizes its negative charge through the oxygen atom on the carbonyl group and the corresponding oxygen atoms on the B ring. Delocalization of this type of negative charge is presented in Figure 3.

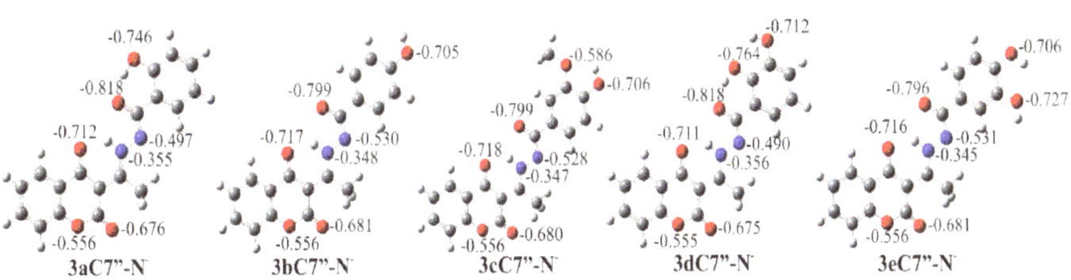

Figure 3. The NBO charge delocalization for anions obtained by proton abstraction from -NH group in position C7″-N for investigated compounds.

3.6. Radical Scavenging Activity against DPPH Radical

The antioxidative capacity of the investigated coumarin-hydroxybenzohydrazide derivatives was examined by the DPPH test, and the results are presented in Table 2. Since the IC_{50} values represent the sample concentration required to inhibit 50% of DPPH radicals, thermodynamic calculation on possible inactivation of DPPH radical through already examined free radical scavenging mechanisms was performed. The potential mechanisms of free radical scavenging were discussed based on a change in enthalpies for reactions following HAT, SET-PT and SPLET mechanisms. The calculations were performed in methanol, simulating the environment in which the in vitro DPPH test was performed. The energy difference between the products and reactants of the reaction was used as the principal to assess whether the reaction can be expected to occur or not. The lowest value of thermodynamic parameters implies the preferability of the mechanistic pathway.

Analysis of the results presented in Table 5 exposes positive ΔH_{IP} values for reaction with all investigated compounds. This means that all the examined reactions of scavenging of DPPH radical via SET-PT are endergonic. The high values of ΔH_{IP} unequivocally indicate that the SET-PT mechanism can be neglected when scavenging of DPPH radical is discussed, which is consistent with the thermodynamic results presented in Table 3. Furthermore, the HAT and SPLET mechanisms will be considered possible mechanisms of scavenging the DPPH radical. Both HAT and SPLET mechanisms can be discussed as potential mechanisms of radical scavenging in methanol, since low positive values (<50 kJ/moL) are obtained. The obtained values of ΔH_{BDE} and ΔH_{PA} for all examined compounds are similar and indicate endogenic reactions, except in the case of the radical, e.g., the anion in the C7″-NH position. Namely, in the C7″-NH position, negative values of ΔH_{PA} are obtained, implying an exergonic reaction. Based on this, it can be expected that DPPH radical in methanol will probably be scavenged following the SPLET mechanistic pathway. This hypothesis should be proven or annulled with the appropriate values of activation energies.

Table 5. Calculated parameters of radical scavenging activity for coumarin hybrids in kJ moL^{-1}.

METHANOL	*HAT* ΔH_{BDE}	*SET-PT* ΔH_{IP}	ΔH_{PDE}	*SPLET* ΔH_{PA}	ΔH_{ETE}
		3a			
C2″-OH	47	201	−142	50	7
C7″-NH	27		−163	2	34
		3b			
C4″-OH	59	184	−134	23	28
C7″-NH	38		−159	−6	29
		3c			
C4″-OH	50	161	−115	38	3
C7″-NH	24		−137	−4	25
		3d			
C3″-OH	34		−118	38	−4
C2″-OH	31	152	−121	29	0.1
C7″-NH	24		−128	−29	53
		3e			
C4″-OH	17		−151	6	9
C3″-OH	15	168	−152	16	−1.8
C7″-NH	25		−143	−8	32

The obtained experimental and theoretical results (Tables 2 and 5) suggest that **3d** is more reactive than **3e** against the DPPH radical. At the same time, IC_{50} values for **3a**, **3b**, and **3c** suggest that their activity is almost non-existent, while ΔH_{PA} values from Table 5 do not infer a significant difference between these three compounds and compound **3e**. This can be explained by the fact that compound **3e**, in comparison to compounds **3a**, **3b**, and **3c**, has one OH group more, which allows a better chance of reacting with the free radical species. The results presented in Table 5 suggest that the mechanism of radical scavenging against DPPH radical is SPLET, following the experimental results.

3.7. Potential Toxicology

To find whether compounds were suitable to be used as antioxidants in food industry, potential toxicity was investigated. For this purpose, the Prediction of Toxicity of Chemicals (ProTox-II) webserver protocol was used [48]. The results of the potential toxicity are presented in Table 6. To closely determine overall toxicity, investigated compounds are divided into six toxicity classes by the median lethal dose (LD_{50}). These toxicity classes are defined according to the globally harmonized system of classification and labelling of chemicals (GHS). Compounds from Class I are highly toxic and fatal if swallowed (LD_{50} < 5 mg/kg). The Class II compounds' LD_{50} is found in the range between 5 and 50 mg/kg, and these compounds are still considered fatal if swallowed. Class III compounds are considered toxic, but with no fatal consequences, with LD_{50} in between 50 mg/kg and 300 mg/kg. Class **IV** compounds are harmful if swallowed, with LD_{50} between 300 mg/kg and 2000 mg/kg. Class **V** consist of compounds that may be harmful if swallowed (2000 mg/kg < $LD_{50} \leq$ 5000 mg/kg), while Class **VI** contains compounds that are considered non-toxic, with LD_{50} > 5000 mg/kg.

Table 6. Predicted toxicity of the investigated compounds **3a–e** with warfarin, esculetin, 4-hydroxy coumarin, quercetin, and ethanol as referent compounds.

Compound:	Predicted Toxicity Class	Predicted LD_{50} (mg/kg)
3a	Class V	3000
3b	Class IV	1460
3c	Class IV	721
3d	Class V	3000
3e	Class IV	2000
Warfarin	Class I	2
Esculetin	Class IV	945
4-hydroxy coumarin	Class IV	2000
Quercetin	Class III	159
Ethanol	Class V	3450

According to the results presented in Table 6, compounds **3a**, **3d**, and **3e** show significantly lower potential toxicity than esculetin and quercetin, which are natural products found in plants with a wide range of commercial uses [49–51]. Additionally, compounds with the OH group in C2″ position, i.e., **3a** and **3d**, showed a significant decrease in potential toxicity in comparison with parent molecule, 4-hydroxy coumarin. Bearing in mind that **3d** has shown excellent antioxidative properties, it is possible to consider it a potential supplement in the food industry, although experimental in vitro/in vivo studies are necessary for the definite conclusion.

4. Conclusions

Green synthesis of coumarin-hydroxybenzohydrazide derivatives by using a mixture of vinegar and ethanol as a catalyst and solvent is reported in this paper. This protocol provides isolation of pure products in moderate yield without any purification. The obtained compounds were structurally characterized by IR and NMR spectroscopy, and examined for their in vitro antioxidant activity. The results of the in vitro DPPH examination showed that two out of five of the investigated compounds, **3d** and **3e**, expressed high antioxidant activity. It should be mentioned that especially compound **3d**, with an activity slightly lower than the reference compounds NDGA and quercetin, acts as an excellent radical scavenger. In addition, the preferred radical scavenging pathways were theoretically investigated by the density functional theory approach. For that purpose, the values of various thermodynamic parameters, such as bond dissociation enthalpy, proton affinity, frontier molecular orbitals, HOMO-LUMO gap, and ionization potential were calculated. In silico calculations showed that HAT and SPLET reaction mechanisms are probable in non-polar and polar solvents, respectively. On the other hand, it was found that the SET-PT was not an operative mechanism in both solvents.

Supplementary Materials: The following are available online at https://www.mdpi.com/article/10.3390/antiox10071106/s1.

Author Contributions: Conceptualization, Z.S.M., M.R.A. and E.H.A.; methodology, Z.S.M., D.M.S., A.Ć. and E.H.A.; software, M.R.A.; validation, Z.S.M., V.S. and Z.D.P.; formal analysis, M.R.A. and J.D.M.; investigation, D.M.S. and E.H.A.; resources, V.S. and Z.D.P.; data curation, V.S. and J.D.M.; writing—original draft preparation, M.R.A., E.H.A.; writing—review and editing, Z.S.M., Z.D.P. and J.D.M.; visualization, D.M.S. and J.D.M.; supervision, Z.S.M.; project administration, D.M.S. and Z.D.P.; funding acquisition, Z.S.M. All authors have read and agreed to the published version of the manuscript.

Funding: This research was funded by the Serbian Ministry of Education, Science, and Technological Development (Agreement Nos 451-03-09/2021-14/200378; 451-03-09/2021-14/200122 and 451-03-9/2021-14/200146).

Institutional Review Board Statement: Not applicable.

Informed Consent Statement: Not applicable.

Data Availability Statement: Data is contained within the article and supplementary material.

Conflicts of Interest: The authors declare no conflict of interest.

References

1. Liu, Y.; Fiskum, G.; Schubert, D. Generation of reactive oxygen species by the mitochondrial electron transport chain. *J. Neurochem.* **2002**, *80*, 780–787. [CrossRef] [PubMed]
2. Martinez-Cayuela, M. Oxygen free radicals and human disease. *Biochimie* **1995**, *77*, 147–161. [CrossRef]
3. Valko, M.; Leibfritz, D.; Moncol, J.; Cronin, M.T.; Mazur, M.; Telser, J. Free radicals and antioxidants in normal physiological functions and human disease. *Int. J. Biochem. Cell Biol.* **2007**, *39*, 44–84. [CrossRef] [PubMed]
4. Sadeer, N.B.; Llorent-Martínez, E.J.; Bene, K.; Mahomoodally, M.F.; Mollica, A.; Sinan, K.I.; Zengin, G. Chemical profiling, antioxidant, enzyme inhibitory and molecular modelling studies on the leaves and stem bark extracts of three African medicinal plants. *J. Pharm. Biomed. Anal.* **2019**, *174*, 19–33. [CrossRef] [PubMed]
5. Mollica, A.; Costante, R.; Fiorito, S.; Genovese, S.; Stefanucci, A.; Mathieu, V.; Epifano, F. Synthesis and anti-cancer activity of naturally occurring 2, 5-diketopiperazines. *Fitoterapia* **2014**, *98*, 91–97. [CrossRef] [PubMed]
6. Uysal, A.; Ozer, O.Y.; Zengin, G.; Stefanucci, A.; Mollica, A.; Picot-Allain, C.M.N.; Mahomoodally, M.F. Multifunctional approaches to provide potential pharmacophores for the pharmacy shelf: *Heracleum sphondylium* L. subsp. ternatum (Velen.) Brummitt. *Comput. Biol. Chem.* **2019**, *78*, 64–73. [CrossRef]
7. Kirsch, G.; Abdelwahab, A.B.; Chaimbault, P. Natural and synthetic coumarins with effects on inflammation. *Molecules* **2016**, *21*, 1322. [CrossRef] [PubMed]
8. Bansal, Y.; Sethi, P.; Bansal, G. Coumarin: A potential nucleus for anti-inflammatory molecules. *Med. Chem. Res.* **2013**, *22*, 3049–3060. [CrossRef]
9. Hassan, M.Z.; Osman, H.; Ali, M.A.; Ahsan, M.J. Therapeutic potential of coumarins as antiviral agents. *Eur. J. Med. Chem.* **2016**, *123*, 236–255. [CrossRef]
10. Thakur, A.; Singla, R.; Jaitak, V. Coumarins as anticancer agents: A review on synthetic strategies, mechanism of action and SAR studies. *Eur. J. Med. Chem.* **2015**, *101*, 476–495. [CrossRef]
11. de Souza, L.G.; Rennó, M.N.; Figueroa-Villar, J.D. Coumarins as cholinesterase inhibitors: A review. *Chem.-Biol. Interact.* **2016**, *254*, 11–23. [CrossRef] [PubMed]
12. A Garro, H.; R Pungitore, C. Coumarins as potential inhibitors of DNA polymerases and reverse transcriptases. Searching new antiretroviral and antitumoral drugs. *Curr. Drug Discov. Technol.* **2015**, *12*, 66–79. [CrossRef] [PubMed]
13. Emami, S.; Dadashpour, S. Current developments of coumarin-based anti-cancer agents in medicinal chemistry. *Eur. J. Med. Chem.* **2015**, *102*, 611–630. [CrossRef]
14. Zhang, L.; Xu, Z. Coumarin-containing hybrids and their anticancer activities. *Eur. J. Med. Chem.* **2019**, *181*, 111587. [CrossRef]
15. Menezes, J.C.; Diederich, M. Translational role of natural coumarins and their derivatives as anticancer agents. *Future Med. Chem.* **2019**, *11*, 1057–1082. [CrossRef]
16. Kotali, A.; Nasiopoulou, D.A.; Harris, P.A.; Helliwell, M.; Joule, J.A. Transformation of a hydroxyl into an acyl group on α-pyrone ring: A novel route to 3, 4-diacylcoumarins. *Tetrahedron* **2012**, *68*, 761–766. [CrossRef]

17. Kotali, A.; Nasiopoulou, D.A.; Tsoleridis, C.A.; Harris, P.A.; Kontogiorgis, C.A.; Hadjipavlou-Litina, D.J. Antioxidant Activity of 3-[N-(Acylhydrazono) ethyl]-4-hydroxy-coumarins. *Molecules* **2016**, *21*, 138. [CrossRef]
18. Govindaiah, P.; Dumala, N.; Mattan, I.; Grover, P.; Prakash, M.J. Design, synthesis, biological and in silico evaluation of coumarin-hydrazone derivatives as tubulin targeted antiproliferative agents. *Bioorg. Chem.* **2019**, *91*, 103143. [CrossRef] [PubMed]
19. Avdović, E.H.; Petrović, I.P.; Stevanović, M.J.; Saso, L.; Dimitrić Marković, J.M.; Filipović, N.D.; Živić, M.Ž.; Cvetić Antić, T.N.; Žižić, M.V.; Todorović, N.V.; et al. Synthesis and Biological Screening of New 4-Hydroxycoumarin Derivatives and Their Palladium (II) Complexes. *Oxidative Med. Cell. Longev.* **2021**, *2021*, 8849568. [CrossRef]
20. Milovanović, V.M.; Petrović, Z.D.; Novaković, S.; Bogdanović, G.A.; Petrović, V.P.; Simijonović, D. Green synthesis of benzamide-dioxoisoindoline derivatives and assessment of their radical scavenging activity–Experimental and theoretical approach. *Tetrahedron* **2020**, *76*, 131456. [CrossRef]
21. Kontogiorgis, C.; Hadjipavlou-Litina, D. Biological evaluation of several coumarin derivatives designed as possible anti-inflammatory/antioxidant agents. *J. Enzym. Inhib. Med. Chem.* **2003**, *18*, 63–69. [CrossRef] [PubMed]
22. Dimić, D.; Milenković, D.; Marković, J.D.; Marković, Z. Antiradical activity of catecholamines and metabolites of dopamine: Theoretical and experimental study. *Phys. Chem. Chem. Phys.* **2017**, *19*, 12970–12980. [CrossRef] [PubMed]
23. Prihantini, A.I.; Tachibana, S.; Itoh, K. Antioxidant active compounds from elaeocarpussylvestris and their relationship between structure and activity. *Procedia Environ. Sci.* **2015**, *28*, 758–768. [CrossRef]
24. Frisch, M.J.; Trucks, G.W.; Schlegel, H.B. *Gaussian 09 (Revision E. 01)*; Gaussian Inc.: Wallingford, CT, USA, 2013.
25. Zhao, Y.; Truhlar, D.G. The M06 suite of density functionals for main group thermochemistry, thermochemical kinetics, noncovalent interactions, excited states, and transition elements: Two new functionals and systematic testing of four M06-class functionals and 12 other functionals. *Theor. Chem. Acc.* **2008**, *120*, 215–241.
26. Walker, M.; Harvey, A.J.; Sen, A.; Dessent, C.E. Performance of M06, M06-2X, and M06-HF density functionals for conformationally flexible anionic clusters: M06 functionals perform better than B3LYP for a model system with dispersion and ionic hydrogen-bonding interactions. *J. Phys. Chem. A* **2013**, *117*, 12590–12600. [CrossRef]
27. Takano, Y.; Houk, K.N. Benchmarking the conductor-like polarizable continuum model (CPCM) for aqueous solvation free energies of neutral and ionic organic molecules. *J. Chem. Theory Comput.* **2005**, *1*, 70–77. [CrossRef] [PubMed]
28. Marković, Z. Study of the mechanisms of antioxidative action of different antioxidants. *J. Serb. Soc. Comput. Mech.* **2016**, *10*, 135–150. [CrossRef]
29. Milenković, D.A.; Dimić, D.S.; Avdović, E.H.; Amić, A.D.; Marković, J.M.D.; Marković, Z.S. Advanced oxidation process of coumarins by hydroxyl radical: Towards the new mechanism leading to less toxic products. *Chem. Eng. J.* **2020**, *395*, 124971. [CrossRef]
30. Galano, A.; Mazzone, G.; Alvarez-Diduk, R.; Marino, T.; Alvarez-Idaboy, J.R.; Russo, N. Food antioxidants: Chemical insights at the molecular level. *Annu. Rev. Food Sci. Technol.* **2016**, *7*, 335–352. [CrossRef]
31. Galano, A. Free radicals induced oxidative stress at a molecular level: The current status, challenges and perspectives of computational chemistry based protocols. *J. Mex. Chem. Soc.* **2015**, *59*, 231–262. [CrossRef]
32. Wang, G.; Xue, Y.; An, L.; Zheng, Y.; Dou, Y.; Zhang, L.; Liu, Y. Theoretical study on the structural and antioxidant properties of some recently synthesised 2, 4, 5-trimethoxy chalcones. *Food Chem.* **2015**, *171*, 89–97. [CrossRef]
33. Wang, L.; Yang, F.; Zhao, X.; Li, Y. Effects of nitro-and amino-group on the antioxidant activity of genistein: A theoretical study. *Food Chem.* **2019**, *275*, 339–345. [CrossRef]
34. Marković, Z.; Tošović, J.; Milenković, D.; Marković, S. Revisiting the solvation enthalpies and free energies of the proton and electron in various solvents. *Comput. Theor. Chem.* **2016**, *1077*, 11–17. [CrossRef]
35. Avdović, E.H.; Stojković, D.L.; Jevtić, V.V.; Kosić, M.; Ristić, B.; Harhaji-Trajković, L.; Vukić, M.; Vuković, N.; Marković, Z.S.; Potočňák, I.; et al. Synthesis, characterization and cytotoxicity of a new palladium (II) complex with a coumarin-derived ligand 3-(1-(3-hydroxypropylamino) ethylidene) chroman-2, 4-dione. Crystal structure of the 3-(1-(3-hydroxypropylamino) ethylidene)-chroman-2, 4-dione. *Inorg. Chim. Acta* **2017**, *466*, 188–196. [CrossRef]
36. Dimić, D.S.; Marković, Z.S.; Saso, L.; Avdović, E.H.; Đorović, J.R.; Petrović, I.P.; Stanisavljević, D.D.; Stevanović, M.J.; Potočňák, I.; Samol'ová, E.; et al. Synthesis and Characterization of 3-(1-((3, 4-Dihydroxyphenethyl) amino) ethylidene)-chroman-2, 4-dione as a Potential Antitumor Agent. *Oxidative Med. Cell. Longev.* **2019**, *2019*, 2069250. [CrossRef] [PubMed]
37. Avdović, E.H.; Milenković, D.; Marković, J.M.D.; Đorović, J.; Vuković, N.; Vukić, M.D.; Jevtić, V.V.; Trifunović, S.R.; Potočňák, I.; Marković, Z. Synthesis, spectroscopic characterization (FT-IR, FT-Raman, and NMR), quantum chemical studies and molecular docking of 3-(1-(phenylamino) ethylidene)-chroman-2, 4-dione. *Spectrochim. Acta Part A Mol. Biomol. Spectrosc.* **2018**, *195*, 31–40. [CrossRef] [PubMed]
38. Vinson, J.A.; Su, X.; Zubik, L.; Bose, P. Phenol antioxidant quantity and quality in foods: Fruits. *J. Agric. Food Chem.* **2001**, *49*, 5315–5321. [CrossRef] [PubMed]
39. Foti, M.C. Use and Abuse of the DPPH• Radical. *J. Agric. Food Chem.* **2015**, *63*, 8765–8776. [CrossRef] [PubMed]
40. Foti, M.C.; Johnson, E.R.; Vinqvist, M.R.; Wright, J.S.; Barclay, L.R.C.; Ingold, K.U. Naphthalene diols: A new class of antioxidants intramolecular hydrogen bonding in catechols, naphthalene diols, and their aryloxyl radicals. *J. Org. Chem.* **2002**, *67*, 5190–5196. [CrossRef]
41. Bendary, E.; Francis, R.R.; Ali, H.M.G.; Sarwat, M.I.; El Hady, S. Antioxidant and structure–activity relationships (SARs) of some phenolic and anilines compounds. *Ann. Agric. Sci.* **2013**, *58*, 173–181. [CrossRef]

42. Simijonović, D.; Petrović, Z.D.; Milovanović, V.M.; Petrović, V.P.; Bogdanović, G.A. A new efficient domino approach for the synthesis of pyrazolyl-phthalazine-diones. Antiradical activity of novel phenolic products. *RSC Adv.* **2018**, *8*, 16663–16673. [CrossRef]
43. Cornard, J.P.; Boudet, A.C.; Merlin, J.C. Theoretical investigation of the molecular structure of the isoquercitrin molecule. *J. Mol. Struct.* **1999**, *508*, 37–49. [CrossRef]
44. Sadasivam, K.; Kumaresan, R. Antioxidant behavior of mearnsetin and myricetin flavonoid compounds—A DFT study. *Spectrochim. Acta Part A Mol. Biomol. Spectrosc.* **2011**, *79*, 282–293. [CrossRef] [PubMed]
45. Vagánek, A.; Rimarčík, J.; Dropková, K.; Lengyel, J.; Klein, E. Reaction enthalpies of OH bonds splitting-off in flavonoids: The role of non-polar and polar solvent. *Comput. Theor. Chem.* **2014**, *1050*, 31–38. [CrossRef]
46. Petrović, Z.D.; Đorović, J.; Simijonović, D.; Petrović, V.P.; Marković, Z. Experimental and theoretical study of antioxidative properties of some salicylaldehyde and vanillic Schiff bases. *RSC Adv.* **2015**, *5*, 24094–24100. [CrossRef]
47. Zheng, Y.; An, L.; Dou, Y.; Liu, Y. Density functional theory study of the structure–antioxidant activity of polyphenolic deoxybenzoins. *Food Chem.* **2014**, *151*, 198–206.
48. Banerjee, P.; Eckert, A.O.; Schrey, A.K.; Preissner, R. ProTox-II: A webserver for the prediction of toxicity of chemicals. *Nucleic Acids Res.* **2018**, *46*, W257–W263. [CrossRef]
49. Gansukh, E.; Nile, A.; Kim, D.H.; Oh, J.W.; Nile, S.H. New insights into antiviral and cytotoxic potential of quercetin and its derivatives–a biochemical perspective. *Food Chem.* **2021**, *334*, 127508. [CrossRef]
50. Keizo, S.; Hiromichi, O.; Shigeru, A. Selective inhibition of platelet lipoxygenase by esculetin. *Biochim. Biophys. Acta (BBA)-Lipids Lipid Metab.* **1982**, *713*, 68–72. [CrossRef]
51. Milanović, Ž.B.; Antonijević, M.R.; Amić, A.D.; Avdović, E.H.; Dimić, D.S.; Milenković, D.A.; Marković, Z.S. Inhibitory activity of quercetin, its metabolite, and standard antiviral drugs towards enzymes essential for SARS-CoV-2: The role of acid–base equilibria. *RSC Adv.* **2021**, *11*, 2838–2847. [CrossRef]

antioxidants

MDPI

Article

Correlations between Total Antioxidant Capacity, Polyphenol and Fatty Acid Content of Native Grape Seed and Pomace of Four Different Grape Varieties in Hungary

Éva Szabó [1], Tamás Marosvölgyi [2,3], Gábor Szilágyi [1], László Kőrösi [4], János Schmidt [1], Kristóf Csepregi [5], László Márk [1,6] and Ágnes Bóna [1,*]

1 Department of Biochemistry and Medical Chemistry, Medical School, University of Pécs, 7624 Pécs, Hungary; szabo.eva.dr@pte.hu (É.S.); gabor.szilagyi@aok.pte.hu (G.S.); janos.schmidt@aok.pte.hu (J.S.); laszlo.mark@aok.pte.hu (L.M.)
2 Department of Pediatrics, Medical School, University of Pécs, 7623 Pécs, Hungary; marosvolgyi.tamas@pte.hu
3 Institute of Bioanalysis, Medical School, University of Pécs, 7624 Pécs, Hungary
4 Research Institute for Viticulture and Oenology, University of Pécs, 7634 Pécs, Hungary; korosi.laszlo@pte.hu
5 Department of Plant Biology, University of Pécs, 7624 Pécs, Hungary; csepregi@gamma.ttk.pte.hu
6 MTA-PTE Human Reproduction Research Group, 7624 Pécs, Hungary
* Correspondence: agnes.bona@aok.pte.hu; Tel.: +36-72-536-276

Abstract: Grape pomace is a valuable source of various bioactive compounds such as plant-derived polyphenols and polyunsaturated fatty acids (PUFAs). The commercial demand of grape skin and seed powders as nutraceuticals is still growing. However, no distinction is currently made between unfermented native grape seed and grape seed pomace powders regarding their antioxidant activities. Our aim was to find the relationship between the polyphenol and fatty acid content as well as the antioxidant capacity of native and fermented grape seeds of four different grape varieties harvested in the Villány wine region. According to our results, none of the three investigated polyphenols (resveratrol, rutin, quercetin) could be detected in native grape seed samples in correlation with their significantly lower total antioxidant capacities compared to fermented seed samples. Pinot Noir (PN) grape seed pomace samples with the highest resveratrol and oil content showed significantly higher total antioxidant capacity than Cabernet Sauvignon (CS), Syrah (S) and Blue Portugal (BP) samples. Based on the statistical analysis, positive correlation was found between the fatty acid content and the resveratrol concentration in the pomace samples of different grape varieties. In contrast, rutin concentrations were negatively proportional to the fatty acid content of the fermented samples. No significant correlation was found considering the quercetin content of the samples. According to our findings, grape pomace seems a more promising source in the production of nutraceuticals, since it contains polyphenols in higher concentration and exerts significantly higher antioxidant activity than native grape seeds.

Keywords: grape seed; pomace; antioxidant activity; polyunsaturated fatty acids; polyphenol; resveratrol; rutin; HPLC; GC

Citation: Szabó, É.; Marosvölgyi, T.; Szilágyi, G.; Kőrösi, L.; Schmidt, J.; Csepregi, K.; Márk, L.; Bóna, Á. Correlations between Total Antioxidant Capacity, Polyphenol and Fatty Acid Content of Native Grape Seed and Pomace of Four Different Grape Varieties in Hungary. *Antioxidants* **2021**, *10*, 1101. https://doi.org/10.3390/antiox10071101

Academic Editors: Irene Dini and Domenico Montesano

Received: 14 June 2021
Accepted: 5 July 2021
Published: 9 July 2021

1. Introduction

The demand for natural plant-based antioxidants promotes the production of wide varieties of nutraceutical products. Bioactive polyphenols present in the skin and seeds of grapes occur in wine and even in grape pomace after winemaking (fermentation) [1]. Grape pomace, as a relatively inexpensive and abundant source of antioxidants, is commercially available in a form of dietary supplements (powders, tablets, capsules), providing an alternative source of polyphenol intake avoiding wine consumption. Interestingly, no difference is made so far between native grape seed and grape pomace products considering their antioxidant effect. However, the latter seems a more promising raw material in the production of nutraceuticals.

The antioxidant activity of plant polyphenols is well known. These bioactive molecules exert their beneficial effects by neutralizing reactive oxygen species (ROS) and chelating prooxidative metal ions [2]. Polyphenols may play a key role in the prevention of degenerative processes, such as cancer [3], diabetes [4], chronic inflammation [5] and aging [6]. In addition, polyphenolic compounds inhibit the oxidation of low-density lipoproteins and prevent platelet aggregation, decreasing the development of cardiovascular and coronary diseases [7]. The level of phenolic compounds in grape seeds depends mostly on the grape variety [8–10], ripeness [11,12], growing region [13,14], year of the harvest [8,15,16] and winemaking technology [17].

The oil content of grape seed pomace is about 5–20 wt%, containing linoleic acid (C18:2n-6, LA) and oleic acid (C18:1n-9, OA) as the most abundant fatty acids [18–20]. The oil content and the fatty acid composition of grape seeds are affected by ripening [21,22], grape variety [18–20,23–27], the year of the harvest [18] and geographical location [28]. The beneficial properties of grape seed oils are associated with polyunsaturated fatty acids (PUFAs). α-Linolenic acid (C18:3n-3, ALA) and LA cannot be synthesized by the human body; they must be ingested with foods. These essential fatty acids decrease the risk of cardiovascular diseases and cancer [18].

Hungary, although a small country in terms of territory, is at the forefront of world wine production: it ranks 14th worldwide and 8th in Europe [29]. In Villány, among other wine-producing regions of Hungary, mainly red wine is produced. Although there are several studies about the oil, fatty acid and polyphenol content as well as the antioxidant capacities of different grape varieties harvested in European countries [30–33], the amount of Hungarian data is very limited. According to present knowledge, no study has been performed until now that focused on the relationship between fatty acid composition, polyphenol content and antioxidant capacity values of native grape seeds and pomace. Our aim was to find a correlation among these data and to investigate the possible relationship between the concentration of PUFAs and the investigated polyphenols (resveratrol, rutin and quercetin), since they exert beneficial physiological effects.

2. Materials and Methods

2.1. Chemicals

GC-grade methanol, *n*-hexane and chloroform as well as the ACS grade K_2CO_3 were purchased from Merck (Darmstadt, Germany). Puriss. p.a.-grade acetyl chloride and ACS grade pyrogallol were purchased from Sigma-Aldrich (Steinheim, Germany).

Resveratrol, rutin and quercetin standards were obtained from Sigma-Aldrich (Steinheim, Germany). The HPLC-grade acetic acid, ethanol, water and methanol were purchased from Merck (Darmstadt, Germany).

Folin–Ciocalteu (FC) reagent, 2,2′-azino-bis(3-ethylbenzothiazoline-6-sulphonic acid) diammonium salt (ABTS), Trolox (6-hydroxy-2,5,7,8-tetramethylchroman-2-carboxylic acid), horseradish peroxidase, and H_2O_2 were purchased from Sigma-Aldrich (Steinheim, Germany). ACS-grade ascorbic acid, Na_2CO_3 and $FeCl_3$ solutions as well as the HPLC-grade 2,4,6-Tris(2-pyridyl)-s-triazine (TPTZ) and gallic acid were obtained from Sigma-Aldrich (Steinheim, Germany).

2.2. Plant Materials

The investigated grape varieties (Pinot Noir, PN; Cabernet Sauvignon, CS; Syrah, S and Blue Portugal, BP) were harvested in 2017 in the Villány wine region (Hungary). The seeds of fermented and unfermented grape berries provided by the Bock winery were removed manually, were washed with distilled water and dried in an exsiccator at 35 °C for 24 h. Dried native and pomace seeds of the four different grape varieties were homogeneously ground by an electric grinder. The samples were stored at −80 °C until further measurements were carried out.

2.3. Total Antioxidant Capacity Determination

The antioxidant activities of grape seeds were measured using three different methods. Since individual phenolic compounds may have different responses to different assays, a complex, multi-aspect approach is necessary to obtain a good assessment of the total antioxidant capacities (TACs) of biological samples [34]. Powdered grape seed samples (60 mg) were placed into Eppendorf tubes. For the extraction, 1.5 mL of a 1:1 mixture of ethanol and water was used. Samples were placed into an ultrasonic water bath (Industrial Ultrasonic Cleaner, YS-JP-020, Shenzhen, China) for 15 min at room temperature and then centrifuged (Heraeus Fresco 17 Microcentrifuge, Thermo Fisher Scientific, Bremen, Germany) at 13,000 rpm for 10 min at room temperature. Supernatants were transferred into Eppendorf tubes and pellets were re-extracted following the above procedure two more times. Pooled supernatants of the three repeated extractions made one sample.

The total antioxidant capacities of grape seed extracts were measured by using a modified Folin–Ciocalteu assay [35]. Diluted FC reagent (0.5 mL, 1:10 in distilled water) was mixed with the seed extracts in microplate wells and incubated at room temperature for 5 min. Na_2CO_3 solution (0.5 mL, 6 wt%) was added to the extracts and the mixture was incubated for 90 min at room temperature. Absorbance was measured at 765 nm (Multiskan FC plate reader, Thermo Fisher Scientific, Waltham, MA, USA). Calibration was made with gallic acid. To compare the results obtained from the different assays, FC antioxidant capacity values were expressed uniformly as mg ASA equivalents per mg dry grape seed.

Ferric-reducing antioxidant potential (FRAP) is a TAC method based on the reduction of ferric TPTZ to ferrous TPTZ complex. The FRAP reagent was made by mixing acetate buffer (300 mM, pH 3.60), TPTZ solution (10 mM TPTZ in distilled water) and $FeCl_3$ solution (20 mM $FeCl_3$ in 40 mM HCl) [34]. Seed extracts of 10 µL were incubated with 190 µL of FRAP reagent in microplate wells at room temperature for 30 min (shaking the samples every 10 min), and the absorbance of the samples was measured at 620 nm by using the plate reader. Calibration was made with ascorbic acid. FRAP data of seed extracts were expressed as mg ASA equivalents per mg dry grape seed.

The Trolox equivalent antioxidant activity (TEAC) of samples was measured according to Majer and Hideg [36] with modifications. The blue colored 2,2′-azino-bis(3-ethylbenzothiazoline-6-sulphonic acid) radical cation ($ABTS^+$) was prepared in a phosphate buffer (50 mM, pH 6.00), by mixing $ABTS^+$ (0.1 mM), horseradish peroxidase (0.0125 µM), and H_2O_2 (1 mM). The solution was incubated for 15 min at room temperature. Seed extracts of 10 µL were added to 190 µL of $ABTS^+$ solution and the loss of blue color was followed at 651 nm. Calibration was made with Trolox. Total antioxidant capacity values were calculated from TEAC results and expressed as mg ascorbic acid (ASA) equivalents per mg dry grape seed in order to make the results of different assays comparable.

2.4. Polyphenol Content Determination by HPLC

The homogeneously grinded grape seed samples (100 mg) were extracted with the 1:1 mixture of ethanol and water (1 mL) for six hours at 80 °C. Extracts were centrifuged at 10,000 rpm for 10 min at room temperature. The supernatants of the samples were filtered through Acrodisc syringe filters (25 mm, 0.2 µm) (Pall Life Sciences, New York, NY, USA) prior to the injection.

The HPLC system used consisted of an Ultimate 3000 SD pump (Dionex Softron GmbH, Germering, Germany), an Ultimate 3000 autosampler, and an Ultimate 3000 RS UV–vis diode array detector. The eluate was monitored at three different wavelengths (306, 350 and 365 nm), respectively, according to the absorption maxima of the investigated polyphenols. Chromatographic separations were achieved using a Kinetex C18 reversed phase column (5 µm, 2.1 mm × 150 mm i.d.) (Phenomenex, Torrance, CA, USA). Chromeleon (Version 6.8, Thermo Fisher Scientific) and Hystar (Version 3.2, Bruker Daltonics, Bremen, Germany) data management software were used to control the equipment and DataAnalysis (Version 4.0, Bruker Daltonics) was used for data evaluation.

A multi-step gradient method was applied, using methanol/water/acetic acid (10/90/1 *v*/*v*/*v*) as solvent A and methanol/water/acetic acid (90/10/1 *v*/*v*/*v*) as solvent B at a flow rate of 0.6 mL/min. The gradient profile was: 0.0–10.0 min, from 0% to 15% B; 10.0–35.0 min, from 15.0% to 35.0% B, 35.0–40.0 min, from 35.0% to 100.0% B followed by a hold of 5 min. The equilibrium time to original conditions was 5 min. The total runtime was 55 min. The injection volume was 5 µL. The column temperature was set to 50 °C. The limit of detection (LOD)/limit of quantification (LOQ) values were 3/10 ng/mL for resveratrol, 6/21 ng/mL for quercetin and 10/34 ng/mL for rutin.

2.5. Oil content Determination

The oil content of the different grape seeds was determined in three parallel samples by modified Bligh and Dyer extraction [37,38].

Shortly, for the lipid extraction, 2.5 mL of methanol and 1.25 mL of chloroform, then 1.25 mL of distilled water and 1.25 mL of chloroform, were added to 200.0 mg of powdered grape seed sample. The different phases were separated by centrifugation at 3000 rpm for 15 min at 4 °C. The lower chloroform phase containing the lipids was transferred into a test tube of known weight by using a Brand macro pipette controller. The weight of the N_2-dried lipids were measured, and the lipid content was calculated. Oil content data are given as g oil in 100 g dried grape seed.

2.6. Fatty Acid Analysis

The lipid extract (about 0.1 mg) was transferred into the extraction tube and 2 mL of methanol/hexane (4/1 *v*/*v*) was added. To prevent auto-oxidation, 0.5% pyrogallol was used. During shaking on a vortex mixer, 200 µL of acetyl chloride was added. For derivatization, the reaction tubes were placed into a heating block for 1 h at 100 °C. After cooling down, 4.8 mL of K_2CO_3 solution (6% *w*/*v*) was added. The samples were centrifuged at 3200 rpm for 10 min at 4 °C. The upper, fatty acid methyl ester containing hexane phase was transferred to vials and analyzed by gas chromatography (GC).

Fatty acid composition was determined by using Agilent 6890N GC, which consisted of autosampler 7683B, and a flame ionization detector (FID). Separation was performed on capillary column DB-23 (60 m × 0.25 mm × 0.25 µm; Agilent J&W Scientific, Folsom, CA, USA) [39].

The Chromeleon (Version 7.1, Thermo Fisher Scientific, Sunnyvale, CA, USA) data management software was used to control the equipment and for data evaluation.

Peak identification was verified by comparison with standard reference mixtures (NuChek Prep Inc., Elysian, MN, USA: GLC-463, 674, 642 and 643). Fatty acid contents were expressed as weight percent of total fatty acids (wt%).

2.7. Statistical Analysis

Statistical analysis was performed with SPSS for Windows 25.0 (SPSS Inc., Chicago, IL, USA); the results of each sample were compared with ANOVA and an independent *T*-test. For correlations between fatty acids, polyphenols and total antioxidant capacity values, Pearson's correlation was used. The level of significance was $p < 0.05$. All measurements were performed in triplicate and data are presented as mean ± SD.

3. Results

3.1. Polyphenol Content

The mg/kg concentrations of the three different polyphenols in grape seed pomace were calculated (Table 1). Considering the resveratrol content of the analyzed grape seed pomace samples, the highest concentrations were detected in the PN samples, while BP, CS and S samples contained significantly lower levels of resveratrol. The levels of rutin were significantly higher in BP pomace seeds than in PN and CS samples. The lowest amount of rutin was found in CS samples. S samples showed higher while CS samples significantly lower quercetin content compared to BP and PN pomace seeds with similar quercetin values. Rutin and quercetin values correlated significantly in the investigated samples (Pearson correlation coefficient: +0.645, $p = 0.024$), while there was no significant correlation between the values of the other polyphenols.

Table 1. Polyphenol content (mg/kg dry weight) of grape seed pomace of four different grape varieties harvested in the Villány wine region.

	Blue Portugal	Syrah	Pinot Noir	Cabernet Sauvignon
resveratrol (mg/kg)	7.11 ± 0.44 ^{ABC}	19.02 ± 0.17 ^{ADE}	37.93 ± 2.78 ^{BDF}	15.04 ± 0.41 ^{CEF}
rutin (mg/kg)	9.37 ± 0.64 ^{aA}	8.81 ± 1.56	7.98 ± 0.52 ^{ab}	6.74 ± 0.37 ^{Ab}
quercetin (mg/kg)	34.17 ± 2.35 ^{A}	38.52 ± 1.11 ^{Ba}	33.09 ± 0.88 ^{Ca}	15.69 ± 0.80 ^{ABC}

Means in a row sharing common Roman superscript denote significant differences between the grape varieties: ^{a,b} $p < 0.05$; ^{A,B,C,D,E,F}: $p \leq 0.001$.

The investigated polyphenols could not be detected in native grape seeds by HPLC (data not shown).

3.2. Total Antioxidant Capacity

The TAC was determined by three different methods (FCR, FRAP and TEAC) and similar results were found. The highest TAC was found in PN samples compared to BP, S and CS samples with significantly lower antioxidant activities both in native grape seeds and grape pomace (Table 2). In native seeds CS, and in pomace samples BP and S showed the lowest antioxidant capacity. Grape seed pomace samples showed significantly higher TAC than native seeds in all investigated grape varieties measured with three different methods ($p < 0.001$, data not shown)

Table 2. Total antioxidant capacity (FCR, FRAP, TEAC) of native grape seed and grape pomace of four different grape varieties harvested in the Villány wine region.

mg ASA in mg Dry Seed; Mean ± SD		Blue Portugal	Syrah	Pinot Noir	Cabernet Sauvignon
FCR	Native	0.021 ± 0.001 ^{aAB}	0.018 ± 0.000 ^{aCD}	0.027 ± 0.000 ^{ACE}	0.010 ± 0.000 ^{BDE}
	Pomace	0.041 ± 0.001 ^{AB}	0.041 ± 0.001)^{CD}	0.063 ± 0.000 ^{ACE}	0.046 ± 0.000 ^{BDE}
FRAP	Native	0.052 ± 0.001 ^{AaB}	0.052 ± 0.001 ^{ACD}	0.064 ± 0.002 ^{aCE}	0.025 ± 0.001 ^{BDE}
	Pomace	0.078 ± 0.001 ^{Aa}	0.077 ± 0.002 ^{Bb}	0.116 ± 0.004 ^{ABC}	0.085 ± 0.001 ^{abC}
TEAC	Native	0.062 ± 0.002 ^{aAB}	0.056 ± 0.002 ^{aCD}	0.083 ± 0.001 ^{ACE}	0.028 ± 0.001 ^{BDE}
	Pomace	0.096 ± 0.002 ^{Aa}	0.094 ± 0.002 ^{Bb}	0.132 ± 0.001 ^{ABC}	0.104 ± 0.002 ^{abC}

FCR: Folin–Ciocalteu reagent, FRAP: ferric-reducing antioxidant potential, TEAC: Trolox equivalent antioxidant activity. The antioxidant capacities (FCR, FRAP, TEAC) of the different grape varieties (native and pomace) were compared with ANOVA and independent T-test. Means in a row sharing common Roman superscript denote significant differences between the grape varieties: ^{a,b} $p < 0.05$; ^{A,B,C,D,E}: $p \leq 0.001$.

3.3. Fatty Acid Composition

The fatty acid composition and oil content of grape seed pomace is shown in Table 3. The highest oil content was found in the PN and CS pomace seeds followed by S seeds, while the lowest levels were detected in BP samples.

Regarding the SFA composition of grape seed pomace samples, C16:0 (palmitic acid) and C18:0 (stearic acid) were detected in the highest values. The highest level of C16:0 was found in the BP samples, while S samples showed higher C18:0 values than the other investigated grape varieties.

Table 3. Oil content (g oil/100 g dried seed) and fatty acid composition (wt%, weight percent of total fatty acids) of grape seed pomace of four different grape varieties harvested in the Villány wine region.

	Blue Portugal	Syrah	Pinot Noir	Cabernet Sauvignon
Oil content	9.54 ± 0.21 ABC	12.13 ± 0.14 ADa	13.91 ± 0.21 BD	13.51 ± 0.26 Ca
Saturated fatty acids (SFA)				
C14:0	0.13 ± 0.01 ab	0.10 ± 0.01 a	0.11 ± 0.01 c	0.09 ± 0.01 bc
C15:0	0.02 ± 0.0	0.01 ± 0.00 a	0.02 ± 0.00 Aa	0.01 ± 0.00 A
C16:0	9.68 ± 0.05 ABC	8.05 ± 0.05 Aa	8.03 ± 0.11 Bb	8.34 ± 0.04 Cab
C17:0	0.07 ± 0.00 a	0.07 ± 0.00 b	0.07 ± 0.00 abc	0.06 ± 0.00 c
C18:0	4.12 ± 0.02 ABC	4.55 ± 0.01 ADE	3.85 ± 0.00 BDF	4.45 ± 0.02 CEF
C20:0	0.17 ± 0.00 AaB	0.20 ± 0.00 AC	0.16 ± 0.00 aCD	0.20 ± 0.01 BD
C22:0	0.06 ± 0.01 a	0.07 ± 0.00	0.07 ± 0.00 ab	0.06 ± 0.01 b
C24:0	0.04 ± 0.00 ABa	0.03 ± 0.00 A	0.03 ± 0.00 Bb	0.04 ± 0.00 ab
Monounsaturated fatty acids (MUFA)				
C16:1n-7	0.20 ± 0.00 ABC	0.09 ± 0.00 AD	0.16 ± 0.00 BDE	0.09 ± 0.00 CE
C18:1n-9	15.40 ± 0.03 ABC	13.37 ± 0.01 AD	16.80 ± 0.15 BDE	13.53 ± 0.13 CE
C18:1n-7	0.82 ± 0.01 AaB	0.65 ± 0.02 AC	0.81 ± 0.00 aCD	0.66 ± 0.02 BD
C20:1n-9	0.14 ± 0.00 Aa	0.16 ± 0.00 A	0.17 ± 0.01 a	0.16 ± 0.01
Polyunsaturated fatty acids (PUFA)				
C18:2n-6	68.43 ± 0.05 AaB	71.95 ± 0.07 ACb	69.01 ± 0.24 aCD	71.61 ± 0.12 BbD
C18:3n-3	0.46 ± 0.00 ABC	0.44 ± 0.00 Aab	0.42 ± 0.00 Ba	0.42 ± 0.01 Cb

Means in a row sharing common Roman superscript denote significant differences between the grape varieties: a,b,c: $p < 0.05$; A,B,C,D,E: $p \leq 0.001$.

OA was the most abundant monounsaturated fatty acid in grape seed pomace. The highest concentration of OA was found in PN pomace samples, followed by CS, while S and BP samples contained significantly lower levels of OA.

Among the identified polyunsaturated fatty acids, LA was present in the highest concentration in S and ALA in BP grape pomace seeds.

3.4. Correlations between Polyphenols and Total Antioxidant Capacities

We found significant positive correlations between resveratrol values and total antioxidant capacities: FCR (Pearson correlation coefficient: +0.935; $p < 0.000$), FRAP (Pearson correlation coefficient: +0.909; $p < 0.000$) and TEAC (Pearson correlation coefficient: +0.900; $p < 0.000$). In contrast, there were no correlations between the antioxidant capacities and the values of the two other investigated polyphenols, rutin and quercetin (data not presented).

3.5. Correlations between Polyphenols, Total Antioxidant Capacities and Fatty Acids

The oil content of grape pomace seeds correlated significantly and positively with the resveratrol concentrations and total antioxidant capacities measured with three different methods (Table 4). A similar relationship was found considering the OA, and C20:1n-9 content of the samples. The saturated fatty acid C16:0 showed positive but no significant correlation with the measured resveratrol values. The total antioxidant capacities (FCR, FRAP, TEAC) were in a significant direct proportion with C16:0 concentrations. In contrast, the rutin content of grape pomace samples showed a significant inverse correlation considering the oil content as well as the C16:0, C18:0, OA and C20:1n-9 content of the samples.

Table 4. Pearson's correlation coefficients of total antioxidant capacities, oil content, fatty acid and polyphenol concentrations in grape seed pomace samples.

	Resveratrol		Rutin		FCR		FRAP		TEAC	
	Pearson	Sig	Pearson	Sig	Pearson	Sig	Pearson	Sig	Pearson	Sig
Oil Content	**0.747**	**0.008**	**−0.686**	**0.014**	**0.678**	**0.015**	**0.643**	**0.024**	**0.654**	**0.021**
C16:0	0.555	0.077	**−0.864**	**0.001**	**0.691**	**0.013**	**0.643**	**0.024**	**0.682**	**0.015**
C18:0	0.273	0.417	**−0.664**	**0.026**	0.262	0.411	0.224	0.484	0.233	0.466
C18:1n-9	**0.800**	**0.003**	**−0.673**	**0.023**	**0.972**	**0.000**	**0.950**	**0.000**	**0.959**	**0.000**
C20:1n-9	**0.691**	**0.019**	**−0.700**	**0.016**	**0.695**	**0.012**	**0.655**	**0.021**	**0.659**	**0.020**

Bold numbers denote significant correlations.

Furthermore, significant positive correlations were found between essential fatty acids (ALA and LA) and the resveratrol concentrations of grape seed pomace samples (Figure 1) ALA and total antioxidant capacities: FCR (Pearson correlation coefficient: +0.705; $p < 0.05$), FRAP (Pearson correlation coefficient: +0.669; $p < 0.05$) and TEAC (Pearson correlation coefficient: +0.676; $p < 0.05$) showed significant positive correlations as well. In contrast, rutin values were in significant inverse proportion with the LA and ALA content of the samples (Figure 1). No significant correlation was found between the antioxidant capacities and LA values; however, there was a trend of positive correlation between LA and FCR (Pearson correlation coefficient: +0.574; $p = 0.051$). There was no significant relationship between fatty acid and quercetin values (data not shown).

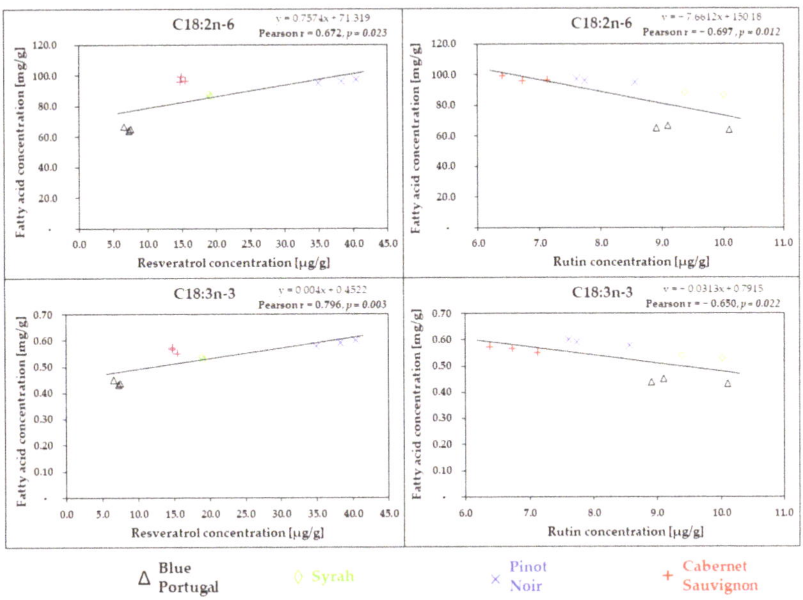

Figure 1. Correlations between the essential fatty acids (C18:3n-3 and C18:2n-6) and polyphenols (resveratrol and rutin) in grape seed pomace of four different grape varieties.

4. Discussion

In this study, the total antioxidant capacity, polyphenol and oil content as well as the fatty acid composition of both native and fermented grape seeds in four Hungarian widely cultivated red grape varieties were investigated.

The TAC values were higher in all four analyzed grape seed pomace samples than in native grape seeds. We found only one similar measurement: in an Argentine study [40], the antioxidant capacities of whole grape and grape pomace (skin and seeds) extracts in three different varieties were compared. According to these measurements, S and Merlot pomace extracts showed higher antioxidant activities than whole grape extracts in contrast with the CS samples. In whole grape extracts, CS, while in pomace samples, S showed the highest antioxidant capacity.

The TAC measured with three different methods was the highest both in fermented and native seeds of PN variety. Interestingly, CS seeds showed the second-highest antioxidant activity among fermented samples while the lowest values were detected in the unfermented seeds of CS. According to Chambre et al. [41] the antioxidant capacity values of PN seed pomace samples were also higher than that of CS samples.

Similarly to our findings, other studies described that the concentrations of individual flavonols and stilbenes, such as quercetin, rutin and resveratrol, vary greatly in the pomace of different grape varieties [15,25,40].

Numerous studies found strong positive correlation between the antioxidant capacity and the total phenolic content of various grape seeds [1,9,16,42–48]. Although the total polyphenol content was not determined in our study, a similar significant positive correlation was identified between antioxidant capacities and resveratrol values.

Based on our measurements, the investigated polyphenols could not be detected in native seeds. Similarly, rutin and resveratrol could not be identified in native grape seeds only in grape peel in a Chinese study [47]. Guchu et al. [49] could not detect quercetin either in the unfermented or in fermented grape seed samples. The stilbene content of grape seeds in the different ripening stages of grapes was analyzed by Dudoit et al. [50]. Based on this study, resveratrol could not be detected in the seeds, but it was present in the skin of grapes. Similar results were found by other researchers regarding to the resveratrol content of native seeds in different grape varieties [51,52]. Furthermore, resveratrol [15,25,53], rutin [25,53] and quercetin [25,40] were identified from grape seed pomace samples by other research groups as well. However, some of the previous studies showed conflicting results about the polyphenol content of native grape seeds and pomace. Farhadi et al. [51] investigated six red grape varieties and found low levels of rutin and quercetin in native seed samples. Furthermore, in a Turkish study [54], high levels of rutin, quercetin and resveratrol were found in native grape seeds. Somkuvar et al. [55] could also detect all of the three investigated polyphenols in the native seeds of CS and S red grape varieties. In a study in Brazil [56], neither rutin nor quercetin derivatives could be detected in pomace seed extracts of CS samples.

Considering the oil content of the analyzed grape pomace seeds, the highest values were measured in PN seed samples, followed by S seeds, while the oil content of BP samples was the lowest (Table 3). However, in a Canadian study [23], CS grape pomace seeds showed the highest (11.17 $\pm$ 0.05 wt%) and PN samples (9.83 $\pm$ 0.05 wt%) the lowest oil content, and the values of S samples were between these two concentrations (10.10 $\pm$ 0.10 wt%). Compared to our results, Brazilian CS grape pomace seed samples show lower (4.83 $\pm$ 0.11 wt%) [25], while Chinese CS samples higher oil content (14.45 $\pm$ 0.03 wt%) [19].

The most abundant fatty acids in grape pomace seed oil are OA and LA. Among the investigated grape pomaces, PN samples had the highest OA content, followed by BP, CS and S. In contrast, one of the previous studies described higher OA content in S grape pomace samples (16.61 wt%) than in PN (15.00 wt%) and CS (12.63 wt%) seeds [23]. Riberio et al. found significantly lower levels of OA (10.01 $\pm$ 0.69 wt%) in CS pomace samples [25]. Regarding to our measurements, the highest LA concentrations were detected in S and CS pomace seed samples, while PN and BP seeds contained lower levels of LA. Wen et al. found similar LA concentrations (73.39 $\pm$ 0.08 wt%) in CS grape seed pomace [19]. In contrast, Ribeiro et al. found LA in significantly lower amounts (54.58 $\pm$ 2.78 wt%) in CS pomace [25]. The results of Beveridge et al. about the LA values of CS (72.77 wt%) and PN (70.00 wt%) grape seed pomace samples correspond well with our findings; however, they detected lower concentrations of LA in S (68.00 wt%) seed samples [23].

Many studies have investigated both the fatty acid composition and polyphenol content of different grape varieties [20,21,43,57–60], but to the best of our knowledge, this is the first study investigating the relationship between these values. In this study, significant positive correlations were found between the concentration of resveratrol and the oil, OA, C20:1n-9, LA and ALA content of the pomace seed samples. This direct proportion suggests a potential connection between the concentrations and beneficial physiological effects of these molecules. PUFAs may decrease the risk of several diseases, such as dyslipidemia [61], type 2 diabetes mellitus [62], cardiovascular disease [63–65], atherosclerosis and metabolic syndrome [66,67]. Resveratrol may also exert beneficial effects in cardiovascular diseases because of its blood pressure lowering [68], anti-inflammatory [69] and NO production-enhancing effects [70]. However, the rutin concentration of the pomace samples correlated significantly and negatively to their OA, C20:1n-9, LA and ALA content. The TAC values of pomace samples correlated significantly and directly to their oil, C16:0, OA, C20:1n-9 and ALA content.

5. Conclusions

According to our measurements, the TAC values of grape pomace seeds are significantly higher than that of native grape seeds independent of grape variety. The distinct difference can be explained by their polyphenol contents. The absence of resveratrol, rutin and quercetin in native grape seed samples suggests that the process of fermentation may exert a strong impact on the polyphenol content of grape seeds. Polyphenols present in grape skin, pulp and stem are presumably enriched in grape seeds during fermentation. In addition, a significant positive correlation was found between the resveratrol concentration and the OA, C20:1n-9, LA and ALA content of grape seed pomace samples. Furthermore, a significant inverse proportion was found between rutin values and the concentration of fatty acids in pomace seed samples. Besides, no relationship was found between the quercetin content and the fatty acid composition of grape pomace samples. According to our findings, grape pomace seems a more valuable source of raw material used in the production of modern nutraceuticals due to its higher polyphenol content and stronger antioxidant activity.

Author Contributions: Conceptualization, L.M.; methodology, Á.B., K.C., L.K., T.M., J.S., G.S.; software, Á.B., K.C., T.M., J.S.; writing, Á.B., K.C., T.M., É.S.; visualization, T.M. and É.S. All authors have read and agreed to the published version of the manuscript.

Funding: This research was funded by ÁOK PD-2015-07, GINOP-2.3.2-15-2016-00021 and the National Research, Development and Innovation Office (grant numbers K-124165 and K-120193).

Institutional Review Board Statement: Not applicable.

Informed Consent Statement: Not applicable.

Data Availability Statement: The data presented in this study are available on request from the corresponding author.

Acknowledgments: We are grateful to Bock Winery (H-7773 Villány, Batthyány u. 15.; Hungary) for all the grape seed samples provided.

Conflicts of Interest: The authors declare no conflict of interest.

References

1. Guaita, M.; Bosso, A. Polyphenolic characterization of grape skins and seeds of four Italian red cultivars at harvest and after fermentative maceration. *Foods* **2019**, *8*, 395. [CrossRef]
2. Bhuiyan, M.N.; Mitsuhashi, S.; Sigetomi, K.; Ubukata, M. Quercetin inhibits advanced glycation end product formation via chelating metal ions, trapping methylglyoxal, and trapping reactive oxygen species. *Biosci. Biotechnol. Biochem.* **2017**, *81*, 882–890. [CrossRef]
3. Bian, Y.; Wei, J.; Zhao, C.; Li, G. Natural polyphenols targeting senescence: A novel prevention and therapy strategy for cancer. *Int. J. Mol. Sci.* **2020**, *21*, 684. [CrossRef]

4. Burton-Freeman, B.; Brzeziński, M.; Park, E.; Sandhu, A.; Xiao, D.; Edirisinghe, I. A selective role of dietary anthocyanins and flavan-3-ols in reducing the risk of type 2 diabetes mellitus: A review of recent evidence. *Nutrients* **2019**, *11*, 841. [CrossRef]
5. Yahfoufi, N.; Alsadi, N.; Jambi, M.; Matar, C. The immunomodulatory and anti-Inflammatory role of polyphenols. *Nutrients* **2018**, *10*, 1618. [CrossRef]
6. Khurana, S.; Venkataraman, K.; Hollingsworth, A.; Piche, M.; Tai, T.C. Polyphenols: Benefits to the cardiovascular system in health and in aging. *Nutrients* **2013**, *5*, 3779–3827. [CrossRef]
7. Mattera, R.; Benvenuto, M.; Giganti, M.G.; Tresoldi, I.; Pluchinotta, F.R.; Bergante, S.; Tettamanti, G.; Masuelli, L.; Manzari, V.; Modesti, A.; et al. Effects of polyphenols on oxidative stress-mediated injury in cardiomyocytes. *Nutrients* **2017**, *9*, 523. [CrossRef] [PubMed]
8. Ky, I.; Lorrain, B.; Kolbas, N.; Crozier, A.; Teissedre, P.L. Wine by-products: Phenolic characterization and antioxidant activity evaluation of grapes and grape pomaces from six different French grape varieties. *Molecules* **2014**, *19*, 482–506. [CrossRef] [PubMed]
9. Liang, Z.; Cheng, L.; Zhong, G.Y.; Liu, R.H. Antioxidant and antiproliferative activities of twenty-four *Vitis vinifera* grapes. *PLoS ONE* **2014**, *9*, e105146. [CrossRef] [PubMed]
10. Zhu, L.; Zhang, Y.; Lu, J. Phenolic contents and compositions in skins of red wine grape cultivars among various genetic backgrounds and originations. *Int. J. Mol. Sci.* **2012**, *13*, 3492–3510. [CrossRef] [PubMed]
11. Benbouguerra, N.; Richard, T.; Saucier, C.; Garcia, F. Voltammetric behavior, flavanol and anthocyanin contents, and antioxidant capacity of grape skins and seeds during ripening (*Vitis vinifera* var. Merlot, Tannat, and Syrah). *Antioxidants* **2020**, *9*, 800. [CrossRef]
12. Gil, M.; Kontoudakis, N.; Gonzalez, E.; Esteruelas, M.; Fort, F.; Canals, J.M.; Zamora, F. Influence of grape maturity and maceration length on color, polyphenolic composition, and polysaccharide content of Cabernet Sauvignon and Tempranillo wines. *J. Agric. Food Chem.* **2012**, *60*, 7988–8001. [CrossRef]
13. Del-Castillo-Alonso, M.A.; Castagna, A.; Csepregi, K.; Hideg, E.; Jakab, G.; Jansen, M.A.; Jug, T.; Llorens, L.; Matai, A.; Martinez-Luscher, J.; et al. Environmental factors correlated with the metabolite profile of *Vitis vinifera* cv. Pinot Noir berry skins along a European latitudinal gradient. *J. Agric. Food Chem.* **2016**, *64*, 8722–8734. [CrossRef] [PubMed]
14. Martinez-Gil, A.M.; Gutierrez-Gamboa, G.; Garde-Cerdan, T.; Perez-Alvarez, E.P.; Moreno-Simunovic, Y. Characterization of phenolic composition in Carignan noir grapes (*Vitis vinifera* L.) from six wine-growing sites in Maule Valley, Chile. *J. Sci. Food Agric.* **2018**, *98*, 274–282. [CrossRef]
15. Kammerer, D.; Claus, A.; Carle, R.; Schieber, A. Polyphenol screening of pomace from red and white grape varieties (*Vitis vinifera* L.) by HPLC-DAD-MS/MS. *J. Agric. Food Chem.* **2004**, *52*, 4360–4367. [CrossRef]
16. Sochorova, L.; Prusova, B.; Jurikova, T.; Mlcek, J.; Adamkova, A.; Baron, M.; Sochor, J. The study of antioxidant components in grape seeds. *Molecules* **2020**, *25*, 3736. [CrossRef]
17. Grieco, F.; Carluccio, M.A.; Giovinazzo, G. Autochthonous *Saccharomyces cerevisiae* starter dultures enhance polyphenols content, antioxidant activity, and anti-inflammatory response of Apulian red wines. *Foods* **2019**, *8*, 453. [CrossRef]
18. Lachman, J.; Hejtmánková, A.; Táborský, J.; Kotíková, Z.; Pivec, V.; Střalková, R.; Vollmannová, A.; Bojňanská, T.; Dědina, M. Evaluation of oil content and fatty acid composition in the seed of grapevine varieties. *LWT Food Sci. Technol.* **2015**, *63*, 620–625. [CrossRef]
19. Wen, X.; Zhu, M.; Hu, R.; Zhao, J.; Chen, Z.; Li, J.; Ni, Y. Characterisation of seed oils from different grape cultivars grown in China. *J. Food Sci. Technol.* **2016**, *53*, 3129–3136. [CrossRef] [PubMed]
20. Zdunic, G.; Godevac, D.; Savikin, K.; Krivokuca, D.; Mihailovic, M.; Przic, Z.; Markovic, N. Grape seed polyphenols and fatty acids of autochthonous Prokupac vine variety from Serbia. *Chem. Biodivers.* **2019**, *16*, e1900053. [CrossRef] [PubMed]
21. Bombai, G.; Pasini, F.; Verardo, V.; Sevindik, O.; di Foggia, M.; Tessarin, P.; Bregoli, A.M.; Caboni, M.F.; Rombola, A.D. Monitoring of compositional changes during berry ripening in grape seed extracts of cv. Sangiovese (*Vitis vinifera* L.). *J. Sci. Food Agric.* **2017**, *97*, 3058–3064. [CrossRef]
22. Rubio, M.; Alvarez-Orti, M.; Alvarruiz, A.; Fernandez, E.; Pardo, J.E. Characterization of oil obtained from grape seeds collected during berry development. *J. Agric. Food Chem.* **2009**, *57*, 2812–2815. [CrossRef]
23. Beveridge, T.H.; Girard, B.; Kopp, T.; Drover, J.C. Yield and composition of grape seed oils extracted by supercritical carbon dioxide and petroleum ether: Varietal effects. *J. Agric. Food Chem.* **2005**, *53*, 1799–1804. [CrossRef]
24. Fernandes, L.; Casal, S.; Cruz, R.; Pereira, J.A.; Ramalhosa, E. Seed oils of ten traditional Portuguese grape varieties with interesting chemical and antioxidant properties. *Food Res. Int.* **2013**, *50*, 161–166. [CrossRef]
25. Ribeiro, L.F.; Ribani, R.H.; Francisco, T.M.; Soares, A.A.; Pontarolo, R.; Haminiuk, C.W. Profile of bioactive compounds from grape pomace (*Vitis vinifera* and *Vitis labrusca*) by spectrophotometric, chromatographic and spectral analyses. *J. Chromatogr. B Anal. Technol. Biomed. Life Sci.* **2015**, *1007*, 72–80. [CrossRef]
26. Sabir, A.; Unver, A.; Kara, Z. The fatty acid and tocopherol constituents of the seed oil extracted from 21 grape varieties (*Vitis* spp.). *J. Sci. Food Agric.* **2012**, *92*, 1982–1987. [CrossRef]
27. Tangolar, S.G.; Ozogul, Y.; Tangolar, S.; Torun, A. Evaluation of fatty acid profiles and mineral content of grape seed oil of some grape genotypes. *Int. J. Food Sci. Nutr.* **2009**, *60*, 32–39. [CrossRef]
28. Bada, J.C.; León-Camacho, M.; Copovi, P.; Alonso, L. Characterization of grape seed oil from wines with protected denomination of origin (PDO) from Spain. *Grasas Aceites* **2015**, *66*, e085. [CrossRef]

29. Roca, P. 2019 Statistical Report on World Vitiviniculture. Available online: https://www.oiv.int/public/medias/6782/oiv-2019-statistical-report-on-world-vitiviniculture.pdf (accessed on 27 June 2021).
30. Biniari, K.; Xenaki, M.; Daskalakis, I.; Rusjan, D.; Bouza, D.; Stavrakaki, M. Polyphenolic compounds and antioxidants of skin and berry grapes of Greek *Vitis vinifera* cultivars in relation to climate conditions. *Food Chem.* **2020**, *307*, 125518. [CrossRef]
31. Fidan, M.; Erez, M.E.; İnal, B.; Pinar, S.M.; Altintaş, S. Antioxidant capacity and phylogenetic analysis of twenty native grape cultivars in Siirt province, Turkey. *Cell Mol. Biol.* **2018**, *64*, 14–18. [CrossRef]
32. Martin, M.E.; Grao-Cruces, E.; Millan-Linares, M.C.; Montserrat-de la Paz, S. Grape (*Vitis vinifera* L.) seed oil: A functional food from the winemaking industry. *Foods* **2020**, *9*, 1360. [CrossRef]
33. Sikuten, I.; Stambuk, P.; Andabaka, Z.; Tomaz, I.; Markovic, Z.; Stupic, D.; Maletic, E.; Kontic, J.K.; Preiner, D. Grapevine as a rich source of polyphenolic compounds. *Molecules* **2020**, *25*, 5604. [CrossRef] [PubMed]
34. Csepregi, K.; Neugart, S.; Schreiner, M.; Hideg, E. Comparative evaluation of total antioxidant capacities of plant polyphenols. *Molecules* **2016**, *21*, 208. [CrossRef]
35. Csepregi, K.; Kocsis, M.; Hideg, E. On the spectrophotometric determination of total phenolic and flavonoid contents. *Acta Biol. Hung.* **2013**, *64*, 500–509. [CrossRef]
36. Majer, P.; Hideg, E. Developmental stage is an important factor that determines the antioxidant responses of young and old grapevine leaves under UV irradiation in a green-house. *Plant Physiol. Biochem.* **2012**, *50*, 15–23. [CrossRef] [PubMed]
37. Bligh, E.G.; Dyer, W.J. A rapid method of total lipid extraction and purification. *Can. J. Biochem. Physiol.* **1959**, *37*, 911–917. [CrossRef]
38. Beermann, C.; Green, A.; Möbius, M.; Schmitt, J.J.; Boehm, G. Lipid class separation by HPLC combined with GC FA analysis: Comparison of seed lipid compositions from different *Brassica napus* L. varieties. *J. Am. Oil Chem. Soc.* **2003**, *80*, 747–753. [CrossRef]
39. Siziba, L.P.; Lorenz, L.; Stahl, B.; Mank, M.; Marosvölgyi, T.; Decsi, T.; Rothenbacher, D.; Genuneit, J. Changes in human milk fatty acid composition during lactation: The Ulm SPATZ health study. *Nutrients* **2019**, *11*, 2842. [CrossRef]
40. Lingua, M.S.; Fabani, M.P.; Wunderlin, D.A.; Baroni, M.V. From grape to wine: Changes in phenolic composition and its influence on antioxidant activity. *Food Chem.* **2016**, *208*, 228–238. [CrossRef]
41. Chambre, D.R.; Tociu, M.; Stanescu, M.D.; Popescu, C. Influence of composition on the thermal behavior of oils extracted from the seeds of some Romanian grapes. *J. Sci. Food Agric.* **2019**, *99*, 6324–6332. [CrossRef]
42. Alonso, A.M.; Guillen, D.A.; Barroso, C.G.; Puertas, B.; Garcia, A. Determination of antioxidant activity of wine byproducts and its correlation with polyphenolic content. *J. Agric. Food Chem.* **2002**, *50*, 5832–5836. [CrossRef]
43. Da Porto, C.; Porretto, E.; Decorti, D. Comparison of ultrasound-assisted extraction with conventional extraction methods of oil and polyphenols from grape (*Vitis vinifera* L.) seeds. *Ultrason. Sonochem.* **2013**, *20*, 1076–1080. [CrossRef]
44. Gonzalez-Centeno, M.R.; Jourdes, M.; Femenia, A.; Simal, S.; Rossello, C.; Teissedre, P.L. Proanthocyanidin composition and antioxidant potential of the stem winemaking byproducts from 10 different grape varieties (*Vitis vinifera* L.). *J. Agric. Food Chem.* **2012**, *60*, 11850–11858. [CrossRef] [PubMed]
45. Lutterodt, H.; Slavin, M.; Whent, M.; Turner, E.; Yu, L.L. Fatty acid composition, oxidative stability, antioxidant and antiproliferative properties of selected cold-pressed grape seed oils and flours. *Food Chem.* **2011**, *128*, 391–399. [CrossRef] [PubMed]
46. Lutz, M.; Jorquera, K.; Cancino, B.; Ruby, R.; Henriquez, C. Phenolics and antioxidant capacity of table grape (*Vitis vinifera* L.) cultivars grown in Chile. *J. Food Sci.* **2011**, *76*, C1088–C1093. [CrossRef] [PubMed]
47. Tang, G.Y.; Zhao, C.N.; Liu, Q.; Feng, X.L.; Xu, X.Y.; Cao, S.Y.; Meng, X.; Li, S.; Gan, R.Y.; Li, H.B. Potential of grape wastes as a natural source of bioactive compounds. *Molecules* **2018**, *23*, 2598. [CrossRef]
48. Xia, L.; Xu, C.; Huang, K.; Lu, J.; Zhang, Y. Evaluation of phenolic compounds, antioxidant and antiproliferative activities of 31 grape cultivars with different genotypes. *J. Food Biochem.* **2019**, *43*, e12626. [CrossRef]
49. Guchu, E.; Ebeler, S.E.; Lee, J.; Mitchell, A.E. Monitoring selected monomeric polyphenol composition in pre- and post-fermentation products of *Vitis vinifera* L. cv. Airén and cv. Grenache noir. *LWT Food Sci. Technol.* **2015**, *60*, 552–562. [CrossRef]
50. Dudoit, A.; Benbouguerra, N.; Richard, T.; Hornedo-Ortega, R.; Valls-Fonayet, J.; Coussot, G.; Saucier, C. α-glucosidase inhibitory activity of Tannat grape phenolic extracts in relation to their ripening stages. *Biomolecules* **2020**, *10*, 1088. [CrossRef]
51. Farhadi, K.; Esmaeilzadeh, F.; Hatami, M.; Forough, M.; Molaie, R. Determination of phenolic compounds content and antioxidant activity in skin, pulp, seed, cane and leaf of five native grape cultivars in West Azerbaijan province, Iran. *Food Chem.* **2016**, *199*, 847–855. [CrossRef]
52. Pastrana-Bonilla, E.; Akoh, C.C.; Sellappan, S.; Krewer, G. Phenolic content and antioxidant capacity of muscadine grapes. *J. Agric. Food Chem.* **2003**, *51*, 5497–5503. [CrossRef]
53. Rockenbach, I.I.; Rodrigues, E.; Gonzaga, L.V.; Caliari, V.; Genovese, M.I.; de Souza Schmidt Gonçalves, A.E.; Fett, R. Phenolic compounds content and antioxidant activity in pomace from selected red grapes (*Vitis vinifera* L. and *Vitis labrusca* L.) widely produced in Brazil. *Food Chem.* **2011**, *127*, 174–179. [CrossRef]
54. Özcan, M.M.; Al Juhaimi, F.; Gülcü, M.; Uslu, N.; Geçgel, Ü. Determination of bioactive compounds and mineral contents of seedless parts and seeds of grapes. *S. Afr. J. Enol. Vitic.* **2017**, *38*, 212–220. [CrossRef]
55. Somkuwar, R.G.; Bhange, M.A.; Oulkar, D.P.; Sharma, A.K.; Ahammed Shabeer, T.P. Estimation of polyphenols by using HPLC–DAD in red and white wine grape varieties grown under tropical conditions of India. *J. Food Sci. Technol.* **2018**, *55*, 4994–5002. [CrossRef]

56. Rockenbach, I.I.; Gonzaga, L.V.; Rizelio, V.M.; de Souza Schmidt Gonçalves, A.E.; Genovese, M.I.; Fett, R. Phenolic compounds and antioxidant activity of seed and skin extracts of red grape (*Vitis vinifera* and *Vitis labrusca*) pomace from Brazilian winemaking. *Food Res. Int.* **2011**, *44*, 897–901. [CrossRef]
57. Agostini, F.; Bertussi, R.A.; Agostini, G.; Atti dos Santos, A.C.; Rossato, M.; Vanderlinde, R. Supercritical extraction from vinification residues: Fatty acids, α-tocopherol, and phenolic compounds in the oil seeds from different varieties of grape. *Sci. World J.* **2012**, *2012*, 1–9. [CrossRef]
58. Al Juhaimi, F.; Geçgel, Ü.; Gülcü, M.; Hamurcu, M.; Özcan, M.M. Bioactive properties, fatty acid composition and mineral contents of grape seed and oils. *S. Afr. J. Enol. Vitic.* **2017**, *38*, 103–108. [CrossRef]
59. Elagamey, A.A.; Abdel-Wahab, M.A.; Shimaa, M.M.E.; Abdel-Mogib, M. Comparative study of morphological characteristics and chemical constituents for seeds of some grape table varieties. *J. Am. Sci.* **2013**, *9*, 447–454. [CrossRef]
60. Jin, Q.; O'Hair, J.; Stewart, A.C.; O'Keefe, S.F.; Neilson, A.P.; Kim, Y.-T.; McGuire, M.; Lee, A.; Wilder, G.; Huang, H. Compositional characterization of different industrial white and red grape pomaces in Virginia and the potential valorization of the major components. *Foods* **2019**, *8*, 667. [CrossRef]
61. Decsi, T.; Csabi, G.; Török, K.; Erhardt, É.; Minda, H.; Burus, I.; Molnár, S.; Molnár, D. Polyunsaturated fatty acids in plasma lipids of obese children with and without metabolic cardiovascular syndrome. *Lipids* **2000**, *35*, 1179–1184. [CrossRef]
62. Wu, J.H.Y.; Marklund, M.; Imamura, F.; Tintle, N.; Ardisson Korat, A.V.; de Goede, J.; Zhou, X.; Yang, W.-S.; de Oliveira Otto, M.C.; Kröger, J.; et al. Omega-6 fatty acid biomarkers and incident type 2 diabetes: Pooled analysis of individual-level data for 39 740 adults from 20 prospective cohort studies. *Lancet Diabetes Endocrinol.* **2017**, *5*, 965–974. [CrossRef]
63. Farvid, M.S.; Ding, M.; Pan, A.; Sun, Q.; Chiuve, S.E.; Steffen, L.M.; Willett, W.C.; Hu, F.B. Dietary linoleic acid and risk of coronary heart disease: A systematic review and meta-analysis of prospective cohort studies. *Circulation* **2014**, *130*, 1568–1578. [CrossRef]
64. Marklund, M.; Wu, J.H.Y.; Imamura, F.; del Gobbo, L.C.; Fretts, A.; de Goede, J.; Shi, P.; Tintle, N.; Wennberg, M.; Aslibekyan, S.; et al. Biomarkers of dietary omega-6 fatty acids and incident cardiovascular disease and mortality. *Circulation* **2019**, *139*, 2422–2436. [CrossRef]
65. Zhuang, P.; Zhang, Y.; He, W.; Chen, X.; Chen, J.; He, L.; Mao, L.; Wu, F.; Jiao, J. Dietary fats in relation to total and cause-specific mortality in a prospective cohort of 521 120 individuals with 16 years of follow-up. *Circ. Res.* **2019**, *124*, 757–768. [CrossRef]
66. Das, U.N. Essential fatty acids—A review. *Curr. Pharm. Biotechnol.* **2006**, *7*, 467–482. [CrossRef] [PubMed]
67. Julibert, A.; Bibiloni, M.d.M.; Mateos, D.; Angullo, E.; Tur, J.A. Dietary fat intake and metabolicsyndrome in older adults. *Nutrients* **2019**, *11*, 1901. [CrossRef]
68. Fogacci, F.; Tocci, G.; Presta, V.; Fratter, A.; Borghi, C.; Cicero, A.F.G. Effect of resveratrol on blood pressure: A systematic review and meta-analysis of randomized, controlled, clinical trials. *Crit. Rev. Food Sci. Nutr.* **2018**, *59*, 1605–1618. [CrossRef] [PubMed]
69. Koushki, M.; Dashatan, N.A.; Meshkani, R. Effect of resveratrol supplementation on inflammatory markers: A systematic review and meta-analysis of randomized controlled trials. *Clin. Ther.* **2018**, *40*, 1180–1192.e5. [CrossRef]
70. Li, H.; Xia, N.; Hasselwander, S.; Daiber, A. Resveratrol and vascular function. *Int. J. Mol. Sci.* **2019**, *20*, 2155. [CrossRef]

antioxidants

MDPI

Review

Spices, Condiments, Extra Virgin Olive Oil and Aromas as Not Only Flavorings, but Precious Allies for Our Wellbeing

Irene Dini * and **Sonia Laneri**

Department of Pharmacy, University of Naples Federico II, Via Domenico Montesano 49, 80131 Naples, Italy; slaneri@unina.it
* Correspondence: irdini@unina.it

Abstract: Spices, condiments and extra virgin olive oil (EVOO) are crucial components of human history and nutrition. They are substances added to foods to improve flavor and taste. Many of them are used not only to flavor foods, but also in traditional medicine and cosmetics. They have antioxidant, antiviral, antibiotic, anticoagulant and antiinflammatory properties and exciting potential for preventing chronic degenerative diseases such as cardiomyopathy and cancer when used in the daily diet. Research and development in this particular field are deeply rooted as the consumer inclination towards natural products is significant. It is essential to let consumers know the beneficial effects of the daily consumption of spices, condiments and extra virgin olive oil so that they can choose them based on effects proven by scientific works and not by the mere illusion that plant products are suitable only because they are natural and not chemicals. The study begins with the definition of spices, condiments and extra virgin olive oil. It continues by describing the pathologies that can be prevented with a spicy diet and it concludes by considering the molecules responsible for the beneficial effects on human health (phytochemical) and their eventual transformation when cooked.

Keywords: spices; condiments; extra-virgin olive oil; antiviral properties; antioxidant properties; nutricosmetic

Citation: Dini, I.; Laneri, S. Spices, Condiments, Extra Virgin Olive Oil and Aromas as Not Only Flavorings, but Precious Allies for Our Wellbeing. *Antioxidants* **2021**, *10*, 868. https://doi.org/10.3390/antiox10060868

Academic Editor: Jesús R. Huertas

Received: 11 May 2021
Accepted: 27 May 2021
Published: 28 May 2021

1. Introduction

Spices and condiments have played an essential role in human nutrition and participated in developing most cultures worldwide. The use of curry was known in 2000 B.C.E. in India. In Egypt and Babylon, spices such as garlic, cumin and coriander were considered magical. The Greeks and Romans used anise, savory, basil, garlic, hyssop, fennel, mustard, capers, cumin, coriander, oregano, myrtle, parsley, verbena in the kitchen, medicine and cosmetics. Marco Polo in the 13th and the European colonization of Africa, America and Asia during the 15th to 17th centuries improved and spread condiments and spices worldwide [1]. Spices and cooking processes contribute to the ethnic identity of food [2]. Ethnic foods have increased their popularity among consumers worldwide since tourism, international trade and immigration raised the possibility of tasting them. Social media and the opportunity to share culinary experiences also contributed [3–6]. Partly driven by the improved popularity of ethnic food consumption, the global seasoning and spices market was USD 136.24 billion in 2019 and its growth rate is probable to grow by 4.8% from 2015 to 2025 steadily. The global seasoning and spices market size was valued at USD 13.77 billion in 2019 and is expected to grow at a compound annual growth rate (CAGR) of 6.3% from 2020 to 2027 [7]. The nutrients and phytochemicals in spices, extra virgin olive oil and flavorings are widely used in traditional medicine, pharmaceuticals, dental preparation, aromatherapy and nutraceuticals [8]. The Dietary Supplement and the Education Act have defined "nutraceuticals" as supplements containing herbs, plant products, metabolites, or extracts singly or combined [9]. Currently, the cosmetics industry uses spices and extra virgin olive oil to prepare food supplements and topical skincare cosmetics to combat blemishes from the inside and outside simultaneously [10–13]. In this study, the nutraceutical potential of spices, herbs and condiments, is revised to demonstrate the fundamental role of safeguarding our health that they have in the diet. Special attention is paid to the transformations that can occur during cooking, the synergistic effects linked to the simultaneous use of more than one seasoning and any toxic effects to maximize the biological impact and avoid any side effects. A brief mention of extra virgin olive oil properties is given since it contributes synergistically with spices, herbs and condiments to the nutraceutical value of the finished dish when used to flavor foods.

2. What Differentiates Spices from Herbs, Condiments, Aromas and Extra Virgin Olive Oil?

According to the Codex Alimentarius, "herbs, spices, seasonings and condiments" are considered food flavoring substances usually from botanical sources, dehydrated, ground or whole, added food to improve aroma and taste. Instead, extra virgin olive oil is contained in the section "fat and oils" [14]. Spices, herbs, salt, salt substitutes, vinegar, seasonings, condiments, mustards, sauces, soups and broths, salads, spreads, soy-based condiments, protein products from sources other than soy, yeast and similar products are a part of "herbs, spices, seasonings, and condiments" section.

2.1. *Spices and Herbs*

Spices are mixtures in powder or paste form, such as chili seasoning and curry paste. They are sometimes dried before use. Spices are obtained from the bark, bulbs (e.g., onion and garlic), fruits (e.g., peppers and star anise), flowers (orange and lavender, seeds (e.g., fennel, coriander, sesame and cumin), roots (e.g., ginger and turmeric), or the entire plant (e.g., cinnamon) [15]. Spice plants are often used as sources of phytochemicals and essential oils (Eos) [16]. Phytochemicals are bioactive substances which quality and content in spices depend on plant variety, part of the plant, pedoclimatic condition, harvest period, drying, type of processing and storage [15]. The Eos in spices are aromatic oily liquids containing pharmacologically active components (principally terpenoid and phenolic compounds). Eos are obtained by steam distillation, cold-soaking, extrusion, or solvent extraction [17].

The "herbs or culinary herbs" are plants with aromatic leaves, stems and flowers such as basil, parsley, rosemary and oregano.

2.2. *Condiments and Seasonings*

Condiments are prepared food flavoring containing spices or spices extractives in single forms (e.g., onion salt, garlic salt) or mixtures of constituents (e.g., mustard, chili sauce) which are added to food during cooking and/or eating [18]. They are available in liquid, semisolid and solid forms. The food category "condiments", in the Codex Alimentarius, does not contain condiment sauces (e.g., mayonnaise, ketchup, mustard) or relishes. Sauces are liquid or semi-liquid products able to enhance the appearance, aroma and flavor of foods. They may or not include spice or spice extracts [18]. Seasoning is the method of adding salts, spices, or herbs to food to improve the flavor. In the last years, condiments, seasonings and spices are used as vehicles for micronutrient fortification since they are cheap and widely consumed by people of all socioeconomic backgrounds [19]. Their colors and flavors mask undesirable organoleptic characteristics from fortification and intensely flavored condiments avoid the overconsumption of a fortified nutrient. Soy sauce, bouillon cubes and fish sauce are used as vehicles to enhance iron consumption [20]. Bouillon cubes are dry broth made by dehydrated meat stock or vegetables. In West Africa, where the bouillon industry is more enhanced than the salt production industry, bouillon cubes are fortified with iodized salt alone or combined with zinc, iron, folic acid, vitamin A, or other B vitamins [21,22].

3. Condiment

3.1. Vinegar

Vinegar is a common condiment worldwide. It is obtained from the transformation of sugars to ethanol by yeasts and ethanol's oxidation to acetic acid by bacteria in cider, wine, malt, grain, spirit, raisin and other fruit [23]. It is possible to buy a spray-dried vinegar powder, which provides maximum vinegar taste and may be rehydrated with water in a ratio of 1:1/vinegar:water. The health benefits of vinegar are due to acetic acid and other organic acids (mainly acetic acid and lactic acid together with malic acid, tartaric acid, citric acid and succinic acid), amino acids, phenolic compounds (e.g., catechin, chlorogenic acid, syringic acid, ferulic acid, protocatechuic acid, caffeic acid, gallic acid and *p*-coumaric acid), flavanols (e.g., epicatechin), flavonols (e.g., rutin), anthocyanidin (e.g., malvidin-3-glucoside), anthocyanin (e.g., pyranoanthocyanin), carotenoids, vitamins (B group and C), minerals (Al, Ca, Cr, Fe, K, Mn, Mg, Na, P, S and Zn), alkaloids and sugars (glucose, arabinose, fructose, xylose and mannitose) able to induce antioxidant, antitumor, antidiabetic, antihypertensive, antiobesity, antiinflammatory and antimicrobial effects [24,25]. The bioactive compounds' composition and concentration depend on the raw material used to produce the vinegar and the production method [26]. Cosmetic products based on vinegar are formulated to counteract the signs of skin aging [27].

3.2. Extra-Virgin Olive Oil

Extra virgin olive oil (EVOO) is described in Section 2 of the Codex Alimentarius: "Standards for Fats and Oils from Vegetable Sources". EVOO is made from the olive tree's fruit by mechanical processes and purified by washing with water, filtering, settling and centrifuging only [28]. It is often debated whether EVOO should be considered a condiment or a functional condiment since it is used to enhance the appearance, aroma and flavor of foods and it is rich in nutrients and natural antioxidants like a functional food. The healthy and nutritional values are ascribable to monounsaturated fatty acids (MUFAs: 16:1; 18:1; 20:1), polyunsaturated fatty acids (PUFAs), squalene, triterpenic acids, phytosterols, dialcohols, polyphenols and tocopherols [29,30]. Oleic acid (55.0–83.0% of the lipid content), palmitic (7.5–20.0% of the lipid content), linoleic (3.5–21.0% of the lipid content), stearic (0.5–5.0% of the lipid content), palmitoleic (0.3–3.5% of the lipid content), linolenic acids are the more representative fatty acids in the EVOO. Instead, myristic, eicosanoic acids, heptadecanoic are present in traces [31]. Olive variety, agronomic conditions and the olives' ripening affect the fatty acid composition and content. The International Olive Council [32], Codex Alimentarius [33] and European regulations [34–36] norm their content in the EVOO. Sterols (or phytosterols) are another group of naturally occurring lipids in EVOO. Their range varies between 800 and 2600 mg/Kg [37]. Three sterols' classes are identified in the EVOO according to the presence of methylic groups in position C4 in rings A. 4-desmethylsterols are sterols without methyl group (e.g., β-sitosterol, Δ5-campesterol and avenasterol), 4-monomethylsterols have one methyl group (e.g., citrostadienol, cycloeucalenol, obtusifoliol and gramisterol) and 4,4′-dimethylsterols have two methyl groups (e.g., α-amyrin, cycloartenol and β-amyrin) (Figure 1). The most abundant sterol is β-sitosterol (75–90%). The sterol' levels diminish during the oil's storage and enhance peroxides [38]. The most concentrated phenolic compounds in the EVOO are lignans (e.g., pinoresinol, hydroxypinoresinol, acetoxypinoresinol), followed by secoiridoids (e.g., ligstroside, ligstroside decarboxymethyl aglycone, oleuropein, oleuropein aglycone mono-aldehyde, etc.), phenolic alcohols (e.g., tyrosol), flavones (e.g., luteolin, apigenin), flavonols (e.g., quercetin-3-rutinoside), anthocyanidin, (e.g., cyanidin glucosides) and phenolic acids (e.g., vanillic acid, ferulic acid, cinnamic acid, hydroxybenzoic acid) [39]. These substances modulate aging-associated processes and have antiviral, antitumor, anti-atherogenic, antihepatotoxic, antiinflammatory, immunomodulatory, anti-autoimmune (i.e., rheumatoid arthritis) and hypoglycemic properties [40]. The phenolic compounds profile and concentration in EVOO vary significantly according to the olive cultivar, pedoclimatic factors (altitude and amount of irrigation), agricultural practices [41], oil extraction methods and storage conditions.

The EVOO quality is linked to olive fruits free of damage and the absence of pesticide residues (e.g., fungicides, insecticides and herbicides). Biological control using *Trichoderma* species or their metabolites are new options to select the EVOO phenolic profile [40] and terpenoid profiles [42]. The nutraceutical importance of phenolics forced researchers to develop reliable analytical methods for their oil dosage [43]. Moreover, EVOO contains tocopherols [31]. They act as free radical scavengers in membranes and lipoproteins and transform fatty acid peroxyl radicals into tocopheroxyl radicals. α-tocopherol regulates signal transduction, apoptosis pathways and transcriptional regulation of the cell cycle [44].

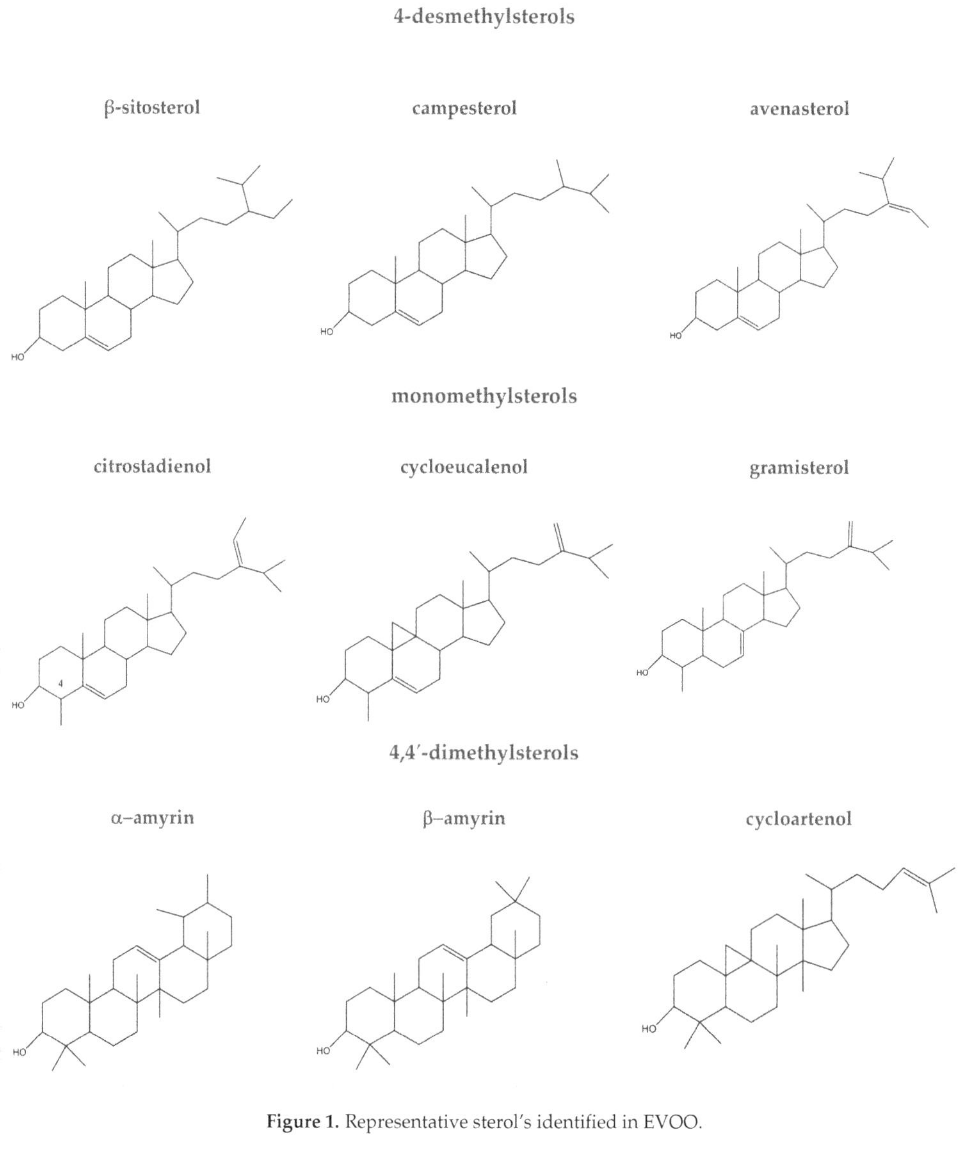

Figure 1. Representative sterol's identified in EVOO.

4. Sauces

4.1. Soy Products Fermented

Soybeans contain some bioactive compounds such as proteins (e.g., β-conglycinin and glycinin), polyunsaturated fatty acids, lecithin, vitamin E, saponin and isoflavones (e.g., genistein, daidzin, daidzein and glycitein) [45,46]. The fermentation processing method enhances the organoleptic and nutritional properties of soybean products. In fermented soybean products, the isoflavone profiles change (isoflavones aglycone-form increase) and improve the phenols' total content, flavonoids [47] and bioactive peptides. Bioactive peptides have shown hypotriglyceridemic, hypocholesterolemic, antiobesity, antidiabetic, hypotensive, anticancer, antioxidant and antiinflammatory in experimental models [48].

4.1.1. Soy Sauce

Soy sauce is a liquid seasoning currently, used in cooking worldwide. Studies suggest that soy sauce contains polysaccharides, protein, MUFA (e.g., 18:1 fatty acids), PUFA (18:2 and 18:3 fatty acids), minerals (calcium, iron, magnesium, phosphorus, potassium, sodium, zinc, copper, selenium), vitamins (C, B1, B2, B3 and B6) [49], melanoidins, isoflavones, phenolic acids, furan ketones, peptides, organic acids and β-carbolines [50]. The polysaccharides from soy sauce regulate iron and lipid absorption in the gastrointestinal tract [51]. Soy sauce exhibits antioxidant, antihyperuricemic, antimicrobial, anticarcinogenic, anticataract and antiplatelet activities [52]. Sometimes medicinal herbs are added in soy sauce to improve the antioxidant profile, e.g., citrus peels rich in flavanons (e.g., rutin and hesperidin), flavones (nobiletin), with anticancer, antioxidation, anti-inflammation and cardiovascular diseases protective properties [52].

4.1.2. Miso

Miso is a seasoning made by fermenting soybeans with kōji (the fungus *Aspergillus oryzae*), salt and sometimes barley, rice, seaweed or other ingredients. It contains vitamins (B2, B12, E), fat, protein, minerals (iron, calcium, sodium), carbohydrates, [53], saponin, lecithin, phenolic acids (ferulic, vanillic, p-OH-benzoic, p-coumaric, syringic) and isoflavonoids (daidzein, genistein) [54,55]. Miso has antioxidant, ACE-inhibitor, stroke preventive, antihypertensive and anticancer properties [56]. Miso improves the absorption of Coenzyme Q10 (CoQ10) supplements. The CoQ10 is a lipid-soluble antioxidant involved in energy production which can improve the symptoms of some geriatric disorders (e.g., glucose metabolism in diabetes, high blood pressure and the symptoms of Parkinson's disease), decrease peripheral oxidative stress and inflammation [57].

4.2. Fish Sauce

Fish sauce is a fermented condiment with a mild fishy flavor, traditionally used in East and Southeast Asian countries [58]. It is obtained by transforming lipids and proteins with enzymes and halophilic microorganisms [59]. Endogenous proteases and proteases produced by microorganisms hydrolyze the proteins in fish into peptides and amino acids [60,61]. The amino acids in fish sauces contribute to the umami taste and have some biological activity among these anti-oxidative, antithrombotic, hypocholesterolemic, antidiabetic and antihypertensive effects are reported [62,63]. Moreover, they inhibit the ACE enzyme (angiotensin I-converting enzyme) [64], able to stabilize blood pressure by transforming angiotensin I to the potent vasoconstrictor angiotensin II and inactivating the vasodilator peptide bradykinin [65]. The fish sauce contains vitamins (A, B1, B2, B3 and B9), fat, protein, minerals (iron, calcium and phosphorus), carbohydrates [66,67] and high levels of docosahexaenoic acid [68] are able to regulate the symptoms of atopic dermatitis [69].

4.3. Tabasco

Tabasco is red pepper hot-sauce sauce obtained by lactic acid fermentation due to autochthonous bacteria [70]. Sodium chloride is added to the peppers to make the pulp microbiologically safe. It selects homofermentative lactic acid bacteria and destroys enterobacteria [71]. After four weeks, vinegar and salt are added to pepper extract to obtain tabasco sauce. Tabasco contains vitamins (C, B1, B2, B3, B5, B6, B9, A and E), fat, protein, minerals (iron, calcium, magnesium, potassium, sodium and phosphorus), carbohydrates [72]. Enzymatic hydrolysis disrupts the cell walls cutting the polysaccharide chains, improve the extract yield and make available bioactive molecules, such as capsaicinoids (capsaicin, dihydrocapsaicin and nordihydrocapsaicin), carotenoids, flavor compounds and polyphenols [73].

5. Spices Commonly Used in Food Preparation

5.1. Curry

Curry is a combination of spices (turmeric, cumin, coriander, paprika, cardamom and other spices) and herbs and its composition varies between regions [74]. It contains fat, protein, minerals (e.g., iron, calcium, sodium), carbohydrates, fiber [75] and phytochemicals such as flavanols (e.g., catechin), flavonols (e.g., quercetin, kaempferol) [76], carbazole (murrayanol, murrayagetin, marmesin-1″-O-rutinoside, mukoenine-A, -B and C, murrastifoline–F, bis–2-hydroxy-3-methyl carbazole, bismahanine, biskoeniquinone-A and bismurrayaquinone A, koenoline, mukoline, mukolidine). Phytochemicals in curry have antioxidant, antidiabetic, cytotoxic, anticancer, immunomodulatory, antiobesity, antihyperlipidemic, hepatoprotective [77] and skincare activities [78].

5.2. Tumeric

Turmeric (*Curcuma longa*) is considered the golden spice of India. It is obtained from the rhizome of a herbaceous plant that belongs to the ginger family *Zingiberaceae*. [79]. Tumeric is widely used as a spice, coloring material, food and preservative in South East Asia, Africa and Brazil. The bright yellow spice is obtained by boiling and drying rhizomes. Turmeric spice has a hot, bitter flavor and a minor fragrance of ginger and orange. It is used to make a curry spice and mustard [80]. The rhizomes contain vitamin C, minerals (e.g., iron, calcium and sodium) [81], flavanols (e.g., catechin), flavonols (e.g., kaempferol and myricetin) [76], curcuminoids (e.g., curcumin, 5-methoxycurcumin, demethoxycurcumin, bis-demthoxycurcumin, cyclocurcumin and dihydrocurcumin), sesquiterpenes (e.g., germacrone, ar-, α, β-turmerones, turmerone, β-bisabolene, zingiberene, α-curcumene, bisacurone, β-sesquiphellandene, curcumenone, procurcumadiol dehydrocurdinone, bisacumol, isoprocurcumenol, curcumenol, epiprocurecumenol, curlone zedoaronediol and turmeronols A and B), steroids (e.g., β-sitosterol, stigmasterol, cholesterol, 2-hydroxymethyl anthraquinone and anthraquinone) and essential oils (e.g., α-phellandrene, cineol, sabinene, sesquiterpenes with turmerones skeleton and borneol) [82]. Tumeric rhizomes are used as stimulants, stomachs and blood purifiers to prevent anorexia, diabetic wounds, hepatic disorders, rheumatism, sinusitis, bronchitis, asthma, skin infections and eye infections [82].

5.3. Fenugreek

Fenugreek (*Trigonella foenum-graecum* Linn.) belongs to the *Fabaceae* family. The leaves, seeds and flowers are used dry. The seeds release a maple–curry–nutty flavor by crashing. Leaves and sprouts have a sweeter taste than the seeds and are eaten as a vegetable and mixed into dough, stews and beans. It contains amino acids (glutamic acid, aspartic acid, leucine, tyrosine, phenyl cysteine and alanine), fatty acids (e.g., mono- and di-galactodiacylglycerols, oleic acid, linolenic acid, linoleic acid, glycolipids, phosphatidylethanolamine and phosphatidylcholine), vitamins (e.g., A, B1, B2, C, niacin, nicotinic acid and folic acid) and minerals (e.g., Fe, P, Ca, Mg, S, Cu, Co, Zn, Mn and Br) [83]. The phytochemical analysis of fenugreek has revealed the presence of furostanols (e.g., protodioscin derivatives) and spirostanols (e.g., dioscin derivatives) saponins, steroids, alkaloids (e.g., trigonelline), flavonols (e.g., quercetin-3-O-rhamnoside), flavons (e.g., vitexin-7-O-glucoside, apigenin-6-C-glucoside, apigenin-6-C-glucoside, apigenin-8-C-glucoside, apigenin-6-C-xyloside-8-C-glucoside, apigenin-6 and 8-C-diglucoside), isoflavonoids (e.g., maackiaian and medicarpin), terpenes and phenolic acid derivatives (e.g., caffeic acid, p-coumaric acid and chlorogenic acid, hymecromone, trigocoumarin, trigoforin, scopoletin and γ-schizandrin) [83]. Pharmaceutical employment of fenugreek is related to diabetes, obesity, hyperlipidemia, inflammation damages, cancer, oxidative stress reparations and improving women's health [83].

5.4. Garlic

Garlic (*Allium sativum*) is an herb of the Liliaceae family. *Allium* is derived from the Celtic word al (burning, pungent). The bulb is widely used as a culinary spice and in traditional medicine [84]. It contains vitamins (e.g., A and C) [85] and some bioactive compounds such as flavanols (e.g., catechin), flavonols (e.g., kaempferol, myricetin and quercetin) [76], organosulfur compounds (e.g., allicin, diallyl sulfide, diallyl disulfide, diallyl trisulfide, S-allyl-cysteine, E/Z-ajoene and alliin), phenolic compounds (e.g., β-resorcylic acid, pyrogallol, gallic acid and protocatechuic acid), saponins (e.g., proto-desgalactotigonin, desgalactotigonin-rhamnose, proto-desgalactotigonin-rhamnose, sativoside B1-rhamnose, voghieroside D1 and sativoside R1) and polysaccharides [86–90]. Garlic has antioxidant, antiinflammatory, antiobesity, antidiabetic, anticancer, cardiovascular protective, immunomodulatory and antibacterial properties [91]. The antioxidant properties of garlic are related to organosulfur compounds, flavonoids and saponins. Garlic improves and regulates the antioxidant enzyme activities (heme oxygenase-1 and the glutamate-cysteine ligase modifier) and the nuclear erythroid 2-related factor 2 (Nrf2-ARE) pathway [92,93]. Garlic could constrain inflammation by impeding inflammatory mediators' action (e.g., nitric oxide, tumor necrosis factor-α and interleukine-1). It decreases nitric oxide production and prostaglandin E-2 by reducing the expression of inducible NO synthase, cyclooxygenase-2 and the transcription of the nuclear factor-kappa B [94,95]. The main immune-modulating components in garlic are polysaccharides. They have an immunomodulatory effect and regulate the expressions of tumor necrosis factor-α, IL-6, IL-10 and interferon-γ in macrophages. Polysaccharides in fresh garlic exhibit a more potent activity on the immune system than fermented garlic since the fructans degrade during processing [96]. Garlic's cardiovascular protective effects are related to inhibition of oxidative stress and lipid peroxidation, control of angiotensin-converting enzymes and NO and H_2S production. Moreover, garlic powder can reduce blood pressure, cholesterol (total and low-density lipoprotein cholesterol) and platelet aggregation [91]. It decreases hypertension by reducing oxidative stress, improving NO and hydrogen sulfide production and inhibiting the angiotensin-converting enzyme [92]. Garlic prevents different cancer pathologies by regulating carcinogen metabolism, decreasing cell growth and proliferation, inducing apoptosis, destructing angiogenesis and preventing invasion and migration [91]. Garlic enhances gastrointestinal functions and relieves gastric ulcers and colitis, by decreasing inflammation, oxidative stress and Helicobacter pylori levels [91]. Finally, fermented garlic reduces obesity by impeding lipogenesis and controlling lipid metabolism [91].

5.5. Ginger

Ginger (*Zingiber officinale*) rhizome is consumed as a fresh paste, dried powder, slices preserved in syrup, crystallized ginger, or tea flavoring. It contains carbohydrates, protein, free amino acids, fatty acids, triglycerides, ash, crude fiber [97–99], minerals (e.g., potassium, copper, magnesium, silicon, manganese), vitamins (e.g., A, E, C, B1, B2, B3, B5, B6, B9 and B12) [100,101], flavanols (e.g., catechin), flavonols (e.g., myricetin) [76], oleoresin (e.g., sesquiterpene hydrocarbons), phenolic compounds (e.g., gingerole, shogoals), diasyleheptanoids (e.g., gingerenone), curcuminoids (e.g., curcumin), alkaloids, carotenoids, tannins, flavonoids, saponins, cardinolides and steroids [100]. Ginger has antioxidant, antiinflammatory, anticancer, hypocholesterolemic, cardio preventive, antibiotic and antimicrobial effects [102].

5.6. Chilli Pepper

Chilli pepper (*Capsicum annuum*) is a well-known domesticated species of the genus *Capsicum*. It contains vitamin C, carotenoids (e.g., β-carotene, antheraxanthin, violaxanthin, zeaxanthin, capsanthin, capsorubin and lutein), capsaicinoids, phenolic acids (e.g., chlorogenic acid, caffeic acid, ferulic acid, coumaric acid), flavonols (e.g., rutin and quercetin) and flavanones (e.g., hesperidin), [103,104]. Health-promoting chilly pepper activities are associated with antioxidants and antiinflammatory activities of carotenoids and phenols [104]. It has chemopreventive, antidiabetic, antiobesity, cardioprotective, hepatoprotective and photoprotective skin properties [104].

6. Herbs Commonly Used in Food Preparation

6.1. Basil

Ocimum basilicum L., belonging to the Lamiaceae family, contains polysaccharides, vitamins, minerals (e.g., magnesium, calcium, iron and zinc), fatty acids (e.g., stearic acid, oleic acid, palmitic acid, linoleic acid, myristic acid, α-linolenic acid, capric acid, lauric acid and arachidonic acid), steroids, phenolic acids (e.g., caffeic acid, vanillic acid, rosmarinic acid, chlorogenic acid and p-hydroxybenzoic), flavonols (e.g., quercetin and rutin), flavones (e.g., apigenin) [105], β-carotene, terpenes, alcohols, aldehydes, ketones esters and ethers [106]. Biological effects of basil include antioxidant, anticancer, antiatherosclerotic, hypolipidemic, antidiabetic, immunity boost and antiaging activities [107].

6.2. Parsley

Parsley (*Petroselinum crispum*) belongs to Apiaceae (Umbelliferae) family. It contains flavones (e.g., diosmetin 7-malonylapiosylglucoside, diosmetin/chrysoeriol 7-malonylapiosylglucoside, diosmetin 7-apiosylglucoside, apigenin dihexoside and apigenin 7-apiosylglucoside), flavonols (e.g., isorhamnetin 3-alonylglucoside-7-glucoside, isorhamnetin 3-malonylglucoside-7-glucoside, isorhamnetin 3-glucoside, apigenin 7-malonylapiosylglucoside, apigenin 7-malonylapiosylglucoside, apigenin 7-malonylapiosylglucoside, apigenin 7-dimalonylapiosylglucoside and apigenin 7-glucoside), flavanones (e.g., hesperetin 7-glucoside), carotenoids (e.g., beta-cryptoxanthin, beta-carotene, lutein and zeaxanthin), coumarins (bergapten, psoralen and xanthotoxin), tannins and triterpenes [108]. Biological activities ascribed to parsley are antidiabetic, antihypertensive, antioxidant, antitumorigenic and gastroprotective activities [109].

6.3. Fennel

Fennel (*Foeniculum vulgare* Mill.) is a perennial, umbelliferous herb. Stems, leaves and shoots are used in culinary traditions. It contains carbohydrates, essential fatty acids (e.g., ω-6 and ω-3) [110], phenylpropanoids (e.g., estragole and trans-anethole), monoterpenes (e.g., α-pinene and α-phellandrene), hydrocinnamic acids (e.g., neochlorogenic acid, chlorogenic acid, criptochlrogenic acid, 5-feruoylquinic acid,1,4-O-dicaffeoylquinic acid and 1,5-O-dicaffeoylquinic acid) and flavonols (e.g., quercetin-3-O-glucuronide and kaempferol-3-O-glucuronide). Fennel's beneficial properties include antithrombotic, anticancer, antioxidant, antiinflammatory, hepatoprotective and antidiabetic properties [111].

6.4. Sage

Sage (*Salvia officinalis* L.) is an aromatic herb belonging to the Lamiaceae family. It contains carbohydrates, protein, total lipids, MUFA, fiber, minerals (e.g., calcium, iron, magnesium, phosphorus, potassium, zinc, manganese, selenium), vitamins (e.g., C, B1, B2, B3, B6, B9 and E) [112], phenolic acids (e.g., gallic acid, chlorogenic acid, caffeic acid, coumaric acid ferulic acid, rosmarinic acid) flavonols (e.g., quercetin, rutin and myricetin), terpene/terpenoids (camphor, borneol, caryophyllene, elemene, cineole, humulene, ledene, thujone and pinene) [113]. The antioxidant and antiinflammatory properties of sage are related to the terpenoids compounds, phenolic acids and flavonoids [114]. The cryptotanshinone (a diterpenoid) produces an agonistic activity on the opioid system [115]. Terpenes and terpenoids are responsible for anticancer activity. The α-humulene and caryophyllene inhibit breast cancer and colorectal cancer tumor cells; manool induces selective cytotoxicity on human glioblastoma and cervical adenocarcinoma; ursolic acid constrains angiogenesis and invasion of melanoma cells; and rosmarinic acid inhibits the growth of the colon, breast, hepatocellular, prostate, small cell lung carcinomas and chronic myeloid leukemia [114].

7. Description of Processes in Which the Phytochemicals and Nutrients in Spices, Condiments and Sauces Can Intervene to Exert Their Beneficial Action

7.1. Oxidative Stress

Oxidative stress is produced by an excess of the reactive oxygen species (ROS, e.g., superoxide, hydroxyl radical, hydrogen peroxide and singlet oxygen) and reactive nitrogen species (RNS, e.g., nitric oxide, peroxynitrite, nitrogen dioxide) [116]. ROS production in living organisms is due to phagocytosis, respiratory chain (endogenous reactions), exposure to UV radiation, air pollutants and other physical and chemical agents. The electrons made in the mitochondrial respiratory chain are transferred to molecular oxygen forming superoxide anion. The nitric oxide improves the generation of superoxide anion-producing peroxynitrite enzyme, leading to the increased oxidation of proteins, carbohydrates and lipids. ROS make membrane lipid peroxidation and determines the loss of membrane fluidity, altering cell homeostasis. Humans have endogenous defense mechanisms, such as superoxide dismutase, glutathione peroxidase and catalase, to protect against ROS-induced damage. Improved ROS production alters the balance between oxidant and antioxidant levels, determining a pro-oxidative condition [117]. Oxidative stress is involved in some diseases, including inflammation, atherosclerosis, type 2 diabetes mellitus and cancer [118]. Biomarkers of oxidative stress are lipoproteins oxidation, lipid hydroperoxides, conjugated dienes, malondialdehyde (MDA), F2-isoprostanes (F2-IsoPs), glutathione, protein carbonyls and activities of antioxidant enzymes [119]. A universal index does not identify oxidative stress since the biomarkers used to define the stress status have different kinetics of production and elimination [120]. The assay methods used to control lipid peroxidation determine the levels of lipid peroxides or the end products of lipid peroxidation. [121]. The standard assays to determine protein modifications measure the nitration of protein tyrosine residues and the carbonyl groups of the oxidized proteins [122,123].

7.2. The Immune Response

The immune system is a biological system that evolved host protection against viruses, bacteria, fungi, parasites and cancer cells [124]. Four functions of the immune system determine host defense. These include barrier production to stop pathogens, identifying and removing the pathogens that pass the barrier by immune cells and immunological memory creation [125]. Physical barriers are skin, respiratory, gastrointestinal tract (including microbiota), nasopharynx, hair and cilia. Immune cells are granulocytes (neutrophils, basophils and eosinophilic), lymphocytes (T-, B- and natural killer-cells) and phagocytes (monocytes, macrophages, dendritic cells and mast cells) [126]. Cellular and humoral responses can be innate and adaptive in the identification and eradication of pathogens. The innate responses are non-specific responses to pathogens that occur when there is no previous exposure or immunization. This action's actors are physical barriers, biochemical mechanisms, inflammatory response, complement system and phagocytes [127]. The speed and effectiveness of these responses are independent of the number of exposures to the pathogen. The adaptive responses are linked to the immunological "memory" and can generate an antigen-specific response. They involve antigen-specific T lymphocytes, which determine the adaptive response or destroy virally infected cells and B lymphocytes can secrete immunoglobulins (antibodies specific against the infecting pathogen) [127]. When the adaptive immune responses occur, the T helper cells (Th1 and Th17) migrate into circulation from lymphoid tissue, penetrate infected sites and make cytokines. The innate and adaptive immune responses control inflammation and the progress of the self and non-self-discrimination. Immature T cell populations express antigen-specific receptors that distinguish self or non-self-macromolecules [128]. In the thymus, T lymphocytes with T cell receptors (TCRs) recognize the self-peptides and major histocompatibility complex (MHC) proteins destroy nonself-macromolecules [129]. Autoimmune diseases happen when central and or induced peripheral tolerance do not work. Old age, obesity and diet determine the most severe symptoms of the disease. Aging can cause the thymus involution that decreases the output of naive T lymphocytes (T CD8+ kill cells directly and T CD4+, T helper cells that secrete cytokines) [130–133], the answer to new antigens and an increase of the inflammatory mediators in the blood (inflammageing) [134]. An excessive inflammatory response determines a loss in acquired immunity [134]. Obesity reduces T lymphocytes, B lymphocytes, natural killer cell activity, the antibody and IFN-γ (Interferon-gamma) production [135–137]. Food bioactive molecules and micronutrients can increase immune functions [138]. Fatty acids, amino acids, vitamins and mineral ions produce leukotrienes, prostaglandins, chemokines, immunoglobulins, cytokines and acute-phase proteins [135,139], valid for the immunity response. Carbazoles and tryptophan-enriched proteins determine antiinflammatory action, activating aryl hydrocarbon receptors. Moreover, diet regulates the microbiota to produce short-chain fatty acids that affect immune responses activating the G-protein–coupled receptors and epigenetic mechanisms [140].

8. Nutrient with Potential Nutraceutical Effects

8.1. Lipids

Lipids are essential energy sources for the human body. Fatty acids are constituents of fats and oils. They are classified into: saturated (without double bond), monounsaturated (with one double bond) and polyunsaturated fatty acids (with some double bond). Polyunsaturated fatty acids (PUFAs) are considered essential acids because humans cannot synthesize them. They are divided into two groups: the omega-3 and omega-6 fatty acids [141]. The free fatty acids (FFAs) serve as energy sources and natural ligands free fatty acid receptors, regulating the secretion of peptide hormones and inflammation. The viruses use the fats to fuse the viral membrane and host cell during replication, endocytosis and exocytosis [142].

8.1.1. Unsaturated Fatty Acids

Monounsaturated Fatty Acids' Health Properties

Monounsaturated fatty acids (MUFAs) are carboxylic acids with hydrocarbon chains having only one double bond. MUFAs inhibit coagulation, improve blood pressure and glucose homeostasis, reduce oxidative states and inflammation and modify plasma lipids, lipoprotein patterns, membrane composition and fluidity of blood cells [29]. In EVOO, the main MUFA by content is the oleic acid (18:1 ω-9), representing 49% to 83% of the total fatty acid. It promotes bile secretion and enhances gastric mucosa protection by decreasing hydrochloric acid secretion [143]. The American Heart Association sets a limit of MUFA consumption at 20% of total energy. The American Diabetes Association and Dietitians of Canada approve almost 25% of energy [144].

Polyunsaturated Fatty Acids Health Properties

PUFAs are carboxylic acids with hydrocarbon chains having more than one double bond. PUFAs are classified into ω-3 and ω-6 families. ω-3 fatty (alpha-linolenic, docosahexaenoic and eicosapentaenoic acids) have the double bonds in the third bond from the methyl end. ω-6 acids (linoleic, arachidonic and gamma acids) have it in the sixth bond from the fatty acid's methyl end [145]. PUFAs are precursors of eicosanoids (mediator signaling molecules). Eicosanoids derived from ω-6 PUFAs are proinflammatory molecules. Eicosanoids from ω-3 PUFAs have anti-inflammatory properties [146]. Arachidonic acid (ARA, n-6), eicosapentaenoic acid (EPA, n-3) and docosahexaenoic acid (DHA, n-3) are lipid immune mediators (SPMs) [147]. Mammals do not synthesize them. Humans transform linoleic acid (LA) and alpha-linolenic acid (ALA) found in foods to n-6 and n-3 long-chain polyunsaturated fatty acids [148]. SPMs contribute to cytokine "kidnapping" and eliminating remnants making the available restoration of structure and tissue homeostasis [149–152]. They decrease inflammation (duration and magnitude) and accelerate reepithelization, tissue regeneration and wound healing [153]. Moreover, EPA and DHA are the precursors of resolvins and ARA of maresins and protectins. Resolvins, maresins and protectins inactivate polymorphonuclear leukocytes and increase leukocytes, which remove remnants from neutrophil apoptosis (efferocytosis). Finally, EPA and DHA activate nuclear factor kappa B (NFkB), the peroxisome proliferator-activated receptor (PPAR) and destabilize membrane lipid rafts.

8.2. Vitamins

8.2.1. Vitamin A

Vitamin A is a fat-soluble retinoid group (retinol, retinyl esters and retinal) [154–156]. Retinol is obtained from animal sources as retinyl palmitate, or it can be synthesized in the intestine starting from beta carotene, a precursor/pro vitamin of vegetable origin. Vitamin A controls the differentiation of epithelial tissue, the imprinting of the B and T cells with gut-homing specificity, the arranging T cells and IgA+ cells into intestinal tissues [157], supports the gut barrier [158–160], reduces the toxic effects of ROS and regulate the membrane fluidity and gap-junctional communication [161,162]. Vitamin A improves epithelial construction (keratinization, stratification, differentiation) and functional maturation of epithelial cells [163]. It is part of the respiratory and intestine apparatus's mucus layer, promoting the antigen non-specific immunity function enhancing mucin secretion [163–165]. Vitamin A promotes the proliferation and regulation of the thymocytes apoptosis [166,167]. It plays a crucial role in controlling the differentiation, maturation and function of macrophages and neutrophils, which respond to pathogen invasion through phagocytosis and activation of natural killer T cells [168,169]. Vitamin A supervises the early differentiation of the natural killer T (CD4+) and dendritic cells (antigen-presenting cells) [170] and synthesizes immunoglobulins [171]. The balance between T helper 1 and T helper 2 lymphocytes is altered when vitamin A is deficient. Retinoic acid is essential for CD8+ T lymphocyte proliferation and antibody generation by B lymphocytes [172].

8.2.2. B-Group Vitamins

B vitamins are a family of water-soluble vitamins able to act as cofactors and coenzymes in metabolic pathways and play roles in maintaining immune homeostasis [173,174]. B vitamins are obtained from the intestinal microbiota and diet [175]. They contribute to gut barrier function controlling the intestinal immune regulation and are involved in intestinal immune regulation. Vitamin B6 regulates T lymphocyte migration, the folic acid and the T cells in the small intestine [176,177]. Vitamin B12 influences the phagocytic and bacterial killing capacity of the neutrophils [125]. Human gut microbes use vitamin B12 as a cofactor for metabolic pathways [126]. Both vitamins maintain or enhance NK cell cytotoxic activity [177–179]. Villarruz-Sulit and Cabaluna (2020) hypothesized that vitamin B supplementation affects the treatment of COVID-19 [179].

8.2.3. Vitamin C

Vitamin C is a water-soluble vitamin. Humans cannot synthesize it since they lack a crucial enzyme in the biosynthetic pathway [180,181]. Vitamin C is an antioxidant vitamin. It can donate electrons [182] and is a cofactor of monooxygenase and dioxygenase (antioxidant enzymes) [183,184]. It is involved in collagen biosynthesis and plays a crucial role in the immune system, including barrier integrity preservation and action on leukocyte function [185–190]. It is a cofactor for two enzymes (lysyl and prolyl hydroxylases) needed to maintain the tertiary structure of collagen I [184]. Moreover, vitamin C protects against ROS-induced damage [157], improves keratinocyte differentiation, lipid synthesis [191] and fibroblast proliferation and migration. It acts as an electron donor [192–195]. Enhances motility/chemotaxis [196–205], phagocytosis, ROS generation [206–213] and microbial killing [196,197,214–217]. Moreover, vitamin C facilitates apoptosis, clearance [212,216,218], decreases necrosis and the formation of the extracellular trap (NETosis), the cell death independent of apoptosis [216,218]. Vitamin C improves differentiation and proliferation [204,205,219–225] of B and T lymphocytes, increases antibody levels [219,224–228], controls the cytokines production [219,229–238] and decreases the histamine levels [198,204,239–245].

8.2.4. Vitamin E

Vitamin E is a collection of eight fat-soluble antioxidants that includes α-, β-, γ- and δ-) tocopherol and (α-, β-, γ- and δ-) tocotrienol derivative [246]. Vitamin E determines both humoral and cell-mediated immune functions [247] and enhances infectious pathogens' susceptibility [248]. It protects membranes from damage caused by free radicals [158,249], regulates NK cell cytotoxic activity [158,179,250,251] and decreases the production of the PGE2 by macrophages [175,252,253]. Vitamin E protects the polyunsaturated fatty acids and immune cells from oxidation [247,254]. It regulates natural killer cell activity, lymphocyte proliferation, specific antibody production following vaccination, the neutrophils' phagocytosis and promotes interaction between CD4+ T lymphocytes and dendritic cells [125].

8.3. *Phytochemicals in Spices with Nutraceutical Properties*

8.3.1. Phenols

Polyphenols are secondary metabolites of plants, involved in defense against attack by pathogens or UV (ultraviolet) radiation. They are classified based on phenol rings in phenolic acids, flavonoids, stilbenes, curcuminoids and lignans.

Flavonoids

The flavonoids consist of two aromatic rings bound together (A and B) and one heterocycle (ring C). They are divided into six subclasses depending on the degree of unsaturation and oxidation of the C ring and the carbon of the C ring on which the B ring is attached: flavones, flavonols, isoflavones, chalcones, anthocyanins, flavanones, flavanols, catechins and proanthocyanidins [255] (Table 1). Flavonoids have antioxidant, anti-mutagenic anti-angiogenic, antibacterial, anti-allergic, antiinflammatory, anticancer, enzyme modulation properties [256–258]. They can directly scavenge ROS, stabilize the free radicals, chelate metal ions with phenolic hydroxyl groups, activate phase II detoxification enzymes and block pro-oxidant enzymes. [259]. The flavonoids' anticancer mechanisms employ the control of the ROS-scavenging enzyme's activities, autophagy, apoptosis and inhibition of the cancer cell proliferation and invasiveness [259]. The flavonoid's antiinflammatory actions involve immune cell regulation, suppressing the chemokines, COX-2, cytokines, proinflammatory transcription factors and kappa kinase/c-Jun amino-terminal kinases [260,261]. FDA (Food and Drug Administration) approved a clinical trial of quercetin against Covid-19 [262]. In silico modeling works have shown that quercetin is one of the top five most potent compounds in a database of 8000 small molecules) able to bind the interface site of the ACE2 receptors and theoretically disrupt the initiating infection process of the SARS-CoV-2 Viral Spike Protein [263]. Apigenin, a 5,7-trihydroxyflavon, employs immune-regulatory activity in an organ-specific manner modulating NF-κB activity in the lungs [264], decreasing the secretion of the mast cell [265], T cells [266], COX-2, IL, TNF and NO [267]. In silico modeling, works have shown that apigenin binds the interface site of the ACE2 receptors and has great potential to act as COVID-19 main proteases inhibitors [268]. Isoflavones are phytoestrogens belonging to the non-steroidal estrogens. They improve the adaptive immune system, inhibiting lymphocyte proliferation, antigen-specific immune activities (T- and B-cells) and allergic responses [269–272]. The isoflavone genistein enhances CD8 T-cells and cytokines' production by T-cells [269,273–276]. Phytoestrogens interact with the T-cell and contribute to cytokine responses compartment enhancing or inhibiting the NF-kB pathway. Kojima et al. have shown that they improve gene expression mediated by ROR γ and α (retinoic-acid-receptor-related orphan receptor) in T-lymphoma cells and enhancing the expression of IL-17 [275]. The phytoestrogens interact with the B-cell compartment. The isoflavones inhibit IgG2a (immunoglobulin G2a) antibodies [270–272], the antigen-specific IgG1 and IgG3 in thyroiditis [272], the expression of IgE [270], the inflammatory immune response by inhibiting the antigen-presentation and functions of dendritic cells (DCs) [270,271]. They modulate the innate immune system inhibiting the production of IFN-γ, TNF-α, IL-9 and IL-13 from CD4+ T-cells [270,271], suppress allergic inflammation-reducing mast cell degranulation [270,272] and control NK cell activity reducing expression of IL-18Rα (IL-18 receptor α) and IFN-γ production in response to IL-12 and IL-18 [277]. Finally, they induce antiinflammatory responses in macrophages. Dia et al. showed that some phytoestrogens (daidzein and genistein) reduce the production of NO (nitric oxide), the expression of iNOS (inducible nitric oxide synthase) and enhance the superoxide dismutase and catalase activities [278].

Table 1. Class of flavonoids and their seasoning sources.

Class of Flavonoids	Chemical Structures	Condiments and/or Spices Sources
Anthocyanin	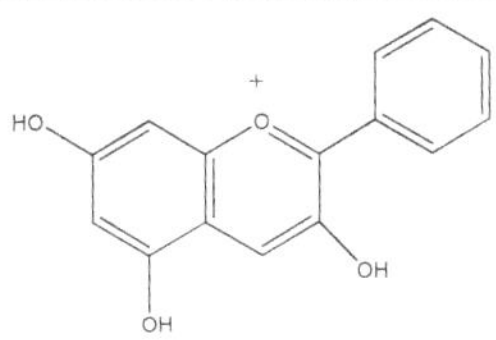	Wine vinegar
Flavanones	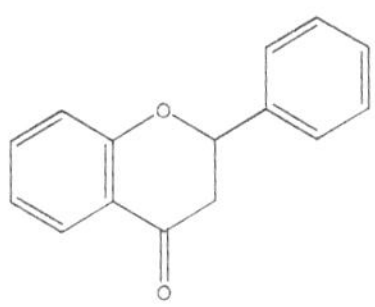	Soy sauce, parsley, chilli pepper.
Flavanols	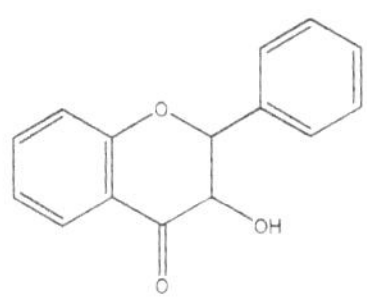	Fenugreek, curry, tumeric, garlic, ginger.
Flavones	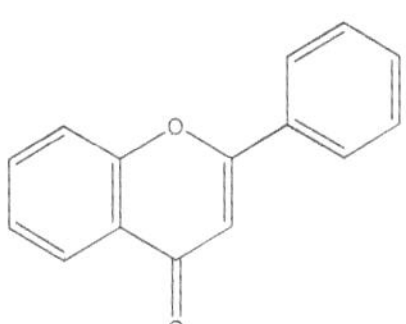	EVOO, soy sauce, fenugreek, basil, parsley.
Flavonols	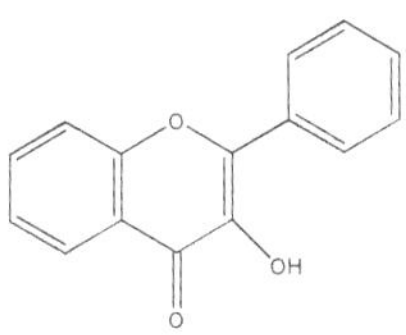	EVOO, fenugreek, curry, tumeric, garlic, ginger, basil, parsley, fennel, chilli pepper, sage.

Table 1. *Cont.*

Class of Flavonoids	Chemical Structures	Condiments and/or Spices Sources
Isoflavonoids	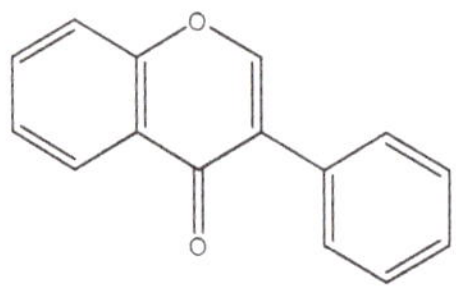	Fenugreek, soy sauce, Miso.
Chalchones	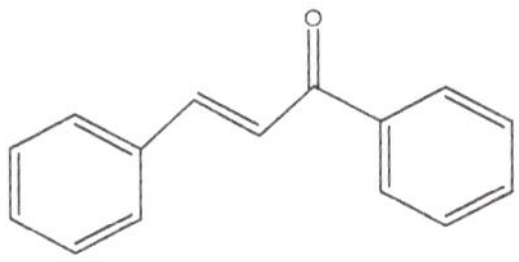	Tumeric

Curcuminoids

Curcuminoids are phenolic compounds with neuroprotective, antioxidant, antitumor, anti-acidogenic, antiinflammatory and radioprotective activities [279]. They have a diarylheptanoic nucleus with varying degrees of oxidation and unsaturation (Figure 2).

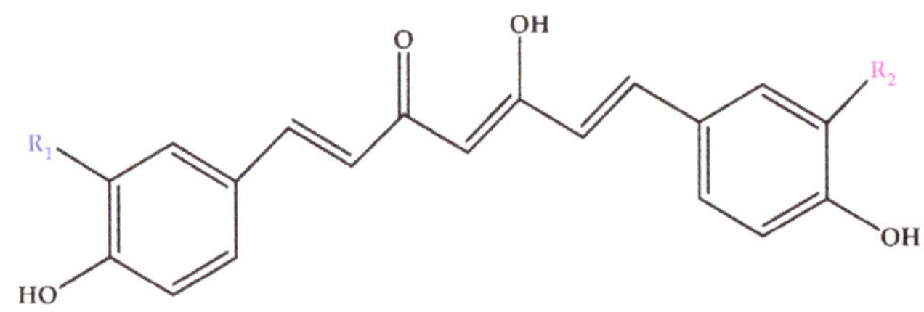

$R_1 = R_2 = OCH_3$ Curcumin

$R_1 = H$; $R_2 = OCH_3$ Desmethoxycurcumin

$R_1 = R_2 = H$ bis-desmethoxycurcumin

Figure 2. Curcuminoids in turmeric spice.

Curcumin, the principal polyphenol of *Curcuma longa* helps the antiinflammatory system to decrease the metabolism of arachidonic acid, lipoxygenase and cyclooxygenase activities, tumor necrosis factor, interleukins cytokines, nuclear factor-κB and steroids production [280]. It supports antioxidant defense mechanisms, such as scavenge hydroxyl radicals and superoxide anions, protection of cells from DNA damage, lipid peroxidation, protein carbonylation, protein oxidation, improvement of the glutathione's levels, stabilization of the superoxide-dismutase, glutathione S-transferase and glutathione peroxidase [281] and chelation of the heavy metals (aluminum, cadmium, copper, manganese, zinc and iron) responsible of the ROS production [282]. Curcumin carries out chemopreventive activities improving the glutathione transferase and NADPH quinone reductase (phase 2 detoxification enzymes), reducing cytochrome formation P450 1A1, a pro-carcinogen activating phase 1 enzyme and arachidonic acid production [283]. It has potential as a therapeutic agent for neurological diseases (e.g., Alzheimer's) since it inhibits amyloid-beta protein aggregation (e.g., α-synuclein, huntingtin, phosphorylated tau, prion proteins, Aβ-oligomers and fibrils) and enhances motor coordination and cognition peroxidase [281]. Finally, curcumin has cardioprotective actions (e.g., antiplatelet and anticoagulant) and improves the activities of detoxifying enzymes (e.g., glutathione-S-transferase) [284].

Capsaicinoids

Capsaicinoids (CAPs) (Figure 3) are compounds responsible for the burning sensation. Thirteen different CAPs are identified. They are characterized by one vanillyl group, a carboxamide group and a variable aliphatic chain (Figure 3). Capsaicinoids have hypocholesterolemic, antioxidant, antiinflammatory, antitumoral, antidiabetic and antiobesity properties [285]. Capsaicin and dihydrocapsaicin have shown hypocholesterolemic action obtained by reducing cholesterol absorption, improving its hepatic conversion to bile acids, the excretion in the feces and inducing expression of hepatic LDL receptors [286]. Antioxidant activity of CAPs is due to inhibition of lipid peroxidation; radical scavenge formation, depletion of total hepatic thiols and hepatic antioxidant enzyme activities (glutathione-reductase, glutathione-transferase, superoxide dismutase and catalase) [286]. CAPs block the arachidonate metabolites' production (PgE2, leukotrienes) and the release of the lysosomal enzymes (elastase, hyaluronidase and collagenase) by macrophages [287]. Antitumoral properties are related to the interaction with microsomal xenobiotic-metabolizing enzymes, inactivation of cytochrome P-450 HE1 (and other isoforms of the P-450 family) block of microsomal monooxygenases interested in carcinogen activation [288]. Capsaicinoids have potential application in diabetes prevention since they improve insulin secretion by activating, in islet β-cells, the transient receptor potential vanilloid subfamily member 1 (TRPV1) [289] decrease the concentration of postprandial blood glucose, enhance insulin secretion and glucose tolerance [290]. Finally, capsaicinoids have an antiobesity effect due to their ability to stimulate human brown adipose tissue growth. The human brown adipose tissue is the primary site of non-shivering thermogenesis (NST). It increases whole-body energy expenditure and regulates energy balance and body fatness [291].

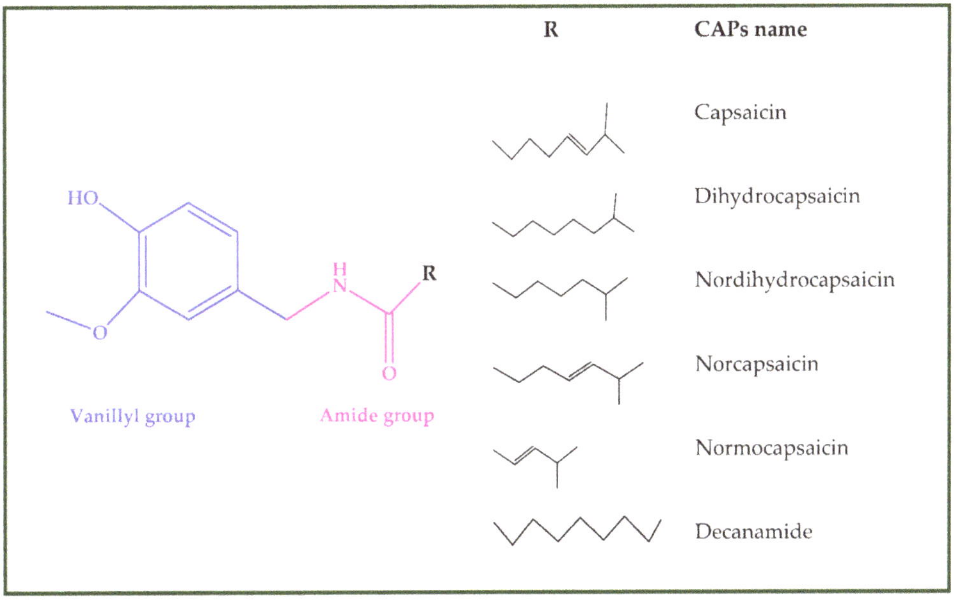

Figure 3. Chemical structures of some capsaicinoids.

8.3.2. Organosulfur Compounds

Organosulfur compounds are biosynthesized for defensive purposes against abiotic stressors by the Allium family's plants (e.g., garlic and onion). Thiosulfates are transformed by pH, temperature and solvent, into alk(en)yl cysteine sulfoxides, mono- di- and tri-sulfides, S-allyl cysteine, (E)- and (Z)-ajoene and vinyl dithiins [292]. Organosulfur compounds act as anticancer molecules improving the immune system and inducing proliferative signals by converting allyl sulfides into sulfane sulfur [293,294]. They stimulate apoptosis, induce xenobiotic-metabolizing enzymes, enhance the detoxification of carcinogens, have a role in cell cycle arrest and prevent nitrosamines and hydrocarbons' metabolism, scavenging free radicals and modulate the enzymes responsible for DNA repair [295,296]. Organosulfur compounds have antiinflammatory activities. The allicin inhibits the proinflammatory cytokines from epithelial digestive cells, blocking TNF-α secretion. Diallyl sulfide (DAS), diallyl tri- (DATS), tetra-sulfides and S-allylcysteine (SAC) decreased inflammatory lipopolysaccharide. The allyl methyl disulfide reduces the formation of the IL-8/IP-10 by the TNF-α in intestinal cells [297].

9. Effect of Spices and Herbs on the Shelf Life of Foods

In past few years, the protective effects of essential oils (Eos) as antimicrobial and antifungal agents in dairy products (e.g., chicken and meat) have been studied. Essential oils are mixtures of organic chemical compounds from the terpenoid family (mainly mono- and sesquiterpenes), phenols, aldehydes and ketones [298]. Eos have different modes of fungal inactivation: chalcones reduce the synthesis of the cell wall polysaccharide 1,3-beta-D-glucan causing fungal cell wall disruption [299] and decrease the conversion of tubulin into microtubules, causing the interruption of the cell division [300]; aldehydes inhibit fungal cell division by reacting with sulfhydryl involved in fungal cell division and interfere with fungal metabolism forming a charge-transfer complex with electron donors in fungal cells [301]; enones and enals stop fungal growth reacting with nucleophiles in fungi [302]; ascaridole decrease hemin making toxic radicals in the presence of Fe^{2+} [303]; carvacrol causes a breakdown of ion gradients distributing into membranes and interferes with the intracellular calcium homeostasis, improving the passive permeability of the cell membrane, modulating the Ca^{2+} permeable transient receptor channels, preventing sarcoplasmic reticulum Ca^{2+} ATPase and activating ryanodine receptors [303].

9.1. Spice Essential Oils in Postharvest Disease Mitigation

The composite solutions of aqueous extract of ginger, garlic and onion improve the shelf life (for about 5–6 days), anti-bacteria, antioxidation and sensory quality of stewed pork [304]. The addition in active packaging of ginger Eos extend the shelf life of poultry meat and meat products since they reduce lipid oxidation and microbiological growth [305,306]. Nanoemulsions of *Thymus daenensis* L. Eos are antibacterial, which can prolong stability and meliorate sensorial attributes in mayonnaise [307]. Garlic and ginger extracts improve the antioxidant activity, antimicrobial ability against some foodborne pathogens (e.g., *Bacillus subtilis* DB 100 host, *Escherichia coli* BA 12296, *Clostridium botulinum* ATCC 3584, *Staphylococcus aureus* NCTC10788 and *Salmonella senftenberg* ATCC 8400) and reduce thiobarbituric acid reactive substances levels, in herring fish fillet [305]. The nanoencapsulated form of the Garlic Eos is more effective than free form when incorporating into an active packaging (chitosan and whey protein films) to extend the shelf life of refrigerated vacuum-packed sausage [308].

9.2. Herb Essential Pils in Postharvest Disease Mitigation

The cumin and clove essential oils reduced Escherichia coli, Listeria monocytogenes, Salmonella, Campylobacter jejuni, Yersinia enterocolitica, Clostridium perfringens, Toxoplasma Gondi and Staphylococcus aureus bacterial cells in processed meat products [309].

Ethanolic extract of the *Cinnamomum zeylanicum* bark has a high antimicrobial effect against *Staphylococcus aureus, Escherichia coli* [310], *Listeria monocytogenes, Salmonella enterica*, food-borne pathogens [311].

Extracts of oregano and clove inhibit the lipid oxidation and reduce the *Listeria monocytogenes, Salmonella enterica* and *Staphylococcus aureus* numbers in cheese at room temperature [311].

The *Ocimum basilicum* L. Eos. have antifungal activity against *Candida albicans*, *Aspergillus niger* [312], *Aspergillus flavus* [313] and *Penicillium nalgiovense* [314]. The basil (*Ocimum basilicum*) leaves extract added to active film extends the shelf life of eggplant up to 16 days. It reserved eggplants' moisture loss, retarded the improvement in total soluble solids, firmness and color changes [315]. Recently, Gundewadi et al. (2018) have shown a desirable inhibitory activity of basil Eos against fungi, *P. chrysogenum* and *A. flavus* when the nanoemulsions of the lipophilic active ingredients are dispersed in an aqueous media and *sapindus* extract serves as a surfactant for nano emulsification purposes [316].

Chitosan packages infused with *Origanum vulgare* essential oil maintain the grapes' quality, physical and sensory attributes in post-harvest storage [317].

Eugenol and thymol extend strawberry shelf life by improving resistance to spoilage, deterioration and enhances their free radical scavenging capacity [318]. Eugenol and thymol reduce weight loss, skin color variation, ripening and decay of grape berries when used in a modified atmosphere package [319].

The fennel Eos addition to biodegradable film (based on polyhydroxybutyrate and polylactic acids) preserves oysters' shelf-life improving the oxygen barrier performance, antioxidant activity and antimicrobial activity against aerobic and anaerobic bacteria [320]. Carvacrol, perillaldehyde and anethole improve total anthocyanins, phenolics and antioxidant activity, in blueberry fruit [321].

Postharvest Eos (e.g., linalool, perillaldehyde, cinnamaldehyde, anethole, cinnamic acid and carvacrol) treatments improve the antioxidant potential in raspberries with perillaldehyde [322].

9.3. Sauce Contribution in Postharvest Disease Mitigation

The soy sauce added to the Tambaqui fillet (processed by the sous vide) improves its shelflife [323].

9.4. Condiment Contribution in Postharvest Disease Mitigation

Vinegar prolongs the shelf life and palatability of common mackerel [324]. The powdered-buffered vinegar and liquid-buffered vinegar decrease the psychrotrophic growth of *Salmonella typhimurium* in ground beef patties [325]. During chilled storage, vinegar added to silver carp inhibit acid phosphatase (related to the freshness and flavor of fish) and alkaline phosphatase and enhances the accumulation of inosine monophosphate and free amino acids [326].

10. Food-to-Food Fortification

Food fortification aims to enhance people's health. The main problem of classic food fortification depends on the economic problems related to food processing, especially in developing countries where the risk of malnutrition is very worrying. Food to food fortification is a new approach to food fortification. It uses the accessible local resource (animal or plant) to fortify another food [327]. Herbs, spices (essential oils, extract powder, fresh, etc.) and sauces are added to dairy products to improve their functional properties [323].

10.1. Spice's Contribution to the Functional Properties of Foods

The addition of *Allium sativum* water extract during fermentation increases lactic acid bacteria in yogurt [328].

The supplement of cinnamon powder into yogurt enhances the antioxidant activity, the total phenolic content and phenolics bioaccessibility into the gastrointestinal tract [329]. The addition of ginger extracts into yogurt improves its antigenotoxic and antioxidant effects [330].

10.2. Herb's Contribution to the Functional Properties of Foods

The supplement of basil into yogurt enhances the content of bioactive peptides and antioxidant properties [331]. The addition of dry basil leaves preserves and functionalizes cheeses. The basil leaves improve the cheese's antioxidant activity, prevent the degradation of the protein (due to basil's antimicrobial activity), fatty acids peroxidation and accelerating moisture loss [332]. The treatment of shredded iceberg lettuce with basil leaves' extract positively affects its total phenolic content, antioxidant potential and no effect on consumer acceptability [333].

The fortify of cheese with dry rosemary and parsley improves cheese's antioxidant properties [334].

The supplement of sage, or thyme and cumin essential oils to butter enhances its oxidative stability during storage time [335,336].

The addition of ginger into ice cream improves its total phenols and antioxidant activity [337].

10.3. Sauces Contribution to the Functional Properties of Foods

Soy sauce reduces lipid oxidation in meat products through the chelating activity of Fe^{2+} [338]. Soy sauce, Miso and fish sauces improve the Z-isomerization of lycopene in processed tomato products, enhancing the functionality of dishes since Z isomer is the most bioavailable [339].

10.4. Condiments Contribution to the Functional Properties of Foods

Extra virgin olive oil improves polyunsaturated fatty acids oxidative stability in algae oil. The EVOO's MUFA decreased the n-3 algae' PUFA and the EVOO's secoiridoid reduced the algae's triglyceride hydrolysis, simplifying their industry's application [340].

11. Effect of Cooking on Spices, Condiments, Extra Virgin Olive Oil and Aromas

11.1. Effect of Cooking on Spices

The radical scavenging potential of ginger, garlic, cinnamon and turmeric depends on the cooking method. Microwaves decrease antioxidant activity. Instead, boiling and steaming increase it. Heat treatment also regulates their antimicrobial activity. Microwave, bake, grill and frying determine to lose their antimicrobial activity. Instead, boiled and steam methods decreased it [341]. The use of pepper onion, garlic, chili pepper, fennel and cumin, before grilling, frying, or roasting the meat prevents the formation of heterocyclic aromatic amines (Has) and polycyclic aromatic hydrocarbons (PAHs), compounds known to be associated with cancer development [342,343]. HAs are made during the cooking of protein-rich foods by the interaction between creatine/creatinine and free amino acids or with hexoses from the Maillard reaction [344]. The incomplete pyrolysis or combustion makes PAHs of organic matter. They are obtained by thermal degradation of fatty acids, triglycerides, steroids and amino acids [345]. The phenolics and organosulfur compounds in spices suppress the reactive species and/or interact on reactions, stopping byproducts' formation [342]. Cooking methods and the fermentation process affect garlic antioxidant capacities. Raw garlic has more antioxidant activity than cooked garlic and black garlic (fermented garlic) has more significant antioxidant activity than crude garlic [346].

11.2. Effects of Cooking on Soy Sauce

Soy sauce preserves lipids from oxidation during cooking [347]. Phenolic compounds (by soybean) and antioxidative Maillard reaction products (by a non-enzymatic browning reaction) chelate ferrous ion a catalyst of lipid oxidation reaction [348].

11.3. Effects of Cooking on EVOO Degradation

EVOO has good thermal resistance in comparison with other vegetable oils. It is due to the high MUFA profile, phenolic composition and vitamin E. The initial olive oil composition, the period of the olive harvest and heating conditions (temperature, cooking process time and food presence) regulate the degradation rate and time required to degrade the antioxidant pool [338]. The absence of refining gave a high acidity to the EVOO and decreased its upper thermal limits in response to the released free fatty acids' lower boiling point. Under frying and roasting conditions (180–190 °C), EVOO performance is better than other vegetable oils. It is quickly degraded under microwave processing, but food decreases the thermo-oxidative effects of microwave heating [338]. EVOO degradation and underwater boiling conditions, mainly depend on heating time, food and loss of phenolic compounds into the water phase. As a result, wherever possible, olive oil must be added more closely to the final cooking process [349].

12. Spices Side Effects

The consumption of some spices can cause side effects. For example, ginger can determine gastrointestinal (e.g., heartburn, diarrhea, bloating, gas, abdominal pain and epigastric distress), cardiovascular and respiratory symptoms [350]. High doses are not recommended in pregnancy, lactation and patient with bleeding disorders since it has anti-platelet property [351]. The most common garlic's side effect is halitosis (bad breath) and body odor, especially when the raw form of the herbs is taken due to allyl methyl sulfide. A rare allergy to garlic was ascribed to protein allinase, which made hypersensitivity responses via immunoglobulin E [352]. Fenugreek ingestion can determine hypoglycemia in diabetics persons, diarrhea, abdominal distention, dyspepsia and flatulence [353]. Tumeric and curcumin can determine dermatitis and urticaria (immunoglobulin E mediated) especially following direct curcumin exposure to the skin or scalp [354]. Higher curcumin doses increase carcinogenesis, enhancing ROS cell levels [355].

13. Conclusions

The spices and condiments play an essential role in our diet that goes far beyond flavoring our dishes. They can be considered supplements of molecules functional to prevent oxidative stress injury, inflammation damage and chronic degenerative diseases that afflict our society, such as cardiovascular diseases and cancer. When multiple spices are used to prepare a dish, there is the possibility that they can have synergistic effects, increasing their health potential. However, more in-depth information on the effects of exposure to their bio components is needed to define intervention strategies to maximize beneficial effects and minimize unwanted side effects.

Author Contributions: I.D. writing—original draft preparation review and editing, S.L. funding acquisition. All authors have read and agreed to the published version of the manuscript.

Funding: This research received no external funding.

Institutional Review Board Statement: Not applicable.

Informed Consent Statement: Not applicable.

Conflicts of Interest: The authors declare no conflict of interest.

References

1. García-Casal, M.N.; Pena-Rosas, J.P.; Malave, H.G. Sauces, spices, and condiments: Definitions, potential benefits, consumption patterns, and global markets. *Ann. N. Y. Acad. Sci.* **2016**, *1379*, 3–16. [CrossRef]
2. Camarena, D.M.; Sanjuán, A.I.; Philippidis, G. Influence of ethnocentrism and neophobia on ethnic food consumption in Spain. *Appetite* **2011**, *57*, 121–130. [CrossRef] [PubMed]
3. Verbeke, W.; López, G.P. Ethnic food attitudes and behaviour among Belgians and Hispanics living in Belgium. *Br. Food J.* **2005**, *107*, 823–840. [CrossRef]
4. Bell, B.; Adhikari, K.; Chambers, E.; Cherdchu, P.; Suwonsichon, T. Ethnic food awareness and perceptions of consumers in Thailand and the United States. *Nutr. Food Sci.* **2011**, *41*, 268–277. [CrossRef]
5. Chung, L.; Chung, S.J.; Kim, J.Y.; Kim, K.O.; O' Mahony, M.; Vickers, Z.; Cha, S.M.; Ishii, R.; Baures, K.; Kim, H.R. Comparing the liking for Korean style salad dressings and beverages between US and Korean consumers: Effects of sensory and non-sensory factors. *Food Qual. Prefer.* **2012**, *26*, 105–108. [CrossRef]
6. Hong, J.H.; Lee, K.W.; Chung, S.J.; Chung, L.; Kim, H.R.; Kim, K.O. Sensory characteristics and cross-cultural comparisons of consumer acceptability for *Gochujang* dressing. *Food Sci. Biotechnol.* **2012**, *21*, 829–837. [CrossRef]

7. Seasoning and Spices Market Size, Share & Trends Analysis Report by Product (Herbs, Salt & Salts Substitutes, Spices), by Application, by Region, and Segment Forecasts, 2020–2027. Available online: https://www.grandviewresearch.com/industry-analysis/seasonings-spices-market#:~{}:text=The%20global%20seasoning%20and%20spices%20market%20size%20was%20estimated%20at,USD%2015.44%20billion%20in%202020 (accessed on 15 July 2020).
8. Bhagya, H.P.; Raveendra, Y.C.; Lalithya, K.A. Multi benificial uses of spices: A brief review. *Acta Sci. Nutr. Health* **2015**, *1*, 3–7.
9. Fanelli, F.; Cozzi, G.; Raiola, A.; Dini, I.; Mulè, G.; Logrieco, A.F.; Ritieni, A. Raisins and Currants as Conventional Nutraceuticals in Italian Market: Natural Occurrence of Ochratoxin A. *J. Food Sci.* **2017**, *82*, 2306–2312. [CrossRef] [PubMed]
10. Dini, I.; Laneri, S. Nutricosmetics: A brief overview. *Phytother. Res.* **2019**, *33*, 3054–3063. [CrossRef]
11. Laneri, S.; Di Lorenzo, R.M.; Bernardi, A.; Sacchi, A.; Dini, I. *Aloe barbadensis*: A Plant of Nutricosmetic Interest. *Nat. Prod. Commun.* **2020**, *15*. [CrossRef]
12. Laneri, S.; Di Lorenzo, R.; Sacchi, A.; Dini, I. Dosage of bioactive molecules in the nutricosmeceutical *Helix aspersa muller* mucus and formulation of new cosmetic cream with moisturizing effect. *Nat. Prod. Commun.* **2019**, *14*, 1–7. [CrossRef]
13. Dini, I. Spices and herbs as therapeutic foods. In *Food Quality: Balancing Health and Disease*; Holban, A.M., Grumezescu, A.M., Eds.; Academic Press: New York, NY, USA; Elservier: London, UK, 2018; pp. 433–469.
14. CODEX STAN 192-1995. 2013. Codex Alimentarius. Codex General Standard for Food Additives. Adopted in 1995. Revision 1997, 1999, 2001, 2003, 2004, 2005, 2006, 2007, 2008, 2009, 2010, 2011, 2012, 2013, 2014, 2015, 2016, 2017, 2018, 2019. Available online: http://www.fao.org/fao-who-codexalimentarius/sh-proxy/en/?lnk=1&url=https%253A%252F%252Fworkspace.fao.org%252Fsites%252Fcodex%252FStandards%252FCXS%2B192-1995%252FCXS_192e.pdf (accessed on 15 July 2020).
15. Diniz do Nascimento, L.; Moraes, A.A.B.d.; Costa, K.S.d.; Pereira Galúcio, J.M.; Taube, P.S.; Costa, C.M.L.; Neves Cruz, J.; de Aguiar Andrade, E.H.; Faria, L.J.G.d. Bioactive Natural Compounds and Antioxidant Activity of Essential Oils from Spice Plants: New Findings and Potential Applications. *Biomolecules* **2020**, *10*, 988. [CrossRef]
16. Raj Singh, L. Golden Shower: A wonder medicinal plant and its processing technology. *Indian J. Appl. Res.* **2019**, *9*, 66–67.
17. Solorzano-Santos, F.; Miranda-Novales, M.G. Essential oils from aromatic herbs as antimicrobial agents. *Curr. Opin. Biotechnol.* **2012**, *23*, 136–141. [CrossRef] [PubMed]
18. Farrell, K.T. *Spices, Condiments and Seasonings*, 2nd ed.; Aspen Publishers: Gainthersburg, MD, USA, 1990; pp. 291–298.
19. Chadare, F.J.; Idohou, R.; Nago, E.; Affonfere, M.; Hounhouigan, D.J.; Agossadou, J.; Kévin, T.; Christel, F.; Sewanou, K.; Azokpota, P.; et al. Conventional and food-to-food fortification: An appraisal of past practices and lessons learned. *Food Sci. Nutr.* **2019**, *7*, 2781–2795. [CrossRef]
20. WHO/FAO. *Guidelines on Food Fortification with Micronutrients*; World Health Organization/Food and Agriculture Organization of the United Nations: Geneva, Switzerland, 2006. Available online: https://www.who.int/nutrition/publications/guide_food_fortification_micronutrients.pdf (accessed on 1 January 2006).
21. Klassen-Wigger, P.; Geraets, M.; Messier, M.C.; Detzel, P.; Lenoble, H.P.; Barclay, D.V. Micronutrient fortification of bouillon cubes in Central and West Africa. In *Food Fortification in a Globalized World*; Mannar, M.G.V., Hurrell, R.F., Eds.; Academic Press: New York, NY, USA, 2018; pp. 363–372.
22. Mkambula, P.; Mbuya, M.N.N.; Rowe, L.A.; Sablah, M.; Friesen, V.M.; Chadha, M.; Osei, A.K.; Ringholz, C.; Vasta, F.C.; Gorstein, J. The Unfinished Agenda for Food Fortification in Low- and Middle-Income Countries: Quantifying Progress, Gaps and Potential Opportunities. *Nutrients* **2020**, *12*, 354. [CrossRef]
23. Dini, I.; Di Lorenzo, R.; Senatore, A.; Coppola, D.; Laneri, S. Validation of Rapid Enzymatic Quantification of Acetic Acid in Vinegar on Automated Spectrophotometric System. *Foods* **2020**, *9*, 761. [CrossRef]
24. Xia, T.; Zhang, B.; Duan, W.; Zhang, J.; Wang, M. Nutrients and bioactive components from vinegar: A fermented and functionalfood. *J. Funct. Foods* **2020**, *64*, 103681. [CrossRef]
25. Cerezo, A.; Cuevas, E.; Winterhalter, P.; Garcia, M.; Troncoso, A. Anthocyanin composition in Cabernet Sauvignon red wine vinegar obtained by submerged acetification. *Food Res. Int. J.* **2010**, *43*, 1577–1584. [CrossRef]
26. Kandylis, P.; Bekatorou, A.; Dimitrellou, D.; Plioni, I.; Giannopoulou, K. Health Promoting Properties of Cereal Vinegars. *Foods* **2021**, *10*, 344. [CrossRef] [PubMed]
27. Koulbanis, C.; Mellul, M.; Candau, D. Cosmetic Composition Containing Vinegar as Active Anti-Ageing Agent, and Its Use in the Treatment of Dermatological Ageing. U.S. Patent 5,560,916, 1996.
28. Servili, M.; Selvaggini, R.; Esposto, S.; Taticchi, A.; Montedoro, G.; Morozzi, G. Health and sensory properties of virgin olive oil hydrophilic phenols: Agronomic and technological aspect of production that affect their occurence in the oil. *J. Chromatogr.* **2004**, *1054*, 113–127. [CrossRef]
29. Piroddi, M.; Albini, A.; Fabiani, R.; Giovannelli, L.; Luceri, C.; Natella, F.; Rosignoli, P.; Rossi, T.; Taticchi, A.; Servili, M.; et al. Nutrigenomics of extra-virgin olive oil: A review. *BioFactors* **2017**, *43*, 17–41. [CrossRef] [PubMed]
30. FoodData Central Search Results. US Department of Agriculture (Agricultural Research Service) Fats and Oils, Oil, Olive, Extra Virgin. Available online: https://fdc.nal.usda.gov/fdc-app.html#/food-details/748608/nutrients (accessed on 16 December 2019).
31. Angerosa, F.; Campestre, C.; Giansante, L. Analysis and authentication. In *Olive Oil: Chemistry and Technology*; Boskou, D., Ed.; AOCS Press: Champaign, IL, USA, 2006; pp. 113–172.

32. Trade Standard Applying to Olive Oil and Olive-Pomace Oil. COI/T 15/NC No 3/Rev 7, May 2013. Available online: https://www.internationaloliveoil.org/wp-content/uploads/2019/11/COI-T.15-NC.-No-3-Rev.-13-2019-Eng.pdf (accessed on 15 November 2019).
33. Codex Alimentarius. Codex Standard for Olive Oils, and Olive Pomace Oils, CODEX STAN 33-1981. Codex Alimentarius, Roma, Itália, Rev, 2. 2003. Available online: https://img.21food.cn/img/biaozhun/20100729/180/11294201.pdf (accessed on 18 January 2021).
34. EC Regulation (1991). No. 2568/91/EEC, 1991. Off J Eur Commun L 248:1–83, July 11. 1991. Available online: https://eur-lex.europa.eu/eli/reg/1991/2568/2015-10-16 (accessed on 11 June 1991).
35. EC Regulation (2012). No. 432/2012 of 16 May 2012 Establishing a List of Permitted Health Claims Made on Foods, Other than Those Referring to the Reduction of Disease Risk and to Children's Development and Health. *O.J.E.U.* L 136:1–40. Available online: http://data.europa.eu/eli/reg/2012/432/oj (accessed on 16 May 2012).
36. EC Regulation (2013). No. 1332/2013 Amending Regulation (EU) No. 36/2012 Concerning Restrictive Measures in View of the Situation in Syria, 13 December 2013. OJ L, 335(14.12). Available online: http://data.europa.eu/eli/reg/2013/1332/oj (accessed on 14 December 2013).
37. Kyçyk, O.; Aguilera, M.P.; Gaforio, J.J. Sterol composition of virgin olive oil of forty-three olive cultivars from the World Collection Olive Germplasm Bank of Cordoba. *J. Sci. Food Agric.* **2016**, *96*, 4143–4450. [CrossRef]
38. Ghanbari, R.; Anwar, F.; Alkharfy, K.M.; Gilani, A.-H.; Saari, N. Valuable Nutrients and Functional Bioactives in Different Parts of Olive (*Olea europaea* L.)—A Review. *Int. J. Mol. Sci.* **2012**, *13*, 3291–3340. [CrossRef] [PubMed]
39. Dini, I.; Graziani, G.; Fedele, F.L.; Sicari, A.; Vinale, F.; Castaldo, L.; Ritieni, A. An environmentally friendly practice used in olive cultivation capable of increasing commercial interest in waste products from oil processing. *Antioxidants* **2020**, *9*, 466. [CrossRef] [PubMed]
40. Dini, I.; Graziani, G.; Gaspari, A.; Fedele, F.L.; Sicari, A.; Vinale, F.; Cavallo, P.; Lorito, M.; Ritieni, A. New strategies in the cultivation of olive trees and repercussions on the nutritional value of the extra virgin olive oil. *Molecules* **2020**, *25*, 2345. [CrossRef] [PubMed]
41. Dini, I.; Graziani, G.; Fedele, F.L.; Sicari, A.; Vinale, F.; Castaldo, L.; Ritieni, A. Effects of *Trichoderma* Biostimulation on the Phenolic Profile of Extra-Virgin Olive Oil and Olive Oil By-Products. *Antioxidants* **2020**, *9*, 284. [CrossRef]
42. Dini, I.; Marra, R.; Cavallo, P.; Pironti, A.; Sepe, I.; Troisi, J.; Scala, G.; Lombari, P.; Vinale, F. *Trichoderma* Strains and Metabolites Selectively Increase the Production of Volatile Organic Compounds (VOCs) in Olive Trees. *Metabolites* **2021**, *11*, 213. [CrossRef]
43. Dini, I.; Seccia, S.; Senatore, A.; Coppola, D.; Morelli, E. Development and Validation of an Analytical Method for Total Polyphenols Quantification in Extra Virgin Olive Oils. *Food Anal. Methods* **2020**, *13*, 457–464. [CrossRef]
44. Azzi, A. Tocopherols, tocotrienols and tocomonoenols: Many similar molecules but only one vitamin E. *Redox Biol.* **2019**, *26*, 101259. [CrossRef] [PubMed]
45. Ohara, M.; Lu, H.; Shiraki, K.; Ishimura, Y.; Uesaka, T.; Katoh, O.; Watanabe, H. Prevention by long-term fermented miso of induction of colonic aberrant crypt foci by azoxymethane in F344 rats. *Oncol. Rep.* **2002**, *9*, 69–73. [CrossRef] [PubMed]
46. Watanabe, H.; Kashimoto, N.; Kajimura, J.; Kamiya, K. A miso (Japanese soybean paste) diet conferred greater protection against hypertension than a sodium chloride diet in Dahl salt-sensitive rats. *Hypertens. Res.* **2006**, *29*, 731–738. [CrossRef]
47. Xu, L.; Du, B.; Xu, B. A systematic, comparative study on the beneficial health components and antioxidant activities of commercially fermented soy products marketed in china. *Food Chem.* **2015**, *174*, 202–213. [CrossRef]
48. Chatterjee, C.; Gleddie, S.; Xiao, C.-W. Soybean Bioactive Peptides and Their Functional Properties. *Nutrients* **2018**, *10*, 1211. [CrossRef] [PubMed]
49. FoodData Central Search Results. US Department of Agriculture (Agricultural Research Service) Sauce, Ready-to-Serve, Soy Souce. Available online: https://fdc.nal.usda.gov/fdc-app.html#/food-details/1100456/nutrients (accessed on 3 October 2020).
50. Li, H.; Zhao, M.; Su, G.; Lin, L.; Wang, Y. Effect of soy sauce on serum uric acid levels in hyperuricemic rats and identification of flazin as a potent xanthine oxidase inhibitor. *J. Agric. Food Chem.* **2016**, *64*, 4725–4734. [CrossRef] [PubMed]
51. Kobayashi, M. Nutritional function of polysaccharides from soy sauce in the gastrointestinal tract. In *Bioactive Food as Dietary Interventions for Liver and Gastrointestinal Disease*; Watson, R.R., Preedy, V.R., Eds.; Elsevier Inc.: San Diego, CA, USA, 2013; pp. 139–147.
52. Peng, M.; Liu, J.; Liu, Z.; Fu, B.; Hu, Y.; Zhou, M.; Fu, C.; Gao, B.; Wang, C.; Li, D.; et al. Effect of citrus peel on phenolic compounds, organic acids and antioxidant activity of soy sauce. *LWT-Food Sci. Technol.* **2018**, *90*, 627–635. [CrossRef]
53. Watanabe, H. Beneficial biological effects of miso with reference to radiation injury, cancer and hypertension. *J. Toxicol. Pathol.* **2013**, *26*, 91–103. [CrossRef]
54. Nakamura, Y.; Tsuji, S.; Tonogai, Y. Determination of the levels of isoflavonoids in soybeans and soy-derived foods and estimation of isoflavonoids in the Japanese daily intake. *J. AOAC Int.* **2000**, *83*, 635–650. [CrossRef] [PubMed]
55. Minamiyama, Y.; Okada, S. Miso: Production, properties and benefits to heath. In *Handbook of Fermented Functional Foods*, 2nd ed.; Farnworth, E.R., Ed.; CRC Pressp: Boca Raton, FL, USA; Taylor & Francis Group: London, UK, 2008; pp. 277–284.
56. Sanlier, N.; Gökcen, B.B.; Sezgin, A.C. Health benefits of fermented foods. *Crit. Rev. Food Sci. Nutr.* **2019**, *59*, 506–527. [CrossRef] [PubMed]
57. Takahashi, M.; Nagata, M.; Kaneko, T.; Suzuki, T. Miso soup consumption enhances the bioavailability of the reduced form of supplemental coenzyme q(10). *J. Nutr. Metab.* **2020**, *2020*, 5349086. [CrossRef]

58. Yoshikawa, S.; Kurihara, H.; Kawai, Y.; Yamazaki, K.; Tanaka, A.; Nishikiori, T.; Ohta, T. Effect of halotolerant starter microorganisms on chemical characteristics of fermented Chum Salmon (*Oncorhynchus keta*) sauce. *J. Agric. Food Chem.* **2010**, *58*, 6410–6417. [CrossRef]
59. Giri, A.; Osako, K.; Ohshima, T. Identification and characterisation of headspace volatiles of fish *miso*, a Japanese fish meat based fermented paste, with special emphasis on effect of fish species and meat washing. *Food Chem.* **2010**, *120*, 621–631. [CrossRef]
60. Gowda, S.G.S.; Narayan, B.; Gopal, S. Bacteriological properties and health-related biochemical components of fermented fish sauce: Anoverview. *Food Rev. Int.* **2016**, *32*, 203–229. [CrossRef]
61. Sun, J.; Yu, X.; Fang, B.; Ma, L.; Xue, C.; Zhang, Z.; Mao, X. Effect of fermentation by Aspergillus oryzaeon the biochemical and sensory proper-ties of anchovy (*Engraulis japonicus*) fish sauce. *Int. J. Food Sci. Technol.* **2016**, *51*, 133–141. [CrossRef]
62. Rhyu, M.-R.; Kim, Y.; Misaka, T. Suppression of hTAS2R16 Signaling by Umami Substances. *Int. J. Mol. Sci.* **2021**, *21*, 7045.
63. Hamzeh, A.; Noisa, P.; Yongsawatdigul, J. Characterization of the antioxidant and ACE-inhibitory activities of Thai fish sauce at different stages of fermentation. *J. Funct. Foods* **2020**, *64*, 103699. [CrossRef]
64. Li, F.; Liu, W.; Yamaki, K.; Liu, Y.; Fang, Y.; Li, Z.; Chem, M.; Wang, C. Angiotensin I-converting enzyme inhibitory effect of chinese soypaste along fermentation and ripening: Contribution of early soybean protein borne peptides and late Maillard reaction products. *Int. J. Food Prop.* **2016**, *19*, 2805–2816. [CrossRef]
65. Lopetcharat, K.; Choi, Y.J.; Park, J.W. Fish sauce products and manufacturing: A review. *Food Rev. Int.* **2001**, *17*, 65–88. [CrossRef]
66. FoodData Central Search Results. U.S. Department of Agriculture (Agricultural Research Service) Sauce, Ready-to-Serve, Fish Sauce. Available online: https://fdc.nal.usda.gov/fdc-app.html#/food-details/1099270/nutrients (accessed on 3 October 2020).
67. Dincer, T.; Cakli, S.; Kilinc, B.; Tolasa, S. Amino acids and fatty acid composition content of fish sauce. *J. Adv. Vet. Anim. Res.* **2010**, *9*, 311–315. [CrossRef]
68. Han, S.-C.; Kang, G.-J.; Ko, Y.-J.; Kang, H.-K.; Moon, S.-W.; Ann, Y.-S.; Yoo, E.-S. Fermented fish oil suppresses T helper 1/2 cell response in a mouse model of atopic dermatitis via generation of CD4+ CD25+ Foxp3+ T cells. *BMC Immunol.* **2012**, *13*, 44. [CrossRef] [PubMed]
69. Di Cagno, R.; Surico, R.F.; Paradiso, A.; De Angelis, M.; Salmon, J.C.; Buchin, S.; De Gara, L.; Gobbetti, M. Effect of autochthonous lactic acid bacteria starters on health-promoting and sensory properties of tomato juices. *Int. J. Food Microb.* **2009**, *128*, 473–483. [CrossRef]
70. Cousin, F.J.; Le Guellec, R.; Schlusselhuber, M.; Dalmasso, M.; Laplace, J.-M.; Cretenet, M. Microorganisms in fermented apple beverages: Current knowledge and future directions. *Microorganisms* **2017**, *5*, E39. [CrossRef]
71. FoodData Central Search Results. US Department of Agriculture (Agricultural Research Service) Sauce, Ready-to-Serve, Pepper, TABASCO. Available online: https://fdc.nal.usda.gov/fdc-app.html#/food-details/174528/nutrients (accessed on 26 October 2020).
72. Farias, V.L.D.; Araújo, Í.M.D.S.; Rocha, R.F.J.D.; Garruti, D.D.S.; Pinto, G.A.S. Enzymatic maceration of Tabasco pepper: Effect on the yield, chemical and sensory aspects of the sauce. *LWT* **2020**, *127*, 109311. [CrossRef]
73. Dhakal, S.; Chao, K.; Schmidt, W.; Qin, J.; Kim, M.; Huang, Q. Detection of Azo Dyes in Curry Powder Using a 1064-nm Dispersive Point-Scan Raman System. *Appl. Sci.* **2018**, *8*, 564. [CrossRef]
74. *FoodData Central Search Results*; US Department of Agriculture (Agricultural Research Service): Washington, DC, USA, 2020. Available online: https://fdc.nal.usda.gov/fdc-app.html#/food-details/170924/nutrients (accessed on 4 January 2019).
75. Ashokkumar, K.; Pandian, A.; Murugan, M.; Dhanya, M.K.; Sathyan, T.; Sivakumar, P.; Raj, S.; Warkentin, T.D. Profiling bioactive flavonoids and carotenoids in select south Indian spices and nuts. *Nat. Prod. Res.* **2020**, *34*, 1306–1310. [CrossRef]
76. Reddy, B.M.; Dhanpal, C.K.; Lakshmi, B.V.S. A review on curr y leaves (*Murray akoenigii): Versatile* multi-potential medicinal plant. *Int. J. Adv. Pharm. Med. Bioallied Sci.* **2018**, *6*, 31–41.
77. Tachibana, Y.; Kikuzaki, H.; Lajis, N.H.; Nakatani, N. Antioxidative activity of carbazoles from *Murraya koenigii* leaves. *J. Agric. Food Chem.* **2001**, *49*, 5589–5594. [CrossRef]
78. Jurenka, J.S. Anti-inflammatory properties of curcumin, a major constituent of *Curcuma longa*: A review of preclinical and clinical research. *Altern. Med. Rev.* **2009**, *14*, 141–153.
79. Irshad, S.; Muazzam, A.; Shahid, Z.; Dalrymple, M.B. *Curcuma longa* (Turmeric): An auspicious spice for antibacterial, phytochemical and antioxidant activities. *Pak. J. Pharm. Sci.* **2018**, *31*, 2689–2696. [PubMed]
80. *FoodData Central Search Results*; US Department of Agriculture (Agricultural Research Service): Washington, DC, USA, 2020. Available online: https://fdc.nal.usda.gov/fdc-app.html#/food-details/1445985/nutrients; (accessed on 26 October 2020).
81. Omosa, L.K.; Midiwo, J.O.; Kuete, V. *Curcuma Longa*; Elsevier: Edinburgh, UK, 2012.
82. Venkata, K.C.; Bagchi, D.; Bishayee, A. A small plant with big benefits: Fenugreek (*Trigonella foenum-graecum* Linn.) for disease prevention and health promotion. *Mol. Nutr. Food Res.* **2017**, *61*, 1–26.
83. Jacob, B.; Narendhirakannan, R.T. Role of medicinal plants in the management of diabetes mellitus: A review. *Biotechnol. J.* **2019**, *9*, 4.
84. *FoodData Central Search Results*; US Department of Agriculture (Agricultural Research Service): Washington, DC, USA, 2019. Available online: https://fdc.nal.usda.gov/fdc-app.html#/food-details/1275176/nutrients; (accessed on 5 January 2019).
85. Diretto, G.; Rubio-Moraga, A.; Argandona, J.; Castillo, P.; Gomez-Gomez, L.; Ahrazem, O. Tissue-specific accumulation of sulfur compounds and saponins in different parts of garlic cloves from purple and white ecotypes. *Molecules* **2017**, *22*, 1359. [CrossRef] [PubMed]

86. Szychowski, K.A.; Rybczynska-Tkaczyk, K.; Gawel-Beben, K.; Swieca, M.; Karas, M.; Jakubczyk, A.; Matysiak, M.; Binduga, U.E.; Gminski, J. Characterization of active compounds of different garlic (*Allium sativum* L.) cultivars. *Pol. J. Food Nutr. Sci.* **2018**, *68*, 73–81. [CrossRef]
87. Bradley, J.M.; Organ, C.L.; Lefer, D.J. Garlic-derived organic polysulfides and myocardial protection. *J. Nutr.* **2016**, *146*, 403S–409S. [CrossRef]
88. Wang, Y.C.; Guan, M.; Zhao, X.; Li, X.L. Effects of garlic polysaccharide on alcoholic liver fibrosis and intestinal microflora in mice. *Pharm. Biol.* **2018**, *56*, 325–332. [CrossRef]
89. Nagella, P.; Thiruvengadam, M.; Ahmad, A.; Yoon, J.Y.; Chung, I.M. Composition of polyphenols and antioxidant activity of garlic bulbs collected from different locations of Korea. *Asian J. Chem.* **2014**, *26*, 897–902. [CrossRef]
90. Shang, A.; Cao, S.-Y.; Xu, X.-Y.; Gan, R.-Y.; Tang, G.-Y.; Corke, H.; Mavumengwana, V.; Li, H.-B. Bioactive Compounds and Biological Functions of Garlic (*Allium sativum* L.). *Foods* **2019**, *8*, 246. [CrossRef] [PubMed]
91. Liu, J.; Guo, W.; Yang, M.L.; Liu, L.X.; Huang, S.X.; Tao, L.; Zhang, F.; Liu, Y.S. Investigation of the dynamic changes in the chemical constituents of chinese "laba" garlic during traditional processing. *RSC Adv.* **2018**, *8*, 41872–41883. [CrossRef]
92. Kang, J.S.; Kim, S.O.; Kim, G.Y.; Hwang, H.J.; Kim, B.W.; Chang, Y.C.; Kim, W.J.; Kim, C.M.; Yoo, Y.H.; Choi, Y.H. An exploration of the antioxidant effects of garlic saponins in mouse-derived C2C12 myoblasts. *Int. J. Mol. Med.* **2016**, *37*, 149–156. [CrossRef] [PubMed]
93. Park, S.Y.; Seetharaman, R.; Ko, M.J.; Kim, D.Y.; Kim, T.H.; Yoon, M.K.; Kwak, J.H.; Lee, S.J.; Bae, Y.S.; Choi, Y.W. Ethyl linoleate from garlic attenuates lipopolysaccharide-induced pro-inflammatory cytokine production by inducing heme oxygenase-1 in RAW 264.7 cells. *Int. Immunopharmacol.* **2014**, *19*, 253–261. [CrossRef] [PubMed]
94. Rabe, S.Z.T.; Ghazanfari, T.; Siadat, Z.; Rastin, M.; Rabe, S.Z.T.; Mahmoudi, M. Anti-inflammatory effect of garlic 14-kDa protein on LPS-stimulated-J774A.1 macrophages. *Immunopharmacol. Immunotoxicol.* **2015**, *37*, 158–164. [CrossRef] [PubMed]
95. Li, M.; Yan, Y.X.; Yu, Q.T.; Deng, Y.; Wu, D.T.; Wang, Y.; Ge, Y.Z.; Li, S.P.; Zhao, J. Comparison of immunomodulatory effects of fresh garlic and black garlic polysaccharides on RAW 264. *7 macrophages. J. Food Sci.* **2017**, *82*, 765–771. [CrossRef] [PubMed]
96. Asdaq, S.M.; Inamdar, M.N. Potential of garlic and its active constituent, S-allyl cysteine, as antihypertensive and cardioprotective in presence of captopril. *Phytomedicine* **2010**, *17*, 1016–1026. [CrossRef]
97. Leung, A.Y. *Chinese Herbal Remedies*; Leug, A.Y., Ed.; Universe Books: New York, NY, USA, 1984; pp. 1–192.
98. Tang, W.; Eisenbrand, G. Chinese Drugs of Plant Origin. In *Chemistry, Pharmacology, and Use in Traditional and Modern Medicine*; Springer: Berlin/Heidelberg, Germany, 1992; pp. 545–547.
99. Shirin Adel, P.R.; Prakash, J. Chemical composition and antioxidant properties of ginger root (Zingiber officinale). *J. Med. Plants Res.* **2010**, *4*, 2674–2679. [CrossRef]
100. *FoodData Central Search Results*; US Department of Agriculture (Agricultural Research Service): Washington, DC, USA, 2017. Available online: https://fdc.nal.usda.gov/fdc-app.html#/food-details/1103672/nutrients; (accessed on 1 January 2017).
101. Bhatt, N.; Waly, M.I.; Essa, M.M.; Ali, A. Ginger: A functional herb. In *Food as Medicine*; Nova Science Publishers Inc.: Hauppauge, NY, USA, 2013; pp. 51–72.
102. De Krishna, A.; Minakshi, D. Functional and therapeutic applications of some important spices. In *The Role of Functional Food Security in Global Health*; Watson, R.R., Singh, R.B., Takahashi, T., Eds.; Academic Press: Cambridge, UK, 2019; pp. 499–510.
103. Mohd Hassan, N.; Yusof, N.A.; Yahaya, A.F.; Mohd Rozali, N.N.; Othman, R. Carotenoids of *Capsicum* Fruits: Pigment Profile and Health-Promoting Functional Attributes. *Antioxidants* **2019**, *8*, 469. [CrossRef]
104. Shahrajabian, M.H.; Sun, W.; Cheng, Q. Chemical components and pharmacological benefits of basil (Ocimum basilicum): A review. *Int. J. Food Propert.* **2020**, *23*, 1961–1970. [CrossRef]
105. Ciesielski, W.; Gąstoł, M.; Kulawik, D.; Oszczęda, Z.; Pisulewska, E.; Tomasik, P. Specific Controlling Essential Oil Composition of Basil (*Ocimum basilicum* L.) Involving Low-Temperature, Low-Pressure Glow Plasma of Low Frequency. *Water* **2020**, *12*, 3332. [CrossRef]
106. Zhan, Y.; An, X.; Wang, S.; Sun, M.; Zhou, H. Basil polysaccharides: A review on extraction, bioactivities and pharmacological applications. *Bioorg. Med. Chem.* **2020**, *28*, 115179. [CrossRef]
107. Slimestad, R.; Fossen, T.; Brede, C. Flavonoids and other phenolics in herbs commonly used in Norwegian commercial kitchens. *Food Chem.* **2020**, *309*, 125678. [CrossRef] [PubMed]
108. Dorman, H.J.; Lantto, T.A.; Raasmaja, A.; Hiltunen, R. Antioxidant, pro-oxidant and cytotoxic properties of parsley. *Food Funct.* **2011**, *2*, 328–337. [CrossRef] [PubMed]
109. Barros, L.; Carvalho, A.M.; Ferreira, I.C.F.R. The nutritional composition of fennel (*Foeniculum vulgare*): Shoots, leaves, stems and inflorescences. *LWT Food Sci. Technol.* **2010**, *45*, 814–818. [CrossRef]
110. Ferioli, F.; Giambanelli, E.; D'Antuono, L.F. Fennel (*Foeniculum vulgare* Mill. Subsp. *piperitum*) florets, a traditional culinary spice in Italy: Evaluation of phenolics and volatiles in local populations, and comparison with the composition of other plant parts. *J. Sci. Food Agric.* **2017**, *97*, 5369–5380. [CrossRef]
111. FoodData Central Search Results. US Department of Agriculture (Agricultural Research Service) Sage. Available online: https://fdc.nal.usda.gov/fdc-app.html#/food-details/170935/nutrients (accessed on 1 April 2019).
112. Ben Khedher, M.R.; Khedher, S.B.; Chaieb, I.; Tounsi, S.; Hammami, M. Chemical composition and biological activities of Salvia officinalis essential oil from Tunisia. *EXCLI J.* **2017**, *16*, 160–173.

113. Ghorbani, A.; Esmaeilizadeh, M. Pharmacological properties of *Salvia officinalis* and its components. *J. Tradit. Complement. Med.* **2017**, *7*, 433–440. [CrossRef] [PubMed]
114. De Caro, C.; Raucci, F.; Saviano, A.; Cristiano, C.; Casillo, G.M.; Di Lorenzo, R.; Sacchi, A.; Laneri, S.; Dini, I.; De Vita, S.; et al. Pharmacological and molecular docking assessment of cryptotanshinone as natural-derived analgesic compound. *Biomed. Pharmacother.* **2020**, *126*, 110042. [CrossRef] [PubMed]
115. Radi, R.; Beckman, J.S.; Bush, K.M.; Freeman, B.A. Peroxynitrite, a stealthy biological oxidant. *Int. J. Biol. Chem.* **2013**, *288*, 26464–26472. [CrossRef]
116. Rehman, K.; Akash, M.S.H. Mechanism of generation of oxidative stress and pathophysiology of type 2 diabetes mellitus: How are they interlinked? *J. Cell Biochem.* **2017**, *118*, 3577–3585. [CrossRef] [PubMed]
117. Lobo, V.; Patil, A.; Phatak, A.; Chandra, N. Free radicals, antioxidants and functional foods: Impact on human health. *Pharmacogn. Rev.* **2010**, *4*, 118–126. [CrossRef]
118. Ito, F.; Sono, Y.; Ito, T. Measurement and Clinical Significance of Lipid Peroxidation as a Biomarker of Oxidative Stress: Oxidative Stress in Diabetes, Atherosclerosis, and Chronic Inflammation. *Antioxidants* **2019**, *8*, 72. [CrossRef]
119. Lichtenberg, D.; Pinchuk, I. Oxidative stress, the term and the concept. *Biochem. Biophys. Res. Commun.* **2015**, *461*, 441–444. [CrossRef]
120. Dotan, Y.; Lichtenberg, D.; Pinchuk, I. Lipid peroxidation cannot be used as a universal criterion of oxidative stress. *Prog. Lipid Res.* **2004**, *43*, 200–227. [CrossRef]
121. Guo, W.; Adachi, T.; Matsui, R.; Xu, S.; Jiang, B.; Zou, M.H.; Kirber, M.; Lieberthal, W.; Cohen, R.A. Quantitative assessment of tyrosine nitration of manganese superoxide dismutase in angiotensin II-infused rat kidney. *Am. J. Physiol. Heart Circ. Physiol.* **2003**, *285*, H1396–H1403. [CrossRef]
122. Khan, J.; Brennand, D.M.; Bradley, N.; Gao, B.; Bruckdorfer, R.; Jacobs, M. 3-Nitrotyrosine in the proteins of human plasma determined by an ELISA method. *Biochem. J.* **1998**, *330*, 795–801. [CrossRef]
123. Parkin, J.; Cohen, B. An overview of the immune system. *Lancet* **2001**, *357*, 1777–1789. [CrossRef]
124. Calder, P.C. Nutrition, immunity and COVID-19. *BMJ Nutr. Prev. Health* **2020**, *3*, 74–92. [CrossRef] [PubMed]
125. Gombart, A.F.; Pierre, A.; Maggini, S. A Review of Micronutrients and the Immune System–Working in Harmony to Reduce the Risk of Infection. *Nutrients* **2020**, *12*, 236. [CrossRef] [PubMed]
126. *Janeway's Immunobiology*, 9th ed; Murphy, K.; Weaver, C. (Eds.) Garland Science/Taylor & Francis: New York, NY, USA, 2016; pp. 28–32.
127. Hengartner, H.; Odermatt, B.; Schneider, R.; Schreyer, M.; Walle, G.; MacDonald, H.R.; Zinkernagel, R.M. Deletion of self-reactive T cells before entry into the thymus medulla. *Nature* **1988**, *336*, 388–390. [CrossRef] [PubMed]
128. Jenkinson, E.J.; Kingston, R.; Smith, C.A.; Williams, G.T.; Owen, J.J. Antigen-induced apoptosis in developing T cells: A mechanism for negative selection of the T cell receptor repertoire. *Eur. J. Immunol.* **1989**, *19*, 2175–2177. [CrossRef] [PubMed]
129. Pawelec, G.; Larbi, A.; Derhovanessian, E. Senescence of the human immune system. *J. Comp. Pathol.* **2010**, *142*, S39–S44. [CrossRef]
130. Pera, A.; Campos, C.; Lopez, N.; Hassouneh, F.; Alonso, C.; Tarazona, R.; Solana, R. Immunosenescence: Implications for response to infection and vaccination in older people. *Maturitas* **2015**, *82*, 50–55. [CrossRef] [PubMed]
131. Agarwal, S.; Busse, P.J. Innate and adaptive immunosenescence. *Ann. Allergy Asthma Immunol.* **2010**, *104*, 183–190. [CrossRef] [PubMed]
132. Montgomery, R.R.; Shaw, A.C. Paradoxical changes in innate immunity in aging: Recent progress and new directions. *J. Leukoc. Biol.* **2015**, *98*, 937–943. [CrossRef]
133. Ventura, M.T.; Casciaro, M.; Gangemi, S.; Buquicchio, R. Immunosenescence in aging: Between immune cells depletion and cytokines up-regulation. *Clin. Mol. Allergy* **2017**, *15*, 21–29. [CrossRef] [PubMed]
134. Calder, P.C.; Bosco, N.; Bourdet-Sicard, R.; Capuron, L.; Delzenne, N.; Doré, J.; Franceschi, C.; Lehtinen, M.J.; Recker, T.; Salvioli, S.; et al. Health relevance of the modification of low-grade inflammation in ageing (inflammageing) and the role of nutrition. *Ageing Res. Rev.* **2017**, *40*, 95–119. [CrossRef] [PubMed]
135. Bennett, M.; Gilroy, D.W. Lipid mediators in inflammation. In *Myeloid cells in health and disease: A synthesis*; Wiley: Hoboken, NJ, USA, 2017; pp. 343–366.
136. Honce, R.; Schultz-Cherry, S. Impact of obesity on influenza A virus pathogenesis, immune response, and evolution. *Front. Immunol.* **2019**, *10*, 1071. [CrossRef] [PubMed]
137. Frasca, D.; Diaz, A.; Romero, M.; Blomberg, B.B. Ageing and obesity similarly impair antibody responses. *Clin. Exp. Immunol.* **2017**, *187*, 64–70. [CrossRef]
138. O'Shea, D.; Hogan, A.E. Dysregulation of natural killer cells in obesity. *Cancers* **2019**, *11*, 573. [CrossRef]
139. Huttunen, R.; Syrjanen, J. Obesity and the risk and outcome of infection. *Int. J. Obes.* **2013**, *37*, 333–340. [CrossRef]
140. Bogden, J.D.; Oleske, J.M. The essential trace minerals, immunity, and progression of HIV-1 infection. *Nutr. Res.* **2007**, *27*, 69–77. [CrossRef]
141. Kris-Etherton, P.M.; Harris, W.S.; Appel, L.J. Fish consumption, fish oil, omega-3 fatty acids, and cardiovascular disease. *Circulation* **2002**, *106*, 2747–2757. [CrossRef] [PubMed]
142. Huang, C.; Wang, Y.; Li, X.; Ren, L.; Zhao, J.; Hu, Y.; Zhang, L.; Fan, G.; Xu, J.; Gu, X.; et al. Clinical features of patients infected with 2019 novel coronavirus in Wuhan, China. *Lancet* **2020**, *395*, 497–506. [CrossRef]

143. Bermudez, B.; Lopez, S.; Ortega, A.; Varela, L.M.; Pacheco, Y.M. Oleic acid in olive oil: From a metabolic framework toward a clinical perspective. *Curr. Pharm. Des.* **2011**, *17*, 831–843. [CrossRef]
144. Schwingshack, L.; Hoffmann, G. Monounsaturated fatty acids and risk of cardiovascular disease: Synopsis of the evidence available from systematic reviews and meta-analyses. *Nutrients* **2012**, *4*, 1989–2007. [CrossRef]
145. *Principles of Biochemistry*; Lehninger, A.L.; Nelson, D.L.; Cox, M.M. (Eds.) W.H. Freeman and Company, Macmillan Higher Education: New York, NY, USA, 2005; Volume 1, pp. 1119–1120.
146. Patterson, E.; Wall, R.; Fitzgerald, G.F.; Ross, R.P.; Stanton, C. Health implications of high dietary omega-6 polyunsaturated fatty acids. *J. Nutr. Metab.* **2012**, *2012*, 539426. [CrossRef] [PubMed]
147. Food and Nutrition Tips during Self-Quarantine. Available online: https://www.euro.who.int/en/health-topics/health-emergencies/coronavirus-covid-19/publications-and-technical-guidance/food-and-nutrition-tips-during-self-quarantine (accessed on 6 January 2021).
148. Abu-Farha, M.; Thanaraj, T.A.; Qaddoumi, M.G.; Hashem, A.; Abubaker, J.; Al-Mulla, F. The Role of Lipid Metabolism in COVID-19 Virus Infection and as a Drug Target. *Int. J. Mol. Sci.* **2020**, *21*, 3544. [CrossRef]
149. Waitzberg, D.L.; Torrinhas, R.S. Fish oil lipid emulsions and immune response: What clinicians need to know. *Nutr. Clin. Pract.* **2009**, *24*, 487–499. [CrossRef]
150. Saini, R.K.; Keum, Y.S. Omega-3 and omega-6 polyunsaturated fatty acids: Dietary sources, metabolism, and significance—A review. *Life Sci.* **2018**, *203*, 255–267. [CrossRef]
151. Fredman, G.; Serhan, C.N. Specialized proresolving mediator targets for RvE1 and RvD1 in peripheral blood and mechanisms of resolution. *Biochem. J.* **2011**, *437*, 185–197. [CrossRef]
152. Spite, M.; Norling, L.V.; Summers, L.; Yang, R.; Cooper, D.; Petasis, N.A.; Flower, J.; Perretti, M.; Serhan, C.N. Resolvin D2 is a potent regulator of leukocytes and controls microbial sepsis. *Nature* **2009**, *461*, 1287–1291. [CrossRef]
153. Hasturk, H.; Kantarci, A.; Goguet-Surmenian, E.; Blackwood, A.; Andry, C.; Serhan, C.N.; Van Dyke, T.E. Resolvin E1 regulates inflammation at the cellular and tissue level and restores tissue homeostasis in vivo. *J. Immunol.* **2007**, *179*, 7021–7029. [CrossRef] [PubMed]
154. Titos, E.; Rius, B.; González-Périz, A.; López-Vicario, C.; Morán-Salvador, E.; Martínez-Clemente, M.; Arroyo, V.; Claria, J. Resolvin D1 and Its Precursor Docosahexaenoic Acid Promote Resolution of Adipose Tissue Inflammation by Eliciting Macrophage Polarization toward an M2-Like. Phenotype. *J. Immunol.* **2011**, *187*, 5408–5418. [CrossRef] [PubMed]
155. Lythgoe, M.P.; Middleton, P. Ongoing Clinical Trials for the Management of the COVID-19 Pandemic. *Trends Pharmacol. Sci.* **2020**, *41*, 363–382. [CrossRef]
156. Huo, Q.; Li, B.; Cheng, L.; Wu, T.; You, P.; Shen, S.; Li, Y.; He, Y.; Tian, W.; Li, R.; et al. Dietary Supplementation of Lysophospholipids Affects Feed Digestion in Lambs. *Animals* **2019**, *9*, 805. [CrossRef] [PubMed]
157. Johnson, E.J.; Russell, R.M. Beta-Carotene. In *Encyclopedia of Dietary Supplements*, 2nd ed.; Coates, P.M., Betz, J.M., Blackman, M.R., Eds.; Informa Healthcare: London, UK; New York, NY, USA, 2010; pp. 115–120.
158. Ross, C.A.; Vitamin, A. *Encyclopedia of Dietary Supplements*, 2nd ed.; Coates, P.M., Betz, J.M., Blackman, M.R., Eds.; Informa Healthcare: London, UK; New York, NY, USA, 2010; pp. 778–791.
159. Abdelhamid, L.; Luo, X.M. Retinoic Acid, Leaky Gut, and Autoimmune Diseases. *Nutrients* **2018**, *10*, 1016. [CrossRef] [PubMed]
160. Maggini, S.; Beveridge, S.; Sorbara, J.P.; Senatore, G. Feeding the immune system: The role of micronutrients in restoring resistance to infections. *CAB Rev.* **2008**, *3*, 1–21. [CrossRef]
161. Biesalski, H.K. Nutrition meets the microbiome: Micronutrients and the microbiota. *Ann. N. Y. Acad. Sci.* **2016**, *1372*, 53–64. [CrossRef]
162. Levy, M.; Thaiss, C.A.; Elinav, E. Metabolites: Messengers between the microbiota and the immune system. *Genes Dev.* **2016**, *30*, 1589–1597. [CrossRef] [PubMed]
163. Sirisinha, S. The pleiotropic role of vitamin A in regulating mucosal immunity. *Asian Pac. J. Allergy Immunol.* **2015**, *33*, 71–89.
164. Chew, B.P.; Park, J.S. Carotenoid action on the immune response. *J. Nutr.* **2004**, *134*, 257S–261S. [CrossRef]
165. Solomons, N.W.; Vitamin, A. *Present Knowledge in Nutrition*, 9th ed.; Bowman, B., Russell, R., Eds.; International Life Sciences Institute Nutrition Foundation: Washington, DC, USA, 2006; pp. 157–183.
166. Mccullough, F.S.; Northropclewes, C.A.; Thurnham, D.I. The effect of vitamin A on epithelial integrity. *Proc. Nutr. Soc.* **1999**, *58*, 289–293. [CrossRef]
167. Wang, J.L.; Swartz-Basile, D.A.; Rubin, D.C.; Levin, M.S. Retinoic acid stimulates early cellular proliferation in the adapting remnant rat small intestine after partial resection. *J. Nutr.* **1997**, *127*, 1297–1303. [CrossRef]
168. Riabroy, N.; Tanumihardjo, S.A. Oral doses of retinyl ester track chylomicron uptake and distribution of vitamin A in a male piglet model for newborn infants. *J. Nutr.* **2014**, *144*, 1188–1195. [CrossRef] [PubMed]
169. Kiss, I.; Rühl, R.; Szegezdi, E.; Fritzsche, B.; Toth, B.; Pongrácz, J.; Perlmann, T.; Fésüs, L.; Szondy, Z. Retinoid receptor-activating ligands are produced within the mouse thymus during postnatal development. *Eur. J. Immunol.* **2008**, *38*, 147–155. [CrossRef]
170. Kuwata, T.; Wang, I.M.; Tamura, T.; Ponnamperuma, R.M.; Levine, R.; Holmes, K.L.; Morse, H.C.; De Luca, L.M.; Ozato, K. Vitamin A deficiency in mice causes a systemic expansion of myeloid cells. *Blood* **2000**, *95*, 3349–3356. [CrossRef] [PubMed]
171. Chang, H.K.; Hou, W.S. Retinoic acid modulates interferon- production by hepatic natural killer T cells via phosphatase 2A and the extracellular signal-regulated kinase pathway. *J. Interferon Cytokine Res.* **2015**, *35*, 200–212. [CrossRef] [PubMed]
172. Wynn, T.A.; Vannella, K.M. Macrophages in Tissue Repair, Regeneration, and Fibrosis. *Immunity* **2016**, *44*, 450–462. [CrossRef]

173. Ross, A.C. Vitamin A deficiency and retinoid repletion regulate the antibody response to bacterial antigens and the maintenance of natural killer cells. *J. Clin. Immunol. Immunopathol.* **1996**, *80*, S63–S72. [CrossRef]
174. Hu, N.; Li, Q.B.; Zou, S. Effect of vitamin A as an adjuvant therapy for pneumonia in children: A Meta-analysis. *Zhongguo Dang dai er ke za zhi = Chin. J. Contemp. Pediatrics* **2018**, *20*, 146–153.
175. Chowdhury, A.I. Role and Effects of Micronutrients Supplementation in Immune System and SARS-Cov-2(COVID-19). *Asian Pac. J. Allergy Immun.* **2020**, *4*, 47–55.
176. Hosomi, K.; Kunisawa, J. The specific roles of vitamins in the regulation of immunosurveillance and maintenance of immunologic homeostasis in the gut. *Immune Netw.* **2017**, *17*, 13–19. [CrossRef] [PubMed]
177. Ross, A. Vitamin A and Carotenoids. In *Modern Nutrition in Health and Disease*, 10th ed.; Shils, M., Shike, M., Ross, A., Caballero, B., Cousins, R., Eds.; Lippincott Williams & Wilkins: Baltimore, MD, USA, 2006; pp. 351–375.
178. Yoshii, K.; Hosomi, K.; Sawane, K.; Kunisawa, J. Metabolism of Dietary and Microbial Vitamin B Family in the Regulation of Host Immunity. *Front. Nutr.* **2019**, *6*, 48. [CrossRef]
179. Luna, G.; Alping, P.; Burman, J.; Fink, K.; Fogdell-Hahn, A.; Gunnarsson, M.; Hillert, J.; Langer-Gould, A.; Lycke, J.; Nilsson, P.; et al. Infection Risks Among Patients with Multiple Sclerosis Treated With Fingolimod, Natalizumab, Rituximab, and Injectable Therapies. *JAMA Neurol.* **2019**, *77*, 2. [CrossRef]
180. Villarruz-Sulit, M.V.; Cabaluna, I.T. Should, B Vitamins be used in the treatment of COVID-19? *Asia Pac. Cent. Evid. Based Healthcare* **2020**, *1*, 1–2.
181. Kant, A.K.; Block, G. Dietary vitamin B-6 intake and food sources in the US population: NHANES II,1976–1980. *Am. J. Clin. Nutr.* **1990**, *52*, 707–716. [CrossRef] [PubMed]
182. Watanabe, F. Vitamin B12 sources and bioavailability. *Exp. Biol. Med.* **2007**, *232*, 1266–1274. [CrossRef]
183. Burns, J.J. Missing step in man, monkey and guinea pig required for the biosynthesis of L-ascorbic acid. *Nature* **1957**, *180*, 553. [CrossRef]
184. Nishikimi, M.; Fukuyama, R.; Minoshima, S.; Shimizu, N.; Yagi, K. Cloning and chromosomal mapping of the human nonfunctional gene for L-gulono-gamma-lactone oxidase, the enzyme for L-ascorbic acid biosynthesis missing in man. *J. Biol. Chem.* **1994**, *269*, 13685–13688. [CrossRef]
185. Carr, A.; Frei, B. Does vitamin C act as a pro-oxidant under physiological conditions? *FASEB J.* **1999**, *13*, 1007–1024. [CrossRef]
186. Mandl, J.; Szarka, A.; Banhegyi, G. Vitamin C: Update on physiology and pharmacology. *Br. J. Pharmacol.* **2009**, *157*, 1097–1110. [CrossRef] [PubMed]
187. Englard, S.; Seifter, S. The biochemical functions of ascorbic acid. *Annu. Rev. Nutr.* **1986**, *6*, 365–406. [CrossRef]
188. Kivirikko, K.I.; Myllyla, R.; Pihlajaniemi, T. Protein hydroxylation: Prolyl 4-hydroxylase, an enzyme with four cosubstrates and a multifunctional subunit. *FASEB J.* **1989**, *3*, 1609–1617. [CrossRef]
189. Geesin, J.C.; Darr, D.; Kaufman, R.; Murad, S.; Pinnell, S.R. Ascorbic acid specifically increases type I and type III procollagen messenger RNA levels in human skin fibroblast. *J. Investig. Dermatol.* **1988**, *90*, 420–424. [CrossRef] [PubMed]
190. Kishimoto, Y.; Saito, N.; Kurita, K.; Shimokado, K.; Maruyama, N.; Ishigami, A. Ascorbic acid enhances the expression of type 1 and type 4 collagen and SVCT2 in cultured human skin fibroblasts. *Biochem. Biophys. Res. Commun.* **2013**, *430*, 579–584. [CrossRef]
191. Nusgens, B.V.; Humbert, P.; Rougier, A.; Colige, A.C.; Haftek, M.; Lambert, C.A.; Richard, A.; Creidi, P.; Lapiere, C.M. Topically applied vitamin C enhances the mRNA level of collagens I and III, their processing enzymes and tissue inhibitor of matrix metalloproteinase 1 in the human dermis. *J. Investig. Dermatol.* **2001**, *116*, 853–859. [CrossRef] [PubMed]
192. Tajima, S.; Pinnell, S.R. Ascorbic acid preferentially enhances type I and III collagen gene transcription in human skin fibroblasts. *J. Dermatol. Sci.* **1996**, *11*, 250–253. [CrossRef]
193. Davidson, J.M.; Lu Valle, P.A.; Zoia, O.; Quaglino, D., Jr.; Giro, M. Ascorbate differentially regulates elastin and collagen biosynthesis in vascular smooth muscle cells and skin fibroblasts by pretranslational mechanisms. *J. Biol. Chem.* **1997**, *272*, 345–352. [CrossRef]
194. Carr, A.; Maggini, S. Vitamin C and immune function. *Nutrients* **2017**, *9*, 1211. [CrossRef] [PubMed]
195. Stankova, L.; Gerhardt, N.B.; Nagel, L.; Bigley, R.H. Ascorbate and phagocyte function. *Infect. Immunol.* **1975**, *12*, 252–256. [CrossRef] [PubMed]
196. Winterbourn, C.C.; Vissers, M.C. Changes in ascorbate levels on stimulation of human neutrophils. *Biochim. Biophys. Acta* **1983**, *763*, 175–179. [CrossRef]
197. Parker, A.; Cuddihy, S.L.; Son, T.G.; Vissers, M.C.; Winterbourn, C.C. Roles of superoxide and myeloperoxidase in ascorbate oxidation in stimulated neutrophils and H(2)O(2)-treated HL60 cells. *Free Radic. Biol. Med.* **2011**, *51*, 1399–1405. [CrossRef] [PubMed]
198. Oberritter, H.; Glatthaar, B.; Moser, U.; Schmidt, K.H. Effect of functional stimulation on ascorbate content in phagocytes under physiological and pathological conditions. *Int. Arch. Allergy Appl. Immunol.* **1986**, *81*, 46–50. [CrossRef]
199. Goldschmidt, M.C. Reduced bactericidal activity in neutrophils from scorbutic animals and the effect of ascorbic acid on these target bacteria in vivo and in vitro. *Am. J. Clin. Nutr.* **1991**, *54*, 1214S–1220S. [CrossRef]
200. Goldschmidt, M.C.; Masin, W.J.; Brown, L.R.; Wyde, P.R. The effect of ascorbic acid deficiency on leukocyte phagocytosis and killing of actinomyces viscosus. *Int. J. Vitam. Nutr. Res.* **1988**, *58*, 326–334.
201. Johnston, C.S.; Huang, S.N. Effect of ascorbic acid nutriture on blood histamine and neutrophil chemotaxis in guinea pigs. *J. Nutr.* **1991**, *121*, 126–130. [CrossRef] [PubMed]

202. Rebora, A.; Dallegri, F.; Patrone, F. Neutrophil dysfunction and repeated infections: Influence of levamisole and ascorbic acid. *Br. J. Dermatol.* **1980**, *102*, 49–56. [CrossRef]
203. Patrone, F.; Dallegri, F.; Bonvini, E.; Minervini, F.; Sacchetti, C. Disorders of neutrophil function in children with recurrent pyogenic infections. *Med. Microbiol. Immunol.* **1982**, *171*, 113–122. [CrossRef]
204. Boura, P.; Tsapas, G.; Papadopoulou, A.; Magoula, I.; Kountouras, G. Monocyte locomotion in anergic chronic brucellosis patients: The in vivo effect of ascorbic acid. *Immunopharmacol. Immunotoxicol.* **1989**, *11*, 119–129. [CrossRef]
205. Anderson, R.; Theron, A. Effects of ascorbate on leucocytes: Part III. In vitro and in vivo stimulation of abnormal neutrophil motility by ascorbate. *S. Afr. Med. J.* **1979**, *56*, 429–433.
206. Johnston, C.S.; Martin, L.J.; Cai, X. Antihistamine effect of supplemental ascorbic acid and neutrophil chemotaxis. *J. Am. Coll. Nutr.* **1992**, *11*, 172–176. [CrossRef]
207. Anderson, R.; Oosthuizen, R.; Maritz, R.; Theron, A.; Van Rensburg, A.J. The effects of increasing weekly doses of ascorbate on certain cellular and humoral immune functions in normal volunteers. *Am. J. Clin. Nutr.* **1980**, *33*, 71–76. [CrossRef]
208. Anderson, R. Ascorbate-mediated stimulation of neutrophil motility and lymphocyte transformation by inhibition of the peroxidase/H_2O_2/halide system in vitro and in vivo. *Am. J. Clin. Nutr.* **1981**, *34*, 1906–1911. [CrossRef]
209. Ganguly, R.; Durieux, M.F.; Waldman, R.H. Macrophage function in vitamin C-deficient guinea pigs. *Am. J. Clin. Nutr.* **1976**, *29*, 762–765. [CrossRef]
210. Corberand, J.; Nguyen, F.; Fraysse, B.; Enjalbert, L. Malignant external otitis and polymorphonuclear leukocyte migration impairment. Improvement with ascorbic acid. *Arch. Otolaryngol.* **1982**, *108*, 122–124. [CrossRef]
211. Levy, R.; Schlaeffer, F. Successful treatment of a patient with recurrent furunculosis by vitamin C: Improvement of clinical course and of impaired neutrophil functions. *Int. J. Dermatol.* **1993**, *32*, 832–834. [CrossRef] [PubMed]
212. Levy, R.; Shriker, O.; Porath, A.; Riesenberg, K.; Schlaeffer, F. Vitamin C for the treatment of recurrent furunculosis in patients with imparied neutrophil functions. *J. Infect. Dis.* **1996**, *173*, 1502–1505. [CrossRef]
213. Nungester, W.J.; Ames, A.M. The relationship between ascorbic acid and phagocytic activity. *J. Infect. Dis.* **1948**, *83*, 50–54. [CrossRef]
214. Shilotri, P.G. Phagocytosis and leukocyte enzymes in ascorbic acid deficient guinea pigs. *J. Nutr.* **1977**, *107*, 1513–1516. [CrossRef] [PubMed]
215. Collins, J.F. *Modern Nutrition in Health and Disease*, 11th ed.; Ross, A.C., Caballero, B., Cousins, R.J., Tucker, K.L., Ziegler, T.R., Eds.; Lippincott Williams & Wilkins: Baltimore, MD, USA, 2014; pp. 206–216.
216. Prohaska, J.R. Copper. In *Present Knowledge in Nutrition*, 10th ed.; Erdman, J.W., Macdonald, I.A., Zeisel, S.H., Eds.; Wiley-Blackwell: Washington, DC, USA, 2012; pp. 540–553.
217. Shilotri, P.G. Glycolytic, hexose monophosphate shunt and bactericidal activities of leukocytes in ascorbic acid deficient guinea pigs. *J. Nutr.* **1977**, *107*, 1507–1512. [CrossRef]
218. Sharma, P.; Raghavan, S.A.; Saini, R.; Dikshit, M. Ascorbate-mediated enhancement of reactive oxygen species generation from polymorphonuclear leukocytes: Modulatory effect of nitric oxide. *J. Leukoc. Biol.* **2004**, *75*, 1070–1078. [CrossRef]
219. Rebora, A.; Crovato, F.; Dallegri, F.; Patrone, F. Repeated staphylococcal pyoderma in two siblings with defective neutrophil bacterial killing. *Dermatologica* **1980**, *160*, 106–112. [CrossRef]
220. Vissers, M.C.; Wilkie, R.P. Ascorbate deficiency results in impaired neutrophil apoptosis and clearance and is associated with up-regulation of hypoxia-inducible factor 1alpha. *J. Leukoc. Biol.* **2007**, *81*, 1236–1244. [CrossRef]
221. Fisher, B.J.; Kraskauskas, D.; Martin, E.J.; Farkas, D.; Wegelin, J.A.; Brophy, D.; Ward, K.R.; Voelkel, N.F.; Fowler, A.A., III; Natarajan, R. Mechanisms of attenuation of abdominal sepsis induced acute lung injury by ascorbic acid. *Am. J. Physiol. Lung Cell. Mol. Physiol.* **2012**, *303*, L20–L32. [CrossRef]
222. Huijskens, M.J.; Walczak, M.; Koller, N.; Briede, J.J.; Senden-Gijsbers, B.L.; Schnijderberg, M.C.; Bos, G.M.; Germeraad, W.T. Technical advance: Ascorbic acid induces development of double-positive T cells from human hematopoietic stem cells in the absence of stromal cells. *J. Leukoc. Biol.* **2014**, *96*, 1165–1175. [CrossRef] [PubMed]
223. Molina, N.; Morandi, A.C.; Bolin, A.P.; Otton, R. Comparative effect of fucoxanthin and vitamin C on oxidative and functional parameters of human lymphocytes. *Int. Immunopharmacol.* **2014**, *22*, 41–50. [CrossRef] [PubMed]
224. Tanaka, M.; Muto, N.; Gohda, E.; Yamamoto, I. Enhancement by ascorbic acid 2-glucoside or repeated additions of ascorbate of mitogen-induced IgM and IgG productions by human peripheral blood lymphocytes. *Jpn. J. Pharmacol.* **1994**, *66*, 451–456. [CrossRef] [PubMed]
225. Manning, J.; Mitchell, B.; Appadurai, D.A.; Shakya, A.; Pierce, L.J.; Wang, H.; Nganga, V.; Swanson, P.C.; May, J.M.; Tantin, D.; et al. Vitamin C promotes maturation of T-cells. *Antioxid. Redox Signal.* **2013**, *19*, 2054–2067. [CrossRef]
226. Kennes, B.; Dumont, I.; Brohee, D.; Hubert, C.; Neve, P. Effect of vitamin C supplements on cell-mediated immunity in old people. *Gerontology* **1983**, *29*, 305–310. [CrossRef]
227. Anderson, R.; Hay, I.; van Wyk, H.; Oosthuizen, R.; Theron, A. The effect of ascorbate on cellular humoral immunity in asthmatic children. *S. Afr. Med. J.* **1980**, *58*, 974–977. [PubMed]
228. Fraser, R.C.; Pavlovic, S.; Kurahara, C.G.; Murata, A.; Peterson, N.S.; Taylor, K.B.; Feigen, G.A. The effect of variations in vitamin C intake on the cellular immune response of guinea pigs. *Am. J. Clin. Nutr.* **1980**, *33*, 839–847. [CrossRef] [PubMed]

229. Feigen, G.A.; Smith, B.H.; Dix, C.E.; Flynn, C.J.; Peterson, N.S.; Rosenberg, L.T.; Pavlovic, S.; Leibovitz, B. Enhancement of antibody production and protection against systemic anaphylaxis by large doses of vitamin C. *Res. Commun. Chem. Pathol. Pharmacol.* **1982**, *38*, 313–333. [CrossRef]
230. Prinz, W.; Bloch, J.; Gilich, G.; Mitchell, G. A systematic study of the effect of vitamin C supplementation on the humoral immune response in ascorbate-dependent mammals. I. The antibody response to sheep red blood cells (a T-dependent antigen) in guinea pigs. *Int. J. Vitam. Nutr. Res.* **1980**, *50*, 294–300. [PubMed]
231. Prinz, W.; Bortz, R.; Bregin, B.; Hersch, M. The effect of ascorbic acid supplementation on some parameters of the human immunological defence system. *Int. J. Vitam. Nutr. Res.* **1977**, *47*, 248–257.
232. Mohammed, B.M.; Fisher, B.J.; Kraskauskas, D.; Farkas, D.; Brophy, D.F.; Fowler, A.A.; Natarajan, R. Vitamin C: A novel regulator of neutrophil extracellular trap formation. *Nutrients* **2013**, *5*, 3131–3151. [CrossRef]
233. Chen, Y.; Luo, G.; Yuan, J.; Wang, Y.; Yang, X.; Wang, X.; Li, G.; Liu, Z.; Zhong, N. Vitamin C mitigates oxidative stress and tumor necrosis factor-alpha in severe community-acquired pneumonia and LPS-induced macrophages. *Mediat. Inflamm.* **2014**, *14*, 426740. [CrossRef]
234. Jeng, K.C.; Yang, C.S.; Siu, W.Y.; Tsai, Y.S.; Liao, W.J.; Kuo, J.S. Supplementation with vitamins C and E enhances cytokine production by peripheral blood mononuclear cells in healthy adults. *Am. J. Clin. Nutr.* **1996**, *64*, 960–965. [CrossRef]
235. Kim, Y.; Kim, H.; Bae, S.; Choi, J.; Lim, S.Y.; Lee, N.; Kong, J.M.; Hwang, Y.I.; Kang, J.S.; Lee, W.J. Vitamin C is an essential factor on the antiviral immune responses through the production of interferon-a/b at the initial stage of influenza A virus (H3N2) infection. *Immune Netw.* **2013**, *13*, 70–74. [CrossRef] [PubMed]
236. Gao, Y.L.; Lu, B.; Zhai, J.H.; Liu, Y.C.; Qi, H.X.; Yao, Y.; Chai, Y.F.; Shou, S.T. The parenteral vitamin C improves sepsis and sepsis-induced multiple organ dysfunction syndrome via preventing cellular immunosuppression. *Mediat. Inflamm.* **2017**, *2017*, 4024672. [CrossRef]
237. Portugal, C.C.; Socodato, R.; Canedo, T.; Silva, C.M.; Martins, T.; Coreixas, V.S.; Loiola, E.C.; Gess, B.; Rohr, D.; Santiago, A.R.; et al. Caveolin-1-mediated internalization of the vitamin C transporter SVCT2 in microglia triggers an inflammatory phenotype. *Sci. Signal.* **2017**, *10*, eaal2005. [CrossRef]
238. Dahl, H.; Degre, M. The effect of ascorbic acid on production of human interferon and the antiviral activity in vitro. *Acta Pathol. Microbiol. Scand. B* **1976**, *84*, 280–284. [CrossRef] [PubMed]
239. Karpinska, T.; Kawecki, Z.; Kandefer-Szerszen, M. The influence of ultraviolet irradiation, L-ascorbic acid and calcium chloride on the induction of interferon in human embryo fibroblasts. *Arch. Immunol. Ther. Exp.* **1982**, *30*, 33–37.
240. Siegel, B.V. Enhancement of interferon production by poly(rI)-poly(rC) in mouse cell cultures by ascorbic acid. *Nature* **1975**, *254*, 531–532. [CrossRef] [PubMed]
241. Canali, R.; Natarelli, L.; Leoni, G.; Azzini, E.; Comitato, R.; Sancak, O.; Barella, L.; Virgili, F. Vitamin C supplementation modulates gene expression in peripheral blood mononuclear cells specifically upon an inflammatory stimulus: A pilot study in healthy subjects. *Genes Nutr.* **2014**, *9*, 390. [CrossRef] [PubMed]
242. Dawson, W.; West, G.B. The influence of ascorbic acid on histamine metabolism in guinea-pigs. *Br. J. Pharmacol. Chemother.* **1965**, *24*, 725–734. [CrossRef]
243. Nandi, B.K.; Subramanian, N.; Majumder, A.K.; Chatterjee, I.B. Effect of ascorbic acid on detoxification of histamine under stress conditions. *Biochem. Pharmacol.* **1974**, *23*, 643–647. [CrossRef]
244. Subramanian, N.; Nandi, B.K.; Majumder, A.K.; Chatterjee, I.B. Role of L-ascorbic acid on detoxification of histamine. *Biochem. Pharmacol.* **1973**, *22*, 1671–1673. [CrossRef]
245. Chatterjee, I.B.; Gupta, S.D.; Majumder, A.K.; Nandi, B.K.; Subramanian, N. Effect of ascorbic acid on histamine metabolism in scorbutic guinea-pigs. *J. Physiol.* **1975**, *251*, 271–279. [CrossRef] [PubMed]
246. Clemetson, C.A. Histamine and ascorbic acid in human blood. *J. Nutr.* **1980**, *110*, 662–668. [CrossRef]
247. Phokela, S.S.; Peleg, S.; Moya, F.R.; Alcorn, J.L. Regulation of human pulmonary surfactant protein gene expression by 1alpha,25-dihydroxyvitamin D3. *Am. J. Physiol. Lung Cell Mol. Physiol.* **2005**, *289*, 617–626. [CrossRef]
248. Ovesen, L.; Brot, C.; Jakobsen, J. Food contents and biological activity of 25-hydroxyvitamin D: A vitamin D metabolite to be reckoned with? *Ann. Nutr. Metab.* **2003**, *47*, 107–113. [CrossRef] [PubMed]
249. Traber, M.G.; Vitamin, E. *Modern Nutrition in Health and Disease*, 10th ed.; Shils, M.E., Shike, M., Ross, A.C., Caballero, B., Cousins, R., Eds.; Lippincott Williams & Wilkins: Baltimore, MD, USA, 2006; pp. 396–411.
250. Mihajlovic, M.; Fedecostante, M.; Oost, M.J.; Steenhuis, S.K.P.; Lentjes, E.; Maitimu-Smeele, I.; Janssen, M.J.; Hilbrands, L.B.; Masereeuw, R. Role of Vitamin D in Maintaining Renal Epithelial Barrier Function in Uremic Conditions. *Int. J. Mol. Sci.* **2017**, *18*, 2531. [CrossRef] [PubMed]
251. Suzuki, H.; Kunisawa, J. Vitamin-mediated immune regulation in the development of inflammatory diseases. *Endocr. Metab. Immune Disord Drug Targets.* **2015**, *15*, 212–215. [CrossRef] [PubMed]
252. Yin, Z.; Pintea, V.; Lin, Y.; Hammock, B.D.; Watsky, M.A. Vitamin D enhances corneal epithelial barrier function. *Investig. Ophthalmol. Vis. Sci.* **2011**, *52*, 7359–7364. [CrossRef]
253. Wu, D.; Meydani, S.N. Vitamin E, immunity, and infection. In *Nutrition, Immunity, and Infection*; CRC Press: Boca Raton, FL, USA, 2017; pp. 197–212.
254. Chavance, M.; Herbeth, B.; Fournier, C.; Janot, C.; Venhes, G. Vitamin status, immunity and infections in an elderly population. *Eur. J. Clin. Nutr.* **1989**, *43*, 827–835. [PubMed]

255. Wu, D.; Meydani, S.N. Age-associated changes in immune function: Impact of vitamin E intervention and the underlying mechanisms. *Endocr. Metab. Immune. Disord. Drug Targets* **2014**, *14*, 283–289. [CrossRef]
256. Panche, A.N.; Diwan, A.D.; Chandra, S.R. Flavonoids: An overview. *J. Nutr. Sci.* **2016**, *5*, e47. [CrossRef]
257. Ravishankar, D.; Rajora, A.K.; Greco, F.; Osborn, H.M. Flavonoids as prospective compounds for anti-cancer therapy. *Int. J. Biochem. Cell Biol.* **2013**, *45*, 2821–2831. [CrossRef]
258. Hodek, P.; Trefil, P.; Stiborová, M. Flavonoids-potent and versatile biologically active compounds interacting with cytochromes P450. *Chem. Biol. Interact.* **2002**, *139*, 1–21. [CrossRef]
259. Liu, Y.; Jing, Y.Y.; Zeng, C.Y.; Li, C.G.; Xu, L.H.; Yan, L.; Bai, W.-J.; Zha, Q.-B.; Ouyang, D.-Y.; He, X.-H. Scutellarin suppresses NLRP3 inflammasome activation in macrophages and protects mice against bacterial sepsis. *Front Pharmacol.* **2018**, *8*, 975. [CrossRef]
260. Dini, I.; Falanga, D.; Di Lorenzo, R.; Tito, A.; Carotenuto, G.; Zappelli, C.; Grumetto, L.; Sacchi, A.; Laneri, S.; Apone, F. An Extract from *Ficus carica* Cell Cultures Works as an Anti-Stress Ingredient for the Skin. *Antioxidants* **2021**, *10*, 515. [CrossRef]
261. Kopustinskiene, D.M.; Jakstas, V.; Savickas, A.; Bernatoniene, J. Flavonoids as Anticancer Agents. *Nutrients* **2020**, *12*, 457. [CrossRef] [PubMed]
262. Gupta, S.C.; Kunnumakkara, A.B.; Aggarwal, S.; Aggarwal, B.B. Inflammation, a Double-Edge Sword for Cancer and Other Age-Related Diseases. *Front. Immunol.* **2018**, *9*, 2160. [CrossRef]
263. Sargiacomo, C.; Sotgia, F.; Lisanti, M.P. COVID-19 and chronological aging: Senolytics and other anti-aging drugs for the treatment or prevention of corona virus infection? *Aging* **2020**, *12*, 6511–6517. [CrossRef] [PubMed]
264. Smith, M.S.; Smith, J.C. Repurposing therapeutics for COVID-19: Supercomputer-based docking to the SARS-CoV-2 Viral Spike Protein and Viral Spike Protein-Human ACE2 Interface. *ChemRxiv.* **2020**. [CrossRef]
265. Ginwala, R.; Bhavsar, R.; Chigbu, D.G.I.; Jain, P.; Khan, Z.K. Potential Role of Flavonoids in Treating Chronic Inflammatory Diseases with a Special Focus on the Anti-Inflammatory Activity of Apigenin. *Antioxidants* **2019**, *8*, 35. [CrossRef]
266. Chirumbolo, S. The role of quercetin, flavonols and flavones in modulating inflammatory cell function. *Inflamm. Allergy Drug Targets* **2010**, *9*, 263–285. [CrossRef] [PubMed]
267. Verbeek, R.; Plomp, A.C.; van Tol, E.A.; van Noort, J.M. The flavones luteolin and apigenin inhibit in vitro antigen-specific proliferation and interferon-gamma production by murine and human autoimmune T cells. *Biochem. Pharmacol.* **2004**, *68*, 621–629. [CrossRef] [PubMed]
268. Alschuler, L.; Weil, A.; Horwitz, R.; Stamets, P.; Chiasson, A.M.; Crocker, R.; Maizes, V. Integrative considerations during the COVID-19 pandemic. *Explore* **2020**, S1550–S8307. [CrossRef]
269. Guo, T.L.; Mc Cay, J.A.; Zhang, L.X.; Brown, R.D.; You, L.; Karrow, N.A. Genistein modulations immune responses and increases host resistance to B16F10 tumor in adult female B6cF1 mice. *J Nutr.* **2001**, *131*, 3251–3258. [CrossRef] [PubMed]
270. Sakai, T.; Kogiso, M. Soy isoflavones and immunity. *J. Investig. Med.* **2008**, *55*, 167–173. [CrossRef]
271. Masilamani, M.; Wei, J.; Bhatt, S.; Paul, M.; Yakir, S.; Sampson, H.A. Soybean isoflavones regulate dendritic cell function and suppress allergic sensitization to peanut. *J. Allergy Clin. Immunol.* **2011**, *128*, 1242–1250. [CrossRef]
272. Wei, J.; Bhatt, S.; Chang, L.M.; Sampson, H.A.; Masilamani, M. Isoflavones, genistein and daidzein, regulate mucosal immune response by suppressing dendritic cell function. *PLoS ONE.* **2012**, *7*, e47979. [CrossRef]
273. Smith, B.N.; Dilger, R.N. Immunomodulatory potential of dietary soybean-derived isoflavones and saponins in pigs. *J. Anim. Sci.* **2018**, *96*, 1288–1304. [CrossRef] [PubMed]
274. Jin, X.; Wang, S.; Zhao, X.; Jin, Q.; Fan, C.; Li, J.; Shan, Z.; Teng, W. Coumestrol inhibits autoantibody production through modulating Th1 response in experimental autoimmune thyroiditis. *Oncotarget* **2016**, *7*, 52797–52809. [CrossRef]
275. Sirotkin, A.V.; Harrath, A.H. Phytoestrogens and their effects. *Eur. J. Pharmacol.* **2014**, *741*, 230–236. [CrossRef] [PubMed]
276. Park, J.; Kim, S.H.; Cho, D.; Kim, T.S. Formononetin, a phyto-oestrogen, its metabolites up-regulate interleukin-4 production in activated T cells via increased AP-1 DNA binding activity. *Immunology* **2005**, *116*, 71–81. [CrossRef] [PubMed]
277. Kojima, H.; Takeda, Y.; Muromoto, R.; Takahashi, M.; Hirao, T.; Takeuchi, S.; Jetten, A.M.; Matzuda, T. Isoflavones enhance interleukin-17 gene expression via retinoic acid receptor-related orphan receptors alpha and gamma. *Toxicology* **2015**, *329*, 32–39. [CrossRef]
278. Abron, J.D.; Singh, N.P.; Price, R.L.; Nagarkatti, M.; Nagarkatti, P.S.; Singh, U.P. Genistein induces macrophage polarization and systemic cytokine to ameliorate experimental colitis. *PLoS ONE* **2018**, *13*, e0199631. [CrossRef]
279. Mace, T.A.; Ware, M.B.; King, S.A.; Loftus, S.; Farren, M.R.; McMichael, E.; Scoville, S.; Geraghty, C.; Young, G.; Carson, W.E.; et al. Soy isoflavones and their metabolites modulate cytokine-induced natural killer cell function. *Sci. Rep.* **2019**, *9*, 5068. [CrossRef]
280. Dia, V.P.; Berhow, M.A.; Gonzalez De Mejia, E. Bowman-Birk inhibitor and genistein among soy compounds that synergistically inhibit nitric oxide and prostaglandin E2 pathways in lipopolysaccharide-induced macrophages. *J. Agric. Food Chem.* **2008**, *56*, 11707–11717. [CrossRef] [PubMed]
281. Amalraj, A.; Pius, A.; Gopi, S.; Gopi, S. Biological activities of curcuminoids, other biomolecules from turmeric and their derivatives—A review. *J. Tradit. Complement. Med.* **2017**, *7*, 205–233. [CrossRef]
282. Kohli, K.; Ali, J.; Ansari, M.J.; Raheman, Z. Curcumin: A natural antiinflammatory agent. *Indian J. Pharmacol.* **2005**, *37*, 141–147. [CrossRef]
283. Maiti, P.; Dunbar, G.L. Use of Curcumin, a Natural Polyphenol for Targeting Molecular Pathways in Treating Age-Related Neurodegenerative Diseases. *Int. J. Mol. Sci.* **2018**, *19*, 1637. [CrossRef]

284. Wanninger, S.; Lorenz, V.; Subhan, A.; Edelmann, F.T. Metal complexes of curcumin—Synthetic strategies, structures and medicinal applications. *Chem. Soc. Rev.* **2015**, *44*, 4986–5002. [CrossRef]
285. Itokawa, H.; Shi, Q.; Akiyama, T.; Morris-Natschke, S.L.; Lee, K.-H. Recent advances in the investigation of curcuminoids. *Chin. Med.* **2008**, *3*, 11. [CrossRef] [PubMed]
286. Chan, K.K.; Bin Hamid, M.S.; Webster, R.D. Quantification of capsaicinoids in chillies by solid-phase extraction coupled with voltammetry. *Food Chem.* **2018**, *265*, 152–158. [CrossRef]
287. Srinivasan, K. Biological Activities of Red Pepper (Capsicum annuum) and Its Pungent Principle Capsaicin: A Review. *Crit. Rev. Food Sci. Nutr.* **2016**, *56*, 1488–1500. [CrossRef] [PubMed]
288. Joe, B.; Lokesh, B.R. Effect of curcumin and capsaicin on arachidonic acid metabolism and lysosomal enzyme secretion by rat peritoneal macrophages. *Lipids* **1997**, *32*, 1173–1180. [CrossRef]
289. Anandakumar, P.; Kamaraj, S.; Jagan, S.; Ramakrishnan, G.; Asokkumar, S.; Naveenkumar, C.; Raghunandhakumar, S.; Devaki, T. Capsaicin inhibits benzo(a)pyrene-induced lung carcinogenesis in an in vivo mouse model. *Inflamm. Res.* **2012**, *61*, 1169–1175. [CrossRef] [PubMed]
290. Zhang, L.L.; Yan Liu, D.; Ma, L.Q.; Luo, Z.D.; Cao, T.B.; Zhong, J.; Yan, Z.C.; Wang, L.J.; Zhao, Z.G.; Zhu, S.J.; et al. Activation of transient receptor potential vanilloid type-1 channel prevents adipogenesis and obesity. *Circ. Res.* **2007**, *100*, 1063–1070. [CrossRef]
291. Zhang, S.; Ma, X.; Zhang, L.; Sun, H.; Liu, X. Capsaicin reduces blood glucose by increasing insulin levels and glycogen content better than capsiate in streptozotocin-induced diabetic rats. *J. Agric. Food Chem.* **2017**, *65*, 2323–2330. [CrossRef]
292. Saito, M.; Matsushita, M.; Yoneshiro, T.; Okamatsu-Ogura, Y. Brown Adipose Tissue, Diet-Induced Thermogenesis, and Thermogenic Food Ingredients: From Mice to Men. *Front. Endocrinol.* **2020**, *11*, 222. [CrossRef]
293. Ramirez, D.A.; Locatelli, D.A.; González, R.E.; Cavagnaro, P.F.; Camargo, A.B. Analytical methods for bioactive sulfur compounds in *Allium*: An integrated review and future directions. *J. Food Comp. Anal.* **2017**, *61*, 4–19. [CrossRef]
294. Pinto, J.T.; Krasnikov, B.F.; Cooper, A.J.L. Redox-sensitive proteins are potential targets of garlic-derived mercaptocysteine derivatives. *J. Nutr.* **2006**, *136*, 835S–841S. [CrossRef] [PubMed]
295. Wang, H.C.; Pao, J.; Lin, S.Y.; Sheen, L.Y. Molecular mechanisms of garlic-derived allyl sulfides in the inhibition of skin cancer progression. *Ann. N. Y. Acad. Sci.* **2012**, *1271*, 44–52. [CrossRef] [PubMed]
296. Nuutila, A.M.; Puupponen-Pimiä, R.; Aarni, M.; Oksman-Caldentey, K.-M. Comparison of antioxidant activities of onion and garlic extracts by inhibition of lipid peroxidation and radical scavenging activity. *Food Chem.* **2003**, *81*, 485–493. [CrossRef]
297. Das, A.; Banik, N.L.; Ray, S.K. Garlic compounds generate reactive oxygen species leading to activation of stress kinases and cysteine proteases for apoptosis in human glioblastoma T98G and U87MG cells. *Cancer* **2007**, *110*, 1083–1095. [CrossRef] [PubMed]
298. Burt, S. Essential oils: Their antibacterial properties and potential applications in foods-a review. *Int. J. Food Microbiol.* **2004**, *94*, 223–253. [CrossRef]
299. Sivakumar, P.M.; Kumar, T.M.; Doble, M. Antifungal activity, mechanism and QSAR studies on chalcones. *Chem. Biol. Drug Des.* **2009**, *74*, 68–79. [CrossRef]
300. Rozmer, Z.; Perjési, P. Naturally occurring chalcones and their biological activities. *Phytochem. Rev.* **2016**, *15*, 87–120. [CrossRef]
301. Kurita, N.; Miyaji, M.; Kurane, R.; Takahara, Y.; Ichimura, K. Antifungal activity and molecular orbital energies of aldehyde compounds from oils of higher plants. *Agric. Biol. Chem.* **1979**, *43*, 2365–2371.
302. Babu, K.S.; Li, X.C.; Jacob, M.R.; Zhang, Q.; Khan, S.; Ferreia, D.; Clark, A.M. Synthesis, antifungal activity, and structure-activity relationships of coruscanone A. analogs. *J. Med. Chem.* **2006**, *49*, 7877–7886. [CrossRef]
303. Monzote, L.; Stamberg, W.; Staniek, K.; Gille, L. Toxic effects of carvacrol, caryophyllene oxide, and ascaridole from essential oil of Chenopodium ambrosioides on mitochondria. *Toxicol. Appl. Pharmacol.* **2009**, *240*, 337–347. [CrossRef]
304. Cao, Y.; Gu, W.; Zhang, J.; Chu, Y.; Ye, X.; Hu, Y.; Chen, J. Effects of chitosan, aqueous extract of ginger, onion and garlic on quality and shelf life of stewed-pork during refrigerated storage. *Food Chem.* **2013**, *141*, 1655–1660. [CrossRef]
305. Souza, V.G.L.; Pires, J.R.A.; Vieira, É.T.; Coelhoso, I.M.; Duarte, M.P.; Fernando, A.L. Shelf Life Assessment of Fresh Poultry Meat Packaged in Novel Bionanocomposite of Chitosan/Montmorillonite Incorporated with Ginger Essential Oil. *Coatings* **2018**, *8*, 177. [CrossRef]
306. Khaledian, Y.; Pajohi-Alamoti, M.; Bazargani-Gilani, B. Development ofcellulose nanofibers coating incorporated with ginger essential oiland citric acid to extend the shelf life of ready-to-cook barbecuechicken. *J. Food Process. Preserv.* **2019**, *43*, e14114. [CrossRef]
307. Mansouri, S.; Pajohi-Alamoti, M.; Aghajani, N.; Bazargani-Gilani, B.; Nourian, A. Stability and antibacterial activity of Thymus daenensis L. essential oil nanoemulsion in mayonnaise. *J. Sci. Food Agric.* **2020**. [CrossRef] [PubMed]
308. Esmaeili, H.; Cheraghi, N.; Khanjari, A.; Rezaeigolestani, M.; Basti, A.A.; Kamkar, A.; Aghaee, E.M. Incorporation of nanoencapsulated garlic essential oil into edible films: A novel approach for extending shelf life of vacuum-packed sausages. *Meat Sci.* **2020**, *166*, 108135. [CrossRef] [PubMed]
309. Hernández-Ochoa, L.; Aguirre-Prieto, Y.B.; Nevárez-Moorillón, G.V.; Gutierrez-Mendez, N.; Salas-Munoz, E. Use of essential oils and extracts from spices in meat protection. *J. Food Sci. Technol.* **2014**, *51*, 957–963. [CrossRef]
310. Salma, U.; Saha, S.K.; Sultana, S.; Ahmed, S.M.; Haque, S.D.; Mostaqim, S. The Antibacterial Activity of Ethanolic Extract of Cinnamon (*Cinnamomum zeylanicum*) against two Food Borne Pathogens: *Staphylococcus aureus* And *Escherichia coli*. *Mymensingh Med. J.* **2019**, *28*, 767–772. [PubMed]

311. Shan, B.; Cai, Y.Z.; Brooks, J.D.; Corke, H. Potential application of spice and herb extracts as natural preservatives in cheese. *J. Med. Food* **2011**, *14*, 284–290. [CrossRef]
312. Badea, G.; Bors, A.G.; Lacatusu, I.; Oprea, O.; Ungureanu, C.; Stan, R.; Meghea, A. Influence of basil oil extract on the antioxidant and antifungal activities of nanostructured carriers loaded with nystatin. *Comptes Rendus Chim.* **2015**, *18*, 668–677. [CrossRef]
313. El-Soud, N.H.A.; Deabes, M.; . El-Kassem, L.A.; Khalil, M. Chemical composition and antifungal activity of *Ocimum basilicum* L. essential oil. *Maced. J. Med. Sci.* **2015**, *3*, 374–379. [CrossRef]
314. Saggiorato, A.G.; Gaio, I.; Treichel, H.; de Oliveira, D.; Cichoski, A.J.; Cansian, R.L. Antifungal activity of basil essential oil (Ocimum basilicum L.): Evaluation in vitro and on an Italian-type sausage surface. *Food Bioproc. Technol.* **2012**, *5*, 378–384. [CrossRef]
315. Kumar, R.; Ghoshal, G.; Goyal, M. Effect of basil leaves extract on modified moth bean starch active film for eggplant surface coating. *LWT. Food Sci.Technol.* **2021**, *145*, 111380. [CrossRef]
316. Gundewadi, G.; Sarkar, D.J.; Rudra, S.G.; Singh, D. Preparation of basil oil nanoemulsion using Sapindus mukorossi pericarp extract: Physico-chemical properties and antifungal activity against food spoilage pathogens. *Ind. Crop. Prod.* **2018**, *125*, 95–104. [CrossRef]
317. Santos, N.S.T.; Aguiar, A.J.A.A.; de Oliveira, C.E.V.; de Sales, C.V.; Silva, S.M.; Silva, R.S.; Stamford, T.C.M.; de Souza, E.L. Efficacy of the application of a coating composed of chitosan and Origanum vulgare L. essential oil to control *Rhizopus stolonifer* and *Aspergillus niger* in grapes (*Vitis labrusca* L.). *Food Microbiol.* **2012**, *32*, 345–353. [CrossRef]
318. Wang, C.Y.; Wang, S.Y.; Yin, J.J.; Parry, J.; Yu, L.L. Enhancing Antioxidant, Antiproliferation, and free radical scavenging activities in strawberries with essential oils. *J. Agric. Food Chem.* **2007**, *55*, 6527–6532. [CrossRef]
319. Valero, D.; Valverde, J.M.; Martı´nez-Romero, D.; Guille´n, F.; Castillo, S.; Serrano, M. The combination of modified atmosphere packaging with eugenol or thymol to maintain quality, safety and functional properties of table grapes. *Postharvest Biol. Technol.* **2006**, *41*, 317–327. [CrossRef]
320. Miao, L.; Walton, W.C.; Wang, L.; Li, L.; Wang, Y. Characterization of polylactic acids-polyhydroxybuty rate based packaging film with fennel oil, and its application on oysters. *Food Packag. Shelf Life.* **2019**, *22*, 1–8. [CrossRef]
321. Wang, C.Y.; Wang, S.Y.; Chen, C. Increasing antioxidant activity and reducing decay of blueberries by essential oils. *J. Agric. Food. Chem.* **2008**, *56*, 3587–3592. [CrossRef]
322. Jin, P.; Wang, S.Y.; Gao, H.; Chen, H.; Zheng, Y.; Wang, C.Y. Effect of cultural system and essential oil treatment on antioxidant capacity in raspberries. *Food Chem.* **2012**, *132*, 399–405. [CrossRef] [PubMed]
323. Kato, H.C.A.; Peixoto Joele, M.R.S.; Sousa, C.L.; Ribeiro, S.C.A.; Lourenço, L.F.H. Evaluation of the shelf life of tambaqui fillet processed by the sous vide method. *J. Aquat. Food Prod. Technol.* **2017**, *26*, 1144–1156. [CrossRef]
324. Toyohara, M.; Murata, M.; Ando, M.; Kubota, S.; Sakaguchi, M.; Toyohara, H. Texture changes associated with insolubilization of sarcoplasmic proteins during salt-vinegar curing of fish. *Food Sci. J.* **1999**, *64*, 804–807. [CrossRef]
325. Stelzleni, A.M.; Ponrajan, A.; Harrison, M.A. Effects of buffered vinegar and sodium dodecyl sulfate plus levulinic acid on Salmonella Typhimurium survival, shelf-life, and sensory characteristics of ground beef patties. *Meat Sci.* **2013**, *95*, 1–7. [CrossRef] [PubMed]
326. Shi, C.; Cui, J.; Na, Q.; Luo, Y.; Han, L.; Hang, W. Effect of ginger extract and vinegar on atp metabolites, imp-related enzyme activity, reducing sugars and phosphorylated sugars in silver carp during postslaughter storage. *Int. J. Food Sci. Technol.* **2017**, *52*, 413–423. [CrossRef]
327. Uvere, P.O.; Onyekwere, E.U.; Ngoddy, P.O. Production of maize-bambara groundnut complementary foods fortified pre-fermentation with processed foods rich in calcium, iron, zinc and provitamin A. *J. Sci. Food Agric.* **2010**, *90*, 566–573. [CrossRef]
328. Bakrm, S.A.; Salihin, B.A. Effects of inclusion of Allium sativum and Cinnamomumverum in milk on the growth and activity of lactic acid bacteria during yoghurt fermentation Amer-Euras. *J. Agric. Environ. Sci.* **2013**, *13*, 1448–1457.
329. Helal, A.; Tagliazucchi, D. Impact of in-vitro gastro-pancreatic digestion on polyphenols and cinnamaldehyde bioaccessibility and antioxidant activity in stirred cinnamon-fortified yoghurt. *LWT Food Sci. Technol.* **2018**, *89*, 164–170. [CrossRef]
330. Srivastava, P.; Prasad, S.G.M.; Mohd, N.A.; Prasad, M. Analysis of antioxidant activity of herbal yoghurt prepared from different milk. *J. Pharma. Innov.* **2015**, *4*, 18–20.
331. Amirdivani, S.; Baba, A.S. Changes in yoghurt fermentation characteristics, and antioxidant potential and in vitro inhibition of angiotensin-1 converting enzyme upon the inclusion of peppermint, dill and basil. *LWT Food Sci. Technol.* **2011**, *44*, 1458–1464. [CrossRef]
332. Carocho, M.; Barros, L.; Barreira, J.C.; Calhelha, R.C.; Sokovi´c, M.; Fernández-Ruiz, V.; Buelga, C.S.; Morales, P.; Ferreira, I.C. Basil as functional and preserving ingredient in "serra da estrela" cheese. *Food Chem.* **2016**, *207*, 51–59. [CrossRef]
333. Sikora, M.; Złotek, U.; Kordowska-Wiater, M.; Świeca, M. Effect of Basil Leaves and Wheat Bran Water Extracts on Antioxidant Capacity, Sensory Properties and Microbiological Quality of Shredded Iceberg Lettuce during Storage. *Antioxidants* **2020**, *9*, 355. [CrossRef] [PubMed]
334. Josipović, R.; Kneževi, Z.M.; Frece, J.; Markov, K.; Kazazić, S.; Mrvčić, J. Improved properties and microbiological safety of novel cottage cheese containing spices. *Food Technol. Biotechnol.* **2015**, *53*, 454–462. [CrossRef]
335. Najgebauer, L.D.; Grega, T.; Sady, M. The quality and storage stability of butter made from sour cream with addition of dried sage and rosemary. *Biotechnol. Anim. Husb.* **2009**, *25*, 753–761.

336. Farag, R.S.; Ali, M.N.; Taga, S.H. Use of some essential oils as natural preservatives for butter. *J. Am. Oil Chem. Soc.* **1990**, *67*, 188–191. [CrossRef]
337. Pinto, S.V.; Patel, A.M.; Jana, A.H. Evaluation of different forms of ginger as flavouring in herbal ice cream. *Int. J. Food Sci. Technol. Nutr.* **2009**, *3*, 73–83.
338. Ham, Y.-K.; Hwang, K.-E.; Song, D.-H.; Choi, J.-H.; Choi, Y.-S.; Kim, H.-W. Relationship between the antioxidant capacity of soy sauces and its impact on lipid oxidation of beef patties. *Meat Sci.* **2019**, *158*, 107907. [CrossRef]
339. Honda, M.; Kageyama, H.; Hibino, T.; Takemura, R.; Goto, M.; Fukaya, T. Enhanced Z-isomerization of tomato lycopene through the optimal combination of food ingredients. *Sci Rep.* **2019**, *9*, 7979. [CrossRef]
340. González-Hedström, D.; Granado, M.; Inarejos-García, A.M. Protective effects of extra virgin olive oil against storage-induced omega 3 fatty acid oxidation of algae oil. *NFS J.* **2020**, *21*, 9–15. [CrossRef]
341. Zhang, Y.; Wang, Y.; Zhang, F.; Wang, K.; Liu, G.; Yang, M.; Zhang, D. Allyl methyl disulfide inhibits IL-8 and IP-10 secretion in intestinal epithelial cells via the NF-kappa B signaling pathway. *Inter. Immunopharmac.* **2015**, *27*, 156–163. [CrossRef]
342. Bordoloi, R.; Ahmed, N.S.; Dora, K.C.; Chowdhury, S. Effect of cooking on antioxidant and antimicrobial property of spices. *Biochem. Cell. Arch.* **2017**, *17*, 361–365.
343. Neves, T.M.; da Cunha, D.T.; de Rosso, V.V.; Domene, S.M.Á. Effects of seasoning on the formation of heterocyclic amines and polycyclic aromatic hydrocarbons in meats: A meta-analysis. *Compr. Rev. Food Sci. Food Saf.* **2021**, *20*, 526–541. [CrossRef]
344. IARC. *Red Meat and Processed Meat*; International Agency for Research on Cancer: Lyon, France, 2015; 114p. Available online: https://hal.inrae.fr/hal-02787204/document (accessed on 15 March 2018).
345. Shabbir, A.M.; Raza, A.; Faqir, M.A.; Khan, M.R.; Suleria, H.A.R. Effect of thermal treatment on meat proteins with special reference to heterocyclic aromatic amines (HAAs). *Crit. Rev. Food Sci. Nutr.* **2015**, *55*, 82–93. [CrossRef]
346. Janoszka, B.; Nowak, A.; Szumska, M.; Śnieżek, E.; Tyrpień-Golder, K. Human exposure to biologically active heterocyclic aromatic amines arising from thermal processing of protein rich food. *Wiadomosci Lek.* **2019**, *72*, 1542–1550. [CrossRef]
347. Liang, T.F.; Wei, F.F.; Lu, Y.; Kodani, Y.; Nakada, M.; Miyakawa, T.; Tanokura, M. Comprehensive nmr analysis of compositional changes of black garlic during thermal processing. *J. Agric. Food Chem.* **2015**, *63*, 683–691. [CrossRef]
348. Mashilipa, C.; Wang, Q.; Slevin, M.; Ahmed, N. Antiglycation and antioxidant properties of soy sauces. *Med. Food J.* **2011**, *14*, 1647–1653. [CrossRef]
349. Santos, C.S.P.; Cruz, R.; Cunha, S.C.; Casal, S. Effect of cooking on olive oil quality attributes. *Food Res. Int.* **2013**, *54*, 2016–2024. [CrossRef]
350. Anh, N.H.; Kim, S.J.; Long, N.P.; Min, J.E.; Yoon, Y.C.; Lee, E.G.; Kim, M.; Kim, T.J.; Yang, Y.Y.; Son, E.Y.; et al. Ginger on Human Health: A Comprehensive Systematic Review of 109 Randomized Controlled Trials. *Nutrients* **2020**, *12*, 157. [CrossRef] [PubMed]
351. Sellami, M.; Slimeni, O.; Pokrywka, A.; Kuvačić, G.; Hayes, L.D.; Milic, M.; Padulo, J. Herbal medicine for sports: A review. *J. Int. Soc. Sports Nutr.* **2018**, *15*, 14. [CrossRef] [PubMed]
352. Onuora, C.; Ofili, C.T.; Salawu, S.; Elimian, I.; Shehu, H. Therapeutic Effects of Garlic: A Review. *Sci. J. Biol. Life Sci.* **2019**, *1*, 2019.
353. Ouzir, M.; El Bairi, K.; Amzazi, S. Toxicological properties of fenugreek (Trigonella foenum graecum). *Food Chem. Toxicol.* **2016**, *96*, 145–154. [CrossRef] [PubMed]
354. Liddle, M.; Hull, C.; Liu, C.; Powell, D. Contact urticaria from curcumin. *Dermatitis* **2006**, *17*, 196–197. [CrossRef] [PubMed]
355. Lopez-Lazaro, M. Anticancer and carcinogenic properties of curcumin: Considerations for its clinical development as a cancer chemopreventive and chemotherapeutic agent. *Mol. Nutr. Food Res.* **2008**, *52* (Suppl. 1), S103–S127. [CrossRef] [PubMed]

antioxidants

MDPI

Article

Antioxidant and Inflammatory Gene Expression Profiles of Bovine Peripheral Blood Mononuclear Cells in Response to *Arthrospira platensis* before and after LPS Challenge

Magdalena Keller †, Elisa Manzocchi †, Deborah Rentsch, Rosamaria Lugarà and Katrin Giller *

Animal Nutrition, ETH Zurich, Universitaetstrasse 2, 8092 Zurich, Switzerland; magdalena.keller@usys.ethz.ch (M.K.); elisa.manzocchi@usys.ethz.ch (E.M.); deborah.rentsch@usys.ethz.ch (D.R.); rosamaria.lugara@usys.ethz.ch (R.L.)
* Correspondence: katrin.giller@usys.ethz.ch; Tel.: +41-52-354-9209
† Authors contributed equally to the manuscript.

Abstract: Oxidative stress and inflammatory diseases are closely related processes that need to be controlled to ensure the desirable high performance of livestock. The microalga spirulina has shown antioxidant and anti-inflammatory properties in monogastric species. To investigate potential beneficial effects in ruminants, we replaced soybean meal (SOY) in the diets of dairy cows and fattening bulls by spirulina (SPI) and analyzed plasma concentrations of antioxidants (β-carotene, α-tocopherol, polyphenols) and serum total antioxidant capacity. Following in vitro stimulation with lipopolysaccharide (LPS), peripheral blood mononuclear cells (PBMCs) were isolated for expression analysis of inflammation- and antioxidant-defense-related genes. Plasma β-carotene concentration was higher in SPI, compared to SOY cows, but did not differ in bulls. Plasma total phenol concentration was significantly higher in SPI, compared to SOY bulls, but not in cows. Stimulation of bovine PBMCs with LPS increased the expression of most cytokines and some antioxidant enzymes. Gene expression of PBMCs derived from SPI animals, compared to SOY animals, hardly differed. Our results indicate that in ruminants, spirulina might not have potent antioxidant and anti-inflammatory properties. Future studies should evaluate the microbial degradation of spirulina and its bioactive compounds in the rumen to provide further data on potential beneficial health effects in ruminants.

Keywords: microalgae; spirulina; inflammation; lipopolysaccharide; dairy cows; fattening bulls; leukocytes

Citation: Keller, M.; Manzocchi, E.; Rentsch, D.; Lugarà, R.; Giller, K. Antioxidant and Inflammatory Gene Expression Profiles of Bovine Peripheral Blood Mononuclear Cells in Response to *Arthrospira platensis* before and after LPS Challenge. *Antioxidants* **2021**, *10*, 814. https://doi.org/10.3390/antiox10050814

Academic Editors: Irene Dini and Domenico Montesano

Received: 26 April 2021
Accepted: 17 May 2021
Published: 20 May 2021

1. Introduction

Oxidative stress and inflammatory diseases are common problems in high-yielding livestock and impair their health and productivity, thus resulting in economic losses for the farmers [1]. Dairy cows are most susceptible to oxidative stress and inflammation during the transition from pregnancy to lactation, resulting from substantial metabolic and physiological adaptations [1,2]. This contributes to the occurrence of inflammatory diseases such as mastitis and metritis. In fattening cattle, one of the major inflammatory disorders is bovine respiratory disease, recognized as the leading cause of death in feedlots in the USA [3]. Consequently, understanding and preventing oxidative stress and the associated inflammatory processes in livestock is of utmost importance.

Oxidative stress is characterized by an imbalance of reactive oxygen species (ROS) and antioxidants, including non-enzymatic antioxidants (e.g., carotenoids, tocopherols, polyphenols, and glutathione) and antioxidant enzymes (e.g., superoxide dismutase (SOD), catalase (CAT), glutathione peroxidase (GPX)) [4]. The accumulation of ROS can severely damage cellular macromolecules and continuously promote ROS production. Literature provides evidence for a vicious cycle with oxidative stress resulting in compromised inflammatory reactions, which, in turn, result in additional ROS production at the site of inflammation [5–8]. Inflammatory processes involve the actions of peripheral blood mononuclear cells (PBMCs), which comprise lymphocytes, natural killer (NK) cells, and monocytes, with the latter being able to differentiate to macrophages upon activation. Stimulation of the Toll-like receptor 4 (TLR4) on the surface of PBMCs by the bacterial endotoxin lipopolysaccharide (LPS) activates the transcription factor nuclear factor kappa B (NFKB). After translocation into the nucleus, NFKB induces the transcription of cytokines and chemokines such as tumor necrosis factor alpha (*TNFA*), interleukin 1 beta (*IL1B*), and *IL8* (also known as C-X-C motif chemokine ligand 8 (*CXCL8*)). The proinflammatory cytokines TNFA and IL1B are important in the early onset of inflammation and defense. They direct leucocytes to the site of inflammation and can also cause systemic effects. Production of large amounts of TNFA can damage the organism and lead to a septic shock. Interferon gamma (IFNG) is produced by T-cells and NK cells and activates macrophages, which release ROS and thus contribute to oxidative stress at the site of inflammation [9]. The induction of prostaglandin–endoperoxide synthase 2 (PTGS2), also known as cyclooxygenase 2, leads to prostaglandin synthesis at the site of inflammation, which further modulates the inflammatory process [10]. In a negative feedback manner, activated macrophages produce IL10. It inhibits the activated macrophages and thus counteracts the inflammatory process, also by inhibiting cytokine production.

The close relationship between oxidative stress and inflammation has been described in the literature [11,12] and is further emphasized by the concomitant occurrence of oxidative stress and inflammatory production diseases such as mastitis and metritis in dairy cows during the periparturient period [13]. In contrast, antioxidant supplementation has been associated with a decreased incidence of mastitis [14,15]. Thus, to control common production diseases in livestock, both inflammatory processes and oxidative stress need to be addressed. In order to limit the globally increasing application of antibiotics in livestock production [16], it is desirable to find alternatives to antibiotics that can be applied both for treatment and prevention of production diseases. Such alternatives are searched for, among others, in the group of bioactive feeds containing compounds that exhibit antioxidant and anti-inflammatory properties and might therefore be advantageous for the prevention of oxidative stress and inflammation in livestock [17].

The filamentous blue-green microalga *Arthrospira platensis* (spirulina) belongs to the group of cyanobacteria and occurs naturally in alkaline lakes [18]. Spirulina is known for its high protein content (60–70%), which resulted in its consideration as a valuable protein source in livestock nutrition to replace the often critically discussed soybean meal [19,20]. This applicability has already been confirmed in different studies with ruminants that showed a maintained or even improved performance when soybean meal was replaced by spirulina [21–25]. In addition to its recognition as a sustainable protein source, spirulina has recently gained attention as a source of health-promoting bioactive compounds. These include carotenoids, tocopherols, polyphenols, γ-linoleic acid (GLA), and phycocyanin, all contributing to the observed antioxidant and anti-inflammatory effects of spirulina (reviewed by [26]). Spirulina has been shown to induce the activity of antioxidant enzymes such as SOD, GPx, and glutathione-S-transferase (GST), increase glutathione levels, decrease lipid oxidation, and consequently reduce oxidative stress in humans, rats, chickens, and fish [27–31]. Several studies have also reported an immunomodulatory effect of spirulina. Spirulina increased the accumulation of NK cells in tissues, induced phagocytosis by macrophages, modulated the production of key inflammatory molecules such as IL1B and IFNG in chickens, humans, and murine macrophages [32–34]. A close correlation has been reported between the antioxidant and anti-inflammatory activities of spirulina [35]. The proposed underlying mechanism of action is the inhibition of NFKB activation, a pivotal mediator of inflammatory processes, by spirulina's bioactive compounds [36,37].

The antioxidant and anti-inflammatory properties of spirulina have been described in monogastrics such as rodents, humans, and chickens [28,33,38,39]. However, to the best of our knowledge, the interplay of spirulina, oxidative stress, and inflammation has not yet been investigated in ruminants. Since spirulina and its bioactive compounds may be significantly affected by the microbial metabolism in the rumen, the effects on oxidative stress and inflammation in ruminants might differ from those observed in monogastrics. This is especially interesting because spirulina is evaluated as an alternative protein source to replace soy for improved sustainability of both dairy cow and beef cattle nutrition and might be applied in large quantities in ruminant diets in the future.

Consequently, the aim of the present study was to assess a potential effect of dietary spirulina, compared to soybean meal, on basal and LPS-induced inflammation- and oxidative-stress-related gene expression of PBMCs obtained from dairy cows and fattening bulls. We hypothesized that (I) spirulina increases blood concentration of antioxidant molecules such as β-carotene, α-tocopherol, and polyphenols, as well as the total antioxidant capacity; (II) basal antioxidant and inflammatory gene expression is more beneficial in spirulina-fed, compared to soy-fed, animals; and (III) the gene expression response to in vitro challenge with increasing concentrations of LPS is less pronounced in spirulina-fed, compared to soy-fed, animals.

2. Materials and Methods

2.1. Experiment 1—Fattening Bulls

The experiment was approved by the Cantonal Veterinary Office of Zurich, Switzerland (license no. ZH129/18), and was conducted at the AgroVet-Strickhof research station (Lindau, Switzerland). A total of 12 Limousin-sired crossbred bulls with an initial age of 130 ± 8 (mean ± SD) days and an average body weight of 164 ± 9 kg were randomly divided into two experimental groups (n = 6 per group), balanced for initial body weight, sire, and breed of the dam (see also [40]). All animals were fed the same basal diet consisting of 50% grass silage, 30% maize silage, and 20% concentrate for five months. The main protein source in the concentrate differed between the two experimental groups: the control group received soybean meal (SOY), whereas the other group received the microalga spirulina (SPI) as part of the concentrate (Table 1). The spirulina had been cultivated in a Spirulina–Ogawa–Terui medium in production ponds [41]. After harvesting, vacuum belt filtering, and spray drying, the spirulina was sifted and blended (Institut für Getreideverarbeitung, Nuthetal, Germany). The crude protein contents of soybean meal and spirulina were 536 and 710 g/kg DM, respectively. The forage mixture was supplemented with 12 g/kg DM of a vitaminized mineral mix providing, per kg: 138 g Ca; 46 g P; 36 g Mg; 167 g Na; 5 g Zn; 3 g Mn; 1 g Cu; vitamin A, 625,000 IE, vitamin D_3, 62,500 IE, vitamin E, 1 mg, and *Saccharomyces cerevisiae* (*NCYC Sc* 47), 333 colony-forming units. Additionally, animals had free access to a NaCl-containing licking block. The concentrates were formulated to be isocaloric and isonitrogenous. Diets were designed according to the Swiss feeding recommendations [42]. After five months of experimental feeding, blood samples were collected from the jugular vein after disinfecting the puncture site with 70% ethanol using serum, Na-heparinized, and EDTA plasma tubes, in that order, thereby flushing the needle before collecting the sample for PBMC isolation (Na-heparinized tube) to avoid contamination and activation of immune cells due to the skin perforation. The serum and EDTA plasma tubes were centrifuged at 2100× g for 10 min at 4 °C. The respective supernatants were collected and frozen at −80 °C until further analysis.

2.2. Experiment 2—Dairy Cows

The experiment was approved by the Cantonal Veterinary Office of Zurich, Switzerland (license no. ZH125/18), and was conducted at the AgroVet-Strickhof research station. A total of 12 multiparous, late-lactating dairy cows (Brown Swiss and Holstein) were randomly divided into two groups (n = 6 per group), balanced for breed, parity, days in milk, and milk yield (see also [24]). All animals were fed a hay-based diet with the addition of one of two isocaloric and isonitrogenous concentrates for 31 days. The control group received soybean meal (SOY), whereas the other group received the microalga spirulina (SPI) as part of the concentrate (Table 1). The spirulina was produced as described for experiment 1 but from a different batch. The crude protein contents of soybean meal and spirulina were 428 and 629 g/kg DM, respectively. In addition, the cows received 50 g/d of NaCL and 120 g/d of a vitaminized mineral mix. This mix contained per kilogram: 80 g Ca; 160 g P; 50 g Mg; 45 g Na; 4 g Zn; 2 g Mn; 0.5 g Cu; 0.02 g Se; 0.02 g I; 0.015 g Co, as well as, per kilogram, vitamin A, 1,200,000 IU; vitamin D3, 200,000 IU; vitamin E, 2 g. Diets were designed to cover requirements for maintenance and milk yield according to the Swiss feeding recommendations [42]. After 31 days of experimental feeding, blood samples were collected as described for experiment 1.

Table 1. Components of the experimental concentrate used in experiment 1 (fattening bulls) and the experimental total mixed ration used in experiment 2 (dairy cows), their chemical composition and contents of extractable phenols, β-carotene, and α-tocopherol.

	Experiment 1—Fattening Bulls (Concentrate)		Experiment 2—Dairy Cows (Total Mixed Ration)	
	Control	Spirulina	Control	Spirulina
Ingredients (g/kg DM)				
Hay	-	-	740	740
Sugar beet pulp	-	-	130	130
Soybean meal	277	-	60	-
Spirulina	-	198	-	50
Wheat flakes	-	-	40	50
Molasses	20	20	30	30
Wheat	450	518	-	-
Maize grain	237	250	-	-
Wheat bran	11	-	-	-
Tallow-lard mixture	5	5	-	-
Composition (%)				
DM (% of wet weight)	88.2	88.1	65.4	52.4
OM	96.8	97.1	85.6	86.1
CP	22.8	21.4	14.6	14.2
NDF	46.4	n.a.	46.9	43.7
ADF	10.5	n.a.	30.5	27.5
ADL	n.a.	n.a.	5.22	4.10
EE	2.60	3.71	1.64	2.36
Total extractable phenols (mg/100 g DM)	n.a.	n.a.	957	942
β-carotene (mg/kg DM)	n.a.	n.a.	7.6	24.3
α-tocopherol (mg/kg DM)	n.a.	n.a.	31.4	37.7

ADF: acid detergent fiber; ADL: acid detergent lignin; CP: crude protein; DM: dry matter; EE: ether extract; n.a.: not analyzed; NDF: neutral detergent fiber; OM: organic matter; -: not contained in the diet.

2.3. *Plasma and Serum Analyses of Antioxidants and Total Antioxidant Capacity*

The EDTA plasma samples were used to analyze the concentrations of β-carotene, α-tocopherol, and total phenols. For β-carotene and α-tocopherol, the plasma was deproteinized with pure water and ethanol (1:2) and extracted with hexane. For the analysis of β-carotene, the hexane was evaporated. The dried residue was dissolved in dichloromethane and dimethyl sulfoxide (1:4). The β-carotene content was analyzed with a UV/VIS detector and an HPLC system (Chromaster, Merck-Hitachi, Darmstadt, Germany) by applying the European standard method [43]. The α-tocopherol content was analyzed with a fluorescence detector and the same HPLC system according to Kälber et al. [44]. The total phenols from plasma were analyzed according to Sinz et al. [45]. Total antioxidant capacity (TAC) was measured in serum samples using the OxiSelectTM Total Antioxidant Capacity Assay Kit (Cell Biolabs, San Diego, CA, USA) according to the manufacturer's protocol.

2.4. Short-Term Whole Blood Stimulation with LPS and Isolation of PBMCs

In the laboratory, the blood samples collected in Na-heparinized tubes in both experiments were carefully mixed by inversion. Per animal, six endotoxin-free cell culture tubes (VWR, Dietikon, Switzerland) were filled with 2 mL of blood (method adapted from [46]). The samples were treated with two different concentrations of LPS (purified LPS from Escherichia coli O111:B4; Merck, Buchs, Switzerland) in duplicate. As a control (LPS0), 20 μL of RPMI 1640 medium (Merck) were added to two tubes per animal. The remaining four tubes per animal were incubated in duplicate with 10 and 100 ng LPS/mL blood (LPS10 and LPS100), respectively. Dilutions of LPS were made in RPMI 1640 medium, and a total volume of 20 μL was added to the respective cell culture tubes. All tubes were incubated for 2 h in a water bath at 38 °C. After the incubation, all tubes were placed on ice to stop the reaction. Subsequently, samples were centrifuged (10 min, 1500× *g*, 4 °C), and the supernatant was removed. The red blood cells were lysed by resuspending the pellet in ice-cold ACK lysing buffer (Gibco, Thermo Fisher, CA, USA). Ice-cold phosphate-buffered saline (PBS) was added, and the remaining PBMCs were pelleted for 5 min at 300× *g* and 4 °C. After two more washings with ice-cold PBS and complete removal of the supernatant, the white cell pellets were frozen at −80 °C until further processing.

2.5. RNA Isolation, Quality Control, and cDNA Synthesis

To isolate total RNA from the PBMCs, TRIzol (Invitrogen, Thermo Fisher Scientific, Reinach, Switzerland) was added to the frozen cell pellets before thawing on ice according to the manufacturer's protocol. Samples were homogenized and precipitated using chloroform. The RNeasy Micro Kit (Qiagen, Venlo, the Netherlands) was used to purify the RNA according to the manufacturer's instructions. The RNA concentration was determined using a NanoDrop One (Thermo Fisher, MA, USA). The integrity of the RNA was determined using the Bioanalyzer (Agilent Technologies, CA, USA). The RNA integrity number (RIN) varied between 7 and 10. In total, 250 ng of RNA were used for reverse transcription with the GoScript™ Reverse Transcription System (Promega, Duebendorf, Switzerland) at the following conditions: 5 min at 25 °C, 60 min at 42 °C, 15 min at 70 °C. The cDNA was stored at −20 °C until further analysis. For two LPS0 samples from SOY dairy cows, not enough RNA was obtained, hence $n = 4$ in this group.

2.6. Gene Expression Analysis

The quantitative real-time polymerase chain reaction was performed using the KAPA Sybr Fast Mix (Kapa Biosystems, Wilmington, NC, USA) according to the manufacturer's instructions on a LightCycler instrument (System LightCycler® 96, Roche, Basel, Switzerland; used for inflammation-related genes) and a CFX384 Real-Time PCR Detection System (Bio-Rad, Munich, Germany; used for oxidative-stress related genes) with 40 cycles of amplification in three steps (10 s at 95 °C, 20 s at 60 °C, 1 s at 72 °C). The cycle of quantification (Cq) values resulted from applying a single threshold. Four reference genes were tested (actin beta (ACTB), H3.3 histone A (H3-3A), ubiquitin B (UBB), tyrosine 3-monooxygease/tryptophan 5-monooxygenase activation protein zeta (YWHAZ)). According to the geNorm output, the relative expression level (ΔCq) for each target gene was obtained by scaling the target gene Cq of each individual sample to the geometric mean of the Cqs of UBB and YWHAZ. Relative expression of target genes was related to the respective SOY LPS0 group. Primers (Table 2) were ordered at Microsynth (Balgach, Switzerland).

Table 2. Sequences of the primers used for qPCR in both experiments.

Gene	Accession Number	Forward Primer (5′–3′)	Reverse Primer (5′–3′)	Amplicon Length
ACTB	NM_173979.3	GAT CTG GCA CCA CAC CTT CT	AGA GAC AGC ACA GCC TGG AT	174
BAX	NM_173894.1	GCC CTT TTG CTT CAG GGT TT	ACA GCT GCG ATC ATC CTC TG	179
CAT	NM_001035386.2	CTG GGA CCC AAC TAT CTC CA	AAG TGG GTC CTG TGT TCC AG	179
CXCL8	NM_173925.2	GTT GCT CTC TTG GCA GCT TT	GGT GGA AAG GTG TGG AAT GT	118
GPX1	NM_174076.3	CCT GAC ATT GAA ACC CTG CT	TCA TGA GGA GCT GTG GTC TG	220
GSR	NM_001114190.2	TGT CAT TGT TGG TGC TGG TT	AGC GTT CTC CAG CTC TTC AG	154
H3-3A	NM_001014389.2	GTA CTG TGG CAC TCC GTG AA	GAT AGG CCT CAC TTG CCT CC	168
IFNG	NM_174086.1	TTC TTG AAT GGC AGC TCT GA	TTC TCT TCC GCT TTC TGA GG	154
IL1B	NM_174093.1	TGA CCT GAG GAG CAT CCT TT	AGA GGA GGT GGA GAG CCT TC	179
IL10	NM_174088.1	GTG AAC TCA CTG GGG GAG	ACC GCC TTG CTC TTG TTT T	92
RELA	NM_001080242.2	GCC TGT CCT CTC TCA CCC CAT CTT TG	ACA CCT CGA TGT CCT CTT TCT GCA CC	152
NFKB2	NM_001102101.1	ATC TGA GCA TTG TGC GAC TG	CTT CAG GTT TGA GGC TCC AG	131
NQO1	NM_001034535.1	ATG AAG GAG GCT GCC ATA GA	CTG CAG CTT CCA GCT TCT TT	223
NFE2L2	NM_001011678.2	CTC CAG CCA GTT GAC AGT GA	GTT GTG CTT TCA GGG TGG TT	225
PTGS2	NM_174445.2	TCG AGG TGT ATG TAT GAG TGT A	GTG CTG GGC AAA GAA TGC AA	487
SOD1	NM_174615.2	AGA GGC ATG TTG GAG ACC TG	CAG CGT TGC CAG TCT TTG TA	189
SOD2	NM_201527.2	ATT GCT GGA AGC CAT CAA AC	AGC AGG GGG ATA AGA CCT GT	192
TLR4	NM_174198.6	GAC CCT TGC GTA CAG GTT GT	GGT CCA GCA TCT TGG TTG AT	103
TNF	NM_173966.3	GCC CTC TGG TTC AAA CAC TC	AGA TGA GGT AAG CCC GTC A	191
UBB	NM_174133.2	AGA TCC AGG ATA AGG AAG GCA T	GCT CCA CCT CCA GGG TGA T	198
YWHAZ	NM_174814.2	AGG CTG AGC GAT ATG ATG AC	GAC CCT CCA AGA TGA CCT AC	140

2.7. Statistical Analyses

The statistical analysis was performed using SPSS version 26 (SPSS GmbH Software, Munich, Germany) separately for both experiments. The Shapiro–Wilk test was used to test for normal distribution of data and residuals. Plasma and serum parameters were analyzed using Student's *t*-test. Data analysis for qPCR results was performed on ΔCq values using a general linear model with diet group (SOY, SPI), LPS concentration (LPS0, LPS10, LPS100), and their interaction as fixed factors, while the individual animal was included as a categorical control variable. Differences in means were controlled for multiple comparisons using the Games–Howell post hoc test. Results are presented as boxplots of the relative gene expression with Spear style whiskers (min to max). Statistical significance was defined as $p \leq 0.05$, and $p \leq 0.1$ was considered a trend.

3. Results

3.1. Blood Concentrations of Antioxidants and Total Antioxidant Capacity

The plasma concentration of β-carotene was higher ($p < 0.001$) in SPI, compared to SOY dairy cows, whereas no difference was observed in fattening bulls (Figure 1A). In dairy cows, the concentrations of plasma total phenols and serum TAC did not significantly differ among SOY and SPI animals (Figure 1B,C). However, plasma total phenol concentration was higher ($p = 0.045$) and a trend ($p = 0.074$) was observed toward a higher serum TAC in SPI, compared to SOY bulls (Figure 1B,C). In both fattening bulls and dairy cows, plasma α-tocopherol was not affected by the diet.

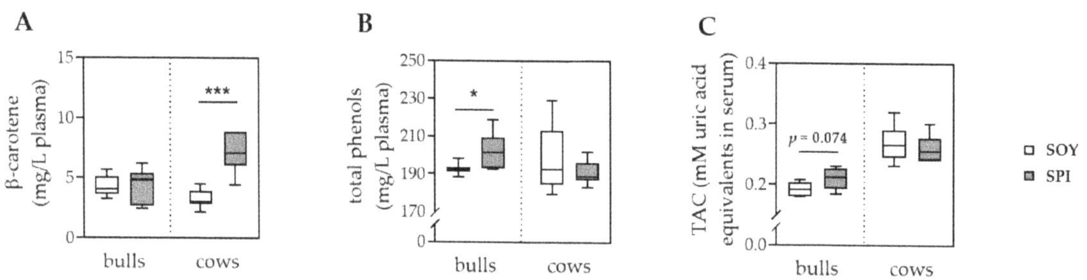

Figure 1. Antioxidants and antioxidant capacity in blood of fattening bulls and dairy cows fed with soybean meal (white boxes) or spirulina (gray boxes) as protein source. The concentrations of β-carotene (**A**) and total phenols (**B**) were analyzed in plasma, whereas the total antioxidant capacity (TAC, **C**) was determined in serum. Results are presented in boxplots with Spear style whiskers (min to max). * $p < 0.05$; *** $p < 0.001$.

3.2. Expression of Antioxidant Genes in Peripheral Blood Mononuclear Cells

Stimulation of PBMCs with LPS10 ($p = 0.008$), but not LPS100 ($p = 0.077$), decreased *GPX1* expression in SPI bulls, whereas no significant difference was observed in SOY bulls (Figure 2A). In dairy cows, *GPX1* expression did not significantly differ between groups (Figure 2F). For *CAT* expression, a pattern contrary to that of *GPX1* was observed, with no significant effect of diet and LPS in bull PBMCs (Figure 2B) and a decreased expression in PBMCs obtained from SPI (LPS100 versus LPS0, $p = 0.049$) but not from SOY cows (Figure 2G). In bull PBMCs, the expression of *SOD2* was increased in LPS10 and LPS100 (both $p < 0.001$), compared to LPS0 in both diet groups (Figure 2C). In PBMCs from dairy cows, a significant increase in *SOD2* expression was observed only in the SOY group (LPS10 versus LPS0, $p = 0.049$) since the observed increase in the SPI group (LPS100 versus LPS0, $p = 0.071$) could be considered only a trend (Figure 2H). The expression of NAD(P)H quinone oxidoreductase 1 (*NQO1*) was higher ($p = 0.040$) in SOY LPS100, compared to SOY LPS0 bull PBMCs, whereas LPS stimulation did not significantly affect *NQO1* expression in SPI bull PBMCs (Figure 2D). In dairy cow PBMCs, neither diet nor LPS stimulation affected *NQO1* expression (Figure 2I). The glutathione reductase (*GSR*) expression was lower (all $p < 0.05$) in PBMCs from bulls stimulated with LPS10 and LPS100, compared to LPS0 independent of diet group (Figure 2E). No significant difference in *GSR* expression was observed in PBMCs from dairy cows (Figure 2J). In PBMCs from both experiments, the mRNA expression of the transcription factor nuclear factor erythroid-derived 2 like 2 (*NFE2L2*) and the antioxidant enzyme SOD1 was not significantly affected by diet or LPS stimulation.

3.3. Expression of Inflammation-Related Genes in Peripheral Blood Mononuclear Cells

In PBMCs obtained from fattening bulls, the expression of *TLR4* was higher in LPS10 ($p = 0.031$) and LPS100 ($p < 0.001$), compared to LPS0 in SOY bulls, whereas no significant difference of *TLR4* expression was observed in SPI bulls (Figure 3A). The *TLR4* expression did not significantly differ in PBMCs from dairy cows in any of the diet groups (Figure 3J). The expression of *NFKB2* was upregulated in LPS10 and LPS100 (all $p < 0.01$), compared to LPS0 PBMCs in bulls and cows independent of diet (Figure 3B,K). In fattening bulls, the mRNA expression of *TNF*, *IL1B*, *CXCL8*, *IL10*, and *PTGS2* was higher (all $p < 0.01$) for LPS10, compared to LPS0 PBMCs in both diet groups (Figure 3C–F,H). Stimulation with 100 ng/mL of LPS did not further increase the expression of any of these genes, as compared to 10 ng/mL of LPS. The same results (all $p < 0.05$) were observed in PBMCs from dairy cows (Figure 3L,M,O), except for *CXCL8* and *PTGS2*. The *CXCL8* expression did not significantly differ in PBMCs from dairy cows (Figure 3N). For *PTGS2*, an increase in expression in the SPI group was observed only with LPS100 ($p = 0.003$) but not with LPS10 ($p = 0.137$), compared to LPS0, whereas in the SOY group, the expression was higher in both LPS10 and LPS100, compared to LPS0 (Figure 3Q). The expression of *IFNG* in PBMCs from bulls was higher in SOY ($p = 0.029$) and SPI ($p < 0.001$) LPS100, compared to LPS0, whereas no increase was observed with LPS10 (Figure 3G). In cows, *IFNG* expression remained unaffected (Figure 3P). The BCL-2-associated X protein (*BAX*) expression was lower in LPS10 ($p = 0.022$) and LPS100 ($p = 0.033$), compared to LPS0 in SOY bulls, while in SPI bulls, no significant decrease in *BAX* expression was observed following LPS stimulation (Figure 3I). In cows, a contrasting effect was observed with a decrease in *BAX* expression in the SPI group with LPS10, compared to LPS0 ($p = 0.036$), and no significant effect ($p = 0.169$) in the SOY group (Figure 3R). In PBMCs from both bulls and cows, expression of the NFKB subunit *RELA* did not significantly differ due to diet or LPS stimulation.

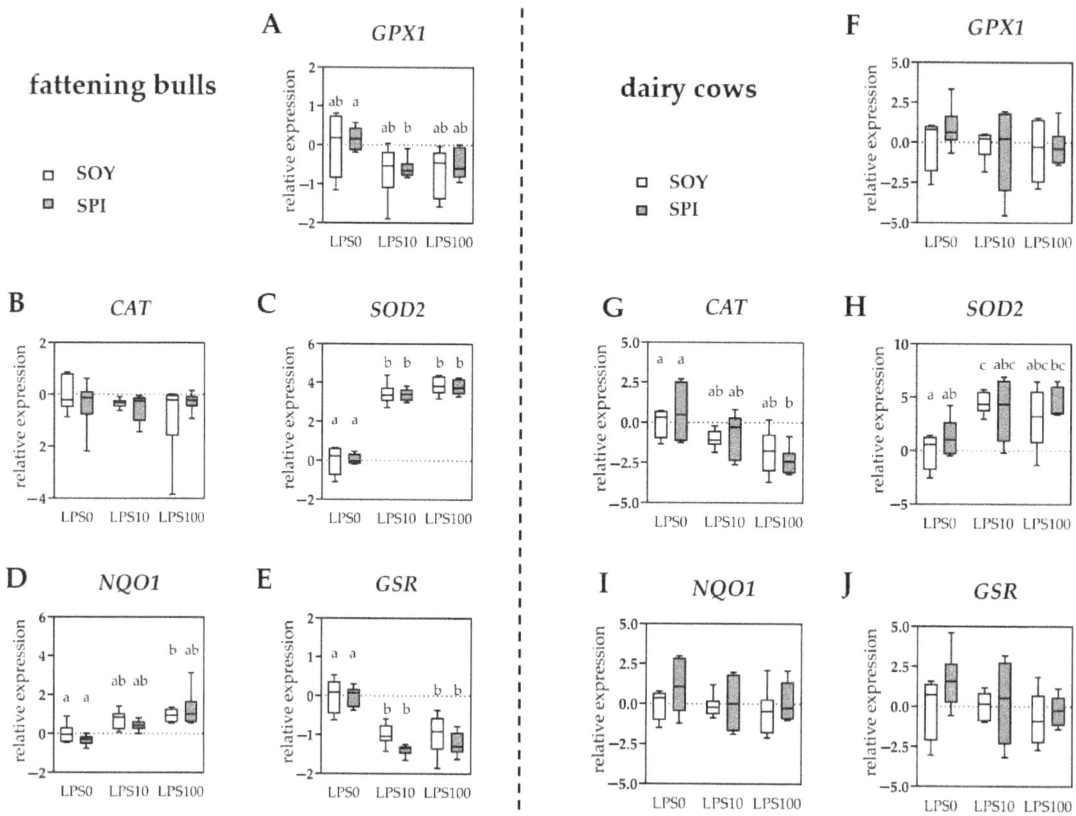

Figure 2. Relative expression of antioxidant genes in peripheral blood mononuclear cells isolated from fattening bulls (**A–E**) and dairy cows (**F–J**) fed with soybean meal (white boxes) or spirulina (gray boxes) as protein source. Individual results were normalized to the mean of the respective SOY LPS0 group. Results are presented in boxplots with Spear style whiskers (min to max). Different lowercase letters indicate significant differences ($p \leq 0.05$).

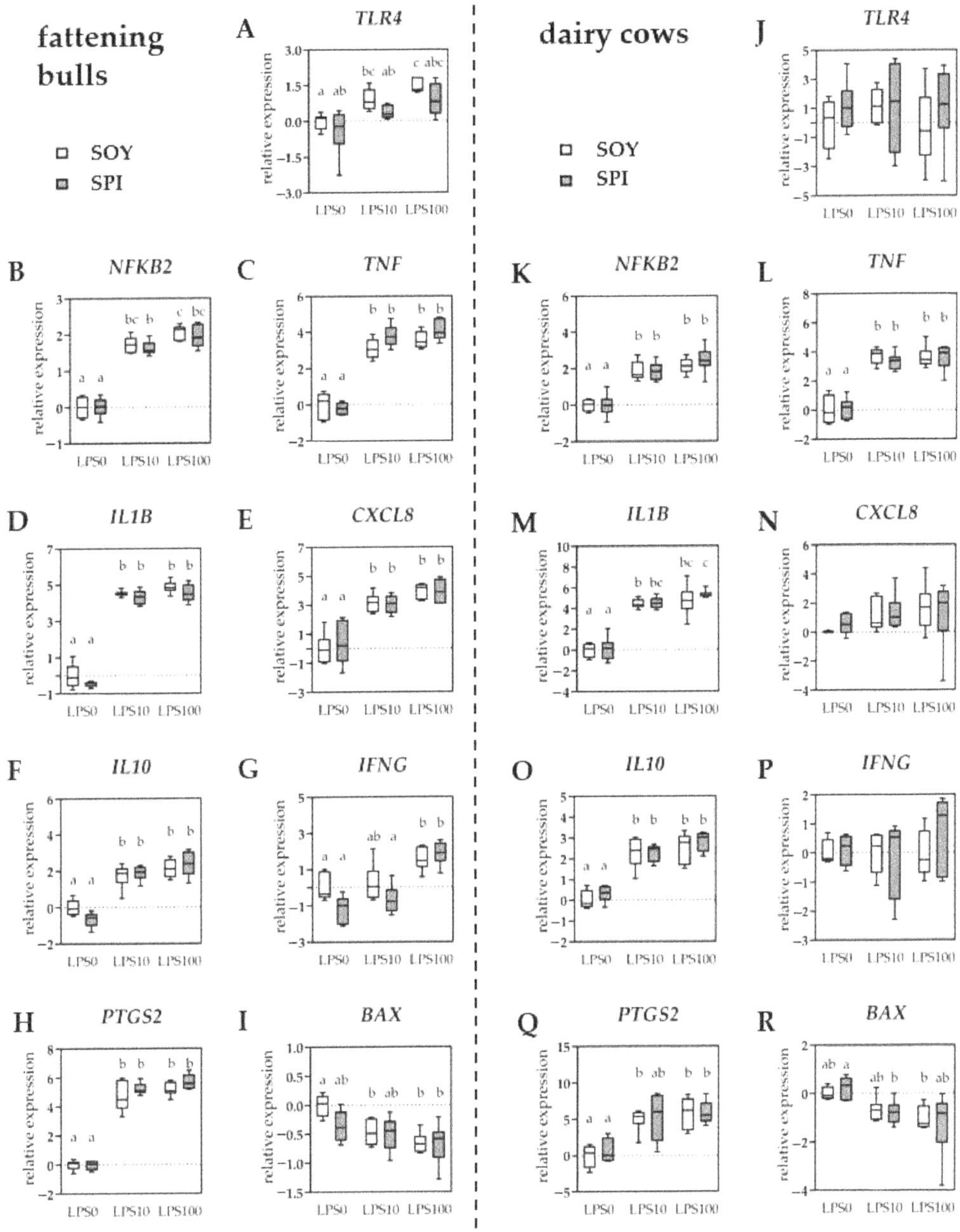

Figure 3. Relative expression of inflammation-related genes in peripheral blood mononuclear cells isolated from fattening bulls (**A–I**) and dairy cows (**J–R**) fed with soybean meal (white boxes) or spirulina (gray boxes) as protein source. Individual results were normalized to the mean of the respective SOY LPS0 group. Results are presented in boxplots with Spear style whiskers (min to max). Different lowercase letters indicate significant differences ($p \leq 0.05$).

4. Discussion

The microalga spirulina has been reported to possess antioxidant and anti-inflammatory properties in a wide range of in vivo studies performed in monogastric species [28,31,33,38,39]. These effects are in part attributed to spirulina's contents of antioxidant and anti-inflammatory bioactive compounds including β-carotene, α-tocopherol, and polyphenols. Oral intake of spirulina has been demonstrated to increase plasma concentrations of β-carotene in rats [47]. In line with this, the β-carotene plasma concentration was almost doubled in SPI, compared to SOY dairy cows of the present study. This can be explained by the threefold higher β-carotene concentration in the SPI, compared to the SOY dairy cow diet (24.3 vs. 7.6 mg/kg DM). In contrast, no difference in plasma β-carotene concentration was observed in the fattening bulls. A potential explanation might be provided by the differing basal diets. The dairy cows received a basal diet consisting mainly of hay (74%), whereas the basal diet of the fattening bulls consisted of 50% grass silage. The concentration of β-carotene in the fattening bull diet was unfortunately not analyzed, but it is known that grass silage contains two- to threefold higher β-carotene proportions compared to hay [48]. Therefore, the β-carotene intake from the basal diet was likely higher in fattening bulls than in dairy cows. Additional dietary β-carotene from spirulina might thus have been absorbed more effectively in hay-fed dairy cows than in grass-silage-fed fattening bulls. This might also have been necessary since the excretion of β-carotene with the milk was higher in SPI, compared to SOY dairy cows [24], whereas fattening bulls do not experience this "loss" of β-carotene. Furthermore, other sex-specific effects on β-carotene metabolism might have played a role in the present findings since β-carotene metabolism is influenced by sex hormones [49]. The influence of culture conditions on spirulina composition can be ruled out in the present study because the spirulina used in both experiments was derived from the same batch. Consequently, also the differential effect of spirulina feeding on phenol plasma concentration seems to be related to the differing basal diet composition (similar total extractable phenol proportions in the dairy cow diets (SOY: 957 vs. SPI: 942 mg/100 g DM); not analyzed in fattening bull diets) and/or differences in bioavailability and metabolism of phenols between dairy cows and fattening bulls. Additionally, the time between the last intake of experimental feeds from the morning feed portion and blood sampling could have played a role since it was not controlled for and peak plasma levels of dietary phenols are reached after 1 to 2 h after ingestion and then rapidly decrease [50]. The observed trend for increased TAC in fattening bulls, accompanied by the significantly higher total phenol plasma concentration, is reasonably in line with studies suggesting that phenols are a major contributor to TAC [51,52]. Importantly, the different β-carotene plasma concentrations in dairy cows did not affect serum TAC. The similar plasma α-tocopherol concentrations in SPI and SOY animals is in line with the similar dietary proportions of α-tocopherol (SOY: 31.4 vs. SPI: 37.7 mg/kg DM in dairy cow diets; not analyzed in fattening bull diets) as well as previous results that did not observe differences in plasma α-tocopherol concentrations after supplementing rats with spirulina extracts [53].

Despite the observed differences in β-carotene and phenol plasma concentrations in SPI, compared to SOY dairy cows and fattening bulls, respectively, the expression of antioxidant enzymes hardly differed. This is in contrast with the literature that reports activation of NFE2L2 signaling by carotenoids and polyphenols, ultimately resulting in induced expression of antioxidant enzymes [54,55]. However, only a few studies regarding the potential antioxidant and anti-inflammatory effects of phenols in cattle are available to date, and their results are not consistent (summarized by [17]). Notably, the expression of the redox-sensitive transcription factor NFE2L2 and consequently that of its downstream antioxidant enzymes showed much higher variability in dairy cows, compared to fattening bulls, pointing toward larger interindividual differences, likely an effect of the varying metabolic activity during lactation. Studies with higher numbers of animals per group may be required to obtain more robust results. The immunomodulatory potential of spirulina with effects on the expression of pro-inflammatory molecules such as *IL1B*, *IL2*, *IL8*, *TNFA*, their transcription factor *NFKB*, as well as *PTGS2* in different monogastric species was summarized by Ravi et al. [56]. In vitro, incubation of human PBMCs with spirulina increased IFNG, IL1B, and IL4 secretion [33]. Mao et al. [34] observed a decrease in IL4 levels by spirulina in humans suffering from allergic rhinitis and oral intake of spirulina augmented the production of IFNG in humans [57]. In consequence, literature data on the direction of inflammation-related effects of spirulina are not consistent, but immunomodulatory effects have frequently been observed. In the present study, inflammation-related gene expression was hardly affected by spirulina supplementation in dairy cows and fattening bulls. The few differences observed were limited to an increased expression of *TLR4* in LPS10 and LPS100, compared to LPS0 treatments in SOY but not SPI bulls, as well as decreased expression of the proapoptotic *BAX* in LPS10-treated, compared to LPS0-treated, PBMCs in SPI but not SOY dairy cows. A numeric increase in *TLR4* expression with LPS stimulation was, however, also observed in SPI bulls and was apparently sufficient to significantly increase expression of *NFKB2* and several cytokines (*TNF, IL1B, CXCL8, IL10,* and *IFNG*). Due to the observed differences in β-carotene and phenol plasma concentrations, the similar expression of inflammation-related genes in SPI and SOY animals was surprising. Supplementation with β-carotene in vitro in murine macrophages and in vivo in mice prevented intracellular ROS accumulation and inhibited the expression of inflammatory genes such as *PTGS2, IL1B,* and *TNFA* upon LPS stimulation [36,58]. In dairy cows, Wang et al. [59] observed lower incidences of mastitis after supplementation with β-carotene. An underlying mechanism for the beneficial effects of β-carotene observed in monogastrics seems to be the blockage of the NFKB p65 subunit (RELA) nuclear translocation [36]. Since we analyzed only gene expression but not nuclear protein proportions of RELA, we cannot state if this mechanism was at all induced in our dairy cows. Additionally, the higher plasma phenol concentration in fattening bulls did not affect the expression of inflammation-related genes in isolated PBMCs.

Taken together, despite the broad evidence of antioxidant and immunomodulatory effects of spirulina in a variety of monogastric species, the present study could not confirm these effects on the level of antioxidant blood parameters and gene expression in PBMCs obtained from ruminants both in the native and the LPS-challenged state. Ruminants efficiently digest especially the carbohydrate fraction of spirulina, but about 20% of dietary spirulina still bypasses the rumen [60,61]. Currently, not much information exists about the rumen degradability of spirulina's bioactive compounds during ruminal fermentation. It seems, however, likely that the different digestive processes in monogastrics and ruminants result in a differing bioavailability of spirulina's bioactive compounds. This might, in turn, also differently affect the beneficial effects on oxidative and inflammatory processes in the animals' metabolism. Our results indicate that in ruminants, plasma β-carotene, α-tocopherol, and phenols are likely not involved in mediating antioxidant and anti-inflammatory gene expression in PBMCs: first, some of these antioxidants were present in similar concentrations in plasma in both diet groups, and second, even when concentrations differed, expression of antioxidant and inflammatory genes was not significantly affected.

It has previously been suggested that the deep blue chromophore–protein complex phycocyanin, forming up to 40% of spirulina's protein, may be the major antioxidant compound in spirulina [26,62]. Phycocyanin was reported to inhibit radical generation and lipid peroxidation in vitro and in vivo [35,63]. The chromophore phycocyanobilin was shown to possess an antioxidant activity very similar to that of phycocyanin, suggesting that phycocyanobilin is responsible for most of the antioxidant effect of phycocyanin [62]. However, a decrease in the antioxidant activity of phycocyanin after trypsin-mediated hydrolysis suggests that also the apoprotein significantly contributes to the antioxidant properties of phycocyanin [62]. In addition, phycocyanin possesses immunomodulatory properties by regulating cytokine expression and selectively inhibiting PTGS2, whereas results concerning anti-inflammatory effects following LPS-stimulation of macrophages remain contradictory [64–66]. The beneficial antioxidant effects of phycocyanin have been investigated so far only in monogastrics. Interestingly, lactic acid fermentation of spirulina has been shown to release phenols and phycocyanin and to improve the antioxidant activity in vitro [67]. This effect was, however, diminished when fermentation exceeded 36 h. It can thus be speculated that by hydrolyzing the cell wall of spirulina, rumen fermentation could in a first step enhance the release and thus the bioavailability of bioactive compounds including phycocyanin. However, prolonged exposure to fermentation processes presumably leads to microbial degradation of the protein compound of phycocyanin and potentially also the chromophore. Thus, the compound contributing significantly to spirulina's antioxidant and immunomodulatory properties may be lost in ruminants.

Some additional aspects need to be considered when evaluating the present results, namely, (a) a direct comparison of the outcomes of the two experiments in the present study has to be regarded with care because of substantial differences in basal diets, duration, and proportion of spirulina supplementation, animal sex, age, and production stage. Further studies are required to directly compare the effects of spirulina in male and, preferably non-lactating, female animals of similar age; (b) the LPS concentrations applied in the present study were rather high. In most cases, the inflammatory response was not further increased by LPS100, compared to LPS10, as indicated by similar cytokine expression. Therefore, it can be assumed that LPS10 already induced the maximum inflammatory response. Presumably, the antioxidant and anti-inflammatory effects of spirulina were not potent enough to counteract the induction of the maximum inflammatory response. It can only be assumed that when using a lower concentration of LPS, representing a milder inflammatory state, potential attenuating effects of dietary spirulina intake on cytokine expression might have been observed. This should be further investigated in additional ex vivo studies; (c) in the case of the dairy cows, it needs to be considered that the animals were all in the late-lactating stage. Dairy cows are most susceptible to oxidative stress and inflammation during the transition period when substantial metabolic and physiological adaptations occur [1,2]. Therefore, it would be interesting to observe the effects when subjecting transition cows to spirulina supplementation; (d) the analysis of gene expression alone does not provide complete information about dietary effects. Instead of a change in gene expression, the activity, e.g., of antioxidant enzymes may be affected by spirulina as has been previously reported [30]. Due to the minor differences in serum TAC, an increase in antioxidant enzyme activity seems, however, unlikely; (e) as a consequence of intense hepatic xenobiotic metabolism, changes in gene expression might have been limited to liver tissue and hepatic immune cells and were consequently not visible in the circulating immune cells; (f) the type and concentration of LPS and the type of leukocytes play a role in the results [64–66]. Using PBMCs in the present study as well as a different type and concentration of LPS compared to other studies may have contributed to differing results; (g) interestingly, the antioxidant effect of spirulina seems to taper off when exceeding a specific threshold concentration [68]. This threshold, however, is not yet determined and might also differ between different species. The dietary proportion of 4–5% of spirulina in DM fed in the present study resulted in spirulina intakes of 0.8 g/kg body weight (fattening bulls) and 1.3 g/kg body weight (dairy cows). This proportion is slightly higher than what has been applied in studies with rats (0.5–1 g/kg body weight [28,69]) and much higher than what was applied in a study with humans (<0.1 g/kg body weight [30]). If spirulina's antioxidant and anti-inflammatory properties become only active at a specific low intake level, as indicated by the results of Ding et al. [68], this might at least in part explain the absence of effects in the present study. Further studies should evaluate if such threshold concentrations exist, and if so, if and how they differ between different species and interact with rumen metabolism.

5. Conclusions

The effects of dietary spirulina on antioxidant plasma concentrations in ruminants differed in the experiments in fattening bulls and dairy cows of the present study. Our hypothesis I was mostly disproved since the only significant increases in blood antioxidants were observed for β-carotene (in dairy cows) and total phenol (in fattening bulls). Basal mRNA expression of antioxidant and inflammatory genes in PBMCs did not differ between the experimental groups, thus disproving hypothesis II. The partially higher plasma antioxidant concentrations observed after spirulina feeding were clearly not effective in attenuating the oxidative and inflammatory effect of increasing LPS concentrations, as demonstrated by similar changes in the expression of antioxidant enzymes and cytokines in PBMCs from SPI and SOY animals, which disproves hypothesis III. The results of the present study obtained from two independent feeding experiments indicate that in ruminants, spirulina might not have antioxidant and anti-inflammatory properties similar to those observed in monogastric species. Rumen degradation of key bioactive compounds such as phycocyanin might explain the species-related differences in spirulina-mediated effects, which requires further investigation. In conclusion, further research is required to understand the rumen degradation, bioavailability, metabolism, and actions of spirulina and its bioactive compounds in ruminants and to provide further data on potential beneficial health effects in fattening bulls and dairy cows.

Author Contributions: Conceptualization, M.K., E.M. and K.G.; methodology, R.L. and K.G.; validation, K.G.; formal analysis, D.R. and K.G.; investigation, M.K., E.M., D.R., R.L. and K.G.; data curation, M.K., E.M., D.R. and K.G.; writing—original draft preparation, K.G.; writing—review and editing, M.K., E.M., D.R., R.L. and K.G.; visualization, K.G.; supervision, R.L. and K.G.; project administration, K.G.; funding acquisition, M.K., E.M. and K.G. All authors have read and agreed to the published version of the manuscript.

Funding: The financial support of the Federation of Migros Cooperatives, Micarna SA, ETH Zurich Foundation, and World Food System Centre of ETH Zurich, as well as the H. Wilhelm Schaumann Foundation (Hamburg, Germany) is greatly acknowledged.

Institutional Review Board Statement: The study was approved by the Cantonal Veterinary Office of Zurich, Switzerland (license no. ZH129/18 and ZH125/18).

Informed Consent Statement: Not applicable.

Data Availability Statement: Data will be made available on request by the corresponding author for reasonable purposes.

Acknowledgments: We are grateful to the staff of AgroVet-Strickhof and ETH Zurich for their assistance and support during the animal experiment and in the lab, especially P. Bucher, C. Kunz, S. Meese, M. Mergani, and H. Renfer and his team. Access to the lab equipment used for RNA isolation and qPCR provided by S. Bauersachs (University of Zurich) and S.E. Ulbrich (ETH Zurich) is greatly acknowledged.

Conflicts of Interest: The authors declare no conflict of interest.

References

1. Sordillo, L.; Aitken, S. Impact of oxidative stress on the health and immune function of dairy cattle. *Vet. Immunol. Immunopathol.* **2009**, *128*, 104–109. [CrossRef] [PubMed]
2. Bradford, B.J.; Yuan, K.; Farney, J.K.; Mamedova, L.K.; Carpenter, A.J. Invited review: Inflammation during the transition to lactation: New adventures with an old flame. *J. Dairy Sci.* **2015**, *98*, 6631–6650. [CrossRef]
3. United States Department of Agriculture. *Part III: Health Management and Biosecurity in U.S. Feedlots, 1999*; #N336.1200; USDA, APHIS, VS, CEAH, National Animal Health Monitoring System: Fort Collins, CO, USA, 2000.
4. Sies, H. Biochemistry of Oxidative Stress. *Angew. Chem. Int. Ed. Engl.* **1986**, *25*, 1058–1071. [CrossRef]
5. Splettstoesser, W.D.; Schuff-Werner, P. Oxidative stress in phagocytes—"The enemy within". *Microsc. Res. Tech.* **2002**, *57*, 441–455. [CrossRef] [PubMed]
6. Victor, V.M.; Rocha, M.; De la Fuente, M. Immune cells: Free radicals and antioxidants in sepsis. *Int. Immunopharmacol.* **2004**, *4*, 327–347. [CrossRef] [PubMed]
7. Mittal, M.; Siddiqui, M.R.; Tran, K.; Reddy, S.P.; Malik, A.B. Reactive oxygen species in inflammation and tissue injury. *Antioxid. Redox Signal.* **2014**, *20*, 1126–1167. [CrossRef]
8. Ramos-Tovar, E.; Muriel, P. Molecular mechanisms that link oxidative stress, inflammation, and fibrosis in the liver. *Antioxidants* **2020**, *9*, 1279. [CrossRef] [PubMed]
9. Boehm, U.; Klamp, T.; Groot, M.; Howard, J.C. Cellular responses to interferon-gamma. *Annu. Rev. Immunol.* **1997**, *15*, 749–795. [CrossRef]
10. Fournier, D.B.; Gordon, G.B. COX-2 and colon cancer: Potential targets for chemoprevention. *J. Cell Biochem. Suppl.* **2000**, *77*, 97–102. [CrossRef]

11. Sordillo, L.M. Factors affecting mammary gland immunity and mastitis susceptibility. *Livest. Prod. Sci.* **2005**, *98*, 89–99. [CrossRef]
12. Muri, J.; Kopf, M. Redox regulation of immunometabolism. *Nat. Rev. Immunol.* **2020**, 1–19. [CrossRef]
13. Miller, J.K.; Brzezinska-Slebodzinska, E.; Madsen, F.C. Oxidative stress, antioxidants, and animal function. *J. Dairy Sci.* **1993**, *76*, 2812–2823. [CrossRef]
14. Smith, K.L.; Harrison, J.H.; Hancock, D.D.; Todhunter, D.A.; Conrad, H.R. Effect of vitamin E and selenium supplementation on incidence of clinical mastitis and duration of clinical symptoms. *J. Dairy Sci.* **1984**, *67*, 1293–1300. [CrossRef]
15. Weiss, W.P.; Hogan, J.S.; Todhunter, D.A.; Smith, K.L. Effect of vitamin E supplementation in diets with a low concentration of selenium on mammary gland health of dairy cows. *J. Dairy Sci.* **1997**, *80*, 1728–1737. [CrossRef]
16. European Medicines Agency. *Sales of Veterinary Antimicrobial Agents in 19 EU/EEA Countries in 2010*; EMA/88728/2012; European Medicines Agency: Amsterdam, The Netherlands, 2012.
17. Gessner, D.K.; Ringseis, R.; Eder, K. Potential of plant polyphenols to combat oxidative stress and inflammatory processes in farm animals. *J. Anim. Physiol. Anim. Nutr.* **2017**, *101*, 605–628. [CrossRef]
18. Belay, A.; Ota, Y.; Miyakawa, K.; Shimamatsu, H. Current knowledge on potential health benefits of spirulina. *Appl. Phycol.* **1993**, *5*, 235–241. [CrossRef]
19. Wild, K.J.; Steingaß, H.; Rodehutscord, M. Variability in nutrient composition and in vitro crude protein digestibility of 16 microalgae products. *J. Anim. Physiol. Anim. Nutr.* **2018**, *102*, 1306–1319. [CrossRef]
20. Da Silva, V.P.; van der Werf, H.M.G.; Spies, A.; Soares, S.R. Variability in environmental impacts of Brazilian soybean according to crop production and transport scenarios. *J. Environ. Manag.* **2010**, *91*, 1831–1839. [CrossRef]
21. Costa, D.F.A.; Quigley, S.P.; Isherwood, P.; McLennan, S.R.; Poppi, D.P. Supplementation of cattle fed tropical grasses with microalgae increases microbial protein production and average daily gain. *J. Anim. Sci.* **2016**, *94*, 2047–2058. [CrossRef]
22. Garcés, C.N.; Vela, D.; Mullo, A.; Cabezas, V.; Alvear, A.; Ponce, C.H. Spirulina supplementation during the transition period by grazing dairy cattle at tropical highland conditions. *Trop. Anim. Health Prod.* **2019**, *51*, 477–480. [CrossRef]
23. Kulpys, J.; Paulauskas, E.; Pilipavičius, V.; Stankevičius, R. Influence of *cyanobacteria Arthrospira* (Spirulina) *platensis* biomass additives towards the body condition of lactation cows and biochemical milk indexes. *Agron. Res.* **2009**, *7*, 823–835.
24. Manzocchi, E.; Guggenbühl, B.; Kreuzer, M.; Giller, K. Effects of the substitution of soybean meal by spirulina in a hay-based diet for dairy cows on milk composition and sensory perception. *J. Dairy Sci.* **2020**, *103*, 11349–11362. [CrossRef] [PubMed]
25. Kurrig, M.; Kreuzer, M.; Reidy, B.; Scheurer, A.; Giller, K. Impact of substituting soybean meal by alternative protein sources in a grass and maize-silage based diet on performance, carcass and meat quality of growing bulls. In Proceedings of the 74th conference of the Society of Nutrition Physiology, Göttingen, Germany, 3–5 March 2020; Kampf, D., Ausmeier, S., Eds.; DLG-Verlag GmbH: Frankfurt am Main, Germany, 2020; Volume 29, p. 84.
26. Wu, Q.; Liu, L.; Miron, A.; Klímová, B.; Wan, D.; Kuča, K. The antioxidant, immunomodulatory, and anti-inflammatory activities of Spirulina: An overview. *Arch. Toxicol.* **2016**, *90*, 1817–1840. [CrossRef] [PubMed]
27. Kalafati, M.; Jamurtas, A.Z.; Nikolaidis, M.G.; Paschalis, V.; Theodorou, A.A.; Sakellariou, G.K.; Koutedakis, Y.; Kouretas, D. Ergogenic and antioxidant effects of spirulina supplementation in humans. *Med. Sci. Sports Exerc.* **2010**, *2*, 142–151. [CrossRef] [PubMed]
28. Abdel-Daim, M.M.; Abuzead, S.M.; Halawa, S.M. Protective role of *Spirulina platensis* against acute deltamethrin-induced toxicity in rats. *PLoS ONE* **2013**, *8*, e72991. [CrossRef]
29. Abdelkhalek, N.K.; Ghazy, E.W.; Abdel-Daim, M.M. Pharmacodynamic interaction of *Spirulina platensis* and deltamethrin in freshwater fish *Nile tilapia, Oreochromis niloticus*: Impact on lipid peroxidation and oxidative stress. *Environ. Sci. Pollut. Res. Int.* **2015**, *22*, 3023–3031. [CrossRef]
30. Ismail, M.; Hossain, M.F.; Tanu, A.R.; Shekhar, H.U. Effect of spirulina intervention on oxidative stress, antioxidant status, and lipid profile in chronic obstructive pulmonary disease patients. *BioMed Res. Int.* **2015**, *2015*, 486120. [CrossRef]
31. Tufarelli, V.; Baghban-Kanani, P.; Azimi-Youvalari, S.; Hosseintabar-Ghasemabad, B.; Slozhenkina, M.; Gorlov, I.; Seidavi, A.; Ayaşan, T.; Laudadio, V. Effects of Horsetail (*Equisetum arvense*) and Spirulina (*Spirulina platensis*) Dietary Supplementation on Laying Hens Productivity and Oxidative Status. *Animals* **2021**, *11*, 335. [CrossRef]
32. Qureshi, M.A.; Ali, R.A. Spirulina platensis exposure enhances macrophage phagocytic function in cats. *Immunopharmacol. Immunotoxicol.* **1996**, *18*, 457–463. [CrossRef]
33. Mao, T.K.; Van de Water, J.; Gershwin, M.E. Effect of Spirulina on the secretion of cytokines from peripheral blood mononuclear cells. *J. Med. Food* **2000**, *3*, 135–141. [CrossRef] [PubMed]
34. Mao, T.K.; Van de Water, J.; Gershwin, M.E. Effects of a Spirulina based dietary supplement on cytokine production from allergic rhinitis patients. *J. Med. Food* **2005**, *8*, 27–30. [CrossRef] [PubMed]
35. Pak, W.; Takayama, F.; Mine, M.; Nakamoto, K.; Kodo, Y.; Mankura, M.; Egashira, T.; Kawasaki, H.; Mori, A. Anti-oxidative and anti-inflammatory effects of spirulina on rat model of non-alcoholic steatohepatitis. *J. Clin. Biochem. Nutr.* **2012**, *51*, 227–234. [CrossRef] [PubMed]
36. Bai, S.K.; Lee, S.J.; Na, H.J.; Ha, K.S.; Han, J.A.; Lee, H.; Kwon, Y.G.; Chung, C.K.; Kim, Y.M. Beta-Carotene inhibits inflammatory gene expression in lipopolysaccharide-stimulated macrophages by suppressing redox-based NF-kappaB activation. *Exp. Mol. Med.* **2005**, *37*, 323–334. [CrossRef]

37. Riss, J.; Décordé, K.; Sutra, T.; Delage, M.; Baccou, J.C.; Jouy, N.; Brune, J.P.; Oréal, H.; Cristol, J.P.; Rouanet, J.M. Phycobiliprotein C-phycocyanin from Spirulina platensis is powerfully responsible for reducing oxidative stress and NADPH oxidase expression induced by an atherogenic diet in hamsters. *J. Agric. Food Chem.* **2007**, *55*, 7962–7967. [CrossRef]
38. Qureshi, M.A.; Garlich, J.D.; Kidd, M.T. Dietary Spirulina platensis enhances humoral and cell-mediated immune functions in chickens. *Immunopharmacol. Immunotoxicol.* **1996**, *18*, 465–476. [CrossRef]
39. Gad, A.S.; Khadrawy, Y.A.; El-Nekeety, A.A.; Mohamed, S.R.; Hassan, N.S.; Abdel-Wahhab, M.A. Antioxidant activity and hepatoprotective effects of whey protein and Spirulina in rats. *Nutrition* **2011**, *27*, 582–589. [CrossRef]
40. Keller, M.; Reidy, B.; Scheurer, A.; Eggerschwiler, L.; Morel, I.; Giller, K. Soybean meal can be replaced by faba bean, pumpkin seed cake, spirulina or be completely omitted in a forage-based diet of fattening bulls to achieve comparable performance, carcass and meat quality. *Animals* **2021**. under review.
41. Ogawa, T.; Terui, G. Studies on the growth of *Spirulina platensis*. On the pure culture of *Spirulina platensis*. *J. Ferment. Technol.* **1970**, *48*, 361–367.
42. Agroscope. Feeding Recommendations for Ruminants. Available online: https://www.agroscope.admin.ch/agroscope/de/home/services/dienste/futtermittel/fuetterungsempfehlunge-wiederkaeuer.html (accessed on 24 January 2020). (In German)
43. Beuth Publishing DIN. *Foodstuffs—Determination of Vitamin A by High Performance Liquid Chromatography—Part 2: Measurement of β-Carotene*; Beuth Publishing DIN: Berlin, Deutschland, 2000; German version EN 12823-2:2000. [CrossRef]
44. Kälber, T.; Meier, J.S.; Kreuzer, M.; Leiber, F. Flowering catch crops used as forage plants for dairy cows: Influence on fatty acids and tocopherols in milk. *J. Dairy Sci.* **2011**, *94*, 1477–1489. [CrossRef]
45. Sinz, S.; Liesegang, A.; Kreuzer, M.; Marquardt, S. Do supplements of *Acacia mearnsii* and grapeseed extracts alone or in combination alleviate metabolic nitrogen load and manure nitrogen emissions of lambs fed a high crude protein diet? *Arch. Anim. Nutr.* **2019**, *73*, 306–323. [CrossRef] [PubMed]
46. Mann, S.; Sipka, A.; Leal Yepes, F.A.; Nydam, D.V.; Overton, T.R.; Wakshlag, J.J. Nutrient-sensing kinase signaling in bovine immune cells is altered during the postpartum nutrient deficit: A possible role in transition cow inflammatory response. *J. Dairy Sci.* **2018**, *101*, 9360–9370. [CrossRef]
47. Rao, A.R.; Raghunath Reddy, R.L.; Baskaran, V.; Sarada, R.; Ravishankar, G.A. Characterization of microalgal carotenoids by mass spectrometry and their bioavailability and antioxidant properties elucidated in rat model. *J. Agric. Food Chem.* **2010**, *58*, 8553–8559. [CrossRef]
48. Chauveau-Duriot, B.; Thomas, D.; Portelli, J.; Doreau, M. Carotenoids content in forages: Variation during conservation. *Renc. Rech. Rumin.* **2005**, *12*, 117.
49. Van Helden, Y.G.; Godschalk, R.W.; von Lintig, J.; Lietz, G.; Landrier, J.F.; Bonet, M.L.; van Schooten, F.J.; Keijer, J. Gene expression response of mouse lung, liver and white adipose tissue to β-carotene supplementation, knockout of Bcmo1 and sex. *Mol. Nutr. Food Res.* **2011**, *55*, 1466–1474. [CrossRef] [PubMed]
50. Wein, S.; Beyer, B.; Zimmermann, B.F.; Blank, R.H.; Wolffram, S. Bioavailability of Quercetin from Onion Extracts after Intraruminal Application in Cows. *J. Agric. Food Chem.* **2018**, *66*, 10188–10192. [CrossRef] [PubMed]
51. Jimenez-Escrig, A.; Jimenez-Jimenez, I.; Pulido, R.; Saura-Calixto, F. Antioxidant activity of fresh and processed edible seaweeds. *J. Sci. Food Agric.* **2001**, *81*, 530–534. [CrossRef]
52. Zheng, W.; Wang, S.Y. Antioxidant activity and phenolic compounds in selected herbs. *J. Agric. Food Chem.* **2001**, *49*, 5165–5170. [CrossRef]
53. García-Martínez, D.; Rupérez, F.J.; Ugarte, P.; Barbas, C. Tocopherol fate in plasma and liver of streptozotocin-treated rats that orally received antioxidants and Spirulina extracts. *Int. J. Vitam. Nutr. Res.* **2007**, *77*, 263–271. [CrossRef]
54. Ben-Dor, A.; Steiner, M.; Gheber, L.; Danilenko, M.; Dubi, N.; Linnewiel, K.; Zick, A.; Sharoni, Y.; Levy, J. Carotenoids activate the antioxidant response element transcription system. *Mol. Cancer Ther.* **2005**, *4*, 177–186.
55. Martínez-Huélamo, M.; Rodríguez-Morató, J.; Boronat, A.; de la Torre, R. Modulation of Nrf2 by olive oil and wine polyphenols and neuroprotection. *Antioxidants* **2017**, *6*, 73. [CrossRef]
56. Ravi, M.; De, S.L.; Azharuddin, S.; Paul, S.F.D. The beneficial effects of spirulina focusing on its immunomodulatory and antioxidant properties. *Nutr. Diet. Suppl.* **2010**, *2*, 73–83. [CrossRef]
57. Hirahashi, T.; Matsumoto, M.; Hazeki, K.; Saeki, Y.; Ui, M.; Seya, T. Activation of the human innate immune system by Spirulina: Augmentation of interferon production and NK cytotoxicity by oral administration of hot water extract of Spirulina platensis. *Int. Immunopharmacol.* **2002**, *2*, 423–434. [CrossRef]
58. Katsuura, S.; Imamura, T.; Bando, N.; Yamanishi, R. Beta-Carotene and beta-cryptoxanthin but not lutein evoke redox and immune changes in RAW264 murine macrophages. *Mol. Nutr. Food Res.* **2009**, *53*, 1396–1405. [CrossRef]
59. Wang, J.Y.; Owen, F.G.; Larson, L.L. Effect of beta-carotene supplementation on reproductive performance of lactating Holstein cows. *J. Dairy Sci.* **1988**, *71*, 181–186. [CrossRef]
60. Gouveia, L.; Batista, A.P.; Sousa, I.; Raymundo, A.; Bandarra, N.M. Microalgae in Novel Food Products. In *Food Chemistry Research Developments*; Papadopoulos, K.N., Ed.; Nova Science Publishers: New York, NY, USA, 2008; pp. 1–37.
61. Panjaitan, T.; Quigley, S.P.; McLennan, S.R.; Poppi, D.P. Effect of the concentration of Spirulina (*Spirulina platensis*) algae in the drinking water on water intake by cattle and the proportion of algae bypassing the rumen. *Anim. Prod. Sci.* **2010**, *50*, 405–409. [CrossRef]

62. Zhou, Z.P.; Liu, L.N.; Chen, X.L.; Wang, J.X.; Chen, M.; Zhang, Y.Z. Factors that effect antioxidant activity of C-phycocyanins from *Spirulina platensis*. *J. Food Biochem.* **2005**, *29*, 313–322. [CrossRef]
63. Bermejo-Bescós, P.; Piñero-Estrada, E.; Villar del Fresno, A.M. Neuroprotection by Spirulina platensis protean extract and phycocyanin against iron-induced toxicity in SH-SY5Y neuroblastoma cells. *Toxicol. In Vitro* **2008**, *22*, 1496–1502. [CrossRef]
64. Reddy, M.C.; Subhashini, J.; Mahipal, S.V.; Bhat, V.B.; Reddy, P.S.; Kiranmai, G.; Madyastha, K.M.; Reddanna, P. C-phycocyanin, a selective cyclooxygenase-2 inhibitor, induces apoptosisin lipopolysaccharide-stimulated RAW 264.7 macrophages. *Biochem. Biophys. Res. Commun.* **2003**, *304*, 385–392. [CrossRef]
65. Cherng, S.C.; Cheng, S.N.; Tarn, A.; Chou, T.C. Anti-inflammatory activity of C-phycocyanin in lipopolysaccharide-stimulated RAW 264.7 macrophages. *Life Sci.* **2007**, *81*, 1431–1435. [CrossRef]
66. Chen, H.W.; Yang, T.S.; Chen, M.J.; Chang, Y.C.; Wang, E.I.C.; Ho, C.L.; Lai, Y.J.; Yu, C.C.; Chou, J.C.; Chao, L.K.P.; et al. Purification and immunomodulating activity of C-phycocyanin from *Spirulina platensis* cultured using power plant flue gas. *Process Biochem.* **2014**, *49*, 1337–1344. [CrossRef]
67. De Marco Castro, E.; Shannon, E.; Abu-Ghannam, N. Effect of fermentation on enhancing the nutraceutical properties of *Arthrospira platensis* (Spirulina). *Fermentation* **2019**, *5*, 28. [CrossRef]
68. Ding, J.; Jin, A.; Shi, L.; Huo, T. Effect of spirulina on antioxidation ability of liver during CCl4 induced chronic liver injury in mice. *J. Lake Sci.* **2004**, *16*, 343–348. [CrossRef]
69. Abdel-Daim, M.M.; Farouk, S.M.; Madkour, F.F.; Azab, S.S. Anti-inflammatory and immunomodulatory effects of *Spirulina platensis* in comparison to *Dunaliella salina* in acetic acid-induced rat experimental colitis. *Immunopharmacol. Immunotoxicol.* **2015**, *37*, 126–139. [CrossRef] [PubMed]

antioxidants

Review

Recent Molecular Mechanisms and Beneficial Effects of Phytochemicals and Plant-Based Whole Foods in Reducing LDL-C and Preventing Cardiovascular Disease

Salman Ul Islam [1,†], Muhammad Bilal Ahmed [1,†], Haseeb Ahsan [1,2,†] and Young-Sup Lee [1,*]

1 School of Life Sciences, BK21 FOUR KNU Creative BioResearch Group, Kyungpook National University, Daegu 41566, Korea; salman2013@knu.ac.kr (S.U.I.); muhammad786@knu.ac.kr (M.B.A.); haseeb2020@knu.ac.kr (H.A.)

2 Department of Pharmacy, Faculty of Life and Environmental Sciences, University of Peshawar, Peshawar 25120, Pakistan

* Correspondence: yselee@knu.ac.kr; Tel.: +82-53-950-6353; Fax: +82-53-943-2762

† These authors contributed equally to this work.

Abstract: Abnormal lipid metabolism leads to the development of hyperlipidemia, a common cause of multiple chronic disorders, including cardiovascular disease (CVD), obesity, diabetes, and cerebrovascular disease. Low-density lipoprotein cholesterol (LDL-C) currently remains the primary target for treatment of hyperlipidemia. Despite the advancement of treatment and prevention of hyperlipidemia, medications used to manage hyperlipidemia are limited to allopathic drugs, which present certain limitations and adverse effects. Increasing evidence indicates that utilization of phytochemicals and plant-based whole foods is an alternative and promising strategy to prevent hyperlipidemia and CVD. The current review focuses on phytochemicals and their pharmacological mode of actions for the regulation of LDL-C and prevention of CVD. The important molecular mechanisms illustrated in detail in this review include elevation of reverse cholesterol transport, inhibition of intestinal cholesterol absorption, acceleration of cholesterol excretion in the liver, and reduction of cholesterol synthesis. Moreover, the beneficial effects of plant-based whole foods, such as fresh fruits, vegetables, dried nuts, flax seeds, whole grains, peas, beans, vegan diets, and dietary fibers in LDL-C reduction and cardiovascular health are summarized. This review concludes that phytochemicals and plant-based whole foods can reduce LDL-C levels and lower the risk for CVD.

Keywords: plant-based foods; LDL; CVD; lipid oxidation; dietary fiber; cholesterol; hyperlipidemia

Citation: Islam, S.U.; Ahmed, M.B.; Ahsan, H.; Lee, Y.-S. Recent Molecular Mechanisms and Beneficial Effects of Phytochemicals and Plant-Based Whole Foods in Reducing LDL-C and Preventing Cardiovascular Disease. *Antioxidants* **2021**, *10*, 784. https://doi.org/10.3390/antiox10050784

Academic Editors: Irene Dini and Domenico Montesano

Received: 25 March 2021
Accepted: 12 May 2021
Published: 15 May 2021

1. Introduction

Cholesterol is circulated in the human body by five major types of lipoproteins: high-density lipoprotein (HDL), low-density lipoprotein (LDL), intermediate-density lipoprotein (IDL), very low-density lipoprotein (VLDL), and chylomicrons [1]. The metabolism and plasma levels of cholesterol are mostly regulated by the liver. During the first step of LDL formation, intrahepatic cholesterol, either via gut absorption or de novo synthesis, is repackaged by the liver along with phospholipids, triglycerides (TGs), and proteins into VLDL particles, which then enter the general blood circulation and are converted into more cholesterol-enriched species, first IDL and then LDL, by lipoprotein lipase and cholesteryl ester transfer protein (Figure 1A). The concentrations of these circulating lipoprotein species are then regulated by the liver primarily by clearance through LDL receptors on the hepatocyte surface [2,3].

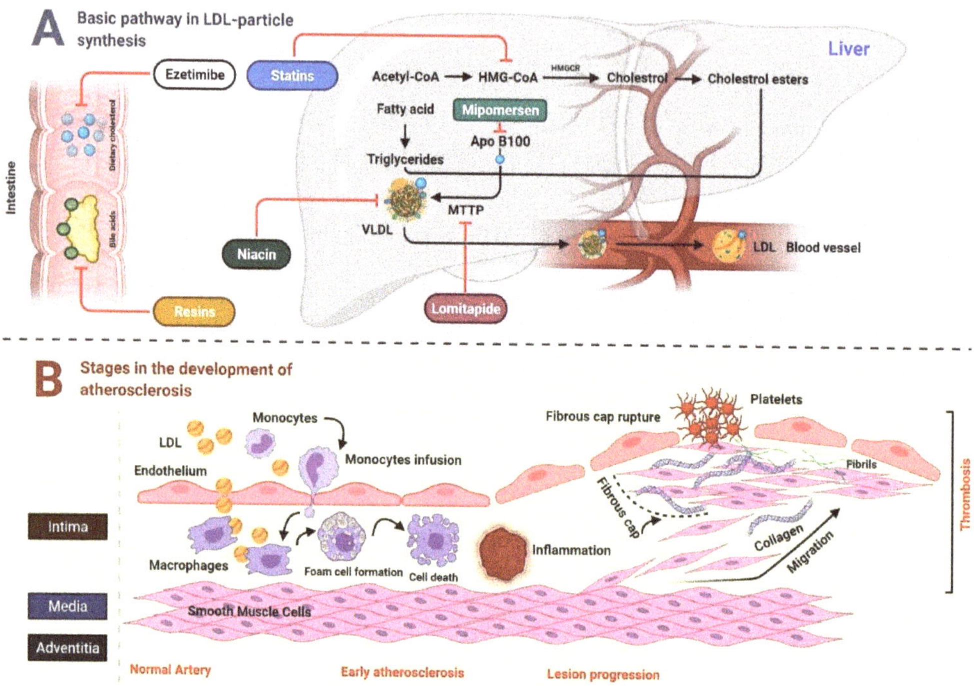

Figure 1. Connection between LDL and CVD.

Figure 1A Bile acids and dietary cholesterol are absorbed from the lower and upper small intestine, respectively. Cholesterol absorption inhibitor, for example, ezetimibe, and bile acid sequestrants (resins) disrupt these pathways, subsequently reducing the intrahepatic cholesterol pool. The synthesis of cholesterol occurs in the liver through a multistep process that starts with acetyl-CoA. HMGCR is the rate-limiting enzyme, whose action is blocked by statins. TG are produced by esterification of fatty acids on a 3-carbon glycerol backbone. TG and esterified cholesterol are assembled by microsomal triglyceride transfer protein MTTP into nascent VLDL particles with ApoB100 on their surface, whereas MTTP and ApoB100 remain the targets for lomitapide and mipomersen, respectively. In the blood, endovascular lipases process VLDL particles to LDL particles, which are catabolized by the LDLR mainly on liver cells. Niacin inhibits the transport of VLDL. Figure 1B Key steps in the development of atherosclerosis include early atherosclerosis, lesion progression, and thrombosis. LDL play a crucial role in the development of atherosclerosis. During early atherosclerosis, monocytes are capture by the endothelial cells of the inner layer of arterial wall. The endothelial permeability assists the LDL particles in migrating into the arterial wall. Monocytes become mature, and are transformed to macrophages, which uptake the LDL particles yielding to fat-laden foam cells. Early atherosclerosis is followed by lesion progression where the smooth muscle cells move from the middle layer of the arterial wall into the tunica intima. The last step is the thrombosis which is characterized by the rapturing of the fibrous cap of a plaque and establishment of contact of blood coagulation components with the thrombogenic plaque.

2. Correlation of LDL Cholesterol with CVD

Several investigations provide strong evidence that LDL cholesterol (LDL-C) is a potent cardiovascular risk factor [4]. Early studies such the "Multiple Risk Factor Intervention Trial" measured total cholesterol instead of LDL-C, indicating a strong correlation between cholesterol and cardiovascular mortality [5]. However, this relationship can be assigned to LDL-C because LDL contains a major part of total cholesterol. Multiple investigations have confirmed LDL-C to be the most atherogenic lipoprotein. Studies have shown that circulating LDL particles penetrate the endothelium of arterial walls and are oxidized. Then, these oxidized LDL particles induce inflammation of the overlying endothelium and surrounding smooth muscle cells [6] (Figure 1B). Persistent elevations in circulating LDL-C levels have been directly linked to the progression from early-stage fatty streaks to advanced-stage, lipid-rich plaques. For instance, LDL receptor-deficient mice, which fail to clear LDL from the blood, have excessive LDL-C, which promoted the development of severe atherosclerosis [7], whereas mice with virtually no LDL-C did not develop atherosclerosis irrespective of diet and other risk factors for coronary heart disease (CHD) [8].

An epidemiological study demonstrated LDL-C as an independent predictor of CVD risk, as LDL-C levels > 160 mg/dL are associated with > 1.5-fold greater risk of CHD than levels < 130 mg/dL [9]. However, besides the role of LDL-C as a risk marker, researchers have also established it as a true risk factor based on investigations where inhibition of LDL-C via β-hydroxy-β-methylglutaryl coenzyme A (HMG-CoA) reductase inhibitors decreased cardiovascular events [10]. These findings have been verified by multiple large randomized controlled trials of LDL lowering like the MRC/BHF Heart Protection Study in 20,536 UK adults [11]. Most of the time, these trials focused to investigate the actions of statins, and were further supported by large meta-analyses. For instance, the Prospective Pravastatin Pooling Project (PPP) pooled the data from the West of Scotland Coronary Prevention Study (WOSCOPS), the Cholesterol and Recurrent Events trial (CARE), and the Long-term Intervention with Pravastatin in Ischemic Disease study (LIPID), providing over 100,000 person-years of follow-up [12]. Likewise, the prospective meta-analysis of the Cholesterol Treatment Trialists' (CTT) Collaboration pooled the data from 14 randomized statin trials, containing over 90,000 individuals [13]. These trials offer exceptional statistical power for proving the potency and safety of statin therapy for a multitude of patient subgroups and endpoints.

It has been reported that statin treatment reduced the five-year incidence of major coronary events, stroke, and coronary revascularization by about one-fifth per mmol/L reduction in LDL-C [14]. Another meta-analysis from the CTT Collaboration analyzed the efficacy and safety of more intensive versus standard LDL-C lowering by statin therapy. The data were collected from 170,000 participants in a total of 26 randomized trials, which demonstrated that further decrease in LDL-C (0.51 mmol/L at one year vs. standard therapy) reduced the incidence of major coronary events by 15% [15]. Based on this information, guidelines have been established suggesting different target levels of LDL-C for different subgroups of patients. Almost all cardiovascular guidelines point to the evidence for LDL-C being both a prime cause of CHD, and a primary target of therapy [16]. Moreover, although many single-nucleotide polymorphisms (SNPs) of genes associated with increased LDL-C levels, including LDL receptor (LDLR), apolipoprotein E (ApoE), proprotein convertase subtilisin/kexin type 9 (PCSK9), and apolipoprotein B (ApoB), have been correlated with an increased risk of CVD, specific SNPs of these same genes have been associated with decreased LDL-C levels and lower risks of CVD [17–20].

At present, hyperlipidemia is primarily treated with allopathic antihyperlipidemic drugs. However, due to intolerance and adverse effects associated with these medicines, plant-based foods are important alternatives [21,22]. Plant-based foods contain various bioactive phytochemicals that can decrease LDL levels through multiple hyperlipidemia-related biological pathways. Consumption of plant-based foods has emerged as a promising and potentially cost-effective approach to decrease LDL levels while also adhering to the concept of "green" healthcare [23,24]. The following sections describe the underlying mechanisms of phytochemicals to reduce the cholesterol levels and prevent CVD.

3. Major Cholesterol Regulatory Mechanisms of Phytochemicals

3.1. Acceleration of Reverse Cholesterol Transport

Reverse cholesterol transport (RCT) is a crucial pathway that removes excess cholesterol from peripheral tissues and delivers them to the liver [25,26]. The RCT comprises of three main processes: cholesterol efflux, where excess cholesterol is removed from cells; modulation of lipoprotein, where HDL gains structural and functional changes; hepatic lipid uptake, where HDL delivers cholesterol to the liver, which is finally excreted into bile and feces [27]. In vivo investigations have demonstrated that promotion of RCT might decrease CVD and atherosclerotic plaque burden [28].

3.1.1. Cholesterol Efflux

Cholesterol efflux is referred to the removal of excess cholesterol from macrophages. Studies have demonstrated the ATP-binding cassette transporter A1 (ABCA1) and ATP-binding cassette transporter G1 (ABCG1) to be the most important transporters contributing to regulate cholesterol efflux from cells. ABCA1 is responsible for the efflux to lipid-free apolipoprotein A-I (ApoA-I), whereas ABCG1 regulates efflux to mature HDL [29–31]. It has been reported that promotion of cholesterol efflux effectively inhibited the formation of foam cells and subsequent atherosclerosis caused by dyslipidemia [32,33].

Multiple investigations have suggested that phytochemicals such resveratrol [34], puerarin [35], leonurine [36], luteolin [37], andrographolide [38], leoligin [39], chrysin [40], and allicin [41] could enhance cholesterol efflux to HDL through ABCA1 or ABCG1. A Chawla et al. [42] reported that the PPARγ-LXR-ABCA1 pathway contributed to cholesterol efflux in macrophages. It was demonstrated that most of the above-mentioned phytochemicals increased the expressions of ABCA1 or ABCG1 through PPARγ or LXR. Moreover, previous studies have reported that quercetin-induced ABCA1 levels and cholesterol efflux were mediated by activation of TAK1-MKK3/6-p38 signaling cascade [43–45].

3.1.2. Modulation of Lipoprotein

Besides cholesterol efflux, inhibiting lipid uptake in macrophages is another mechanism to inhibit foam cell formation, which eventually leads to suppress atherosclerotic plaque formation. CD36 (cluster of differentiation 36) and scavenger receptor class A (SR-A) are mainly responsible for uptake of lipoprotein-derived cholesterol by macrophages [46]. Several mechanisms have been described for phytochemicals through which they induce intracellular cholesterol efflux. For instance, a study reported that icariin, an active flavonol diglycoside, downregulated the CD36 expressions level through p38MAPK signaling pathway [47]. Additionally, paeonol was shown to repress the CD36 at both mRNA and protein levels by inhibiting the nuclear translocation of C—Jun [48]. Puerarin blocked the TLR4/NFκB signaling and decreased the expressions of CD36 [49]. Likewise, rographolide [38], and salvianolic acid B [50] were reported to inhibit CD36. An investigation reported that ginsenoside-Rd blocked the activity of SR-A, which caused reduction of oxidized LDL uptake and cholesterol aggregation in macrophages [51].

After removal from cells, free cholesterol is converted to cholesteryl esters by lecithin: cholesterol acyltransferase (LCAT) to form mature HDL [52]. Relevant investigations have been conducted on phytochemical is this area. Researchers have demonstrated that curcumin [53] and naringin [54] increased the RCT via LCAT and exerted anti-atherosclerosis effects.

It has been reported that cholesterol ester transporter (CETP) transfers cholesterol esters (CEs) from HDL towards ApoB-containing lipoproteins, resulting in reduced concentration of HDL and ApoA-I, while elevating the concentration of CE in VLDL and remnants [55]. As CETP elevates the concentration of VLDL and LDL-C, its specific knockdown can reduce atherosclerotic CVD [56]. It has been reported that anthocyanins could effectively inhibit the activity of CETP in humans [57].

3.1.3. Hepatic Lipid Uptake

As already mentioned, that cholesterol metabolism is mostly regulated by the liver, where it takes up LDL and HDL-CE particles by LDLR and scavenger receptor class B type I (SR-BI), respectively. The LDLR binds to LDL on the cell surface. PCSK9 has been shown to post-transcriptionally downregulate the LDLR by binding to the receptor's epidermal growth factor repeat A (EGF-A) on the cell surface and shuttling it to the lysosomes for degradation [58].

Multiple phytochemicals have been shown to alleviate atherogenesis by modulating the activity of LDLR and PCSK9. For instance, berberine was shown to upregulate the hepatic expression of LDLR and sterol regulatory element-binding protein 2 (SREBP-2), whereas it downregulated the expression of hepatocyte nuclear factor 1 [59,60]. Another investigation reported that berberine exerted anti-lipid effects by regulating hepatic LDLR and PCSK9 via the ERK signaling pathway [61]. Piseth Nhoek et al. [62] reported that flavonoid compounds, 3,7,2′-trihydroxy-5-methoxy-flavanone and skullcapflavone II, isolated from the roots of *Scutellaria baicalensis*, downregulated the PCSK9 at mRNA level via sterol regulatory element-binding protein-1 (SREBP-1). Furthermore, other phytochemicals such as curcumin [63], and tanshinone IIA [64], also modulated the activity of LDLR via downregulation of PCSK9.

Lipoprotein lipase (LPL) is the rate-limiting enzyme in the circulation of cholesterol metabolism, hydrolyzing the TG core of circulating TG-rich lipoproteins, VLDL, and chylomicrons [65]. In addition, hepatic lipase is required for the hydrolysis of triglyceride-rich lipoproteins [66]. A study reported that paeoniflorin regulated the GALNT2-ANGPTL3-LPL signaling pathway to diminish dyslipidemia in mice [67]. Yan Zhang and coworkers reported that osthole (an active constituent obtained from the fruit of *Cnidium monnieri* (L) Cusson) decreased the TC and TG in rat serum, and this effect was related with the elevated activities of LPL and hepatic lipase [68] (Figure 2).

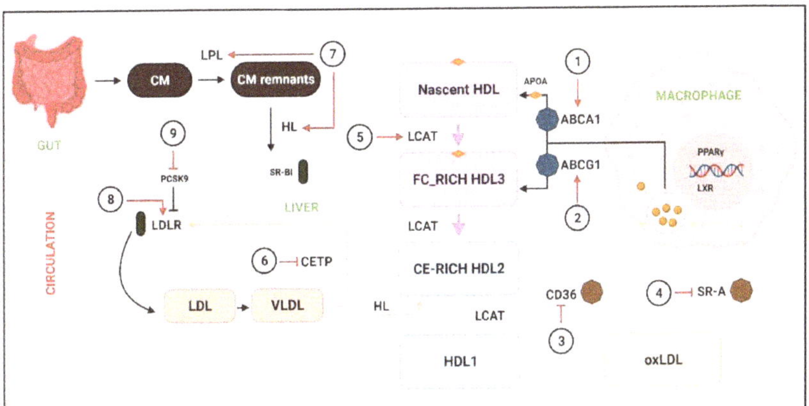

Figure 2. Acceleration of RCT by various phytochemicals.

RCT is responsible to facilitate the transport of excess cholesterol from peripheral tissues to the liver, where it is redistributed to other tissues or eliminated from the body through the gallbladder. The transport of cholesterol into macrophages occurs through CD36 and SR-A, whereas extra cholesterol is effluxed through ABCA1 and ABCG1 to the nascent HDL and free FC-rich HDL3, respectively. LCAT converts nascent HDL, FC-rich HDL3, and CE-rich HDL2 to HDL1, whereas CETP catalyzes HDL2 to VLDL. Next, HDL2 and VLDL are taken up by SR-B1 and LDLR, respectively. (1, 2) The activity of ABCA1 and ABCG1 is promoted by resveratrol [34], puerarin [35], leonurine [36], luteolin [37], andrographolide [38], leoligin [39], and chrysin [40]. (3) Downregulation of CD36 by icariin [47], paeonol [48], and puerarin [49], salvianolic acid B [50], and rographolide [38].

(4) Inhibition of SR-A by ginsenoside-Rd [51]. (5) LCAT levels are elevated by naringin [54] and curcumin [53]. (6) CETP is downregulated by crocin anthocyanins [57]. (7) LPL and HL are stimulated by paeoniflorin [67] and osthole [68]. (8) LDLR expression is increased by berberine [59,60]. (9) PCSK9 expression are downregulated by 3,7,2′-trihydroxy-5-methoxy-flavanone, skullcapflavone II [62], curcumin [63], and tanshinone IIA [64]. CM = chylomicron; CM remnants = chylomicron remnants; HL = hepatic lipase; FC = free cholesterol; CE = cholesterol ester; oxLDL = Oxidized low-density lipoprotein; LXR = Liver X receptor.

3.2. *Inhibition of Intestinal Cholesterol Absorption*

Absorption of cholesterol refers to the transfer of intraluminal cholesterol into enterocytes or thoracic duct lymph [69]. The food substances and bile enter from the intestinal lumen into enterocytes via the transmembrane protein Niemann-Pick C1 like1 (NPC1L1) [70]. Inside the enterocytes, free cholesterol is esterified to CEs by an enzyme called acyl CoA: cholesterol acyltransferase-2 (ACAT)-2 in the endoplasmic reticulum [71]. Afterward, CEs and TG, under the action of MTTP, form chylomicrons, which are then secreted into the lymphatic system [72]. Studies have reported that inhibition of intestinal cholesterol absorption effectively lowered the plasma LDL-C level [73] and reduced the risk of CVD [74]. Thus, it is necessary to block excessive absorption of cholesterol from the diet and bile [75].

3.2.1. Cholesterol Uptake Inhibition

Phytochemicals show blood lipid-lowering effects and inhibit cholesterol uptake mainly by targeting NPC1L1. It has been shown that downregulation of NPC1L1 considerably reduces intestinal cholesterol absorption [70], which is modulated by SREBP-2 [76] and LXR [77]. Curcumin has been shown to block cholesterol uptake by binding to the NPC1L1-related transporter [78,79]. A study reported the curcumin response elements to be present in the region between −291 and +56 of NPC1L1 promoter [80]. Additionally, curcumin was shown to inhibit NPC1L1 pathways by activating the SREBP2 transcription factor [80]. Jun Zou and Dan Feng reported that lycopene, the predominant carotenoid in tomatoes, blocked intestinal cholesterol absorption by blockade of the LXRα pathway [81]. Another study reported that ankaflavin and monascin suppressed the protein levels of NPC1L1 associated with small intestine tissue lipid absorption [82].

3.2.2. Enhancement of Cholesterol Esterification

Studies have reported that ACAT2 exhibits a strong connection with the plasma cholesterol levels and catalyzes the formation of cholesteryl ester in enterocytes [71,83]. Phytochemicals such as oleanolic acid and ursolic acid have been reported to reduce cholesterol levels by inhibiting the activity of ACAT [84,85]. Additionally, downregulation of MTTP is associated with decreased ApoB secretion and chylomicron assemblage. Flavonoids, such as hesperetin [86], quercetin [87], taxifolin [88], tangeretin [89], and naringenin [86] have shown MTTP inhibitory activities. Ioanna Vallianou and Margarita Hadzopoulou-Cladaras reported that camphene, in response to a decrease in the intracellular cholesterol exerts, upregulated the expression of SREBP-1 and blocked the activity of MTTP [90]. It was shown that Tanshinone IIA repressed the MTTP's transcripts and stimulated cellular ApoB proteasomal degradation [91]. Likewise, nobiletin [92], tangeretin [92], and lignin [93] have been shown to reduce the secretion of ApoB. Collectively, phytochemicals inhibit excessive cholesterol absorption and reduce blood lipid levels by downregulating NPC1L1, MTTP, ACAT2, ApoB, LXR, and SREBP (Figure 3).

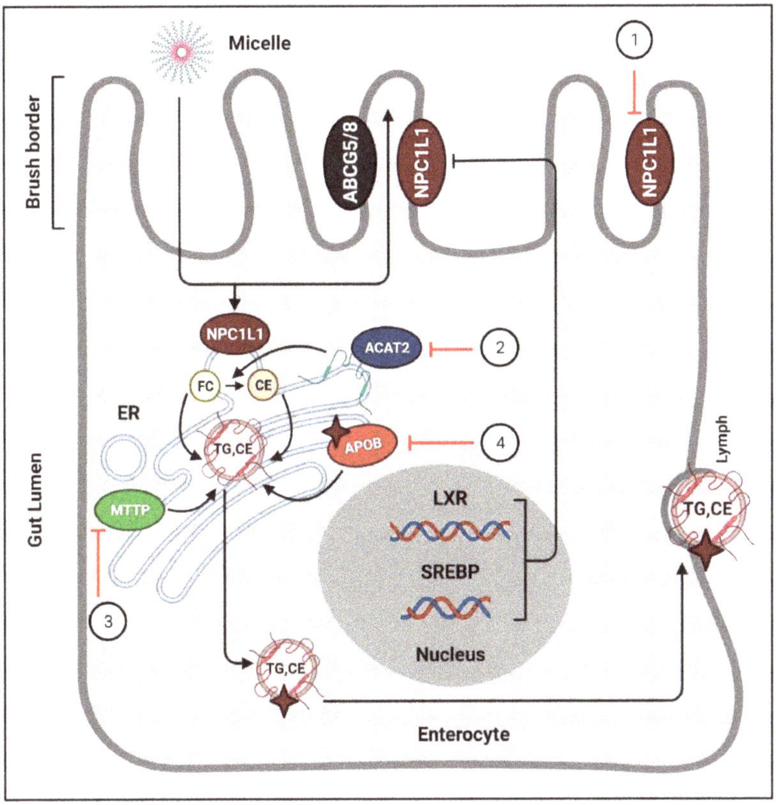

Figure 3. Phytochemicals block the absorption of intestinal cholesterol.

In intestinal lumen, cholesterol incorporates into bile salt micelles and diffuses to the brush border membrane of enterocytes via NPC1L1. Inside the intestinal cells, free cholesterols are esterified by ACAT-2, and afterwards they enter the lymphatic system in the form of chylomicrons. Phytochemicals exert their actions to regulate these processes. (1) NPC1L1 is suppressed by curcumin [78,79], lycopene [81], monascin, and ankaflavin [82]. (2) ACAT2 is targeted by oleanolic acid, and ursolic acid [84,85]. (3) The activity of MTTP is blocked by hesperetin [86], quercetin [87], taxifolin [88], tangeretin [89], and naringenin [86]. (4) Tanshinone IIA stimulates the proteasomal degradation of cellular ApoB [91]. ER = endoplasmic reticulum; FC = free cholesterol; CE = cholesterol ester; ABCG5/8 = ATP-binding cassette sub-family G member 5/8; TG = triglycerides

3.3. Promotion of Cholesterol Excretion in the Liver

Hepatic cholesterol, after conversion to bile acids, is removed from the body through biliary secretion [94]. Cholesterol 7 alpha-hydroxylase (CYP7A1) is the first and rate-limiting enzyme in the bile acid synthesis pathway, which performs a crucial role in maintaining cholesterol homeostasis [95,96]. The activity of CYP7A1 has been shown to be promoted by phytochemicals such as catechins and gypenosides [97,98]. Furthermore, utilization and excretion of cholesterol was improved by palmatine (main alkaloids in *Coptis chinensis*) and jatrorrhizine (extracted from *Rhizoma coptidis*) through upregulation of CYP7A1 mRNA [99,100]. Another phytochemical called columbamine, obtained from *Rhizoma coptidis,* indirectly transactivated CYP7A1 by the stimulation of hepatocyte nuclear factor 4-alpha and fetoprotein transcription factor, resulting in enhanced cholesterol catabolism and bile acids secretion [101]. A study demonstrated that punicalagin and ellagic acid extracted from pomegranate enhanced cholesterol metabolism in human hepatocytes by activating the PPARγ-CYP7A1 signaling [102] (Figure 2).

3.4. Inhibition of Cholesterol Synthesis

The synthesis of cholesterol is regulated through an elegant system of feedback inhibition that senses intracellular cholesterol and eventually regulates several proteins participating in cholesterol homeostasis [103,104]. Squalene synthase (SQS) and HMGCR remain the crucial enzymes involved in cholesterol homeostasis, and the genes of these enzymes are regulated by SREBP-2 [105]. Likewise, AMP-activated protein kinase (AMPK) remains a key sensor in the regulation of lipid metabolism [106]. It was reported that alteration of AMPK restricted the rate of HMG-CoA expression, which, in turn, regulated the synthesis of cholesterol [107].

Studies have described several phytochemicals such as curcumin [108], leoligin [109], ODP-Ia [110], puerarin [111], and geraniol [112], which could suppress the synthesis of cholesterol through inhibition of HMGCR. Moreover, certain phytochemicals have been shown to inhibit synthesis of cholesterol by activating the AMPK [62–70]. SREBPs contribute to the intake of cholesterol and exert regulatory effect on genes encoding HMGCR [113]. Emodin has been reported to inhibit the transcription of SREBP-2 and subsequently suppress the biosynthesis of cholesterol [114]. T Grand-Perret et al. [115] reported that activation of SCAP/SREBP signaling pathway remarkably inhibited cholesterol biosynthesis. Another study reported that (−)-epicatechin and tetramethylpyrazine blocked the SCAP/SREBP-1c pathway, which resulted in the amelioration of atherosclerosis and lipid metabolism disorders [116,117]. Hassan Hajjaj and coworkers [118] reported that 26-oxygenosterol, obtained from *Ganoderma lucidum*, downregulated lanosterol 14 alpha-demethylase, which is responsible for the conversion of 24, 25-dihydrolanosterol into cholesterol. A study demonstrated SQS to be an attractive target for antihyperlipidemic drugs due to its contribution in cholesterol synthesis [119]. Yankun Chen et al. [120] reported that cynarin inhibited SQS and led to decrease the TG levels (Figure 4).

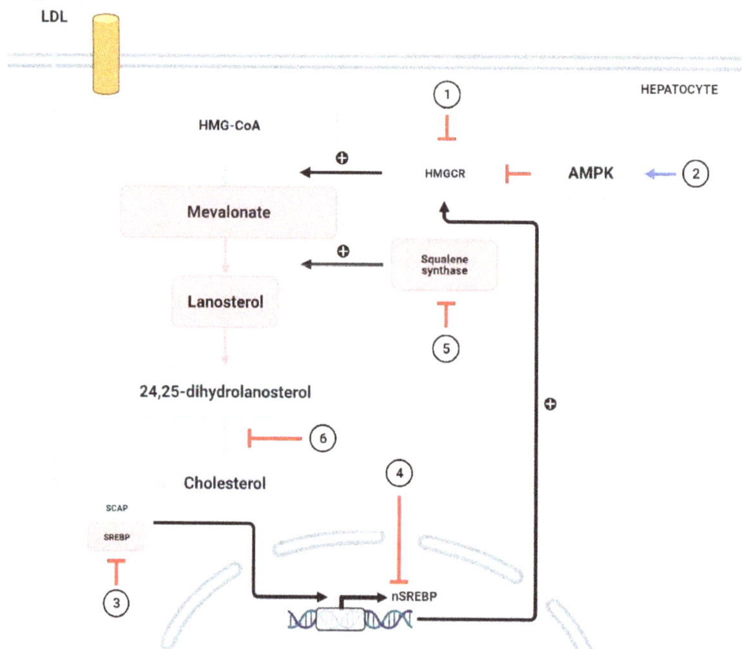

Figure 4. Phytochemicals inhibit the synthesis of cholesterol.

(**1**) HMGCR is blocked by curcumin [108], leoligin [109], ODP-Ia [110], puerarin [111], and geraniol [112]. (**2**) Activation of AMPK by phytochemicals such as curcumin, tanshinone IIA, and paeoniflorin leads to inhibition of cholesterol synthesis [62–70]. (**3**,**4**) Emodin inhibits the transcription of SREBP-2 [114], whereas (−)-epicatechin and tetramethylpyrazine block the SCAP/SREBP-1c pathway [116,117]. (**5**) 26-oxygenosterol promotes the activity of SQS and subsequently block cholesterol synthesis [118]. (**6**) Cynarin exerts inhibitory action in the process where 24,25-sdiydrolanosterol forms cholesterol [120]. SCAP = sterol regulatory element binding protein (SREBP) cleavage-activating protein.

4. Plant-Based Whole Foods Reducing LDL-C and Contributing to Prevent CVD

Low intake of fruits and vegetables was reportedly responsible for ~25.5 million premature deaths globally in 2013 [121]. Vegetables and fruits are good sources of various beneficial substances, such as dietary fiber, minerals, vitamins, and antioxidant entities, which can collectively reduce the risk of chronic disorders and total mortality and exert beneficial effects on the gut microbiota [122–124]. The intake of vegetables, fruits, and dietary fiber has been reported to have positive effects on serum cholesterol levels and platelet aggregation [125]. The results of a meta-analysis found that a daily increase of 200 g of fruits and vegetables reduced the relative risks of CHD, stroke, and CVD by 8–16%, 13–18%, and 8–13%, respectively [126]. Moreover, daily consumption of ~500 g of fruits and vegetables has been correlated with a 22% lower risk of CVD than with a dietary intake of 0–40 g/day [126].

4.1. Grapes (*Vitis vinifera*)

Grapes are rich sources of phenolic compounds, with an appreciable anthocyanin content of ~46% [127]. Although anthocyanins convey antioxidant, anti-inflammatory, antihypertensive, and antiplatelet activities [128], some studies have reported conflicting results regarding the effects on lipid profiles [129,130].

The effects of grape juice on dyslipidemia were studied using a mouse model homozygous for the absence of the LDLR gene (LDLR$^{-/-}$) and fed a hyperlipidemic diet. During this study, 30 male mice (12 weeks old) were assigned to one of three groups (10 mice/group): the HL group, which received a high-fat diet; the HLU group, fed a high-fat diet and grape juice (2 g/kg/day), and the HLS group, which received a high-fat diet along with simvastatin (20 mg/kg/day). Blood pressure, lipid levels, glycemic and insulinemic profiles, and C-reactive protein levels were determined. It was noted that the 60-day outcomes of the HLU and HLS group were similar, as the addition of grape juice diminished dyslipidemia and effectively elevated HDL-C levels. Moreover, left ventricular hypertrophy and arterial hypertension was prevented in the HLU group. These results suggest that dietary grape juice can potentially prevent CVD [131].

Various clinical trials have found that grape polyphenols are effective against cholesterolemia. For instance, van Mierlo et al. [132] reported that, compared with a placebo, the intake of grape polyphenols (800 mg/day) for 2 weeks led to decreased TC and TG levels. Similar outcomes were obtained by another investigation on 60 healthy volunteers who received 700 mg/day of a polyphenol-rich grape extract supplement for one month [133]. A study of 44 pre-or postmenopausal women found that dietary supplementation of lyophilized grape powder (39 g/day for 4 weeks) effectively reduced serum levels of LDL-C, ApoE, ApoB, and TGs [134]. Furthermore, administration of grape polyphenols for three weeks and consumption of red wine for one month was reported to reduce LDL-C levels and the risk of CVD [135,136].

4.2. Cranberries (*Vaccinium macrocarpon*)

Cranberries are a rich source of flavonoids (flavanols, flavan3-ols, and anthocyanins) and phenolic acids (ellagic, benzoic, and hydroxycinnamic acids), which contribute to reducing the risk of CVD through antioxidant, anti-inflammatory, and antithrombotic mechanisms [137]. Wilson et al. first reported the LDL-protective properties of cranberry juice (pressed berries), as 0.10% cranberry juice suppressed the formation of thiobarbituric acid reactive substances via Cu^{2+}-induced oxidation of LDL [138]. Another study reported that dietary intake of 2.8 mg/g of cold-pressed cranberry seed oil inhibited LDL oxidation [139]. Besides improving the resistance of LDL to oxidation, cranberry extract has also been shown to increase cholesterol uptake by HepG2 cells and to enhance the synthesis of LDL receptors, which resulted in accelerated cholesterol excretion in vivo [140]. In another study of the effects of cranberry juice powder on blood cholesterol levels, pigs with familial hypercholesterolemic (FH) were fed a diet supplemented with 47 g/day of citric acid and 57 g/day of fructose for two weeks. On day 15, 150 g/day of cranberry juice powder was added and continued for four weeks. Total blood cholesterol, HDL, and LDL levels were observed weekly. At baseline, LDL levels in the FH pigs were 11-fold greater than in normal pigs (428 vs. 37 mg/dL, respectively), whereas total blood cholesterol was sevenfold greater (458 vs. 67 mg/dL, respectively). At the end of the investigation, the LDL levels decreased to 94 mg/dL and total blood cholesterol to 92 mg/dL in the FH pigs. These findings indicate that cranberry juice powder can decrease cholesterol levels in hypercholesterolemic individuals [141].

4.3. Pomegranate (*Punica granatum*)

Pomegranates contain several potent antioxidants (anthocyanins and tannins), which act as effective anti-atherogenic agents [142]. Daily consumption of pomegranate juice has been shown to reduce serum LDL-C and TG levels and to increase HDL-C levels [143]. The considerable amounts of steroidal compounds in pomegranate seed oil were reported to decrease cholesterol levels [144]. A study of the effect of concentrated pomegranate juice on cholesterol profiles of type-2 diabetes patients with hyperlipidemia reported that daily intake of 40 g of pomegranate for 8 weeks effectively reduced TC and LDL-C levels, as well as the TC/HDL-C and LDL-C/HDL-C ratios [145]. Moreover, pomegranate juice was found to reduce LDL accumulation and increase HDL levels by 20% in humans, whereas a 90% decrease in LDL levels was noted in mice [146]. Al-Moraie et al. [147] reported that consumption of 1–5 mL/kg of pomegranate juice for 28 days effectively decreased LDL-C, VLDL-C, TC, and TG levels while elevating the expression of antioxidant enzymes and HDL-C levels. A study investigating the correlation of punicalagin (the main polyphenol in pomegranate) with ApoB100 that surrounds LDL particles showed that punicalagin bound to ApoB100 at low concentrations (0.25–4 μM) and stimulated LDL influx to macrophages (up to 2.5-fold) in a dose-dependent manner. The study further demonstrated that LDL influx to macrophages occurred specifically via the LDL receptor. The most important fact demonstrated by this investigation was that the interaction of punicalagin with LDL led specifically to LDL influx to the macrophages without their conversion into foam cells. The study concluded that upon binding to ApoB100, punicalagin induced LDL influx to macrophages, thereby decreasing circulating cholesterol levels [148].

4.4. Apple (*Malus domestica*)

The effect of apple polyphenols on blood lipid profile has been the focus of numerous studies. A study, while investigating the cholesterol-lowering effect of five different apple species (annurca apple, red delicious, Granny Smith, fuji, and golden delicious) in mildly hypercholesterolaemic healthy subjects, showed that annurca apples exerted the most significant effects; allowing a reduction in TC and LDL-C levels by 8.3% and 14.5%, respectively, while increased HDL-C level by 15.2% [149]. Another study observed the effects of eating two apples/day for four months on blood cholesterol levels. The results showed that blood cholesterol was lowered by 14.5% as compared with the untreated group, whereas HDL-C levels were increased by 15% [150]. However, other than a change in oxidized LDL levels, there was no reduction in blood cholesterol [151]. Another study of lyophilized apples found that administration of 0.21–1.43 g of polyphenols daily for one month did not improve the cardiovascular health of obese patients [152]. Similar results were obtained from another investigation where 300 g golden delicious apple per day for eight weeks increased the serum levels of VLDL and TG, but had no effect on TC, LDL-C, HDL-C, LDL/HDL ratio, and ApoB [153]. Hence, dietary supplementation of apple as whole fruit led to inconsistent results. Although the polyphenolic content of apples was found to lower blood cholesterol levels and LDL oxidation, these data are insufficient to conclude that apples, as a dietary nutraceutical, can lower plasma cholesterol [154,155].

4.5. Dried Nuts

Nuts are rich in polyunsaturated fatty acids, phytosterols, polyphenolics, and fiber [156]. Many studies have reported that daily consumption of nuts can improve cardiovascular health [157–159]. Almonds, pistachios, and walnuts are the most highly consumed nuts worldwide [160]. In a previous meta-analysis, of all edible nuts, pistachios were found to improve blood lipid profiles and effectively reduce serum TG, TC, and lipoprotein levels. Walnuts were second to pistachios in lowering TGs and cholesterol. Controlled levels of LDL, TGs, and TC are important serum markers of cardiovascular health [60,128,161]. Almonds were found to reduce LDL levels more effectively compared with other markers. Mechanisms responsible for the lipid-lowering abilities of nuts include reduced absorption of dietary cholesterol and increased bile production. Nuts are also reported to have an inhibitory effect against HMG-CoA, which is required for biosynthesis of cholesterol via acetyl CoA. Multiple bioactive constituents (phytosterols and fiber) may also individually convey cardiovascular benefits [162–164]. Ellagitannins and lutein in nuts have also been shown to reduce blood lipid levels [165,166]. At the cellular level, nuts have been reported to influence the expression levels of several miRNAs associated with lipid metabolism and uptake [167,168].

Several meta-analyses of the cardiovascular benefits of nuts have reported similar results. However, there have been conflicting reports, as one study claimed that walnuts and pistachios do not lower serum TG levels [169]. As a possible explanation for these conflicting findings, some studies evaluated nut-enriched foods containing other ingredients, such as skimmed milk and components of Mediterranean diets, which could influence the effects of nuts [169–172]. Nonetheless, the superiority of pistachios in lowering blood cholesterol is reportedly due to the greater content of β-carotene, γ-tocopherol and lutein [156].

The reported effects of nut-enriched diets on body weight varies among studies, as some reported a slight reduction in adipose tissue content in response to daily intake of nut-enriched foods [173], whereas others found that the high fat and calories of nuts can cause weight gain [160]. In a meta-analysis of 34 studies, only one reported significant weight loss, whereas the others found no significant correlation between a nut-enriched diet and body weight [174,175].

When compared with lipid-lowering drugs, such as statins, the effect of nuts on cholesterol seems to be quite modest. Moreover, the duration of most of the studies were relatively short (<six months), whereas dyslipidemia disorders are chronic. Hence, well-designed studies of larger populations for greater durations are required to accurately evaluate the effects of nut-enriched diets on cardiovascular health. Another concern is the reliance on self-reporting the frequency and quantity of nut-enriched diets, which may be misleading. Furthermore, many studies failed to report daily intake of nuts on regular basis. In addition, nuts, such as pecans, peanuts, and Brazil nuts, were often not included in the studies. Most meta-analyses did not conduct sensitivity analysis due to limited data. Hence, based on the available published studies with sensitivity analysis, diets enriched with pistachios and walnuts have more favorable lipid-lowering effects than other nut-enriched diets. However, the quality of evidence is debatable due to the several shortcomings addressed earlier [176].

4.6. Fruits of *Opuntia* spp.

Many studies have reported that consumption of various fruits of *Opuntia* spp. can significantly lower TC [177,178]. A recent study also reported the lipid-lowering effects of juice consumption at 150 mL/day for two weeks consecutively in both healthy and hyperlipidemic populations [179]. The processes thought to be involved in this effect was a reduction in fat absorption in the intestine, increased bile synthesis and secretion, and increased density of LDL receptors at cholesterol uptake sites. The high fiber content of fruits has been suggested to be the primary lipid-lowering ingredient. Reduced enterohepatic recirculation of bile is another factor responsible for indirect reduction in blood cholesterol levels [180]. Other studies have suggested that pectin derived from fruit also promotes the production of bile by increasing the biosynthesis of chenodeoxycholic acid and increased uptake of LDL from the blood. Other studies have reported such effects without reporting the quantity of daily fiber intake. Hence, due to the phytochemical constituents and fiber content, *Opuntia* spp. are good candidates for managing CVD [177].

4.7. Flax Seeds

Flax seeds are rich in dietary fiber, which consists of pentose- and hexose-based hydrophilic polymers, such as arabinoxylans, galactose, ketose (fructose), pectin, and omega-3 fatty acids, which form high consistency solutions in the gut [181]. Beverages containing flax fibers have been reported to reduce fasting levels of TC and LDL by 12% and 15%, respectively [182]. Another study reported that consumption of roasted flax seed powder for three months significantly reduced serum levels of TG, TC, VLDL, and LDL [183].

Lignans isolated from the flax seeds are being extensively studied for their hypocholesterolemic effects. A study was conducted for two months, in patients with high blood cholesterol, to observe the effects of administering the dietary secoisolariciresinol diglucoside (SDG) from flax seeds on lipid profile. The results exhibited that dose of 600 mg SDG was sufficient to reduce TC and LDL up to 24%. Authors concluded that flax seed lignans had significant anti hypercholestrolemic effects [184]. Another study was conducted which involved ingestion of whole flax and sunflower seeds by a special population of hypercholestrolemic postmenopausal women. Patients were given 38 g of one diet for six weeks followed by switching of diet for another six weeks. The washout period between two dietary regimens was a two-week interval. Flax seed diet was able to reduce TC and LDL up to 6.9% and 14.7% respectively. Marked reduction in the concentrations of lipoprotein A (7.4%) was also observed. Reduction in TC and LDL due to sunflower seeds was lower than flax seeds. The authors were of the view that the above effects of whole seeds of both plants were due to the presence of linoleic acids, and fibers present in adequate amounts [185]. Another study investigated the effects of ground flax seeds on the lipid profile of patients taking statins to control blood cholesterol. The randomised double blind study conducted for 12 months involved the administration of 30g ground seeds to 58 patients. A significant reduction in LDL (15%) was observed after one month of diet therapy. 11% reduction in TC was observed after 6 mon of flax seeds intake. The authors also reported fading of flax seeds effects on cholesterol lowering after 6 months of treatment. Combining flax seeds diet with statins caused a consistent reduction in LDL by 8.5% after 12 months [186].

4.8. Whole Grains

Whole grains, such as rice, corn, barley, and rye, are rich sources of fiber. An analysis of 64 studies [183] concluded that regular intake of oat reduced serum TG and LDL levels by up to 19% and 23%, respectively. Among these 64 studies, some also described the favorable HDL-elevating effects of oat. Another review of 24 studies reported that intake of whole grain foods also lowered serum TC and LDL levels, with the most dominant effect on TC [183].

Yet another study conducted on 12754 individuals to observe their dietary habits with respect to adequate intake of whole grains (more than 3 oz per day) and their daily dose of statins. The researchers reported that one fourth of total individuals were regular in consuming whole grains and statins for the total duration of study (12 months). They also reported that concomitant use of dietary and pharmacological means of cholesterol control was better than either of the two regimens used alone. Therefore, inclusion of whole grains in adequate amounts lessens the odds of patients in developing more severe forms of cardiovascular emergencies [187]. Just as two or more servings of refined sugars and carbohydrates have been associated with greater risk of developing CVDs, daily servings of whole grains (two or more) have been associated with 10–20% lower risk of developing CVDs [188]. Another study assessed the effect of whole grains by analysing the data from 24 studies. Whole grain intake caused reduction in triglycerides and cholesterol levels without affecting HDL [189]. A study assessed the effects of a diet containing whole grains, fruits and vegetables on CVD risk and weight gain. A total of 75 overweight women were randomized and given one of the above-mentioned diets for 10 weeks. The results, when examined after 10 weeks, showed that women who took whole grains for 10 weeks displayed better lipid control profile (lowered LDL), lower hypertension, and greater weight reduction than those who were given fruits and vegetables [190]. Another study tried to analyse the effect of specific grains against whole grains in reducing CVD risk. A prospective study conducted on 2329 old age (late 50s) individuals with previous history of myocardial infarction (MI). Examination of eating habits of these individuals followed by risk estimation using Cox proportional hazard model revealed that whole grains are effective in lowering the risk of future CVDs. Among individual grains, rye and oats were found to be specifically more effective than other individual grains in lowering risk of MI and other CVDs [191].

4.9. Soy Components

Soy proteins are one of the plant based food components approved by FDA for their health claim regarding their cholesterol reducing effects. Researchers studied 46 trials after the approval to judge if the health claim still holds value. Of 46, 41 had data on LDL-C while 43 had data on TC. Analysis showed that 25 mg daily dose of soy proteins for 6 weeks caused reduction in LDL-C and TC by 4.76 and 6.41 mg/dL respectively. This reduction amounted to 3–4% reduced LDL-C by consumption of soy proteins, in adults [192]. Combining phospholipids, soy fibers and proteins revealed a synergistic effect on lowering blood cholesterol than soy proteins alone [193]. Another study conducted in population of postmenopausal women, reported that soy phytosterols (4 g) taken in combination with fibers and soy proteins induce greater antihyperlipidemic effects than soy proteins alone [194].

Soy isoflavones have been reported to regulate lipogenesis, beta oxidation of fatty acids, and lipolysis by inhibiting the Akt/mTORC1 pathway [99]. In addition, isoflavones were reported to reduce serum TC and LDL levels by 1.7% and 3.6%, respectively, in hypercholesterolemic patients with a lesser effect in normal individuals [195]. Another study found greater lipid-lowering effects in earlier periods of treatment with overall reductions in serum TC, LDL, and TG levels of 3.7%, 5.25%, and 7.27%, respectively. Moreover, soy isoflavones were reported to increase the amount of beneficial HDL by 3.03% [196]. A meta-analysis confirmed that whole soya from four weeks to one year reduced more serum LDL compared with processed soya-extracts [197].

Consumption of 2 g/day of soybean leaf extract for 12 weeks was reported to reduce serum TG levels in overweight subjects with mildly high blood glucose levels [110], whereas another study showed that 70 mg/day for three months decreased serum TG levels in postmenopausal women [198].

Soy isoflavones include daidzein, genistein, and glycitein, which occur as glycosides. The linkage is broken by digestive enzymes to liberate the active aglycone portion of glycosides. A study of the lipid-lowering effects of daidzein (soy isoflavone) at 40 and 80 mg/day for six months found that the effects on TG levels were dose-dependent [199]. In the colon, daidzein is converted to S-equol. In a previous study, 10 mg/day of S-equol was found to reduce blood LDL levels and the cardio-ankle vascular index [200]. Although daidzein is metabolized in the colon, 30–50% of the Caucasian population had dissimilar results among the different groups. Likewise, genistein at 54 mg/day for 12 months increased HDL levels from 46.4 mg/dL to 56.8 mg/dL (22.4% increase) and decreased LDL levels from 108.8 mg/dL to 78.7 mg/dL (27.66% decrease) in postmenopausal women [201].

4.10. Vegetarian Diet

Vegetarian diets excludes meat, poultry and fish. Vegan diet, in addition to the exclusions of vegetarian diet, also excludes eggs and dairy products. Health benefits of vegetarian diets are due to the presence of adequate quantities of $n-6$ fatty acids, fiber, carotenoids, folate, ascorbic acid, vitamin E and magnesium. On the other hand, these diets lack sufficient protein, vitamin A, B12 and zinc. Vegan foods are low in calcium too due to absence of dairy products [202]. A scientific opinion suggested that a person does not need to become vegan or vegetarian to improve their cardiovascular health but slight changes in dietary components and habits might do the needful [203].

Reducing consumption of meat is thought to be correlated with improved cardiovascular health [204]. Vegetarian and Mediterranean-type diets have been recommended as health imparting diets by the American College of Cardiology/American Heart Association. The diets of Africans are largely plant-based, which have been associated with a lower prevalence of CVD [205]. A study conducted in the US reported a clear correlation between meat consumption and CVD [206]. Another significant study showed that a vegetarian diet reduces the risk of CVD by 32%, as vegetables have anti anti-inflammatory and antioxidant effects due to the prevalence of carotenoids, flavonoids, and other polyphenols [207]. A main drawback of such diets is the lower availability of proteins and certain minerals, especially zinc and calcium. Some researchers have argued that the presence of trypsin inhibitors (pulses, tomato), lyso-alanine (in processed vegetables), and glucosinolates in plant-based foods can also cause harm to the body due to the addition of non-vegetarian diets; thus, a slight reduction in consumption of vegetarian diets is recommended [208–210].

A meta-analytic study of 11 trials was conducted to observe the effects of vegetarian diets on lipid profile. Random effect model was used for determination of net changes in lipid profile. Vegetarian diets caused significant reduction in TC and LDL-C without having significant effect on lowering triglycerides [211]. A meta analytic and prospective study reported that vegetarian diets continued for 5 years or more reduced the risk of mortality from CHD by 24% [212]. A study conducted in Taiwanese patients to compare the effects of vegetarian diet against an omnivorous diet on CVDs reported that vegetarian diet was more effective in lowering TC and LDL-C. No effect was observed on HDL-C and TGs. However, homocysteine levels (indicative of folate and B12 deficiency) were higher in the vegetarian population [213].

5. Dietary Fiber

Fibers are carbohydrate polysaccharides that are not hydrolyzed by digestive enzymes in gastrointestinal tract. Low fiber intake has been associated with a higher risk of CVD. In contrast, fiber-rich foods, such as fruits and vegetables, have been associated with a low risk for CVD [214]. This correlation is attributed to the increased bile secretion induced by fibers, lower fatty acid biosynthesis, increased insulin sensitivity at target tissues, feeling satiated due to increased indigestible bulk in the intestine, lower cravings, and improved bowel habits [215,216]. Beta glucan has been shown to increase the consistency of stomach and intestinal contents and reduce enterohepatic recirculation of bile via adsorption of bile acids by viscose fiber dispersion and inhibition of micelle formation for lipid emulsification, which leads to indirect reduction in blood cholesterol levels (Figure 5). Moreover, pectin and guar gum have been reported to possess such effects [217]. A previous meta-analysis of 15 studies with a large combined cohort of ~1 million reported a correlation between mortality and fiber consumption and found that daily intake of ~30 g/day of dietary fiber reduced the risk of mortality due to CVDs by 23% [218]. The source of fiber also affects its physiological function [219], as soluble fibers have cholesterol-lowering effects, whereas insoluble fibers can prevent sudden increases in blood glucose levels [220]. Beta glucan is a soluble fiber found in oat and barley. Many studies have associated the intake of 3 g beta glucan per day with lower serum TG, cholesterol, and LDL levels (5–10%) [186]. In a murine model of hypertension, dietary fiber was found to play a role in the downregulation of early growth response 1, which is a transcription factor that regulates many genes and pathways involved in the progression of CVD [219,221]. Another meta-analysis reported that intake of 3 g/day of beta glucan contained in fibrous foods decreased LDL levels with no effect on HDL and TGs [222]. Another study of 2295 individuals over a period of ~5 years found that fibrous foods, such as pulses, fruits, and vegetables, reduced the risk of CVD [223].

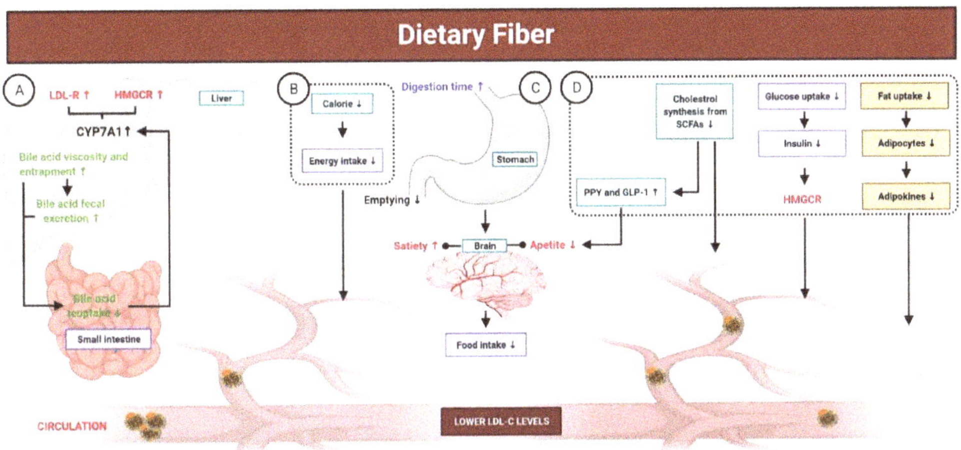

Figure 5. Mechanisms of dietary fiber in the gut. Dietary fibers are carbohydrate polysaccharides; not hydrolyzed by digestive enzymes in gastrointestinal tract and are associated with a low risk for CVD. (**A**) Dietary fibers elevate the fecal excretion of bile acid, decrease its re-uptake in the small intestine, and inhibit bile acid permeation. A reduced enterohepatic pool of bile acid stimulates the CYP7A (the rate-limiting enzyme involved in the production of bile acid), which, in turn, enhances liver uptake of LDL-C from blood through the upregulation of LDL-R and HMGCR. (**B**,**C**) Dietary fibers exhibit fewer calories and their consumption results in a prolonged digestion time with delayed gastric emptying. Additionally, it leads to an increase in bulk-forming and satiety, as well as viscosity-induced reduced absorption of cholesterol, which eventually lower the concentration of LDL-C. (**D**) Reduced cholesterol synthesis from SCFAs, produced by dietary fibers fermentation in the intestine, contributes to decrease the concentration of LDL-C. Certain SCFAs (propionate) enhances the release of PYY and GLP-1, both of which contribute to reduce LDL-C concentration. Additionally, dietary fibers intake leads to reduced fat uptake, altering the production of adipokines (TNF-α, resistin, and leptin), which play key roles in lipid metabolism and improving cholesterol concentration. The viscosity of dietary fibers reduces the intestinal absorption of glucose, leading to decreased secretion of insulin. Insulin is responsible for stimulating HMGCR; thus, lower insulin could decrease the concentration of LDL-C.

5.1. Epidemiological Evidence on Dietary Fiber and Health

Risk factors for CVD include obesity, diabetes, and hypertension. A prospective cohort study reported that consumption of 7 g/day of total fiber reduced the risk of CVD by 9% [224]. The same amount has been reported to decrease the risk of type-2 diabetes up to 6%, especially with whole grain foods [225]. Additionally, the study also revealed that three servings of whole grain bread reduced the risk of diabetes by 32% [225]. The ability of fibers (beta glucan) to reduce hypertension has also been observed in healthy individuals, although the reduction has been modest, as systolic pressure was reduced by 2·9 mmol Hg and diastolic pressure by 1·5 mm Hg [226]. Another systematic review deduced from various trials that the fibers content in a cereal bowl of porridge with an oat snack reduced blood cholesterol; 0·15 mmol in unclassified participants and 0·20 mmol in participants with hypercholesterolaemia [227].

Butyrate is used as an energy source by various bacteria for survival. Other SCFAs also affect fatty acid biosynthesis and oxidation. Once absorbed into the portal circulation, propionate and acetate can reduce hepatic synthesis of cholesterol and lipids [228]. Moreover, the receptor-mediated activation of SCFAs causes change in the activity of protein kinase A, which affects lipid metabolism. Propionate has also been reported to cause the release of peptide YY and glucagon-like peptide 1, which affects the metabolism and muscle uptake of free fatty acids and reduces the severity of hepatic steatosis. Adipokines are chemical messengers responsible for the metabolism of body fat and adipocytes. Fiber consumption has been reported to reduce the production of adipokines, which may result in lower production of fatty tissue. These chemical messengers are also involved in glycemic metabolic control [229,230].

5.2. Fiber Supplements

It has been estimated that risk of CHD in a population reduces by 2% for every 1% decrease in cholesterol [231]. Estimates indicate that in the two decades (1980–2000), health conscious behavior led to significant reductions in cholesterol intake that caused 33% reduction in mortality due to CHD in US [3,232]. Intake of dietary fibers in adequate amount is found in 5% adults. Therefore, supplements are required to provide concentrated form of fiber. Among many marketed products, psyllium, β-glucan and inulin are famous as natural supplements. Semisynthetic processed fibers include wheat dextrin, cyclodextrin and methylcellulose [233]. β-glucan containing foods can reduce the cholesterol levels of consumers. Viscous fibers such as those containing β-glucan have greater tendency to reduce cholesterol absorption. A study was conducted on 345 individuals to assess the effect of processed β-glucan supplement on LDL-C. Results depicted that administration of 3 g high viscosity β-glucan (achieved via low heat and pressurized processing) caused 5.5% reduction in LDL-C against placebo (wheat fiber) [234]. Cyclodextrin, which is produced from corn, is reported to reduce TC by 5%, LDL by 6.7%, and ApoB by 5.6%. This reduction in lipids occurs due to the presence of cyclodextrin in the diet without changing any other dietary component, although this effect was more pronounced in patients with high serum TG and cholesterol levels [235]. A study of the effect of 2 g/day of cyclodextrin on post prandial TG levels found that the dose was sufficient to reduce blood TG levels after a fatty meal [236].

6. Concluding Remarks and Future Directions

Scientists believe that CVD can be prevented through adoption of healthy lifestyle habits and optimal improvements in risk factors including high levels of LDL-C. Phytochemicals and plant-based whole foods can achieve better LDL-C control, and their addition is highly recommended to preventive strategies either for all subjects or more focused to patients with CVD. Our review focused the beneficial effects of phytochemicals and plant-based whole foods in the management of hyperlipidemia and CVD prevention.

Phytochemicals regulate the process of lipid metabolism and multi-target intervention, which occur in cholesterol biosynthesis, absorption, transport, and elimination. NPC1L1, HMGCR, PCSK9, and CYP7A1 remain the key molecules involved in these processes. In addition, during these processes, transcription factors such as LXR, SREBP, PPARα, and PPARγ actively participate and mediate the lipid-lowering actions of phytochemicals. The outcomes regarding the utilization of phytochemicals are promising and emphasize the potential indication of these phytochemicals in different categories of patients. The phytochemicals described in this review are diverse, including polyphenols (pomegranate), alkaloids (berberine), flavonoids (taxifolin, quercetin), and saponins (ginsenoside). These phytochemicals are mostly safe and well tolerated. Collectively, the phytochemicals can reduce the levels of LDL-C and prevent CVD by regulating different metabolic pathways.

Plant-based whole foods are known for their LDL-C lowering potential and cardiovascular health benefits [237,238]. These foods include vegetables, fruits, legumes, nuts, whole grains, dietary fibers, and seeds [157,237]. Plant-based whole foods are abundant in dietary fibers, unsaturated fatty acids, plant proteins, multiple micronutrients like vitamins, and phytochemicals like polyphenols and phytosterols [237]. These foods influence the development of CVD either directly or indirectly by different underlying mechanisms. For example, replacing saturated fats with unsaturated fatty acids in the diet is known to reduce LDL-C [239,240]. It has been reported in multiple observational studies and randomized controlled trials, that replacing saturated fatty acids with vegetable oil polyunsaturated fatty acids decreases the risk of CVD [239].

Dietary fibers, especially viscous soluble fibers like beta-glucan, inhibit the intestinal absorption of cholesterol and re-absorption of bile acids, producing SCFAs in the colon, which may influence hepatic cholesterol synthesis, and eventually result in LDL-C-lowering effect [241]. A significant correlation has been established between dietary fibers intake and lower risk of all-cause mortality [218] and mortality from CVD as well as CHD [218,241]. Phytosterols have also been shown to block the intestinal cholesterol absorption and reduce the concentrations of circulating LDL-C [242]. Siying S Li et al. [98] showed that replacing animal protein with plant protein resulted in lower LDL-C.

Of note, all plant-based whole foods are not necessarily effective in lowering CVD risk because not all plant-based whole foods exhibit beneficial CV effects [237,243]. As we mentioned earlier that golden delicious apple per day for eight weeks elevated the serum levels of VLDL and TG, and did not show considerable effect on TC, LDL-C, HDL-C, LDL/HDL ratio, and ApoB [153]. Similarly, foods pattern with more refined grains have been linked with a higher CVD risk [244]. Therefore, the quality of plant-based whole foods and food components are of prime importance.

Although plant-based foods are generally perceived positively because of their health benefits, there are also multiple barriers that hinder the switch to, and maintenance of a plant-based diet. Common hurdles encompass health concerns that plant-based foods may lack specific nutrients, enjoyment of eating meat and animal-based foods, and reluctance to change dietary behavior [245,246]. Similarly, there are certain other limitations and unresolved issues. For example, several bioactive components of plant-based foods have not yet been characterized; thus, further intensive investigations are needed. Moreover, the dosage of plant-based foods greatly varies, which subsequently leads to inconsistent outcomes. It is worth mentioning that plant-based foods can reduce LDL-C when utilized in defined effective nontoxic quantities. Hence, further studies are needed to optimize the doses of processed products, extracts, and compounds obtained from plant-based foods to enhance efficacy and minimize toxicity and adverse effects. Moreover, to obtain the maximum advantages of plant-based foods for the management of LDL-C and CVD, current knowledge about bioavailability and toxicity in human must be improved. The antihyperlipidemic effects of plant-based foods need to be investigated at the cellular and molecular levels in high-quality studies with systemic and in-depth analyses. Furthermore, large-scale clinical trials are also required to assure the hypolipidemic effects. Despite these shortcoming, gaps, and limitations that have to be addressed, plant-based foods are an effective strategy to manage LDL-C and CVD.

Author Contributions: Y.-S.L.: Conceptualization, Supervision; S.U.I. and H.A.: Writing—Original Draft; M.B.A.: Designed all the figures. All authors have read and agreed to the published version of the manuscript.

Funding: This research was funded by a National Research Foundation of Korea (NRF) grant from the Korean government (MSIT) (NRF- 2019R1A2C1003003).

Conflicts of Interest: The authors declare no conflict of interest, financial or otherwise.

Abbreviations

Acetyl CoA	Acetyl coenzyme A
Akt	Protein kinase B
ApoB	Apolipoprotein B
ApoB100	Apolipoprotein B100
ApoE	Apolipoprotein E
CHD	Coronary heart disease
CVD	Cardiovascular disease
CYP7A1	Cytochrome P450 Family 7 Subfamily A Member 1
ERK	Extracellular Signal-Regulated Kinase
GLP-1	Glucagon-like peptide-1
HDL	High-density lipoprotein
HDL-C	HDL cholesterol
HMG-CoA	β-hydroxy-β-methylglutaryl coenzyme A
HMGCR	β-hydroxy-β-methylglutaryl coenzyme A reductase
IDL	Intermediate-density lipoprotein
LDL	Low-density lipoprotein
LDL-C	LDL Cholesterol
LDLR	Low-density lipoprotein receptor
$LDLR^{-/-}$	Deleted low-density lipoprotein receptor
LXR	Liver X receptor
MKK3	Mitogen-activated protein kinase kinase 3
mTORC1	Mammalian target of rapamycin complex 1 or mechanistic target of rapamycin complex 1
MTTP	Microsomal triglyceride transfer protein large subunit
PPARα	Peroxisome proliferator-activated receptor alpha
PPARβ	Peroxisome proliferator-activated receptor beta
PPARγ	Peroxisome proliferator-activated receptor gamma
PPY	Peptide YY
ROS	Reactive oxygen species
SCFAs	Short-chain fatty acids
S-equol	7-Hydroxy-3-(4′-hydroxyphenyl)-chroman
SNP	Single-nucleotide polymorphism
SOD	Superoxide dismutase
SREBP-1c	Sterol regulatory element-binding protein-1c
TAK1	Mitogen-activated protein kinase kinase kinase 7
TC	Total cholesterol
TGTC	TriglyceridesTotal cholesterol
VLDLTG	Very low-density lipoproteinTriglycerides
VLDL	Very low-density lipoprotein

References

1. Saeed, A.; Feofanova, E.V.; Yu, B.; Sun, W.; Virani, S.S.; Nambi, V.; Coresh, J.; Guild, C.S.; Boerwinkle, E.; Ballantyne, C.M.; et al. Remnant-Like Particle Cholesterol, Low-Density Lipoprotein Triglycerides, and Incident Cardiovascular Disease. *J. Am. Coll. Cardiol.* **2018**, *72*, 156–169. [CrossRef]
2. Ramasamy, I. Recent advances in physiological lipoprotein metabolism. *Clin. Chem. Lab. Med.* **2014**, *52*, 1695–1727. [CrossRef] [PubMed]
3. Wadhera, R.K.; Steen, D.L.; Khan, I.; Giugliano, R.P.; Foody, J.M. A review of low-density lipoprotein cholesterol, treatment strategies, and its impact on cardiovascular disease morbidity and mortality. *J. Clin. Lipidol.* **2016**, *10*, 472–489. [CrossRef] [PubMed]
4. Yu, D.; Liao, J.K. Emerging views of statin pleiotropy and cholesterol lowering. *Cardiovasc. Res.* **2021**. [CrossRef] [PubMed]
5. Stamler, J.; Wentworth, D.; Neaton, J.D. Is relationship between serum cholesterol and risk of premature death from coronary heart disease continuous and graded? Findings in 356,222 primary screenees of the Multiple Risk Factor Intervention Trial (MRFIT). *JAMA* **1986**, *256*, 2823–2828. [CrossRef] [PubMed]
6. Borén, J.; Chapman, M.J.; Krauss, R.M.; Packard, C.J.; Bentzon, J.F.; Binder, C.J.; Daemen, M.J.; Demer, L.L.; Hegele, R.A.; Nicholls, S.J.; et al. Low-density lipoproteins cause atherosclerotic cardiovascular disease: Pathophysiological, genetic, and therapeutic insights: A consensus statement from the European Atherosclerosis Society Consensus Panel. *Eur. Heart J.* **2020**, *41*, 2313–2330. [CrossRef]

7. Véniant, M.M.; Sullivan, M.A.; Kim, S.K.; Ambroziak, P.; Chu, A.; Wilson, M.D.; Hellerstein, M.K.; Rudel, L.L.; Walzem, R.L.; Young, S.G. Defining the atherogenicity of large and small lipoproteins containing apolipoprotein B100. *J. Clin. Investig.* **2000**, *106*, 1501–1510. [CrossRef]
8. Lieu, H.D.; Withycombe, S.K.; Walker, Q.; Rong, J.X.; Walzem, R.L.; Wong, J.S.; Hamilton, R.L.; Fisher, E.A.; Young, S.G. Eliminating Atherogenesis in Mice by Switching Off Hepatic Lipoprotein Secretion. *Circulation* **2003**, *107*, 1315–1321. [CrossRef]
9. Wilson, P.W.F.; D'Agostino, R.B.; Levy, D.; Belanger, A.M.; Silbershatz, H.; Kannel, W.B. Prediction of Coronary Heart Disease Using Risk Factor Categories. *Circulation* **1998**, *97*, 1837–1847. [CrossRef] [PubMed]
10. Iqbal, D.; Khan, M.S.; Khan, M.S.; Ahmad, S.; Hussain, M.S.; Ali, M. Bioactivity guided fractionation and hypolipidemic property of a novel HMG-CoA reductase inhibitor from Ficus virens Ait. *Lipids Health Dis.* **2015**, *14*, 1–15. [CrossRef]
11. Group HPSC. MRC/BHF Heart Protection Study of antioxidant vitamin supplementation in 20 536 high-risk individuals: A randomised placebo-controlled trial. *Lancet* **2002**, *360*, 23–33. [CrossRef]
12. Sacks, F.M.; Tonkin, A.M.; Shepherd, J.; Braunwald, E.; Cobbe, S.; Hawkins, C.M.; Keech, A.; Packard, C.; Simes, J.; Byington, R.; et al. Effect of pravastatin on coronary disease events in subgroups defined by coronary risk factors: The Prospective Pravastatin Pooling Project. *Circulation* **2000**, *102*, 1893–1900. [CrossRef] [PubMed]
13. Unit ES. Efficacy and safety of cholesterol-lowering treatment: Prospective meta-analysis of data from 90 056 participants in 14 randomised trials of statins. *Lancet* **2005**, *366*, 1267–1278.
14. Poess, J.; Boehm, M.; Laufs, U. Are the guidelines correct? Should all patients with coronary heart disease or diabetes be treated with a statin? *Med. Klin. Munich Ger. 1983* **2009**, *104*, 74–78.
15. Baigent, C.; Blackwell, L.; Emberson, J.; Holland, L.E.; Reith, C.; Bhala, N.; Peto, R.; Barnes, E.H.; Keech, A.; Simes, J.; et al. *Efficacy and Safety of More Intensive Lowering of LDL Cholesterol: A Meta-Analysis of Data from 170,000 Participants in 26 Randomised Trials*; Elsevier: Amsterdam, The Netherlands, 2010.
16. National Cholesterol Education Program (U.S.) Expert Panel on Detection, Evaluation, and Treatment of High Blood Cholesterol in Adults (Adult Treatment Panel III): "Third Report of the National Cholesterol Education Program (NCEP) Expert Panel on Detection, Evaluation, and Treatment of High Blood Cholesterol in Adults (Adults Treatment Panel III) Final report". *Circulation* **2002**, *106*, 3143–3421.
17. Willer, C.J.; Sanna, S.; Jackson, A.U.; Scuteri, A.; Bonnycastle, L.L.; Clarke, R.; Heath, S.C.; Timpson, N.J.; Najjar, S.S.; Stringham, H.M.; et al. Newly identified loci that influence lipid concentrations and risk of coronary artery disease. *Nat. Genet.* **2008**, *40*, 161–169. [CrossRef] [PubMed]
18. Kathiresan, S.; Melander, O.; Anevski, D.; Guiducci, C.; Burtt, N.P.; Roos, C.; Hirschhorn, J.N.; Berglund, G.; Hedblad, B.; Groop, L.; et al. Polymorphisms Associated with Cholesterol and Risk of Cardiovascular Events. *N. Engl. J. Med.* **2008**, *358*, 1240–1249. [CrossRef]
19. Investigators MIGC. Inactivating mutations in NPC1L1 and protection from coronary heart disease. *N. Engl. J. Med.* **2014**, *371*, 2072–2082. [CrossRef]
20. Blood, I.; Crosby, J.; Peloso, G.M.; Auer, P.L.; Crosslin, D.R.; Stitziel, N.O.; Lange, L.A.; Lu, Y. TG, HDL Working Group of the Exome Sequencing Project NH, Lung, Institute B. Loss-of-function mutations in APOC3, triglycerides, and coronary disease. *N. Engl. J. Med.* **2014**, *371*, 22–31.
21. Kumar, A.; Mosa, K.A.; Ji, L.; Kage, U.; Dhokane, D.; Karre, S.; Madalageri, D.; Pathania, N. Metabolomics-assisted biotechnological interventions for developing plant-based functional foods and nutraceuticals. *Crit. Rev. Food Sci. Nutr.* **2018**, *58*, 1791–1807. [CrossRef]
22. Mahamuni, S.P.; Khose, R.D.; Menaa, F.; Badole, S.L. Therapeutic approaches to drug targets in hyperlipidemia. *BioMedicine* **2012**, *2*, 137–146. [CrossRef]
23. Hlaing, T.T.; Park, A. Hyperlipidaemia. *Medicine* **2013**, *41*, 607–609. [CrossRef]
24. George, V.C.; Dellaire, G.; Rupasinghe, H.V. Plant flavonoids in cancer chemoprevention: Role in genome stability. *J. Nutr. Biochem.* **2017**, *45*, 1–14. [CrossRef] [PubMed]
25. Rosenson, R.S.; Brewer, H.B., Jr.; Davidson, W.S.; Fayad, Z.A.; Fuster, V.; Goldstein, J.; Hellerstein, M.; Jiang, X.C.; Phillips, M.C.; Rader, D.J.; et al. Cholesterol efflux and atheroprotection: Advancing the concept of reverse cholesterol transport. *Circulation* **2012**, *125*, 1905–1919. [CrossRef]
26. Rader, D.J.; Alexander, E.T.; Weibel, G.L.; Billheimer, J.; Rothblat, G.H. The role of reverse cholesterol transport in animals and humans and relationship to atherosclerosis. *J. Lipid Res.* **2009**, *50*, S189–S194. [CrossRef] [PubMed]
27. Favari, E.; Chroni, A.; Tietge, U.J.; Zanotti, I.; Bernini, F. Cholesterol Efflux and Reverse Cholesterol Transport. *High Density Lipoproteins* **2015**, *224*, 181–206. [CrossRef]
28. Feig, J.E.; Hewing, B.; Smith, J.D.; Hazen, S.L.; Fisher, E.A. High-density lipoprotein and atherosclerosis regression: Evidence from preclinical and clinical studies. *Circ. Res.* **2014**, *114*, 205–213. [CrossRef] [PubMed]
29. Yvan-Charvet, L.; Wang, N.; Tall, A.R. Role of HDL, ABCA1, and ABCG1 Transporters in Cholesterol Efflux and Immune Responses. *Arter. Thromb. Vasc. Biol.* **2010**, *30*, 139–143. [CrossRef]
30. Du, X.-M.; Kim, M.-J.; Hou, L.; Le Goff, W.; Chapman, M.J.; Van Eck, M.; Curtiss, L.K.; Burnett, J.R.; Cartland, S.P.; Quinn, C.M.; et al. HDL Particle Size Is a Critical Determinant of ABCA1-Mediated Macrophage Cellular Cholesterol Export. *Circ. Res.* **2015**, *116*, 1133–1142. [CrossRef]

31. Yue, P.; Chen, Z.; Nassir, F.; Bernal-Mizrachi, C.; Finck, B.; Azhar, S.; Abumrad, N.A. Enhanced hepatic apoA-I secretion and peripheral efflux of cholesterol and phospholipid in CD36 null mice. *PLoS ONE* **2010**, *5*, e9906. [CrossRef]
32. Mody, P.; Joshi, P.H.; Khera, A.; Ayers, C.R.; Rohatgi, A. Beyond coronary calcification, family history, and C-reactive protein: Cholesterol efflux capacity and cardiovascular risk prediction. *J. Am. Coll. Cardiol.* **2016**, *67*, 2480–2487. [CrossRef] [PubMed]
33. Ye, D.; Lammers, B.; Zhao, Y.; Meurs, I.; Van Berkel, T.J.C.; Van Eck, M. ATP-binding cassette transporters A1 and G1, HDL metabolism, cholesterol efflux, and inflammation: Important targets for the treatment of atherosclerosis. *Curr. Drug Targets* **2011**, *12*, 647–660. [CrossRef] [PubMed]
34. Ye, G.; Chen, G.; Gao, H.; Lin, Y.; Liao, X.; Zhang, H.; Liu, X.; Chi, Y.; Huang, Q.; Zhu, H.; et al. Resveratrol inhibits lipid accumulation in the intestine of atherosclerotic mice and macrophages. *J. Cell. Mol. Med.* **2019**, *23*, 4313–4325. [CrossRef] [PubMed]
35. Li, C.H.; Gong, D.; Chen, L.Y.; Zhang, M.; Xia, X.D.; Cheng, H.P.; Huang, C.; Zhao, Z.W.; Zheng, X.L.; Tang, X.E.; et al. Puerarin promotes ABCA1-mediated cholesterol efflux and decreases cellular lipid accumulation in THP-1 macrophages. *Eur. J. Pharmacol.* **2017**, *811*, 74–86. [CrossRef] [PubMed]
36. Jiang, T.; Ren, K.; Chen, Q.; Li, H.; Yao, R.; Hu, H.; Lv, Y.-C.; Zhao, G.-J. Leonurine Prevents Atherosclerosis Via Promoting the Expression of ABCA1 and ABCG1 in a Pparγ/Lxrα Signaling Pathway-Dependent Manner. *Cell. Physiol. Biochem.* **2017**, *43*, 1703–1717. [CrossRef] [PubMed]
37. Francisco, V.; Figueirinha, A.; Costa, G.; Liberal, J.; Ferreira, I.; Lopes, M.C.; García-Rodríguez, C.; Cruz, M.T.; Batista, M.T. The Flavone Luteolin Inhibits Liver X Receptor Activation. *J. Nat. Prod.* **2016**, *79*, 1423–1428. [CrossRef]
38. Lin, H.-C.; Lii, C.-K.; Chen, H.-C.; Lin, A.-H.; Yang, Y.-C.; Chen, H.-W. Andrographolide Inhibits Oxidized LDL-Induced Cholesterol Accumulation and Foam Cell Formation in Macrophages. *Am. J. Chin. Med.* **2018**, *46*, 87–106. [CrossRef]
39. Wang, L.; Ladurner, A.; Latkolik, S.; Schwaiger, S.; Linder, T.; Hošek, J.; Palme, V.; Schilcher, N.; Polanský, O.; Heiss, E.H.; et al. Leoligin, the Major Lignan from Edelweiss (Leontopodium nivale subsp. alpinum), Promotes Cholesterol Efflux from THP-1 Macrophages. *J. Nat. Prod.* **2016**, *79*, 1651–1657. [CrossRef]
40. Wang, S.; Zhang, X.; Liu, M.; Luan, H.; Ji, Y.; Guo, P.; Wu, C. Chrysin inhibits foam cell formation through promoting cholesterol efflux from RAW264.7 macrophages. *Pharm. Biol.* **2015**, *53*, 1481–1487. [CrossRef]
41. Lin, X.-L.; Hu, H.-J.; Liu, Y.-B.; Hu, X.-M.; Fan, X.-J.; Zou, W.-W.; Pan, Y.-Q.; Zhou, W.-Q.; Peng, M.-W.; Gu, C.-H. Allicin induces the upregulation of ABCA1 expression via PPARγ/LXRα signaling in THP-1 macrophage-derived foam cells. *Int. J. Mol. Med.* **2017**, *39*, 1452–1460. [CrossRef]
42. Chawla, A.; Boisvert, W.A.; Lee, C.-H.; Laffitte, B.A.; Barak, Y.; Joseph, S.B.; Liao, D.; Nagy, L.; Edwards, P.A.; Curtiss, L.K.; et al. A PPARγ-LXR-ABCA1 Pathway in Macrophages Is Involved in Cholesterol Efflux and Atherogenesis. *Mol. Cell* **2001**, *7*, 161–171. [CrossRef]
43. Ren, K.; Jiang, T.; Zhao, G.-J. Quercetin induces the selective uptake of HDL-cholesterol via promoting SR-BI expression and the activation of the PPARγ/LXRα pathway. *Food Funct.* **2018**, *9*, 624–635. [CrossRef] [PubMed]
44. Sun, L.; Li, E.; Wang, F.; Wang, T.; Qin, Z.; Niu, S.; Qiu, C. Quercetin increases macrophage cholesterol efflux to inhibit foam cell formation through activating PPARγ-ABCA1 pathway. *Int. J. Clin. Exp. Pathol.* **2015**, *8*, 10854.
45. Chang, Y.-C.; Lee, T.-S.; Chiang, A.-N. Quercetin enhances ABCA1 expression and cholesterol efflux through a p38-dependent pathway in macrophages. *J. Lipid Res.* **2012**, *53*, 1840–1850. [CrossRef] [PubMed]
46. Li, X.-Y.; Kong, L.-X.; Li, J.; He, H.-X.; Zhou, Y.-D. Kaempferol suppresses lipid accumulation in macrophages through the downregulation of cluster of differentiation 36 and the upregulation of scavenger receptor class B type I and ATP-binding cassette transporters A1 and G1. *Int. J. Mol. Med.* **2013**, *31*, 331–338. [CrossRef] [PubMed]
47. Yang, H.; Yan, L.; Qian, P.; Duan, H.; Wu, J.; Li, B.; Wang, S.; Wang, S. Icariin Inhibits Foam Cell Formation by Down-Regulating the Expression of CD36 and Up-Regulating the Expression of SR-BI. *J. Cell. Biochem.* **2015**, *116*, 580–588. [CrossRef] [PubMed]
48. Li, X.; Zhou, Y.; Yu, C.; Yang, H.; Zhang, C.; Ye, Y.; Xiao, S. Paeonol suppresses lipid accumulation in macrophages via upregulation of the ATP-binding cassette transporter A1 and downregulation of the cluster of differentiation 36. *Int. J. Oncol.* **2015**, *46*, 764–774. [CrossRef] [PubMed]
49. Zhang, H.; Zhai, Z.; Zhou, H.; Li, Y.; Li, X.; Lin, Y.; Li, W.; Shi, Y.; Zhou, M.S. Puerarin inhibits oxLDL-induced macrophage activation and foam cell formation in human THP1 macrophage. *BioMed Res. Int.* **2015**, *2015*. [CrossRef]
50. Bao, Y.; Wang, L.; Xu, Y.; Yang, Y.; Wang, L.; Si, S.; Cho, S.; Hong, B. Salvianolic acid B inhibits macrophage uptake of modified low density lipoprotein (mLDL) in a scavenger receptor CD36-dependent manner. *Atherosclerosis* **2012**, *223*, 152–159. [CrossRef]
51. Li, J.; Xie, Z.-Z.; Tang, Y.-B.; Zhou, J.-G.; Guan, Y.-Y. Ginsenoside-Rd, a purified component from panax notoginseng saponins, prevents atherosclerosis in apoE knockout mice. *Eur. J. Pharmacol.* **2011**, *652*, 104–110. [CrossRef]
52. Acuña-Aravena, M.; Cohen, D.E. Lipoprotein Metabolism and Cholesterol Balance. *Liver Biol. Pathobiol.* **2020**, *2020*, 255–267. [CrossRef]
53. Ganjali, S.; Blesso, C.N.; Banach, M.; Pirro, M.; Majeed, M.; Sahebkar, A. Effects of curcumin on HDL functionality. *Pharmacol. Res.* **2017**, *119*, 208–218. [CrossRef]
54. Rotimi, S.O.; Adelani, I.B.; Bankole, G.E.; Rotimi, O.A. Naringin enhances reverse cholesterol transport in high fat/low streptozocin induced diabetic rats. *Biomed. Pharmacother.* **2018**, *101*, 430–437. [CrossRef] [PubMed]

55. Chapman, M.J.; Le Goff, W.; Guerin, M.; Kontush, A. Cholesteryl ester transfer protein: At the heart of the action of lipid-modulating therapy with statins, fibrates, niacin, and cholesteryl ester transfer protein inhibitors. *Eur. Heart J.* **2010**, *31*, 149–164. [CrossRef]
56. Shrestha, S.; Wu, B.J.; Guiney, L.; Barter, P.J.; Rye, K.-A. Cholesteryl ester transfer protein and its inhibitors. *J. Lipid Res.* **2018**, *59*, 772–783. [CrossRef]
57. Qin, Y.; Xia, M.; Ma, J.; Hao, Y.; Liu, J.; Mou, H.; Cao, L.; Ling, W. Anthocyanin supplementation improves serum LDL- and HDL-cholesterol concentrations associated with the inhibition of cholesteryl ester transfer protein in dyslipidemic subjects. *Am. J. Clin. Nutr.* **2009**, *90*, 485–492. [CrossRef] [PubMed]
58. Lagace, T.A. PCSK9 and LDLR degradation: Regulatory mechanisms in circulation and in cells. *Curr. Opin. Lipidol.* **2014**, *25*, 387. [CrossRef]
59. Li, H.; Dong, B.; Park, S.W.; Lee, H.-S.; Chen, W.; Liu, J. Hepatocyte Nuclear Factor 1α Plays a Critical Role in PCSK9 Gene Transcription and Regulation by the Natural Hypocholesterolemic Compound Berberine. *J. Biol. Chem.* **2009**, *284*, 28885–28895. [CrossRef]
60. Jia, Y.-J.; Xu, R.-X.; Sun, J.; Tang, Y.; Li, J.-J. Enhanced circulating PCSK9 concentration by berberine through SREBP-2 pathway in high fat diet-fed rats. *J. Transl. Med.* **2014**, *12*, 103. [CrossRef]
61. Cao, S.; Xu, P.; Yan, J.; Liu, H.; Liu, L.; Cheng, L.; Qiu, F.; Kang, N. Berberrubine and its analog, hydroxypropyl-berberrubine, regulate LDLR and PCSK9 expression via the ERK signal pathway to exert cholesterol-lowering effects in human hepatoma HepG2 cells. *J. Cell. Biochem.* **2019**, *120*, 1340–1349. [CrossRef]
62. Nhoek, P.; Chae, H.-S.; Masagalli, J.N.; Mailar, K.; Pel, P.; Kim, Y.-M.; Choi, W.J.; Chin, Y.-W. Discovery of Flavonoids from Scutellaria baicalensis with Inhibitory Activity against PCSK 9 Expression: Isolation, Synthesis and Their Biological Evaluation. *Molecules* **2018**, *23*, 504. [CrossRef] [PubMed]
63. Tai, M.-H.; Chen, P.-K.; Chen, P.-Y.; Wu, M.-J.; Ho, C.-T.; Yen, J.-H. Curcumin enhances cell-surface LDLR level and promotes LDL uptake through downregulation of PCSK9 gene expression in HepG2 cells. *Mol. Nutr. Food Res.* **2014**, *58*, 2133–2145. [CrossRef] [PubMed]
64. Chen, H.-C.; Chen, P.-Y.; Wu, M.-J.; Tai, M.-H.; Yen, J.-H. Tanshinone IIA Modulates Low Density Lipoprotein Uptake via Down-Regulation of PCSK9 Gene Expression in HepG2 Cells. *PLoS ONE* **2016**, *11*, e0162414. [CrossRef] [PubMed]
65. Wang, H.; Eckel, R.H. Lipoprotein lipase: From gene to obesity. *Am. J. Physiol. Metab.* **2009**, *297*, E271–E288. [CrossRef] [PubMed]
66. Bos, G.; Snijder, M.B.; Nijpels, G.; Dekker, J.M.; Stehouwer, C.D.; Bouter, L.M.; Heine, R.J.; Jansen, H. Opposite Contributions of Trunk and Leg Fat Mass with Plasma Lipase Activities: The Hoorn Study. *Obes. Res.* **2005**, *13*, 1817–1823. [CrossRef]
67. Xiao, H.-B.; Liang, L.; Luo, Z.-F.; Sun, Z.-L. Paeoniflorin regulates GALNT2-ANGPTL3-LPL pathway to attenuate dyslipidemia in mice. *Eur. J. Pharmacol.* **2018**, *836*, 122–128. [CrossRef]
68. Zhang, Y.; Xie, M.L.; Zhu, L.J.; Gu, Z.L. Therapeutic effect of osthole on hyperlipidemic fatty liver in rats 3. *Acta Pharmacol. Sin.* **2007**, *28*, 398–403. [CrossRef]
69. Sesorova, I.S.; Dimov, I.D.; Kashin, A.D.; Sesorov, V.V.; Karelina, N.R.; Zdorikova, M.A.; Beznoussenko, G.V.; Mironov, A.A. Cellular and sub-cellular mechanisms of lipid transport from gut to lymph. *Tissue Cell* **2021**, *72*, 101529. [CrossRef]
70. Altmann, S.W.; Davis, H.R.; Zhu, L.-J.; Yao, X.; Hoos, L.M.; Tetzloff, G.; Iyer, S.P.N.; Maguire, M.; Golovko, A.; Zeng, M.; et al. Niemann-Pick C1 Like 1 Protein Is Critical for Intestinal Cholesterol Absorption. *Science* **2004**, *303*, 1201–1204. [CrossRef]
71. Lee, R.G.; Willingham, M.C.; Davis, M.A.; Skinner, K.A.; Rudel, L.L. Differential expression of ACAT1 and ACAT2 among cells within liver, intestine, kidney, and adrenal of nonhuman primates. *J. Lipid Res.* **2000**, *41*, 1991–2001. [CrossRef]
72. Sirwi, A.; Hussain, M.M. Lipid transfer proteins in the assembly of apoB-containing lipoproteins. *J. Lipid Res.* **2018**, *59*, 1094–1102. [CrossRef] [PubMed]
73. Davidson, M.H.; Voogt, J.; Luchoomun, J.; Decaris, J.; Killion, S.; Boban, D.; Glass, A.; Mohammad, H.; Lu, Y.; Villegas, D.; et al. Inhibition of intestinal cholesterol absorption with ezetimibe increases components of reverse cholesterol transport in humans. *Atherosclerosis* **2013**, *230*, 322–329. [CrossRef] [PubMed]
74. Pirillo, A.; Catapano, A.L.; Norata, G.D. Niemann-Pick C1-Like 1 (NPC1L1) inhibition and cardiovascular diseases. *Curr. Med. Chem.* **2016**, *23*, 983–999. [CrossRef] [PubMed]
75. Wang, D.Q.-H. Regulation of Intestinal Cholesterol Absorption. *Annu. Rev. Physiol.* **2007**, *69*, 221–248. [CrossRef]
76. Alrefai, W.A.; Annaba, F.; Sarwar, Z.; Dwivedi, A.; Saksena, S.; Singla, A.; Dudeja, P.K.; Gill, R.K. Modulation of human Niemann-Pick C1-like 1 gene expression by sterol: Role of sterol regulatory element binding protein 2. *Am. J. Physiol.-Gastrointest. Liver Physiol.* **2007**, *292*, G369–G376. [CrossRef]
77. Duval, C.; Touche, V.; Tailleux, A.; Fruchart, J.-C.; Fievet, C.; Clavey, V.; Staels, B.; Lestavel, S. Niemann–Pick C1 like 1 gene expression is down-regulated by LXR activators in the intestine. *Biochem. Biophys. Res. Commun.* **2006**, *340*, 1259–1263. [CrossRef]
78. Feng, D.; Ohlsson, L.; Duan, R.-D. Curcumin inhibits cholesterol uptake in Caco-2 cells by down-regulation of NPC1L1 expression. *Lipids Health Dis.* **2010**, *9*, 40. [CrossRef]
79. Feng, D.; Zou, J.; Zhang, S.; Li, X.; Lu, M. Hypocholesterolemic Activity of Curcumin Is Mediated by Down-regulating the Expression of Niemann-Pick C1-like 1 in Hamsters. *J. Agric. Food Chem.* **2017**, *65*, 276–280. [CrossRef]
80. Kumar, P.; Malhotra, P.; Ma, K.; Singla, A.; Hedroug, O.; Saksena, S.; Dudeja, P.K.; Gill, R.K.; Alrefai, W.A. SREBP2 mediates the modulation of intestinal NPC1L1 expression by curcumin. *Am. J. Physiol.-Gastrointest. Liver Physiol.* **2011**, *301*, G148–G155. [CrossRef]

81. Zou, J.; Feng, D. Lycopene reduces cholesterol absorption through the downregulation of Niemann-Pick C1-like 1 in Caco-2 cells. *Mol. Nutr. Food Res.* **2015**, *59*, 2225–2230. [CrossRef]
82. Lee, C.-L.; Wen, J.-Y.; Hsu, Y.-W.; Pan, T.-M. Monascus-Fermented Yellow Pigments Monascin and Ankaflavin Showed Antiobesity Effect via the Suppression of Differentiation and Lipogenesis in Obese Rats Fed a High-Fat Diet. *J. Agric. Food Chem.* **2013**, *61*, 1493–1500. [CrossRef] [PubMed]
83. Afonso, M.S.; Machado, R.M.; Lavrador, M.S.; Quintao, E.C.R.; Moore, K.J.; Lottenberg, A.M. Molecular Pathways Underlying Cholesterol Homeostasis. *Nutrients* **2018**, *10*, 760. [CrossRef] [PubMed]
84. Lin, Y.; Vermeer, M.A.; Trautwein, E.A. Triterpenic Acids Present in Hawthorn Lower Plasma Cholesterol by Inhibiting Intestinal ACAT Activity in Hamsters. *Evid.-Based Complement. Altern. Med.* **2011**, *2011*, 1–9. [CrossRef] [PubMed]
85. Wang, Y.; Yi, X.; Ghanam, K.; Zhang, S.; Zhao, T.; Zhu, X. Berberine decreases cholesterol levels in rats through multiple mechanisms, including inhibition of cholesterol absorption. *Metabolism* **2014**, *63*, 1167–1177. [CrossRef] [PubMed]
86. Wilcox, L.J.; Borradaile, N.M.; De Dreu, L.E.; Huff, M.W. Secretion of hepatocyte apoB is inhibited by the flavonoids, naringenin and hesperetin, via reduced activity and expression of ACAT2 and MTP. *J. Lipid Res.* **2001**, *42*, 725–734. [CrossRef]
87. Casaschi, A.; Wang, Q.; Richards, A.; Theriault, A. Intestinal apolipoprotein B secretion is inhibited by the flavonoid quercetin: Potential role of microsomal triglyceride transfer protein and diacylglycerol acyltransferase. *Lipids* **2002**, *37*, 647–652. [CrossRef] [PubMed]
88. Casaschi, A.; Rubio, B.K.; Maiyoh, G.K.; Theriault, A.G. Inhibitory activity of diacylglycerol acyltransferase (DGAT) and microsomal triglyceride transfer protein (MTP) by the flavonoid, taxifolin, in HepG2 cells: Potential role in the regulation of apolipoprotein B secretion. *Atherosclerosis* **2004**, *176*, 247–253. [CrossRef] [PubMed]
89. Kurowska, E.M.; Manthey, J.A.; Casaschi, A.; Theriault, A.G. Modulation of hepG2 cell net apolipoprotein B secretion by the citrus polymethoxyflavone, tangeretin. *Lipids* **2004**, *39*, 143–151. [CrossRef]
90. Vallianou, I.; Hadzopoulou-Cladaras, M. Camphene, a Plant Derived Monoterpene, Exerts Its Hypolipidemic Action by Affecting SREBP-1 and MTP Expression. *PLoS ONE* **2016**, *11*, e0147117. [CrossRef]
91. Kang, Y.-J.; Jin, U.-H.; Chang, H.-W.; Son, J.-K.; Lee, S.H.; Son, K.-H.; Chang, Y.-C.; Lee, Y.-C.; Kim, C.-H. Inhibition of microsomal triglyceride transfer protein expression and atherogenic risk factor apolipoprotein B100 secretion by tanshinone IIA in HepG2 cells. *Phytotherapy Res. Int. J. Devoted Pharmacol. Toxicol. Eval. Nat. Prod. Deriv.* **2008**, *22*, 1640–1645. [CrossRef] [PubMed]
92. Lin, Y.; Vermeer, M.A.; Bos, W.; Van Buren, L.; Schuurbiers, E.; Miret-Catalan, S.; Trautwein, E.A. Molecular Structures of Citrus Flavonoids Determine Their Effects on Lipid Metabolism in HepG2 Cells by Primarily Suppressing ApoB Secretion. *J. Agric. Food Chem.* **2011**, *59*, 4496–4503. [CrossRef] [PubMed]
93. Norikura, T.; Mukai, Y.; Fujita, S.; Mikame, K.; Funaoka, M.; Sato, S. Lignophenols Decrease Oleate-Induced Apolipoprotein-B Secretion in HepG2 Cells. *Basic Clin. Pharmacol. Toxicol.* **2010**, *107*, 813–817. [CrossRef] [PubMed]
94. Wang, D.Q.H.; Portincasa, P.; Tso, P. Transintestinal cholesterol excretion: A secondary, nonbiliary pathway contributing to reverse cholesterol transport. *Hepatology* **2017**, *66*, 1337–1340. [CrossRef] [PubMed]
95. Pullinger, C.R.; Eng, C.; Salen, G.; Shefer, S.; Batta, A.K.; Erickson, S.K.; Verhagen, A.; Rivera, C.R.; Mulvihill, S.J.; Malloy, M.J.; et al. Human cholesterol 7α-hydroxylase (CYP7A1) deficiency has a hypercholesterolemic phenotype. *J. Clin. Investig.* **2002**, *110*, 109–117. [CrossRef] [PubMed]
96. Li, T.; Matozel, M.; Boehme, S.; Kong, B.; Nilsson, L.-M.; Guo, G.; Ellis, E.; Chiang, J.Y.L. Overexpression of cholesterol 7α-hydroxylase promotes hepatic bile acid synthesis and secretion and maintains cholesterol homeostasis. *Hepatology* **2010**, *53*, 996–1006. [CrossRef]
97. Lee, M.-S.; Park, J.-Y.; Freake, H.; Kwun, I.-S.; Kim, Y. Green tea catechin enhances cholesterol 7α-hydroxylase gene expression in HepG2 cells. *Br. J. Nutr.* **2008**, *99*, 1182–1185. [CrossRef] [PubMed]
98. Lu, Y.; Du, Y.; Qin, L.; Wu, D.; Wang, W.; Ling, L.; Ma, F.; Ling, H.; Yang, L.; Wang, C.; et al. Gypenosides Altered Hepatic Bile Acids Homeostasis in Mice Treated with High Fat Diet. *Evid.-Based Complement. Altern. Med.* **2018**, *2018*, 1–10. [CrossRef]
99. Ning, N.; He, K.; Wang, Y.; Zou, Z.; Wu, H.; Li, X.; Ye, X. Hypolipidemic Effect and Mechanism of Palmatine from Coptis chinensis in Hamsters Fed High-Fat diet. *Phytother. Res.* **2015**, *29*, 668–673. [CrossRef]
100. Wu, H.; He, K.; Wang, Y.; Xue, D.; Ning, N.; Zou, Z.; Ye, X.; Li, X.; Wang, D.; Pang, J. The antihypercholesterolemic effect of jatrorrhizine isolated from Rhizoma Coptidis. *Phytomedicine* **2014**, *21*, 1373–1381. [CrossRef]
101. Wang, Y.; Han, Y.; Chai, F.; Xiang, H.; Huang, T.; Kou, S.; Han, B.; Gong, X.; Ye, X. The antihypercholesterolemic effect of columbamine from Rhizoma Coptidis in HFHC-diet induced hamsters through HNF-4α/FTF-mediated CYP7A1 activation. *Fitoterapia* **2016**, *115*, 111–121. [CrossRef]
102. Lv, O.; Wang, L.; Li, J.; Ma, Q.; Zhao, W. Effects of pomegranate peel polyphenols on lipid accumulation and cholesterol metabolic transformation in L-02 human hepatic cells via the PPARγ-ABCA1/CYP7A1 pathway. *Food Funct.* **2016**, *7*, 4976–4983. [CrossRef]
103. Luo, J.; Yang, H.; Song, B.-L. Mechanisms and regulation of cholesterol homeostasis. *Nat. Rev. Mol. Cell Biol.* **2020**, *21*, 225–245. [CrossRef]
104. Maxfield, F.R.; Tabas, I. Role of cholesterol and lipid organization in disease. *Nature* **2005**, *438*, 612–621. [CrossRef]
105. Brown, M.S.; Radhakrishnan, A.; Goldstein, J.L. Retrospective on Cholesterol Homeostasis: The Central Role of Scap. *Annu. Rev. Biochem.* **2018**, *87*, 783–807. [CrossRef]
106. Steinberg, G.R.; Kemp, B.E. AMPK in Health and Disease. *Physiol. Rev.* **2009**, *89*, 1025–1078. [CrossRef]

107. Hardie, D.G.; Carling, D. The AMP-Activated Protein Kinase. Fuel Gauge of the Mammalian Cell? *Eur. J. Biochem.* **1997**, *246*, 259–273. [CrossRef]
108. Shin, S.-K.; Ha, T.-Y.; McGregor, R.A.; Choi, M.-S. Long-term curcumin administration protects against atherosclerosis via hepatic regulation of lipoprotein cholesterol metabolism. *Mol. Nutr. Food Res.* **2011**, *55*, 1829–1840. [CrossRef] [PubMed]
109. Scharinger, B.; Messner, B.; Türkcan, A.; Schuster, D.; Vuorinen, A.; Pitterl, F.; Heinz, K.; Arnhard, K.; Laufer, G.; Grimm, M.; et al. Leoligin, the major lignan from Edelweiss, inhibits 3-hydroxy-3-methyl-glutaryl-CoA reductase and reduces cholesterol levels in ApoE−/− mice. *J. Mol. Cell. Cardiol.* **2016**, *99*, 35–46. [CrossRef] [PubMed]
110. Zhao, L.-Y.; Huang, W.; Yuan, Q.-X.; Cheng, J.; Huang, Z.-C.; Ouyang, L.-J.; Zeng, F.-H. Hypolipidaemic effects and mechanisms of the main component of Opuntia dillenii Haw. polysaccharides in high-fat emulsion-induced hyperlipidaemic rats. *Food Chem.* **2012**, *134*, 964–971. [CrossRef]
111. Chung, M.J.; Sung, N.-J.; Park, C.-S.; Kweon, D.-K.; Mantovani, A.; Moon, T.-W.; Lee, S.-J.; Park, K.-H. Antioxidative and hypocholesterolemic activities of water-soluble puerarin glycosides in HepG2 cells and in C57 BL/6J mice. *Eur. J. Pharmacol.* **2008**, *578*, 159–170. [CrossRef]
112. Galle, M.; Kladniew, B.R.; Castro, M.A.; Villegas, S.M.; Lacunza, E.; Polo, M.; De Bravo, M.G.; Crespo, R. Modulation by geraniol of gene expression involved in lipid metabolism leading to a reduction of serum-cholesterol and triglyceride levels. *Phytomedicine* **2015**, *22*, 696–704. [CrossRef] [PubMed]
113. Ma, S.; Sun, W.; Gao, L.; Liu, S. Therapeutic targets of hypercholesterolemia: HMGCR and LDLR. *Diabetes Metab. Syndr. Obes. Targets Ther.* **2019**, *12*, 1543–1553. [CrossRef]
114. Kim, Y.-S.; Lee, Y.-M.; Oh, T.-I.; Shin, D.H.; Kim, G.-H.; Kan, S.-Y.; Kang, H.; Kim, J.H.; Kim, B.M.; Yim, W.J.; et al. Emodin Sensitizes Hepatocellular Carcinoma Cells to the Anti-Cancer Effect of Sorafenib through Suppression of Cholesterol Metabolism. *Int. J. Mol. Sci.* **2018**, *19*, 3127. [CrossRef] [PubMed]
115. Grand-Perret, T.; Bouillot, A.; Perrot, A.; Commans, S.; Walker, M.; Issandou, M. SCAP ligands are potent new lipid-lowering drugs. *Nat. Med.* **2001**, *7*, 1332–1338. [CrossRef] [PubMed]
116. Zhang, Y.; Ren, P.; Kang, Q.; Liu, W.; Li, S.; Li, P.; Liu, H.; Shang, J.; Zhang, L.; Gong, Y.; et al. Effect of Tetramethylpyrazine on Atherosclerosis and SCAP/SREBP-1c Signaling Pathway in ApoE−/−Mice Fed with a High-Fat Diet. *Evid.-Based Complement. Altern. Med.* **2017**, *2017*, 1–8. [CrossRef]
117. Cheng, H.; Xu, N.; Zhao, W.; Su, J.; Liang, M.; Xie, Z.; Wu, X.; Li, Q. (−)-Epicatechin regulates blood lipids and attenuates hepatic steatosis in rats fed high-fat diet. *Mol. Nutr. Food Res.* **2017**, *61*. [CrossRef]
118. Hajjaj, H.; Macé, C.; Roberts, M.; Niederberger, P.; Fay, L.B. Effect of 26-Oxygenosterols from Ganoderma lucidum and Their Activity as Cholesterol Synthesis Inhibitors. *Appl. Environ. Microbiol.* **2005**, *71*, 3653–3658. [CrossRef]
119. Davidson, M.H. Squalene synthase inhibition: A novel target for the management of dyslipidemia. *Curr. Atheroscler. Rep.* **2007**, *9*, 78–80. [CrossRef]
120. Chen, Y.; Chen, X.; Luo, G.; Zhang, X.; Lu, F.; Qiao, L.; He, W.; Li, G.; Zhang, Y. Discovery of Potential Inhibitors of Squalene Synthase from Traditional Chinese Medicine Based on Virtual Screening and In Vitro Evaluation of Lipid-Lowering Effect. *Molecules* **2018**, *23*, 1040. [CrossRef]
121. Abubakar, I.; Tillmann, T.; Banerjee, A. Global, regional, and national age-sex specific all-cause and cause-specific mortality for 240 causes of death, 1990–2013: A systematic analysis for the Global Burden of Disease Study 2013. *Lancet* **2015**, *385*, 117–171.
122. Bøhn, S.K.; Myhrstad, M.C.; Thoresen, M.; Holden, M.; Karlsen, A.; Tunheim, S.H.; Erlund, I.; Svendsen, M.; Seljeflot, I.; Moskaug, J. Ø; et al. Blood cell gene expression associated with cellular stress defense is modulated by antioxidant-rich food in a randomised controlled clinical trial of male smokers. *BMC Med.* **2010**, *8*, 54. [CrossRef] [PubMed]
123. Anderson, J.W.; Baird, P.; Davis, R.H.; Ferreri, S.; Knudtson, M.; Koraym, A.; Waters, V.; Williams, C.L. Health benefits of dietary fiber. *Nutr. Rev.* **2009**, *67*, 188–205. [CrossRef] [PubMed]
124. Islam, S.U.; Ahmed, M.B.; Ahsan, H.; Islam, M.; Shehzad, A.; Sonn, J.K.; Lee, Y.S. An Update on the Role of Dietary Phytochemicals in Human Skin Cancer: New Insights into Molecular Mechanisms. *Antioxidants* **2020**, *9*, 916. [CrossRef]
125. Alissa, E.M.; Ferns, G.A. Dietary Fruits and Vegetables and Cardiovascular Diseases Risk. *Crit. Rev. Food Sci. Nutr.* **2017**, *57*, 1950–1962. [CrossRef] [PubMed]
126. Aune, D.; Giovannucci, E.; Boffetta, P.; Fadnes, L.T.; Keum, N.; Norat, T.; Greenwood, D.C.; Riboli, E.; Vatten, L.J.; Tonstad, S. Fruit and vegetable intake and the risk of cardiovascular disease, total cancer and all-cause mortality—A systematic review and dose-response meta-analysis of prospective studies. *Int. J. Epidemiol.* **2017**, *46*, 1029–1056. [CrossRef]
127. Carmona-Jiménez, Y.; Palma, M.; Guillén-Sánchez, D.A.; García-Moreno, M.V. Study of the Cluster Thinning Grape as a Source of Phenolic Compounds and Evaluation of Its Antioxidant Potential. *Biomolecules* **2021**, *11*, 227. [CrossRef]
128. Cardoso, L.M.; Viana Leite, J.P.; Gouveia Peluzio, M.D.C. Biological effects of anthocyanins on the atherosclerotic process. *Rev. Colomb. Cienc. Químico-Farm.* **2011**, *40*, 116–138.
129. Castilla, P.; Echarri, R.; Dávalos, A.; Cerrato, F.; Ortega, H.; Teruel, J.L.; Lucas, M.F.; Gómez-Coronado, D.; Ortuño, J.; Lasunción, M.A. Concentrated red grape juice exerts antioxidant, hypolipidemic, and antiinflammatory effects in both hemodialysis patients and healthy subjects. *Am. J. Clin. Nutr.* **2006**, *84*, 252–262. [CrossRef]
130. Vaisman, N.; Niv, E. Daily consumption of red grape cell powder in a dietary dose improves cardiovascular parameters: A double blind, placebo-controlled, randomized study. *Int. J. Food Sci. Nutr.* **2015**, *66*, 342–349. [CrossRef]

131. Martins, Â.M.; Silva Sarto, D.A.Q.; Caproni, K.D.P.; Silva, J.; Silva, J.; Souza, P.S.; Dos Santos, L.; Ureña, M.J.E.; Souza Carvalho, M.D.G.D.; Vilas Boas, B.M.; et al. Grape juice attenuates left ventricular hypertrophy in dyslipidemic mice. *PLoS ONE* **2020**, *15*, e0238163. [CrossRef]
132. Van Mierlo, L.A.; Zock, P.L.; van der Knaap, H.C.; Draijer, R. Grape polyphenols do not affect vascular function in healthy men. *J. Nutr.* **2010**, *140*, 1769–1773. [CrossRef]
133. Yubero, N.; Sanz-Buenhombre, M.; Guadarrama, A.; Villanueva, S.; Carrión, J.M.; Larrarte, E.; Moro, C. LDL cholesterol-lowering effects of grape extract used as a dietary supplement on healthy volunteers. *Int. J. Food Sci. Nutr.* **2013**, *64*, 400–406. [CrossRef] [PubMed]
134. Zern, T.L.; Wood, R.J.; Greene, C.; West, K.L.; Liu, Y.; Aggarwal, D.; Shachter, N.S.; Fernandez, M.L. Grape Polyphenols Exert a Cardioprotective Effect in Pre- and Postmenopausal Women by Lowering Plasma Lipids and Reducing Oxidative Stress. *J. Nutr.* **2005**, *135*, 1911–1917. [CrossRef] [PubMed]
135. Zunino, S.J.; Peerson, J.M.; Freytag, T.L.; Breksa, A.P.; Bonnel, E.L.; Woodhouse, L.R.; Storms, D.H. Dietary grape powder increases IL-1β and IL-6 production by lipopolysaccharide-activated monocytes and reduces plasma concentrations of large LDL and large LDL-cholesterol particles in obese humans. *Br. J. Nutr.* **2014**, *112*, 369–380. [CrossRef]
136. Chiva-Blanch, G.; Urpi-Sarda, M.; Ros, E.; Arranz, S.; Valderas-Martínez, P.; Casas, R.; Sacanella, E.; Llorach, R.; Lamuela-Raventos, R.M.; Andres-Lacueva, C.; et al. Dealcoholized Red Wine Decreases Systolic and Diastolic Blood Pressure and Increases Plasma Nitric Oxide. *Circ. Res.* **2012**, *111*, 1065–1068. [CrossRef] [PubMed]
137. Pourmasoumi, M.; Hadi, A.; Najafgholizadeh, A.; Joukar, F.; Mansour-Ghanaei, F. The effects of cranberry on cardiovascular metabolic risk factors: A systematic review and meta-analysis. *Clin. Nutr.* **2020**, *39*, 774–788. [CrossRef] [PubMed]
138. Wilson, T.; Porcari, J.P.; Harbin, D. Cranberry extract inhibits low density lipoprotein oxidation. *Life Sci.* **1998**, *62*, A381–A386. [CrossRef]
139. Yu, L.L.; Zhou, K.K.; Parry, J. Antioxidant properties of cold-pressed black caraway, carrot, cranberry, and hemp seed oils. *Food Chem.* **2005**, *91*, 723–729. [CrossRef]
140. Chu, Y.-F.; Liu, R.H. Cranberries inhibit LDL oxidation and induce LDL receptor expression in hepatocytes. *Life Sci.* **2005**, *77*, 1892–1901. [CrossRef]
141. Reed, J. Cranberry Flavonoids, Atherosclerosis and Cardiovascular Health. *Crit. Rev. Food Sci. Nutr.* **2002**, *42*, 301–316. [CrossRef]
142. Aviram, M.; Rosenblat, M. Pomegranate for Your Cardiovascular Health. *Rambam Maimonides Med. J.* **2013**, *4*, e0013. [CrossRef] [PubMed]
143. Fuhrman, B.; Volkova, N.; Aviram, M. Pomegranate juice inhibits oxidized LDL uptake and cholesterol biosynthesis in macrophages. *J. Nutr. Biochem.* **2005**, *16*, 570–576. [CrossRef] [PubMed]
144. Boroushaki, M.T.; Mollazadeh, H.; Afshari, A.R. Pomegranate seed oil: A comprehensive review on its therapeutic effects. *Int. J. Pharm. Sci. Res.* **2016**, *7*, 430.
145. Esmaillzadeh, A.; Tahbaz, F.; Gaieni, I.; Alavi-Majd, H.; Azadbakht, L. Cholesterol-Lowering Effect of Concentrated Pomegranate Juice Consumption in Type II Diabetic Patients with Hyperlipidemia. *Int. J. Vitam. Nutr. Res.* **2006**, *76*, 147–151. [CrossRef]
146. Aviram, M.; Dornfeld, L.; Rosenblat, M.; Volkova, N.; Kaplan, M.; Coleman, R.; Hayek, T.; Presser, D.; Fuhrman, B. Pomegranate juice consumption reduces oxidative stress, atherogenic modifications to LDL, and platelet aggregation: Studies in humans and in atherosclerotic apolipoprotein E–deficient mice. *Am. J. Clin. Nutr.* **2000**, *71*, 1062–1076. [CrossRef]
147. Al-Moraie, M.M.; Arafat, R.A.; Al-Rasheedi, A.A. Effect of pomegranate juice on lipid profile and antioxidant enzymes in hypercholesterolemic rats. *Life Sci. J.* **2013**, *10*, 2717–2728.
148. Atrahimovich, D.; Khatib, S.; Sela, S.; Vaya, J.; Samson, A.O. Punicalagin Induces Serum Low-Density Lipoprotein Influx to Macrophages. *Oxidative Med. Cell. Longev.* **2016**, *2016*, 1–9. [CrossRef]
149. Tenore, G.C.; Caruso, D.; Buonomo, G.; D'Urso, E.; D'Avino, M.; Campiglia, P.; Marinelli, L.; Novellino, E. Annurca (*Malus pumilaMiller* cv. Annurca) apple as a functional food for the contribution to a healthy balance of plasma cholesterol levels: Results of a randomized clinical trial. *J. Sci. Food Agric.* **2017**, *97*, 2107–2115. [CrossRef]
150. Nagasako-Akazome, Y.; Kanda, T.; Ohtake, Y.; Shimasaki, H.; Kobayashi, T. Apple Polyphenols Influence Cholesterol Metabolism in Healthy Subjects with Relatively High Body Mass Index. *J. Oleo Sci.* **2007**, *56*, 417–428. [CrossRef]
151. Hyson, D.; Studebaker-Hallman, D.; Davis, P.A.; Gershwin, M.E. Apple Juice Consumption Reduces Plasma Low-Density Lipoprotein Oxidation in Healthy Men and Women. *J. Med. Food* **2000**, *3*, 159–166. [CrossRef]
152. Auclair, S.; Chironi, G.; Milenkovic, D.; Hollman, P.; Renard, C.; Mégnien, J.; Gariepy, J.; Paul, J.-L.; Simon, A.; Scalbert, A. The regular consumption of a polyphenol-rich apple does not influence endothelial function: A randomised double-blind trial in hypercholesterolemic adults. *Eur. J. Clin. Nutr.* **2010**, *64*, 1158–1165. [CrossRef] [PubMed]
153. Vafa, M.R.; Haghighatjoo, E.; Shidfar, F.; Afshari, S.; Gohari, M.R.; Ziaee, A. Effects of Apple Consumption on Lipid Profile of Hyperlipidemic and Overweight Men. *Int. J. Prev. Med.* **2011**, *2*, 94–100.
154. Ravn-Haren, G.; Dragsted, L.O.; Buch-Andersen, T.; Jensen, E.N.; Jensen, R.I.; Németh-Balogh, M.; Paulovicsová, B.; Bergström, A.; Wilcks, A.; Licht, T.R.; et al. Intake of whole apples or clear apple juice has contrasting effects on plasma lipids in healthy volunteers. *Eur. J. Nutr.* **2013**, *52*, 1875–1889. [CrossRef] [PubMed]
155. Barth, S.W.; Koch, T.C.L.; Watzl, B.; Dietrich, H.; Will, F.; Bub, A. Moderate effects of apple juice consumption on obesity-related markers in obese men: Impact of diet–gene interaction on body fat content. *Eur. J. Nutr.* **2011**, *51*, 841–850. [CrossRef]

156. Kim, Y.; Keogh, J.B.; Clifton, P.M. Benefits of Nut Consumption on Insulin Resistance and Cardiovascular Risk Factors: Multiple Potential Mechanisms of Actions. *Nutrients* **2017**, *9*, 1271. [CrossRef]
157. Bechthold, A.; Boeing, H.; Schwedhelm, C.; Hoffmann, G.; Knüppel, S.; Iqbal, K.; De Henauw, S.; Michels, N.; Devleesschauwer, B.; Schlesinger, S.; et al. Food groups and risk of coronary heart disease, stroke and heart failure: A systematic review and dose-response meta-analysis of prospective studies. *Crit. Rev. Food Sci. Nutr.* **2019**, *59*, 1071–1090. [CrossRef]
158. Afshin, A.; Micha, R.; Khatibzadeh, S.; Mozaffarian, D. Consumption of nuts and legumes and risk of incident ischemic heart disease, stroke, and diabetes: A systematic review and meta-analysis. *Am. J. Clin. Nutr.* **2014**, *100*, 278–288. [CrossRef]
159. Zhou, D.; Yu, H.; He, F.; Reilly, K.H.; Zhang, J.; Li, S.; Zhang, T.; Wang, B.; Ding, Y.; Xi, B. Nut consumption in relation to cardiovascular disease risk and type 2 diabetes: A systematic review and meta-analysis of prospective studies. *Am. J. Clin. Nutr.* **2014**, *100*, 270–277. [CrossRef]
160. Freeman, A.M.; Morris, P.B.; Barnard, N.; Esselstyn, C.B.; Ros, E.; Agatston, A.; Devries, S.; O'Keefe, J.; Miller, M.; Ornish, D.; et al. Trending Cardiovascular Nutrition Controversies. *J. Am. Coll. Cardiol.* **2017**, *69*, 1172–1187. [CrossRef]
161. Holmes, M.V.; Asselbergs, F.W.; Palmer, T.M.; Drenos, F.; Lanktree, M.B.; Nelson, C.P.; Dale, C.E.; Padmanabhan, S.; Finan, C.; Swerdlow, D.I.; et al. Mendelian randomization of blood lipids for coronary heart disease. *Eur. Heart J.* **2015**, *36*, 539–550. [CrossRef]
162. Griel, A.E.; Kris-Etherton, P.M. Tree nuts and the lipid profile: A review of clinical studies. *Br. J. Nutr.* **2006**, *96*, S68–S78. [CrossRef] [PubMed]
163. Zhao, J.V.; Schooling, C.M. Effect of linoleic acid on ischemic heart disease and its risk factors: A Mendelian randomization study. *BMC Med.* **2019**, *17*, 1–8. [CrossRef]
164. Trautwein, E.A.; McKay, S. The Role of Specific Components of a Plant-Based Diet in Management of Dyslipidemia and the Impact on Cardiovascular Risk. *Nutrients* **2020**, *12*, 2671. [CrossRef]
165. Xu, X.-R.; Zou, Z.-Y.; Xiao, X.; Huang, Y.-M.; Wang, X.; Lin, X.-M. Effects of Lutein Supplement on Serum Inflammatory Cytokines, ApoE and Lipid Profiles in Early Atherosclerosis Population. *J. Atheroscler. Thromb.* **2013**, *20*, 170–177. [CrossRef]
166. García-Conesa, M.T.; Chambers, K.; Combet, E.; Pinto, P.; Garcia-Aloy, M.; Andres-Lacueva, C.; De Pascual-Teresa, S.; Mena, P.; Ristic, A.K.; Hollands, W.J.; et al. Meta-Analysis of the Effects of Foods and Derived Products Containing Ellagitannins and Anthocyanins on Cardiometabolic Biomarkers: Analysis of Factors Influencing Variability of the Individual Responses. *Int. J. Mol. Sci.* **2018**, *19*, 694. [CrossRef]
167. Hernández-Alonso, P.; Giardina, S.; Salas-Salvadó, J.; Arcelin, P.; Bulló, M. Chronic pistachio intake modulates circulating microRNAs related to glucose metabolism and insulin resistance in prediabetic subjects. *Eur. J. Nutr.* **2016**, *56*, 2181–2191. [CrossRef]
168. Ortega, F.J.; Cardona-Alvarado, M.I.; Mercader, J.M.; Moreno-Navarrete, J.M.; Moreno, M.; Sabater, M.; Fuentes-Batllevell, N.; Ramírez-Chávez, E.; Ricart, W.; Molina-Torres, J.; et al. Circulating profiling reveals the effect of a polyunsaturated fatty acid-enriched diet on common microRNAs. *J. Nutr. Biochem.* **2015**, *26*, 1095–1101. [CrossRef]
169. Del Gobbo, L.C.; Falk, M.C.; Feldman, R.; Lewis, K.; Mozaffarian, D. Effects of tree nuts on blood lipids, apolipoproteins, and blood pressure: Systematic review, meta-analysis, and dose-response of 61 controlled intervention trials. *Am. J. Clin. Nutr.* **2015**, *102*, 1347–1356. [CrossRef]
170. Guasch-Ferré, M.; Li, J.; Hu, F.B.; Salas-Salvadó, J.; Tobias, D.K. Effects of walnut consumption on blood lipids and other cardiovascular risk factors: An updated meta-analysis and systematic review of controlled trials. *Am. J. Clin. Nutr.* **2018**, *108*, 174–187. [CrossRef]
171. Musa-Veloso, K.; Paulionis, L.; Poon, T.; Lee, H.Y. The effects of almond consumption on fasting blood lipid levels: A systematic review and meta-analysis of randomised controlled trials. *J. Nutr. Sci.* **2016**, *5*. [CrossRef]
172. Njike, V.Y.; Ayettey, R.; Petraro, P.; Treu, J.A.; Katz, D.L. Walnut ingestion in adults at risk for diabetes: Effects on body composition, diet quality, and cardiac risk measures. *BMJ Open Diabetes Res. Care* **2015**, *3*, e000115. [CrossRef] [PubMed]
173. Flores-Mateo, G.; Rojas-Rueda, D.; Basora, J.; Ros, E.; Salas-Salvadó, J. Nut intake and adiposity: Meta-analysis of clinical trials. *Am. J. Clin. Nutr.* **2013**, *97*, 1346–1355. [CrossRef] [PubMed]
174. Schwingshackl, L.; Schwarzer, G.; Rücker, G.; Meerpohl, J.J. Perspective: Network Meta-analysis Reaches Nutrition Research: Current Status, Scientific Concepts, and Future Directions. *Adv. Nutr.* **2019**, *10*, 739–754. [CrossRef]
175. Hutton, B.; Salanti, G.; Caldwell, D.M.; Chaimani, A.; Schmid, C.H.; Cameron, C.; Ioannidis, J.P.; Straus, S.E.; Thorlund, K.; Jansen, J.P.; et al. The PRISMA Extension Statement for Reporting of Systematic Reviews Incorporating Network Meta-analyses of Health Care Interventions: Checklist and Explanations. *Ann. Intern. Med.* **2015**, *162*, 777–784. [CrossRef] [PubMed]
176. Liu, K.; Hui, S.; Wang, B.; Kaliannan, K.; Guo, X.; Liang, L. Comparative effects of different types of tree nut consumption on blood lipids: A network meta-analysis of clinical trials. *Am. J. Clin. Nutr.* **2020**, *111*, 219–227. [CrossRef] [PubMed]
177. Gouws, C.; Mortazavi, R.; Mellor, D.; McKune, A.; Naumovski, N. The effects of Prickly Pear fruit and cladode (*Opuntia* spp.) consumption on blood lipids: A systematic review. *Complement. Ther. Med.* **2020**, *50*, 102384. [CrossRef] [PubMed]
178. Palumbo, B.; Efthimiou, Y.; Stamatopoulos, J.; Oguogho, A.; Budinsky, A.; Palumbo, R.; Sinzinger, H. Prickly pear induces upregulation of liver LDL binding in familial heterozygous hypercholesterolemia. *Nucl. Med. Rev.* **2003**, *6*, 35–39.
179. Khouloud, A.; Abedelmalek, S.; Chtourou, H.; Souissi, N. The effect of Opuntia ficus-indica juice supplementation on oxidative stress, cardiovascular parameters, and biochemical markers following yo-yo Intermittent recovery test. *Food Sci. Nutr.* **2017**, *6*, 259–268. [CrossRef]

180. Wolfram, R.M.; Kritz, H.; Efthimiou, Y.; Stomatopoulos, J.; Sinzinger, H. Effect of prickly pear (Opuntia robusta) on glucose- and lipid-metabolism in non-diabetics with hyperlipidemia–a pilot study. *Wien. Klin. Wochenschr.* **2002**, *114*, 840–846.
181. Goh, K.K.T.; Pinder, D.N.; Hall, C.E.; Hemar, Y. Rheological and Light Scattering Properties of Flaxseed Polysaccharide Aqueous Solutions. *Biomacromolecules* **2006**, *7*, 3098–3103. [CrossRef]
182. Kristensen, M.; Jensen, M.G.; Aarestrup, J.; Petersen, K.E.; Søndergaard, L.; Mikkelsen, M.S.; Astrup, A. Flaxseed dietary fibers lower cholesterol and increase fecal fat excretion, but magnitude of effect depend on food type. *Nutr. Metab.* **2012**, *9*, 8. [CrossRef] [PubMed]
183. Surampudi, P.; Enkhmaa, B.; Anuurad, E.; Berglund, L. Lipid Lowering with Soluble Dietary Fiber. *Curr. Atheroscler. Rep.* **2016**, *18*, 75. [CrossRef] [PubMed]
184. Zhang, W.; Wang, X.; Liu, Y.; Tian, H.; Flickinger, B.; Empie, M.W.; Sun, S.Z. Dietary flaxseed lignan extract lowers plasma cholesterol and glucose concentrations in hypercholesterolaemic subjects. *Br. J. Nutr.* **2008**, *99*, 1301–1309. [CrossRef]
185. Arjmandi, B.H.; Khan, D.A.; Juma, S.; Drum, M.L.; Venkatesh, S.; Sohn, E.; Wei, L.; Derman, R. Whole flaxseed consumption lowers serum LDL-cholesterol and lipoprotein(a) concentrations in postmenopausal women. *Nutr. Res.* **1998**, *18*, 1203–1214. [CrossRef]
186. Edel, A.L.; Rodriguez-Leyva, D.; Maddaford, T.G.; Caligiuri, S.P.; Austria, J.A.; Weighell, W.; Guzman, R.; Aliani, M.; Pierce, G.N. Dietary Flaxseed Independently Lowers Circulating Cholesterol and Lowers It beyond the Effects of Cholesterol-Lowering Medications Alone in Patients with Peripheral Artery Disease. *J. Nutr.* **2015**, *145*, 749–757. [CrossRef] [PubMed]
187. Vaidean, G.D.; Manczuk, M.; Vansal, S.S.; Griffith, J. The cholesterol-lowering effect of statins is potentiated by whole grains intake. The Polish Norwegian Study (PONS). *Eur. J. Intern. Med.* **2018**, *50*, 47–51. [CrossRef]
188. Temple, N.J. Fat, Sugar, Whole Grains and Heart Disease: 50 Years of Confusion. *Nutrients* **2018**, *10*, 39. [CrossRef] [PubMed]
189. Hollænder, P.L.; Ross, A.B.; Kristensen, M. Whole-grain and blood lipid changes in apparently healthy adults: A systematic review and meta-analysis of randomized controlled studies. *Am. J. Clin. Nutr.* **2015**, *102*, 556–572. [CrossRef] [PubMed]
190. Fatahi, S.; Daneshzad, E.; Kord-Varkaneh, H.; Bellissimo, N.; Brett, N.R.; Azadbakht, L. Impact of Diets Rich in Whole Grains and Fruits and Vegetables on Cardiovascular Risk Factors in Overweight and Obese Women: A Randomized Clinical Feeding Trial. *J. Am. Coll. Nutr.* **2018**, *37*, 568–577. [CrossRef] [PubMed]
191. Helnæs, A.; Kyrø, C.; Andersen, I.; Lacoppidan, S.; Overvad, K.; Christensen, J.; Tjønneland, A.; Olsen, A. Intake of whole grains is associated with lower risk of myocardial infarction: The Danish Diet, Cancer and Health Cohort. *Am. J. Clin. Nutr.* **2016**, *103*, 999–1007. [CrossRef]
192. Blanco Mejia, S.; Messina, M.; Li, S.S.; Viguiliouk, E.; Chiavaroli, L.; Khan, T.A.; Srichaikul, K.; Mirrahimi, A.; Sievenpiper, J.L.; Kris-Etherton, P.; et al. A meta-analysis of 46 studies identified by the FDA demonstrates that soy protein decreases circulating LDL and total cholesterol concentrations in adults. *J. Nutr.* **2019**, *149*, 968–981. [CrossRef] [PubMed]
193. Høie, L.H.; Morgenstern, E.C.; Grünwald, J.; Graubaum, H.J.; Busch, R.; Lüder, W.; Zunft, H.J. A double-blind placebo-controlled clinical trial compares the cholesterollowering effects of two different soy protein preparations in hypercholesterolemic subjects. *Eur. J. Nutr.* **2005**, *44*, 65–71. [CrossRef] [PubMed]
194. Lukaczer, D.; DeAnn, J.L.; Lerman, R.H.; Darland, G.; Schiltz, B.; Tripp, M.; Bland, J.S. Effect of a low glycemic index diet with soy protein and phytosterols on CVD risk factors in postmenopausal women. *Nutrition* **2006**, *22*, 104–113. [CrossRef]
195. Taku, K.; Umegaki, K.; Sato, Y.; Taki, Y.; Endoh, K.; Watanabe, S. Soy isoflavones lower serum total and LDL cholesterol in humans: A meta-analysis of 11 randomized controlled trials. *Am. J. Clin. Nutr.* **2007**, *85*, 1148–1156. [CrossRef]
196. Zhan, S.; Ho, S.C. Meta-analysis of the effects of soy protein containing isoflavones on the lipid profile. *Am. J. Clin. Nutr.* **2005**, *81*, 397–408. [CrossRef]
197. Tokede, O.A.; Onabanjo, T.A.; Yansane, A.; Gaziano, J.M.; Djoussé, L. Soya products and serum lipids: A meta-analysis of randomised controlled trials. *Br. J. Nutr.* **2015**, *114*, 831–843. [CrossRef] [PubMed]
198. Kim, J.; Lee, H.; Lee, O.; Lee, K.-H.; Lee, Y.-B.; Young, K.D.; Jeong, Y.H.; Choue, R. Isoflavone supplementation influenced levels of triglyceride and luteunizing hormone in Korean postmenopausal women. *Arch. Pharmacal Res.* **2013**, *36*, 306–313. [CrossRef]
199. Qin, Y.; Shu, F.; Zeng, Y.; Meng, X.; Wang, B.; Diao, L.; Wang, L.; Wan, J.; Zhu, J.; Wang, J.; et al. Daidzein Supplementation Decreases Serum Triglyceride and Uric Acid Concentrations in Hypercholesterolemic Adults with the Effect on Triglycerides Being Greater in Those with the GA Compared with the GG Genotype of ESR-β Rsa I. *J. Nutr.* **2014**, *144*, 49–54. [CrossRef]
200. Usui, T.; Tochiya, M.; Sasaki, Y.; Muranaka, K.; Yamakage, H.; Himeno, A.; Shimatsu, A.; Inaguma, A.; Ueno, T.; Uchiyama, S.; et al. Effects of naturalS-equol supplements on overweight or obesity and metabolic syndrome in the Japanese, based on sex and equol status. *Clin. Endocrinol.* **2013**, *78*, 365–372. [CrossRef]
201. Squadrito, F.; Marini, H.; Bitto, A.; Altavilla, D.; Polito, F.; Adamo, E.B.; D'Anna, R.; Arcoraci, V.; Burnett, B.P.; Minutoli, L.; et al. Genistein in the Metabolic Syndrome: Results of a Randomized Clinical Trial. *J. Clin. Endocrinol. Metab.* **2013**, *98*, 3366–3374. [CrossRef]
202. Key, T.J.; Appleby, P.N.; Rosell, M.S. Health effects of vegetarian and vegan diets. *Proc. Nutr. Soc.* **2006**, *65*, 35–41. [CrossRef] [PubMed]
203. Ashen, M.D. Vegetarian diets in cardiovascular prevention. *Curr. Treat. Options Cardiovasc. Med.* **2013**, *15*, 735–745. [CrossRef] [PubMed]

204. Anderson, T.J.; Grégoire, J.; Pearson, G.J.; Barry, A.R.; Couture, P.; Dawes, M.; Francis, G.A.; Genest, J., Jr.; Grover, S.; Gupta, M.; et al. 2016 Canadian Cardiovascular Society Guidelines for the Management of Dyslipidemia for the Prevention of Cardiovascular Disease in the Adult. *Can. J. Cardiol.* **2016**, *32*, 1263–1282. [CrossRef] [PubMed]
205. Sterling, S.R.; Bowen, S.-A. The potential for plant-based diets to promote health among blacks living in the United States. *Nutrients* **2019**, *11*, 2915. [CrossRef] [PubMed]
206. Larsson, S.C.; Orsini, N. Red Meat and Processed Meat Consumption and All-Cause Mortality: A Meta-Analysis. *Am. J. Epidemiol.* **2014**, *179*, 282–289. [CrossRef]
207. Kwok, C.S.; Umar, S.; Myint, P.K.; Mamas, M.A.; Loke, Y.K. Vegetarian diet, Seventh Day Adventists and risk of cardiovascular mortality: A systematic review and meta-analysis. *Int. J. Cardiol.* **2014**, *176*, 680–686. [CrossRef]
208. Migliaccio, S.; Brasacchio, C.; Pivari, F.; Salzano, C.; Barrea, L.; Muscogiuri, G.; Savastano, S.; Colao, A. What is the best diet for cardiovascular wellness? A comparison of different nutritional models. *Int. J. Obes. Suppl.* **2020**, *10*, 50–61. [CrossRef] [PubMed]
209. Crowe, F.L.; Appleby, P.N.; Travis, R.C.; Key, T.J. Risk of hospitalization or death from ischemic heart disease among British vegetarians and nonvegetarians: Results from the EPIC-Oxford cohort study. *Am. J. Clin. Nutr.* **2013**, *97*, 597–603. [CrossRef]
210. Liu, R.H. Dietary Bioactive Compounds and Their Health Implications. *J. Food Sci.* **2013**, *78*, A18–A25. [CrossRef]
211. Wang, F.; Zheng, J.; Yang, B.; Jiang, J.; Fu, Y.; Li, D. Effects of Vegetarian Diets on Blood Lipids: A Systematic Review and Meta-Analysis of Randomized Controlled Trials. *J. Am. Heart Assoc.* **2015**, *4*, e002408. [CrossRef]
212. Key, T.J.; Fraser, G.E.; Thorogood, M.; Appleby, P.N.; Beral, V.; Reeves, G.; Burr, M.L.; Chang-Claude, J.; Frentzel-Beyme, R.; Kuzma, J.W.; et al. Mortality in vegetarians and nonvegetarians: Detailed findings from a collaborative analysis of 5 prospective studies. *Am. J. Clin. Nutr.* **1999**, *70*, 516s–524s. [CrossRef] [PubMed]
213. Chen, C.; Lin, Y.; Lin, T.; Lin, C.; Chen, B.; Lin, C. Total cardiovascular risk profile of Taiwanese vegetarians. *Eur. J. Clin. Nutr.* **2008**, *62*, 138–144. [CrossRef]
214. Jones, J.R.; Lineback, D.M.; Levine, M.J. Dietary reference intakes: Implications for fiber labeling and consumption: A summary of the International Life Sciences Institute North America Fiber Workshop, June 1–2, 2004, Washington, DC. *Nutr. Rev.* **2006**, *64*, 31–38. [CrossRef]
215. Lia, A.; Hallmans, G.; Sandberg, A.S.; Sundberg, B.; Aman, P.; Andersson, H. Oat beta-glucan increases bile acid excretion and a fiber-rich barley fraction increases cholesterol excretion in ileostomy subjects. *Am. J. Clin. Nutr.* **1995**, *62*, 1245–1251. [CrossRef] [PubMed]
216. Brown, L.; Rosner, B.; Willett, W.W.; Sacks, F.M. Cholesterol-lowering effects of dietary fiber: A meta-analysis. *Am. J. Clin. Nutr.* **1999**, *69*, 30–42. [CrossRef] [PubMed]
217. Evans, C.E.; Greenwood, D.C.; Threapleton, D.E.; Cleghorn, C.L.; Nykjaer, C.; Woodhead, C.E.; Gale, C.P.; Burley, V.J. Effects of dietary fibre type on blood pressure: A systematic review and meta-analysis of randomized controlled trials of healthy individuals. *J. Hypertens.* **2015**, *33*, 897–911. [CrossRef]
218. Kim, Y.; Je, Y. Dietary fibre intake and mortality from cardiovascular disease and all cancers: A meta-analysis of prospective cohort studies. *Arch. Cardiovasc. Dis.* **2016**, *109*, 39–54. [CrossRef]
219. McKeown, N.M.; Meigs, J.B.; Liu, S.; Saltzman, E.; Wilson, P.W.; Jacques, P.F. Carbohydrate Nutrition, Insulin Resistance, and the Prevalence of the Metabolic Syndrome in the Framingham Offspring Cohort. *Diabetes Care* **2004**, *27*, 538–546. [CrossRef]
220. Bazzano, L.A.; Thompson, A.M.; Tees, M.T.; Nguyen, C.H.; Winham, D.M. Non-soy legume consumption lowers cholesterol levels: A meta-analysis of randomized controlled trials. *Nutr. Metab. Cardiovasc. Dis.* **2011**, *21*, 94–103. [CrossRef]
221. Marques, F.Z.; Nelson, E.; Chu, P.-Y.; Horlock, D.; Fiedler, A.; Ziemann, M.; Tan, J.K.; Kuruppu, S.; Rajapakse, N.W.; El-Osta, A.; et al. High-Fiber Diet and Acetate Supplementation Change the Gut Microbiota and Prevent the Development of Hypertension and Heart Failure in Hypertensive Mice. *Circulation* **2017**, *135*, 964–977. [CrossRef]
222. Whitehead, A.; Beck, E.J.; Tosh, S.; Wolever, T.M. Cholesterol-lowering effects of oat β-glucan: A meta-analysis of randomized controlled trials. *Am. J. Clin. Nutr.* **2014**, *100*, 1413–1421. [CrossRef] [PubMed]
223. Mirmiran, P.; Bahadoran, Z.; Khalili Moghadam, S.; Zadeh Vakili, A.; Azizi, F. A Prospective Study of Different Types of Dietary Fiber and Risk of Cardiovascular Disease: Tehran Lipid and Glucose Study. *Nutrients* **2016**, *8*, 686. [CrossRef] [PubMed]
224. Threapleton, D.E.; Greenwood, D.C.; Evans, C.E.; Cleghorn, C.L.; Nykjaer, C.; Woodhead, C.; Cade, J.E.; Gale, C.P.; Burley, V.J. Dietary fibre intake and risk of cardiovascular disease: Systematic review and meta-analysis. *BMJ* **2013**, *347*, f6879. [CrossRef]
225. Greenwood, D.C.; Threapleton, D.E.; Evans, C.E.; Cleghorn, C.L.; Nykjaer, C.; Woodhead, C.; Burley, V.J. Glycemic Index, Glycemic Load, Carbohydrates, and Type 2 Diabetes: Systematic review and dose-response meta-analysis of prospective studies. *Diabetes Care* **2013**, *36*, 4166–4171. [CrossRef] [PubMed]
226. Khan, K.; Jovanovski, E.; Ho, H.V.T.; Marques, A.C.R.; Zurbau, A.; Mejia, S.B.; Sievenpiper, J.L.; Vuksan, V. The effect of viscous soluble fiber on blood pressure: A systematic review and meta-analysis of randomized controlled trials. *Nutr. Metab. Cardiovasc. Dis.* **2018**, *28*, 3–13. [CrossRef]
227. Evans, C.E.L. Dietary fibre and cardiovascular health: A review of current evidence and policy. *Proc. Nutr. Soc.* **2020**, *79*, 61–67. [CrossRef] [PubMed]
228. Den Besten, G.; Van Eunen, K.; Groen, A.K.; Venema, K.; Reijngoud, D.-J.; Bakker, B.M. The role of short-chain fatty acids in the interplay between diet, gut microbiota, and host energy metabolism. *J. Lipid Res.* **2013**, *54*, 2325–2340. [CrossRef]
229. Ban, S.J.; Rico, C.W.; Um, I.C.; Kang, M.Y. Comparative evaluation of the hypolipidemic effects of hydroxyethyl methylcellulose (HEMC) and hydroxypropyl methylcellulose (HPMC) in high fat-fed mice. *Food Chem. Toxicol.* **2012**, *50*, 130–134. [CrossRef]

230. Liu, X.; Yang, F.; Song, T.; Zeng, A.; Wang, Q.; Sun, Z.; Shen, J. Therapeutic Effect of Carboxymethylated and Quanternized Chitosan on Insulin Resistance in High-Fat-Diet-Induced Rats and 3T3-L1 Adipocytes. *J. Biomater. Sci. Polym. Ed.* **2012**, *23*, 1271–1284. [CrossRef]
231. Lipid Research Clinics Coronary Primary Prevention Trial Results. II. The relationship of reduction in incidence of coronary heart disease to cholesterol lowering. *JAMA* **1984**, *251*, 365–374. [CrossRef]
232. Mathews, R.; Kamil, A.; Chu, Y. Global review of heart health claims for oat beta-glucan products. *Nutr. Rev.* **2020**, *78*, 78–97. [CrossRef]
233. McRorie, J.; Fahey, G.; Wallace, T. Fiber supplements and clinically meaningful health benefits: Identifying the physiochemical characteristics of fiber that drive specific physiologic effects. In *The CRC Handbook on Dietary Supplements in Health Promotion*; CRC Press: Boca Raton, FL, USA, 2015; pp. 161–206.
234. Wolever, T.M.; Tosh, S.M.; Gibbs, A.L.; Brand-Miller, J.; Duncan, A.M.; Hart, V.; Lamarche, B.; Thomson, B.A.; Duss, R.; Wood, P.J. Physicochemical properties of oat β-glucan influence its ability to reduce serum LDL cholesterol in humans: A randomized clinical trial. *Am. J. Clin. Nutr.* **2010**, *92*, 723–732. [CrossRef] [PubMed]
235. Comerford, K.B.; Artiss, J.D.; Jen, K.-L.C.; Karakas, S.E. The Beneficial Effects α-Cyclodextrin on Blood Lipids and Weight Loss in Healthy Humans. *Obesity* **2011**, *19*, 1200–1204. [CrossRef] [PubMed]
236. Jarosz, P.A.; Fletcher, E.; Elserafy, E.; Artiss, J.D.; Jen, K.-L.C. The Effect of α-Cyclodextrin on postprandial lipid and glycemic responses to a fat-containing meal. *Metabolism* **2013**, *62*, 1443–1447. [CrossRef]
237. Hemler, E.C.; Hu, F.B. Plant-Based Diets for Cardiovascular Disease Prevention: All Plant Foods Are Not Created Equal. *Curr. Atheroscler. Rep.* **2019**, *21*, 18. [CrossRef]
238. Zhou, D.-D.; Luo, M.; Shang, A.; Mao, Q.-Q.; Li, B.-Y.; Gan, R.-Y.; Li, H.-B. Antioxidant Food Components for the Prevention and Treatment of Cardiovascular Diseases: Effects, Mechanisms, and Clinical Studies. *Oxidative Med. Cell. Longev.* **2021**, *2021*, 1–17. [CrossRef]
239. Zock, P.L.; Blom, W.A.; Nettleton, J.A.; Hornstra, G. Progressing Insights into the Role of Dietary Fats in the Prevention of Cardiovascular Disease. *Curr. Cardiol. Rep.* **2016**, *18*, 1–13. [CrossRef]
240. Schwingshackl, L.; Bogensberger, B.; Benčič, A.; Knüppel, S.; Boeing, H.; Hoffmann, G. Effects of oils and solid fats on blood lipids: A systematic review and network meta-analysis. *J. Lipid Res.* **2018**, *59*, 1771–1782. [CrossRef]
241. Theuwissen, E.; Mensink, R.P. Water-soluble dietary fibers and cardiovascular disease. *Physiol. Behav.* **2008**, *94*, 285–292. [CrossRef]
242. Gylling, H.; Plat, J.; Turley, S.D.; Ginsberg, H.N.; Ellegård, L.; Jessup, W.; Jones, P.J.H.; Lütjohann, D.; März, W.; Masana, L.; et al. Plant sterols and plant stanols in the management of dyslipidaemia and prevention of cardiovascular disease. *Atherosclerosis* **2014**, *232*, 346–360. [CrossRef]
243. Satija, A.; Bhupathiraju, S.N.; Spiegelman, D.; Chiuve, S.E.; Manson, J.E.; Willett, W.; Rexrode, K.M.; Rimm, E.B.; Hu, F.B. Healthful and Unhealthful Plant-Based Diets and the Risk of Coronary Heart Disease in U.S. Adults. *J. Am. Coll. Cardiol.* **2017**, *70*, 411–422. [CrossRef]
244. Willett, W.; Rockström, J.; Loken, B.; Springmann, M.; Lang, T.; Vermeulen, S.; Garnett, T.; Tilman, D.; DeClerck, F.; Wood, A.; et al. Food in the Anthropocene: The EAT–Lancet Commission on healthy diets from sustainable food systems. *Lancet* **2019**, *393*, 447–492. [CrossRef]
245. Corrin, T.; Papadopoulos, A. Understanding the attitudes and perceptions of vegetarian and plant-based diets to shape future health promotion programs. *Appetite* **2017**, *109*, 40–47. [CrossRef]
246. Fehér, A.; Gazdecki, M.; Véha, M.; Szakály, M.; Szakály, Z. A Comprehensive Review of the Benefits of and the Barriers to the Switch to a Plant-Based Diet. *Sustainability* **2020**, *12*, 4136. [CrossRef]

antioxidants

MDPI

Article

Production and Characterization of Medium-Sized and Short Antioxidant Peptides from Soy Flour-Simulated Gastrointestinal Hydrolysate

Chiara Cavaliere [1], Angela Michela Immacolata Montone [2,3], Sara Elsa Aita [1], Rosanna Capparelli [4], Andrea Cerrato [1], Paola Cuomo [4], Aldo Laganà [1,5], Carmela Maria Montone [1,*], Susy Piovesana [1] and Anna Laura Capriotti [1]

1 Department of Chemistry, Università di Roma "La Sapienza", Piazzale Aldo Moro 5, 00185 Rome, Italy; chiara.cavaliere@uniroma1.it (C.C.); saraelsa.aita@uniroma1.it (S.E.A.); andrea.cerrato@uniroma1.it (A.C.); aldo.lagana@uniroma1.it (A.L.); susy.piovesana@uniroma1.it (S.P.); annalaura.capriotti@uniroma1.it (A.L.C.)
2 Istituto Zooprofilattico Sperimentale del Mezzogiorno, Via Salute 2, Portici, 80055 Naples, Italy; angela.montone@izsmportici.it
3 Department of Industrial Engineering, Università degli Studi di Salerno, Via Giovanni Paolo II 132, 84084 Fisciano, Italy
4 Department of Agriculture Sciences, University of Naples "Federico II", Via Università 100, Portici, 80055 Naples, Italy; capparel@unina.it (R.C.); paola.cuomo@unina.it (P.C.)
5 CNR NANOTEC, Campus Ecotekne, University of Salento, Via Monteroni, 73100 Lecce, Italy
* Correspondence: carmelamaria.montone@uniroma1.it

Citation: Cavaliere, C.; Montone, A.M.I.; Aita, S.E.; Capparelli, R.; Cerrato, A.; Cuomo, P.; Laganà, A.; Montone, C.M.; Piovesana, S.; Capriotti, A.L. Production and Characterization of Medium-Sized and Short Antioxidant Peptides from Soy Flour-Simulated Gastrointestinal Hydrolysate. *Antioxidants* **2021**, *10*, 734. https://doi.org/10.3390/antiox10050734

Academic Editor: Simone Carradori

Received: 1 April 2021
Accepted: 30 April 2021
Published: 6 May 2021

Abstract: Soybeans (*Glycine max*) are an excellent source of dietary proteins and peptides with potential biological activities, such as antihypertensive, anti-cholesterol, and antioxidant activity; moreover, they could prevent cancer. Also, soy contains all the essential amino acids for nutrition; therefore, it represents an alternative to animal proteins. The goal of this paper was the comprehensive characterization of medium-sized and short peptides (two to four amino acids) obtained from simulated gastrointestinal digestion. Two different analytical approaches were employed for peptide characterization, namely a common peptidomic analysis for medium-sized peptides and a suspect screening analysis for short peptides, employing an inclusion list of exact m/z values of all possible amino acid combinations. Moreover, fractionation by preparative reversed-phase liquid chromatography was employed to simplify the starting protein hydrolysate. Six fractions were collected and tested for antioxidative activity by an innovative antioxidant assay on human gastric adenocarcinoma AGS cell lines. The two most active fractions (2 and 3) were then characterized by a peptidomic approach and database search, as well as by a suspect screening approach, in order to identify potential antioxidant amino acid sequences. Some of the peptides identified in these two fractions have been already reported in the literature for their antioxidant activity.

Keywords: antioxidative activity; bioactive peptides; peptidomics; mass spectrometry; soybean

1. Introduction

In the last few decades, the opportunity to prevent, alleviate, or even treat some diseases by assuming functional food or bioactive compounds obtained from food is gaining increasing interest for both researchers and consumers. Therefore, in this field, efforts are aimed at the discovery of new bioactive compounds and new physiological functions of known compounds, as well as the identification of the most suitable and sustainable foods to extract these valuable substances at the industrial level.

It is known that certain amino acid sequences can have one or more biological functions; for this reason, many studies are focused on the identification of bioactive peptides, which can be naturally present in food or can be obtained from the parent protein in which they are encrypted [1,2]. These bioactive peptides are generally 2–20 amino acids long, even if some bioactivities have been attributed to longer sequences (up to 43 amino acids, as in lunasin).

Bioactive peptides can be released from the parent protein during food processing (such as ripening, fermentation, and cooking), storage, or gastrointestinal digestion [2,3]. For research and industrial applications, several methodological approaches are available to obtain potential bioactive peptides encrypted in proteins, including chemical, physical, and biological ones. Among these, treatment with enzymes is the most suitable one for preserving functional and nutritional values of protein hydrolysate [4]. Nonetheless, the high cost, as well as production of bitter-tasting hydrolysates, are some of the main drawbacks in the employment of enzymes. To better simulate the physiological conditions in which peptides are formed from proteins, human digestive enzymes, which are found in the stomach, intestines, and pancreas (e.g., pepsin, trypsin, chymotrypsin, and pancreatin) can be used [5].

Since dietary proteins represent a cheap and valuable source of bioactive peptides, several foods of both animal and plant origin [2,6,7] have been investigated for this aim, including milk [3,8–10], meat, fish [11,12], egg, cereals, and soybean [1,3,7,13].

Soybean (*Glycine max*) has been cultivated for five millennia in Asian countries where, together with its derived products (e.g., soy milk, miso, tofu, etc.), it represents an important source of proteins (ca. 40% of the content) and peptides. Since the last century, soybean cultivation has become widespread in western countries, too [4]. Soy proteins contain all the essential amino acids, and therefore they represent a valid alternative to food of animal origin; furthermore, several biological functions and bioactivities have been attributed to their derived peptides [13], such as cancer and cardiovascular disease prevention, antihypertensive activity [14], hypocholesterolemic effect, and antioxidant properties [1,4,7,15,16]. In particular, the antioxidant activity, i.e., the defense against the free radical damage [17], could be useful in contrasting several pathologies, as well as in nutraceutical and cosmetics applications.

In the present work, we wanted to exploit the antioxidative activity of soybean peptides. Generally, antioxidant sequences are 3–6 amino acid long [6], which are easily metabolized and absorbed in the gastrointestinal tract. Therefore, we focused our attention on short and medium-sized peptides obtained by simulated gastrointestinal digestion of soy protein extracts. First, the soy hydrolysate was separated into six fractions, which were tested for antioxidative activity using an innovative, intracellular reactive oxygen species (ROS) detection assay; in this way, cytotoxic fractions were identified and discarded. Then the peptides, contained in two fractions showing the highest antioxidative properties, were identified using two different strategies for medium-sized and short peptides, respectively. Medium-sized peptides were analyzed by a shotgun proteomic approach, using nano-ultra-high-performance liquid chromatography coupled to tandem mass spectrometry (nanoUHPLC-MS/MS), with identification based on a database search using Proteome Discover software. For short peptides (2–4 amino acid long), the analysis was carried out by UHPLC-MS/MS, following the suspect screening strategy, and their identification was assisted by Compound Discoverer software. Indeed, short peptides have not been fully elucidated in soy yet, mainly because their identification is challenging and cannot be obtained using conventional proteomics approaches and informatics tools. Among the identified peptides, seven were already validated for antioxidative activity and are reported in a public database of food bioactive peptides.

2. Materials and Methods

2.1. Materials

All chemicals, reagents, and organic solvents of the highest grade available were purchased from Sigma-Aldrich (St. Louis, MO, USA) unless otherwise stated. Trifluoroacetic acid (TFA) was supplied by Romil Ltd. (Cambridge, UK). Ultrapure water was prepared by an Arium 611 VF system from Sartorius (Göttingen, Germany). Mass grade solvents used for medium-sized peptides were purchased from VWR International (Milan, Italy).

Optima LC-MS grade water and acetonitrile (ACN), used for short peptide analysis, were purchased from Thermo Fisher Scientific (Waltham, MA, USA).

2.2. Protein Extraction

Soy flour samples were purchased in a local market. An aliquot of 2.7 g of soy flour was extracted with 15 mL of a cold buffer consisting of 6 mol L^{-1} urea, 10 mmol L^{-1} Tris (hydroxymethyl) aminomethane hydrochloride (Tris-HCl, pH 8), 75 mmol L^{-1} NaCl. The sample was vortexed for 50 min, then centrifuged at 9400× *g* at 4 °C for 30 min. All protein samples were quantified by the bicinchoninic acid (BCA) assay using bovine serum albumin (BSA) as standard, and stored at −80 °C until digestion.

Protein Digestion

A simulated in vitro gastrointestinal digestion was performed as previously described [18], with some modifications. The pH of the protein extract was adjusted to 2.0 with 1 mol L^{-1} HCl to mimic the stomach environment, then pepsin was used in an enzyme-to-protein ratio of 1:20. The solution was incubated at 37 °C for 1 h under static conditions. Then, the pH was adjusted to 7.5 with 1 mol L^{-1} NaOH to stop pepsin digestion and create the best environment for pancreatin hydrolysis; pancreatin was added with an enzyme-to-protein ratio of 1:10. After 2 h, α-chymotrypsin was added with an enzyme-to-protein ratio of 1:20. The solution was incubated overnight at 37 °C. Enzymatic hydrolysis was stopped by decreasing the pH to 2.0 with TFA. The volume of the digested extract was reduced to 30 µL by an IKA RV 8 rotary evaporator (IKA-Werke GmbH & Co. KG, Staufen, Germany) for subsequent analyses.

2.3. Purification of Antioxidant Peptides from Soybeans

The hydrolyzed peptides were purified by preparative liquid chromatography (LC) using the Xbridge BEH C18 OBD Prep, 19 mm id × 250 mm column (particle size = 5 µm; Waters, Milford, MA, USA) as previously described [19,20], with some modifications. The chromatographic system was a Shimadzu Prominence LC-20A, including a CBM-20A controller, two LC-20 AP preparative pumps, and a DGU-20A3R inline degasser. An SPD-20A UV detector with a preparative cell (0.5 mm) was used. The FRC-10A (Shimadzu, Kyoto, Japan) auto-collector was employed. Data acquisition was performed by LabSolution software version 5.53 (Shimadzu, Kyoto, Japan). The detector was set at 214 nm.

Chromatography was operated at a flow-rate of 17 mL min^{-1}, using H_2O with 10 mmol L^{-1} ammonium formate at pH 10 as phase A, and methanol/H_2O (90/10, *v*/*v*, with 10 mmol L^{-1} ammonium formate, pH 10) as phase B. The chromatographic gradient was as follows: B was increased from 0% to 50% in 20 min, then B was brought to 95% in 4 min and kept constant for 6 min. The column was equilibrated at starting condition for 6 min. Six fractions were collected every 5 min (as shown in Figure 1). Each peptide fraction was evaporated with a rotary evaporator and stored at −80 °C for subsequent analysis.

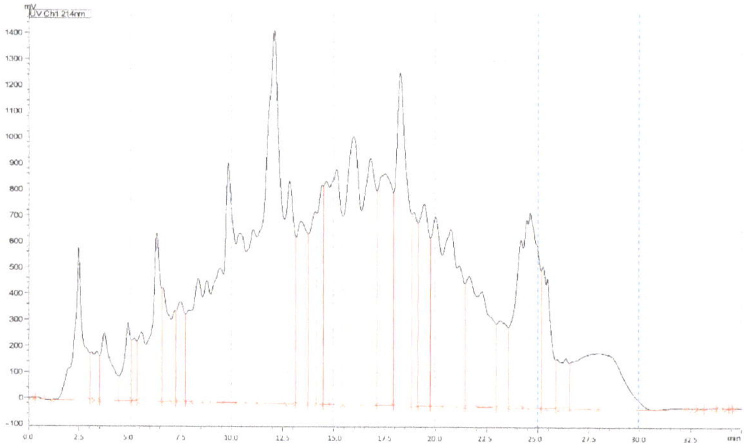

Figure 1. LC-UV chromatogram of hydrolyzed soybean extracts showing the applied fractionation.

2.4. *Bioactivity Assays*

2.4.1. Cell Culture

Human gastric adenocarcinoma cell line AGS (ATCC CRL-1739) was grown in DMEM-F12 medium, supplemented with 10% fetal bovine serum (FBS), 1% glutamine, and 1% penicillin + streptomycin (all from Microtech, Vero Beach, FL, USA) and cultured in T-75 flasks in a humidified incubator containing 5% CO_2 at 37 °C.

2.4.2. Intracellular Reactive Oxygen Species Measurement

AGS cells were split at 80–90% of confluency, seeded (0.5×10^6) in 35 mm culture dishes, and incubated at 37 °C in a 5% CO_2 atmosphere overnight. After cell attachment, ROS detection assay was assayed using dihydrorhodamine 123 (DHR, Sigma Aldrich, St. Louis, MO, USA), as described by Cuomo et al. [21]. Briefly, cells were loaded with DHR (10 μmol L^{-1} for 20 min), and were treated with (1) 1 mmol L^{-1} H_2O_2 for 30 min; (2) 1 mg mL^{-1} soybean peptide fractions for 1 h, and (3) 1 mg mL^{-1} soybean peptide fractions for 1 h and 1 mmol L^{-1} H_2O_2 for 30 min. After treatments, DAPI (Thermo Fischer Scientific, Waltham, MA, USA) was used as a nuclear counterstain, and cells were analyzed with a Zeiss Axioskop 2 Hal100 fluorescence microscope equipped with a digital camera (Nikon, Tokyo, Japan). The excitation and emission wavelengths were 488 and 515 nm, respectively. Images were digitally acquired with exposure times of 100–400 ms, and processed for fluorescence determination with ImageJ software. The fluorescence signal of the intracellular ROS was normalized on the DAPI fluorescence signal.

2.5. *Analysis of Medium-Sized Peptides by nanoUHPLC-MS/MS*

The chromatographic system was an Ultimate 3000 nanoUHPLC (Thermo Scientific, Bremen, Germany) coupled to an Orbitrap Elite mass spectrometer (Thermo Scientific). Pierce LTQ Velos ESI Positive Ion Calibration Solution (Thermo Fisher Scientific) was used to calibrate the instrument once a week, and the mass accuracy was <1.5 ppm.

Medium-sized peptides were analyzed as described in a previous work [22], with some modifications. A 10 μL sample was injected and preconcentrated on a μ-precolumn (300 μm id × 5 mm, Acclaim PepMap 100 C18, particle size 5 μm, pore size 100 Å; Thermo Fisher Scientific) at a 10 μL min^{-1} flow-rate of H_2O/ACN 99:1 (*v*/*v*) containing 0.1% (*v*/*v*) TFA. Then the peptide mixture was separated on an EASY-Spray column (Thermo Fisher Scientific, 75 μm id × 15 cm, PepMap C18, 3 μm particles, 100 Å pore size) operated at 300 nL min^{-1} and 35 °C.

The mobile phase was constituted by (1) H_2O and (2) ACN, both with 0.1% formic acid. The chromatographic gradient, referred to as B, was 1% for 5 min, 1–5% in 2 min,

5–35% in 90 min, and 35–90% in 3 min. Finally, the column was rinsed at 90% B for 10 min and equilibrated at 1% B for 19 min.

Full-scan mass spectra were acquired in the 300–2000 m/z range at 30,000 resolution (full width at half maximum, FWHM, at m/z 400). Tandem mass spectra were acquired at 15,000 resolution (FWHM, at m/z 400) in the top 10 data-dependent acquisition (DDA) mode, with the rejection of singly charged ions and unassigned charge states. Precursor ions were fragmented by higher-energy collisional dissociation (HCD), with 35% normalized collision energy (NCE) and a 2 m/z isolation window. Dynamic exclusion was enabled with a repeat count of 1 and a repeat duration of 30 s, with an exclusion duration of 20 s. For each sample, three technical replicates were performed. Raw data files were acquired by Xcalibur software (version 2.2, Thermo Fisher Scientific).

2.6. *Analysis of Short Peptides by UHPLC-MS/MS*

The short peptides were analyzed by reverse-phase (RP) chromatography, using the Vanquish H UHPLC system coupled to an Q Exactive mass spectrometer (Thermo Fisher Scientific, Bremen, Germany) and a heated electrospray source (ESI) [23]. The instrument was calibrated every 48 h using a Pierce LTQ Velos ESI Positive Ion Calibration Solution (Thermo Fisher Scientific); mass accuracy was <1 ppm. The peptide mixtures were separated on a Kinetex XB-C18 chromatographic column (2.1 mm id × 100 mm, particle size 2.6 μm; Phenomenex, Torrance, CA, USA) kept at 40 °C. The mobile phase consisted of (1) H_2O and (2) ACN, both containing 0.1% TFA (v/v); the flow-rate was 0.4 mL min^{-1}. The gradient, referred to as B, was 1% for 2 min, 1–35% in 20 min, and 35–99% in 3 min. Finally, B was maintained at 99% for 3 min and then lowered again to 1%. The MS acquisition was performed in positive ionization mode in the range m/z 150–750 with a resolution (FWHM, m/z 200) of 70,000, while MS/MS spectra were recorded in top 5 DDA mode, using 35% NCE for HCD and a resolution of 35,000 (FWHM, m/z 200). For the DDA (4980 unique masses), an inclusion list with exact m/z values for individually charged precursor ions was used. The inclusion lists were prepared using MatLab R2018, as previously described [24].

For each sample, three technical replicates were carried out. Raw data files were acquired by Xcalibur software (version 2.2 SP1.48; Thermo Fisher Scientific).

2.7. *Peptide Identification*

2.7.1. Medium-Sized Peptide Identification

The raw files of the digested fractions were submitted to the Proteome Discoverer software (version 1.3; Thermo Scientific) with the Mascot search engine (v.2.3; Matrix Science, London, UK) for the identification of peptides/proteins.

The research was performed against the proteome of *Glycine max* (Soybean) downloaded from Uniprot (http://www.uniprot.org/ accessed on 11 January 2020, 85022 sequences). Nonspecific digestion was chosen, and neither fixed nor variable changes were set. The mass tolerances for the precursor and the product ions were set at 10 ppm and 0.05 Da, respectively. A decoy function was used for false discovery rate calculation, which was set at 1%.

2.7.2. Short Peptide Identification

The identification of short peptides in fractionated gastrointestinal digests was performed following a data processing workflow implemented on Compound Discoverer (v. 3.1, Thermo Fisher Scientific, Bremen, Germany) [25]. The masses were extracted from the raw files based on customized parameters, aligning the signals, removing empty or missing MS/MS spectrum signals, and using short lists of complete peptides to match a possible composition. Manual validation of MS/MS spectra was also aided by the Compound Class Assessment Tool, which allowed to automatically match typical product ions from amino acids at the N-terminus, C-terminus, and in the middle of the sequence, and assign them to 20 classes of compounds (one for each natural amino acid). Provisional identification of short peptides was obtained according to diagnostic fragmentation spectra, aided by mMass, which allows for in silico fragmentation of peptides [26].

3. Results and Discussion

3.1. Sample Preparation

3.1.1. Extraction

Three different extraction protocols were tested in terms of protein recovery, also aiming to minimize interference content in the extract. Furthermore, the total time required for sample preparation was also considered for a possible industrial scale-up of the process. Experimental details of protocols I and II are reported in Appendix A, whereas protocol III was selected for this work, and is reported in the experimental section. Briefly, protocol I [27] was based on the use of an extraction buffer containing sodium dodecyl sulfate (SDS), whereas glass beads and a buffer containing sodium deoxycholate (SDC) were used in protocol II [28] to enhance the lysis of cell walls. For protocol III, a buffer containing urea was used.

A BCA assay was employed to evaluate the protein extraction recovery of the three protocols, and results are shown in Figure 2. Protocol I gave the worst extraction efficiency, and was the most time-consuming; therefore, it is not suitable for a possible industrial scale-up. Protocol II gave the best result; however, the employment of glass beads does not allow us to extract large flour samples. Protocol III provided similar results and was the fastest, and for these reasons was selected.

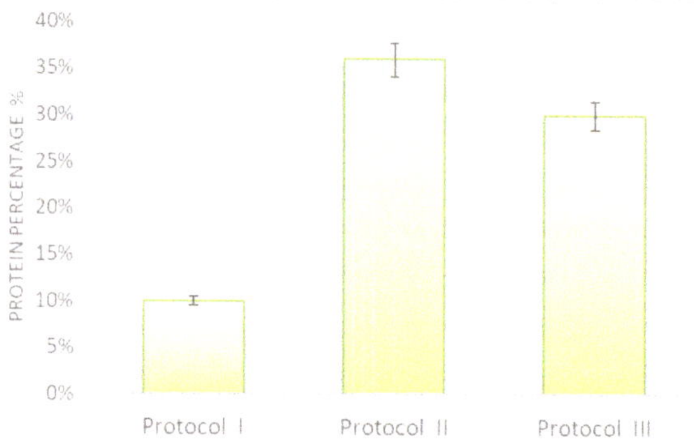

Figure 2. Protein recovery percentage obtained by applying three different extraction protocols. Evaluation was carried out by BCA assay, using BSA as a standard.

In protocol III, the employment of urea allows us to not only denature proteins but also to solubilize and extract them, as urea is a mild chaotropic agent. Moreover, all the other buffer constituents were compatible with the following analysis steps.

3.1.2. Gastrointestinal Digestion

In the gut, endogenous proteases, such as pepsin, trypsin, and chymotrypsin, hydrolyze proteins into peptides, which will be further processed by peptidases in the intestinal tract [29]. Therefore, to reproduce gastrointestinal digestion, a sequence of different enzymes is generally used, mainly pepsin, trypsin, chymotrypsin, and papain [17].

To obtain putative bioactive peptides from soy by simulated gastrointestinal digestion, different enzymes were employed, mainly alcalase [16,30,31]; however, other proteases, also in combination, were used as well, including bromelain, papain, pepsin, flavourzyme, neutrase, protemax, and transglutaminase [4].

In this work, we sequentially used pepsin, papain, and chymotrypsin. During hydrolysis, temperature, pH, and enzyme-to-substrate ratio were carefully checked to obtain reproducible results (data not shown).

3.2. Soybean-Derived, Hydrolyzed Peptides in Oxidative Stress In Vitro

Recently, the antioxidant role of soybean, and specifically of peptides derived from hydrolysed soybean proteins, has been reported [15,32,33]. ROS are molecules physiologically produced during cell metabolism, participating in cell proliferation and survival. Nevertheless, external stimuli, i.e., environmental factors, xenobiotics, and microbial infection, may contribute to ROS accumulation, causing an imbalance between their production and removal by cellular antioxidant systems [34]. This can result in oxidative stress, leading to cell damage.

Using a DHR cell-permeable fluorogenic probe, which is useful to detect ROS production, the ability of bioactive soybean peptides was investigated to prevent ROS formation in H_2O_2 stimulated cells. Among the six peptide fractions, we tested the antioxidant activity of fractions 2 and 3. The remaining fractions were excluded because of their cytotoxic effect (data not shown). Indeed, the choice of using a gastric cell line was determined by the need for evaluating the potential cytotoxic effect of the derived peptide fractions, justifying its potential use as functional food. As shown in Figure 3, the cells pre-treated with both fractions 2 and 3 displayed reduced ROS levels compared to H_2O_2-treated cells. Interestingly, compared to fraction 3, fraction 2 exhibited a higher protective effect against oxidative stress induced by H_2O_2, almost returning the cells to the control group. These results suggest that soybean peptides have important antioxidant activities, with particular attention to peptides contained in fraction 2.

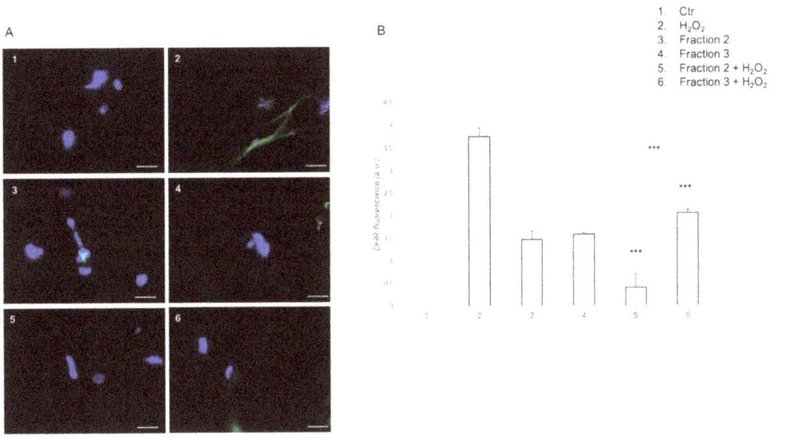

Figure 3. Soybean peptides prevents H_2O_2-induced, intracellular ROS formation. (**A**) DHR-loaded cells were pre-treated for 1 h with peptide fractions 2 or 3 (1 mg mL^{-1}), then incubated for further 30 min with H_2O_2 (1 mmol L^{-1}), and finally observed and analyzed with a fluorescence microscope (scale bar: 100 μm); (**B**) Quantification of the mean fluorescence of individual cells. Results are expressed as arbitrary units and represent the average ± SD calculated from three independent experiments, each performed in duplicate. Statistical analysis was performed with ANOVA followed by Bonferroni's test. *** $p < 0.001$, comparing 5 with 2 or 6 and 6 with 2.

Compared to classical assays used to evaluate the antioxidant activity of soy hydrolysates, such as 2,2-diphenyl-1-picrylhydrazyl (DPPH) [16], oxygen radical absorbance capacity (ORAC) [31], lipid peroxidation inhibitory activity [30], and the employment of an intracellular ROS measurement allowed also to recognize the cytotoxic effect of some fractions.

3.3. Identification of Peptides in Fractions with High Antioxidative Activity

The antioxidant function of soy peptides has been attributed to multiple mechanisms, including hydrogen donation, scavenging of hydroxyl radicals, transition-metal ion chelation, and active-oxygen quenching; therefore, it is widely accepted that there is a synergistic effect [4]. Nonetheless, it is still interesting to characterize the most bioactive species, also to better investigate the amino acid sequences responsible for the antioxidative activity. Therefore, the peptides contained in the two most active fractions, namely 2 and 3, were fully characterized and searched in BIOPEP-UMW database for confirmation.

3.3.1. Analysis of Short Peptides

The identification of short peptides is challenging for several reasons, as already reported [24]. Briefly, the main issues consist of the low possibility of identifying short sequences (<5 amino acids) by a proteomic database search [35], also due to the high occurrence of isobaric peptides and the scarce ionization efficiency of short peptides, which mainly form monoprotonated molecules in positive ESI. Moreover, scarce fragmentation data can be obtained. Therefore, a previously developed, specific method was employed for short peptide identification. Following a HRMS-based suspect screening approach, a list containing the exact masses of precursors relative to all the possible combinations of the 20 natural amino acids (from two to four, resulting in 168,400 unique combinations) was used.

A total of 132 unique amino acid sequences were identified in fractions 2 and 3. However, under the operating conditions, it was not possible to discriminate the isobaric leucine and isoleucine by tandem mass spectra; therefore, the occurrence of the two amino acids within the identified sequence was retained as equally probable (see Table S1 of Supplementary Materials). Consequently, 203 possible sequences were searched in BIOPEP-UWM database [36], where, on a total of 4216 food bioactive peptides, there are 689 validated sequences for antioxidative activity.

Table 1 reports the short peptides identified in fractions 2 and 3 of the soybean hydrolysate and matching antioxidative sequences reported in BIOPEP-UWM database. The sequences IR and LK, obtained from ovotransferrin protein, showed radical-scavenging activity by ORAC assay [37]. The dipeptide AW was obtained from the marine bivalve *Mactra veneriformis*, and it showed hydroxyl, DPPH, and superoxide radical scavenging activities [38]. The dipeptide EL from casein positively responded to SOSA and DPPH assays [39]. The two last sequences, LH and ADF, were identified in soybean [40] and one of its derived products [41], respectively, and both were tested for antioxidative activity against the peroxidation of linoleic acid.

Table 1. Short peptides matching with BIOPEP-UWM sequences validated for antioxidative activity. Peak areas are average values of all the replicates.

Sequence	Peak Area Fraction 2 ($\times 10^6$)	Peak Area Fraction 3 ($\times 10^6$)	Peptide Source
Ile–Arg (IR)	13	815	Egg white ovotransferrin [37]
Leu–Lys (LK)	692	0.4	Egg white ovotransferrin [37]
Ala–Trp (AW)	1	332	Marine bivalve [38]
Glu–Leu (EL)	261	2	Milk casein [39]
Leu–His (LH)	2	92	Soybean [40]
Ala–Asp–Phe (ADF)	124	0.9	Okara [41]

Most of the other identified di-, tri-, and tetrapeptides are reported in the BIOPEP-UWM database with some biological functions, and the dipeptidyl peptidase IV inhibitor and ACE inhibitor are the most frequently occurring.

The list of the identified short sequences was also submitted to PeptideRanker [42], a server for the prediction of bioactive peptides based on a novel N-to-1 neural network, which assigns a generic bioactivity probability rank. Although 0.5 is the threshold value for labeling a peptide as bioactive, a 0.8 threshold is suggested to reduce false-positive predictions. The rank assignment of the 203 short unique amino acid sequences is reported in Table S2 of the Supplementary Materials: 32 and 15 short peptides had a rank above 0.5 and 0.8, respectively. The latter included 10 dipeptides, 3 tripeptides, and 2 tetrapeptides. Of the six antioxidant peptides found in BIOPEP-UWM, only AW and ADF had a high rank, namely 0.9669 and 0.8062, respectively.

Concerning the amino acid composition of bioactive peptides, the presence of hydrophobic amino acid residues, such as W, F, P, G, K, I, and V, at both the N-terminus and C-terminus, are relevant for peptide antioxidative activity, as well as H and R at the C-terminus position [43]. In particular, the presence of H has been associated with metal-ion chelator, active-oxygen quencher, and hydroxyl radical scavenger properties [4,7]. Another contribution to the antioxidative activity (by free radical quenching) is provided by the excess electrons of the negatively charged residues E and D. Also, the relative positions of these amino acids can determine or enhance their activity [44].

All the bioactive short peptides reported in Table 1 present at least one of these characteristic residues. Moreover, all the identified sequences of our soybean hydrolysate with a prediction rank >0.8 assigned by PeptideRanker contain at least one of these amino acids.

3.3.2. Analysis of Medium-Sized Peptides

As expected, most of the medium-sized peptides are derived from glycinin, which is the major seed storage protein of soybean, β-conglycinin, and their isoforms and subunits. Indeed, these two proteins account for up to 80–90% of the total soy protein content. The peptide identification in fractions 2 and 3 are reported in Tables S3 and S4 of the Supplementary Materials, respectively.

Table 2 shows the only medium-sized peptide, identified in fraction 3, that matched with a validated antioxidant sequence in the BIOPEP-UWM database.

Table 2. Medium-sized peptide identified in fraction 3 that matched with a BIOPEP-UWM sequence validated for antioxidative activity.

Sequence	Peptide Source	PeptideRanker Score
VNPESQQGSPR	Soy [17]	0.155249

PeptideRanker was used to predict the general bioactivity of the medium-sized peptides as well. The peptide reported in Table 2 obtained a very low prediction score. In general, among the 136 sequences identified in fraction 2, 33 and 19 peptides had bioactivity prediction scores >0.5 and >0.8, respectively (see Table S5 in Supplementary Materials). Concerning fraction 3, only 3 and 1 sequences out of a total of 66 presented a score >0.5 and >0.8, respectively (see Table S6 of Supplementary Materials).

3.3.3. Further Considerations on Identified Peptides

The characteristics of peptides identified in the two active fractions were further analyzed by their grand average of hydropathy (GRAVY) index score, which is usually used to estimate the hydrophobicity or hydrophilicity of a protein. The resulting index, ranging between −2 and +2, will be positive for hydrophobic amino acid sequences. Figure 4A,B reports the GRAVY value distributions for short and medium-sized peptides. As is known, medium-sized peptides are more hydrophilic than short peptides.

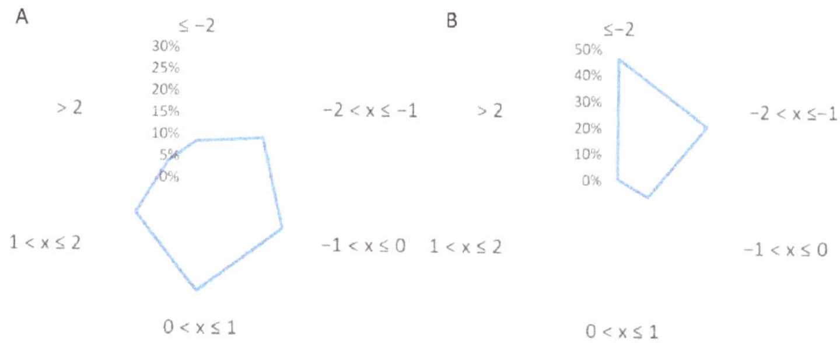

Figure 4. Gravy index score distribution for (**A**) short peptides and (**B**) medium-sized peptides identified in fractions 2 and 3 of soybean hydrolysate.

Generally, antioxidant peptides, besides containing the amino acids reported above, have a low molecular mass and a GRAVY score ranging between −0.5 and +0.5 [45]. About 50% of the short peptides identified in our research also meet the last criterion, as can be seen in Figure 4A. Regarding medium-size peptides, this requisite is possessed only by a small percentage (Figure 4B).

4. Conclusions

In this work, peptide fractions obtained by the simulated gastrointestinal digestion of soybean protein extracts were evaluated for antioxidative activity by an intracellular test. Only the most active and non-cytotoxic fractions were fully characterized for their short and medium-sized peptide content. The analysis of the two peptide typologies required two different analytical approaches and two diverse software packages for their identification. In particular, the challenge in the identification of short peptides is the impossibility of following the classical informatics tools generally employed in proteomics analysis.

Among the various peptide sequences identified, five dipeptides, one tripeptide, and one medium-sized peptide had already been reported to possess antioxidative activity in the most up-to-date database of food bioactive peptides. In particular, one dipeptide, the tripeptide, and the medium-sized peptide were already identified in soy or a soy by-product. A further purification of the two active fractions (containing almost the same peptides) could help in identifying the peptides with the highest antioxidant function. This could be particularly interesting for the short peptides, which are less frequently investigated for the known issues related to their analysis, and which have not been fully elucidated in soy yet. Nevertheless, it is important to keep in mind that the amino acid sequences could exert their bioactivity only in combination or with a synergic effect. Furthermore, the short peptides could be tested for assessing other bioactivities.

Nowadays, soy cultivation is widespread in several countries, mainly for feed production. Nonetheless, the economic benefits derived from its intensive cultivation are controversial, because of the negative ecological impact on soil, forests, and other plant cultivation. For this reason, the possibility of maximizing the exploitation of this plant and its by-products by extracting the most active amino acid sequences for several industrial applications, including nutraceuticals, could support its sustainable employment.

Supplementary Materials: The following are available online at https://www.mdpi.com/article/10.3390/antiox10050734/s1, Table S1: Short peptide identification, Table S2: Short peptide bioactivity prediction by PeptideRanker, Table S3: Medium-sized peptides identified in fraction 2, Table S4: Medium-sized peptides identified in fraction 3, Table S5: Medium-sized peptide (fraction 2) bioactivity prediction by PeptideRanker, Table S6: Medium-sized peptides (fraction 3) bioactivity prediction by PeptideRanker.

Author Contributions: Conceptualization, A.L.C. and A.L.; methodology, P.C., S.E.A. and A.M.I.M.; software, C.M.M., S.P. and P.C.; validation, A.C. and R.C.; formal analysis, C.C. and A.L.C.; investigation, C.M.M., A.C. and P.C.; resources, C.C. and R.C.; data curation, P.C. and C.M.M.; writing—

original draft preparation, C.C., R.C. and C.M.M.; writing—review and editing, S.P.; visualization, A.M.I.M. and S.E.A.; supervision, A.L.C. and R.C.; project administration, A.L. and R.C.; funding acquisition, A.L.C. All authors have read and agreed to the published version of the manuscript.

Funding: This research received no external funding.

Institutional Review Board Statement: Not applicable.

Informed Consent Statement: Not applicable.

Data Availability Statement: Data is contained within the article.

Conflicts of Interest: The authors declare no conflict of interest.

Appendix A. Comparison of Three Protein Extraction Protocols

Protocol I: soy flour (2.7 g) was added to 1 mL of extraction buffer, i.e., 50 mmol L^{-1} Tris (pH 8.8), 15 mmol L^{-1} KCl, 20 mmol L^{-1} dithiothreitol (DTT), and 1% (*w*/*v*) sodium dodecyl sulfate (SDS) [27]. Then, the sample was placed on ice for 1 h, and vortexed for 1 min every 15 min. Insoluble matter was removed by centrifugation at 4 °C for 15 min at 11,000× *g*. The supernatant was collected, and the proteins were precipitated overnight at −20 °C by adding four volumes of acetone. After centrifugation at 11,000× *g* at 4 °C for 30 min, the pellet was dried and dissolved in 6 mol L^{-1} of urea and 10 mmol L^{-1} Tris-HCl (pH 8).

Protocol II: proteins were extracted from soy flour with the aid of glass beads [28]. A 25 mg soy sample was added to 150 mg glass microbeads and 1 mL lysis buffer (60 mmol L^{-1} Tris pH 9, 15 mmol L^{-1} KCl, 20 mmol L^{-1} DTT, 2% (*w*/*v*) SDC). The sample was placed in an ultrasonic bath for 4 h, under agitation for 1 h every 30 min. Then, the sample was incubated at 100 °C for 30 min; centrifugation at 4 °C at 11,000× *g* for 15 min allowed us to remove insoluble matter. The supernatant was collected as proteins precipitated, as described above.

Protocol III: this is reported in the experimental section, and was used for this work.

References

1. Wang, W.; de Mejia, E.G. A New Frontier in Soy Bioactive Peptides that May Prevent Age-related Chronic Diseases. *Compr. Rev. Food Sci. Food Saf.* **2005**, *4*, 63–78. [CrossRef] [PubMed]
2. Daroit, D.J.; Brandelli, A. In vivo bioactivities of food protein-derived peptides—A current review. *Curr. Opin. Food Sci.* **2021**, *39*, 120–129. [CrossRef]
3. Fernández-Tomé, S.; Hernández-Ledesma, B. Gastrointestinal Digestion of Food Proteins under the Effects of Released Bioactive Peptides on Digestive Health. *Mol. Nutr. Food Res.* **2020**, *64*, 2000401. [CrossRef] [PubMed]
4. Ashaolu, T.J. Health Applications of Soy Protein Hydrolysates. *Int. J. Pept. Res. Ther.* **2020**, *26*, 2333–2343. [CrossRef]
5. Capriotti, A.L.; Caruso, G.; Cavaliere, C.; Samperi, R.; Ventura, S.; Zenezini Chiozzi, R.; Laganà, A. Identification of potential bioactive peptides generated by simulated gastrointestinal digestion of soybean seeds and soy milk proteins. *J. Food Compos. Anal.* **2015**, *44*, 205–213. [CrossRef]
6. Piovesana, S.; Capriotti, A.L.; Cavaliere, C.; La Barbera, G.; Montone, C.M.; Zenezini Chiozzi, R.; Laganà, A. Recent trends and analytical challenges in plant bioactive peptide separation, identification and validation. *Anal. Bioanal. Chem.* **2018**, *410*, 3425–3444. [CrossRef]
7. Matemu, A.; Nakamura, S.; Katayama, S. Health Benefits of Antioxidative Peptides Derived from Legume Proteins with a High Amino Acid Score. *Antioxidants* **2021**, *10*, 316. [CrossRef]
8. Zenezini Chiozzi, R.; Capriotti, A.L.; Cavaliere, C.; La Barbera, G.; Piovesana, S.; Samperi, R.; Laganà, A. Purification and identification of endogenous antioxidant and ACE-inhibitory peptides from donkey milk by multidimensional liquid chromatography and nanoHPLC-high resolution mass spectrometry. *Anal. Bioanal. Chem.* **2016**, *408*, 5657–5666. [CrossRef] [PubMed]
9. Capriotti, A.L.; Cavaliere, C.; Piovesana, S.; Samperi, R.; Laganà, A. Recent trends in the analysis of bioactive peptides in milk and dairy products. *Anal. Bioanal. Chem.* **2016**, *408*, 2677–2685. [CrossRef]
10. Piovesana, S.; Capriotti, A.L.; Cavaliere, C.; La Barbera, G.; Samperi, R.; Zenezini Chiozzi, R.; Laganà, A. Peptidome characterization and bioactivity analysis of donkey milk. *J. Proteomics* **2015**, *119*, 21–29. [CrossRef]
11. Capriotti, A.L.; Cavaliere, C.; Foglia, P.; Piovesana, S.; Samperi, R.; Zenezini Chiozzi, R.; Laganà, A. Development of an analytical strategy for the identification of potential bioactive peptides generated by in vitro tryptic digestion of fish muscle proteins. *Anal. Bioanal. Chem.* **2015**, *407*, 845–854. [CrossRef]

12. Halim, N.R.A.R.A.; Yusof, H.M.M.; Sarbon, N.M.M. Functional and bioactive properties of fish protein hydolysates and peptides: A comprehensive review. *Trends Food Sci. Technol.* **2016**, *51*, 24–33. [CrossRef]
13. Singh, B.P.; Vij, S.; Hati, S. Functional significance of bioactive peptides derived from soybean. *Peptides* **2014**, *54*, 171–179. [CrossRef]
14. Xu, Z.; Wu, C.; Sun-Waterhouse, D.; Zhao, T.; Waterhouse, G.I.N.; Zhao, M.; Su, G. Identification of post-digestion angiotensin-I converting enzyme (ACE) inhibitory peptides from soybean protein Isolate: Their production conditions and in silico molecular docking with ACE. *Food Chem.* **2021**, *345*, 128855. [CrossRef]
15. Zhang, Q.; Tong, X.; Li, Y.; Wang, H.; Wang, Z.; Qi, B.; Sui, X.; Jiang, L. Purification and Characterization of Antioxidant Peptides from Alcalase-Hydrolyzed Soybean (*Glycine max* L.) Hydrolysate and Their Cytoprotective Effects in Human Intestinal Caco-2 Cells. *J. Agric. Food Chem.* **2019**, *67*, 5772–5781. [CrossRef] [PubMed]
16. Liu, T.X.; Zhao, M. Physical and chemical modification of SPI as a potential means to enhance small peptide contents and antioxidant activity found in hydrolysates. *Innov. Food Sci. Emerg. Technol.* **2010**, *11*, 677–683. [CrossRef]
17. Tonolo, F.; Moretto, L.; Grinzato, A.; Fiorese, F.; Folda, A.; Scalcon, V.; Ferro, S.; Arrigoni, G.; Bellamio, M.; Feller, E.; et al. Fermented Soy-Derived Bioactive Peptides Selected by a Molecular Docking Approach Show Antioxidant Properties Involving the Keap1/Nrf2 Pathway. *Antioxidants* **2020**, *9*, 1306. [CrossRef] [PubMed]
18. Cerrato, A.; Capriotti, A.L.; Capuano, F.; Cavaliere, C.; Montone, A.M.I.; Montone, C.M.; Piovesana, S.; Zenezini Chiozzi, R.; Laganà, A. Identification and Antimicrobial Activity of Medium-Sized and Short Peptides from Yellowfin Tuna (*Thunnus albacares*) Simulated Gastrointestinal Digestion. *Foods* **2020**, *9*, 1185. [CrossRef] [PubMed]
19. Caliceti, C.; Capriotti, A.L.; Calabria, D.; Bonvicini, F.; Zenezini Chiozzi, R.; Montone, C.M.; Piovesana, S.; Zangheri, M.; Mirasoli, M.; Simoni, P.; et al. Peptides from Cauliflower By-Products, Obtained by an Efficient, Ecosustainable, and Semi-Industrial Method, Exert Protective Effects on Endothelial Function. *Oxid. Med. Cell. Longev.* **2019**, *2019*, 1–13. [CrossRef] [PubMed]
20. Montone, C.M.; Zenezini Chiozzi, R.; Marchetti, N.; Cerrato, A.; Antonelli, M.; Capriotti, A.L.; Cavaliere, C.; Piovesana, S.; Laganà, A. Peptidomic approach for the identification of peptides with potential antioxidant and anti-hyperthensive effects derived from Asparagus by-products. *Molecules* **2019**, *24*, 3627. [CrossRef]
21. Cuomo, P.; Papaianni, M.; Fulgione, A.; Guerra, F.; Capparelli, R.; Medaglia, C. An Innovative Approach to Control H. pylori-Induced Persistent Inflammation and Colonization. *Microorganisms* **2020**, *8*, 1214. [CrossRef]
22. Cerrato, A.; Aita, S.E.; Cavaliere, C.; Laganà, A.; Montone, C.M.; Piovesana, S.; Zenezini Chiozzi, R.; Capriotti, A.L. Comprehensive identification of native medium-sized and short bioactive peptides in sea bass muscle. *Food Chem.* **2021**, *343*, 128443. [CrossRef]
23. Montone, C.M.; Capriotti, A.L.; Cerrato, A.; Antonelli, M.; La Barbera, G.; Piovesana, S.; Laganà, A.; Cavaliere, C. Identification of bioactive short peptides in cow milk by high-performance liquid chromatography on C18 and porous graphitic carbon coupled to high-resolution mass spectrometry. *Anal. Bioanal. Chem.* **2019**, *411*, 3395–3404. [CrossRef] [PubMed]
24. Piovesana, S.; Montone, C.M.; Cavaliere, C.; Crescenzi, C.; La Barbera, G.; Laganà, A.; Capriotti, A.L. Sensitive untargeted identification of short hydrophilic peptides by high performance liquid chromatography on porous graphitic carbon coupled to high resolution mass spectrometry. *J. Chromatogr. A* **2019**, *1590*, 73–79. [CrossRef]
25. Cerrato, A.; Aita, S.E.; Capriotti, A.L.; Cavaliere, C.; Montone, C.M.; Laganà, A.; Piovesana, S. A new opening for the tricky untargeted investigation of natural and modified short peptides. *Talanta* **2020**, *219*, 121262. [CrossRef]
26. Niedermeyer, T.H.J.; Strohalm, M. mMass as a Software Tool for the Annotation of Cyclic Peptide Tandem Mass Spectra. *PLoS ONE* **2012**, *7*, e44913. [CrossRef]
27. Capriotti, A.L.; Cavaliere, C.; Piovesana, S.; Stampachiacchiere, S.; Ventura, S.; Zenezini Chiozzi, R.; Laganà, A. Characterization of quinoa seed proteome combining different protein precipitation techniques: Improvement of knowledge of nonmodel plant proteomics. *J. Sep. Sci.* **2015**, *38*, 1017–1025. [CrossRef] [PubMed]
28. Montone, C.M.; Capriotti, A.L.; Cavaliere, C.; La Barbera, G.; Piovesana, S.; Zenezini Chiozzi, R.; Laganà, A. Peptidomic strategy for purification and identification of potential ACE-inhibitory and antioxidant peptides in Tetradesmus obliquus microalgae. *Anal. Bioanal. Chem.* **2018**, *410*, 3573–3586. [CrossRef]
29. Yu, S.; Bech Thoegersen, J.; Kragh, K.M. Comparative study of protease hydrolysis reaction demonstrating Normalized Peptide Bond Cleavage Frequency and Protease Substrate Broadness Index. *PLoS ONE* **2020**, *15*, e0239080. [CrossRef] [PubMed]
30. Park, S.Y.; Lee, J.-S.S.; Baek, H.-H.H.; Lee, H.G. Purification and characterization of antioxidant peptides from soy protein hydrolysate. *J. Food Biochem.* **2010**, *34*, 120–132. [CrossRef]
31. Darmawan, R.; Bringe, N.A.; de Mejia, E.G. Antioxidant Capacity of Alcalase Hydrolysates and Protein Profiles of Two Conventional and Seven Low Glycinin Soybean Cultivars. *Plant Foods Hum. Nutr.* **2010**, *65*, 233–240. [CrossRef] [PubMed]
32. Jiménez-Escrig, A.; Alaiz, M.; Vioque, J.; Rupérez, P. Health-promoting activities of ultra-filtered okara protein hydrolysates released by in vitro gastrointestinal digestion: Identification of active peptide from soybean lipoxygenase. *Eur. Food Res. Technol.* **2010**, *230*, 655–663. [CrossRef]
33. Yi, G.; ud Din, J.; Zhao, F.; Liu, X. Effect of soybean peptides against hydrogen peroxide induced oxidative stress in HepG2 cells via Nrf2 signaling. *Food Funct.* **2020**, *11*, 2725–2737. [CrossRef]
34. Pizzino, G.; Irrera, N.; Cucinotta, M.; Pallio, G.; Mannino, F.; Arcoraci, V.; Squadrito, F.; Altavilla, D.; Bitto, A. Oxidative Stress: Harms and Benefits for Human Health. *Oxid. Med. Cell. Longev.* **2017**, *2017*, 1–13. [CrossRef]

35. Koskinen, V.R.; Emery, P.A.; Creasy, D.M.; Cottrell, J.S. Hierarchical clustering of shotgun proteomics data. *Mol. Cell. Proteomics* **2011**, *10*, M110.003822. [CrossRef]
36. Minkiewicz, P.; Iwaniak, A.; Darewicz, M. BIOPEP-UWM Database of Bioactive Peptides: Current Opportunities. *Int. J. Mol. Sci.* **2019**, *20*, 5978. [CrossRef]
37. Huang, W.Y.; Majumder, K.; Wu, J. Oxygen radical absorbance capacity of peptides from egg white protein ovotransferrin and their interaction with phytochemicals. *Food Chem.* **2010**, *123*, 635–641. [CrossRef]
38. Liu, R.; Zheng, W.; Li, J.; Wang, L.; Wu, H.; Wang, X.; Shi, L. Rapid identification of bioactive peptides with antioxidant activity from the enzymatic hydrolysate of Mactra veneriformis by UHPLC–Q-TOF mass spectrometry. *Food Chem.* **2015**, *167*, 484–489. [CrossRef] [PubMed]
39. Suetsuna, K.; Ukeda, H.; Ochi, H. Isolation and characterization of free radical scavenging activities peptides derived from casein. *J. Nutr. Biochem.* **2000**, *11*, 128–131. [CrossRef]
40. Chen, H.M.; Muramoto, K.; Yamauchi, F.; Nokihara, K. Antioxidant Activity of Designed Peptides Based on the Antioxidative Peptide Isolated from Digests of a Soybean Protein. *J. Agric. Food Chem.* **1996**, *44*, 2619–2623. [CrossRef]
41. Yokomizo, A.; Takenaka, Y.; Takenaka, T. Antioxidative Activity of Peptides Prepared from Okara Protein. *Food Sci. Technol. Res.* **2002**, *8*, 357–359. [CrossRef]
42. Mooney, C.; Haslam, N.J.; Pollastri, G.; Shields, D.C. Towards the Improved Discovery and Design of Functional Peptides: Common Features of Diverse Classes Permit Generalized Prediction of Bioactivity. *PLoS ONE* **2012**, *7*, e45012. [CrossRef] [PubMed]
43. Torres-Fuentes, C.; del Mar Contreras, M.; Recio, I.; Alaiz, M.; Vioque, J. Identification and characterization of antioxidant peptides from chickpea protein hydrolysates. *Food Chem.* **2015**, *180*, 194–202. [CrossRef]
44. Zou, T.-B.; He, T.-P.; Li, H.-B.; Tang, H.-W.; Xia, E.-Q. The Structure-Activity Relationship of the Antioxidant Peptides from Natural Proteins. *Molecules* **2016**, *21*, 72. [CrossRef]
45. Ji, D.; Udenigwe, C.C.; Agyei, D. Antioxidant peptides encrypted in flaxseed proteome: An in silico assessment. *Food Sci. Hum. Wellness* **2019**, *8*, 306–314. [CrossRef]

antioxidants

MDPI

Article

Effects of Drying Methods on Antioxidant, Anti-Inflammatory, and Anticancer Potentials of Phenolic Acids in Lovage Elicited by Jasmonic Acid and Yeast Extract

Urszula Złotek [1,*], Sławomir Lewicki [2,3], Anna Markiewicz [2], Urszula Szymanowska [1] and Anna Jakubczyk [1]

[1] Department of Biochemistry and Food Chemistry, University of Life Sciences in Lublin, Skromna 8, 20-704 Lublin, Poland; urszula.szymanowska@up.lublin.pl (U.S.); anna.jakubczyk@up.lublin.pl (A.J.)

[2] Department of Microwave Safety, Military Institute of Hygiene and Epidemiology, 01-163 Warsaw, Poland; slawomir.lewicki@wihe.pl (S.L.); anna.kucza7@gmail.com (A.M.)

[3] Department of Medicine, Faculty of Medical Sciences and Health Sciences, Kazimierz Pulaski University of Technology and Humanities, 26-600 Radom, Poland

* Correspondence: urszula.zlotek@up.lublin.pl; Tel.: +48-81-462-33-28

Abstract: The study presents the effect of drying methods (traditional, convection, microwave, and freeze-drying) on the content and bioactivity (determined as antioxidative, anti-inflammatory, and antiproliferative potential) of potentially bioavailable fractions of phenolic acids contained in lovage elicited with jasmonic acid (JA) and yeast extract (YE) and in untreated control leaves. The highest amount of syringic acid was recorded in the convectionally dried lovage samples, while ethanolic extracts from lyophilized lovage had the highest content of protocatechuic and caffeic acids. The drying method significantly influenced the tested properties only in some cases. The traditional drying resulted in lower antioxidant potential, while convectional drying caused a reduction of the lipoxygenase inhibition ability of the samples after simulated digestion. Samples containing the control and elicited lovage leaves dried with convectional and traditional methods exhibited the highest cytotoxicity against a prostate cancer epithelial cell line.

Keywords: lovage; elicitation; phenolic acids; antioxidant activity; potential anti-inflammatory potential; anticancer properties

Citation: Złotek, U.; Lewicki, S.; Markiewicz, A.; Szymanowska, U.; Jakubczyk, A. Effects of Drying Methods on Antioxidant, Anti-Inflammatory, and Anticancer Potentials of Phenolic Acids in Lovage Elicited by Jasmonic Acid and Yeast Extract. *Antioxidants* **2021**, *10*, 662. https://doi.org/10.3390/antiox10050662

Academic Editor: Irene Dini and Domenico Montesano

Received: 26 March 2021
Accepted: 22 April 2021
Published: 24 April 2021

Publisher's Note: MDPI stays neutral with regard to jurisdictional claims in published maps and institutional affiliations.

1. Introduction

The food industry has recently become increasingly interested in herbs due to their organoleptic, pro-health, and preservative properties. The *Apiaceae* family is one of the groups of herbal plants comprising many herbs with culinary and medical importance. Herbs from this family are rich in bioactive compounds that are characterized by many health-promoting properties, i.e., antioxidant, antibacterial, hepatoprotective, vasorelaxant, anti-inflammatory, and antitumor activities [1]. Lovage, i.e., one of the important plants from the *Apiaceae* family, is especially popular in cooking, mainly for its aromatic properties [2]. Additionally, lovage has many bioactive properties, which predispose it to be used in other applications, e.g., in medicine, as indicated in some publications. This herb has been used in folk medicine mainly due to its carminative, spasmolytic, and diuretic effects, but there is ample information regarding other pro-health properties of lovage, e.g., antioxidant, antibacterial, hepatoprotective, and anticancer activities [2,3]. When considering this fact and the growing consumers' interest in natural spices related to the growing awareness of the impact of food on human health, enhancement of the health potential of food of plant origin is still an important issue.

Previous studies have indicated that elicitation can be an effective and safe method for the improvement of the health-promoting properties of selected herbs, including lovage [3–5]. Elicitation with jasmonic acid and yeast extract can improve some biological activities via an enhancement of the production of phenolic acids in fresh lovage leaves, as shown in our previous study [3,5].

In turn, there is no information regarding the effect of the drying methods on preservation of these properties in elicited herbs. Drying, which is commonly used in food technology, is a very valuable method for the enhancement of the storability of herbs after harvest. However, it can cause changes in the organoleptic quality, which has been analyzed in many studies [6,7]. Drying may also affect the health properties of herbs, but there is very limited information on this subject (especially for elicited herbs).

The aim of the present study was to evaluate the influence of drying methods on the content and bioactivity (determined as antioxidative, anti-inflammatory, and antiproliferative potential) of potentially bioavailable fractions of phenolic acids that are contained in lovage leaves elicited with jasmonic acid and yeast extracts.

2. Materials and Methods

2.1. Growth Conditions

Lovage (*Levisticum officinale* Koch. cv. Elsbetha) leaves growing as a control and elicited with 10 µM of jasmonic acid (JA) and 0.1% yeast extract (YE) were the plant materials that were used in this study. The growth conditions and elicitation method were described in our previous manuscript [5]. The concentrations of the elicitors were selected based on our previous study [5], which indicated that 0.1% yeast extract (YE) and 10 µM of jasmonic acid (JA) proved to be the most effective concentrations of the elicitors for lovage plants. The plants were collected twenty-five days after the elicitation. Next, the plant materials were assigned into four groups, which were subjected to the different drying conditions, as in Figure 1.

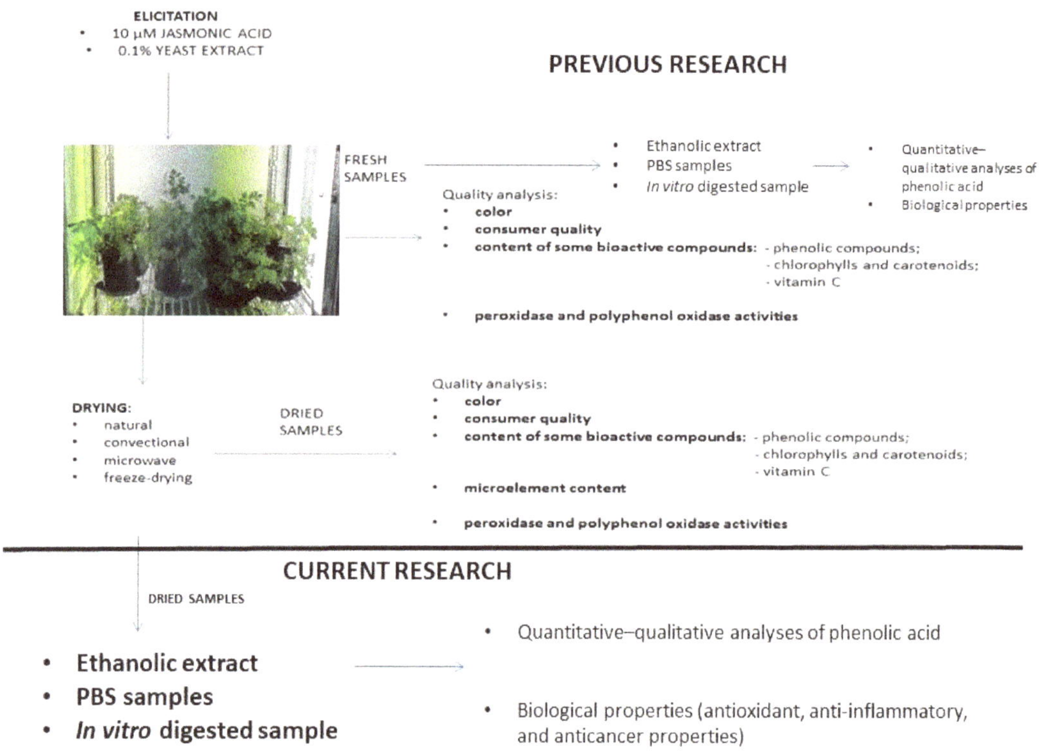

Figure 1. Scheme of the experimental procedure.

2.2. Drying Techniques

Fresh herbs were dried using four techniques: natural (traditional) drying (in a darkened room, a temperature of 20 °C to 22 °C, for approximately seven days), convection drying (in a drying oven at 40 °C for approximately 5 h), microwave drying (in a laboratory microwave dryer, microwave power 360 W at 20 °C, for ca. 5 min.), and freeze-drying (in a lyophilizer at a temperature of −49 °C and pressure of 0.045 mbar) [8,9].

2.3. Preparation of Extracts

2.3.1. Ethanolic Extracts

The ethanolic extracts were prepared for further analysis (0.3 g dw dissolved in 15 mL of 70% (*v/v*) acidified ethanol (0.1% HCl) were subjected to sonication at 30 °C for 1 h and centrifugation at 9000× *g* for 30 min.).

2.3.2. PBS Extracts

For the preparation of buffer extracts, dried leaf tissue (0.5 g) was homogenized, extracted for 30 min with 20 mL of PBS buffer (phosphate buffered saline, pH 7.4), and then centrifuged at 9000× *g* for 20 min Next, the residues were extracted again with 20 mL of PBS buffer. The supernatants were combined and adjusted to a final volume of 50 mL of with PBS buffer.

2.3.3. In Vitro Digestion

In vitro digestion was performed, as described previously by Gawlik-Dziki et al. [10]. The dried leaf tissue (0.5 g) was homogenized in a Stomacher Laboratory Blender for 1 min to simulate mastication in the presence of 5 mL of a simulated saliva solution containing 7 mM $NaHCO_3$, 0.35 mM NaCl (pH 6.75), and α-amylase (E.C. 3.2.1.1., 200 U per mL of enzyme activity). Subsequently, the mixture was stirred for 10 min at 37 °C in darkness. For gastric digestion, the solution was adjusted to pH 2.5 with 1 M HCl and 15 mL of 300 U/mL of pepsin (from porcine stomach mucosa, pepsin A, EC 3.4.23.1) in 0.03 mol/L NaCl, pH −1.2, were added. The reaction was carried out for 60 min at 37 °C. Next, the solution was adjusted to pH 7 with 1 M NaOH and 15 mL of a mixture of a 0.7% solution of pancreatin and a 2.5% solution of bile extract were added. The incubation was carried out for 120 min at 37 °C in darkness. Thereafter, the samples were centrifuged and the supernatants (gastrointestinally digested extracts; GD extracts) were used for further analysis.

2.4. Analysis of Phenolic Acids

Quantitative–qualitative analyses of phenolic acid extracts were performed while using a VarianProStar HPLC system separation module (Varian, Palo Alto, CA, USA) equipped with a Varian ChromSpher C18 reverse-phase column (250 mm × 4.6 mm) and a ProStar diode array detector. The column thermostat was set at 40 °C. All of the analyses were carried out with 4.5% acetic acid (solvent A) and 50% acetonitrile (solvent B) mobile phases at a flow rate of 0.8 mL min^{-1}. At the end of the gradient, the column was washed with 50% acetonitrile and then equilibrated to the initial condition for 10 min. Quantitative determinations were conducted by calculation of an external standard, using calibration curves of the standards. Detection was performed at 270 nm and 370 nm. Phenolic acids that were present in the sample were identified by comparing the retention times and UV-visible absorption spectra with those of the standard compounds. Phenolics were expressed as micrograms per gram dry weight (DW) [11].

2.5. Antioxidant Activities

2.5.1. Free Radical Scavenging Assay

The free radical scavenging activity was determined using 2,2′-azino-bis[3 -ethylbenz othiazoline-6-sulphonic acid ($ABTS^{•+}$) as a source of free radicals as in Re et al. [12]. The antioxidant activity was expressed as μmol of Trolox per gram of dry weight (DW) (TEAC, Trolox equivalent antioxidant activity).

2.5.2. Ferric Reducing Antioxidant Power

Ferric reducing antioxidant power (RP) was determined according to the methods that were described by Oyaizu [13]. The reducing power was expressed as a Trolox equivalent (TE) in μmol of Trolox per gram of dry weight (DW).

2.5.3. Chelating Power

The chelating power (CHP) was determined using the method that was developed by Guo et al. [14]. The percentage of inhibition of the ferrozine-Fe^{2+} complex formation was calculated using the formula:

$$\% \text{ inhibition} = [1 - A_A/A_C] \times 100$$

where:

A_C—the absorbance of the control (the solvent instead of the extract), A_A—absorbance of the sample

The chelating power was expressed as an EDTA (Ethylenediaminetetraacetic acid) equivalent in μg EDTA per g of dry weight (DW).

2.6. *Determination of Potential Anti-Inflammatory Properties*

2.6.1. Lipoxygenase Inhibitory Activity

The method described by Szymanowska et al. [15] and adapted to the BioTek Microplate Reader was used to analyze the effect of the different concentrations of the control and the elicited lovage extracts (ethanolic, PBS and obtained after the simulated digestion) on lipoxygenase activity. Based on the relationship between the concentration of the analyzed extract and the degree of inhibition of LOX activity, a curve was drawn, from which the EC_{50} value, i.e., an extract concentration (mg DW/mL) providing 50% inhibition, was determined.

2.6.2. Cyclooxygenase-2 Inhibitory Activity

A COX Activity Assay kit from Cayman Chemical Company was used to determine the inhibition of cyclooxygenase-2 activity by the analyzed extracts. The determination was carried out according to the procedure that was specified by the manufacturer of the kit. Based on the degree of inhibition of COX-2 activity by various concentrations of the analyzed extracts, the EC_{50} value, i.e., a concentration (mg DW/mL) providing 50% enzyme inhibition, was determined.

2.7. *Anticancer Properties*

The studies were performed on two cancer cell lines: gastric epithelial cell line NCI-N87 (ATCC® CRL5822 ™) and prostate cancer line VCaP (ATCC® CRL-2876 ™). Healthy prostate epithelial cells HPrEC (ATCC® PCS-440-010™) were used as a control. All of the cell lines were purchased from the American Type Culture Collection (ATCC), University Boulevard, Manassas, VA, USA.

2.7.1. Cell Culture

The studied cells were cultured in the same conditions, as described in our previous publication [3].

2.7.2. MTT and NR Tests

The MTT and NR assays were performed, as described previously [3,16].

2.7.3. Cell Viability and Type of Cell Death

The study was conducted according to the procedure that was specified by Leśniak et al. [17] and described in detail by [3].

2.7.4. Cell Cycle

After trypsinization (0.05% trypsin, 3–7 min, 37 °C), the cells were seeded onto a 48-cell culture plate. After 24 h, the cells were washed with the PBS solution and appropriate culture medium with the studied dried lovage leaf extracts at the concentrations of 0.1 and 20 mg/mL was added. After 24 h, the cells were trypsinized (0.05% trypsin, 3 min, 37 °C), centrifuged (300× *g*, 5 min), washed twice with the PBS solution, and then centrifuged again (300× *g*, 5 min). The cell pellet was then resuspended in 1 mL of a cold 70% ethanol/PBS solution (4 °C) and left in a freezer (−20 °C) for permeabilization of the cell membrane until the time of analysis (min. one day, max. 30 days). On the day of analysis, the cells were centrifuged (300× *g*, 5 min) and suspended in 300 µL of stain buffer (PI/RNase Staining Buffer, BD Warsaw, Poland). Subsequently, iodide propidium (PI, Bioscene Warsaw, Poland) was added for 15 min at room temperature. Cell cycle analysis was performed using a flow cytometer (FACS Calibur, BD, San Jose, CA, USA) (25,000 cells stopped the acquisition) and calculations were conducted using FCS Express program (De Novo Software, Pasadena, CA, USA). The results are presented as mean ±SD of three independent experiments (n = 6).

2.8. Statistical Analysis

All of the the determinations were performed in triplicate. Statistical analysis was performed using STATISTICA 7.0 software for a comparison of means in ANOVA with post-hoc Tukey's HSD (honestly significant difference) test at the significance level $p < 0.05$.

The data obtained from the experiment on cancer cells were checked for the normality of distribution (Shapiro–Wilk test). The level of statistical significance was calculated: in the case of normal distribution, one-way ANOVA with Bonferroni correction and Student's t test were used; otherwise, non-parametric one-way ANOVA with correction Kruskal–Wallis and Mann–Whitney tests were employed. The data were analyzed using the GraphPad Prism program (version 5, GraphPad Software, Inc., La Jolla, CA, USA) at a significance level of $p < 0.05$. The data were also evaluated using Pearson's correlation coefficients to identify the relationships between phenolic acid content and selected biological activities of the studied samples.

3. Results

3.1. Phenolic Acid Analysis

The drying method influenced the content of some phenolic acids, e.g., the highest amount of syringic acid was recorded in the convectionally dried lovage samples, while the ethanolic extracts from lyophilized lovage had the highest content of protocatechuic and caffeic acids. Ferulic acid was only identified in the ethanolic extracts from dried lovage, with its highest amount being detected in the microwave-dried lovage samples (Table 1).

The chemical (ethanolic) extracts were characterized by the highest content of caffeic and sinapic acids, while an opposite trend was observed in the case of *p*-hydroxybenzoic acid—the highest concentration was detected in the in vitro digested samples, as in Table 1.

The elicitation with JA caused a statistically significant increase in the content of most of the analyzed phenolic acids. In all samples of lovage dried with the three methods (freeze-drying, microwave drying, and traditional drying), elicitation with jasmonic acid resulted in a statistically significant increase in the content of protocatechuic, caffeic, syringic, ferulic, and salicylic acids. In addition, lovage dried with all of these methods was characterized by increased content of sinapic acid, as seen in Table 1.

Table 1. Qualitative and quantitative analysis of phenolic acids in the ethanolic, buffer (PBS (phosphate buffered saline)), and gastrointestinally digested (GD) extracts from control and elicited dried lovage leaves.

Sample	Compounds [mg/g DW]							
	Protocatechuic Acid	*p*-hydroxybenzoic Acid	Caffeic Acid	Syringic Acid	Vanillic Acid	Ferulic Acid	Sinapic Acid	Salicylic Acid
Freeze-Dried Samples								
Ethanolic extracts								
CL-E	159.59 ± 0.42 dC	nd	177.32 ± 0.98 eE	15.34 ± 0.02 aA	7.71 ± 0.02 bAB	10.40 ± 0.06 aA	102.14 ± 0.52 aC	492.75 ± 1.37 aC
JAL-E	413.56 ± 0.78 iE	nd	440.86 ± 1.43 jG	38.55 ± 0.23 eC	nd	15.84 ± 0.01 cC	260.91 ± 0.93 hF	1149.10 ± 6.23 hF
YEL-E	264.84 ± 0.41 hD	nd	312.03 ± 0.62 iF	22.02 ± 0.12 cB	6.19 ± 0.03 aA	13.52 ± 0.05 bB	232.78 ± 0.49 fgE	1036.70 ± 8.10 gE
PBS extracts								
CL-P	nd	15.65 ± 0.06 cdC	10.22 ± 0.01 aA	nd	12.58 ± 0.49 cD	nd	84.73 ± 4.20 cB	nd
JAL-P	9.56 ± 0.08 aA	13.64 ± 1.92 cB	12.45 ± 0.16 abB	nd	10.42 ± 0.13 bC	nd	110.77 ± 7.57 dC	33.65 ± 0.44 aA
YEL-P	nd	9.45 ± 0.15 aA	nd	nd	8.42 ± 1.58 aB	nd	61.44 ± 3.21 bA	nd
GD extracts								
CL-GD	31.27 ± 9.91 abcB	150.22 ± 12.05 bD	18.73 ± 4.47 abcC	14.13 ± 3.67 abcA	16.88 ± 0.76 bcE	nd	99.14 ± 7.89 abBC	185.15 ± 34.42 abB
JAL-GD	126.67 ± 29.42 eC	273.40 ± 22.48 deE	138.38 ± 2.19 eD	54.39 ± 7.79 eD	26.63 ± 3.12 deF	nd	303.54 ± 40.57 fG	693.13 ± 74.54 eD
YEL-GD	nd	290.60 ± 0.21 eF	nd	nd	18.40 ± 0.02 cE	nd	188.43 ± 1.05 deD	nd
Microwave-Dried Samples								
Ethanolic extracts								
CM-E	111.43 ± 0.34 bD	nd	148.59 ± 3.46 dD	34.12 ± 0.20 dC	10.68 ± 0.07 cA	20.82 ± 0.06 dA	186.76 ± 2.82 bcD	654.54 ± 16.19 cF
JAM-E	235.26 ± 6.39 gE	nd	257.94 ± 7.48 hF	52.38 ± 0.69 hE	15.91 ± 0.57 fC	33.94 ± 2.17 hC	210.74 ± 8.11 dE	968.39 ± 74.54 fH
YEM-E	130.39 ± 6.43 cD	nd	173.08 ± 5.51 eE	44.66 ± 1.46 fD	12.32 ± 0.12 dB	25.96 ± 3.38 gB	223.97 ± 13.16 efE	781.96 ± 55.85 eG
PBS extracts								
CM-P	46.45 ± 0.21 dB	nd	58.92 ± 0.33 dC	15.17 ± 0.11 cB	15.93 ± 0.14 dC	nd	108.55 ± 0.56 dB	146.09 ± 1.19 eB
JAM-P	240.68 ± 0.65 hE	nd	290.26 ± 0.61 iG	71.93 ± 0.19 hF	30.65 ± 0.08 fD	nd	190.61 ± 0.12 fD	557.83 ± 1.30 hE
YEM-P	22.88 ± 0.16 cA	37.42 ± 0.25 gA	18.80 ± 0.06 cA	10.01 ± 0.04 bA	12.04 ± 0.25 cB	nd	86.61 ± 0.22 cA	117.31 ± 0.23 cA
GD extracts								
CM-GD	74.52 ± 16.44 dC	210.29 ± 19.73 cC	36.61 ± 6.50 dB	29.62 ± 5.29 dC	16.72 ± 1.60 bcC	nd	159.91 ± 24.89 cdC	409.11 ± 91.18 cD
JAM-GD	253.35 ± 27.51 fE	272.01 ± 2.78 deD	153.74 ± 14.10 fD	102.83 ± 10.04 fG	30.36 ± 1.09 eD	nd	278.95 ± 1.77 fF	1019.41 ± 23.70 fH
YEM-GD	50.05 ± 16.66 abcdC	109.75 ± 26.71 aB	27.97 ± 7.26 bcdAB	20.18 ± 5.99 abcdBC	14.05 ± 1.92 bC	nd	93.20 ± 26.45 abAB	197.93 ± 84.82 abC

Table 1. *Cont.*

Sample	Compounds [mg/g DW]							
	Protocatechuic Acid	*p*-hydroxybenzoic Acid	Caffeic Acid	Syringic Acid	Vanillic Acid	Ferulic Acid	Sinapic Acid	Salicylic Acid
Traditionally Dried Samples								
Ethanolic extracts								
CT-E	67.93 ± 0.25 aB	61.63 ± 0.06 dB	73.88 ± 0.33 bE	20.54 ± 0.30 bC	16.35 ± 0.10 fC	16.67 ± 0.05 cA	183.75 ± 0.29 bD	544.54 ± 0.92 bE
JAT-E	109.24 ± 0.61 bC	97.05 ± 0.19 fC	119.42 ± 0.82 cG	46.42 ± 0.04 gE	18.73 ± 0.08 iE	21.49 ± 0.02 deB	261.88 ± 0.32 hF	770.86 ± 0.55 eF
YET-E	62.99 ± 0.45 aB	63.64 ± 0.14 eB	65.42 ± 0.14 aD	19.50 ± 0.06 bC	16.03 ± 0.12 fC	16.48 ± 0.01 cA	194.22 ± 0.07 bcD	536.23 ± 0.98 abE
PBS extracts								
CT-P	20.10 ± 0.22 cA	25.47 ± 0.12 eA	20.55 ± 0.13 cB	9.82 ± 0.07 bB	9.41 ± 0.13 abB	nd	48.32 ± 0.03 aA	74.72 ± 0.60 bA
JAT-P	106.57 ± 0.33 fC	99.79 ± 0.06 hC	96.23 ± 0.15 gF	48.25 ± 0.10 gF	29.18 ± 0.21 eG	nd	228.75 ± 0.03 gE	438.91 ± 1.48 gD
YET-P	15.59 ± 0.46 bA	27.46 ± 0.09 fA	15.24 ± 0.11 bA	6.73 ± 0.04 aA	8.61 ± 0.04 aA	nd	63.59 ± 0.20 bB	74.10 ± 0.08 bA
GD extracts								
CT-GD	21.49 ± 3.90 abA	273.23 ± 16.57 deD	16.14 ± 0.40 abA	10.21 ± 0.02 abB	18.75 ± 1.50 c	nd	171.10 ± 33.59 dD	298.74 ± 49.77 bcC
JAT-GD	62.93 ± 0.46 cdB	296.99 ± 1.38 eE	36.94 ± 0.27 dC	25.10 ± 0.19 cdD	26.18 ± 0.14 dF	nd	226.70 ± 0.89 eE	546.00 ± 4.45 dE
YET-GD	26.81 ± 0.47 abA	174.08 ± 0.18 bD	17.81 ± 0.42 bAB	11.74 ± 0.10 abB	17.41 ± 0.02 bcD	nd	115.97 ± 0.22 bcC	189.18 ± 0.10 abB
Convectionally Dried Samples								
Ethanolic extracts								
CC-E	186.76 ± 0.52 fE	27.05 ± 0.16 aC	202.31 ± 0.66 gG	63.05 ± 0.07 jH	17.20 ± 0.17 gE	24.19 ± 0.10 fgC	197.18 ± 0.63 cG	812.49 ± 1.04 eH
JAC-E	171.51 ± 0.29 eE	40.76 ± 0.16 bD	188.60 ± 0.38 fF	58.11 ± 0.26 iG	14.79 ± 0.09 eC	23.29 ± 0.03 efB	215.38 ± 0.19 deH	775.73 ± 0.59 eG
YEC-E	129.45 ± 0.14 cD	42.06 ± 0.02 cD	153.35 ± 0.29 dE	51.98 ± 0.11 hF	17.71 ± 0.05 hE	22.52 ± 0.01 defA	236.07 ± 0.20 gI	712.59 ± 0.26 dF
PBS extracts								
CC-P	65.00 ± 0.59 eC	11.36 ± 0.03 bA	68.90 ± 0.50 fD	26.33 ± 0.13 eD	11.90 ± 0.04 cB	nd	57.19 ± 0.04 bB	127.35 ± 0.72 dB
JAC-P	63.31 ± 1.37 eBC	14.55 ± 0.07 cdB	65.20 ± 1.16 eD	23.59 ± 0.23 dC	9.60 ± 0.06 abA	nd	64.71 ± 1.26 bC	160.75 ± 1.45 fC
YEC-P	126.44 ± 2.94 gD	nd	157.84 ± 3.83 hE	42.82 ± 0.66 fE	16.89 ± 0.01 dD	nd	174.35 ± 5.69 eE	440.20 ± 8.90 gE
GD extracts								
CC-GD	52.58 ± 11.53 bcdB	162.02 ± 21.63 bF	32.07 ± 3.97 cdC	24.36 ± 3.61 cdCD	17.16 ± 2.83 bcE	nd	121.51 ± 11.87 bcD	279.64 ± 55.88 bD
JAC-GD	50.07 ± 0.42 abcdB	240.28 ± 0.27 cdG	23.97 ± 0.06 abcdB	20.58 ± 0.13 bcdB	24.69 ± 0.05 dF	nd	187.68 ± 0.13 deF	287.81 ± 1.49 bcD
YEC-GD	17.09 ± 0.05 aA	79.95 ± 0.08 aE	12.92 ± 0.09 aA	8.65 ± 0.03 aA	10.17 ± 0.02 aA	nd	55.19 ± 0.14 aA	86.44 ± 0.19 aA

Abbreviations: C, control; JA, plants elicited with 10 µM of jasmonic acid; YE, plants elicited with 0.1% yeast extract; CL-E, JAL-E, YEL-E —ethanolic extracts from freeze-dried samples; CL-P, JAL-P, YEL-P—PBS extracts from freeze-dried samples; CL-GD, JAL-GD, YEL-GD gastrointestinally digested extracts from freeze-dried samples; CM-E, JAM-E, YEM-E— ethanolic extracts from microwave-dried samples; CM-P, JAM-P, YEM-P -PBS extracts from microwave-dried samples; CM-GD, JAM-GD, YEM-GD—gastrointestinally digested extracts from microwave-dried samples; CT-E, JAT-E, YET-E— ethanolic extracts from traditionally dried samples; CT-P, JAT-P, YET-P –PBS extracts from traditionally dried samples; CT-GD, JAT-GD, YET-GD—gastrointestinally digested extracts from traditionally dried samples; CC-E, JAC-E, YEC-E— ethanolic extracts from convectionally dried samples; CC-P, JAC-P, YEC-P –PBS extracts from convectionally dried samples; CC-GD, JAC-GD, YEC-GD—gastrointestinally digested extracts from convectionally dried samples. The means (±SD) in the columns followed by different lowercase letters are significantly different for the same type of extract but different drying methods. Means (±SD) in the columns followed by different capital letters are significantly different for the same of drying methods but different extracts ($p \leq 0.05$).

In some cases, elicitation with the yeast extract also resulted in an increase in the content of some identified compounds. This can be noticed in particular in the ethanolic extracts from the freeze-dried and microwave-dried lovage. In the ethanolic extracts from the freeze-dried lovage, the elicitation with YE caused a 0.65-fold, 0.75-fold, 0.30-fold, over two-fold, and over two-fold increase in the content of protocatechuic acid, caffeic acid, ferulic acid, sinapic acid, and salicylic acid, respectively. In the case of the microwave-dried lovage leaves, the content of caffeic, syringic, vanillic, ferulic, sinapic, and salicylic acids in the ethanolic extracts of herbs treated with the yeast extract increased by 16.4%, 30.8%, 15.3%, 24.6%, 19.9%, and 19.4%, respectively, as in Table 1. Additionally, in the case of the convectionally dried herbs, the elicitation with YE resulted in an increase in the content of protocatechuic acid, caffeic acid, syringic acid, vanillic acid, sinapic, acid, and salicylic acid in the PBS extract, as in Table 1.

3.2. Antioxidant Activities

The lovage drying method did not significantly affect the antioxidant properties of the tested extracts from the control and elicited herbs, with the exception of the reducing power (in this case, the traditional drying resulted in substantially lower antioxidant potential of the samples after simulated digestion), as summarized in Table 2. It should be noted that a significant enhancement of all antioxidant properties of the GD extracts from the control and elicited lovage was detected, regardless of the drying method used (compared to the ethanolic and PBS-extracts), as seen in Table 2. Additionally, in some cases, the elicitation with JA and YE improved the antioxidant properties of the potentially bioavailable fraction and the PBS extracts of phenolic acids from dried lovage leaves. The highest antiradical properties of the potentially bioavailable fraction of phenolic compounds estimated against ABTS were detected in the samples of convectionally dried lovage treated with JA (the activities increased by 58.52% when compared to the control), as in Table 2. Additionally, the reducing power of samples after in vitro digestion was statistically significantly increased by the elicitation in some cases. The GD extract from JA-elicited freeze-dried lovage was characterized by 17.34% higher RP compared to the control, while the YE elicitation increased this ability by 23.46% in the GD samples of convectionally dried lovage (Table 2).

The PBS-extractable phenolic compounds from the JA-elicited lovage showed a statistically significantly higher ability to neutralize free radicals ABTS (in the case of traditionally and microwave-dried herbs). The same result was obtained in the case of the, microwave-dried, traditionally, and convectionally dried YE-elicited lovage samples (Table 2).

However, the elicitation with JA and YE did not have a positive effect on the ability to chelate iron ions by the tested dried lovage extracts, as in Table 2. The statistical analysis indicated the content of *p*-hydroxybenzoic acid was correlated significantly and positively with the antioxidant activities—($R^2 = 0.63$; $R^2 = 0.72$; $R^2 = 0.76$ for ABTS, RP and CHP, respectively)—Table 3.

3.3. Lipoxygenase and Cyclooxygenase 2 Inhibition

The extracts from dried lovage leaves exhibited potential anti-inflammatory activity reflected in the inhibition of pro-inflammatory enzymes, such as lipoxygenase and cyclooxygenase 2, as shown in Table 2. Convectional drying, as compared to the other drying methods, resulted in a reduction of the LOX inhibition ability by the PBS and GD samples of the control and elicited lovage. The COX2 inhibition by the ethanolic extracts was significantly lower in the case of the convectionally and microwave dried samples, as shown in Table 2. It should be noted that the potentially bioavailable fractions of phenolic acids (GD extracts) were characterized by a significantly higher ability to inhibit COX2 in comparison to the PBS extract, while such a relationship for the ability to inhibit LOX activity was only observed in the convectionally dried lovage samples (Table 2). However, the elicitation with JA and YE did not have a positive effect on these properties of the potentially bioavailable phenolic acids from dried lovage leaves—only the samples from convectionally dried plant material subjected to in vitro digestion and JA and YE elicitation exhibited higher COX2 inhibitory ability (ca. 40% and 53%, respectively) than the control, as seen in Table 2.

Table 2. The effect of elicitation and drying methods on the antioxidant and potentially anti-inflammatory properties of the ethanolic, buffer (PBS (phosphate buffered saline)), and gastrointestinally digested (GD) extracts from dried lovage leaves.

Extract Type	Sample	Antioxidant Activities			Potential Anti-Inflammatory Activities	
		ABTS [μMTrolox/g DW]	RP [mg Trolox/g DW]	CHP [mg EDTA/g DW]	LOX Inhibition EC_{50} [mg DW/mL]	COX2 Inhibition EC_{50} [mg DW/mL]
Freeze-Dried Samples						
Ethanolic extracts	CL-E	1.87 ± 0.21 bdAB	19.97 ± 5.98 bcB	3.18 ± 0.35 gB	0.115 ± 0.004 dD	0.113 ± 0.0004 deDE
	JAL-E	1.99 ± 0.10 dB	18.35 ± 5.73 bcB	3.19 ± 0.40 gB	0.145 ± 0.004 eE	0.021 ± 0.000 aA
	YEL-E	1.53 ± 0.17 bcA	7.90 ± 2.94 aA	2.28 ± 0.32 cefA	0.208 ± 0.015 fDE	0.048 ± 0.0001 bB
PBS extracts	CL-P	3.98 ± 0.79 cdeC	22.03 ± 4.52 dB	6.50 ± 0.34 abC	0.044 ± 0.001 aA	0.075 ± 0.004 cC
	JAL-P	4.25 ± 0.62 eC	19.64 ± 0.72 bcdB	5.95 ± 0.63 aC	0.068 ± 0.003 cC	0.122 ± 0.006 eE
	YEL-P	4.17 ± 0.98 deC	15.73 ± 1.32 abcdAB	6.68 ± 0.50 abC	0.063 ± 0.0005 cC	0.103 ± 0.001 dD
GD extracts	CL-GD	6.82 ± 0.36 eE	98.81 ± 4.83 bcdC	14.25 ± 0.49 bD	0.059 ± 0.004 abcABC	0.019 ± 0.0001 aA
	JAL-GD	6.40 ± 0.41 deDE	115.93 ± 6.04 cdeC	14.48 ± 0.35 bD	0.061 ± 0.002 bcBC	0.038 ± 0.003 bB
	YEL-GD	5.71 ± 0.32 dD	108.29 ± 4.36 cdC	14.77 ± 0.14 bD	0.047 ± 0.004 abAB	0.013 ± 0.0005 aA
Microwave-Dried Samples						
Ethanolic extracts	CM-E	0.92 ± 0.16 aA	14.17 ± 0.93 abA	2.41 ± 0.17 efA	0.296 ± 0.032 eE	0.209 ± 0.011 fF
	JAM-E	1.46 ± 0.16 bB	17.88 ± 1.25 bcA	0.98 ± 0.12 aA	0.164 ± 0.008 cC	0.147 ± 0.012 dD
	YEM-E	0.95 ± 0.12 aAB	9.42 ± 0.46 aA	2.25 ± 0.16 cefA	0.224 ± 0.013 dD	0.171 ± 0.0003 eE
PBS extracts	CM-P	3.05 ± 0.95 bcdC	20.90 ± 1.13 cdA	6.92 ± 0.71 abB	0.072 ± 0.011 abAB	0.044 ± 0.0004 bB
	JAM-P	4.59 ± 0.57 eD	20.90 ± 8.03 cdA	7.30 ± 1.15 bB	0.097 ± 0.002 bB	0.058 ± 0.002 bcBC
	YEM-P	4.66 ± 0.24 eD	20.31 ± 6.81 bcdA	7.03 ± 0.33 abB	0.072 ± 0.0001 abAB	0.061 ± 0.003 cC
GD extracts	CM-GD	4.18 ± 0.22 bcD	115.46 ± 4.47 cdeC	13.97 ± 1.29 bD	0.068 ± 0.004 abAB	0.012 ± 0.0001 aA
	JAM-GD	4.06 ± 0.37 bcD	131.72 ± 6.39 eD	12.11 ± 1.86 aC	0.059 ± 0.002 aA	0.013 ± 0.0006 aA
	YEM-GD	4.12 ± 0.15 bcD	101.66 ± 9.20 bcdB	14.03 ± 0.53 bD	0.069 ± 0.001 abAB	0.013 ± 0.0002 aA
Traditionally Dried Samples						
Ethanolic extracts	CT-E	1.40 ± 0.17 bA	25.98 ± 2.06 cC	1.62 ± 0.40 abcA	0.221 ± 0.028 bB	0.051 ± 0.001 dD
	JAT-E	1.82 ± 0.16 cdAB	20.70 ± 4.23 bcBC	2.32 ± 0.14 cefAB	0.198 ± 0.0005 bB	0.024 ± 0.001 bB
	YET-E	2.05 ± 0.11 cBC	15.20 ± 2.01 abAB	2.75 ± 0.49 fgB	0.205 ± 0.002 bB	0.031 ± 0.001 cC
PBS extracts	CT-P	1.89 ± 0.14 aB	8.83 ± 3.93 aA	7.29 ± 0.35 bC	0.051 ± 0.008 aA	0.055 ± 0.004 eD
	JAT-P	4.53 ± 0.34 eF	11.88 ± 3.56 abcdA	7.08 ± 0.12 abC	0.049 ± 0.004 aA	0.057 ± 0.002 eD
	YET-P	3.00 ± 0.20 abcD	10.15 ± 2.06 abA	6.72 ± 0.19 abC	0.060 ± 0.0001 aA	0.055 ± 0.001 deD
GD extracts	CT-GD	4.64 ± 0.61 cF	37.69 ± 3.43 aD	14.10 ± 0.91 bD	0.049 ± 0.002 aA	0.013 ± 0.0007 aA
	JAT-GD	4.03 ± 0.25 bcF	42.60 ± 2.51 aD	13.69 ± 0.84 abD	0.043 ± 0.003 aA	0.019 ± 0.002 abAB
	YET-GD	3.70 ± 0.24 bE	41.34 ± 3.84 aD	13.44 ± 0.69 abD	0.051 ± 0.002 aA	0.030 ± 0.001 cC
Convectionally Dried Samples						
Ethanolic extracts	CC-E	1.99 ± 0.20 dAB	25.98 ± 2.06 cA	2.50 ± 0.30 efgA	0.175 ± 0.011 cC	0.083 ± 0.0003 cC
	JAC-E	2.00 ± 0.15 dAB	20.70 ± 4.23 bcA	1.86 ± 0.38 bceA	0.179 ± 0.001 cC	0.167 ± 0.005 eE
	YEC-E	1.46 ± 0.09 bA	15.20 ± 2.01 abA	1.36 ± 0.84 abA	0.212 ± 0.003 dD	0.176 ± 0.004 eE
PBS extracts	CC-P	2.43 ± 0.50 abBC	11.02 ± 3.60 abcA	6.49 ± 1.14 abB	0.146 ± 0.001 bB	0.099 ± 0.010 dD
	JAC-P	2.40 ± 0.44 abBC	13.14 ± 4.19 abcdA	6.20 ± 0.73 abB	0.211 ± 0.012 dD	0.091 ± 0.001 cdD
	YEC-P	3.97 ± 0.50 cdeD	13.60 ± 4.43 abcdA	7.13 ± 0.27 abB	0.22 ± 0.001 dD	0.093 ± 0.0003 cdD
GD extracts	CC-GD	3.93 ± 0.65 bcD	98.54 ± 24.95 bcB	13.81 ± 0.32 bC	0.131 ± 0.010 abAB	0.03 ± 0.002 bB
	JAC-GD	6.23 ± 0.35 deE	80.82 ± 0.34 bB	13.51 ± 0.43 abC	0.122 ± 0.006 aA	0.018 ± 0.001 abAB
	YEC-GD	2.84 ± 0.50 aC	121.30 ± 11.08 deC	14.28 ± 0.56 bC	0.117 ± 0.005 aA	0.016 ± 0.001 aA

Abbreviations: C, control; JA, plants elicited with 10 μM of jasmonic acid; YE, plants elicited with 0.1% yeast extract; CL-E, JAL-E, YEL-E —ethanolic extracts from freeze-dried samples; CL-P, JAL-P, YEL-P—PBS extracts from freeze-dried samples; CL-GD, JAL-GD, YEL-GD gastrointestinally digested extracts from freeze-dried samples; CM-E, JAM-E, YEM-E— ethanolic extracts from microwave-dried samples; CM-P, JAM-P, YEM-P -PBS extracts from microwave-dried samples; CM-GD, JAM-GD, YEM-GD—gastrointestinally digested extracts from microwave-dried samples; CT-E, JAT-E, YET-E— ethanolic extracts from traditionally dried samples; CT-P, JAT-P, YET-P –PBS extracts from traditionally dried samples; CT-GD, JAT-GD, YET-GD—gastrointestinally digested extracts from traditionally dried samples; CC-E, JAC-E, YEC-E—ethanolic extracts from convectionally dried samples; CC-P, JAC-P, YEC-P –PBS extracts from convectionally dried samples; CC-GD, JAC-GD, YEC-GD—gastrointestinally digested extracts from convectionally dried samples. Means (±SD) in the columns followed by different lowercase letters are significantly different for the same type of extract but different drying methods. The means (±SD) in the columns followed by different capital letters are significantly different for the same of drying methods, but different extracts ($p \leq 0.05$).

Table 3. Pearson's correlation coefficients for phenolic acid content and in vitro bioactivities.

	Protocatechuic Acid	p-hydroxybenzoic Acid	Caffeic Acid	Syringic Acid	Vanillic Acid	Ferulic Acid	Sinapic Acid	Salicylic Acid
ABTS	−0.32	0.63	−040	−0.11	0.50	−0.69	−0.04	−0.40
RP	−0.14	0.72	−0.29	0.08	0.36	−0.45	0.08	−0.10
CHP	−0.43	0.76	−0.54	−0.22	0.38	−0.78	−0.19	−0.47
LOX inhibition *	0.37	−0.48	0.47	0.29	−0.26	0.66	0.28	0.50
COX2 inhibition *	0.14	−0.60	0.27	0.18	−0.27	0.57	0.01	0.19
inhibition of proliferation of NCI-N87 cell line	0.13	0.27	0.14	0.06	0.11	0.45	0.43	0.40
inhibition of viability of NCI-N87 cell line	0.34	0.12	0.38	0.16	−0.01	0.62	0.54	0.59
inhibition of proliferation of VCaP cell line	0.07	0.37	0.09	0.08	0.18	0.36	0.47	0.33
inhibition of viability of VCaP cell line	0.18	0.27	0.21	0.14	0.13	0.52	0.52	0.45
inhibition of proliferation of HPrEC cell line	0.36	0.11	0.38	0.02	−0.14	0.39	0.41	0.49
inhibition of viability of HPrEC cell line	0.37	0.14	0.40	0.21	−0.02	0.63	0.56	0.62

* a negative correlation between the phenolic acid content and the EC50 values indicates a positive relationship between phenolic content and LOX/COX2 inhibition (low EC50 values indicate high ability to inhibit LOX/COX2).

It should be also noticed that, in the case of the ethanolic extracts, the elicitation with JA and YE caused an increase in the ability to inhibit COX2 (except for the ethanolic extracts from the convection-dried lovage leaves, where an opposite trend was observed). The greatest increase in this activity (over five-fold) was observed in the case of extracts from the JA-treated freeze-dried lovage (Table 2).

In the case of the ethanolic extracts from the microwave-dried and traditionally dried lovage leaves, the elicitation with JA and YE also resulted in an increase in the ability to inhibit LOX, as shown in Table 2.

The analysis of correlations only indicated a significant positive relationship between LOX and COX2 inhibition and phenolic content for the p-hydroxybenzoic acid content (Table 3).

3.4. *Influence of Analyzed Extracts on Healthy and Cancer Cell Lines*

The PBS extracts of the control and elicited lovage leaves dried with all of the tested methods in the tested concentrations (0.1–20 mg dw/mL) had no effect on the number of cells (in MTT and NR assays), viability (no impact on apoptosis or necrosis), or percentages of cell cycle phases (G1, S, or G2) in the gastric cancer NCI-N87 (ATCC® CRL5822 ™), prostate cancer VCaP (ATCC® CRL-2876 ™), and healthy prostate HPrEC (ATCC® PCS-440-010 ™) cell lines, as compared to cells that were not exposed to the extracts (Table 4). In contrast, the ethanolic extracts and in vitro digested samples exerted a significant influence on these properties (Table 4 and Supplement S1–S4).

The ethanolic extracts of the studied samples (control and elicited) caused a significant reduction of the number of healthy prostate epithelial HPrEC cells, but only at the concentrations of 10 and 20 mg/mL (Figures S1–S3, S37–S39 and S55–S57). It should be noted that this effect was significantly lower in the case of samples that were obtained from the microwave-dried lovage (CM-E, JAM-E, YEM-E samples), as in Table 4; Figures S19–S21. At the highest concentration, all of the ethanolic extracts caused apoptosis. With the exception of the microwave-dried lovage samples, this was associated with changes in the cell cycle (an increased percentage of cells in the S phase with a reduction of cells in the G1 phase). It should also be noted that there were no significant changes in the influence on the healthy prostate epithelial HPrEC cells between the control sample and samples elicited with jasmonic acid (JAL-E) or with the 0.1% yeast extract (Figures S1–S3, S37–S39 and S55–S57). A similar effect on the healthy prostate epithelial HPrEC cells was also observed in the case of the in vitro digested samples; however, the effect depended on the method of drying (Table 4; Figures S4–S6, S22–S24, S40–S42 and S58–S60). In the case of the in vitro digested samples from the lyophilized and convectionally dried lovage, a significantly lower effect on the cytotoxicity parameters (proliferation and viability) was observed when compared to the ethanolic extracts, but the effect of the changes in the cell cycle was preserved, as seen in Table 4. The strongest negative effect on the healthy cells (comparable to that of the ethanolic extracts) was noted in the case of the digested samples from the traditionally dried lovage (CT-GD, JAT-GD, YET-GD), but these samples caused opposite changes in the cell cycle (a reduction of the percentage of S-phase cells), as in

Figures S40–S42. In turn, the GD samples from the microwave-dried lovage showed no cytotoxic effect on the HPrEC cells in the proliferation assays—there was only a slight, but significant, increase in the percentage of apoptotic cells after the JAM-GD treatment (21.5% vs. 16.6% when compared to the non-treated cells, $p = 0.0112$). However, at the highest concentration (20 mg/mL), a significantly higher percentage of G2 phase cells was noted (Figures S22–S24).

Table 4. In vitro effects of the highest concentration of the analyzed extracts after 72-h treatment on the cancer gastric epithelial NCI-N87 (ATCC® CRL5822 ™), prostate cancer VCaP (ATCC® CRL-2876 ™), and healthy prostate epithelial HPrEC cell lines (ATCC® PCS-440-010™).

Sample	NCI-N87 (ATCC® CRL5822 ™)			VCaP—(ATCC® CRL-2876 ™)			HPrEC (ATCC® PCS-440-010 ™)		
	Proliferation	Viability	Cell Cycle	Proliferation	Viability	Cell Cycle	Proliferation	Viability	Cell Cycle
Freeze-Dried Samples									
Ethanolic extracts									
CL-E	↓↓	↓↓↓	↑G1↓G2	↓	↓↓	n.c.	↓↓↓↓	↓↓↓	↑S↓G1
JAL-E	↓↓	↓↓↓	↑G1↓G2	↓	↓↓	↑G2	↓↓↓↓	↓↓↓	↑S↓G1
YEL-E	↓	↓↓↓	↑G1↓S	↓	↓	↑G2	↓↓↓↓	↓↓↓	↑S↓G1
PBS extracts									
CL-P	n.c.	n.c.	n.c.	n.c.	n.c.	n.c.	n.c.	n.c.	n.c.
JAL-P	n.c.	n.c.	n.c.	n.c.	n.c.	n.c.	n.c.	n.c.	n.c.
YEL-P	n.c.	n.c.	n.c.	n.c.	n.c.	n.c.	n.c.	n.c.	n.c.
GD extracts									
CL-GD	↓	↓↓	n.c.	↓	↓	↑G1↓S	n.c.	↓	↑S↓G1,G2
JAL-GD	↓	↓↓	n.c.	↓↓	↓↓	↑G1↓S,G2	↓	↓↓	↑S↓G2
YEL-GD	↓	↓	n.c.	↓↓	↓↓	↑G1↓S,G2	↓	↓↓	↑S
Microwave-Dried Samples									
Ethanolic extracts									
CM-E	↓↓	↓↓↓	↑S↓G2	↓↓↓	↓↓	↑G1↓S	↓	↓↓↓	n.c.
JAM-E	↓↓	↓↓↓	↑G1	↓	↓↓	↑G1↓S	↓	↓↓↓	n.c.
YEM-E	↓↓	↓↓↓	n.c.	n.c.	↓	↑G1↓S	↓	↓↓↓	↑G2
PBS extracts									
CM-P	n.c.	n.c.	n.c.	n.c.	n.c.	n.c.	n.c.	n.c.	n.c.
JAM-P	n.c.	n.c.	n.c.	n.c.	n.c.	n.c.	n.c.	n.c.	n.c.
YEM-P	n.c.	n.c.	n.c.	n.c.	n.c.	n.c.	n.c.	n.c.	n.c.
GD extracts									
CM-GD	n.c.	n.c.	↑S↓G1	n.c.	n.c.	↑G1	n.c.	n.c.	↑G2
JAM-GD	n.c.	n.c.	↑S↓G1	↑	n.c.	n.c.	n.c.	↓	↑G2
YEM-GD	↑	n.c.	n.c.	n.c.	n.c.	↑G1	n.c.	n.c.	↑G2
Traditionally Dried Samples									
Ethanolic extracts									
CT-E	↓↓↓	↓↓↓	↓G2	↓↓	↓↓↓	↓G2	↓↓↓	↓↓↓	↑G2↓G1
JAT-E	↓↓↓	↓↓↓	↓G2	↓↓	↓↓↓	n.c.	↓↓↓	↓↓↓	↑G2↓G1
YET-E	↓↓↓	↓↓↓	↑G2	↓↓	↓↓↓	↑G1↓G2	↓↓↓	↓↓↓	↑G2↓G1
PBS extracts									
CT-P	n.c.	n.c.	n.c.	n.c.	n.c.	n.c.	n.c.	n.c.	n.c.
JAT-P	n.c.	n.c.	n.c.	n.c.	n.c.	n.c.	n.c.	n.c.	n.c.
YET-P	n.c.	n.c.	n.c.	n.c.	n.c.	n.c.	n.c.	n.c.	n.c.
GD extracts									
CT-GD	↓↓↓	↓↓↓	n.c.	↓↓	↓↓↓	↑S	↓↓↓	↓↓↓	↓S
JAT-GD	↓↓↓	↓↓↓	↓G1	↓↓	↓↓↓	↑S	↓↓↓	↓↓↓	↓S
YET-GD	↓↓↓	↓↓↓	↓G1	↓↓	↓↓↓	↓G2	↓↓↓	↓↓↓	↓S
Convectionally Dried Samples									
Ethanolic extracts									
CC-E	↓↓	↓↓↓	n.c.	↓↓	↓↓↓	↑G1↓S,G2	↓↓	↓↓↓	↑S↓G2
JAC-E	↓↓	↓↓↓	n.c.	↓↓	↓↓↓	↑G1↓S,G2	↓↓↓	↓↓↓	↑S↓G2
YEC-E	↓↓	↓↓↓	n.c.	↓↓	↓↓↓	↑G1↓S	↓↓	↓↓↓	↑S↓G2

Table 4. *Cont.*

Sample	NCI-N87 (ATCC® CRL5822 ™)			VCaP—(ATCC® CRL-2876 ™)			HPrEC (ATCC® PCS-440-010 ™)		
	Proliferation	Viability	Cell Cycle	Proliferation	Viability	Cell Cycle	Proliferation	Viability	Cell Cycle
				PBS extracts					
CC-P	n.c.	n.c.	n.c.	n.c.	n.c.	n.c.	n.c.	n.c.	n.c.
JAC-P	n.c.	n.c.	n.c.	n.c.	n.c.	n.c.	n.c.	n.c.	n.c.
YEC-P	n.c.	n.c.	n.c.	n.c.	n.c.	n.c.	n.c.	n.c.	n.c.
				GD extracts					
CC-GD	↓↓↓	↓↓	↑S	↓↓	↓	n.c.	↓↓↓	↓↓↓	↑G2↓S
JAC-GD	↓↓↓	↓↓↓	↑S	↓↓	↓↓	n.c.	↓↓	↓	↑G2
YEC-GD	n.c.	n.c.	↑S	↑↑	n.c.	↑G1	↓	↓	↑G1↓S

Abbreviations: C, control; JA, plants elicited with 10 μM of jasmonic acid; YE, plants elicited with 0.1% yeast extract; CL-E, JAL-E, YEL-E —ethanolic extracts from freeze-dried samples; CL-P, JAL-P, YEL-P—PBS extracts from freeze-dried samples; CL-GD, JAL-GD, YEL-GD gastrointestinally digested extracts from freeze-dried samples; CM-E, JAM-E, YEM-E— ethanolic extracts from microwave-dried samples; CM-P, JAM-P, YEM-P -PBS extracts from microwave-dried samples; CM-GD, JAM-GD, YEM-GD—gastrointestinally digested extracts from microwave-dried samples; CT-E, JAT-E, YET-E— ethanolic extracts from traditionally dried samples; CT-P, JAT-P, YET-P –PBS extracts from traditionally dried samples; CT-GD, JAT-GD, YET-GD—gastrointestinally digested extracts from traditionally dried samples; CC-E, JAC-E, YEC-E— ethanolic extracts from convectionally dried samples; CC-P, JAC-P, YEC-P –PBS extracts from convectionally dried samples; CC-GD, JAC-GD, YEC-GD—gastrointestinally digested extracts from convectionally dried samples. n.c.—no change; ↓—decrease; ↑—increase; G1, S, G2—cell cycle phases; number of ↓ or ↑—level of changes: ↓↓↓↓—the highest, ↓—the lowest.

The ethanolic extracts of the studied samples (control and elicited) at the concentrations of 10 and 20 mg/mL caused a decrease in the number of cancer gastric epithelial NCI-N87 cells measured by the MTT and NR assays (Figures S7–S9, S25–S27, S43–S45 and S61–S63). The strongest effect was observed for ethanolic extracts from the traditionally dried lovage (the largest decrease in the number of cells as well as the cytotoxic effect when samples at the lower concentration of 5 mg/mL were used). The highest concentration of ethanolic extracts (20 mg/mL) affected the viability of the cells by increasing apoptosis (over 90% cells—extracts from the lyophilized lovage (CL-E, JAL-E, YEL-E), 95% cells—extracts from the microwave- and traditionally-dried samples (CM-E, JAM-E, YEM-E, CT-E, JAT-E, YET-E), and 98% cells—extracts from the convectionally dried samples- CC-E, JAC-E, YEC-E). Additionally, a reduction of the number of necrotic cells and, in some cases, also changes in the cell cycle were observed (Table 4). The increased percentage of G1 cells that were associated with the reduction of the percentage of G2 cells observed after the addition of the CL-E and JAL-E samples may be a result of the apoptosis process. It is noteworthy that, after the addition of ethanolic extracts from the convectionally dried lovage, there were no changes in the cell cycle in the cancer gastric epithelial NCI-N87 cells (Table 4). Generally, there were no significant changes in the influence on NCI-N87 cells between the control and elicited samples, with some exceptions. After the application of the highest concentrations of the JAL-E and JAM-E extracts, a significant reduction in the percentage of apoptotic cells was observed when compared to extracts from the control lovage (Figures S7–S8 and S25–S26).

The effect of the samples after in vitro digestion on the cancer gastric epithelial NCI-N87 cells was generally significantly lower than that of the ethanolic extracts, except for the GD samples from the traditionally dried lovage—CT-GD, JAT-GD, and YET-GD. In this case, the cytotoxicity of the GD-samples against the NCI-N87 cells was as high as that of the ethanolic extracts, as seen in Table 4, Figures S10–S12, S28–S30, S46–S48 and S64–S66. It should also be emphasized that, in the case of the GD samples from the microwave-dried lovage (CM-GD, JAM-GD, YEM-GD), there was no significant cytotoxic effect on the NCI-N87 cells, as in Table 4, Figures S28–S30. Additionally, the elicitation significantly changed these properties in some cases. Lower levels of apoptosis were detected after the addition of the JAL-GD and YEL-GD samples (40.7 and 15.1% of cells, respectively) in comparison to the CL-GD sample (46.8% of cells), as shown in Figures S10–S12. Additionally, the GD sample from the convectionally dried lovage elicited with the 0.1% yeast extract had no cytotoxic effect on the cancer gastric epithelial NCI-N87 cells (Table 4, Figure S66).

The prostate cancer epithelial VCaP cells turned out to be the least sensitive to the ethanolic and gastrointestinally digested extracts from both the control and elicited freeze-dried lovage leaves, as seen in Figures S13–S18. No cytotoxic effect was shown by the MTT assay in the case of the ethanolic extracts from the freeze-dried lovage, whereas the NR assay revealed a cytotoxic effect even for the 1 and 5 mg/mL concentrations of the samples. Additionally, in the case of the ethanolic extracts from the microwave-dried lovage, the cytotoxic effect in the NR test was observed, even for the low concentrations of the extracts used (0.1 and 1 mg/mL). It has to be underlined that these results indicate that the traditionally and convectionally dried samples from the control and elicited lovage plants exhibited the highest cytotoxicity against the VCaP cell line, in comparison to the samples from plants that were dried with the other methods (Table 4). This was especially visible in the influence on apoptosis: all extracts—from the control and elicited lovage dried with these methods caused a large increase in the level of apoptosis—over 80% of apoptotic cells, as seen in Figures S49–S51 and S67–S69.

The cytotoxic effect of the GD samples from the convectionally dried lovage and the control lyophilized lovage sample on the prostate cancer epithelial VCaP cells was relatively lower than in the case of the ethanolic extracts. It should also be added that the GD-samples from the microwave-dried lovage (control and elicited) did not exert a cytotoxic effect on the VCaP cells (Table 4, Figures S16–S18, S34–S36, S52–S54 and S70–S72). Interestingly, the GD samples from the elicited freeze-dried lovage (JAL-GD and YEL-GD) showed much higher cytotoxicity towards the VCaP cells than the GD samples from the control lovage, as shown in Figures S16–S18. A higher apoptosis level was found after treatment with the YEL-GD extract (72.8%) than JAL-GD (65.8%). In both of the elicited samples, a lower necrosis percentage (more than 65% reduction) and a lower mean level of apoptosis (40.7 and 15.1% of cells, respectively) were observed. There were also significant changes in the cell cycle phases. Generally, the percentage of cells in the G1 phase was increased and the number of S and G2 phase cells was reduced. Interestingly, the results for both GD samples from the elicited lyophilized plants (JAL-GD and YEL-GD) also produced changes in the cell cycle phases at the lowest concentration studied (0.1 mg of extracts/mL). Elicitation with jasmonic acid also increased the cytotoxicity against the VCaP cells exhibited by the in vitro digested convection-dried lovage samples (Figures S70–S71). The percentage of apoptotic cells amounted to 50.1% after the JAC-GD exposure, but it was only 12.3% in the case of application of the control lovage samples (CC-GD). In turn, the YEC-GD samples showed no cytotoxicity against the prostate cancer epithelial VCaP cells (Figure S72).

The ability to inhibit the proliferation and viability of the studied cell lines was positively correlated with the content of almost all identified phenolic acids, but a statistically significant positive correlation was only observed in the case of ferulic acid, sinapic acid, and salicylic acid (Table 3).

4. Discussion

Our recent research has confirmed that JA and YE elicitation exerts positive effects on phenolic acid biosynthesis, and it has revealed some pro-health properties of the potentially bioavailable fraction of phenolic acids from fresh lovage leaves [3]. However, the food industry very often needs dried lovage leaves. Therefore, the influence of drying methods on these properties of elicited lovage leaves is a very important issue, but no such research has been conducted so far. Drying can significantly extend the shelf life of herbs. Unfortunately, it may also cause some negative changes in the organoleptic or pro-health quality of herbal plants. [7,18]. Previous research indicated that drying caused significant losses of some bioactive compounds, e.g., vitamin C and plant pigments (carotenoids and chlorophylls). Additionally, it should be noted that phenolic compounds that were present in elicited and control lovage leaves turned out to be the least sensitive to degradation during the drying process [9]. Therefore, this group of bioactive compounds will determine the health-enhancing properties of dried lovage. In contrast, in the research conducted by Lim and Murtijaya [8], drying with three methods (oven drying, sun drying, and microwave drying) resulted in a significant decline in total phenolics in *Phyllanthus amarus*. A significant decrease in the content of polyphenols in *Cistus creticus* leaves after convection and microwave drying was reported earlier [7]. Noteworthy, these studies were focused on "chemical extracts". In the present study, an attempt was made to investigate the potential bioavailability and health-promoting activity of phenolic acids contained in lovage leaves that were dried using various methods and subjected to the elicitation process with jasmonic acid and yeast extract. During digestion in the human body, phenolic compounds can be released from the food matrix or can be complexed with digestive enzymes, as indicated in scientific literature. This may result in the formation of some metabolites whose pro-health properties may differ from those of the initial material [19,20]. The bioavailability of phenolic compounds can be determined by several types of factors that are associated with the chemical structure of phenolics, interaction with the food matrix, and digestion conditions [19,21]. Seven phenolic acids (protocatechuic, *p*-hydroxybenzoic, syringic, vanillic, sinapic, salicylic, and caffeic acids) were detected in the PBS and GD samples from dried lovage leaves, as shown in Table 1. Additionally, ferulic acid was identified in the ethanolic extracts, as in Table 1. These results correspond to earlier studies on fresh leaves of JA- and YE-elicited lovage [3,5].

Our previous study indicated that elicitation with jasmonic acid increased the content of phenolic acids in PBS extracts from fresh lovage leaves; however, in the case of potentially available fractions of phenolic acids (GD samples), the treatment only increased the content of protocatechuic acid and *p*-hydroxybenzoic acid [3]. The highest levels of protocatechuic, caffeic, syringic, and salicylic acids were recorded in the GD samples of microwave-dried lovage (both elicited and control), when compared to the other drying methods, as shown by the analysis of the influence of various drying methods (traditional, convection, microwave, and freeze-drying) on the content of the potentially bioavailable fraction of phenolic acids (Table 1). In turn, the freeze-drying process yielded the highest levels of potentially bioavailable *p*-hydroxybenzoic acid and sinapic acid. Similarly, in studies on other herbs where three drying methods (conventional, microwave, and freeze-drying) were used, the best extraction results were observed in the case of microwave drying and freeze-drying of melissa and thyme, respectively [22]. Importantly, these results only apply to "chemical extracts" [22]. In the case of dried lovage leaves, the jasmonic acid elicitation caused an increase in all phenolic acids identified in the potentially available fractions of phenolic compounds (Table 1). Surprisingly, only in the case of lyophilization, the YE-elicited dried lovage leaves were characterized by a higher level of hydroxybenzoic and sinapic acids in the potentially bioavailable fraction of phenolic acids, in comparison to the control, as seen in Table 1. Therefore, it can be assumed that, especially in the case of the JA-elicited lovage leaves, drying had a positive effect on the release of phenolic compounds during the simulated digestion process. There are no similar data on the effect of drying on elicited herbs.

Drying that is associated with the use of high temperatures can degrade phenolic compounds. However, other studies also suggest that high temperature can induce some modifications that can positively influence the bioavailability of phenolics, e.g., degradation or modification of food matrix factors that may contribute to increasing the extractability of phenolic compounds during digestion [23]. In a study that was conducted by Kamiloglu et al. [24], oven-drying had a positive effect on the bioaccessibility of phenolic acids in tomato. Similarly, sun-drying enhanced the bioavailability of caffeic acid derivatives (50–60% increase) in figs, but negatively influenced the bioavailability of flavonoids [25]. Therefore, drying has a positive effect on bioavailability, especially in the case of phenolic acids, as confirmed in this paper.

Hydroxybenzoic acid is quite common in plants, but low amounts thereof are identified in food of plant origin, mainly because it is usually associated with cell wall structures, i.e., tannins or lignins [26]. The simulated digestion caused a significant increase in the content of this compound in the GD extracts (Table 1). This was probably related to the release of this acid from plant cell wall structures during the in vitro digestion process.

Many studies have indicated that phenolic compounds that are present in herbs largely determine their biological properties, i.e., the antioxidant and anti-inflammatory potential [27,28]; thus, it can be assumed that both the drying method and in vitro digestion of lovage leaves may have influenced these properties. The drying method had an impact only on the reducing power of the extract from lovage leaves dried with the traditional method, as summarized in Table 2. Similarly, vacuum drying did not have a significant effect on the antiradical activity of basil and peppermint in a study that was conducted by Dogan and Tornuk [29]. However, the drying methods significantly influenced the antioxidant activity measured as ferric reducing antioxidant power and oxygen radical absorbance capacity of dried herbs, such as rosemary, oregano, marjoram, sage, basil, and thyme, as demonstrated by Hossain et al. [30]. Among the three drying methods used (air-, freeze-, and vacuum oven-drying), the air-dried herbs had the best antioxidant properties. Similarly, drying at low temperature, e.g., sun-drying and 40 °C oven-drying, increased the total polyphenol contents and antioxidant capacity of peppermint, rosemary, and motherwort in a study that was conducted by Yi and Wetzstein [31]. Through the release of bioactive compounds (e.g., phenolics) from the food matrix, digestion can positively influence many biological properties, e.g., antioxidant, anti-inflammatory, antimicrobial, antimutagenic, hypoglycemic, and cardioprotective activities, as indicated in some research [32,33]. In the present study, the simulated digestion caused a significant enhancement of all the tested antioxidant activities and potential anti-inflammatory (expressed as COX2 inhibition) properties of the extracts from the control and elicited lovage, as in Table 2. However, an increase in LOX inhibition after the in vitro digestion was only detected in samples from convectionally dried lovage leaves (Table 2). These results partially correspond with the previous study on the biological properties of fresh lovage leaves after in vitro digestion. It was shown that the simulated digestion significantly improved some biological activities, in particular, the ability to inhibit the LOX and COX 2 [3]. This may suggest that drying had a negative effect on the bioavailability of compounds that are responsible for LOX inhibition.

Antioxidative and anti-inflammatory properties are also largely responsible for the anti-cancer properties of bioactive compounds [34]. A study that was conducted by Spréa et al. [2] showed cytotoxic potential of extract from lovage leaves against liver cancer cells. Similarly, in a study conducted by Bogucka-Kocka et al. [35], hydroalcoholic extract from lovage exhibited the cytotoxic potential against seven leukemia human cell lines and two normal cell lines. Noteworthy, these studies were conducted with the use of only decoction (water) and hydroethanolic extracts, and did not evaluate the potentially bioavailable fraction of lovage bioactive compounds.

In the present studies, the ethanolic extracts and GD samples from the lovage leaves (dried using all of the tested methods) had a cytotoxic effect against the gastric epithelial NCI-N87 cell line and the prostate VCaP cancer line, while samples before in vitro digestion (PBS) did not show such an effect (Supplementary Materials). Additionally, the ethanolic extracts showed generally greater cytotoxicity against the cancer cells and the healthy cell line, in comparison to the potentially available samples (GD samples), as in Table 4. This result was not confirmed by previous investigations, in which in vitro digested basil samples showed greater cytotoxicity against cancer cells than ethanolic extracts [33].

The drying method only significantly influenced the tested properties in some cases—the traditional and convectional dried lovage leaves showed significantly higher cytotoxicity against the prostate cancer epithelial cell line, in comparison to the other drying methods (Table 4).

Additionally, in some cases, elicitation had an effect on these properties of the analyzed herb. For example, samples from freeze-dried plants elicited with jasmonic acid (JAL-GD) and 0.1% yeast extract (YEL-GD) exhibited higher cytotoxicity against prostate cancer epithelial cells than the control plants, as in Supplementary Materials, Figures S13–S18. Additionally, in the case of the in vitro digested samples from convectionally dried plants, the highest cytotoxicity against the prostate cancer cell line was exhibited by samples from jasmonic acid-elicited lovage, whereas an opposite effect was observed in the yeast extract-elicited lovage samples. The information about the influence of elicitation on the anticancer properties of herbal extracts is very limited in the available literature. In the studies that were conducted by Złotek et al. [33], the elicitation of basil with arachidonic acid did not cause significant changes in the cytotoxic properties of the potentially bioavailable fraction of bioactive compounds against a cancer cell line.

In the literature, some phenolic acids, e.g., ferulic acid, *p*-hydroxybenzoic acid, vanillic acid, caffeic acid, and coumaric acid, are indicated as compounds with anti-cancer activity [34,36]. The correlation analysis that was carried out in this study showed that the ability of the tested extracts to inhibit the proliferation and viability of the tested cell lines was mainly related to the presence of ferulic, sinapic, and salicylic acids in the tested extracts (Table 3). A study conducted by Elansary et al. [36] partially confirmed our observation, which suggested that ferulic acid, i.e., the major phenolic compound found in *Catalpa speciosa* bark extract, is highly associated with anticancer activity.

5. Conclusions

In conclusion, dried lovage (control and elicited with jasmonic acid and yeast extract) has significant advantages in terms of health-promoting properties, which is shown in the study of the potential bioavailability of phenolic acids and the pro-health properties of the bioavailable fraction of phenolic acids. The elicitation (especially with JA) resulted in a statistically significant increase in the content of most of the analyzed phenolic acids, which, in some cases, resulted in an increase in bioactive properties (especially antioxidant and antiproliferative activity). The content of *p*-hydroxybenzoic acid was correlated with the antioxidant activity and with the ability of the studied extracts to inhibit LOX and COX2, while the antiproliferative activity seems to be mainly related to ferulic, sinapic, and salicylic acids. The present research proves that the drying of elicited herbs does not deteriorate their health-promoting properties and the use of elicitation to modify the biological activity of herbal plants does not exclude the use of such plants in a dried form, which is a valuable indication for the food industry.

Supplementary Materials: The following are available online at https://www.mdpi.com/article/10.3390/antiox10050662/s1.

Author Contributions: Conceptualization, U.Z., S.L.; methodology, U.Z., U.S., S.L.; formal analysis, U.Z.; investigation, U.Z., S.L., U.S., A.M., A.J.; data curation, U.Z., S.L., U.S.; writing—original draft preparation, U.Z., S.L.; writing—review and editing, U.Z., S.L., U.S., A.J.; visualization, U.Z., S.L., U.S.; supervision, U.Z.; project administration, U.Z.; funding acquisition, U.Z. All authors have read and agreed to the published version of the manuscript.

Funding: This research was funded by the NATIONAL SCIENCE CENTRE, Poland (NCN) [Grant SONATA 12 No 2016/23/D/NZ9/00553] and The APC was funded by the NATIONAL SCIENCE CENTRE, Poland (NCN) [Grant SONATA 12 No 2016/23/D/NZ9/00553].

Institutional Review Board Statement: Not applicable.

Informed Consent Statement: Not applicable.

Data Availability Statement: Not applicable.

Conflicts of Interest: The authors declare no conflict of interest.

References

1. Amiri, M.S.; Joharchi, M.R. Ethnobotanical knowledge of Apiaceae family in Iran: A review. *J. Phytomed.* **2016**, *6*, 621–635.
2. Spréa, R.M.; Fernandes, Â.; Calhelha, R.C.; Pereira, C.; Pires, T.C.S.P.; Alves, M.J.; Canan, C.; Barros, L.; Amaral, J.S.; Ferreira, I.C.F.R. Chemical and bioactive characterization of the aromatic plant: Levisticum officinale W.D.J. Koch: A comprehensive study. *Food Funct.* **2020**, *11*, 1292–1303. [CrossRef]
3. Jakubczyk, A.; Złotek, U.; Szymanowska, U.; Jęderka, K.; Rybczyńska-Tkaczyk, K.; Lewicki, S. In vitro Antioxidant, Anti-inflammatory, Anti-metabolic Syndrome, Antimicrobial, and Anticancer Effect of Phenolic Acids Isolated from Fresh Lovage Leaves [Levisticum o ffi cinale Koch] Elicited with Jasmonic Acid and Yeast Extract. *Antioxidants* **2020**, *9*, 554. [CrossRef]

4. Złotek, U.; Szymanowska, U.; Karaś, M.; Świeca, M. Antioxidative and anti-inflammatory potential of phenolics from purple basil (*Ocimum basilicum* L.) leaves induced by jasmonic, arachidonic and β-aminobutyric acid elicitation. *Int. J. Food Sci. Technol.* **2016**, *51*, 163–170. [CrossRef]
5. Złotek, U.; Szymanowska, U.; Pecio, Ł.; Kozachok, S.; Jakubczyk, A. Antioxidative and Potentially Anti-inflammatory Activity of Phenolics from Lovage Leaves Levisticum officinale Koch Elicited with Jasmonic Acid and Yeast Extract. *Molecules* **2019**, *24*, 1441. [CrossRef]
6. Śledź, M.; Nowacka, M.; Wiktor, A.; Witrowa-Rajchert, D. Selected chemical and physico-chemical properties of microwave-convective dried herbs. *Food Bioprod. Process.* **2013**, *91*, 421–428. [CrossRef]
7. Stępień, A.E.; Gorzelany, J.; Matłok, N.; Lech, K.; Figiel, A. The effect of drying methods on the energy consumption, bioactive potential and colour of dried leaves of Pink Rock Rose (*Cistus creticus*). *J. Food Sci. Technol.* **2019**, *56*, 2386–2394. [CrossRef]
8. Lim, Y.Y.; Murtijaya, J. Antioxidant properties of Phyllanthus amarus extracts as affected by different drying methods. *LWT-Food Sci. Technol.* **2007**, *40*, 1664–1669. [CrossRef]
9. Złotek, U.; Szymanowska, U.; Rybczyńska-Tkaczyk, K.; Jakubczyk, A. Effect of jasmonic acid, yeast extract elicitation, and drying methods on the main bioactive compounds and consumer quality of lovage (*Levisticum officinale Koch*). *Foods* **2020**, *9*, 323. [CrossRef] [PubMed]
10. Gawlik-Dziki, U.; Dziki, D.; Baraniak, B.; Lin, R. The effect of simulated digestion in vitro on bioactivity of wheat bread with Tartary buckwheat flavones addition. *LWT-Food Sci. Technol.* **2009**, *42*, 137–143. [CrossRef]
11. Świeca, M.; Baraniak, B. Nutritional and antioxidant potential of lentil sprouts affected by elicitation with temperature stress. *J. Agric. Food Chem.* **2014**, *62*, 3306–3313. [CrossRef]
12. Re, R.; Pellegrini, N.; Proteggente, A.; Pannala, A.; Yang, M.; Rice-Evans, C. Antioxidant Activity Applying an Improved Abts Radical Cation Decolorization Assay. *Free Radic. Biol. Med.* **1999**, *26*, 1231–1237. [CrossRef]
13. Oyaizu, M. Studies on products of browning reaction—Antioxidative activities of products of browning reaction prepared from glucosamine. *Jpn. J. Nutr.* **1986**, *44*, 307–315. [CrossRef]
14. Guo, J.T.; Lee, H.L.; Chiang, S.H.; Lin, H.I.; Chang, C.Y. Antioxidant properties of the extracts from different parts of broccoli in Taiwan. *J. Food Drug Anal.* **2001**, *9*, 96–101.
15. Szymanowska, U.; Jakubczyk, A.; Baraniak, B.; Kur, A. Characterisation of lipoxygenase from pea seeds (*Pisum sativum* var. Telephone L.). *Food Chem.* **2009**, *116*, 906–910. [CrossRef]
16. Rokicki, D.; Zdanowski, R.; Lewicki, S.; Leśniak, M.; Suska, M.; Wojdat, E.; Skopińska-Rózewska, E.; Skopiński, P. Inhibition of proliferation, migration and invasiveness of endothelial murine cells culture induced by resveratrol. *Cent. Eur. J. Immunol.* **2014**, *39*, 449–454. [CrossRef] [PubMed]
17. Leśniak, M.; Zdanowski, R.; Suska, M.; Brewczyńska, A.; Stankiewicz, W.; Kloc, M.; Kubiak, J.Z.; Lewicki, S. Effects of Hexachlorophene, a Chemical Accumulating in Adipose Tissue, on Mouse and Human Mesenchymal Stem Cells. *Tissue Eng. Regen. Med.* **2018**, *15*, 211–222. [CrossRef]
18. Mbondo, N.N.; Owino, W.O.; Ambuko, J.; Sila, D.N. Effect of drying methods on the retention of bioactive compounds in African eggplant. *Food Sci. Nutr.* **2018**, *6*, 814–823. [CrossRef] [PubMed]
19. Jakobek, L. Interactions of polyphenols with carbohydrates, lipids and proteins. *Food Chem.* **2015**, *175*, 556–567. [CrossRef] [PubMed]
20. Karaś, M.; Jakubczyk, A.; Szymanowska, U.; Złotek, U.; Zielińska, E. Digestion and bioavailability of bioactive phytochemicals. *Int. J. Food Sci. Technol.* **2017**, *52*, 291–305. [CrossRef]
21. Mandalari, G.; Vardakou, M.; Faulks, R.; Bisignano, C.; Martorana, M.; Smeriglio, A.; Trombetta, D. Food matrix effects of polyphenol bioaccessibility from almond skin during simulated human digestion. *Nutrients* **2016**, *8*, 568. [CrossRef] [PubMed]
22. Jimenez-Garcia, S.N.; Vazquez-Cruz, M.A.; Ramirez-Gomez, X.S.; Beltran-Campos, V.; Contreras-Medina, L.M.; Garcia-Trejo, J.F.; Feregrino-Pérez, A.A. Changes in the content of phenolic compounds and biological activity in traditional Mexican herbal infusions with different drying methods. *Molecules* **2020**, *25*, 1601. [CrossRef]
23. Ribas-Agustí, A.; Martín-Belloso, O.; Soliva-Fortuny, R.; Elez-Martínez, P. Food processing strategies to enhance phenolic compounds bioaccessibility and bioavailability in plant-based foods. *Crit. Rev. Food Sci. Nutr.* **2018**, *58*, 2531–2548. [CrossRef] [PubMed]
24. Kamiloglu, S.; Demirci, M.; Selen, S.; Toydemir, G.; Boyacioglu, D.; Capanoglu, E. Home processing of tomatoes (*Solanum lycopersicum*): Effects on in vitro bioaccessibility of total lycopene, phenolics, flavonoids, and antioxidant capacity. *J. Sci. Food Agric.* **2014**, *94*, 2225–2233. [CrossRef]
25. Kamiloglu, S.; Capanoglu, E. Investigating the in vitro bioaccessibility of polyphenols in fresh and sun-dried figs (*Ficus carica* L.). *Int. J. Food Sci. Technol.* **2013**, *48*, 2621–2629. [CrossRef]
26. Shahidi, F.; Varatharajan, V.; Oh, W.Y.; Peng, H. Phenolic compounds in agri-food by-products, their bioavailability and health effects. *J. Food Bioact.* **2019**, *5*, 57–119. [CrossRef]
27. Sytar, O.; Hemmerich, I.; Zivcak, M.; Rauh, C.; Brestic, M. Comparative analysis of bioactive phenolic compounds composition from 26 medicinal plants. *Saudi J. Biol. Sci.* **2018**, *25*, 631–641. [CrossRef]
28. Dzoyem, J.P.; Kuete, V.; McGaw, L.J.; Eloff, J.N. The 15-lipoxygenase inhibitory, antioxidant, antimycobacterial activity and cytotoxicity of fourteen ethnomedicinally used African spices and culinary herbs. *J. Ethnopharmacol.* **2014**, *156*, 1–8. [CrossRef]

29. Dogan, K.; Tornuk, F. Improvement of bioavailability of bioactive compounds of medicinal herbs by drying and fermentation with Lactobacillus plantarum. *Funct. Foods Health Dis.* **2019**, *9*, 735–748. [CrossRef]
30. Hossain, M.B.; Barry-Ryan, C.; Martin-Diana, A.B.; Brunton, N.P. Effect of drying method on the antioxidant capacity of six Lamiaceae herbs. *Food Chem.* **2010**, *123*, 85–91. [CrossRef]
31. Yi, W.; Wetzstein, H.Y. Effects of drying and extraction conditions on the biochemical activity of selected herbs. *HortScience* **2011**, *46*, 70–73. [CrossRef]
32. Gawlik-Dziki, U.; Świeca, M.; Dziki, D.; Sęczyk, Ł.; Złotek, U.; Rózyło, R.; Kaszuba, K.; Ryszawy, D.; Czyz, J. Anticancer and antioxidant activity of bread enriched with broccoli sprouts. *BioMed Res. Int.* **2014**, *2014*, 608053. [CrossRef] [PubMed]
33. Złotek, U.; Szychowski, K.A.; Świeca, M. Potential in vitro antioxidant, anti-inflammatory, antidiabetic, and anticancer effect of arachidonic acid-elicited basil leaves. *J. Funct. Foods* **2017**, *36*, 290–299. [CrossRef]
34. Rosa, L.; Silva, N.; Soares, N.; Monteiro, M.; Teodoro, A. Anticancer Properties of Phenolic Acids in Colon Cancer—A Review. *J. Nutr. Food Sci.* **2016**, *6*, 1–7.
35. Bogucka-Kocka, A.; Smolarz, H.D.; Kocki, J. Apoptotic activities of ethanol extracts from some Apiaceae on human leukaemia cell lines. *Fitoterapia* **2008**, *79*, 487–497. [CrossRef] [PubMed]
36. Elansary, H.O.; Szopa, A.; Kubica, P.; Al-Mana, F.A.; Mahmoud, E.A.; Zin El-Abedin, T.K.A.; Mattar, M.A.; Ekiert, H. Phenolic Compounds of Catalpa speciosa, Taxus cuspidate, and Magnolia acuminata have Antioxidant and Anticancer Activity. *Molecules* **2019**, *24*, 412. [CrossRef]

antioxidants MDPI

Article

Untargeted Metabolomics of Korean Fermented Brown Rice Using UHPLC Q-TOF MS/MS Reveal an Abundance of Potential Dietary Antioxidative and Stress-Reducing Compounds

Akanksha Tyagi, Su-Jung Yeon, Eric Banan-Mwine Daliri, Xiuqin Chen, Ramachandran Chelliah and Deog-Hwan Oh *

Department of Food Science and Biotechnology, College of Agriculture and Life Sciences, Kangwon National University, Chuncheon 200-701, Korea; akanksha@kangwon.ac.kr (A.T.); sujung0811@gmail.com (S.-J.Y.); ericdaliri@kangwon.ac.kr (E.B.-M.D.); cxq20135331@gmail.com (X.C.); ramachandran865@gmail.com (R.C.)
* Correspondence: deoghwa@kangwon.ac.kr; Fax: +82-33-2595565

Citation: Tyagi, A.; Yeon, S.-J.; Daliri, E.B.-M.; Chen, X.; Chelliah, R.; Oh, D.-H. Untargeted Metabolomics of Korean Fermented Brown Rice Using UHPLC Q-TOF MS/MS Reveal an Abundance of Potential Dietary Antioxidative and Stress-Reducing Compounds. *Antioxidants* **2021**, *10*, 626. https://doi.org/10.3390/antiox10040626

Academic Editors: Antonio Segura-Carretero and Irene Dimi

Received: 17 February 2021
Accepted: 15 April 2021
Published: 19 April 2021

Abstract: Free radical-induced oxidative stress is the root cause of many diseases, such as diabetes, stress and cardiovascular diseases. The objective of this research was to screen GABA levels, antioxidant activities and bioactive compounds in brown rice. In this study, we first fermented brown rice with different lactic acid bacteria (LABs), and the best LAB was selected based on the levels of GABA in the fermentate. *Lactobacillus reuterii* generated the highest levels of GABA after fermentation. To ascertain whether germination can improve the GABA levels of brown rice, we compared the levels of GABA in raw brown rice (Raw), germinated brown rice (Germ), fermented brown rice (Ferm) and fermented-germinated brown rice (G+F) to identify the best approach. Then, antioxidant activities were investigated for Raw BR, Germ BR, Ferm BR and G+F BR. Antioxidant activity was calculated using a 2,2-diphenyl-1-picryl hydrazile radical assay, 2,2-azino-bis-(3-ethylene benzothiozoline-6-sulfonic acid) radical assay and ferric-reducing antioxidant power. In Ferm BR, DPPH (114.40 ± 0.66), ABTS (130.52 ± 0.97) and FRAP (111.16 ± 1.83) mg Trolox equivalent 100 g, dry weight (DW), were observed as the highest among all samples. Total phenolic content (97.13 ± 0.59) and total flavonoids contents (79.62 ± 1.33) mg GAE/100 g and catechin equivalent/100 g, DW, were also found to be highest in fermented BR. Furthermore, an untargeted metabolomics approach using ultra-high-performance liquid tandem chromatography quadrupole time of flight mass spectrometry revealed the abundance of bioactive compounds in fermented BR, such as GABA, tryptophan, coumaric acid, L-ascorbic acid, linoleic acid, β-carotenol, eugenol, 6-gingerol, etc., as well as bioactive peptides which could contribute to the health-promoting properties of *L. reuterii* fermented brown rice.

Keywords: brown rice; fermentation; germination; antioxidants; stress; bioactive compounds; untargeted metabolomics; functional food; health benefits

1. Introduction

Rice (*Oryza sativa*) is one of the most important and stable human food crops in the world. Rice grain is a major source of carbohydrate and protein, as well as other important nutrients, for billions of people worldwide, particularly in developing countries [1]. In recent years, brown rice (BR) has gained growing attention due to its obvious advantages over white rice for the content of bioactive compounds [2], as it is known that the wrong diet may lead to an imbalance between free radical formation and radical scavenging ability, causing oxidative stress [3].

A lower reactive oxygen species (ROS) concentration is important for normal cellular signaling, although higher concentrations and long-term exposure to ROS cause damage to cellular macromolecules such as DNA, lipids and proteins, eventually leading to necrosis and apoptotic cell death [4]. The normal and proper functioning of the central nervous system (CNS) depends entirely on the chemical integrity of the brain. It is well known that the brain absorbs a significant amount of oxygen and is very high in lipid content and is vulnerable to oxidative stress [5]. High oxygen consumption contributes to an unsustainable output of ROS. Besides, neuronal membranes are found to be rich in polyunsaturated fatty acids, which are particularly susceptible to ROS [5].

Nowadays, antioxidants from natural sources such as fruits, cereals and vegetables have become a profitable alternative to prevent oxidative stress. A variety of studies have identified various bioactive compounds in brown rice, such as phenolic acids, flavonoids, γ-oryzanol, aminobutyric acid (GABA), α-tocopherol and γ -tocotrienol, which contribute to the health-promoting properties of brown rice [6]. Additionally, gamma-aminobutyric acid (GABA) is a recognized non-protein amino acid. GABA is commonly recognized as one of the main brain neurotransmitters, and ingestion of GABA has been recognized as affecting many important physiological functions, including enhancing brain function, postponing intelligence loss [7] and relief of nervous stress [8].

Although polyphenols or bioactive compounds are not available only in free form, insoluble or bound polyphenols are part of the cell wall, while free polyphenols can be contained within the cell wall of the plant [9]. Bound polyphenols are connected by ester, ether or glycoside linkages to the cell wall components that increase the mechanical strength of the cell wall. The use of bioconversion processes like fermentation and germination has been established as an efficient method for releasing bound bioactive compounds (phenols, flavonoids, organic acids, etc.). Enzymes produced in these processes (hydrolyzing enzymes) or biological process enzymes can break down the bound bioactive compounds or the linkages between the cell wall components [10]. Lactic acid bacteria have been involved in fermenting food materials for a long time, to release polyphenols or bioactive compounds [11]. Moreover, germination also activates many dormant enzymes which induce the degradation of molecules by synthesis and respiration of new constituents of cells [11]. The potential antioxidant and beneficial properties of health-associated phenolic compounds found in brown rice varieties make them potential targets for functional food markets. Many fermented brown rice and germination brown rice products are also well recognized in the community because of their various beneficial effects on health.

Besides, metabolomics techniques like gas chromatography-mass spectrometry (GC-MS), ^{1}H-nuclear magnetic resonance (^{1}H-NMR) and liquid chromatography-mass spectrometry (LC-MS) have been used to classify food metabolites [12]. Among different commonly used techniques, LC-MS is the most widely used in metabolomics studies because of its sensitivity to detection, high resolution and non-derivatization of samples [13]. Based on the same LC-MS approach, ultra-high-performance liquid tandem chromatography quadrupole flight mass spectrometry (UHPLC-QTOF/MS) is a new approach in chromatography to evaluate and quantify further metabolites with more sensitivity [14].

Recently, there has been a significant rise in interest in the use of nutritious foods, rich in essential amino acids, polyphenols and bioactive compounds like GABA. Therefore, the present study sought to investigate antioxidant properties, including phenolic compounds, flavonoids and organic acids alongside the amino acid composition of brown rice using UHPLC-ESI-QTOF-MS/MS to explore the ability of bio convergence processes to be used as natural antioxidants and therapeutics for the development of functional foods.

2. Materials and Methods

2.1. Rice Samples

BR sample (*Oryza sativa* L. Variety Japonica) was purchased from the local market, Chuncheon, Gangwon-do, South Korea. Raw BR and after-processing (germination) BR was ground into a fine powder using an electric mill and sifted through mesh 40. Samples were stored at −20 °C before further extraction.

2.2. Chemicals and Cultures

All chemical reagents were of analytical grade. Acetonitrile, ethanol, acetone, methanol, sodium carbonate, sodium hydroxide, anhydrous sodium acetate, hydrochloric acid, potassium persulfate, acetic acid and sulfuric acid were purchased from Daejung Chemicals and Metals Co., Ltd., South Korea. The phenolic standards and other chemicals like Folin–Ciocalteu reagent, Trolox (6-hydroxy-2,5,7,8-tetramethylchroman-2-carboxylic acid), DPPH (2,2-diphenyl-1-picrylhydrazyl); ABTS (2,2′-Azino-bis (3-ethylbenzothiazoline-6-sulfonic acid), TPTZ (2,4,6-Tris(2-pyridyl)-s-triazine) and gallic, ferulic, caffeic, o-Coumaric and p-Coumaric acids were obtained from Sigma, South Korea.

Lactobacillus reuterii AKT1 and all other strains used in our study were obtained from the Department of Food Science and Biotechnology, Kangwon National University, Korea, to be used for fermentation. The bacteria stock culture was stored at −80 °C, in MRS broth (Difco), containing 20% glycerol (*v*/*v*).

2.3. Sample Preparation

2.3.1. Brown Rice Germination

Brown rice (BR) germination was carried out by the method of [15] Cáceres et al. (2017), with some modification. Briefly, BR seeds (50 g) were washed 3–4 times with distilled water, 0.2% of sodium hypochlorite solution (1:5 *w*/*v*) as a disinfectant for 20 min at room temperature, and then rinsed thoroughly with distilled water. Drained BR was then soaked in distilled water for 12 h at 28 ± 1 °C. The soaking solution was changed every 6 h. After soaking, germination was carried out at a constant temperature (DAIHAN LABTECH. Co., Ltd., Incubator, Gyeonggi-do, Korea) at 30 ± 1 °C and 85% relative humidity in the dark for 48 h. Following germination, the germinated brown rice (GBR) was dried in a blast drying oven (OF-22GW, JEIO Tech Instrument Co., Ltd., Seoul, Korea) at 55 °C for 6 h (Figure S1). After drying, the GBR was ground into a fine powder using an electric mill and sifted through mesh 40.

2.3.2. Brown Rice Fermentation

Ground brown rice powder (10 gm/10 mL) was mixed with distilled water, sterilized in an autoclave at 121 °C for 15 min and then left until cool. Then, approximately 5 mL of 2×10^7 cfu/mL spores suspension^{-1} of 10 different lactic acid bacteria strains (*P. pentosaceus* (FMC1), *L. fermentum* (FMF2), *L. fermentum* (AKT2), *L. rhamnosus* (FMR1), *L. rhamnosus* (FMR2), *L. brevis* (FMB1), *L. brevis* ATCC (STANDARD), *L. plantarum* (FMP1), *L. plantarum* (FMP2) and *L. reuterii* (AKT1)) was obtained from actively growing slants in sterile water, which was then inoculated into sterilized (autoclaved) BR samples and incubated at 37 °C with 150 rpm agitation for 48 h. After 48 h of fermentation, the media were then centrifuged at 10,000× *g* for 10 min and the supernatant was freeze-dried and stored at −20 °C until further studies.

2.3.3. Preparation of Ethanolic Extracts

Extraction was carried out using a procedure of Shao-Hua et al. (2012) [16], with some modification. Brown rice powder (1 gm) was mixed with 20 mL of 50% ethanol (1:20 *w*/*v*) for 4 h in an electric shaker (RK-2D, DAIHAN scientific, Korea) at 50 °C. After that, extracts were centrifuged (Union 32R plus, Hanil Science Industrial, Korea) and the supernatant was collected at 4000× *g* for 10 min and the residue was re-extracted twice under the same conditions. The supernatants were filtered (0.20 µm Whatman™ syringe filter, Lk Lab Korea), evaporated at 50 °C and stored at −20 °C for further use. A stock solution of the sample was prepared with a concentration of 1 mg/mL. This was the stock that was used for the entire analysis.

2.4. Detection of Gamma-Aminobutyric Acid (GABA)

GABA content of the 50% ethanolic extract of raw rice, fermented rice, germinated brown rice (GBR) and germination combined with fermentation was determined as reported by Tiebing Liu et al. (2015) [17], with some modification. A Poroshell HPH-C18 column (4.6 × 100 mm, 2.7 μm) using pre-column derivatization with dansyl chloride by HPLC was used for GABA detection. The derivatization method was as follows: 0.2 mL of GABA standard solution, as well as rice samples (1 mg/mL), were mixed with the same volume, 0.2 mL, of dansyl chloride solution in a 1 mL brown volumetric flask and shaken. Then, samples were placed in a 50 °C water bath (VS-1205W, Vision Scientific Co. Ltd., Korea) for 50 min away from light, cooled to room temperature and diluted to a volume of 1.0 mL with methanol. Then, the solution was filtered through filters of 0.45 μm. The pH of the samples was adjusted to 8–8.5 before derivatization by adding 0.1 mol/L sodium bicarbonate.

The HPLC conditions were as follows: HPLC (Agilent Technologies, Baudrats, Germany) with an Infinity Lab Poroshell HPH-C18 column (4.6 × 100 mm, 2.7 μm) fitted with a guard column (Agilent Technologies, CA, USA); a flow rate of 1 mL/min; a column temperature of 30 °C; phase A, methanol; mobile phase B, water; and an amount of injection of approximately 2 μL.

2.5. *Total Phenolic Content (TPC)*

TPC was measured using the Folin–Ciocalteu colorimetric method mentioned formerly by Chang et al., (2018) [18] with some modification. Briefly, for a short duration of 6 min, 100 μL of sample extract or standard solution of gallic acid was treated with 200 μL of Folin–Ciocalteu reagent. The mixture was then alkalized with 1 mL of 700 Mm Na_2CO_3. After being kept dark for 90 min, the absorbance was measured at 760 nm using the SpectraMax i3 plate reader (Molecular Devices Korea, LLC). The total phenolic content was determined from the standard gallic acid curve and expressed as milligrams of gallic acid equivalent per 100 g of sample (mg GAE/100 g).

2.6. *Total Flavonoid Content (TFC)*

The total flavonoid content (TFC) of ethanol extracts was calculated using the 24-well microplate method as defined by Bah et al. (2016) [19], with some modifications. Briefly, 200 μL of our sample extracts were added to 1 mL of distilled water and 75 μL of $NaNO_2$ (50 g L^{-1}). After 5 min of incubation, 75 μL of $AlCl_3$ (100 g L^{-1}) was applied to the reaction mixture. Later, after 6 min, 600 μL of distilled water and 500 μL of 1 M NaOH were added. The absorbance was read at 510 nm by using the SpectraMax i3 plate reader (Molecular Devices Korea, LLC). Catechin was used as a baseline and the results were expressed as milligram catechin equivalents per 100 g of sample (mg CE/100 g).

2.7. *Antioxidant Assays*

2.7.1. DPPH Radical Scavenging Activity

DPPH radical scavenging activity assay was determined following the method of Chen et al. (2016) [20], with some modification. Briefly, a 24-well microplate was used with 200 μL sample extracts, which were mixed with 2 mL of freshly formulated 100 μM DPPH radical solution (4 mg DPPH in 100 mL 95 % *v*/*v* methanol). After 60 min of incubation at room temperature in the dark, the absorbance was measured at 515 nm using a spectrophotometer. The Trolox concentration plot with DPPH radical scavenging activity was used as a standard curve. DPPH values were expressed as mg Trolox equivalent (TE) per 100 g of sample DW.

2.7.2. ABTS Radical Scavenging Activity

ABTS activity was measured as described by Ofosu et al. (2020) [21] with some modification. ABTS radical cation reagent was formed by the reaction of 7 mmol/L ABTS stock solution with 2.45 mmol/L potassium persulfate stock solution (1:1, *v*/*v*). The mixture was kept in the dark for 12–16 h at room temperature before use. The $ABTS^{\bullet+}$ reagent was then mixed with methanol and the absorbance was modified to 0.700 ± 0.020 at 734 nm. In 1 mL of ABTS radical solution, 100 μL of properly diluted crude extracts or standards were mixed. After 30 min of incubation at room temperature in the dark, the spectrophotometer (SpectraMax i3 plate reader (Molecular Devices Korea, LLC) was used to measure the absorbance at 734 nm. Trolox concentration with ABTS radical scavenging rate was used as a reference curve. ABTS values were expressed as mg Trolox equivalent (TE) per 100 g of sample DW.

2.7.3. Ferric-Reducing Antioxidant Power (FRAP)

Ferric-reducing antioxidant strength was analyzed using the method as documented by Zeng et al. (2019) [22], with some improvements. Briefly, 0.1 mL of extract was combined with a FRAP reagent of 3.9 mL (FRAP reagent was prepared by using 50 mL acetate buffer (0.3 M, pH 3.6), 5 mL Tripyridyl triazine (TPTZ) solution (10 mM TPTZ in 40 mM HCl) and 5 mL $FeCl_3 \cdot 6H_2O$ (20 mM)), and set at 37 °C for 10 min, then the absorbance was performed at 593 nm. The findings were expressed as mg TE/100 g, DW.

2.8. UHPLC Q-TOF-MS/MS Identification

Ultra-high-performance liquid tandem chromatography quadrupole flight mass spectrometry (UHPLC Q-TOF-MS/MS) was used for the identification of bioactive compounds in rice. A rice sample (1 mg/mL) prepared after ethanol extraction was used for the analysis. The samples were centrifuged at 12,000× *g* for 15 min at 4 °C. Aliquots (1 mL) of the supernatants were filtered through 0.25 m pore size Millex syringe filters (Merck KGaA, Darmstadt, Germany) and transferred to LC-MS vials. The mass spectrometric analysis was conducted in both positive (ESI+) and negative (ESI-) ion modes. UHPLC (SCIEX ExionLC AD machine, MA, USA) fitted with a controller, AD pump, degasser, AD autosampler, AD column oven and photodiode array (PDA) detector (ExionLC), coupled with a quadrupole time-of-flight mass spectrometer (Q-TOF-MS) (X500R QTOF), was used for UHPLC and mass spectrometric analyses (LC-MS^2). The protocol of Daliri et al. (2020) [23] was used in the detection method with some modifications. Briefly, the analytical column used was a 100 × 3 mm Accucore column C-18 (Thermo Fisher Scientific, Waltham, MA, USA). The sample (10 μL) was injected by an autosampler and eluted through a binary mobile phase column consisting of solution A, which was water containing 0.1% formic acid, and solution B, which was methanol. A flow rate of 0.4 mL/min was used with a linear gradient which was programmed for 25 min as follows: 0–3.81 min, 9 to 14% B; 3.81–4.85 min, 14 to 15% B; 4.85–5.89 min, 15% B; 5.89–8.32 min, 15 to 17% B; 8.32–9.71 min, 17 to 19% B; 9.71–10.40 min, 19% B; 10.40–12.48 min, 19 to 26% B; 12.48–13.17 min, 26 to 28% B; 13.17–14.21 min, 28 to 35% B; 14.21–15.95 min, 35 to 40% B; 15.95–16.64 min, 40 to 48% B; 16.64–18.37 min, 48 to 53% B; 18.37–22.53 min, 53 to 70% B; 22.53–22.88 min, 70 to 9% B; and 22.88–25.00 min, 9% B. For positive ion mode, the capillary temperature and voltage were set at 320 °C and 40 V, respectively, the sheath gas flow was set at 40 arb. units and an aux gas flow of 8 arb. units, spray voltage at 3.6 kV and tube lens voltage at 120 V. In negative ion mode, capillary temperature and voltage were set at 320 °C and −40 V, respectively, sheath gas flow at 40 arb. units and gas flow to 8 arb. units, spray voltage up to 2.7 kV and tube lens voltage up to −120 V. In both positive and negative ion modes, the study was carried out using the Fourier transform mass spectrometry (FTMS) full scan ion mode, applying a mass scan range of 115–1000 m/z and a resolution of 30,000 full width at half maximum (FWHM) while the spectra were acquired in centroid mode. The scanning time under these conditions was approximately 1 s.

2.9. Statistical Analysis

The data was analyzed using Graphpad Prism 8.0. Differences in mean values between differently treated brown rice samples phenolic extracts were calculated using one-way variance analysis (ANOVA) followed by a Tukey and Duncan test at $p < 0.05$ significance stage using the SPSS program and Graphpad Prism 8.0. The findings were described as average ± standard deviation (SD).

Compound identification was achieved by using empirical formula finders like Pubchem (https://pubchem.ncbi.nlm.nih.gov/) (access on 9 April 2021) or ChemSpider (http://www.chemspider.com/) (access on 9 April 2021). ClustVis program (http://biit.cs.ut.ee/clustvis/) (access on 17 April 2021) and Paleontological Statistics (PAST) was used in multivariate statistical analyses, including main component analysis (PCA) and heat maps [24]. PCA was used to compare the changes among brown rice samples. Heat maps were drawn using peak areas of samples by clustVis [25].

3. Results and Discussion

3.1. Detection of Gamma-Aminobutyric Acid (GABA)

GABA, the main inhibitory neurotransmitter, has multiple physiological functions such as being a blood pressure regulator, cardiovascular disease defender and hormonal and cell regulator. It is also associated with brain and mental disorders [26]. Therefore, we first screened different fermentation bacteria (LABs) for GABA concentration (Figure 1A), as it is linked to stress-related disorders. Among 10 lactobacillus strains, *L. reuterii* (AKT1) showed the highest GABA concentration in HPLC detection (Table S1). All the results are expressed as the mean $\pm$ SD of triplicate analyses, and additional statistical analyses were carried out using they Tukey and Duncan tests, $p \leq 0.05$. These results show *L. reuterii* as the most efficient bacteria in GABA production during fermentation among all other LABs used for screening. In raw rice, GABA content was found to be 1.61 $\pm$ 0.001 μg/mL, whereas in *L. reuterii* fermented brown rice it was 27.03 $\pm$ 0.055 μg/mL. In this research, we found a higher GABA concentration than reported by Cáceres et al. and Le et al. [27,28].

After screening fermentation bacteria, we compared fermentation (Ferm) with germinated brown rice (Germ) and the germination combined with fermentation approach (G+F). In germinated samples, the GABA content was 34.43 $\pm$ 0.026, and in combination (G+F) samples it was found to be 22.91 $\pm$ 0.028 (Figure 1B). Germination was found to be higher than fermentation, so further analysis was carried out to find out more about the bioactive compounds produced during different bio convergence approaches in brown rice. We performed germination for 24, 48 and 72 h, and 48 h germination was found to be the best among GABA screenings. We found no difference in GABA concentration after 48 h and an increase in GABA content after 24 h, so 48 h was selected for further analysis (24 and 72 h data not shown). Fermentation was also carried out for 48 h to compare the same duration of fermentation and germination.

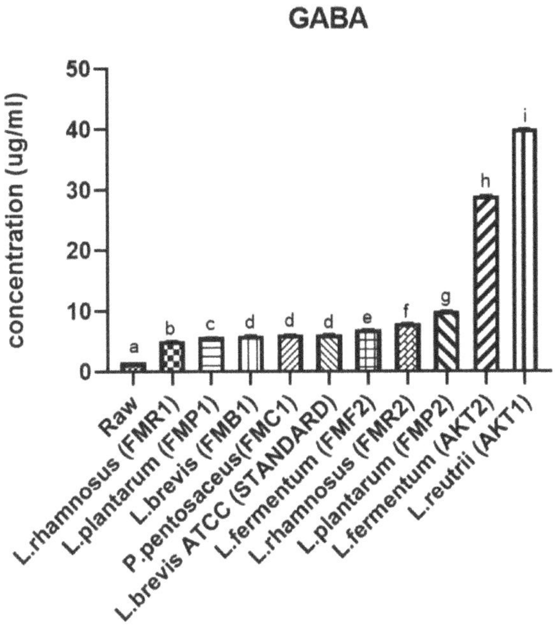

A

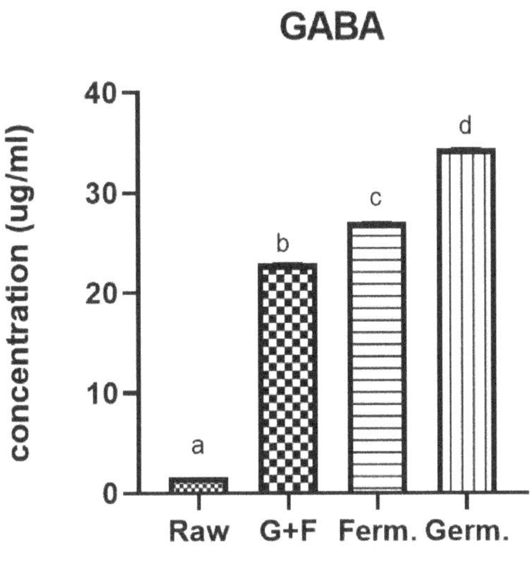

B

Figure 1. (**A**) Screening of different *Lactobacillus* strains (48 h fermented). (**B**) Comparison of GABA content among raw, germination (Germ), fermentation (Ferm) and germination combined with fermentation (G+F) treatments. All values are expressed as the mean ± SD of triplicate experiments. The sample concentration used was 1 mg/mL, and a–i superscripts with different letters indicate a significant difference, while sample superscript letters indicate no significant difference ($p < 0.05$) found by using SPSS and Graphpad Prism 8.0 programs.

3.2. Total Phenolic Content and Total Flavonoid Content of Ethanol Extracts

Phenolics are a class of phytochemicals that have one or more hydroxyl groups of aromatic rings, and phenolic content has been linked to the antioxidant effects of grains [29]. Phenolic-rich foods have high antioxidant potential and may counteract the negative effects of pro-inflammatory and pro-oxidative foods. The TPC and TFC of all four samples are shown in Table 1 as the mean ± SD of triplicate analyses with statistically significant differences ($p \leq 0.05$). TPC was found to be lowest in raw brown rice samples, 16.75 ± 0.75 mg GAE/100 g, DW, and highest in *L. reuterii* AKT1 fermented brown rice samples, 97.13 ± 0.59 mg GAE/100 g, DW. Hydrolysis by enzymes in germination and fermentation usually enhances the total phenolic content and this was also observed in this study. TPC increased with germination (74.70 ± 0.72 mg GAE/100 g, DW) and fermentation (97.13 ± 0.59 mg GAE/100 g, DW) as compared with the raw BR sample. This study shows a higher TPC content of brown rice than reported by [30,31].

Table 1. Total antioxidants, DPPH, ABTS and FRAP, total phenolic content (TPC) and total flavonoid content (TFC) of different rice samples.

S.NO	Sample	DPPH (mg Trolox Equivalent 100 g, DW)	ABTS (mg Trolox Equivalent 100 g, DW)	FRAP (mg Trolox Equivalent 100 g, DW)	TPC (mg Gallic Acid Equivalent 100 g, DW)	TFC (mg Catechin Equivalent 100 g, DW)
1	Raw BR	25.653 ± 0.98 [a]	21.86 ± 0.93 [a]	19.21 ± 2.64 [a]	16.75 ± 0.75 [a]	14.42 ± 0.80 [a]
2	Germinated BR	89.602 ± 1 [b]	98.37 ± 1.38 [b]	88.04 ± 1.27 [b]	74.70 ± 0.72 [b]	62.88 ± 2.62 [b]
3	Fermented BR	114.40 ± 0.66 [d]	130.52 ± 0.97 [d]	111.16 ± 1.83 [d]	97.13 ± 0.59 [d]	79.62 ± 1.33 [d]
4	G+F BR	105.99 ± 0.59 [c]	110.11 ± 0.26 [c]	96.61 ± 2.29 [c]	89.94 ± 0.25 [c]	69.85 ± 0.80 [c]

BR—brown rice, G+F—germination combined with fermentation. Results are expressed as mean ± SD of triplicate analyses. Different alphabetical letters ([a–d]) in each column represent statistically significant differences (Tukey and Duncan test, $p \leq 0.05$). DW—dry weight sample.

Flavonoids are a group of phenolic compounds consisting of two aromatic rings (A and B rings) bound together by a three-carbon structure generally found in an oxygenated heterocyclic ring (C ring). They have potent antioxidant activity and are associated with a reduced risk of chronic diseases. Total flavonoid content was found to be higher in fermented brown rice, 79.62 ± 1.33 mg catechin equivalent 100 g, DW, followed by the combined approach (G+F BR), germination and raw BR (Table 1). TFC content was lower than TPC in brown rice samples. The total TFC in this study was found to be higher than reported by Huang et al. [16], similarly to TPC, but both TPC and TFC were found to be lower than reported by Gong et al. [32]. These differences in brown rice values from different researchers may be due to genotype, cultivation landscape and climate conditions [33]. Besides, it is worth noting that the phenolic content can be significantly influenced by different extraction solvents and procedures.

3.3. Antioxidant Assay (DPPH, ABTS, FRAP)

Antioxidants are exogenous or endogenous molecules that reduce any form of oxidative/nitrosative stress or its consequences. Phenolic compounds are generally considered to be desired components for human health, due to their antioxidant activity. Various mechanisms, such as free radical scavenging, reducing capacity, metal ion chelation and lipid peroxidation inhibition, have been studied to explain how rice extracts could be used as effective antioxidants [34]. In recent years, germination and fermentation processes have also been considered to be an effective approach for enhancing antioxidant activities in cereals. In the present study, DPPH, ABTS and FRAP assays were used to evaluate the antioxidant activity of different treated brown rice samples. The antioxidant values of DPPH, ABTS and FRAP of phenolic extracts of BR are presented in Table 1.

One of the most frequently used assays in the determination of antioxidant activity in natural products is the measurement of the scavenging activity of the 2,2′-diphenyl-1-

picrylhydrazyl (DPPH) radical by spectrophotometry. The DPPH radical scavenging activity was highest in the fermented brown rice sample (*L. reuterii* AKT1) (114.40 ± 0.66 mg Trolox equivalent 100 g, DW), followed by G+F BR and germinated brown rice samples (105.99 ± 0.59 and 89.60 ± 1 mg Trolox equivalent 100 g, DW). The lowest values were found in raw BR (25.653 ± 0.98 mg Trolox equivalent 100 g, DW).

Similarly, in in-plant materials and grains, ABTS is an important method for the determination of radical scavenging activity. In addition, the FRAP assay was originally designed to assess plasma antioxidant ability, but it has also been commonly used in a wide variety of pure compounds and biological samples to determine antioxidant capacity. The test operates on the principles of antioxidant donation of electrons. It measures changes in absorption as a result of the formation of blue iron (II) from colorless iron oxide (III). In our research, the same trend with DPPH was observed in ABTS and FRAP assays. ABTS activity was measured to be highest in fermented BR (130.52 ± 0.97 mg Trolox equivalent 100 g, DW), followed by G+F and germination (Table 1). In ABTS the lowest activity was from raw BR. Similarly, ferric-reducing antioxidant power (FRAP) was the highest in fermented BR (111.16 ± 1.83 mg Trolox equivalent 100 g, DW) followed by G+F and germination. A correlation was observed among the trends in all antioxidant assays. These findings were found to be higher than in earlier reports by Lin et al. (2019) [35] and Ilowefah et al. (2017) [30] in fermented BR and Cho et al. (2018) [36] for germinated BR. Besides, other studies have reported the high antioxidant activity of brown rice, such as Rao et al. [37].

3.4. *UHPLC Q-TOF-MS/MS Detection*

Ultra-high-performance liquid tandem chromatography quadrupole flight mass spectrometry (UHPLC-Q-TOF-MS/MS) detection is considered the gold standard technique for the precise detection and quantification of a wide variety of components. Therefore, in this study, we used this detection technique for the identification of bioactive compounds in brown rice.

3.4.1. Amino Acid Level in Brown Rice

In the growth and development of organisms, amino acids play an important role and can also improve the taste of food. In this study, a total of eighteen amino acids were detected in brown rice (Figure 2A, Table S2) and differently treated samples showed statistically significant differences from each other after comparing levels of amino acids (ANOVA Table S2 $p \leq 0.05$). Raw brown rice contained the least number of amino acids, whereas germination and fermentation lead to an increase in amino acid content. The levels of amino acids were detected to be highest in the *L. reuterii* AKT1 fermented BR sample, which is potentially because during fermentation microorganisms produce enzymes that lead to the formation of several metabolites and bioactive compounds from the food matrix [38]. Additionally, an increase was also observed in the germination process due to the activation of many dormant enzymes which enhance bioactive compounds in rice, whereas in the raw sample, a low level might be the reason for more bound amino acids with parent molecules. In the fermented sample, levels of some essential amino acids like methionine, histidine, valine, tryptophan, leucine and lysine increased drastically after fermentation, but not phenylalanine and threonine. Additionally, certain conditionally essential amino acids were also found to be higher in fermented brown rice samples (ornithine, serine, arginine, proline, tyrosine and glutamine) (Figure 2A). Moreover, the G+F approach also enhanced the level of amino acids as compared with the germination-only approach. Principal component analysis (PCA) was applied on amino acids to see the diversity and correlation among samples, and the raw BR profile was found to be correlated with the germinated sample but divergent from the fermented BR and G+F BR samples, whereas after comparing PC1 with PC3, the G+F BR sample showed some correlation with the fermented (*L. reuterii*) BR sample, as shown in the PCA graph (Figure 2B) (Table S2). PCA analysis showed the same trend as was observed in the heat map analysis.

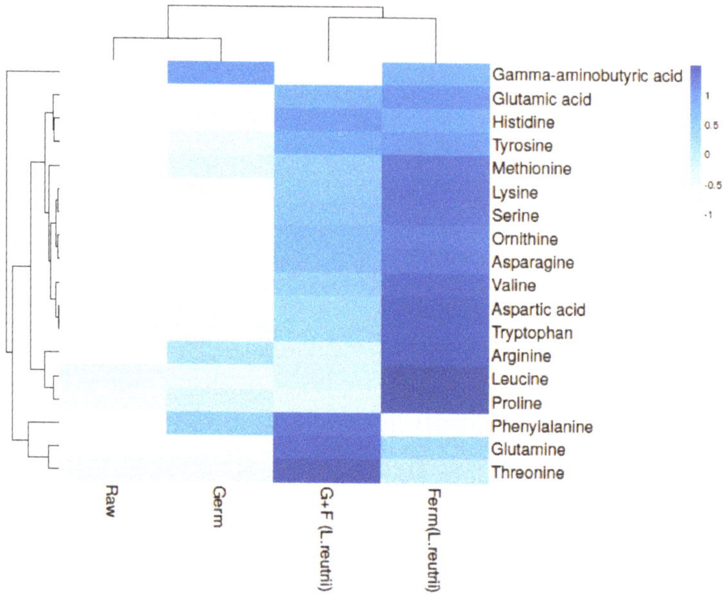

(A)

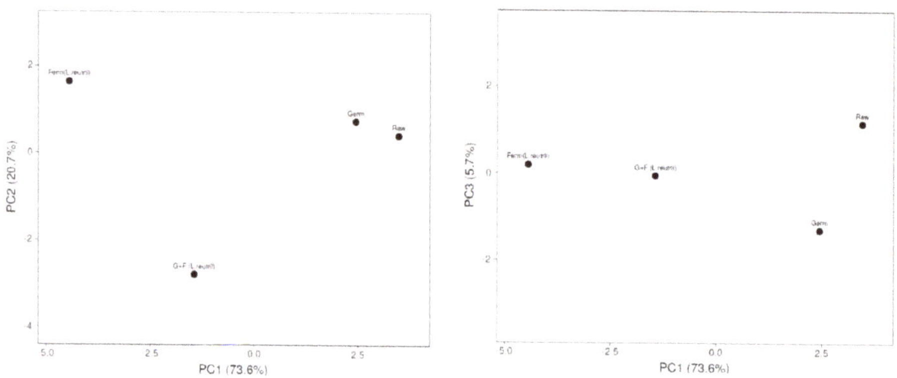

(B)

Figure 2. Levels of amino acids in raw, germinated, fermented (*L. reuterii* AKT1) and G+F (*L. reuterii* AKT1) samples. (**A**) The heat map shows different levels of amino acids in samples based on shades of blue. (**B**) Identification of principal component analysis (PCA) of Raw, Germ, Ferm (*L. reuterii*) and G+F samples was shown by comparing PC1 with PC2 and PC 1 with PC3. Germ-germinated brown rice, Ferm- fermented brown rice, G+F-germinated + fermented brown rice.

3.4.2. Phenolic Compounds in Brown Rice Samples

Fourteen phenolic compounds were identified in brown rice samples by UPLC-ESI-Q-TOF-MS/MS. Phenolic compounds have been suggested to account for the health benefits of rice intake in the prevention of chronic diseases, including cardiovascular disease, type II diabetes, obesity, oxidative stress, stress-related disorders and cancer [39]. The contents of phenolic compounds in different processed brown rice samples are presented in Table S3 and significant differences were observed after comparing the levels of each sample ($p \leq 0.05$). Heat map analysis was used for clustering phenolic compounds based on their concentrations (Figure 3A) where the color scheme from blue to red shows concentration in decreasing order. The results show that the highest phenolic compounds were detected in the fermented (*L. reuterii* AKT1) brown rice sample, followed by G+F (*L. reuterii* AKT1), germinated and raw. Phenolic compounds usually occur in an esterified form linked to the cereal wall matrix in the cereals, as they are not readily available [40]. Fermentation is considered to be a possible strategy not only to increase AAs but also to release insoluble bound phenolic compounds and thus to improve the poor bioavailability of grain phenolics [41]. In the present study, many phenolic compounds such as β-carotenol [42], eugenol [43], apigenin [44], 6-gingerol [45], cinnamic acid [46] and vanillic acid [47] were detected in high concentrations in fermented brown rice, and all of these have been reported as strong antioxidants, as well as able to reduce stress-related behaviors. Principal component analysis (PCA) was applied for better interpretations, and to avoid multicollinearity (Figure 3B) (Table S3). In PCA, fermented (*L. reuterii* AKT1) brown rice was found to be divergent from all other samples, as it contained the highest concentration of phenolic compounds.

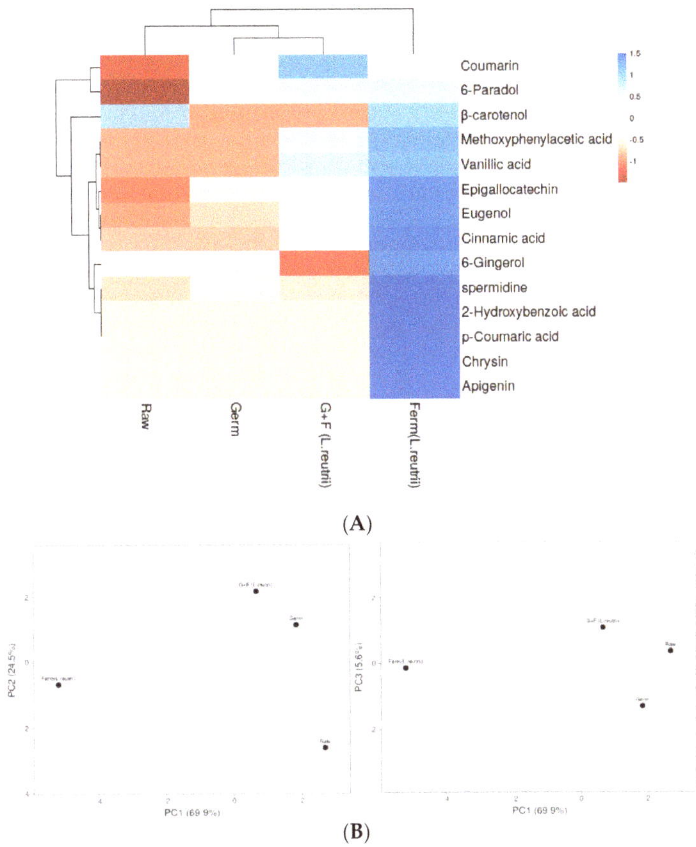

Figure 3. Levels of phenolic compounds in raw, germinated, fermented (*L. reuterii* AKT1) and G+F (*L. reuterii* AKT1) samples. (**A**) The heat map shows different levels of phenolic compounds in samples based on colors from blue to red, representing the level of phenolic compounds in decreasing order. (**B**) Identification of principal component analysis (PCA) of Raw, Germ, Ferm (*L. reuterii*) and G+F samples was shown by comparing PC1 with PC2 and PC1 with PC3. Germ- germinated brown rice, Ferm- fermented brown rice, G+F-germinated + fermented brown rice.

Comparing different approaches, we agree with the present reported studies that fermentation increases phenolic compounds as well as the bioavailability and bio-accessibility of cereal grains like brown rice [48].

3.4.3. Organic Acid Level in Brown Rice Samples

In the present study, nineteen organic acids, including lipoic acid, pyrazinoic acid, anofinic acid, malonic acid, arabinonic acid, nicotinic acid, homocitric acid, p-coumaric acid, ascorbic acid, malyngic acid, hydroxybutyric acid, gluconic acid, isophthalic acid, succinic acid, neuraminic acid, malic acid, itaconic acid, butanoic acid and glutaric acid, were detected in the tested brown rice samples (Table S4). Additionally, as shown in Table S4, organic acid levels were found to be significantly different in differently treated samples ($p \leq 0.05$). Heat map analysis was used for clustering brown rice samples based on their organic acid concentration in differently treated samples (Figure 4A). The levels of organic acids were enhanced after germination and fermentation treatments, leading to the highest organic acids content in the *L. reuterii* AKT1 fermented brown rice sample (Table S4). Organic acids like ascorbic acid [49], p-coumaric acid [50], hydroxybutyric acid [51] and lipoic acid [52] were widely reported as strong antioxidants, as well as in the improvement of stress-related disorders in various studies. Fermentation triggers the structural degradation of the cell wall via the metabolic activities of microbes, leading to the synthesis of different bioactive compounds [53]. The functions of proteases, amylases and xylanases derived from fermenting microorganisms and cereal grains also contribute to grain modification and to the distortion of chemical bonds that, subsequently, release bound bioactive compounds. Additionally, during germination, due to the respiration and synthesis of new cell constituents, the degradation of certain molecules leads to bioactive components that enhance the functional and nutritional properties of germinated rice [54]. Principal component analysis (PCA) was applied to group organic acids that were more strongly correlated to avoid multicollinearity for the better elucidation of results (Figure 4B) (Table S4). The organic acid profile of fermented brown rice was found to be divergent from the other three samples, similar to the heat map analysis. Whereas raw BR and germinated BR samples were found to be more strongly correlated with each other in PC1 with PC2 analysis, G+F samples were found to be somehow correlated with Germ BR in PC1 with PC3 analysis.

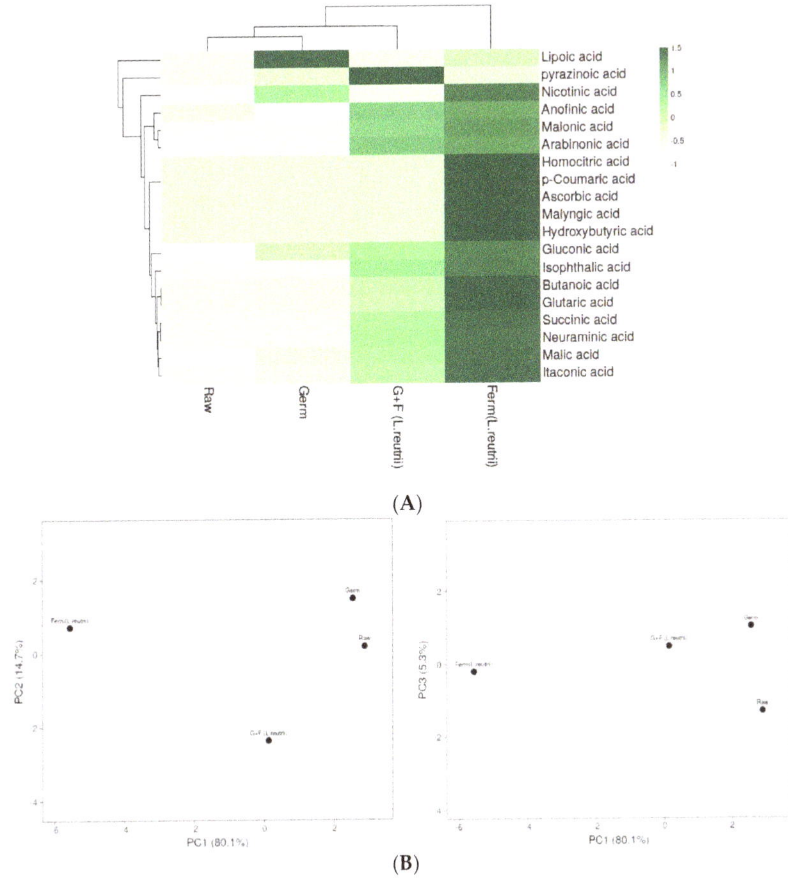

Figure 4. Levels of organic acids in Raw, Germinated, Fermented (*L. reuterii* AKT1) and G+F (*L. reuterii* AKT1) samples. (**A**) The heat map shows different levels of organic acids in samples based on colours as green to white represents the level of organic acids in decreasing order. (**B**) Identification of principal component analysis (PCA) of Raw, Germ, Ferm (*L. reuterii* AKT1) and G+F samples was shown by comparing PC1 with PC2 and PC1 with PC3. Germ- germinated brown rice, Ferm- fermented brown rice, G+F- germinated + fermented brown rice.

3.4.4. Fatty Acid Composition in Brown Rice Samples

In the present study, seventeen fatty acids were detected in differently treated brown rice samples (Table S5), and as shown in Table S5, fatty acid levels were found to be significantly different in all four differently treated samples ($p \leq 0.05$). The results show that the highest levels of fatty acids were found in fermented brown rice. In particular, fermentation has been indicated as a tool to enhance the nutritional value of foods, both in terms of the increased bioavailability of bioactive compounds and in terms of the production of health-promoting end-products. Among the latter, short-chain fatty acids (SCFAs) have emerged as some of the most studied compounds in the last decade, due to their proven benefit. Heat map analysis was used for separating fatty acids based on different concentration, represented from blue to green in decreasing order (Figure 5A). Fatty acids like stearic acid [55], linoleic acid [56], valeric acid [57], lauric acid [58] and azelaic acid [59] were already reported in the literature as strong antioxidants and stress-reducing compounds. Principal component analysis (PCA) was used to understand the correlation among different groups, and we found that the fermented brown rice sample (Ferm *L. reuterii* AKT1) was divergent from the other three samples. Moreover, germinated and G+F BR were found to be correlated (Figure 5B) (Table S5).

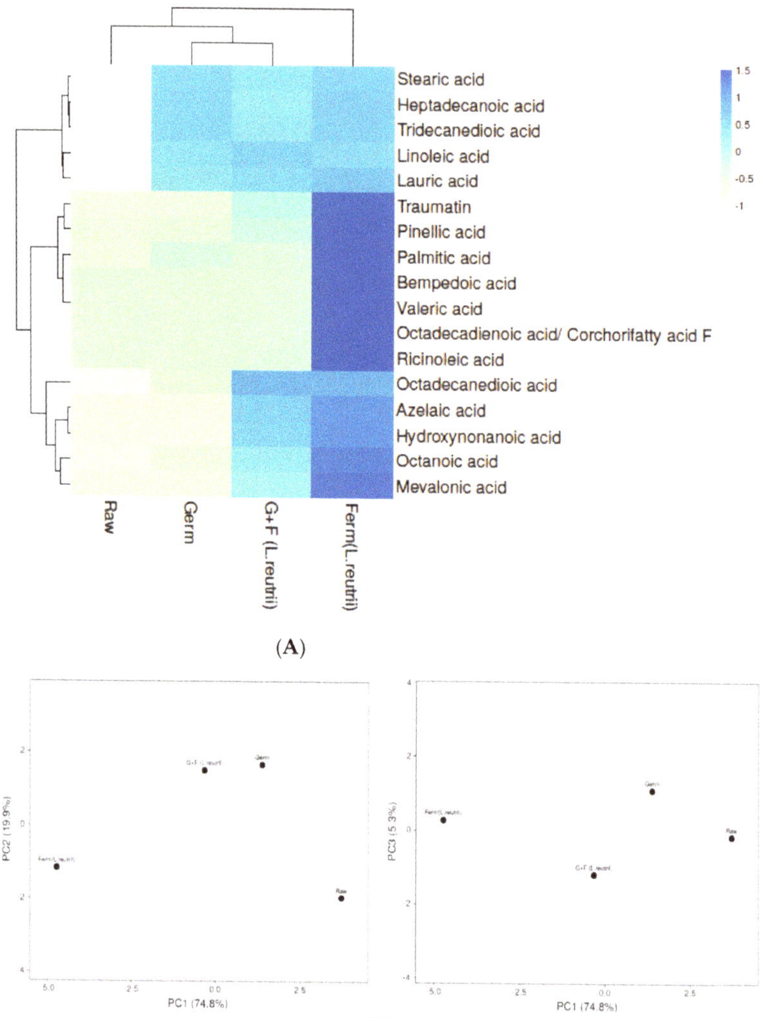

Figure 5. Levels of fatty acid in raw, germinated, fermented (*L. reuterii* AKT1) and G+F (*L. reuterii* AKT1) samples. (**A**) The heat map shows different levels of fatty acid in samples based on colors, as blue to green represents the level of fatty acids in decreasing order. (**B**) Identification of principal component analysis (PCA) of Raw, Germ, Ferm (*L. reuterii* AKT1) and G+F was shown by comparing PC1 with PC2 as well as PC1 with PC3. Germ- germinated brown rice, Ferm- fermented brown rice, G+F- germinated + fermented brown rice.

Additionally, to understand the correlations and diversity among differently treated samples, as well as for the better elucidation of results, PCA was also performed by combining all the analyses together (amino, phenolic, organic and fatty acids) (Figure S2). In joint PCA, fermented BR was found to be divergent from the other three samples, similar to that which we have observed in all separate analyses. Additionally, Germ BR and Raw BR were found to be correlated with each other.

Hence, by comparing different approaches in our study, it is clear that fermentation increases most of the bioactive compounds in cereal grains like brown rice [60].

3.4.5. Peptides Identified in Brown Rice Samples

An alternative, less expensive method for producing bioactive peptides from protein substrates is microbial fermentation. The main advantage of using microbes instead of enzymes is that appropriately used microorganisms will not only break down proteins into peptides and free amino acids, but can also remove anti-nutritional or hyper-allergenic factors that may be present (such as phytate, stachyose, saponins, etc.) [61,62]. In the present study, most of the peptides were reported in fermented brown rice samples (*L. reuterii* AKT1) (Table 2). Therefore, we analyzed the peptides produced in the different fermented and germinated samples and a total of 10 known peptides were found, and their potential functions were determined by comparing them with similar peptides already reported in the literature (Table 2). Most of the peptides identified in our study were not known in the databases so it is difficult to predict their functions. These bioactive peptides reported in our study in the fermented samples contribute to the strong antioxidative and stress-reducing abilities. In the meantime, functional analysis is required to confirm the antioxidative and stress-related abilities of peptides in the fermented sample containing sequences of already reported peptides.

Table 2. Identification of peptides in different processed brown rice samples.

S.No	Sample Name	Retention Time	Peak Area	Adduct/Charge	Precursor Mass	Found at Mass	Formula Finder Result	Peptide	Role	Reference
1	Raw	0.9	3.92×10^2	[M+H]+	229.155	229.1556	$C_{11}H_{20}N_2O_3$	Leucylproline (Leu-Pro)	Reduction in Stress/Depression And Antioxidant	[63]
	Germ	1.4	2.34×10^5	[M+H]+	229.155	229.1553				
	Ferm	1.41	5.18×10^3	[M+H]+	229.155	229.1552				
	G+F	1.4	1.02×10^5	[M+H]+	229.155	229.1552				
2	Raw	18.52	2.98×10^4	[M+Na]+	525.287	525.2886	$C_{14}H_{26}N_4O_6$	Gln-Ser-Leu Glutaminyl-seryl-leucine	Antioxidant	[64]
	Germ	18.51	1.52×10^5	[M+Na]+	525.287	525.2892				
	Ferm	18.51	2.31×10^5	[M+Na]+	525.287	525.2891				
	G+F	18.51	1.60×10^5	[M+Na]+	525.287	525.2890				
3	Raw	24.01	1.15×10^4	[M+NH4]+	445.288	445.2884	$C_{18}H_{33}N_7O_5$	Gly-Pro-Arg-Val	Sleep inducing peptide	[65]
	Germ	24.01	9.03×10^3	[M+Na]+	450.244	450.2448				
	Ferm	24.01	9.88×10^6	[M+Na]+	450.244	450.2446				
	G+F	23.98	1.18×10^5	[M+Na]+	450.244	450.2441				
4	Raw	ND	ND	ND	ND	ND	$C_{12}H_{25}N_5O_3$	Arg-leu	Antioxidant	[66]
	Germ	ND	ND	ND	ND	ND				
	Ferm	1.29	1.89×10^3	[M+H]+	288.203	288.2027				
	G+F	ND	ND	ND	ND	ND				
5	Raw	ND	ND	[M−H]−	ND	ND	$C_{11}H_{20}N_2O_5$	Glu-Leu	Antioxidant	[67]
	Germ	ND	ND	[M−H]−	ND	ND				
	Ferm	10.08	3.79×10^3	[M−H]−	259.131	259.1294				
	G+F	0	0	[M−H]−	0	0				
6	Raw	ND	ND	[M+CH3OH+H]+	ND	ND	$C_7H_{12}N_2O_3$	gly-pro	Stress reduction	[68]
	Germ	1.26	9.30×10^4	[M+CH3OH+H]+	205.118	205.1187				
	Ferm	1.18	1.51×10^3	[M+CH3OH+H]+	205.118	205.1193				
	G+F	1.39	6.27×10^3	[M+CH3OH+H]+	205.118	205.1193				
7	Raw	ND	ND	[M+H]+	ND	ND	$C_{10}H_{17}N_3O_6S$	L.gamma.glu-L.cystein (Glu-tathione)Peptide	Antioxidant	[69]
	Germ	ND	ND	[M+H]+	ND	ND				
	Ferm	1.82	1.08×10^4	[M+H]+	437.163	437.1693				
	G+F	ND	ND	[M+H]+	ND	ND				
8	Raw	ND	ND	[M+H]+	ND	ND	$C_{14}H_{20}N_4O_7S$	Gly-Gly-Gly	Antioxidant	[70]
	Germ	ND	ND	[M+H]+	ND	ND				
	Ferm	40.08	1.08×10^4	[M+H]+	389.113	389.1127				
	G+F	ND	ND	[M+H]+	ND	ND				
9	Raw	16.03	1.23×10^4	[M−H]−	186.115	186.1144	$C_9H_{17}NO_3$	Pivagabine (N-pivaloyl-γ-aminobutyric acid)	Stress and Anxiety	[71]
	Germ	16.02	8.71×10^3	[M−H]−	186.115	186.1139				
	Ferm	16.04	6.83×10^5	[M−H]−	186.115	186.1138				
	G+F	16.04	3.15×10^5	[M−H]−	186.115	186.1138				
10	Raw	47.27	6.72×10^5	[M−H]−	459.196	459.1957	$C_{16}H_{28}N_8O_8$	Gly-Arg-Gly-Asp-Gly	Peripheral Nerve Regeneration and Antioxidant	[72]
	Germ	47.26	7.26×10^5	[M−H]−	459.196	459.196				
	Ferm	47.28	8.73×10^5	[M−H]−	459.196	459.1959				
	G+F	47.27	9.41×10^5	[M−H]−	459.196	459.1958				

Germ- germinated brown rice, Ferm- fermented brown rice, G+F- germinated + fermented brown rice.

4. Conclusions

In our research, we identified that fermented brown rice exhibited high antioxidant activities as well as a significant amount of phenolic compounds and flavonoids, as compared to raw, germination and germination combined with fermentation samples. This might be because, when germination and fermentation were combined, biological enzymes utilized bioactive compounds as their energy sources. Additionally, using untargeted metabolomics, we found that *L. reuterii* fermented brown rice enhanced the production of essential amino acids and organic acids. The fermented samples also enhanced phenolic compounds and fatty acids, and were found to be rich in antioxidant and stress-reducing peptides. These findings have shown that fermented *L. reuterii* AKT1 brown rice consumption can provide optimal health benefits, as it is rich in antioxidant phytonutrients. Moreover, these findings also make this sample a promising material for the development of functional food with great antioxidant and stress-reducing abilities. However, further studies concerning the ability of *L. reuterii* AKT1 fermented brown rice to reduce stress-related disorders require in-vivo studies for further confirmation of their health-promoting effects and functional food development.

Supplementary Materials: The following are available online at https://www.mdpi.com/article/10.3390/antiox10040626/s1. Figure S1: Germinated brown rice. Figure S2: Principal component analysis (PCA) of Raw, Germ, Ferm (*L. reuterii* AKT1) and G+F were shown by comparing component 1(PC1) with component 2 (PC2). Table S1: GABA concentration detected by HPLC in Raw, different lactic acid bacteria fermented brown rice, germinated, and germination combined with fermentation (G+F). Table S2: Amino acids detected in different processed brown rice samples (raw, germinated, fermented (*L. reuterii* AKT1), and germinated+ fermented (*L. reuterii* AKT1). Table S3: Phenolic componds detected in different processed brown rice samples (raw, germinated, fermented (*L. reuterii* AKT1), and germinated+ fermented (*L. reuterii* AKT1). Table S4: Organic acids detected in different processed brown rice samples (raw, germinated, fermented (*L. reuterii* AKT1), and germinated+ fermented (*L. reuterii* AKT1). Table S5: Fatty acids detected in different processed brown rice samples (raw, germinated, fermented (*L. reuterii* AKT1), and germinated+ fermented (*L. reuterii* AKT1).

Author Contributions: Conceptualization, A.T. and D.-H.O.; data curation, A.T. and S.-J.Y.; formal analysis, A.T. and E.B.-M.D.; funding acquisition, D.-H.O.; investigation, X.C. and R.C.; methodology, X.C.; software; R.C.; supervision, D.-H.O.; writing—original draft, A.T.; writing—review and editing, A.T., S.-J.Y. and E.B.-M.D. All authors have read and agreed to the published version of the manuscript.

Funding: This research work was supported by a grant from the Brain Korea (BK) 21 Plus Project (Grant No. 22A20153713433), funded by the Korean Government, Republic of Korea.

Institutional Review Board Statement: Not required.

Informed Consent Statement: Not required.

Data Availability Statement: Not required.

Conflicts of Interest: The authors declare that they have no conflict of interest. The funders had no role in study design, data collection and analysis, decision to publish, or preparation of the manuscript.

Abbreviations

BR	Raw brown rice
Germ	Germinated brown rice
Ferm/Ferm (*L. reuterii*)	Fermented brown rice
G+F/G+F (*L. reuterii*)	Germinated combined with fermented brown rice

References

1. Huang, Y.; Tong, C.; Xu, F.; Chen, Y.; Zhang, C.; Bao, J. Variation in mineral elements in grains of 20 brown rice accessions in two environments. *Food Chem.* **2016**, *192*, 873–878. [CrossRef] [PubMed]
2. Saleh, A.S.; Wang, P.; Wang, N.; Yang, L.; Xiao, Z. Brown rice versus white rice: Nutritional quality, potential health benefits, development of food products, and preservation technologies. *Compr. Rev. Food Sci. Food Saf.* **2019**, *18*, 1070–1096. [CrossRef]
3. Tyagi, A.; Daliri, E.B.-M.; Kwami Ofosu, F.; Yeon, S.-J.; Oh, D.-H. Food-Derived Opioid Peptides in Human Health: A Review. *Int. J. Mol. Sci.* **2020**, *21*, 8825. [CrossRef] [PubMed]
4. Auten, R.L.; Davis, J.M. Oxygen toxicity and reactive oxygen species: The devil is in the details. *Pediatric Res.* **2009**, *66*, 121–127. [CrossRef] [PubMed]
5. Singh, A.; Kukreti, R.; Saso, L.; Kukreti, S. Oxidative stress: A key modulator in neurodegenerative diseases. *Molecules* **2019**, *24*, 1583. [CrossRef] [PubMed]
6. Pang, Y.; Ahmed, S.; Xu, Y.; Beta, T.; Zhu, Z.; Shao, Y.; Bao, J. Bound phenolic compounds and antioxidant properties of whole grain and bran of white, red and black rice. *Food Chem.* **2018**, *240*, 212–221. [CrossRef]
7. Palmer, L.M.; Schulz, J.M.; Murphy, S.C.; Ledergerber, D.; Murayama, M.; Larkum, M.E. The cellular basis of GABAB-mediated interhemispheric inhibition. *Science* **2012**, *335*, 989–993. [CrossRef] [PubMed]
8. Abdou, A.M.; Higashiguchi, S.; Horie, K.; Kim, M.; Hatta, H.; Yokogoshi, H. Relaxation and immunity enhancement effects of γ-aminobutyric acid (GABA) administration in humans. *Biofactors* **2006**, *26*, 201–208. [CrossRef] [PubMed]
9. Beckman, C.H. Phenolic-storing cells: Keys to programmed cell death and periderm formation in wilt disease resistance and in general defence responses in plants? *Physiol. Mol. Plant Pathol.* **2000**, *57*, 101–110. [CrossRef]
10. Madeira Junior, J.V.; Teixeira, C.B.; Macedo, G.A. Biotransformation and bioconversion of phenolic compounds obtainment: An overview. *Crit. Rev. Biotechnol.* **2015**, *35*, 75–81. [CrossRef]
11. Khosravi, A.; Razavi, S.H. The role of bioconversion processes to enhance bioaccessibility of polyphenols in rice. *Food Biosci.* **2020**, *35*, 100605. [CrossRef]
12. Zhao, S.; Zhao, J.; Bu, D.; Sun, P.; Wang, J.; Dong, Z. Metabolomics analysis reveals large effect of roughage types on rumen microbial metabolic profile in dairy cows. *Lett. Appl. Microbiol.* **2014**, *59*, 79–85. [CrossRef]
13. Goldansaz, S.A.; Guo, A.C.; Sajed, T.; Steele, M.A.; Plastow, G.S.; Wishart, D.S. Livestock metabolomics and the livestock metabolome: A systematic review. *PLoS ONE* **2017**, *12*, e0177675. [CrossRef]
14. Yang, Y.; Dong, G.; Wang, Z.; Wang, J.; Zhang, Z.; Liu, J. Rumen and plasma metabolomics profiling by UHPLC-QTOF/MS revealed metabolic alterations associated with a high-corn diet in beef steers. *PLoS ONE* **2018**, *13*, e0208031. [CrossRef] [PubMed]
15. Cáceres, P.J.; Peñas, E.; Martinez-Villaluenga, C.; Amigo, L.; Frias, J. Enhancement of biologically active compounds in germinated brown rice and the effect of sun-drying. *J. Cereal Sci.* **2017**, *73*, 1–9. [CrossRef]
16. Huang, S.-H.; Ng, L.-T. Quantification of polyphenolic content and bioactive constituents of some commercial rice varieties in Taiwan. *J. Food Compos. Anal.* **2012**, *26*, 122–127. [CrossRef]
17. Liu, T.; Li, B.; Zhou, Y.; Chen, J.; Tu, H. HPLC determination of γ-aminobutyric acid in Chinese rice wine using pre-column derivatization. *J. Inst. Brew.* **2015**, *121*, 163–166. [CrossRef]
18. Chang, X.; Ye, Y.; Pan, J.; Lin, Z.; Qiu, J.; Guo, X.; Lu, Y. Comparative assessment of phytochemical profiles and antioxidant activities in selected five varieties of wampee (*Clausena lansium*) fruits. *Int. J. Food Sci. Technol.* **2018**, *53*, 2680–2686. [CrossRef]
19. Apea-Bah, F.B.; Minnaar, A.; Bester, M.J.; Duodu, K.G. Sorghum–cowpea composite porridge as a functional food, Part II: Antioxidant properties as affected by simulated in vitro gastrointestinal digestion. *Food Chem.* **2016**, *197*, 307–315. [CrossRef]
20. Chen, Z.; Yu, L.; Wang, X.; Gu, Z.; Beta, T. Changes of phenolic profiles and antioxidant activity in canaryseed (*Phalaris canariensis* L.) during germination. *Food Chem.* **2016**, *194*, 608–618. [CrossRef] [PubMed]
21. Ofosu, F.K.; Elahi, F.; Daliri, E.B.-M.; Chelliah, R.; Ham, H.J.; Kim, J.-H.; Han, S.-I.; Hur, J.H.; Oh, D.-H. Phenolic Profile, Antioxidant, and Antidiabetic Potential Exerted by Millet Grain Varieties. *Antioxidants* **2020**, *9*, 254. [CrossRef]
22. Zeng, Z.; Hu, X.; McClements, D.J.; Luo, S.; Liu, C.; Gong, E.; Huang, K. Hydrothermal stability of phenolic extracts of brown rice. *Food Chem.* **2019**, *271*, 114–121. [CrossRef] [PubMed]
23. Daliri, E.B.-M.; Ofosu, F.K.; Chelliah, R.; Kim, J.-H.; Kim, J.-R.; Yoo, D.; Oh, D.-H. Untargeted metabolomics of fermented rice using UHPLC Q-TOF MS/MS reveals an abundance of potential antihypertensive compounds. *Foods* **2020**, *9*, 1007. [CrossRef] [PubMed]
24. Metsalu, T.; Vilo, J. ClustVis: A web tool for visualizing clustering of multivariate data using Principal Component Analysis and heatmap. *Nucleic Acids Res.* **2015**, *43*, W566–W570. [CrossRef] [PubMed]
25. Pinkard, B.R.; Gorman, D.J.; Rasmussen, E.G.; Maheshwari, V.; Kramlich, J.C.; Reinhall, P.G.; Novosselov, I.V. Raman spectroscopic data from Formic Acid Decomposition in subcritical and supercritical water. *Data Brief* **2020**, *29*, 105312. [CrossRef] [PubMed]
26. Diana, M.; Quílez, J.; Rafecas, M. Gamma-aminobutyric acid as a bioactive compound in foods: A review. *J. Funct. Foods* **2014**, *10*, 407–420. [CrossRef]
27. Le, P.H.; Verscheure, L.; Le, T.T.; Verheust, Y.; Raes, K. Implementation of HPLC Analysis for γ-Aminobutyric Acid (GABA) in Fermented Food Matrices. *Food Anal. Methods* **2020**, *13*, 1190–1201. [CrossRef]
28. Cáceres, P.J.; Peñas, E.; Martínez-Villaluenga, C.; García-Mora, P.; Frías, J. Development of a multifunctional yogurt-like product from germinated brown rice. *LWT* **2019**, *99*, 306–312. [CrossRef]

29. Van Hung, P. Phenolic Compounds of Cereals and Their Antioxidant Capacity. *Crit. Rev. Food Sci. Nutr.* **2016**, *56*, 25–35. [CrossRef]
30. Ilowefah, M.; Bakar, J.; Ghazali, H.M.; Muhammad, K. Enhancement of Nutritional and Antioxidant Properties of Brown Rice Flour Through Solid-State Yeast Fermentation. *Cereal Chem.* **2017**, *94*, 519–523. [CrossRef]
31. Gong, E.S.; Liu, C.; Li, B.; Zhou, W.; Chen, H.; Li, T.; Wu, J.; Zeng, Z.; Wang, Y.; Si, X.; et al. Phytochemical profiles of rice and their cellular antioxidant activity against ABAP induced oxidative stress in human hepatocellular carcinoma HepG2 cells. *Food Chem.* **2020**, *318*, 126484. [CrossRef] [PubMed]
32. Gong, E.S.; Luo, S.J.; Li, T.; Liu, C.M.; Zhang, G.W.; Chen, J.; Zeng, Z.C.; Liu, R.H. Phytochemical profiles and antioxidant activity of brown rice varieties. *Food Chem.* **2017**, *227*, 432–443. [CrossRef] [PubMed]
33. Rao, S.; Santhakumar, A.B.; Chinkwo, K.A.; Wu, G.; Johnson, S.K.; Blanchard, C.L. Characterization of phenolic compounds and antioxidant activity in sorghum grains. *J. Cereal Sci.* **2018**, *84*, 103–111. [CrossRef]
34. Ghasemzadeh, A.; Jaafar, H.Z.; Juraimi, A.S.; Tayebi-Meigooni, A. Comparative evaluation of different extraction techniques and solvents for the assay of phytochemicals and antioxidant activity of hashemi rice bran. *Molecules* **2015**, *20*, 10822–10838. [CrossRef] [PubMed]
35. Chih-Cheng, L. Antioxidant Properties and Antibacterial Activity of Fermented *Monascus purpureus* Extracts. 2019. Available online: http://120.106.195.12/handle/310904600Q/18260 (accessed on 1 April 2021).
36. Cho, D.-H.; Lim, S.-T. Changes in phenolic acid composition and associated enzyme activity in shoot and kernel fractions of brown rice during germination. *Food Chem.* **2018**, *256*, 163–170. [CrossRef]
37. Rao, S.; Santhakumar, A.B.; Chinkwo, K.; Snell, P.; Oli, P.; Blanchard, C.L. Rice phenolic compounds and their response to variability in growing conditions. *Cereal Chem.* **2020**, *97*, 1045–1055. [CrossRef]
38. Adebo, O.A.; Njobeh, P.B.; Adebiyi, J.A.; Gbashi, S.; Phoku, J.Z.; Kayitesi, E. Fermented pulse-based food products in developing nations as functional foods and ingredients. In *Functional Food—Improve Health through Adequate Food*; Hueda, M.C., Ed.; INTECH: Rijeka, Croatia, 2017; pp. 77–109.
39. Okarter, N.; Liu, R.H. Health benefits of whole grain phytochemicals. *Crit. Rev. Food Sci. Nutr.* **2010**, *50*, 193–208. [CrossRef]
40. Dai, T.; Chen, J.; McClements, D.J.; Hu, P.; Ye, X.; Liu, C.; Li, T. Protein–polyphenol interactions enhance the antioxidant capacity of phenolics: Analysis of rice glutelin–procyanidin dimer interactions. *Food Funct.* **2019**, *10*, 765–774. [CrossRef]
41. Annunziata, G.; Arnone, A.; Ciampaglia, R.; Tenore, G.C.; Novellino, E. Fermentation of Foods and Beverages as a Tool for Increasing Availability of Bioactive Compounds. Focus on Short-Chain Fatty Acids. *Foods* **2020**, *9*, 999. [CrossRef]
42. Padmanabhan, P.; Cheema, A.; Paliyath, G. Solanaceous Fruits Including Tomato, Eggplant, and Peppers. In *Encyclopedia of Food and Health*; Caballero, B., Finglas, P.M., Toldrá, F., Eds.; Academic Press: Oxford, UK, 2016; pp. 24–32.
43. Al-Trad, B.; Alkhateeb, H.; Alsmadi, W.; Al-Zoubi, M. Eugenol ameliorates insulin resistance, oxidative stress and inflammation in high fat-diet/streptozotocin-induced diabetic rat. *Life Sci.* **2019**, *216*, 183–188. [CrossRef] [PubMed]
44. Zamani, F.; Samiei, F.; Mousavi, Z.; Azari, M.R.; Seydi, E.; Pourahmad, J. Apigenin ameliorates oxidative stress and mitochondrial damage induced by multiwall carbon nanotubes in rat kidney mitochondria. *J. Biochem. Mol. Toxicol.* **2021**, e22762. [CrossRef]
45. Hegazy, A.M.S.; Mosaed, M.M.; Elshafey, S.H.; Bayomy, N.A. 6-gingerol ameliorates gentamicin induced renal cortex oxidative stress and apoptosis in adult male albino rats. *Tissue Cell* **2016**, *48*, 208–216. [CrossRef] [PubMed]
46. Martirosyan, D.; Mirmiranpour, H.; Ashoori, M.R. Synergistic effect of laser irradiation and cinnamic acid as a functional food on oxidative stress in type 2 diabetes mellitus. *Bioact. Compd. Health Dis.* **2020**, *3*, 154–165. [CrossRef]
47. Ahmadi, N.; Safari, S.; Mirazi, N.; Karimi, S.A.; Komaki, A. Effects of vanillic acid on Aβ1-40-induced oxidative stress and learning and memory deficit in male rats. *Brain Res. Bull.* **2021**, *170*, 264–273. [CrossRef] [PubMed]
48. Janarny, G.; Gunathilake, K.D.P.P. Changes in rice bran bioactives, their bioactivity, bioaccessibility and bioavailability with solid-state fermentation by Rhizopus oryzae. *Biocatal. Agric. Biotechnol.* **2020**, *23*, 101510. [CrossRef]
49. Moritz, B.; Schmitz, A.E.; Rodrigues, A.L.S.; Dafre, A.L.; Cunha, M.P. The role of vitamin C in stress-related disorders. *J. Nutr. Biochem.* **2020**, *85*, 108459. [CrossRef] [PubMed]
50. Godarzi, S.M.; Gorji, A.V.; Gholizadeh, B.; Mard, S.A.; Mansouri, E. Antioxidant effect of p-coumaric acid on interleukin 1-β and tumor necrosis factor-α in rats with renal ischemic reperfusion. *Nefrologia* **2020**, *40*, 311–319. [CrossRef] [PubMed]
51. Ting Wong, C.G.; Bottiglieri, T.; Snead, O.C., III. GABA, γ-hydroxybutyric acid, and neurological disease. *Ann. Neurol. Off. J. Am. Neurol. Assoc. Child Neurol. Soc.* **2003**, *54*, S3–S12. [CrossRef] [PubMed]
52. Andreeva-Gateva, P.; Traikov, L.; Sabit, Z.; Bakalov, D.; Tafradjiiska-Hadjiolova, R. Antioxidant Effect of Alpha-Lipoic Acid in 6-Hydroxydopamine Unilateral Intrastriatal Injected Rats. *Antioxidants* **2020**, *9*, 122. [CrossRef]
53. Wang, Y.; Compaoré-Sérémé, D.; Sawadogo-Lingani, H.; Coda, R.; Katina, K.; Maina, N.H. Influence of dextran synthesized in situ on the rheological, technological and nutritional properties of whole grain pearl millet bread. *Food Chem.* **2019**, *285*, 221–230. [CrossRef] [PubMed]
54. Kaur, M.; Asthir, B.; Mahajan, G. Variation in antioxidants, bioactive compounds and antioxidant capacity in germinated and ungerminated grains of ten rice cultivars. *Rice Sci.* **2017**, *24*, 349–359. [CrossRef]
55. Wang, Z.J.; Liang, C.L.; Li, G.M.; Yu, C.Y.; Yin, M. Stearic acid protects primary cultured cortical neurons against oxidative stress 4. *Acta Pharmacol. Sin.* **2007**, *28*, 315–326. [CrossRef] [PubMed]

56. Queiroz, M.P.; Lima, M.d.S.; de Melo, M.F.F.T.; Bertozzo, C.C.d.M.S.; de Araújo, D.F.; Guerra, G.C.B.; Queiroga, R.d.C.R.d.E.; Soares, J.K.B. Maternal suppplementation with conjugated linoleic acid reduce anxiety and lipid peroxidation in the offspring brain. *J. Affect. Disord.* **2019**, *243*, 75–82. [CrossRef] [PubMed]
57. Jayaraj, R.L.; Beiram, R.; Azimullah, S.; Mf, N.M.; Ojha, S.K.; Adem, A.; Jalal, F.Y. Valeric Acid Protects Dopaminergic Neurons by Suppressing Oxidative Stress, Neuroinflammation and Modulating Autophagy Pathways. *Int. J. Mol. Sci.* **2020**, *21*, 7670. [CrossRef] [PubMed]
58. Zaidi, A.A.; Khan, M.A.; Shahreyar, Z.A.; Ahmed, H. Lauric acid: Its role in behavioral modulation, neuro-inflammatory and oxidative stress markers in haloperidol induced Parkinson's disease. *Pak. J. Pharm. Sci.* **2020**, *33*, 755–763.
59. Zhang, D.; Luo, Z.; Jin, Y.; Chen, Y.; Yang, T.; Yang, Q.; Wu, B.; Shang, Y.; Liu, X.; Wei, Y.; et al. Azelaic Acid Exerts Antileukemia Effects against Acute Myeloid Leukemia by Regulating the Prdxs/ROS Signaling Pathway. *Oxid. Med. Cell. Longev.* **2020**, *2020*, 1295984. [CrossRef] [PubMed]
60. Ritthibut, N.; Oh, S.-J.; Lim, S.-T. Enhancement of bioactivity of rice bran by solid-state fermentation with Aspergillus strains. *LWT* **2021**, *135*, 110273. [CrossRef]
61. Rizzello, C.G.; Lorusso, A.; Russo, V.; Pinto, D.; Marzani, B.; Gobbetti, M. Improving the antioxidant properties of quinoa flour through fermentation with selected autochthonous lactic acid bacteria. *Int. J. Food Microbiol.* **2017**, *241*, 252–261. [CrossRef]
62. Tamam, B.; Syah, D.; Suhartono, M.T.; Kusuma, W.A.; Tachibana, S.; Lioe, H.N. Proteomic study of bioactive peptides from tempe. *J. Biosci. Bioeng.* **2019**, *128*, 241–248. [CrossRef] [PubMed]
63. Zhang, F.; Jia, Z.; Gao, P.; Kong, H.; Li, X.; Lu, X.; Wu, Y.; Xu, G. Metabonomics study of urine and plasma in depression and excess fatigue rats by ultra fast liquid chromatography coupled with ion trap-time of flight mass spectrometry. *Mol. Biosyst.* **2010**, *6*, 852–861. [CrossRef] [PubMed]
64. Adjimani, J.P.; Asare, P. Antioxidant and free radical scavenging activity of iron chelators. *Toxicol. Rep.* **2015**, *2*, 721–728. [CrossRef] [PubMed]
65. Akhrem, A.; Galaktionov, S.; Golubovich, V.; Kirnarskiĭ, L. Theoretical conformational analysis of delta-sleep inducing peptide. *Biofizika* **1982**, *27*, 324–325. [PubMed]
66. Pandey, M.; Kapila, R.; Kapila, S. Osteoanabolic activity of whey-derived anti-oxidative (MHIRL and YVEEL) and angiotensin-converting enzyme inhibitory (YLLF, ALPMHIR, IPA and WLAHK) bioactive peptides. *Peptides* **2018**, *99*, 1–7. [CrossRef] [PubMed]
67. Suetsuna, K.; Ukeda, H.; Ochi, H. Isolation and characterization of free radical scavenging activities peptides derived from casein. *J. Nutr. Biochem.* **2000**, *11*, 128–131. [CrossRef]
68. Tomova, T.; Zamoshchina, T.; Svetlik, M.; Sedokova, M.; Gostyukhina, A.; Fatyushina, A. Gly-Pro and Adaptive Reactions in Multicomponent Stress. *Neurosci. Behav. Physiol.* **2020**, *50*, 1083–1089. [CrossRef]
69. Marí, M.; de Gregorio, E.; de Dios, C.; Roca-Agujetas, V.; Cucarull, B.; Tutusaus, A.; Morales, A.; Colell, A. Mitochondrial Glutathione: Recent Insights and Role in Disease. *Antioxidants* **2020**, *9*, 909. [CrossRef]
70. Yin, Z.; Wu, Y.; Chen, Y.; Qie, X.; Zeng, M.; Wang, Z.; Qin, F.; Chen, J.; He, Z. Analysis of the interaction between cyanidin-3-O-glucoside and casein hydrolysates and its effect on the antioxidant ability of the complexes. *Food Chem.* **2020**, *340*, 127915. [CrossRef]
71. Genazzani, A.D.; Despini, G.; Chierchia, E.; Benedetti, C.; Prati, A. Pharmacological and integrative treatment of stress-induced hypothalamic amenorrhea. In *Frontiers in Gynecological Endocrinology*; Springer: Berlin/Heidelberg, Germany, 2016; pp. 69–84.
72. Yan, Q.; Yin, Y.; Li, B. Use new PLGL-RGD-NGF nerve conduits for promoting peripheral nerve regeneration. *Biomed. Eng. Online* **2012**, *11*, 36. [CrossRef]

antioxidants

MDPI

Article

An Extract from *Ficus carica* Cell Cultures Works as an Anti-Stress Ingredient for the Skin

Irene Dini [1,*], Danila Falanga [2], Ritamaria Di Lorenzo [1], Annalisa Tito [2], Gennaro Carotenuto [2], Claudia Zappelli [3], Lucia Grumetto [1], Antonia Sacchi [1], Sonia Laneri [1,*] and Fabio Apone [2,3]

1 Department of Pharmacy, University of Naples Federico II, Via Domenico Montesano 49, 80131 Napoli, Italy; dilorenzo@ritamaria.unina.it (R.D.L.); lucia.grumetto@unina.it (L.G.); antonia.sacchi@unina.it (A.S.)

2 Arterra Bioscience SpA, Via Benedetto Brin 69, 80142 Napoli, Italy; danila@arterrabio.it (D.F.); annalisa@arterrabio.it (A.T.); gennaro@arterrabio.it (G.C.); fapone@arterrabio.it (F.A.)

3 Vitalab Srl, Via Benedetto Brin 69, 80142 Napoli, Italy; claudiazappelli@vitalabactive.com

* Correspondence: irdini@unina.it (I.D.); slaneri@unina.it (S.L.)

Abstract: Psychological stress activates catecholamine production, determines oxidation processes, and alters the lipid barrier functions in the skin. Scientific evidence associated with the detoxifying effect of fruits and vegetables, the growing awareness of the long-term issues related to the use of chemical-filled cosmetics, the aging of the population, and the increase in living standards are the factors responsible for the growth of food-derived ingredients in the cosmetics market. A *Ficus carica* cell suspension culture extract (FcHEx) was tested in vitro (on keratinocytes cells) and in vivo to evaluate its ability to manage the stress-hormone-induced damage in skin. The FcHEx reduced the epinephrine (−43% and −24% at the concentrations of 0.002% and 0.006%, respectively), interleukin 6 (−38% and −36% at the concentrations of 0.002% and 0.006%, respectively), lipid peroxide (−25%), and protein carbonylation (−50%) productions; FcHEx also induced ceramide synthesis (+150%) and ameliorated the lipid barrier performance. The in vivo experiments confirmed the in vitro test results. Transepidermal water loss (TEWL; −12.2%), sebum flow (−46.6% after two weeks and −73.8% after four weeks; on the forehead −56.4% after two weeks and −80.1% after four weeks), and skin lightness (+1.9% after two weeks and +2.7% after four weeks) defined the extract's effects on the skin barrier. The extract of the *Ficus carica* cell suspension cultures reduced the transepidermal water loss, the sebum production, the desquamation, and facial skin turning to a pale color from acute stress, suggesting its role as an ingredient to fight the signs of psychological stress in the skin.

Keywords: *Ficus carica*; oxidative stress protection; lipid peroxidation; stress hormones; epidermal skin barrier; nutricosmetics

Citation: Dini, I.; Falanga, D.; Di Lorenzo, R.; Tito, A.; Carotenuto, G.; Zappelli, C.; Grumetto, L.; Sacchi, A.; Laneri, S.; Apone, F. An Extract from *Ficus carica* Cell Cultures Works as an Anti-Stress Ingredient for the Skin. *Antioxidants* **2021**, *10*, 515. https://doi.org/10.3390/antiox10040515

Academic Editor: David Arráez-Román

Received: 8 February 2021
Accepted: 22 March 2021
Published: 25 March 2021

1. Introduction

The focus on "cosmetics from food" has been a significant trend in recent years. The integration of natural ingredients from food into dietary supplements and cosmetic products is a reality that has already been commercialized. Fruits contain anti-aging, antioxidant, and anti-inflammatory molecules, and consequently, have a high potential in cosmetics [1–3]. The skin is the most exposed organ of our body such that its health is threatened by a large variety of environmental factors (temperature shifts, sun radiation, tobacco smoke, ozone, chemical pollutants), which is known as the "exposome." The concept of the exposome includes the totality of factors to which an individual gets exposed from conception to death; this term entered our vocabulary quite recently [4]. Besides the exposome, additional factors have been recognized as skin-aging potentiators, acting separately or in combination with the exposome [5], such as psychological stress, lack of sleep, and inappropriate nutrition. Among these, psychological stress represents one of the most occurring issues nowadays. It can be defined as the condition in which environmental demands exceed the individual's ability to cope, resulting in behavioral and mood changes [6]. Psychological stress determines the release of corticotropin releasting hormone (CRH), adrenocorticotropin (ACTH), glucocorticoids, catecholamines, epinephrine (adrenaline), and norepinephrine (noradrenaline) into the bloodstream in response to the unfavorable conditions [7]. If the stress response is persistent, those hormones may trigger adverse physiological events and exacerbate the stressful condition [8]. The skin and brain are strictly connected: the brain delivers signals to the skin that influences its stress response, mainly through the hypothalamic–pituitary–adrenal (HPA) axis, while the skin cells, in turn, produce a CRH, ACTH, epinephrine, and their receptors, triggering autocrine pathways and regulating the endocrine and immune systems [9]. Catecholamines stimulate keratinocyte differentiation by inducing keratin, involucrin, and transglutaminase [10], and exacerbate atopic dermatitis in an inflamed state via the continuous induction of a cytokine cascade [11]. Moreover, catecholamines, which are overproduced during psychological stress, impair the skin barrier's recovery, reducing the accumulation of epidermal lipids—mainly cholesterol, fatty acids, and ceramides—by suppressing their synthetic pathways [12]. Chronic psychological stress strongly increases reactive oxygen species in mice skin, where these effects dramatically accelerate skin aging [13]. Oxidative stress in the skin is strictly related to the formation of age spots too. The accumulation of oxidized lipids and proteins culminates in the formation of lipofuscin particles, which are responsible, together with melanin, for dark spots [14] that the hydrolase enzymes hardly degrade. In this study, we evaluated the activity of an extract from *Ficus carica* cell suspension cultures to alleviate skin damage caused by psychological stress using in vitro and in vivo analyses. Plant cell cultures are becoming a popular source of cosmetic ingredients with proven efficacy [15], where their characteristics of safety and sustainability have been very much requested by the skincare market [16].

2. Materials and Methods

2.1. *In Vitro Studies*

2.1.1. Plant Cell Culturing and Extract Preparation

A certified fig plant (*Ficus carica*) was obtained from a local nursery ("Vivai Milone," Lamezia Terme, Calabria Region, Italy). The leaves were surface sterilized with 70% *(v/v)* EtOH (Sigma Aldrich, St. Louis, MO, USA), followed by a treatment with 1% bleach (supplemented with Tween 20 (Sigma Aldrich, St. Louis, MO, USA)), and finally rinsed with sterile distilled water. Then, the leaves were excised into 0.5–1.0 cm segments and cultured on solidified Gamborg B5 medium (Merk, Darmstadt, Germany) containing 500 mg/L myo-inositol, 30 g/L sucrose, phytohormones, and 8 g/L phytoagar. The explants were subcultured every four weeks onto a fresh medium for three months. Once a compact and friable callus was obtained, the cells were transferred to the liquid Gamborg B5 medium, supplemented with 500 mg/L myo-inositol, 30 g/L sucrose, 1 mg/L adenine, and phytohormones. The suspensions were shaken on a rotary shaker at 27 °C, with a 16-h daily photoperiod. Once the liquid suspension cultures of about 200 g/L were obtained, the cells were collected, lysed in a water-based buffer, and lyophilized to prepare a water-soluble extract. The powder (*Ficus carica* hydrosoluble extract (FcHEx)) was dissolved in water or cell culture media at appropriate testing concentrations.

2.1.2. Skin Cells and Explants

Immortalized Human Keratinocytes (HaCaT), purchased from Addexbio Technologies (San Diego, CA, USA), were maintained in Dulbecco's Modified Eagle Medium (DMEM; Sigma Aldrich, St. Louis, MO, USA) that was supplemented with 10% fetal bovine serum (FBS; Sigma Aldrich, St. Louis, MO, USA) in 95% air, 5% CO_2, and humidified atmosphere at 37 °C.

Skin explants were obtained from the skin of healthy female donors (aged 35–49) following mastectomy or breast reduction procedures at the Villa Cinzia surgery center (Naples, IT). The use of skin tissue was done according to the Declaration of Helsinki and all patients had given their written informed consent. Skin punch biopsies (8 mm) were obtained from skin explants and cultured in 24-transwell plates in DMEM/FBS plus antibiotics in air-liquid conditions at 37 °C in 5% CO_2 humidified air.

2.1.3. Gene Expression Analysis

A total of 2×10^5 HaCaT were seeded in six-well plates, incubated for 16 h, and then treated with the FcHEx or ICI-118,551 (Sigma Aldrich, St. Louis, MO, USA) for 4 h in the presence of 10 μM of epinephrine (Sigma Aldrich, St. Louis, MO, USA). At the end of the incubation, total RNA was extracted with the GenElute Mammalian Total RNA Purification Kit (Sigma Aldrich, St. Louis, MO, USA) and treated with DNAse I (Sigma Aldrich, St. Louis, MO, USA) at 37 °C for 30 min. Reverse transcription was performed using the RevertAid™ First Strand cDNA Synthesis Kit (Thermo Fisher Scientific, Dallas, TX, USA). RT-PCR was performed with the Quantum RNA™ kit (Thermo Fisher, Waltham, MA, USA) containing specific primers to amplify 18S rRNA, along with competimers that reduced the amplified 18S rRNA product within the range to be used as an endogenous standard. The amplification reactions were performed using the Mastercycler™ ProS (Eppendorf, Milan, Italy) with the following general scheme: 2 min at 94 °C, followed by 35 cycles of 94 °C for 30 s, 50 °C for 30 s, and 72 °C for 30 s, with a 10 min final extension at 72 °C. The PCR products were loaded onto 1.5% agarose gel (Merk, Darmstadt, Germany), and the amplification bands were visualized and quantified with the Geliance 200 Imaging system (Perkin Elmer, Waltham, MA, USA). The amplification band corresponding to the analyzed gene was normalized relative to the amplification band corresponding to the 18S and reported as a percentage of untreated controls (fixed at 100%). The RT-PCRs were done in triplicate, where the average results are reported in the graphs. The sequences of the used primers were as follows: interleukin 6 (IL-6) forward (Fw)—GTCCTGATCCAGTTCCTGCAG, IL-6 reverse (Rv)—CTACACTTTCCAAGAAATGATC (gene ID: 3569), β-glucocerebrosidase (GBA) Fw—AGTTGCACAACTTCAGC, and GBA Rv—GTCCAGGTACCAATGTAC (gene ID: 2629).

2.1.4. Determination of the Lipid Peroxides

A total of 1.8×10^4 HaCaT were seeded in 96-well plates and then treated with the FcHEx or ICI-118,551 for 24 h. After incubation, the cells were washed in phosphate-buffered saline (PBS; Sigma Aldrich, St. Louis, MO, USA) and then incubated with the dye C11-Bodipy (Thermo Fisher, Waltham, MA, USA) at 37 °C for 30 min. A total of 50 μM of epinephrine was used to induce lipid peroxidation, and the fluorescence was read at 510/580 using the Victor3 plate reader system (Perkin Elmer, Waltham, MA, USA).

2.1.5. Lipofuscin Measurements

A total of 8×10^3 HaCaT were seeded in a 96-well plate and then treated with the FcHEx for 24 h. A total of 50 nM of epinephrine was added to the cells and incubated for an additional 48 h. The cells were then washed in PBS and incubated with Sudan Black (Sigma Aldrich, St. Louis, MO, USA) (0.7% dilution in EtOH 70%) for 5 min. The cells were lysed in 1 M of NaOH (Sigma Aldrich, St. Louis, MO, USA) for 20 min at 70 °C to solubilize the dye. The quantity of solubilized dye was determined by reading the absorbance at 595 nm in the Victor3 plate reader system.

2.1.6. Epidermal Lipid Measurements

A total of 1.0×10^4 HaCaT were seeded in 96-well plates, grown for 48 h, and then treated with the FcHEx or ICI-118,551 as a positive control for an additional 24 h. The cells were incubated with 10 μM of epinephrine for 24 h, and the lipid accumulation was detected via the addition of the AdipoRed assay Reagent (Lonza, Basel, Switzerland). The fluorescence at 485 nm was read using the Victor3 plate reader system, and the content of the epidermal lipid was normalized relative to the cell density determined by crystal violet staining.

2.1.7. Measuring the Carbonylated Proteins in Cells and Skin Explants

A total of 1.5×10^4 HaCaT were seeded in 96-well plates and then incubated with the FcHEx or ICI-118,551 in the presence of epinephrine 50 μM for 24 h. The cells were then washed in PBS and fixed in 4% paraformaldehyde (PFA; Sigma Aldrich, St. Louis, MO, USA). After washing in PBS + 0.05% Tween 20, the samples were incubated with 5 mM 2,4-Dinitrophenyl-hydrazine (DNPH; Sigma Aldrich, St. Louis, MO, USA) in 2N HCl (Sigma Aldrich, St. Louis, MO, USA) for 1 h at room temperature. The carbonylated products were detected via ELISA using a specific antibody against DNP (sc69697, Santa Cruz Biotechnology, Heidelberg, Germany).

The skin punches were treated with the FcHEx in the presence of 56 nM of epinephrine for 24 h. After incubation, the punches were fixed with PFA for 6 h, washed in PBS, and then incubated in 15% and 30% sucrose (Sigma Aldrich, St. Louis, MO, USA). The punches were embedded in optimal cutting temperature (OCT) medium (Tissue Tek, Sakura Finetek USA, Inc., Torrance, CA, USA), frozen in dry ice, and stored −80 °C. Cryosections (10 μm) were obtained using the CM1520 cryostat (Leica Microsystems, Wetzlar, Germany). The slides were incubated with 5 mM DNPH in 2N HCl for one hour at room temperature, washed in PBS/EtOH (1:1), and then in PBS/Tween 20. The slides were incubated in a blocking buffer (3% bovine serum albumin (BSA) in PBS) for 30 min, washed with PBS/Tween 20, incubated with the primary antibody anti-DNP (1:50 dilution), and then with the secondary conjugated antibody Alexa Fluor 488 (Life Technologies, Carlsbad, CA, USA). Nuclei of the cells were stained using 4′,6-Diamidino-2-phenylindole (DAPI; Sigma Aldrich, St. Louis, MO, USA). The signals were detected using fluorescent microscopy.

2.1.8. cAMP Measurements

The HaCaT cells were washed in PBS, detached from the flask with a non-enzymatic dissociation solution (Sigma-Aldrich, St. Louis, MO, USA), and resuspended at the concentration of 10^6/mL in a stimulation buffer (BSA 0.1%, isobutylmethylxanthine (IBMX) 0.5% in PBS) containing Alexa Fluor 647-labeled anti-cAMP antibody (Thermo Fisher Scientific, Dallas, TX, USA), according to the protocol described by the provider of the Lance Kit (Perkin Elmer, Waltham, MA, USA). A total of 1.2×10^4 cells were distributed in aliquots in 384-well plates and treated with either the FcHEx or ICI-118,551 in the presence of 10 μM of epinephrine. In parallel, a standard curve for cAMP (Thermo Fisher Scientific, Dallas, TX, USA) was prepared by diluting known concentrations of cAMP in the stimulation buffer in the presence of an anti-cAMP antibody. The cells were incubated for 1 h at room temperature, and then a detection mix containing Europium-W8044-labeled streptavidin and biotin-labeled cAMP (Thermo Fisher Scientific, Dallas, TX, USA) was added to each well. The cAMP concentration was measured by exciting at 320 nm and recording at 615 and 665 nm using an EnVision instrument (Perkin Elmer, Waltham, MA, USA).

2.1.9. Determination of the Ficin Amount and Activity

The ficin amount was determined using ELISA: scalar dilutions of the extract were incubated on a 96-well plate at 4 °C for 16 h, and then the quantity of ficin was calculated using a specific antibody (AS09550, Agrisera, Vännäs, Sweden). Scalar dilutions of the extract were incubated with 0.43 mM of the substrate pGLU-PHE-LEU-pNitroanilide (Sigma Aldrich, St. Louis, MO, USA) in a pH 6.5 phosphate buffer (Sigma Aldrich, St. Louis, MO, USA) to measure the ficin activity. Purified ficin was used as standard starting from 0.015 enzymatic units. The color developed was measured by reading the absorbance at 405nm with the Victor3 multiplate reader system.

2.1.10. Statistical Analysis

The in vitro experiments were performed in triplicates and repeated at least three times. The data are expressed as mean ± standard deviation (SD) of three independent experiments. The statistical tests were performed with the aid of the GraphPad Prism version 9 software (GraphPad Software, San Diego, CA, USA). A two-way analysis of variance (ANOVA) was used, followed by a Tukey's multiple comparisons post-test; a p-value lower than 0.05 was considered statistically significant. The number of asterisks in the graphs indicate the level of significance (*** p-value < 0.001; ** $0.001 < p < 0.01$; * $0.01 < p < 0.05$).

2.2. *In Vivo Studies*

The studies were carried out according to the principles of the 1964 Helsinki Declaration European Community revisions (fourth revision, called Somerset West, South Africa [17], and the May 2008 Colipa Guidelines for the estimation of a cosmetic product's efficiency [18]. The study was conducted on forty healthy volunteers, aged between 20 and 27 years, going through psychological stress. They were preparing for a 15-credit exam. All the volunteer students successfully passed the exam (scores of 27/30–30/30). The measures were performed two weeks before the examination (T_0), the day of the exam (acute stress time—T_{2w}), and a successive period of two weeks (recovery time—T_{4w}). These subjects provided informed consent and received an objective evaluation using standardized numerical scales, i.e., they were subjected to a Complete Skin Investigation (CSI; Courage + Khazaka Electronic GmbH, Cologne, Germany version for windows ®10) analysis for the evaluation of the degree of skin and the type of skin. In this study, twenty volunteers used an active formulation and twenty used a placebo in a double-blind trial.

Each volunteer was examined in a closed room, which was kept at a controlled temperature and humidity (20 ± 2 °C, 40 ± 5% relative humidity (RH)) at the same time of day after a stationing time of about 30 minutes.

The probes were previously calibrated in the same room, and the measurements were carried out at pre-established times (T_0, T_{2w}, T_{4w}). The panelists applied about 2 mg/cm^2 of cream on the face and eye area, which had been previously cleaned, avoiding direct contact with the eyes and gently massaging until completely absorbed.

At the beginning (T_0), after 2 weeks (T_{2w}), and at the end of the study period (after 4 weeks, T_{4w}), the following instrumental evaluations were carried out:

Tewameter® TM300 (Courage + Khazaka Electronic GmbH, Mathias-Brüggen-Str. 91 50829, Cologne, Germany) measured the transepidermal water loss.

Sebufix® F 16 and Corneofix® F 20 applied to a Visioscope PC35® camera (Courage + Khazaka Electronic GmbH, Cologne, Germany) measured the sebum secretion. The Sebufix® F 16 is a special foil that absorbs the skin surface's sebum due to its micropores and shows them as spots of different sizes. The Visioscope PC35® camera monitored the qualitative sebum production in real-time. The number, size, and area covered with spots (mm^2) were evaluated for five increasing spot sizes, together with the slope of the sebum development during the measurement time on the cheeks and foreheads.

Corneofix® F 20 (Courage + Khazaka Electronic GmbH, Cologne, Germany) special adhesive tapes collected flakes of dead cells, known as corneocytes. The number, size, and thickness of the corneocytes is correlated with the exfoliation of the stratum corneum. When mounted on the Visioscope PC35® camera, the exfoliation can be evaluated using its software.

Visioface 1000D (Courage + Khazaka Electronic GmbH, Cologne, Germany) for images confirmed the pore counting and type.

The skin lightness evaluation was undertaken using a Colorimeter CL 400 (Courage + Khazaka Electronic GmbH, Cologne, Germany). The Colorimeter CL 400 probe sends out white LED light arranged circularly to illuminate the skin uniformly. The emitted light is scattered in all directions. Parts of it travel through the layers and some of it scatters out of the skin. The light reflected from the skin is measured in the probe and expressed accordingly. The probe's raw data are corrected with a particular color matrix to be as close as possible to norm values. The skin's color is expressed as an XYZ-value (tristimulus) and can be converted into RGB values (red/green/blue) or L*a*b.

A Skin-pH-Meter PH 905 (Courage + Khazaka Electronic GmbH, Mathias-Brüggen-Str. 91 50829, Cologne, Germany) measured the cute pH. The measurements were based on a high-quality combined electrode, where both a glass H^+-ion-sensitive electrode and a reference electrode were placed in one housing. It was connected to a probe handle containing the measurement electronics.

The data analyses were obtained using CK-MPA-Multi-Probe-Adapter FBQ software version for windows ®10 (Courage + Khazaka Electronic GmbH, Cologne, Germany).

2.2.1. Study Population

Forty healthy volunteers, female (n = 38) and male (n = 2), from all social categories, were enrolled according to these inclusion criteria:

- Healthy Caucasian subjects.
- Aged between 20 and 27 years and going through a period of psychological stress two weeks before the examination T_0, the day of the examination (T_{2w}—stress time), and 2 weeks after the examination (T_{4w}—recovery time).
- Sex: female and male.
- Phototype: II–IV Fitzpatrick scale.
- Subjects who have read and signed the informed consent form written by the investigators.
- Subjects with stressed skin, as assessed via CSI analysis.
- Subjects who did not apply products other than the one studied on the test area and no product within seven days before the test.
- Subjects who agreed to follow the study rules and the planned check-ups.
- Subjects who agreed not to expose themselves to UV for the duration of the study.
- Exclusion criteria:
- Pregnant or breastfeeding women.
- Subjects with anamnesis of cutaneous hyper-reactivity or intolerance reactions to cosmetic products/ingredients.
- Subjects with diseases in the period immediately preceding the current study.
- Subjects undergoing topical or systemic treatment with any drug that may affect the outcome of the test or subjects affected by skin diseases (eczema, psoriasis, lesions).
- Subjects treated with topical retinoids in the previous six months at the start of the study or with systemic retinoids in the previous 12 months.
- Subjects who performed treatments with topical products based on alpha and beta-hydroxy acids in the 45 days before the start of the study.

2.2.2. Cream Composition

The cream with *Ficus carica* cell suspension culture extract (FcHEx) contained two phases:

- Phase O (oil phase) polyglyceryl-3 methylglucose 5.0%, cetyl alcohol 2%, and cetearyl alcohol 3%.
- Phase W (water phase) containing water (88.8%), the FcHEx (0.5%), sodium benzoate 0.5%, potassium sorbate, and perfume (0.1%).

The placebo cream contained all the components without the FcHEx.

All components that were used to produce the creams were bought from ACEF (Fiorenzuola D'Arda, Italy) and Parfum by Farotti Essenze (Rimini, Italy). The two creams were produced by energetically shaking the two phases at 70 °C using a Silverson L5M-A Laboratory Mixer (SBL, Shanghai, China), cooled in an ice bath, and adding the remaining components at room temperature. Finally, the pH (5.1–5.2) and viscosity (28.179–29.287 mPa; L4, 20 rpm) were tested using a Crison GPL20 pH-Meter (Crison, Barcelona, Spain) and a Visco Basic Plus rheometer (Fungilab, Barcelona, Spain).

2.2.3. Data Analysis and Statistics

The measurement averages were considered at the times T_0, T_{2w}, and T_{4w}. Student's t-test was used to assess whether the differences between the averages were significant, where the significance level was set to $p \leq 0.05$. The mean value and standard deviation were calculated for the initial, intermediate, and final instrumental values using a spreadsheet by setting:

T_0 = average value before the study started;
T_{2w} = average value after 2 weeks of treatment;
T_{4w} = average value after 4 weeks of treatment;
$T_{2w} - T_0$ = average value variation after 2 weeks;
$T_{4w} - T_0$ = average value variation after 4 weeks.

This difference is also reported as a relative percentage of variation (Δ%).

3. Results

3.1. In Vitro Studies

3.1.1. Extract Preparation

A hydrosoluble extract was obtained from the *Ficus carica* liquid suspension culture by disrupting the cells in a saline buffer and taking the water-soluble supernatant (FcHEx). *Ficus carica* cell cultures are rich in polyphenols (flavones, flavanones, catechins, anthocyanidins, and phenolic acids) [19]. The concentration of ficin, a sulfhydryl protease with a papain-like activity that is abundant in the fig latex [20], was measured in the FcHEx using a specific antibody in ELISA, as its presence was already reported in *F. carica* cell cultures [21]. The concentration of ficin was about 1.48 µg/mg of dried extract, which was higher than that previously reported. The ficin activity was measured using an enzymatic assay and was found to be 0.9 enzymatic units/µg of extract. The extract concentrations to be used in the cellular tests were calculated via the MTT (3-(4,5-dimethylthiazol-2-yl)-2,5-diphenyltetrazolium bromide) cytotoxicity assay on the keratinocyte cell lines (HaCaT), treated with increasing concentrations of the FcHEx (from 0.01 to 0.0002% *w/v*). No significant cytotoxicity was recorded at all for the used concentrations (data not shown). For convenience, the two doses of 0.006% and 0.002% of the extract were chosen for all the following experiments.

3.1.2. Activity of the FcHEx on Epinephrine Signaling

Keratinocytes were stimulated with epinephrine alone or in the presence of the FcHEx at the two concentrations, and the level of the second messenger cAMP was measured to investigate the capacity of the FcHEx to affect the epinephrine signal cascade in the skin cells. The ICI-118,551 was used as a positive control, as it inhibits the β2 adrenergic receptors [22]. As shown in Figure 1, the 10 µM of epinephrine stimulated the cAMP synthesis by about 40-fold, but this increase was attenuated by the FcHEx (about 43% and 24% at the concentrations of 0.002% and 0.006%, respectively). As expected, at a concentration of 10 µM, the ICI-118,551 totally abolished the epinephrine stimulation.

3.1.3. Activity of the FcHEx on the Inflammatory Cytokine IL-6

The IL-6 gene expression was measured in keratinocytes treated with the FcHEx after stimulation with 10 µM of epinephrine, as it activates the synthesis of inflammatory cytokines through cAMP [23]. As shown in Figure 2, the IL-6 gene expression increased by about 50% in stimulated cells, but this effect was significantly reduced by the extract (38.3% and 35.6% at the concentrations of 0.002% and 0.006%, respectively) and the positive control ICI-118,551.

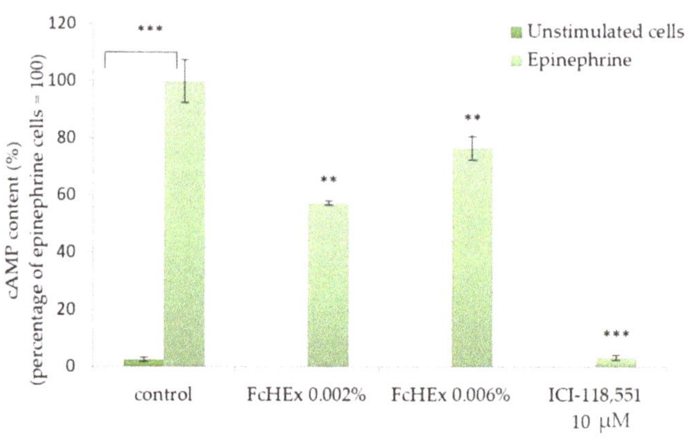

Control	FcHEx 0.002%	FcHEx 0.006%	ICI-118,551 10 μM
2.61 ± 0.75%			
100 ± 7.50%	57.28 ± 0.78%	76.67 ± 4.16%	3.51 ± 0.87%

Figure 1. Effect of the *Ficus carica* cell suspension culture extract (FcHEx) on the cAMP synthesis in skin keratinocytes. The keratinocytes were stimulated with 10 μM of epinephrine alone or with the FcHEx or with a positive control for 30 min and then processed to determine the cAMP content. The reported values represent the averages of three independent experiments and are expressed as the percentages of cells treated with epinephrine only, which was set to 100%. The asterisks indicate statistically significant values (*** p-value was between 0.0001 and 0.001; ** p-value was between 0.001 and 0.01).

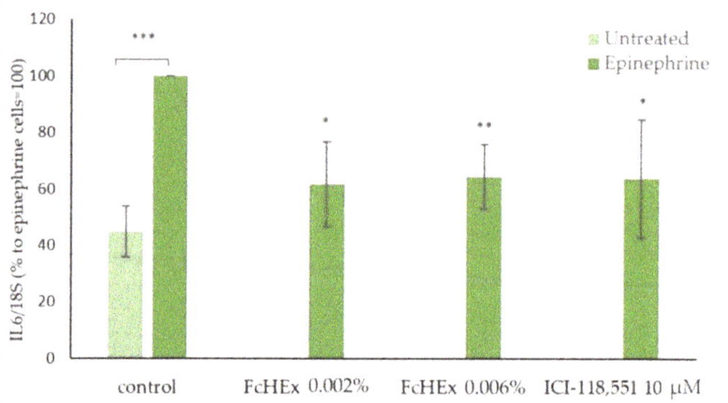

Control	FcHEx 0.002%	FcHEx 0.006%	ICI-118,551 10 μM
44.99 ± 0.001%			
100%	61.74 ± 0.02%	64.38 ± 0.01%	63.74 ± 0.04%

Figure 2. Effect of the FcHEx on the IL-6 gene expression in the skin keratinocytes. The keratinocytes were stimulated with 10 μM of epinephrine alone or with the FcHEx or the positive control for 4 h and then processed for RT-PCR analysis. The reported values represent the averages of three independent experiments and are expressed as the percentages of cells treated with epinephrine only, which was set to 100%. The asterisks indicate statistically significant values (*** p-value was between 0.0001 and 0.001; ** p-value wa between 0.001 and 0.01; * p-value was between 0.01 and 0.05).

3.1.4. Lipid Peroxide Measurements

The FcHEx's ability to reduce the oxidative stress induced by epinephrine was investigated by measuring the lipid peroxide production in keratinocytes. As shown in Figure 3, 50 µM of epinephrine increased the lipid peroxide production by 50%. Pretreatment with the FcHEx (at both the used concentrations) decreased this event by 25%.

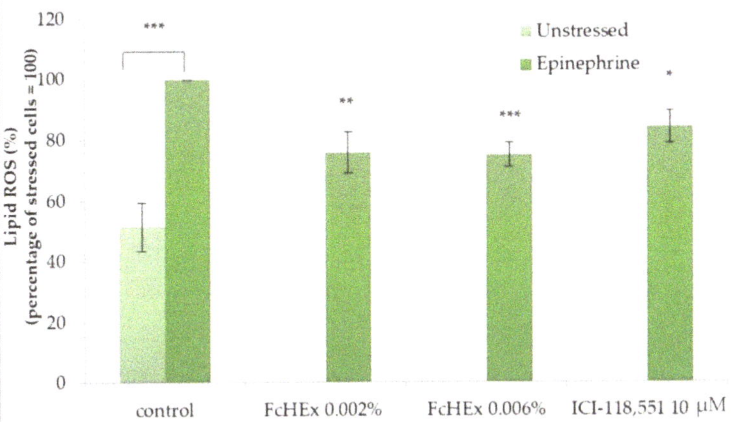

Control	FcHEx 0.002%	FcHEx 0.006%	ICI-118,551 10 µM
51.46 ± 7.93%			
100	75.75 ± 6.73%	74.95 ± 4.02%	84.04 ± 5.45%

Figure 3. Effect of the FcHEx on the lipid peroxidation in skin keratinocytes. The keratinocytes were treated with the FcHEX or the positive control for 24 h, stressed with 50 µM of epinephrine, and then analyzed for lipid peroxide content by using C11-Bodipy dye. The reported values represent the average of three independent experiments and are expressed as the percentages of cells treated with epinephrine only, which was set to 100%. The asterisks indicate statistically significant values (*** *p*-value was between 0.0001 and 0.001; ** *p*-value was between 0.001 and 0.01; * *p*-value was between 0.01 and 0.05).

3.1.5. Carbonylated Protein Determination

The concentration of carbonylated proteins was measured in the keratinocytes stressed with epinephrine 50 µM and then treated with the FcHEx using a specific antibody to verify whether the lipid peroxide inhibition was associated with a reduction in protein oxidation. Epinephrine induced an increase in protein carbonylation by 40%, which was significantly inhibited by the extract (about 50%) and the inhibitor ICI-118,551 (Figure 4).

The protection against protein carbonylation of the FcHEx was evaluated in the skin explants as well. Three punches from biopsy explants were topically treated for each condition. The induction of protein carbonylation was achieved with 56 nM of epinephrine, which corresponded to the hormone dose found in psychologically stressed individuals' blood plasma [24]. After the treatment with epinephrine, the level of carbonylated proteins increased by about 45%, while the treatment with the FcHEx 0.002% reduced this effect by 50%, analogously to the inhibitor drug (Figure 5).

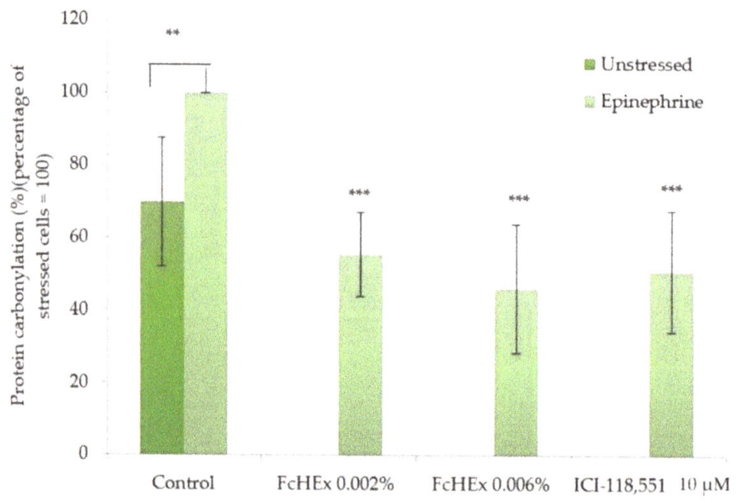

Control	FcHEx 0.002%	FcHEx 0.006%	ICI-118.551 10 µM
69.93 ± 0.02%			
100%	55.49 ± 0.0005%	46.13 ± 0.001%	50.95 ± 0.105%

Figure 4. Effect of the FcHEx on the protein carbonylation in the skin keratinocytes. The keratinocytes were treated with the FcHEX or the positive control in the presence of 50 µM epinephrine for 24 h, fixed with formaldehyde, stained with DNPH (2,4-dinitrophenylhydrazine), and incubated with the specific antibody. The reported values represent the averages of three independent experiments and are expressed as the percentages of cells treated with epinephrine only, which was set to 100%. The asterisks indicate statistically significant values (*** *p*-value was between 0.0001 and 0.001; ** *p*-value was between 0.001 and 0.01).

3.1.6. Lipofuscin Measurements

The cells were stressed with epinephrine in the absence and presence of the extract, and the concentration of lipofuscin particles was measured as the accumulation of oxidated compounds, either lipids or proteins, which can often determine lipofuscin production. The treatment with epinephrine significantly increased the lipofuscin accumulation; in contrast, the treatment with both concentrations of the FcHEx decreased significantly (Figure 6).

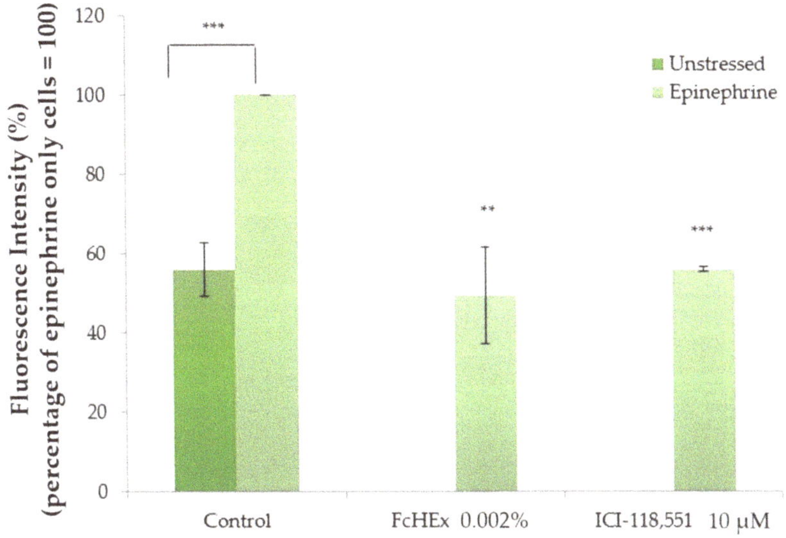

Control	FcHEx 0.002%	ICI-118.551 10 μM
55.89 ± 10.49%		
100%	49.31 ± 0.001%	55.93 ± 12.07%

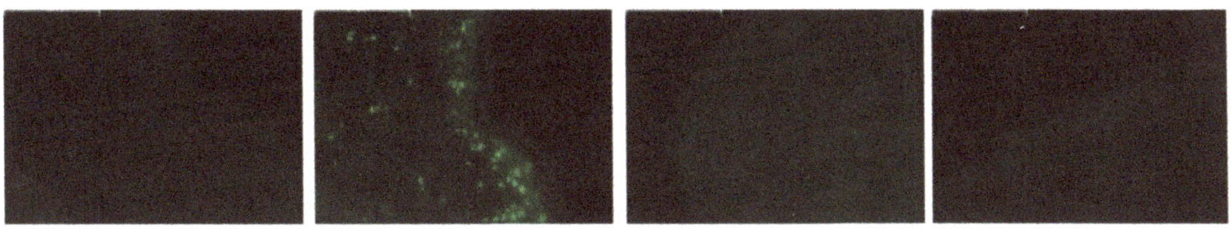

Figure 5. Effect of the FcHEx on the protein carbonylation in the human skin explants. Human skin punches, which were derived from aesthetic surgical biopsies, were treated with the FcHEx or the positive control in the presence of 56 nM of epinephrine. The reported values represent the averages of three independent experiments and are expressed as the percentages of skin explants treated with epinephrine only, which was set to 100%. The asterisks indicate statistically significant values (*** p-value was between 0.0001 and 0.001; ** p-value was between 0.001 and 0.01).

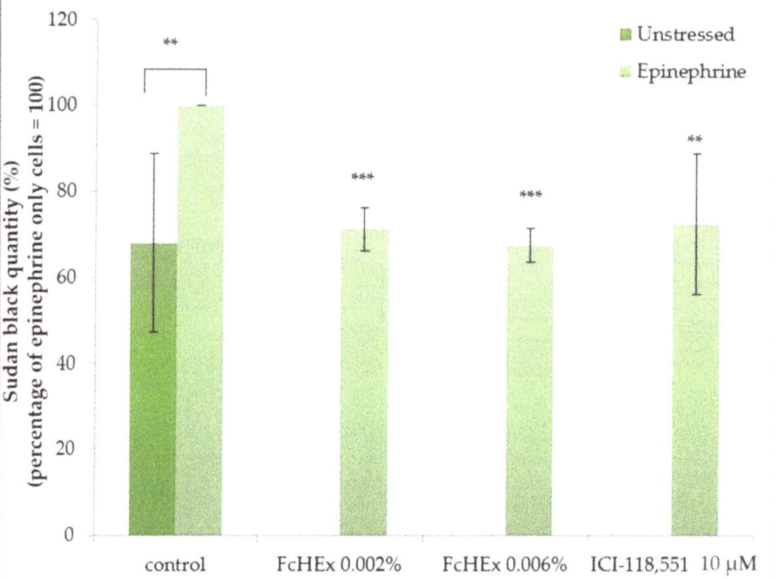

Control	FcHEx 0.002%	FcHEx 0.006%	ICI-118,551 10 μM
67.99 ± 20.78%			
100%	71.15 ± 4.97%	67.43 ± 3.87%	72.49 ± 16.4%

Figure 6. Effect of the FcHEx on the lipofuscin accumulation in the skin keratinocytes. The keratinocytes were treated with the FcHEx or the positive control for 24 h, then 50 nM of epinephrine was added to the cells incubated for an additional 48 h. At the end of the incubation, the cells were stained and the absorbance at 595 nM was measured. The obtained values were normalized relative to the protein content determined using a Bradford assay. The reported values represent the averages of three independent experiments and are expressed as the percentages of cells treated with epinephrine only, which was set to 100%. The asterisks indicate statistically significant values (*** p-value was between 0.0001 and 0.001; ** p-value was between 0.001 and 0.01).

3.1.7. Activity of the FcHEx on the Ceramide Production

The gene expression of the GBA enzyme, which is responsible for the hydrolysis of glucocerebrosides into ceramides, and the amount of newly synthesized ceramides were measured to investigate the FcHEx's ability to preserve the epidermal skin barrier against epinephrine stress. The treatment with the FcHEx increased the GBA expression by almost 150% in cells stressed with epinephrine (Figure 7).

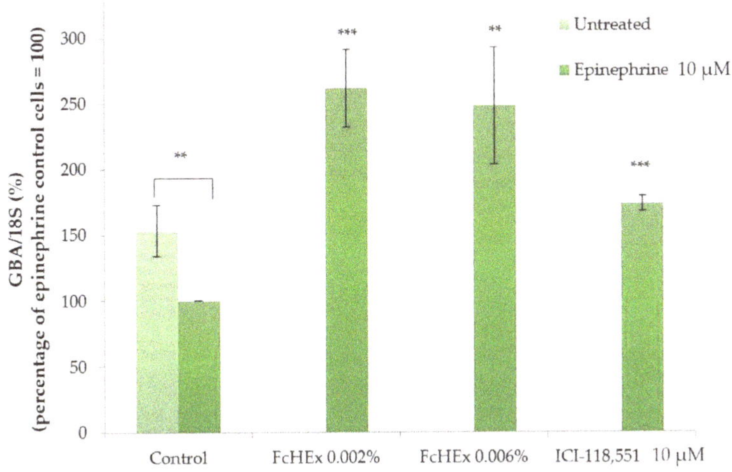

Control	FcHEx 0.002%	FcHEx 0.006%	ICI-118,551 10 μM
153.21 ± 19.49%			
100%	261.45 ± 29.45%	248.4 ± 44.63%	173.06 ± 5.74%

Figure 7. Effect of the FcHEx on the β-glucocerebrosidase (GBA) enzyme gene expression in the skin keratinocytes. The keratinocytes were stimulated with 10 μM of epinephrine alone or with the FcHEx or the positive control for 4 h and then processed for RT-PCR analysis. The reported values represent the averages of three independent experiments and are expressed as the percentages of cells treated with epinephrine only, which was set to 100%. The asterisks indicate statistically significant values (*** *p*-value was between 0.0001 to 0.001; ** *p*-value was between 0.001 to 0.01).

The concentration of ceramides was measured using Nile red staining in cells stressed with epinephrine and treated with the FcHEx. The ceramide content decreased in the epinephrine-treated keratocytes. In contrast, it significantly increased when the FcHEx was administered at both the used concentrations (Figure 8).

3.2. *In Vivo Studies*

3.2.1. Transepidermal Water Loss (TEWL) Evaluation

Stress is known to affect skin barrier permeability and negatively alter TEWL values. An evaporimeter evaluated the TEWL values were expressed in grams per square hectometer. As shown in Figure 9, the cream containing the extract counteracted the TEWL increase. During acute stress, i.e., after two weeks, a peak in the TEWL values in volunteers applying the placebo was observed, while the extract abolished this increase. After four weeks, the extract reduced the TEWL values by 12.2% during the recovery time, showing barrier recovery efficacy, while the placebo showed no significant effect.

3.2.2. Sebum Production Determination

The sebum secretion was monitored using a Sebufix® F 16 and a Corneofix® F 20 applied to a Visioscope PC35® camera. After two and four weeks of use, a statistically significant decrease in the sebum concentration on the cheek and forehead in the volunteers treated with cream containing the FcHEx was recorded. More precisely, sebum production on the cheek decreased by 46.6% after two weeks (during the acute stress time) and 73.8% after four weeks (recovery time). On the forehead, the sebum concentration decreased by 56.4% after two weeks and 80.1% after four weeks (Figure 10).

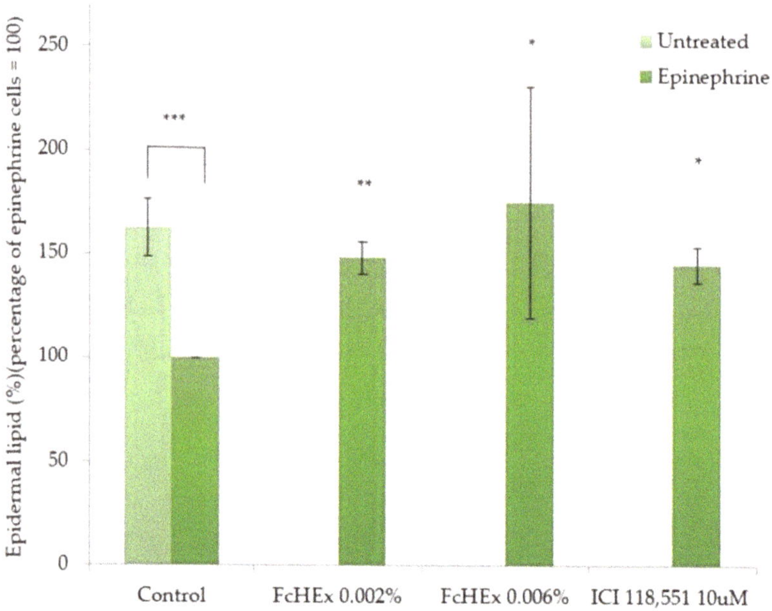

Control	FcHEx 0.002%	FcHEx 0.006%	ICI 118,551 10 uM
162.60 ± 22.34%			
100%	148.25 ± 23.08%	175.33 ± 64.40%	144.15 ± 37.45%

Figure 8. Effect of the FcHEx on the epidermal lipid content in the skin keratinocytes. The HaCaT cells were treated with the FcHEx or the positive control for 24 h, then incubated with epinephrine 10 μM for an additional 24 h. The lipid content was measured using an Adipored assay. The values were normalized relative to the cell density determined using crystal violet staining. The reported values represent the averages of three independent experiments and are expressed as the percentages of cells treated with epinephrine only, which was set to 100%. The asterisks indicate statistically significant values (*** *p*-value was between 0.0001 and 0.001; ** *p*-value was between 0.001 and 0.01; * *p*-value was between 0.01 and 0.05).

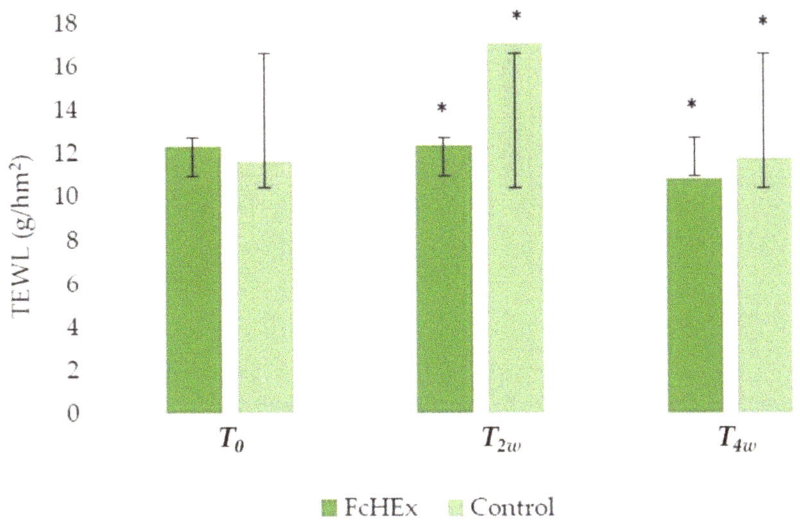

	T_0	T_{2w}	T_{4w}
FcHEx (g/hm²)	12.32 ± 2.45	12.39 ± 3.09	10.82 ± 2.34
Control (g/hm²)	11.63 ± 4.01	17.08 ± 6.77	11.76 ± 3.16

Figure 9. Transepidermal water loss evaluation (TEWL) evaluation. The TEWL trend values at time T_0 (before the application), T_{2w} (after two weeks—acute stress), and T_{4W} (after four weeks—recovery effect). All the measurements for T_{2w} vs. T_0 and T_{4w} vs. T_0 for the FcHEx and the control were statistically significant. The asterisks indicate statistically significant values (* p-value was between 0.01 and 0.05).

3.2.3. Skin Exfoliation

The skin exfoliation was evaluated using a Corneofix® F 20. After two and four weeks of use, during the acute stress and recovery times, a significant decrease in skin renewal in the volunteers treated with cream containing the FcHEx was recorded (T_{2w} = −13%, T_{4w} = −45.7%), and an increase was shown in the volunteers treated with the placebo (T_{2w} = +84.3%, T_{4w} = +9.7%) (Figure 11).

3.2.4. Skin Lightness

The skin lightness was evaluated using a Colorimeter CL 400. The L* value of the volunteers increased in both those who used the cream containing the FcHEx (T_{2w} = 1.9%, T_{4w} = 2.7%) and in the placebo users (T_{2w} = 4.3%, T_{4w} = 0.8%), but the volunteers that used the placebo showed paler faces than the group used the active cream (Figure 12).

3.2.5. Skin pH

A skin pH-meter measured the pH on the skin surface. The cream containing the FcHEx maintained the volunteers' face pH in a normal range (5.5–6) during the acute stress time. In contrast, the placebo increased the pH value during the acute stress time (at T_{2w} pH = 6.05; Figure 13).

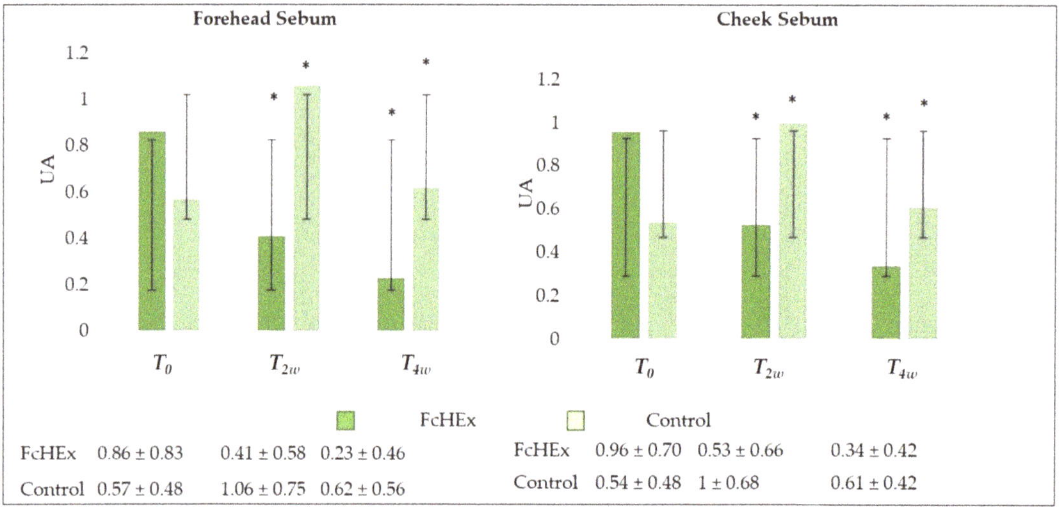

Figure 10. Sebum production evaluation on the forehead and cheek of volunteers. The sebum production trend values at time T_0 (before the application), T_{2w} (after two weeks—acute stress), T_{4w} (after four weeks—recovery effect). All the measurements for T_{2w} vs. T_0 and T_{4w} vs. T_0 for the FcHEx and the control were statistically significant. The asterisks indicate statistically significant values (* p-value was between 0.01 and 0.05).

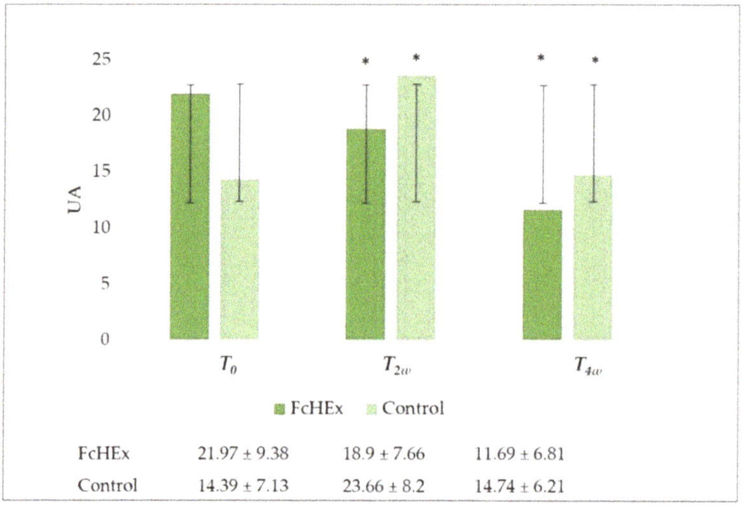

Figure 11. Skin exfoliation evaluation. The corneum exfoliation trend value at time T_0 (before the application), T_{2w} (after two weeks—acute stress), and T_{4w} (after four weeks—recovery effect). All the measurements for T_{2w} vs. T_0 and T_{4w} vs. T_0 for the FcHEx and the control were statistically significant. The asterisks indicate statistically significant values (* p-value was between 0.01 and 0.05).

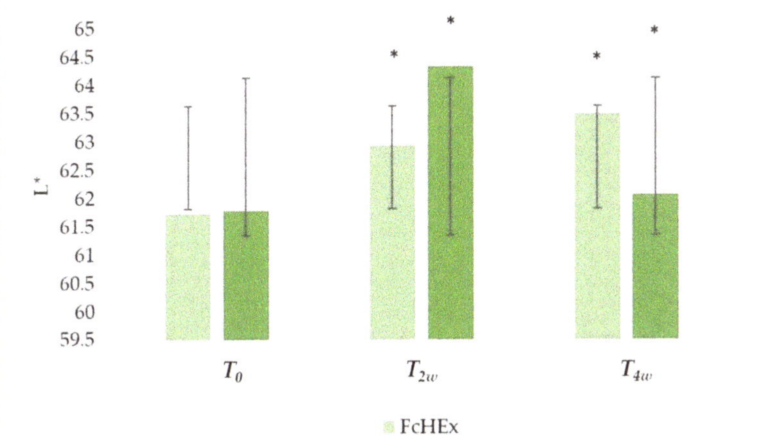

	T_0	T_{2w}	T_{4w}
FcHEx	61.74 ± 2.95	62.95 ± 2.60	63.5 ± 2.39
Control	61.8 ± 2.80	64.34 ± 3.88	62.08 ± 3.21

Figure 12. Skin lightness evaluation. Skin lightness trend values at time T_0 (before the application), T_{2w} (after two weeks—acute stress), T_{4w} (after four weeks—recovery effect). All the measures T_{2W} vs. T_0 and T_{4w} vs. T_0 for the FcHEx and the control were statistically significant. The asterisks indicate statistically significant values (* p-value was between 0.01 and 0.05).

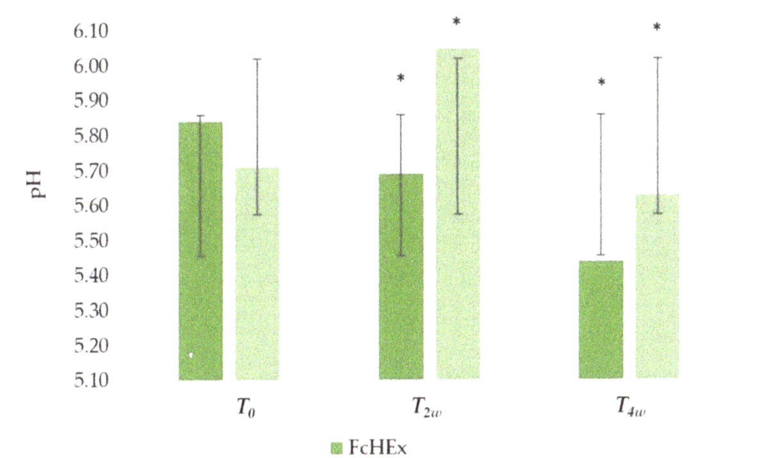

	T_0	T_{2w}	T_{4w}
FcHEx	5.84 ± 0.31	5.69 ± 0.15	5.44 ± 0.16
Control	5.71 ± 0.40	6.05 ± 0.61	5.63 ± 0.37

Figure 13. Skin pH evaluation. The skin pH trend values at time T_0 (time of application), T_{2w} (after two weeks—acute stress), and T_{4w} (after four weeks—recovery effect). All the measures for T_{2w} vs. T_0 and T_{4w} vs. T_0 for the FcHEx and the control were statistically significant. The asterisks indicate statistically significant values (* p-value was between 0.01 and 0.05).

4. Discussion

Psychological stress is the process by which the environmental requirements exceed a person's capacity to adapt, leading to emotional, behavioral, and physiological changes [6]. Chronic stress-induced augmentation in the release of glucocorticoids, epinephrine, and norepinephrine can accelerate aging [25], bacterial colonization, and human peptide protection [26,27]. Stressful conditions, such as those of a chronically ill patient undergoing treatment, might decrease the leukocyte telomere length and telomerase activity, accelerating cellular senescence and decreasing life expectancy [28]. These changes depend on β2-adrenoceptor activation by catecholamines, causing DNA damage and p53 suppression [29]. Oxidation and inflammation processes enhance neutrophil elastase and matrix metalloproteinase-8 (MMP-8) production, which leads to collagen and elastic fiber degradation and wrinkle formation [13]. Research on new cosmetic materials to relieve psychological stress diseases of the skin has been actively pursued. Natural ingredients are generally safer to use than synthetic materials [30]. In this study, a hydrosoluble extract of *Ficus carica* cell suspension cultures (FcHEx), a known source of ficin and phenolic compounds [31], was characterized for its capacity to protect skin cells against the oxidative damage caused by psychological stress. In vitro and in vivo experiments were performed to demonstrate the positive action of the FcHEx that can revert oxidative damage on lipids and proteins due to epinephrine's stress-induced effect. Skin cells and neurons share a common embryonic origin, as both express proteins that are involved in conserved signaling pathways. Psychological stress triggers the autonomic nervous system to release catecholamines, such as epinephrine and norepinephrine, which can significantly affect many cell types' metabolisms and functions, including those of the skin [9]. Epinephrine interacts directly with neutrophils, reducing these cell responses to various cAMP-mediated proinflammatory stimuli [32]. The activation of the epinephrine downstream signals was measured in keratinocytes cells to investigate the capacity of the FcHEx to affect the epinephrine signal cascade in skin cells. In keratinocytes, the inflammatory process was induced via epinephrine administration, and the FcHEx significantly reduced the epinephrine stimulation. The cAMP-mediated inflammatory cytokine gene expression was analyzed, and the FcHEx administration reduced the interleukin 6 production by 50%. Moreover, the lipid peroxides production in the keratinocytes was used to test the antioxidant properties of the FcHEx, which showed a significant antioxidant property, decreasing ROS production by 25%. The lipofuscin test confirmed these data as the treatment with both the FcHEx doses entirely abolished the production of the lipofuscin that was mediated by epinephrine. The inhibition of the carbonylated protein concentration by 50% in the keratinocytes showed that the antioxidant properties of the FeHEx were also linked to an inhibition of protein oxidation. Experiments conducted in skin explants confirmed the protective effect of the FcHEx against protein carbonylation, suggesting that the antioxidant action of the FcHEx was mainly linked to an inhibition of protein oxidation, which is the main threat to skin functionality and integrity [33]. These results encouraged us to study the potential role of the FcHEx in the development of the skin hydrolipidic film, which is able to moisturize and stop excessive water loss of the skin [34] and protect the epidermal lipid barrier from microbe attacks and physical injuries, supporting skin integrity. HaCaT cells were treated with the FcHEx to verify its action on the gene expression of GBA, an enzyme that is able to catalyze ceramide production [35]. The results showed that the GBA expression increased in a dose-dependent manner (50% and 75% at the concentrations of 0.002% and 0.006%, respectively). Furthermore, the content of ceramides by Nile red staining in cells stressed with epinephrine and treated with the FcHEx was measured to confirm the gene expression results.

The biological activity of the FcHEx was confirmed via a clinical evaluation on a group of 40 human volunteers, aged between 20 and 27 years, going through a period of psychological stress. For 28 days, twice daily, the volunteers were asked to apply either a cream containing the FcHEx at the concentration of 0.5% or a placebo cream without the ingredient on the whole face. Parameters such as TEWL, sebum flow, skin exfoliation, and lightness, were evaluated. The measurements were carried out during a high-stress condition, namely, a mid-year examination, and after two weeks from the high-stress condition (recovery time). The TEWL value in the volunteers treated with placebo was higher (48.8%) than the volunteers treated with the cream containing the FcHEx (active cream −0.2%), indicating that the extract restored the regular epidermal permeability barrier homeostasis, which was perturbed by the psychological stress. Indeed, the loss of transepidermal water is an indicator of skin barrier integrity since there is a water concentration gradient inside the stratum corneum, which is continuously spread from the body to the skin's surface and the environment. Thus, low TEWL values are representative of an unaltered skin barrier [36]. Generally, during a student's psychological stress period due to an approaching examination, acne-related symptoms increase [37], and cellular turnover accelerates, depending on the stress levels [38]. The cream containing the *Ficus carica* extract decreased sebum production and inhibited exfoliation in the clinical tests. Desquamation involves the degradation of corneodesmosomes by enzymes that depend on water and pH for their activity. In the skin, the exfoliation is highest at neutral-to-low alkaline pH levels and decreases at acidic pH levels [39]. The active cream normalized and maintained all volunteers' face pH in a normal range during the acute stress, which usually rises in stress conditions. A neutral pH impedes the secretion of polar lipids (glucosylceramides) into mature lamellar bilayers (ceramides), enhancing proteolytic activities and increasing corneodesmolysis aberrations in corneocyte cohesion [40]. Moreover, the skin pH influences the peroxidase activity of the ficin, which is one of the components of the FcHEx that is responsible for the antioxidant activity of the extract [41]. The ficin peroxidase activity was around 60% or higher at a pH of 4.5–5.0, the pH of normal skin [42]. Finally, whether the skin turned pale was evaluated to test the cream's effect on the sympathetic axis (adrenaline and noradrenaline) activated by the acute stress [38]. A small increase in L* was observed in volunteers to whom the cream was applied, suggesting a good efficacy of the extract to prevent the facial skin turning to a pale color due to psychological stress.

5. Conclusions

In vitro and in vivo tests demonstrated that the extract from *Ficus carica* cell cultures alleviated skin damage caused by psychological stress. The FcHEx worked as an anti-stress ingredient that alleviated the negative consequences of stress hormone activity on the skin, such as inflammation, oxidation, alteration of the skin barrier, and skin turning to a pale color. All the data support the conclusion that the *Ficus carica* cell culture extract has good potential to be employed in skincare products, particularly those formulated for dry and stressed skins.

Author Contributions: I.D., writing—original draft; D.F., investigation; R.D.L., investigation; A.T., investigation; G.C., investigation; C.Z., investigation; L.G., investigation; A.S., investigation; S.L., data curation; F.A., data curation. All authors have read and agreed to the published version of the manuscript.

Funding: This research received no external funding.

Institutional Review Board Statement: Ethical review and approval were waived for this study, due to the experimentation involved in the cosmetic products testing.

Informed Consent Statement: Informed consent was obtained from all subjects involved in the study.

Data Availability Statement: Data is contained within the article or supplementary material.

Conflicts of Interest: The authors declare no conflict of interest.

References

1. Dini, I.; Laneri, S. Nutricosmetics: A brief overview. *Phytother. Res.* **2019**, *33*, 3054–3063. [CrossRef]
2. Laneri, S.; Di Lorenzo, R.M.; Bernardi, A.; Sacchi, A.; Dini, I. Aloe barbadensis: A Plant of Nutricosmetic Interest. *Nat. Prod. Comm.* **2020**, *15*. [CrossRef]

3. Laneri, S.; Di Lorenzo, R.; Sacchi, A.; Dini, I. Dosage of bioactive molecules in the nutricosmeceutical Helix aspersa muller mucus and formulation of new cosmetic cream with moisturizing effect. *Nat. Prod. Comm.* **2019**, *14*, 1–7. [CrossRef]
4. Wild, C.P. Complementing the genome with an "exposome": The outstanding challenge of environmental exposure measurement in molecular epidemiology. *Cancer Epidemiol. Biomark. Prev.* **2005**, *14*, 1847–1850. [CrossRef]
5. Krutmann, J.; Bouloc, A.; Sore, G.; Bernard, B.A.; Passeron, T. The skin aging exposome. *Dermatol. Sci. J.* **2017**, *85*, 152–161. [CrossRef] [PubMed]
6. Vileikyte, L. Stress and wound healing. *Clin. Dermatol.* **2007**, *25*, 49–55. [CrossRef]
7. Papadimitriou, A.; Priftis, K.N. Regulation of the hypothalamic-pituitary-adrenal axis. *Neuroimmunomodulation* **2009**, *16*, 265–271. [CrossRef] [PubMed]
8. Cohen, S.; Janicki-Deverts, D.; Miller, G.E. Psychological stress and disease. *JAMA* **2007**, *298*, 1685–1687. [CrossRef]
9. Chen, Y.; Lyga, J. Brain-skin connection: Stress, inflammation and skin aging. *Inflamm. Allergy Drug Targets* **2014**, *13*, 177–190. [CrossRef]
10. Mammone, T.; Marenus, K.; Maes, D.; Lockshin, R.A. The induction of terminal differentiation markers by the cAMP pathway in human HaCaT keratinocytes. *Skin Pharmacol. Physiol.* **1998**, *11*, 152–160. [CrossRef] [PubMed]
11. Grando, S.A.; Pittelkow, M.R.; Schallreuter, K.U. Adrenergic and cholinergic control in the biology of epidermis: Physiological and clinical significance. *J. Investig. Dermatol.* **2006**, *126*, 1948–1965. [CrossRef]
12. Choi, E.H.; Brown, B.E.; Crumrine, D.; Chang, S.; Man, M.Q.; Elias, P.M.; Feingold, K.R. Mechanisms by which psychologic stress alters cutaneous permeability barrier homeostasis and stratum corneum integrity. *J. Investig. Dermatol.* **2005**, *124*, 587–595. [CrossRef] [PubMed]
13. Romana-Souza, B.; Santos Lima-Cezar, G.; Monte-Alto-Costa, A. Psychological stress-induced catecholamines accelerates cutaneous aging in mice. *Mech. Ageing Dev.* **2015**, *152*, 63–73. [CrossRef] [PubMed]
14. Santacruz-Perez, C.; Paulo Newton Tonolli, P.; Ravagnani, F.G.; Baptista, F.S. Photochemistry of Lipofuscin and the Interplay of UVA and Visible Light in Skin Photosensitivity, Photochemistry and Photophysics. In *Fundamentals to Applications*; Saha, S., Mondal, S., Eds.; IntechOpen: London, UK, 2018.
15. Laneri, S.; Dini, I.; Tito, A.; Di Lorenzo, R.; Bimonte, M.; Tortora, A.; Zappelli, C.; Angelillo, M.; Bernardi, A.; Sacchi, A.; et al. Plant cell culture extract of Cirsium eriophorum with skin pore refiner activity by modulating sebum production and inflammatory response. *Phytother. Res.* **2021**, *35*, 530–540. [CrossRef] [PubMed]
16. Apone, F.; Tito, A.; Arciello, S.; Carotenuto, G.; Colucci, M.G. Plant tissue cultures as sources of ingredients for skin care applications. *Ann. Plant Rev.* **2020**, *3*, 135–150.
17. Carlson, R.V.; Boyd, K.M.; Webb, D.J. The revision of the Declaration of Helsinki: Past, present and future. *Br. J. Clin. Pharmacol.* **2004**, *57*, 695–713. [CrossRef]
18. Renner, G.; Audebert, F.; Burfeindt, J.; Calvet, B.; Caratas-Perifan, M.; Leal, M.E.; Gorni, R.; Long, A.; Meredith, E.; O'Sullivan, Ú.; et al. Cosmetics Europe guidelines on the management of undesirable effects and reporting of serious undesirable effects from cosmetics in the European Union. *Cosmetics* **2017**, *4*, 1. [CrossRef]
19. Tsalokostas, G. Using Tissue Culture as an Alternative Source of Polyphenols Produced by *Ficus carica* L. Ph.D. Thesis, Faculty in Biology, City University of New York, New York, NY, USA, 2009.
20. Gagaoua, M.; Boucherba, N.; Bouanane-Darenfed, A.; Ziane, F.; Nait-Rabah, S.; Hafid, K.; Boudechicha, H.-R. Three-phase partitioning as an efficient method for the purification and recovery of ficin from Mediterranean fig (*Ficus carica* L.) latex. *Sep. Purif. Technol.* **2014**, *132*, 461–467. [CrossRef]
21. Nassar, A.H.; Newbury, H.J. Ficin production by callus cultures of *Ficus carica*. *Plant Phys. J.* **1987**, *131*, 171–179. [CrossRef]
22. Cuesta, A.M.; Albiñana, V.; Gallardo-Vara, E.; Recio-Poveda, L.; de Rojas, P.I.; de Las-Heras, K.V.G.; Aguirre, D.T.; Botella, L.M. The b_2-adrenergic receptor antagonist ICI-118,551 blocks the constitutively activated HIF signaling in hemangioblastomas from von Hippel-Lindau disease. *Sci. Rep.* **2019**, *9*, 10062. [CrossRef] [PubMed]
23. Zumwalt, J.W.; Thunstrom, B.J.; Spangelo, B.L. Interleukin-1beta and catecholamines synergistically stimulate interleukin-6 release from rat C6 glioma cells in vitro: A potential role for lysophosphatidylcholine. *Endocrinology* **1999**, *140*, 888–896. [CrossRef] [PubMed]
24. Wortsman, J.; Frank, S.; Cryer, P.E. Adre-nomedullary response to maximal stress in humans. *Am. J. Med.* **1984**, *77*, 779–784. [CrossRef]
25. Dunn, J.H.; Koo, J. Psychological Stress and skin aging: A review of possible mechanisms and potential therapies. *Dermat. Online J.* **2013**, *19*, 18561.
26. Scudiero, O.; Brancaccio, M.; Mennitti, C.; Laneri, S.; Lombardo, B.; De Biasi, M.G.; De Gregorio, E.; Pagliuca, C.; Colicchio, R.; Salvatore, P.; et al. Human Defensins: A Novel Approach in the Fight against Skin Colonizing Staphylococcus aureus. *Antibiotics* **2020**, *9*, 198. [CrossRef]
27. Pero, R.; Angrisano, T.; Brancaccio, M.; Falanga, A.; Lombardi, L.; Natale, F.; Laneri, S.; Lombardo, B.; Galdiero, S.; Scudiero, O. Beta-defensins and analogs in Helicobacter pylori infections: mRNA expression levels, DNA methylation, and antibacterial activity. *PLoS ONE* **2019**, *14*, e0222295. [CrossRef]
28. Damjanovic, A.K.; Yang, Y.; Glaser, R.; Kiecolt-Glaser, J.K.; Nguyen, H.; Laskowski, B.; Beversdorf, D.Q.; Zou, Y.; Weng, N.P. Accelerated telomere erosion is associated with a declining immune function of caregivers of Alzheimer's disease patients. *Immun. J.* **2007**, *179*, 4249–4254. [CrossRef]

29. Hara, M.R.; Kovacs, J.J.; Whalen, E.J.; Rajagopal, S.; Strachan, R.T.; Grant, W.; Towers, A.J.; Williams, B.; Lam, C.M.; Xiao, K.; et al. A stress response pathway regulates DNA damage through beta2-adrenoreceptors and beta-arrestin-1. *Nature* **2011**, *477*, 349–353. [CrossRef]
30. Minich, D.M.; Bland, J.S. A review of the clinical efficacy and safety of cruciferous vegetable phytochemicals. *Nutr. Rev.* **2007**, *65*, 259–267. [CrossRef]
31. Cormier, F.; Charest, C.; Dufresne, C. Partial purification and properties of proteases from fig (*Ficus carica*) callus cultures. *Biotechnol. Lett.* **1989**, *11*, 797–802. [CrossRef]
32. Moore, A.R.; Willoughby, D. A The role of cAMP regulation in controlling inflammation. *Clin. Exp. Immunol.* **1995**, *101*, 387–389. [CrossRef]
33. Rinnerthaler, M.; Bischof, J.; Streubel, M.K.; Trost, A.; Richter, K. Oxidative Stress in Aging Human Skin. *Biomolecules* **2015**, *5*, 545–589. [CrossRef]
34. Proksch, E.; Brandner, J.M.; Jensen, J.M. The skin: An indispensable barrier. *Exp. Dermatol.* **2008**, *17*, 1063–1972. [CrossRef]
35. Feingold, K.R. The role of epidermal lipids in cutaneous permeability barrier homeostasis. *J. Lipid Res.* **2007**, *48*, 2531–2546. [CrossRef] [PubMed]
36. Garg, A.; Chren, M.M.; Sands, L.P.; Matsui, M.S.; Marenus, K.D.; Feingold, K.R.; Elias, P.M. Psychological stress perturbs epidermal permeability barrier homeostasis: Implications for the pathogenesis of stress-associated skin disorders. *Arch Dermatol.* **2001**, *137*, 53–59. [CrossRef]
37. Chiu, A.; Chon, S.Y.; Kimball, A.B. The response of skin disease to stress: Changes in acne severity vulgaris as affected by examination stress. *Arch Dermatol.* **2003**, *139*, 897–900. [CrossRef]
38. Peters, E.M. Stressed skin a molecular psychosomatic update on stress-causes and effects in dermatologic diseases. *J. Dtsch. Dermatol. Ges.* **2016**, *14*, 233–252. [CrossRef]
39. Loden, M. Role of topical emollients and moisturizers in the treatment of dry skin barrier disorders. In *Dry Skin and Moisturizers: Chemistry and Function*, 2nd ed.; Loden, M., Mailbach, H., Eds.; CRC Press: Boca Raton, FL, USA, 2006; pp. 771–778.
40. Fluhr, J.W.; Elias, P.M. Stratum corneum pH: Formation and function of the acid mantle. *Exog. Dermatol.* **2002**, *1*, 163–175. [CrossRef]
41. Cho, U.M.; Choi, D.H.; Yoo, D.S. Inhibitory Effect of Ficin Derived from Fig Latex on Inflammation and Melanin Production in Skin Cells. *Biotechnol. Bioproc. E* **2019**, *24*, 288–297. [CrossRef]
42. Devaraj, K.B.; Kumar, P.R.; Prakash, V. Purification, characterization, and solvent-induced thermal stabilization of ficin from Ficus carica. *J. Agric. Food Chem.* **2008**, *56*, 11417–11423. [CrossRef]

antioxidants

MDPI

Article

8-Week Supplementation of 2S-Hesperidin Modulates Antioxidant and Inflammatory Status after Exercise until Exhaustion in Amateur Cyclists

Francisco Javier Martínez-Noguera [1,*], Cristian Marín-Pagán [1], Jorge Carlos-Vivas [2] and Pedro E. Alcaraz [1]

1 Research Center for High Performance Sport, Catholic University of Murcia, Campus de los Jerónimos N° 135, UCAM, 30107 Murcia, Spain; cmarin@ucam.edu (C.M.-P.); palcaraz@ucam.edu (P.E.A.)

2 Health, Economy, Motricity and Education Research Group (HEME), Faculty of Sport Sciences, University of Extremadura, Avda. de Elvas, s/n., 06006 Badajoz, Spain; jorge.carlosvivas@gmail.com

* Correspondence: fjmartinez3@ucam.edu; Tel.: +34-968-278-566

Abstract: Both acute and chronic ingestion of 2S-hesperidin have shown antioxidant and anti-inflammatory effects in animal studies, but so far, no one has studied this effect of chronic ingestion in humans. The main objective was to evaluate whether an 8-week intake of 2S-hesperidin had the ability to modulate antioxidant-oxidant and inflammatory status in amateur cyclists. A parallel, randomized, double-blind, placebo-controlled trial study was carried out with two groups (500 mg/d 2S-hesperidin; $n = 20$ and 500 mg/d placebo; $n = 20$). An incremental test was performed to determine the working zones in a rectangular test, which was used to analyze for changes in antioxidant and inflammatory biomarkers. After 2S-hesperidin ingestion, we found in the rectangular test: (1) an increase in superoxide dismutase (SOD) after the exercise phase until exhaustion ($p = 0.045$) and the acute recovery phase ($p = 0.004$), (2) a decrease in the area under the oxidized glutathione curve (GSSG) ($p = 0.016$), and (3) a decrease in monocyte chemoattractant protein 1 (MCP1) after the acute recovery phase ($p = 0.004$), post-intervention. Chronic 2S-hesperidin supplementation increased endogenous antioxidant capacity (↑SOD) after maximal effort and decreased oxidative stress (↓AUC-GSSG) during the rectangular test, decreasing inflammation (↓MCP1) after the acute recovery phase.

Keywords: polyphenols; flavonids; endogenous antioxidant enzymes; reduced glutathione; oxidized glutathione; catalase; superoxide dismutase; interleukin 6; tumor necrosis factor; endurance sports

Citation: Martínez-Noguera, F.J.; Marín-Pagán, C.; Carlos-Vivas, J.; Alcaraz, P.E. 8-Week Supplementation of 2S-Hesperidin Modulates Antioxidant and Inflammatory Status after Exercise until Exhaustion in Amateur Cyclists. *Antioxidants* **2021**, *10*, 432. https://doi.org/10.3390/antiox10030432

Academic Editors: Irene Dini and Domenico Montesano

Received: 3 February 2021
Accepted: 8 March 2021
Published: 11 March 2021

1. Introduction

Flavonoids are bioactive substances found mainly in fruits and vegetables, with more than 15,000 molecules identified within this family [1]. However, one of the most well-known is hesperidin, which is a flavonoid present at high concentrations in citrus fruits, being the main one in sweet orange (*Citrus sinensis*). Hesperidin may be found in two isomeric forms, 2S- and 2R-, where the 2S isomer is predominant in nature [2]. When hesperidin reaches the intestine, bacterial flora converts it into hesperetin (aglycon), which is effectively absorbed, being the main metabolite of hesperidin [3]. Previous studies have shown the positive effects of hesperidin on some diseases (neurological, cardiovascular, insulin sensitization) due to its antioxidant and anti-inflammatory properties [4,5]. Moreover, the intake of hesperidin (in orange juice) has been shown to modulate leukocyte gene expression, boosting its antioxidant and inflammatory profile, and therefore showing a nutrigenomic effect [6]. On the other hand, the ability of 2S-hesperidin to improve performance has been observed [7]. It should be noted that there are other important factors that can modulate the effect of flavonoids like hesperidin, such as intestinal flora transformations, absorption and bioavailability [8].

The antioxidant effect of hesperidin is mainly related to its radical scavenging capabilities, as well as the increase in antioxidant cellular defense catalase (CAT), superoxide dismutase (SOD), reduced glutathione (GSH) and oxidized glutathione (GSSG) via the nuclear respiratory factor 2 (NRF2) signaling pathway [4]. On the other hand, the hesperidin anti-inflammatory effect is produced by a decrease in inflammatory markers, such as nuclear factor kappa B (NF-κB), interleukin 6 (IL6), tumor necrosis α (TNFα) and inducible nitric oxide synthase (iNOS) [4].

Regarding the potential of hesperidin on physical performance, a recent study reported that the acute intake of 500 mg of 2S-hesperidin significantly improved anaerobic performance [9]. In the same study, they also found small non-significant changes in CAT, SOD, GSH and the GSSG/GSH ratio compared to a placebo during a rectangular test (with different intensities) in amateur cyclists. Similarly, a study performed in rats observed that 2S-hesperidin (200 mg/kg, three days per week during five weeks) showed a protective effect on the oxidative stress induced by an exhausting exercise [10]. Hesperidin supplementation prevented the increase in reactive oxygen species (ROS) production and avoided a decrease in SOD and catalase activities, while leading to a higher physical performance. In the same way, 6 weeks of hesperetin (main metabolite of hesperidin) supplementation (50 $mg \cdot kg^{-1} \cdot d^{-1}$) significantly increased the GSH/GSSG ratio and improved running performance (exercise time) in aged mice [11]. In addition, a recent study found that eight weeks' intake of 2S-hesperidin improved performance at the threshold of estimated functional power and maximum power in an incremental test until exhaustion compared to a placebo in amateur cyclists [7]. Other polyphenols have shown hesperidin-like effects. For example, the intake of 100 mL per day for six weeks of acai berry-based juice (↑ anthocyanins) increased the levels of GSH and CAT post-exercise and after 1 h of recovery, without changes in SOD and exercise performance (300 m running times) in junior athletes [12].

With regards to exercise, it is known that almost 0.15% of the oxygen consumed is converted into ROS, which can be detrimental to muscle and mitochondrial function [13]. In sports physiology, it is hypothesized that rapid increases in ROS during intensive exercise may be a contributor to fatigue [14]. Based on recent findings, a new theory proposes that antioxidant supplementation (vitamins A, C, E, thiols, ubiquinones and flavonoids) may delay fatigue [15]. However, this mitigation of ROS generation may disrupt cellular signaling involved in training adaptations [16]. ROS are intracellular messengers and activators of transcription factors that promote the expression of genes related to training adaptations and performance improvement [16]. Thus, antioxidant supplementation could decrease ROS production and delay fatigue, but in turn it may slow down the physiological adaptations of training [17,18]. Due to the current controversy on this topic, further investigations are required to evaluate if the intake of antioxidant polyphenols, such as hesperidin, could improve endogenous antioxidant status without negatively affecting performance.

Currently, studies have shown that acute [9] and chronic [7] intake of 2S-hesperidin in amateur cyclists improve anaerobic and aerobic performance, respectively. However, no research has explained the metabolic, biochemical and molecular mechanisms by which 2S-hesperidin intake improves performance. We hypothesized that the chronic intake of 2S-hesperidin would improve amateur cyclists' antioxidant status, evaluated through markers such as CAT, SOD, GSSG, GSH and hemoxigenase 1(HO1), but decrease inflammatory markers, such as IL6, TNFα, monocyte chemoattractant protein-1 (MCP1) and C reactive protein (CRP). However, the implications of long-term or prolonged use are unknown. Therefore, this study aimed to evaluate the effect of eight weeks of 2S-hesperidin supplementation (500 mg/day) on the antioxidant-oxidant (CAT, SOD, GSH, GSSG, HO1 and TBARS) and anti-inflammatory (IL6, TNFα, MCP1 and CRP) state in amateur cyclists before and at the end of the rectangular test and after the resting phase.

2. Methodology

2.1. Study Design

A randomized, double-blind, parallel clinical trial was conducted. Forty subjects were divided into 2 groups: 2S-heperidin (n = 20) and placebo (n = 20). Subjects were randomized into groups using the Randomizer software. Participants consumed two 250 mg capsules of either Placebo (microcellulose, 500 mg) or 2S-hesperidin (500 mg Cardiose®, produced by HealthTech BioActives (HTBA), Murcia, Spain) at breakfast for 8 weeks. The Cardiose® supplement consisted of a natural orange extract that, due to its unique manufacturing process, retains most of the natural isomeric form of hesperidin (NLT 85% 2S-hesperidin). The placebo supplements were similar in appearance to the 2S-hesperidin capsule. Cyclists were instructed to continue their usual diet and training program. The usual total training distance was balanced between the groups (Table 1).

Table 1. Baseline general characteristics and training variables of the cyclists.

	2S-Hesperidin	Placebo	p-Value
Age (years)	35.0 (9.20)	32.6 (8.90)	0.407
Body mass (kg)	71.0 (6.98)	70.4 (6.06)	0.773
Height (cm)	175.3 (6.20)	176.5 (6.10)	0.541
BMI ($kg{\cdot}m^{-2}$)	23.1 (1.53)	22.6 (1.43)	0.292
BF (%)	8.9 (1.63)	9.0 (1.64)	0.803
VO_{2MAX} ($L{\cdot}min^{-1}$)	3.99 (0.36)	3.98 (0.63)	0.971
VO_{2MAX} ($mL{\cdot}kg^{-1}{\cdot}min^{-1}$)	57.5 (6.97)	57.9 (9.53)	0.88
HR_{MAX} (bpm)	184.9 (11.11)	183.2 (8.68)	0.593
VT1 (%)	50.9 (5.63)	50.0 (4.78)	0.61
VT2 (%)	84.9 (5.85)	84.1 (5.70)	0.644
Training variables	**2S-Hesperidin**	**Placebo**	**p-value**
Total distance (km)	1121.12 (534.99)	1082.43 (810.46)	0.868
HR_{AVG} (bpm)	144.76 (8.88)	137.48 (13.11)	0.067
W_{AVG} (W)	174.86 (15.79)	163.47 (32.49)	0.435
RPE	6.34 (0.82)	6.33 (1.16)	0.975

Values are expressed as mean (SD). BMI = body mass index; BF = body fat; VO_{2max} = maximum oxygen volume; VT1 = ventilatory threshold 1 (aerobic); VT2 = ventilatory threshold 2 (anaerobic); Total distance = of all the training sessions carried out during the study period; HRavg = average heart rate of all the training sessions carried out during the study period; Wavg = average power output of all training sessions during the study period and RPE = rating of perceived exertion of all training sessions during the study.

2.2. Participants

Forty healthy male, amateur cyclists completed the study (Table 1). Subjects met the following inclusion criteria: 18–55 years old, BMI of 19–25.5 $kg{\cdot}m^{-2}$, at least 3 years of cycling experience, and training for 6–12 $h{\cdot}wk^{-1}$. Amateur cyclists were excluded if: (a) regular smoking or alcohol drinking, (b) metabolic, cardiorespiratory or digestive pathology or abnormality, (c) injury in the previous 6 months, (d) supplements or medication in the previous 2 weeks and (e) abnormal values in blood test parameters. Before the start of the study, participants were informed about the procedures, and signed informed consent was obtained. The study was conducted following the Declaration of Helsinki guidelines for research on human subjects [19] and was approved by the Ethics Committee of the Catholic University of Murcia (CE091802), registered in ClinicalTrials.gov (Identifier: NCT04597983).

2.3. Procedures

Participants visited the laboratory on five different occasions. Visit 1 consisted of a medical examination and blood extraction to determine health status. On visits 2 and 4, a 24-h diet recall was conducted, followed by an incremental test until exhaustion on a cycle ergometer to estimate the rectangular test zones. On visits 3 and 5, the 24-h diet recall was repeated, and participants performed a rectangular test on the cycle ergometer

(Figure 1) (Table 2). Before each testing session (visits 2, 3, 4 and 5), a standardized breakfast composed of 95.2 g of carbohydrates (68%), 18.9 g of protein (14%) and 11.3 g of lipids (18%) was prescribed by a sports nutritionist. Intake of both treatments began at visit 1 under the supervision of an investigator and finished at visit 5. Subjects in both groups were instructed not to consume foods with a high content of citrus flavonoids (grapefruit, lemons, or oranges) for 5 days prior to and during the study. This was verified by diet recall records.

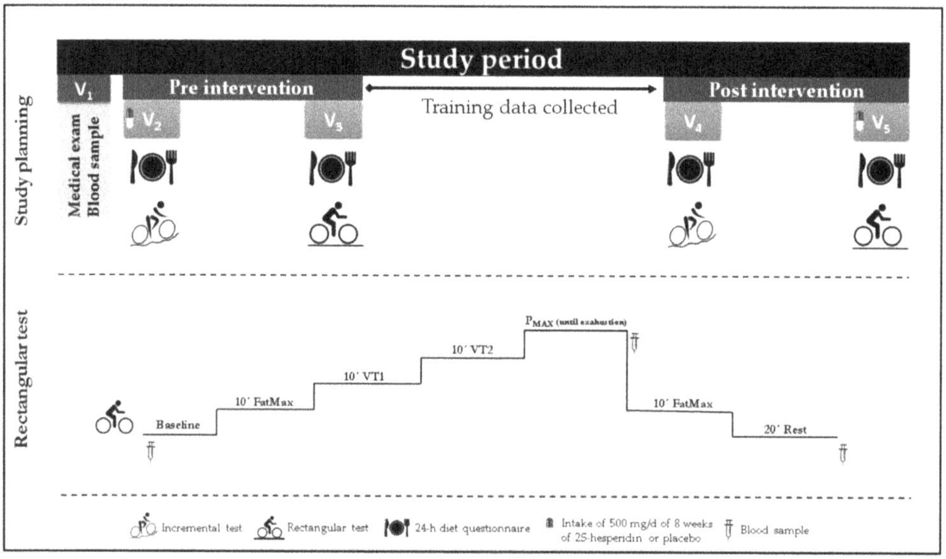

Figure 1. Study planning and rectangular test protocol.

Table 2. Between-group comparisons of dietary intake of cyclists.

	Pre-Intervention			Post-Intervention		
	2S-Hesperidin	Placebo	*p*-Value	2S-Hesperidin	Placebo	*p*-Value
Kcal	2163.6 (519.02)	2100.2 (515.77)	0.708	1974.1 (377.97)	2133.5 (437.98)	0.237
Kcal/BM	31.1 (9.34)	30.2 (8.71)	0.768	27.9 (6.53)	30.3 (6.46)	0.249
CHO (g)	245.7 (73.46)	222.0 (69.68)	0.312	216.6 (63.47)	248.3 (58.15)	0.117
CHO/BM	3.5 (1.31)	3.2 (1.14)	0.416	3.1 (1.08)	3.5 (0.94)	0.173
PRO (g)	113.5 (25.21)	115.2 (25.37)	0.837	109.0 (23.05)	101.5 (23.67)	0.332
PRO/BM	1.6 (0.41)	1.7 (0.48)	0.778	1.5 (0.35)	1.5 (0.42)	0.596
LP (g)	80.8 (27.24)	83.5 (23.65)	0.739	71.5 (17.61)	71.6 (18.89)	0.985
LP/BM	1.2 (0.45)	1.2 (0.37)	0.758	1.0 (0.27)	1.0 (0.29)	0.823

Values are expressed as mean (SD). Kcal = kilocalories; CHO = carbohydrates; PRO = protein; LP = lipids; BM = body mass. The mean values correspond to the average of all 24-h diet recall data collected at pre-intervention (visits 2, 3 and 4) and post-intervention (visits 5, 6 and 7).

2.4. Testing

2.4.1. Medical Exam

A medical examination was conducted by the research center's medical doctor and consisted of medical and health history, resting electrocardiogram and examination (auscultation, blood pressure, etc.). These evaluations confirmed that the volunteer was healthy enough to be enrolled in the study.

2.4.2. Maximal Test

Incremental step with final ramp test was performed on a cycle ergometer using a metabolic cart (Metalyzer 3B. Leipzig, Germany) to determine the maximal fat oxidation zone (FatMax), ventilatory thresholds 1 (VT1) and 2 (VT2) and maximal oxygen consumption (VO_{2max}). Participants started cycling at 35 W for 2 min, increasing by 35 W for every 2 min until RER > 1.05, and then initiating the final ramp (+ 35 $W{\cdot}min^{-1}$) until exhaustion. To ensure they reached VO_{2max}, at least 2 of the following criteria had to be fulfilled: plateau in final VO_2 values (increase $\leq$ 2.0 $mL{\cdot}kg^{-1}{\cdot}min^{-1}$ in the 2 last loads), reaching maximal theoretical HR ((220–age)·0.95), RER $\geq$ 1.15 and lactate $\geq$ 8.0 mmoL/L. Ventilatory thresholds were obtained using the ventilatory equivalents method described by Wasserman [20].

2.4.3. Rectangular Test

Rectangular test procedures are shown in Figure 1. This test was performed on a cycle ergometer using power output values achieved during the maximal test at different intensity zones (FatMax, VT1, VT2 and maximum power). Participants exercised continuously as follows: 10 min at FatMax, 10 min at VT1, 10 min at VT2, at maximum power until exhaustion (post-P_{MAX}) and 30 min rest (post-REC). There were no rest periods between phases.

2.4.4. Blood Samples

Venous blood (arm antecubital area) was collected into one 3 mL ethylenediaminetetraacetic acid (EDTA) tube for hemogram and another 3.5 mL polyethene terephthalate (PET) tube was collected by a nurse for overall health analysis (visit 1). Red blood cell count was carried out in an automated Cell-Dyn 3700 analyzer (Abbott Diagnostics, Chicago, IL, USA) using internal (Cell-Dyn 22) and external (Program of Excellence for Medical Laboratories-PEML) controls. Values of erythrocytes, haemoglobin, haematocrit and haematimetric indexes were estimated.

Additionally, venous blood samples were collected in the baseline, after the maximum power stage (post-P_{MAX}) and during the resting phase (post-REC), to measure antioxidant and anti-inflammatory parameters (visits 3 and 5) (Figure 1). During every extraction point, 6 tubes of 3 mL of EDTA were obtained. Blood samples were centrifuged at 3500 rpm in 4 °C for 10 min and sent to the laboratory for later analysis.

2.4.5. Urine Samples

The main hesperidin metabolites were analyzed in the urine of participants. Urine samples were collected for 24 h before V2 and V5 visits from each participant, before and after the supplementation, and were frozen in liquid nitrogen after collection and thawed for its analysis. For analysis, 50 µL of urine was mixed with 100 µL of water with 1% formic acid containing the internal standard. Then, the mixture was injected into LC-MS/MS (UHPLC 1290 Infinity II Series coupled to a QqQ/MS 6490 Series Agilent Technologies, Sta. Clara, CA, USA). Metabolites were quantified by external standard calibration using *rac*-Hesperetin-d3 as the internal standard.

2.5. Antioxidant and Inflammatory State Markers

The following parameters were selected to measure the antioxidant and inflammatory status.

2.5.1. TBARS (Lipoperoxidation Biomarker)

Thiobarbituric acid reactive substances (TBARS) are a by-product of the oxidative degradation of lipids by reactive oxygen species (lipid peroxidation), which is commonly used as an oxidative stress marker [21]. TBARS assay involves the reaction of malondialdehyde (MDA), a product of lipid peroxidation, with thiobarbituric acid (TBA) under high temperature and acidic conditions to form an MDA-TBA complex that can be measured colorimetrically [22]. The coefficient of variation between replicas had to be less than or equal to 4.6% (Supplementary File 1).

2.5.2. Catalase (CAT)

CAT activity was determined using a UV-VIS spectrophotometer. This was expressed in sec^{-1} per gram of hemoglobin [23]. The coefficient of variation between replicas had to be less than or equal to 4.9% (Supplementary File 1).

2.5.3. Superoxide Dismutase (SOD)

SOD activity was measured using an SD125 Ransod kit (Randox Ltd. Crumlin, United Kingdom) [24]. The coefficient of variation between replicas had to be less than or equal to 5.1% (Supplementary File 1).

2.5.4. Glutathione Reduced (GSH) and Oxidized (GSSG)

GSH was analyzed by the glutathione-S-transferase assay described by Akerboom and Sies [25]. Glutathione oxidized form and glutathione disulfide (GSSG) were determined in a similar way to GSH as shown above, as described by Asensi [26]. The coefficient of variation for GSH between replicas must be less than or equal to 4.1% (Supplementary File 1).

2.5.5. Hemoxygenase 1 (HO1)

A commercial kit was used based on the Enzyme-Linked ImmunoSorbent Assay (ELISA) method (Shanghai BlueGeneBiotech Co., Ltd., Shanghai, China) with a detection limit of 0.1 ng/mL, according to the manufacturer's instructions. The coefficient of variation between replicas must be less than or equal to 4.9% (Supplementary File 1).

2.5.6. Measurement of Cytokines IL6, TNFα and MCP1

These assays employed the quantitative sandwich enzyme immunoassay technique (DRG Instruments GmbH, Marburg, Germany), according to the manufacturer's instructions. A monoclonal antibody specific for IL6, TNFα and MCP1 was precoated onto a microplate. Standards and samples were placed into the wells, and any IL6, TNFα and MCP1 present were bounded by the immobilized antibody. After washing away any unbounded substances, an enzyme-linked polyclonal antibody specific for IL6, TNFα and MCP1 was added to the wells. Following a wash to remove any unbounded antibody–enzyme reagent, a substrate solution was added to the wells, and color developed in proportion to the amount of IL6, TNFα and MCP1 bounded in the initial step. The color development was stopped, and the intensity of the color was measured. The coefficient of variation for IL6, TNFα and MCP1 between replicas must be less than or equal to 4.4, 6.4 and 4.7%, respectively (Supplementary File 1).

2.5.7. C reactive Protein (CRP)

For CRP-ultrasensitive (PCR-Turbilátex, Spinreact, Girona, Spain) detection, a turbidimetric test was used for the quantification of low serum CRP levels, according to the manufacturer's instructions. Latex particles coated with anti-human CRP antibodies were agglutinated by CRP that was present in the subject's sample. The agglutination process caused an absorbance change proportional to the CRP concentration of the sample, and by comparison with a CRP calibrator of known concentration, the CRP content in the analyzed sample was determined. The coefficient of variation between replicas had to be less than or equal to 4.7%.

2.6. Statistical Analyses

Data analysis was conducted using IBM Social Sciences software (SPSS, version 21.0, Chicago, IL, USA). Descriptive statistics are presented as mean and standard deviation (SD). Levene's and Shapiro–Wilk tests were applied to check the homogeneity and normality of the data, respectively. A group × time × moment ANOVA was conducted to analyze within-group and between-group differences in all dependent variables and for every time-point of measurement (baseline, post-P_{MAX} and post-REC) and in both moments (pre-test and post-test). In addition, the area under the curve (AUC), resulting from the integration of the three time-points of measurement taken during the rectangular test, was calculated for each variable. The AUC was used to analyze pre-post differences both within groups and between groups. The within-group differences in the AUC were analyzed by repeated-measures *t*-test, and between-group comparisons in the AUC were conducted by applying an independent samples T-test. Cohen's d effect size (ES) (95% confidence interval) was calculated for all comparisons. Threshold values for ES statistics were as follows: > 0.2 small, > 0.5 moderate, > 0.8 large [27]. Significant differences were considered for $p \leq 0.05$.

3. Results

3.1. Biomarkers of Antioxidants and Oxidants Endogenous

Obtained values for CAT, SOD, GSSG, GSH, GSSG/GSH and HO1 during the rectangular test, pre- and post-intervention, are presented in Table 3. For each parameter, within group changes at each time point (baseline, Post-P_{MAX} and Post- REC) during supplementation were evaluated. A significant increase in SOD activity was found for the 2S-hesperidin group in post-P_{MAX} (15.5%) and post-REC (16.3%), while the placebo showed a significant increase in SOD at baseline (18.1%), intragroup pre-post-intervention (Figure 2). In addition, a similar increase in SOD in the AUC was found in 2S-hesperidin (14.1%) and placebo (11.9%) in the intragroup statistical analysis, without significant differences between groups (Figure 3).

Additionally, a trend towards a decrease with a moderate size effect in GSSG levels at post-P_{MAX} (−17.7%) was found in 2S-hesperidin in the post-intervention intragroup statistical analysis. In contrast, a significant decrease with a large size effect in GSSG was observed in the placebo at post-P_{MAX} (−15.1%) after the intervention (Figure 2). When comparing baseline post-intervention between groups, 2S-hesperidin had lower GSSG values (−20.1%) than the placebo (Figure 2). After the analysis of the AUC intragroup, there was a decrease in GSSG (−14.6%) only in 2S-hesperidin, without differences between groups (Figure 3).

Table 3. Changes in enzymes and peptides endogenous antioxidants before, during and after rectangular test comparing pre- and post-intervention.

		2S-HESPERIDIN				PLACEBO				Between-Group Comparison			
		Baseline	**Post- P_{MAX}**	**Post- REC**	**AUC**	**Baseline**	**Post- P_{MAX}**	**Post- REC**	**AUC**	**ΔBaseline**	**ΔPost-P_{MAX}**	**ΔPost-REC**	**ΔAUC**
CAT	Pre-Int	23.89 (5.14)	38.66 (11.4)	23.89 (3.75)	63.29 (14.2)	25.15 (4.54)	39.14 (8.51)	24.34 (3.85)	63.88 (11.20)				
	Post-Int	25.17 (4.95)	35.57 (9.66)	25.5 (7.65)	61.91 (12.62)	24.78 (4.02)	35.45 (11.12)	25.01 (4.41)	60.34 (12.74)				
	p-value	0.272	0.148	0.263	0.526	0.757	0.058	0.633	0.188	0.316	0.76	0.636	0.527
	Effect size	0.24	0.26	0.41	0.09	0.08	0.42	0.17	0.30	0.32	0.07	0.15	0.2
SOD	Pre-Int	1509 (435.05)	1442 (282.29)	1541 (280.16)	2971 (584.34)	1566 (284.36)	1573 (274.41)	1563 (303.05)	3138 (479.34)				
	Post-Int	1698 (238.55)	1666 (246.110	1792 (231.2)	3391 (308.32)	1849 (364.77)	1727 (491.32)	1718 (412.24)	3511 (754.07)				
	p-value	0.11	**0.045**	**0.004**	**0.011**	**0.009**	0.124	0.057	**0.023**	0.449	0.699	0.402	0.826
	Effect size	0.42	0.76	0.86	0.69	0.95	0.54	0.49	0.75	0.2	0.16	0.27	0.07
GSSG	Pre-Int	0.357 (0.19)	0.322 (0.07)	0.314 (0.060	0.657 (0.14)	0.363 (0.13)	0.369 (0.07)	0.34 (0.1)	0.720 (0.11)				
	Post-Int	0.278 (0.06)	0.265 (0.07)	0.316 (0.1)	0.561 (0.06)	0.348 (0.14)	0.313 (0.12)	0.328 (0.13)	0.650 (0.21)				
	p-value	0.08	**0.058**	0.963	**0.016**	0.734	**0.049**	0.719	0.219	0.326	0.979	0.772	0.627
	Effect size	0.39	0.75	0.02	0.64	0.11	0.82	0.11	0.59	0.31	0.01	0.09	0.12
GSH	Pre-Int	26.37 (3.14)	26.41 (3.31)	26.93 (3.66)	53.07 (4.25)	26.99 (2.46)	26.61 (2.40)	26.98 (2.66)	53.60 (3.30)				
	Post-Int	23.88 (3.09)	23.59 (2.86)	25.00 (4.61)	48.03 (5.57)	24.76 (3.66)	25.58 (3.78)	25.38 (2.80)	50.65 (5.30)				
	p-value	**0.014**	**0.029**	0.081	**0.011**	**0.042**	0.335	0.110	**0.027**	0.862	0.246	0.812	0.343
	Effect size	0.76	0.82	0.51	1.14	0.87	0.41	0.58	0.86	0.06	0.37	0.08	0.3
GSSG/ GSH	Pre-Int	1.16 (0.32)	1.25 (0.32)	1.17 (0.27)	2.41 (0.37)	1.29 (0.48)	1.36 (0.41)	1.27 (0.35)	2.65 (0.61)				
	Post-Int	1.20 (0.30)	1.15 (0.25)	1.31 (0.49)	2.40 (0.42)	1.43 (0.53)	1.25 (0.50)	1.28 (0.42)	2.61 (0.75)				
	p-value	0.770	0.390	0.272	0.950	0.338	0.361	0.956	0.082	0.608	0.938	0.431	0.903
	Effect size	0.11	0.29	0.52	0.02	0.27	0.26	0.02	0.06	0.16	0.02	0.25	0.04

Table 3. *Cont.*

		2S-HESPERIDIN				PLACEBO				Between-Group Comparison			
		Baseline	Post- P_{MAX}	Post- REC	AUC	Baseline	Post- P_{MAX}	Post- REC	AUC	ΔBaseline	ΔPost-P_{MAX}	ΔPost-REC	ΔAUC
HO1	Pre-Int	0.539 (0.24)	0.473 (0.20)	0.513 (0.22)	0.998 (0.41)	0.626 (0.23)	0.551 (0.28)	0.589 (0.27)	1.158 (0.51)				
	Post-Int	0.650 (0.26)	0.581 (0.31)	0.585 (0.20)	1.198 (0.51)	0.740 (0.34)	0.678 (0.26)	0.705 (0.30)	1.400 (0.55)				
	P-value	0.135	0.172	0.281	0.081	0.059	0.066	**0.027**	**0.038**	0.972	0.849	0.548	0.789
	Effect size	0.44	0.53	0.32	0.47	0.47	0.43	0.41	0.46	0.01	0.06	0.19	0.09
TBARS	Pre-Int	2.55 (0.42)	2.80 (0.38)	2.41 (0.48)	5.27 (0.68)	2.64 (0.37)	2.92 (0.54)	2.75 (0.41)	5.62 (0.78)				
	Post-Int	2.79 (0.52)	2.88 (0.49)	2.54 (0.33)	5.54 (0.69)	2.73 (0.50)	2.80 (0.49)	2.57 (0.50)	5.45 (0.77)				
	P-value	**0.052**	0.519	0.393	0.246	0.527	0.493	0.216	0.551	0.399	0.406	0.134	0.227
	Effect size	0.55	0.22	0.25	0.38	0.22	0.21	0.42	0.21	0.27	0.27	0.48	0.39

Values are expressed as mean (SD). Abbreviations: AUC = area under curve; CAT = catalase; SOD = superoxide dismutase; GSH = reduced glutathione; GSSG = oxidized glutathione; % GSSG/GSH = oxidized/reduced glutathione ratio; HO1 = hemoxygenase 1; TBARS = thiobarbituric acid reactive substances and SD = standard deviation. Group comparison = *p*-value comparison of Δ pre-post intervention between groups at different times (baseline, post-P_{MAX} and post-REC) of rectangular test.

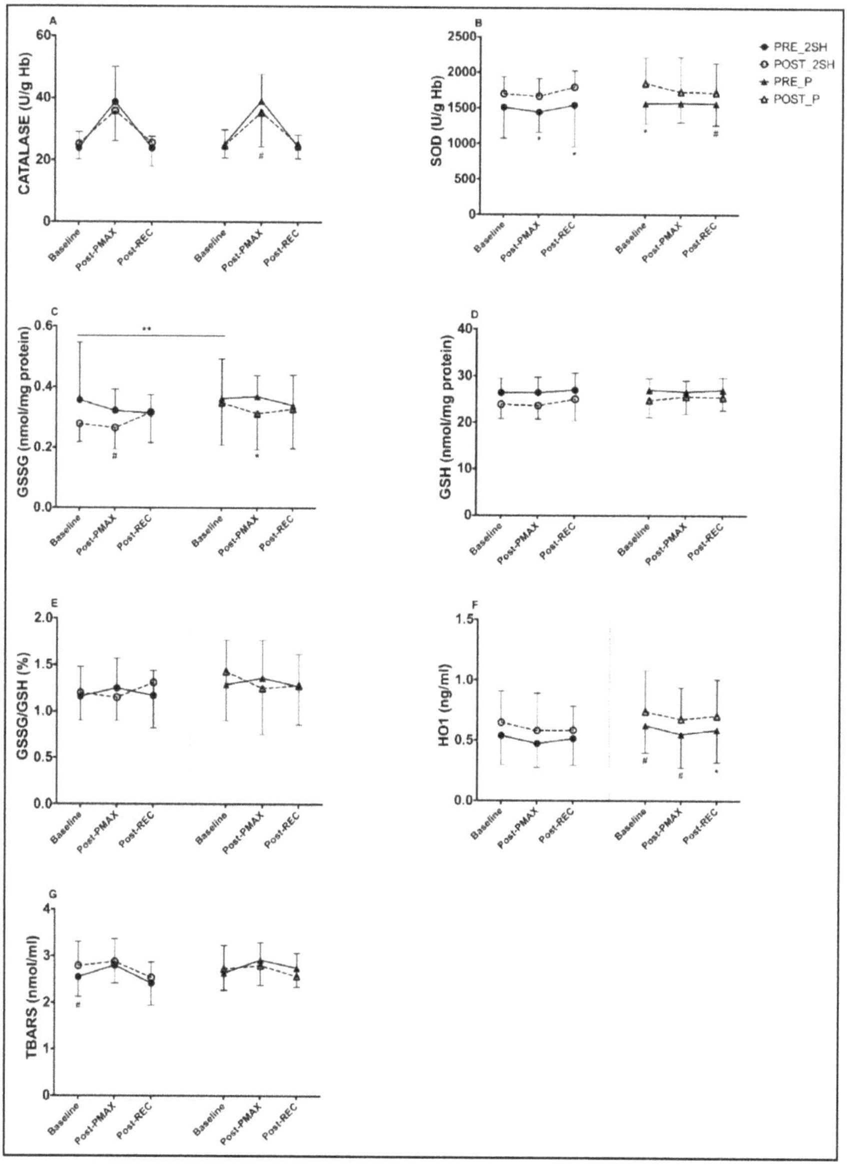

Figure 2. Differences between pre- and post-intervention intragroup in antioxidant and oxidant parameters at different points of the rectangular test (**A–G**). (**C**), a significant difference (p = 0.04) appears, comparing baseline of the second rectangular test between groups. * $p < 0.05$. # p = 0.05–0.06. ** $p < 0.05$ between post-intervention time points of rectangular test between groups (2S-hesperidin vs. placebo).

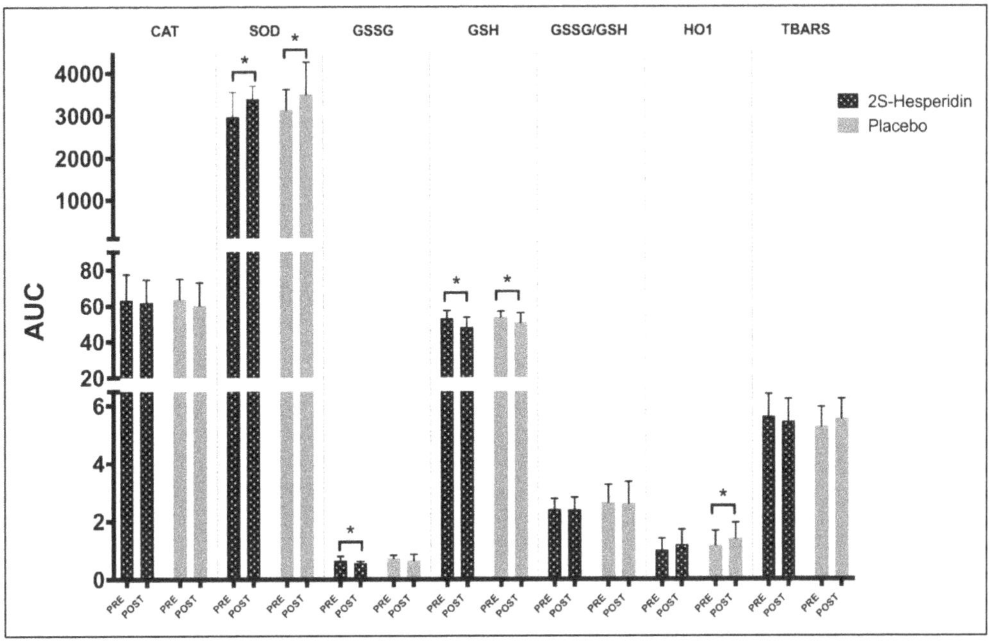

Figure 3. Intragroup differences between AUC (pre- and post-intervention) in antioxidant parameters. * $p < 0.05$. There were no significant differences between groups in AUC.

For GSH, a decrease was reported at baseline (−9.4%) and post-P_{MAX} (−10.7%) in 2S-hesperdin after the intervention. On the other hand, a significant decline was found at baseline (−8.3%) in the placebo (Figure 2). Intragroup AUC analysis of GSH showed a decrease in 2S-hesperidin (−9.5%) and the placebo (5.5%), without differences between groups (Figure 3).

After the intragroup analysis, HO1 significantly increased at post-REC (19.7%) in the placebo, while there was a non-significant increase with a moderate size effect in HO1 at post-P_{MAX} (22.8%) in 2S-hesperidin (Figure 2). Intragroup AUC analysis showed an increase in HO1 in the placebo (20.9%) without any differences between groups (Figure 3).

When we analyzed the intragroup TBARS data, we found a trend towards an increase with a moderate size effect at baseline (9.4%) in 2S-hesperidin, without significant differences between groups.

When results for each parameter at each time point during supplementation were compared between groups, no significant changes were found in any antioxidant-oxidant parameter.

3.2. *Inflammatory Biomarkers*

Table 4 shows the obtained values for inflammatory biomarkers IL6, TNFα, MCP1 and CRP during the rectangular test, pre- and post-intervention. Within-group changes for each parameter and time point (baseline, Post-P_{MAX} and Post- REC) during supplementation have been evaluated. The placebo group showed a significant decrease in IL6 at Post-P_{MAX} (−35.7%), without significant changes in 2S-hesperidin (Figure 4), in the intra-group comparison pre- and post-intervention. However, after the intragroup analysis, we reported a decline in the AUC of IL6 (−33.0%) in the placebo after the supplementation period, without differences between groups (Figure 5).

Table 4. Changes in inflammatory status markers before, during and after rectangular test comparing pre- and post-intervention.

		2S-Hesperidin				Placebo				Between-Group Comparison			
		Pre	Post- P_{MAX}	Post- REC	AUC	Pre	Post- P_{MAX}	Post- REC	AUC	ΔBaseline	ΔPost-P_{MAX}	ΔPost-REC	ΔAUC
IL6	Pre-Int	2.46 (3.78)	3.05 (4.92)	4.85 (6.55)	6.71 (9.94)	5.41 (9.26)	7.17 (12.02)	9.44 (10.62)	14.59 (20.67)				
	Post-Int	2.04 (2.32)	2.57 (2.94)	2.01 (2.61)	4.59 (5.01)	4.35 (8.65)	4.61 (7.78)	5.96 (11.98)	9.77 (17.69)				
	p-value	0.537	0.695	0.128	0.255	0.129	**0.045**	0.065	**0.021**	0.514	0.243	0.807	0.31
	Effect size	0.11	0.10	0.42	0.20	0.11	0.20	0.31	0.22	0.21	0.39	0.08	0.33
TNFα	Pre-Int	8.06 (1.51)	8.43 (2.07)	7.97 (2.08)	16.44 (3.17)	8.71 (1.62)	8.90 (1.83)	7.94 (1.35)	17.22 (2.76)				
	Post-Int	7.58 (1.70)	8.17 (1.98)	7.44 (2.00)	15.68 (3.40)	7.55 (1.89)	7.61 (2.46)	7.41 (1.94)	15.09 (3.6)				
	P-value	0.271	0.466	0.148	0.338	**0.015**	**0.038**	0.127	**0.021**	0.252	0.21	0.999	0.239
	Effect size	0.31	0.12	0.25	0.23	0.69	0.67	0.38	0.74	0.37	0.4	0	0.38
MCP1	Pre-Int	568 (177.05)	614 (240.06)	631 (184.77)	1214 (403.32)	714 (324.12)	766 (359.13)	656 (291.75)	1452 (617.47)				
	Post-Int	453 (137.74)	543 (174.45)	466 (131.81)	1002 (299.31)	550 (228.61)	657 (268.48)	571 (231.280	1218 (484.67)				
	P-value	**0.007**	0.096	**0.004**	**0.002**	**<0.001**	**0.026**	0.108	**0.021**	0.388	0.567	0.275	0.845
	Effect size	0.63	0.29	0.86	0.5	0.49	0.29	0.28	0.36	0.28	0.18	0.35	0.06
CRP	Pre-Int	1.200 (2.48)	1.268 (2.47)	1.063 (2.00)	2.399 (4.71)	1.487 (2.56)	1.579 (2.57)	2.049 (3.45)	3.347 (5.12)				
	Post-Int	0.724 −0.49	0.798 −0.54	0.669 (0.45)	1.494 (1.01)	1.517 (2.50)	1.539 (2.49)	1.404 (2.49)	3.000 (4.98)				
	p-value	0.521	0.523	0.634	0.398	0.981	0.943	0.402	0.832	0.606	0.658	0.819	0.774
	Effect size	0.18	0.18	0.19	0.18	0.01	0.01	0.18	0.07	0.16	0.14	0.07	0.09

Values are expressed as mean (SD). Abbreviations: AUC = area under curve; IL-6 = interleukin 6; TNFα = tumor necrosis factor α; MCP-1 = monocytes chemoattractant protein 1 and CRP = C-reactive protein. Group comparison = p-value comparison of Δ pre-post intervention between groups at different times (baseline, post-P_{MAX} and post-REC) of rectangular test.

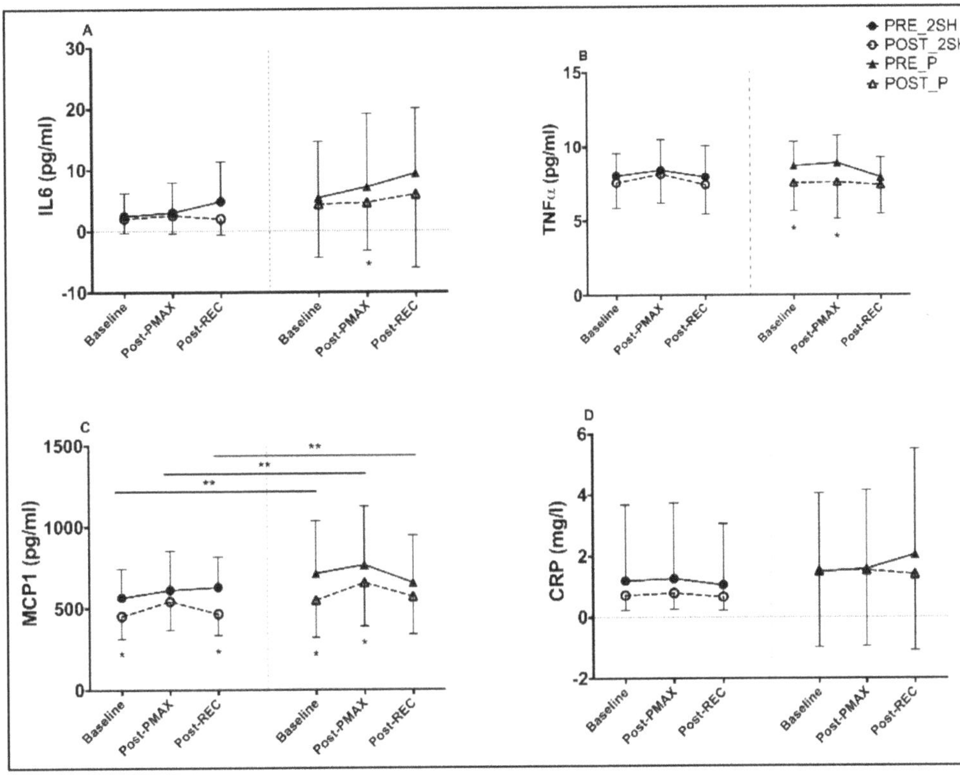

Figure 4. Differences between pre- and post-intervention intragroup and intergroup in inflammatory parameters at different points of the rectangular test (**A–D**). (**C**) a significant difference appears, comparing baseline ($p = 0.043$), Post-P_{MAX} ($p = 0.026$) and Post-REC ($p = 0.045$) of the second rectangular test between groups. * $p < 0.05$. ** $p < 0.05$ between post-intervention time points of rectangular test between groups (2S-hesperidin vs. placebo).

Regarding TNFα, a significant drop in levels at baseline (−13.3%) and post-P_{MAX} (−14.5%) was found in the placebo (Figure 4). In addition, intragroup AUC analysis of TNFα found a decrease (−12.4%) in the placebo without differences between groups (Figure 5).

Significant decreases were observed in MCP1 at baseline (−20.2%) and post-REC (−26.1%) in 2S-hesperidin. In the placebo, significant decreases were also observed in MCP1 at baseline (−23.0%) and post-P_{MAX} (−14.2%) in the post-intervention intragroup statistical analysis (Figure 4). When comparing MCP1 at different times (baseline, Post-P_{MAX} and Post-REC) of the rectangular test post-intervention between groups, 2S-hesperidin had lower MCP1 values (−17.6%, −17.4% and −18.4%, respectively) than the placebo (Figure 4). In addition, a similar decrease in the AUC was found in 2S-hesperidin (−17.5%) and the placebo (16.1%) in intragroup statistical analysis, but in the case of 2S-hesperidin, a moderate size effect was observed, without significant differences between groups (Figure 5).

No significant within-group changes were reported for CRP in any group (Figure 4).

When results for each parameter at each time point during supplementation were compared between groups, no significant changes were found in any inflammatory parameter.

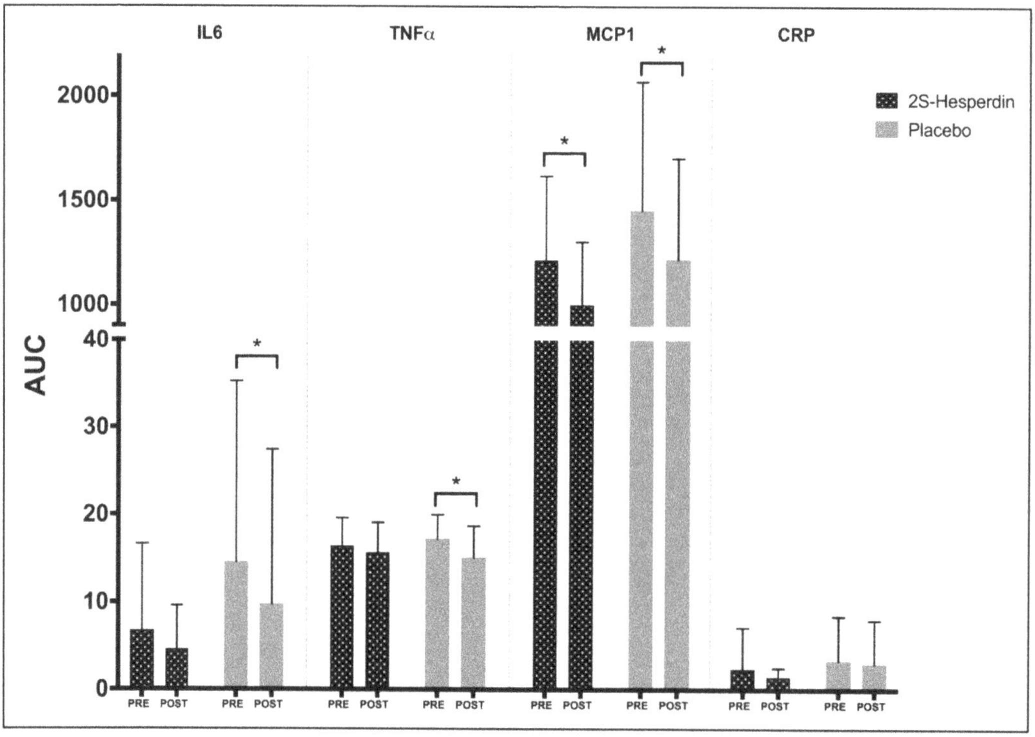

Figure 5. Intragroup differences between AUC (pre- and post-intervention) in inflammatory parameters. * $p < 0.05$. There were no significant differences between groups in AUC.

3.3. Hesperidin Metabolites Urine

Different hesperidin metabolites, mainly hesperetin glucuronides and sulfates, were analyzed in urine after Cardiose® intake. The main metabolite detected was hesperetin-3-glucuronide, representing 78.9 ± 5.0% ($n = 20$) of the total, while hesperetin-7-glucuronide and hesperetin-7-sulfate made up 6.9 ± 2.9% ($n = 20$) and 14.7 ± 4.1% ($n = 20$) of the excreted metabolites. Despite the similarities in the excreted metabolites' profiles, a large interindividual variability was observed in the number of excreted hesperidin metabolites ranging from 2.3 to 37.5 µmol.

4. Discussion

This study evaluated the effect of 8-week supplementation with 500 mg/d of 2S-hesperidin or placebo on antioxidant and inflammatory status in amateur cyclists during a rectangular cycle-ergometer test. To the best of our knowledge, this is the first study that examines the effect of chronic 2S-hesperidin intake on the antioxidant and inflammatory status of athletes at baseline, during and after exercise. In the rectangular test, oxidative status improved (↓GSSG AUC) after the 2S-hesperidin intervention, but not with the placebo. In addition, significant improvements in antioxidant capacity (↑SOD) after maximal exercise (Post-P_{MAX}) and inflammatory status after the acute recovery phase (↓MCP1) were found in the 2S-hesperidin group compared to the placebo (baseline and Post-REC).

4.1. Changes in Endogenous Antioxidant Markers

SOD activity is usually increased after training, as an exercise-mediated adaptation [28]. In contrast, a previous study showed no increases in SOD activity in untrained individuals after an 8-week moderate training program (35-min aerobic cycle, 3 times/week) [29]. Conversely, we observed a maintenance of SOD from baseline to Post-P_{MAX} and an increase to Post-REC in the post-intervention rectangular test, with significant increases in Post-P_{MAX} and Post-REC pre-post-intervention in 2S-hesperidin. However, an increase in SOD activity levels, evaluated as the AUC, was observed for both groups during intervention. In amateur cyclists, the acute intake of 2S-hesperidin (single dose; 500 mg) led to no significant decrease in SOD at baseline [9]. In animals, 2S-hesperidin supplementation (200 mg/kg for three days per week), along with a 5-week training program, led to no significant changes in SOD activity in rats after an exhaustive exercise test [10]. Additionally, 2S-hesperidin significantly increased SOD activity in heart tissues, which was attenuated by doxorubicin (induced cardiac toxicity) treatment [30]. The 2S-hesperidin antioxidant capacity enhancement may be explained by the antioxidant characteristics of this molecule, related to hydroxyl groups in its B-ring [31]. In addition, Parhiz et al. found that hesperidin had significant radical scavenging activity and prevented H_2O_2-induced oxidative damage on the cellular membranes of red blood cells, with radical scavenging activities comparable to ascorbic acid and trolox (a vitamin E derivative) [32]. Furthermore, 2S-hesperidin shows a neutralizing effect on non-enzymatic lipid peroxidation and superoxide, hydroxyl, peroxynitrite, and nitric oxide radicals [4,31], leading to a lower depletion of antioxidant enzymes and allowing the maintenance of high antioxidant levels, even after exercise-induced oxidative stress.

Another mechanism that has been reported in vitro and in animal models, but has yet to be confirmed in humans, is the improvement of the antioxidant status through a nutrigenetic effect. Hesperidin has shown increased regulation of respiratory nuclear factor 2 (NRF2) [33]. NRF2 is a basic leucine zipper transcription factor that binds and activates the antioxidant response element in the promoters of many antioxidant and detoxification genes encoding proteins, such as SOD, glutathione, thioredoxin and HO1, and thus it promotes the regulation of the intracellular redox environment [34]. Interestingly, flavonoids have been proposed as inductors of the expression of genes related to enzymes of the endogenous antioxidant system through the activation of the NRF2 transcription pathway [35]. The higher SOD activity at the end of maximal effort and after a short recovery period after the intervention indicates that the chronic intake of 2S-hesperidin improves antioxidant capacity at maximal effort and in the acute phase of recovery in amateur cyclists.

The oxidation of GSH to GSSG is a sensitive marker of oxidative stress [36]. In addition, a GSH decrease and GSSG and GSSG/GSH ratio increases have been observed in professional cyclists after competitions [37]. When comparing both groups at baseline after the intervention, the 2S-hesperidin group had lower GSSG values than the placebo, indicating lower levels of oxidative stress. This is in line with the decrease found in the AUC (GSSG) in 2S-hesperidin, indicating a decrease in oxidative stress when considering the whole rectangular test, which may be related to detraining adaptation. In fact, lower training volumes and intensities are associated with lower levels of GSH and GSSG in professional cyclists [38], which was also found Post-P_{MAX} in the placebo. Therefore, this decrease in both groups is due to a lower exposure to high levels of free radicals leading to a maladaptation in the glutathione antioxidant system. The main advantage of incorporating the AUC in this study is that it allows us to precisely define the duration and magnitude of the variable being evaluated, which cannot be done in a point-by-point comparison [39]. Despite the fact that there are no previous studies in humans evaluating the effects of chronic hesperidin intake on GSH and GSSG, instead, non-significant decreases in GSH, GSSH and the GSSG/GSH ratio were observed after a repeated sprint test in amateur cyclists after a single-dose of 2S-hesperidin (500 mg) [9]. In the same way, pathological animal models have shown the positive effect of hesperidin supplementation on these glutathione markers (↑GSH and ↓GSSG) [40,41].

It is known that regulation exists between GSH and GSSG by the enzymes glutathione reductase (GR) and GPx to maintain a balance between both molecules and avoid an increase in ROS [42]. The changes observed in GSSG in the experimental group could be due to the modulation of GPx and GR activity, which was not measured in this study. In addition, another factor that may influence the GSSG/GSH ratio is the levels of nicotinamide adenine dinucleotide phosphate (NADPH), which are used with an indispensable cofactor by the GR and GPx enzymes to synthesize the GSH and GSSG forms [43]. In this context, NADPH donates two electrons to reduce GSSG to GSH by GR; the recycled GSH can then be used to reduce H_2O_2 to water by GPxs [44]. In addition, increased glucose-6-phosphatase dehydrogenase (G6PD) (a major source of cytosolic NADPH) activity by genetic or pharmacological means has been seen to raise cellular stores of NAPDH and GSH, promote the detoxification of ROS, and increase cell viability in primary vascular endothelial and smooth muscle cells in vitro [45]. Increased G6PD activity is positively correlated with increased GR activity, where hesperidin was capable of restoring the activity of G6PD in rats [46]. In addition, Salvemini et al. [47] reported that a three-fold increase in G6PD activity resulted in a two-fold increase in GSH levels, as well as a very significant increase in resistance to oxidative stress. As we can see, the GSSG/GSH ratio can be modulated by different components involved in the endogenous antioxidant system, which makes it difficult to explain its changes. Therefore, the chronic intake of 2S-hesperidin could decrease GSSG levels (evidenced by ↓ AUC), indicating a better antioxidant state in the rectangular test, but specifically immediately after exercise. This would facilitate faster post-training recovery or competition for cyclists.

In relation to HO1, in the placebo there was an increase Post-REC with an increasing trend in baseline and Post-P_{MAX} post-intervention; however, after 2S-hesperidin supplementation, no significant change was seen, but there was a moderate effect in Post-P_{MAX}. The high variability in the HO1 data in 2S-hesperidin may have been the consequence of no significant changes being observed. What is clear is that amateur-level cycling for 8 weeks improves HO1 levels.

Although there is no clear pattern of improvement in antioxidant markers in 2S-hesperidin, there is an improvement in certain components (↑SOD, ↓GSSG) of the endogenous antioxidant system measured in this study and at key times during recovery. However, further studies are needed to provide clarity on this issue.

4.2. Changes in Inflammatory Markers

The production of ROS at the mitochondria of the working muscle stimulates the production of myokines or pro-inflammatory cytokines [48]. IL6 (inflammatory cytokine) plasma levels can increase up to 100-fold after exercise, and circulating muscle-derived IL6 levels are closely related to the duration and intensity of exercise [49]. To our knowledge, no studies have evaluated the effects of 2S-hesperidin intake on inflammatory markers in humans. In this study, IL6 levels increased during the first and second rectangular test from baseline to Post-P_{MAX} in both groups, but there were different trends from Post-P_{MAX} to Post-REC in the second rectangular test (↓2S-hesperidin and ↑placebo). A significant decrease in IL6 during the recovery stage was observed in the placebo, post-intervention. Other flavonoids, such as cocoa-derived flavanols, have also failed to inhibit the increase in IL6 after intense exercise (75% of peak power output for 30 min) in cyclists [50]. We believe that the high variability in the IL6 data was a factor that did not allow us to find significant intra- and intergroup differences. In addition, IL6 values in the placebo were quantitatively higher than those of 2S-hesperidin, which may favor a significant decrease in IL6 after the reduction in the training load performed by cyclists from post-season to pre-season, as was the period in which the study was conducted (from the end of September to the end of December). As IL6 is known to stimulate the expression of TNFα [51], a decrease in IL6 levels in the placebo would lead to a decrease in TNFα levels (baseline and Post-P_{MAX}). Since there is less training load, there is less induction of oxidative stress [38] and, consequently, less stimulation of the inflammatory system.

However, numerous in vitro studies (inflammatory models) have shown the ability of hesperidin to lower IL6 levels and TNFα [52–55]. A recent study in trained animals showed that hesperidin intake (200 mg/kg for three days per week) during 5 weeks prevented an increase in IL6 levels in peritoneal macrophages after an exhausting exercise [56]. Interestingly, in this study, a significant increase in IL6 after an exhausting exercise, from pre-training to post-intervention, was observed in the placebo group. Hesperidin intake has also led to a decrease in IL6 in a rheumatoid arthritis rat model [57]. In rats, the intake of alcohol to induce a gastric ulcer increased the expression of cyclooxygenase-2 mRNA and decreased GPx, SOD, and CAT, but the intake of hesperidin reversed these changes, improving the antioxidant and inflammatory status [58]. In addition, in a model of Alzheimer's disease in mice, treatment with hesperidin (40 mg/kg, 90 days intragastric) increased HO1 and decreased levels of TNFα, CRP, NF-κB and MCP1, suppressing oxidative stress and inflammation [59].

A hypothesis has been generated at the nutrigenomic level of how the intake of hesperidin can improve the inflammatory state, related to the activation of the Akt/NRF2 axis and the inhibition of NF-κB [59], with the latter being a transcription factor well known for its role in the innate immune response and a transcriptional activator of inflammatory mediators such as cytokines [60]. On the other hand, NRF2 is not only important for redox signaling, but also for the attenuation of the inflammatory mediator synthesis [59]. In this sense, the impairment of NRF2 signaling by ultraviolet B (UVB) was reversed by the topical application of hesperidin methyl chalcone, which inhibited the production of the cytokines TNFα, IL-1β, IL6, and IL-10 that had been induced by UVB irradiation in hairless mice [61]. This suggests that there is a connection between the antioxidant and inflammatory status and their signaling pathways. In our case, the group ingesting 2S-hesperidin did not experience a significant decrease in TNFα.

MCP1 is another inflammatory cytokine that increases after exercise in plasma [62]. In our study, lower MPC-1 levels during the whole exercise (AUC) were observed after supplementation in both groups. This decrease was statically significant at baseline and during the recovery phase for the 2S-hesperidin supplemented group. In addition, when comparing between groups at different post-intervention test times, the 2S-hesperidin group had lower levels compared to the placebo. In previous studies with an acute lung damage model, both in vitro and in vivo, hesperidin has shown immunomodulatory effects, down-regulating the expression of MCP1 as well as other pro-inflammatory cytokines, such as IL6 and TNFα [52,63]. Precisely, treatment with hesperetin-7-O-glucuronide (5 mg kg^{-1}) has been observed to decrease the MCP1 mRNA expression in rat aortic endothelial cells [63]. On the other hand, the oral administration of 100 or 200 mg/kg of hesperidin three times a week for four weeks in rats produced a decrease in the pro-inflammatory cytokines interferon-gamma-γ and MCP1 in the lymphocyte of the mesenteric lymph node [10]. Additionally, polyphenols and hesperidin can modulate gut microbial composition or functionality, which modulate the release of microbial-derived metabolites [64]. In addition, hesperidin has the ability to inhibit the growth of harmful bacteria, such as Escherichia coli, Pseudomonas aeruginosa, Prevotella spp., Porphyromonas gingivalis and Fusobacterium nucleatum, among others [3]. In particular, hesperidin can increase the abundance of Faecalibacterium prausnitzii, which inhibits NF-κB activation and consequently attenuates the inflammatory response [65]. The inhibitory capacity of hesperidin in some bacteria may modify the composition of the intestinal microbiota acting as an immunomodulator and anti-inflammatory (↓IL-1β, TNFα, and IL6), with a direct relationship between the two effects [3]. In contrast, the effects of quercetin (flavonoid) intake (1g/day) for 3 weeks in trained cyclists were evaluated by Nieman et al. [62]. In this study, no significant changes in MCP1 plasma levels were observed after a 3-week supplementation and a 3-day period in which subjects cycled for 3 h/day at ~57% maximal work rate. Muscle biopsies showed a within-group significant post-exercise increase in muscle cytokine mRNA expression for IL6 and TNFα, but without differences between the quercetin and placebo groups [62]. No anti-inflammatory effect was observed after the intake of quercetin. On the other hand, the placebo group showed a decrease in MCP1 at baseline and Post-P_{MAX} at post-intervention, possibly related to the decrease in TNF-α, as a positive correlation between MCP-1 and TNF-α concentrations after short-term exercise training has been previously demonstrated [66], indicating a relationship between these two cytokines.

Although there is no clear pattern of improvement in inflammatory markers in 2S-hesperidin, there is an enhancement in MCP1 (baseline and Post-REC) compared to the placebo in the second rectangular test at all points and at key times during recovery. However, further studies are needed to bring clarity to this question.

These 2S-hesperidin properties such as antioxidant and anti-inflammatory properties may be related: a decrease in oxidative stress during the exercise maximum intensity could modulate the inflammatory state in the acute phase of recovery. As has been shown in the studies presented in this publication, there is a close relationship between antioxidant and inflammatory status and their signaling pathways. Redox balance can be altered during periods of high intensity physical exercise and low rest periods, leading to a chronic oxidative stress state [15]. Moreover, high oxidative stress levels can inhibit exercise physiological adaptations, reducing performance and leading to overtraining [15]. Therefore, an optimal redox homeostasis is essential for a proper muscle physiological function (i.e., antioxidant status, biochemistry, signaling, bioenergetics and muscle contraction).

The effects of antioxidant supplementation on performance are a controversial topic, which still needs additional research. On one hand, it has been pointed out that the use of antioxidant substances may help to maintain optimal ROS levels in the muscle, avoiding possible decreases in performance [15]. On the other hand, it has been hypothesized that chronic antioxidant intake can hinder training adaptations, negatively affecting performance [17]. Different studies show that antioxidant intake does not prevent the exercise-induced activation of redox-sensitive signaling pathways [67]. A recent publication summarized the performance measurements that were carried out in this same intervention trial, along with the antioxidant and inflammatory marker results reported in this paper [7]. In this trial, amateur cyclists' supplementation (8 weeks) with 2S-hesperidin (500 mg/day) led to an increase in power production at estimated functional threshold power (2.3% = 6.40 W; p = 0.049) and maximum power (1.9% = 7.40 W; p = 0.049) during an incremental test after the intervention [7]. Thus, 2S-hesperidin does not appear to interfere with training-induced adaptations, improving performance while avoiding oxidative stress and inflammation.

The study described in this paper has some limitations. One limitation was the short recovery time after the rectangular test (20 min after exercise), in which changes in antioxidant and inflammatory markers were evaluated. Measurements 24 and 48 h after exercise would have provided valuable additional information; however, funding constraints made it impossible. Additionally, a larger sample would have given more statistical power to the reported results due to the high individual variability in some markers. Given the few studies carried out in this field, future research could shed light on the effectiveness of 2S-hesperidin as an ergogenic aid with antioxidant and inflammatory effects.

Differences with current results may be related to the different stage of the season in which studies were done in the sample used, and the different aerobic and anaerobic demand profiles of the used tests. In the same way, one of the factors influencing the variability of 2S-hesperidin effects in different studies may be its pharmacokinetics, and the resulting exposure of the body to hesperidin metabolites. It has been described that the concentration of 2S-hesperidin metabolites in plasma reaches its maximum peak 5–7 h after intake, being almost completely eliminated after 24 h. In the urine, the maximum peak of metabolites is usually found at 24 h of 2S-hesperidin intake, and its total excretion occurs after 48 h [68].

5. Conclusions

Supplementation with 2S-hesperidin (500mg/d) for 8 weeks improves the post-rectangular test antioxidant (↑SOD and ↓AUC-GSSG) and inflammatory status during the acute phase of post-exercise recovery (↓MCP1). This modulation in antioxidant and inflammatory markers can help cyclists to improve their recovery after intense efforts or long exercise sessions that, due to their characteristics, led to an increase in inflammation and oxidative stress. Unlike other polyphenols, 2S-hesperidin supplementation does not appear to interrupt adaptations produced by training in amateur cyclists, enhancing their performance [7].

Supplementary Materials: The following are available online at https://www.mdpi.com/2076-3921/10/3/432/s1, Supplementary File 1. Methodology oxidative, antioxidant and inflammatory markers.

Author Contributions: Conceptualization, F.J.M.-N., C.M.-P. and P.E.A.; methodology, F.J.M.-N., C.M.-P. and P.E.A.; formal analysis, F.J.M.-N., C.M.-P. and J.C.-V.; investigation, F.J.M.-N., C.M.-P. and J.C.-V.; resources, F.J.M.-N., C.M.-P. and J.C.-V.; data curation, F.J.M.-N., C.M.-P. and J.C.-V.; writing—original draft preparation, F.J.M.-N.; writing—review and editing, F.J.M.-N., C.M.-P. and J.C.-V.; visualization, C.M.-P.; supervision, C.M.-P. and P.E.A.; project administration, C.M.-P. and P.E.A.; funding acquisition, P.E.A. All authors have read and agreed to the published version of the manuscript.

Funding: The authors declare that this study has been financed by HTBA (Murcia, Spain), who kindly provided the product Cardiose®, but they did not participate in the experimental design, data collection, data analysis, interpretation of the data, writing of the manuscript, or in the decision to publish the results.

Institutional Review Board Statement: The study was conducted according to the guidelines of the Declaration of Helsinki, and approved by the Ethics Committee of the Catholic University of Murcia (CE091802).

Informed Consent Statement: Informed consent was obtained from all subjects involved in the study.

Data Availability Statement: All data is contained within the article.

Acknowledgments: This study was supported by the Research Center for High Performance Sport of the Catholic University of Murcia and HTBA (Murcia, Spain). We would like to acknowledge Linda H. Chung for her help in this project. We also thank Iris Samarra, Antoni del Pino and Nuria Canela, from the Metabolomics facility of the Centre for Omic Sciences (COS) Joint Unit of the Universitat Rovira i Virgili-Eurecat, for their contribution to the urine analysis. The results of the current study do not constitute endorsement of the product by the authors or the journal.

Conflicts of Interest: The authors declare no conflict of interest.

References

1. Xiao, J. Dietary flavonoid aglycones and their glycosides: Which show better biological significance? *Crit. Rev. Food Sci. Nutr.* **2017**, *57*, 1874–1905. [CrossRef] [PubMed]
2. Asztemborska, M.; Zukowski, J. Determination of diastereomerization barrier of some flavanones by high-performance liquid chromatography methods. *J. Chromatogr. A* **2006**, *1134*, 95–100. [CrossRef]
3. Mas-Capdevila, A.; Teichenne, J.; Domenech-Coca, C.; Caimari, A.; Del Bas, J.M.; Escoté, X.; Crescenti, A. Effect of Hesperidin on Cardiovascular Disease Risk Factors: The Role of Intestinal Microbiota on Hesperidin Bioavailability. *Nutrients* **2020**, *12*, 1488. [CrossRef] [PubMed]
4. Parhiz, H.; Roohbakhsh, A.; Soltani, F.; Rezaee, R.; Iranshahi, M. Antioxidant and anti-inflammatory properties of the citrus flavonoids hesperidin and hesperetin: An updated review of their molecular mechanisms and experimental models. *Phytother. Res.* **2015**, *29*, 323–331. [CrossRef]
5. Barreca, D.; Gattuso, G.; Bellocco, E.; Calderaro, A.; Trombetta, D.; Smeriglio, A.; Lagana, G.; Daglia, M.; Meneghini, S.; Nabavi, S.M. Flavanones: Citrus phytochemical with health-promoting properties. *Biofactors* **2017**, *43*, 495–506. [CrossRef]
6. Milenkovic, D.; Deval, C.; Dubray, C.; Mazur, A.; Morand, C. Hesperidin displays relevant role in the nutrigenomic effect of orange juice on blood leukocytes in human volunteers: A randomized controlled cross-over study. *PLoS ONE* **2011**, *6*, e26669. [CrossRef] [PubMed]
7. Martínez-Noguera, F.J.; Marín-Pagán, C.; Carlos-Vivas, J.; Alcaraz, P.E. Effects of 8 Weeks of 2S-Hesperidin Supplementation on Performance in Amateur Cyclists. *Nutrients* **2020**, *12*, 3911. [CrossRef]
8. Kawabata, K.; Yoshioka, Y.; Terao, J. Role of intestinal microbiota in the bioavailability and physiological functions of dietary polyphenols. *Molecules* **2019**, *24*, 370. [CrossRef] [PubMed]

9. Martinez-Noguera, F.J.; Marin-Pagan, C.; Carlos-Vivas, J.; Rubio-Arias, J.A.; Alcaraz, P.E. Acute Effects of Hesperidin in Oxidant/Antioxidant State Markers and Performance in Amateur Cyclists. *Nutrients* **2019**, *11*, 1898. [CrossRef] [PubMed]
10. Estruel-Amades, S.; Massot-Cladera, M.; Garcia-Cerdà, P.; Pérez-Cano, F.J.; Franch, À.; Castell, M.; Camps-Bossacoma, M. Protective effect of hesperidin on the oxidative stress induced by an exhausting exercise in intensively trained rats. *Nutrients* **2019**, *11*, 783. [CrossRef] [PubMed]
11. Biesemann, N.; Ried, J.S.; Ding-Pfennigdorff, D.; Dietrich, A.; Rudolph, C.; Hahn, S.; Hennerici, W.; Asbrand, C.; Leeuw, T.; Strübing, C. High throughput screening of mitochondrial bioenergetics in human differentiated myotubes identifies novel enhancers of muscle performance in aged mice. *Sci. Rep.* **2018**, *8*, 9408. [CrossRef] [PubMed]
12. Sadowska-Krępa, E.; Kłapcińska, B.; Podgórski, T.; Szade, B.; Tyl, K.; Hadzik, A. Effects of supplementation with acai (Euterpe oleracea Mart.) berry-based juice blend on the blood antioxidant defence capacity and lipid profile in junior hurdlers. A pilot study. *Biol. Sport* **2015**, *32*, 161–168. [CrossRef] [PubMed]
13. Martínez-Noguera, F.J.; Alcaraz, P.E.; Ortolano-Ríos, R.; Dufour, S.P.; Marín-Pagán, C. Differences between Professional and Amateur Cyclists in Endogenous Antioxidant System Profile. *Antioxidants* **2021**, *10*, 282. [CrossRef] [PubMed]
14. Gomez-Cabrera, M.-C.; Borrás, C.; Pallardó, F.V.; Sastre, J.; Ji, L.L.; Viña, J. Decreasing xanthine oxidase-mediated oxidative stress prevents useful cellular adaptations to exercise in rats. *J. Physiol.* **2005**, *567*, 113–120. [CrossRef] [PubMed]
15. Bentley, D.J.; Ackerman, J.; Clifford, T.; Slattery, K.S. Acute and Chronic Effects of Antioxidant Supplementation on Exercise Performance. In *Antioxidants in Sport Nutrition*; Lamprecht, M., Ed.; CRC Press/Taylor & Francis(c) 2015 by Taylor & Francis Group, LLC: Boca Raton, FL, USA, 2015.
16. Merry, T.L.; Ristow, M. Do antioxidant supplements interfere with skeletal muscle adaptation to exercise training? *J. Physiol.* **2016**, *594*, 5135–5147. [CrossRef] [PubMed]
17. Watson, T.A.; MacDonald-Wicks, L.K.; Garg, M.L. Oxidative stress and antioxidants in athletes undertaking regular exercise training. *Int. J. Sport Nutr. Exerc. Metab.* **2005**, *15*, 131–146. [CrossRef]
18. Palazzetti, S.; Richard, M.J.; Favier, A.; Margaritis, I. Overloaded training increases exercise-induced oxidative stress and damage. *Can. J. Appl. Physiol.* **2003**, *28*, 588–604. [CrossRef] [PubMed]
19. World Medical Association. WMA Declaration of Helsinki-Ethical principles for medical research involving human subjects. *JAMA* **2013**, *310*, 2191–2194. [CrossRef]
20. Wasserman, K.; Beaver, W.L.; Whipp, B.J. Gas exchange theory and the lactic acidosis (anaerobic) threshold. *Circulation* **1990**, *81*, 14–30.
21. Schisterman, E.F.; Faraggi, D.; Browne, R.; Freudenheim, J.; Dorn, J.; Muti, P.; Armstrong, D.; Reiser, B.; Trevisan, M. TBARS and cardiovascular disease in a population-based sample. *J. Cardiovasc. Risk* **2001**, *8*, 219–225. [CrossRef] [PubMed]
22. Lykkesfeldt, J. Determination of malondialdehyde as dithiobarbituric acid adduct in biological samples by HPLC with fluorescence detection: Comparison with ultraviolet-visible spectrophotometry. *Clin. Chem.* **2001**, *47*, 1725–1727. [CrossRef] [PubMed]
23. Aebi, H. Catalase in vitro. *Methods Enzymol.* **1984**, *105*, 121–126. [PubMed]
24. Randox Laboratories Ltd. *Radicales Libres*; Randox Laboratories Ltd.: Crumlin, UK, 1996; pp. 1–166.
25. Akerboom, T.P.; Sies, H. Assay of glutathione, glutathione disulfide, and glutathione mixed disulfides in biological samples. *Methods Enzymol.* **1981**, *77*, 373–382. [CrossRef]
26. Asensi, M.; Sastre, J.; Pallardo, F.V.; Estrela, J.M.; Vina, J. Determination of oxidized glutathione in blood: High-performance liquid chromatography. *Methods Enzymol.* **1994**, *234*, 367–371. [CrossRef] [PubMed]
27. Cohen, J. *Statistical Power Analysis for the Social Sciences*; Laurence Erlbaum Associates: Hillsdale, MI, USA, 1988.
28. Urso, M.L.; Clarkson, P.M. Oxidative stress, exercise, and antioxidant supplementation. *Toxicology* **2003**, *189*, 41–54. [CrossRef]
29. Tiidus, P.M.; Pushkarenko, J.; Houston, M.E. Lack of antioxidant adaptation to short-term aerobic training in human muscle. *Am. J. Physiol* **1996**, *271*, R832–R836. [CrossRef] [PubMed]
30. Abdel-Raheem, I.T.; Abdel-Ghany, A.A. Hesperidin alleviates doxorubicin-induced cardiotoxicity in rats. *J. Egypt Natl. Cancer Inst.* **2009**, *21*, 175–184.
31. Suarez, J.; Herrera, M.D.; Marhuenda, E. In vitro scavenger and antioxidant properties of hesperidin and neohesperidin dihydrochalcone. *Phytomedicine* **1998**, *5*, 469–473. [CrossRef]
32. Kalpana, K.B.; Srinivasan, M.; Menon, V.P. Evaluation of antioxidant activity of hesperidin and its protective effect on H2O2 induced oxidative damage on pBR322 DNA and RBC cellular membrane. *Mol. Cell. Biochem.* **2009**, *323*, 21–29. [CrossRef]
33. Elavarasan, J.; Velusamy, P.; Ganesan, T.; Ramakrishnan, S.K.; Rajasekaran, D.; Periandavan, K. Hesperidin-mediated expression of Nrf2 and upregulation of antioxidant status in senescent rat heart. *J. Pharm. Pharmacol.* **2012**, *64*, 1472–1482. [CrossRef]
34. Friling, R.S.; Bensimon, A.; Tichauer, Y.; Daniel, V. Xenobiotic-inducible expression of murine glutathione S-transferase Ya subunit gene is controlled by an electrophile-responsive element. *Proc. Natl. Acad. Sci. USA* **1990**, *87*, 6258–6262. [CrossRef]
35. Arredondo, F.; Echeverry, C.; Abin-Carriquiry, J.A.; Blasina, F.; Antúnez, K.; Jones, D.P.; Go, Y.M.; Liang, Y.L.; Dajas, F. After cellular internalization, quercetin causes Nrf2 nuclear translocation, increases glutathione levels, and prevents neuronal death against an oxidative insult. *Free Radic. Biol. Med.* **2010**, *49*, 738–747. [CrossRef]
36. Sen, C.K. Glutathione homeostasis in response to exercise training and nutritional supplements. In *Stress Adaptation, Prophylaxis and Treatment*; Springer: Berlin/Heidelberg, Germany, 1999; pp. 31–42.
37. Cordova, A.; Sureda, A.; Albina, M.L.; Linares, V.; Belles, M.; Sanchez, D.J. Oxidative stress markers after a race in professional cyclists. *Int J. Sport Nutr. Exerc. Metab.* **2015**, *25*, 171–178. [CrossRef] [PubMed]

38. Leonardo-Mendonça, R.C.; Concepción-Huertas, M.; Guerra-Hernández, E.; Zabala, M.; Escames, G.; Acuña-Castroviejo, D. Redox status and antioxidant response in professional cyclists during training. *Eur. J. Sport Sci.* **2014**, *14*, 830–838. [CrossRef] [PubMed]
39. Talluri, R.; Shete, S. Using the weighted area under the net benefit curve for decision curve analysis. *BMC Med. Inform. Decis. Mak.* **2016**, *16*, 94. [CrossRef]
40. Zhang, G.; Zhu, J.; Zhou, Y.; Wei, Y.; Xi, L.; Qin, H.; Rao, Z.; Han, M.; Ma, Y.; Wu, X.A. Hesperidin Alleviates Oxidative Stress and Upregulates the Multidrug Resistance Protein 2 in Isoniazid and Rifampicin-Induced Liver Injury in Rats. *J. Biochem. Mol. Toxicol.* **2016**, *30*, 342–349. [CrossRef]
41. Selvaraj, P.; Pugalendi, K.V. Hesperidin, a flavanone glycoside, on lipid peroxidation and antioxidant status in experimental myocardial ischemic rats. *Redox Rep.* **2010**, *15*, 217–223. [CrossRef] [PubMed]
42. Owen, J.B.; Butterfield, D.A. Measurement of oxidized/reduced glutathione ratio. *Methods Mol. Biol.* **2010**, *648*, 269–277. [CrossRef] [PubMed]
43. Bradshaw, P.C. Cytoplasmic and Mitochondrial NADPH-Coupled Redox Systems in the Regulation of Aging. *Nutrients* **2019**, *11*, 504. [CrossRef] [PubMed]
44. Buettner, G.R.; Wagner, B.A.; Rodgers, V.G. Quantitative redox biology: An approach to understand the role of reactive species in defining the cellular redox environment. *Cell Biochem. Biophys.* **2013**, *67*, 477–483. [CrossRef]
45. Leopold, J.A.; Zhang, Y.Y.; Scribner, A.W.; Stanton, R.C.; Loscalzo, J. Glucose-6-phosphate dehydrogenase overexpression decreases endothelial cell oxidant stress and increases bioavailable nitric oxide. *Arter. Thromb. Vasc. Biol.* **2003**, *23*, 411–417. [CrossRef]
46. Préville, X.; Salvemini, F.; Giraud, S.; Chaufour, S.; Paul, C.; Stepien, G.; Ursini, M.V.; Arrigo, A.P. Mammalian small stress proteins protect against oxidative stress through their ability to increase glucose-6-phosphate dehydrogenase activity and by maintaining optimal cellular detoxifying machinery. *Exp. Cell Res.* **1999**, *247*, 61–78. [CrossRef]
47. Salvemini, F.; Franzé, A.; Iervolino, A.; Filosa, S.; Salzano, S.; Ursini, M.V. Enhanced glutathione levels and oxidoresistance mediated by increased glucose-6-phosphate dehydrogenase expression. *J. Biol. Chem.* **1999**, *274*, 2750–2757. [CrossRef] [PubMed]
48. Scheele, C.; Nielsen, S.; Pedersen, B.K. ROS and myokines promote muscle adaptation to exercise. *Trends Endocrinol. Metab.* **2009**, *20*, 95–99. [CrossRef] [PubMed]
49. Ost, M.; Coleman, V.; Kasch, J.; Klaus, S. Regulation of myokine expression: Role of exercise and cellular stress. *Free Radic. Biol. Med.* **2016**, *98*, 78–89. [CrossRef] [PubMed]
50. Decroix, L.; Tonoli, C.; Soares, D.D.; Descat, A.; Drittij-Reijnders, M.J.; Weseler, A.R.; Bast, A.; Stahl, W.; Heyman, E.; Meeusen, R. Acute cocoa Flavanols intake has minimal effects on exercise-induced oxidative stress and nitric oxide production in healthy cyclists: A randomized controlled trial. *J. Int. Soc. Sports Nutr.* **2017**, *14*, 28. [CrossRef]
51. Petersen, A.M.; Pedersen, B.K. The anti-inflammatory effect of exercise. *J. Appl. Physiol.* **2005**, *98*, 1154–1162. [CrossRef]
52. Yeh, C.-C.; Kao, S.-J.; Lin, C.-C.; Wang, S.-D.; Liu, C.-J.; Kao, S.-T. The immunomodulation of endotoxin-induced acute lung injury by hesperidin in vivo and in vitro. *Life Sci.* **2007**, *80*, 1821–1831. [CrossRef]
53. Choi, I.Y.; Kim, S.J.; Jeong, H.J.; Park, S.H.; Song, Y.S.; Lee, J.H.; Kang, T.H.; Park, J.H.; Hwang, G.S.; Lee, E.J.; et al. Hesperidin inhibits expression of hypoxia inducible factor-1 alpha and inflammatory cytokine production from mast cells. *Mol. Cell Biochem.* **2007**, *305*, 153–161. [CrossRef]
54. Nizamutdinova, I.T.; Jeong, J.J.; Xu, G.H.; Lee, S.-H.; Kang, S.S.; Kim, Y.S.; Chang, K.C.; Kim, H.J. Hesperidin, hesperidin methyl chalone and phellopterin from Poncirus trifoliata (Rutaceae) differentially regulate the expression of adhesion molecules in tumor necrosis factor-alpha-stimulated human umbilical vein endothelial cells. *Int. Immunopharmacol.* **2008**, *8*, 670–678. [CrossRef]
55. Moon, P.-D.; Kim, H.-M. Antiinflammatory Effects of Traditional Korean Medicine, JinPi-tang and Its Active Ingredient, Hesperidin in HaCaT cells. *Phytother. Res.* **2012**, *26*, 657–662. [CrossRef]
56. Ruiz-Iglesias, P.; Estruel-Amades, S.; Camps-Bossacoma, M.; Massot-Cladera, M.; Franch, À.; Pérez-Cano, F.J.; Castell, M. Influence of Hesperidin on Systemic Immunity of Rats Following an Intensive Training and Exhausting Exercise. *Nutrients* **2020**, *12*, 1219. [CrossRef]
57. Li, R.; Li, J.; Cai, L.; Hu, C.M.; Zhang, L. Suppression of adjuvant arthritis by hesperidin in rats and its mechanisms. *J. Pharm. Pharmacol.* **2008**, *60*, 221–228. [CrossRef] [PubMed]
58. Selmi, S.; Rtibi, K.; Grami, D.; Sebai, H.; Marzouki, L. Protective effects of orange (Citrus sinensis L.) peel aqueous extract and hesperidin on oxidative stress and peptic ulcer induced by alcohol in rat. *Lipids Health Dis.* **2017**, *16*, 152. [CrossRef]
59. Hong, Y.; An, Z. Hesperidin attenuates learning and memory deficits in APP/PS1 mice through activation of Akt/Nrf2 signaling and inhibition of RAGE/NF-κB signaling. *Arch. Pharm Res.* **2018**, *41*, 655–663. [CrossRef] [PubMed]
60. Nennig, S.E.; Schank, J.R. The Role of NFkB in Drug Addiction: Beyond Inflammation. *Alcohol Alcohol.* **2017**, *52*, 172–179. [CrossRef]
61. De Oliveira, C.A.; Peres, D.D.A.; Rugno, C.M.; Kojima, M.; Pinto, C.A.S.d.O.; Consiglieri, V.O.; Kaneko, T.M.; Rosado, C.; Mota, J.; Velasco, M.V.R.; et al. Functional photostability and cutaneous compatibility of bioactive UVA sun care products. *J. Photochem. Photobiol. B Biol.* **2015**, *148*, 154–159. [CrossRef] [PubMed]
62. Nieman, D.C.; Henson, D.A.; Davis, J.M.; Angela Murphy, E.; Jenkins, D.P.; Gross, S.J.; Carmichael, M.D.; Quindry, J.C.; Dumke, C.L.; Utter, A.C.; et al. Quercetin's influence on exercise-induced changes in plasma cytokines and muscle and leukocyte cytokine mRNA. *J. Appl. Physiol.* **2007**, *103*, 1728–1735. [CrossRef] [PubMed]

63. Yamamoto, M.; Jokura, H.; Hashizume, K.; Ominami, H.; Shibuya, Y.; Suzuki, A.; Hase, T.; Shimotoyodome, A. Hesperidin metabolite hesperetin-7-O-glucuronide, but not hesperetin-3′-O-glucuronide, exerts hypotensive, vasodilatory, and anti-inflammatory activities. *Food Funct.* **2013**, *4*, 1346–1351. [CrossRef] [PubMed]
64. Cardona, F.; Andrés-Lacueva, C.; Tulipani, S.; Tinahones, F.J.; Queipo-Ortuño, M.I. Benefits of polyphenols on gut microbiota and implications in human health. *J. Nutr. Biochem.* **2013**, *24*, 1415–1422. [CrossRef]
65. Corrêa, T.A.F.; Rogero, M.M.; Hassimotto, N.M.A.; Lajolo, F.M. The Two-Way Polyphenols-Microbiota Interactions and Their Effects on Obesity and Related Metabolic Diseases. *Front. Nutr.* **2019**, *6*, 188. [CrossRef] [PubMed]
66. Middelbeek, R.J.W.; Motiani, P.; Brandt, N.; Nigro, P.; Zheng, J.; Virtanen, K.A.; Kalliokoski, K.K.; Hannukainen, J.C.; Goodyear, L.J. Exercise intensity regulates cytokine and klotho responses in men. *Nutr. Diabetes* **2021**, *11*, 5. [CrossRef]
67. Petersen, A.C.; McKenna, M.J.; Medved, I.; Murphy, K.T.; Brown, M.J.; Della Gatta, P.; Cameron-Smith, D. Infusion with the antioxidant N-acetylcysteine attenuates early adaptive responses to exercise in human skeletal muscle. *Acta Physiol.* **2012**, *204*, 382–392. [CrossRef] [PubMed]
68. Manach, C.; Morand, C.; Gil-Izquierdo, A.; Bouteloup-Demange, C.; Rémésy, C. Bioavailability in humans of the flavanones hesperidin and narirutin after the ingestion of two doses of orange juice. *Eur. J. Clin. Nutr.* **2003**, *57*, 235–242. [CrossRef] [PubMed]

antioxidants

MDPI

Article

A Comparative Study of the Antioxidative Effects of *Helichrysum italicum* and *Helichrysum arenarium* Infusions

Katja Kramberger [1], Zala Jenko Pražnikar [1], Alenka Baruca Arbeiter [2], Ana Petelin [1], Dunja Bandelj [2] and Saša Kenig [1,*]

[1] Faculty of Health Sciences, University of Primorska, 6310 Izola, Slovenia; katja.kramberger@fvz.upr.si (K.K.); zala.praznikar@upr.si (Z.J.P.); ana.petelin@fvz.upr.si (A.P.)

[2] Faculty of Mathematics, Natural Sciences and Information Technologies, University of Primorska, 6000 Koper, Slovenia; alenka.arbeiter@upr.si (A.B.A.); dunja.bandelj@upr.si (D.B.)

* Correspondence: sasa.kenig@fvz.upr.si; Tel.: +386-5-66-35-801

Abstract: *Helichrysum arenarium* (L.) Moench (abbrev. as HA) has a long tradition in European ethnomedicine and its inflorescences are approved as a herbal medicinal product. In the Mediterranean part of Europe, *Helichrysum italicum* (Roth) G. Don (abbrev. as HI) is more common. Since infusions from both plants are traditionally used, we aimed to compare their antioxidative potential using in vitro assays. Two morphologically distinct HI plants, HIa and HIb, were compared to a commercially available HA product. Genetic analysis using microsatellites confirmed a clear differentiation between HI and HA and suggested that HIb was a hybrid resulting from spontaneous hybridization from unknown HI subspecies. High-performance liquid chromatography–mass spectrometry analysis showed the highest amounts of hydroxycinnamic acids and total arzanol derivatives in HIa, whereas HIb was richest in monohydroxybenzoic acids, caffeic acids, and coumarins, and HA contained the highest amounts of flavonoids, especially flavanones. HIa exhibited the highest radical scavenging activity; it was more efficient in protecting different cell lines from induced oxidative stress and in inducing oxidative stress-related genes superoxide dismutase 1, catalase, and glutathione reductase 1. The antioxidative potential of HI was not only dependent on the morphological type of the plant but also on the harvest date, revealing important information for obtaining the best possible product. Considering the superior properties of HI compared to HA, the evaluation of HI as a medicinal plant could be recommended.

Keywords: *Helichrysum*; medicinal plants; infusions; phenolic compounds; antioxidative potential

Citation: Kramberger, K.; Jenko Pražnikar, Z.; Baruca Arbeiter, A.; Petelin, A.; Bandelj, D.; Kenig, S. A Comparative Study of the Antioxidative Effects of *Helichrysum italicum* and *Helichrysum arenarium* Infusions. *Antioxidants* **2021**, *10*, 380. https://doi.org/10.3390/antiox10030380

Academic Editors: Irene Dini and Domenico Montesano

Received: 14 February 2021
Accepted: 26 February 2021
Published: 3 March 2021

1. Introduction

Helichrysum arenarium (L.) Moench (hereinafter abbreviated as HA) has a long tradition in European ethnomedicine as a medicinal plant with many known activities; the EU monograph supports its use for digestive disorders, feeling of fullness and bloating, and, importantly, no side effects are reported [1]. An assessment report of the European Medicines Agency (EMA) Committee on Herbal Medicinal Products (HMPC) supports its stimulatory role in bile flow and in the release of cholesterol into the bile. HA has also been used in chronic liver inflammatory disease and as a scavenger of free radicals [2]. HA inflorescences are the only herbal medicinal product from the *Helichrysum* genus (*Asteraceae* family) approved by the HMPC and currently available on the market as such.

While HA can be found mainly in Poland, Russia, Lithuania, and Latvia [3], in the Mediterranean region of Slovenia and in other regions with similar climatic conditions, such as coastal Croatia, Italy, Spain, Portugal, and Corsica, *Helichrysum italicum* (Roth) G. Don (abbreviated as HI) is more common, both as a wild and as a cultivated plant [4]. Similar to HA, HI is widely used in traditional medicine but has not yet been recognized and evaluated by the HMPC. Moreover, scientific data are mostly limited either to topical applications in wound healing or to the investigation of individual compounds in in vitro models [5,6]. The antimicrobial activity of ethanol, methanol, and acetone extracts and essential oil is also well documented [7–9].

Lately, much attention has been paid to natural antioxidants and their ability to delay the progression of chronic diseases related to the formation and the activity of reactive oxygen species (ROS) and other free radicals [10,11]. Among these natural products, various herbal extracts containing numerous plant secondary metabolites exhibit the ability to scavenge ROS. Herbs are very often used to prepare recreational tea and are thereby important sources of antioxidants in different cultures [12]. Although the English term "tea" refers to the infusion made from the leaves of a tea plant (*Camellia sinensis* (L.) Kuntze), in colloquial language it also refers to the wide variety of locally grown herbs used in different regions of the world for recreational tea. Among the most important ones in Europe are Roman chamomile (*Chamaemelum nobile* (L.) All.), red raspberry (*Rubus ideus* L.), and lemon balm (*Melissa officinalis* L.) whereas in southern Europe, chamomile (*Matricaria chamomilla* L.) and lemon thyme (*Thymus pulegioides* L.) are frequently used [12]. Due to the intense fragrance of HI, its plantations are primarily dedicated for use in the cosmetic industry. However, in traditional medicine, HI is also used in the diet, mainly for the preparation of infusions to treat respiratory and digestive conditions [9]. Recently, it was confirmed by Kramberger, et al [13] that hot water extracts prepared from dried HI aerial parts contain high amounts of bioactive compounds and have comparable antioxidative activity to the ethanol and methanol extracts.

Several subspecies (ssp.) of HI are known, but genetic or morphological description of the plant material used is often lacking in studies [9]. While it has been shown that the method of extract preparation crucially contributes to the heterogeneity of extracts, genetic background is also a major factor influencing the profile of the different bioactive compounds [13]. This suggests that comprehensive plant characterization would be important to obtain the best health beneficial effects of infusions or a recreational tea.

Therefore, we analyzed the content of the main bioactive compounds and the antioxidative properties, tested with cellular and non-cellular in vitro methods, of infusions prepared from two morphologically distinct HI plants (HIa and HIb) in comparison to the commercially available HA. In addition to evaluating HI and comparing it with a recognized medicinal plant, we also aimed to provide information relevant for the farmers to select appropriate plant material for their plantations to obtain the most advantageous health beneficial effects of their crop.

2. Materials and Methods

2.1. Plant Material

The analyzed plant material included two *Helichrysum* species: dried flower heads of commercially available *Helichrysum arenarium* (L.) Moench tea (sample HA), purchased from Flora Ltd., Rogatec, Slovenia, and *Helichrysum italicum* (Roth) G. Don plants (samples HIa and HIb) cultivated in two Slovenian plantations. Within the *H. italicum* group, two morphological types were identified according to their morphological differences: plant *H. italicum* a (sample HIa, characteristic of common plants) and plant *H. italicum* b (sample HIb, characteristic of one sample). Based on taxonomically relevant characters [14], morphological type HIb differs from type HIa, mainly by a shorter leaf length, a clear presence of axillary leaf fascicles, a leaf margin undulation, and a lower number of capitula per synflorescence. The HIa plants originated from a commercial, private plantation in Dragonja (45°27′05″ N 13°41′31″ E), Slovenia. The HIb plants were collected from the ex situ experimental collection of the University of Primorska, established in 2018 near Ankaran (45°34′19″ N 13°46′32″ E), Slovenia. Plants from both plantations were grown under similar ecological conditions of a sub-Mediterranean climate. Herbarium specimens of both HIa and HIb plants were deposited at the University of Primorska, Faculty of Mathematics, Natural Sciences and Information Technologies, Slovenia, under accession numbers HIa_UP20 and HIb_UP20.

2.2. DNA Extraction and Microsatellite Analysis

Total genomic DNA was extracted from fresh leaves of cultivated HIa HIb and from dried flower heads of HA using the modified cetyltrimethyl ammonium bromide-polyvinylpyrrolidone (CTAB-PVP) protocol [15]. DNA extracts from plants from both collection fields and of commercial tea Flora Ltd. were deposited in the genetic laboratory of the University of Primorska, Slovenia under the accession numbers HIaUP_1-30 (*H. italicum* plant a), HIbUP_1 (*H. italicum* plant b), and HAUP_1-7 (*H. arenarium*).

The concentration of DNA was determined by Qubit^{TM} 3 (Thermo Fisher Scientific, Waltham, MA, USA). A set of eight newly developed microsatellite primers (HiUP-01, HiUP-02, HiUP-09, HiUP-13, HiUP-16, HiUP-18, HiUP-22, HiUP-24) designed by Baruca Arbeiter, et al. [15] were selected for genetic analysis. Amplification was carried out using a DNA Engine Thermal Cycler 200 (Bio-Rad Laboratories, Hercules, CA, USA) containing 1x AllTaq PCR Buffer, 2 mM $MgCl_2$, dNTPs (0.2 mM of each dNTP), 1x Q-Solution, 1.25 U of AllTaq DNA polymerase (Qiagen, Hilden, Germany), 0.2 μM of each locus-specific primer (synthesized by IDT), 0.25 μM of universal primer M13(-21) labelled with 6-FAM, VIC, PET or NED (Applied Biosystems, Foster City, CA, USA), and 40 ng of template DNA. The two-step amplification profile consisted of an initial denaturation at 94 °C for 5 min, followed by 30 cycles (step 1) of 94 °C for 30 s, 56 °C for 45 s and 72 °C for 45 s, followed by eight cycles (step 2) of 94 °C for 30 s, 53 °C for 45 s and 72 °C for 45 s. Final extension was carried out at 72 °C for 8 min [16]. Fragment analysis was conducted using SeqStudio^{TM} Genetic Analyzer (Applied Biosystems, Foster City, CA, USA), and the allele calling was done with GeneMapper version 4.1 (Applied Biosystems, Foster City, CA, USA).

2.3. Infusion Preparation

For chemical analysis, HIa and HIb aerial parts were harvested from the 24th to 29th of June 2019 and air-dried. For analysis of the optimal harvest period, the material was harvested in May and June 2020. Plant material (HIa, HIb and HA flower heads) was milled to 2–3 mm fine particles and stored at room temperature in a dry and dark place until further use. For the analysis of separate plant parts, samples of HIa were separated to green parts (stems and leaves) and flowers. Fresh infusions were prepared from the same stock material for all cell culture experiments and for chemical analysis by infusing 0.5 g of dried plant material in 100 mL of boiling water. After a 10-min incubation, infusions were filtered through Whatman paper and passed through a 0.2 μm cellulose acetate membrane filter (Macherey-Nagel GmbH & Co KG, Düren, Germany). For the chemical analysis, samples were screened in the same dilution.

2.4. Chemical Analysis

High-performance liquid chromatography–mass spectrometry (HPLC-MS) analysis was performed using an Agilent 1260 Infinity II HPLC system (Agilent Technologies, Santa Clara, CA, USA) equipped with a diode array detector (DAD, model G7115A) and coupled to an Agilent 6530 Accurate-Mass Quadrupole Time-of-Flight (Q-TOF) MS system equipped with an Agilent Jet Stream dual electrospray ionization (ESI) source. The HPLC system included a binary pump (model G7112B), Agilent 1260 Autosampler (model G7129A) and a Poroshell 120, EC-C18, 2.1 × 150 mm, 2.7 μm column (693775-902, Agilent Technologies, Santa Clara, CA, USA). Separation was obtained with a linear gradient of (A) water + 0.1% formic acid (v/v) and (B) acetonitrile/methanol (50:50, v/v), starting at 3.0% B and increased to 100.0% B in 15 min and held for 5 min (flow rate 0.30 mL/min, column temperature 50 °C, injection volume 1 μL). The MS scans were performed under the following conditions: gas temperature 250 °C, drying gas flow 8 L/min, nebulizer 35 psig, sheath gas temperature 375 °C, sheath gas flow 11 L/min, capillary voltage 1000 V, and fragmentor voltage 150 V. Mass spectra were recorded as centroid data for m/z 100–1000 in MS mode and m/z 40–1000 in MS/MS mode, with an acquisition rate of 14.0 spectra/sec. The automated MS/MS data-dependent acquisition was done for ions detected in the full scan above 2000 counts with a cycle time of 0.5 s, a quadrupole isolation width in narrow

~1.3 Da, using fixed collision energies of 10, 20, and 40 eV, and a maximum of three selected precursor ions per cycle. The instrument was tuned in low mass range (1700 *m/z*) and in extended dynamic range (2 GHz) mode. Under those conditions, the instrument was expected to provide experimental data with accuracy within ±3 ppm. Agilent MassHunter Data Acquisition and Qualitative Analysis Workflows (version B.08.00) software were used to acquire and process the data, respectively. The compound identification procedure is described in detail by Kramberger, et al. [13]. Semiquantitative comparison of different *H. italicum* samples was made by comparing abundances obtained from extracted ion chromatograms of individually identified compounds. Identified compounds were sorted by the chemical classes and subclasses and their abundances were summed.

2.5. Radical Scavenging Activity Assay

1,1-diphenyl-2-picrylhydrazyl (DPPH) reagent was dissolved in methanol, and samples of infusions were diluted in water. Ascorbic acid was used as a positive control. Samples were mixed with 0.2 mM DPPH methanol solution and incubated for 1 h at room temperature in the dark. To determine the reduction of DPPH, radical absorbance at 515 nm was measured on spectrophotometer Infinite F200 (Tecan Group Ltd., Zürich, Switzerland). The radical scavenging activity was calculated as a percentage of DPPH discoloration using the equation:

$$(\text{A (sample with DPPH)} - \text{A (sample)})/(\text{A (DPPH)} - \text{A (solvent)}) * 100, \quad (1)$$

The presented results are for the 12.5% *v/v* concentration, which was used for all infusions in the linear range of the test. Each test was performed in five parallels and repeated twice.

2.6. Cell Cultures

Caco-2 human colorectal adenocarcinoma cells (ATCC® HTB37™) and primary colon fibroblasts CCD112CoN (ATCC® CRL1541™) were purchased from American Type Culture Collection (ATCC, Manassas, VA, USA) and cultured in Dulbecco's Modified Eagle's high glucose medium (DMEM) supplemented with 20% fetal bovine serum (FBS). U937 histiocytic lymphoma cells were purchased from ATCC (ATCC® CRL-1593.2™) and grown in Roswell Park Memorial Institute (RPMI) culture media supplemented with 10% FBS. All cultures were maintained at 37 °C in a humidified atmosphere containing 5% CO_2.

For all cell culture experiments, infusions were mixed directly with cell culture media, and the concentrations are expressed in *v/v* concentrations.

2.7. Cell ViabilityAassay

Cell viability after exposure to *Helichrysum* infusions was determined using PrestoBlue™ reagent (Invitrogen™, Carlsbad, CA, USA) according to the manufacturer's instructions. Cells were treated with infusions diluted in cell culture media for 24 h. PrestoBlue™ was added and after 30 min, fluorescence was measured on a spectrophotometer Infinite F200 (Tecan Group Ltd., Zürich, Switzerland) at excitation/emission (ex/em) of 535/595 nm.

2.8. Intracellular ROS Level

Intracellular oxidative stress levels were evaluated using the 2′,7′-Dichlorofluorescin Diacetate (DCF-DA) assay [17]. Caco-2 and CCD112CoN cells were plated on black, clear-bottom 96-well plates at concentrations of 10.000 and 5.000 cells per well, respectively, and left to adhere to the plates for 24 h. U937 suspension cells were plated at a concentration of 10.000 cells per well. Cells were pretreated with different noncytotoxic dilutions of infusions for 24 h, whereas untreated samples served as controls. Cells were then washed in phosphate-buffered saline (PBS) and exposed to DCFH-DA to a final concentration of 50 µM and freshly prepared tert-butyl hydroperoxide solution (t-BOOH, 250 µM in PBS). The resulting increase in fluorescence was determined immediately and after 30 min incubation at room temperature using a spectrophotometer, Infinite F200 (Tecan Group Ltd., Zürich, Switzerland), at ex/em of 485/535 nm. The fluorescence increase is expressed as relative fluorescence, where the fluorescence of the control sample is set to 1.

2.9. Gene Expression Analysis

RNA was isolated from infusion-treated and control cells using TRIzol™ reagent (Thermo Fisher Scientific, Waltham, MA, USA) following the manufacturer's instructions. One μg of RNA was reverse transcribed to cDNA with a cDNA Archive kit (Applied Biosystems, Foster City, CA, USA), and for the quantitative RT-PCR reactions, QuantStudio® 3 Real-Time PCR System (Thermo Fisher Scientific, Waltham, MA, USA), SYBR Green master mix and 40 ng of cDNA template were used. The primer sequences 48762945c1 for superoxide dismutase 1 (SD1), 260436906c3 for catalase (CAT), and 305410788c1 for glutathione reductase (GR) were selected from PrimerBank [18] and used in a 0.5 μM final concentration. 18S rRNA was used as an internal control. Reaction conditions were 50 °C for 2 min, 95 °C for 10 min, and 40 cycles of 95 °C for 15 s and 60 °C for 1 min. Melting curves were inspected to ensure primer specificity. The results were analyzed using the ΔΔCt algorithm and are presented as the fold-change compared to non-treated cells.

2.10. Statistical Analysis

Values are presented as the mean ± standard deviation and were analyzed using statistical software (SPSS 23.0; IBM, Tokyo, Japan). Differences between groups were evaluated with one-way analysis of variance (ANOVA) followed by Dunnett's test for multiple comparisons. Statistical significance was defined as $p < 0.05$.

3. Results

3.1. Plant Identification and Characterization

In the literature, there is often not enough information about the plant material used in the study. It is well known that both environmental factors and genotype can influence the chemical composition of a plant and consequently its biological effects. Therefore, we used genetic analysis to characterize both the commercially available product HA and the two morphological types of HI (HIa and HIb) grown in the commercial plantation and experimental collection, respectively. Newly developed microsatellite markers for HI [15] were successfully amplified in all analyzed plant materials. Examination of the eight microsatellite markers revealed 64 alleles in HI (HIa and HIb) and 58 alleles in HA. The number of amplified alleles identified in both species was similar, which indicates comparable polymorphism detected in the two species. A total of 89 alleles were detected, with the number of alleles per locus ranging from eight for loci HiUP-13 and HiUP-16 to 17 for locus HiUP-18. Microsatellite analysis identified a large number of alleles specific to each of the *Helichrysum* species. Eight microsatellite loci exhibited 31 (35%) HI specific (unique) alleles (Nu) and 25 (28%) HA unique alleles (Nu). The alleles of seven microsatellite markers (HiUP-02, HiUP-09, HiUP-13, HiUP-16, HiUP-18, HiUP-22, HiUP-24) shared by HI and HA (Ns) were 33 (37%). Only at locus HiUP-01 were no shared alleles found, and microsatellite profiles were unique for each of the species. In addition, microsatellite analysis revealed no substantial genetic differences between HIa and HIb genotypes, despite clear morphological differences between these two morphological types. Morphological differences between HIa and HIb, the total number of amplified alleles for HI and HA, and the proportion of shared alleles between HI and HA are presented in Figure 1.

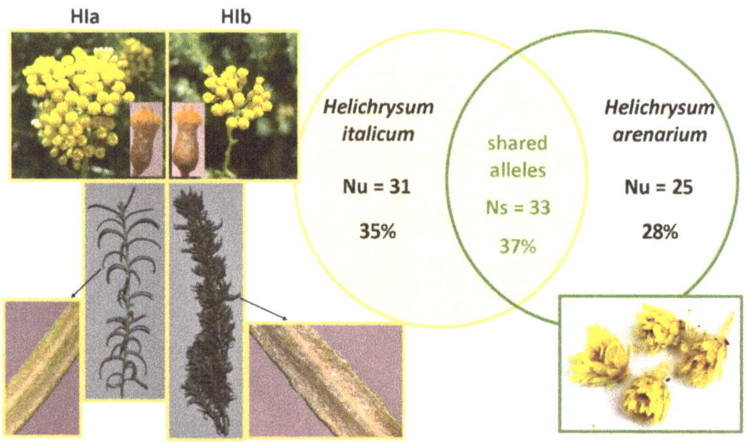

Figure 1. Plant material characterization. Taxonomically relevant differences among two identified HI morphological types HIa and HIb, the number (Nu) and the proportion of taxon-specific (unique) alleles for HI (Nu = 31; 35%) and HA (Nu = 25; 28%), and the number (Ns) and proportion of shared alleles among HI and HA (Ns = 33; 37%).

The main chemical compounds which could be responsible for the biological effects, such as antioxidative potential and induction of gene expression, were analyzed (Table 1). Chemical composition comparison was done based only on the targeted compounds identified in our previous study [13]. The analysis is semiquantitative and therefore the relative abundances were used for the comparison between infusions [19]. The sum of the identified phenolic compounds was highest in HIa, whereas in HIb it was two-fold lower, and the lowest in HA. Looking at separate classes and subclasses of the identified compounds, we found hydroxycinnamic acids and total arzanol derivatives and other pyrones to be the most abundant in HIa. The representatives of hydroxycinnamic acids which were particularly abundant were caffeoylquinic and dicaffeoylquinic acids, while caffeic acid and its derivatives were more abundant in HIb. HIa was also the richest in cyclic polyols, isobenzofuranones, and neolignans. Only in this infusion were kaempferol and quercetin derivatives detected, and there were no flavanones. Subclasses of monohydroxybenzoic acids and coumarins were present in the largest quantities in HIb, and unlike the other two infusions, HIb also contained acetophenones and tremetones. The HA infusion was the richest in total flavonoids, especially flavanones. The abundances of caffeoylquinic acid, total hydroxybenzoic acids, and cyclic polyols in HA were not substantially lower than those in HIa. On the other hand, some flavonoid subclasses such as flavonols and flavones in HA were not detected. Furthermore, a difference from HI was observed in the absence of other hydroxycinnamic acids, pyrones, and coumarins.

Table 1. HPLC–MS analysis of the infusions. Semiquantitative analysis was preformed to compare the abundances of the main components in HIa, Hib, and HA. The relative abundances are presented for each infusion. Calculations were performed based on abundance summation of individual compounds belonging to each subclass and class. Names of subclasses are written in italics.

Compound Name	Infusion		
	HIa	HIb	HA
Total Identified Phenolic Compounds [1]	**2071068**	**974921**	**681606**
Total hydroxycinnamic acids	1104310	472357	189173
Caffeic acid and its derivatives	*174492*	*218381*	*31144*

Table 1. *Cont.*

Compound Name	Infusion		
	HIa	HIb	HA
Other hydroxycinnamic acids and their derivatives	*43618*	*44407*	*0*
Caffeoylquinic acids	*443858*	*197797*	*150534*
Other monoesters	*59450*	*0*	*4167*
Dicaffeoylquinic acids	*382892*	*11772*	*3328*
Total hydroxybenzoic acids	89445	92466	64752
Monohydroxybenzoic acids	*43026*	*72816*	*42956*
Dihydroxybenzoic acids	*46419*	*19650*	*21796*
Total flavonoids	92004	24180	227018
Quercetin derivatives	*25573*	*0*	*0*
Myricetin derivatives	*10056*	*5223*	*0*
Kaempferol derivatives	*47970*	*0*	*0*
Flavanones	*0*	*18957*	*227018*
Flavones	*8405*	*0*	*0*
Total arzanol derivatives and other pyrones	392448	275723	7521
Pyrones	*384837*	*256991*	*0*
Arzanol and its derivatives	*7611*	*18732*	*7521*
Total other phenolic compounds	392861	110195	193142
Cyclic polyols	*180368*	*54374*	*168071*
Isobenzofuranones	*156472*	*4638*	*22703*
Neolignans	*48243*	*15502*	*2368*
Acetophenones	*0*	*3750*	*0*
Tremetones	*0*	*14525*	*0*
Coumarins	*7778*	*17406*	*0*

[1] Sum of total abundances of all compound classes (hydroxycinnamic acids, hydroxybenzoic acids, flavonoids, arzanol derivatives, other pyrones, and other phenolic compounds) listed in the table.

3.2. Radical Scavenging Activity

The DPPH test was used to determine the in vitro antioxidative potential of infusions. Green parts of HI exhibited higher radical scavenging activity compared to flowers alone (Figure 2A). Due to this reason and regardless of the fact that commercially available product contained only flowers, we decided to use whole aerial parts of HIa and HIb to fully exploit their putative health benefits. A comparison of infusions prepared from different species and morphological types revealed the highest radical scavenging activity of HIa, which was in the tested concentration of 12.5% v/v comparable to the activity of 25 μM ascorbic acid (AA) and reached 43.6% inhibition of DPPH (Figure 2B). The infusion of HIb at the same concentration inhibited 21.4% of DPPH, and the infusion of HA inhibited only 4.4%. As it was observed that the two HI types on a particular date were not always at the same growth stage and their scent was variably strong, in 2020, plant material was collected from both types on several dates during May and June. DPPH revealed only mild variability of radical scavenging activity for HIa. On the contrary, for HIb, the activity markedly varied and was on all tested dates significantly lower than that of HIa, with the exception of June the 12th, when the activities were comparable (Figure 2C).

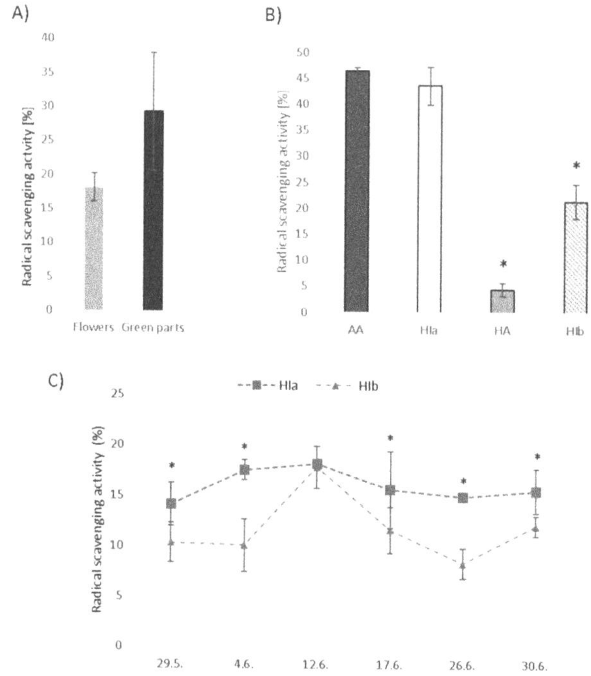

Figure 2. Radical scavenging activity measured by the DPPH test in (**A**) infusions prepared from flowers and green parts of HIa, (**B**) HIa, HIb, and HA infusions in comparison to 25 μM ascorbic acid (AA) and (**C**) HIa and HIb collected on different dates, all at a 12.5% *v*/*v* concentration. Mean ± SD is presented; * $p < 0.05$.

3.3. Cytotoxicity and the Protective Effect from Oxidative Stress Induction

Due to higher radical scavenging activity, HIa was selected for the analysis in cell models in order to compare its effects to commercially available HA. The cytotoxicity of both infusions was inspected for cancerous colon cells Caco-2, primary colon cells CCD112CoN, and the lymphoma U937 cell line with PrestoBlue™ assay, and the results are presented in Figure 3. Infusions prepared from HIa and HA were for U937 cells toxic at 5% *v*/*v* concentration, whereas for Caco-2 and CCD112CoN, cells of one of the two were toxic already at a 1% *v*/*v* concentration. The HIa infusion at 1% *v*/*v* concentration was toxic for Caco-2 cells but not for the primary colon cells. For further experiments, non-cytotoxic test concentrations were used.

To determine the protective effect of infusions against oxidative stress induction, DCFH-DA assay was employed. In Caco-2 cells, HIa at the highest tested concentration (0.5% *v*/*v*) reduced the level of reactive oxygen species concentrations to 84.2 ± 8%. The effect of HIa was even more pronounced in primary colon CCD112CoN cells, where it increased with concentration and reached a reduction of 70.5 ± 4.5% at a 0.5% (*v*/*v*) concentration of HIa infusion. HA had no significant effect on cancerous or on primary colon cells (Figure 4A,B). HIa was also more efficient than HA in reducing the levels of ROS in U937 cells; at the highest tested concentration of 2% (*v*/*v*), it reduced ROS to 64.1 ± 1%. HA at lower concentrations reduced the ROS level to 88.3 ± 1.6%, but the effect disappeared when a higher infusion concentration was used (Figure 4C).

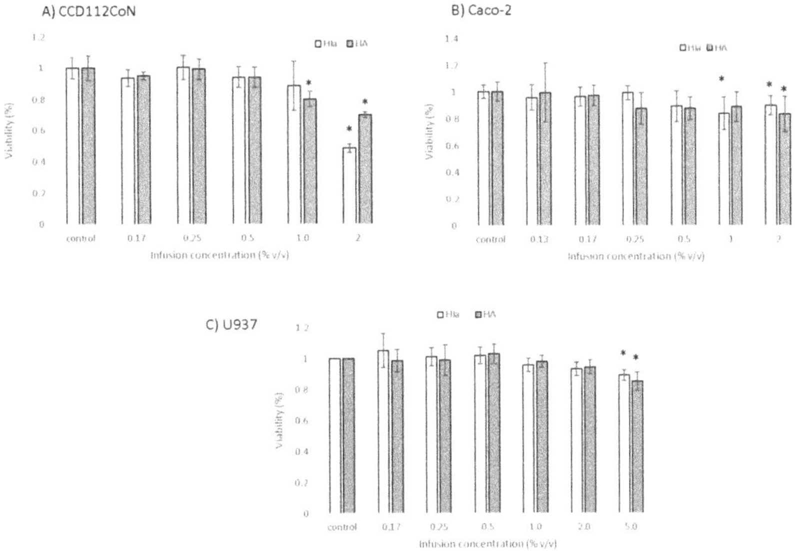

Figure 3. Cytotoxicity of HIa and HA infusions as determined by the PrestoBlue Assay in (**A**) CCD112CoN cells, (**B**) Caco-2 cells, and (**C**) U937 cells. The mean ± SD of three separate experiments is presented; * $p < 0.05$.

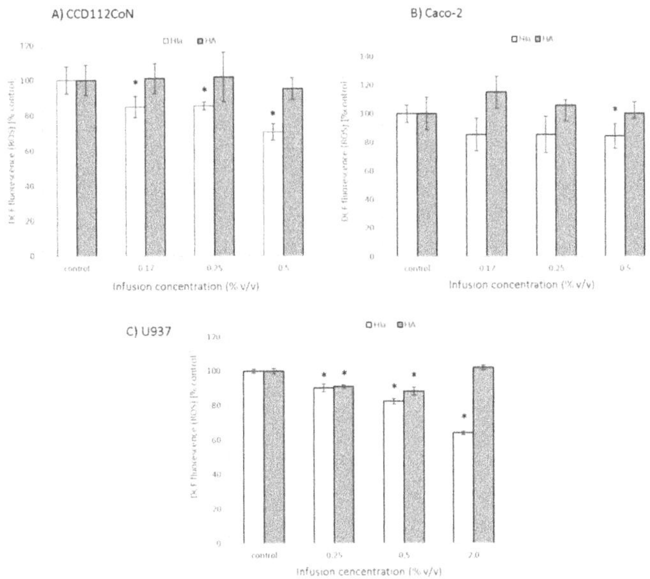

Figure 4. Protective effect of infusions against oxidative stress induction, determined with DCFH-DA assay in (**A**) CCD112CoN, (**B**) Caco-2, and (**C**) U937 cells. The mean ± SD of two separate experiments is presented; * $p < 0.05$.

3.4. The Expression of Genes Related to Oxidative Stress

To explore the mechanism of the antioxidative activity of *Helichrysum* infusions, the expression of three genes related to oxidative stress (SD1, GR1, and CAT) was determined by RT–PCR in infusion-treated cells. Similar to the determined protective effects, the greatest differences between the two infusions were detected in the primary colon cells (Figure 5). Here, HIa caused a significant 2–2.5-fold upregulation of all three enzymes; HA, on the contrary, downregulated all three genes. In Caco-2 cells, both infusions comparably upregulated the expression of SD1 (1.4-fold). In U937 cells, the HIa infusion caused a slight but significant increase in the expression of SD1 and GR1 and a decrease in the expression of CAT, whereas HA decreased the expression of both GR1 and CAT (Figure 5).

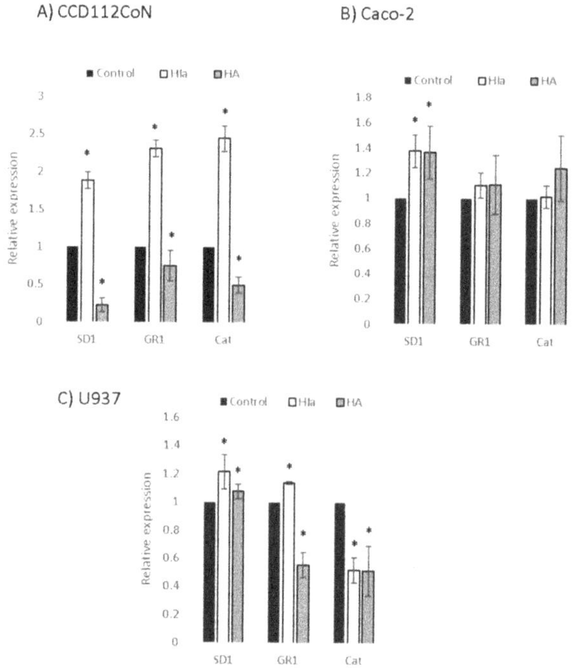

Figure 5. Expression levels of antioxidative enzymes. RT–PCR was performed to analyze the expressions of superoxide dismutase 1 (SD1), glutathione reductase 1 (GR1), and catalase (Cat) in (**A**) CD112CoN, (**B**) Caco-2, and (**C**) U937 cells. 18S rRNA was used as an endogenous control; the relative gene expression of non-treated cells was set to one, * $p < 0.05$.

4. Discussion

Genetic analysis with microsatellites confirmed a clear differentiation among HI and HA species. In both *Helichrysum* species, unique alleles, characteristic of a defined species were found, confirming the usefulness of eight HiUP microsatellites in the identification process and traceability of the plant material. In addition, HI and HA shared more than one third of all detected alleles, indicating a common genetic background. Despite remarkable differences in the morphology and chemical profiles of plants HIa and HIb, these two forms shared identical alleles. Nevertheless, HIb was characterized by unique allele combinations allowing clear differentiation among the two morphs. The plant HIb is probably a hybrid, generated through spontaneous hybridization from unknown HI subspecies, which usually occur in the nature, where different subspecies of HI overlap [14,20]. Differences in the chemical composition, genetic structure, and antioxidative effects of *Helichrysum* species confirmed that accurate authentication is very important to ensure the most suitable and safe use of medicinal plants.

Infusions prepared from HI exhibited superior antioxidative properties to the infusion from a commercially available and approved medicinal plant of HA. This was true for both morphologically distinct HI plants used here; in vitro tests revealed the highest radical scavenging activity for HIa, followed by HIb, and HA. Furthermore, HIa was able to protect primary colon cells, colon cancerous cells, and lymphocytes from induced oxidative stress in a dose-dependent manner. HA also exerted a protective effect, but only in lymphocytes, and not at all tested concentrations. It can be pointed out that HIa was toxic to tumor Caco-2 cells at a lower concentration than in primary colon cells, which is an additional benefit, but for any claims about antitumorigenic effects, further studies should be performed.

The observations discussed above can be attributed to the chemical composition of the tested infusions. The total content of all identified phenolic compounds was substantial in all tested infusions, but there were considerable differences in the amounts of individual compounds. The content of total arzanol derivatives and other pyrones in the HIa infusion was approximately 50-fold higher than that in the HA infusion. Arzanol is one of the best investigated compounds found in HI and was previously found to exert strong antioxidative effects such as inhibition of lipid peroxidation [21]. The content of total hydroxycinnamic acids, for which the antioxidative potential was confirmed for many representatives, such as free caffeic acid [22], caffeoylquinic acids, and dicaffeoylquinic acids [23], was also the highest in HIa. The same was true for total flavonols, of which kaempferol derivatives, most likely also responsible for the somewhat more bitter taste, were the most abundant. HIa was the only infusion containing quercetin and its derivatives, which have a strong antioxidant activity for use in inflammatory conditions, mood disorders, and circulatory disfunction [24]. The total amount of flavonoids was, on the contrary, higher in HA, mostly due to the high content of flavanones. The latter was also expected, as apigenin, naringenin, and their derivatives were previously identified as major antioxidative constituents of HA methanol extract and tea made from inflorescences [25,26]. Previous studies on organic solvent-extracts of HA reported the presence of hydroxycinnamic, hydroxybenzoic acids, kaempferol, and low levels of quercetin [25,27]. While we detected representatives of the first two classes in the infusion of HA, kaempferol and quercetin were not detected. A chemical profile of HI infusions was also similar to that in previous reports; previously identified major polar constituents in methanol extract of HI flowers were caffeic acid, chlorogenic acid, quercetin derivatives, caffeoylquinic and dicaffeoylquinic acids, and tiliroside [28]. In line with our results, pyrone derivatives, such as mycropyrone, arzanol, and its derivatives were also reported [29]. On the contrary, flavanones such as naringenin were not detected in the HIa infusion, but were previously found in the methanolic extracts of the same species [30]. It is, however, important to stress that most previous chemical composition analyses were focused on organic solvent extracts, and the extraction technique largely influences the chemical composition [13,30].

Commercially available tea of HA contains only dried flowers, as only these are recognized as a herbal drug by the HMPC. We show here that for HIa, the antioxidative properties of green parts (stems and leaves) are superior to those of flowers, which led to a decision to use all aerial parts for the cell-culture experiments. Use of all aerial parts is also reported in ethnomedicine; fumes from flowers and leaves were reported to help with sleeplessness and headache and in decoctions against stomach-ache [31]. Infusions made from all aerial parts were used against cold [32] and to treat allergies [33]. Since green parts of HA were not available, we cannot exclude that these parts may also contain bioactive compounds which could contribute to the higher antioxidative potential of HA.

Based on the chemical analysis, it can be concluded that recreational tea prepared from HI or HA can be an important non-caloric source of health beneficial phenolic compounds. Despite the fact that some of these compounds are widely available in the diet, the intake of flavonols, for example, ranges only from 9 to 36.2 mg/day in the USA [34] and is comparatively low in Sweden and Norway [35]. The absence of caffeine, for those who wish to avoid its intake, makes *Helichrysum* infusions a good alternative to some other popular antioxidant-rich beverages, such as green tea.

To investigate whether the differential protective effects were exclusively due to the antioxidative properties of the infusions alone or also due to the differential response of particular cells to the treatments, the expressions of antioxidative enzymes were analyzed. It was shown for *Helichrysum plicatum* DC. that the extracts can increase the activities of antioxidative enzymes SD, GPx, and CAT when the concentration is not too high [36]. The effect on the expression of genes relevant in the response against oxidative stress may partly explain the mechanism through which HI reduces induced oxidative stress. In primary colon cells, where the protective effect was the most efficient, HIa infusion caused the most significant upregulation of SD1. In lymphocytes, both SD1 and GR1 upregulation in HIa-treated cells was consistent with the observed protective effect. In Caco-2 cells, no difference in the expression levels of any tested genes was observed, pointing to other possible mechanisms. A higher potential of the infusion itself is one possible explanation, whereas changes in the expression of other genes or in the production of endogenous antioxidants are also possible.

The principle of hormesis, a well-known phenomenon, where a mild stress caused by a low dose of a particular compound induces adaptive responses in cells, preparing them for forthcoming stressful events [37,38], could also partially explain our observations. Such a principle was previosly confirmed for several stressors, such as oxygen, nitric oxide, carbon monoxide, and physical exercise [37]. Among well-investigated stressors are plant extracts, such as extracts of Ginkgo biloba [39] and cruciferous vegetables [40], and/or single phenolic compounds [41,42]. In the present study, the protective effect against induced stress was indeed observed in cells pretreated with the infusions and the upregulation of the genes, coding for the antioxidative enzymes, could be one of such adaptations. The main idea of hormetic activity is that a stressor in a low dose induces a mild stress to which cells respond with beneficial overcompensation to the damaging stimulus, whereas at higher doses, the damaging effects predominate [38]. The fact that the HA infusion in U937 cells was protective at a lower but not a higher concentration may point to such a mechanism.

For the initial experiments, plant material was collected in early June 2019. In the following weeks, we noticed that the scent of the plants varied. Even though the scent depends on volatile compounds which are not responsible for the antioxidative properties of infusions, these differences could indicate that the content of bioactive compounds changes over time. The same has been observed for other plants [43]. This was confirmed when plant material was collected at different time points in 2020. While HIa exerted a stable antioxidative potential, HIb had an obvious peak in mid-June. The result suggests that the harvesting time has an important influence on the quality of the harvested material. However, the optimal harvest period cannot be suggested based solely on our data, as it may depend not only on the species/subspecies but also on the specific weather conditions. Observing the antioxidative properties over a longer period of time in several seasons with special attention on the phenophase of the plant would provide the most useful and transferable information.

5. Conclusions

Taken together, our results provide scientific confirmation of the health beneficial value of traditionally used HI infusions and answer, at least in part, the frequently expressed call for such data in the literature. The superior antioxidative potential of HI compared to HA observed here may suggest that HI could also be evaluated as a medicinal plant, and further in vivo studies can be recommended. At the same time, we show that the two distinct morphological types are different in terms of chemical composition and consequent functional properties. This points to the importance of thorough plant characterization prior to plantation set-up. Additionally, the harvesting time also importantly contributes to the antioxidative effects of the infusions, which suggests this parameter is crucial and should be considered to obtain the best possible product and thus fully exploit its commercial potential.

Author Contributions: Conceptualization, S.K., Z.J.P. and D.B.; methodology, K.K., A.B.A. and S.K.; investigation, S.K., K.K., A.B.A., A.P. and Z.J.P.; resources, Z.J.P. and D.B.; data curation, K.K. and S.K.; writing—original draft preparation, S.K.; writing—review and editing, K.K., A.B.A., D.B. and Z.J.P.; visualization, S.K. and A.P.; supervision, S.K. and Z.J.P.; funding acquisition, Z.J.P. and D.B. All authors have read and agreed to the published version of the manuscript.

Funding: This research was financially supported by the Slovenian Research Agency (research program P1-0386, post-doctoral project Z4-1875), European Agricultural Fund for Rural Development LAG Istra (project 33152-247/217/15), and by the Republic of Slovenia Ministry of Education, Science and sport and European Union from European Social Funds (project SMILJ).

Institutional Review Board Statement: Not applicable.

Informed Consent Statement: Not applicable.

Data Availability Statement: The data presented in this study are available on request from the corresponding author.

Acknowledgments: The authors would like to thank Jana Bergant for permission to collect plant material from her plantation. The authors acknowledge the European Commission for funding InnoRenew CoE (grant agreement #739574), under the H2020 Widespread-Teaming program and Republic of Slovenia (investment funding of the Republic of Slovenia and the European Union's European Regional Development Fund). The authors would like to thank the InnoRenew CoE for providing the laboratory equipment for the sample preparation and the chemical analysis.

Conflicts of Interest: The authors declare no conflict of interest. The funders had no role in the design of the study; in the collection, analyses, or interpretation of data; in the writing of the manuscript, or in the decision to publish the results.

References

1. Committee on Herbal Medicinal Products (HMPC). *European Union Herbal Monograph on Helichrysum arenarium (L.)*; Flos: Merano, Italy, 2016.
2. Committee on Herbal Medicinal Products (HMPC). *Assessment Report on Helichrysum arenarium (L.)*; Flos: Moench, Italy, 2016.
3. Pljevljakušić, D.; Bigović, D.; Janković, T.; Jelačić, S.; Šavikin, K. Sandy Everlasting (*Helichrysum arenarium* (L.) Moench): Botanical, Chemical and Biological Properties. *Front. Plant Sci.* **2018**, *9*, 1123. [CrossRef] [PubMed]
4. Ninčević, T.; Grdiša, M.; Šatović, Z.; Jug-Dujaković, M. *Helichrysum italicum* (Roth) G. Don: Taxonomy, Biological Activity, Biochemical and Genetic Diversity. *Ind. Crop. Prod.* **2019**, *138*, 111487. [CrossRef]
5. Sala, A.; Recio, M.; Giner, R.M.; Máñez, S.; Tournier, H.; Schinella, G.; Ríos, J.-L. Anti-Inflammatory and Antioxidant Properties of *Helichrysum italicum*. *J. Pharm. Pharmacol.* **2002**, *54*, 365–371. [CrossRef]
6. Taglialatela-Scafati, O.; Pollastro, F.; Chianese, G.; Minassi, A.; Gibbons, S.; Arunotayanun, W.; Mabebie, B.; Ballero, M.; Appendino, G. Antimicrobial Phenolics and Unusual Glycerides from *Helichrysum italicum* Subsp. Microphyllum. *J. Nat. Prod.* **2013**, *76*, 346–353. [CrossRef] [PubMed]
7. Oliva, A.; Garzoli, S.; Sabatino, M.; Tadić, V.; Costantini, S.; Ragno, R.; Božović, M. Chemical Composition and Antimicrobial Activity of Essential Oil of *Helichrysum italicum* (Roth) G. Don Fil. (Asteraceae) from Montenegro. *Nat. Prod. Res.* **2019**, *34*, 445–448. [CrossRef] [PubMed]
8. D'Abrosca, B.; Buommino, E.; Caputo, P.; Scognamiglio, M.; Chambery, A.; Donnarumma, G.; Fiorentino, A. Phytochemical Study of *Helichrysum italicum* (Roth) G. Don: Spectroscopic Elucidation of Unusual Amino-Phlorogucinols and Antimicrobial Assessment of Secondary Metabolites from Medium-Polar Extract. *Phytochemistry* **2016**, *132*, 86–94. [CrossRef]
9. Antunes Viegas, D.; Palmeira-de-Oliveira, A.; Salgueiro, L.; Martinez-de-Oliveira, J.; Palmeira-de-Oliveira, R. *Helichrysum italicum*: From Traditional Use to Scientific Data. *J. Ethnopharmacol.* **2014**, *151*, 54–65. [CrossRef]
10. Griffiths, K.; Aggarwal, B.B.; Singh, R.B.; Buttar, H.S.; Wilson, D.; De Meester, F. Food Antioxidants and Their Anti-Inflammatory Properties: A Potential Role in Cardiovascular Diseases and Cancer Prevention. *Diseases* **2016**, *4*, 28. [CrossRef] [PubMed]
11. Ciulla, M.; Marinelli, L.; Cacciatore, I.; Stefano, A.D. Role of Dietary Supplements in the Management of Parkinson's Disease. *Biomolecules* **2019**, *9*, 271. [CrossRef] [PubMed]
12. Sõukand, R.; Quave, C.L.; Pieroni, A.; Pardo-de-Santayana, M.; Tardío, J.; Kalle, R.; Łuczaj, Ł.; Svanberg, I.; Kolosova, V.; Aceituno-Mata, L.; et al. Plants Used for Making Recreational Tea in Europe: A Review Based on Specific Research Sites. *J. Ethnobiol. Ethnomed.* **2013**, *9*, 58. [CrossRef]
13. Kramberger, K.; Barlič-Maganja, D.; Bandelj, D.; Baruca Arbeiter, A.; Peeters, K.; Miklavčič Višnjevec, A.; Jenko Pražnikar, Z. HPLC-DAD-ESI-QTOF-MS Determination of Bioactive Compounds and Antioxidant Activity Comparison of the Hydroalcoholic and Water Extracts from Two *Helichrysum italicum* Species. *Metabolites* **2020**, *10*, 403. [CrossRef]
14. Herrando Moraira, S.; Blanco Moreno, J.M.; Sáez, L.; Galbany Casals, M. Re-Evaluation of *Helichrysum italicum* Complex (Compositae: Gnaphalieae): A New Species from Majorca (Balearic Islands). *Collect. Bot.* **2016**, *35*, e009.

15. Baruca Arbeiter, A.; Hladnik, M.; Jakše, J.; Bandelj, D. First Set of Microsatellite Markers for Immortelle (*Helichrysum italicum* (Roth) G. Don): A Step towards the Selection of the Most Promising Genotypes for Cultivation. *Ind. Crop. Prod.* **2021**, *162*, 113298. [CrossRef]
16. Schuelke, M. An Economic Method for the Fluorescent Labeling of PCR Fragments. *Nat. Biotechnol.* **2000**, *18*, 233–234. [CrossRef] [PubMed]
17. Wang, H.-F.; Wang, Y.-K.; Yih, K.-H. DPPH Free-radical Scavenging Ability, Total Phenolic Content, and Chemical Composition Analysis of Forty-five Kinds of Essential Oils. *Int. J. Cosmet. Sci.* **2009**, *31*, 475–476. [CrossRef]
18. Spandidos, A.; Wang, X.; Wang, H.; Seed, B. PrimerBank: A Resource of Human and Mouse PCR Primer Pairs for Gene Expression Detection and Quantification. *Nucleic Acids Res.* **2010**, *38*, D792–D799. [CrossRef]
19. Pereira, C.G.; Barreira, L.; Bijttebier, S.; Pieters, L.; Neves, V.; Rodrigues, M.J.; Rivas, R.; Varela, J.; Custódio, L. Chemical Profiling of Infusions and Decoctions of *Helichrysum italicum* Subsp. Picardii by UHPLC-PDA-MS and in Vitro Biological Activities Comparatively with Green Tea (*Camellia sinensis*) and Rooibos Tisane (*Aspalathus linearis*). *J. Pharm. Biomed. Anal.* **2017**, *145*, 593–603. [CrossRef]
20. Galbany-Casals, M.; Blanco-Moreno, J.M.; Garcia-Jacas, N.; Breitwieser, I.; Smissen, R.D. Genetic Variation in Mediterranean *Helichrysum italicum* (Asteraceae; Gnaphalieae): Do Disjunct Populations of Subsp. Microphyllum Have a Common Origin? *Plant Biol.* **2011**, *13*, 678–687. [CrossRef]
21. Rosa, A.; Pollastro, F.; Atzeri, A.; Appendino, G.; Melis, M.P.; Deiana, M.; Incani, A.; Loru, D.; Dessì, M.A. Protective Role of Arzanol against Lipid Peroxidation in Biological Systems. *Chem. Phys. Lipids* **2011**, *164*, 24–32. [CrossRef] [PubMed]
22. Spagnol, C.M.; Assis, R.P.; Brunetti, I.L.; Isaac, V.L.B.; Salgado, H.R.N.; Corrêa, M.A. In Vitro Methods to Determine the Antioxidant Activity of Caffeic Acid. *Spectrochim. Acta Part A Mol. Biomol. Spectrosc.* **2019**, *219*, 358–366. [CrossRef] [PubMed]
23. Jeng, T.L.; Lai, C.C.; Liao, T.C.; Lin, S.Y.; Sung, J.M. Effects of Drying on Caffeoylquinic Acid Derivative Content and Antioxidant Capacity of Sweet Potato Leaves. *J. Food Drug Anal.* **2015**, *23*, 701–708. [CrossRef]
24. D'Andrea, G. Quercetin: A Flavonol with Multifaceted Therapeutic Applications? *Fitoterapia* **2015**, *106*, 256–271. [CrossRef]
25. Gradinaru, A.C.; Silion, M.; Trifan, A.; Miron, A.; Aprotosoaie, A.C. *Helichrysum arenarium* Subsp. Arenarium: Phenolic Composition and Antibacterial Activity against Lower Respiratory Tract Pathogens. *Nat. Prod. Res.* **2014**, *28*, 2076–2080. [CrossRef]
26. Czinner, E.; Kéry, A.; Hagymási, K.; Blázovics, A.; Lugasi, A.; Szöke, E.; Lemberkovics, E. Biologically Active Compounds of *Helichrysum arenarium* (L.) Moench. *Eur. J. Drug Metab. Pharmacokinet.* **1999**, *24*, 309–313. [CrossRef] [PubMed]
27. Sroka, Z.; Kuta, I.; Cisowski, W.; Dryś, A. Antiradical Activity of Hydrolyzed and Non-Hydrolyzed Extracts from Helichrysi Inflorescentia and Its Phenolic Contents. *Z. Nat. C* **2004**, *59*, 363–367. [CrossRef]
28. Mari, A.; Napolitano, A.; Masullo, M.; Pizza, C.; Piacente, S. Identification and Quantitative Determination of the Polar Constituents in *Helichrysum italicum* Flowers and Derived Food Supplements. *J. Pharm. Biomed. Anal.* **2014**, *96*, 249–255. [CrossRef] [PubMed]
29. Werner, J.; Ebrahim, W.; Özkaya, F.C.; Mándi, A.; Kurtán, T.; El-Neketi, M.; Liu, Z.; Proksch, P. Pyrone Derivatives from *Helichrysum italicum*. *Fitoterapia* **2019**, *133*, 80–84. [CrossRef]
30. Maksimovic, S.; Tadic, V.; Skala, D.; Zizovic, I. Separation of Phytochemicals from *Helichrysum italicum*: An Analysis of Different Isolation Techniques and Biological Activity of Prepared Extracts. *Phytochemistry* **2017**, *138*, 9–28. [CrossRef] [PubMed]
31. Cornara, L.; La Rocca, A.; Marsili, S.; Mariotti, M.G. Traditional Uses of Plants in the Eastern Riviera (Liguria, Italy). *J. Ethnopharmacol.* **2009**, *125*, 16–30. [CrossRef]
32. Pieroni, A. Medicinal Plants and Food Medicines in the Folk Traditions of the Upper Lucca Province, Italy. *J. Ethnopharmacol.* **2000**, *70*, 235–273. [CrossRef]
33. Bruni, A.; Ballero, M.; Poli, F. Quantitative Ethnopharmacological Study of the Campidano Valley and Urzulei District, Sardinia, Italy. *J. Ethnopharmacol.* **1997**, *57*, 97–124. [CrossRef]
34. Dabeek, W.M.; Marra, M.V. Dietary Quercetin and Kaempferol: Bioavailability and Potential Cardiovascular-Related Bioactivity in Humans. *Nutrients* **2019**, *11*, 2288. [CrossRef]
35. Zamora-Ros, R.; Cayssials, V.; Jenab, M.; Rothwell, J.A.; Fedirko, V.; Aleksandrova, K.; Tjønneland, A.; Kyrø, C.; Overvad, K.; Boutron-Ruault, M.-C.; et al. Dietary Intake of Total Polyphenol and Polyphenol Classes and the Risk of Colorectal Cancer in the European Prospective Investigation into Cancer and Nutrition (EPIC) Cohort. *Eur. J. Epidemiol.* **2018**, *33*, 1063–1075. [CrossRef] [PubMed]
36. Apaydin Yildirim, B.; Kordali, S.; Terim Kapakin, K.A.; Yildirim, F.; Aktas Senocak, E.; Altun, S. Effect of *Helichrysum plicatum* DC. Subsp. Plicatum Ethanol Extract on Gentamicin-Induced Nephrotoxicity in Rats. *J. Zhejiang Univ. Sci. B* **2017**, *18*, 501–511. [CrossRef] [PubMed]
37. Mattson, M.P. Hormesis and Disease Resistance: Activation of Cellular Stress Response Pathways. *Hum. Exp. Toxicol.* **2008**, *27*, 155–162. [CrossRef] [PubMed]
38. Calabrese, V.; Santoro, A.; Monti, D.; Crupi, R.; Di Paola, R.; Latteri, S.; Cuzzocrea, S.; Zappia, M.; Giordano, J.; Calabrese, E.J.; et al. Aging and Parkinson's Disease: Inflammaging, Neuroinflammation and Biological Remodeling as Key Factors in Pathogenesis. *Free Radic. Biol. Med.* **2018**, *115*, 80–91. [CrossRef]
39. Calabrese, E.J.; Calabrese, V.; Tsatsakis, A.; Giordano, J.J. Hormesis and Ginkgo Biloba (GB): Numerous Biological Effects of GB Are Mediated via Hormesis. *Ageing Res. Rev.* **2020**, *64*, 101019. [CrossRef] [PubMed]

40. Ping, Z.; Liu, W.; Kang, Z.; Cai, J.; Wang, Q.; Cheng, N.; Wang, S.; Wang, S.; Zhang, J.H.; Sun, X. Sulforaphane Protects Brains against Hypoxic–Ischemic Injury through Induction of Nrf2-Dependent Phase 2 Enzyme. *Brain Res.* **2010**, *1343*, 178–185. [CrossRef] [PubMed]
41. Siracusa, R.; Scuto, M.; Fusco, R.; Trovato, A.; Ontario, M.L.; Crea, R.; Di Paola, R.; Cuzzocrea, S.; Calabrese, V. Anti-Inflammatory and Anti-Oxidant Activity of Hidrox® in Rotenone-Induced Parkinson's Disease in Mice. *Antioxidants* **2020**, *9*, 824. [CrossRef] [PubMed]
42. Martel, J.; Ojcius, D.M.; Ko, Y.-F.; Ke, P.-Y.; Wu, C.-Y.; Peng, H.-H.; Young, J.D. Hormetic Effects of Phytochemicals on Health and Longevity. *Trends Endocrinol. Metab.* **2019**, *30*, 335–346. [CrossRef] [PubMed]
43. Duda, S.C. Changes in Major Bioactive Compounds with Antioxidant Activity of Agastache Foeniculum, Lavandula Angustifolia, Melissa Officinalis and Nepeta Cataria: Effect of Harvest Time and Plant Species. *Ind. Crop. Prod.* **2015**, *77*, 499–507. [CrossRef]

antioxidants

MDPI

Article

Onion Peel: Turning a Food Waste into a Resource

Rita Celano [1,†], Teresa Docimo [2,†], Anna Lisa Piccinelli [1,*], Patrizia Gazzerro [1], Marina Tucci [2], Rosa Di Sanzo [3], Sonia Carabetta [3], Luca Campone [4], Mariateresa Russo [3,*] and Luca Rastrelli [1]

1 Department of Pharmacy, University of Salerno, Via Giovanni Paolo II 132, 84084 Fisciano, Italy; rcelano@unisa.it (R.C.); pgazzerro@unisa.it (P.G.); rastrelli@unisa.it (L.R.)

2 Institute of Bioscience and BioResources, National Research Council, Via Università 100, 80055 Portici, Italy; teresa.docimo@ibbr.cnr.it (T.D.); marina.tucci@ibbr.cnr.it (M.T.)

3 Department of Agriculture Science, Food Chemistry, Safety and Sensoromic Laboratory (FoCuSS Lab), University of Reggio Calabria, Via Salita Melissari, 89124 Reggio Calabria, Italy; rosa.disanzo@unirc.it (R.D.S.); sonia.carabetta@unirc.it (S.C.)

4 Department of Biotechnology and Biosciences, University of Milano-Bicocca, Piazza Della Scienza 2, 20126 Milan, Italy; luca.campone@unimib.it

* Correspondence: apiccinelli@unisa.it (A.L.P.); mariateresa.russo@unirc.it (M.R.)

† These authors contributed equally to this work.

Abstract: Food waste is a serious problem for food processing industries, especially when it represents a loss of a valuable source of nutrients and phytochemicals. Increasing consumer demand for processed food poses the problem of minimizing waste by conversion into useful products. In this regard, onion (*Allium cepa*) waste consisting mainly of onion skin is rich in bioactive phenolic compounds. Here, we characterized the flavonoid profiles and biological activities of onion skin wastes of two traditional varieties with protected geographical indication (PGI), the red "Rossa di Tropea" and the coppery "Ramata di Montoro", typically cultivated in a niche area in southern Italy. The phytochemical profiles of exhaustive extracts, characterized by ultra-high-performance liquid chromatography coupled with ultraviolet (UV) detection and high-resolution mass spectrometry, revealed that flavonols and anthocyanins were the characteristic metabolite classes of onion skins. Quercetin, quercetin glucosides and their dimer and trimer derivatives, and, among anthocyanins, cyanidin 3-glucoside, were the most abundant bioactive compounds. The potential of onion skins was evaluated by testing several biological activities: ABTS/oxygen radical absorbance capacity (ORAC) and in vitro alpha-glucosidase assays were performed to evaluate the antioxidant and anti-diabetic properties of the extracts and of their main compounds, respectively, and proliferative activity was evaluated by MTT assay on human fibroblasts. In the present study, by observing various biological properties of "Rossa di Tropea" and "Ramata di Montoro" onion-dried skins, we clearly indicated that this agricultural waste can provide bioactive molecules for multiple applications, from industrial to nutraceutical and cosmetical sectors.

Keywords: sustainable agriculture; onion skin; traditional varieties; flavonols; dietary antioxidants,

Citation: Celano, R.; Docimo, T.; Piccinelli, A.L.; Gazzerro, P.; Tucci, M.; Di Sanzo, R.; Carabetta, S.; Campone, L.; Russo, M.; Rastrelli, L. Onion Peel: Turning a Food Waste into a Resource. *Antioxidants* **2021**, *10*, 304. https://doi.org/10.3390/antiox10020304

Academic Editor: Francesca Giampieri

Received: 13 January 2021
Accepted: 11 February 2021
Published: 16 February 2021

1. Introduction

The increasing progression of non-communicable diseases (NCD), such as cardiovascular diseases, cancer, and diabetes, as the main causes of deaths worldwide has prompted an improvement of dietary habits [1]. It has been shown that the daily intake of phenolics-rich food or polyphenols-enriched food supplements is highly related to NCD disease prevention. The Mediterranean diet is usually associated with health benefits and favorable life expectancy, due to the daily consumption of different types of vegetables and fruit as a rich source of antioxidants [2,3]. This is particularly true for some key aliments of the Mediterranean diet, such as vegetables belonging to the *Allium* genus, such as onion, garlic, scallion, and chives, whose protective role towards the prevention of several chronic disorders has been widely proved [4]. The multilayer tissue of onion bulbs is a rich source of bioactive compounds belonging to two main chemical groups: the alk(en)yl cysteine sulfoxides and flavonoids. The first group of molecules accounts for the characteristic odor and taste of onion, thus their composition and content determine unique flavors in the different species/varieties [5,6]. Regarding color, different colors of skin and bulbs, ranging from red to yellow, are strictly dependent on flavonoid composition, where the yellow color is mostly due to quercetin and derivatives [7], while red is due to anthocyanins [8]. As recently reported, onion bulbs are ranked among the best sources of dietary flavonoids, mostly belonging to flavonols and anthocyanins [9]. Quercetin glycosides are the predominant flavonols in all onion cultivars whereas the anthocyanins are mainly present in red onions, where they account for approximately 10% of the total flavonoid content of fresh weight [9,10].

Over the past 20 years, world production of onion has raised at least by 25%; [11] reports a production of about 47 million tons every year, becoming the second most important horticultural crop. This is also because lifestyle changes increased the demand for fresh-cut ready-to-use vegetables, including onion. Increased market of processed onion has also led to a higher waste accumulation, and the main European onion producers, Spain, Netherlands, and the UK, reached a waste production of about 50,000 t [12–14]. Therefore, to face the high production of food products ingredients and waste, there is the need to find sustainable agricultural processes and develop sustainable solutions for recovering key natural products. The onion skin, being not edible, is fast removed before processing and sale, but is unsuitable as fodder or landfill disposal because of its composition and smell, thus representing the major waste of onion processing. Therefore, sustainable production of *Allium* requires valorizing onion skin by turning this food waste into "food by-product". By-products from food processing often contain valuable molecules, which can be used as functional ingredients in food, cosmetic, and pharmaceutical industries [15]. In this regard, chemical characterization and knowledge of biological activities of onion skin is necessary for the development of optimal and efficient systems of resource recovery.

The growing demand for food with improved nutritional value and increased sustainability due to the value of its by-products, has increased the interest toward the selection of improved agronomic species and/or local varieties producing food and food ingredients with enriched content of bioactive phytomolecules and improved health beneficial properties [8,10,16,17]. In this regard, the cultivation of superior varieties also offers the opportunity to recover high value by-products. The dried skin of onion is a rich source of flavonols with a high percent of aglycone forms, whereas the content of anthocyanins is limited to red-colored varieties. Most studies have been focused on chemical and biological characterization of edible onion bulbs whereas less attention has been addressed to onion skins because of its inedible nature. [18] reported the characterization of several compounds and of the antioxidant activity of brown and red onion skins from Spanish cultivars, highlighting the brown onion outer layer as the better waste source to valorize. Another study on Californian onions [19], characterized flavonoids in the outer layer of several commercially important onion varieties, highlighting that Californian varieties are richest in flavonoids than some unmentioned Italian varieties. However, literature on onion skins of Italian varieties, is still very scant and especially regarding red skin onion cultivars. Indeed, two commercially important traditional Italian PGI onion varieties, known as "Rossa di Tropea"" and "Ramata di Montoro" are of great interest. The red "Rossa di Tropea", from Calabria and the coppery "Ramata di Montoro" from Campania local varieties are outstanding food products with several peculiar organoleptic properties conferring distinctive tastes. The Calabrian variety has a typical red-colored envelope and a unique sweetness, whereas the "Ramata di Montoro" has a coppery skin and a delicate and persisting aroma. [8] reported flavonoids and sugar content in the bulb and skin of "Rossa di Tropea" onion, underlying the highest flavonol concentration in this local cultivar than other red Italian onions [6]. Nevertheless, being the high content of anthocyanins concentrated in the inedible outer layer of "Rossa di Tropea" onion, the health promoting potential was not associated with a possible dietary intake. More recently, [10] reported the identification and quantification of polyphenols of "Rossa di Tropea" and "Ramata di Montoro" Italian varieties, showing "Rossa di Tropea" onion bulbs having higher content of bioactive polyphenols, and both exhibiting a strong protective role from oxidative damage thus highlighting a possible synergistic role of the constituents of onion bulbs.

To valorize further the "Rossa di Tropea" and "Ramata di Montoro" onion cultivation, we focused on onion skins to envisage their potential nutraceutical and pharmaceutical use. In this study, we report the chemical characterization and screening of the biological activities of onion skin extracts from these two southern cultivars. We performed a quali-quantitative characterization by ultra-high-performance liquid chromatography (UHPLC) coupled with UV detection and high-resolution mass spectrometry (UHPLC-UV-HRMS) of polyphenols from "Ramata di Montoro" and "Rossa di Tropea" dried skin onions along with an accurate chemical identification of several distinctive compounds by nuclear magnetic resonance (NMR) techniques. Moreover, since phytochemicals can be used in different industrial applications, including the food industry, for the development of functional or enriched foods and food additives, the health industry, for medicines and pharmaceuticals, and/or functional ingredients for cosmetics, we analyzed the biological activities of the onion skin extracts and of their main constituents by evaluation of their antioxidant activities by ABTS and oxygen radical absorbance capacity (ORAC) methods. In addition, we analyzed the potential of the extracts as modulators of postprandial glucose blood levels by testing their activity as alpha-glucosidase inhibitors, and evaluated cell viability by MTT assay on human fibroblasts to assess their safe use in a cosmetic dermatological formulation. by characterizing onion waste, our study aims to promote its use for the production of various crucial bioactive components as step toward sustainable production.

2. Materials and Methods

2.1. Reagents and Materials

Ultrapure water (18 MΩ) was obtained using a Milli-Q purification system (Millipore, Bedford, TX, USA). MS-grade water (H_2O), and acetonitrile (CH_3CN) were supplied by Romil (Cambridge, UK). Analytical grade methanol (MeOH), CH_3CN, ethyl acetate, n-buthanol (n-BuOH), chloroform, formic acid (HCOOH) and absolute ethanol (EtOH) were supplied by Carlo Erba Reagents (Milan, Italy). Deuterated methanol (99.8%, CD3OD), dimethyl sulfoxide (DMSO), sephadex LH20, MS-grade formic acid (HCOOH), quercetin, isorhamnethin, kaempferol, (±)-6-hydroxy-2,5,7,8-tetramethylchromane-2-carboxylic acid (Trolox) and 2,2′-azinobis (3-ethylben-zothiazoline-6-sulphonic acid) diammonium salt (ABTS), sodium fluorescein and 2,2′-azobis(2-methylpropionamidine) dihydrochloride were provided by Merk Chemical (Milan, Italy). Cyanidine-3-*O*-glucoside (CyG) was purchased from Extrasynthese (Lyon, France).

2.2. Plant Material and Exhaustive Extraction

Outer dry protective layers of brown skin onion bulbs (*Allium cepa* L.) of "Ramata di Montoro" were supplied by the company "Gaia Società Semplice Agricola" (Montoro, Avellino, Italy) located in the production area of onion PGI "Ramata di Montoro". Outer dry protective layers of red skin onion bulbs (*Allium cepa* L.) of PGI "Rossa di Tropea" were provided from Calabrian farmers which organically cultivated and harvested "Rossa di Tropea" onions in Calabria region (Tropea, Vibo Valentia, Italy). Samples were collected after the harvest and naturally dried then before being grounded and sieved according to [16] (Figure S1). The powders were used for exhaustive extraction.

Exhaustive extraction was performed using ultrasound-assisted solid-liquid extraction (UAE) at 25 °C in a thermostat-controlled ultrasound bath (Labsonic LBS2, Treviglio, Italy) at the frequency of 20.0 kHz. Dried *Allium cepa* skins from "Rossa di Tropea" and "Ramata di Montoro" onion varieties were extracted using aqueous EtOH (70% v/v) and a matrix/solvent ratio of 1:20 (3 × 30 min). At each extraction cycle, after centrifugation 10 min at 9000 g, the extracts of each variety were pooled, filtered (Whatman No. 1 filter) and freeze-dried (freeze dryer Alpha 1–2 LD, Christ, Germany), after the removal of the organic solvent under vacuum at 40 °C in a rotary evaporator (Rotavapor R-200, Buchi Italia s.r.l, Cornaredo, Italy). Extraction yields of 11.6% and 22.6% (g extract per 100 g dry skin, DM) were obtained for "Ramata di Montoro" and "Rossa di Tropea" onion skins, respectively. Exhaustive "Ramata di Montoro" and "Rossa di Tropea" extracts were then used for further chemical characterization and biological tests.

2.3. UHPLC-HRMS^n Analysis

Identification of phenolic compounds in the exhaustive skin extracts was carried out using a similar approach and same equipment as reported in [16]. The chromatographic analyses were performed using a PLATINblue UHPLC chromatographic system (Knauer GmbH, Berlin, Germany), consisting of two high-pressure pumps, an autosampler, a column temperature control system and a photodiode array detector, coupled to high-resolution mass spectrometer LTQ Orbitrap (ThermoFisher Scientific, Milan, Italy) equipped with an heated electrospray ionization (HESI) source. Chromatographic separation was performed using a Kinetex C18 column (2.1 × 50 mm, 1.7 µm; Phenomenex, Torrance, CA, USA), thermostated at 30 °C, and a binary gradient H_2O (A) and CH_3CN (B) both containing 0.1% HCOOH, at a flow of 600 µL / min. The elution gradient used is the following: 2% B, 0–3 min, 2–13% B, 3–5 min, 13% B, 5–9 min, 13–18% B, 9–12 min, 18% B, 12–13 min, 18–30% B 13–17 min, 30–50% B, 17–21 min, 50–98% B, 21–22 min, 98% B, 22–27 min. Injection volume was set at 5 µL. The mass spectrometer was used in negative and positive ionization mode, high purity nitrogen (N_2) was used as sheath gas and auxiliary gas, 30 and 10 arbitrary units, respectively. High purity helium (He) was used as collision gas. Mass spectrometer parameters were as follows: source voltage 4.0 kV; capillary voltage, −33 V; tube lens voltage, −41.4 V; capillary temperature, 300 °C; MS spectra were acquired by full range acquisition covering 140–1500 m/z. For fragmentation study, a data dependent scan was performed, and the normalized collision energy of the collision induced dissociation (CID) cell was set at 30 eV and the isolation width of precursor ions was set at 2.0 m/z. The resolution was 60,000 and 7500 for the full mass and the data dependent MS scan, respectively. Phenolic compounds were characterized according to the corresponding mass spectra, accurate mass, characteristic fragmentation, and retention time. Xcalibur software (version 2.2) was used for instrument control, data acquisition, and data analysis.

2.4. Isolation and Identification of Onion Skin Flavonols

Major flavonols of onion skin were isolated from "Ramata di Montoro" onion skin using a purification procedure guided by UHPLC-UV analysis (Section 2.5). The exhaustive skin extract was suspended in distilled water and partitioned with ethyl acetate (200 mL × 5) and n-BuOH (150 mL × 3). For each extraction solvent, the organic phases were pooled and vacuum dried with a rotary evaporator at 40 °C (yields of 59% and 15% for ethyl acetate and n-BuOH portions, respectively). Subsequently, a portion of ethyl acetate extract (about 3 g) was fractionated over a Sephadex LH 20 column (60 g, 1 m × 3 cm i.d), using methanol as eluent at a flow of 0.5 mL min^{-1}. Fractions (8 mL each) were collected and analyzed by TLC (Si−gel, chloroform/methanol/water, 80: 20: 2, v/v) and highlighted with Ce $(SO_4)/H_2SO_4$. Fractions with similar Rf were combined into 12 major fractions (I–XII) and analyzed by UHPLC-UV. The fractions containing target flavonols were further purified by semipreparative high-performance liquid chromatography with refractive index detector (HPLC−IR) on a Luna C18 column (250 × 10 mm i.d., 10 µm, Phenomenex, Bologna, Italy). In detail, fraction IV was separated using $MeOH/H_2O$ 6:4 v/v as mobile phase (flow rate of 1.8 mL min^{-1}) to yield compound **11** (quercetin 4′ glucoside, QG). Fractions V, X and XII were separated with $MeOH/H_2O$ 70:30 v/v (flow rate of 1.8 mL min^{-1}) to yield compound **13** (quercetin, Q), **21** (quercetin dimer, Q2) and **22** (quercetin trimer, Q3), respectively. Compounds **19** (Q2Gb) and **18** (Q2Ga) were isolated from fractions VII and VIII using as mobile phase $MeOH/H_2O$ 65:35 v/v (flow rate of 2 mL min^{-1}). Fraction II contained compounds **2** (Qox) with a degree of purity suitable for NMR analysis. The butanol extract was directly purified by HPLC-IR using MeOH/H_2O 40:60 v/v as mobile phase (flow of 3.0 mL min^{-1}) to yield compound **6** (quercetin-3, 4′-*O*-diglucoside, QdG). The degree of purity of the isolated compounds was evaluated by HPLC-DAD analysis (220–600 nm). A Bruker DRX 600 spectrometer operating at 599.19 MHz for ^1H and 150.86 MHz for ^{13}C was used for the NMR spectra. For the measurement of the ^1H NMR and ^{13}C NMR spectra in CD_3OD the relative signal of CHD_2OD (δH = 3.31, δC = 49.05) was used as standard. NMR data of isolated compounds were compared with literature data [20] to establish their structures.

2.5. Quantitative Analysis by UHPLC-UV

Quantitative UHPLC-UV analyses of onion skin extracts were performed using a Dionex Ultimate 3000 UHPLC chromatographic system (ThermoFisher, Milan, Italy), consisting of two ternary pumps, an autosampler, a column temperature control system and a UV/Vis detector. Chromatographic separation was performed using UHPLC-HRMS assay conditions. UV spectra were acquired in the range of 200–600 nm, and three wavelengths, 295, 365, and 520 nm were selected for the detection of target analytes based on their maximum absorbance. External standard method was employed to determine the levels of main skin compounds. Stock solutions of commercial standards (Q and CyG) and isolated standards (Q-dG, QG, Q2 and Q3) were prepared in DMSO at a concentration of 5 mM, and stored at −20 °C. The calibration levels were prepared from the stock solutions by appropriate serial dilutions with MeOH/H_20 1:1, v/v, and analyzed in triplicate. The linearities of the calibration curves were evaluated in the concentration range of 6.25–100 μM for Q, QG and Q-dG, 25–100 μM for Q2Ga, Q2Gb, Q2, Q3, and CyG. The regression curves were tested with the analysis of variance (ANOVA) and linear model was found appropriate over the tested concentration range (R^2 values > 0.998). The samples were analyzed after appropriate dilutions, to fall within the dynamic calibration range.

2.6. Antioxidant Assays

The antioxidant activities (AOAs) of onion skin exhaustive extracts and their major components (Q, QG, Q2, Q3, and CyG) were evaluated using ABTS scavenging capacity and ORAC assays according to [21–23] and using the procedures adapted for use in 96-well plates as reported in [24]. A microplate spectrophotometer reader Multiskan Go (Thermo Scientific) and a multimode plate reader EnSpire 2300 (E, PerkinElmer, Waltham, MA, USA) were employed for ABTS and ORAC assays, respectively. ABTS and ORAC assay results were expressed as Trolox equivalent antioxidant capacity (TEAC) per g of extracts (μmol TE/g) or per μmol of pure compound (μmol TE/μmol). In the ABTS assay, the curves of Trolox, pure compounds and onion skin extracts were obtained by plotting concentration (mg mL^{-1} for onion skin extracts and mM for Trolox and standards) against the average % Inhibition of radical absorbances (($Abs_{control} - Abs_{sample})/Abs_{control} \times 100$). Concentrations corresponding to % Inhibition of 50 (50% I) were extrapolated from curves and TEAC was calculated as:

Trolox $concentration_{50\%I}$/Sample $concentration_{50\%I}$ × g extract or μM pure compounds.

In ORAC assay, the net area under the FL decay curve (AUC) vs concentration curves were considered, and TEAC values were calculated as:

Trolox $concentration_{net\ AUC}$/Sample $concentration_{net\ AUC}$ × g extract or μM pure compounds.

2.7. Alpha-Glucosidase Inhibitory Activity

The assay was carried as reported by [25]. 10mg of dried EtOH extract from "Rossa di Tropea" and "Ramata di Montoro" onion skins, were dissolved in EtOH and appropriately diluted in 0.05 M phosphate buffer. A mixture containing 25 μL of extract (final concentrations of 25–150 μg/mL), 25 μL of α-glucosidase (0.2 U ml^{-1}, from Baker's yeast), and 175 μL of phosphate buffer (50 mM, pH 6.8) was left to stand for 10 min at room temperature. The reaction was started by the addition of 25 μL of 23.2 mM *p*-nitrophenyl-α-

D-glucopyranoside (2.5 mM) and incubated for 15 min at 37 °C. The assay was conducted in a 96-well plate, and the absorbance was determined at 405 nm using a Bio-Rad microplate reader (Bio-Rad, Richmond, CA, USA). The amount of released *p*-nitrophenol was determined spectrophotometrically, measuring the absorbance of the solution at 405 nm. The percent inhibition was calculated as follows:

$$\text{Abs } 100\% - \text{Abs } 0\%) - (\text{Abs sample} - \text{Abs blank})/(\text{Abs } 100\% - \text{Abs } 0\%) \times 100$$

where: Abs 100% is the absorbance of 100% enzyme activity (reaction mixture containing only the enzyme and the substrate); Abs 0% is the absorbance of 0% enzyme activity (reaction mixture containing the substrate and without enzyme); Abs sample is the absorbance of the reaction mixture, containing the substrate, enzyme, and the tested extract; Abs blank is the absorbance of the reaction mixture without the enzyme, containing only the substrate and tested extract.

2.8. *Cell Culture and Cell Viability Assay*

Human dermal fibroblast (HDFA) cell line was obtained from Gibco Life Corporation (ThermoFischer Scientific, Milan, Italy) and maintained in M106 medium supplemented with Low Serum Growth Supplement (LSGS), 100 mg/L streptomycin and penicillin 100 IU/mL at 37 °C in a humidified atmosphere of 5% CO_2. Stock solutions (20 mg/mL) of extracts in DMSO were stored in the dark at 4 °C until use. Appropriate dilutions were prepared in culture medium immediately prior to use. In all experiments, the final concentration of DMSO did not exceed 0.15% (*v*/*v*).

To evaluate cell viability the colorimetric MTT (3-(4,5 di-methylthiazol-2-yl)-2,5-diphenyltetrazolium bromide) metabolic activity assay was performed as described previously. Briefly, 6×10^3 cells/well were seeded in 96-well plates and exposed to increasing concentrations of extracts and purified compounds for 44 h. MTT stock solution (5 mg/mL in PBS, Sigma) was added to each well and incubated for 4 h at 37 °C in humidified 5% CO_2 atmosphere. The formazan crystals were solubilized with acidic isopropanol (0.1 N HCl in absolute isopropanol). MTT conversion to formazan by metabolically viable cells was monitored by spectrophotometer at an optical density of 595 nm. Each data point represents the average of three separate experiments in triplicate.

2.9. *Statistical Analysis*

Statistical analyses were performed with Sigma Plot The criterion for statistical significance was $p < 0.05$. All values are expressed as means of at least 3 values ± SD.

3. Results and Discussion

3.1. *UHPLC-DAD-HRMS/MS Analysis of "Rossa di Tropea" and "Ramata di Montoro" Onion Skins*

The qualitative profiles of both onion-dried skin extracts were determined by UHPLC-HRMS/MS analysis, taking advantage from the our previous qualitative findings on composition of supercritical fluid extraction (SFE) extract of "Ramata di Montoro" onion skin [16]. In this study, the previous chromatographic conditions were adapted to optimize the detection of anthocyanins in the "Rossa di Tropea" type, particularly 1% formic acid was added to the mobile phases and sample solutions either to improve the chromatographic resolution and to increase the UV absorbance of anthocyanins. Moreover, MS analyses were performed in positive and negative ionization modes to extend the ability of UHPLC-HRMS/MS method to detection and characterization of anthocyanins. Figure 1 shows the UV (365 and 520 nm) profiles of "Ramata di Montoro" and "Rossa di Tropea" onion skin extracts under optimal chromatographic condition. Metabolite assignment were made by comparing retention time, UV/Vis and HRMS/MS data of detected compounds, whenever available, or interpreting MS data combined with chemo-taxonomic data reported in the literature or databases. Twenty-one (**1–22**) major peaks were detected, and their retention times, λ_{max} values and HRMS data are listed in Table 1.

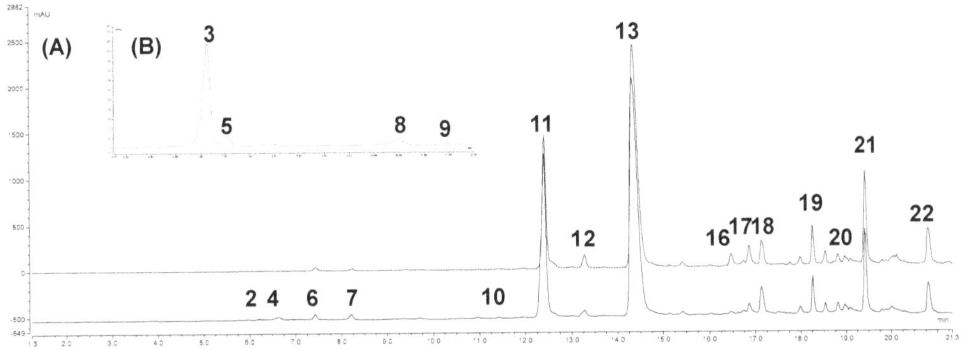

Figure 1. (**A**) UHPLC-UV profile (365 nm) of "Rossa di Tropea" onion skin (solid line) and "Ramata di Montoro" onion skin extract (dashed line); (**B**) UHPLC-UV profile (520 nm) of "Rossa di Tropea" onion skin.

"Ramata di Montoro" and "Rossa di Tropea" onion skin extracts showed a comparable qualitative profile except for the exclusive presence of anthocyanins (**3**, **5**, **8** and **9**) in "Rossa di Tropea". In both extracts, most of the detected compounds were flavonols, in particular quercetin derivatives, and the isorhamnetin glycosides as minority compounds. No difference in flavonol profiles was observed between the two onion varieties (Figure 1). Both extracts contained flavonol glycosides, with QG (**11**) as major flavonol glycoside. In addition to QG, two mono-*O*-hexoside derivatives (**10** and **11**) and two di-O-hexosides derivatives of quercetin (**4** and **6**) were identified in onion skin extracts (Table 1). Also, isorhamnetin glycosides, namely isorhamnetin-*O*-dihexoside (**7**) and isorhamnetin-*O*-hexoside (**12**), were detected as minority compounds (Table 1). These results are in agreement with data on the flavonols composition of *Allium cepa* [4,8–10,17,26]. Besides flavonol glycosides, quercetin oxidation products were also detected in the onion skins. Quercetin is the most abundant flavonol of the onion skin, according to previous studies [9,16,18], and it can undergo oxidation by enzymes contained in plant tissues, such as polyphenol oxidase and peroxidase, generating various oxidation products [27,28]. The oxidized derivatives of quercetin were detected only in the onion skin [16,20,29] and they are considered aging and senescence products [30,31]. It has been hypothesized that they contribute to the browning of the outer onion scales [30,32]. Generally, polyphenol oxidized products in plant foods, are considered unwanted products, as they are responsible for organoleptic changes. However, in some cases (black tea and coffee), the oxidation processes can generate nutritionally and functionally interesting products, worthy of attention and further study [30,31]. In both skin extracts several oxidized derivatives of quercetin were detected: one monomer (**2**), four dimers (**18**, **19**, **20** and **21**) of which three glycosylates and one trimer (**22**) (Table 1). These compounds were tentatively identified by searching the databases for the structure corresponding to the molecular formula calculated from the accurate mass, and comparing their MS/MS spectra with data reported in [16]. In this study, the proposed structures were confirmed by isolation and NMR analysis (Section 3.2). Other compounds found in both onion skins were protocatechuic acid (**1**), isorhamnethin (**17**) and kaempferol (**16**), identified by comparison with reference standards, and two isomers, **14** and **15**, at *m/z* 453.0455 and with calculated molecular formula $C_{22}H_{14}O_{11}$ (Table 1). According to [20], they might correspond to isomers of quercetin linked to a residue of protocatechuic acid. However, their MS/MS spectra do not correlate with these structures and therefore their identification could not be confirmed.

Table 1. UHPLC-HRMS/MS data of detected compounds in "Ramata di Montoro" (M) and "Rossa di Tropea" (T) onion-dried skin extracts.

N°	Compound	Molecular Formula	RT(UV) (min)	Measured (m/z) [M − H]⁻ (m/z)	Error (ppm)	Product Ion MS/MS	Measured (m/z) [M + H]⁺ (m/z)	Error (ppm)	Product Ion MS/MS	Onion Cultivar
1	Protocatechuic acid	$C_7H_6O_4$	1.0	153.1235	1.6	/	/	/	/	M, T
2	2-(3,4-Dihydroxybenzoyl)-2,4,6-trihydroxy-3(2H)-benzofuranone (Qox)	$C_{15}H_{10}O_8$	6.0	317.0303	3.3	299; 191; 207; 273	/	/	/	M, T
3	Cyanidin 3-glucoside (CyG)	$C_{21}H_{21}O_{11}$	6.2	/	/	/	449.1077	−0.3	287	T
4	Quercetin dihexoside	$C_{27}H_{30}O_{17}$	6.6	625.1405	0.9	463; 301	627.1557	0.3	465; 303	M, T
5	Cyanidin 3-laminaribioside	$C_{27}H_{31}O_{16}$	7.0	/	/	/	611.1609	0.3	287	T
6	Quercetin 3,4′-diglucoside (QdG)	$C_{27}H_{30}O_{17}$	7.4	625.1410	−0.3	463; 301	627.1559	0.6	465; 303	M, T
7	Isorhamnetin dihexoside	$C_{28}H_{32}O_{17}$	8.2	639.1571	0.04	477; 315	641.1716	0.5	317; 479	M, T
8	Cyanidin 3-malonilglucoside	$C_{24}H_{23}O_{14}$	10.5	/	/	/	535.1084	0.3	287	T
9	Cyanidin 3-malonillaminaribioside	$C_{30}H_{33}O_{19}$	11.4	/	/	/	697.1609	0.2	287	T
10	Quercetin-3-glucoside	$C_{21}H_{20}O_{12}$	11.6	463.0874	0.6	301	465.1027	−0.2	303	M, T
11	Quercetin-4′-glucoside (QG)	$C_{21}H_{20}O_{12}$	12.4	463.0873	0.8	301	465.1025	−0.6	303	M, T
12	Isorhamnetin-O-hexoside	$C_{22}H_{22}O_{12}$	13.3	477.1031	0.7	315	479.1186	0.5	317	M, T
13	Quercetin (Q)	$C_{15}H_{10}O_7$	14.3	301.0350	0.6	179; 151	303.0496	−1.0	285; 257; 229;	M, T
14	Protocatecoyl quercetin	$C_{22}H_{14}O_{11}$	14.8	453.0454	0.4	299	455.0610	0.1	437; 301;	M, T
15	Protocatecoyl quercetin	$C_{22}H_{14}O_{11}$	14.9	453.0455	0.6	299	455.00609	0.1	437; 301;	M, T
16	Kaempferol	$C_{15}H_{10}O_6$	16.3	285.0399	2.1	/	287.0548	−0.6	/	M, T
17	Isorhamnetin	$C_{16}H_{12}O_7$	16.9	315.0503	1.2	300; 257	317.0656	0.0	302, 285; 257	M, T
18	Quercetin dimer 4′-glucoside (Q2Ga)	$C_{36}H_{28}O_{19}$	17.2	763.1140	−0.2	611; 449;	765.1300	0.3	603; 451	M, T
19	Quercetin dimer 4′-glucoside (Q2Gb)	$C_{36}H_{28}O_{19}$	18.3	763.1139	−0.3	611; 600; 299	765.1298	0.1	603; 585	M, T
20	Quercetin dimer hexoside	$C_{36}H_{28}O_{19}$	18.8	763.1139	−0.2	611; 600; 299	765.1299	0.2	603; 585	M, T
21	Quercetin dimer (Q2)	$C_{30}H_{18}O_{14}$	19.4	601.0617	0.8	449; 299	603.0772	0.5	585; 313; 303	M, T
22	Quercetin trimer (Q3)	$C_{45}H_{26}O_{21}$	20.8	901.0881	−0.2	299; 449; 599; 601	903.1044	0.5	885; 751; 585; 613	M, T

Anthocyanins are the distinctive feature of red onion varieties and they are responsible for their characteristic red/purple color [9,33]. They are highly concentrated in the skin and outer fleshy layer, whereas in the edible tissue they are limited to a single layer of cells of the epidermal tissue [8]. The most frequently reported anthocyanins in red onion are cyanidin derivatives [9]. Few studies are reported in the literature about the anthocyanins detected in onion skin [8,34,35], but most of the available data refer to the whole onion bulb [9,10]. In "Rossa di Tropea" onion skin extracts, four cyanidin derivatives (**3**, **5**, **8** and **9**) were detected. Compounds **3**, **5**, **8** and **9** showed characteristic UV spectra of anthocyanins with an absorption maximum 520 nm [36]. As is well known, anthocyanins and their derivatives ionized better in the positive mode. Therefore, in HRMS/MS spectra, these compounds were characterized by the product ion at m/z 287.0550 (ppm 1.02 to 2.52) corresponding to cyanidin aglycone, generated by the loss of one hexoside and one laminaribioside unit for **3** and **5**, respectively, and the malonylhexoside and malonyllaminaribioside unit, for **8** and **9**, respectively. Compound **3** was unequivocally identified as CyG by comparison with reference standards. According to literature data, **5**, **8**, and **9** were tentatively identified as cyanidin 3-laminaribioside, cyanidin 3-malonylglucoside (CymG) and cyanidin 3-malonyllaminaribioside, respectively [34,35]. In general, the obtained results are in agreement with literature data [9]. Cyanidin 3-glucoside and cyanidin 3-laminaribioside are considered the main anthocyanins identified in red onions and the acetylated anthocyanins with malonic acid are the predominant pigments detected in the anthocyanin profile of red onion cultivars [9]. The literature data regarding the phenolic profile of "Rossa di Tropea" red onion are scarce [8,10]. As reported, [8] identified also delphinidin 3-laminaribioside in addition to cyanidin 3-malonylglucoside and cyanidin 3-malonyllaminaribioside in onion skin. In a more recent study conducted on the bulb of the "Rossa di Tropea" red onion, our same anthocyanin profile was reported [10].

3.2. Isolation of Major Flavonols of Onion Skin

To confirm the flavonol structures proposed by UHPLC-DAD-HRMS/MS analysis of the "Ramata di Montoro" and "Rossa di Tropea" onion skins (glycosylation site of the glycosidic derivatives and the structure of the oxidation products), and to obtain pure compounds for the quantitative analysis and the evaluation of biological properties, a preparative procedure for isolation of major flavonols was undertaken.

The exhaustive extract of "Ramata di Montoro" onion skin was first fractionated in ethyl acetate and BuOH soluble portions by liquid-liquid partition. Their UHPLC-UV analyses (Figure S2) indicated that ethyl acetate fraction contains all target compounds (**11**, **13**, **18**, **19**, **21** and **22**), except the diglycoside derivatives of quercetin (**6**), present in n-BuOH extract and purified from it by preparative reversed-phase HPLC-IR. The ethyl acetate extract was then purified using gel permeation chromatography followed by reversed-phase HPLC-IR. This procedure (Figure S3) allowed to isolate the compounds QG (**11**), Q (**13**), Q2Ga (**19**), Q2Gb (**18**), Q2 (**21**), Q3 (**22**) and QdG (**6**) from ethyl acetate extract, with a good degree of purity (>90%, UHPLC-DAD). The structures of isolated compounds were elucidated by 1D- and 2D-NMR experiments, confirming the structures hypothesized by the HRMS/MS data (Figure S3).

3.3. Quantitative Analysis of "Ramata di Montoro" and "Rossa di Tropea" Onion Skins

The content of major components (QG, Q, Q2Ga, Q2Gb, Q2, Q3, CyG and CymG) of "Ramata di Montoro" and "Rossa di Tropea" onion skins was estimated by UHPLC-UV analysis, to evaluate their abundance and the richness of these by-products. Quantitative determination was performed using calibration curves of corresponding pure compounds, and the data, expressed both as content in onion skin (dry matter) and as concentration of extracts, are listed in Table 2.

Table 2. Flavonoid content levels in "Ramata di Montoro" and "Rossa di Tropea" onion skins and their exhaustive extract.

Compounds	"Ramata di Montoro" Onion		"Rossa di Tropea" Onion	
	Exhaustive Extract (mg/g)	Onion Skin (mg/100 gDM)	Exhaustive Extract (mg/g)	Onion Skin (mg/100 gDM)
CyG	nd	nd	16.2 ± 0.2 [c]	365.2 ± 4.5 [c]
CymG [a]	nd	nd	1.5 ± 0.1 [a]	34.5 ± 2.3 [a]
QdG	5.5 ± 0.9 [a]	64.0 ± 10.5 [a]	8.8 ± 0.2 [b]	198.2 ± 4.5 [b]
QG	44.6 ± 5.2 [e]	517.7 ± 60.3 [e]	71.2 ± 3.8 [g]	1602.9 ± 85.5 [g]
Q	30.1± 2.5 [d]	349.5 ± 29.1 [d]	52.4 ± 2.9 [f]	1180.0 ± 65.3 [f]
Q2	22.9 ± 2.8 [c]	265.5 ± 45.4 [c]	27.1 ± 4.1 [d]	608.9 ± 18.8 [d]
Q2Ga	14.6 ± 2.5 [b]	168.9 ± 28.9 [b]	19.4 ± 0.6 [c]	436.6 ± 13.5 [c]
Q2Gb	16.4 ± 1.6 [b]	190.3 ± 18.5 [b]	17.8 ± 2.6 [c]	399.6 ± 58.3 [c]
Q3	29.0 ± 3.7 [d]	336.7 ± 42.9 [d]	35.1 ± 9.0 [e]	790.4 ± 17.3 [e]

Values are means of three replicates ± SD. nd = not detected. [a–g] Expressed as equivalents of CyG. Different superscript letters within each column indicate significant differences among samples ($p < 0.05$, 95% confidence level).

In accordance to literature data [10], "Rossa di Tropea" onion skin showed an overall flavonoid content about three times higher than "Ramata di Montoro" variety, 4.7 and 1.5 g/100 g of skin, respectively, also excluding the contribution of anthocyanins (4.4 g/100 g DM) (Table 2). Flavonols were the major bioactive constituents of both studied onion skins, and the estimated amount of "Rossa di Tropea" and "Ramata di Montoro" flavonols fits in the range reported in the literature for skin of different onion varieties, 0.1–4.6 g/100 g DM [18,19]. The most abundant flavonols were QG, representing 34% of the total target compounds in both varieties, and Q (25 and 23% of total in "Rossa di Tropea" and "Ramata di Montoro", respectively). Our quantitative data agree with the literature data referring to flavonol composition of onion skins [18–20,37]. Oxidized derivatives Q2 and Q3 were also relatively abundant (14–17 and 18–22% of total flavonols, respectively). No previous data on their distribution in onion skin are available. Regarding anthocyanins, detected at significant levels only in "Rossa di Tropea" onion skin, CyG resulted the most abundant red pigment and contributed sufficiently to the total flavonoid content (about 8%).

These results highlighted the potential of this food by-product as a rich and economical source of bioactive compounds.

3.4. *Antioxidant Activity of "Rossa di Tropea" and "Ramata di Montoro" Onion Skin Extracts*

Quali-quantitative profiles of "Rossa di Tropea" and "Ramata di Montoro" onion skins highlight that these waste products are an excellent source of Q and its derivatives and of CyG for "Rossa di Tropea" onion skin (Table 2). As is well known, Q represents one of the compounds with the greatest antioxidant activity present in nature [38,39], and the anthocyanins are natural colorant with interesting antioxidant properties [8,40–42]. Thus, in this study, the in vitro antioxidant activities of "Rossa di Tropea" and "Ramata di Montoro" onion skin extracts and of their major bioactive compounds were evaluated, to provide a preliminary assessment of the potential of these onion by-products as source of antioxidants and obtain further information regarding the oxidized quercetin derivatives.

When the antioxidant activity is measured by an individual test, the results can reflect only the chemical reactivity under the specific conditions applied for the assay. Therefore, to have a full view on the antioxidant properties of "Rossa di Tropea" and "Ramata di Montoro" onion skins and their bioactive constituents, two different antioxidant assays were used: ABTS (SET/HAT mechanisms) and ORAC (HAT mechanism) assays [43–45]. The results expressed in terms of TEAC (μmol TE/g and μmol TE/μmol for extract and pure compounds, respectively) are reported in Table 3. The two onion skin extracts showed similar trends in both assays, with "Rossa di Tropea" exhibiting higher antioxidant properties than "Ramata di Montoro". Among the tested compounds, Q and CyG resulted the most powerful antioxidants of onion skins, with comparable TEAC values in both assays (Table 3). These data, along with the differences observed in the chemical composition of the onion skins of the two traditional varieties (Table 2), indicate that CyG contributes appreciably to the antioxidant activity of "Rossa di Tropea" extract, as well as the higher content of Q. About the quercetin derivatives, QG and QdG displayed much lower antioxidant activities than their aglycone (Table 3), whereas the oxidated derivatives Q2 and Q3 showed valuable antioxidant properties even if slightly lower than Q. These results are in accordance with previous studies on the flavonoid structure-antioxidant activity relationships [18–20,37]. As is common knowledge, the strong antioxidant activity of quercetin depends on the number and position of the hydroxyl groups, especially *o*-dihydroxy (3′-4′), 3- and 5-OH substituents and the 2,3-double bond conjugated with keto function [8,39,46,47]. Therefore, in the case of QG and QdG the glycosylation of hydroxyl groups in position 4′ and/or 3 reduces noticeably their antioxidant properties. Conversely, the oxidative products Q2 and Q3 retain *o*-dihydroxy substituent in the B-ring and consequently the structural criteria underlying the radical scavenging potential. In the same way, the degree and position of hydroxylation, glycosylation and methoxylation in the B-ring of anthocyanins affect their stability and reactivity and thereby antioxidant activity [42]. In the case of cyanidin derivatives, [48] have observed that the glycosylation of cyanidin to cyanidin 3-glucoside increased the antioxidant activity [48]. According to the obtained results we can conclude that both onion skins are a rich source of antioxidants exploitable in different applications as food additives and/or functional ingredients for food supplements, nutraceuticals, and cosmetic products.

Table 3. Antioxidant Activity of "Rossa di Tropea" and "Ramata di Montoro" onion skin extracts and main compounds.

	ORAC	ABTS
	μmol TE/g ± SD	μmol TE/g ± SD
"Rossa di Tropea"	7.82 ± 0.72 [a]	11.32 ± 1.40 [a]
"Ramata di Montoro"	4.13 ± 0.29 [b]	5.77 ± 0.88 [b]
Compounds	μmol TE/ μmol ± SD	μmol TE/ μmol ± SD
CyG	5.55 ± 1.88	7.21 ± 2.03
QdG	1.01 ± 0.09	2.14 ± 0.55
QG	1.68 ± 1.93	3.60 ± 0.11
Q	5.17 ± 1.76	9.80 ± 0.64
Q2	4.47 ± 3.93	5.58 ± 0.14
Q3	2.42 ± 1.86	6.99 ± 0.15

Values are means of three replicates ± SD. Different superscript letters within each column indicate significant differences among samples ($p < 0.05$, 95% confidence level).

3.5. Onion Skin Extracts as Source of α-Glucosidase Inhibitors

Several studies report that a strong correlation exists between antioxidant power of plant extracts and α-glucosidase inhibitory activity, and especially onion bulbs have been widely tested for their biological activities including anti-diabetic capabilities [49]. Instead, few studies investigated the anti-diabetic potential of onion skins although they represent the richest source of phenolics. Indeed, [50] reported a higher alpha-glucosidase inhibitory activity of the skin extract than of the onion pulp one for a Korean onion variety, about 90% vs 40%, respectively. Therefore, the potential of "Ramata di Montoro" and "Rossa di Tropea" onion skin extracts for the in vitro α-glucosidase inhibitory activity were assayed toward *Saccharomyces cerevisiae* alpha-glucosidase enzyme. First, to understand whether the extracts could show a different inhibitory capability according to their concentration, a range of concentrations from 5 to 100 μg of extracts were tested (Figure 2). Our data indicated that "Ramata di Montoro" extracts showed an almost 100% capability to inhibit alpha-glucosidase enzyme at each tested concentration, whereas "Rossa di Tropea" extract had the highest inhibitory effect of about 90% only at the lowest tested concentration and its inhibitory power was almost inversely related to the concentration increase.

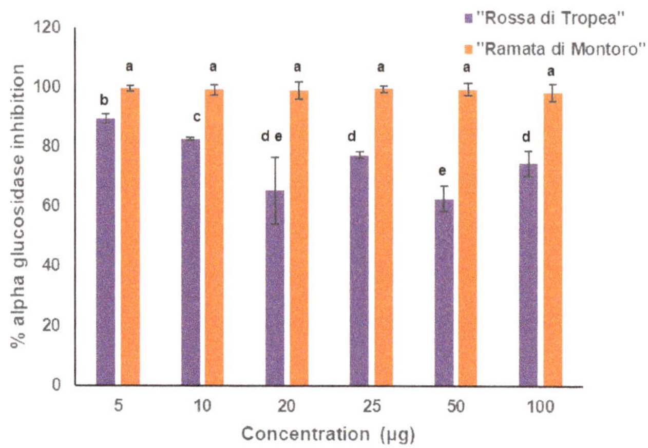

Figure 2. α-glucosidase inhibitory activities of "Ramata di Montoro" and "Rossa di Tropea" extracts at different concentrations. Value represents the mean of three replicates. Different letters denote significant differences among tested concentration within each extract by analysis of variance [ANOVA]. Statistical significance was defined as $p < 0.05$, using Tukey's post hoc test for mean separation.

To estimate the correlation between major flavonoids of "Ramata di Montoro" and "Rossa di Tropea" onion skins and the high α-glucosidase inhibitory activity observed, the α-glucosidase inhibitory activity of the extracts was also compared to pure compounds in a range of concentrations mimicking the actual content when 25 ug of each extract are tested. As shown in Table 4, at each tested concentration CyG, QdG, and QG showed a very low activity, whereas Q, Q2, Q3 showed an almost total enzyme inhibition within the range of tested concentrations. This is not surprising since several studies have reported that quercetin glycosylation strongly reduces its inhibitory effect [51]. Similarly, CyG from "Rossa di Tropea" onion skin extract showed a very low inhibitory potential in agreement with Chen et al. [52], who recently showed that among the major Cinnamon anthocyanins, cyanidin behaves as powerful α-glucosidase inhibitor, while cyanidin 3-rutinoside and cyanidin 3-glucoside did not show inhibitory activity on α-glucosidase. As reported by [53], the increase in hydrophilicity by glycosylation, especially in the 3 position, determines a significant decline of the inhibitory effect, most likely because the free 3-OH is crucial for blocking enzyme active site, as also reported by [54]. Consequently, it is clear that glucosides of cyanidin and quercetin are negatively affecting glucosidase inhibition. Therefore, "Ramata di Montoro" onion skin extract, being devoid of CyG and having also lower content of quercetin glucoside, exhibits a stronger inhibitory activity than "Rossa di Tropea" onion skin extract. In our assay, it can be argued that the use α-glucosidase from yeast *Saccharomyces cerevisiae* could be misleading because the origin of α-glucosidase can influence the activity of the potential inhibitors. Nevertheless, our results agree with data performed on rat intestinal α-glucosidase thus supporting that also in yeast the tested compounds have a constant inhibitory effect. Moreover, α-glucosidase inhibition mechanism of flavonoids and in particular of quercetin, as common plant flavonoid, and main compound in our extracts, has been often investigated [52,55–57]. All these studies, reported quercetin acting with as mixed close to non-competitive inhibition type, thus suggesting that this flavonoid can bind both free enzyme and enzyme-substrate complex, whereas 3-OH glycosylated derivatives can vary their inhibitory mechanism toward a competitive inhibition [55–58]. Overall, our results confirmed that these extracts, act as potent α-glucosidase inhibitors, and suggest that both "Ramata di Montoro" and "Rossa di Tropea" onion skins extracts have the potential to efficaciously contribute as dietary supplements for both the postprandial glycemia control and diabetes-related cellular oxidative stress; nevertheless, further studies will be performed to elucidate the structure-activity relationships of flavonoids inhibitory power in our onion skin extracts.

Table 4. α-glucosidase inhibitory activity of major compounds of "Ramata di Montoro" and "Rossa di Tropea" onion skin extracts.

Compounds	% Alpha-Glucosidase Inhibition				Concentration Range (C1–C4) *
	C1	C2	C3	C4	
CyG	8.08 ± 1.00	8.58 ± 0.59	7.34 ± 0.41	7.17 ± 0.35	0.15–0.40 μg
QdG	7.30 ± 0.51	6.27 ± 2.03	7.12 ± 0.85	6.98 ± 0.15	0.05–0.25 μg
QG	4.61 ± 0.08	4.55 ± 0.28	5.63 ± 1.22	3.88 ± 0.20	0.75–2.0 μg
Q	99.25 ± 0.32	99.53 ± 0.03	99.46 ± 0.03	99.76 ± 0.43	0.50–1.50 μg
Q2	95.59 ± 1.86	99.07 ± 2.64	97.84 ± 1.80	98.70 ± 2.69	0.15–0.75 μg
Q3	99.27 ± 0.56	99.58 ± 0.07	99.41 ± 0.34	99.45 ± 0.18	0.30–0.90 μg

* C1–C4 are the concentration ranges tested for % alpha-glucosidase inhibitory activity evaluation of the main compounds of "Rossa di Tropea" and "Ramata di Montoro" skin extracts. C1 and C4 correspond to the lowest and highest tested concentrations, respectively. Concentrations have been selected in a range of representative concentrations for each compound when 25 ug of both extracts are tested.

3.6. Onion Skin Extracts Stimulate Cell Viability

Fibroblasts along with keratinocytes, serve to maintain integrity of normal skin. Upon wounding, skin barrier is disrupted, and cells migrate toward the damaged area to regenerate skin by remodeling of the extracellular matrix upon proteolytic degradation and providing novel biosynthesis and deposition of matrix components. To restore cellular integrity is necessary that cells possess high viability and are well proliferating. To test weather onion skin extracts could influence cell proliferation MTT assay was performed.

MTT assay revealed that no cytotoxic effects were detected when cells were cultured with different high concentrations (0.250 mg ml^{-1} and 0.125 mg ml^{-1}) of both onion skin extracts. However, in cells cultured with higher concentration of onion waste extracts (0.250 mg ml^{-1}) the cell viability was significantly induced after 48 h of culturing, as compared to control untreated cells, being more effective at higher concentration) (Figure 3).

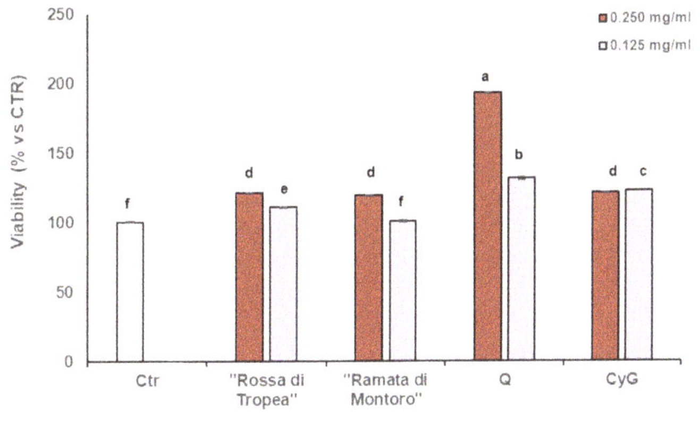

Figure 3. Effects of "Rossa di Tropea" and "Ramata di Montoro" extracts and main compounds on HDFA cell growth. Cells were treated for 48 h with DMSO (Ctrl) or with the indicated concentrations of different extracts. Viability was assessed by MTT assay and was expressed as a percentage of control. Values represent means ± SD. Different letters denote significant differences among tested concentration within each extract by analysis of variance [ANOVA]. Statistical significance was defined as $p < 0.05$, using Tukey's post hoc test for mean separation.

To test the effects of single compounds we selected two of the main constituents, Q and CyG, the latter only present only in "Rossa di Tropea" onion skin extract. Both pure compounds were able to increase cell viability; nevertheless, CyG at both tested concentrations showed a similar effect, not dose-dependent, whereas Q showed the highest proliferative effect to be strictly dose-dependent, namely a higher dose was able to induce a higher proliferative effect. Interestingly, "Rossa di Tropea" extract has comparable content of Q than "Ramata di Montoro", but the latter is devoid of CyG and derivatives. Taken together our result indicate that both the extracts at higher concentration possess the same proliferative effect but at lower concentration a higher activity can be detected in "Rossa di Tropea" extract, therefore the main effect can be ascribed to Q that at higher extract dilution remains quantitatively high and effective in "Rossa di Tropea" onion skin extract. Absence of cytotoxic effects and indeed the increase in fibroblast proliferation exerted by both onion extracts is a good indication of the potential of the extracts and active ingredients to induce tissue regeneration and facilitate the progression of wound healing.

Our results confirmed the relationship between antioxidant activity due to the presence of bioactive phenolic compounds and cell proliferation, as reported by other studies [59,60]. In particular, this study further highlights that a proactive use of waste to ameliorate regenerative and antioxidant processes can be associated with a wide range of final usage.

4. Conclusions

The results presented in this work provide a detailed analysis of skin flavonoid composition of two economically important varieties of *Allium cepa* cultivated in the South of Italy, thus enabling the valorization of value-added products in an otherwise waste product. We are aware that in order to create a sustainable biorefinery approach, mass and energy balances need to be modeled; nevertheless, in this study the use of sun-dried onion matrix, standardized extraction methods with solvent recycling, and low-energy extraction, such as SFE-CO_2/co-solvent and UAE procedures, ensure the limiting of energetic/environmental burdens. In addition, the screening of biological activities conducted on "Rossa di Tropea" and "Ramata di Montoro" onion skins revealed multiple potential applications of this waste products in several industrial uses as nutraceuticals, food preserving agents, and pharmaceuticals. With a remarkable bioactivity, phytochemicals present in these onion skin extracts can be further investigated for numerous other uses, ranging from sustainable "green" management of agricultural pests to formulation of smart packaging, or formulation of functional food ingredients. Therefore, the multiple re-use of onion waste by-products can be the driving force for the development of sustainable, cost-effective, and efficient technologies for reduction of agricultural and food waste. Hence, our findings can be considered a starting contribution to optimize the sustainable use of this natural resource and make the onion production chain more efficient and sustainable.

Supplementary Materials: The following data are available online at https://www.mdpi.com/2076-3921/10/2/304/s1, Figure S1: Picture depicting Rossa di Tropea" and "Ramata di Montoro" varieties and relative onion skins Figure S2: UHPLC-UV profile (365 nm) of ethyl acetate extract (*black*) and n-BuOH extract (*blue*), Figure S3. Preparative procedure for the isolation of the main components of "Ramata di Montoro" exhaustive onion skin extract, Figure S4: Chemical structures of isolated compounds.

Author Contributions: Conceptualization, R.C., T.D., A.L.P., M.R. and L.R.; Data curation, R.C., T.D. and R.D.S.; Formal analysis, R.C., T.D. and L.C.; Funding acquisition, M.T., M.R. and L.R.; Investigation, R.C., T.D., P.G. and S.C.; Methodology, R.C., T.D., A.L.P., P.G. and L.C.; Project administration, M.R.; Resources, M.R. and L.R.; Software, R.C. and T.D.; Supervision, A.L.P. and M.R.; Validation, R.C., T.D., A.L.P. and S.C.; Writing—Original draft, R.C., T.D., A.L.P., P.G., M.T., R.D.S. and M.R.; Writing—Review and editing, R.C., T.D., A.L.P. and M.R. All authors have read and agreed to the published version of the manuscript.

Funding: This research was supported by the following project: OroNero (CUP: J88C17000130006) POR Calabria 204-2020.

Institutional Review Board Statement: Not applicable.

Informed Consent Statement: Not applicable.

Data Availability Statement: The authors confirm that the data supporting the findings of this study are available within the article and its Supplementary Materials.

Acknowledgments: The paper involved the contribution of researchers from Italian Society of Food Chemistry (ITACHEMFOOD).

Conflicts of Interest: The authors declare no conflict of interest.

References

1. Marmot, M. Diet, cancer, and NCD prevention. *Lancet Oncol.* **2018**, *19*, 863–864. [CrossRef]
2. Dinu, M.; Pagliai, G.; Casini, A.; Sofi, F. Mediterranean diet and multiple health outcomes: An umbrella review of meta-analyses of observational studies and randomised trials. *Eur. J. Clin. Nutr.* **2018**. [CrossRef]
3. Castro-Barquero, S.; Lamuela-Raventós, R.M.; Doménech, M.; Estruch, R. Relationship between Mediterranean Dietary Polyphenol Intake and Obesity. *Nutrients* **2018**, *10*, 1523. [CrossRef]
4. Griffiths, G.; Trueman, L.; Crowther, T.; Thomas, B.; Smith, B. Onions—A global benefit to health. *Phyther. Res.* **2002**, *16*, 603–615. [CrossRef]
5. Russo, M.; Serra, D.; Suraci, F.; Di, R.; Fuda, S. The potential of e-nose aroma profiling for identifying the geographical origin of licorice (*Glycyrrhiza glabra* L.) roots. *FOOD Chem.* **2014**, *165*, 467–474. [CrossRef]
6. Russo, M.; Cefaly, V.; Di Sanzo, R.; Carabetta, S.; Postorino, S.; Serra, D. Characterization of different "tropea red onion" (*Allium cepa* L.) ecotypes by aroma precursors, aroma profiles and polyphenolic composition. *Acta Hortic.* **2012**, 197–203. [CrossRef]
7. Beretta, V.H.; Bannoud, F.; Insani, M.; Galmarini, C.R.; Cavagnaro, P.F. Variability in spectrophotometric pyruvate analyses for predicting onion pungency and nutraceutical value. *Food Chem.* **2017**, *224*, 201–206. [CrossRef]
8. Rice-Evans, C.A.; Miller, N.J.; Paganga, G. Structure-antioxidant activity relationships of flavonoids and phenolic acids. *Free Radic. Biol. Med.* **1996**, *20*, 933–956. [CrossRef]
9. Slimestad, R.; Fossen, T.; Vågen, I.M. Onions: A source of unique dietary flavonoids. *J. Agric. Food Chem.* **2007**, *55*, 10067–10080. [CrossRef]
10. Tedesco, I.; Carbone, V.; Spagnuolo, C.; Minasi, P.; Russo, G.L. Identification and quantification of flavonoids from two southern italian cultivars of *Allium cepa* L., Tropea (Red Onion) and Montoro (Copper Onion), and their capacity to protect human erythrocytes from oxidative stress. *J. Agric. Food Chem.* **2015**, *63*, 5229–5238. [CrossRef] [PubMed]
11. FAOSTAT. FAO Statistics Division 2018. Available online: http//www.fao.org/faostat/en/#data (accessed on 21 September 2020).
12. Roldán, E.; Sánchez-Moreno, C.; de Ancos, B.; Cano, M.P. Characterisation of onion (*Allium cepa* L.) by-products as food ingredients with antioxidant and antibrowning properties. *Food Chem.* **2008**, *108*, 907–916. [CrossRef]
13. Sharma, K.; Mahato, N.; Nile, S.H.; Lee, E.T.; Lee, Y.R. Economical and environmentally-friendly approaches for usage of onion (*Allium cepa* L.) waste. *Food Funct.* **2016**, *7*, 3354–3369. [CrossRef]
14. Santiago, B.; Arias Calvo, A.; Gullón, B.; Feijoo, G.; Moreira, M.T.; González-García, S. Production of flavonol quercetin and fructooligosaccharides from onion (*Allium cepa* L.) waste: An environmental life cycle approach. *Chem. Eng. J.* **2020**. [CrossRef]
15. Gontard, N.; Sonesson, U.; Birkved, M.; Majone, M.; Bolzonella, D.; Celli, A.; Angellier-Coussy, H.; Jang, G.W.; Verniquet, A.; Broeze, J.; et al. A research challenge vision regarding management of agricultural waste in a circular bio-based economy. *Crit. Rev. Environ. Sci. Technol.* **2018**. [CrossRef]
16. Marotti, M.; Piccaglia, R. Characterization of flavonoids in different cultivars of onion (*Allium cepa* L.). *J. Food Sci.* **2002**, *67*, 1229–1232. [CrossRef]
17. Campone, L.; Celano, R.; Piccinelli, A.L.; Pagano, I.; Carabetta, S.; Di Sanzo, R.; Russo, M.; Ibañez, E.; Cifuentes, A.; Rastrelli, L. Response surface methodology to optimize supercritical carbon dioxide/co-solvent extraction of brown onion skin by-product as source of nutraceutical compounds. *Food Chem.* **2018**, *269*, 495–502. [CrossRef]
18. Benítez, V.; Mollá, E.; Martín-Cabrejas, M.A.; Aguilera, Y.; López-Andréu, F.J.; Cools, K.; Terry, L.A.; Esteban, R.M. Characterization of Industrial Onion Wastes (*Allium cepa* L.): Dietary Fibre and Bioactive Compounds. *Plant Foods Hum. Nutr.* **2011**, *66*, 48–57. [CrossRef]
19. Lee, J.; Mitchell, A.E. Quercetin and isorhamnetin glycosides in onion (*Allium cepa* L.): Varietal comparison, physical distribution, coproduct evaluation, and long-term storage stability. *J. Agric. Food Chem.* **2011**, *59*, 857–863. [CrossRef]
20. Ly, T.N.; Hazama, C.; Shimoyamada, M.; Ando, H.; Kato, K.; Yamauchi, R. Antioxidative Compounds from the Outer Scales of Onion. *J. Agric. Food Chem.* **2005**, *53*, 8183–8189. [CrossRef]
21. Re, R.; Pellegrini, N.; Proteggente, A.; Pannala, A.; Yang, M.; Rice-Evans, C. Antioxidant activity applying an improved ABTS radical cation decolorization assay. *Free Radic. Biol. Med.* **1999**, *26*, 1231–1237. [CrossRef]
22. Sánchez-Camargo, A.P.; Mendiola, J.A.; Valdés, A.; Castro-Puyana, M.; García-Cañas, V.; Cifuentes, A.; Herrero, M.; Ibáñez, E. Supercritical antisolvent fractionation of rosemary extracts obtained by pressurized liquid extraction to enhance their antiproliferative activity. *J. Supercrit. Fluids* **2016**, *107*, 581–589. [CrossRef]
23. Ou, B.; Chang, T.; Huang, D.; Prior, R.L. Determination of total antioxidant capacity by oxygen radical absorbance capacity (ORAC) using fluorescein as the fluorescence probe: First Action 2012.23. *J. AOAC Int.* **2013**, *96*, 1372–1376. [CrossRef] [PubMed]
24. Celano, R.; Piccinelli, A.L.; Pagano, I.; Roscigno, G.; Campone, L.; De Falco, E.; Russo, M.; Rastrelli, L. Oil distillation wastewaters from aromatic herbs as new natural source of antioxidant compounds. *Food Res. Int.* **2017**, *99*, 298–307. [CrossRef] [PubMed]

25. Pratap Chandran, R.; Nishanth Kumar, S.; Manju, S.; Abdul Kader, S.; Dileep Kumar, B.S. In Vitro α-glucosidase inhibition, antioxidant, anticancer, and antimycobacterial properties of ethyl acetate extract of *Aegle tamilnadensis* Abdul Kader (Rutaceae) leaf. *Appl. Biochem. Biotechnol.* **2015**, *175*, 1247–1261.
26. Bonaccorsi, P.; Caristi, C.; Gargiulli, C.; Leuzzi, U. Flavonol glucoside profile of Southern Italian red onion (*Allium cepa* L.). *J. Agric. Food Chem.* **2005**, *53*, 2733–2740. [CrossRef]
27. Makris, D.P.; Rossiter, J.T. An investigation on structural aspects influencing product formation in enzymic and chemical oxidation of quercetin and related flavonols. *Food Chem.* **2002**, *77*, 177–185. [CrossRef]
28. Makris, D.P.; Rossiter, J.T. Hydroxyl Free Radical-Mediated Oxidative Degradation of Quercetin and Morin: A Preliminary Investigation. *J. Food Compos. Anal.* **2002**, *15*, 103–113. [CrossRef]
29. Osman, A.; Makris, D.P.; Kefalas, P. Investigation on biocatalytic properties of a peroxidase-active homogenate from onion solid wastes: An insight into quercetin oxidation mechanism. *Process Biochem.* **2008**, *43*, 861–867. [CrossRef]
30. Ramos, F.A.; Takaishi, Y.; Shirotori, M.; Kawaguchi, Y.; Tsuchiya, K.; Shibata, H.; Higuti, T.; Tadokoro, T.; Takeuchi, M. Antibacterial and antioxidant activities of quercetin oxidation products from yellow onion (*Allium cepa*) skin. *J. Agric. Food Chem.* **2006**, *54*, 3551–3557. [CrossRef]
31. Gülşen, A.; Makris, D.P.; Kefalas, P. Biomimetic oxidation of quercetin: Isolation of a naturally occurring quercetin heterodimer and evaluation of its In Vitro antioxidant properties. *Food Res. Int.* **2007**, *40*, 7–14. [CrossRef]
32. Takahama, U.; Oniki, T. Flavonoids and Some Other Phenolics as Substrates of Peroxidase: Physiological Significance of the Redox Reactions. *J. Plant Res.* **2000**, *113*, 301–309. [CrossRef]
33. Metrani, R.; Singh, J.; Acharya, P.; Jayaprakasha, G.K.; Patil, B.S. Comparative Metabolomics Profiling of Polyphenols, Nutrients and Antioxidant Activities of Two Red Onion (*Allium cepa* L.) Cultivars. *Plants* **2020**, *9*, 1077. [CrossRef]
34. Donner, H.; Gao, L.; Mazza, G. Separation and characterization of simple and malonylated anthocyanins in red onions, *Allium cepa* L. *Food Res. Int.* **1997**, *30*, 637–643. [CrossRef]
35. Sharif, A.; Saim, N.; Jasmani, H.; Ahmad, W.Y.W. Effects of Solvent and Temperature on the Extraction of Colorant from Onion (*Allium cepa*) Skin using Pressurized Liquid Extraction. *Asian J. Appl. Sci.* **2010**, *3*, 262–268. [CrossRef]
36. Steimer, S.; Sjöberg, P.J.R. Anthocyanin characterization utilizing liquid chromatography combined with advanced mass spectrometric detection. *J. Agric. Food Chem.* **2011**, *59*, 2988–2996. [CrossRef] [PubMed]
37. Yang, X.N.; Park, M.J.; Ha, I.J.; Moon, J.S.; Kang, Y.-H. Functional Components and Antioxidant Effects of Colored Onions. *Curr. Res. Agric. Life Sci.* **2015**, *33*, 69–73.
38. Lesjak, M.; Beara, I.; Simin, N.; Pintać, D.; Majkić, T.; Bekvalac, K.; Orčić, D.; Mimica-Dukić, N. Antioxidant and anti-inflammatory activities of quercetin and its derivatives. *J. Funct. Foods* **2018**, *40*, 68–75. [CrossRef]
39. Shahidi, F.; Ambigaipalan, P. Phenolics and polyphenolics in foods, beverages and spices: Antioxidant activity and health effects—A review. *J. Funct. Foods* **2015**, *18*, 820–897. [CrossRef]
40. Chen, L.; Teng, H.; Xie, Z.; Cao, H.; Cheang, W.S.; Skalicka-Woniak, K.; Georgiev, M.I.; Xiao, J. Modifications of dietary flavonoids towards improved bioactivity: An update on structure-activity relationship. *Crit. Rev. Food Sci. Nutr.* **2018**, *58*, 513–527. [CrossRef]
41. López, J.G.-E. Flavonoids in Health and Disease. *Curr. Med. Chem.* **2019**, *26*, 6972–6975. [CrossRef]
42. Kähkönen, M.P.; Heinonen, M. Antioxidant activity of anthocyanins and their aglycons. *J. Agric. Food Chem.* **2003**, *51*, 628–633. [CrossRef]
43. Zulueta, A.; Esteve, M.J.; Frígola, A. ORAC and TEAC assays comparison to measure the antioxidant capacity of food products. *Food Chem.* **2009**, *114*, 310–316. [CrossRef]
44. Jacobo-Velázquez, D.A.; Cisneros-Zevallos, L. Correlations of Antioxidant Activity against Phenolic Content Revisited: A New Approach in Data Analysis for Food and Medicinal Plants. *J. Food Sci.* **2009**, *74*, R107–R113. [CrossRef]
45. Prior, R.L.; Wu, X.; Schaich, K. Standardized Methods for the Determination of Antioxidant Capacity and Phenolics in Foods and Dietary Supplements. *J. Agric. Food Chem.* **2005**, *53*, 4290–4302. [CrossRef]
46. Miller, N.J.; Rice-Evans, C.A. Spectrophotometric determination of antioxidant activity. *Redox Rep.* **1996**, *2*, 161–171. [CrossRef]
47. Rice-evans, C.A.; Miller, N.J.; Bolwell, P.G.; Bramley, P.M.; Pridham, J.B. The relative antioxidant activities of plant-derived polyphenolic flavonoids. *Free Radic. Res.* **1995**. [CrossRef]
48. Wang, H.; Cao, G.; Prior, R.L. Oxygen Radical Absorbing Capacity of Anthocyanins. *J. Agric. Food Chem.* **1997**, *45*, 304–309. [CrossRef]
49. Wu, H.; Xu, B. Inhibitory effects of onion against α-glucosidase activity and its correlation with phenolic antioxidants. *Int. J. Food Prop.* **2014**, *17*, 599–609. [CrossRef]
50. Kim, M.H.; Jo, S.H.; Jang, H.D.; Lee, M.S.; Kwon, Y.I. Antioxidant activity and α-glucosidase inhibitory potential of onion (*Allium cepa* L.) extracts. *Food Sci. Biotechnol.* **2010**, *19*, 159–164. [CrossRef]
51. Şöhretoğlu, D.; Sari, S. Flavonoids as alpha-glucosidase inhibitors: Mechanistic approaches merged with enzyme kinetics and molecular modelling. *Phytochem. Rev.* **2020**, *19*, 1081–1092. [CrossRef]
52. Chen, J.; Wu, S.; Zhang, Q.; Yin, Z.; Zhang, L. α-Glucosidase inhibitory effect of anthocyanins from Cinnamomum camphora fruit: Inhibition kinetics and mechanistic insights through In Vitro and in silico studies. *Int. J. Biol. Macromol.* **2020**, *143*, 696–703. [CrossRef] [PubMed]
53. Thanakosai, W.; Phuwapraisirisan, P. First identification of α-glucosidase inhibitors from okra (*Abelmoschus esculentus*) seeds. *Nat. Prod. Commun.* **2013**, *8*, 1085–1088. [CrossRef]

54. Mohamed, G.A. Alliuocide G, a new flavonoid with potent α-amylase inhibitory activity from *Allium cepa* L. *Arkivoc* **2008**, *2008*, 202–209. [CrossRef]
55. Tadera, K.; Minami, Y.; Takamatsu, K.; Matsuoka, T. Inhibition of alpha-Glucosidase and alpha-Amylase by Flavonoids. *J. Nutr. Sci. Vitaminol.* **2006**, *52*, 149–153. [CrossRef]
56. Iio, M.; Yoshioka, A.; Imayoshi, Y.; Koriyama, C.; Moriyama, A. Effect of flavonoids on a-glucosidase and ß-fructosidase from yeast. *Agric. Biol. Chem.* **1984**, *48*, 1559–1563.
57. Nile, A.; Gansukh, E.; Park, G.S.; Kim, D.H.; Hariram Nile, S. Novel insights on the multi-functional properties of flavonol glucosides from red onion (*Allium cepa* L.) solid waste–In Vitro and in silico approach. *Food Chem.* **2021**. [CrossRef] [PubMed]
58. Li, Y.Q.; Zhou, F.C.; Gao, F.; Bian, J.S.; Shan, F. Comparative evaluation of quercetin, isoquercetin and rutin as inhibitors of α-glucosidase. *J. Agric. Food Chem.* **2009**. [CrossRef] [PubMed]
59. Bahramsoltani, R.; Farzaei, M.H.; Rahimi, R. Medicinal plants and their natural components as future drugs for the treatment of burn wounds: An integrative review. *Arch. Dermatol. Res.* **2014**. [CrossRef]
60. Addis, R.; Cruciani, S.; Santaniello, S.; Bellu, E.; Sarais, G.; Ventura, C.; Maioli, M.; Pintore, G. Fibroblast proliferation and migration in wound healing by phytochemicals: Evidence for a novel synergic outcome. *Int. J. Med. Sci.* **2020**. [CrossRef]

antioxidants

MDPI

Article

Antitussive, Antioxidant, and Anti-Inflammatory Effects of a Walnut (*Juglans regia* L.) Septum Extract Rich in Bioactive Compounds

Ionel Fizeșan [1,†], Marius Emil Rusu [2,†], Carmen Georgiu [3,*], Anca Pop [1], Maria-Georgia Ștefan [1], Dana-Maria Muntean [2], Simona Mirel [4], Oliviu Vostinaru [5,*], Béla Kiss [1] and Daniela-Saveta Popa [1]

1 Department of Toxicology, Faculty of Pharmacy, Iuliu Hatieganu University of Medicine and Pharmacy, 8 Victor Babes, 400012 Cluj-Napoca, Romania; ionel.fizesan@umfcluj.ro (I.F.); anca.pop@umfcluj.ro (A.P.); stefan.georgia@umfcluj.ro (M.-G.Ș.); kbela@umfcluj.ro (B.K.); dpopa@umfcluj.ro (D.-S.P.)
2 Department of Pharmaceutical Technology and Biopharmaceutics, Faculty of Pharmacy, Iuliu Hatieganu University of Medicine and Pharmacy, 8 Victor Babes, 400012 Cluj-Napoca, Romania; rusu.marius@umfcluj.ro (M.E.R.); dana.muntean@umfcluj.ro (D.-M.M.)
3 Department of Pathological Anatomy, Faculty of Medicine, Iuliu Hatieganu University of Medicine and Pharmacy, 8 Victor Babes, 400012 Cluj-Napoca, Romania
4 Department of Medical Devices, Faculty of Pharmacy, Iuliu Hatieganu University of Medicine and Pharmacy, 8 Victor Babes, 400012 Cluj-Napoca, Romania; smirel@umfcluj.ro
5 Department of Pharmacology, Physiology and Physiopathology, Faculty of Pharmacy, Iuliu Hatieganu University of Medicine and Pharmacy, 8 Victor Babes, 400012 Cluj-Napoca, Romania
* Correspondence: cgeorgiu@umfcluj.ro (C.G.); oliviu.vostinaru@umfcluj.ro (O.V.); Tel.: +40-741-185163 (O.V.)
† These authors contributed equally to this work.

Abstract: The antitussive, antioxidant, and anti-inflammatory effects of a walnut (*Juglans regia* L.) septum extract (WSE), rich in bioactive compounds were investigated using the citric acid aerosol-induced cough experimental model in rodents. Wistar male rats were treated orally for three days with distilled water (control), codeine (reference), and WSE in graded doses. On the third day, all rats were exposed to citric acid aerosols, the number of coughs being recorded. Each animal was sacrificed after exposure, and blood and lung tissue samples were collected for histopathological analysis and the assessment of oxidative stress and inflammatory biomarkers. The results of the experiment showed a significant antitussive effect of WSE, superior to codeine. This activity could be due to cellular protective effect and anti-inflammatory effect via the stimulation of the antioxidant enzyme system and the decrease of IL-6 and CXC-R1 concentration in the lung tissue of WSE-treated animals. The antioxidant and anti-inflammatory effects of WSE were confirmed by biochemical assays and histopathological analysis. This is the first scientific study reporting the antitussive effect of walnut septum, a new potential source of non-opioid antitussive drug candidates, and a valuable bioactive by-product that could be used in the treatment of respiratory diseases.

Keywords: walnut; by-products; antitussive; ROS; NOx; IL-6; CXC-R1; histopathological analysis

Citation: Fizeșan, I.; Rusu, M.E.; Georgiu, C.; Pop, A.; Ștefan, M.-G.; Muntean, D.-M.; Mirel, S.; Vostinaru, O.; Kiss, B.; Popa, D.-S. Antitussive, Antioxidant, and Anti-Inflammatory Effects of a Walnut (*Juglans regia* L.) Septum Extract Rich in Bioactive Compounds. *Antioxidants* **2021**, *10*, 119. https://doi.org/10.3390/antiox10010119

Received: 27 December 2020
Accepted: 12 January 2021
Published: 15 January 2021

1. Introduction

Natural products such as endogenous or exogenous metabolites of animals and plants, played a significant role in the discovery of medicines as more than half of the medicines in use were developed from natural products [1]. Famous examples are aspirin, a product derived from salicylic acid, present in the bark of several species of genus *Salix* [2], codeine, a naturally occurring alkaloid found in plants of *Papaver* spp. [3], quinine, a component of the bark of trees in the genus *Cinchona* [4], or penicillin, the first antibiotic compound isolated from the fungus *Penicillium* [5]. The isolation and use of bioactive compounds from natural products can mitigate the high cost of chemical synthetic drugs in the development of new medications.

Cough is a defensive reflex often present in pulmonary diseases including asthma, chronic bronchitis, pulmonary neoplasm, and upper respiratory tract infections. Chronic cough is associated with significant morbidity such as exacerbation of asthmatic symptoms, breathlessness, rib fractures, pneumothorax, or syncope. The commonly used treatment can include administration of antitussive medication, codeine being one of the most prominent. Codeine has been prescribed as an analgesic or antitussive for many decades, but clinical evidence emphasized the potential danger of this compound. Codeine is metabolized into morphine in the liver, which is partially responsible for its analgesic and antitussive effects. However, the individual patient response to codeine differs greatly in children, with documented cases of unanticipated respiratory depression and death. Various regulatory bodies, including the US Food and Drug Administration and European Medicines Agency issued warnings regarding the occurrence of the negative effects of codeine in this age group. Alternative therapies are needed to prevent future opioid problems especially in pediatric patients [6].

It was already revealed that some macro- and micronutrients contained in tree nuts have favorable health benefits via different mechanisms of action [7,8]. Walnuts are considered nutraceuticals, as they contain proteins and essential unsaturated fatty acids, as well as tocopherols, sterols, and polyphenols with recognized antioxidant, anti-inflammatory, and antibacterial properties [9]. In addition to kernel, walnut by-products including leaves, green husk or septum represent important sources of nutraceutical compounds [10,11].

Walnut septum (WS), rich in polyphenolic compounds, demonstrated hypoglycemic effect, antioxidant and antimicrobial activity, hematopoiesis-stimulating potential, or anti-aging effects, both in vitro and in vivo [12,13]. A previous study on Wistar rats demonstrated that WS extract (WSE) had no subacute or acute toxic effects at doses of 1000 mg/kg body weight (b.w.) [14].

In folk medicine, WS is used along with hard walnut shell in decoctions to obtain an antitussive effect, but no scientific research supporting this claim has been published so far. Hence, in this study we planned to investigate the antitussive effects of a WSE obtained in optimum extraction conditions with rich polyphenolic composition [15] using a citric acid aerosol-induced cough rodent model [16,17]. This research could lead to the development of new non-opioid antitussive drug candidates of natural origin, potentially useful in a variety of respiratory diseases.

2. Materials and Methods

2.1. Reagents

All reagents and standards were of analytical grade. Acetone, Folin–Ciocâlteu (FC) reagent, hydrochloric acid, hematoxylin-eosin, paraffin were acquired from Merck (Darmstadt, Germany). Sodium carbonate, sodium chloride, sodium nitrite, acetic acid, disodium hydrogen phosphate, potassium dihydrogen phosphate, bovine serum albumin (BSA), Coomassie Brilliant Blue G (CBB), tris(hydroxymethyl)aminomethane (Tris), 2,2′-azinobis-(3-ethylbenzothiazoline-6-sulfonate) (ABTS), 6-hydroxy-2,5,7,8-tetramethyl chroman-2-carboxylic acid (Trolox), 2′,7′-dichloro-dihydro-fluorescein diacetate (DCFH-DA), phosphate-buffered saline (PBS), vanadium (III) chloride, sulphanilic acid, and alpha-naphthylamine were bought from Sigma-Aldrich (Schnelldorf, Germany). The normal saline solution (0.9% sodium chloride) was from B. Braun Melsungen AG (Melsungen, Germany). Neutral buffered formalin was obtained from Chempur (Lodz, Poland), codeine phosphate from Terapia S.A. (Cluj-Napoca, Romania). The water used in our study was ultrapure obtained from a Milli-Q ultrapure water system (Bucharest, Romania).

2.2. Preparation of the Walnut Septum Extract

Walnut septum was obtained from walnuts harvested in Maramureș County, Romania, in the fall of 2018. The previously identified optimal extraction conditions [15] were used to make the WSE with the highest total phenolic content (TPC). Briefly, WS was weighed (0.5 g) and mixed with 5 mL water: acetone (50:50, v/v) as extraction solvent. Turbo-extraction was applied as extraction method, first using an Ultra-Turrax homogenizer (T 18; IKA Labortechnik, Staufen, Germany) for 2 min (1 min at 9500 rpm and 1 min at 13,500 rpm), then a Vortex RX-3 (Velp Scientifica, Usmate, Italy) for another 2 min. The homogenate was centrifuged (Hettich, Micro 22R, Andreas Hettich GmbH & Co., Tuttlingen, Germany) at 3000 rpm for 15 min, sustaining the 40 °C extraction temperature. The supernatant was separated and kept in the dark in open trays (24 h) for acetone evaporation.

The WSE was further standardized in TPC by FC spectrophotometric assay according to a method previously described [18]. In brief, 20 µL of each sample were mixed with 100 µL of FC reagent in a 96 well plate. After 3 min, 80 µL of 7.5% sodium carbonate solution were added and the plate was incubated in the dark at room temperature for 30 min. The absorbance was measured at 760 nm against a solvent blank using a Synergy HT Multi-Detection Microplate Reader (BioTek Instruments, Inc., Winooski, VT, USA), and the TPC was expressed as mg gallic acid equivalents (GAE)/g.

2.3. *Animals and Experimental Protocol*

The experimental protocol was reviewed and approved by the Commission of Ethics of the University of Medicine and Pharmacy from Cluj-Napoca (decision no. 97/09.03.2020) and the Veterinary and Food Safety Department from Cluj-Napoca (decision no. 218/26.05.2020). The experiment was conducted in accordance with the internationally accepted principles (Directive 2010/63/EC) on the protection of laboratory animals used for scientific purposes [19]. In line with the 3Rs (Replacement, Reduction and Refinement) of humane animal research, the minimum number of animals was used for the data to attain statistical significance. As the experiment was designed to cause as little pain and suffering as possible, the laboratory animals tolerated very well the administered substances with no changes in their behavior or state of health being detected.

Healthy male Wistar rats (n = 24), 3 months old, were acquired from the Practical Skills and Experimental Medicine Centre of the University of Medicine and Pharmacy from Cluj-Napoca, Romania. The animals were housed in cages (Tecniplast, Italy) following a 12-h/12-h light/dark cycle, with environment temperature and relative humidity maintained at 22 ± 2 °C and 45 ± 10%, respectively, and access to standard pelleted feed (Cantacuzino Institute, Bucharest, Romania) and filtered water throughout the investigation.

2.3.1. Antitussive Test

We selected 24 rats with body weight of 227.75 ± 14.60 g (mean ± SD) which were randomly divided into four groups (n = 6). The animals were treated for three consecutive days by gavage: group 1 (CN)—distilled water (negative control); group 2 (Cd)—codeine phosphate, 3 mg/kg b.w./day (positive control); group 3 (WSE)—10 mL WSE (containing 134 mg GAE)/kg b.w./day; group 4 (WSE 1:2)—5 mL WSE (containing 67 mg GAE)/kg b.w./day). One hour after the last oral treatment, each rat was placed in an air-tight transparent exposure chamber and exposed to citric acid aerosols (17.5%) using an ultrasonic nebulizer (Laica MD6026, Laica Spa, Italy). The antitussive effect of WSE in rats was assessed by the number and frequency of coughs caused by citric acid aerosols, cough latency after exposure being also evaluated.

The exposure period to citric acid was 4 min, followed by a further observation period of 4 min. The total number of coughs was determined over a total period of 8 min. Individual coughs were recorded with a video data acquisition platform and detected by trained staff. Each of the animals was exposed to citric acid aerosols only once.

The antitussive effect was expressed as the percentage of inhibition of number of coughs and was calculated using the equation:

$$\% \text{ of inhibition} = 100 - (\text{Ct} \times 100)/\text{C}_0$$

where C_0 was the number of coughs in the negative control group and Ct was the number of coughs in the treatment group [20].

2.3.2. Biological Samples

Each animal used in the antitussive test was subjected to isoflurane general anesthesia 6 h after the citric acid aerosol exposure and blood was collected from the retro-orbital sinus in the absence of anticoagulant. Afterwards, all rats were sacrificed by cervical spine dislocation and autopsied. Lungs were collected from all animals. Immediately, a lung lobe from three rats in each group was fixed in 10% neutral buffered formalin for histopathological analyses. Another lobe was frozen in liquid nitrogen and preserved at −80 °C. Lung tissue samples were weighed and homogenized with 50 mM Tris buffer 50 mM (pH 7.4) (1:5, *w*/*v*), in two stages. A crude tissue homogenate was obtained in the first step using a manual Potter-Elvehjem (Sigma-Aldrich, Schnelldorf, Germany) tissue grinder, which was then sonicated by means of an ultrasonic homogenizer (T 18; IKA Labortechnik, Staufen, Germany). The obtained tissue homogenates were used for the assay of nitric oxide (NO), reactive oxygen species (ROS), IL-6, CXC-R1, and CXC-R2. Serum was separated from blood by centrifugation after coagulation and served to determine total antioxidant capacity (TAC).

2.4. *Determination of the Total Protein Content*

The Bradford method was applied for determination of total protein content using BSA (2 mg/mL) as standard for calibration and CBB as color reagent [21]. The tissue samples were homogenized with 50 mM Tris buffer (pH 7.4) (1:10, *w*/*v*) and the absorbance was measured at 595 nm using a Jasco V-530 UV/Vis spectrophotometer (Jasco, Japan). The total protein content was used to normalize the results for oxidative stress and inflammatory biomarkers.

2.5. *Oxidative Stress Biomarkers*

2.5.1. Reactive Oxygen Species

The cellular redox processes are commonly monitored using DCFH-DA. This assay was selected to measure the ability of WSE to protect against oxidative stress in the rat tissues employing a method described earlier [22]. Briefly, 10 µL of homogenate, 180 µL of PBS, and 10 µL of 1 mM DCFH-DA were mixed in a microplate well (sample processed in triplicate). The blank control was achieved by adding 190 µL PBS to 10 µL homogenate. The conversion of DCFH-DA to the fluorescent compound 2′,7′ dichlorofluorescein (DCF) was analyzed using a Synergy 2 Multi-Mode Microplate Reader at λexc = 484 nm and λem = 530 nm, and the results were conveyed as arbitrary units (AU)/mg protein.

2.5.2. Nitric Oxide Level

The quantification of total NO (nitrites and nitrates) was performed using a slightly modified previously described assay [23]. The homogenates were deproteinized by adding an equal volume of acetonitrile, vortexed for 1 min and centrifuged at 5000 rpm and 4 °C for 10 min. To reduce nitrates to nitrites, a volume of 50 µL VCl_3 (0.5% prepared in 0.5 M HCl) was added to the supernatant, followed by 50 µL Griess reagent obtained extempore from Griess I (0.33% sulphanilic acid in 15% acetic acid) and Griess II (0.066% alpha-naphthylamine in 15% acetic acid) (1:1, *v*/*v*). The absorbance of the solution was measured at 540 nm after incubation at 37 °C for 30 min. A scale of sodium nitrite standards (concentrations from 0 to 152 nmol/mL) was employed and the results were expressed as nmol/mg protein.

2.5.3. Total Antioxidant Capacity by Trolox Equivalent Antioxidant Capacity (TEAC) Assay

The total antioxidant capacity (TAC) was assessed by a spectrophotometric method previously described [24]. The assay is based on the capacity of antioxidants in the solution to decolorize the blue-green ABTS radical cation according to their concentrations and antioxidant capacities. Antioxidants present in the sample accelerate the bleaching rate to a degree proportional to their concentrations. This reaction can be monitored spectrophotometrically at 660 nm and the bleaching rate is inversely related with the TAC of the sample.

For the calibration curve, Trolox was used, a water-soluble analog of vitamin E, which is widely used as a traditional standard for TEAC measurement assays, and the results are expressed as mmol Trolox equivalent/L.

2.6. Inflammatory Biomarkers

The anti-inflammatory potential of the extract was evaluated by measuring the levels of three inflammatory markers, IL-6, CXC-R1 and CXC-R2 in lung homogenates using commercially available ELISA Kits according to the manufacturer's instructions and further normalized to the protein content measured using the Bradford method.

2.7. Histopathological Analysis

The lung tissues from each group were fixed in formaldehyde for 24 h and embedded in paraffin blocks. Sections of 5 µm-thickness were obtained using a microtome (Microtec, Rotary Microtome CUT 4060, Germany), displayed on albumin treated glass slides and stained with hematoxylin-eosin (HE) for histopathological evaluation [25]. The tissue samples were processed in an autostainer (Leica TP 1020, Leica Biosystems, Wetzlar, Germany). The stained lung sections were assayed for possible toxic effects. The slides were examined using a microscope (Leica DM750, Leica Biosystems, Wetzlar, Germany) connected to a digital camera (Leica ICC50 ND, Leica Biosystems, Wetzlar, Germany), and analyzed by Leica Application Suite (LAS) V4.12 software Leica Biosystems, Wetzlar, Germany). For the objective evaluation of the histological changes, a morphometric procedure of image analysis, using the open-source platform for biological-image analysis ImageJ software (http://imagej.nih.gov/ij/), was performed on the scanned histopathological slides (Slide Converter, 3DHISTECH Ltd., Budapest, Hungary).

2.8. Statistical Analysis

The data are presented as mean values ± standard error mean (SEM) and, unless stated otherwise, the normally distributed result sets were evaluated using one-way analysis of variance (ANOVA). Data analyses and graphical representation were performed in SigmaPlot 11.0 computer software (Systat Software GmbH, Erkrath, Germany). Outcomes with p values < 0.05 were considered statistically significant.

3. Results

3.1. Antitussive Effect

The citric acid aerosol-induced cough model in rats was carried out by exposing individual animals to citric acid aerosols (17.5%) for 4 min. In the control group, citric acid treatment induced a mean of 31.8 coughs in 8 min, with a latency of 34 sec (Figure 1), accompanied by nasal irritation and minor bleeding around the nostrils. In the positive control group, codeine phosphate at 3 mg/kg b.w./day significantly reduced the number of coughs to a mean of 15.4 and increased the latency to 144 sec, compared to control ($p < 0.01$). Minimal nasal irritation and bleeding was also observed in the codeine-treated group.

As shown in Figure 1 and Table 1, the tested extract demonstrated dose-dependent, potent antitussive activity, comparable with codeine. In the WSE group, for a dose of 10 mL/kg b.w./day, corresponding to 134 mg GAE/kg b.w./day, WSE significantly reduced cough frequency ($p < 0.01$) with a mean of 10.3 coughs in 8 min and prolonged the latency to 81 sec ($p < 0.05$). In the WSE 1:2 group, at 67 mg GAE/kg b.w./day, the cough frequency decreased by 20% and the latency increased to 56 sec. In addition, the animals in both groups treated with WSE showed no signs of nasal irritation and nose bleeding.

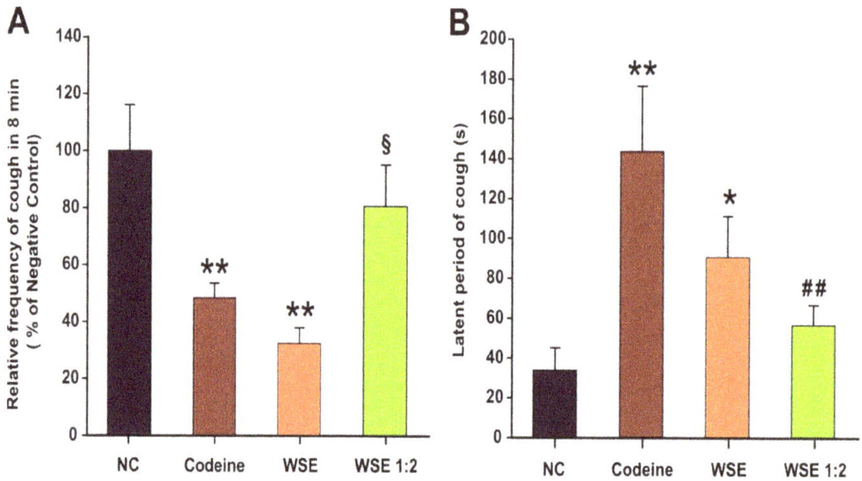

Figure 1. The antitussive activities of codeine, WSE, and WSE 1:2. Relative frequency of cough in 8 min (**A**) and Latent period of cough (**B**). Data are presented as the mean ± SEM ($n = 6$). * $p < 0.05$ and ** $p < 0.01$ compared to NC; ## $p < 0.01$ compared to Codeine); § $p < 0.05$ between the two concentration of WSE. Statistical significance differences in datasets were determined with one-way ANOVA with the Holm-Sidak post hoc test. NC—negative control; WSE—walnut septum extract; WSE 1:2—walnut septum extract diluted 1:2.

Table 1. Effects of different treatments on the citric acid-induced cough in rats.

Group	Dose (mg/kg b.w./day)	Cough Latency [1] (sec)	No. of Coughs [1]	Inhibition (%)
NC	0	34 (15–70)	31.8 (18–44)	-
Codeine	3	144 (75–200)	15.4 (11–20)	51.57
WSE	134 *	81 (60–190)	10.3 (5–15)	67.50
WSE 1:2	67 *	56 (33–100)	25.6 (11–40)	19.28

[1] Values expressed as mean (minimum and maximum values) ($n = 6$); NC—negative control; WSE—walnut septum extract; WSE 1:2—walnut septum extract diluted 1:2. * Gallic acid equivalents.

3.2. Oxidative Stress and Inflammatory Biomarkers

The oxidative stress and inflammatory status were assessed in lung tissue homogenates to outline a pharmacological explanation for the previously observed antitussive effects. Oxidative stress was measured using DCFH-DA, a dye that can measure ROS production in the cytoplasm as well as in intracellular organelles [26]. As shown in Figure 2B, prior treatment with WSE for 3 days, did not significantly influenced the level of total NO. However, a decrease in the quantity of ROS in lung tissue was observed using the DCFH-DA assay (Figure 2A). The decrease was dose-dependent and statistically significant in both WSE-treated groups revealing that WSE could replenish the antioxidant defense systems and reduce the risk of oxidative damage. In the case of codeine, no statistical differences were observed (Figure 2A). Regarding the TAC in serum, a statistical significance between the treatments was not observed (Figure 3). Although the latter results were not statistically significant, longer periods of supplementation might change the outcomes for this biomarker.

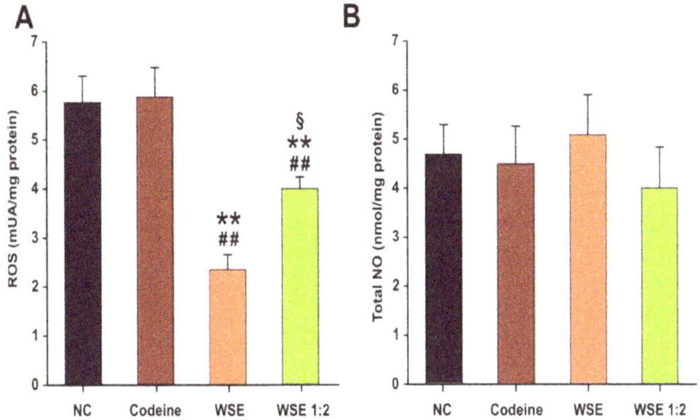

Figure 2. The effects of the treatments on reactive oxygen species (**A**) and total nitric oxide (**B**) in lung tissue homogenates. Values are expressed as mean ± SEM (n = 6). ** $p < 0.01$ compared to NC; ## $p < 0.01$ compared to positive control (Codeine); § $p < 0.05$ between the two concentration of WSE. Statistical significance differences in datasets were determined with one-way ANOVA with the Holm-Sidak post hoc test. NC—negative control; NO—nitric oxide; ROS—reactive oxygen species; WSE—walnut septum extract; WSE 1:2—walnut septum extract diluted 1:2.

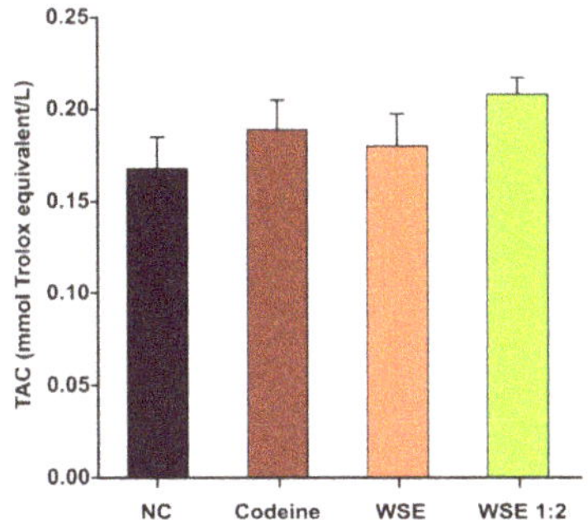

Figure 3. The effects of the treatments on total antioxidant capacity (TAC) in serum. Values are expressed as mean ± SEM (n = 6). Statistical significance differences in datasets were determined with one-way ANOVA with the Holm-Sidak post hoc test. NC—negative control; WSE—walnut septum extract; WSE 1:2—walnut septum extract diluted 1:2.

Previous studies in humans indicated that patients with non-asthmatic chronic cough present airway inflammation with an increased sputum neutrophilia and increased levels of IL-6, IL-8, and tumor necrosis factor α (TNF-α). In this regard, the anti-inflammatory potential of the walnut septum extracts was evaluated by measuring the level of IL-6 and two receptors (CXC-R1 and CXC-R2) for IL-8 in the lung tissue homogenates. In the case of IL-6, a clear decrease in the levels of this pleiotropic pro-inflammatory cytokine was observed for codeine and WSE at both concentrations. Codeine decreased the IL-6 levels by approximately 62.5%, while WSE and WSE 1:2 decreased the levels by approximately 54% and 43%, respectively (Figure 4A). Regarding the expression of the CXC receptors, the lung tissue from the Wistar rats was more abundant in the CXC-R1 subtype (Figure 4B,C). Similar to what was observed in the case of IL-6, pre-treatment for 3 days with codeine and WSE decreased the concentration of CXC-R1 in the lung tissue homogenates (Figure 4B). The decrease was statistically significant in the case of codeine and the highest tested dose of WSE. No statically significant differences were observed in the case of CXC-R2 (Figure 4C).

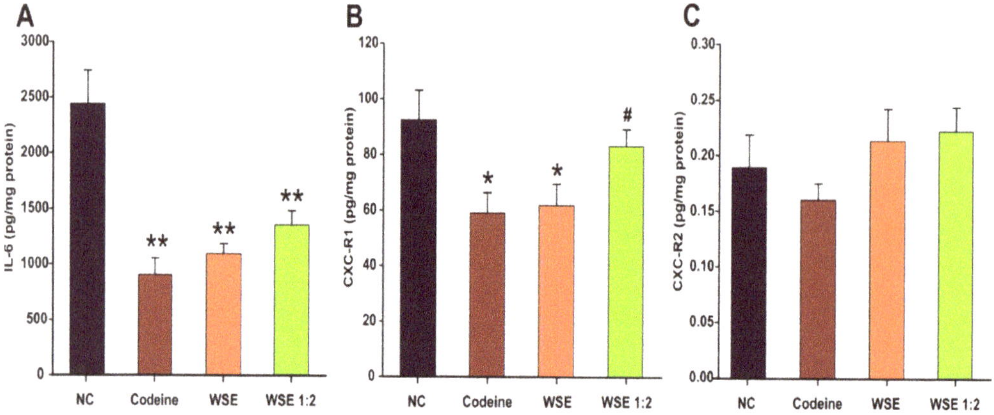

Figure 4. The effects of the treatments on the secretion of pro-inflammatory cytokines and receptors, IL-6 (**A**), CXC-R1 (**B**) and CXC-R2 (**C**) in lung tissue. Values are expressed as mean ± SEM ($n = 6$). * $p < 0.05$ and ** $p < 0.01$ compared to NC; # $p < 0.05$ compared to positive control (Codeine). Statistical significance differences in datasets were determined with one-way ANOVA with the Holm-Sidak post hoc test. CXC-R1—chemokine receptor type 1; CXC-R2—chemokine receptor type 2; IL-6—Interleukin 6; NC—negative control; WSE—walnut septum extract; WSE 1:2—walnut septum extract diluted 1:2.

3.3. Histopathological Analysis

After the rats were sacrificed, the isolated lung tissue, carefully manipulated, without crushing, was fixed in 9% formaldehyde, subsequently included in paraffin, and stained with HE solution. Although the time between exposure to citric acid and sacrification was relatively short, histological changes in the bronchopulmonary parenchyma were observed (Figure 5), some of them severe, such as intra-alveolar hemorrhage (control group).

In the control group, the tracheobronchial tract showed focal erosions of the surface epithelium and a serohemorrhagic intraluminal exudate (Figure 5A). The lung parenchyma showed areas of normal appearance (Figure 5B), as well as areas with atelectatic alveoli along with emphysematous alveoli (Figure 5C), thus reducing the surface area available for gas exchange. Areas with thickening of the alveolar septa (Figure 5D) were also observed due to a hyperemia in the alveolar capillaries, a slight interstitial edema and a septal hypercellularity (Figure 5E). This hypercellularity can be partially explained by the lack of alveolar dilation, the alveolar spaces remaining focal small, round, therefore lined with unflattened cubic pneumocytes (Figure 5F), but also by a real increase in the number of cells in the alveolar septa (Figure 5G), with some macrophages and inflammatory cells, one case showing significant lymphocytic infiltrate (Figure 5H). Severe lesions were also observed in this group, such as extensive intra-alveolar hemorrhage (Figure 5I,J). Except the intra-alveolar hemorrhage, all the lesions previously described in the control group were also observed in the other groups exposed to citric acid, but with different intensity (Figure 5J–L).

Differences between the four groups were observed regarding the alveolar spaces, but without reaching statistical significance. In the control group, the alveolar spaces had a mean value of 11.39 ± 2.46% of the total area. The most extensive area of alveolar spaces was noticed in the group treated with WSE, with a mean value of 14.94 ± 2.64%, followed by the group treated with codeine, with a mean value of 13.17 ± 1.98% and WSE 1:2, with a mean value of 12.80 ± 3.30% of the total area.

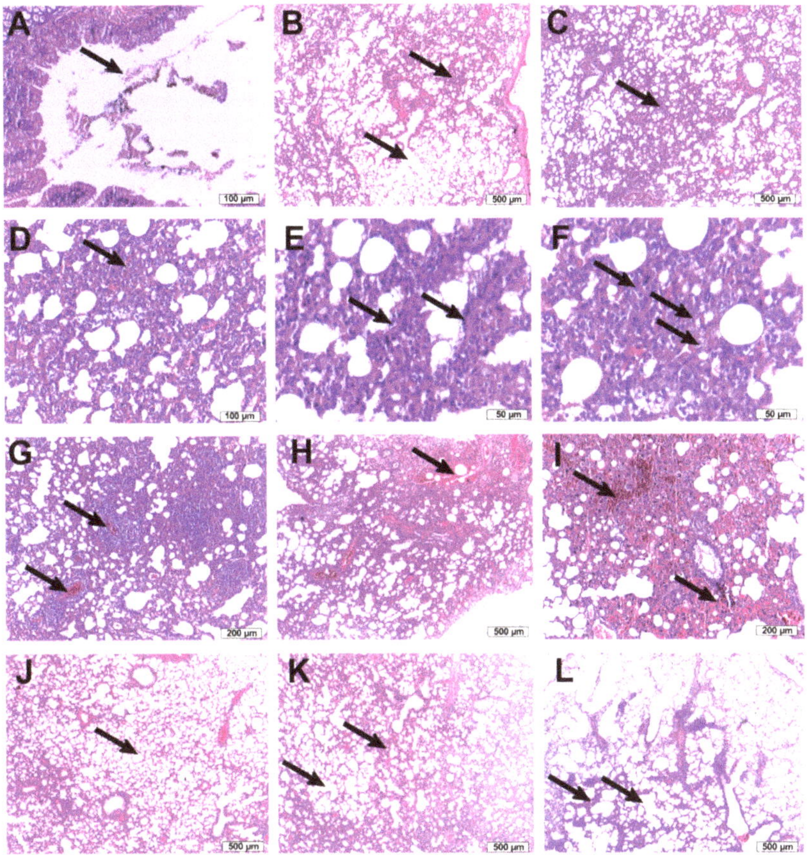

Figure 5. Lung tissue histopathological analysis: Negative control group (**A–I**): (**A**) serohemorrhagic bronchial intraluminal exudate; (**B**) collapsed alveolar septa and the emphysematous alveoli; (**C**) thickened alveolar septa; (**D**) hyperhemic alveolar capillaries with septal hypercellularity; (**E**) unflattened cubic pneumocytes; (**F**) septal macrophages and inflammatory cells; (**G**) lymphocytic infiltrate; (**H**) extensive intra-alveolar hemorrhage; (**I**) intra-alveolar red cells (hemorrhage); WSE group: (**J**) normal appearance of the respiratory parenchyma; WSE 1:2 group: (**L**) the thickened alveolar septa and emphysematous alveoli; Codeine group: (**K**) the thickened alveolar septa and emphysematous alveoli.

4. Discussion

In folk medicine, the wooden diaphragm inside the walnut kernel has been known as an antitussive agent but, to the best of our knowledge, there is no scientific evidence to support this effect. The phytochemical analysis of walnut septum revealed a high content of bioactive compounds, especially antioxidants from the polyphenol class, such as flavonoids, ellagic acid derivatives and gallotannins, with comparable results between the studies [15,27,28].

In the current study, the antitussive, antioxidant, and anti-inflammatory potentials of an optimized WSE characterized by a high quantity of bioactive compounds were investigated in a citric acid-induced cough rodent model. The phytochemical profile of this extract was previously analyzed by liquid chromatography-mass spectrometry (LC-MS) and liquid chromatography coupled with tandem mass spectrometry (LC-MS/MS) methods (Table 2).

Table 2. The phytochemical profile of tested WSE (determined in previously performed analytical studies on lyophilized extract).

Method (Ref.)	Chromatographic Conditions	Detection Mode	Bioactive Compound	*m/z*	mg/100 g
LC-MS [15]	Zorbax SB-C18; methanol: 0.1% acetic acid (*v*/*v*) and binary gradient at 48 °C; flow rate: 1 mL/min; injection volume: 5 µL	UV: 330 nm for 0–17 min (to detect phenolic acids); 370 nm for 18–38 min (to detect flavonoids and their aglycones) MS: ESI [a], negative mode, SIM [b] mode	Caftaric acid	311	<LOD
			Gentisic acid	179	<LOQ
			Caffeic acid	179	<LOD
			Chlorogenic acid	353	<LOD
			p-Coumaric acid	163	<LOQ
			Ferulic acid	193	<LOQ
			Sinapic acid	223	<LOD
			Hyperoside	463	6.73
			Isoquercitrin	463	10.36
			Rutoside	609	<LOD
			Myricetol	317	<LOD
			Fisetin	285	<LOD
			Quercitrin	447	107.31
			Quercetin	301	<LOQ
			Patuletin	331	<LOD
			Luteolin	285	<LOD
			Kaempferol	285	<LOD
			Apigenin	279	<LOD
LC-MS [15]	Zorbax SB-C18; methanol: 0.1% acetic acid (*v*/*v*) and binary gradient at 48 °C; flow rate: 1 mL/min; injection volume: 5 µL	MS: ESI, negative mode, SIM mode	Epicatechin	289	1.25
			Catechin	289	59.76
			Gallic acid	169	7.96
			Syringic acid	197	0.52
			Protocatechuic acid	153	0.99
			Vanillic acid	167	0.56
LC-MS/MS [15]	Zorbax SB-C18; methanol: acetonitrile (10:90, *v*/*v*) and isocratic elution, at 40 °C; flow rate: 1 mL/min; injection volume: 5 µL	MS: APCI [c], positive mode, MRM [d] mode	β-Sitosterol	397 **	3101.8
			Stigmasterol	395	<LOQ
			Campesterol	383	29.21
			Ergosterol	379	<LOD
LC-MS/MS [29]	Zorbax SB-C18; water: methanol (7:93, *v*/*v*) and isocratic elution, at 40 °C; flow rate: 1 mL/min; injection volume: 10 µL	MS: APCI, negative mode, MRM mode	α-Tocopherol	401→386	3.35
			β/γ-Tocopherols	415→400	1.73
			δ-Tocopherol	429→163	1.47

LOD—Limit of detection (0.1 µg/mL); LOQ—Limit of quantification (0.5 µg/mL); ** [$M-H_2O+H^+$] > Main ions from spectrum (β-Sitosterol: 397 > 160.9, 174.9, 188.9, 202.9, 214.9, 243, 257, 287.1, 315.2; Stigmasterol: 395 > 255, 297, 283, 311, 241, 201; Campesterol: 383 > 147, 149, 161, 175, 189, 203, 215, 229, 243, 257; Ergosterol: 379 > 158.9, 184.9, 199, 213, 225, 239, 253, 295, 309, 323); [a] ESI—electrospray ionization; [b] SIM—selected ion monitoring; [c] APCI—atmospheric pressure chemical ionization; [d] MRM—multiple reactions monitoring.

In our experiment, the citric acid-induced cough rodent model was obtained by exposing the animals to citric acid aerosols (17.5%) using an ultrasonic nebulizer in a closed chamber. In previous research, this model was successfully used to assess the antitussive effects of various chemical compounds or complex plant extracts from *Eriobotrya japonica* [30], *Crocus sativus* [31], and *Stemona tuberosa* [32]. Prior to exposure to the irritant, the animals were pre-treated by gavage for three consecutive days with two different doses of WSE (WSE and WSE 1:2) and a positive control (Codeine), to test a possible antitussive effect. The positive control, Codeine, significantly decreased the cough frequency by approximately 50%. Interestingly, pre-exposure to the WSE at the highest dose (134 mg GAE/kg b.w./day) resulted in a comparable effect or even superior to codeine. The antitussive effect of WSE was dose-dependent: the number of coughs decreased by approximately 68% and 20% in case of WSE and WSE 1:2, respectively, versus negative control group. In addition to lowering the cough frequency, codeine treatment prolonged the latency of coughs four-fold. Treatment with the highest tested dose of WSE significantly increased the latency of coughs, a tendency to increase this latency time being also observed for the lowest tested dose of WSE (Figure 1B). For WSE pretreated animals, the protective effects against irritative and hemorrhagic phenomena induced by citric acid aerosols at the nasal level were obvious. This effect could be based on several mechanisms. Previous studies showed that WS extracts increase erythrocyte membrane resistance to lysis, most likely by stimulating glutathione-dependent enzyme systems [33].

Airway inflammation characterized by an increased number of immune cells in sputum and increased secretion of pro-inflammatory cytokines in airway tissues is a hallmark in patients with chronic cough. In this regard we evaluated the ability of WSE to reduce pulmonary inflammation induced by exposure to citric acid. Due to the intricate interplay between oxidative stress and inflammation, as these two processes can propagate and potentiate each other, reactive oxygen species were also measured in lung tissue homogenate. Moreover, oxidative stress was shown to contribute to airway obstruction by impairing the activation of β2-adrenoreceptors, antioxidants possessing the capacity to restore the sensitivity of these receptors to bronchodilators [34].

Pre-treatment with WSE exerted an antioxidant effect in lung tissue, decreasing ROS in a dose-dependent manner. At the highest tested dose, the ROS concentration decreased by approximately 60%, while at the lowest tested dose the decrease was approximately 30%. The ability of the WSE to mitigate oxidative stress was previously reported in an in vitro cellular model [29]. Exposure to non-cytotoxic concentrations of WSE mitigated ROS in a dose-dependent manner in different cell types (A549, T47D-Kbluc, MCF-7 and human gingival fibroblasts cells). Moreover, WSE displayed antioxidant capacity in non-stimulated and stimulated (exposure to H_2O_2) conditions in these in vitro models [29]. Although phytochemicals present in WSE were reported to directly neutralize ROS in in vitro studies, a short dietary supplementation may not result in a sufficient accumulation within a specific tissue to directly mitigate ROS and diminish DCFH oxidation. As in other studies [35], a more plausible explanation would be that reduced ROS concentration is related to the modulation of endogenous antioxidant defense systems. The in vivo antioxidant capacity of WSE was also reported in a D-gal-induced aging model in rats where administration of this extract for 8 weeks statistically decreased the ROS levels in brain and liver tissues [36]. The results obtained indicate that WSE possess in vitro and in vivo antioxidant potential, thus being in accordance with previously published other studies on walnut-derived extracts [37,38].

Similar to other polyphenol rich plant matrices, the observed antioxidant capacity of WSE could be explained by the content in bioactive compounds. Phytochemicals present in WSE were shown in several studies to possess antioxidant potential [39] by interacting and activating the Nrf2/ARE cellular pathway [40] that initiates the expression and synthesis of endogenous antioxidants such as glutathione [41]. Quercetin and the highest quantified quercetin glycosides in WSE, quercitrin (quercetin-3-O-rhamnoside) and isoquercitrin (quercetin 3-β-D-glucoside), were reported to possess antioxidant potential in in vitro and in vivo studies [42]. Choi et al. reported that co-exposure to quercitrin and isoquercitrin can mitigate ROS in human hepatocytes exposed to prooxidants such as 2,2′-Azobis(2-amidinopropane) dihydrochloride (AAPH), Cu^{2+} and H_2O_2 [43]. In addition to their direct action against ROS, quercetin and quercetin glycosides were shown to up-regulate the Nrf2/ARE pathway by repression of the Keap1 protein that inhibits the activation of this pathway [44]. A similar antioxidative potential was reported for hyperoside (quercetin 3-galactoside), exposure of human lens epithelial cells to hyperoside resulting in an increased mRNA and protein expression of heme-oxygenase-1 [45]. Other biomolecules found in WS that can be involved in the antioxidant and anti-inflammatory activities are the ellagitannins. In the gastrointestinal tract, ellagic acid, released from ellagitannins, is metabolized by the colonic microorganisms to urolithins, important modulators of oxidative stress and inflammatory action [46].

Endogenous NO plays a delicate role in homeostasis maintaining of varied cellular functions, including in the respiratory system. The local concentrations of NO are highly dynamic as the synthesis is regulated by independent enzymatic pathways. These are either constitutively expressed or induced at the gene-transcriptional level by cytokines, chemokines, and mediators. Depending on achieved local concentrations, NO can present protective respiratory effects by promoting smooth muscle relaxation and mitigating airway hyperresponsiveness to bronchoconstrictors or deleterious effects by promoting pro-inflammatory events [47]. Interestingly, NO has been shown to have a modulatory effect on inflammation, decreasing the secretion of pro-inflammatory cytokines in human alveolar macrophages challenged with bacterial lipopolysaccharides (LPS), while not altering the basal cytokine levels [48].

In the current study, pre-exposure of animals to WSE influenced NO in lung tissue homogenates of animals exposed to citric acid but not to a statistically significant extent. However, it should be noted that changes in NO were mostly reported in asthmatic patients that present eosinophilic inflammation, while no correlation between exhaled NO and other respiratory pathologies characterized by cough were demonstrated [49]. In a previous study in which we evaluated the anti-aging potential of WSE, no changes in this parameter were observed in liver and brain tissues of naturally aged animals [36]. Alternatively, these results could be due to the fact that whole lung tissues were used for the assay, and local changes of NO concentration in superficial tissue could have been masked by the bulk of the tissue. Regarding other previously reported data, a polysaccharide extract from *Opilia celtidifolia* was shown to possess bronchodilator effects in a citric acid-induced cough model, an effect partially explained by the increase in the NO production [50]. Conversely, in an ovalbumin-induced asthma model, a decrease in NO production accompanied by a down-regulation in the expression of iNOS was observed after exposure to an Orai 1 antagonist with antitussive and bronchodilator effects [51].

The anti-inflammatory potential of the WSE was further evaluated by measuring the level of IL-6 and two receptors (CXC-R1 and CXC-R2) for IL-8 in the lung tissue homogenates. Pre-treatment with WSE resulted in a decreased concentration only in IL-6 and CXC-R1. Additional studies showed that not all pro-inflammatory cytokines could be reduced by the treatment [52]. Regarding the effects observed for IL-6, we obtained similar results in an in vitro study on human gingival fibroblasts, where exposure to non-cytotoxic doses of WSE decreased the extracellular release of IL-6 in a dose-dependent manner [15]. Besides having antioxidant potential, phytochemicals present in *Juglans regia* were shown to possess an anti-inflammatory potential by inhibiting the activation of the NF-κB pathway that regulates the expression of pro-inflammatory cytokines and chemokines [53]. The complementary nature of the NF-κB and Nrf2 signaling pathways was previously reported not only for phytochemicals, but also in the case of various toxicants such as diesel exhaust particles, a concept known as the three-tiered oxidative stress paradigm [54]. The activation of the Nrf2 pathway by the polyphenols and other bioactive compounds in WSE can lead to an anti-inflammatory effect mediated by the inhibition of the NF-κB nuclear translocation [55].

In a recent study, Muzaffer et al. reported that a methanolic extract of male flowers of walnut reduced the release of pro-inflammatory cytokines IL-1 and IL-6 and TNF-α after exposure of human keratinocytes to ultraviolet B rays [38]. Moreover, different extracts from the same species, decreased the expression of pro-inflammatory adhesion molecules in isolated human endothelial cells [56]. Previous studies also demonstrated the importance of IL-8 in the development of airway diseases. Its receptors CXC-R1 and CXC-R2 are present on airway smooth muscles, being involved in the process of cell contraction with subsequent bronchial hyper-responsiveness [57]. Our data showed a decreased concentration of CXC-R1 in the lung tissue homogenates from WSE-treated animals. Thus, by reducing the risk of a chemically triggered bronchospasm, WSE could prove beneficial effects not only for cough suppression, but in larger categories of patients with airway diseases, further research being necessary to ascertain the therapeutical importance of these findings.

The anti-inflammatory effects observed in vitro are also supported by in vivo studies conducted on animals. Exposure to walnut kernel extract reduced inflammation and lung injuries in rats exposed to cigarette smoke extract [58], while exposure of mice to an extract from leaves resulted in antinociceptive and anti-inflammatory effects [59]. Moreover, Pang et al. reported that an extract isolated from the roots of *Pseudostellaria heterophylla* manifested antitussive activity via attenuation of airway inflammation by adjustment of multi-cytokine levels [52]. Similar studies on *Zataria multiflora* [60] indicated that the phytochemical compounds present in this species present a protective antitussive effect by decreasing the level of systemic cytokines such as IL-6 and IL-8 [61].

The histopathological analysis revealed small but remarkable differences between the groups. Initially, the areas represented by the thick broncho-vascular tracts were excluded, the remaining alveolar parenchyma areas being subsequently evaluated. At the level of the alveolar parenchyma, the area represented by the air spaces was quantified, as an indirect indicator of the damage/thickening of the alveolar septa or of the accumulation of fluid or blood in the alveolar spaces.

The area occupied by the alveolar spaces was the largest in the WSE group, followed by codeine and WSE 1:2 groups, while the control group presented the smallest alveolar space area relative to the total area. Therefore, even after a short treatment period, the WSE showed a positive dose-dependent effect on the alveolar space area compared to both control and codeine-treated groups. This effect could be explained by the composition of the WSE, rich in polyphenols and tocopherols. These bioactive phytochemicals revealed, as shown before, antioxidant and anti-inflammatory activities.

In our experimental model, cough was induced by an aerosolized irritative agent, citric acid, which stimulated cough-evoking chemoreceptors from the airways of exposed animals. According to previously published research, an additional inflammatory process can alter the activation profile of these receptors with a subsequent heightened sensitivity and increased transmission of stimulus to the central formations involved in cough control [62]. Thus, a reduction of inflammatory processes in the airways could represent a peripheral mechanism involved in the cough suppressant effect of WSE. However, due to the complexity of chemical composition of the extract and of physiological processes involved in cough regulation, other mechanisms could be involved in the antitussive effect of WSE.

5. Conclusions

This study evaluated the antitussive, antioxidant, and anti-inflammatory effects of a WSE rich in bioactive compounds, using a citric acid-induced cough model in rats. Walnut septum showed antitussive and anti-inflammatory activities and displayed potent antitussive effects.

The antitussive action of WS could be due to a cellular protective effect, by increasing the resistance of cell membranes and erythrocytes during exposure to citric acid aerosols, and an anti-inflammatory effect. These effects could be a consequence of the stimulation of the antioxidant enzyme systems and the reduction of IL-6 and CXC-R1 concentration in the lung tissue homogenates from WSE-treated animals.

Our findings provide initial scientific evidence that walnut septum could be a novel source of non-opioid antitussive drug candidates. More in-depth studies are needed in the future to clarify the molecular mechanisms responsible for the antitussive effect of walnut septum.

Author Contributions: Conceptualization, M.E.R., D.-S.P.; methodology, I.F., M.E.R., D.-S.P.; validation, I.F., O.V., D.-S.P.; formal analysis, I.F., M.E.R., C.G., A.P., M.-G.È., D.-M.M., S.M., O.V., B.K., D.-S.P.; investigation, I.F., M.E.R., C.G., A.P., O.V., D.-S.P.; writing original draft preparation, I.F., M.E.R., C.G., O.V., D.-S.P.; writing, reviewing, and editing, I.F., M.E.R., C.G., A.P., M.-G.È., D.-M.M., S.M., O.V., B.K., D.-S.P.; supervision, I.F., O.V., D.-S.P. All authors have read and agreed to the published version of the manuscript.

Funding: This research received no external funding.

Institutional Review Board Statement: The experimental protocol was reviewed and approved by the Commission of Ethics of the University of Medicine and Pharmacy from Cluj-Napoca (decision no. 97/09.03.2020) and the Veterinary and Food Safety Department from Cluj-Napoca (decision no. 218/26.05.2020).

Informed Consent Statement: Not applicable.

Conflicts of Interest: The authors declare no conflict of interest.

Abbreviations

ABTS	2,2′-azinobis-(3-ethylbenzothiazoline-6-sulfonate)
AU	arbitrary units
BSA	bovine serum albumin
b.w.	body weight
CBB	Coomassie Briliant Blue G
CXC-R1	chemokine receptor type 1
CXC-R2	chemokine receptor type 2
DCFH-DA	2′,7′-dichlorodihydrofluorescein diacetate
FC	Folin-Ciocâlteu
GAE	gallic acid equivalents
HE	hematoxylin and eosin
IL-6	interleukin 6
IL-8	interleukin 8
LC-MS	liquid chromatography-mass spectrometry
LC-MS/MS	liquid chromatography coupled with tandem mass spectrometry
LOD	Limit of detection
LOQ	Limit of quantification
NF-κB	Nuclear Factor kappa B
NO	nitric oxide
Nrf2/ARE	Nuclear factor erythroid 2-related factor 2/antioxidant response element
PBS	phosphate-buffered saline
ROS	reactive oxygen species
SD	standard deviation
SEM	standard error mean
TAC	total antioxidant capacity
TEAC	Trolox equivalent antioxidant capacity
TNF-α	tumor necrosis factor α
TPC	total phenolic content
Tris	tris(hydroxymethyl)aminomethane
Trolox	6-hydroxy-2,5,7,8-tetramethyl chroman-2-carboxylic acid
WS	walnut septum
WSE	walnut septum extract

References

1. Xu, Y.; Wang, F.; Guo, H.; Wang, S.; Ni, S.; Zhou, Y.; Wang, Z.; Bao, H.; Wang, Y. Antitussive and Anti-inflammatory Dual-active Agents Developed from Natural Product Lead Compound 1-Methylhydantoin. *Molecules* **2019**, *24*, 2355. [CrossRef] [PubMed]
2. Oketch-Rabah, H.A.; Marles, R.J.; Jordan, S.A.; Low Dog, T. United States Pharmacopeia Safety Review of Willow Bark. *Planta Med.* **2019**, *85*, 1192–1202. [CrossRef] [PubMed]
3. Dicpinigaitis, P.V.; Morice, A.H.; Birring, S.S.; McGarvey, L.; Smith, J.A.; Canning, B.J.; Page, C.P. Antitussive Drugs—Past, Present, and Future. *Pharmacol. Rev.* **2014**, *66*, 468–512. [CrossRef]
4. Achan, J.; Talisuna, A.O.; Erhart, A.; Yeka, A.; Tibenderana, J.K.; Baliraine, F.N.; Rosenthal, P.J.; D'Alessandro, U. Quinine, an old anti-malarial drug in a modern world: Role in the treatment of malaria. *Malar. J.* **2011**, *10*, 144. [CrossRef] [PubMed]
5. García-Estrada, C.; Martín, J.; Cueto, L.; Barreiro, C. Omics Approaches Applied to Penicillium chrysogenum and Penicillin Production: Revealing the Secrets of Improved Productivity. *Genes* **2020**, *11*, 712. [CrossRef]
6. Tobias, J.D.; Green, T.P.; Coté, C.J. Codeine: Time to Say "No". *Pediatrics* **2016**, *138*, e20162396. [CrossRef]
7. Rusu, M.E.; Mocan, A.; Ferreira, I.C.F.R.; Popa, D.-S. Health Benefits of Nut Consumption in Middle-Aged and Elderly Population. *Antioxidants* **2019**, *8*, 302. [CrossRef]

8. Rusu, M.E.; Simedrea, R.; Gheldiu, A.-M.; Mocan, A.; Vlase, L.; Popa, D.-S.; Ferreira, I.C.F.R. Benefits of tree nut consumption on aging and age-related diseases: Mechanisms of actions. *Trends Food Sci. Technol.* **2019**, *88*, 104–120. [CrossRef]
9. Alasalvar, C.; Bolling, B. Review of nut phytochemicals, fat-soluble bioactives, antioxidant components and health effects. *Br. J. Nutr.* **2015**, *113*, S68–S78. [CrossRef]
10. Santos, A.; Barros, L.; Calhelha, R.C.; Dueñas, M.; Carvalho, A.M.; Santos-Buelga, C.; Ferreira, I.C.F.R. Leaves and decoction of Juglans regia L.: Different performances regarding bioactive compounds and in vitro antioxidant and antitumor effects. *Ind. Crops Prod.* **2013**, *51*, 430–436. [CrossRef]
11. Vieira, V.; Pereira, C.; Abreu, R.M.V.; Calhelha, R.C.; Alves, J.M.; Coutinho, J.A.P.; Ferreira, O.; Barros, L.; Ferreira, I.C.F.R. Hydroethanolic extract of Juglans regia L. green husks: A source of bioactive phytochemicals. *Food Chem. Toxicol.* **2020**, *137*, 111189. [CrossRef] [PubMed]
12. Dehghani, F.; Mashhoody, T.; Panjehshahin, M. Effect of aqueous extract of walnut septum on blood glucose and pancreatic structure in streptozotocin-induced diabetic mouse. *Iran J. Pharmacol. Ther.* **2012**, *11*, 10–14.
13. Ramishvili, L.; Gordeziani, M.; Tavdishvili, E.; Bedineishvili, N.; Dzidziguri, D.; Kotrikadze, N. The effect of extract of greek walnut (*Juglans regia* L.) septa on some functional characteristics of erythrocytes. *Georg. Med. News* **2016**, *261*, 51–57.
14. Ravanbakhsh, A.; Mahdavi, M.; Jalilzade-Amin, G.; Javadi, S.; Maham, M.; Mohammadnejad, D.; Rashidi, M.R. Acute and subchronic toxicity study of the median septum of Juglans regia in Wistar rats. *Adv. Pharm. Bull.* **2016**, *6*, 541–549. [CrossRef]
15. Rusu, M.E.; Gheldiu, A.-M.; Mocan, A.; Moldovan, C.; Popa, D.-S.; Tomuta, I.; Vlase, L. Process Optimization for Improved Phenolic Compounds Recovery from Walnut (*Juglans regia* L.) Septum: Phytochemical Profile and Biological Activities. *Molecules* **2018**, *23*, 2814. [CrossRef]
16. Reynoso, M.; Brodkiewics, I.; Villagra, J.; Balderrama Coca, M.; Sanchez Riera, A.; Vera, N. Evaluation of antitussive and expectorant potential of Ziziphus Mistol Fruits (Mistol). *Int. J. Pharm. Sci. Res.* **2017**, *8*, 3726–3733. [CrossRef]
17. Yang, L.; Jiang, H.; Wang, S.; Hou, A.; Man, W.; Zhang, J.; Guo, X.; Yang, B.; Kuang, H.; Wang, Q. Discovering the Major Antitussive, Expectorant, and Anti-Inflammatory Bioactive Constituents in Tussilago Farfara L. Based on the Spectrum—Effect Relationship Combined with Chemometrics. *Molecules* **2020**, *25*, 620. [CrossRef]
18. Rusu, M.E.; Fizeșan, I.; Pop, A.; Gheldiu, A.-M.; Mocan, A.; Crișan, G.; Vlase, L.; Loghin, F.; Popa, D.-S.; Tomuta, I. Enhanced Recovery of Antioxidant Compounds from Hazelnut (*Corylus avellana* L.) Involucre Based on Extraction Optimization: Phytochemical Profile and Biological Activities. *Antioxidants* **2019**, *8*, 460. [CrossRef]
19. Directive 2010/63/EU of the European Parliament and of the Council. 2010. Available online: https://eur-lex.europa.eu/legal-content/EN/TXT/PDF/?uri=CELEX:02010L0063-20190626&from=EN (accessed on 18 November 2020).
20. Song, K.J.; Shin, Y.J.; Lee, K.R.; Lee, E.J.; Suh, Y.S.; Kim, K.S. Expectorant and Antitussive Effect of Hedera helix and Rhizoma coptidis extracts mixture. *Yonsei Med. J.* **2015**, *56*, 819–824. [CrossRef]
21. Gheldiu, A.-M.; Popa, D.-S.; Loghin, F.; Vlase, L. Oxidative Metabolism of Estrone Modified by Genistein and Bisphenol A in Rat Liver Microsomes. *Biomed. Environ. Sci.* **2015**, *28*, 834–838. [CrossRef]
22. Garg, G.; Singh, S.; Singh, A.K.; Rizvi, S.I. Antiaging Effect of Metformin on Brain in Naturally Aged and Accelerated Senescence Model of Rat. *Rejuvenation Res.* **2017**, *20*, 173–182. [CrossRef] [PubMed]
23. Boşca, A.B.; Dinte, E.; Colosi, H.; Ilea, A.; Câmpian, R.-S.; Uifălean, A.; Pârvu, A.E. Curcumin effect on nitro-oxidative stress in ligature-induced rat periodontitis. *Rom. Biotechnol. Lett.* **2015**, *20*, 10708–10717.
24. Erel, O. A novel automated direct measurement method for total antioxidant capacity using a new generation, more stable ABTS radical cation. *Clin. Biochem.* **2004**, *37*, 277–285. [CrossRef] [PubMed]
25. Grozav, A.; Miclaus, V.; Vostinaru, O.; Ghibu, S.; Berce, C.; Rotar, I.; Mogosan, C.; Therrien, B.; Loghin, F.; Popa, D. Acute toxicity evaluation of a thiazolo arene ruthenium (II) complex in rats. *Regul. Toxicol. Pharmacol.* **2016**, *80*, 233–240. [CrossRef]
26. Gopalakrishnan, A.; Ji, L.L.; Cirelli, C. Sleep Deprivation and Cellular Responses to Oxidative Stress. *Sleep* **2004**, *27*, 27–35. [CrossRef]
27. Li, L.; Song, L.; Sun, X.; Yan, S.; Huang, W.; Liu, P. Characterisation of phenolics in fruit septum of Juglans regia Linn. by ultra performance liquid chromatography coupled with Orbitrap mass spectrometer. *Food Chem.* **2019**, *286*, 669–677. [CrossRef]
28. Genovese, C.; Cambria, M.T.; D'Angeli, F.; Addamo, A.P.; Malfa, G.A.; Siracusa, L.; Pulvirenti, L.; Anfuso, C.D.; Lupo, G.; Salmeri, M. The double effect of walnut septum extract (*Juglans regia* L.) counteracts A172 glioblastoma cell survival and bacterial growth. *Int. J. Oncol.* **2020**, *57*, 1129–1144. [CrossRef]
29. Rusu, M.E.; Fizesan, I.; Pop, A.; Mocan, A.; Gheldiu, A.-M.; Babota, M.; Vodnar, D.C.; Jurj, A.; Berindan-Neagoe, I.; Vlase, L.; et al. Walnut (*Juglans regia* L.) Septum: Assessment of Bioactive Molecules and In Vitro Biological Effects. *Molecules* **2020**, *25*, 2187. [CrossRef]
30. Ju, J.; Zhou, L.; Lin, G.; Liu, D.; Wang, L.; Yang, J. Studies on constituents of triterpene acids from Eriobotrya japonica and their anti-inflammatory and antitussive effects. *J. Chin. Pharm. Sci.* **2003**, *38*, 752–757.
31. Hosseinzadeh, H.; Ghenaati, J. Evaluation of the antitussive effect of stigma and petals of saffron (*Crocus sativus*) and its components, safranal and crocin in guinea pigs. *Fitoterapia* **2006**, *77*, 446–448. [CrossRef]
32. Chung, H.-S.; Hon, P.-M.; Lin, G.; But, P.P.-H.; Dong, H. Antitussive activity of Stemona alkaloids from Stemona tuberosa. *Planta Med.* **2003**, *69*, 914–920. [CrossRef] [PubMed]
33. Dzidziguri, D.; Rukhadze, M.; Modebadze, I.; Bakuradze, E.; Kurtanidze, M.; Giqoshvili, V. The study of the immune corrective properties of greek walnut (*Juglans regia* L.) septa on the experimental model of leukopenia. *Georg. Med. News* **2016**, *252*, 84–89.

34. Dal Negro, R.W.; Wedzicha, J.A.; Iversen, M.; Fontana, G.; Page, C.; Cicero, A.F.; Pozzi, E.; Calverley, P.M.A. Effect of erdosteine on the rate and duration of COPD exacerbations: The RESTORE study. *Eur. Respir. J.* **2017**, *50*, 1700711. [CrossRef] [PubMed]
35. Fu, Y.; Ji, L. Chronic Ginseng Consumption Attenuates Age-Associated Oxidative Stress in Rats. *J. Nutr.* **2003**, *133*, 3603–3609. [CrossRef]
36. Rusu, M.E.; Georgiu, C.; Pop, A.; Mocan, A.; Kiss, B.; Vostinaru, O.; Fizesan, I.; Stefan, M.-G.; Gheldiu, A.-M.; Mates, L.; et al. Antioxidant Effects of Walnut (*Juglans regia* L.) Kernel and Walnut Septum Extract in a D-Galactose-Induced Aging Model and in Naturally Aged Rats. *Antioxidants* **2020**, *9*, 424. [CrossRef]
37. Muthaiyah, B.; Essa, M.M.; Chauhan, V.; Chauhan, A. Protective effects of walnut extract against amyloid beta peptide-induced cell death and oxidative stress in PC12 cells. *Neurochem. Res.* **2011**, *36*, 2096–2103. [CrossRef]
38. Muzaffer, U.; Paul, V.; Prasad, N.R.; Karthikeyan, R.; Agilan, B. Protective effect of *Juglans regia* L. against ultraviolet B radiation induced inflammatory responses in human epidermal keratinocytes. *Phytomedicine* **2018**, *42*, 100–111. [CrossRef]
39. Zhang, H.; Tsao, R. Dietary polyphenols, oxidative stress and antioxidant and anti-inflammatory effects. *Curr. Opin. Food Sci.* **2016**, *8*, 33–42. [CrossRef]
40. Erlank, H.; Elmann, A.; Kohen, R.; Kanner, J. Polyphenols activate Nrf2 in astrocytes via H_2O_2, semiquinones, and quinones. *Free Radic. Biol. Med.* **2011**, *51*, 2319–2327. [CrossRef]
41. Zarogoulidis, P.; Cheva, A.; Zarampouka, K.; Huang, H.; Li, C.; Huang, Y.; Katsikogiannis, N.; Zarogoulidis, K. Tocopherols and tocotrienols as anticancer treatment for lung cancer: Future nutrition. *J. Thorac. Dis.* **2013**, *5*, 349–352. [CrossRef]
42. Boots, A.W.; Haenen, G.R.; Bast, A. Health effects of quercetin: From antioxidant to nutraceutical. *Eur. J. Pharmacol.* **2008**, *585*, 325–337. [CrossRef] [PubMed]
43. Choi, S.-J.; Tai, B.H.; Cuong, N.M.; Kim, Y.-H.; Jang, H.-D. Antioxidative and anti-inflammatory effect of quercetin and its glycosides isolated from mampat (*Cratoxylum formosum*). *Food Sci. Biotechnol.* **2012**, *21*, 587–595. [CrossRef]
44. Tanigawa, S.; Fujii, M.; Hou, D.-X. Action of Nrf2 and Keap1 in ARE-mediated NQO1 expression by quercetin. *Free Radic. Biol. Med.* **2007**, *42*, 1690–1703. [CrossRef] [PubMed]
45. Park, J.; Han, X.; Piao, M.; Oh, M.; Fernando, P.; Kang, K.; Ryu, Y.; Jung, U.; Kim, I.; Hyun, J. Hyperoside Induces Endogenous Antioxidant System to Alleviate Oxidative Stress. *J. Cancer Prev.* **2016**, *21*, 41–47. [CrossRef] [PubMed]
46. Djedjibegovic, J.; Marjanovic, A.; Panieri, E.; Saso, L. Ellagic Acid-Derived Urolithins as Modulators of Oxidative Stress. *Oxid. Med. Cell. Longev.* **2020**, *2020*, 5194508. [CrossRef] [PubMed]
47. Ricciardolo, F.L.M.; Sterk, P.J.; Gaston, B.; Folkerts, G. Nitric oxide in health and disease of the respiratory system. *Physiol. Rev.* **2004**, *84*, 731–765. [CrossRef] [PubMed]
48. Thomassen, M.J.; Buhrow, L.T.; Connors, M.J.; Takao Kaneko, F.; Erzurum, S.C.; Kavuru, M.S. Nitric oxide inhibits inflammatory cytokine production by human alveolar macrophages. *Am. J. Respir. Cell Mol. Biol.* **1997**, *17*, 279–283. [CrossRef]
49. Hoyte, F.C.L.; Gross, L.M.; Katial, R.K. Exhaled nitric oxide: An update. *Immunol. Allergy Clin.* **2018**, *38*, 573–585. [CrossRef]
50. Šutovská, M.; Fraňová, S.; Sadloňová, V.; Grønhaug, T.E.; Diallo, D.; Paulsen, B.S.; Capek, P. The relationship between dose-dependent antitussive and bronchodilatory effects of Opilia celtidifolia polysaccharide and nitric oxide in guinea pigs. *Int. J. Biol. Macromol.* **2010**, *47*, 508–513. [CrossRef]
51. Sutovská, M.; Kocmálová, M.; Adamkov, M.; Výbohová, D.; Mikolka, P.; Mokrá, D.; Hatok, J.; Antošová, M.; Fraňová, S. The long-term administration of Orai 1 antagonist possesses antitussive, bronchodilatory and anti-inflammatory effects in experimental asthma model. *Gen. Physiol. Biophys.* **2013**, *32*, 251–259. [CrossRef]
52. Pang, W.; Lin, S.; Dai, Q.; Zhang, H.; Hu, J. Antitussive Activity of Pseudostellaria heterophylla (Miq.) Pax Extracts and Improvement in Lung Function via Adjustment of Multi-Cytokine Levels. *Molecules* **2011**, *16*, 3360–3370. [CrossRef]
53. Wardyn, J.D.; Ponsford, A.H.; Sanderson, C.M. Dissecting molecular cross-talk between Nrf2 and NF-κB response pathways. *Biochem. Soc. Trans.* **2015**, *43*, 621–626. [CrossRef]
54. Fizeșan, I.; Chary, A.; Cambier, S.; Moschini, E.; Serchi, T.; Nelissen, I.; Kiss, B.; Pop, A.; Loghin, F.; Gutleb, A.C. Responsiveness assessment of a 3D tetra-culture alveolar model exposed to diesel exhaust particulate matter. *Toxicol. Vitr.* **2018**, *53*, 67–79. [CrossRef]
55. Ahmed, S.; Luo, L.; Namani, A.; Wang, X.; Tang, X. Nrf2 signaling pathway: Pivotal roles in inflammation. *Biochim. Biophys. Acta Mol. Basis Dis.* **2017**, *1863*, 585–597. [CrossRef]
56. Papoutsi, Z.; Kassi, E.; Chinou, I.; Halabalaki, M.; Skaltsounis, L.; Moutsatsou, P. Walnut extract (*Juglans regia* L.) and its component ellagic acid exhibit anti-inflammatory activity in human aorta endothelial cells and osteoblastic activity in the cell line KS483. *Br. J. Nutr.* **2008**, *99*, 715–722. [CrossRef]
57. Govindaraju, V.; Michoud, M.-C.; Al-Chalabi, M.; Ferraro, P.; Powell, W.S.; Martin, J.G. Interleukin-8: Novel roles in human airway smooth muscle cell contraction and migration. *Am. J. Physiol. Cell Physiol.* **2006**, *291*, C957–C965. [CrossRef]
58. Qamar, W.; Sultana, S. Polyphenols from *Juglans regia* L.(walnut) kernel modulate cigarette smoke extract induced acute inflammation, oxidative stress and lung injury in Wistar rats. *Hum. Exp. Toxicol.* **2011**, *30*, 499–506. [CrossRef]
59. Hosseinzadeh, H.; Zarei, H.; Taghiabadi, E. Antinociceptive, anti-inflammatory and acute toxicity effects of *Juglans regia* L. leaves in mice. *Iran. Red Crescent Med. J.* **2011**, *13*, 27–33.
60. Boskabady, M.H.; Gholami Mhtaj, L. Effect of the Zataria multiflora on systemic inflammation of experimental animals model of COPD. *Biomed. Res. Int.* **2014**, *2014*, 802189. [CrossRef]

61. Khazdair, M.R.; Ghorani, V.; Alavinezhad, A.; Boskabady, M.H. Effect of Zataria multiflora on serum cytokine levels and pulmonary function tests in sulfur mustard-induced lung disorders: A randomized double-blind clinical trial. *J. Ethnopharmacol.* **2020**, *248*, 112325. [CrossRef]
62. Keller, J.A.; McGovern, A.E.; Mazzone, S.B. Translating cough mechanisms into better cough suppressants. *Chest* **2017**, *152*, 833–841. [CrossRef] [PubMed]

antioxidants

MDPI

Article

Quercetin and Egg Metallome

Evangelos Zoidis [1,*], Athanasios C. Pappas [1], Michael Goliomytis [2], Panagiotis E. Simitzis [2], Kyriaki Sotirakoglou [3], Savvina Tavrizelou [1], George Danezis [4,5] and Constantinos A. Georgiou [4,5]

1 Laboratory of Nutritional Physiology and Feeding, Department of Animal Science, Agricultural University of Athens, 11855 Athens, Greece; apappas@aua.gr (A.C.P.); stud217094@aua.gr (S.T.)
2 Laboratory of Animal Breeding and Husbandry, Department of Animal Science, Agricultural University of Athens, 11855 Athens, Greece; mgolio@aua.gr (M.G.); pansimitzis@aua.gr (P.E.S.)
3 Laboratory of Mathematics and Statistics, Department of Natural Resources Management and Agricultural Engineering, Agricultural University of Athens, 11855 Athens, Greece; sotirakoglou@aua.gr
4 Chemistry Laboratory, Department of Food Science and Human Nutrition, Agricultural University of Athens, 11855 Athens, Greece; gdanezis@aua.gr (G.D.); cag@aua.gr (C.A.G.)
5 FoodOmics.GR Research Infrastructure, Agricultural University of Athens, 11855 Athens, Greece
* Correspondence: ezoidis@aua.gr; Tel.:+30-21-0529-4415

Abstract: The objective of the present study was to investigate the effect of the natural flavonoid quercetin dietary supplementation on the alteration of egg metallome by applying the basic principles of elemental metabolomics. One hundred and ninety-two laying hens were allocated into 4 treatment groups: the control (C) group that was fed with a commercial basal diet and the other experimental groups that were offered the same diet further supplemented with quercetin at 200, 400 and 800 mg per kg of feed (Q2, Q4 and Q8 group, respectively) for 28 days. The diets contained the same vitamin and mineral premix, thus all birds received the same amount of elements since no differences on feed intake existed. The egg elemental profile consisted of As, Ca, Cd, Co, Cr, Cu, Fe, Mg, Mn, Mo, Ni, Pb, Sb, Se, Sr, V, Zn and was determined using inductively coupled plasma mass spectrometry (ICP-MS). Quercetin supplementation altered the elemental profile. Most notably, quercetin altered the element concentrations predominantly in egg shell and albumen. It increased the concentration of Sb while reduced that of Cr and Se in both egg shell and albumen. Moreover, it increased As, Cd in albumen and V in yolk, while compared to the control, reduced As, Cd, Cr, Cu and V and also raised Ca, Fe, Mg and Ni in egg shell. The presence of quercetin led to differentiation of the deposition of certain trace minerals in egg compartments compared to that of hens fed a basal diet, possibly indicating that tailor made eggs for specific nutritional and health requirements could be created in the future.

Keywords: egg; flavonoids; inductively coupled plasma mass spectrometry (ICP-MS); metallome; quercetin

Citation: Zoidis, E.; Pappas, A.C.; Goliomytis, M.; Simitzis, P.E.; Sotirakoglou, K.; Tavrizelou, S.; Danezis, G.; Georgiou, C.A. Quercetin and Egg Metallome. *Antioxidants* **2021**, *10*, 80. https://doi.org/10.3390/antiox10010080

Received: 14 December 2020
Accepted: 3 January 2021
Published: 9 January 2021

1. Introduction

Consumers are seeking for safe foods that also possess high nutritional value. Much attention has been paid to develop animal products with physiological functions that promote human health. Products of animal origin enriched with bioactive compounds seem to have improved quality characteristics [1,2] Furthermore, fortified animal products may protect consumers against oxidation and spoilage originating from bacteria [1,2]. Several functional foods have been created including but not limited to eggs enriched with ω-3 fatty acids, vitamin E, selenium and vitamin D.

Polyphenols present potential beneficial effects in humans and animals. Flavonoids, a class of polyphenols, are antioxidants of natural origin and they possess anti-inflammatory, antioxidant, hepatoprotective, antibacterial and anticarcinogenic properties in several animal models [3] Most notably, poultry production and health have been affected positively by quercetin [3,4]. Particularly, quercetin strengthens immune system via lymphocyte and macrophage stimulation and antibody production [4,5]. It has been shown to possess anti-inflammatory activity in animals [6,7]) and immunoregulatory effects through enhancing IgY antibody production, weight of lymphoid organs (spleen, thymus and bursa) and natural killer cell activity thereby meliorating productive performance [8]. Flavonoids supplementation in hens' diet can also improve nutritional, sensorial and microbiological properties of eggs [1].

Molecular formula of quercetin is $C_{15}H_{10}O_7$. Quercetin belongs to flavonols, subclass of flavonoids and it is widely found in different varieties of vegetables and fruits such as onions, apples and their by-products [9,10]. Quercetin properties on human health are recognized by EFSA [11]. Thus, quercetin contributes to cardiovascular, nervous system, brain, liver and kidney health. Furthermore, it supports healthy ageing and may help strengthen body's defenses [9,11]. Quercetin has antioxidant activities which are important in the control of autoimmune diseases such as pulmonary hypertension syndrome (PHS) in broilers [12]. It plays an important role in improving bone mineral density and volume, suggesting that it influences the micro-architecture of bone tissue [13]. Moreover, quercetin presents further biological actions such as antibacterial, hepatoprotective, growth promoter, antiallergic, antiviral (influenza A virus and rhinovirus), and antithrombotic activities in various animal models [4,14].

Flavonoids could affect metal homeostasis and cellular oxidative status due to dietary flavonoids interactions with trace minerals and flavonoids effect on metallothionein level [15]. Flavonoids display antioxidant characteristics via chelation with transition metals. Metal-flavonoids chelates are more effective free radical scavengers than the parent flavonoids and protect from oxidative stress [7]. Complexes of flavonoids play an important role in limiting metal bioavailability and suppressing metal toxicity [7].

Quercetin occurs as a glycoside (with linked sugars) or as an aglycone (without linked sugars) [9] and, as mentioned above, binds heavy metals and its chelating characteristics could result in lower metal toxicity. Thus, supplementation with quercetin could be regarded as a feasible approach for heavy metal poisoning [4]. Quercetin-metal complexes also present higher anticancer and antibacterial activities than that of free quercetin [16]. In addition, quercetin is chelated with iron, a crucial trace element for many essential biochemical processes. At the same time, quercetin has various roles in absorption of iron, hepcidin (a key regulator of iron metabolism and mediator of anemia of inflammation) regulation and cellular iron uptake and release. Quercetin's iron chelation has direct scavenging action against ROS (reactive oxygen species) and lessens iron overload induced by various pathologies [16]. Furthermore, quercetin's complex with Zn interacts with DNA and exerts antioxidative and anti-tumour activities [17]. As it has been shown in diabetic rats, iron-quercetin complex could reverse oxidative stress and iron deficiency mostly caused by the diabetes but concurrently it induces an imbalance in redistribution of other essential metals [18].

Therefore, it would be interesting to investigate possible alteration of the elemental fingerprint due to quercetin supplementation in various doses. Not only nutritive elements but also toxic elements where studied because as previously mentioned quercetin has presented interactions with many metals. This study is part of a project [19] where quercetin was supplemented to laying hens. Possible alterations in the elemental profile could further enhance the nutritional value (nutritive elements) and safety aspects (toxic elements) of these eggs.

2. Materials and Methods

2.1. Animals and Experimental Design

Previously, detailed description of management of birds used in the study was provided [19]. Briefly, a total of 192 Lohmann Brown-Classic laying hens (70 weeks old), from the flock kept in the facilities of the experimental station of Agricultural University of Athens, were randomly allocated into 4 treatment groups. Each treatment group consisted of 6 replicate enriched cages with 8 hens each. A commercial basal diet was provided to one group namely control (C) whereas, the same diet with quercetin (MP Biomedicals,

LLC, Illkirch, France, 97%) added at 200, 400 and 800 mg per kg of feed was provided to other three groups Q2, Q4 and Q8 group, respectively. The duration of the study was 28 days. All diets were in mash form in order to uniformly mix quercetin in the basal diet. The composition of the basal diet is presented in [19]. Birds of all treatments fed the same diet with the exception of added level of quercetin. The same amount of elements was provided to all birds since the diets contained the same vitamin and mineral premix and no differences on feed intake were noted [19]. Throughout the experimental period, the provision of water was ad libitum. The light regimen was 16 h of continuous light a day. At the end of the trial, after 28 days, a number of eggs, 7 per treatment group, was collected (28 eggs total). Eggs were analyzed for trace and macro-element content. The present study followed the guidelines of the directive 2010/63/EU of the European Parliament and of the Council on the protection of animals used for scientific purposes. The Research Ethics Committee of the Faculty of Animal Science of the Agricultural University of Athens approved the study (code number 38/07022017).

2.2. Determination of Egg Trace and Macro Elements

Fifteen elements were determined, namely As, Ca, Cd, Co, Cr, Cu, Fe, Mg, Mn, Ni, Sb, Se, Sr, V and Zn. Various elements were included with several roles, for example Ca as major structural component, Ca and Mg as responsible for activation or signalling, Co, Cr, Cu, Fe, Mn, Ni, Se, V and Zn as components of enzymes or hormones, toxic elements (As and Cd) and other that their functionality is not clearly defined (Sb and Sr). As previously described [20,21], elemental profile was determined through inductively coupled plasma mass spectrometry, ICP-MS (Perkin Elmer, Elan 9000, SCIEX, Toronto, ON, Canada).

Chemicals used were nitric acid (Suprapur®, 65% *w/v*, Merck, Darmstadt, Germany) and ICP-MS certified multi-element standards (Inorganic Ventures, NJ, USA). Ultrapure water with a resistance of 18.2 $M\Omega\ cm^{-1}$ obtained from a MilliQ plus system (Millipore, Saint Quentin Yvelines, France) was used in all procedures.

Sample digestion was performed with a microwave-assisted digestion system (CEM, Mars X-Press, Matthews, NC, USA). One grammar (1 g wet weight) of albumen or yolk and 0.1 g of shell were weighted in an analytical balance in a polypropylene tube. Then, 10 mL of HNO_3 were added to pre-digest samples for 30 min. The resulting sample suspension was transferred quantitatively in the microwave digestion PTFE (polytetrafluoroethylene) vessel. The samples were heated in the microwave accelerated digestion system according the following program: the power was ramped during 20 min from 100 W to 1200 W and held for 15 min. The temperature reached a maximum of 200 °C and followed by a cool-down cycle for 15 min. PTFE vessels were sealed throughout the aforementioned cycle to avoid volatilization losses. Although all samples were completely brought to solution, to disregard any small particle passing optical inspection entering the ICP-MS, solutions were filtered with polyester disposable syringe filters 0.20 μm/15 mm (Chromafil, Macherey-Nagel, Düren, Germany). Before injection in the ICP-MS, sample solutions were diluted, as required, with ultrapure water.

Operating conditions of the ICP-MS were as follows: nebulizer gas flow of 0.91 $L\ min^{-1}$, ICP RF (radio-frequency) power of 1050 W, lens voltage of 8.5 V and pulse stage voltage of 950 V. Calibration curves were prepared from standard solutions of high purity standards.

2.3. Statistical Analysis

Data were analyzed using the Statgraphics Centurion statistical package (version 16.1) and are presented as means ± pooled standard errors. Treatment effects on element concentrations were explored using one-way analysis of variance (ANOVA) followed by Duncan's multiple range test. Kolmogorov-Smirnov test revealed that all variables followed the normal distribution. Principal component analysis (PCA) was also applied in order to reduce the dimensionality of the data and investigate the relationships between the trace and macro elements. Moreover, data of albumen, yolk and shell were subjected to discriminant analysis to investigate samples distinguishment to the four dietary treatments and establishment of those elements capable to distinguish and classify the samples. Selection of discriminant variables was done using Wilk's lambda (λ) criterion. For all tests, the significance level was set at 5%.

3. Results

3.1. Concentration of Egg Elements

Concentrations of selected trace- and macro-elements in egg albumen are illustrated in Table 1. Addition of 200 mg quercetin per kg of feed, increased the concentration of Cd (34%) compared to control treatment. Higher levels of quercetin (400 mg/kg) significantly increased the concentration of As (18%) and decreased that of Cr (9%) and Se (23%) compared to that of control. Most notably, feed supplementation with quercetin at a concentration of 800 mg/kg, increased, compared to control, the concentration of As (24%), Cd (32%) and Sb (140-fold) and reduced the concentrations of Cr (15%) and Se (34%) (Table 1).

Table 1. The effect of dietary supplementation with quercetin on selected trace- and macro-elements concentrations in laying hens' egg albumen ($n = 7$).

Elements (ppb)	Treatment					
	C	Q2	Q4	Q8	SEM	p-Value
As	4.27 ^{a}	4.40 ^{a}	5.06 ^{b}	5.31 ^{b}	0.113	<0.001
Ca	7.15×10^4	6.78×10^4	5.59×10^4	5.19×10^4	7.41×10^3	0.214
Cd	2.77 ^{a}	3.73 ^{b}	3.14 ^{ab}	3.67 ^{b}	0.233	0.022
Co	1.31	1.60	1.33	1.33	0.099	0.147
Cr	89.5 ^{c}	87.2 ^{c}	81.2 ^{b}	76.3 ^{a}	1.386	<0.001
Cu	276.7	244.3	246.7	248.7	15.81	0.446
Fe	5.21×10^3	7.31×10^3	6.65×10^3	5.50×10^3	806.6	0.246
Mg	1.06×10^5	0.908×10^5	1.02×10^5	0.959×10^5	4.74×10^3	0.154
Mn	58.5	65.4	61.7	62.7	9.014	0.959
Ni	40.7	37.0	37.9	40.8	2.207	0.519
Sb	0.019 ^{a}	0.014 ^{a}	0.585 ^{a}	2.69 ^{b}	0.426	0.001
Se	88.6 ^{c}	84.2 ^{bc}	67.7 ^{ab}	58.0 ^{a}	6.672	0.011
Sr	150.3	152.9	132.4	146.4	9.593	0.454
V	78.8	81.6	82.9	84.3	1.631	0.129
Zn	1.68×10^3	1.41×10^3	1.18×10^3	1.49×10^3	143.8	0.129

C: no additive, Q2, Q4 and Q8: 200, 400 and 800 mg quercetin per kg of feed, respectively. ^{a,b,c} means in a row sharing no common superscript differ significantly ($p < 0.05$).

Egg yolk elements' concentrations are shown in Table 2. Addition of quercetin significantly increased V levels (ca. 7.5–9%), compared to control. Moreover, quercetin added to the diet at 400 and 800 mg/kg markedly increased, compared to control, the concentration of Sb 7.5-fold and 8-fold, respectively (Table 2).

Egg shell elements' concentrations are presented in Table 3. Addition of 200 mg quercetin per kg of diet significantly decreased the concentration, compared to control, of As (14%) and increased the concentration of Fe (9.5%) and Ni (7.5%). Quercetin added to the diet at 400 mg/kg decreased, compared to control, the concentrations of six elements namely As (40%), Cd (22%), Cr (7%), Cu (30%), Se (42%) and V (7%) while increased the levels of Ca (7.5%), Fe (12%), Mg (14%), Ni (13%) and Sb (2-fold). Accordingly, addition of 800 mg quercetin per kg of diet, significantly decreased the concentrations, compared to control, of six elements namely As (56%), Cd (46%), Cr (10%), Cu (25%), Se (43%) and V (14%) while significantly increased the levels of Fe (8%), Ni (11%) and Sb (2.8-fold) (Table 3).

Table 2. The effect of dietary supplementation with quercetin on selected trace- and macro-elements concentrations in laying hens' egg yolk ($n = 7$).

Elements (ppb)	Treatment					
	C	Q2	Q4	Q8	SEM	*p*-value
As	3.73	3.61	3.69	3.71	0.110	0.887
Ca	1.03×10^6	1.11×10^6	1.13×10^6	1.11×10^6	3.73×10^4	0.291
Cd	3.93	5.47	4.61	4.31	0.437	0.108
Co	3.30	3.44	3.73	3.41	0.134	0.164
Cr	148.1	155.7	152.6	150.7	4.072	0.614
Cu	1436	1473	1578	1509	43.46	0.153
Fe	6.89×10^4	7.20×10^4	7.06×10^4	6.55×10^4	3.86×10^3	0.672
Mg	1.25×10^5	1.35×10^5	1.29×10^5	1.33×10^5	3.77×10^3	0.307
Mn	1141	1169	1221	1179	79.01	0.914
Ni	66.7	70.1	67.9	68.0	2.220	0.754
Sb	0.077^{a}	0.408^{ab}	0.582^{b}	0.613^{b}	0.116	0.012
Se	232.9	316.4	236.8	301.5	29.76	0.125
Sr	696.9	752.9	725.0	697.4	34.28	0.618
V	104.7^{a}	114.2^{b}	114.4^{b}	112.7^{b}	2.173	0.016
Zn	4.00×10^4	4.15×10^4	4.22×10^4	4.20×10^4	1.21×10^3	0.577

C: no additive, Q2, Q4 and Q8: 200, 400 and 800 mg quercetin per kg of feed, respectively. [a,b] means in a row sharing no common superscript differ significantly ($p < 0.05$).

Table 3. The effect of dietary supplementation with quercetin on selected trace- and macro-elements concentrations in laying hens' egg shell ($n = 7$).

Elements (ppb)	Treatment					
	C	Q2	Q4	Q8	SEM	*p*-value
As	16.63^{d}	14.21^{c}	9.84^{b}	7.24^{a}	0.857	<0.001
Ca	$2.27 \times 10^{8\,a}$	$2.39 \times 10^{8\,ab}$	$2.44 \times 10^{8\,b}$	$2.39 \times 10^{8\,ab}$	4.00×10^6	0.040
Cd	70.2^{c}	63.8^{bc}	54.5^{b}	37.7^{a}	4.098	<0.001
Co	270.6	281.7	288.3	270.6	5.699	0.097
Cr	692.9^{b}	663.1^{ab}	642.9^{a}	625.2^{a}	14.99	0.024
Cu	1218^{b}	996.0^{ab}	843.3^{a}	905.1^{a}	80.16	0.016
Fe	$1.79 \times 10^{6\,a}$	$1.96 \times 10^{6\,b}$	$2.00 \times 10^{6\,b}$	$1.93 \times 10^{6\,b}$	4.20×10^4	0.033
Mg	$2.87 \times 10^{6\,a}$	$2.99 \times 10^{6\,a}$	$3.28 \times 10^{6\,b}$	$3.06 \times 10^{6\,a}$	8.66×10^4	0.020
Mn	491.1	632.1	514.6	488.6	48.70	0.148
Ni	$9.49 \times 10^{3\,a}$	$1.02 \times 10^{4\,b}$	$1.07 \times 10^{4\,b}$	$1.06 \times 10^{4\,b}$	192.3	<0.001
Sb	33.86^{a}	47.70^{a}	72.26^{b}	94.88^{c}	6.907	<0.001
Se	273.9^{b}	221.9^{ab}	156.8^{a}	154.2^{a}	27.78	0.015
Sr	1.05×10^5	1.11×10^5	1.10×10^5	1.07×10^5	3.42×10^3	0.506
V	663.9^{c}	654.7^{c}	618.4^{b}	572.2^{a}	9.019	<0.001
Zn	6.95×10^3	9.02×10^3	6.51×10^3	6.11×10^3	888.4	0.124

C: no additive, Q2, Q4 and Q8: 200, 400 and 800 mg quercetin per kg of feed, respectively. [a,b,c] means in a row sharing no common superscript differ significantly ($p < 0.05$).

3.2. Principal Components Analysis

The concentrations of the 15 selected trace and macro elements in egg albumen, yolk and shell were subjected to principal components analysis (PCA) in order to investigate the relationships between the elements and detect those elements capable of distinguishing samples (Figure 1). PCA resulted in two principal components, which accounted for 92.40% of the total variability. Trace and macro elements in albumen, yolk and shell samples are presented in Figure 1, as a function of both first and second principal components. The first principal component (PC1) explained 69.64% of the total variability and was defined by the elements As, Ca, Cd, Co, Cr, Fe, Mg, Ni, Sb, Sr and V. The aforementioned elements were located away from the axis origin, suggesting that they were well represented by PC1 and closed together indicating a strong positive correlation. Egg shell samples were clustered near them and therefore, they had higher contents of these elements compared to the samples from egg albumen and yolk, which were clustered on the negative side of PC1. The second principal component (PC2) explained another 22.76% of the total variability and was defined by Cu, Mn, Se and Zn, which were placed closed together on the positive side of PC2, indicating a positive correlation. Samples collected from egg yolk were clustered near them, indicating higher contents of these elements compared to the samples from egg albumen and shell. Thus, a very clear separation of albumen, yolk and shell samples was observed.

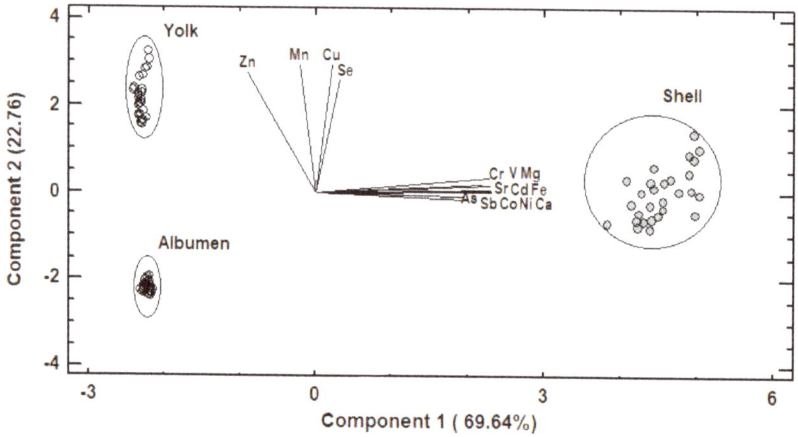

Figure 1. Principal components analysis. Loadings of trace and macro elements for the two first principal components and score plot of albumen, yolk or shell samples.

3.3. Discriminant Analysis

Trace and macro elements data of egg albumen were subjected to discriminant analysis in order to investigate samples distinguishment according to the four dietary treatments and establishment of those elements capable to distinguish the samples. In Figure 2, it can be seen a discriminant plot of the first two discriminant functions, even though one discriminant function was statistically significant ($p < 0.001$). All the samples were successfully clustered by the dietary treatment and a 100% correct classification was observed. Control samples were placed in the bottom right-hand corner of the plot, far away from all the treated samples. On the contrary, samples from treatment Q8 were clustered on the left side of the plot, having the longest distance from control samples and indicating that the most differentiations in the selected trace and macro elements were observed among these dietary treatments. Samples from treatments Q2 and Q4 were clustered in the middle of the plot and therefore samples from these treatments had intermediate concentrations of the selected elements. In addition, a stepwise discriminant analysis revealed that As, Cd and Cr were mainly responsible for the observed discrimination. Another discriminant plot can be seen in Figure 3, categorizing samples according to the dietary treatment, based on selected trace and macro elements of egg yolk. Samples were successfully clustered by the dietary treatment, although some overlapping between samples from treatments Q4 and Q8 was observed. One discriminant function was statistically significant ($p < 0.001$) for distinguishing the samples and the elements Sb, Se and V contributed the most for this discrimination. The percentage of the samples that were classified into the correct group according to the dietary treatment was 92.9%. Further discrimination, based on selected trace and macro elements of egg shell, was attempted (Figure 4). In this case, control samples were also placed far away from samples of treatment Q8, as in albumen's discriminant plot and samples from treatments Q2 and Q4 were also clustered in the middle of the plot. One discriminant function was statistically significant ($p < 0.001$) and a 96.4% correct classification was succeeded. Furthermore, a stepwise discriminant analysis revealed that As, Fe, Sb, Se and V were mainly responsible for distinguishing the observations.

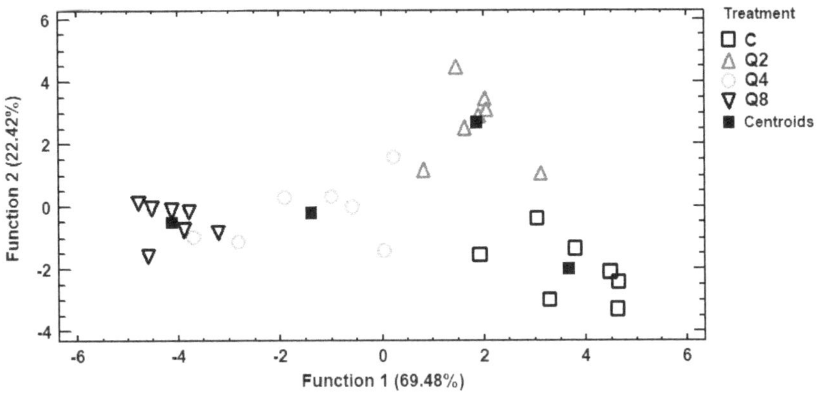

Figure 2. Discriminant plot separating egg albumen samples, according to the dietary treatment, based on trace and macro elements.

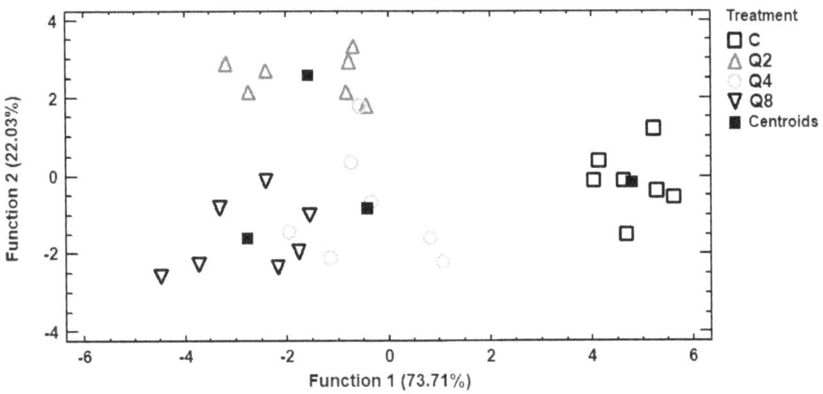

Figure 3. Discriminant plot separating egg yolk samples, according to the dietary treatment, based on trace and macro elements.

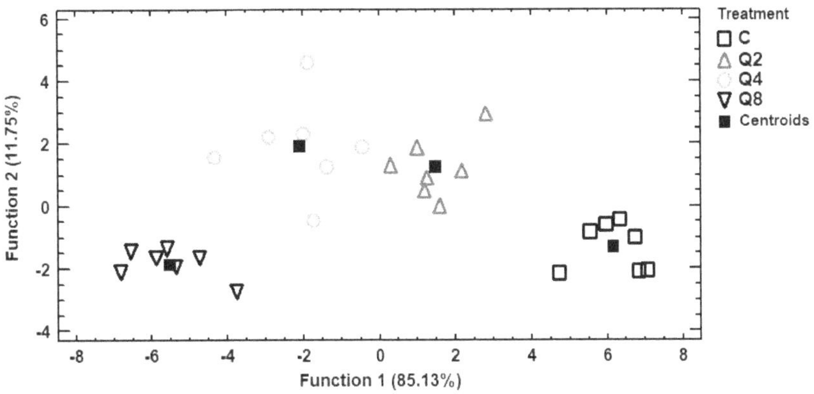

Figure 4. Discriminant plot separating egg shell samples, according to the dietary treatment, based on trace and macro elements.

4. Discussion

4.1. Flavonoids in Poultry Nutrition

Quercetin has been shown to be a great in vitro antioxidant [22] that exerts anti-inflammatory effects in rodents [23] and humans [24] and antitumor effects in rodents [25]. Various studies in poultry have indicated that quercetin and its glycosides are metabolized and absorbed in a similar way to mammals [26]. Utilization of quercetin by laying hens enhances their antioxidant status [27,28], reduces yolk [3] and serum cholesterol [27] levels, and modifies the intestinal environment as it decreases cecal microflora populations of total aerobes and coliforms while increases the populations of *Bifidobacteria* [29].

Our earlier results, in agreement with other studies [3], showed that supplementation of hens' diet with various levels of quercetin for four weeks promoted oxidative stability in a dose dependent manner and was beneficial in improving egg quality [19]. Naringin and hesperidin can be used to alter the elemental profile of the egg by increasing or decreasing the concentration of certain elements in the egg compared to the concentration of those elements in eggs of hens fed a control diet or a control diet supplemented with extra vitamin E [30].

In the present study, the highest contents for the majority of elements were measured in the egg shell compared to yolk and albumen. Moreover, yolk had higher levels than albumen indicating that basic elements are stored in the yolk, where most egg minerals are deposited, and from where they can be effectively transferred via the yolk sac to the developing embryo. Elements accumulated in the albumen may also be accessible during embryogenesis [31,32].

The bioavailability of Zn, Fe, Ca, Mg and *P* from the eggshell was examined in an in vitro study studying the role of citrus flavonoids and reported enhanced bioavailability of these elements by citrus flavonoids [33]. A balance between absorption and excretion of elements exists, mineral homeostasis, and the key role in this balance is assigned to specialized proteins such as Cu- and Zn transporters and Mg channels to maintain [34]. Metallothioneins, that help mammals against heavy metals toxicity and contribute to the homeostasis of certain essential metals, are cysteine-rich protein of low molecular weight, localized to the membrane of the Golgi apparatus that exhibit high binding capacity for both physiological (such as Cu, Se, Zn) and xenobiotic (such as Ag, As, Cd, Hg,) essential and toxic elements [35,36]. Previous studies have indicated a protective effect of quercetin against oxidative damage of erythrocyte membrane which was attributed to its iron chelating activity [37].

4.2. Modulation of Egg Metallome

In the present study, supplementation of layers diet with quercetin mostly affected the trace- and macro elements of the egg shell compared to control fed layers. Quercetin supplementation increased the concentrations of five elements (Ca, Fe, Mg, Ni and Sb) while decreased the levels of six elements namely As, Cd, Cr, Cu, Se and V. Solely the concentration of Sb was uniformly increased with quercetin in albumen, yolk and shell compared to control fed layers. Quercetin seemed to affect the homeostasis of trace elements in hen eggs as reflected by changes of egg's elemental profile in the three egg compartments.

The creation of complexes of flavonoids with elements may be related to certain effects of flavonoids on trace element homeostasis. Quercetin and other flavonoids have received considerable attention as dietary constituents in the recent years [38]. Numerous studies have indicated that flavonoids possess a wide range of biological activities, such as antiviral, antioxidant, anticancer, antibacterial, and anti-inflammatory. Flavonoids are best known as radical scavengers. Beneficial effects are related to the capacity of accepting free radicals and to exhibiting complexation properties with metal ions [39]. Several studies including but not limited to experimental and theoretical ones have reported that the flavonoid complexation is related to the hydroxyl group on carbon 3 or 5 and the adjacent 4-carbonylgroup in the C ring (Scheme 1) [40].

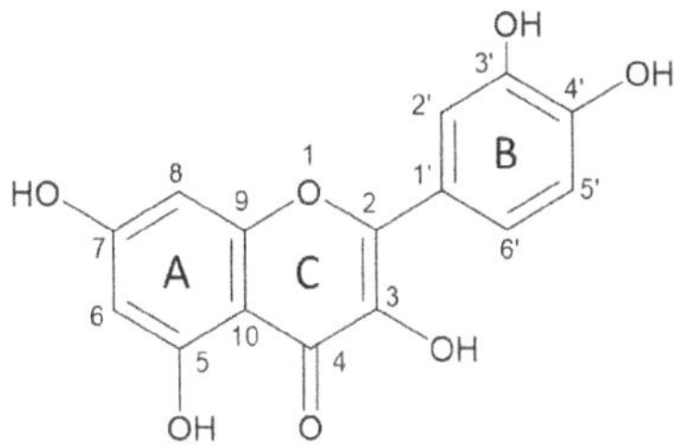

Scheme 1. Structure of quercetin.

The results of the current study work are in line with earlier results on rodents that reported a relation between tissue availability of several elements and absorption and membrane transport of some metal ions which can be affected by the formation of chelates and complexes with organic ligands [41]. The present study revealed that deposition of certain elements was higher in eggs from quercetin fed layers compared to those from the control treatment indicating the chelating capacity of quercetin. Previous studies on humans, animal models or plants have shown the formation of such complexes. Markedly, in a study with postmenopausal women, supplementation with Ca in combination with a flavonoid improved bone Ca retention by 5.5% [42]. However, this was not evident in the case of hens in the present study. In addition, in obese rats fed high-fat or high-carbohydrate diets it was noticed that flavonoid supplementation exerted a vasodilatation effect probably by activating large conductance Ca^{2+}-activated K^{+} currents in a concentration-dependent manner [43].

Antioxidant but also chelating activities of flavonoids in in vivo biological systems are not straightforward and depend on a variety of factors including: (i) the active concentrations in the target tissue compartment, which are very low, (ii) the absorption efficiency, which is relatively low, (iii) metabolite formation after absorption which could reduce their activities and (iv) the surrounding flavonoid environment which can strongly determine in which extend polyphenols exert their activities [44]. Clearly, further research is required to elucidate the metal complexation properties of different dietary quercetin and/or quercetin metabolites levels in egg.

In the present study, the presence of quercetin resulted in higher Fe concentration in the shell. Flavonoids have been shown to protect against induced oxidation and having strong antioxidant activity and Fe chelating properties. In glutathione depleted mouse erythrocytes, the role of quercetin as an antioxidant against induced oxidation was examined and it was showed markable protection against lipid peroxidation by Fe chelation and erythrocytes penetration [32]. Similarly, Deng et al. [45] examined five flavonoids namely quercetin, hesperidin, naringin, baicalin, and rutin and indicated that their antioxidant function originates from chelating Fe ions and from scavenging peroxyl radicals. Differences between flavonoids may be due to their structure and the site of chelation. It has been reported [46] that the majority of flavonoids display a yield of redox reactions greater with Cu than with Fe, as a result of the lower redox potential of Cu^{2+}/Cu^{+} compared with that of Fe^{3+}/Fe^{2+}. Furthermore, it has been shown that quercetin protected cells against oxidative damages caused by Fe overloading. The anti-carcinogenic effect of quercetin has been linked with the chelating effect of Fe as well as its antioxidant effect. It has been reported that quercetin-metal complexes might possess even greater anticancer and antibacterial activities than free quercetin [9,16]. Nevertheless, our results cannot explain the relationship among quercetin, Fe and potential antioxidant effects since there was no positive correlation between quercetin administration levels and Fe concentrations in egg yolk and egg albumen.

Addition of quercetin, in this study, manifested in altered concentrations of Mg and Cd in specific egg fractions. Earlier studies with Mg and flavonoids indicated formation of quercetin–magnesium complexes. More specifically, the interaction of $MgSO_4.7H_2O$ with quercetin was tested and it was reported that the Mg ions significantly modify the chemical properties of quercetin [47]. Though, regarding egg yolk and albumen, Mg levels were not changed by quercetin supplementation. Studies conducted in plants (*Avicennia marina* roots) reported increased absorption rate of Cd under flavonoid amendment [48]. In this aspect, regarding egg albumen, addition of quercetin in the highest dose (Q8) increased Cd levels, opposite to Cd levels in egg shell that were decreased and to egg yolk that remained unaffected. Thus, obtaining clear conclusions regarding mitigation or increase of toxicity is difficult. Nevertheless, further research is required to detect a direct relationship between dietary quercetin levels and metal complexing properties in this flavonoid.

In the present study, the presence of 400 and 800 mg quercetin per kg of diet resulted in lower Cr, Cu and Se concentrations in egg albumen and shell. It has been reported that some flavonoids can also decrease the absorption of trace elements, when included in a diet. For example, addition of grape seed extracts impeded the absorption of Zn and Fe in a dose-dependent manner in intestinal cells [44,49]. Different factors including but not limited to the type of compounds, the dosage and the combination with other compounds affect assimilation and absorption of trace metals by polyphenol compounds [50].

Development of eggs fortified in certain trace elements has been proposed through feed supplementation [30]. Notwithstanding, bioavailability restrictions have led to the usage of appropriate forms of supplements. These limitations could be transcended through chelation with flavonoids rich in phenolic groups. Homeostasis of macro- and trace element has significant features that can be studied by elemental metabolomics through determination of the complete elemental profile [51].

Earlier studies [26] revealed that quercetin and its glycosides can be absorbed and metabolized by chicken. In avian species, similar to mammals, quercetin following absorption can be subjected to glucuronidation, sulfation, and methylation. However, it was reported that incorporation of quercetin or its glycoside in the diet did not increase the antioxidant capacity in plasma or tissues, suggesting that the absorption of quercetin was rather weak or the concentrations of quercetin metabolites in plasma or tissues were not sufficient to significantly increase the antioxidant capacity assayed by the ferric reducing ability of plasma (FRAP) [26].

The present study revealed that complex interactions exist that cannot be solely attributed to chelation of elements. Feeding trials at different levels of flavonoids may further illuminate interactions. Such kind of future studies with fine tuning of the elemental supplementation and simultaneous adjustment of flavonoid levels may lead to creation of tailor-made eggs, to address specific needs like Fe enriched diets [52]. Although data presented here derived from a flavonoid used in its pure chemical form, various agricultural by-products of the olive oil-, fruit juice- and wine industry are rich in phenolics and flavonoids could be also used.

The egg shell metallome alteration due to quercetin dietary supplementation found in the present study may be considered favorable from the avian embryonic development perspective. Macro- and trace elements, essential for embryonic growth such as Ca, Fe and Mg, were increased in the eggshells of quercetin fed hens in comparison to control fed ones. At the same time, elements that are considered toxic at high levels such as Cd, Cr, and Cu were decreased. The eggshell is a macro- and trace nutrient pool for the developing avian embryo and therefore its composition affects embryo development. Further research is required to elucidate how eggshell metallome may affect embryonic and post-hatch development.

5. Conclusions

In conclusion, quercetin supplemented to the diet of layers can affect the elemental profile of egg because a differentiation in the deposition of certain elements in the egg constituents, compared to that of hens fed a control diet, was noted. Quercetin added at 200, 400 and 800 mg per kg of hen feed increased the concentration of Sb (2-140-fold) while reduced that of Cr (7–15%) and Se (34–43%) in both egg shell and albumen. Moreover, it increased As (18–24%), Cd (32–34%) in albumen and V (7.5–9%) in yolk, while compared to the control, reduced As (14–56%), Cd (22–46%), Cr (7–10%), Cu (25–30%) and V (7–14%) and also raised Ca (7.5%), Fe (8–9.5%), Mg (14%) and Ni (7.5–13%) in egg shell.

Author Contributions: Conceptualization, E.Z., A.C.P., M.G., P.E.S., C.A.G.; statistical analyses, K.S.; validation, A.C.P., E.Z., G.D., S.T.; laboratory analyses, G.D., E.Z., A.C.P., M.G., P.E.S., S.T.; resources, E.Z., A.C.P., C.A.G.; writing—original draft preparation E.Z., G.D., A.C.P., M.G., P.E.S.; writing—review and editing, E.Z., A.C.P., M.G., P.E.S., K.S., G.D., C.A.G.; visualization, A.C.P., E.Z., M.G., P.E.S., K.S., G.D.; supervision and project administration, C.A.G. All authors have read and agreed to the published version of the manuscript.

Funding: C.A.G. and G.D. acknowledge support by the project "FoodOmicsGR Comprehensive Characterisation of Foods" (MIS 5029057) which is implemented under the Action "Reinforcement of the Research and Innovation Infrastructure", funded by the Operational Program "Competitiveness, Entrepreneurship and Innovation" (NSRF 2014-2020) and co-financed by Greece and the European Union (European Regional Development Fund).

Institutional Review Board Statement: The present study followed the guidelines of the directive 2010/63/EU of the European Parliament and of the Council on the protection of animals used for scientific purposes. The Research Ethics Committee of the Faculty of Animal Science of the Agricultural University of Athens approved the study (code number 38/07022017).

Informed Consent Statement: Not applicable.

Data Availability Statement: Data is contained within the article.

Co-financed by Greece and the European Union

Conflicts of Interest: The authors declare no conflict of interest.

References

1. Kamboh, A.A.; Leghari, R.A.; Khan, M.A.; Kaka, U.; Naseer, M.; Sazili, A.Q.; Malhi, K.K. Flavonoids supplementation-An ideal approach to improve quality of poultry products. *Worlds Poult. Sci. J.* **2019**, *75*, 115–126. [CrossRef]
2. Batiha, G.-S.; Beshbishy, A.M.; Ikram, M.; Mulla, Z.S.; El-Hack, M.E.A.; Taha, A.E.; Algammal, A.M.; Elewa, Y.H.A. The pharmacological activity, biochemical properties, and pharmacokinetics of the major natural polyphenolic flavonoid: Quercetin. *Foods* **2020**, *9*, 374. [CrossRef] [PubMed]
3. Liu, Y.; Li, Y.; Liu, H.N.; Suo, Y.L.; Hu, L.L.; Feng, X.A.; Zhang, L.; Jin, F. Effect of quercetin on performance and egg quality during the late laying period of hens, Br. *Poul. Sci.* **2013**, *54*, 510–514. [CrossRef] [PubMed]
4. Saeed, M.; Naveed, M.; Arain, M.A.; Arif, M. Quercetin: Nutritional and beneficial effects in poultry. *Worlds Poult. Sci. J.* **2017**, *73*, 355–364. [CrossRef]
5. Hager-Theodorides, A.L.; Goliomytis, M.; Delis, S.; Deligeorgis, S. Effects of dietary supplementation with quercetin on broiler immunological characteristics. *Anim. Feed Sci. Technol.* **2014**, *198*, 224–230. [CrossRef]
6. Comalada, M.; Ballester, I.; Bailon, E.; Xaus, J.; Galvez, J.; De Medina, F.S.; Zarzuelo, A. Inhibition of pro-inflammatory markers in primary bone marrow-derived mousemacrophages by naturally occurring flavonoids: Analysis of the structure-activity relationship. *Biochem. Pharmacol.* **2006**, *72*, 1010–1021. [CrossRef]
7. Malešev, D.; Kuntić, V. Investigation of metal-flavonoid chelates and the determination of flavonoids via metal-flavonoid complexing reactions. *J. Serb. Chem. Soc.* **2007**, *72*, 921–939. [CrossRef]
8. Manach, C.; Texier, O.; Morand, C.; Crespy, V.; Régérat, F.; Demigné, C.; Rémésy, C. Comparison of the bioavailability of quercetin and catechin in rats. *Free Radic. Biol. Med.* **1999**, *27*, 1259–1266. [CrossRef]
9. Ulusoy, H.G.; Sanlier, N. A minireview of quercetin: From its metabolism to possible mechanisms of its biological activities. *Crit. Rev. Food Sci. Nutr.* **2019**, 1–14. [CrossRef]
10. Lee, S.Y.; Lee, S.J.; Yim, D.G.; Hur, S.J. Changes in the content and bioavailability of onion quercetin and grape resveratrol during in vitro human digestion. *Foods* **2020**, *9*, 694. [CrossRef]

11. EFSA Panel on Dietetic Products, Nutrition and Allergies (NDA). Scientific Opinion on the substantiation of health claims related to quercetin and protection of DNA, proteins and lipids from oxidative damage (ID 1647), "cardiovascular system" (ID 1844), "mental state and performance" (ID 1845), and "liver, kidneys" (ID 1846) pursuant to Article 13 (1) of Regulation (EC) No 1924/2006. *EFSA J.* **2011**, *9*, 2067. [CrossRef]
12. Iqbal, M.; Cawthon, D.; Beers, K.; Wideman, R.F., Jr.; Bottje, W.G. Antioxidant enzyme activities and mitochondrial fatty acids in pulmonary hypertension syndrome (PHS) in broilers. *Poult. Sci.* **2002**, *81*, 252–260. [CrossRef] [PubMed]
13. Tsuji, M.; Yamamoto, H.; Sato, T.; Mizuha, Y.; Kawai, Y.; Taketani, Y.; Kato, S.; Terao, J.; Inakuma, T.; Takeda, E. Dietary quercetin inhibits bone loss without effect on the uterus in ovariectomized mice. *J. Bone Miner. Metabol.* **2009**, *27*, 673–681. [CrossRef] [PubMed]
14. Ganesan, S.; Faris, A.N.; Comstock, A.T.; Wang, Q.; Nanua, S.; Hershenson, M.B.; Sajjan, U.S. Quercetin inhibits rhinovirus replication in vitro and in vivo. *Antivir. Res.* **2012**, *94*, 258–271. [CrossRef]
15. Kuo, S.M.; Leavitt, P.S.; Lin, C.P. Dietary flavonoids interact with trace metals and affect metallothionein level in human intestinal cells. *Biol. Trace Elem. Res.* **1998**, *62*, 135–153. [CrossRef]
16. Xiao, L.; Luo, G.; Tang, Y.; Yao, P. Quercetin and iron metabolism: What we know and what we need to know. *Food Chem. Toxicol.* **2018**, *114*, 190–203. [CrossRef]
17. Zhou, J.; Wang, L.; Wang, J.; Tang, N. Antioxidative and anti-tumour activities of solid quercetin metal (II) complexes. *Transit. Metal Chem.* **2001**, *26*, 57–63. [CrossRef]
18. Berroukeche, F.; Mokhtari-Soulimane, N.; Imessaoudene, A.; Cherrak, A.; Ronot, P.; Boos, A.; Belhandouz, A.; Merzouk, H.; Elhabiri, M. Oral supplementation effect of iron and its complex form with quercetin on oxidant status and on redistribution of essential metals in organs of streptozotocin diabetic rats. *Rom. J. Diabetes Nutr. Metab. Dis.* **2019**, *26*, 39–53. [CrossRef]
19. Simitzis, P.; Spanou, D.; Glastra, N.; Goliomytis, M. Impact of dietary quercetin on laying hen performance, egg quality and yolk oxidative stability. *Anim. Feed Sci. Technol.* **2018**, *239*, 27–32. [CrossRef]
20. Pappas, A.C.; Zoidis, E.; Georgiou, C.A.; Demiris, N.; Surai, P.F.; Fegeros, K. Influence of organic selenium supplementation on the accumulation of toxic and essential trace elements involved in the antioxidant system of chicken. *Food Addit. Contam.* **2011**, *28*, 446–454. [CrossRef]
21. Georgiou, C.A.; Danezis, G.P. *Comprehensive Analytical Chemistry*; Pico, Y., Ed.; Elsevier: Amsterdam, The Netherlands, 2015; Volume 68, Chapter 3; pp. 131–243. [CrossRef]
22. Boots, A.W.; Haenen, G.R.M.M.; Bast, A. Health effects of quercetin: From antioxidant to nutraceutical. *Eur. J. Pharmacol.* **2008**, *585*, 325–337. [CrossRef] [PubMed]
23. Huang, R.Y.; Yu, Y.L.; Cheng, W.C.; OuYang, C.N.; Fu, E.; Chu, C.L. Immunosuppressive effect of quercetin on dendritic cell activation and function. *J. Immunol.* **2010**, *184*, 6815–6821. [CrossRef] [PubMed]
24. Sternberg, Z.; Chadha, K.; Lieberman, A.; Hojnacki, D.; Drake, A.; Zamboni, P.; Rocco, P.; Grazioli, E.; Weinstock-Guttman, B.; Munschauer, F. Quercetin and interferon-beta modulate immune response(s) in peripheral blood mononuclear cells isolated from multiple sclerosis patients. *J. Neuroimmunol.* **2008**, *205*, 142–147. [CrossRef] [PubMed]
25. Schlachterman, A.; Valle, F.; Wall, K.M.; Azios, N.G.; Castillo, L.; Morell, L.; Washington, A.V.; Cubano, L.A.; Dharmawardhane, S.F. Combined resveratrol, quercetin, and catechin treatment reduces breast tumor growth in a nude mouse model. *Transl. Oncol.* **2008**, *1*, 19–27. [CrossRef] [PubMed]
26. Rupasinghe, H.P.V.; Ronalds, C.M.; Rathgeber, B.; Robinson, R.A. Absorption and tissue distribution of dietary quercetin and quercetin glycosides of apple skin in broiler chickens. *J. Sci. Food Agric.* **2010**, *90*, 1172–1178. [CrossRef] [PubMed]
27. Iskender, H.; Yenice, G.; Dokumacioglu, E.; Kaynar, O.; Hayirli, A.; Kaya, A. The effects of dietary flavonoid supplementation on the antioxidant status of laying hens. *Braz. J. Poult. Sci.* **2016**, *18*, 663–668. [CrossRef]
28. Zhang, S.; Kim, I.H. Effect of quercetin (flavonoid) supplementation on growth performance, meat stability, and immunological response in broiler chickens. *Livest. Sci.* **2020**, 104286. [CrossRef]
29. Liu, H.N.; Liu, Y.; Hu, L.L.; Suo, Y.L.; Zhang, L.; Jin, F.; Feng, X.A.; Teng, N.; Li, Y. Effects of dietary supplementation of quercetin on performance, egg quality, cecal microflora populations, and antioxidant status in laying hens. *Poult. Sci.* **2014**, *93*, 347–353. [CrossRef]
30. Pappas, A.C.; Zoidis, E.; Goliomytis, M.; Simitzis, P.E.; Sotirakoglou, K.; Charismiadou, M.A.; Nikitas, C.; Danezis, G.; Deligeorgis, S.G.; Georgiou, C.A. Elemental Metabolomics: Modulation of egg metallome with flavonoids, an exploratory study. *Antioxidants* **2019**, *8*, 361. [CrossRef]
31. Richards, M.P. Trace mineral metabolism in the avian embryo. *Poult. Sci.* **1997**, *76*, 152–164. [CrossRef]
32. Miles, R.D. Trace minerals and avian embryo development. *Ciênc. Anim. Bras.* **2000**, *2*, 1–10.
33. Siddique, S.; Firdous, S.; Durrani, A.I.; Khan, S.J.; Saeed, A. Hesperidin, a citrus flavonoid, increases the bioavailability of micronutrients of Gallus domesticus (chicken) eggshell: In vitro study. *Chem. Spec. Bioavailab.* **2016**, *28*, 88–94. [CrossRef]
34. Hashimoto, A.; Kambe, T. Mg, Zn and Cu transport proteins: A brief overview from physiological and molecular perspectives. *J. Nutr. Sci. Vitaminol.* **2015**, *61*, 116–118. [CrossRef] [PubMed]
35. Klaassen, C.D.; Liu, J.; Choudhuri, S. Metallothionein: An intracellular protein to protect against cadmium toxicity. *Annu. Rev. Pharmacol. Toxicol.* **1999**, *39*, 267–294. [CrossRef] [PubMed]
36. Ganapathy, V.; Ganapathy, M.E.; Leibach, F.H. Intestinal transport of peptides and amino acids. *Curr. Top. Membr.* **2000**, *50*, 379–412. [CrossRef]

37. Ferrali, M.; Signorini, C.; Caciotti, B.; Sugherini, L.; Ciccoli, L.; Giachetti, D.; Comporti, M. Protection against oxidative damage of erythrocyte membrane by the flavonoid quercetin and its relation to iron chelating activity. *FEBS Lett.* **1997**, *416*, 123–129. [CrossRef]
38. Dolatabadi, J.E.N. Molecular aspects on the interaction of quercetin and its metal complexes with DNA. *Int. J. Biol. Macromol.* **2011**, *48*, 227–233. [CrossRef]
39. Jabeen, E.; Janjua, N.K.; Ahmed, S.; Murtaza, I.; Ali, T.; Hameed, S. Radical scavenging propensity of Cu^{2+}, Fe^{3+} complexes of flavonoids and in-vivo radical scavenging by Fe^{3+}-primuletin. *Spectrochim. Acta Part A* **2017**, *171*, 432–438. [CrossRef]
40. Dehghan, G.; Dolatabadi, J.E.N.; Jouyban, A.; Zeynali, K.A.; Ahmadi, S.M.; Kashanian, S. Spectroscopic studies on the interaction of quercetin–terbium(iii) complex with calf thymus DNA. *DNA Cell. Biol.* **2011**, *30*, 195–201. [CrossRef]
41. Jaccob, A.A.; Hussain, S.A.; Hussain, S.A. Effects of long-term use of flavonoids on the absorption and tissue distribution of orally administered doses of trace elements in rats. *Pharmacol. Pharm.* **2012**, *3*, 474–480. [CrossRef]
42. Martin, B.R.; McCabe, G.P.; McCabe, L.; Jackson, G.S.; Horcajada, M.N.; Offord-Cavin, E.; Peacock, M.; Weaver, C.M. Effect of hesperidin with and without a calcium (Calcilock) supplement on bone health in postmenopausal women. *J. Clin. Endocrinol. Metab.* **2016**, *101*, 923–927. [CrossRef] [PubMed]
43. Saponara, S.; Testai, L.; Iozzi, D.; Martinotti, E.; Martelli, A.; Chericoni, S.; Sgaragli, G.; Fusi, F.; Calderone, V. (+/−)-Naringenin as large conductance Ca^2+-activated K^+ (BKCa) channel opener in vascular smooth muscle cells. *Br. J. Pharmacol.* **2006**, *149*, 1013–1021. [CrossRef] [PubMed]
44. Surai, P.F. Polyphenol compounds in the chicken/animal diet: From the past to the future. *J. Anim. Physiol. Anim. Nutr.* **2014**, *98*, 19–31. [CrossRef] [PubMed]
45. Deng, W.; Fang, X.; Wu, J. Flavonoids function as antioxidants: By scavenging reactive oxygen species or by chelating iron? *Radiat. Phys. Chem.* **1997**, *50*, 271–276. [CrossRef]
46. Mira, L.; Fernandez, M.T.; Santos, M.; Rocha, R.; Helena Florêncio, M.H.; Jennings, K.R. Interactions of flavonoids with iron and copper ions: A mechanism for their antioxidant activity. *Free Radical Res.* **2002**, *36*, 1199–1208. [CrossRef]
47. Ghosh, N.; Chakraborty, T.; Mallick, S.; Mana, S.; Singha, D.; Ghosh, B.; Roy, S. Synthesis, characterization and study of antioxidant activity of quercetin–magnesium complex. *Spectrochim. Acta Part A* **2015**, *151*, 807–813. [CrossRef] [PubMed]
48. Li, J.; Lu, H.; Liu, J.; Hong, H.; Yan, C. The influence of flavonoid amendment on the absorption of cadmium in Avicennia marina roots. *Ecotoxicol. Environ. Saf.* **2015**, *120*, 1–6. [CrossRef]
49. Kim, E.Y.; Pai, T.K.; Han, O. Effect of bioactive dietary polyphenols on zinc transport across the intestinal Caco-2cell monolayers. *J. Agric. Food Chem.* **2011**, *59*, 3606–3612. [CrossRef]
50. Martel, F.; Monteiro, R.; Calhau, C. Effect of polyphenols on the intestinal and placental transport of some bioactive compounds. *Nutr. Res. Rev.* **2010**, *23*, 47–64. [CrossRef]
51. Zhang, P.; Georgiou, C.A.; Brusic, V. Elemental metabolomics. *Brief. Bioinform.* **2017**, *19*, 524–536. [CrossRef]
52. Aspuru, K.; Villa, C.; Bermejo, F.; Herrero, P.; López, S.G. Optimal management of iron deficiency anemia due to poor dietary intake. *Int. J. Gen. Med.* **2011**, *4*, 741–750. [CrossRef] [PubMed]

antioxidants

MDPI

Article

Simultaneous Quantification of Antioxidants Paraxanthine and Caffeine in Human Saliva by Electrochemical Sensing for CYP1A2 Phenotyping

Rozalia-Maria Anastasiadi [1,*], Federico Berti [2], Silvia Colomban [3], Claudio Tavagnacco [2], Luciano Navarini [3] and Marina Resmini [1,*]

1 Department of Chemistry, Queen Mary University of London, Mile End Road, London E1 4NS, UK
2 Department of Chemical and Pharmaceutical Sciences, University of Trieste, via L. Giorgieri 1, 34127 Trieste, Italy; fberti@units.it (F.B.); c.tavagnacco@units.it (C.T.)
3 Aromalab, illycaffè S.p.A., Area Science Park, Localita' Padriciano 99, 34149 Trieste, Italy; silvia.colomban@illy.com (S.C.); luciano.navarini@illy.com (L.N.)
* Correspondence: rozalia-maria.anastasiadi@utc.fr (R.-M.A.); m.resmini@qmul.ac.uk (M.R.)

Abstract: The enzyme CYP1A2 is responsible for the metabolism of numerous antioxidants in the body, including caffeine, which is transformed into paraxanthine, its main primary metabolite. Both molecules are known for their antioxidant and pro-oxidant characteristics, and the paraxanthine-to-caffeine molar ratio is a widely accepted metric for CYP1A2 phenotyping, to optimize dose–response effects in individual patients. We developed a simple, cheap and fast electrochemical based method for the simultaneous quantification of paraxanthine and caffeine in human saliva, by differential pulse voltammetry, using an anodically pretreated glassy carbon electrode. Cyclic voltammetry experiments revealed for the first time that the oxidation of paraxanthine is diffusion controlled with an irreversible peak at ca. +1.24 V (vs. Ag/AgCl) in a 0.1 M H_2SO_4 solution, and that the mechanism occurs via the transfer of two electrons and two protons. The simultaneous quantification of paraxanthine and caffeine was demonstrated in 0.1 M H_2SO_4 and spiked human saliva samples. In the latter case, limits of detection of 2.89 μM for paraxanthine and 5.80 μM for caffeine were obtained, respectively. The sensor is reliable, providing a relative standard deviation within 7% (n = 6). Potential applicability of the sensing platform was demonstrated by running a small scale trial on five healthy volunteers, with simultaneous quantification by differential pulse voltammetry (DPV) of paraxanthine and caffeine in saliva samples collected at 1, 3 and 6 h postdose administration. The results were validated by ultra-high pressure liquid chromatography and shown to have a high correlation factor (r = 0.994).

Keywords: paraxanthine; caffeine; CYP1A2 phenotyping; antioxidants; human saliva; differential pulse voltammetry

Citation: Anastasiadi, R.-M.; Berti, F.; Colomban, S.; Tavagnacco, C.; Navarini, L.; Resmini, M. Simultaneous Quantification of Antioxidants Paraxanthine and Caffeine in Human Saliva by Electrochemical Sensing for CYP1A2 Phenotyping. *Antioxidants* **2021**, *10*, 10. https://dx.doi.org/10.3390/antiox10010010

Received: 19 November 2020
Accepted: 21 December 2020
Published: 24 December 2020

Publisher's Note: MDPI stays neutral with regard to jurisdictional claims in published maps and institutional affiliations.

1. Introduction

Caffeine is probably one of the most studied methylxanthines to date and is also one of the early examples of functional food ingredients, with multiple studies reporting on the significant health benefits associated with caffeine (CAF) consumption [1]. The roles of CAF as central nervous stimulant, antioxidant [2,3], pro-oxidant [4], cognitive enhancer [5] and neuroprotectant [6,7] are all well documented. An interesting physiological effect of CAF derives by acting as an adenosine analogue, blocking the adenosine receptors [8], and resulting in increased release of hormones, with an overall neuroprotective effect [9]. This has also sparked interest towards the development of new adenosine and caffeine analogues [10]. Furthermore, the consumption of CAF has recently been shown to reduce oxidative DNA damage and prevent lipid peroxidation [11], leading to suggestions that its consumption could positively impact diseases such as type 2 diabetes, depression, and liver, Parkinson's [12] and Alzheimer's disease [13,14].

In the human body, CAF is metabolised more than 90% by the enzyme CYP1A2 and selectively catalysed to paraxanthine (PX), its major primary metabolite (81.5%), and theobromine (TB) and theophylline (TP) are found in 10.8% and 5.4%, respectively (Figure 1) [8].

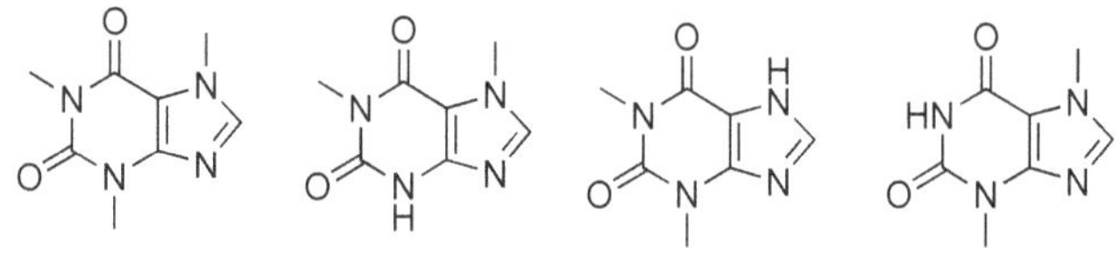

Figure 1. Chemical structures of caffeine and its primary metabolites.

These metabolites share many of the physiological activities of CAF including its antioxidant and pro-oxidant properties, with PX and TP showing a higher affinity towards the adenosine receptors compared to CAF. Even though TB and TP are found in food sources, PX is present in the body only as a result of the metabolism of CAF [15].

The human CYP1A2 enzyme is also responsible for the metabolism of more than 200 clinically important drugs including antidepressants, antipsychotics, anticancer and analgesic drugs, with many of them also having antioxidant properties [16–20]. However, the activity of CYP1A2 shows significant interindividual variability (10- to 200-fold) that seriously influences drugs' metabolism and efficacy [21–23]. This highlights the importance of being able to evaluate the metabolic activity of CYP1A2 as a useful tool to optimise the drug dose–response effect and effectively monitor drug therapy [24,25]. This of course also impacts the metabolism of caffeine and any pharmaceutically active analogues. Considerable progress has been made in predicting drug pharmacokinetics and response in individuals by CYP1A2 phenotyping, but its application in a routine clinical setting remains a challenge [26]. Currently, CAF represents the only fully validated probe for CYP1A2 phenotyping [25,27]. The quantification of the PX-to-CAF (PX/CAF) molar ratio, at a single time-point, 4 to 8 hours upon oral administration of a 100–200 mg CAF dose is an accepted standard in clinical practice [24,25,27,28], although it currently relies on the use of highly specialised equipment, such as high pressure liquid chromatography. In terms of biological matrix, blood, plasma or urine have been traditionally used for CYP1A2 phenotyping; however, the blood and plasma required specialised personnel for sample collection, while drugs' concentrations in urine tend to have a much higher variability and are seldom considered to be reliable.

Saliva has been investigated as an alternative sampling matrix for CYP1A2 phenotyping since it reflects 65–85% of plasma kinetics and shows close correlation with the immunoreactive CYP1A2 liver intrinsic activity [24,29,30]. This makes saliva a more viable approach, being non-invasive, patient-friendly, easier to acquire and it also allows the sampling of patients outside of a clinical setting, and for this reason was selected in this work [31,32]. To date, CAF and PX have been quantified simultaneously using high-performance liquid chromatography (HPLC) coupled with an ultraviolet (UV) detector [33–37], or, to a minor extent, with a mass spectrometer (MS) [38,39]. Although these techniques provide indubitable selectivity and sensitivity, they require costly and heavy equipment, and trained personnel, in addition to being non-environmentally friendly given the large amounts of solvents used. There is an urgency for alternative robust, cheaper, portable and easy-to-operate techniques with high throughput capabilities.

Electrochemical sensors are a valid alternative that allow point-of-care testing and nowadays are attracting increased interest for routine drug analysis of biomarkers [40]. Electrochemical techniques can be used to study the redox behaviour and therefore the antioxidant capacity, as well as quantification [41,42]. Carbon-based electrode materials are ubiquitous in electroanalytical laboratories due to their broad applicability for numerous molecules, wide potential window, low cost, low background current, chemical inertness and feasibility for modification, with glassy carbon (GC) being one of the most commonly used electrodes for analytical purposes, with a good chemical stability and satisfactory potential window [43,44]. Recent efforts on further improving their analytical performance have been reported using a number of modifying materials (e.g., carbon nanomaterials [45], imprinted polymers [46] and enzymes [47]) and their combinations. Despite the improvements on the sensitivity as a result of modifications, their fabrication requires additional steps and higher costs, the former potentially leading to lower reproducibility, short-term stability and challenges when used with biofluids. Despite the significant amount of literature data on modified electrodes [48], the use of bare electrodes still represents the simplest possible approach for electrochemical sensing given the commercial availability and its long-term stability.

Nowadays, electrochemical sensing using human saliva as matrix is attracting a lot of interest [49,50]. A number of studies have been reported on the detection in human saliva of important biomarkers such as uric acid [51,52], cortisol [53] and selegiline [54] using a variety of solid and screen printed carbon electrodes. Although some work has been published on the detection of CAF and other dimethylxanthines (PX, TP and TB) in biofluids such as urine and blood [55,56], beverages [57], food [58] and pharmaceutical formulations [59], the use of saliva as matrix has yet to be evaluated. Using human saliva as matrix brings a major challenge associated with the complexity of its composition and the combination of large and small molecules present; this is further complicated by the interpersonal variability observed among individuals, due to genetics, nutrition and habits [32,60].

Here, we report, for the first time, the simultaneous quantification of antioxidants PX and CAF in human saliva by differential pulse voltammetry (DPV) in aqueous media, employing an anodically pretreated GC electrode. The oxidation behaviour of PX was also investigated by cyclic voltammetry (CV) on the GC electrode. The experimental parameters of the DPV method were optimised using equimolar mixtures of PX and CAF, their analytical performance was evaluated and the assay was validated using artificially spiked human saliva. Qualitative assessment using a boron-doped diamond (BDD) and a screen-printed carbon paste (SP-CP) electrode was also performed for comparison purposes. Preliminary pharmacokinetic studies of CAF were carried out on five healthy human volunteers in order to provide initial evidence of the suitability of the sensing platform. The PX/CAF metabolic ratio was determined at various time intervals by DPV, validating the data with ultra-high pressure liquid chromatography coupled with ultraviolet detector (UHPLC-UV). The developed electrochemical system provides a new insight on designing efficient and robust platforms for enzyme metabolic phenotyping as personalised medicine tools.

2. Materials and Methods

2.1. Chemicals

Caffeine (CAF, ≥99%), paraxanthine (PX, ≥98%), theophylline (TP, ≥99%), theobromine (TB, ≥98%) and boric acid (≥95%), purchased from Sigma-Aldrich Chemie GmbH (Steinheim, Germany), were used as received. Sulphuric acid (H_2SO_4, 95–98%), phosphoric acid (H_3PO_4, ≥85%), nitric acid (HNO_3, ≥65%), acetic acid (≥99%), formic acid (≥95%), sodium hydroxide (NaOH, ≥99%), ethanol, acetonitrile and ethyl acetate were obtained from Sigma-Aldrich with analytical grade purity. Potassium chloride (KCl, ≥3 M) solution for Ag/AgCl reference electrodes was purchased from Sigma-Aldrich. PROPLUS tablets, with a declared amount of 50 mg CAF per tablet, were obtained from a local pharmacy in London, UK. Ultra-pure deionised water (resistivity not lower than 18 MΩ cm at 25 °C) obtained from a Milli-Q unit (Millipore) was used to prepare all the solutions.

2.2. Apparatus

Voltammetric measurements were conducted in a conventional glass cell using an μSTAT 400 potentiostat controlled by DropView 8400 software (DropSens, Oviedo, Spain). A three-electrode setup was used, consisting of a working glassy carbon (GC) electrode with an active surface of 5 mm diameter, a platinum counter electrode and a Ag/AgCl (3 M KCl) reference electrode, chemically isolated from the test solution with a porous Vycor frit (Metrohm Autolab B.V., Utrecht, The Netherlands). A boron-doped diamond (3 mm diameter, BDD, Windsor Scientific Ltd., Slough, UK) and disposable screen-printed carbon paste (SP-CP, model DRP-110, 4 mm diameter, DropSens, Oviedo, Spain) were also used as working electrodes.

The UHPLC system consisted of a 1290 Infinity LC system equipped with an autosampler, a degasser, a quaternary pump, a column thermostat and a diode array detector (DAD) operating at 273 nm. A Kinetex XB-C18 2.1 mm × 100 mm, 2.6 µm column (Phenomenex Torrance, CA, USA) was used at room temperature. For the analysis of the chromatograms, LC OpenLab was used. All equipment was purchased from Agilent Technologies (Waldbronn, Germany).

2.3. Preparation of Working Standards

Stock solutions of CAF (5 mM), PX (5 mM), TP (5 mM) and TB (2.5 mM) were prepared by dissolving the appropriate amount in Milli-Q water or 0.1 M H_2SO_4 for construction of the calibration curves. All stock solutions were prepared and stored in volumetric flasks. Supporting electrolyte solutions (H_2SO_4, H_3PO_4 and HNO_3) were prepared at concentrations in the range 0.01–0.8 M. Britton–Robinson buffer (BR) solutions were prepared at a 0.1 M concentration, in different pHs (2–10), by mixing acetic acid (40 mM), boric acid (40 mM) and H_3PO_4 (40 mM) in Milli-Q water and the pH was adjusted using a 0.2 M NaOH solution. All stock solutions were stored at 4 °C for a maximum of 4 weeks.

2.4. Electrochemical Measurements

The GC and BDD electrodes were mechanically cleaned using an aqueous slurry of 0.05 µm alumina oxide powder on a polishing cloth pad to a mirror-like finish and rinsed thoroughly with Milli-Q water. Before use, all electrodes were electrochemically conditioned, in a 0.1 M H_2SO_4 solution, using cyclic voltammetry (CV) between −1.0 V and +1.8 V at a scan rate of 100 mV s^{-1} (5 cycles).

CV was used for evaluation of the electrochemical behavior of PX and CAF as an equimolar mixture in different supporting electrolytes and for assessment of the oxidation behavior (pH and scan rate studies) of PX, in a 0.1 M H_2SO_4 solution, on the GC electrode. Differential pulse voltammetry (DPV) was employed for assessing the analytical performance of PX and CAF and for their quantitative assessment in human saliva upon optimisation of the DPV operating parameters such as pretreatment potential and time, scan rate and step potential. Standard calibration curves were constructed by successive addition of PX and CAF solutions, individually or simultaneously, into the electrochemical cell containing 4 mL of 0.1 M H_2SO_4. The calibration curve of the individual analytes was obtained by increasing the concentration of either PX or CAF from 1 µM to 300 µM while the other analyte remained fixed at 10 µM (Figures S1 and S2). A combined calibration curve was obtained by simultaneously increasing both analytes from 1 µM to 200 µM (1, 2, 5, 8, 20, 40, 60, 80, 120, 150, 180 and 200 µM), using equimolar mixtures. The peak area (charge, Q) of the analytes was estimated by straight lines connecting the minima and maxima before and after the peak, respectively, without applying any background correction. All measurements were performed in triplicates and the average charge values were plotted against the concentration.

2.5. Assay Validation for Real Application

Matrix-matched calibration curves were constructed using saliva samples obtained from a healthy volunteer who abstained from methylxanthine-containing products for 60 h. Nine saliva samples (10 mL) were collected, one was used as blank and the rest were spiked with for PX (1–10 µM) and CAF (5–25 µM). The samples were treated as previously described, which involved the separation of the analytes from all the proteins and biomolecules present in the saliva samples, and the final solutions were analysed with the optimised DPV method. The first scan was only used each time as the following ones resulted in considerably lower intensities, presumably due to partial blocking of the working surface with co-extracted biocomponents from saliva. Between measurements, the working electrode was repolished to mirror finish. The calibration was repeated three times at three different days and the average charge (Q) values were plotted against the final concentrations. This calibration curve was used for quantification of the PX and CAF in human saliva and CYP1A2 phenotyping. The same samples were also measured by UHPLC-UV.

2.6. Clinical Study

This study was approved by Queen Mary Research Ethics committee (Reference QMERC2019/39) and, upon authorisation, a clinical trial with five volunteers was conducted at the Department of Chemical and Pharmaceutical Sciences at the University of Trieste (Italy). Consent forms were signed by all participants and the guidelines for conducting experiments using human biofluids were respected. The application of the proposed method was demonstrated by the analysis of saliva samples from five healthy volunteers (two females and three males), aged between 21 and 29, that did not take regular medication. The females declared that they were not pregnant or did not take oral contraceptives. The volunteers abstained from methylxanthine-containing beverages and food products (coffee, tea, chocolate, cola, etc.) for 24 h prior the experiment to avoid any specific interference. On the day of the experiment, a single 200 mg CAF dose was orally administrated to the subjects. Saliva samples (10 mL) were collected predose, and 1, 3 and 6 h postdose. The experiments were carried out within the same working day of the saliva collection.

2.7. Data Analysis and Statistical Evaluation

The calibration curves (slope and intercept) were obtained and statistically analysed at a 95% confidence interval using OriginPro 8.0 (OriginLab, Northampton, MA, USA) software. Limit of detection (LOD) and quantification (LOQ) were calculated as three and ten times the standard deviation of the intercept divided by the slope of the corresponding curve, respectively. All voltammetric measurements were performed in triplicate ($n = 3$), except of the intraday repeatability study ($n = 6$), which was expressed in terms of relative standard deviation (RSD). The paired t-test was applied to evaluate the significance of the statistical difference for the results obtained by DPV and UHPLC-UV.

3. Results

3.1. Influence of Supporting Electrolyte on Voltammetric Behavior of PX and CAF

The electrochemical response of an equimolar mixture of PX and CAF (0.2 mM) was investigated by carrying out CV experiments (100 mV s^{-1}) using an anodically pretreated GC electrode, evaluating H_2SO_4, HNO_3 and H_3PO_4 (0.01 to 0.8 M concentration range), as well as Britton–Robinson (BR) buffers (pH 2–10) as supporting electrolytes (Figure S3). Well-defined signals, with high intensity, together with low background current were observed when using electrolytes at pH < 2, consistent with literature data reported for CAF, TP and TB [61,62]. When using BR buffers at pH > 8, a higher background current was observed, caused by discharge of the electrolyte solution (Figure S3). In Figure 2a, the cyclic voltammograms for a mixture with both PX and CAF are shown, displaying the highest signal-to-noise ratio (S/N) in the various supporting electrolytes. These were obtained using, as electrolytes, H_2SO_4, HNO_3 and H_3PO_4 at 0.1 M and BR buffer at pH 2. Among these, 0.1 M H_2SO_4 showed the best peak resolution and was selected as the electrolyte of choice for the next set of experiments. The individual cyclic voltammograms of PX and CAF solutions at 0.5 mM in 0.1 M H_2SO_4, within the potential window from 0.0 V and +1.8 V at a scan rate of 100 mV s^{-1}, are shown in Figure 2b. Both analytes display oxidation peaks at different potentials: +1.24 V for PX and +1.46 V for CAF (vs. Ag/AgCl). In the reverse scan using the GC electrode, no reduction peak was observed, suggesting that the electron transfer during the oxidation of PX and CAF is completely irreversible. Although similar behavior has been reported for CAF [57,63], to the best of our knowledge this is the first report of the electrochemical characterisation of PX.

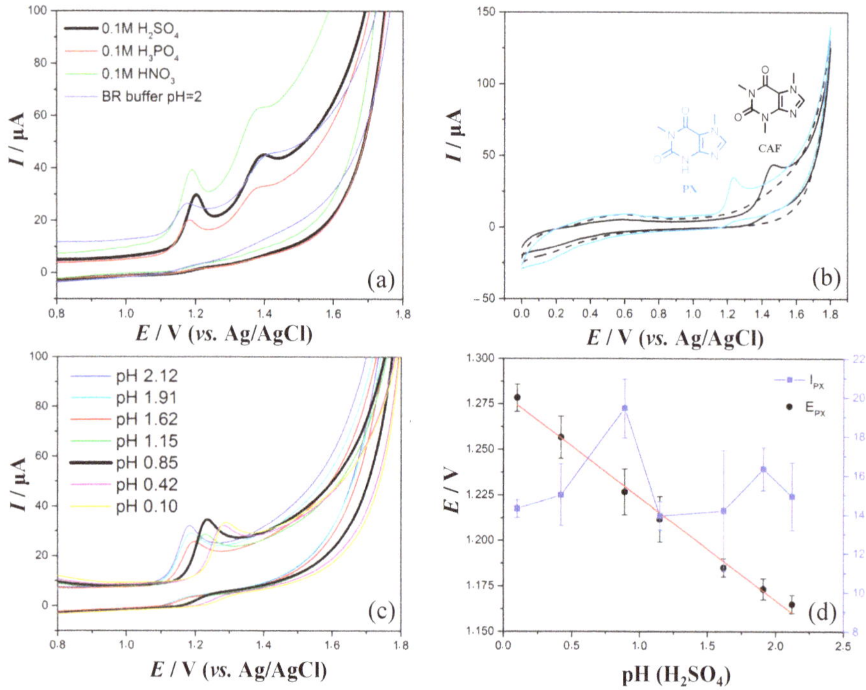

Figure 2. (**a**) Cyclic voltammograms of an equimolar mixture (200 µM) of paraxanthine (PX) and caffeine (CAF) in the optimal concentration for the various supporting electrolytes on the glassy carbon (GC) electrode with a scan rate of 100 mV s^{-1}; (**b**) cyclic voltammograms of blank 0.1 M H_2SO_4 (black dashed line), 500 µM CAF (solid blue line) and 500 µM PX (solid blue line) in 0.1 M H_2SO_4 on the GC electrode with a scan rate of 100 mV s^{-1}. The chemical structures of the paraxanthine (blue) and caffeine (black) are also given in the figure; (**c**) cyclic voltammograms of 500 µM PX at various pH values of H_2SO_4 on the GC electrode with a scan rate of 100 mV s^{-1}. The lower lines are the reverse scan of the corresponding upper lines of the same color; (**d**) the effect of pH on the peak potential (*E*) and peak current (*I*) of PX.

3.2. Oxidation Mechanism of Paraxanthine on the GC Electrode

CV experiments can yield significant kinetic and mechanistic information related to the electron transfer mechanism of an oxidation reaction, by investigating the effect of the pH and the scan rate on the peak current (*I*) and peak potential (*E*). First, the influence of the pH on the peak current and potential of 0.5 mM PX was studied by CV, at a scan rate of 100 mV s^{-1}, using H_2SO_4 in the pH range 0.10 to 2.12 (Figure 2c).

The peak's current value was not altered significantly as a function of pH in the range investigated (Figure 2d), when the experimental error is taken into account. For the oxidation reaction of PX on the GC electrode, a shift in peak potential of 155 mV to lower values, as illustrated in Figure 2d, was observed when the pH of the electrolyte was increased from 0.10 to 2.12. This provides evidence of the participation of protons in the reaction. However, no significant displacement of the potential was observed as PX is present in its protonated form in the investigated pH range. The relationship between potential and pH was found to be linear and expressed by the following equation:

$$E\ (V) = 1.28 - 0.056 \times pH \quad (1)$$

The negative slope of the potential vs. pH data indicates the release of protons during the oxidation reaction of PX and its value (-0.056 V pH^{-1}) was recognised to be very close to the theoretical value of 0.059 V pH^{-1} for a Nernstian process, revealing that the oxidation of PX on the GC electrode occurs via the overall release of two protons and two electrons [64].

Interestingly, the oxidation mechanism of TP, a structural analogue of PX (Figure 1), has also been studied by voltammetry [65–67] at various pHs, which can play a critical role in its mechanism [68], and the results were confirmed using other techniques, such as pulse radiolysis [68,69] followed by high resolution mass spectroscopy [68,70,71], UV/Vis spectrophotometry [65,69,72] and computational methods [65,70]. The reported data were used as comparison. The oxidation of TP is a two-electron and two-proton process at pH < 3, with data suggesting the formation of a radical cation as intermediate, resulting in the formation of 1,3-dimethyluric acid. The pK_a of H on N3 for PX is 8.5, very similar to the H on N7 of TP ($pK_a = 8.6$) [73,74]. Furthermore, the two molecules also show identical oxidising potential (Figure S13) [62,72]. Given the similarities between PX with TP and the CV data obtained for PX, it can be reasonably concluded that the oxidation of PX in pH < 2 also leads to the formation of 1,7-dimethyluric acid (Scheme 1).

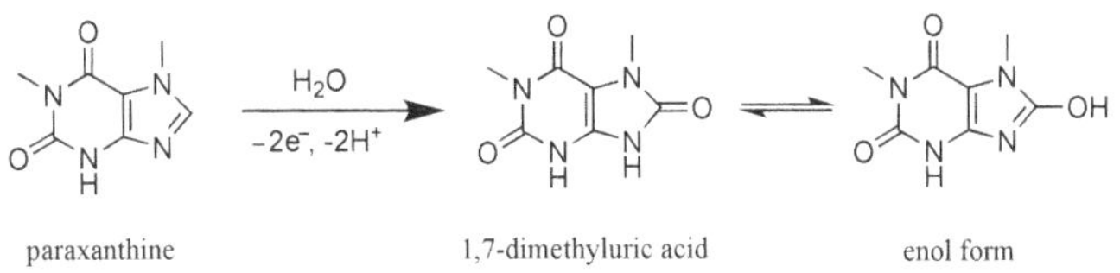

Scheme 1. Overall proposed oxidation mechanism of paraxanthine in 0.1 M H_2SO_4 on the GC electrode.

CV was then used to investigate the effect of scan rate (v) on the oxidation of PX (400 μM) in 0.1 M H_2SO_4 in the range of 5–800 mV s^{-1} (Figure 3). The oxidising potential shifted to higher values, from 1.195 to 1.270 V (Table S1), with increased scan rate, as expected for an electron transfer, followed by a chemical reaction (EC reaction). A plot of the $\log I$ vs. $\log v$ data (Figure 3b) showed a good linear fitting with a slope of 0.51, close to the calculated value of 0.50, for a process entirely driven by diffusion [75,76]. This is further confirmed by the linearity of the peak current (I) vs. $v^{1/2}$ data, with $R^2 = 0.994$ (Figure 3c), and the low intercept value (0.82 μA).

In the case of a totally irreversible reaction with a linear dependence of I with $v^{1/2}$, the electron transfer coefficient (α) of the rate determining step for a multi-step electron process can be obtained from the Tafel plot (Figure S4), E vs. $\log I$, and can be related to the Tafel slope (Equation (S1)) [77]. In the case of PX, the calculated electron transfer coefficient (α) was 0.7, which suggests a one-electron process when compared with the theoretical value (0.5 ± 0.2) [78].

The number of electrons involved in the rate determining step was confirmed from the plot of E vs. $\ln v$ (Figure S5, Equation (S2)) [77], resulting in $\alpha n = 0.7$ and therefore indicating that $n = 1$. The data confirm that the oxidation of PX is a multi-step electron-transfer process with a rate determining step, defining the electrode's kinetics, involving a single electron. The next step focused on the quantitative assessment of PX and CAF and DPV was identified as the most sensitive voltammetric technique.

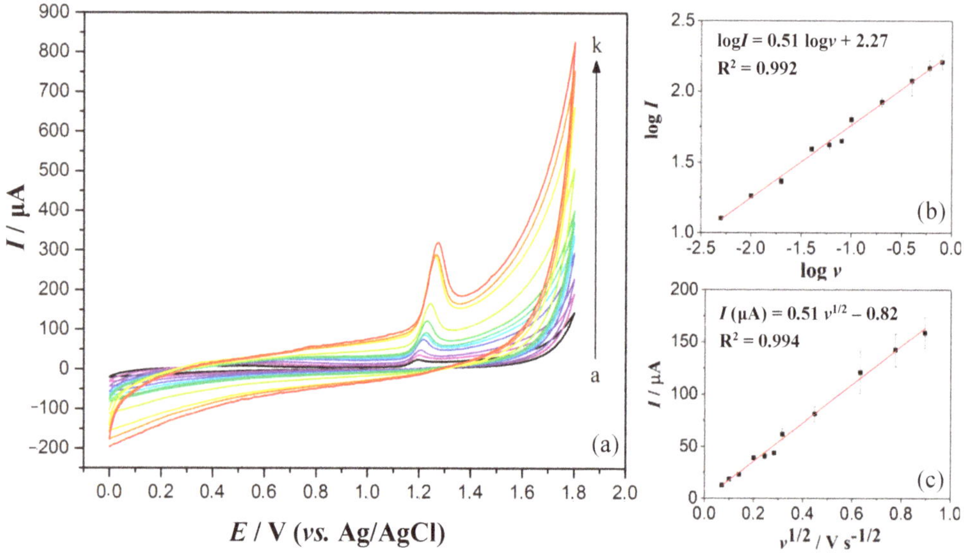

Figure 3. (**a**) Cyclic voltammograms of 400 µM PX at various scan rates (from a to k): 5, 10, 20, 40, 60, 80, 100, 200, 400, 600 and 800 mV s^{-1} in 0.1 M H_2SO_4 on the GC electrode; (**b**) log$I$ against log$v$; (**c**) I against $v^{1/2}$.

3.3. Optimisation of DPV Operating Parameters

The use of DPV for analytical purposes required optimal peak separation between PX and CAF, with the lowest possible background noise, in the range of concentrations found in saliva samples (1–30 µM). Optimisation was carried out using an equimolar mixture of 25 µM PX and CAF in 0.1 M H_2SO_4. The pretreatment potential (E_{pret}) and time (t_{pret}), scan rate (v) and step potential (E_{step}) were considered essential parameters for evaluation. The effect of electrochemical pretreatment was investigated by application of various anodic (+1 V and +2 V) and cathodic potentials (−2 V and −1 V), and pretreatment times (30, 60 and 120 s) at a fixed step potential of 5 mV and scan rate of 10 mV s^{-1} (Figures S4 and S5). The measurements for the anodically pretreated electrode displayed an increased sensitivity compared to the cathodically pretreated one and the untreated electrode, with +1 V giving the best results in terms of peak separation and height (Figure S6). Changes in the pretreatment time did not improve the signals and it was therefore kept at the lowest value of 30 s (Figure S7). Instead, an increase in the current response was observed with a decrease in the scan rate (Figure S8), and therefore the latter was set at 5 mV s^{-1}. Lastly, with a fixed scan rate of 5 mV s^{-1}, pretreatment potential +1 V and time of 30 s, the step potential was investigated in the range 1 to 6 mV (Figure S9). Optimal resolution and intensity of the peaks were obtained at a step potential of 5 mV. Thus, the optimised DPV operating parameters were identified as: E_{pret} of +1 V, t_{pret} of 30 s, E_{step} of 5 mV and v of 5 mV s^{-1}. These experimental conditions were used for the quantification of PX and CAF.

3.4. Quantification of PX and CAF in 0.1 M H_2SO_4 by DPV

All calibration curves were obtained by measuring the charge (peak area, Q) as a function of concentration and the data were confirmed by UHPLC-UV. For both analytes, the DPV measurements show a well-defined peak at potential +1.15 V for PX and +1.36 V for CAF in a 0.1 M H_2SO_4 solution. The potential difference of 210 mV between the two peaks was sufficiently large to allow simultaneous quantification of the two compounds. The obtained DP voltammograms and the corresponding calibration curves for DPV and UHPLC-UV are shown in Figure 4.

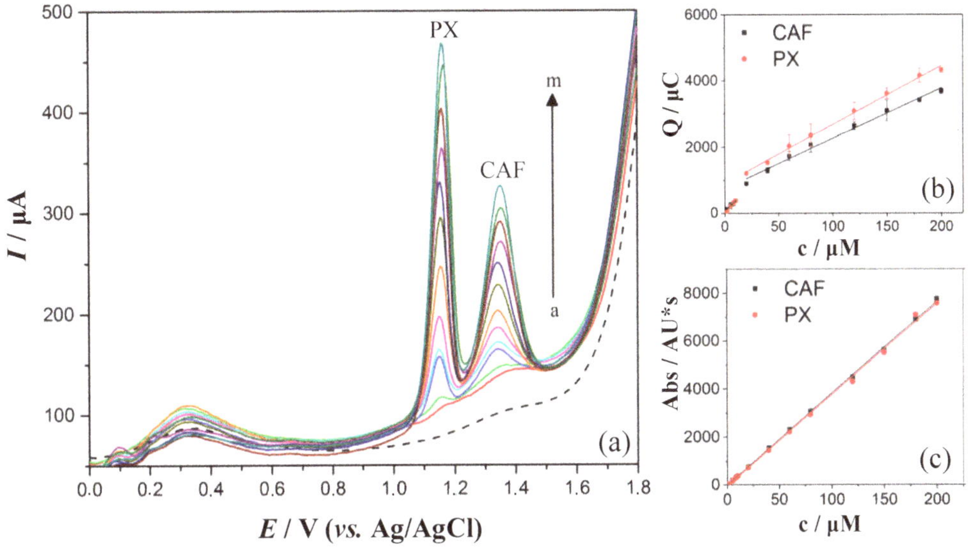

Figure 4. (**a**) DP voltammograms of equimolar mixtures of paraxanthine (PX, E_{PX} = +1.15 V) and caffeine (CAF, E_{CAF} = +1.36 V): blank, 1, 2, 5, 8, 20, 40, 60, 80, 120, 150, 180 and 200 μM in 0.1 M H_2SO_4 on the GC electrode at a scan rate of 5 mV s^{-1}; corresponding calibration curves by (**b**) differential pulse voltammetry (DPV) and (**c**) UHPLC-UV.

The peak charges of PX and CAF at +1.16 V and +1.36 V (vs. Ag/AgCl), respectively, increased proportionally with their concentrations up to 200 μM. Higher concentrations were not evaluated, as these would likely result in partial merging of the peaks, considerably limiting the accurate quantification of the analytes. All analytical parameters for their individual and simultaneous quantification by DPV and UHPLC-UV can be found in the supplementary information (Tables S2 and S3). In the case of solutions containing both analytes, linearity of response was observed up to 200 μM using UHPLC-UV. In the case of DPV, two linear segments were observed, the first one up to 10 μM (Figure S10) and the second one between 11 and 200 μM. This previously reported occurrence is likely due to a slight adsorption of the compounds on the surface of the GC electrode, at higher analyte concentrations, and can result in lower sensitivity [79–81]. The LOD was found to be 0.59 μM for PX and 0.96 μM for CAF when analysed simultaneously; these values are consistent with literature data for CAF, TP and TB using bare carbon electrodes [61,62,82].

The intraday repeatability of the method gave a relative standard deviation (RSD) of 5.4% and 4.7% for an equimolar mixture of 10 μM PX and CAF, respectively, suggesting good reproducibility of the methodology.

3.5. Selectivity Study

Prior to the application of the proposed DPV method for CYP1A2 phenotyping, its selectivity was evaluated. The influence of potential interfering molecules that may be present in saliva was examined with TP and TB, the other two primary metabolites of CAF, which are considered to be the most significant interferents for the target application. The DP voltammograms (Figure S11) showed that TB has identical oxidising potential as CAF, and TP identical to PX, consistent with previous literature data [55,58,61,62].

This could have an impact on the simultaneous quantification of PX and CAF in biological samples. However, literature data report that when CAF is completely metabolised, TB and TP are present in 10.8 and 5.4%, respectively, while PX is present in around 82% [8]. Therefore, their contribution to the peaks of CAF and PX can be considered minimal, within the time frame of 6 h used for this assay, with little effect on the final molar ratios. Other common compounds that can be found in saliva are glucose [47,83], ascorbic acid [84,85] and uric acid [51,85], but they are reported to be oxidised at considerably lower potentials between +0.2 V and +0.6 V and therefore were not considered to impact this study.

3.6. Quantification of PX and CAF in Spiked Human Saliva

Literature data report that following consumption of a 100–200 mg CAF dose, the concentrations of PX and CAF found in human saliva 1 h to 6 h postdose are up to 8 μM for PX and 30 μM for CAF [25,34]. This suggested the need for a pre-concentration step of saliva samples that was achieved by a liquid–liquid extraction, providing recovery of around 70–78% for both PX and CAF [33,34,86] and resulting in a five-fold increase in concentration. A new calibration curve in the saliva matrix was obtained with both DPV and UHPLC (Figure 5) using saliva samples from a volunteer and spiking them with PX and CAF solutions. As all the samples are treated to separate the analytes from the saliva biomolecules, the impact of variability of the salivary samples was significantly reduced. Following the established protocol, PX and CAF showed an oxidising potential of +1.25 V and +1.45 V, respectively. A positive shift of around 90 mV of the peak potential for both analytes was observed compared to previous values (Figure 4); the results suggest that the presence of biocomponents co-extracted from saliva impact the oxidation process. All analytical parameters for the simultaneous quantification of PX and CAF in the saliva samples by DPV and UHPLC are presented in Tables 1 and 2, respectively.

As expected, the LODs by UHPLC were lower than the ones obtained by DPV; however, the simplicity and practicality of DPV mitigates this. The intraday repeatability of the DPV method for binary mixtures of spiked PX (9 μM) and CAF (20 μM) in a saliva sample resulted in RSDs of 7.0% and 4.3%, respectively.

It is of interest to note that experiments using a boron doped diamond (BDD) electrode and screen-printed carbon paste (SP-CP) electrode in both supporting electrolyte solutions or spiked saliva samples showed a significantly lower performance (Figures S12–S15).

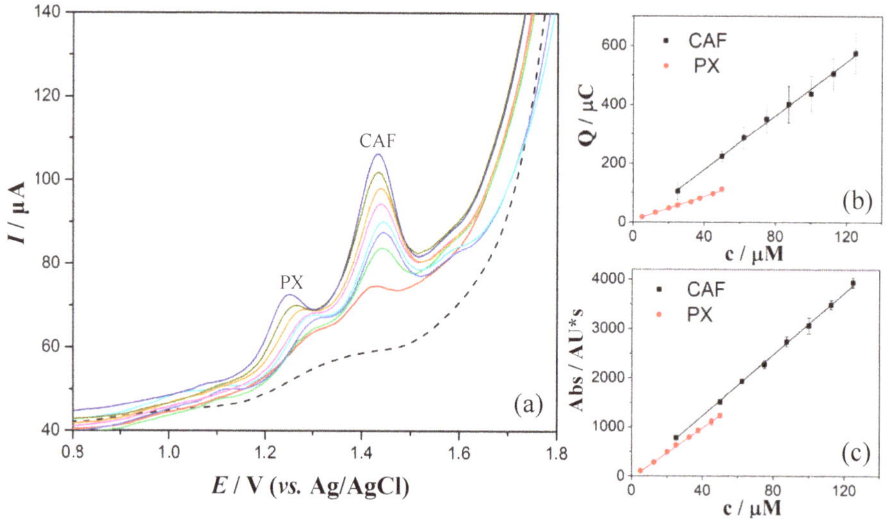

Figure 5. (**a**) DP voltammograms of paraxanthine (PX, E_{PX} = +1.25 V) and caffeine (CAF, E_{CAF} = +1.45 V) extracted from spiked saliva on the GC electrode at a scan rate of 5 mV s^{-1}; PX and CAF concentrations assuming quantitative extraction: blank saliva (black dashed line), 25 μM and 5 μM (red), 50 μM and 12.5 μM (green), 62.5 μM and 20 μM (blue), 75 μM and 25 μM (light blue), 87.5 μM and 32.5 μM (pink), 100 μM and 37.5 μM (yellow), 112.5 μM and 45 μM (dark yellow), 125 μM and 50 μM (blue); corresponding calibration curves by (**b**) DPV and (**c**) UHPLC-UV.

Table 1. Analytical parameters for the simultaneous quantification of paraxanthine (PX) and caffeine (CAF) in spiked saliva by DPV.

Analytical Parameter	DPV	
	PX	CAF
Sensitivity (μC/μM)	2.02 ± 0.06	4.55 ± 0.10
Intercept (μC)	7.37 ± 1.95	−3.67 ± 8.80
R^2	0.994	0.996
LOD (μM)	2.89	5.80
Linear range (μM)	10–60	19–125

Table 2. Analytical parameters for the simultaneous quantification of paraxanthine (PX) and caffeine (CAF) in spiked saliva by UHPLC-UV.

Analytical Parameter	UHPLC-UV	
	PX	CAF
Sensitivity (AU/μM)	25.01 ± 0.28	31.41 ± 0.41
Intercept (AU)	−14.01 ± 9.01	−39.01 ± 35.01
R^2	0.999	0.999
LOD (μM)	1.07	3.34
Linear range (μM)	4–60	11–125

3.7. Evaluation of Method with Five Volunteers

In order to assess the potential of this electrochemical sensing platform, CYP1A2 phenotyping was carried out by monitoring the PX/CAF molar ratio in saliva from five healthy regular coffee drinkers (two females and three males) over time (4 time points; 0–6 h). For the purpose of this study, non-genetic characteristics known to influence the caffeine metabolism, like smoker vs. non-smoker, were not taken into account. A single dose of 200 mg CAF was orally administered to five healthy subjects and saliva samples were collected predose, and 1, 3 and 6 h postdose. The saliva samples were treated as described in the experimental section and analysed with the optimised DPV method and the data were confirmed by UHPLC-UV. Figure 6a shows representative DP voltammograms at the various time intervals from a male volunteer for a single day experiment.

Typical UHPLC-UV chromatograms obtained for the various time points are shown in Figure S16 (ESI). The concentration of PX and CAF in each sample was calculated using the matrix-matched calibration curve. The concentration profiles (Figure 6a and Figure S17) for PX and CAF and their corresponding molar ratios were obtained for three experiments in three different days per subject, employing the proposed DPV method and UHPLC-UV (Table S4). The concentration profiles with time for a male volunteer are shown in Figure 6b. For all tested subjects, a linear decrease in CAF and increase in PX concentration were observed over time, consistent with the first-order pharmacokinetics of CAF in humans [8,87]. Furthermore, for all tested subjects, a linear increase in the PX/CAF molar ratios over time was observed with both methods, as illustrated in Figure 6c, indicating a proportional increase in the concentration of PX with the decrease in CAF. The molar ratios obtained 1 h postdose by DPV showed the lowest correlation with UHPLC-UV (Table S4), as a result of difficulties in resolving the baseline. However, for CYP1A2 phenotyping, the PX/CAF molar ratio at a single time-point is required. Previous studies in the literature have recommended the use of PX/CAF molar ratios in saliva 4 h and 6 h postdose as a convenient sampling time-point for implementation in a clinical setting [28,34]. In this work, the strongest correlation between the two methods, accurately reflecting the CYP1A2 activity, was found to be 6 h postdose, where the PX/CAF molar ratios ranged from 0.34 to 0.54, consistent with the values reported using HPLC-UV by Jordan et al. and Perera et al. [33,34].

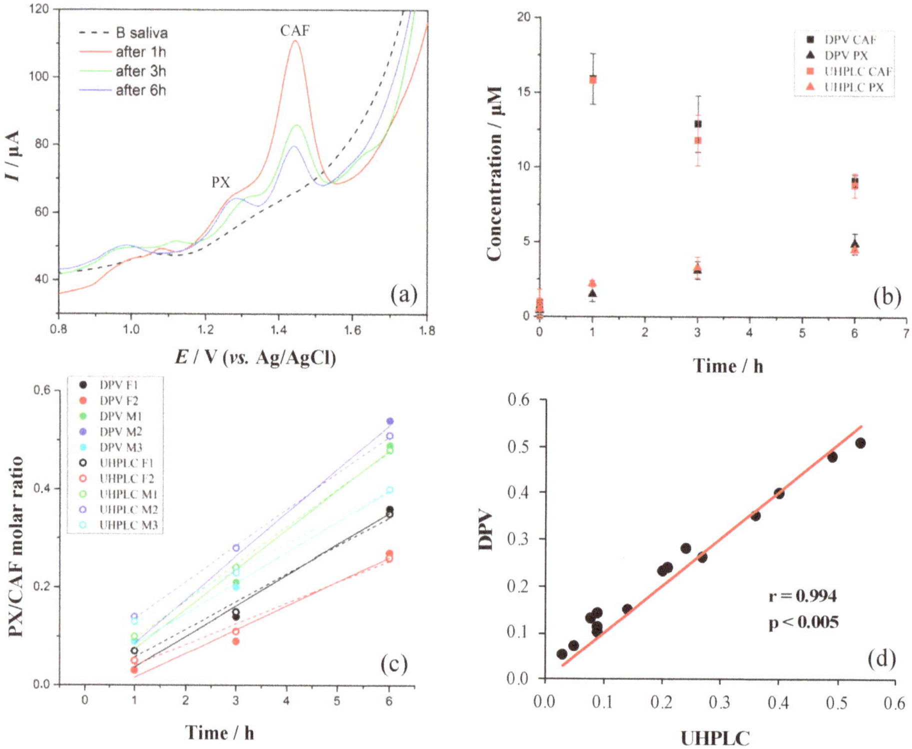

Figure 6. (**a**) DP voltammograms containing paraxanthine (PX, E_{PX} = +1.24 V) and caffeine (CAF, E_{CAF} = +1.44 V), isolated from saliva, in 0.1 M H_2SO_4 on the GC electrode, with a scan rate of 5 mV s^{-1}: predose as blank saliva (B saliva), 1, 3 and 6 h upon administration of a single 200 mg CAF dose from a male volunteer; (**b**) CAF and PX concentration-time profiles by DPV and UHPLC for a male volunteer in three different days upon administration of a single 200 mg CAF dose with SD of the three measurements as error bars; (**c**) PX/CAF molar ratios vs. time (h) by DPV (filled circles and solid lines) and UHPLC-UV (empty circles and dashed lines) for all five subjects (F1 and F2 for females and M1, M2 and M3 for males); (**d**) Correlation plot of PX/CAF molar ratios obtained by DPV vs. UHPLC.

The data demonstrate a high correlation (Pearson correlation coefficient r = 0.994) between the PX/CAF molar ratios values obtained with DPV and UHPLC (Figure 6d). The paired t-test was applied, and the calculated t value (1.74) was lower than the critical value (2.20) at a 95% confidence level, rejecting the null hypothesis. This indicates that the estimated molar ratios by DPV are not statistically significantly different to those measured by UHPLC-UV. This demonstrates that the proposed protocol effectively mitigates the interfering matrix effects of human saliva for the selective determination of the analytes by DPV.

The proposed sensing system based on DPV is fast, requiring only 6.5 min for the evaluation of each sample. Moreover, the proposed electrochemical system is a portable sensor that is considerably cheaper and easier to operate compared with UHPLC. Although the LOD for the UHPLC-UV is lower compared to DPV, the concentrations of PX detected in human saliva are sufficiently high for estimating the CYP1A2 activity.

4. Conclusions

In this work, we developed a simple and fast analytical tool for the simultaneous quantification of antioxidants CAF and PX in human saliva by DPV on an anodically pretreated GC electrode, for applications in CYP1A2 phenotyping. Voltammetry allowed us to study, for the first time, the oxidation behaviour of PX in aqueous media, revealing that its oxidation occurs via the transfer of two electrons and two protons, similarly to the mechanism reported for TP but very different from CAF. The DPV method was validated via UHPLC-UV, and matrix-matched calibration curves were used for the quantification of PX and CAF in saliva. A limit of detection of 2.89 μM for PX and 5.80 μM for CAF was obtained. Potential applicability of the sensing platform was demonstrated by running a small scale trial on five healthy volunteers, with simultaneous quantification by DPV of PX and CAF in saliva samples collected at 1, 3 and 6 h postdose administration. The results were validated by UHPLC-UV, and were shown to have a high correlation factor (r = 0.994).

The new electrochemical sensor allows the selective and simultaneous quantification of PX and CAF in human saliva with an estimated 20-min total analysis time (from sample collection to voltammogram integration), providing a more accessible tool compared to the standard chromatographic techniques. The precision and accuracy demonstrated makes this approach an attractive alternative and therapeutically useful for monitoring probe drugs and their metabolites in saliva and potentially other biofluids (e.g., sweat and plasma). A higher sensitivity could be achieved by modifying the working electrode, but this would require time-consuming steps, additional costs and potentially lead to lower reproducibility. As a next step, further studies using larger numbers of volunteers will be required to evaluate how the electrochemical sensing platform handles interpatient variations. Our work highlights a range of opportunities for the development of point-of-care devices based on electrochemical sensing for phenotyping drug-metabolising enzymes with clinical significance in the area of personalised medicine.

Supplementary Materials: The following are available online at https://www.mdpi.com/2076-3921/10/1/10/s1. Figure S1: (a) DP voltammograms for increasing concentrations of caffeine keeping paraxanthine constant at 10 μM. Caffeine concentrations: (a) 0, (b) 1, (c) 4, (d) 8, (e) 20, (f) 40, (g) 80, (h) 100, (i) 130, (j) 160, (k) 200, (l) 250 and (m) 300 μM in 0.1 M H_2SO_4 on the GC electrode at a scan rate of 5 mV s^{-1}; corresponding calibration curves by (b) DPV and (c) UHPLC-UV; Figure S2: (a) Differential pulse voltammograms for increasing concentrations of paraxanthine keeping caffeine constant at 10 μM. Paraxanthine concentrations: (a) 0, (b) 1, (c) 2, (d) 5, (e) 8, (f) 20, (g) 40, (h) 60, (i) 80, (j) 120, (k) 150, (l) 180 and (m) 200 μM in 0.1 M H_2SO_4 on the glassy carbon electrode at a scan rate of 5 mV s^{-1}; corresponding calibration curves by (b) DPV and (c) UHPLC-UV; Figure S3: Cyclic voltammograms of an equimolar mixture (0.2 mM) of paraxanthine and caffeine from 0.8 to 1.8 V in various concentrations of H_2SO_4, HNO_3, H_3PO_4 and Britton–Robinson (BR) buffers on the glassy carbon electrode at a fixed scan rate of 100 mV s^{-1}; Figure S4: Tafel plot for paraxanthine: oxidising potential (E) against the logarithm of the current (log$I$); Figure S5: Oxidising potential (E) against the natural logarithm of the scan rate (ln$v$) for paraxanthine on the glassy carbon electrode; Figure S6: Differential pulse voltammograms of an equimolar mixture (25 μM) of paraxanthine and caffeine in 0.1 M H_2SO_4 on the glassy carbon electrode without pretreatment and different pretreatment potentials (E_{PRET}) with a fixed step potential of 5 mV and scan rate of 10 mV s^{-1}; Figure S7: Differential pulse voltammograms of an equimolar (25 μM) mixture of paraxanthine and caffeine in 0.1 M H_2SO_4 on a glassy carbon electrode at various pretreatment times: 30, 60 and 120 s with fixed pretreatment potential of +1 V and step potential of 0.005 V at a scan rate of 10 mV s^{-1}; Figure S8: Differential pulse voltammograms of an equimolar (25 μM) mixture of paraxanthine and caffeine in 0.1 M H_2SO_4 on the glassy carbon electrode at various scan rates: 5, 10, 20, 30, 40, 60 and 100 mV s^{-1}; Figure S9: Differential pulse voltammograms of an equimolar (25 μM) mixture of paraxanthine and caffeine in 0.1 M H_2SO_4 on the glassy carbon electrode with a pretreatment potential of +1 V and pretreatment time of 30 s for various step potentials (1, 2, 5 and 6 mV) at a scan rate of 5 mV s^{-1}; Figure S10: Calibration curve for the simultaneous quantification of equimolar mixtures of paraxanthine (PX) and caffeine (CAF) by differential pulse voltammetry in 0.1 M H_2SO_4 on the GC electrode: first linear segment (1–10 μM); Figure S11: Differential pulse voltammograms of an equimolar mixture (50 μM) of caffeine (CAF), paraxanthine (PX), theophylline (TP) and theobromine (TB) in 0.1 M H_2SO_4 on the GC electrode, at a scan rate (v) of 5 mV s^{-1}; Figure S12: Differential pulse voltammograms of 0.1 M H_2SO_4 (black dashed line) and an equimolar (50 μM) mixture of paraxanthine (E_{PX} = +1.23 V) and caffeine (E_{CAF} = +1.40 V, red line) on the boron-doped diamond electrode with differential pulse voltammetry parameters:

pretreatment potential (E_{PRET}) of +2 V, pretreatment time (t_{PRET}) of 30 s, step potential (E_{STEP}) of 0.005 V, pulse potential (E_{PULS}) of 0.1 V, pulse time (t_{PULS}) of 10 ms and scan rate (v) of 100 mV s^{-1}; Figure S13: Differential pulse voltammograms for 0.01 M H_2SO_4 (black dashed line), treated blank saliva (red line), treated spiked water (green line) and treated spiked saliva (blue line) with equimolar concentrations (100 μM) caffeine and paraxanthine on the boron doped diamond electrode; Figure S14: Differential pulse voltammograms of 0.1 M H_3PO_4 (black dashed line) and an equimolar mixture (100 μM) of paraxanthine (E_{PX} = +0.82 V) and caffeine (E_{CAF} = +1.01 V, red line) on a screen-printed carbon paste electrode with differential pulse voltammetry parameters: step potential (E_{STEP}) of 0.004 V, pulse potential (E_{PULS}) of 0.15 V, pulse time (t_{PULS}) of 10 ms and scan rate (v) of 10 mV s^{-1}; Figure S15: Differential pulse voltammograms (DPV) for 0.01 M H_2SO_4 (black dashed line), extracted caffeine and paraxanthine from saliva 2 h postdose (red line) and 4h postdose (green line); on the screen-printed carbon paste electrode with DPV parameters: step potential of 0.004 V, pulse potential of 0.15 V, pulse time of 10 ms and scan rate of 10 mV s^{-1}; Figure S16: UHPLC-UV chromatograms from a volunteer (abstinence from caffeine for 24 h) for a one day experiment obtained upon administration of 200 mg caffeine oral dose: predose (blue), 1 h (red), 3 h (green) and 6 h (pink) postdose; Figure S17: Caffeine (CAF) and paraxanthine (PX) concentration-time profiles by DPV and UHPLC-UV for the rest four tested subjects (F1 and F2 for females and M1 and M2 for males) in three different days upon administration of a single 200 mg CAF dose with standard deviation of the three measurements as error bars. Table S1: Averaged values of peak potential (E) and peak current (I) for the various scan rates for 400 μM paraxanthine (PX) in 0.1 M H_2SO_4 by cyclic voltammetry on the glassy carbon electrode; Table S2: Analytical parameters for the individual and simultaneous determination of paraxanthine and caffeine by differential pulse voltammetry in 0.1 M H_2SO_4 on the glassy carbon electrode; Table S3: Analytical parameters for individual and simultaneous determination of paraxanthine and caffeine by UHPLC-UV; Table S4: Human saliva sample analysis for five volunteers at three different days using the proposed DPV method and UHPLC-UV as reference method. Equations (S1) and (S2) were used to calculate the electron transfer coefficient (α) and the number of electrons involved in the rate determining step of the oxidation of paraxanthine.

Author Contributions: Conceptualization, R.-M.A., L.N. and M.R.; Data curation, R.-M.A. and F.B.; Formal analysis, M.R.; Funding acquisition, F.B., L.N. and M.R.; Investigation, R.-M.A.; Methodology, R.-M.A., S.C. and C.T.; Supervision, F.B., L.N. and M.R.; Writing—original draft, R.-M.A. and M.R.; Writing—review and editing, R.-M.A., C.T. and M.R. All authors have read and agreed to the published version of the manuscript.

Funding: This work was financially supported by the European Union's Horizon 2020 research and innovation program under the Marie Sklodowska-Curie Grant Agreement no. 642014 (IPCOS).

Institutional Review Board Statement: The above study was conditionally approved by the Queen Mary Ethics of Research Committee (Review Panel E) on the 1 May 2019; approval was given by Chair's Action on the 11 June 2019; with full approval ratified by the Facilitator on the 12 June 2019, on receipt of paperwork (Reference QMERC2019/39).

Informed Consent Statement: Informed consent was obtained from all subjects involved in the study.

Data Availability Statement: The data presented in this study are available in https://www.mdpi.com/2076-3921/10/1/10/s1.

Acknowledgments: The authors would like to thank Miroslav Stredansky (Biorealis s.r.o., Bratislava, Slovakia), Luca Redivo and Lubomir Švorc (University of Bratislava, Bratislava, Slovakia) for providing helpful comments and discussions.

Conflicts of Interest: The authors declare no conflict of interest.

Abbreviations

IUPAC: International Union of Pure and Applied Chemistry; LOD, limit of detection; LOQ, limit of quantification; RSD, relative standard deviation; GC, glassy carbon; BDD, boron-doped diamond; SP-CP, screen-printed carbon paste; CV, cyclic voltammetry; DPV, differential pulse voltammetry; CYP, P450 cytochromes; PX, paraxanthine; CAF, caffeine; TP, theophylline; TB, theobromine; BR, Britton–Robinson; UHPLC, ultra-high performance liquid chromatography; UV, ultraviolet; MS, mass spectrometry.

References

1. Monteiro, J.P.; Alves, M.G.; Oliveira, P.F.; Silva, B.M. Structure-bioactivity relationships of methylxanthines: Trying to make sense of all the promises and the drawbacks. *Molecules* **2016**, *21*, 974. [CrossRef] [PubMed]
2. Shi, X.; Dalal, N.S.; Jain, A.C. Antioxidant behaviour of caffeine: Efficient scavenging of hydroxyl radicals. *Food Chem. Toxicol.* **1991**, *29*, 1–6. [CrossRef]
3. Liang, N.; Kitts, D.D. Antioxidant property of coffee components: Assessment of methods that define mechanism of action. *Molecules* **2014**, *19*, 19180–19208. [CrossRef] [PubMed]
4. Li, H.; Roxo, M.; Cheng, X.; Zhang, S.; Cheng, H.; Wink, M. Pro-oxidant and lifespan extension effects of caffeine and related methylxanthines in Caenorhabditis elegans. *Food Chem. X* **2019**, *1*, 1–9. [CrossRef]
5. Nehlig, A. Is caffeine a cognitive enhancer? *J. Alzheimer's Dis.* **2010**, *20*, 585–594. [CrossRef] [PubMed]
6. Ullah, F.; Ali, T.; Ullah, N.; Kim, M.O. Caffeine prevents d-galactose-induced cognitive deficits, oxidative stress, neuroinflammation and neurodegeneration in the adult rat brain. *Neurochem. Int.* **2015**, *90*, 114–124. [CrossRef]
7. Ikram, M.; Park, T.J.; Ali, T.; Kim, M.O. Antioxidant and Neuroprotective Effects of Caffeine against Alzheimer's and Parkinson's Disease: Insight into the Role of Nrf-2 and A2AR Signaling. *Antioxidants* **2020**, *9*, 902. [CrossRef]
8. Fredholm, B.B.; Ashihara, H.; Misako, K.; Crozier, A.; Francis, S.H.; Arnaud, M.J.; Sekhar, K.R. *Methylxanthines*; Springer International Publishing: Heidelberg, Germany, 2011; Volume 200, ISBN 9783642134425.
9. Doré, A.S.; Robertson, N.; Errey, J.C.; Ng, I.; Hollenstein, K.; Tehan, B.; Hurrell, E.; Bennett, K.; Congreve, M.; Magnani, F.; et al. Structure of the adenosine A 2A receptor in complex with ZM241385 and the xanthines XAC and caffeine. *Structure* **2011**, *19*, 1283–1293. [CrossRef]
10. Redivo, L.; Anastasiadi, R.M.; Pividori, M.; Berti, F.; Peressi, M.; Di Tommaso, D.; Resmini, M. Prediction of self-assembly of adenosine analogues in solution: A computational approach validated by isothermal titration calorimetry. *Phys. Chem. Chem. Phys.* **2019**, *21*, 4258–4267. [CrossRef]
11. Chavez-Valdez, R.; Wills-Karp, M.; Ahlawat, R.; Cristofalo, E.A.; Nathan, A.; Gauda, E.B. Caffeine modulates TNF-α production by cord blood monocytes: The role of adenosine receptors. *Pediatr. Res.* **2009**, *65*, 203–208. [CrossRef]
12. Postuma, R.B.; Lang, A.E.; Munhoz, R.P.; Charland, K.; Pelletier, A.; Moscovich, M.; Filla, L.; Zanatta, D.; Romenets, S.R.; Altman, R.; et al. Caffeine for treatment of Parkinson disease: A randomized controlled trial. *Neurology* **2012**, *79*, 651–658. [CrossRef] [PubMed]
13. Arendash, G.W.; Cao, C. Caffeine and coffee as therapeutics against Alzheimer's disease. *J. Alzheimer's Dis.* **2010**, *20*, 117–126. [CrossRef] [PubMed]
14. Maia, L.; De Mendonça, A. Does caffeine intake protect from Alzheimer's disease? *Eur. J. Neurol.* **2002**, *9*, 377–382. [CrossRef] [PubMed]
15. Stavric, B. Methylxanthines: Toxicity to humans. 3. Theobromine, paraxanthine and the combined effects of methylxanthines. *Food Chem. Toxicol.* **1988**, *26*, 725–733. [CrossRef]
16. Zhou, S.-F.; Wang, B.; Yang, L.-P.; Liu, J.-P. Structure, function, regulation and polymorphism and the clinical significance of human cytochrome P450 1A2. *Drug Metab. Rev.* **2010**, *42*, 268–354. [CrossRef] [PubMed]
17. Rendic, S. Summary of Information on Human Cyp Enzymes: Human P450 Metabolism Data. *Drug Metab. Rev.* **2002**, *34*, 83–448. [CrossRef]
18. Dalla Libera, A.; Scutari, G.; Boscolo, R.; Rigobello, M.P.; Bindoli, A. Antioxidant properties of clozapine and related neuroleptics. *Free Radic. Res.* **1998**, *29*, 151–157. [CrossRef]
19. Brinholi, F.F.; de Farias, C.C.; Bonifácio, K.L.; Higachi, L.; Casagrande, R.; Moreira, E.G.; Barbosa, D.S. Clozapine and olanzapine are better antioxidants than haloperidol, Quetiapine, Risperidone and ziprasidone in in vitro models. *Biomed. Pharmacother.* **2016**, *81*, 411–415. [CrossRef]
20. Sala, G.; Arosio, A.; Conti, E.; Beretta, S.; Lunetta, C.; Riva, N.; Ferrarese, C.; Tremolizzo, L. Riluzole selective antioxidant effects in cell models expressing amyotrophic lateral sclerosis endophenotypes. *Clin. Psychopharmacol. Neurosci.* **2019**, *17*, 438–442. [CrossRef]
21. Zanger, U.M.; Schwab, M. Cytochrome P450 enzymes in drug metabolism: Regulation of gene expression, enzyme activities, and impact of genetic variation. *Pharmacol. Ther.* **2013**, *138*, 103–141. [CrossRef]
22. Dobrinas, M.; Cornuz, J.; Oneda, B.; Kohler Serra, M.; Puhl, M.; Eap, C.B. Impact of smoking, smoking cessation, and genetic polymorphisms on CYP1A2 activity and inducibility. *Clin. Pharmacol. Ther.* **2011**, *90*, 117–125. [CrossRef] [PubMed]
23. Ingelman-Sundberg, M.; Oscarson, M.; McLellan, R.A. Polymorphic human cytochrome P450 enzymes: An opportunity for individualized drug treatment. *Trends Pharmacol. Sci.* **1999**, *20*, 342–349. [CrossRef]
24. De Kesel, P.M.M.; Lambert, W.E.; Stove, C.P. Alternative Sampling Strategies for Cytochrome P450 Phenotyping. *Clin. Pharmacokinet.* **2016**, *55*, 169–184. [CrossRef] [PubMed]
25. Fuhr, U.; Jetter, A.; Kirchheiner, J. Appropriate phenotyping procedures for drug metabolizing enzymes and transporters in humans and their simultaneous use in the "cocktail" approach. *Clin. Pharmacol. Ther.* **2007**, *81*, 270–283. [CrossRef]
26. Holmes, E.; Wilson, I.D.; Nicholson, J.K. Metabolic Phenotyping in Health and Disease. *Cell* **2008**, *134*, 714–717. [CrossRef]
27. Faber, M.S.; Jetter, A.; Fuhr, U. Assessment of CYP1A2 activity in clinical practice: Why, how, and when? *Basic Clin. Pharmacol. Toxicol.* **2005**, *97*, 125–134. [CrossRef]

28. Doude Van Troostwijk, L.J.A.E.; Koopmans, R.P.; Guchelaar, H.J. Two novel methods for the determination of CYP1A2 activity using the paraxanthine/caffeine ratio. *Fundam. Clin. Pharmacol.* **2003**, *17*, 355–362. [CrossRef]
29. Nehlig, A. Interindividual differences in caffeine metabolism and factors driving caffeine consumption. *Pharmacol. Rev.* **2018**, *70*, 384–411. [CrossRef]
30. Schievano, E.; Finotello, C.; Navarini, L.; Mammi, S. Quantification of caffeine in human saliva by nuclear magnetic resonance as an alternative method for cytochrome CYP1A2 phenotyping. *Talanta* **2015**, *140*, 36–41. [CrossRef]
31. Caporossi, L.; Santoro, A.; Papaleo, B. Saliva as an analytical matrix: State of the art and application for biomonitoring. *Biomarkers* **2010**, *15*, 475–487. [CrossRef]
32. Chiappin, S.; Antonelli, G.; Gatti, R.; De Palo, E.F. Saliva specimen: A new laboratory tool for diagnostic and basic investigation. *Clin. Chim. Acta* **2007**, *383*, 30–40. [CrossRef] [PubMed]
33. Jordan, N.Y.; Mimpen, J.Y.; van den Bogaard, W.J.M.; Flesch, F.M.; van de Meent, M.H.M.; Torano, J.S. Analysis of caffeine and paraxanthine in human saliva with ultra-high-performance liquid chromatography for CYP1A2 phenotyping. *J. Chromatogr. B Anal. Technol. Biomed. Life Sci.* **2015**, *995–996*, 70–73. [CrossRef] [PubMed]
34. Perera, V.; Gross, A.S.; McLachlan, A.J. Caffeine and paraxanthine HPLC assay for CYP1A2 phenotype assessment using saliva and plasma. *Biomed. Chromatogr.* **2010**, *24*, 1136–1144. [CrossRef] [PubMed]
35. Fuhr, U.; Rost, K.L. Simple and reliable CYP1A2 phenotyping by the paraxanthine/caffeine ratio in plasma and saliva. *Pharmacogenetics* **1994**, *4*, 109–116. [CrossRef] [PubMed]
36. Begas, E.; Kouvaras, E.; Tsakalof, A.K.; Bounitsi, M.; Asprodini, E.K. Development and validation of a reversed-phase HPLC method for CYP1A2 phenotyping by use of a caffeine metabolite ratio in saliva. *Biomed. Chromatogr.* **2015**, *29*, 1657–1663. [CrossRef]
37. Lopez-Sanchez, R.D.C.; Lara-Diaz, V.J.; Aranda-Gutierrez, A.; Martinez-Cardona, J.A.; Hernandez, J.A. HPLC Method for Quantification of Caffeine and Its Three Major Metabolites in Human Plasma Using Fetal Bovine Serum Matrix to Evaluate Prenatal Drug Exposure. *J. Anal. Methods Chem.* **2018**, *2018*. [CrossRef]
38. Martínez-López, S.; Sarriá, B.; Baeza, G.; Mateos, R.; Bravo-Clemente, L. Pharmacokinetics of caffeine and its metabolites in plasma and urine after consuming a soluble green/roasted coffee blend by healthy subjects. *Food Res. Int.* **2014**, *64*, 125–133. [CrossRef]
39. Yin, O.Q.P.; Lam, S.S.L.; Lo, C.M.Y.; Chow, M.S.S. Rapid determination of five probe drugs and their metabolites in human plasma and urine by liquid chromatography/tandem mass spectrometry: Application to cytochrome P450 phenotyping studies. *Rapid Commun. Mass Spectrom.* **2004**, *18*, 2921–2933. [CrossRef]
40. Idili, A.; Parolo, C.; Ortega, G.; Plaxco, K.W. Calibration-Free Measurement of Phenylalanine Levels in the Blood Using an Electrochemical Aptamer-Based Sensor Suitable for Point-of-Care Applications. *ACS Sens.* **2019**, *4*, 3227–3233. [CrossRef]
41. Benbouguerra, N.; Richard, T.; Saucier, C.; Garcia, F. Voltammetric Behavior, Flavanol and Anthocyanin Contents, and Antioxidant Capacity of Grape Skins and Seeds during Ripening (Vitis vinifera var. Merlot, Tannat, and Syrah). *Antioxidants* **2020**, *9*, 800. [CrossRef]
42. Jara-Palacios, M.J.; Gonçalves, S.; Heredia, F.J.; Hernanz, D.; Romano, A. Extraction of antioxidants from winemaking byproducts: Effect of the solvent on phenolic composition, antioxidant and anti-cholinesterase activities, and electrochemical behaviour. *Antioxidants* **2020**, *9*, 675. [CrossRef] [PubMed]
43. McCreery, R.L. ChemInform Abstract: Advanced Carbon Electrode Materials for Molecular Electrochemistry. *Chem. Rev.* **2008**, *108*, 2646–2687. [CrossRef] [PubMed]
44. Scholz, F. Voltammetric techniques of analysis: The essentials. *ChemTexts* **2015**, *1*, 1–24. [CrossRef]
45. Karikalan, N.; Karthik, R.; Chen, S.M.; Chen, H.A. A voltammetric determination of caffeic acid in red wines based on the nitrogen doped carbon modified glassy carbon electrode. *Sci. Rep.* **2017**, *7*, 1–10. [CrossRef] [PubMed]
46. Udomsap, D.; Branger, C.; Culioli, G.; Dollet, P.; Brisset, H. A versatile electrochemical sensing receptor based on a molecularly imprinted polymer. *Chem. Commun.* **2014**, *50*, 7488–7491. [CrossRef]
47. Li, J.; Lin, X. Glucose biosensor based on immobilization of glucose oxidase in poly(o-aminophenol) film on polypyrrole-Pt nanocomposite modified glassy carbon electrode. *Biosens. Bioelectron.* **2007**, *22*, 2898–2905. [CrossRef] [PubMed]
48. Zoski, C. (Ed.) *Handbook of Electrochemistry*; Elsevier: Amsterdam, The Netherlands, 2007; ISBN 9780444519580.
49. Ngamchuea, K.; Chaisiwamongkhol, K.; Batchelor-Mcauley, C.; Compton, R.G. Chemical analysis in saliva and the search for salivary biomarkers-a tutorial review. *Analyst* **2018**, *143*, 81–99. [CrossRef]
50. Campuzano, S.; Yánez-Sedeño, P.; Pingarrón, J.M. Electrochemical bioaffinity sensors for salivary biomarkers detection. *TrAC Trends Anal. Chem.* **2017**, *86*, 14–24. [CrossRef]
51. Ngamchuea, K.; Batchelor-McAuley, C.; Compton, R.G. Understanding electroanalytical measurements in authentic human saliva leading to the detection of salivary uric acid. *Sens. Actuators B Chem.* **2018**, *262*, 404–410. [CrossRef]
52. Jesny, S.; Girish Kumar, K. Electrocatalytic resolution of guanine, adenine and cytosine along with uric acid on poly (4-amino-3-hydroxy naphthalene-1-sulfonic acid) modified glassy carbon electrode. *J. Electroanal. Chem.* **2017**, *801*, 153–161. [CrossRef]
53. Dhull, N.; Kaur, G.; Gupta, V.; Tomar, M. Highly sensitive and non-invasive electrochemical immunosensor for salivary cortisol detection. *Sens. Actuators B Chem.* **2019**, *293*, 281–288. [CrossRef]
54. Dos Santos, W.T.P.; Amin, H.M.A.; Compton, R.G. A nano-carbon electrode optimized for adsorptive stripping voltammetry: Application to detection of the stimulant selegiline in authentic saliva. *Sens. Actuators B Chem.* **2019**, *279*, 433–439. [CrossRef]

55. Kesavan, S.; Gowthaman, N.S.K.; Alwarappan, S.; John, S.A. Real time detection of adenosine and theophylline in urine and blood samples using graphene modified electrode. *Sens. Actuators B Chem.* **2019**, *278*, 46–54. [CrossRef]
56. Güney, S.; Cebeci, F. Selective electrochemical sensor for theophylline based on an electrode modified with imprinted sol-gel film immobilized on carbon nanoparticle layer. *Sens. Actuators B Chem.* **2015**, *208*, 307–314. [CrossRef]
57. Carolina Torres, A.; Barsan, M.M.; Brett, C.M.A. Simple electrochemical sensor for caffeine based on carbon and Nafion-modified carbon electrodes. *Food Chem.* **2014**, *149*, 215–220. [CrossRef]
58. Santos, W.D.J.R.; Santhiago, M.; Yoshida, I.V.P.; Kubota, L.T. Electrochemical sensor based on imprinted sol-gel and nanomaterial for determination of caffeine. *Sens. Actuators B Chem.* **2012**, *166–167*, 739–745. [CrossRef]
59. Wang, Y.; Ding, Y.; Li, L.; Hu, P. Nitrogen-doped carbon nanotubes decorated poly (L-Cysteine) as a novel, ultrasensitive electrochemical sensor for simultaneous determination of theophylline and caffeine. *Talanta* **2018**, *178*, 449–457. [CrossRef]
60. Chojnowska, S.; Baran, T.; Wilińska, I.; Sienicka, P.; Cabaj-Wiater, I.; Knaś, M. Human saliva as a diagnostic material. *Adv. Med. Sci.* **2018**, *63*, 185–191. [CrossRef]
61. Spătaru, N.; Sarada, B.V.; Tryk, D.A.; Fujishima, A. Anodic voltammetry of xanthine, theophylline, theobromine and caffeine at conductive diamond electrodes and its analytical application. *Electroanalysis* **2002**, *14*, 721–728. [CrossRef]
62. Cinková, K.; Zbojeková, N.; Vojs, M.; Marton, M.; Samphao, A. Švorc Electroanalytical application of a boron-doped diamond electrode for sensitive voltammetric determination of theophylline in pharmaceutical dosages and human urine. *Anal. Methods* **2015**, *7*, 6755–6763. [CrossRef]
63. Jesny, S.; Girish Kumar, K. Non-enzymatic Electrochemical Sensor for the Simultaneous Determination of Xanthine, its Methyl Derivatives Theophylline and Caffeine as well as its Metabolite Uric Acid. *Electroanalysis* **2017**, *29*, 1828–1837. [CrossRef]
64. Compton, R.G.; Banks, C.E. *Understanding Voltammetry*, 2nd ed.; World Scientific: Singapore, 2010; ISBN 9781848165878.
65. Petrucci, R.; Zollo, G.; Curulli, A.; Marrosu, G. A new insight into the oxidative mechanism of caffeine and related methylxanthines in aprotic medium: May caffeine be really considered as an antioxidant? *Biochim. Biophys. Acta Gen. Subj.* **2018**, *1862*, 1781–1789. [CrossRef] [PubMed]
66. Hansen, B.H.; Dryhurst, G. Electrochemical oxidation of theophylline at the pyrolytic graphite electrode. *J. Electroanal. Chem.* **1971**, *32*, 405–414. [CrossRef]
67. Wang, T.; Randviir, E.P.; Banks, C.E. Detection of theophylline utilising portable electrochemical sensors. *Analyst* **2014**, *139*, 2000–2003. [CrossRef] [PubMed]
68. Sunil Paul, M.M.; Aravind, U.K.; Pramod, G.; Saha, A.; Aravindakumar, C.T. Hydroxyl radical induced oxidation of theophylline in water: A kinetic and mechanistic study. *Org. Biomol. Chem.* **2014**, *12*, 5611–5620. [CrossRef] [PubMed]
69. Vinchurkar, M.S.; Rao, B.S.M.; Mohan, H.; Mittal, J.P. Kinetics, spectral and redox behaviour of OH adducts of methylxanthines: A radiation chemical study 1. *J. Chem. Soc. Perkin Trans.* **1999**, *2*, 609–618. [CrossRef]
70. Santos, P.M.P.; Silva, S.A.G.; Justino, G.C.; Vieira, A.J.S.C. Demethylation of theophylline (1,3-dimethylxanthine) to 1-methylxanthine: The first step of an antioxidising cascade. *Redox Rep.* **2010**, *15*, 138–144. [CrossRef]
71. Sun, S.; Jiang, J.; Pang, S.; Ma, J.; Xue, M.; Li, J.; Liu, Y.; Yuan, Y. Oxidation of theophylline by Ferrate (VI) and formation of disinfection byproducts during subsequent chlorination. *Sep. Purif. Technol.* **2018**, *201*, 283–290. [CrossRef]
72. Zhu, Y.H.; Zhang, Z.L.; Pang, D.W. Electrochemical oxidation of theophylline at multi-wall carbon nanotube modified glassy carbon electrodes. *J. Electroanal. Chem.* **2005**, *581*, 303–309. [CrossRef]
73. Geremia, K.L.; Seybold, P.G. Computational estimation of the acidities of purines and indoles. *J. Mol. Model.* **2019**, *25*. [CrossRef]
74. Santos, P.M.P.; Telo, J.P.; Vieira, A.J.S.C. Structure and redox properties of radicals derived from one-electron oxidised methylxanthines. *Redox Rep.* **2008**, *13*, 123–133. [CrossRef]
75. Bard, A.J.; Faulkner, L.R. *Electrochemical Methods: Fundamentals and Applications*; Wiley: New York, NY, USA, 2000; ISBN 9788181892638.
76. Compton, R.G.; Laborda, E.; Ward, K.R. *Understanding Voltammetry: Simulation of Electrode Processes*; World Scientific: Singapore, 2013; ISBN 9781783263257.
77. Guidelli, R.; Compton, R.G.; Feliu, J.M.; Gileadi, E.; Lipkowski, J.; Schmickler, W.; Trasatti, S. Defining the transfer coefficient in electrochemistry: An assessment (IUPAC Technical Report). *Pure Appl. Chem.* **2014**, *86*, 245–258. [CrossRef]
78. Batchelor-Mcauley, C.; Kätelhön, E.; Barnes, E.O.; Compton, R.G.; Laborda, E.; Molina, A. Recent Advances in Voltammetry. *ChemistryOpen* **2015**, *4*, 224–260. [CrossRef] [PubMed]
79. Samiec, P.; Švorc, L'.; Stanković, D.M.; Vojs, M.; Marton, M.; Navrátilová, Z. Mercury-free and modification-free electroanalytical approach towards bromazepam and alprazolam sensing: A facile and efficient assay for their quantification in pharmaceuticals using boron-doped diamond electrodes. *Sens. Actuators B Chem.* **2017**, *245*, 963–971. [CrossRef]
80. Pysarevska, S.; Dubenska, L.; Plotycya, S.; Švorc, L'. A state-of-the-art approach for facile and reliable determination of benzocaine in pharmaceuticals and biological samples based on the use of miniaturized boron-doped diamond electrochemical sensor. *Sens. Actuators B Chem.* **2018**, *270*, 9–17. [CrossRef]
81. Ojani, R.; Alinezhad, A.; Abedi, Z. A highly sensitive electrochemical sensor for simultaneous detection of uric acid, xanthine and hypoxanthine based on poly(L-methionine) modified glassy carbon electrode. *Sens. Actuators B Chem.* **2013**, *188*, 621–630. [CrossRef]

82. Švorc, L'.; Haššo, M.; Sarakhman, O.; Kianičková, K.; Stanković, D.M.; Otřísal, P. A progressive electrochemical sensor for food quality control: Reliable determination of theobromine in chocolate products using a miniaturized boron-doped diamond electrode. *Microchem. J.* **2018**, *142*, 297–304. [CrossRef]
83. Raymundo-Pereira, P.A.; Shimizu, F.M.; Coelho, D.; Piazzeta, M.H.O.; Gobbi, A.L.; Machado, S.A.S.; Oliveira, O.N. A Nanostructured Bifunctional platform for Sensing of Glucose Biomarker in Artificial Saliva: Synergy in hybrid Pt/Au surfaces. *Biosens. Bioelectron.* **2016**, *86*, 369–376. [CrossRef]
84. Gupta, V.K.; Jain, A.K.; Shoora, S.K. Multiwall carbon nanotube modified glassy carbon electrode as voltammetric sensor for the simultaneous determination of ascorbic acid and caffeine. *Electrochim. Acta* **2013**, *93*, 248–253. [CrossRef]
85. Ren, W.; Luo, H.Q.; Li, N.B. Simultaneous voltammetric measurement of ascorbic acid, epinephrine and uric acid at a glassy carbon electrode modified with caffeic acid. *Biosens. Bioelectron.* **2006**, *21*, 1086–1092. [CrossRef]
86. Perera, V.; Gross, A.S.; Xu, H.; McLachlan, A.J. Pharmacokinetics of caffeine in plasma and saliva, and the influence of caffeine abstinence on CYP1A2 metrics. *J. Pharm. Pharmacol.* **2011**, *63*, 1161–1168. [CrossRef] [PubMed]
87. Newton, R.; Broughton, L.J.; Lind, M.J.; Morrison, P.J.; Rogers, H.J.; Bradbrook, I.D. Plasma and Salivary Pharmacokinetics of Caffeine in Man. *Eur. J. Clin. Pharmacol.* **1981**, *21*, 45–52. [CrossRef] [PubMed]

antioxidants

MDPI

Article

Does Herbal and/or Zinc Dietary Supplementation Improve the Antioxidant and Mineral Status of Lambs with Parasite Infection?

Klaudia Čobanová [1,*], Zora Váradyová [1], Ľubomíra Grešáková [1], Katarína Kucková [1], Dominika Mravčáková [1] and Marián Várady [2,*]

1 Centre of Biosciences of the Slovak Academy of Sciences, Institute of Animal Physiology, 040 01 Košice, Slovakia; varadyz@saske.sk (Z.V.); gresakl@saske.sk (Ľ.G.); kuckova@saske.sk (K.K.); mravcakova@saske.sk (D.M.)

2 Institute of Parasitology, Slovak Academy of Sciences, 040 01 Košice, Slovakia

* Correspondence: boldik@saske.sk (K.Č.); varady@saske.sk (M.V.)

Received: 16 November 2020; Accepted: 21 November 2020; Published: 24 November 2020

Abstract: This study was conducted to evaluate the effect of feed supplementation with a medicinal herbs mixture (Hmix) and organic zinc (Zn), alone or in combination, on the antioxidant responses and mineral status of lambs infected with the gastrointestinal nematode parasite *Haemonchus contortus*. A total of 24 experimentally infected lambs were randomly allocated to 1 of 4 dietary treatments ($n = 6$). The diets included an unsupplemented control diet (CON) and the CON further supplemented with Hmix, Zn, or both Hmix + Zn. Antioxidant enzymes activities, lipid peroxidation, total antioxidant capacity (TAC) and microelement (Zn, Cu, Fe, Mn) concentrations were analyzed in serum, liver, kidney, and intestinal mucosa. Zinc treatment elevated the superoxide dismutase activities in the duodenal mucosa and ileal TAC. Intake of Hmix resulted in higher kidney and ileal catalase activity and also influenced the TAC of the liver and intestinal mucosa. The inclusion of Hmix or Zn alone into the diet increased glutathione peroxidase activity in the blood, liver and duodenal mucosa. Tissue mineral uptake was not affected by herbal supplementation. Organic Zn intake increased the serum and liver Zn levels and influenced the Cu concentration in duodenal mucosa. Dietary supplementation with Hmix and/or Zn might promote the antioxidant status of lambs infected with *Haemonchus* spp.

Keywords: herbal treatment; organic zinc; lamb; *Haemonchus contortus*; antioxidant enzymes; lipid peroxidation; mineral status

---

1. Introduction

Parasitic gastrointestinal nematodes lead to an adverse impact on health and productivity in livestock, thus causing huge economic losses worldwide. The blood-sucking parasite *Haemonchus contortus* is in particular the predominant pathogenic endoparasite of small ruminants [1]. Nowadays, various trace elements or herbal nutraceuticals which have anthelmintic properties and can be successfully serve as an alternative to chemical control of parasites have been identified [2–4].

The presence of the parasite may induce the formation of a large number of reactive molecules derived from oxygen, such as superoxide radicals, hydroxyl radicals, and hydrogen peroxide [5]. The production of reactive species by phagocytes is a defensive mechanism of the host against the parasites but it promotes oxidative stress, which may also cause serious tissue damage to animals when an imbalance between free radicals' formation and the capacity of the antioxidant defense system occurs [5,6]. The use of medicinal herbs containing bioactive compounds and trace elements may boost a host's antioxidant response to combat oxidative stress and prevent or attenuate the pathological transition caused by parasitic infection [2,7,8].

Due to their redox properties and chemical structure, flavonoids and other phenolic compounds exert antioxidant activity by means of several mechanisms precisely described mainly in vitro [9]. However, the antioxidant effect of polyphenols/flavonoids in biological systems may depend on the efficiency of their absorption and extensive metabolic transformation, the active concentration of flavonoids and their metabolites accumulated in the target tissues as well as on surrounding chemical environment, which varies according to tissues and physiological conditions [10]. For these reasons, the effect of medicinal herbs, rich in plant secondary metabolites with bioactive properties, on the gut antioxidant status and animal health in general needs further elucidation.

Several trace elements are integral components of various antioxidant enzymes and play an essential role as components of antioxidant defense against free radical-induced tissue damage [11]. The presence of gastrointestinal nematode infection can seriously disturb the mineral metabolism of ruminants. In haemonchosis, iron metabolism can be strongly affected, particularly due to anemia in infected animals, and parasitism has a depressive effect on blood copper level in sheep [4,12]. Microelements, such as zinc, copper, and selenium, are essential for the development of functional immunity against parasites [13].

Zinc belongs to the essential trace elements that act as a co-factor for enzymes important for ensuring the proper functioning of the antioxidant defense system. Physiological or adequate levels of zinc protect cells against oxidative damage through the stabilization of cell membranes, inhibition of pro-oxidant enzymes, induction of antioxidant system response and metallothionein synthesis, all of which play a central role in oxidoreductive cellular metabolism [14,15]. Pivoto et al. [7] reported that zinc edetate administered subcutaneously could contribute to the alleviation of the oxidative stress induced by *H. contortus* in lambs.

Some researchers have reported that supplementation with organic Zn sources (amino acid complexes) may improve the growth rate of lambs and increase apparent absorption and retention of zinc, indicating higher bioavailability from organic sources also due to higher zinc concentration in serum and tissues as compared to inorganic source [16,17]. Van Valin et al. [18] found similar zinc retention in lambs regardless of the supplemental Zn source or dietary Zn concentration and pointed out the importance of elucidating dietary factors affecting zinc bioavailability in ruminants. Several studies concerning the interactions between various bioactive plant compounds and mineral absorption in the intestine or their deposition have been reported, with different results depending on the polyphenols used and the microelement examined [19,20]. Plant polyphenolic compounds can form complexes with metal ions which are stable over a wide pH range and throughout the entire gastrointestinal tract but which can inhibit mineral absorption and their bioavailability in ruminants [21].

Information on the antioxidant response of lambs suffering from nematode infection to feed supplementation with herbal nutraceuticals and/or zinc has thus been limited, and to the best of our knowledge, no studies have been undertaken to examine whether the use of dietary additives alone or in combination can affect microelements tissue uptake. Considering the occurrence of oxidative stress in lambs experimentally infected with the gastrointestinal nematode *H. contortus*, this article aims to assess the efficacy of medicinal herbs and organic zinc in improving animal antioxidant status and in influencing microelement (Zn, Cu, Fe, and Mn) levels in the liver, kidney, and intestinal mucosa.

2. Materials and Methods

2.1. Experimental Design and Feeding

All animals were handled in accordance with the guidelines for animal experiments set out in the European Community Directive (2010/63/EU). The study design and all procedures were approved by the Ethics Committee of the Institute of Parasitology of the Slovak Academy of Sciences in accordance with national legislation in Slovakia—Animal Welfare Act No. 23/2009. The animals used in this study were part of comprehensive research aimed at the impact of herbal nutraceuticals and/or zinc for controlling haemonchosis in experimentally infected lambs, and some parts, including the experimental conditions, have been previously described in detail [22].

A total of 24 female lambs (Improved Valachian), 3–4 months old and weighing an average 15.12 ± 1.58 kg, were housed in common stalls with free access to water. At the onset of the experiment, the animals had 7 days to adapt to the environment and consumed meadow hay (ad libitum) and the same commercial concentrate (500 g dry matter (DM)/day/animal). The concentrate was composed of 370 g/kg of wheat bran, 200 g/kg of soybean meal, 230 g/kg of rolled oats and 200 g/kg of maize meal, and the daily ration was divided into two portions offered at ca. 7:00 AM and 3:00 PM. The chemical composition of the meadow hay, concentrate (with and without Zn supplement) and herbal mix are summarized in Table 1. After the adaptation period, all animals were checked for parasite infection using the McMaster method [23]. Subsequently, parasite-free lambs were experimentally infected with approximately 5000 third-stage larvae of *Haemonchus contortus* administered orally to each lamb. For infection, we used the MHco1 strain of *H. contortus*, which is susceptible to all the main classes of anthelmintics. The infection was successful in all infected animals, as was previously shown by Váradyová et al. [22]. Thereafter, all infected lambs were randomly assigned to 1 of 4 dietary treatment groups of six animals in each. The number of animals used in the experiment was assigned according to VICH GL13 guidelines proposed by the European Medicines Agency. The diets included the unsupplemented control diet (CON) and this diet further supplemented with either a Herbal mixture (Hmix), Zn-chelate of glycine hydrate (Zn), or both additives (Hmix + Zn). The animals were housed in collective stalls according to the dietary treatment (n = 6/stall) and the experimental period lasted 70 days (during summer).

Table 1. Chemical composition of the meadow hay, concentrates and herbal mix.

Analyzed Composition	MH	C	C + Zn	Hmix
Dry matter (g/kg)	900	878	876	905
Neutral-detergent fiber (g/kg DM)	651	136	254	532
Acid detergent fiber (g/kg DM)	556	83	93	452
Crude protein (g/kg DM)	163	309	352	207
Nitrogen (g/kg DM)	27	49	56	33
Ash (g/kg DM)	91	29	30	84
Fat (g/kg DM)	21	13	12	26
Microelements Content (mg/kg DM)				
Zinc	45.1	25.4	86.0	26.4
Iron	147.0	68.3	69.8	414.3
Copper	7.0	7.1	8.1	9.9
Manganese	77.7	24.1	25.8	45.1
Phytochemical Content (g/kg DM)				
Phenolic acids	n.d.	n.d.	n.d.	3.55
Flavonoids	n.d.	n.d.	n.d.	9.96
Diterpenes	n.d.	n.d.	n.d.	4.89

MH: Meadow hay, C: Concentrate, C + Zn: Concentrate and zinc chelate of glycine hydrate, Hmix: Herbal mix, DM: Dry matter, n.d.: Not determined.

The herbal mixture (Hmix) used in our study was a non-commercial product composed of herbs typical for Central Europe, which were chosen based on the information about their phytotherapeutic properties from traditional ethnomedicine practice. The Hmix was prepared as a mixture of 9 different dry herbs obtained from commercial sources (AGROKARPATY, Plavnica, Slovak Republic, and BYLINY Mikeš s.r.o, Čičenice, Czech Republic). The mix of medicinal herbs consisted of 11.8% each of *Althaea officinalis* L. (root), *Petasites hybridus* L. (root), *Inula helenium* L. (root), *Plantago lanceolata* L. (leaf), *Rosmarinus officinalis* L. (leaf), *Solidago virgaurea* L. (stem), *Fumaria officinalis* L. (stem), *Hyssopus officinalis*

L. (stem), and 5.6% *Foeniculum vulgare* Mill. (seed). Quantitative analysis of the bioactive compounds in Hmix identified three main groups: Flavonoids (54%) with a high concentration of quercetin, verbascoside and luteolin; diterpenes (27%) with a high concentration of carnosic acid and carnosol; and phenolic acids (19%) with a high concentration of rosmarinic acid and chlorogenic acid [22]. The lambs were fed the Hmix in the amount of 100 g DM/day/animal which was mixed daily with commercial concentrate (with or without Zn supplement) during the experimental period.

The organic zinc source (Zn-chelate of glycine hydrate, Glycinoplex-Zn 26%; Phytobiotics Futterzusatzstoffe GmbH, Eltville, Germany) was directly mixed with the commercial concentrate to provide an additional 60 mg Zn/kg and the concentration was analytically confirmed in triplicates (86.0 mg Zn/kg DM). Consumption of the experimental diets was visually controlled after each feeding and no dietary refusal was recorded. Throughout the whole experiment, the animals had free access to fresh potable water and a specially-prepared mineral lick without zinc offered to each lamb once a week. The mineral lick consisted of (g/kg) 16.2 Ca, 316 Na, 32 Mg, 0.7 Cu, 2.5 Mn, 0.06 Co, 0.02 I, and 0.01 Se.

2.2. Sample Collection

At the end of the experiment, the jugular blood samples were collected from fasting lambs into heparinized tubes in the early morning, just before the sacrifice of each animal. Serum samples were obtained from blood collected into 10-mL serum-separate tubes (Sarstedt AG &Co, Nümbrecht, Germany) and the serum was then separated by centrifugation at 1200× *g* for 10 min at room temperature. The blood and sera samples were stored at −80 °C until analysis. All animals were euthanized (abattoir of the Centre of Biosciences of SAS, Institute of Animal Physiology, Košice, Slovakia, No. SK U 06018) using an overdose of pentobarbital (Dolethal, Vetoquinol, Towcester, Northamptonshire UK). The samples of liver, kidneys and small intestine were excised immediately after death. The duodenal, jejunal and ileal sections were thoroughly rinsed in an ice-cold saline solution (0.9% NaCl, pH 7.0), and the mucosa of these intestinal segments was scraped off using a glass slide onto the ice. All tissue samples were collected at the same site of relevant tissue in all the animals, cleaned thoroughly of adhering tissues, quickly frozen and stored individually in plastic bags at −80 °C until they were analyzed.

2.3. Sample Analysis

2.3.1. Chemical Composition of Feed

Feed samples (meadow hay, concentrate, concentrate + Zn and Hmix) were analyzed in triplicate for dry matter (No. 967.03), nitrogen (No. 968.06), crude protein (No. 990.03), ash (No. 942.05), fat (No. 983.23), acid-detergent fiber and neutral-detergent fiber contents according to standard procedures [24,25], as previously described [22]. The bioactive compounds of the Hmix presented in Table 1 were analyzed by ultra-high-resolution mass spectrometry using a Dionex UltiMate 3000RS system (Thermo Scientific, Darmstadt, Germany) with a charged aerosol detector connected to a high-resolution quadrupole time-of-flight mass spectrometer (HR/Q-TOF/MS, Impact II, Bruker Daltonic GmbH, Bremen, Germany) [22].

2.3.2. Preparation of Tissue Homogenates

For the determination of antioxidant parameters, samples from each tissue were separated into three parts. Each tissue piece was weighed into a test tube and homogenized separately using a homogenizer (Polytron® PT 1600 E, Kinematica AG, Switzerland). One part was homogenized in a cold 50 mM phosphate buffer (pH 7.0) containing 1 mM EDTA to give a 5% homogenate (*w*/*v*). After centrifugation at 10,000× *g* for 20 min at 4 °C (Boeco U-320 R, Hamburg, Germany), the supernatants were separated and used for the analysis of glutathione peroxidase (GPx) activity, catalase (CAT) activity, and total antioxidant capacity (TAC). For determination of superoxide dismutase (SOD) and

Cu/Zn SOD activity, homogenization was carried out in an ice-cold 10 mM TRIS buffer containing 0.25 M sucrose (pH 7.4) to make a 10% homogenate (*w*/*v*). The homogenization was performed at full speed for 30 s for 8 cycles. The homogenates were then centrifuged at 10,000× *g* for 30 min at 4 °C, and the resulting supernatant was used for enzyme assays. The last portion of the tissue sample, used for estimating lipid oxidation products by means of the TBARS (thiobarbituric acid reactive substances) method, was homogenized for 15 s at high speed in deionized distilled water (DDW) to make a 10% homogenate (*w*/*v*), and butylated hydroxytoluene (BHT, 7.2%) dissolved in ethanol (95%) was added before homogenization [26].

2.3.3. Antioxidant Enzyme Activity Assays

Total superoxide dismutase (SOD, EC 1.15.1.1) activity was determined by spectrophotometric assay according to the procedure of Marklund and Marklund [27]. This method is based on the ability of SOD to inhibit the pyrogallol autoxidation at alkaline pH. Briefly, an aliquot of tissue extract was mixed with Tris-cacodylic buffer (50 mM, pH 8.2) containing diethylenetriamine pentaacetic acid (1 mM) and a pyrogallol solution (5 mM) and subsequently the absorbance was monitored at 420 nm for 3 min. using a UV/visible spectrophotometer (Shimadzu UV-2550, Kyoto, Japan). The activity was expressed as the amount of enzyme that inhibits the oxidation of pyrogallol by 50%, which is equal to 1 unit per mg of protein. The activity of Cu/Zn SOD was assayed by inhibition of cytosolic SOD with 1 mM KCN under the same experimental conditions.

The activity of glutathione peroxidase (GPx, EC 1.11.1.9) in tissues was performed by measuring the oxidation of NADPH according to Paglia and Valentine [28]. The assays were carried out at 25 °C and hydrogen peroxide was used as the substrate. The absorbance was read at a wavelength of 340 nm and enzymatic activity was presented as U/g protein. One unit of GPx activity was defined as the amount of sample needed to oxidize 1 µmol of NADPH per min at 25 °C. The activity of blood GPx and hemoglobin (Hb) content was analyzed using commercial kits from Randox, UK.

The activity of catalase (CAT, EC 1.11.1.6) was determined according to the method of Aebi [29], which involves monitoring the rate of disappearance of H_2O_2 in the presence of tissue homogenate at 240 nm. The tissue extract was reacted with freshly prepared 10 mM H_2O_2 in 50 mM phosphate buffer (pH 7.0). One unit of CAT activity is defined as the decomposition of 1 µmol of H_2O_2 per minute at room temperature. Results are expressed in units per mg of tissue protein.

2.3.4. Lipid Oxidation and Total Antioxidant Capacity Determination

A modified fluorometric TBARS method was used to determine lipid oxidation in the serum and tissues, as described by Jo and Ahn [26]. Samples (250 µL) of serum or tissue homogenates were mixed with 100 µL of sodium dodecyl sulfate (8.1%), 750 µL of HCl (0.5 M), 750 µL of thiobarbituric acid (TBA, 20 mM), 25 µL of BHT (7.2 %) and 250 µL of DDW in test tubes and incubated for 15 min in a water bath at 90 °C. After cooling for 10 min, 2.5 mL of n-butanol/pyridine solution (15:1, *v*/*v*) and 0.5 mL of DDW was added and the samples were centrifuged at 3000× *g* for 15 min. The fluorescence of the upper layer was read at 520-nm excitation and 550-nm emission in a fluorometer (Shimadzu RF-1501, Kyoto, Japan). A standard curve was prepared using 1,1,3,3-tetramethoxypropane (malondialdehyde-bis), and the extent of lipid peroxidation was expressed as µmol of malondialdehyde (MDA) formed per L of serum or nmol MDA per g of tissue protein.

The total antioxidant capacities (TAC) of the serum, liver, kidney and intestinal mucosa (supernatant obtained as described above) were performed by ferric reducing antioxidant power (FRAP) assay using the method of Benzie and Strain [30], with some modifications. The Fe^{3+} reduction rate of the tissue sample was measured in a FRAP reagent prepared by mixing 0.3 M acetate buffer with pH 3.6, 10 mM 2,4,6-tris(2-pyridyl)-*s*-triazine and 20 mM $FeCl_3.6H_2O$ (10:1:1) at 37 °C and 593 nm. Absorbance was measured after 30 minutes and the results were calibrated with the FRAP value of the standard ($FeSO_4.7H_2O$) and expressed in µmol Fe^{2+} per g of tissue protein or mmol Fe^{2+} per liter of sera. Total protein in the tissue homogenates was assayed by the method of Bradford [31], with bovine serum albumin as a standard.

2.3.5. Trace Elements Measurement

All tissue and feed samples were dried (105 °C for 48 h), ground and digested with a concentrated HNO_3 (65%) and H_2O_2 (30%) mixture (3:1) using closed pressure vessels in a MWS 4 Speedwave microwave (Berghof Company, Eningen, Germany). The serum samples were only diluted with 0.05% Triton X-100 solution [32]. Trace element (Zn, Cu, Fe and Mn) concentrations were analyzed by flame atomic absorption spectrometry in an air-acetylene flame using a double beam atomic absorption spectrophotometer (AA-7000 Series, Shimadzu Co., Kyoto, Japan) with deuterium background correction. The certificate reference material of lyophilized human plasma ClinCheck Control (Recipe, Munich, Germany), bovine liver BCR-185R and pig kidney ERM-BB186 (Institute for Reference Materials and Measurements, Geel, Belgium) were routinely run in each analysis to verify the precision of the analysis. Mineral concentration in all tissue samples and dietary components was expressed as mg/kg DM and in serum as mg/L.

2.3.6. Statistical Analysis

All statistical analyses were performing using the GraphPad Prism software (GraphPad Prism version 8.4.2., GraphPad Software, San Diego, CA, USA). Differences between diets with and without additives were analyzed by two-way analysis of variance (ANOVA). Experimental data were analyzed as a 2 × 2 factorial in a randomized complete block design representing two main factors: Hmix (with and without) and zinc (with and without), followed by the Tukey post-hoc test for pairwise multiple comparisons, where appropriate. The Least Significant Difference test (Fisher's LSD) was applied post-hoc to determine significant differences among the treatments in case of a significant interaction (Hmix × Zn). For each analysis, the individual lamb was considered as an experimental unit. The data presented are the mean values and pooled standard errors of the mean (SEM). Differences were considered significant when $p \leq 0.05$.

3. Results

3.1. *Antioxidant Enzyme Activity*

The data showing the antioxidant enzyme activities in tissues of infected lambs are presented in Table 2. No effect of Hmix and/or Zn supplementation on the liver total SOD activity was detected, while the Hmix × Zn interaction ($p = 0.014$) affected Cu/Zn SOD activity in the liver, with increased activity in the Hmix group compared to CON ($p < 0.01$). The activity of both SODs did not change in the kidney tissue, but a tendency towards increased SOD activities (SOD, $p = 0.052$; Cu/Zn SOD, $p = 0.063$) in the Hmix treatments was determined.

The effect of Zn intake was observed in the activity of total SOD and Cu/Zn SOD ($p = 0.002$, $p = 0.001$, respectively) in the duodenal mucosa, with the highest activity recorded in lambs fed Zn alone over those fed the CON diet. Neither the total SOD nor Cu/Zn SOD activity in jejunal and ileal mucosa was influenced by the experimental diets.

An interaction of Hmix × Zn was identified for GPx activity in the liver ($p = 0.004$), duodenal mucosa ($p < 0.001$) and blood ($p = 0.020$), and increased GPx activity was found in lambs fed diets with Hmix or Zn alone as compared to those receiving the CON diet or the Hmix and Zn combination (Table 2 and Table 4). GPx activity in the kidney, jejunal and ileal mucosa was not affected by the treatment.

Feeding the Hmix diets significantly increased CAT activity in the kidney ($p = 0.025$) and ileal mucosa ($p = 0.05$), while no effect of dietary treatments on CAT activity in the liver, duodenal, and jejunal mucosa was observed.

Table 2. The effect of dietary supplementation with the herbal mixture and/or organic zinc on the activity of antioxidant enzymes in the tissues of infected lambs (n = 6).

Enzyme Activity	Dietary Treatment				SEM	p-Value		
	CON	Hmix	Zn	Hmix + Zn		Hmix	Zn	Hmix × Zn
Total SOD (U/mg protein)								
Liver	58.23	71.31	67.60	65.98	2.035	0.148	0.601	0.068
Kidney	42.20	47.57	46.44	48.19	0.918	0.052	0.173	0.304
Duodenum	5.87 ^{a}	6.19 ^{ab}	8.49 ^{b}	7.81 ^{ab}	0.366	0.771	**0.002**	0.423
Jejunum	5.82	7.10	7.13	7.45	0.264	0.119	0.106	0.337
Ileum	4.92	6.39	5.88	5.98	0.236	0.093	0.545	0.141
Cu/Zn SOD (U/mg protein)								
Liver	50.16 ^{A}	66.76 ^{B}	59.77 ^{AB}	57.34 ^{AB}	2.065	0.060	0.979	**0.014**
Kidney	31.83	36.64	34.96	35.73	0.681	0.063	0.443	0.170
Duodenum	3.52 ^{a}	3.68 ^{a}	5.36 ^{b}	4.58 ^{ab}	0.231	0.403	**0.001**	0.221
Jejunum	3.32	4.34	4.50	4.60	0.206	0.151	0.072	0.240
Ileum	3.23	3.90	3.46	3.38	0.136	0.280	0.592	0.173
GPx (U/g protein)								
Liver	27.96 ^{AB}	31.00 ^{B}	32.90 ^{B}	24.90 ^{A}	1.013	0.159	0.734	**0.004**
Kidney	37.83	42.22	39.95	39.78	1.111	0.366	0.945	0.331
Duodenum	27.35 ^{A}	32.02 ^{B}	32.90 ^{B}	28.54 ^{A}	0.636	0.862	0.258	**<0.001**
Jejunum	25.26	26.88	26.09	23.72	0.792	0.818	0.479	0.233
Ileum	25.53	27.96	26.44	28.00	0.955	0.328	0.814	0.830
CAT (U/mg protein)								
Liver	45.70	54.92	56.50	58.70	2.672	0.293	0.184	0.514
Kidney	53.86	58.17	51.89	66.46	2.162	**0.025**	0.428	0.203
Duodenum	2.41	2.61	2.50	1.82	0.124	0.296	0.139	0.067
Jejunum	1.55	1.76	1.87	1.96	0.093	0.414	0.181	0.761
Ileum	1.59	1.63	1.30	1.65	0.054	**0.050**	0.172	0.127

CON: Control, Hmix: Herbal mix, Zn: Zinc chelate of glycine hydrate, SOD: Total superoxide dismutase activity, Cu/Zn SOD: Cu/Zn Superoxide dismutase, GPx: Glutathione peroxidase activity, CAT: Catalase. ^{a,b} Means with different superscript letters in a row are significantly different ($p < 0.05$) using Tukey's post hoc test. ^{A,B} Means with different superscript letters in a row are significantly different ($p < 0.05$) using Fisher's Least Significant Difference (LSD) post hoc test. Bold values denote statistical significance at $p < 0.05$.

3.2. Lipid Oxidation and Total Antioxidant Capacity

As shown in Table 3, there were no effects of Hmix and Zn supplementation or interaction between feed additives on lipid oxidation in the tissues. The kidney tissue of lambs fed the diets supplemented with Zn showed a tendency towards lower MDA values ($p = 0.055$). There was an interaction between Hmix and Zn supplementation ($p = 0.01$) on lipid oxidation in the serum (Table 4). The MDA values were decreased in serum of lambs fed diets with Hmix ($p < 0.01$) or Zn ($p < 0.05$) alone or their combination ($p < 0.05$) as compared to those receiving the non-supplemented diet.

Table 3. The effect of dietary supplementation with the herbal mixture and/or organic zinc on lipid oxidation and total antioxidant capacity in the tissues of infected lambs ($n = 6$).

Antioxidant Parameters	Dietary Treatment				SEM	*p*-Value		
	CON	Hmix	Zn	Hmix + Zn		Hmix	Zn	Hmix × Zn
MDA (nmol/g protein)								
Liver	73.48	72.01	77.32	74.50	3.403	0.770	0.666	0.927
Kidney	55.08	48.91	47.09	44.30	1.663	0.164	0.055	0.590
Duodenum	48.38	43.11	48.03	45.47	1.971	0.353	0.809	0.745
Jejunum	82.30	72.39	84.40	91.39	3.806	0.849	0.179	0.278
Ileum	66.52	65.13	67.95	64.61	1.751	0.532	0.904	0.796
TAC (μmol/g protein)								
Liver	46.73	50.47	49.00	54.93	1.243	**0.049**	0.160	0.639
Kidney	41.31	40.58	39.08	45.97	1.276	0.229	0.531	0.140
Duodenum	23.02 ^{A}	27.30 ^{B}	25.46 ^{B}	25.68 ^{B}	0.492	**0.011**	0.617	**0.020**
Jejunum	24.23	27.45	27.25	27.88	0.564	0.074	0.108	0.220
Ileum	24.03 ^{a}	25.40 ^{ab}	27.44 ^{b}	30.04 ^{c}	0.553	**0.004**	**<0.001**	0.329

CON: Control, Hmix: Herbal mix, Zn: Zinc chelate of glycine hydrate, MDA: Malondialdehyde, TAC: Total antioxidant capacity. ^{a,b,c} Means with different superscript letters in a row are significantly different ($p < 0.05$) using Tukey's post hoc test. ^{A,B} Means with different superscript letters in a row are significantly different ($p < 0.05$) using Fisher's Least Significant Difference (LSD) post hoc test. Bold values denote statistical significance at $p < 0.05$.

Table 4. Antioxidant activity and microelements (Zn, Cu, Fe) concentration in serum of infected lambs supplemented with the herbal mixture and/or organic zinc ($n = 6$).

Serum Parameters	Dietary Treatment				SEM	*p*-Value		
	CON	Hmix	Zn	Hmix + Zn		Hmix	Zn	Hmix × Zn
GPx (U/g Hb)	381.08 ^{A}	464.21 ^{B}	476.86 ^{B}	419.44 ^{AB}	15.162	0.648	0.369	**0.020**
MDA (μmol/L)	0.279 ^{B}	0.187 ^{A}	0.203 ^{A}	0.221 ^{A}	0.011	0.074	0.298	**0.010**
TAC (mmol/L)	0.336	0.349	0.341	0.355	0.005	0.199	0.579	0.921
Zinc (mg/L)	0.740 ^{a}	0.810 ^{ab}	0.850 ^{ab}	0.903 ^{b}	0.021	0.109	**0.012**	0.823
Copper (mg/L)	0.815	0.893	0.857	0.845	0.037	0.675	0.967	0.572
Iron (mg/L)	1.418	1.940	1.858	1.920	0.107	0.180	0.329	0.286

CON: Control, Hmix: Herbal mix, Zn: Zinc chelate of glycine hydrate, GPx: Glutathione peroxidase activity, Hb: Hemoglobin, MDA: Malondialdehyde, TAC: Total antioxidant capacity. ^{a,b} Means with different superscript letters in a row are significantly different ($p < 0.05$) using Tukey's post hoc test. ^{A,B} Means with different superscript letters in a row are significantly different ($p < 0.05$) using Fisher's Least Significant Difference (LSD) post hoc test. Bold values denote statistical significance at $p < 0.05$.

The total antioxidant capacity of the liver ($p = 0.049$) and the duodenal mucosa ($p = 0.011$) was affected by Hmix intake. Duodenal TAC was also significantly affected by the Hmix × Zn interaction ($p = 0.020$) and was significantly increased in all supplemented groups compared with the CON group. Antioxidant capacity of the ileal mucosa was affected by both the Hmix ($p = 0.004$) and Zn ($p < 0.001$), with the highest values occuring in lambs fed a combination of Hmix and Zn compared to the CON group ($p < 0.001$) and both experimental treatments ($p < 0.05$). No significant effects on the TAC of serum, kidney, and jejunal mucosa were observed due to treatment (Tables 3 and 4).

3.3. Mineral Profile

Serum Zn concentration was affected by Zn supplementation ($p = 0.012$), with the highest Zn level occurring in the Hmix + Zn treatment compared to the CON group ($p < 0.05$). There were no effects of Hmix and Zn supplements on the levels of Cu and Fe in the serum (Table 4).

Mineral (Zn, Mn, Cu, Fe) concentrations in tissues are presented in Table 5. Intake of the diets enriched with Zn increased the concentration of Zn in the liver, with the highest values found in lambs fed diets with the Hmix and Zn combination over those fed the Hmix alone ($p < 0.05$). The Zn concentration in the other tissues of the lambs was not affected by the dietary treatment.

Table 5. Microelements concentration in the tissues of infected lambs supplemented with the herbal mixture and/or organic zinc ($n = 6$).

Mineral Concentration	Dietary Treatment				SEM	*p*-Value		
	CON	Hmix	Zn	Hmix + Zn		Hmix	Zn	Hmix × Zn
Zinc (mg/kg DM)								
Liver	118.34 ^{ab}	102.32 ^{a}	122.84 ^{ab}	129.20 ^{b}	3.653	0.471	**0.026**	0.104
Kidney	105.23	106.42	104.16	106.16	1.481	0.617	0.835	0.897
Duodenum	98.61	97.72	96.27	98.54	0.700	0.635	0.599	0.282
Jejunum	98.48	96.76	98.18	100.35	1.687	0.950	0.647	0.589
Ileum	98.26	98.51	95.93	97.94	0.824	0.512	0.400	0.609
Copper (mg/kg DM)								
Liver	295.88	350.36	295.90	322.67	17.420	0.270	0.704	0.704
Kidney	16.44 ^{A}	17.41 ^{AB}	18.56 ^{B}	17.06 ^{A}	0.286	0.613	0.099	**0.024**
Duodenum	7.03	6.84	8.08	7.64	0.220	0.462	**0.039**	0.777
Jejunum	7.44	9.04	8.15	8.04	0.242	0.116	0.749	0.074
Ileum	11.71	12.20	11.85	11.81	0.182	0.563	0.751	0.493
Iron (mg/kg DM)								
Liver	81.62	93.78	87.40	94.11	2.810	0.106	0.591	0.631
Kidney	104.28	107.23	103.06	107.70	2.671	0.505	0.948	0.881
Duodenum	143.38 ^{AB}	178.28 ^{AB}	215.41 ^{B}	103.97 ^{A}	13.395	0.109	0.961	**0.004**
Jejunum	59.54	65.22	64.16	64.76	1.713	0.383	0.561	0.480
Ileum	56.96	64.29	61.45	61.22	1.352	0.195	0.792	0.168
Manganese (mg/kg DM)								
Liver	11.43	10.35	11.48	11.42	0.294	0.347	0.354	0.403
Kidney	4.12	4.26	4.46	4.53	0.078	0.495	0.063	0.819
Duodenum	22.81	19.24	17.40	18.44	1.071	0.555	0.155	0.286
Jejunum	6.08	6.57	5.59	6.22	0.225	0.231	0.361	0.874
Ileum	7.33	7.80	6.53	7.98	0.263	0.071	0.544	0.345

CON: Control, Hmix: Herbal mix, Zn: Zinc chelate of glycine hydrate, DM: Dry matter. ^{a,b} Means with different superscript letters in a row are significantly different ($p < 0.05$) using Tukey's post hoc test. ^{A,B} Means with different superscript letters in a row are significantly different ($p < 0.05$) using Fisher's Least Significant Difference (LSD) post hoc test. Bold values denote statistical significance at $p < 0.05$.

An interaction ($p = 0.024$) between Hmix and Zn was recorded for the Cu concentration in the kidney, with an increased Cu level in lambs fed diets with Zn alone in comparison with those fed the CON diet ($p < 0.01$) or the one supplemented with the Hmix and Zn combination ($p < 0.05$). The dietary inclusion of Zn resulted in an elevated Cu concentration in the duodenal mucosa ($p = 0.039$). The Cu concentrations in the liver and intestinal mucosa of the jejunum and ileum were not influenced by the Zn and Hmix treatment.

In the duodenal mucosa, there was observed a Hmix × Zn interaction ($p = 0.004$) for the concentration of Fe, with increased values in lambs fed the diet with Zn alone compared to those receiving the Hmix and Zn combination ($p < 0.01$). Other tissue Fe concentrations were not changed by the dietary treatment. The Mn concentration in tissues did not differ among the dietary treatments; however, there was a trend of increased Mn level in the kidneys of lambs fed the Zn diets ($p = 0.063$).

4. Discussion

Our previous study in lambs infected with the gastrointestinal nematode *H. contortus* indicate the anthelmintic potential of the medicinal herbs and zinc due to a reduction in the fecal egg count and abomasal worm burden; in addition, the clinical signs of haemonchosis, such as anemia, were improved by treatment. However, no significant effects on body weight and mean live-weight gain of infected lambs were observed throughout the experiment due to Hmix and/or Zn treatment [22].

The occurrence of oxidative stress in lambs experimentally infected by *H. contortus* has been documented mainly due to the detection of lipid peroxidation, decreased GPx activity and TAC in plasma or serum [2,33] as well as a decreased level of reduced glutathione and SOD activity in erythrocytes [34] or elevated plasma lactate dehydrogenase level [35]. Data on the antioxidant status in tissues such as liver, kidney and intestinal mucosa of nematode-infected lambs treated with herbal nutraceuticals and organic Zn are currently lacking.

Zn is a co-factor for the enzymatic activity of cytosolic Cu/Zn superoxide dismutase (Cu/Zn-SOD), but in this study dietary Zn supplementation did not affect the activity of this Zn-containing metalloenzyme in tissues, except for the duodenal mucosa. In contrast, Pal et al. [36] reported a positive correlation between plasma and liver Cu/Zn SOD activity and Zn levels in ewes fed an organic mineral source. Approximately 75–85% of all plasma zinc is bound to albumin, which acts as a transport protein for essential metal ions to allow their systemic distribution [37]. Serum albumin, which is a major Zn-carrier, is reduced in a host due to infection with the blood-sucking nematode *H. contortus* [38]. Therefore, zinc transport and distribution seem to be disturbed in lambs suffering from nematode infection and may influence Zn delivery to cells of target tissues for the synthesis of cytosolic zinc-containing Cu/Zn SOD. Supplementation of diets with Zn increased the activity of total SOD and Cu/Zn SOD only in the duodenal mucosa of lambs, even though duodenal Zn concentration was not affected by the treatment. On the other hand, Zn intake increased the duodenal Cu concentration; therefore, our results point out the relationship between Cu/Zn SOD activity and Cu level in the intestinal mucosa. Regarding the functioning of this antioxidant enzyme, it is important to consider that Cu and Zn work together, and actually, the ratio of both elements rather than the concentration of either of these trace elements helps the enzyme function properly [39].

Several mechanisms of flavonoids antioxidant activity have been described in vitro, including activation of antioxidant enzymes, direct scavenging of reactive oxygen species, metal chelating activity, inhibition of oxidases, reduction of α-tocopheryl radicals, increase in uric acid levels, and antioxidant properties of low molecular antioxidants [9]. The bioavailability of dietary phenolic compounds in ruminants is still controversial and their possible antioxidant mechanism in the animal tissues is unclear. Several studies suggest that only some dietary phenolic compounds are bioavailable in ruminants. Moñino et al. [40] observed several of the polyphenols (i.e., rosmarinic acid, carnosol and carnosic acid) in the muscle of lambs suckling from ewes fed with a rosemary-enriched concentrate, whereas, other phenols from rosemary were not detected in the lamb tissues. According to López-Andrés et al. [41], the improvement of the antioxidant capacity of liver and plasma in lambs fed with grass phenolic compounds was not related to the direct transfer of phenolic compounds from herbage to the animal tissues. Dietary polyphenols undergo extensive biotransformation in the small intestine and in hepatic metabolism; therefore, their antioxidant effects might be expected within the gastrointestinal tract, where polyphenols may come into direct contact with cells before subsequent absorption and metabolism [10]. In our study, the intake of Hmix affected the catalase activity in the ileal mucosa and kidney. Moreover, feed supplementation with Hmix alone resulted in higher Cu/Zn SOD activity in the liver as well as increased GPx activity in the duodenal mucosa and blood when compared to unsupplemented lambs. Although tissue lipid peroxidation was not affected by dietary treatment, lower MDA values than in the control lambs were observed in the serum of all supplemented groups. Polyphenols could indirectly upregulate the antioxidant defense system, probably via activation of various transcription factors, including nuclear factor NrF2 (nuclear factor erythroid 2-related factor) and NF-κB (nuclear factor kappa B) [42,43]. Our previous study [44] showed a strong positive correlation between the total polyphenols content and antioxidant capacity of some of the medicinal plants (wormwood, chamomile, fumitory and mallow) and the dietary substrate containing a mix of these herbs possessed strong ruminal antioxidant capacity. Furthermore, strong antioxidant activity has been documented for the biologically active compounds that were the most abundant in the Hmix, such as quercetin, luteolin, carnosic acid, carnosol and rosmarinic acid [22,45–47] and probably contributed significantly to the antioxidant effect of the Hmix. It was reported that quercetin treatment can increase the activity of antioxidant enzymes (GPx, SOD, CAT, and glutathione reductase) and reduce lipid peroxidation in tissues in vivo, suggesting that the flavonoid quercetin enhances the antioxidant defense system [48,49]. For our future studies, determining the bioaccessibility of the main bioactive compounds found in herb mixture as well as their potential bioactivity and antioxidative effects ought to be addressed.

To further assess the antioxidant potential of both dietary supplements, the total antioxidant capacity of serum and tissues was determined in this experiment. Higher TAC values were observed in the liver, duodenal and ileal mucosa of lambs fed Hmix, and a similar pattern was found in the intestinal mucosa of lambs supplemented with Zn. The antioxidants that react in the FRAP assay include ascorbic acid, α-tocopherol, uric acid, bilirubin, and polyphenolic compounds, such as catechins and other flavonoids [50]. Based on the assay principle as well as the mechanism of polyphenols antioxidant action mentioned above, we assume that increased TAC in the tissues of animals treated with Hmix can likely be associated with the ability of bioactive polyphenolic compounds of herbs to increase non-enzymatic antioxidant levels. Similarly, zinc induces the synthesis of antioxidants, such as metallothioneins (MTs), which can protect cells against oxidative damage, and also influences glutathione formation by affecting the expression of glutamate–cysteine ligase [11,14]. Therefore, Zn acts directly on the neutralization of free radicals by glutathione or indirectly as a glutathione peroxidase cofactor [14]. Intake of the diet with the inclusion of either Zn or Hmix alone increased GPx activity in the blood, liver and duodenal mucosa when compared with control lambs or those fed the diet containing both Hmix and Zn. The results of the present study indicate that in terms of the antioxidant activity of GPx, the sole intake of Hmix or Zn may offer greater potential for the elimination of reactive oxygen metabolites than a combination of both additives.

Various studies have been carried out in animals [51,52] and cell culture systems [53] to determine the effect of bioactive plant compounds on mineral absorption. Stef and Gergen [51] reported generally moderate or poor correlation between total phenols and the accumulation of minerals (Zn, Cu, Fe, Mn) in the liver and muscles of chickens receiving medicinal herbs rich in a different type of polyphenols (flavonoids, phenol acids, benzoic acid derivate, phenylpropanoids derivate, and condensed tannins). Supplementation of the diet with polyphenol-rich plant products for 4 weeks had a marginal effect on the status of Zn, Fe, and Cu in piglets with a nutritionally adequate supply of those microelements [52]. Afsana et al. [54] demonstrated no effect of tannic acid supplementation on Zn, Cu, and Mn absorption in rats, but Fe absorption was reduced. There is great variability in the extent to which polyphenol-metal chelation may affect mineral bioavailability. It has been shown that the content and type of polyphenols present in the feed may have a specific influence on the accumulation of microelements in tissue [51]. In the present work, the tissue concentration of Zn, Cu, Fe and Mn was not affected by the intake of Hmix, indicating that medicinal plants containing flavonoids (54%), diterpenes (27%), and phenolic acids (19%) had no negative effect on microelements uptake in the tissues and serum of infected lambs.

Feed supplementation with organic Zn increased serum and liver levels of Zn, which play a crucial role in host immune response to a nematode infection [3,55]. Schafer et al. [3] reported a reduction of Cu and Zn levels in the liver of lambs infected with *H. contortus*; however, no significant decrease was observed in experimentally infected lamb parenterally treated with a combination of Zn and Cu. Gastrointestinal nematodes in sheep interfere with copper metabolism by causing a pH increase in the abomasal and duodenal digesta [56]. However, it seems that the mechanism of parasite interference with copper by blocking its absorption from the gastrointestinal tract is not the only mechanism by which the parasitism negatively affects Cu metabolism in the sheep [12]. In the present study, Zn supplementation increased Cu content in the duodenal mucosa and also in the kidney tissue. It is important to note the competition between copper and zinc for binding to a common transporter in enterocytes; however, this occurs especially when elevated levels of Zn are present in the diet [57]. Iskandar et al. [58] reported that improved Cu status in rats supplemented with a moderately high level of Zn correlated with increased duodenal mRNA expression of several Zn- trafficking proteins (i.e., MT-1, ZnT-1, ZnT-2, and ZnT-4). Therefore, it is possible that increased expression of these Zn-trafficking proteins may alter the distribution of Zn within absorptive enterocytes in a manner that alleviates the block in the activity of a Cu transporter. The results of the present study showed that Zn from an organic source (Zn chelate of glycine hydrate) added to the diet at the level up to the maximum authorized Zn content in complete feed (120 mg Zn/kg) [59] did not interfere with Cu absorption in the duodenal mucosa. The use of organic Zn sources rather than inorganic salts may improve the mineral supply due to their higher bioavailability, mainly because of their different route of absorption and lower interaction with other minerals and dietary components in the gastrointestinal tract [57]. Our finding is in line with the results of an experiment carried out on rabbits and chicken, which showed that the antagonism between Zn and Cu was not found when organic Zn was added to the diets [60,61].

Zinc status has a marked impact on intestinal iron absorption, metabolism and modulation of Fe homeostasis. With adequate Zn dietary intake, intestinal iron absorption and transcellular transport are stimulated via induction of divalent metal transporter-1 (DMT1) and ferroportin (FPN) expression, respectively [62]. Moreover, regulation of Fe absorption and mobilization by zinc might help to keep the cellular redox status in balance, because iron is known as a pro-oxidant while zinc is seen as an antioxidant [63]. Several dietary factors, such as polyphenols, have been shown to affect iron absorption [64]. In our study, Fe concentration in the duodenal mucosa was significantly affected by the interaction and was elevated in lambs supplemented with Zn alone compared to those fed the combination of Zn and medicinal herbs. Dietary polyphenols possess antioxidant properties, including chelation of metals, such as iron; for this reason, it is important to study whether the regular intake of bioactive polyphenolic compounds may impair iron absorption and homeostasis. Flavonoids can affect iron status by modulating the expression and activity of proteins involved in the systemic regulation of iron metabolism and uptake [64]. Several studies have established that bioactive dietary polyphenolic compounds significantly reduce iron absorption in the duodenum, but the precise mechanism of action is unknown [53,65,66]. Lesjak et al. [66] reported that quercetin treatment can influence intestinal iron uptake by inhibition of iron absorption and by decreasing the expression of duodenal DMT1 and FPN. However, Mazhar et al. [67] showed that various polyphenols regulate ferroportin expression and iron homeostasis differently, which might result from differences in their structures. Furthermore, it is required to elucidate the precise mechanism of action of herbal bioactive compounds on mineral absorption in order to improve the efficient utilization of trace elements from various sources (organic/inorganic) in ruminants with a gastrointestinal infection.

5. Conclusions

In conclusion, the inclusion of medicinal herbs into the diets of infected lambs influenced their antioxidant status due to increasing the kidney and ileal catalase activity, while zinc supplementation elevated the duodenal superoxide dismutase activity. Dietary intake of both additives alone or in combination positively affected the total antioxidant capacity of tissues. The present study showed that herbal supplementation did not significantly influence the tissue uptake of trace elements (Zn, Cu, Fe, and Mn). Feeding diets enriched with organic zinc improved the serum and liver zinc levels and increased copper concentration in the duodenal mucosa of lambs with nematode infection. Our results indicate that feed supplementation with medicinal herbs containing bioactive compounds and organic zinc may attenuate the adverse effects of parasite infection by stimulating endogenous antioxidant defense systems of small ruminants.

Author Contributions: Conceptualization, Z.V., M.V., and K.Č.; methodology, K.Č. and Ľ.G.; formal analysis, K.Č., Ľ.G., and K.K.; resources, M.V. and K.Č.; data curation, Ľ.G. and Z.V.; writing—original draft preparation, K.Č.; writing—review and editing, Z.V., D.M., and M.V.; visualization, K.Č.; supervision, M.V.; project administration, M.V. and Z.V.; funding acquisition, M.V. and K.Č. All authors have read and agreed to the published version of the manuscript.

Funding: This work was supported by the Slovak Research and Development Agency under contract number APVV-17-0297 and APVV-18-0131.

Acknowledgments: We thank David Lee McLean for providing language support that greatly improved the manuscript. The authors are grateful to V. Venglovská, R. Geročová, P. Jerga, and G. Benkovský for their excellent laboratory and technical assistance.

Conflicts of Interest: The authors declare no conflict of interest.

References

1. Hoste, H.; Torres-Acosta, J.F.J.; Quijada, J.; Chan-Perez, I.; Dakheel, M.M.; Kommuru, D.S.; Mueller-Harvey, I.; Terrill, T.H. Interactions between nutrition and infections with *Haemonchus contortus* and related gastrointestinal nematodes in small ruminants. In *Advances in Parasitology, Haemonchus Contortus and Haemonchosis—Past, Present and Future Trends*, 1st ed.; Gasser, R., Samson-Himmelstjerna, G., Eds.; Elsevier Ltd.: London, UK, 2016; Volume 93, pp. 239–351. ISBN 978-0-12-810395-1.
2. Leal, M.L.R.; Pivoto, F.L.; Fausto, G.C.; Aires, A.R.; Grando, T.H.; Roos, D.H.; Sudati, J.H.; Wagner, C.; Costa, M.M.; Molento, M.B.; et al. Copper and selenium: Auxiliary measure to control infection by *Haemonchus contortus* in lambs. *Exp. Parasitol.* **2014**, *144*, 39–43. [CrossRef] [PubMed]
3. Schafer, A.S.; Leal, M.L.R.; Molento, M.B.; Aires, A.R.; Duarte, M.M.M.F.; Carvalho, F.B.; Tonin, A.A.; Schmidt, L.; Flores, E.M.M.; Fran a, R.T.; et al. Immune response of lambs experimentally infected with *Haemonchus contortus* and parenterally treated with a combination of zinc and copper. *Small Rumin. Res.* **2015**, *123*, 183–188. [CrossRef]
4. Mravčáková, D.; Váradyová, Z.; Kopčáková, A.; Čobanová, K.; Grešáková, Ľ.; Kišidayová, S.; Babják, M.; Urda Dolinská, M.; Dvorožňáková, M.; Königová, A.; et al. Natural chemotherapeutic alternatives for controlling of haemonchosis in sheep. *BMC Vet. Res.* **2019**, *15*, 302. [CrossRef] [PubMed]
5. Kotze, A.C. Catalase induction protects *Haemonchus contortus* against hydrogen peroxide in vitro. *Int. J. Parasitol.* **2003**, *33*, 393–400. [CrossRef]
6. Celi, P. The role of oxidative stress in small ruminants' health and production. *Revista Brasileira de Zootecnia* **2010**, *39*, 348–363. [CrossRef]
7. Pivoto, F.L.; Torbitz, V.D.; Aires, A.R.; da Rocha, J.F.X.; Severo, M.M.; Grando, T.H.; Peiter, M.; Moresco, R.N.; da Rocha, J.B.T.; Leal, M.L.R. Oxidative stress by *Haemonchus contortus* in lambs: Influence of treatment with zinc edetate. *Res. Vet. Sci.* **2015**, *102*, 22–24. [CrossRef]
8. Poutaraud, A.; Michelot-Antalik, A.; Plantureux, S. Grasslands: A source of secondary metabolites for livestock health. *J. Agric. Food Chem.* **2017**, *65*, 6535–6553. [CrossRef] [PubMed]
9. Procházková, D.; Boušová, I.; Wilhelmová, N. Antioxidant and prooxidant properties of flavonoids. *Fitoterapia* **2011**, *82*, 513–523. [CrossRef]
10. Surai, P.F. Polyphenol compounds in the chicken/animal diet: From the past to the future. *J. Anim. Physiol. Anim. Nutr.* **2014**, *98*, 19–31. [CrossRef]
11. Lee, S.R. Critical role of zinc as either an antioxidant or a prooxidant in cellular system. *Oxid. Med. Cell. Longev.* **2018**. [CrossRef]
12. Adogwa, A.; Mutani, A.; Ramnanan, A.; Ezeokoli, C. The effect of gastrointestinal parasitism on blood copper and hemoglobin levels in sheep. *Can. Vet. J.* **2005**, *46*, 1017–1021. [PubMed]
13. Hughes, S.; Kelly, P. Interactions of malnutrition and immune impairment, with specific reference to immunity against parasites. *Parasite Immunol.* **2006**, *28*, 577–588. [CrossRef] [PubMed]
14. Marreiro, D.N.; Cruz, K.J.C.; Morais, J.B.S.; Beserra, J.B.; Severo, J.S.; Oliveira, A.R.S. Zinc and Oxidative Stress: Current Mechanisms. *Antioxidants* **2017**, *6*, 24. [CrossRef] [PubMed]
15. Prasad, A.S.; Bao, B. Molecular Mechanisms of Zinc as a Pro-antioxidant Mediator: Clinical Therapeutic Implications. *Antioxidants* **2019**, *8*, 164. [CrossRef] [PubMed]
16. Garg, A.K.; Vishal, M.; Dass, R.S. Effect of organic zinc supplementation on growth, nutrient utilization and mineral profile in lambs. *Anim. Feed Sci. Technol.* **2008**, *144*, 82–96. [CrossRef]
17. Kinal, S.; Slupczynska, M. The bioavailability of different chemical forms of zinc in fattening lambs. *Arch. Tierzucht* **2011**, *54*, 391–398. [CrossRef]
18. VanValin, K.R.; Genther-Schroeder, O.N.; Carmichael, R.N.; Blank, C.P.; Deters, E.L.; Hartman, S.J.; Niedermayer, E.K.; Laudert, S.B.; Hansen, S.L. Influence of dietary zinc concentration and supplemental zinc source on nutrient digestibility, zinc absorption, and retention in sheep. *J. Anim. Sci.* **2018**, *96*, 5336–5344. [CrossRef]
19. Patra, A.K. Interactions of plant bioactives with nutrient transport system in gut of livestock. *Indian J. Anim. Health* **2018**, *57*, 125–136. [CrossRef]

20. Pappas, A.C.; Zoidis, E.; Goliomytis, M.; Simitzis, P.E.; Sotirakoglou, K.; Charismiadou, M.A.; Nikitas, C.; Danezis, G.; Deligeorgis, S.G.; Georgiou, C.A. Elemental Metabolomics: Modulation of Egg Metallome with Flavonoids, an Exploratory Study. *Antioxidants* **2019**, *8*, 361. [CrossRef]
21. Naumann, H.D.; Tedeschi, L.O.; Zeller, W.E.; Huntley, N.F. The role of condensed tannins in ruminant animal production: Advances, limitations and future directions. *Rev. Bras. Zootec.* **2017**, *46*, 929–949. [CrossRef]
22. Váradyová, Z.; Mravčáková, D.; Babják, M.; Bryszak, M.; Grešáková, Ľ.; Čobanová, K.; Kišidayová, S.; Plachá, I.; Königová, A.; Cieslak, A.; et al. Effect of herbal nutraceuticals and/or zinc against *Haemonchus contortus* in lambs experimentally infected. *BMC Vet. Res.* **2018**, *14*, 78. [CrossRef] [PubMed]
23. Coles, G.C.; Bauer, C.; Borgsteede, F.H.M.; Geerts, S.; Klei, T.R.; Taylor, M.A.; Waller, P.J. World Association for the Advancement of Veterinary Parasitology (W.A.A.V.P) methods for the detection of anthelmintic resistance in nematodes of veterinary importance. *Vet. Parasitol.* **1992**, *44*, 35–44. [CrossRef]
24. AOAC (Association of Official Analytical Chemists). *Official Methods of Analysis of AOAC International*, 15th ed.; AOAC: Arlington, VA, USA, 1990.
25. Van Soest, P.J.; Robertson, J.B.; Lewis, B.A. Methods for dietary fiber, neutral detergent fiber, and nonstarch polysaccharides in relation to animal nutrition. *J. Dairy Sci.* **1991**, *74*, 3583–3597. [CrossRef]
26. Jo, C.; Ahn, D.U. Fluorometric analysis of 2-thiobarbituric acid reactive substances in turkey. *Poult. Sci.* **1998**, *77*, 475–480. [CrossRef] [PubMed]
27. Marklund, S.; Marklund, G. Involvement of the superoxide anion radical in the autoxidation of pyrogallol and a convenient assay for superoxide dismutase. *Eur. J. Biochem.* **1974**, *47*, 469–474. [CrossRef] [PubMed]
28. Paglia, D.E.; Valentine, W.N. Studies on the quantitative and qualitative characterization of erythrocyte glutathione peroxidase. *J. Lab. Clin. Med.* **1967**, *70*, 158–160.
29. Aebi, H. Catalase in vitro. *Methods Enzymol.* **1984**, *105*, 121–126.
30. Benzie, I.F.F.; Strain, J.J. The Ferric Reducing Ability of Plasma (FRAP) as a Measure of "Antioxidant Power": The FRAP Assay. *Anal. Biochem.* **1996**, *239*, 70–76. [CrossRef]
31. Bradford, M.M. A rapid and sensitive method for the quantitation of microgram quantities of protein utilizing the principle of protein-dye binding. *Anal. Biochem.* **1976**, *72*, 248–254. [CrossRef]
32. Gresakova, L.; Venglovska, K.; Cobanova, K. Dietary manganese source does not affect Mn, Zn and Cu tissue deposition and the activity of manganese-containing enzymes in lambs. *J. Trace Elem. Med. Biol.* **2016**, *38*, 138–143. [CrossRef]
33. Váradyová, Z.; Kišidayová, S.; Čobanová, K.; Grešáková, Ľ.; Babják, M.; Königová, A.; Urda Dolinská, M.; Várady, M. The impact of a mixture of medicinal herbs on ruminal fermentation, parasitological status and hematological parameters of the lambs experimentally infected with *Haemonchus contortus*. *Small Rumin. Res.* **2017**, *151*, 124–132. [CrossRef]
34. Kamel, H.H.; Mostafa, A.M.; Al-Salahy, M.B.; Walaa, M.S.; Wahba, A.A. Protein carbonyl, oxidative stress, anemia, total free amino acids and sheep haemonchosis relationship. *J. Egypt. Soc. Parasitol.* **2018**, *48*, 21–30.
35. Machado, V.; Da Silva, A.S.; Schafer, A.S.; Aires, A.R.; Tonin, A.A.; Oliveira, C.B.; Hermes, C.L.; Almeida, T.C.; Moresco, R.N.; Stefani, L.M.; et al. Relationship between oxidative stress and pathological findings in abomasum of infected lambs by *Haemonchus contortus*. *Pathol. Res. Pract.* **2014**, *210*, 812–817. [CrossRef] [PubMed]
36. Pal, D.T.; Gowda, N.K.S.; Prasad, C.S.; Amarnath, R.; Bharadwaj, U.; Suresh Babu, G.; Sampath, K.T. Effect of copper- and zinc-methionine supplementation on bioavailability, mineral status and tissue concentrations of copper and zinc in ewes. *J. Trace Elem. Med. Biol.* **2010**, *24*, 89–94. [CrossRef] [PubMed]
37. Lu, J.; Stewart, A.J.; Sadler, P.J.; Pinheiro, T.J.T.; Blindauer, C.A. Albumin as a zinc carrier: Properties of its high-affinity zinc-binding site. *Biochem. Soc. Trans.* **2008**, *36*, 1317–1321. [CrossRef] [PubMed]
38. Pathak, A.K.; Dutta, N.; Banerjee, P.S.; Pattanaik, A.K.; Sharma, K. Influence of dietary supplementation of condensed tannins through leaf meal mixture on intake, nutrient utilization and performance of *Haemonchus contortus* infected sheep. *Asian Australas. J. Anim. Sci.* **2013**, *26*, 1446–1458. [CrossRef] [PubMed]
39. Osredkar, J.; Sustar, N. Copper and Zinc, Biological Role and Significance of Copper/Zinc Imbalance. *J. Clinic. Toxicol.* **2011**, *S3*, 001. [CrossRef]
40. Moñino, I.; Martínez, C.; Sotomayor, J.A.; Lafuente, A.; Jordán, M.J. Polyphenolic transmission to Segureño lamb meat from ewes' diet supplemented with the distillate from rosemary (*Rosmarinus officinalis*) leaves. *J. Agric. Food Chem.* **2008**, *56*, 3363–3367. [CrossRef] [PubMed]

41. López-Andrés, P.; Luciano, G.; Vasta, V.; Gibson, T.M.; Scerra, M.; Biondi, L.; Priolo, A.; Mueller-Harvey, I. Antioxidant effects of ryegrass phenolics in lamb liver and plasma. *Animal* **2014**, *8*, 51–57. [CrossRef] [PubMed]
42. Hu, R.; He, Y.; Arowolo, M.A.; Wu, S.; He, J. Polyphenols as Potential Attenuators of Heat Stress in Poultry Production. *Antioxidants* **2019**, *8*, 67. [CrossRef]
43. Surai, P.F. Antioxidants in Poultry Nutrition and Reproduction: An Update. *Antioxidants* **2020**, *9*, 105. [CrossRef] [PubMed]
44. Petrič, D.; Mravčáková, D.; Kucková, K.; Čobanová, K.; Kišidayová, S.; Cieslak, A.; Slusarczyk, S.; Váradyová, Z. Effect of dry medicinal plants (wormwood, chamomile, fumitory and mallow) on in vitro ruminal antioxidant capacity and fermentation patterns of sheep. *J. Anim. Physiol. Anim. Nutr.* **2020**, *104*, 1219–1232. [CrossRef] [PubMed]
45. Dimitrios, B. Sources of natural phenolic antioxidants. *Trends Food Sci. Technol.* **2006**, *17*, 505–512. [CrossRef]
46. Wojdylo, A.; Oszmianski, J.; Czemerys, R. Antioxidant activity and phenolic compounds in 32 selected herbs. *Food Chem.* **2007**, *105*, 940–949. [CrossRef]
47. Yashin, A.; Yashin, Y.; Xia, X.; Nemzer, B. Antioxidant Activity of Spices and Their Impact on Human Health: A Review. *Antioxidants* **2017**, *6*, 70. [CrossRef] [PubMed]
48. Liu, H.; Guo, X.; Chu, Y.; Lu, S. Heart protective effects and mechanism of quercetin preconditioning on anti-myocardial ischemia reperfusion (IR) injuries in rats. *Gene* **2014**, *545*, 149–155. [CrossRef] [PubMed]
49. Xu, D.; Hu, M.-J.; Wang, Y.-Q.; Cui, Y.-L. Antioxidant Activities of Quercetin and Its Complexes for Medicinal Application. *Molecules* **2019**, *24*, 1123. [CrossRef]
50. Benzie, I.F.F.; Devaki, M. The ferric reducing/antioxidant power (FRAP) assay for non-enzymatic antioxidant capacity: Concepts, procedures, limitations and applications. In *Measurement of Antioxidant Activity & Capacity: Recent Trends and Applications*, 1st ed.; Apak, R., Capanoglu, E., Shahidi, F., Eds.; John Wiley & Sons Ltd.: Oxford, UK, 2018; pp. 77–106. ISBN 978-1-119-13535-7.
51. Stef, D.S.; Gergen, I. Effect of mineral-enriched diet and medicinal herbs on Fe, Mn, Zn, and Cu uptake in chicken. *Chem. Cent. J.* **2012**, *6*, 19. [CrossRef]
52. Fiesel, A.; Ehrmann, M.; Geßner, D.K.; Most, E.; Eder, K. Effect of polyphenol-rich plant products from grape or hop as feed supplements on iron, zinc and copper status in piglets. *Arch. Anim. Nutr.* **2015**, *69*, 276–284. [CrossRef]
53. Kim, E.-Y.; Ham, S.-K.; Shigenaga, M.K.; Han, O. Bioactive dietary polyphenolic compounds reduce nonheme iron transport across human intestinal cell monolayers. *J. Nutr.* **2008**, *138*, 1647–1651. [CrossRef]
54. Afsana, K.; Shiga, K.; Ishizuka, S.; Hara, H. Reducing effect of ingesting tannic acid on the absorption of iron, but not of zinc, copper and manganese by rats. *Biosci. Biotechnol. Biochem.* **2004**, *68*, 584–592. [CrossRef] [PubMed]
55. Scott, M.E.; Koski, K.G. Zinc deficiency impairs immune responses against parasitic nematode infections at intestinal and systemic sites. *J. Nutr.* **2000**, *130*, S1412–S1420. [CrossRef] [PubMed]
56. Bang, K.S.; Familton, A.S.; Sykes, A.R. Effect of ostertagiasis on copper status in sheep: A study involving use of copper oxide wire particles. *Res. Vet. Sci.* **1990**, *49*, 306–314. [CrossRef]
57. Goff, J.P. Invited review: Mineral absorption mechanisms, mineral interactions that affect acid-base and antioxidant status, and diet considerations to improve mineral status. *J. Dairy Sci.* **2018**, *101*, 2763–2813. [CrossRef]
58. Iskandar, M.; Swist, E.; Trick, K.D.; Wang, B.; Abbé, M.R.L.; Bertinato, J. Copper chaperone for Cu/Zn superoxide dismutase is a sensitive biomarker of mild copper deficiency induced by moderately high intakes of zinc. *Nutr. J.* **2005**, *4*, 35. [CrossRef]
59. The European Commision (EC). Commision Implementing Regulation (EU) 2016/1095. *Off. J. Eur. Union* **2016**, *59*, 7–27.
60. Ao, T.; Pierce, J.L.; Power, R.; Pescatore, A.J.; Cantor, A.H.; Dawson, K.A.; Ford, M.J. Effect of feeding different forms of zinc and copper on the performance and tissue mineral content of chicks. *Poult. Sci.* **2009**, *88*, 2171–2175. [CrossRef]
61. Čobanová, K.; Chrastinová, L'.; Chrenková, M.; Polačiková, M.; Formelová, Z.; Ivanišinová, O.; Ryzner, M.; Grešáková, L'. The effect of different dietary zinc sources on mineral deposition and antioxidant indices in rabbit tissues. *World Rabbit Sci.* **2018**, *26*, 241–248. [CrossRef]

62. Kondaiah, P.; Yaduvanshi, P.S.; Sharp, P.A.; Pullakhandam, R. Iron and Zinc Homeostasis and Interactions: Does Enteric Zinc Excretion Cross-Talk with Intestinal Iron Absorption? *Nutrients* **2019**, *11*, 1885. [CrossRef]
63. Kilari, S.; Pullakhandam, R.; Nair, K.M. Zinc inhibits oxidative stress-induced iron signaling and apoptosis in Caco-2 cells. *Free Radic. Biol. Med.* **2010**, *48*, 961–968. [CrossRef]
64. Lesjak, M.; Srai, S.K. Role of Dietary Flavonoids in Iron Homeostasis. *Pharmaceuticals* **2019**, *12*, 119. [CrossRef] [PubMed]
65. Lesjak, M.; Hoque, R.; Balesaria, S.; Skinner, V.; Debnam, E.S.; Srai, S.K.S.; Sharp, P.A. Quercetin inhibits intestinal iron absorption and ferroportin transporter expression in vivo and *in vitro*. *PLoS ONE* **2014**, *9*, 1–10. [CrossRef] [PubMed]
66. Lesjak, M.; Balesaria, S.; Skinner, V.; Debnam, E.S.; Srai, S.K.S.; Sharp, P.A. Quercetin inhibits intestinal non-haem iron absorption by regulating iron metabolism genes in the tissues. *Eur. J. Nutr.* **2019**, *58*, 743–753. [CrossRef] [PubMed]
67. Mazhar, M.; Faizi, S.; Gul, A.; Kabir, N.; Simjee, S.U. Effects of naturally occurring flavonoids on ferroportin expression in the spleen in iron deficiency anemia in vivo. *RSC Adv.* **2017**, *7*, 23238–23245. [CrossRef]

antioxidants

MDPI

Article

Changes in Phenolics and Fatty Acids Composition and Related Gene Expression during the Development from Seed to Leaves of Three Cultivated Cardoon Genotypes

Giulia Graziani [1,*,†], Teresa Docimo [2,†], Monica De Palma [2], Francesca Sparvoli [3], Luana Izzo [1], Marina Tucci [2,*] and Alberto Ritieni [1,4]

1 Department of Pharmacy, University of Naples Federico II, Via Domenico Montesano 49, 80131 Naples, Italy; luana.izzo@unina.it (L.I.); ritialb@unina.it (A.R.)
2 Institute of Bioscience and Bioresources, Consiglio Nazionale delle Ricerche, via Università 133, 80055 Portici, Italy; teresa.docimo@ibbr.cnr.it (T.D.); monica.depalma@ibbr.cnr.it (M.D.P.)
3 Institute of Agricultural Biology and Biotechnology, Consiglio Nazionale delle Ricerche, Via E. Bassini 15, 20133 Milan, Italy; sparvoli@ibba.cnr.it
4 Unesco Chair for Health Education and Sustainable Development, 80131 Naples, Italy
* Correspondence: giulia.graziani@unina.it (G.G.); mtucci@unina.it (M.T.)
† These authors contributed equally to this work.

Received: 19 October 2020; Accepted: 5 November 2020; Published: 8 November 2020

Abstract: Cultivated cardoon (*Cynara cardunculus* var. *altilis*) has long been used as a food and medicine remedy and nowadays is considered a functional food. Its leaf bioactive compounds are mostly represented by chlorogenic acids and coumaroyl derivatives, known for their nutritional value and bioactivity. Having antioxidant and hepatoprotective properties, these molecules are used for medicinal purposes. Apart from the phenolic compounds in green tissues, cultivated cardoon is also used for the seed oil, having a composition suitable for the human diet, but also valuable as feedstock for the production of biofuel and biodegradable bioplastics. Given the wide spectrum of valuable cardoon molecules and their numerous industrial applications, a detailed characterization of different organs and tissues for their metabolic profiles as well as an extensive transcriptional analysis of associated key biosynthetic genes were performed to provide a deeper insight into metabolites biosynthesis and accumulation sites. This study aimed to provide a comprehensive analysis of the phenylpropanoids profile through UHPLC-Q-Orbitrap HRMS analysis, of fatty acids content through GC-MS analysis, along with quantitative transcriptional analyses by qRT-PCR of hydroxycinnamoyl-quinate transferase (*HQT*), stearic acid desaturase (*SAD*), and fatty acid desaturase (*FAD*) genes in seeds, hypocotyls, cotyledons and leaves of the cardoon genotypes "Spagnolo", "Bianco Avorio", and "Gigante". Both oil yield and total phenols accumulation in all the tissues and organs indicated higher production in "Bianco Avorio" and "Spagnolo" than in "Gigante". Antioxidant activity evaluation by DPPH, ABTS, and FRAP assays mirrored total phenols content. Overall, this study provides a detailed analysis of tissue composition of cardoon, enabling to elucidate value-added product accumulation and distribution during plant development and hence contributing to better address and optimize the sustainable use of this natural resource. Besides, our metabolic and transcriptional screening could be useful to guide the selection of superior genotypes.

Keywords: cardoon; multipurpose plant; chlorogenic acid; fatty acids; antioxidant

1. Introduction

Cynara cardunculus L. var. *altilis* DC., the cultivated cardoon, belongs to the Asteraceae family and, with its sister species globe artichoke (*Cynara cardunculus* var. *scolymus*) and their common ancestor wild cardoon (*Cynara cardunculus* var. *sylvestris*), participates in the small *Cynara* genus. Cardoon and artichoke originate from the Mediterranean area [1], where they are mostly cultivated. Artichoke is famous as a food crop for its edible immature inflorescences (heads) and is also produced in America and Asia, while cardoon, although having a similar composition, has a restricted traditional food use in Mediterranean countries, though its cultivation for industrial purposes is increasing. Indeed, cultivated cardoon is gaining interest as a multipurpose crop, due to the versatility of its main products, which offer a wide spectrum of applications [2]. Due to its high biomass and related content of cellulose, emicellulose, and lignin, cardoon is starting to be exploited for energy production and green chemistry [3–6]. Concomitantly, cardoon extracts maintain a prominent role in the nutraceutical and pharmaceutical sectors, since numerous bioactive molecules are biosynthesized and accumulated in different parts of the plant. In particular, most of the health-beneficial compounds are polyphenols, which are present in all plant organs, though in higher amounts in leaves and seeds [7].

Cardoon biosynthesizes specialized metabolites, which are distinctive of Asteraceae [8], namely flavonoids, such as apigenin and luteolin, and hydroxycinnamic derivatives such as mono- and di-caffeoylquinic acids (CGA) [9–11]. Extraction of these cardoon bioactive compounds has been reported in many works [7,12,13] and several in vitro and in vivo studies have demonstrated the health-promoting effects related to its polyphenols. Among these polyphenols, chlorogenic acid (CGA) represents the major bioactive component because of its abundance and numerous bioactive functions, especially as a free-radical scavenger and antioxidant. These functions seem to be associated with a well-known dual role as a protectant against oxidative injury caused by free radicals [14,15] and as a substrate for both chemical and enzymatic browning reactions [9]. Apart from CGA, most cardoon phenolics confer several medicinal properties to cardoon leaf extracts, such as cholesterol-lowering, hepatoprotective, and diuretic activities, but also antifungal and antibacterial properties [16–20]. The biosynthesis of general phenylpropanoids and CGA proceeds through several sequential enzymatic reactions, starting from the amino-acid phenylalanine that, being the substrate of phenylalanine-ammonia-lyase (PAL), is the entry point for the formation of a plethora of phenylpropanoids. Afterward, cinnamate 4-hydroxylase (C4H) and 4-coumarate-CoA ligase generate *p*-coumaroyl-CoA, which is the precursor of both CGA and flavonoids. Three possible routes of CGA biosynthesis have been proposed in several plant species [21,22]. The route characterized in artichoke and other Asteraceae [23,24] proceeds through the catalytic activity of hydroxycinnamoyl-CoA quinate transferase (*HQT*) on caffeoyl-CoA and quinic acid to form CGA. Several *HQT* genes with catalytic activity towards *p*-coumaroyl-CoA have been characterized in Asteraceae species, and it cannot be excluded that several other *HQT* isoforms with a possible tissue-specific activity and specific regulation [25] are still undiscovered. In this regard, the study of the spatio-temporal expression pattern of *HQT* along with dynamic changes in CGA accumulation and distribution can provide insights into possible biosynthetic correlations and any developmental regulatory factors acting on the accumulation of this specialized metabolite.

Indeed, cardoon use is not limited to phenylpropanoids, but also oils. As oleaginous species, cardoon produces seed oils, which have been investigated both for human nutrition [26] and as a potential renewable source for the production of biofuels and bioplastics [27]. In plants, fatty acids and derived molecules represent an energy source but are also fundamental in many metabolic and signaling processes, other than being structural components of membranes and thus directly involved in defenses responses [28,29]. In human and animal nutrition, fatty acids such as oleic and linoleic are essential major nutrients since mammals cannot synthesize but only uptake them through the diet. In plants, fatty acids are synthesized in the plastids starting from acetyl-CoA, and, after passing into the cytosol, are modified and assembled in lipids in the endoplasmic reticulum (ER) [30,31]. The desaturation of stearic acid (C18:0) to oleic acid (C18:1) is catalyzed by stearoyl-acyl carrier protein desaturase (SAD). Further desaturation of oleic to linoleic acid (18:2) is catalyzed by FAD2 in the ER and FAD6 in the plastid [32]. As oleic (C18:1Δ9) and linoleic acids (C18:2Δ9,12) are the main determinants for oil quality, their content is very important to assess the nutritional properties of edible oils or direct their technological applications [33,34]. Oleic and linoleic acids have both the capability of lowering total serum cholesterols, but oleic acid, having one double bond less than linoleic, has higher oxidative stability. For this reason, high oleic acid content has also a great potential for industrial uses, for example for making products such as biodiesel and lubricants, needing high oxidative stability. In this regard, the knowledge of active genes in oleic and linoleic biosynthesis is important for selection of MUFA- (mono-unsaturated fatty acids) and/or PUFA- (poly unsaturated fatty acids) enriched tissues and genotypes.

In our study, we followed the distribution and accumulation of phenylpropanoids and oils during the growth and development of cultivated cardoon from seeds to leaf formation. We characterized the metabolic composition of seeds, hypocotyls, cotyledons, and leaves of the three cultivated cardoon genotypes "Bianco Avorio", "Spagnolo", and "Gigante" by GC and high-resolution mass spectrometric analysis, along with the tissue and developmental expression of key genes in the biosynthesis of chlorogenic acid and oleic and linoleic acid by qRT-PCR. To minimize the effects of environment and growth conditions, that are known to influence the level and composition of these metabolites [35], we grew the three cardoon genotypes in controlled greenhouse conditions, thus allowing to specifically highlight genotype differences and assist the selection of the most suitable tissues and genotypes for different purposes. Moreover, to characterize tissues and genotypes based on the biological properties, we analyzed the antioxidant activities by ABTS, DPPH, and FRAP for all the tissues.

2. Materials and Methods

2.1. Reagents and Materials

Polyphenolic standards (purity > 98%), including luteolin, apigenin, diosmin, apigenin-8-C-glucoside, naringenin, quercetin-3-*O*-glucoside, quercetin, kaempferol, myricetin, naringin, kaempferol-3-*O*-glucoside, luteolin-7-*O*-glucoside, *p*-coumaric acid, chlorogenic acid, and 4-hydroxybenzoic acid were purchased from Sigma-Aldrich Chemical Co. (St. Louis, MO, USA). The reference standards were accurately weighed and dissolved in methanol to prepare stock solutions (1 mg mL^{-1}). For the construction of calibration curves, a multi-compound working solution was prepared by appropriate dilution of stock solutions with methanol–water (80:20 *v/v*) to make concentrations in the range 0.02–5 $\mu g \cdot mL^{-1}$. For the antioxidant tests, gallic acid, 6-hydroxy-2,5,7,8-tetramethylchromane-2-carboxylic acid (Trolox), 1,1-diphenyl-2-picrylhydrazyl (DPPH), 2,3,5-triphenyltetrazolium chloride (TPTZ), anhydrous ferric chloride, hydrochloric acid, potassium persulfate ($K_2S_2O_8$), 2,2′-azino-bis(3-ethylbenzothiazoline-6-sulfonic acid) diammonium salt (ABTS) and sodium acetate were purchased from Sigma-Aldrich (MO, United States). Methanol (MeOH), ethanol, and water (LC-MS grade) were acquired from Carlo Erba reagents (Milan, Italy), whereas formic acid (98–100%) was purchased from Fluka (Milan, Italy).

2.2. Plant Growth and Sample Collection

Seeds of *C. cardunculus* L. var. *altilis* DC. "Gigante" and "Spagnolo" genotypes were provided by ARCA 2010 s.c.a.r.l., Acerra (Italy), and "Bianco Avorio" seeds were provided by La Semiorto Sementi, Lavorate di Sarno (Italy). Seeds were placed in germination pots in the greenhouse of the Institute of Bioscience and Bioresources, Portici, (Italy), and grown for the collection of the different tissues at different vegetative stages (Supplementary Figure S1). Part of the germinated seeds were used for the collection of hypocotyls and cotyledons. Hypocotyls were cut with a razor blade, approx. 5 mm below the cotyledons, about 7–10 days post-germination. Other germinated seeds were used for the collection of cotyledons, and leaves were collected 30 days after emergence. Seeds, hypocotyls, cotyledons, and leaves were flash frozen at −80 °C and stored for RNA extraction and biochemical characterization.

2.3. RNA Extraction and qRT-PCR

Seeds and tissues of cultivated cardoon were harvested in biological triplicates and ground in liquid nitrogen. Total RNA was extracted from 100 mg of tissue using the RNeasy kit (Qiagen, CA, USA). After verification of RNA integrity and NanoDrop ND-8000 spectrophotometer (Fisher Scientific, DE, USA) quantification, 1 µg of total RNA was reversed transcribed using the QuantiTec Reverse Transcription Kit (Quiagen, CA, USA), according to the manufacturer's instructions. Real-time qRT-PCR was performed with Platinum® SYBR® Green qPCR SuperMix (Applied Biosystem, CA, USA) in an ABI7900 HT (Life Technologies, CA, USA). Each PCR reaction (20 µL) contained 10 µL real-time qRT-PCR Mix, 4 µL of a 1:25 dilution of cDNA, and 0.25 µM of each specific primer designed on sequences retrieved from NCBI. Thermal cycling conditions were as follows: 50 °C for 2 min, 95 °C for 2 min, followed by 40 cycles of 15 s at 95 °C and 30 s at 60 °C. The dissociation stage was performed for each primer pairs and a single melting curve for each analyzed gene indicated the amplification of a single band. All reactions were performed on biological triplicates and technical duplicates and the $2^{-\Delta\Delta CT}$ method [36,37] was used for fold change measurements. For the relative expression of each gene in each genotype, the tissue with the lowest expression level was used as an internal calibrator by setting its expression equal to 1. *C. cardunculus* actin gene, as previously reported [38] was used to normalize gene expression. In Supplementary Table S1, the accession number and sequences of primers used for real-time qRT-PCR are reported.

2.4. Ultrasound-Assisted Extraction of Polyphenolic Compounds

Lyophilized samples were extracted using the method reported in the literature [39] with few modifications. In particular, 3 g of dried sample were extracted with 30 mL of ethanol/water (50:50 *v/v*) by sonication at room temperature for 30 min. Samples were centrifuged to 4000 rpm at 4 °C, filtered through 0.45 mm nylon syringe membranes, and then used for high-resolution mass spectrometry analysis and antioxidant activity assay. The ultrasonic plant material extraction procedure was repeated three times.

2.5. Antioxidant Activity: ABTS Assay

The antioxidant capacity assay was conducted based on the method described by [40]. Briefly, 44 µL of aqueous 2.45 mM potassium persulfate were added to aqueous 7mM ABTS and incubated in the dark at room temperature (23 °C) for 12–16 h. After dilution (1:88) with ethanol, the absorbance of this $ABTS^+$ working solution should be 0.700 ± 0.050 at 734 nm. The assay was performed by adding 0.1 mL of filtered and suitably diluted sample to 1 mL of $ABTS^+$ working solution and the absorbance was monitored at 734 nm after 2.5 min. Results were expressed as Trolox® equivalent antioxidant capacity (TEAC, mmol Trolox® equivalents Kg^{-1} dry weight of plant). All determinations were performed in triplicate.

2.6. Antioxidant Activity: DPPH Assay

The DPPH radical-scavenging activity was determined using the method proposed by [41] with minor modifications. Briefly, the DPPH methanolic solution 100 µM (1 mL) was added to polyphenol extract (0.2 mL), and the decrease in absorbance of the resulting solution was monitored at 517 nm after 10 min. The results were corrected for dilution and expressed as Trolox® equivalent antioxidant capacity (TEAC, mmol Trolox® equivalents Kg^{-1} dry weight of plant). All determinations were performed in triplicate.

2.7. Antioxidant Activity: FRAP Assay

The FRAP assay was conducted according to the method reported by [42] with slight adaptations. Briefly, the FRAP reagent contained a solution of 10 μM TPTZ in 40 μM HCl, 20 μM of aqueous $FeCl_3$ and acetate buffer (300 μM, pH 3.6) at 1:1:10 (*v/v/v*). The FRAP reagent (300 μL) and sample solutions (10 μL) were mixed and the absorbance was monitored at 593 nm after 10 min. The results were expressed in mmol Trolox® Kg^{-1} dry weight (dw). The results were corrected for dilution and expressed as Trolox® equivalent antioxidant capacity (TEAC, mmol Trolox® equivalents Kg^{-1} dry weight of plant). All determinations were performed in triplicate.

2.8. HRMS Orbitrap Analysis of Bioactive Polyphenols

An Ultra-High-Pressure Liquid Chromatograph (UHPLC, Dionex UltiMate 3000, Thermo Fisher Scientific, Ma, USA) coupled with a Q-Exactive Orbitrap mass spectrometer (UHPLC, Thermo Fischer Scientific, Ma, USA) was used to investigate the quali-quantitative profile of polyphenolic compounds applying conditions reported in our previous work [43]. A Kinetex 2.6 μm Biphenyl (100 × 2.1 mm, Phenomenex) column was applied for chromatographic separation of polyphenols with a column temperature set at 25 °C. The mobile phase consisted of water containing 0.1% of formic acid (eluent A) and methanol containing 0.1% of formic acid (eluent B). Polyphenolic compounds were eluted using the following gradient program: 0–1.3 min 5% B, 1.3–9.3 min 5–100% B, 9.3–11.3 min 100% B, 11.3–13.3 min 100–5% B, 13.3–20 min 5% B. The flow rate was 0.2 mL min^{-1} and the injection volume was 2 μL. The mass spectrometer was operated in negative ion mode (ESI^-) setting two scan events (Full ion MS and All ion fragmentation, AIF) for all compounds of interest. Full scan data were acquired setting a resolving power of 35,000 FWHM (full width at half maximum) at *m/z* 200. The key parameters were as follows: spray voltage −2.8 kV, sheath gas flow rate 35 arbitrary units, auxiliary-gas flow rate, 10 arbitrary units, capillary temperature 310 °C, auxiliary gas heater temperature 350 °C, S-lens RF level 50. For the scan event of AIF, the resolving power was set at 17,500 FWHM, the collision energies were 10, 20, and 45 eV, and the scan range was *m/z* 80–1200. Data acquisition and processing were performed with Quan/Qual Browser Xcalibur software, v. 3.1.66.10 (Xcalibur, Thermo Fisher Scientific, (Thermo Fisher Scientific, Waltham, MA, USA). Parameters of UHPLC-HRMS Orbitrap validation are reported in Supplementary Table S2.

2.9. Oil Extraction

The oil from the samples was extracted using solvent extraction according to the procedure described in the literature [44]. In particular, 100 mL of hexane were added to 10 g of powdered sample and shaken in a conical flask for 24 h. Then the solution was filtered under reduced pressure through filter paper (Whatman No. 1) and concentrated to remove completely hexane on a rotary evaporator (Rotavapor RE 120; Büchi, Flavil, Sweden). The final weight of solid material left after evaporation was recorded and used for the fatty acid profile study.

2.10. Fatty Acids Analysis

Preparation of fatty acid methyl ester derivatives (FAMEs) was carried out dissolving 0.1 g of oil sample in 1 mL hexane and 500 μL of methanolic KOH (2N). Afterward, the mixture was vortexed for 2 min at room temperature and the upper layer (hexane) was directly injected into GCMS. The fatty acid composition was investigated by GC-MS (Agilent Technologies, Santa Clara, CA, USA) using a capillary column, Rxi-5MS 5% Phenyl 95% Dimethylpolysiloxane (29.7 m × 0.25 mm i.d., 0.25 μm) (Restek, Bellefonte, PA, USA). One μL of the sample was injected in splitless mode into the injection port. Helium (grade 6.0) was used as carrier gas with a flow rate of 1.4 mL min^{-1}. The temperature program of the column was as follows: the initial oven temperature = 50 °C for 1.5 min then raising from 50 °C to 80 °C with a program ramp rate of 2.5 °C/min, the column was kept at 80 °C for 1 min then up to 300 °C at 10 °C min^{-1}, making a total run time of 36 min. The injector temperature was 280 °C.

The mass spectrometer was tuned according to the manufacturer's recommendation using perfluorotributylamine (PFTBA). The raw GC- TOF-MS data were processed by the ChromaTOF software version 5.03.09.0 (LECO Corporation, Saint Joseph, MI, USA) that deconvolutes mass spectra and performs peak identification using the NIST11 library (NIST, MD, USA). Results were expressed as mass response areas in relative percentages.

2.11. Statistical Analysis

Statistical analyses were performed using SigmaStat version 11.0 software (Sigma Stat, Statcon, Witzenhausen, Germany, SigmaStat for Windows). Mean and standard deviation were calculated on three biological and two technical replicates for all the biochemical and qRT-PCR data. Statistical differences were evaluated through a one-way analysis of variance (ANOVA). Tukey's post hoc test was used for mean separation and the statistical significance of the comparisons was defined as $p < 0.05$.

3. Results and Discussion

3.1. Polyphenol Profiling of Cardoon Genotypes

Qualitative and quantitative profiles of polyphenolic compounds were obtained using an ultra-high-performance liquid chromatography method coupled with electrospray ionization hybrid linear trap quadrupole orbitrap mass spectrometry (UHPLC-Q-Orbitrap-HRMS). Data were analyzed using Qual Browser Xcalibur 3.0 (Thermo Fisher Scientific), and identification of individual compounds was supported by retention times, exact mass spectra data, and MS/MS spectra. Individual phenolic compounds were quantified using the calibration curves of the respective reference compounds as described in Section 2.1.

The bibliographic data indicated that the chemical profile of cardoon depends on the variety [45,46] and geographical origin [47]. The UHPLC technique coupled to Orbitrap-HRMS showed that the different analyzed cardoon tissues and organs contained the same phenolic qualitative profile, but their relative abundances were considerably different between samples and genotypes. A typical full-scan MS chromatogram of a cardoon sample (cotyledons) is reported in Supplementary Figure S2. The studied cardoon parts, namely seeds, hypocotyls, cotyledons, and leaves, showed significant differences in phenolic compounds content, with leaves having the highest total phenolic content and reaching an average value of 3.99 mg g^{-1} on a dry weight (dw) basis. Table 1 summarizes the peak characteristics of fifteen detected phenolic compounds. Two phenolic acids (peaks 1 and 3), nine non-anthocyanin flavonoids (peaks 4, 5, 6, 7, 8, 9, 10, 11, 12, 13, 14, and 15), and one caffeoylquinic acid (peak 2) were identified in *C. cardunculus* samples.

Flavonoids are represented by the greatest number of compounds, among which the most abundant were apigenin-8-C-glucoside (vitexin), luteolin glucoside, naringin, and luteolin, which is in accordance with the review by [48]. Peak 1 and 3 were positively identified as 4-hydroxy benzoic and coumaric acid, respectively, by comparing their retention time and mass characteristics with commercial standards. Regarding peak 2, caffeoylquinic acid, ([M − H]$^-$ at *m/z* 353) was positively identified as 5-O-caffeoylquinic acid, yielding the base peak at *m/z* 191 and a secondary ion at *m/z* 179, characteristic of 5-acylchlorogenic acids, as reported by [49]. Peak 4 ([M − H]$^-$ at *m/z* 463) released a unique MS^2 unit at *m/z* 301 (−162 μ), corresponding to the loss of a glucosyl unit, being tentatively identified as quercetin 3-*O*-glucoside. Peak 5 was identified as apigenin 8-C-glucoside, presenting pseudomolecular ion [M − H]$^-$ at *m/z* 431 and an MS^2 fragment at *m/z* 269, corresponding to the loss of glucosyl unit. Peaks 8 and 10 were identified respectively as luteolin and kaempferol glucosides both showing pseudomolecular ion [M − H]$^-$ at *m/z* 447 (MS^2 unit at *m/z* 285). Peak 6 was identified as diosmin, characterized by the typical mass spectra in negative mode with the pseudomolecular ion [M − H]$^-$ at *m/z* 607. Peaks 7, 9, 11, 12, 13, 14, and 15 were identified as naringin, myricetin, naringenin, quercetin, luteolin, kaempferol, and apigenin, presenting pseudomolecular ions $[M - H]^-$ at *m/z* 579, 317, 271, 301, 285, 285, and 269, respectively.

The qualitative and quantitative profile of identified compounds (Table 2) varied between seeds, hypocotyls, cotyledons, and leaves of the three *C. cardunculus* analyzed genotypes. Vitexin (apigenin-8-C-glucoside) concentration in tissues and genotypes showed an increase during plant development in the three genotypes, reaching a maximum in leaves with the highest values in "Bianco Avorio" (2107.906 μg g^{-1}) (dw), followed by "Spagnolo" (2100.344 μg g^{-1}) (dw) (Table 2 and Supplementary Table S3). Similar behavior was also observed for luteolin 7-*O*-glucoside, quercetin 3-glucoside, luteolin, and kaempferol (Supplementary Table S3), possibly suggesting a developmental regulation of their production. In all tissues and genotypes, chlorogenic acid was the most represented compound, reaching concentrations between 786.031 and 3735.790 μg g^{-1} (dw). Leaves accumulated high levels of chlorogenic acid in all genotypes, especially in "Spagnolo", though the highest concentration for "Gigante" and "Bianco Avorio" was detected in hypocotyls (Supplementary Table S3). Phenylpropanoids represent essential constitutive and inducible defenses and are crucial for the safe growth and development of the plant. Therefore, these metabolites are usually allocated to protect tissues more exposed to biotic and abiotic stressors [50]. Consistently, the accumulation of total phenols was found to be high in leaves. The high accumulation of CQA precursors and other phenylpropanoids detected in hypocotyls and, to a lower extent, in cotyledons may be associated with the physiological function of these tissues. In particular, hypocotyls could be actively involved in metabolites transport to new growing tissues, actively translocating phenolics to new leaves, whereas the cotyledons could go through metabolic events to mobilize storage reserves [51]. In our tissues and genotypes, we did not detect di-caffeoyl quinic acids. Therefore, in our conditions di-caffeoyl quinic acid isomers, if any, were below the detection limit. We hypothesize that the accumulation of these compounds might depend on the physiological state of the plant and be developmentally regulated, as already reported [12]. These interesting findings remain to be further elucidated and will be the subject of future investigations.

3.2. *Qualitative and Quantitative Profile of (Seeds, Hypocotyls, Cotyledons, and Leaves) Oil*

Regarding the oil content, the results showed that the highest oil concentration is reached in the seeds, with an average level of 25.2%, in agreement with previous reports [52] but without substantial differences between the three analyzed genotypes (Figure 1). The lowest average level was found in the leaves for all genotypes, with the exception of "Gigante", for which the lowest average oil content was detected in cotyledons.

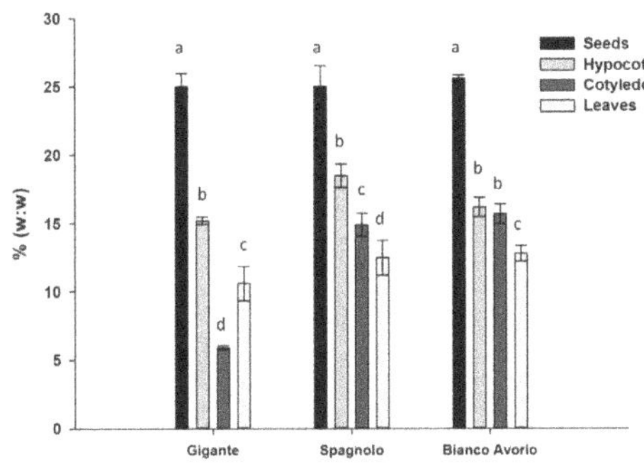

Figure 1. Percent oil content (*w:w*) in the n-exan extract in cotyledons, leaves, hypocotyls, and seeds of cardoon genotypes. Each value represents the mean of three biological and two technical replicates. Different letters denote significant differences among tissues by analysis of variance [ANOVA]. Statistical significance was defined as $p < 0.05$, using Tukey's post hoc test for mean separation.

Table 1. Retention time and exact mass spectra data of cultivated cardoon polyphenols investigated by UHPLC-HRMS Orbitrap.

Peak Number	Polyphenol	Retention Time (min)	Chemical Formula	Theoretical Mass (*m/z*)	Measured Mass (*m/z*)	Accuracy (ppm)
13	Luteolin	10.99	$C_{15}H_{10}O_6$	285.04046	285.04071	0.88
15	Apigenin	11.45	$C_{15}H_{10}O_5$	269.04555	269.04605	1.86
6	Diosmin	9.96	$C_{28}H_{32}O_{15}$	607.16684	607.16821	2.26
5	Apigenin-8-C-glucoside	9.89	$C_{21}H_{20}O_{10}$	431.09837	431.09973	3.15
11	(+/−) Naringenin	10.58	$C_{15}H_{12}O_5$	271.06120	271.06165	1.66
4	Quercitin-3-*O*-glucoside	9.71	$C_{21}H_{19}O_{12}$	463.08820	463.08926	2.29
12	Quercetin	10.71	$C_{15}H_{10}O_7$	301.03538	301.03540	0.07
14	Kaempferol	11.31	$C_{15}H_{10}O_6$	285.04046	285.04077	1.09
9	Myricetin	10.06	$C_{15}H_{10}O_8$	317.03029	317.03119	2.84
7	Naringin	10.01	$C_{27}H_{32}O_{14}$	579.54123	579.54162	0.67
10	Kaempferol-3-*O*-glucoside	10.11	$C_{21}H_{20}O_{11}$	447.09328	447.09396	1.52
8	Luteolin-7-*O*-glucoside	10.05	$C_{21}H_{20}O_{12}$	447.09328	447.09311	−0.38
3	*p*-Coumaric acid	8.66	$C_9H_8O_3$	163.04007	163.04056	3.00
2	Chlorogenic acid	7.85	$C_{16}H_{18}O_9$	353.08781	353.08856	2.12
1	4-Hydroxybenzoic acid	6.76	$C_7H_6O_3$	137.02442	137.02489	3.43

Table 2. Polyphenols content in seeds, hypocotyls, cotyledons, and leaves of "Gigante", "Spagnolo" and "Bianco Avorio" genotypes detected by HRMS-Orbitrap. Values are expressed in mg g^{-1} (dw). Each value represents the mean of three biological and two technical replicates. Different letters denote a significant difference between genotypes within each tissue by analysis of variance [ANOVA]. Statistical significance was defined as $p < 0.05$, using Tukey's post hoc test for mean separation. nd: not detected.

Phenolic Compounds	SEEDS			HYPOCOTYLS			COTYLEDONS			LEAVES		
	Gigante	Spagnolo	B. Avorio	Gigante	Spagnolo	B. Avorio	Gigante	Spagnolo	B. Avorio	Gigante	Spagnolo	B. Avorio
4-hydroxy benzoic acid	2.750a	2.510b	2.040c	5.620a	3.760c	3.910b	12.592b	8.795c	16.771a	1.221a	0.661b	0.329c
Vitexin	107.300a	91.400b	83.700c	121.700b	111.710c	250.110a	350.432c	876.529b	1182.892a	528.396c	2100.344b	2107.906a
luteolin-7-*O*-glucoside	1.470b	2.760a	0.910c	1.610c	8.610b	12.700a	18.092c	34.299a	24.086b	81.922c	122.687a	88.773b
naringin	10.900c	11.700b	14.400a	5.710c	51.700b	67.100a	14.599b	438.433a	436.022a	0.251b	0.357a	0.038c
chlorogenic acid	1036.310c	1201.990b	1430.460a	3461.130b	1006.780c	3735.790a	786.031a	769.301a	463.390b	1468.968b	2467.679a	2637.733a
coumaric acid	1.790a	1.840a	1.820a	3.620a	3.410a	3.720a	0.180a	0.080b	0.080b	0.080c	0.560a	0.200b
quercetin-3-glucoside	0.090c	0.270b	1.160a	0.700b	0.180c	1.180a	1.587b	2.670a	1.970b	3.386a	3.618a	2.050b
diosmin	0.050a	nd	nd	nd	nd	nd	nd	nd	nd	nd	nd	nd
kaempferol-3-*O*-glucoside	2.180a	2.140a	2.200a	4.910ab	4.460b	5.000a	0.420b	0.780a	0.860a	9.960a	9.960a	6.720b
myricetin	0.440a	0.440a	0.450a	0.880a	0.890a	0.880a	2.470c	3.110b	8.470a	6.884a	6.063b	5.493c
naringenin	1.730a	1.610b	1.680ab	3.130a	3.110a	3.110a	0.120a	nd	0.040b	0.160c	0.320b	0.500a
luteolin	0.320a	0.020b	0.020b	0.040b	0.200a	nd	11.700c	37.137a	31.464b	56.040c	84.640a	72.671b
kaempferol	0.340a	0.020b	0.020b	0.040b	0.240a	nd	3.600c	4.300b	7.900a	4.800c	7.360a	5.780b
quercetin	0.820	nd	nd	nd	1.640	nd	nd	0.020a	0.020a	36.580a	21.300b	14.160c
apigenin	0.240a	0.040b	0.020c	nd	0.080a	0.040b	nd	4.580b	7.040a	2.260c	3.710b	8.080a
Total polyphenols	1166.730c	1316.740b	1538.880a	3609.090b	1196.770c	4083.540a	1201.823b	2180.034a	2181.005a	2200.908b	4829.259a	4950.433a

The fatty acid profile of the oil represents the main factor that determines the choice of using the oil either for nutritional or industrial purposes. [53] found that cardoon oil is a rich source of unsaturated fatty acids such as linoleic and oleic acids (44.5% and 42.6%, respectively), whereas saturated fatty acids such as palmitic and stearic acid were detected in lower amounts (9.8% and 3.1%, respectively). Since the oil composition is deeply influenced by genotype, climatic factors, and geographical origin [54], growing the three cardoon genotypes in the same controlled greenhouse conditions should have allowed to highlight only differences due to genetic variability. Our results are in agreement with those reported by [53,55,56], highlighting a considerable variation in fatty acid profile between the analyzed genotypes, as shown in Table 3.

Palmitic acid was the most representative fatty acid in all the analyzed tissues and genotypes (Table 3 and Supplementary Table S4). In contrast to what was reported in the literature [57], in all the studied tissues and genotypes we did not detect stearic acid. Oleic acid content was found to vary from 6.54% to 19.27% and reached the highest amount in the oil extracted from the seeds of the genotype "Gigante", but also in the other genotypes its content was higher in seeds than in all the other analyzed tissues (Table 3 and Supplementary Table S4). The content of linoleic acid varied from 9.20% to 45.43%, with hypocotyls of "Bianco Avorio" showing the highest concentration. Although the hypocotyls of the three genotypes accumulated the highest content of linoleic acid, it is worth noting that a considerable amount of linoleic acid was detected also in "Spagnolo" leaves (Table 3 and Supplementary Table S4). Finally, the content of palmitic acid varied from 34.33 to 63.32% and the highest amount was observed for the oil extracted from the cotyledons of the genotype "Gigante". The fatty acids profile differences had a significant impact on the ratio of polyunsaturated/saturated fatty acids, ranging from 0.38 to 1.42, and highlighted the highest ratio for the oil extracted from hypocotyls of all genotypes analyzed. Surprisingly, the same was found for the oil extracted from the leaves of the genotype "Spagnolo". This value is an important parameter in determining the nutritional value and functional properties of food products and oils as oils with a polyunsaturated/saturated fatty acids ratio greater than one are considered valuable edible oils.

3.3. Transcriptional Analysis of Key Biosynthetic Genes in Chlorogenic Acid and Monounsaturated Fatty Acids

To gain insight into the metabolic activities of cardoon tissues and organs, we analyzed transcripts abundance for *HQT* (hydroxycinnamoyl quinate transferase; Figure 2), as a key gene in the biosynthesis of chlorogenic acid, and for *SAD* (stearoyl acid desaturase) and *FAD2* (fatty acid desaturase) genes (Figure 3), involved in oleic and linoleic acid formation, respectively. We identified the genes coding for enzymes involved in the biosynthesis of CGAs and mono-unsaturated fatty acids from deposited sequences. *HQT* was previously isolated and characterized in *C. cardunculus*. var *altilis* and globe artichoke [23,24], whereas for putative *SAD* and *FAD2* genes, sequences were retrieved from BLAST searches in the *C. cardunculus* database (PRJNA453787) and chosen based on the high nucleotidic and proteic identity to orthologous genes in closely related Asteraceae and other oleaginous species. Regarding *SAD*, we found only two cardoon sequences, highly similar to each other, and coding for proteins closely related to characterized *Lactuca sativa* (XM 023873428.1) and *Olea europaea* var. *sylvestris* (XM 023023317.1) *SAD2*. As regards *FAD* genes, two main oleate desaturase genes occur in oleaginous plants, namely the microsomal and plastidial types [32]. In detail, two microsomial genes are characterized by tissue specificity, the seed type *FAD2.1* and the house-keeping *FAD2.2*, along with the plastidial one, known as *FAD6*. We selected three *FAD2* representatives: *CcFAD2.1* (70.08% to GmFAD2-1, AAB00859, and 77.89% to *Ha*FAD2-1, AAL68981.1), *CcFAD2.2* (90.00% to *Cp*FAD2-2, CAA76157 and 90.82% to *Ha*Fad2.2, AAL68982.1), and *FAD6* (92.88% to *Aa*FAD6, PWA37615 and 83% to *Tc*FAD6, GEU42198). Although gene expression analyses were carried out for only one gene within the phenylpropanoid pathway and for *SAD* and three *FAD2* isoforms for fatty acids biosynthesis, the selected genes represent key steps in the respective pathways. Gene transcripts were detected for all the tissues analyzed, and transcript abundances varied in tissues and genotypes underlining a tight regulation for the accumulation of these metabolites.

Table 3. GC/MS analysis of fatty acids composition in cardoon genotypes (expressed as % of the total fatty acid composition). Each value represents the mean of three biological and two technical replicates. Different letters denote a significant difference between genotypes within each tissue by analysis of variance [ANOVA]. Statistical significance was defined as $p < 0.05$, using Tukey's post hoc test for mean separation. n.d.: not detected.

Fatty Acids	Seeds			Hypocotyls			Cotyledons			Leaves		
	Gigante	Spagnolo	B. Avorio	Gigante	Spagnolo	B. Avorio	Gigante	Spagnolo	B. Avorio	Gigante	Spagnolo	B. Avorio
Pentadeconoic (C15:0)	0.123c	0.283a	0.173b	0.172b	0.182a	0.143c	1.852a	0.578c	0.711b	0.985a	0.441c	0.611b
Palmitic (C16:0)	59.868a	44.265b	56.708a	49.361a	47.722a	40.791b	63.320a	44.594c	50.348b	44.484b	34.330c	61.643a
Margaric (C17:0)	0.694b	1.25a	0.797b	0.109a	0.117a	0.098b	2.253a	0.639b	0.642b	0.882a	0.571b	0.608b
Nonadecanoic (C19:0)	0.013b	0.024a	0.01b4	n.d.	n.d.	n.d.	n.d.	0.252	n.d.	0.289	n.d.	0.187
Arachidic (C20:0)	4.446ab	4.427a	3.4b	n.d.	n.d.	0.659	n.d.	8.176a	5.157b	7.635a	3.852c	5.309b
Behenic (C22:0)	0.634a	0.635a	0.489b	0.271b	0.431a	0.315b	1.976b	3.767a	1.459c	4.162a	2.103c	3.463b
Lignoceric (C24:0)	0.457a	0.246b	0.432a	0.326	n.d.	n.d.	2.508b	5.408a	0.979c	5.219b	n.d.	5.706a
Cerotic (C26:0)	0.052	n.d.	n.d.	n.d.	n.d.	n.d.	0.465a	4.414a	n.d.	3.977	n.d.	n.d.
Melissic (C30:0)	0.015	0.03	n.d.	n.d.	n.d.	0.282	n.d.	n.d.	0.557	n.d.	n.d.	n.d.
Palmitoleic (C16:1)	0.372b	0.816a	0.348b	n.d.	0.135	n.d.	0.607a	n.d.	0.439b	1.452a	0.420c	0.516b
Hexadicadienoic (C16:2)	0.006b	1.389a	0.009b	n.d.	n.d.	n.d.	n.d.	n.d.	n.d.	n.d.	0.088	n.d.
Hexadicatrienoic (C16:3)	n.d.	n.d.	n.d.	n.d.	n.d.	n.d.	9.774	n.d.	n.d.	n.d.	n.d.	n.d.
Oleic (C18:1)	19.271a	18.395b	15.828c	6.54c	7.869b	12.278a	n.d.	9.737a	8.240b	10.524b	15.519a	6.592c
Linoleic (C18:2)	13.684c	27.34a	21.212b	37.892b	38.954b	45.434a	8.935c	12.997b	24.714a	12.806b	37.460a	9.197c
Linolenic (C18:3)	0.355a	0.441a	0.021b	5.259a	4.588b	n.d.	8.312ab	9.438a	6.644b	7.584a	5.215c	6.169b
Nonadecenoic (C19:1)	0.01	0.018	n.d.	n.d.	n.d.	n.d.	n.d.	n.d.	n.d.	n.d.	n.d.	n.d.
Gadoleic (C20:1)	n.d.	0.441b	0.566a	0.07	n.d.	n.d.	n.d.	n.d.	0.110	n.d.	n.d.	n.d.

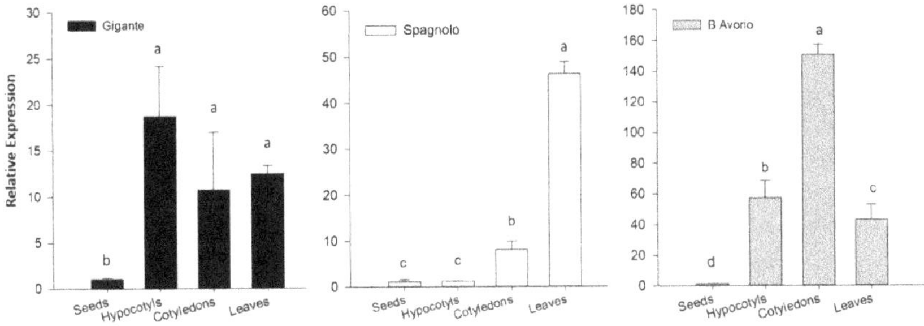

Figure 2. Transcriptional analysis of the *CcHQT* gene by qRT-PCR in seeds, hypocotyls, cotyledons and leaves of "Gigante", "Spagnolo" and "Bianco Avorio" cultivated cardoon genotypes. Results are expressed as fold changes relatively to seeds, used as internal calibrator of gene expression for each genotype. Each value represents the mean ± SD of three biological and two technical replicates. Different letters denote a significant difference between tissues by analysis of variance [ANOVA]. Statistical significance was defined as $p < 0.05$, using the Tukey's post hoc test for mean separation.

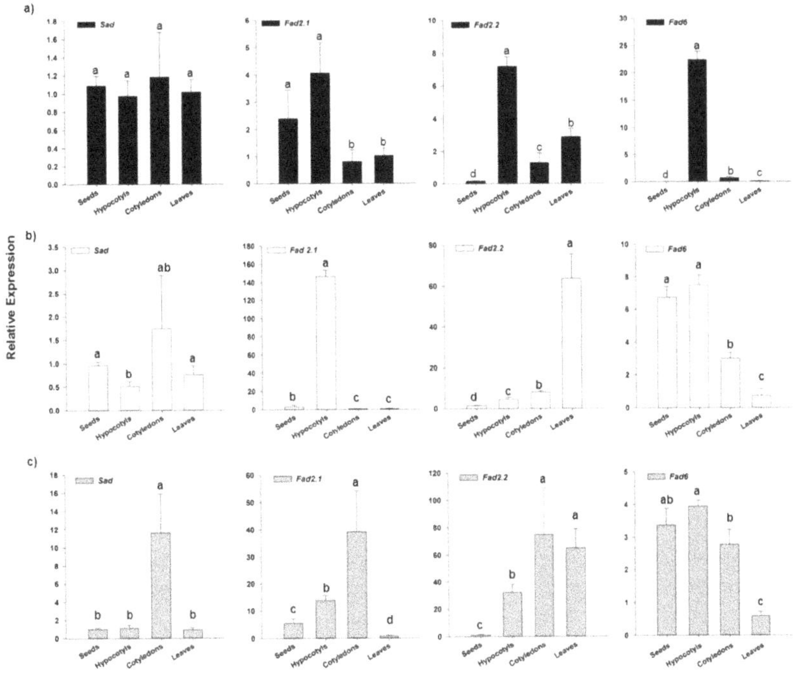

Figure 3. Transcriptional profiling of *SAD*, *FAD2.1*, *FAD2.2* and *FAD6* genes by qRT-PCR in seeds, hypocotyls, cotyledons and leaves of cardoon genotypes. (**a**) "Gigante", (**b**) "Spagnolo", (**c**) "Bianco Avorio". Results are expressed as fold changes relatively to the tissue with the lowest expression level, used as internal calibrator for each gene within each genotype. Each value represents the mean ± SD of three biological and two technical replicates. For each gene, different letters denote a significant difference between tissues within each genotype by analysis of variance [ANOVA]. Statistical significance was defined as $p < 0.05$, using the Tukey's post hoc test for mean separation.

3.3.1. *HQT* Expression Analysis

In details, transcriptional analysis of *HQT* gene expression (Figure 2) was consistent with CQA accumulation among the genotypes (Table 2), which was higher in "Bianco Avorio" than "Gigante" and "Spagnolo". Nevertheless, within each genotype, *HQT* expression did not reflect CQA accumulation for most of the analyzed tissues, in particular seeds and cotyledons. In all genotypes, *HQT* expression was barely detectable in seeds and much higher in cotyledons, while CGA levels were comparable to other tissues in seeds and the lowest in cotyledons (Figure 2 and Table 2). Due to their quiescent state, mature seeds might have a reduced transcriptional activity not temporally related with CGA accumulation, whose biosynthesis is highly regulated during seed development in order to guarantee a safe storage of energy in form of chlorogenic acid at seed maturity. Instead, in cotyledons, the high *HQT* transcriptional activity might be needed to provide high levels of CGA, which is degraded to provide lignin-precursors necessary for cotyledonary cell wall formation [58]. Regarding *HQT* expression in the other tissues we cannot exclude that additional pathways can contribute to CGA synthesis, as reported in other species [59]. Moreover, expression analysis can be only indicative of the transcriptional dynamics linked to CQA accumulation, which is mostly dependent on the enzymatic activity, as reported in other species [59,60]. We believe that *HQT* expression is also the result of a fine regulatory scheme, since this key gene is rate limiting and we cannot exclude that a feedback regulation is involved in its transcriptional rate, as recently reported [61].

3.3.2. *SAD* and *FAD2* Expression Analysis

Transcriptional analysis of key biosynthetic genes in oleic and linoleic acid formation revealed that *SAD* expression was low in seeds of all three genotypes. In "Gigante" and "Spagnolo" it was low also in the other analyzed tissues, whereas in "Bianco Avorio" *SAD* transcription was remarkable in cotyledons but not consistent with quantitative oleic acid detection (Figure 3). The weak correlation between oleic acid accumulation and *SAD* expression might be related to the developmental seed stage investigated in this study. In mature seed, oleic acid might have reached an accumulation threshold, driving a feedback inhibition of *SAD* expression in favor of oleic to linoleic conversion. This hypothesis is consistent with the higher content of linoleic acid in almost all the analyzed tissues and genotypes. Regarding the expression analysis of the FAD isogenes, we found that the expression levels of the three *FAD* genes were higher in hypocotyls than in the other tissues in the "Gigante" genotype. Consistently, high linoleic acid accumulation was detected in hypocotyls of this genotype (Table 3), possibly resulting from the contribution of all the *FAD* isoforms. In "Spagnolo", *FAD2.1* and *FAD2.2* showed a clear tissue-specificity, being highly expressed in hypocotyls and leaves, respectively, and we might hypothesize that each isoform contributed to the higher linoleic acid content of these two tissues in this genotype. As for *FAD6*, transcripts accumulation was higher in "Spagnolo" seeds and hypocotyls than in cotyledons and leaves. Regarding "Bianco Avorio", high *FAD2.1* expression was detected in cotyledons, where also *FAD2.2* accumulated at high levels, similarly to leaves, whereas *FAD6* was higher in seeds and hypocotyls. The distribution of *FAD* transcripts, in particular for *FAD2.2*, was in accordance with published literature in other oleaginous plants, indicating it as constitutive gene [59]. Indeed, *FAD2.2* was expressed in all tissues, except for dormant seeds, and, according to transcriptional data, could play a main role in linoleic acid formation in "Spagnolo" and "Bianco Avorio". Nevertheless, functional studies are necessary to ascertain the role of each gene in this species and whether it might be affected by genotype differences. Overall, our results reflect the harmonic interplay between *SAD* and *FAD* genes, the lower level of *SAD* being seemingly compensated by the higher expression of *FAD* genes, which finally regulates the MUFA/ PUFA ratio in all the cardoon tissues.

3.4. *Antioxidant Activity of Polyphenolic Extracts*

The results obtained for antioxidant activity of cultivated cardoon samples by the ABTS, FRAP and DPPH assays are presented in Table 4 and expressed as TEAC (mmol kg^{-1} dw). The measurement of the antioxidant activity of food is carried out using several methods since it is influenced by various factors. The most commonly used for the evaluation of the antioxidant potential are DPPH radical scavenging, ABTS decolorization and FRAP assays due to their simplicity, stability, accuracy and reproducibility [62].

Table 4. Antioxidant activity in seeds, hypocotyls, cotyledons and leaves of three cultivated cardoon genotypes. Each value represents the mean ± SD of three biological and two technical replicates. Different letters denote a significant difference between genotypes within each tissue by analysis of variance [ANOVA] Statistical significance was defined as $p < 0.05$, using the Tukey's post hoc test for mean separation.

GENOTYPE		DPPH	ABTS	FRAP
		mmol trolox kg^{-1}	**mmol trolox kg^{-1}**	**mmol trolox kg^{-1}**
Seeds	GIGANTE	49.492 ± 0.275c	87.237 ± 3.312c	49.833 ± 1.283c
	SPAGNOLO	143.871 ± 0.942a	135.413 ± 1.119b	59.848 ± 3.340b
	BIANCO AVORIO	137.946 ± 0.265b	148.037 ± 1.127a	107.728 ± 1.303a
Hypocotyls	GIGANTE	57.149 ± 0.377b	667.365 ± 1.036a	72.000 ± 0.954b
	SPAGNOLO	37.543 ± 0.311c	317.206 ± 2.071c	57.939 ± 1.043c
	BIANCO AVORIO	60.866 ± 0.106a	631.482 ± 8.285b	79.758 ± 2.694a
Cotyledons	GIGANTE	19.192 ± 0.123c	487.277 ± 2.111b	47.776 ± 0.987c
	SPAGNOLO	63.951 ± 0.543a	412.113 ± 0.112c	79.148 ± 2.876a
	BIANCO AVORIO	55.126 ± 0.156b	543.012 ± 0.765a	57.122 ± 1.832b
Leaves	GIGANTE	27.167 ± 0.551c	510.315 ± 0.765c	62.450 ± 0.432c
	SPAGNOLO	81.543 ± 0.111b	654.216 ± 0.981b	97.659 ± 0.876a
	BIANCO AVORIO	91.856 ± 0.116a	771.412 ± 0.665a	91.675 ± 0.761b

As for seeds, FRAP values ranged from 49.83 ± 1.28 to 107.73 ± 1.30 mmol kg^{-1} dw, whereas ABTS values ranged from 87.24 ± 53.31 to 148.04 ± 1.13 mmol kg^{-1} dw and DPPH values ranged from 49.49 ± 0.27 to 143.87 ± 0.94 mmol kg^{-1} dw. Antioxidant activity showed significant differences between the tested genotypes, with "Bianco Avorio" having the overall highest antioxidant activity. Our results are in accordance with the results obtained by [11], reporting significant differences between wild and cultivated cardoon genotypes, as well as between hydromethanolic extracts and its residues. [63], similarly to our data, estimated that seed extracts of wild cardoons showed a high DPPH$^{\bullet}$ scavenging activity and a considerable ABTS$^{\bullet+}$ radical scavenging capacity. Moreover, the correlation coefficient between phenolic contents and TEAC was highly significant especially in the case of ABTS and FRAP (r^2 = 0.82 and 0.94, respectively), indicating that polyphenolics may play an important role in free radical cations scavenging and ferric reducing antioxidant power.

Similar results were obtained for hypocotyls, where the correlation coefficients between phenolic contents and TEAC values were highly significant (r^2 = 1, 0.94, and 0.95, respectively for DPPH, ABTS and FRAP). The results obtained showed a remarkable antioxidant activity of the genotype "Bianco Avorio" also for the hypocotyls and the highest antioxidant activity of polyphenols especially as free radical cation scavengers, being the antioxidant activity evaluated with the ABTS assay the highest. For cardoon hypocotyls, in fact, FRAP values ranged from 57.94 ± 1.043 to 79.76.73 ± 2.69 mmol kg^{-1} dw, ABTS values ranged from 317.21 ± 2.071 to 667.37 ± 1. mmol kg^{-1} dw and finally DPPH values ranged from 37.54 ± 0.311 to 60.87 ± 0.106 mmol kg^{-1} dw. These results highlight a significant variability in nutraceutical potential between genotypes.

For cotyledons, the results showed a significant variability in antioxidant potential between genotypes (Table 4). FRAP values ranged from 47.78 ± 0.987 to 79.148 ± 2.876 mmol kg^{-1} dw, whereas ABTS values ranged from 412.113 ± 0.112 to 543.012 ± 0.765 mmol kg^{-1} dw and finally DPPH values ranged from 19.192 ± 0.123 to 63.951 ± 0.543 mmol kg^{-1} dw. The highest antioxidant capacities were recorded for the "Spagnolo" genotype using the DPPH and FRAP methods while, in analogy with the results described so far, "Bianco Avorio" showed the highest scavenger capacity against the ABTS radical cations. In literature, the deposition of phenolic polymers was reported for Coffea arabica cotyledons during development, suggesting that lignification may occur via the utilization of the stored 5-O-caffeoylquinic acid. This result would explain the lower antioxidant activity of the cotyledons also associated with a low level of chlorogenic acid. Finally, for cardoon leaves, FRAP values ranged from 62.45 ± 0.432 to 91.675 ± 0.761 mmol kg^{-1} dw, whereas ABTS values ranged from 510.315 ± 0.766 to 771.412 ± 0.665 mmol kg^{-1} dw and finally DPPH values ranged from 27.167 ± 0.551 to 91.856 ± 0.116 mmol kg^{-1} dw.

The nutraceutical properties of leaves of artichoke and cardoon are linked to their specific chemical composition. Caffeoylquinic acids, followed by flavonoids, are the main phenolic compounds in leaves, among which chlorogenic acid is the most abundant. The same order of activity was found in the DPPH and FRAP assays, but the absolute values were lower than those found in the ABTS assay. According to literature data [64] ABTS assay is the most sensitive assay, as the working solution is soluble in aqueous and organic solvents, at several pH values, and the quenching reaction of free radicals is faster than other methods. The antioxidant capacity of cardoon leaf extracts was reported [12] and different artichoke genotypes were studied in terms of their antioxidant capacity [65]. Antioxidant activity measured on the leaf polyphenolic extracts was remarkably higher than that reported in literature, on the other hand little change in experimental conditions can widely influence the results making the comparison only qualitative. For leaves, the correlation coefficients between phenolic contents and TEAC values were mildly significant (r^2 = 0.98, 0.83 and 0.70, respectively for DPPH, ABTS and FRAP), with the genotype "Bianco Avorio" showing the highest antioxidant activity especially as ABTS radical scavenger.

4. Conclusions

In conclusion, the results presented in this work provide a detailed analysis of tissues composition of cardoon, enabling to elucidate value-added products accumulation and distribution during plant development and hence contributing to optimize the sustainable use of this natural resource. Moreover, transcriptional and metabolic screening could be useful for a careful selection of the genotype.

Supplementary Materials: The following are available online at http://www.mdpi.com/2076-3921/9/11/1096/s1.

Author Contributions: Conceptualization, G.G., T.D.; methodology, G.G., T.D., M.D.P. and L.I.; software, G.G., T.D.; validation, G.G., T.D. and M.D.P.; formal analysis, T.D., G.G. and M.D.P.; investigation, G.G., T.D., L.I. and M.D.P.; resources, M.T. and A.R.; data curation, G.G. and T.D.; writing—original draft preparation, G.G., T.D., M.D.P., M.T. and R.T.; writing—review and editing, G.G., T.D., M.D.P., M.T., A.R.; supervision, A.R. and M.T.; project administration, A.R., F.S. and M.T.; funding acquisition, A.R., F.S. and M.T. All authors have read and agreed to the published version of the manuscript.

Funding: This research was supported by the following project: BOBCAT Biotecnologie per la produzione sostenibile di bio-materiali e prodotti specialistici utilizzando colture cellulari di cardo come bioraffineria. Fondazione Cariplo (2018_0955). In addition, part of this work was also carried out with the support of the CNR project NUTR-AGE (FOE-2019, DSB.AD004.271, Subtask 2.1.

Conflicts of Interest: The authors declare no conflict of interest.

References

1. Bianco, V.V. Present situation and future potential of artichoke in the mediterranean basin. *Acta Hortic.* **2005**, *681*, 39–58. [CrossRef]
2. Pappalardo, H.D.; Toscano, V.; Puglia, G.D.; Genovese, C.; Raccuia, S.A. *Cynara cardunculus* L. as a Multipurpose Crop for Plant Secondary Metabolites Production in Marginal Stressed Lands. *Front. Plant Sci.* **2020**, *11*, 240. [CrossRef] [PubMed]
3. Ciancolini, A.; Alignan, M.; Pagnotta, M.A.; Vilarem, G.; Crinò, P. Selection of Italian cardoon genotypes as industrial crop for biomass and polyphenol production. *Ind. Crop. Prod.* **2013**, *51*, 145–151. [CrossRef]
4. Foti, S.; Mauromicale, G.; Raccuia, S.A.; Fallico, B.; Fanella, F. Possible alternative utilization of Cynara spp. I. Biomass, grain yield and chemical composition of grain. *Ind. Crops Prod.* **1999**, *10*, 219–228. [CrossRef]

5. Petropoulos, S.; Fernandes, Â.; Pereira, C.; Tzortzakis, N.; Vaz, J.; Soković, M.; Barros, L.; Ferreira, I.C. Bioactivities, chemical composition and nutritional value of *Cynara cardunculus* L. seeds. *Food Chem.* **2019**, *289*, 404–412. [CrossRef] [PubMed]
6. Ierna, A.; Mauro, R.P.; Mauromicale, G.; Catania, S.; Lancia, S.V.; Industriale, Z.; Palma, B.; Catania, I. Biomass, grain and energy yield in *Cynara cardunculus* L. as affected by fertilization, genotype and harvest time. *Biomass Bioenergy* **2011**, *36*, 404–410. [CrossRef]
7. Falleh, H.; Ksouri, R.; Chaieb, K.; Karray-Bouraoui, N.; Trabelsi, N.; Boulaaba, M.; Abdelly, C. Phenolic composition of Cynara cardunculus L. organs, and their biological activities. *Comptes Rendus Biol.* **2008**, *331*, 372–379. [CrossRef]
8. Calabria, L.M.; Emerenciano, V.P.; Ferreira, M.J.P.; Scotti, M.T.; Mabry, T.J. A Phylogenetic Analysis of Tribes of the Asteraceae Based on Phytochemical Data. *Nat. Prod. Commun.* **2007**, *2*, 277–285. [CrossRef]
9. Lattanzio, V.; Cardinali, A.; Di Venere, D.; Linsalata, V.; Palmieri, S. Browning phenomena in stored artichoke (*Cynara scolymus* L.) heads: Enzymic or chemical reactions? *Food Chem.* **1994**, *50*, 1–7. [CrossRef]
10. Williamson, G.; Holst, B. Dietary reference intake (DRI) value for dietary polyphenols: Are we heading in the right direction? *Br. J. Nutr.* **2008**, *99*, S55–S58. [CrossRef]
11. Durazzo, A.; Foddai, M.; Temperini, A.; Azzini, E.; Venneria, E.; Lucarini, M.; Finotti, E.; Maiani, G.; Crinò, P.; Saccardo, F.; et al. Antioxidant Properties of Seeds from Lines of Artichoke, Cultivated Cardoon and Wild Cardoon. *Antioxidants* **2013**, *2*, 52–61. [CrossRef] [PubMed]
12. Gouveia, S.C.; Castilho, P.C. Phenolic composition and antioxidant capacity of cultivated artichoke, Madeira cardoon and artichoke-based dietary supplements. *Food Res. Int.* **2012**, *48*, 712–724. [CrossRef]
13. Mandim, F.; Petropoulos, S.A.; Dias, M.I.; Pinela, J.; Kostic, M.; Soković, M.; Santos-Buelga, C.; Ferreira, I.C.F.R.; Barros, L. Seasonal variation in bioactive properties and phenolic composition of cardoon (*Cynara cardunculus var. altilis*) bracts. *Food Chem.* **2020**, *336*, 127744. [CrossRef]
14. Racchi, M.; Daglia, M.; Lanni, C.; Papetti, A.; Govoni, S.; Gazzani, G. Antiradical Activity of Water Soluble Components in Common Diet Vegetables. *J. Agric. Food Chem.* **2002**, *50*, 1272–1277. [CrossRef]
15. Miller, N.J.; Rice-Evans, C.A. Spectrophotometric determination of antioxidant activity. *Redox Rep.* **1996**, *2*, 161–171. [CrossRef]
16. Cho, A.S.; Jeon, S.M.; Kim, M.J.; Yeo, J.; Seo, K.I.; Choi, M.S.; Lee, M.K. Chlorogenic acid exhibits anti-obesity property and improves lipid metabolism in high-fat diet-induced-obese mice. *Food Chem. Toxicol.* **2010**, *48*, 937–943. [CrossRef]
17. Dos Santos, M.D.; Almeida, M.C.; Lopes, N.P.; de Souza, G.E.P. Evaluation of the anti-inflammatory, analgesic and antipyretic activities of the natural polyphenol chlorogenic acid. *Biol. Pharm. Bull.* **2006**, *29*, 2236–2240. [CrossRef] [PubMed]
18. Naveed, M.; Hejazi, V.; Abbas, M.; Kamboh, A.A.; Khan, G.J.; Shumzaid, M.; Ahmad, F.; Babazadeh, D.; FangFang, X.; Modarresi-Ghazani, F.; et al. Chlorogenic acid (CGA): A pharmacological review and call for further research. *Biomed. Pharmacother.* **2018**, *97*, 67–74. [CrossRef]
19. Santana-Gálvez, J.; Cisneros-Zevallos, L.; Jacobo-Velázquez, D. Chlorogenic Acid: Recent Advances on Its Dual Role as a Food Additive and a Nutraceutical against Metabolic Syndrome. *Molecules* **2017**, *22*, 358. [CrossRef]
20. Lou, Z.; Wang, H.; Zhu, S.; Ma, C.; Wang, Z. Antibacterial Activity and Mechanism of Action of Chlorogenic Acid. *J. Food Sci.* **2011**, *76*, M398–M403. [CrossRef]
21. Niggeweg, R.; Michael, A.J.; Martin, C. Engineering plants with increased levels of the antioxidant chlorogenic acid. *Nat. Biotechnol.* **2004**, *22*, 746–754. [CrossRef]
22. Mahesh, V.; Million-Rousseau, R.; Ullmann, P.; Chabrillange, N.; Bustamante, J.; Mondolot, L.; Morant, M.; Noirot, M.; Hamon, S.; de Kochko, A.; et al. Functional characterization of two p-coumaroyl ester 3′-hydroxylase genes from coffee tree: Evidence of a candidate for chlorogenic acid biosynthesis. *Plant Mol. Biol.* **2007**, *64*, 145–159. [CrossRef]
23. Sonnante, G.; D'Amore, R.; Blanco, E.; Pierri, C.L.; De Palma, M.; Luo, J.; Tucci, M.; Martin, C. Novel hydroxycinnamoyl-coenzyme A quinate transferase genes from artichoke are involved in the synthesis of chlorogenic acid. *Plant Physiol.* **2010**, *153*, 1224–1238. [CrossRef] [PubMed]
24. Comino, C.; Hehn, A.; Moglia, A.; Menin, B.; Bourgaud, F.; Lanteri, S.; Portis, E. The isolation and mapping of a novel hydroxycinnamoyltransferase in the globe artichoke chlorogenic acid pathway. *BMC Plant Biol.* **2009**, *9*, 30. [CrossRef]

25. Cheevarungnapakul, K.; Khaksar, G.; Panpetch, P.; Boonjing, P.; Sirikantaramas, S. Identification and Functional Characterization of Genes Involved in the Biosynthesis of Caffeoylquinic Acids in Sunflower (*Helianthus annuus* L.). *Front. Plant Sci.* **2019**, *10*, 968. [CrossRef]
26. Raccuia, S.A.; Melilli, M.G. Biomass and grain oil yields in *Cynara cardunculus* L. genotypes grown in a Mediterranean environment. *Field Crop. Res.* **2007**, *101*, 187–197. [CrossRef]
27. Gominho, J.; Lourenc, A.; Palma, P.; Lourenc, M.E.; Curt, M.D.; Fernández, J.; Pereira, H. Large scale cultivation of *Cynara cardunculus L.* For biomass production—A case study. *Ind. Crops Prod.* **2011**, *33*, 1–6. [CrossRef]
28. Meï, C.; Michaud, M.; Cussac, M.; Albrieux, C.; Gros, V.; Maréchal, E.; Block, M.A.; Jouhet, J.; Rébeillé, F. Levels of polyunsaturated fatty acids correlate with growth rate in plant cell cultures. *Sci. Rep.* **2015**, *5*, 1–9. [CrossRef]
29. Upchurch, R.G. Fatty acid unsaturation, mobilization, and regulation in the response of plants to stress. *Biotechnol. Lett.* **2008**, *30*, 967–977. [CrossRef]
30. Somerville, C.; Browse, J. Plant Lipids: Metabolism, Mutants, and Membranes. *Science* **1991**, *252*, 80–87. [CrossRef]
31. Browse, J.; Somerville, C. Glycerolipid Synthesis: Biochemistry and Regulation. *Annu. Rev. Plant Biol.* **1991**, *42*, 467–506. [CrossRef]
32. Dar, A.A.; Choudhury, A.R.; Kancharla, P.K.; Arumugam, N. The FAD2 Gene in Plants: Occurrence, Regulation, and Role. *Front. Plant Sci.* **2017**, *8*, 1–16. [CrossRef]
33. Kumar, A.; Sharma, A.; Upadhyaya, K.C. Vegetable Oil: Nutritional and Industrial Perspective. *Curr. Genom.* **2016**, *17*, 230–240. [CrossRef]
34. Orsavova, J.; Misurcova, L.; Ambrozova, J.; Vicha, R.; Mlcek, J. Fatty Acids Composition of Vegetable Oils and Its Contribution to Dietary Energy Intake and Dependence of Cardiovascular Mortality on Dietary Intake of Fatty Acids. *Int. J. Mol. Sci.* **2015**, *16*, 12871–12890. [CrossRef] [PubMed]
35. Clé, C.; Hill, L.M.; Niggeweg, R.; Martin, C.R.; Guisez, Y.; Prinsen, E.; Jansen, M.A.K.K. Modulation of chlorogenic acid biosynthesis in *Solanum lycopersicum*; consequences for phenolic accumulation and UV-tolerance. *Phytochemistry* **2008**, *69*, 2149–2156. [CrossRef]
36. Pfaffl, M.W. A new mathematical model for relative quantification in real-time RT-PCR. *Nucleic Acids Res.* **2001**, *29*, e45. [CrossRef]
37. Pfaffl, M. Quantification strategies in real-time PCR. *AZ Quant. PCR* **2004**, *1*, 87–112.
38. Docimo, T.; De Stefano, R.; Cappetta, E.; Lisa Piccinelli, A.; Celano, R.; De Palma, M.; Tucci, M. Physiological, biochemical, and metabolic responses to short and prolonged saline stress in two cultivated cardoon genotypes. *Plants* **2020**, *9*, 554. [CrossRef]
39. Tungmunnithum, D.; Garros, L.; Drouet, S.; Renouard, S.; Lainé, E.; Hano, C. Green Ultrasound Assisted Extraction of trans Rosmarinic Acid from *Plectranthus scutellarioides* (L.) R.Br. Leaves. *Plants* **2019**, *8*, 50. [CrossRef]
40. Re, R.; Pellegrini, N.; Proteggente, A.; Pannala, A.; Yang, M.; Rice-Evans, C. Antioxidant activity applying an improved ABTS radical cation decolorization assay. *Free Radic. Biol. Med.* **1999**, *26*, 1231–1237. [CrossRef]
41. Brand-Williams, W.; Cuvelier, M.E.; Berset, C. Use of a free radical method to evaluate antioxidant activity. *LWT-Food Sci. Technol.* **1995**, *28*, 25–30. [CrossRef]
42. Benzie, I.F.F.; Strain, J.J. The ferric reducing ability of plasma as a measure of antioxodant. *Anal. Biochem.* **1996**, *239*, 70–76. [CrossRef]
43. Graziani, G.; Ritieni, A.; Cirillo, A.; Cice, D.; Di Vaio, C. Effects of Biostimulants on Annurca Fruit Quality and Potential Nutraceutical Compounds at Harvest and during Storage. *Plants* **2020**, *9*, 775. [CrossRef]
44. Kostić, M.D.; Joković, N.M.; Stamenković, O.S.; Rajković, K.M.; Milić, P.S.; Veljković, V.B. Optimization of hempseed oil extraction by n-hexane. *Ind. Crop. Prod.* **2013**, *48*, 133–143. [CrossRef]
45. Pandino, G.; Lombardo, S.; Mauromicale, G.; Williamson, G. Phenolic acids and flavonoids in leaf and floral stem of cultivated and wild *Cynara cardunculus* L. genotypes. *Food Chem.* **2011**, *126*, 417–422. [CrossRef]
46. Dias, M.I.; Barros, L.; Barreira, J.C.M.; Alves, M.J.; Barracosa, P.; Ferreira, I.C.F.R. Phenolic profile and bioactivity of cardoon (*Cynara cardunculus* L.) inflorescence parts: Selecting the best genotype for food applications. *Food Chem.* **2018**, *268*, 196–202. [CrossRef] [PubMed]

47. Ramos, P.A.B.; Santos, S.A.O.; Guerra, Â.R.; Guerreiro, O.; Freire, C.S.R.; Rocha, S.M.; Duarte, M.F.; Silvestre, A.J.D. Phenolic composition and antioxidant activity of different morphological parts of *Cynara cardunculus* L. var. altilis (DC). *Ind. Crop. Prod.* **2014**, *61*, 460–471. [CrossRef]
48. Lattanzio, V.; Kroon, P.A.; Linsalata, V.; Cardinali, A. Globe artichoke: A functional food and source of nutraceutical ingredients. *J. Funct. Foods* **2009**, *1*, 131–144. [CrossRef]
49. Clifford, M.N.; Johnston, K.L.; Knight, S.; Kuhnert, N. Hierarchical scheme for LC-MSn identification of chlorogenic acids. *J. Agric. Food Chem.* **2003**, *51*, 2900–2911. [CrossRef] [PubMed]
50. Tattini, M.; Loreto, F.; Fini, A.; Guidi, L.; Brunetti, C.; Velikova, V.; Gori, A.; Ferrini, F. Isoprenoids and phenylpropanoids are part of the antioxidant defense orchestrated daily by drought-stressed Platanus × acerifolia plants during Mediterranean summers. *New Phytol.* **2015**, *207*, 613–626. [CrossRef]
51. Penfield, S.; Graham, S.; Graham, I.A. Storage reserve mobilization in germinating oilseeds: Arabidopsis as a model system. *Biochem. Soc. Trans.* **2005**, *33*, 380–383. [CrossRef]
52. Gominho, J.; Curt, M.D.; Lourenço, A.; Fernández, J.; Pereira, H. *Cynara cardunculus* L. as a biomass and multi-purpose crop: A review of 30 years of research. *Biomass Bioenergy* **2018**, *109*, 257–275. [CrossRef]
53. Raccuia, S.A.; Piscioneri, I.; Sharma, N.; Melilli, M.G. Genetic variability in *Cynara cardunculus* L. domestic and wild types for grain oil production and fatty acids composition. *Biomass Bioenergy* **2011**, *35*, 3167–3173. [CrossRef]
54. Francisco, M.; Elena Cartea, M.; Maria Butron, A.; Sotelo, T.; Velasco, P. Environmental and Genetic Effects on Yield and Secondary Metabolite Production in Brassica rapa Crops. *J. Agric. Food Chem.* **2012**, *60*, 5507–5514. [CrossRef]
55. Maccarone, E.; Fallico, B.; Fanella, F.; Mauromicale, G.; Raccuia, S.A. Possible alternative utilization of Cynara spp. II. Chemical characterization of their grain oil. *Ind. Crops Prod.* **1999**, *10*, 229–237. [CrossRef]
56. Curt, M.D.; Sánchez, G.; Fernández, J. The potential of *Cynara cardunculus* L. For seed oil production in a perennial cultivation system. *Biomass Bioenergy* **2002**, *23*, 33–46. [CrossRef]
57. Angelova, V.; Nemska, M.P.; Krustev, L. Chemical composition of cardoon (*Cynara cardunculus L.*) Grown in south bulgaria. *AGROFOR Int. J.* **2019**, *4*, 100–110.
58. Volpi E Silva, N.; Mazzafera, P.; Cesarino, I. Should I stay or should I go: Are chlorogenic acids mobilized towards lignin biosynthesis? *Phytochemistry* **2019**, *166*, 112063. [CrossRef]
59. Payyavula, R.S.; Navarre, D.; Kuhl, J.C.; Pantoja, A.; Pillai, S.S. Differential effects of environment on potato phenylpropanoid and carotenoid expression. *BMC Plant Biol.* **2012**, *12*, 39. [CrossRef] [PubMed]
60. Payyavula, R.S.; Shakya, R.; Sengoda, V.G.; Munyaneza, J.E.; Swamy, P.; Navarre, D.A. Synthesis and regulation of chlorogenic acid in potato: Rerouting phenylpropanoid flux in HQT-silenced lines. *Plant Biotechnol. J.* **2015**, *13*, 551–564. [CrossRef]
61. Li, Y.; Kong, D.; Bai, M.; He, H.; Wang, H.; Wu, H. Correlation of the temporal and spatial expression patterns of HQT with the biosynthesis and accumulation of chlorogenic acid in Lonicera japonica flowers. *Hortic. Res.* **2019**, *6*, 73. [CrossRef]
62. Vijaya Kumar Reddy, C.; Sreeramulu, D.; Raghunath, M. Antioxidant activity of fresh and dry fruits commonly consumed in India. *Food Res. Int.* **2010**, *43*, 285–288. [CrossRef]
63. Kollia, E.; Markaki, P.; Zoumpoulakis, P.; Proestos, C. Antioxidant activity of *Cynara scolymus* L. and *Cynara cardunculus* L. extracts obtained by different extraction techniques. *Nat. Prod. Res.* **2017**, *31*, 1163–1167. [CrossRef]
64. Arnao, M.B. Some methodological problems in the determination of antioxidant activity using chromogen radicals: A practical case. *Trends Food Sci. Technol.* **2000**, *11*, 419–421. [CrossRef]
65. Colla, G.; Rouphael, Y.; Cardarelli, M.; Svecova, E.; Rea, E.; Lucini, L. Effects of saline stress on mineral composition, phenolic acids and flavonoids in leaves of artichoke and cardoon genotypes grown in floating system. *J. Sci. Food Agric.* **2013**, *93*, 1119–1127. [CrossRef]

Publisher's Note: MDPI stays neutral with regard to jurisdictional claims in published maps and institutional affiliations.

antioxidants

MDPI

Article

Chemico-Biological Characterization of Torpedino Di Fondi® Tomato Fruits: A Comparison with San Marzano Cultivar at Two Ripeness Stages

Cinzia Ingallina [1,†], Alessandro Maccelli [1,†], Mattia Spano [1], Giacomo Di Matteo [1], Antonella Di Sotto [2], Anna Maria Giusti [3], Giuliana Vinci [4], Silvia Di Giacomo [2], Mattia Rapa [4], Salvatore Ciano [4], Caterina Fraschetti [1], Antonello Filippi [1], Giovanna Simonetti [5], Carlos Cordeiro [6], Marta Sousa Silva [6], Maria Elisa Crestoni [1,*], Anatoly P. Sobolev [7,*], Simonetta Fornarini [1] and Luisa Mannina [1]

1 Dipartimento di Chimica e Tecnologie del Farmaco, Sapienza Università di Roma, P. le Aldo Moro 5, 00185 Rome, Italy; cinzia.ingallina@uniroma1.it (C.I.); alessandro.maccelli@uniroma1.it (A.M.); mattia.spano@uniroma1.it (M.S.); giacomo.dimatteo@uniroma1.it (G.D.M.); caterina.fraschetti@uniroma1.it (C.F.); antonello.filippi@uniroma1.it (A.F.); simonetta.fornarini@uniroma1.it (S.F.); luisa.mannina@uniroma1.it (L.M.)

2 Dipartimento di Fisiologia e Farmacologia "V. Ersparmer", Sapienza Università di Roma, P. le Aldo Moro 5, 00185 Rome, Italy; antonella.disotto@uniroma1.it (A.D.S.); silvia.digiacomo@uniroma1.it (S.D.G.)

3 Dipartimento di Medicina Sperimentale Sapienza, Università di Roma, P. le Aldo Moro 5, 00185 Rome, Italy; annamaria.giusti@uniroma1.it

4 Dipartimento di Management, Laboratorio di Merceologia, Sapienza Università di Roma, Via del Castro Laurenziano 9, 00161 Rome, Italy; giuliana.vinci@uniroma1.it (G.V.); mattia.rapa@uniroma1.it (M.R.); salvatore.ciano@uniroma1.it (S.C.)

5 Dipartimento di Biologia Ambientale, Sapienza Università di Roma, P. le Aldo Moro 5, 00185 Rome, Italy; giovanna.simonetti@uniroma1.it

6 Laboratório de FT-ICR e Espectrometria de Massa Estrutural, Faculdade de Ciências da Universidade de Lisboa, Campo-Grande, 1749-016 Lisboa, Portugal; cacordeiro@fc.ul.pt (C.C.); mfsilva@fc.ul.pt (M.S.S.)

7 Istituto per i Sistemi Biologici, Laboratorio di Risonanza Magnetica "Annalaura Segre", CNR, 00015 Rome, Italy

* Correspondence: mariaelisa.crestoni@uniroma1.it (M.E.C.); anatoly.sobolev@cnr.it (A.P.S.); Tel.: +39-06-4991-3596 (M.E.C.); +39-06-9067-2385 (A.P.S.)

† These authors gave an equal contribution to this work.

Received: 23 September 2020; Accepted: 15 October 2020; Published: 21 October 2020

Abstract: Torpedino di Fondi (TF) is a hybrid tomato landrace developed in Sicily and recently introduced in the south Lazio area along with the classical San Marzano (SM) cultivar. The present study aimed at characterizing TF tomatoes at both pink and red ripening stages, and at comparing them with traditional SM tomatoes. A multidisciplinary approach consisting of morphological, chemical (FT-ICR MS, NMR, HPLC, and spectrophotometric methods), and biological (antioxidant and antifungal in vitro activity) analyses was applied. Morphological analysis confirmed the mini-San Marzano nature and the peculiar crunchy and solid consistency of TF fruits. Pink TF tomatoes displayed the highest content of hydrophilic antioxidants, like total polyphenols (0.192 mg/g), tannins (0.013 mg/g), flavonoids (0.204 mg/g), and chlorophylls a (0.344 mg/g) and b (0.161 mg/g), whereas red TF fruits were characterized by the highest levels of fructose (3000 mg/100 g), glucose (2000 mg/100 g), tryptophan (2.7 mg/100 g), phenylalanine (13 mg/100 g), alanine (25 mg/100 g), and total tri-unsaturated fatty acids (13% mol). Red SM fruits revealed the greatest content of lipophilic antioxidants, with 1234 mg/g of total carotenoids. In agreement with phenolics content, TF cultivar showed the greatest antioxidant activity. Lastly, red TF inhibited *Candida* species (*albicans*, *glabrata* and *krusei*) growth.

Keywords: tomatoes; NMR spectroscopy; FT-ICR mass spectrometry; ripening stage; phenolics; antioxidant activity; metabolomics; phytochemicals

1. Introduction

San Marzano (SM) is a traditional tomato landrace grown in south Italy suitable for both fresh consumption and processing. The distinct SM organoleptic properties made this variety a worldwide model for tomato quality traits, although the scarcity of genetic resistance against pathogens represents a critical SM weakness [1]. Throughout the years, both natural and in vitro selections have led to new SM tomatoes with peculiar accessions; however, several ecotypes not suitable for local environments gradually disappeared. The term "San Marzano" refers to a population of tomatoes with a wide range of characteristics [2]. In this context, a new SM cultivar, namely, Torpedino di Fondi (TF) has been recently introduced in the south Lazio area. Developed in Sicily (Licata and Vittoria), TF is characterized by a peculiar sweetness and palatability and, due to its smaller size and weight compared to SM, it is defined as mini-San Marzano.

Different analytical methodologies, such as NMR, MS, GC-MS, and HPLC, have been applied to characterize different SM cultivars from chemical [2–7], sensorial [4,5], and genomic [1,4,8] points of view. However, to the best of our knowledge, TF tomatoes have not been characterized yet.

It is well established that plants and vegetable foodstuffs represent a unique reservoir of nutrients and phytochemicals with health implications [9]. Tomatoes and tomato-based food have proven to possess a wide variety of bioactive compounds that are beneficial for human well-being; among these, dietary antioxidants like carotenoids, polyphenols, and vitamins are the most abundant in tomato fruits. Indeed, carotenoids have shown to play an important role in reducing the incidence of some chronic diseases, like cancer and cardiovascular diseases [10], whereas polyphenols in tomatoes have proved to prevent the oxidative damage [11–14] in human cells. Recently, the influence of the cultivation system on the polyphenols content has been found to depend mostly on variety and year than the cultivation and drying methods [15,16].

Moreover, a recent study provided evidence on the activity of ethanolic extracts of *Solanum lycopersicum* at concentration of 6.25 mg/mL against *C. albicans*, *C. guilliermondii*, and *C. lusitaniae* isolated from HIV positive patients [17]. Conversely, extracts from tomato crop remains at the end of the cultivation cycle displayed a low antifungal activity against the microfungi *Aspergillus* and *Penicillium species* [18]. Interestingly, an antimicrobial snaking peptide (SN2) obtained from *Solanum lycopersicum* tested as a recombinant peptide in *E. coli* exhibited strong fungicidal bioactivity ascribed to biomembrane perforation [19]. *Candida* spp. is present in the gut, but an overproduction may lead to serious health problems. Some diseases, such as Crohn's disease and ulcerative colitis, are associated with an overgrowth of *Candida* in the gastrointestinal tract [20]. *Candida* overgrowth can be prevented by healthy foods.

The aim of the present study was to fully characterize for the first time TF tomatoes at pink and red ripeness stages, both considered ideal for fresh consumption, through the investigation of morphological characteristics, metabolite profile (carbohydrates, amino acids, organic acids, polyphenols, pigments, sterols, fatty acids), and the evaluation of antioxidant and antifungal (towards *Candida* spp.) properties of tomato extracts. In this study, a comparison with traditional SM tomatoes was also carried out at the same experimental conditions.

The present multidisciplinary analytical approach here employed was already successfully applied to other matrices such as sweet pepper [21], celery [22], extra-virgin olive oil [23], hemp inflorescences [24], but never to tomato fruits. In this study, the powerful combination of high-resolution NMR spectroscopy and FT-ICR MS, not yet widely exploited and largely complementing each other, allowed to obtain a broad untargeted chemical profile, whereas HPLC and spectrophotometric targeted methodologies enabled the content of biogenic amines, polyphenols, and pigments, respectively, to be quantified.

Herein, a biological evaluation of the extracts was finally carried out in terms of antioxidant properties, antifungal activity, and enzyme inhibition. The combined results of radical-scavenging activity, formation of advanced glycation final product (AGE), and the cytoprotective activity towards the oxidative damage induced by tert-butyl hydroperoxide solution (tBuOOH) has allowed to estimate the TF and SM tomatoes antioxidant properties. Moreover, in vitro antifungal activity of the tomatoes extracts towards four *C. albicans*, three *C. glabrata*, and two *C. krusei* strains was assayed.

2. Materials and Methods

2.1. Plant Material

Fresh fruits of *Solanum lycopersicum* L. TF variety were grown and collected by Mafalda SRL (41.342622, 13.420856), whereas SM fresh fruits were grown and collected by San Leone Agricultural Cooperative (41.293929, 13.397638) sited both in Fondi (Latina, Italy). Fondi is characterized by a Mediterranean climate with an average air temperature (T) = 19.5 °C and humidity = 54.1% during the growing season. Irrigation and plant protection, as well as the weed control were carried out following local practices. Samples were harvested at two different ripening stages according to market demand, namely pink (P) stage (from 30% to 60% of not green tomato skin) and red stage (R) (about 90% of not green tomato skin) showing a red colour (Figure S1). Peduncles were removed, some fresh fruits were subjected to morphological analysis and the extraction procedure, while other samples were instantly stored at −80 °C.

2.2. Chemicals

Rutin, quercetin, 2,2-diphenyl-1-picrylhydrazyl (DPPH), 2,2′-azino-bis(3-thylbenzothiazoline-6-sulfonic acid) diammonium salt (ABTS), 2,2′-azobis(2-methylpropionamidine) dihydrochloride (AAPH), trolox, polyvinylpyrrolidone (PVP), tert-butyl hydroperoxide solution (tBuOOH; 900 mg mL^{-1}), ferrozine, hydroxylamine hydrochloride, iron(III) chloride ($FeCl_3 \times 6H_2O$), iron(II) sulfate heptahydrate ($FeSO_4 \times 7H_2O$), potassium hexacyanoferrate(III), iron(II) chloride ($FeCl_2 \times 4H_2O$), magnesium oxide, doxorubicin, 3-(4,5-dimethylthiazol-2-yl)-2,5-diphenyltetrazolium bromide (MTT), 2,7-dichlorofluorescein diacetate (DCFH-DA), and the solvents (HPLC-MS purity grade) were purchased from Sigma-Aldrich (Milan, Italy). Methanol (HPLC-grade), formic acid (99%), perchloric acid (70%), acetone (analytical-grade), chloroform, acetonitrile (HPLC-grade) were obtained from Carlo Erba Reagenti (Milan, Italy). Double-distilled water was obtained using a Millipore Milli-Q Plus water treatment system (Millipore Bedford Corp., Bedford, MA). Sodium carbonate (Na_2CO_3; 99.999% purity), Folin–Ciocalteu's phenol reagent, tannic acid (Ph. Eur. purity) and aluminum chloride hexahydrate ($AlCl_3 \times 6H_2O$; Ph. Eur. purity) were purchased from Merck (Darmstadt, Germany). Deuterated water (D_2O) 99.97 atom% of deuterium, methanol-D4 99.80 atom% of deuterium, chloroform-D 99.80 atom% of deuterium + 0.03% tetramethylsilane (TMS), and 3-(trimethylsilyl)-propionic-2,2,3,3-d_4 acid sodium salt (TSP) were purchased from Euriso-Top (Saclay, France).

2.3. Morphological Analysis

Ten fresh fruits for both tomato varieties and redness stages were subjected to morphological analysis, in order to describe their size (length and diameter), weight, and shape. For each fruit, a careful separation of different components, including peel (i.e., exocarp of the fruit), pulp (i.e., mesocarp of the fruit), seeds (removed with the internal juice), and juice, was performed. The peel was gently separated from the pulp by using a scalpel. The detached components were examined and weighted, and their amount in the whole fruit was determined.

Pigments Characterization

The total carotenoids and chlorophylls analysis in peel and pulp of SM and TF tomatoes samples were performed according to Solovchenko and co-workers [25] with some modifications. The peel was cut from the surface of the fruits, carefully freed from the pulp, and successively weighted. Both peel and pulp from each sample were twice washed with distilled water for 1 min and dried with filter paper. To remove the cuticular lipids, peel fraction was washed with 2 mL portion of chloroform for 1 min. Pigments were extracted according to the Folch method [26]. The samples were homogenized with mortar and pestle in 6 mL of chloroform-methanol (2:1, *v/v*) and, to prevent chlorophyll pheophytinization, 30 mg of MgO were added before the homogenization. The homogenate was passed through a paper filter and after an amount of distilled water equal to 1/5 of the extract volume was added. Finally, this mixture was centrifuged in a glass tube test for 20 min at 2469× *g* for 20 min at 10 °C to complete separation of chloroform fraction from hydroalcoholic one. Absorption spectra of the chloroform phase were recorded with a Beckman Coulter DU 800 instruments, in the range of 350–800 nm with a spectral resolution of 0.5 nm, at a temperature of 20 °C. The concentrations of chlorophyll a and b as well as total carotenoids (*mg/g* of sample) were determined according to Wellburn [27].

2.4. Extraction Procedures

Fifteen fresh whole fruits from pink TF (TF_P), red TF (TF_R), pink SM (SM_P), and red SM (SM_R) were frozen and ground in liquid nitrogen to obtain a homogeneous pool and subjected to the Bligh–Dyer extraction method, which allows to extract both water-soluble and liposoluble metabolites in a quantitative manner.

In details, about 1.0 g of samples (peel, pulp, and seeds) of each variety was added sequentially with 3 mL methanol/chloroform (2:1 *v/v*) mixture, 1 mL of chloroform, and 1.2 mL of distilled water. After each addition the sample was carefully shacked. The emulsion was maintained at 4 °C for 40 min. The sample was then centrifuged (4200× *g* for 15 min at 4 °C) and the upper (hydroalcoholic) and lower (organic) phases were carefully separated. The pellets were re-extracted using half of the solvent volumes (in the same conditions described above) and the separated fractions were pooled. Both hydroalcoholic and organic fractions were filtered with Whatman paper filters and dried under a gentle N_2 flow at room temperature until the solvent was completely evaporated [22]. The dried phases were stored at −20 °C until further analyses. The values of drug to extract ratio (DER) are reported in Table 1.

Table 1. Drug to extract ratio of Bligh–Dyer (hydroalcoholic and organic) extracts of both pink and red fruits from *Solanum lycopersicum* var. Torpedino di Fondi (TF) and San Marzano (SM) [a].

Sample	Drug/Extract Ratio (DER)	
	Hydroalcoholic	Organic
TF_P	20	274
TF_R	9	320
SM_P	20	447

[a] TF_P = pink TF; TF_R = red TF; SM_P = pink SM; SM_R = red SM.

2.5. Metabolite Profile

2.5.1. NMR Analysis

The dried organic fraction of each sample was dissolved in 0.7 mL of a $CDCl_3/CD_3OD$ mixture (2:1 *v/v*) and then placed into a 5 mm NMR tube. Finally, the NMR tube was flame sealed. Conversely, the dried hydroalcoholic phase of each sample was solubilized in 0.7 mL of 400 mM phosphate buffer/D_2O containing 1 mM solution of TSP as internal standard and then transferred into a 5 mm NMR tube. NMR spectra of all hydroalcoholic and organic extracts were recorded at 27 °C on a Bruker AVANCE 600 spectrometer operating at the proton frequency of 600.13 MHz and equipped with a Bruker multinuclear z-gradient 5 mm probe head. ^1H spectra were referenced to methyl group signals of TSP

(δ = 0.00 ppm) in D_2O and to the residual CHD_2 signal of methanol (set to 3.31 ppm) in $CD_3OD/CDCl_3$ mixture. ^1H spectra of hydroalcoholic extracts were acquired with 256 transients with a recycle delay of 5 s. The residual HDO signal was suppressed using a pre-saturation. The experiment was carried out by using 45° pulse of 6.5–7.5 µs, 32 K data points. ^1H spectra of extracts in $CD_3OD/CDCl_3$ were acquired with 256 transients, recycle delay of 5 s, and 90° pulse of 9–11 µs, 32 K data points. The two-dimensional (2D) NMR experiments, such as ^1H-^1H TOCSY, ^1H-^{13}C HSQC, and ^1H-^{13}C HMBC, were carried out under the same experimental conditions previously reported [28]. The integrals of 26 selected signals in hydroalcoholic extract (Table 2) were measured using the Bruker TOPSPIN software and normalized with respect to the resonance at 0.00 ppm, due to methyl group signal of TSP, set to 100. Results were expressed in mg/100 g fresh weight (FW). The quantification of components in organic extracts was described in a previous work [28].

Table 2. Compounds and relative signals ($\delta(^1H)$, ppm) selected for quantitative analysis in the hydroalcoholic and organic extracts.

Ppm	Compounds	Ppm	Compounds
		Hydroalcoholic	
0.96	Leucine	3.25	β-Glucose
0.99	Valine	4.04	Fructose
1.01	Isoleucine	4.31	Malic Acid
1.34	Threonine	4.59	β-Galactose
1.49	Alanine	5.25	α-Glucose
2.30	γ-amino butyric acid (GABA)	6.91	Tyrosine
2.35	Glutamic acid	7.34	Phenylalanine
2.46	Glutamine	7.74	Tryptophan
2.55	Citric Acid	8.14–8.17	Histidine
2.81	Aspartic acid	8.36	Adenosine
2.90	Asparagine	8.46	Formic Acid
3.04	Lysine	8.586	ADP
3.21	Choline	9.13	Trigonelline
		Organic	
0.66	β-Sitosterol	2.73	Di-unsaturated fatty acids (DUFAs)
0.68	Stigmasterol	2.77	Tri-unsaturated fatty acids (TUFAs)
2.30	Total fatty acids (FAs)	5.31	Total unsaturated fatty acids (UFAs)

2.5.2. FT-ICR MS Analysis

A portion (1 mg) of each dried Bligh–Dyer hydroalcoholic (H) and organic (O) fraction of TF and SM cultivars was dissolved in 1 mL (1:1) methanol/water and CH_2Cl_2, respectively. These stock solutions were then vortexed for 3 min, filtered through a 0.45 µm polypropylene Acrodisc (Sigma–Aldrich) syringe filter to remove debris and subsequently diluted in methanol so as to obtain a final concentration of 100 µg L^{-1}, a value chosen to limit ion suppression effects. For each extract, three distinct solutions prepared according to the above procedure were submitted to analysis. In positive mode MS, formic acid (1% *v/v*) was used to assist protonation, while leucine enkephalin (YGGFL, $C_{28}H_{37}N_5O_7$) was added to all samples at a final concentration of 0.5 µg L^{-1} as an internal reference (revealed as $[M+H]^+$ at *m/z* 556.27657 in positive mode and as $[M-H]^-$ at *m/z* 554.26202 in negative mode) to calibrate the spectra by means of the on-line calibration tool (Data Analysis 5.0, Bruker Daltonics). Further internal calibration was achieved by referring to a list of ubiquitous metabolites, including hexose/monosaccharides, citric and palmitic acids, reaching a routine mass accuracy lower than 0.2 ppm. Preliminary mass spectrometric surveys were carried out by using a Bruker BioApex Fourier transform ion cyclotron resonance (FT-ICR) [29] mass spectrometer (Bruker Daltonics GmbH, Bremen, Germany) equipped with an Apollo I electrospray ionization (ESI) source and a 4.7 T superconducting magnet (FT-ICR lab, Sapienza Università di Roma). Ultrahigh-resolution mass spectra were acquired on a Bruker SolariX XR FT-ICR MS endowed with a 7 T superconducting magnet (Magnex Scientific Inc., Yarnton, UK), a ParaCell (Bruker Daltonics GmbH, Bremen, Germany), and an APOLLO II electrospray ionization (ESI) source operated in either the positive (ESI+) or negative ionization mode (ESI-), at Universidade de Lisboa. Samples were directly infused in the ESI source at a flow rate of 120 μL h^{-1}. The nebulizer gas pressure was set at 1.0 bar, the drying gas flow rate at 4.0 L min^{-1} at a temperature of 200 °C, and the capillary exit voltage at 200 V.

All MS spectra were acquired in absorption mode, over a mass range between *m/z* 100 and 3000 (resolution of 650,000 at *m/z* 400), with an acquisition size of 4 mega words, resulting in a free induction decay (FID) of 1.973 s. For each sample, two hundred scans were coadded, corresponding to a run time of 10 min.

Overall, 20 μL of dilute sample solution (corresponding to 2 ng of original dry sample) were used for acquiring one mass spectrum, which not only makes the FT-ICR MS analysis compatible for high sample throughput but also uses relatively small sample amounts.

The list of *m/z* values was exported with a cut-off signal-to-noise ratio (S/N) of 4 and submitted to the free tool MassTRIX [30], taking into account protonated, sodiated, and potassiated (ESI(+)), and deprotonated and chlorinated (ESI(−)) ions, with a maximum deviation range set to ± 1ppm. An accurate check of the isotopic pattern based on the natural abundances of 13C, 15N, 18O, 34S, and 37Cl isotopes, was also performed to minimize false positive results. Only singly charged species were revealed, in both polarity modes. In analyzing each cultivar, peaks with a reproducibility lower than 67% were removed. A large number of unambiguous molecular formulas, for which several isomers are possible, admitting the presence of the elements C, H, O, N, P, and S, could be assigned by both ESI(+) and ESI(−) analyses and were further filtered by application of several chemical constraints as indicated by Kind et al. [31]. Additional information was obtained by acquisition of collision induced dissociation (CID) spectra, though limited to components of adequate abundance, further verified against fragmentation patterns of reference compounds or data inserted into a specialized database. The formulas generated from each sample were then transposed to two-dimensional van Krevelen diagrams, known as elemental ratio analysis, constructed by plotting the molar hydrogen to carbon ratio (H/C) vs. the molar ratio of oxygen to carbon (O/C) for each data point. According to their own characteristic H/C and O/C ratios, main classes of compounds are specifically localized as areas in the plot, thus allowing a depiction of a sample's composition [32].

2.5.3. Phenolic Compounds (Polyphenols, Tannins, and Flavonoids)

Total polyphenols, tannins, and flavonoids per milligram of fresh fruit were determined by spectrophotometric methods according to previous published methods [24]. The total amount of both polyphenols and tannins was expressed as tannic acid equivalents (TAE), while flavonoids were expressed as quercetin equivalents (QE).

2.5.4. Biogenic Amines (BAs) Determination

The BAs determination was carried out as previously described [33]. Briefly, 8 g of tomato extract were added 15 mL 0.6 M $HClO_4$ aqueous solution and 0.5 mL of 1,7diaminoheptane 100 mg mL^{-1} (Internal Standard), then homogenized for 3 min with an Ultra-Turrax and centrifuged at 3000 RPM for 10 min. Supernatant was filtered through a 0.20 μm membrane Millipore filter and sediment was added with 8 mL of $HClO_4$ 0.6 M, mixed, and centrifuged again for 3 min. The second extract was then filtered and added to the first. The final volume was adjusted to 25 mL with $HClO_4$ 0.6M. An aliquot of 1 mL of the final extract was then derivatized by adding 200 μL of NaOH 2 M, 300 μL of saturated $NaHCO_3$ solution, and 2 mL of dansyl chloride solution (10 mg mL^{-1} in acetone). After shaking, samples were left in the dark at 45 °C for 60 min. The final volume was adjusted to 5 mL by adding acetonitrile. The dansylated extract was filtered using 0.22 μm (Polypro Acrodisc, PallGelman Laboratory, USA) filter, injected into the chromatograph, and analyzed with a previous standard method [34]. The determination was carried out twice on 8 samples of each examined tomato cultivar.

2.6. Screening of Biological Activities

2.6.1. Antioxidant Activities

All tests were performed in 96-multiwell microplates away from direct light. To perform the assays, the extracts were assayed at the concentrations of 1, 10, 25, 50, 100, 250, 500, 1000, 1500, 2000, and 5000 µg mL^{-1} in order to achieve a concentration-response curve. The samples were dissolved in 50% or 100% *v/v* EtOH (organic and hydroalcoholic extracts, respectively). The experiments were repeated at least twice, and in every experiment, each concentration was tested in triplicate. Data obtained from at least two experiments were pooled for the statistical analysis.

In each experiment, the vehicle (negative control) and standard antioxidants (positive controls), i.e., trolox (assayed concentrations 0.1, 0.25, 1, 5, 10, 50, and 100 µg mL^{-1}) for the radical scavenger and reducing activity, and quercetin (assayed concentrations 1, 5, 10, 50, 100, and 200 µg mL^{-1}) for the chelating activity, were included too. The absorbance was measured by a microplate reader (Epoch Microplate Spectrophotometer, BioTeK® Instruments Inc., Winooski, VT, USA). Some wells containing only the test samples were also included to determine its possible absorbance.

Scavenging activity towards DPPH and ABTS radicals was determined according to the methods of Di Sotto et al. [35]. Furthermore, the ability of the extracts to indirectly interfere with the ROS-generation through blocking the Fenton reaction was evaluated by testing the iron chelating and reducing activities in the ferrozine assay [35]. Chelation ability was evaluated against both ferrous and ferric ions. The ability of the samples to inhibit the ROS-induced lipid peroxidation was assessed by the ferric thiocyanate method [36].

2.6.2. Advanced Glycation End-Product (AGE) Inhibition

The ability of the tested samples to inhibit the AGE formation was measured through the method of Di Sotto et al. [37]. The phenolics naringenin and rutin were included as standard inhibitors, while the vehicle (50% or 100% *v/v* EtOH for organic and hydroalcoholic extracts respectively) represented the lack of inhibition. The inhibitory activity was calculated as percentage of the control, as follow:

$$(A_{control} - A_{sample}/A_{control}) \times 100 \quad (1)$$

where $A_{control}$ is the fluorescence of the control, whereas A_{sample} is the fluorescence of the sample. Data from at least three replicated experiments (including six replicates for experiment) were pooled for the statistical analysis.

2.6.3. Cytoprotection towards the Oxidative Stress Induced by tBuOOH

Cytoprotective activity of the tested extracts was evaluated towards the oxidative damage induced by tert-butyl hydroperoxide solution (tBuOOH) in HepG2 liver cancer cells (American Type Culture Collection, Milan, Italy). The cells were grown at 37 °C in 5% CO_2 in Dulbecco's modified Eagle's medium, supplemented with fetal bovine serum (10% *v/v*), glutamine (2 mM), streptomycin (100 µg mL^{-1}), and penicillin (100 U mL^{-1}) [38]. All experiments were performed when cells reached the logarithmic growth phase.

Preliminarily, the extracts (1–1000 µg mL^{-1} concentration range) were tested for the mitochondrial cytotoxicity by the 3-(4,5-dimethylthiazol-2-yl)-2,5-diphenyl tetrazolium bromide (MTT) assay [39], in order to define the proper concentrations to be used in the subsequent experiments. Then, 50% or 100% *v/v* EtOH were used as vehicle for organic and hydroalcoholic extracts respectively; the vehicle was nontoxic at final concentration of 1% *v/v* in the medium.

The ability of the tested samples to counteract the oxidative stress induced by tBuOOH was evaluated by measuring the levels of intracellular ROS (reactive oxygen species) through the 2,7-dichlorofluorescein diacetate assay (DCFH-DA) [40]. To this end, 5×10^5 cells were grown into 6-well plates for 24 h, then treated with a nontoxic concentration of the extracts (100 µg mL^{-1})

for 24 h. At the end of incubation, the cells were treated with a low-toxic concentration (about 40% cytotoxicity as found in preliminary experiments) of the pro-oxidant agent tBuOOH (5 mM) for 2 h, then washed twice with Hank's Balanced Salt Solution (HBSS) (1×) and added with DCFH-DA (10 μM; 6 μL). Fluorescence of DCF, obtained by DCFH-DA oxidation, was measured though a BD Accuri™ C6 flow cytometer at an excitation wavelength of 485 nm and emission wavelength of 528 nm. In each experiment, proper treatment with the vehicle control (corresponding to the basal ROS level) and the pro-oxidant agent tBuOOH were included too; furthermore, the extracts alone were assayed to evaluate their effect on the basal ROS levels, released as a consequence of cell metabolism. The oxidation index was obtained by the ratio between the DCF fluorescence of the sample and vehicle control.

2.6.4. In Vitro Metabolic Enzyme Inhibition

The ability of the tested extracts to inhibit in vitro the α-amylase and α-glucosidase enzymes was measured by dinitrosalicilic acid (DNSA) and *p*-nitrophenyl-α-D-glucopiranoside (PNGP) methods described by Di Sotto et al. [37]. Acarbose was included in all the experiments as standard enzyme inhibitor (100% enzyme inhibition), while the vehicle (50% or 100% *v/v* EtOH for organic and hydroalcoholic extracts respectively) represented the maximum enzyme activity. Additional treatments, in which enzyme solution was replaced by buffer solution, were included to evaluate a possible interfering absorbance of the samples. The experiments were performed at least in triplicate and in each experiment about six replicates were prepared. Data obtained from at least two experiments were pooled in the statistical analysis. The inhibitory activity was calculated as percentage of inhibition with respect to the vehicle control.

2.6.5. Antifungal Susceptibility Test

To evaluate the minimal inhibitory concentration (MIC) of the extracts, the broth microdilution method was performed according to a standardized method for yeasts [41].

The assay was carried out with four *C. albicans* strains (ATCC10231, ATCC24433, 3153A, PMC1033), three *C. glabrata* strains (PMC0822, PMC0851, PMC0807), and two *C. krusei* strains (PMC0631, PMC0624). *Candida* spp. strains were grown on Sabouraud dextrose agar at 37 °C for 24 h. Then, cell suspensions of the strains were prepared in RPMI 1640 medium buffered to pH 7.0 with 0.165 mM MOPS. The final concentration of the inoculum was 1×10^3–5×10^3 cells mL^{-1}. The extracts were dissolved in DMSO and diluted 100 times in RPMI-1640 broth. Ten concentrations ranging from 1000 to 1.9 μg mL^{-1} were tested against *Candida* spp. strains in 96-well round-bottom microtitration plates. The antifungal activity is the result of four independent experiments. The MIC_{50}, MIC_{90}, and MIC_{100}, the lowest concentrations of extracts that caused growth inhibitions ≥50%, ≥90%, and 100% respectively, were evaluated. Data were reported as range and geometric mean (GM) of MIC.

2.7. Statistical Analysis

All values are expressed as mean ± standard error (SE). Statistical analysis was performed by GraphPad Prism™ (Version 4.00) software (GraphPad Software, Inc., San Diego, CA, USA). The one-way analysis of variance (one-way ANOVA), followed by a suitable multiple comparison post hoc test (i.e., Bonferroni post-test for comparison among means, while Dunnett's post-test for estimating a difference compared to the control), was used to analyze the difference between treatments. The concentration– response curves were constructed using the "Hill equation":

$$E = Emax/(1 + 10\hat{}(LogEC_{50}/A) \times HillSlope) \quad (2)$$

where E is the effect at a given concentration of agonist, Emax is the maximum activity, EC_{50} is the concentration that produces a 50% of the inhibitory response (namely IC_{50}), A is the agonist concentration in molarity, HillSlope is the slope of the agonist curve. p values < 0.05 were considered as significant. Correlation between two variables was evaluated by the Pearson correlation coefficient and the statistical significance was measured by the two-tailed *t*-test.

3. Results and Discussion

3.1. Morphological and Pigments Analyses

The sampled TF tomatoes showed a shape similar to SM fruits, but were smaller in length and circumference (about two-fold lower) and in weight (about five folds lower), thus supporting their nature of "Mini-San Marzano tomato" (Table 3). Despite a smaller size, peel amount in pink and red TF fruits was two- and ten-fold higher than those of SM at the same ripening stages, respectively (Table S1). Furthermore, at least a doubled peel amount was found in the red TF tomatoes compared to the pink ones, whereas an opposite trend was observed in SM fruits (Table S1).

Table 3. Color, weight, and size (length and diameter) of both pink and red fruits from Solanum lycopersicum TF and SM varieties. Data are displayed as mean ± SE (n = 10) [a].

Sample	Color	Weight (g)	Size (cm)	
			Length	Diameter
TF_P	dark-green	21.7 ± 0.1	6.3 ± 0.1	2.6 ± 0.1
TF_R	bright red	20.6 ± 0.1	5.6 ± 0.1	2.7 ± 0.1
SM_P	pale green	106.1 ± 0.9 **	10.1 ± 0.1 **	4.5 ± 0.1 *
SM_R	light red	110.7 ± 0.6 **,§	11.4 ± 0.1 **	4.4 ± 0.1 *

[a] * $p < 0.05$ and ** $p < 0.01$ denote a statistically significant difference compared to TF tomato at the same ripening stage (t-Student Test). § $p < 0.05$ denotes a statistically significant difference with respect to pink stage within the same cultivar (t-Student Test).

Ripeness also increased the pulp amount in both varieties. The TF pulp content per gram of fruit was at least four-fold higher with respect to SM tomatoes (Table S1). In spite of a lower pulp amount, both pink and red SM (SM_P, SM_R) tomatoes contained high amounts of juice (pH 4.0–4.2). Conversely, the juice content was at least two to six-fold lower in pink and red TF (TF_P, TF_R) tomatoes (Table S1). Altogether, these features support the claimed crunchy and solid consistency of TF tomatoes, likely ascribable to a high peel and pulp content, despite a significant low amount of juice.

A significant difference in the seed number and weight per gram of fruit was observed, which was about four-fold higher in TF compared to SM tomatoes at both ripening stages (Table S1). Conversely, the size and weight of each seed were similar in both varieties, with a slight increase in the SM_R fruits (Table S1). It is widely accepted that seed size and number are strictly linked to the plant reproductive potential. Small seeds were reported to possess lower reproductive capacity, due to a lower endosperm amount, which limits seedling survivorship and competitive ability [42]. On the other hand, a high seed number improves the competitive ability of the plant, due to the increased probability of seedling survivorship. In this context, TF tomatoes seem to possess a higher reproductive competition related to SM, along with similar plant reproductive capacity and survival.

The color of SM tomatoes varied from pale green in the pink fruits to light red in the red ones (Table 3 and Figure S1). The analysis of pigments i.e., chlorophyll a, chlorophyll b, and total carotenoids, carried out on the TF and SM organic extracts from peel and pulp justified these features. The dark-green colour of TF_P tomatoes can be ascribed to the higher content in chlorophylls a and b in peel (+70% and +52% higher amount, respectively) ($p < 0.01$) and in pulp (+10% and +82% higher amount, respectively) ($p < 0.01$) with respect to TF_R fruits (Table 4). On the contrary, carotenoids content (bright red color) was found about five-fold higher in TF_R peel and pulp than in pink ones ($p < 0.001$) (Figure S1).

Table 4. Amounts of chlorophyll a, chlorophyll b, and total carotenoids in the organic extracts of both pink and red fruits from Solanum lycopersicum var. Torpedino di Fondi (TF) compared San Marzano (SM) tomatoes [a].

Sample	Fruit Part	Chlorophyll a	Chlorophyll b	Total Carotenoids	Ratio (a + b)/ Total Carotenoids
			µg/g FW		
TF_P	peel	166 ± 32	67 ± 3	79 ± 7	3.16
	pulp	178 ± 30	94 ± 4	82 ± 9	
TF_R	peel	49 ± 5 §	32 ± 3 §	404 ± 22 §§	0.18
	pulp	18 ± 4 §§	17 ± 1 §§	455 ± 10 §§	
SM_P	peel	63 ± 3 **	33 ± 4 *	26 ± 5 **	3.9
	pulp	31 ± 3 ***	13 ± 3 **	13 ± 3 ***	
SM_R	peel	5 ± 2 *** §	9 ± 1 §	723 ± 4 ***,§§	0.02
	pulp	4 ± 2 §	8 ± 1	511 ± 13 ***,§§	

[a] * $p < 0.05$, ** $p < 0.01$ and *** $p < 0.001$ denote a statistically significant difference compared to TF tomato at the same ripening stage (ANOVA followed by Bonferroni multiple comparison post test). §§ $p < 0.01$ denotes a statistically significant difference with respect to pink stage within the same variety (ANOVA followed by Bonferroni multiple comparison post test).

The difference in chlorophyll a content between pink and red fruits was particularly marked in SM cultivar, where chlorophyll a amount was 90% higher in peel and pulp of SM_P compared to SM_R fruits ($p < 0.01$). SM_P tomatoes displayed both chlorophyll a and b content to be double in peel with respect to pulp, while in SM_R fruits, this difference was not evident (Table 4). The opposite trend was observed with regard to carotenoids content. SM_R peel showed a carotenoid amount 28 times higher with respect to SM_P peel ($p < 0.001$), while in SM_R pulp, a 40 times higher level of total carotenoids with respect to SM_P pulp ($p < 0.001$) was found.

TF fruits were characterized by higher levels of chlorophylls compared to SM ones at both ripening stages ($p < 0.01$). In particular, chlorophyll a and b content was found to be 60% and 80% higher in TF_P peel and pulp fruits than in SM_P peel and pulp, respectively ($p < 0.01$). A similar trend was observed in terms of total carotenoids content, which was 68% higher in TF_P fruits with respect to SM_P ($p < 0.01$). These findings agree with the different shade of green color observed in the TF_P (dark green) compared to SM_P fruits (pale green) (Figure S1). In addition, red fruits of both cultivars were characterized by different pigments proportion. In TF_R peel, chlorophyll levels were 90% higher compared to SM_R peel (+90%, $p < 0.001$), the latter showed 40% higher content of carotenoids than TF_R peel ($p < 0.001$; Table 4).

Carotenoids are mainly responsible for the red color of tomatoes and are involved in the fruit protection against excessive sun irradiation and harmful UV rays. Total carotenoids amount in fruits depends on the ripening stage and other factors, such as cultivar, climate, sun exposure, agronomic practices, irrigation. The ratio of chlorophyll a and b to total carotenoids (a + b/total carotenoids) can be considered as an indicator of tomato ripening stage. During chromoplast development in fruit maturation, the ratio a + b/total carotenoids tends to decrease continuously, thus reaching a value below 1.0 [43]. Our findings confirmed the differences in ripening stages selected for the study, revealing a ratio value of 3.16 in TF_P tomatoes and 0.18 in TF_R ones. In SM fruits, this trend was even wider, in fact, pink fruits showed a ratio value of 3.9, whereas in red fruits, this ratio was 0.02 (far below of 1).

3.2. Metabolite Profiling

FT-ICR MS and NMR untargeted analyses were carried out for a thorough metabolite profile characterization of TF and SM in relation to their pink and red states. The high mass accuracy typically achieved with FT-MS implies that elemental formulas can be determined, pertaining to a large number of metabolites, based on their accurate mass, whereas the NMR capacity of structural determination allows the unambiguous compound identification and quantification.

The ESI FT-ICR MS analysis of both Bligh–Dyer hydroalcoholic and organic fractions of TF_P, TF_R, SM_P, and SM_R fruits has allowed to detect both polar and non-polar metabolites. Each sample was analyzed in both positive and negative ionization mode (Figures S2–S5), detecting up to 1138 molecular formulas; however, a larger number of compounds were detected in positive ionization mode (Table 5). Overall, the TF cultivar was characterized by a smaller number of compounds with respect to SM and the ripeness process promotes a general increase in the number of putatively identified metabolites (Table 5). An overview of all the recorded plausible compounds is available in Tables S2 and S3.

Table 5. Number of chemical formulas detected in hydroalcoholic and organic extracts of pink and red SM and TF by ESI FT-ICR MS.

Sample		Ion Mode	Detected Molecular Formulas	
		San Marzano		
Hydroalcoholic	Pink	ESI(+)	824	935
		ESI(−)	132	
	Red	ESI(+)	808	1031
		ESI(−)	240	
Organic	Pink	ESI(+)	401	508
		ESI(−)	113	
	Red	ESI(+)	865	1138
		ESI(−)	286	
		Torpedino di Fondi		
Hydroalcoholic	Pink	ESI(+)	488	652
		ESI(−)	261	
	Red	ESI(+)	549	751
		ESI(−)	204	
Organic	Pink	ESI(+)	381	586
		ESI(−)	208	
	Red	ESI(+)	800	948
		ESI(−)	151	

Specific data analysis allows to organize the vast amount and complexity of detected formulas to uncover interesting information. Among the detected molecular formulas, the relative frequency distribution was investigated (Figure 1E,F) showing that all tomato extracts contain a majority of CHO species followed by CHON, CHOP and, in smaller amount, CHNOP and CHNOS. In particular, CHO components correspond mainly to polyphenols (more hits in SM_R extracts), steroids (more hits in TF), and fatty acids (more entries in TF), followed by di- and tri-glycerides (more entries in SM_P), terpenoids, organic acids, and arachidonic derivatives (Supplementary Figure S6A). When considering CHON components, they can be ascribed mainly to amino fatty acids, amino-sugars, amines (more hits in red extracts), N-acylamines (more hits in TF_P), followed by amino acids (more entries in pink extracts), solanidines, nucleosides (more hits in SM_P), and vitamins (more hits in TF_R) as shown in Figure S6B.

Van Krevelen diagrams were used to classify the detected molecular formulas in different classes of natural compounds such as lipids, terpenoids, carbohydrates, amino acids, aminosugars, nucleic acids, polyphenols, polyketides, unsaturated hydrocarbons and condensed hydrocarbons (Figure 1A–D). TF and SM tomato extracts showed marked similarities, covering several classes of metabolite families. A relatively higher compound density is present in the area of lipids, terpenoids, and polyketides, followed by components in the areas of amino acids, unsaturated hydrocarbons, polyphenols and (relatively less) in the regions of carbohydrates, aminosugars, nucleic acids, and condensed hydrocarbons (Figure 1A–D).

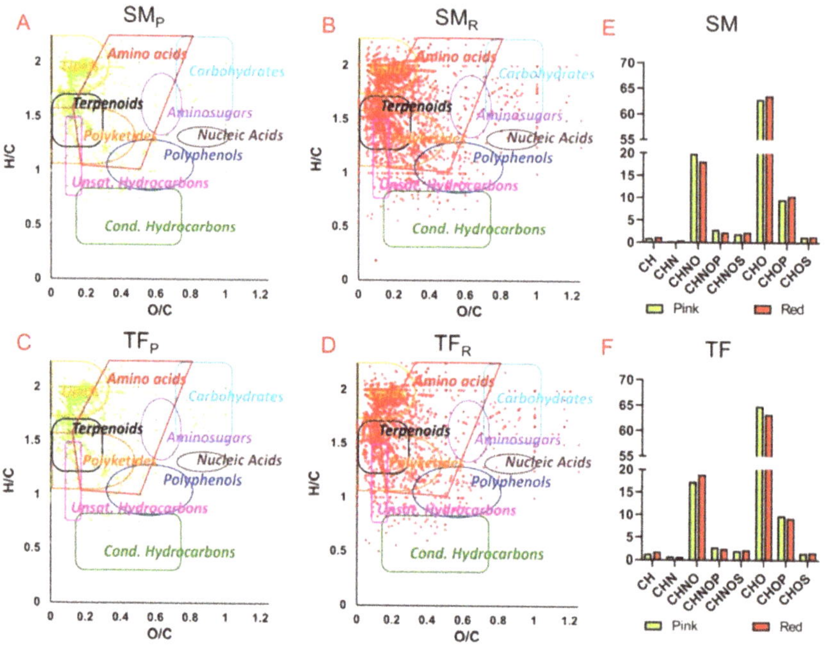

Figure 1. Van Krevelen plot (elemental plot) obtained from the molecular formulas obtained by ESI FT-ICR MS analysis of total hydroalcoholic and organic fractions of: (**A**) SM_P; (**B**) SM_R; (**C**) TF_P; (**D**) TF_R. Histograms of the relative frequency of CH, CHN, CHNO, CHNOP, CHNOS, CHO, CHOP, CHOS compounds: (**E**) in SM_P (green), SM_R (red); (**F**) TF_P (green) and TF_R (red).

Moreover, Venn diagrams (Figure S7) pointed out possible similarities and differences in the metabolic profile of the sampled TF and SM fruits. The combined pattern of hydroalcoholic and organic fractions of pink and red extracts of the two cultivars showed that overall only 19% of the molecular formulas were found to be common, thus suggesting a noticeable extent of chemical diversity, whereas more than 40% of molecular formulas were shared between pink and red samples of each variety.

This wide metabolomics survey supported the untargeted and targeted analyses driving the identification of selected classes of metabolites. In particular, 1D NMR spectra assignment of the TF and SM hydroalcoholic extracts solubilized in D_2O phosphate buffer and organic extracts solubilized in $CDCl_3/CD_3OH$ (Table 2) were obtained by means of literature data [28,44–46]. Furthermore, targeted analytical approaches provided the identification and quantification of total polyphenols, tannins, and flavonoids content and BAs.

Results will be presented and discussed according to compound classes.

3.2.1. Amino Acids and Derivatives

NMR spectra of both red and pink TF and SM hydroalcoholic extracts showed signals of sixteen amino acids, namely leucine, valine, isoleucine, threonine, alanine, GABA, glutamic acid, glutamine, aspartic acid, asparagine, lysine, arginine, tyrosine, phenylalanine, tryptophan, and histidine, as also confirmed by ESI FT-ICR MS. All of them were quantifiable, except arginine. In addition, ESI FT-ICR MS revealed the presence of proline, serine, the non-essential amino acid citrulline, and other amino-acids-related metabolites, like hydroxyproline and phosphoserine. Some peptides were also found. In particular, glutathione was detected in all hydroalcoholic extracts, S-nitrosoglutathione in red fruits, glutathione disulfide in pink fruits.

Alanyl-alanine (in hydroalcoholic SM_P), glycyl-proline (in organic SM_R), glycyl-leucine (in organic TF_R), glutamyl-valine and glutamyl-glutamine (in hydroalcoholic SM_R) were also revealed.

According to NMR scrutiny, TF and SM samples showed some/several similarities (Figure 2A): glutamine was found to be the most abundant amino acid in both cultivars at the pink stage, followed by GABA and glutamic acid, whereas, at the red stage, glutamic acid increased, becoming the most abundant amino acid. The pattern of SM developmental changes in free amino acid content was in agreement with literature data, being glutamic acid characterized by a remarkable increase in all ripe fruits [2–4,45,47–52]. Glutamic acid, aspartate, tryptophan, and alanine rose upon the fruit ripening; asparagine and phenylalanine turned out to be constant; tyrosine, isoleucine, valine, threonine, GABA, and glutamine content decreased from pink to red SM fruits. These findings reflect data reported in literature about the analyses of SM [2,4] and other cultivars [3,45,47–52].

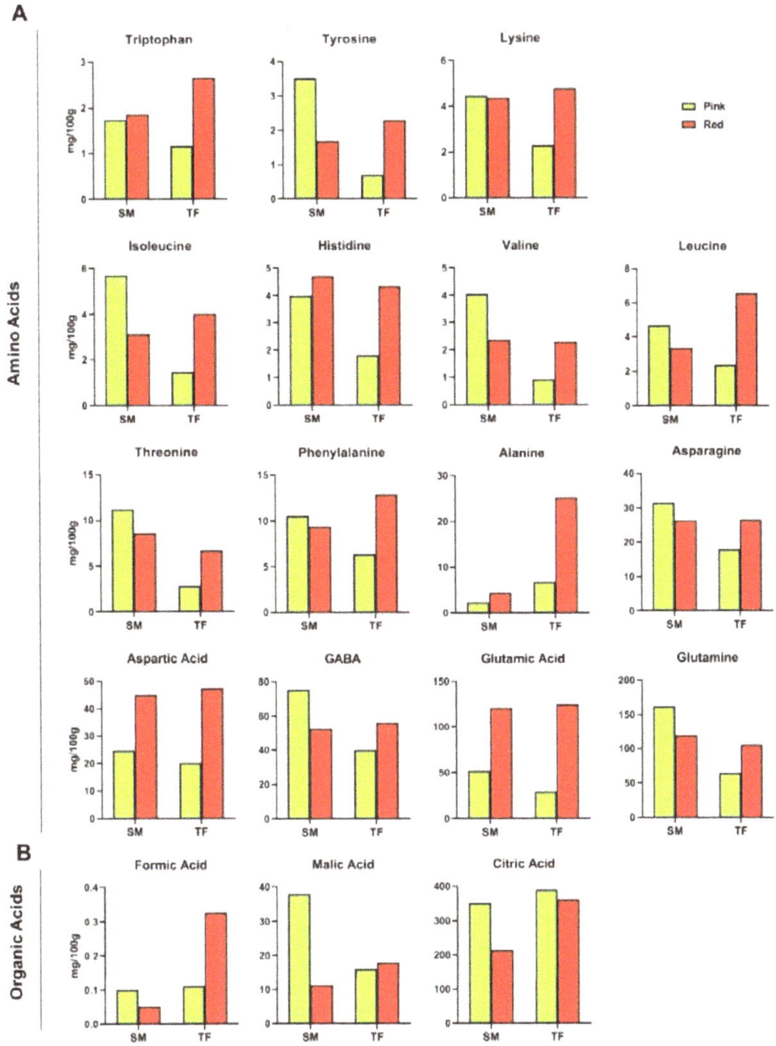

Figure 2. *Cont.*

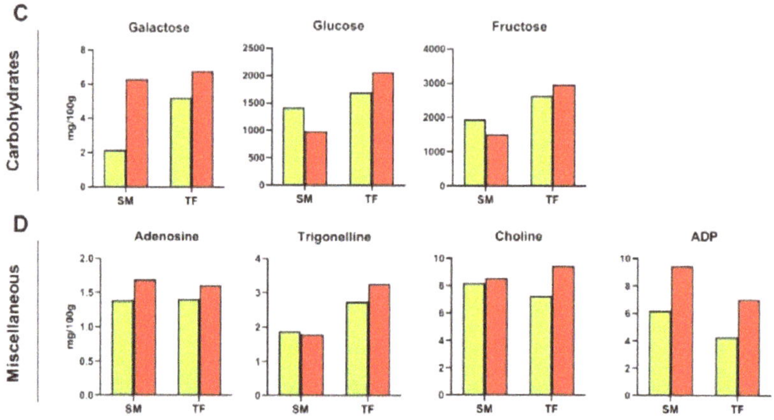

Figure 2. Histograms relative to metabolites identified by H^1-NMR spectroscopy in tomato hydroalcoholic extracts from pink (green) and red (red) SM and TF cultivars: (**A**) Amino acids; (**B**) Organic acids; (**C**) Carbohydrates; (**D**) Other compounds. Data are expressed as mg/100 g FW.

Interestingly, the TF fruit ripening showed a peculiar trend in the amino acid profile, being characterized by an increase in the content of all the amino acids from the pink to red stage. In particular, except for asparagine, GABA, and glutamine, the content of the remaining twelve amino acids rose more than two-fold.

3.2.2. Organic Acids

Ascorbic, citric, chlorogenic, malic and formic acids were detected by NMR analysis. Chlorogenic acid (5-caffeoylquinic acid) was identified only in hydroalcoholic TF_P extracts. The ESI FT-ICR mass spectra in negative mode provided additional peaks corresponding to deprotonated organic acids identified in one or a few samples, like succinic and glutaric (in organic SM_R and hydroalcoholic TF_P extracts), maleic (in hydroalcoholic TF_P extracts), quinic and shikimic (in hydroalcoholic TF_R and TF_P extracts), and lactic (absent only in red hydroalcoholic samples) acids, as the most prominent signals. In addition, sugar esters of caffeic and ferulic acids, caffeoyl- and feruloyl-hexose, were revealed, with both metabolites being present in hydroalcoholic extracts. Although the present method does not allow to recognize which constitutional isomer is formed, glycosylated forms of phenolic acids have been previously identified in methanol extracts of tomato fruit by a HPLC/DAD/MS approach [53].

Histograms reporting NMR data (Figure 2B) showed that formic acid was always present in a minor amount, whereas citric acid represented the main organic acid, contributing to sourness [5] and confirming literature data [2–4,45,47,48]. During the developmental process, both malic and citric acids contents stayed constant in TF fruits and decreased in SM tomatoes. Previous results concerning the organic acid trends are contradictory: Mounet et al. [48] and Jezequel et al. [54] described an increasing citric acid content from the pink to red stage, whereas Perez et al. [45] found out a decrease in citric and malic acids content. These differences might be ascribed to the combination of genetic, pedoclimatic, seasonal, and agronomic factors.

In terms of cultivar, the three organic acids were comparably abundant at the pink stage, except for malic acid which was significantly higher in SM than in TF fruits. At the red stage, formic, malic and citric acids contents were found nearly doubled in TF compared to SM fruits.

3.2.3. Sugars

Fructose, glucose and galactose were the monosaccharides identified in the ^{1}H NMR spectrum of hydroalcoholic extracts. Besides, confirming the widespread incidence of mono and disaccharides, ESI FT-ICR results reported an Amadori compound, fructosyl lysine, in all hydroalcoholic extracts, providing a sensitive marker of early modifications in food nutrients previously described also in unprocessed tomato extracts [55].

Fructose was the most abundant sugar in both cultivars at both ripeness stages, followed by glucose; however, higher levels of both carbohydrates were found in TF compared to SM fruits (Figure 2C). Fructose, glucose, and galactose levels increased in TF cultivar over the ripening period, whereas the opposite trend was observed in SM fruits. Loiudice et al. [2] reported glucose and fructose contents of eleven SM tomato cultivars at three harvesting years: fructose was always found the most abundant sugar in all samples characterized by a mean content of 1.4g/100 g, whereas glucose mean content was 1.2 g/100 g. These data are consistent with the values reported in Figure 2C: fructose was the main sugar in SM with a mean value of 1.7 g/100 g and glucose content was 1.2 g/100 g, with a total amount of 2.9 g/100 g. However, sugars content in hybrid cultivars tended to increase [2], confirming the higher amount of both fructose and glucose in TF, which reached 5 g/100 g as mean value.

3.2.4. Other Compounds

NMR signals of choline, trigonelline, uridine, adenosine, and ADP were identified and, except for uridine, quantified (Figure 2D). A rich variety of miscellaneous compounds were detected by ESI FT-ICR analysis, comprising small amines (serotonin), nucleosides (adenosine, methylthioadenosine, orotidine), and nucleotides (guanosine-, cytidine-, and uridine-monophosphate), sugar alcohols (sorbitol, mannitol), sugar acids (galactonic and glucuronic acids), aminosugars (glucosamine, lactosamine), terpenes (*p*-cymene, caryophyllene, limonene), terpenoid (apiole, oxo-campholide), vitamins and derivatives (ascorbic acid, retinol, α-tocopherol). As expected, both cultivars contained several key secondary metabolites characteristic of tomato fruits, including alkaloids (trigonelline, narciclasine, and catharanthine, only in hydroalcoholic extracts, nicotine and sauroxine, spread in all samples), polyketides such as lycoflexine, mostly in pink fruits, and glycoalkaloids as tomatine and tomatidine, only in pink hydroalcoholic extracts [56,57]. Several phytohormones, recognized as key signaling molecules, were observed in most samples, like derivatives of abscisic acid, only in hydroalcoholic extracts, jasmonic acid, only in red hydroalcoholic fractions, and salicylic acid, mostly in TF_P and SM_R, and several gibberellins, mainly detected in red organic samples [58].

The concentration of choline and adenosine were comparable in SM and TF fruits, slightly increasing at the red stage, whereas trigonelline content was found to be lower in SM with respect to TF at both developmental stages. ADP content severely increased in both cultivars at the red stage, almost doubling its level.

The presence of hydroxyl-substituted fatty acids, metabolites with strong anti-inflammatory and antioxidative effects, confirms the nutraceutical potential of tomato. Hydroxy-stearic acid was observed in all samples except for organic SM_R, hydroxy-linoleic acid, in all extract excluding hydroalcoholic SM_P, and hydroxylinolenic acid, in organic SM_R and all TF extracts [59].

3.2.5. Phenolic Compounds (Polyphenols, Tannins, and Flavonoids)

Spectrophotometric targeted analyses provided the total content of phenolic compounds expressed as polyphenols, tannins, and flavonoids. Highest levels of total polyphenols were found in hydroalcoholic extracts from pink fruits of both cultivars, being TF_P the most enriched sample (triple content compared SM_P); conversely, their levels were reduced of about 14- and 6-fold with ripeness (Table 6). Similarly, a 1.6-fold reduction in total polyphenols was found in TF_R organic extracts with respect to those from TF_P, whereas an opposite trend occurred in the SM variety (Table 6).

Table 6. Amounts of total polyphenols, tannins and flavonoids in Bligh–Dyer hydroalcoholic and organic extracts of both pink and red fruits from *Solanum lycopersicum* var. TF and SM [a].

Sample	ADD HEADING	Polyphenols	Tannins	Flavonoids
		[μg TAEs/g Fruit] [b]		[μg QEs/g Fruit] [c]
TF_P	Hydroalcoholic	155.0 ± 0.1	5.0 ± 0.4	149.7 ± 0.6
	Organic	36.9 ± 0.3	7.7 ± 0.3	54.4 ± 0.4
TF_R	Hydroalcoholic	11.1 ± 0.3 §§§	6.4 ± 0.1 §	101.3 ± 0.8 §§
	Organic	22.9 ± 0.5 §§	0.9 ± 0.1 §§§	12.8 ± 0.5 §§§
SM_P	Hydroalcoholic	55.0 ± 0.4 ***	5.0 ± 0.6	50.0 ± 0.7 ***
	Organic	13.7 ± 0.1 ***	4.7 ± 0.2 **	2.2 ± 0.1 ***
SM_R	Hydroalcoholic	9.1 ± 0.6 §§§	5.8 ± 0.2	91.0 ± 0.3 * §§
	Organic	27.8 ± 0.7 §§§	4.3 ± 0.3 ***	132.2 ± 1.2 *** §§§

[a] * $p < 0.05$, ** $p < 0.01$ and *** $p < 0.001$ denote a statistically significant difference compared to TF tomato at the same stage of ripening (ANOVA followed by Bonferroni multiple comparison post test). § $p < 0.05$, §§ $p < 0.01$ and §§§ $p < 0.001$ denotes a statistically significant difference with respect to pink stage within the same variety (t-Student Test). [b] TAEs, tannic acid equivalents. [c] QEs, quercetin equivalents.

TF_P and SM_P hydroalcoholic extracts contained similar levels of tannins, which slightly increased with ripening (Table 6). Comparing the ripening stages, both SM_P and SM_R organic extracts contained an analogue content of tannins, whereas a marked 8-fold reduction in their levels occurred in TF_R organic extracts compared to TF_P (Table 6).

Flavonoids were mainly concentrated in the hydroalcoholic extracts of both TF and SM tomatoes at both the ripening stages, although high levels were also found in the organic extracts of TF_P and SM_R fruits (Table 6). TF_P hydroalcoholic extract resulted in the most enriched sample in flavonoids, being three times more concentrated than SM_P; similarly, the same trend was observed in pink organic fractions, being TF the cultivar characterized by an almost 30-fold higher flavonoid content compared to SM (Table 6). At the red stage, TF fruits showed a flavonoid reduction of about 1.5- and 4-fold in organic and hydroalcoholic extracts respectively, whereas an opposite trend was registered for SM tomatoes. Indeed, a 2- and 66-fold flavonoid increase was found in the hydroalcoholic and organic SM_R with respect to SM_P (Table 6). This evidence revealed that the highest levels of total polyphenols, tannins, and flavonoids were concentrated in the TF_P tomatoes, although a high flavonoid content was retained in TF_R and SM_R fruits. Our data agree with previous evidence that highlighted a total flavonoid content of 200 μg/g (calculated as quercetin equivalents) in SM_R tomatoes [60]. Conversely, to the best of our knowledge, no comparison data are available in the literature regarding TF landrace.

In addition, ESI FT-ICR MS experiments have revealed flavanols (dihydroxy-methoxy-isoflavanol), flavan-3-ols (epigallocatechin sulfate), flavonoids (apigeniflavan, tetrahydroxyflavanone glucoside), and polyphenol derivatives like catechin-O-glucoside, catechin-O-rutinoside, dihydrokaempferol, trihydroxy-prenyldihydrochalcone glucosyl-coumarate, quercetin glucoside-glucuronide [61].

3.2.6. Sterols

β-Sitosterol and stigmasterol were detected and quantified by NMR analysis in the organic extracts of both cultivars at pink and red developmental stages.

β-Sitosterol content showed a decreasing trend in SM_R fruits, opposite to TF fruits. Conversely, stigmasterol content significantly increased (3-fold higher) over the ripening stage in both cultivars (Figure 3). According to ESI FT-ICR results, cholesterol- and hydroxycholesterol-sulfate were found in all samples, whereas methylstigmasterol was detected in the hydroalcoholic TF_R sample.

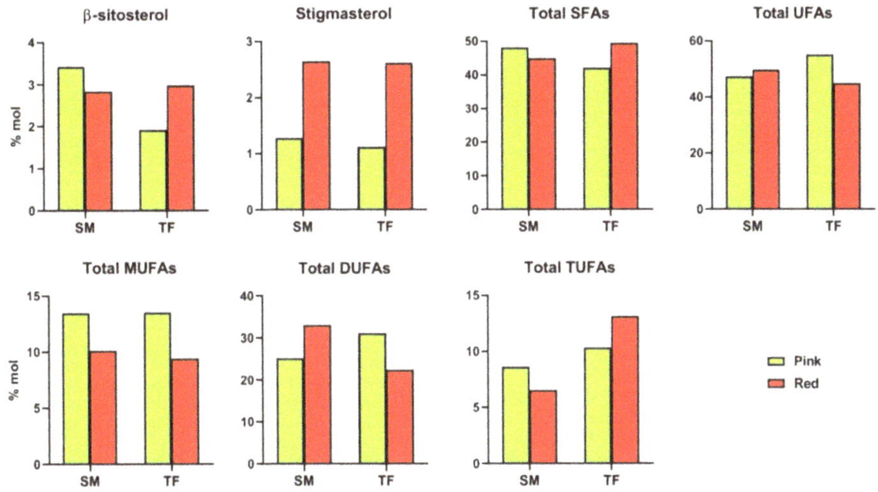

Figure 3. Histograms comparing the concentration (% molar) of metabolites present in tomato organic extracts from pink and red SM and TF. SFA: saturated fatty acids; UFA: unsaturated fatty acids; MUFA: mono-unsaturated fatty acids; DUFA: di-unsaturated.

3.2.7. Fatty Acid Chains

NMR analysis allowed the identification and quantification of the total saturated (SFA) and unsaturated (UFA) fatty acid chains, the latter ones including mono- (MUFA), di- (DUFA) and tri-unsaturated (TUFA) fatty acid chains. The amount of the selected metabolites was comparable in SM and TF cultivars at both ripening stages, except for the case of TUFA, more abundant in TF fruits (Figure 3). Regarding the ripening stages, total UFA content was found to slightly increase in SM fruits and decrease in TF cultivar, conversely total SFA showed the opposite trend.

A drop from pink to red fruits was noticed in MUFA (both SM and TF), DUFA (TF), and TUFA (SM) content, whereas an opposite trend/increase with the ripening stage emerged DUFA (SM) and TUFA (TF) amount.

Despite untargeted investigation, direct infusion ESI FT-ICR MS analysis delivers consistent solution composition and maximum metabolome coverage, ion-suppression effects and differences of signal response are a concern in view of an accurate quantitation. However, careful tuning of experimental conditions has recently allowed a successful quantification of numerous isomeric groups of intact wax esters, where relative ionization efficiency was found to be influenced only by lipid class and saturation degree, while independent on carbon chain length [62]. On this basis, the abundances of molecular formulas classified as free fatty acids (FA) and presenting the expected $(CH_2)_2$ increments were obtained from the lists of organic extracts in the negative ionization mode, relatively richer in lipids (Table S4). Then, these values were summed up within each specific class, namely saturated SFA, containing the series 12:0-20:0, MUFA, with the series 14:1–20:1, DUFA and TUFA, including compounds 18:2, 20:2, 18:3, 20:3, 18:4, 20:4, respectively, to evaluate their relative abundance. Notably, four FAs at *m/z* 227, 255, 277, and 279 have been assigned to myristic (14:0), palmitic (16:0), linolenic (18:3), and linoleic (18:2) acids, respectively, on the basis of their characteristic fragmentation in CID experiments. As shown in Figure 4, an overall similar composition is highlighted in both organic SM_R and TF_R extracts, with the highest percentages found for: (i) 16:0 (ca. 60%) and 18:0 (ca. 28%) among SFAs; (ii) 16:1 (ca. 58%) and 18:1 (ca. 39%) among MUFAs; (iii) 18:2 (ca. 77%) and 20:2 (ca. 17%) among DUFAs; (iv) 18:3 (ca. 37%) and 20:3 (ca. 64%), among TUFAs.

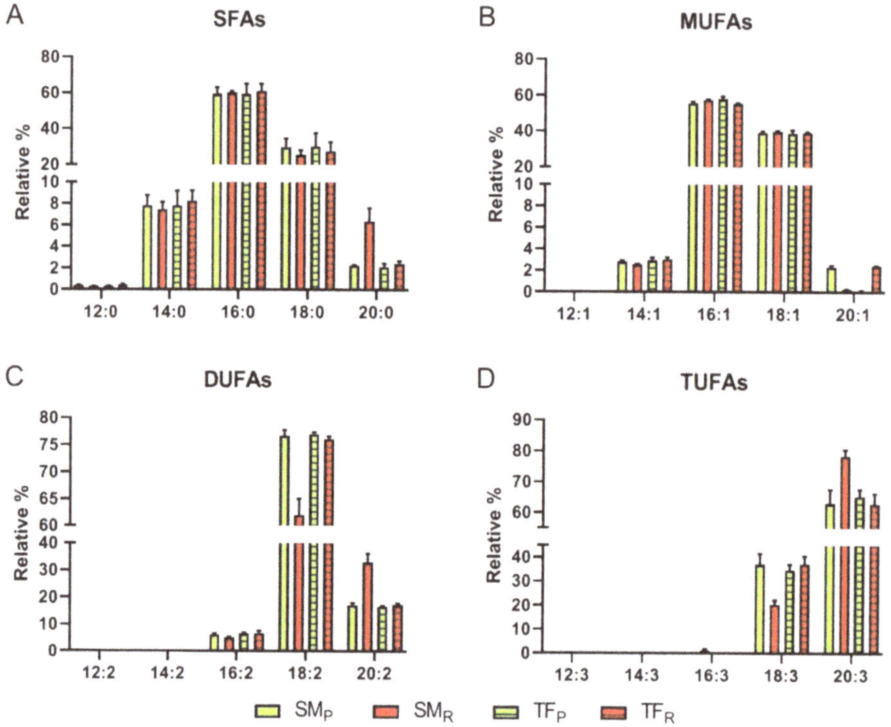

Figure 4. Histograms of the relative abundance distribution within specific classes of FA: saturated (**A**); mono-unsaturated (**B**); di-unsaturated (**C**); tri-unsaturated (**D**) obtained by ESI(−) FT-ICR MS analyses of organic SM_R (red), SM_P (green), TF_R (red), TF_P (green) extracts.

The presence and chemical diversity of long chain fatty acids pointed out in both cultivars is intriguing, since these compounds are known to have a wide range of biological properties, including the promotion of type 2 immune responses [63].

Noteworthy, SM_R organic fraction contains both the lowest percentage of relatively shorter chain di- and tri-unsaturated FA, including 18:2 (ca. 62% vs. the above reported value of 77%) and 18:3 (ca. 21% vs. 37% above reported), and the highest amount of longer chain FA, including 20:2 (ca. 33% vs. 17%) and 20:3 (ca. 78% vs. 64%).

The availability of appropriate unsaturated fatty acids is reported as a significant factor responsible for specific fruit flavor and aroma development, due to the action of lipase enzymes that may release a rich milieu of metabolites from acyl lipids during the ripeness processes.

3.2.8. Biogenic Amines

High performance liquid chromatography allowed to detect seven BAs (Table 7), namely putrescine (PUT), cadaverine (CAD), histidine (HIS), serotonin (SER), spermidine (SPD), and spermine (SPM). All BAs investigated are found in tomato samples, except for tyramine (TYM), an amine with negative health effect, which was not detected in any samples. By contrast, HIS, another BA with negative health effect, was detected in all samples. Histamine presence is regulated in some food, but none in tomatoes. The tolerance presence of histamine in wine and fish is up to 100 μg/g [64]. The maximum HIS level was found in TFR at 1.463 ± 0.015 μg/g, far less than the regulation limits (Commission Regulation (EU) No 1019/2013 of 23 October 2013 amending Annex I to Regulation (EC) No 2073/2005 as regards to histamine in fishery products.) For these reasons, these tomato cultivars appear to be safe for human health. The rest of BAs are usually related to cultivar and storage condition of samples. Indeed, a great variability of BAs concentration was found between SM and TF. In all the tomatoes, SER, an important neurotransmitter, was also detected in high concentration. This BA has positive effects on human health; SER plays an important role in regulating mood, sleep, body temperature, sexuality, and appetite. SER is involved in many neuropsychiatric disorders such as migraine, bipolar disorder; serotonin deficiency causes obsessive-compulsive disorder, repetition, and mania. So, its assumption by diet is highly recommended [65]. About the different level of the BAs in pink with respect to red tomatoes, is possible to highlight that in almost all samples, the highest concentration was found in red fruits (Table 7). This is in accordance with previously studies which demonstrated the accumulation of BAs during ripening of meat [66] or dairy products [67]. Finally, based on the obtained data, BAs could be also used as ripening markers of tomatoes.

Table 7. Biogenic amines determined by HPLC in SM and TF samples, at red and pink ripening stages ± Std. Dev. (μg/g) [a].

ADD HEADING	BPEA	PUT	CAD	HIS	SER	TYM	SPD	SPM
SM_P	0.170 ± 0.001	2.777 ± 0.137	1.127 ± 0.055	0.448 ± 0.027	277.760 ± 5.226	n.d.	0.235 ± 0.010	0.253 ± 0.013
SM_R	0.168 ± 0.001	7.564 ± 0.289	1.888 ± 0.025	1.363 ± 0.037	394.054 ± 12.725	n.d.	0.122 ± 0.006	0.346 ± 0.019
TF_P	0.165 ± 0.001	3.289 ± 0.006	1.287 ± 0.006	0.589 ± 0.032	258.679 ± 7.360	n.d.	0.116 ± 0.001	0.191 ± 0.002
TF_R	0.272 ± 0.050	6.293 ± 0.113	1.533 ± 0.033	1.463 ± 0.015	326.848 ± 8.850	n.d.	0.249 ± 0.010	0.477 ± 0.031

[a] BPEA: β-phenylethylamine, PUT: putrescine; CAD: cadaverine; HIS: histidine; SER: serotonin; TYM: tyramine; SPD: spermidine; SPM: spermine; n.d.: not detected.

3.3. Screening of Biological Activities

3.3.1. Antioxidant Activities

The radical scavenging properties of the tested extracts were evaluated against the synthetic chromogenic DPPH and ABTS radicals. Under our experimental conditions, all extracts (1–5000 μg mL^{-1}) were able to counteract the ABTS radical, despite a weak radical scavenger activity against DPPH, which achieved a lower than 40% inhibition at the highest tested concentration, thus hindering the IC_{50} evaluation (Figure S8). As expected, the positive control trolox (concentration range of 1–100 μg mL^{-1}) was found to be a potent scavenger of both DPPH and ABTS (Figure S8). The measurable IC_{50} values for the extracts and Trolox were displayed in Table 8.

Table 8. Effects of hydroalcoholic and organic extracts from both pink (*p*) and red (R) fruits of *Solanum lycopersicum* var. TF and SM, and standard antioxidant agents in the antioxidant assays [a].

Sample		IC_{50} (CL) (μg mL^{-1}) [a]		
	ADD HEADING	ABTS Radical Scavenging Activity	Ferric Ion Chelating Activity	Ferric Ion Reducing Activity
TF_P	Hydroalcoholic	371.2 (296.6–494.8)	106.6 (69.3–163.6)	-
	Organic	174.9 (119.6–255.8)	-	184.6 (109.3–433.6)
TF_R	Hydroalcoholic	-	208.5 (143.1–303.1)	-
	Organic	573.1 (463.3–708.9)	-	250.0 (152.1–410.5)
SM_P	Hydroalcoholic	629.9 (475.9–720.5)	93.5 (63.9–116.6)	-
	Organic	479.6 (358.7–663.3)	-	397.8 (319.3–438.7)
SM_R	Hydroalcoholic	-	242.7 (233.9–276.6)	-
	Organic	-	-	368.5 (289.3–468.6)
Positive control		2.5 (1.3–5.6) [b]	45.2 (13.1–75.5) [c]	1.5 (1.1–2.0) [b]

[a] CL, confidence limits; - not evaluable since a lower than 40% effect was achieved. [b] trolox; [c] quercetin.

Comparing the tomato varieties at the pink developmental stage, TF_P organic extract displayed the most potent ABTS scavenging activity, being the IC_{50} value about 2-, 3-, and 4-fold less than that of TF_P hydroalcoholic, and both SM_P organic and hydroalcoholic fractions, respectively (Table 8). Similarly, TF_R organic extract was the most effective scavenging sample from red fruits, followed by TF_R hydroalcoholic extract, although with a 3-fold lower potency compared to the TF_P sample (Table 8). Conversely, SM_R organic and hydroalcoholic extracts produced a lower than 50% ABTS inhibition at the highest tested concentrations, thus hindering the evaluation the IC_{50} value (Table 8).

ABTS and DPPH radicals are scavenged by electron- or hydrogen-transfer mechanisms, although with a different specificity and kinetic profile [68]. ABTS usually reacts with both lipophilic and hydrophilic compounds and possesses a poor selectivity in the reaction with hydrogen-atom donors; conversely, DPPH is more selective for small molecules, likely due to the limited steric accessibility of the radical site to larger compounds [68]. ABTS assay has been also reported to better estimate the antioxidant power of fruits and vegetables rich in hydrophilic, lipophilic, and high-pigmented antioxidant compounds compared to DPPH assay [69]. Particularly, carotenoids seem to not react with DPPH, while being able to bleach ABTS [70].

On the base of this evidence, the scavenging abilities of TF and SM extracts towards ABTS radical can be ascribed to the presence of both hydrophilic and lipophilic antioxidant phytochemicals. Among them, the antioxidant contribution of polyphenols, carotenoids, tocopherols, and vitamins C and E to the ABTS scavenging properties of tomato fruits has been previously hypothesized [71]. Further studies are required to clarify their involvement in the radical scavenging activity of TF and SM extracts.

When assessed in ferrozine assay, all samples exhibited a weak chelating activity of ferrous ion; conversely, the hydroalcoholic extracts of TF and SM fruits were able to chelate ferric ions (Table 8), being hydroalcoholic TF_P and SM_P extracts the most potent (IC_{50} values about two-fold lower than that of the corresponding red extracts). The positive control quercetin resulted to be about two- and four-fold more potent than the tested extracts (Table 8).

Despite a marked ferric chelating activity, the hydroalcoholic samples were ineffective as reducing agents; conversely, the organic extracts significantly reduced ferric ions, being that from TF_P the most potent (Table 8). According to the Pearson analysis, a significant correlation occurs between the ABTS scavenger power of TF hydroalcoholic and organic extract and the respective chelating and reducing activities (correlation coefficient r of 0.81 and 0.95, respectively).

Regarding the ferric thiocyanate assay, all the extracts showed an inhibitory activity of linoleic acid peroxidation, being the organic samples from both pink and red tomatoes the most effective ones (Figure 5). Among the tested extracts, TF_P and TF_R organic fractions induced about a 60 and 50% inhibition of lipid peroxidation already after 24 h incubation (Figure 5B), followed by a 50% of SM_P organic fraction and a lower than 40% inhibition of the other samples (Figure 5B).

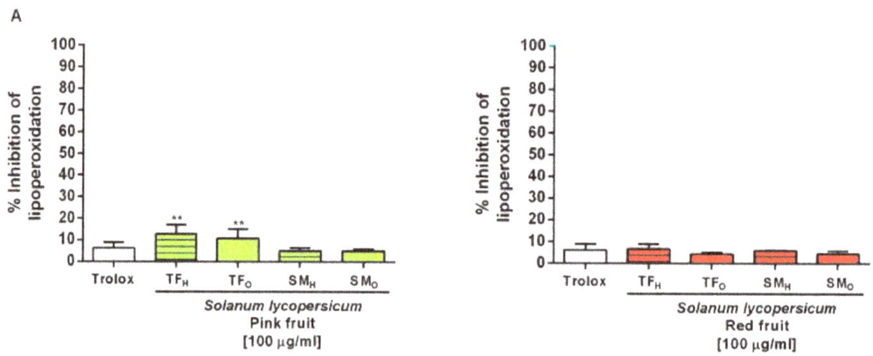

Figure 5. *Cont.*

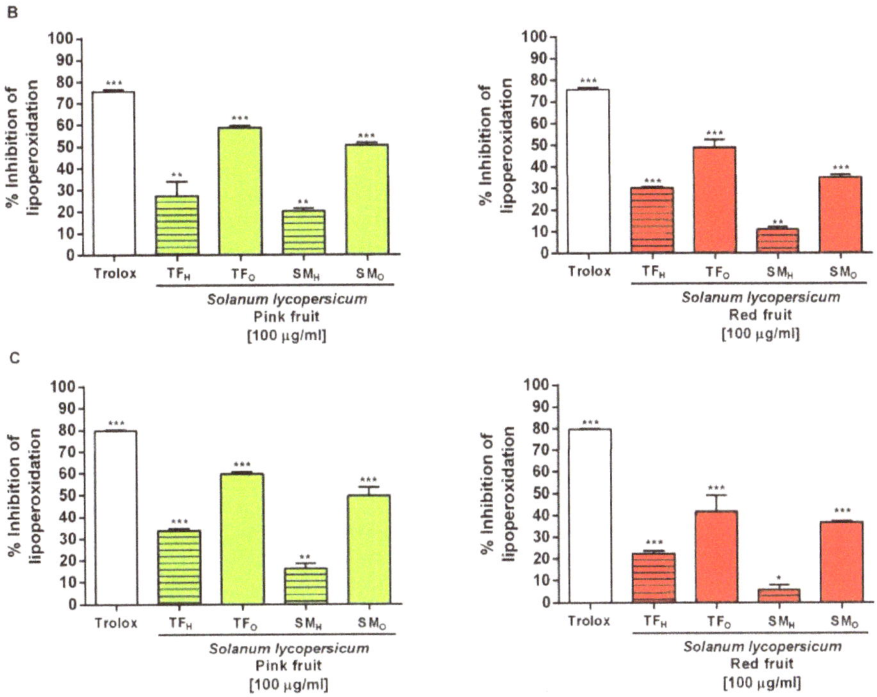

Figure 5. Inhibitory effects of the organic and hydroalcoholic extracts from *Solanum lycopersicum* var. TF and SM tomatoes, at pink (left) and red (right) stages, on linoleic acid peroxidation after different time exposure (**A**) t = 0, (**B**) t = 24 h, and(**C**) t = 48 h. TF organic (TF_O), TF hydroalcoholic (TF_H), SM organic (SM_O), SM hydroalcoholic (SM_H), extracts 100 μg mL^{-1}. * $p < 0.05$, ** $p < 0.01$, and *** $p < 0.01$, represent a statistically significant lipoperoxidation inhibition respect to the basal effect at t = 0 (Anova + Dunnett's multiple comparison post test).

3.3.2. Advanced Glycation End-Product (AGE) Inhibition

Growing evidence highlighted that phenolic compounds are able to prevent the production of advance glycation end products (AGEs), toxic metabolites accumulated under different pathologies, and responsible for the inflammatory and oxidative stress [37]. According to these data, the ability to interfere with AGE formation was assessed as a possible mechanism correlated to the antioxidant and cytoprotective power of TF and SM extracts. Therefore, treatment with AGEs inhibitors is believed to be a potential strategy for preventing diabetes complications.

Under the experimental conditions, despite a null activity of SM samples, both TF_P and TF_R hydroalcoholic fractions produced a concentration-dependent and statistically significant inhibition of the AGE production, although with a potency about 3-fold lower compared to rutin (positive control). The maximum 47% inhibition was achieved at the concentration of 1000 μg mL^{-1} of TF_R hydroalcoholic extract (Figure 6B). Furthermore, that from TF_P produced a maximum 44% inhibition at the highest tested concentration. According to literature [37], phenolic compounds could contribute to the observed effects.

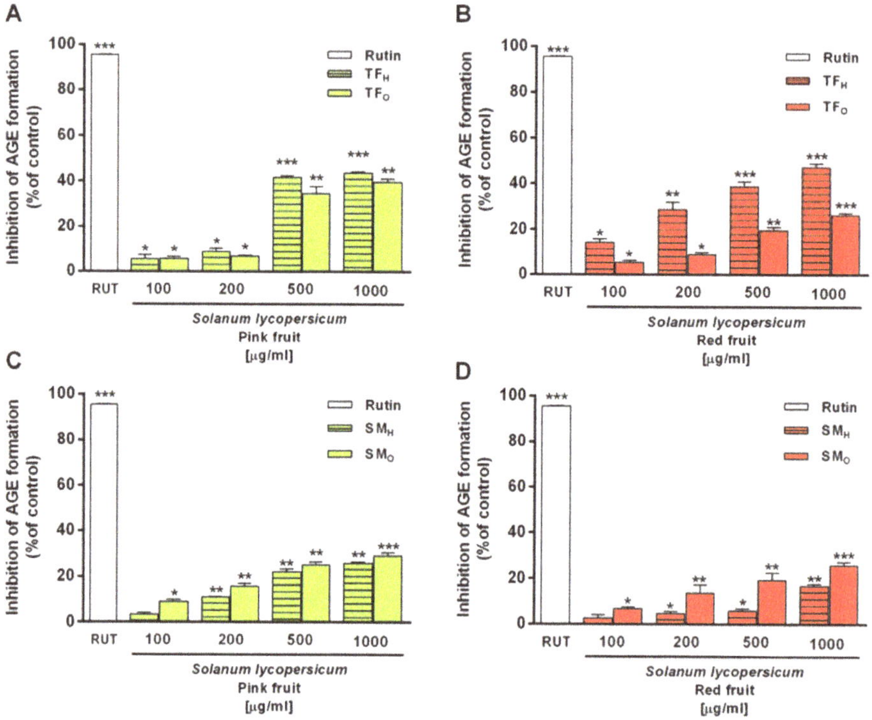

Figure 6. Inhibition of formation of advanced glycation end-products (AGE) induced by organic (O) and hydroalcoholic (H) extracts from *Solanum lycopersicum* var. (**A**) TF_P, (**B**) TF_R, (**C**) SM_P, (**D**) SM_R, and the positive control rutin [200 μg mL^{-1}]. RUT, rutin, TF_H, TF_O, SM_H, SM_O. Each value represents mean ± SEM ($n = 6$). * $p < 0.05$, ** $p < 0.01$, and *** $p < 0.01$, represent a statistically significant AGE inhibition compared to the basal level (Anova + Dunnett's multiple comparison post test).

3.3.3. In Vitro Metabolic Enzyme Inhibition

Taking into account that dietary phenolics are known to decrease the activity of α-amylase and α-glucosidase, thus lowering carbohydrate digestion and absorption [37], the ability of the tested samples was also evaluated to affect the function of both enzymes. Under our experimental conditions, the extracts resulted ineffective towards α-amylase enzyme, whereas a partial α-glucosidase inhibition (maximum 50% inhibition at the highest concentration of 1000 μg mL^{-1}) was found in the presence of the hydroalcoholic extracts of both pink and red TF tomatoes (data not shown), likely ascribable to the highest phenolic content.

3.3.4. Cytoprotection towards the Oxidative Stress Induced by tBuOOH

Preliminarily, the cytotoxicity of selected tomatoes samples on HepG2 cells was evaluated by MTT assay, thus highlighting that the extracts did not affect significantly the cell viability up to the concentration of 100 μg mL^{-1} after 24 h exposure, with early toxicity signs at higher concentrations (data not shown). On the basis of this evidence, the concentration of 100 μg mL^{-1} was used to study the ability of the extracts to inhibit the intracellular oxidative stress induced by tBuOOH after 2 h exposure.

Under our experimental conditions, tBuOOH produced a statistically significant increase of the intracellular ROS-level with respect to the vehicle control, reaching an oxidation index of 2.17 ± 0.04 (Figure 7), while the extracts alone did not affect the ROS levels (data not shown).

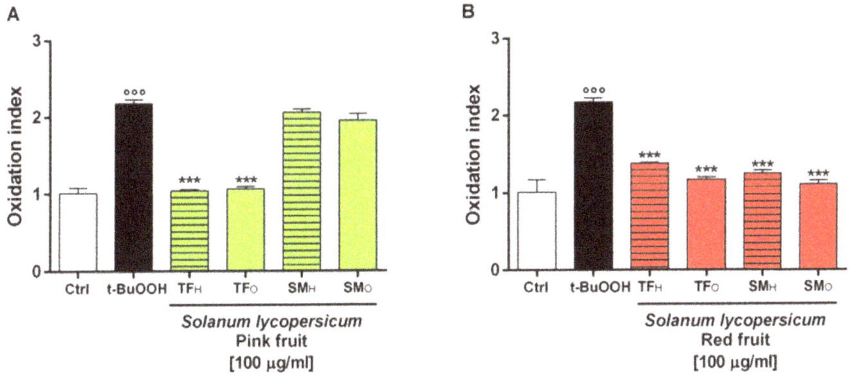

Figure 7. Effect of the organic and hydroalcoholic extracts from the fruits of *Solanum lycopersicum* var. Torpedino di Fondi (TF) and San Marzano (SM) at (**A**) pink and (**B**) red ripening stages on the *t*BuOOH-induced intracellular ROS levels by the DCFH-DA (2,7-dichlorofluorescein diacetate) assay. *t*BuOOH (5 mM) TF_H, TF_O, SM_H, SM_O. The oxidation index was obtained by the ration between the DCF fluorescence of the sample and that of the vehicle control (i.e., EtOH 1% *v/v*). *** $p < 0.001$, represent a statistically significant difference of the treatment with respect to *t*BuOOH (Anova + Dunnett's multiple comparison post test). °°° $p < 0.001$, represent a statistically significant difference of *t*BuOOH vs. Ctrl. $p < 0.001$, represent a statistically significant difference of *t*BuOOH vs. Ctrl.

When the cells were pre-treated overnight with the tested extracts, the pro-oxidant effect of tBuOOH resulted significantly reduced, although with different potency. Among pink tomato samples, both TF hydroalcoholic and organic extracts were able to halve the tBuOOH-induced oxidation (with 48% and 54% inhibition index respectively), thus exhibiting a strong antioxidant activity (Figure 7A). Conversely, the organic and hydroalcoholic extracts from SM_P fruits resulted ineffective, being the oxidation index similar to that of tBuOOH (Figure 7A). All the extracts from red tomatoes displayed marked antioxidant activity, with the oxidation index of tBuOOH reduced from 1.7- to 2-fold. For both varieties, the organic extracts were the most potent samples, achieving the inhibition levels of 46.0 and 49% for TF and SM, respectively (Figure 7B). Analogously, the hydroalcoholic extracts produced antioxidant effects against tBuOOH, although with lower potency, being the inhibition values of 36 and 43% for TF and SM, respectively (Figure 7B).

3.3.5. Antifungal Activity of SM and TF Hydroalcoholic and Organic Extracts

Irving and colleagues have shown antifungal activity of tomato plant extracts against *Candida albicans* ATCC 2091 [72]. In the present study, for the first time has been analyzed the anti-*Candida* activity of different extracts of tomato fruits from TF and SM cultivars against different *Candida* species such as *C. albicans*, *glabrata*, and *krusei*.

Candida is a human commensal in several anatomically distinct sites such as in the gastrointestinal tract. In specific environmental condition, *Candida* can switch to pathogen and can be responsible of some diseases. In the gut, patients with intestinal inflammation have high levels of *Candida* species when compared to healthy individuals [73]. The main *Candida* species isolated from the human gastrointestinal tract are *C. albicans*, *C. glabrata*, and *C. krusei* [74]. Antifungal activity of plants extracts was demonstrated on these species by the broth microdilution method. Between the two varieties tested, TF showed a better activity, against all *Candida* strains (Table S5). Moreover, organic fractions showed the best antifungal activity compared to the hydroalcoholic ones. In particular, organic TF_R and TF_P and hydroalcoholic TF_R and TF_P extracts showed a geometric (GM) MIC_{50} of 707 μg mL^{-1}, 841 μg mL^{-1}, 1361 μg mL^{-1}, 1101 μg mL^{-1}, respectively. Organic SM_R and SM_P, hydroalcoholic SM_R and SM_P fractions showed a GM MIC_{50} of 1236 μg mL^{-1}, 891 μg mL^{-1}, 1442 μg mL^{-1}, 1414 μg mL^{-1}, respectively (Table S5) In particular, TF_R organic extracts showed a GM MIC_{50} of 707 μg mL^{-1}, while the TF_R hydroalcoholic showed a GM MIC_{50} 1442 μg mL^{-1} against all *Candida* strains (Table S5).

Antibacterial and antifungal activity of several plant secondary metabolites and their derivatives such as alkaloids and polyphenols has been reported [75]. The alkaloid such as glycoalkaloid α-tomatine depicted antifungal effects against a variety of fungi [76]. In our results, tomatine was found only in TF_P and not in TF_R, indicating that probably, as hypothesized by some authors, the synergy of several compounds is responsible for the antifungal activity shown.

In conclusion, TF reduces *Candida* cells in the intestinal tract intake of TF, which have a growth inhibiting activity against different *Candida* species, and could be a strategy to restore the intestinal microbiota present in the healthy individual.

4. Conclusions

The combined application of both targeted and untargeted methodologies allowed to outline the chemical profile of both TF, a new hybrid cultivar recently introduced in south Lazio (Italy), and SM tomatoes at two ripening stages. We wish to itemize here in some detail both single important molecules and chemical classes to stimulate an active consideration of these highly complex natural mixtures, rich in compounds that may reveal novel important, hopefully beneficial roles in forthcoming studies. Some metabolites were shared by all extracts, though at different concentration, such as macronutrients like sugars and derivatives (hexose, sorbitol, mannosylglycerate), and amino acids (tryptophan and citrulline), the Amadori adduct fructoselysine, relevant biochemical intermediates (ornithine, chorismic acid, and GABA), terpenes (caryophyllene), nucleobase (adenosine), vitamin precursors and metabolites (diapophytoene, diapolycopene, α-tocopheronolactone), fatty alcohol (panaxytriol), organic acids (citric, chlorogenic, and azelaic acid) and conjugates (caffeic acid 3-glucoside, O-feruloylquinate), sterols (solagenin), free fatty acids (myristic, myristoleic, lauric, palmitic, oleic, linoleic, linolenic, eicosenoic acids). Differently, other metabolites might be considered marker compounds, being detected only in one or a few extracts, like glycyphyllin, maleic and tartaric acids (TF_P), cinnamoyl glucoside (TF_R), the vitamins dihydroretinol, dehydroretinal, retinoic and tocopheronic acids, and the antifungal terpenoid phytuberin (SM_R), the solanine derivative, tomatidinol (SM_P), the vitamin-E precursors, phytol (SM_R, TF_P), and γ-tocotrienol (SM_P, SM_R), the glycoalkaloids tomatine and tomatidine, suberic and ascorbic acids (SM_P, TF_P), quinic, phosphogluconic and shikimic acids (TF_R, TF_P), the polyketide lycoflexine (SM_R, SM_P, TF_P). These characteristic chemical features may concur to the excellent organoleptic properties as well as to antioxidant, antiglycative, and antifungal activities of Torpedino di Fondi, an emerging south Lazio tomato belonging to the mini-San Marzano type. This study may contribute to the unceasing buildup of reliable reference databases useful to guarantee food authenticity and freshness, and to support consumers and further nutraceutical evaluations.

Supplementary Materials: Supplementary materials can be found at http://www.mdpi.com/2076-3921/9/10/1027/s1, Figures S1–S5: TF and SM tomato fruits from Fondi area (Lazio region), Figure S6: Comparison between TFP and TFR and SMP and SMR their ripening stage for the number of possible CHO (A) and CHNO (B) metabolites, Figure S7: Common and uncommon features (in%) in the combined pattern of hydroalcoholic and organic extracts, Table S1: Amount of peel, pulp, seeds and juice in both pink and red Solanum lycopersicum var. TF and SM fruits; Table S2–3: Comprehensive list of metabolites detected in hydroalcoholic and organic fractions of pink and red San Marzano (SM) and Torpedino di Fondi (TF) extracts, Table S4: Overview of the relative abundancies of specific classes of saturated, mono-, di- and poly-unsaturated fatty acids (FA), Table S5: Antifungal activity of extracts against 4 *C.albicans* strains, 3 *C. glabrata* strains and 2 *C. krusei*.

Author Contributions: Conceptualization, C.I., A.M., M.E.C., and L.M.; methodology, A.F., and C.F.; validation, A.P.S., A.D.S., C.C., M.S.S., and A.F.; formal analysis, A.M.G., M.S., G.D.M., M.R., S.C., A.F., C.F., G.S., A.D.S. and S.D.G.; investigation, A.M.G., M.S., G.D.M., M.R., S.C., G.S., A.F., C.F., C.C., M.S.S., A.D.S. and S.D.G.; resources, C.I., A.M., M.S., and G.D.M.; data curation, A.P.S., A.D.S., S.F., C.C., M.S.S. and G.V.; writing—original draft preparation, C.I., A.M., A.D.S. and M.E.C.; writing—review and editing, C.I., A.M. and M.E.C.; visualization, C.I., A.M., and M.E.C.; supervision, S.F., M.E.C., G.V., A.P.S., and L.M.; project administration, L.M.; funding acquisition, S.F., M.E.C., and L.M. All authors have read and agreed to the published version of the manuscript.

Funding: This research was funded by REGIONE LAZIO, "e-ALIERB" Project LR13/2008–Dipartimento di Chimica e Tecnologie del Farmaco and Italian Ministry of Education, Universities and Research-Dipartimenti di Eccellenza-L. 232/2016 and by the EU Horizon 2020 Programme (EU_FT-ICR_MS, under grant number 731077). We also acknowledge the support from the Portuguese Mass Spectrometry Network (LISBOA-01-0145-FEDER-022125) and The Portuguese Foundation for Science and Technology FCT for contract CEECIND/02246/2017 to MSS.

Acknowledgments: Authors wish to thank the Società Cooperativa Agricola "San Leone" (Fondi-Sperlonga, Lazio region, Central Italy) and Mafalda s.r.l. for supplying the fresh raw material. Fellowship of Cinzia Ingallina and Silvia Di Giacomo was financed by "e-ALIERB" Project. Antonella Di Sotto was supported by Enrico and Enrica Sovena Foundation.

Conflicts of Interest: The authors declare no conflict of interest. The funders had no role in the design of the study; in the collection, analyses, or interpretation of data; in the writing of the manuscript, or in the decision to publish the results.

References

1. Rao, R.; Corrado, G.; Bianchi, M.; Di Mauro, A. (GATA)4 DNA fingerprinting identifies morphologically characterized "San Marzano" tomato plants. *Plant Breed.* **2006**, *125*, 173–176. [CrossRef]
2. Loiudice, R.; Impembo, M.; Laratta, B.; Villari, G.; Lo Voi, A.; Siviero, P.; Castaldo, D. Composition of San Marzano tomato varieties. *Food Chem.* **1995**, *53*, 81–89. [CrossRef]
3. Dono, G.; Rambla, J.L.; Frusciante, S.; Granell, A.; Diretto, G.; Mazzucato, A. Color mutations alter the biochemical composition in the san marzano tomato fruit. *Metabolites* **2020**, *10*, 110. [CrossRef]
4. D'Esposito, D.; Ferriello, F.; Molin, A.D.; Diretto, G.; Sacco, A.; Minio, A.; Barone, A.; Di Monaco, R.; Cavella, S.; Tardella, L.; et al. Unraveling the complexity of transcriptomic, metabolomic and quality environmental response of tomato fruit. *BMC Plant Biol.* **2017**, *17*, 1–18. [CrossRef] [PubMed]
5. Ercolano, M.R.; Carli, P.; Soria, A.; Cascone, A.; Fogliano, V.; Frusciante, L.; Barone, A. Biochemical, sensorial and genomic profiling of traditional Italian tomato varieties. *Euphytica* **2008**, *164*, 571–582. [CrossRef]
6. Sobolev, A.P.; Circi, S.; Capitani, D.; Ingallina, C.; Mannina, L. Molecular fingerprinting of food authenticity. *Curr. Opin. Food Sci.* **2017**, *16*. [CrossRef]
7. Sobolev, A.P.; Thomas, F.; Donarski, J.; Ingallina, C.; Circi, S.; Cesare Marincola, F.; Capitani, D.; Mannina, L. Use of NMR applications to tackle future food fraud issues. *Trends Food Sci. Technol.* **2019**, *91*, 347–353. [CrossRef]
8. Dono, G.; Picarella, M.E.; Pons, C.; Santangelo, E.; Monforte, A.; Granell, A.; Mazzucato, A. Characterization of a repertoire of tomato fruit genetic variants in the San marzano genetic background. *Sci. Hortic.* **2020**, *261*, 108927. [CrossRef]
9. Macone, A.; Fontana, M.; Barba, M.; Botta, B.; Nardini, M.; Ghirga, F.; Calcaterra, A.; Pecci, L.; Matarese, R.M. Antioxidant Properties of Aminoethylcysteine Ketimine Decarboxylated Dimer: A Review. *Int. J. Mol. Sci.* **2011**, *12*, 3072–3084. [CrossRef]
10. Burton-Freeman, B.; Reimers, K. Tomato Consumption and Health: Emerging Benefits. *Am. J. Lifestyle Med.* **2011**, *5*, 182–191. [CrossRef]
11. Del Giudice, R.; Raiola, A.; Tenore, G.C.; Frusciante, L.; Barone, A.; Monti, D.M.; Rigano, M.M. Antioxidant bioactive compounds in tomato fruits at different ripening stages and their effects on normal and cancer cells. *J. Funct. Foods* **2015**, *18*, 83–94. [CrossRef]
12. García-Valverde, V.; Navarro-González, I.; García-Alonso, J.; Periago, M.J. Antioxidant Bioactive Compounds in Selected Industrial Processing and Fresh Consumption Tomato Cultivars. *Food Bioprocess Technol.* **2013**, *6*, 391–402. [CrossRef]
13. Slimestad, R.; Verheul, M. Review of flavonoids and other phenolics from fruits of different tomato (*Lycopersicon esculentum Mill.*) cultivars. *J. Sci. Food Agric.* **2009**, *89*, 1255–1270. [CrossRef]
14. Fratianni, F.; Cozzolino, A.; D'Acierno, A.; Nazzaro, F.; Riccardi, R.; Spigno, P. Qualitative Aspects of Some Traditional Landraces of the Tomato "Piennolo" (*Solanum lycopersicum* L.) of the Campania Region, Southern Italy. *Antioxidants* **2020**, *9*, 565. [CrossRef] [PubMed]
15. Anton, D.; Matt, D.; Pedastsaar, P.; Bender, I.; Kazimierczak, R.; Roasto, M.; Kaart, T.; Luik, A.; Püssa, T. Three-Year Comparative Study of Polyphenol Contents and Antioxidant Capacities in Fruits of Tomato (*Lycopersicon esculentum Mill.*) Cultivars Grown under Organic and Conventional Conditions. *J. Agric. Food Chem.* **2014**, *62*, 5173–5180. [CrossRef] [PubMed]

16. Tan, S.; Ke, Z.; Chai, D.; Miao, Y.; Luo, K.; Li, W. Lycopene, polyphenols and antioxidant activities of three characteristic tomato cultivars subjected to two drying methods. *Food Chem.* **2021**, *338*, 128062. [CrossRef] [PubMed]
17. Devadas, S.M.; Giffen, S.R.; Kumar, N.; Lobo, R.; Ballal, M. Activity of Solanum lycopersicum against Candida species isolated from retro-positive patients—An invitro study. *J. Pharm. Sci. Res.* **2017**, *9*, 1233–1236.
18. Añibarro-Ortega, M.; Pinela, J.; Ćirić, A.; Martins, V.; Rocha, F.; Soković, M.D.; Barata, A.M.; Carvalho, A.M.; Barros, L.; Ferreira, I.C.F.R. Valorisation of table tomato crop by-products: Phenolic profiles and in vitro antioxidant and antimicrobial activities. *Food Bioprod. Process.* **2020**, *124*, 307–319. [CrossRef]
19. Herbel, V.; Schäfer, H.; Wink, M. Recombinant Production of Snakin-2 (an Antimicrobial Peptide from Tomato) in *E. coli* and Analysis of Its Bioactivity. *Molecules* **2015**, *20*, 14889–14901. [CrossRef]
20. Vasile Rusu, A.; Alvarez Penedo, B.; Schwarze, A.-K.; Trif, M. The Influence of Candida spp. in Intestinal Microbiota; Diet Therapy, the Emerging Conditions Related to Candida in Athletes and Elderly People. In *Candidiasis [Working Title]*; IntechOpen: London, UK, 2020.
21. Sobolev, A.P.; Mannina, L.; Capitani, D.; Sanzò, G.; Ingallina, C.; Botta, B.; Fornarini, S.; Crestoni, M.E.; Chiavarino, B.; Carradori, S.; et al. A multi-methodological approach in the study of Italian PDO "Cornetto di Pontecorvo" red sweet pepper. *Food Chem.* **2018**, *255*. [CrossRef]
22. Ingallina, C.; Capitani, D.; Mannina, L.; Carradori, S.; Locatelli, M.; Di Sotto, A.; Di Giacomo, S.; Toniolo, C.; Pasqua, G.; Valletta, A.; et al. Phytochemical and biological characterization of Italian "sedano bianco di Sperlonga" Protected Geographical Indication celery ecotype: A multimethodological approach. *Food Chem.* **2020**, *309*, 125649. [CrossRef] [PubMed]
23. Circi, S.; Capitani, D.; Randazzo, A.; Ingallina, C.; Mannina, L.; Sobolev, A.P. Panel test and chemical analyses of commercial olive oils: A comparative study. *Chem. Biol. Technol. Agric.* **2017**, *4*. [CrossRef]
24. Ingallina, C.; Sobolev, A.P.; Circi, S.; Spano, M.; Fraschetti, C.; Filippi, A.; Di Sotto, A.; Di Giacomo, S.; Mazzoccanti, G.; Gasparrini, F.; et al. *Cannabis sativa* L. Inflorescences from Monoecious Cultivars Grown in Central Italy: An Untargeted Chemical Characterization from Early Flowering to Ripening. *Molecules* **2020**, *25*, 1908. [CrossRef]
25. Solovchenko, A.E.; Chivkunova, O.B.; Merzlyak, M.N.; Reshetnikova, I.V. A Spectrophotometric Analysis of Pigments in Apples. *Russ. J. Plant Physiol.* **2001**, *48*, 693–700. [CrossRef]
26. Folch, J.; Lees, M.; Sloane Stanley, G.H. A simple method for the isolation and purification of total lipides from animal tissues. *J. Biol. Chem.* **1957**, *226*, 497–509.
27. Wellburn, A.R. The Spectral Determination of Chlorophylls a and b, as well as Total Carotenoids, Using Various Solvents with Spectrophotometers of Different Resolution. *J. Plant Physiol.* **1994**, *144*, 307–313. [CrossRef]
28. Ingallina, C.; Sobolev, A.P.; Circi, S.; Spano, M.; Giusti, A.M.; Mannina, L. New hybrid tomato cultivars: An NMR-based chemical characterization. *Appl. Sci.* **2020**, *10*, 1887. [CrossRef]
29. Marshall, A.G.; Hendrickson, C.L.; Jackson, G.S. Fourier transform ion cyclotron resonance mass spectrometry: A primer. *Mass Spectrom. Rev.* **1998**, *17*, 1–35. [CrossRef]
30. MassTRIX. Available online: http://masstrix3.helmholtz-muenchen.de/masstrix3/start) (accessed on 20 April 2020).
31. Kind, T.; Fiehn, O. Seven Golden Rules for heuristic filtering of molecular formulas obtained by accurate mass spectrometry. *BMC Bioinform.* **2007**, *8*, 105. [CrossRef] [PubMed]
32. Kim, S.; Kramer, R.W.; Hatcher, P.G. Graphical Method for Analysis of Ultrahigh-Resolution Broadband Mass Spectra of Natural Organic Matter, the Van Krevelen Diagram. *Anal. Chem.* **2003**, *75*, 5336–5344. [CrossRef] [PubMed]
33. Preti, R.; Rapa, M.; Vinci, G. Effect of Steaming and Boiling on the Antioxidant Properties and Biogenic Amines Content in Green Bean (*Phaseolus vulgaris*) Varieties of Different Colours. *J. Food Qual.* **2017**, *2017*, 1–8. [CrossRef]
34. Chiacchierini, E.; Restuccia, D.; Vinci, G. Evaluation of two different extraction methods for chromatographic determination of bioactive amines in tomato products. *Talanta* **2006**, *69*, 548–555. [CrossRef] [PubMed]
35. Di Sotto, A.; Vecchiato, M.; Abete, L.; Toniolo, C.; Giusti, A.M.; Mannina, L.; Locatelli, M.; Nicoletti, M.; Di Giacomo, S. *Capsicum annuum* L. var. Cornetto di Pontecorvo PDO: Polyphenolic profile and in vitro biological activities. *J. Funct. Foods* **2018**, *40*, 679–691. [CrossRef]
36. Di Sotto, A.; Di Giacomo, S.; Toniolo, C.; Nicoletti, M.; Mazzanti, G. *Sisymbrium Officinale* (L.) Scop. and its Polyphenolic Fractions Inhibit the Mutagenicity of Tert-Butylhydroperoxide in Escherichia Coli WP2 uvr AR Strain. *Phyther. Res.* **2016**, *30*, 829–834. [CrossRef]

37. Di Sotto, A.; Locatelli, M.; Macone, A.; Toniolo, C.; Cesa, S.; Carradori, S.; Eufemi, M.; Mazzanti, G.; Di Giacomo, S. Hypoglycemic, Antiglycation, and Cytoprotective Properties of a Phenol-Rich Extract from Waste Peel of *Punica granatum* L. var. Dente di Cavallo DC2. *Molecules* **2019**, *24*, 3103. [CrossRef] [PubMed]
38. Di Sotto, A.; Irannejad, H.; Eufemi, M.; Mancinelli, R.; Abete, L.; Mammola, C.L.; Altieri, F.; Mazzanti, G.; Di Giacomo, S. Potentiation of Low-Dose Doxorubicin Cytotoxicity by Affecting P-Glycoprotein through Caryophyllane Sesquiterpenes in HepG2 Cells: An in Vitro and in Silico Study. *Int. J. Mol. Sci.* **2020**, *21*, 633. [CrossRef]
39. Vitalone, A.; Di Giacomo, S.; Di Sotto, A.; Franchitto, A.; Mammola, C.L.; Mariani, P.; Mastrangelo, S.; Mazzanti, G. Cassia angustifolia Extract Is Not Hepatotoxic in an in vitro and in vivo Study. *Pharmacology* **2011**, *88*, 252–259. [CrossRef]
40. Di Giacomo, S.; Abete, L.; Cocchiola, R.; Mazzanti, G.; Eufemi, M.; Di Sotto, A. Caryophyllane sesquiterpenes inhibit DNA-damage by tobacco smoke in bacterial and mammalian cells. *Food Chem. Toxicol.* **2018**, *111*, 393–404. [CrossRef]
41. Clinical and Laboratory Standards Institute. *Reference Method for Broth Dilution Antifungal Susceptibility Testing of Yeasts CLSI Document M27*, 4th ed.; Clinical and Laboratory Standards Institute: Wayne, PA, USA, 2017; ISBN 1-56238-827-4.
42. Schaal, B.A. Reproductive Capacity and Seed Size in Lupinus Texensis. *Am. J. Bot.* **1980**, *67*, 703–709. [CrossRef]
43. Lichtenthaler, H.K.; Buschmann, C. Chlorophylls and Carotenoids: Measurement and Characterization by UV-VIS Spectroscopy. *Curr. Protoc. Food Anal. Chem.* **2001**, *1*, F4.3.1–F4.3.8. [CrossRef]
44. Sobolev, A.P.; Segre, A.; Lamanna, R. Proton high-field NMR study of tomato juice. *Magn. Reson. Chem.* **2003**, *41*, 237–245. [CrossRef]
45. Pérez, E.M.S.; Iglesias, M.J.; Ortiz, F.L.; Pérez, I.S.; Galera, M.M. Study of the suitability of HRMAS NMR for metabolic profiling of tomatoes: Application to tissue differentiation and fruit ripening. *Food Chem.* **2010**, *122*, 877–887. [CrossRef]
46. Hohmann, M.; Christoph, N.; Wachter, H.; Holzgrabe, U. 1 H NMR Profiling as an Approach To Differentiate Conventionally and Organically Grown Tomatoes. *J. Agric. Food Chem.* **2014**, *62*, 8530–8540. [CrossRef]
47. Carrari, F.; Baxter, C.; Usadel, B.; Urbanczyk-Wochniak, E.; Zanor, M.-I.; Nunes-Nesi, A.; Nikiforova, V.; Centero, D.; Ratzka, A.; Pauly, M.; et al. Integrated Analysis of Metabolite and Transcript Levels Reveals the Metabolic Shifts That Underlie Tomato Fruit Development and Highlight Regulatory Aspects of Metabolic Network Behavior. *Plant Physiol.* **2006**, *142*, 1380–1396. [CrossRef]
48. Mounet, F.; Lemaire-Chamley, M.; Maucourt, M.; Cabasson, C.; Giraudel, J.L.; Deborde, C.; Lessire, R.; Gallusci, P.; Bertrand, A.; Gaudillère, M.; et al. Quantitative metabolic profiles of tomato flesh and seeds during fruit development: Complementary analysis with ANN and PCA. *Metabolomics* **2007**, *3*, 273–288. [CrossRef]
49. Agius, C.; von Tucher, S.; Poppenberger, B.; Rozhon, W. Quantification of Glutamate and Aspartate by Ultra-High Performance Liquid Chromatography. *Molecules* **2018**, *23*, 1389. [CrossRef]
50. Sorrequieta, A.; Ferraro, G.; Boggio, S.B.; Valle, E.M. Free amino acid production during tomato fruit ripening: A focus on l-glutamate. *Amino Acids* **2010**, *38*, 1523–1532. [CrossRef]
51. Boggio, S.B.; Palatnik, J.F.; Heldt, H.W.; Valle, E.M. Changes in amino acid composition and nitrogen metabolizing enzymes in ripening fruits of Lycopersicon esculentum Mill. *Plant Sci.* **2000**, *159*, 125–133. [CrossRef]
52. Ghirga, F.; Quaglio, D.; Ghirga, P.; Berardozzi, S.; Zappia, G.; Botta, B.; Mori, M.; D'Acquarica, I. Occurrence of Enantioselectivity in Nature: The Case of (S)-Norcoclaurine. *Chirality* **2016**, *28*, 169–180. [CrossRef] [PubMed]
53. Gómez-Romero, M.; Segura-Carretero, A.; Fernández-Gutiérrez, A. Metabolite profiling and quantification of phenolic compounds in methanol extracts of tomato fruit. *Phytochemistry* **2010**, *71*, 1848–1864. [CrossRef] [PubMed]
54. Jézéquel, T.; Deborde, C.; Maucourt, M.; Zhendre, V.; Moing, A.; Giraudeau, P. Absolute quantification of metabolites in tomato fruit extracts by fast 2D NMR. *Metabolomics* **2015**, *11*, 1231–1242. [CrossRef]
55. Sanz, M.L.; del Castillo, M.D.; Corzo, N.; Olano, A. Presence of 2-Furoylmethyl Derivatives in Hydrolysates of Processed Tomato Products. *J. Agric. Food Chem.* **2000**, *48*, 468–471. [CrossRef] [PubMed]

56. Bueno, M.J.M.; Díaz-Galiano, F.J.; Rajski, Ł.; Cutillas, V.; Fernández-Alba, A.R. A non-targeted metabolomic approach to identify food markers to support discrimination between organic and conventional tomato crops. *J. Chromatogr. A* **2018**, *1546*, 66–76. [CrossRef]
57. Iijima, Y.; Watanabe, B.; Sasaki, R.; Takenaka, M.; Ono, H.; Sakurai, N.; Umemoto, N.; Suzuki, H.; Shibata, D.; Aoki, K. Steroidal glycoalkaloid profiling and structures of glycoalkaloids in wild tomato fruit. *Phytochemistry* **2013**, *95*, 145–157. [CrossRef]
58. Van Meulebroek, L.; Bussche, J.V.; Steppe, K.; Vanhaecke, L. Ultra-high performance liquid chromatography coupled to high resolution Orbitrap mass spectrometry for metabolomic profiling of the endogenous phytohormonal status of the tomato plant. *J. Chromatogr. A* **2012**, *1260*, 67–80. [CrossRef]
59. Liberati-Čizmek, A.-M.; Biluš, M.; Brkić, A.L.; Barić, I.C.; Bakula, M.; Hozić, A.; Cindrić, M. Analysis of Fatty Acid Esters of Hydroxyl Fatty Acid in Selected Plant Food. *Plant Foods Hum. Nutr.* **2019**, *74*, 235–240. [CrossRef]
60. Tommonaro, G.; de Prisco, R.; Abbamondi, G.R.; Marzocco, S.; Saturnino, C.; Poli, A.; Nicolaus, B. Evaluation of Antioxidant Properties, Total Phenolic Content, and Biological Activities of New Tomato Hybrids of Industrial Interest. *J. Med. Food* **2012**, *15*, 483–489. [CrossRef] [PubMed]
61. Iijima, Y.; Suda, K.; Suzuki, T.; Aoki, K.; Shibata, D. Metabolite Profiling of Chalcones and Flavanones in Tomato Fruit. *J. Jpn. Soc. Hortic. Sci.* **2008**, *77*, 94–102. [CrossRef]
62. Chen, J.; Green, K.B.; Nichols, K.K. Quantitative Profiling of Major Neutral Lipid Classes in Human Meibum by Direct Infusion Electrospray Ionization Mass Spectrometry. *Investig. Opthalmol. Vis. Sci.* **2013**, *54*, 5730. [CrossRef]
63. Berer, K.; Martínez, I.; Walker, A.; Kunkel, B.; Schmitt-Kopplin, P.; Walter, J.; Krishnamoorthy, G. Dietary non-fermentable fiber prevents autoimmune neurological disease by changing gut metabolic and immune status. *Sci. Rep.* **2018**, *8*, 10431. [CrossRef]
64. Guo, Y.-Y.; Yang, Y.-P.; Peng, Q.; Han, Y. Biogenic amines in wine: A review. *Int. J. Food Sci. Technol.* **2015**, *50*, 1523–1532. [CrossRef]
65. Briguglio, M.; Dell'Osso, B.; Panzica, G.; Malgaroli, A.; Banfi, G.; Zanaboni Dina, C.; Galentino, R.; Porta, M. Dietary Neurotransmitters: A Narrative Review on Current Knowledge. *Nutrients* **2018**, *10*, 591. [CrossRef] [PubMed]
66. Suzzi, G. Biogenic amines in dry fermented sausages: A review. *Int. J. Food Microbiol.* **2003**, *88*, 41–54. [CrossRef]
67. Linares, D.M.; del Río, B.; Ladero, V.; Martínez, N.; Fernández, M.; Martín, M.C.; Álvarez, M.A. Factors Influencing Biogenic Amines Accumulation in Dairy Products. *Front. Microbiol.* **2012**, 3. [CrossRef]
68. Prior, R.L.; Wu, X.; Schaich, K. Standardized Methods for the Determination of Antioxidant Capacity and Phenolics in Foods and Dietary Supplements. *J. Agric. Food Chem.* **2005**, *53*, 4290–4302. [CrossRef]
69. Floegel, A.; Kim, D.-O.; Chung, S.-J.; Koo, S.I.; Chun, O.K. Comparison of ABTS/DPPH assays to measure antioxidant capacity in popular antioxidant-rich US foods. *J. Food Compos. Anal.* **2011**, *24*, 1043–1048. [CrossRef]
70. Müller, L.; Fröhlich, K.; Böhm, V. Comparative antioxidant activities of carotenoids measured by ferric reducing antioxidant power (FRAP), ABTS bleaching assay (αTEAC), DPPH assay and peroxyl radical scavenging assay. *Food Chem.* **2011**, *129*, 139–148. [CrossRef]
71. Kotíková, Z.; Lachman, J.; Hejtmánková, A.; Hejtmánková, K. Determination of antioxidant activity and antioxidant content in tomato varieties and evaluation of mutual interactions between antioxidants. *LWT Food Sci. Technol.* **2011**, *44*, 1703–1710. [CrossRef]
72. Irving, G.W.; Fontaine, T.D.; Doolittle, S.P. Partial Antibiotic Spectrum Of Tomatin, an Antibiotic Agent from the Tomato Plant 12. *J. Bacteriol.* **1946**, *52*, 601–607. [CrossRef]
73. Charlet, R.; Bortolus, C.; Sendid, B.; Jawhara, S. Bacteroides thetaiotaomicron and Lactobacillus johnsonii modulate intestinal inflammation and eliminate fungi via enzymatic hydrolysis of the fungal cell wall. *Sci. Rep.* **2020**, *10*, 11510. [CrossRef]
74. Biasoli, M.S.; Tosello, M.E.; Magaró, H.M. Adherence of Candida strains isolated from the human gastrointestinal tract. *Mycoses* **2002**, *45*, 465–469. [CrossRef]

75. Othman, L.; Sleiman, A.; Abdel-Massih, R.M. Antimicrobial Activity of Polyphenols and Alkaloids in Middle Eastern Plants. *Front. Microbiol.* **2019**, *10*. [CrossRef] [PubMed]
76. Khan, H.; Mubarak, M.S.; Amin, S. Antifungal Potential of Alkaloids As An Emerging Therapeutic Target. *Curr. Drug Targets* **2017**, *18*. [CrossRef] [PubMed]

Publisher's Note: MDPI stays neutral with regard to jurisdictional claims in published maps and institutional affiliations.

antioxidants

MDPI

Article

Virgin Olive Oil Extracts Reduce Oxidative Stress and Modulate Cholesterol Metabolism: Comparison between Oils Obtained with Traditional and Innovative Processes

Carmen Lammi [1,*], Nadia Mulinacci [2], Lorenzo Cecchi [2], Maria Bellumori [2], Carlotta Bollati [1], Martina Bartolomei [1], Carlo Franchini [3], Maria Lisa Clodoveo [4], Filomena Corbo [3] and Anna Arnoldi [1]

1 Department of Pharmaceutical Sciences, University of Milan, 20133 Milan, Italy; carlotta.bollati@unimi.it (C.B.); martina.bartolomei@unimi.it (M.B.); anna.arnoldi@unimi.it (A.A.)

2 Department of Neuroscience, Psychology, Drug and Child Health, Pharmaceutical and Nutraceutical Section, University of Florence, 50019 Florence, Italy; nadia.mulinacci@unifi.it (N.M.); lo.cecchi@unifi.it (L.C.); maria.bellumori@unifi.it (M.B.)

3 Department of Pharmacy-Pharmaceutical Sciences, University Aldo Moro Bari, 70125 Bari, Italy; carlo.franchini@uniba.it (C.F.); filomena.corbo@uniba.it (F.C.)

4 Interdisciplinary Department of Medicine, University Aldo Moro Bari, 70125 Bari, Italy; marialisa.clodoveo@uniba.it

* Correspondence: carmen.lammi@unimi.it; Tel.: +39-025-031-9372

Received: 26 July 2020; Accepted: 24 August 2020; Published: 27 August 2020

Abstract: This study was aimed at demonstrating the substantial equivalence of two extra virgin olive oil samples extracted from the same batch of Coratina olives with (OMU) or without (OMN) using ultrasound technology, by performing chemical, biochemical, and cellular investigations. The volatile organic compounds compositions and phenolic profiles were very similar, showing that, while increasing the extraction yields, the innovative process does not change these features. The antioxidant and hypocholesterolemic activities of the extra virgin olive oil (EVOO) phenol extracts were also preserved, since OMU and OMN had equivalent abilities to scavenge the 1,1-diphenyl-2-picrylhydrazyl (DPPH) and 2,2′-azino-bis(3-ethylbenzothiazoline-6-sulfonic acid) diammonium salt (ABTS) radicals in vitro and to protect HepG2 cells from oxidative stress induced by H_2O_2, reducing intracellular reactive oxygen species (ROS) and lipid peroxidation levels. In addition, by inhibiting 3-hydroxy-3-methylglutarylcoenzyme a reductase, both samples modulated the low-density lipoprotein receptor (LDLR) pathway leading to increased LDLR protein levels and activity.

Keywords: antioxidant; HepG2 cells; EVOO extract; IOC methods; LDLR; PCSK9

1. Introduction

New lifestyles, higher incomes, and consumer awareness are creating consumer demand for high-quality, diverse, and innovative food products. One of the goals of the European Union is to guarantee food safety and security in a changing world, under the effects of climate change, resource paucity, and population dynamics. The development of innovative technologies and sustainable business models for food systems is a crucial factor for boosting the competitiveness of the European industry [1]. European research programs supporting new technologies and innovative products are continuously introduced in the food market. However, in respect to most of the other industrial sectors, food consumers are not very favorable to modify their customs and to accept innovations, partly due to a phenomenon known as neophobia that is the rejection of new or unfamiliar foods [2].

Our interdisciplinary research team is applying an innovative and sustainable process method to produce high-quality, cost-effective, and resource-efficient extra virgin olive oil (EVOO), employing an emerging technology based on the simultaneous treatment of olive paste both with ultrasound and heat-exchange [3–5]. The application of ultrasound on olive oil extraction has been studied by many authors in different olive growing areas [3–16]. Ultrasound is sound waves with frequencies higher than the upper audible limit of human hearing (greater than 20 kHz) [17,18]. Due to the mechanical effects of the sound waves within the olive paste, it is possible to eliminate the malaxation, the only batch phase in the continuous extraction process. When the ultrasound wave is propagating in the crushed olive paste, it determines an alternation of positive and negative pressures inside it. When the negative pressure values are below the water vapor pressure in the olive paste, it undergoes a phase change from liquid to gas, forming cavities containing steam and giving rise to the phenomenon of cavitation. Cavitation is a physical phenomenon consisting of the formation of cold vapor bubbles inside a fluid that then implode, producing shock waves. If implosion occurs near the cell wall of the olive fruit, it generates a liquid microjet that breaks the wall, freeing oil and minor compounds [16].

This is a ready-to-market technological solution, combined with practices and management strategies to help olive oil producers to increase the production yields [13], while preserving the healthy properties of the oil, improving the process efficiency, and valorizing the wastes [14–16]. The application of this strategic innovation may occur on a large scale, although only if this type of EVOO is well accepted by the market and a premium price is recognized for its high quality, overcoming the natural neophobia that accompanies every innovation in a traditional food sector, such as EVOO [17,18]. In fact, the consumer may percept a food produced with an innovative process as less natural and less good than a traditional one. In order to improve the diffusion of this strategic innovation in the oil sector, it is thus mandatory to demonstrate the substantial equivalence of the innovative and traditional products.

This investigation was conducted on two EVOO samples extracted from the same batch of fruits of the cultivar Coratina with (OMU) or without (OMN) using the ultrasound technology. The first objective of the study was to show that the two samples had comparable volatile organic compounds (VOCs) profiles, because this profile is more and more used in chemical/statistical models for supporting the panel test in virgin olive oil classification [19]. The VOCs profile was determined using a validated solid-phase microextraction (HS-SPME) followed by gas chromatography (GC) coupled with mass spectrometry (MS) (HS-SPME-GC-MS) method for the reliable quantification of 73 VOCs based on the use of the Multiple Internal Standard Normalization (MISN) and 73 calibration lines built with authentic external standards [19,20].

The second objective was to verify the similarity of the phenolic profiles since phenols provide relevant health benefits. In particular, the European Union has recently approved the health claim that "olive oil polyphenols contribute to the protection of blood lipids from oxidative stress" and that "the claim may be used only for olive oil which contains at least 5 mg of hydroxytyrosol and its derivatives (e.g., oleuropein complex and tyrosol) per 20 g of olive oil" [21,22]. The detailed phenolic profiles were determined by using the International Olive Council (IOC) official method for total phenols as well as a recently validated hydrolytic procedure for total hydroxytyrosol (HT) and tyrosol (Tyr) [21].

Furthermore, another objective was to prove that the new process does not modify or impair the biological properties of the EVOO phenols in term of the antioxidant and hypocholesterolemic properties. The antioxidant activity was evaluated in vitro by employing the 1,1-diphenyl-2-picrylhydrazyl (DPPH) radical assay as well as the 2,2′-azino-bis(3-ethylbenzothiazoline-6-sulfonic acid) diammonium salt (ABTS) assay, whereas the capacity to reduce the level of intracellular radical oxygen species (ROS) and lipid peroxidation was measured in HepG2 where the oxidative stress was induced by H_2O_2. Finally, an in-depth investigation was performed on the capacity of both extracts to modulate cholesterol metabolism, by carrying out molecular and functional experiments. In fact, in a previous work [23], we had shown that OMN is able to modulate, in a favorable way, this metabolic pathway. For this reason, OMN was used as reference extract and the evaluation of the OMU ability to modulate the LDLR-pathway was carried out in order to assess how the ultrasound process affects the biological activity of the new extract.

2. Materials and Methods

2.1. Chemicals

All chemicals employed are from commercial sources. See Supplementary Materials for further details.

2.2. Sonicated Extra Virgin Olive Oil

The oils were produced from olives from the cultivar Coratina cultivated in Apulia in the period 2017–2018 in industrial plants producing olive oil (1500–3000 kg/h) located in Apulia Region (Frantoio MIMI). The extraction line was equipped with a two-phase centrifugal system for oil separation. The mill had a hammer crusher and the malaxers were hermetically closed. The homogeneous batches of olives were divided into samples of 800 kg each. After the selection and the washing, the olives were crushed. The crushed olive paste was then passed into the Sono Heat Exchanger (SHE) and then fell into the malaxer. The SHE was characterized by a work capacity equal to 1500 kg/h and was equipped with 56 transducers (100 watt and 31 kHz) able to transfer a specific energy equal 18,000 kJ/kg [5]. Moreover, the SHE was able to cool and heat the olive paste as a function of the environmental temperature in order to keep the temperature constant. Malaxers were used as buffers to continuously feed the decanter; the time of malaxation was set at 0 min for the sonicated samples and equal to 30 min for the traditional samples. The resulting EVOO was collected, filtered, and stored in dark bottles (at 15 °C) until chemical analysis.

The OMN and OMU samples' production conditions are identical except for the sonication phase present before the crusher in the OMU sonicated samples and absent in the OMN unsonicated samples.

2.3. HS-SPME-GC-MS Analysis of Volatile Organic Compounds

The composition of the EVOO headspaces were characterized using a HS-SPME-GC-MS method, following the same conditions already described in detail in previous studies [19,20]. The peak identification was based on the comparison of the GC-MS parameters with those of authentic standards. The quantitation of each of the 73 identified VOCs was based on the use of 73 calibration lines built using authentic external standards, after normalization of peak areas using 9 suitable internal standards [19,20].

2.4. Analysis of Phenols in EVOO and Defatted Extracts

The extraction of phenols was carried out in triplicate with MeOH:H_2O 80:20 *v/v* according to the IOC method [24] (IOC/T.20/Doc No. 29) and the analyses were carried out with a HP 1100 system (quaternary pump, Diode-Array Detection (DAD) detector, autosampler from Agilent Technologies, Santa Clara, CA, USA). The column was a SphereClone ODS (2), 5 μm, 250 × 4.6 mm; the elution solvents were acidified H_2O by phosphoric acid (pH 2.0), CH_3CN and MeOH and the applied gradient was in accordance with the IOC method [24].; for the quantitative analysis, the area values were collected at 280 nm and the results expressed as mg tyrosol/kg oil, using syringic acid as internal standard (ISTD).

The acid hydrolysis was applied to the extracts obtained as above (the hydrolyzed extracts were not used for the biological tests) and the sum of free and bound hydroxytyrosol and tyrosol was determined at 280 nm [21]. Briefly, the extract (300 μL) was added with 1.0 M H_2SO_4 (300 μL), then heated at 80 °C for 2 h and, after cooling, the solution was diluted with distilled water (400 μL). The analysis was carried out using a column, 150 × 3 mm (5 μm) RP18-Gemini (Phenomenex, CA,

USA); the HPLC-DAD system was HP 1200 (Agilent Technologies, Santa Clara, CA, USA). The flow rate was 0.4 mL/min, the eluent A was H_2O at pH 3.2 by HCOOH, and eluent B was acetonitrile. The linear applied gradient was from A 95% to 70% (5 min), then 5 min to A 50% and other 5 min to A 2% with a final plateau of 5 min. Total time of analysis and equilibration time were 22 min and 10 min respectively. Tyrosol was evaluated using the curve built with pure tyrosol (purity 98%); the hydroxytyrosol amount was evaluated using the same calibration line but applying a corrective factor for not overestimate the concentration: mg OH-tyrosol = mg tyrosol × 0.65 [25].

2.5. *Phenolic Extracts for Testing on Cell Lines*

For each oil, 50 g were exactly weighted and extracted by 150 mL of MeOH:H_2O 80:20 *v*/*v* mixture and vigorously hand-shaken for some minutes, then the extraction was concluded with the aid of an ultrasound bath (10 min). The sample was centrifuged at 5000 rpm for 25 min and the solution recovered and filtered by polyvinylidene fluoride (PVDF) type 0.45 μm filter. The residual lipid was removed adding hexane (75 mL for twice), the defatted hydro-alcoholic solution was dried under vacuum at room temperature. The dry samples were dissolved in ethanol (in a flask of 10 mL) and the solution split in 10 vials in order to obtain 1 mL corresponding to 10 g of EVOO (these samples were analyzed as described in the previous paragraph). The solvent was removed from each vial by a flux of nitrogen to obtain 10 aliquots of the dried extract to be stored over time before the chemical and biological tests. By this way, each vial of OMN contained 5.6 ± 0.12 mg of dry weight and OMU 6.0 ± 0.09 mg of dry weight.

2.6. *Proton Spectra of the Phenolic Extract*

^{1}H-NMR spectra of dry OMN and OMU were acquired as previously suggested [26] to investigate on the aldehydes derived by oleuropein and ligstroside. Briefly, one vial of each extract (obtained as above described from 10 g of EVOO) was dissolved adding 1 mL of $CDCl_3$ and analyzed by 400 MHz instrument (Advance 400 from Bruker, Bremen, Germany).

2.7. *DPPH Assay for Evaluating the In Vitro Radical Scavenging Activity*

The DPPH assay to determine the antioxidant activity in vitro was performed by a standard method with slight modifications [27]. The DPPH solution (0.0125 mM in MeOH, 45 μL) was added to 15 μL of the OMN and OMU EVOO extracts at different concentrations (1.0 and 50.0 μg/mL). The reaction for scavenging the DPPH radicals was performed in the dark at room temperature (RT) and the absorbance was measured at 520 nm after 30 min incubation.

2.8. *ABTS Assay for Evaluating the In Vitro Radical Scavenging Activity*

Aliquots of 10 μL of OMN and OMU EVOO extracts (5.0, 10.0, 50.0 and 100.0 μL/mL) or trolox standards were added to individual wells of the assay plate provided from the kit (ABTS Antioxidant Assay Kit, Zen Bio, NC, USA), using the assay buffer as a negative control. Then, 20 μL of the myoglobin working solution was added to each well. To begin the assay, 100 μL of the ABTS solution per well was added and the absorbance at a wavelength of 405 nm was detected using the kinetic mode, for 30 min, through the Synergy H1 plate reader (Biotek, Bad Friedrichshall, Germany), later the reaction was stopped by adding 50 μL of Stop Solution and the absorbance at 405 nm was detected.

2.9. *Cell Culture Conditions and Treatments*

HepG2 cells, purchased from ATCC (HB-8065, ATCC from LGC Standards, Milan, Italy), were cultured following conditions already optimized [23]. More details are available on Supplementary Materials.

OMN and OMU extracts were tested separately. Briefly, each extract was dissolved in DMSO in order for preparing a stock solution of 50.0 mg/mL, which was diluted in order to reach the final concentration of 25.0 μg/mL in complete growth Doulbecco's Modified Eagle Medium (DMEM). The final 0.05% concentration of DMSO was kept constant either in treated or control cells.

2.10. 3-(4,5-Dimethylthiazol-2-yl)-2,5-Diphenyltetrazolium Bromide (MTT) Assay

A total of 3×10^4 HepG2 cells/well were seeded in 96-well plates and treated with 25.0, 50.0, 100.0 and 200.0 μg/mL of OMN and OMU samples, or vehicle (H_2O) in complete growth media for 48 h at 37 °C under 5% CO_2 atmosphere. MTT experiments have been performed following conditions previously optimized [23]. More details are available in Supplementary Materials.

2.11. Fluorometric Intracellular ROS Assay

For cells preparation, 3×10^4 HepG2 cells/well were seeded on a 96-well plate overnight in growth medium. The day after, the medium was removed, 50 μL/well of Master Reaction Mix was added and the cells were incubated at 5% CO_2, 37 °C for 1 h in the dark. Then, cells were treated with 5 μL of 12 × OMN and OMU EVOO extracts to reach the final concentrations of 1.0, 10.0, 50.0 and 100.0 μg/mL and incubated in the dark at 37 °C for 1 h. To induce ROS, cells were treated with H_2O_2 at a final concentration of 0.5 mM for 30 min a 37 °C in the dark and fluorescence signals (ex./ em. 490/525 nm) were recorded using Synergy H1 fluorescence plate reader (Biotek, Bad Friedrichshall, Germany).

2.12. Lipid Peroxidation (Malondialdehyde Equivalent, MDA Eq) Assay

HepG2 cells (2.5×10^5 cells/well) were incubated with 10 and 25 μg/mL of OMN and OMU samples for 24 h at 37 °C under 5% CO_2 atmosphere. The day after, cells were incubated with 1 mM H_2O_2 or vehicle (H_2O) for 30 min, then collected and homogenized in 150 μL ice-cold MDA lysis buffer containing 1.5 μL of BHT (100×). Samples were centrifuged at 13,000× *g* for 10 min, then they were filtered through a 0.2 μm filter to remove insoluble material. To form the MDA- thiobarbituric acid (TBA) adduct, 300 μL of the TBA solution were added into each vial containing 100 μL of samples and incubated at 95 °C for 60 min, then cooled to RT for 10 min in an ice bath. For analysis, 100 μL of each reaction mixture were pipetted into a 96 well plate and the absorbance was measured at 532 nm using the Synergy H1 fluorescence plate reader (Biotek, Bad Friedrichshall, Germany).

2.13. 3-Hydroxy-3-Methylglutaryl Coenzyme a Reductase (HMGCoAR) Activity Assay

The experiments were carried out following the manufacturer's instructions and conditions previously optimized [28] and more details are available on Supplementary Materials.

2.14. Western Blot Analysis

Western blot experiments have been performed using conditions previously optimized [29]. More details are available on Supplementary Materials.

2.15. In-Cell Western (ICW) Assay

Experiments have been performed using conditions previously optimized [30]. More details are available on Supplementary Materials.

2.16. Assay for the Evaluation of Fluorescent LDL Uptake by HepG2 Cells

Experiments have been performed using conditions previously optimized [31]. More details are available on Supplementary Materials.

2.17. Statistically Analysis

Statistical analyses were carried out by the t-student, One-way and Two-way ANOVA, and GraphPad Prism 8. Values were expressed as means ± standard deviation (s.d.); p-values < 0.05, were considered to be significant.

3. Results & Discussion

This work compares OMN and OMU samples obtained respectively with traditional methods in industrial mills equipped as described in materials and methods and with the use of ultrasound produced through the SHE prototype. The prototype was assembled and tested in the 2019 olive oil campaign and the design and operating characteristics are reported in literature [4,32].

3.1. *Comparison of the VOCs Profile of the EVOO Samples*

The analysis of the VOCs of EVOO samples provide useful information on the product quality, mainly from a sensory standpoint. The positive attributes are mainly associated to the VOCs originated from the lipoxygenase (LOX) pathway, whereas other classes of VOCs are associated to the main sensory defects (i.e., rancid, winey-vinegary, fusty/muddy sediment, musty/humidity/earthy, frostbitten olives/wet wood) [19,20,33,34].

Table 1 shows the amount of the main LOX-related VOCs in the two analyzed samples, but quite significant slight differences between the two samples were only observed for two VOCs (i.e., penten-3-one, E-2-hexenol and hexanal), confirming that the innovative extraction processes did not affect the profile of LOX-related VOCs. Accounting for almost 90% of the LOX-related VOCs content, E-2-hexenal was the most abundant component, according to available literature [20].

Table 1. Profile of the lipoxygenase LOX-related volatile organic compounds VOCs in the two analyzed samples. Results are expressed as mean of triplicate measurements. For each VOC, different letters indicate significant differences between the two samples. EVOO from cultivar Coratina with (**OMU**) or without (**OMN**) using the ultrasound technology

VOC (mg/kg)	OMU	OMN
penten-3-one	0.504 a	0.634 b
E-2-pentenal	0.043 a	0.048 a
penten-3-ol	0.854 a	0.876 a
Z-3-hexenal	<LOD	<LOD
E-2-hexenal	33.889 a	33.160 a
E-2-pentenol	0.058 a	0.054 a
Z-2-pentenol	0.023 a	0.024 a
Z-3-Hexenyl acetate	0.053 a	0.054 a
Z-3-hexenol	0.467 a	0.433 a
E-2-hexenol	0.731 a	0.678 a
Hexanal	0.886 a	1.081 b
Hexyl acetate	0.010 a	0.009 a
Hexanol	0.570 a	0.516 a
ΣLOX	38.090 a	37.569 a

In order to evaluate the impact of the VOCs linked to the negative attributes, it was decided to calculate the marker recently proposed for the rancid defects, consisting of the sum of the content of pentanal, nonanal, and E-2-heptenal [20]. For both samples, the marker was equal to 0.025 mg/kg, much smaller than the limit value of 0.65 mg/kg, above which the sensation of rancid can be perceived [20].

Finally, the application of the chemometric approaches, recently proposed for supporting the Panel Test in Virgin Olive Oil classification [19], confirmed that both samples belong to the EVOO category.

3.2. *Comparison of the Phenolic Profiles of the EVOO Samples*

Similarly, the phenolic profiles of the two oils obtained before hydrolysis (IOC method) were very similar as well as the phenol contents determined after acidic hydrolysis (Table 2).

Table 2. Phenol contents determined before and after acidic hydrolysis. Results are expressed as mean ± s.d. of analysis of three extracts of the same samples.

	Phenol Contents in the Oils (mg/kg)			
	(Before Hydrolysis)		(After Hydrolysis)	
	OMN	OMU	OMN	OMU
Free hydroxytyrosol	17.4 ± 5.1	18.1 ±2.6	169.6 ± 2.6	185.1 ± 13.4
Free tyrosol	15.6 ± 5.4	14.3 ± 3.0	308.6 ± 2.7	325.7 ±23.4
Total Phenols	415.7 ± 7.3	375.1 ± 8.5	478.1 ± 4.7	510.7 ± 36.7

The ^{1}H-NMR spectra of the phenolic extracts allowed the evaluation of the ratios among the characteristic secoiridoids, measured as the aldehydes derived from the transformation during milling of the oleuropein and ligstroside precursors (Figure S1). This analysis showed that the ultrasound treatment did not change the profiles of the main secoiridoidic components of the phenolic fractions (Figure 1A). As for the dry extracts, the concentration of the total simple phenols (Tyr and OH-Tyr) and the total phenolic quantity evaluated before and after hydrolysis (Tyr + OH-Tyr) is shown in Figure 1B that hightlights the strong similarity of the two samples, with only a tiny difference between the total phenols before hydrolysis.

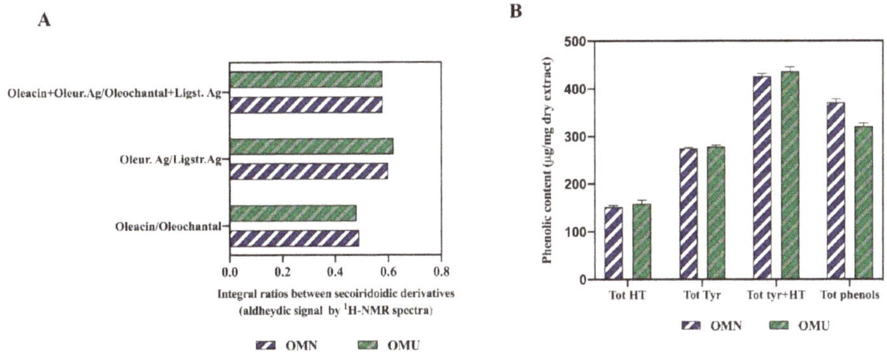

Figure 1. Analysis of secoiridoids. (**A**) Ratios of the integrals corresponding to the signals of secoiridoidic monoaldehydes in ^{1}H-NMR spectra of the dry extracts from OMN and OMU samples. Oleur. Ag, aglycone of oleuropein; Ligstr. Ag. aglycone of ligstroside. (**B**) Total tyrosol + hydroxytyrosol (after hydrolysis) and phenolic content (according to the IOC method) in the dry extracts from OMN and OMU oil; each data represents the mean of the analysis of three extracts of the same samples.

The use of ultrasounds has been already shown to increase oil yields while guaranteeing oil quality preservation. Several couples of EVOOs were analyzed and no significant differences were highlighted for their phenolic profile and volatiles produced by the lipoxygenase cascade [13].

3.3. Comparison of the In Vitro Antioxidant Activity of the Phenolic Extracts by DPPH and ABTS Assays

The correlation between the total phenol contents and the antioxidant activity has been widely studied in olive oil [35]. The antioxidant activity of an EVOO significantly increases in the presence of a high phenol concentrations. The genotype, the cultivation area and climate, and the extraction technique represent the major factors influencing the variability of the levels of phenolic compounds. This variability greatly affects not only the organoleptic, but also the nutraceutical features of each EVOO sample.

Different robust and reproducible methods for measuring the antioxidant capacity of olive oil are successfully and equally applied worldwide [36]. To evaluate the overall antioxidant activity of EVOO extracts, we decided to employ the DPPH and ABTS assays.

The DPPH radical scavenging assay is one of the most commonly used single electron transfer (SET)-based antioxidant procedure due to its ease of performance, rapidness, automation potential, reproducibility, and usability at ambient temperatures [37]. Each extract was tested in the range of concentration from 5.0 to 100 μg/mL. The results clearly suggested that both OMU and OMN are able to scavenge the DPPH radical (Figure 2A). OMU diminished the DPPH radicals by 3.2 ± 5.6%, 35.2 ± 3.9%, 70.9 ± 1.4%, and 67.6 ± 0.4% at 5, 10, 50, and 100 μg/mL, respectively, whereas OMN reduced the DPPH radicals by 10.3 ± 5.4%, 32.4 ± 3.4%, 71.1 ± 1.2%, and 68.7 ± 0.1%, respectively, indicating that both extracts display similar radical scavenging activities. At the low dose of 5 μg/mL, the DPPH reduction was not significant. Starting from the same phenol-rich variety Coratina, the in vitro scavenging activities of the final products do not appear to be affected by the ultrasound process, in agreement with the chemical analyses (Tables 1 and 2).

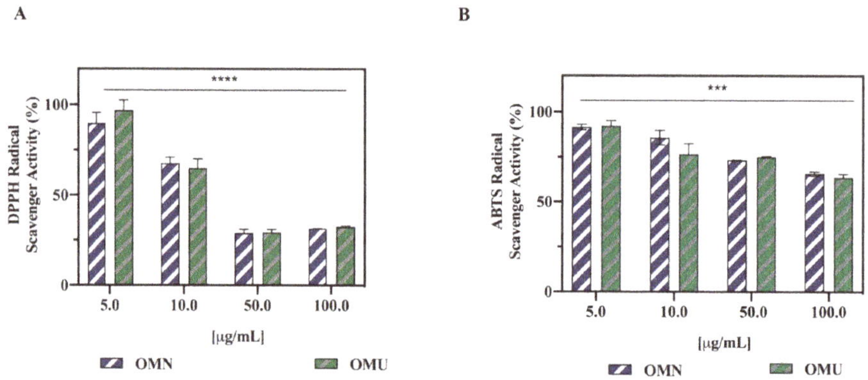

Figure 2. Antioxidant effects of OMN and OMU extracts. (**A**) In vitro radical scavenging activity of OMN and OMU phenol extracts by DPPH assay. (**B**) In vitro radical scavenging activity of OMN and OMU phenol extracts by ABTS assay. Data represent the mean ± s.d. of six determinations performed in triplicate. All the data sets have been analyzed by Two-way ANOVA. In particular, the reductions of DPPH (****) $p < 0.0001$ and ABTS (***) $p < 0.001$ radicals are significant as function of the concentrations, whereas no significant difference have been observed between OMN and OMU extracts.

Hydroxytyrosol (HT) and oleuropein (Ole) are known as peroxy radical scavengers and are usually associated with the antioxidant activities of olive products [38]. In light with this information, additional experiments were performed in order to evaluate the in vitro radical scavenging activities of these compounds by performing the same DPPH assay (Figure S2). HT and Ole appear to be better antioxidants than Tyr (Figure S2), suggesting their active contribution to the scavenging activities of both OMN and OMU phytocomplexes. Similar results have been observed some years ago by Carrasco-Pancorbo and co-workers [39], who hypothesized that the lower antioxidant activity of Tyr compared to HT can be explained by the absence of the ortho-diphenolic group in its structure [40].

In parallel, each extract was assessed by using the ABTS. OMU diminished the ABTS radicals by 8.4 ± 1.4%, 14.2 ± 4%, 26.8 ± 0.1%, and 34.2 ± 1.1% at 5, 10, 50, and 100 μg/mL, respectively (Figure 2B), whereas OMU by 7.7 ± 2.9%, 23.5 ± 5.9%, 25 ± 0.3%, and 34.5 ± 2% at 5, 10, 50, and 100 μg/mL, respectively. Surprisingly, at the low concentration of 5 μg/mL, the ABTS reduction was significant for both extracts (* $p < 0.5$). Both phytocomplexes were therefore active, but with a smaller potency than in the DPPH assay. The differences between the DPPH and ABTS methods may be possibly ascribed to the different solvent used in these assays [41]. In fact, MeOH is used as solvent in the DPPH assay, whereas H_2O is used in ABTS assay [42]. Considering that the phytocomplex had been extracted from the EVOO samples with a MeOH:H_2O 80:20 v/v mixture, it seems possible to affirm that both OMN and OMU extracts are predominantly characterized by methanol soluble compounds. For instance,

HT, which is more soluble in methanol, shows a better antioxidant activity in the DPPH than in the ABTS assay [43].

3.4. Comparison of the Antioxidant Effects on HepG2 Cells

Upon absorption from the gastrointestinal tract, the liver represents not only the main target for phenolic antioxidants, but also the major organ deputed to phenol metabolism [44,45]. Moreover, hepatocyte mitochondria and endoplasmic reticulum are the major sites for the generation of reactive oxygen species (ROS) in various forms of liver diseases [46]. The human hepatic HepG2 cell line was thus chosen here to compare the antioxidant activity of these samples. Indeed, HepG2, a well-differentiated transformed cell line, is a reliable model, easy to culture, well characterized, and widely used for biochemical and nutritional studies where many antioxidants and conditions can be assayed with minor inter-assay variations [47].

Preliminary cell viability experiments were carried out using MTT assay in order to assess the concentrations of the OMU extract that may potentially determine cytotoxicity on HepG2 cells. After a 48 h treatment, no significant cell mortality had been detected up to 100 μg/mL versus untreated cells (C), whereas at 200 μg/mL a 33.1 ± 1.8% cell mortality was detected (Figure 3). In a similar experiment previously performed on OMN, no cytotoxic effect was detected up to 100 μg/mL versus untreated cells (C) [23].

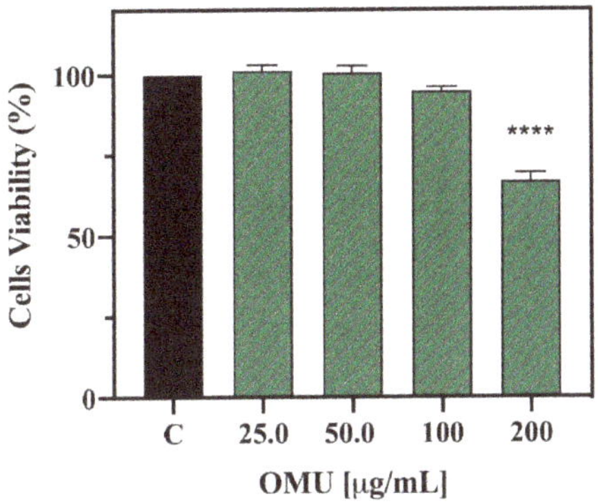

Figure 3. MTT assay. Effect of OMU extract on the HepG2 cell viability. Data represent the mean ± s.d. of three independent experiments performed in triplicate. The statistical significance of C vs OMU 200 μg/mL was analyzed by t-student test. (****) $p < 0.0001$, C: control cells.

In order to evaluate whether OMN and OMU extracts can modulate the H_2O_2-induced ROS production, HepG2 cells were pre-treated with both extracts (in the concentration range 1.0–25.0 μg/mL) at 37 °C overnight. The following day, the cells were treated with H_2O_2 (1.0 mM) at 37 °C for 30 min. Figure 4A,B clearly highlights that HepG2 cells, exposed to H_2O_2 alone, produce a dramatic increment of intracellular ROS levels by 211.4 ± 21.7% versus the control cells (basal value = 100%, $p < 0.5$), which was attenuated by the pre-treatment with OMN and OMU. OMN reduced the H_2O_2-induced intracellular ROS by 198.8 ± 12.5% and 130.3 ± 11.3% at 10.0 and 25.0 μg/mL, respectively ($p < 0.01$) (Figure 4A), whereas OMU by 197.9 ± 11.2, 188.5 ± 5.0, 128.0 ± 0.3 at 1.0, 10.0, and 25.0 μg/mL, respectively ($p < 0.001$) (Figure 4B).

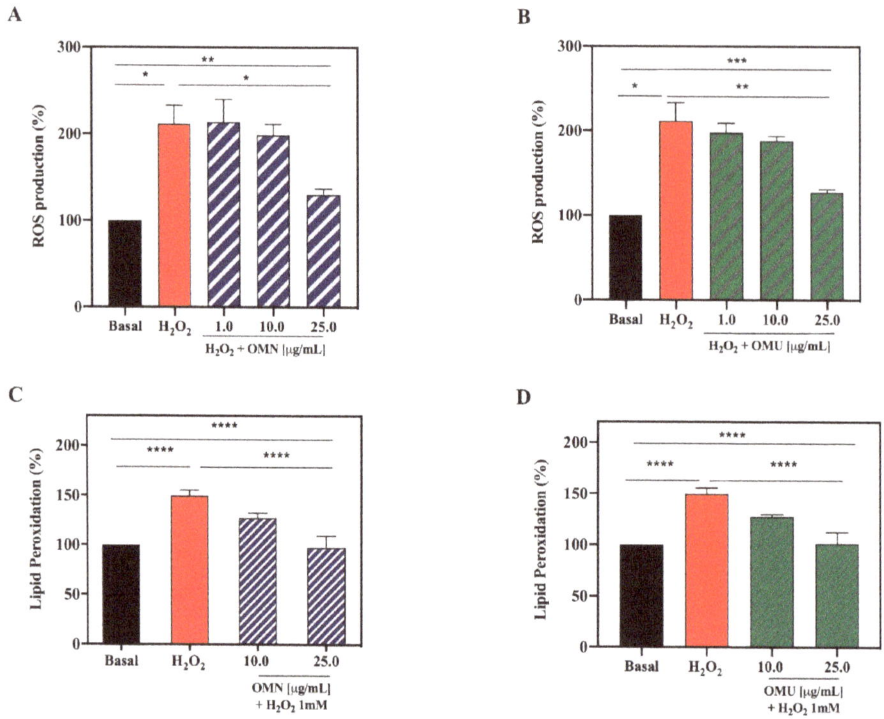

Figure 4. Antioxidant effects of OMN and OMU extracts on HepG2 cells. (**A**) OMN and (**B**) OMU reduce the H_2O_2 (1 mM)-induced ROS levels in HepG2 cells. (**C**) OMN and (**D**) OMU decrease the lipid peroxidation in the same cells after oxidative stress induction by H_2O_2. Data represent the mean ± s.d. of six independent experiments performed in triplicate. All the data sets were analyzed by One-way ANOVA; basal vs H_2O_2 samples were analyzed by t-student test, whereas H_2O_2 vs OMN/OMU + H_2O_2 samples by One-way ANOVA. (*) $p < 0.5$; (**) $p < 0.01$; (***) $p < 0.001$; (****) $p < 0.0001$.

These findings indicate that the pre-treatments with both extracts protect, in a similar way, the HepG2 cells against the increase of intracellular ROS induced by the H_2O_2 addition, thus restoring the ROS levels. These results are in agreement with a literate study on traditional EVOO samples performed using a similar protocol [48]. In addition, other studies have suggested that EVOO extracts impair the ROS generation induced by oxidative stress in other cellular systems, such as intestinal and muscle ones. The observed effects may depend mostly on HT that has been demonstrated to be able to reduce ROS generation induced by tert-butylhydroperoxide (t-BOOH) in HepG2 cells [47].

Lipids of cellular membranes are susceptible to oxidative attack, typically by ROS, resulting in a well-defined chain reaction with the production of end products, such as malondialdehyde (MDA) and related compounds, known as TBA reactive substances (TBARS). MDA is widely used as an index of lipoperoxidation in biological and medical sciences and elevated amounts of this metabolite have been found in various diseases linked to free radical damage [49]. Based on these considerations, the capacity of OMN and OMU to modulate the H_2O_2-induced lipid peroxidation in human hepatic HepG2 cells was assessed measuring the reaction of MDA precursor with the TBA reagent to form a fluorescent product ($\lambda ex = 532/\lambda em = 553$ nm), proportional to the amount of TBARS (MDA equivalents) present. Figure 3 clearly suggests that, in agreement with the observed increase of ROS after the H_2O_2 treatment, a significant increase of the lipid peroxidation at cellular level up to 151.3 ± 6.6% was detected ($p < 0.01$). However, the pre-treatment of HepG2 cells with both EVOO extracts produce a significant reduction of lipid peroxidation even under basal conditions ($p < 0.5$) (Figure 4C,D). OMN decreases the lipid peroxidation up to 127.5 ± 5.3% and 97.8 ± 7.1% at 10 and 25 μg/mL, respectively, whereas OMU up to 127.7 ± 2.1% and 101.4 ± 10.3%, respectively, at the same concentrations. Again, any significant difference was observed as a function of the innovative EVOO extraction method.

Presumably, the reduction of lipid peroxidation induced by oxidative stress is mostly mediated by the contribution of both HT and Tyr in the EVOO extracts. In fact, clear evidence suggests that HT protects the integrity of the HepG2 cellular membrane leading to a reduction of MDA levels after t-BOOH induced oxidative stress [47]. Similarly, Tyr reduces lipid peroxidation in HepG2 cells exposed to acute ethanol treatment [50]. Other evidence confirms that Tyr has a protective effect on membrane integrity in other cellular systems. On the contrary, there are literature indications that the Ole-mediated protective effect against oxidative stress is not associated with a reduction of MDA generation [51].

3.5. Comparison of the Hypocholesterolemic Activity of the OMU and OMN Phenol Extracts

The beneficial effects of EVOO are linked to its ability to reduce the oxidative stress and to limit the lipoprotein oxidation, making the LDL less atherogenic. In fact, in patients with mild hypertension, the intake of EVOO improves the oxidation state of plasmatic LDL, decreases their peroxidation, increases the reduced glutathione, and decreases hypertension [52]. In a trial with 200 volunteers, the intake of EVOO led to an increase in high-density lipoproteins (HDL) and a simultaneous reduction of oxidation state of plasma [53]. Recently, we have disclosed a new molecular mechanism through which EVOO extracts may exhibit cholesterol-lowering activities by a directly modulation of the cholesterol biosynthetic pathway [23].

These innovative and groundbreaking results prompt us to assess the capacity of OMU extract to modulate the HMGCoAR activity (Figure 5). To achieve this goal, in vitro experiments were carried out using the purified recombinant catalytic domain of the enzyme. Testing OMU extract, the residual enzyme activity was 84.2 ± 2.5%, 71.9 ± 1.1%, 56.4 ± 7.9%, and 33.0 ± 2.8% respectively at the concentrations of 10.0, 50.0, 100.0, and 250.0 μg/mL (Figure 5, green bars). Similarly, testing OMN, as the reference extract, the residual in vitro HMGCoAR activity was 82.5 ± 0.7%, 78.5 ± 4.6%, 54.8 ± 1.4%, and 35.6 ± 2.6%, respectively in the same range of concentrations (Figure 5, blue bars). A statistical analysis performed by two-way ANOVA indicated that a significant difference in the residual activity of the enzyme was observed as a function of the tested concentrations for both OMN and OMU (** $p < 0.01$), whereas any significant difference was observed between OMN and OMU extract at each tested dose. These results are also in line with those previously obtained for the reference OMN extract [23]. These results indicate that the new ultrasound process does not impair the molecular mechanism through which EVOO extracts promote hypocholesterolemic effect through the modulation of HMGCoAR activity, an enzyme that is crucial in cholesterol biosynthesis and is also the well-known target of statins [45,54].

In order to assess the role of the main components of the phytocomplex, we investigated in parallel Ole, Tyr and HT alone (Figure S3). Ole and Tyr were ineffective, whereas a slightly inhibition was observed with HT only at the concentration of 100 μM. These results underline the unique feature of the EVOO extracts, excluding the predominant role of any among Ole, Tyr, and HT when tested alone in the modulation of the enzyme activity. Thus, all these results clearly suggest that a synergistic contribution of the total components of phytocomplex is responsible for the inhibition of the HMGCoAR activity.

On this basis, we decided to compare the ability of these extracts to modulate the LDLR pathway. Thus, HepG2 cells were treated with 25.0 μg/mL of each extract (a doses that is about 10 folds lower than the smallest cytotoxic dose and a compromise between it and the effect on the in vitro HMGCoAR activity). The western blot experiments, assessed on cell lysates, showed that LDLR pathway was activated after OMN and OMU treatments (Figure 6A–C). The OMU and OMN extracts increased the protein level of the sterol regulatory element-binding transcription factor (SREBP)-2 (precursor) by 156 ± 6.8% and 134% ± 10.9% ($p < 0.001$), respectively (Figure 6A), which in turn led to an augmentation of LDLR protein levels up to 207 ± 33% and 161 ± 21.7% ($p < 0.01$), respectively (Figure 6B). Upon the activation of the SREBP-2 transcription factor, improvements of HMGCoAR protein levels were also observed, by 158 ± 33% with OMU and by 153 ± 39.8% ($p < 0.05$) with OMN (Figure 6C).

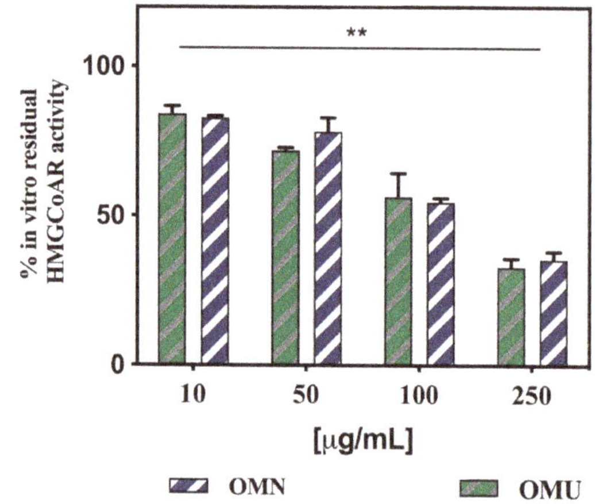

Figure 5. Effect of OMN and OMU extracts on the in vitro activity of HMGCoAR. Data represent the mean ± s.d. of three determinations performed in triplicate. All the data sets were analyzed by two-way ANOVA. In particular, the reduction of enzyme activity is significant as function of the all the tested concentrations (**) $p < 0.01$, whereas no significant difference was observed between OMN and OMU extracts.

The LDLR is a key player in the cholesterol metabolism pathway. In fact, it is regulated at transcriptional, translational, and post-translational levels. In particular, the dynamic trafficking of functional LDLR is finely regulated by a protein named proprotein convertase subtilisin/kexin type 9 (PCSK9). Both SREBP-2 and hepatocyte nuclear factor 1-alpha (HNF1-α) cooperatively transcriptional activate the PCSK9 gene expression, but only SREBP-2 controls the LDLR expression. In this context, statins enhance the level of PCSK9 leading to an improvement of LDLR degradation.

Herein, the contextual modulation of PCSK9 intracellular processing was also investigated in HepG2 cells. OMU and OMN did not modulate the mature PCSK9 protein levels (Figure 6D) and to activate HNF1-α, the PCSK9 transcription factor (Figure 6E). This result reinforces the main outcome of a previous work [23], which has suggested that EVOO phenols are able to improve the cholesterol metabolism pathway without increasing the PCSK9 protein levels, a relevant drawback of statins [55].

In order to compare the ability of the extracts to modulate the levels of the LDLR population localized on the hepatocyte surface, a specific ICW assay was performed in parallel on both extracts tested at 25 µg/mL. This cell-based assay permits the target protein detections in fixed cultured cells. Improvements of the membrane LDLR levels by 197 ± 5.3 % for OMU and by 180 ± 17.2% for OMN extracts were observed (Figure 7A). Also in this case, no significant differences were detected between the two samples.

Finally, functional experiments were carried out to compare the effects of the two extracts on the capacity of HepG2 cells to uptake the LDL from the extracellular environment. The experiment (conducted at 25.0 µg/mL) showed that the ability of HepG2 cells to absorb fluorescent LDL from the extracellular space was increased by 239.3 ± 34.2% by OMU and by 243.9 ± 6.5 % by OMN (Figure 7B). Therefore, even here there are only negligible differences in the capacity of the two extracts to stimulate the uptake of the LDL by HepG2 cells.

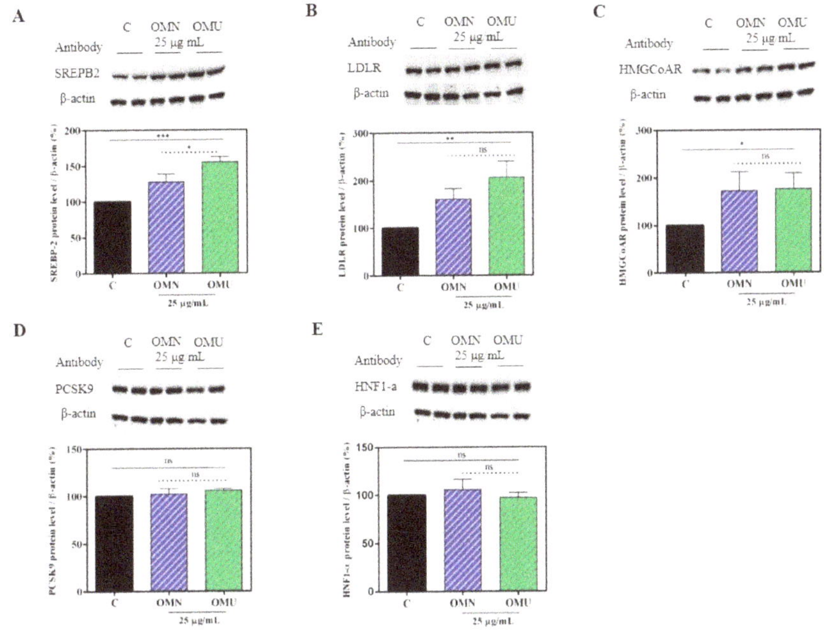

Figure 6. Modulation of cholesterol biosynthesis. (**A**) Western blot of SREBP-2 (precursor); (**B**) western blot of the LDLR; (**C**) western blot of HMGCoAR; (**D**) western blot of PCSK9; (**E**) western blot of HNF1- α. Data represent the mean ± s.d. of eight independent experiments performed in duplicate. All the data sets were analyzed by One-way ANOVA and OMN vs OMU by t-student test. (*) $p < 0.5$; (**) $p < 0.01$; (***) $p < 0.001$; ns: not significant. C: control sample.

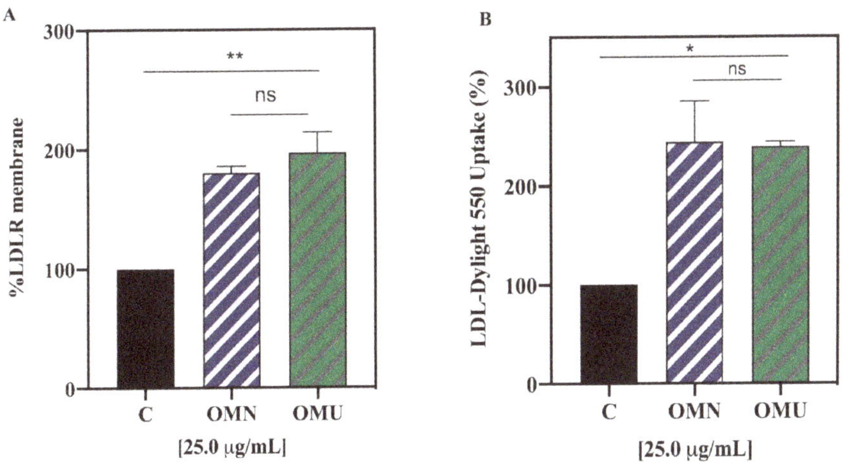

Figure 7. Modulation of the LDLR on HepG2 cell surface and uptake of environmental LDL. (**A**) LDLR protein levels on HepG2 cell surface evaluated by in cell western. (**B**) Uptake of fluorescent LDL from the environment by HepG2 cells. Data represent the mean ± s.d. of five independent experiments performed in triplicate. All the data sets were analyzed by One-way ANOVA and OMN vs OMU by t-student test. (*) $p < 0.5$; (**) $p < 0.001$. ns: not significant, C: control sample.

Indeed, these results show that through the inhibition of the HMGCoAR activity, both OMN and OMU extracts modulate the cellular cholesterol metabolism, leading to the LDLR and HMGCoAR protein level augmentations by increasing the SREBP-2 (precursor) protein levels (Figure 6). In addition, with a similar behavior, both extracts increase the population of LDLR which are localized on the membrane of HepG2 cells, which leads from a functional point of view to an increased ability of HepG2 cells to absorb extracellular LDL, with an in vitro cholesterol-lowering effect (Figure 7). The membrane LDLR activity is also positively regulated by the fact that both OMN and OMU extracts did not produce any variation of PCSK9 protein levels, clearly confirming that similarly to OMN, also OMU shows a distinct and unique behavior in respect to statins.

4. Conclusions

Performing a series of chemical, biochemical and cellular investigations, in this study we have demonstrated the substantial equivalence of the EVOO sample OMU, produced with an innovative process based on the simultaneous treatment of olive paste both with ultrasound and heat-exchange, and the sample OMN produced with the traditional methodology. This means that the increase of the yield has not impaired any relevant feature of the EVOO, neither on the standpoint of the chemical composition and sensory characteristics nor the nutraceutical properties, as far as the antioxidant and hypocholesterolemic activities are involved.

Properly communicated, these results may represent a tool for reducing the neophobia linked to the introduction of a relevant innovation in a traditional food sector such as the EVOO production. The possibility of guaranteeing a price premium to this kind of product may assure a greater environmental and economic sustainability for stakeholders in the supply chain, achieving the European goal of improving simultaneously industry competitiveness, natural resources protection, and citizen health.

Supplementary Materials: The following are available online at http://www.mdpi.com/2076-3921/9/9/798/s1, Description of Material and Methods, Figure S1. ^{1}H-NMR spectra of the phenolic extracts from OMN and OMU. Profiles of the main secoiridoidic components of the phenolic fraction, Figure S2. Antioxidant activity evaluation of (**A**) hydroxytyrosol (HT), (**B**) Oleuropein (Ole), and (**C**) Tyrosol (Tyr) by DPPH assay in the range of concentration 10–250 μM, Figure S3. Effect of HT, Ole, and Tyr on the in vitro HMGCoaR activity.

Author Contributions: Conceptualization, C.L.; Funding acquisition, M.L.C., F.C., C.F., A.A. and N.M.; Investigation, C.L., M.B. (Maria Bellumori), L.C., M.B. (Martina Bartolomei) and C.B.; Supervision, C.L. and N.M.; Writing—original draft, C.L., M.B. (Maria Bellumori) and L.C.; Writing—review & editing, C.L., M.L.C., F.C., C.F., A.A. and N.M. All authors have read and agreed to the published version of the manuscript.

Funding: This research was funded by the following: **1.** The AGER 2 Project, grant n. 2016–0174, AGER Foundation—Olive Tree and Oil: Competitive—Claims of olive oil to improve the market value of the product; **2.** EU project 820587—OLIVE-SOUND-Ultrasound reactor—The solution for a continuous olive oil extraction process H2020-EU.2.1.—INDUSTRIAL LEADERSHIP—EIC-FTI-2018–2020—Fast Track to Innovation (FTI)—European Union's Horizon 2020 research and innovation program under grant agreement No. 820587.

Acknowledgments: We are indebted to Carlo Sirtori Foundation (Milan, Italy) for having provided part of equipment used in this experimentation.

Conflicts of Interest: The authors declare no conflict of interest.

References

1. Clodoveo, M.L.; Dipalmo, T.; Rizzello, C.G.; Corbo, F.; Crupi, P. Emerging technology to develop novel red winemaking practices: An overview. *Innov. Food Sci. Emerg.* **2016**, *38*, 41–56. [CrossRef]
2. Giordano, S.; Clodoveo, M.L.; De Gennaro, B.; Corbo, F. Factors determining neophobia and neophilia with regard to new technologies applied to the food sector: A systematic review. *Int. J. Gastron. Food Sci.* **2018**, *11*, 1–19. [CrossRef]
3. Clodoveo, M.L. An overview of emerging techniques in virgin olive oil extraction process: Strategies in the development of innovative plants. *J. Agric. Eng.* **2013**, *XLIV*, 297–305. [CrossRef]

4. Amirante, R.; Distaso, E.; Tamburrano, P.; Paduano, A.; Pettinicchio, D.; Clodoveo, M.L. Acoustic Cavitation by Means Ultrasounds in The Extra Virgin Olive Oil Extraction Process. In Proceedings of the Ati 2017–72nd Conference of the Italian Thermal Machines Engineering Association 2017, Lecce, Italy, 6–8 September 2017; Volume 126, pp. 82–90. [CrossRef]
5. Amirante, R.; Clodoveo, M.L. Developments in the design and construction of continuous full-scale ultrasonic devices for the EVOO industry. *Eur. J. Lipid Sci. Technol.* **2017**, 119. [CrossRef]
6. Bejaoui, M.A.; Beltran, G.; Aguilera, M.P.; Jimenez, A. Continuous conditioning of olive paste by high power ultrasounds: Response surface methodology to predict temperature and its effect on oil yield and virgin olive oil characteristics. *LWT Food Sci. Technol.* **2016**, *69*, 175–184. [CrossRef]
7. Bejaoui, M.A.; Sanchez-Ortiz, A.; Sanchez, S.; Jimenez, A.; Beltran, G. The high power ultrasound frequency: Effect on the virgin olive oil yield and quality. *J. Food Eng.* **2017**, *207*, 10–17. [CrossRef]
8. Aydar, A.Y.; Bagdatlioglu, N.; Koseoglu, O. Effect of ultrasound on olive oil extraction and optimization of ultrasound-assisted extraction of extra virgin olive oil by response surface methodology (RSM). *Grasas Aceites* **2017**, 68. [CrossRef]
9. Bejaoui, M.A.; Sanchez-Ortiz, A.; Aguilera, M.P.; Ruiz-Moreno, M.J.; Sanchez, S.; Jimenez, A.; Beltran, G. High power ultrasound frequency for olive paste conditioning: Effect on the virgin olive oil bioactive compounds and sensorial characteristics. *Innov. Food Sci. Emerg.* **2018**, *47*, 136–145. [CrossRef]
10. Iqdiam, B.M.; Mostafa, H.; Goodrich-Schneider, R.; Baker, G.L.; Welt, B.; Marshall, M.R. High Power Ultrasound: Impact on Olive Paste Temperature, Malaxation Time, Extraction Efficiency, and Characteristics of Extra Virgin Olive Oil. *Food Bioprocess Technol.* **2018**, *11*, 634–644. [CrossRef]
11. Iqdiam, B.M.; Abuagela, M.O.; Marshall, S.M.; Yagiz, Y.; Goodrich-Schneider, R.; Baker, G.L.; Welt, B.A.; Marshall, M.R. Combining high power ultrasound pre-treatment with malaxation oxygen control to improve quantity and quality of extra virgin olive oil. *J. Food Eng.* **2019**, *244*, 1–10. [CrossRef]
12. Rigane, G.; Yahyaoui, A.; Acar, A.; Mnif, S.; Salem, R.B.; Arslan, D. Change in some quality parameters and oxidative stability of olive oils with regard to ultrasound pretreatment, depitting and water addition. *Biotechnol. Rep. (Amst)* **2020**, *26*, e00442. [CrossRef] [PubMed]
13. Cecchi, L.; Bellumori, M.; Corbo, F.; Milani, G.; Clodoveo, M.L.; Mulinacci, N. Implementation of the Sono-Heat-Exchanger in the Extra Virgin Olive Oil Extraction Process: End-User Validation and Analytical Evaluation. *Molecules* **2019**, *24*, 2379. [CrossRef] [PubMed]
14. Amirante, P.; Clodoveo, M.L.; Tamborrino, A.; Leone, A. A New Designer Malaxer to Improve Thermal Exchange Enhancing Virgin Olive Oil Quality. In Proceedings of the VI International Symposyum Olive Grow, Evora, Portugal, 9–13 September 2008; Volume 949, pp. 455–462.
15. Amirante, R.; Demastro, G.; Distaso, E.; Hassaan, M.A.; Mormando, A.; Pantaleo, A.M.; Tamburrano, P.; Tedone, L.; Clodoveo, M.L. Effects of Ultrasound and Green Synthesis ZnO Nanoparticles on Biogas Production from Olive Pomace. In *Proceedings of the Ati 2018—73rd Conference of the Italian Thermal Machines Engineering Association, Pisa, Italy, 12–14 September 2018*; Elsevier: Amsterdam, The Netherlands, 2018; Volume 148, pp. 940–947. [CrossRef]
16. Clodoveo, M.L.; Dipalmo, T.; Crupi, P.; Durante, V.; Pesce, V.; Maiellaro, I.; Lovece, A.; Mercurio, A.; Laghezza, A.; Corbo, F.; et al. Comparison Between Different Flavored Olive Oil Production Techniques: Healthy Value and Process Efficiency. *Plant Foods Hum. Nutr.* **2016**, *71*, 81–87. [CrossRef] [PubMed]
17. Clodoveo, M.L.; Dipalmo, T.; Schiano, C.; La Notte, D.; Pati, S. What's now, what's new and what's next in virgin olive oil elaboration systems? A perspective on current knowledge and future trends. An overview of emerging techniques in virgin olive oil extraction process: Strategies in the development of innovative plants. *J. Agric. Eng.* **2014**, *45*, 49–59.
18. Roselli, L.; Cicia, G.; Cavallo, C.; Del Giudice, T.; Carlucci, D.; Clodoveo, M.L.; De Gennaro, B.C. Consumers' willingness to buy innovative traditional food products: The case of extra-virgin olive oil extracted by ultrasound. *Food Res. Int.* **2018**, *108*, 482–490. [CrossRef] [PubMed]
19. Cecchi, L.; Migliorini, M.; Giambanelli, E.; Rossetti, A.; Cane, A.; Melani, F.; Mulinacci, N. Headspace Solid-Phase Microextraction-Gas Chromatography-Mass Spectrometry Quantification of the Volatile Profile of More than 1200 Virgin Olive Oils for Supporting the Panel Test in Their Classification: Comparison of Different Chemometric Approaches. *J. Agric. Food Chem.* **2019**, *67*, 9112–9120. [CrossRef] [PubMed]

20. Cecchi, L.; Migliorini, M.; Giambanelli, E.; Rossetti, A.; Cane, A.; Mulinacci, N. New Volatile Molecular Markers of Rancidity in Virgin Olive Oils under Nonaccelerated Oxidative Storage Conditions. *J. Agric. Food Chem.* **2019**, *67*, 13150–13163. [CrossRef]
21. Bellumori, M.; Cecchi, L.; Innocenti, M.; Clodoveo, M.L.; Corbo, F.; Mulinacci, N. The EFSA Health Claim on Olive Oil Polyphenols: Acid Hydrolysis Validation and Total Hydroxytyrosol and Tyrosol Determination in Italian Virgin Olive Oils. *Molecules* **2019**, *24*, 2179. [CrossRef]
22. Roselli, L.; Clodoveo, M.L.; Corbo, F.; De Gennaro, B. Are health claims a useful tool to segment the category of extra-virgin olive oil? Threats and opportunities for the Italian olive oil supply chain. *Trends Food Sci. Technol.* **2017**, *68*, 176–181. [CrossRef]
23. Lammi, C.; Bellumori, M.; Cecchi, L.; Bartolomei, M.; Bollati, C.; Clodoveo, M.L.; Corbo, F.; Arnoldi, A.; Nadia, M. Extra Virgin Olive Oil Phenol Extracts Exert Hypocholesterolemic Effects through the Modulation of the LDLR Pathway: In Vitro and Cellular Mechanism of Action Elucidation. *Nutrients* **2020**, *12*, 1723. [CrossRef]
24. IOC/T.20/Doc No. 29. *Official Method of Analysis. Determination of Biophenols in Olive oil by HPLC*; International Olive Council: Madrid, Spain, 2009.
25. Bellumori, M.; Cecchi, L.; Romani, A.; Mulinacci, N.; Innocenti, M. Recovery and stability over time of phenolic fractions by an industrial filtration system of olive mill wastewaters: A three-year study. *J. Sci. Food Agric.* **2018**, *98*, 2761–2769. [CrossRef]
26. Karkoula, E.; Skantzari, A.; Melliou, E.; Magiatis, P. Direct measurement of oleocanthal and oleacein levels in olive oil by quantitative (1)H NMR. Establishment of a new index for the characterization of extra virgin olive oils. *J. Agric. Food Chem.* **2012**, *60*, 11696–11703. [CrossRef] [PubMed]
27. Lammi, C.; Bollati, C.; Arnoldi, A. Antioxidant activity of soybean peptides on human hepatic HepG2 cells. *J. Food Bioact.* **2019**, *7*. [CrossRef]
28. Aiello, G.; Lammi, C.; Boschin, G.s.; Zanoni, C.; Arnoldi, A. Exploration of Potentially Bioactive Peptides Generated from the Enzymatic Hydrolysis of Hempseed Proteins. *J. Agric. Food Chem.* **2017**, *65*, 10174–10184. [CrossRef]
29. Lammi, C.; Zanoni, C.; Calabresi, L.; Arnoldi, A. Lupin protein exerts cholesterol-lowering effects targeting PCSK9: From clinical evidences to elucidation of the in vitro molecular mechanism using HepG2 cells. *J. Funct. Foods* **2016**, *23*, 230–240. [CrossRef]
30. Lammi, C.; Zanoni, C.; Arnoldi, A. A simple and high-throughput in-cell Western assay using HepG2 cell line for investigating the potential hypocholesterolemic effects of food components and nutraceutics. *Food Chem.* **2015**, *169*, 59–64. [CrossRef]
31. Zanoni, C.; Aiello, G.; Arnoldi, A.; Lammi, C. Investigations on the hypocholesterolaemic activity of LILPKHSDAD and LTFPGSAED, two peptides from lupin beta-conglutin: Focus on LDLR and PCSK9 pathways. *J. Funct. Foods* **2017**, *32*, 1–8. [CrossRef]
32. Clodoveo, M.L.; Moramarco, V.; Paduano, A.; Sacchi, R.; Di Palmo, T.; Crupi, P.; Corbo, F.; Pesce, V.; Distaso, E.; Tamburrano, P.; et al. Engineering design and prototype development of a full scale ultrasound system for virgin olive oil by means of numerical and experimental analysis. *Ultrason. Sonochem.* **2017**, *37*, 169–181. [CrossRef] [PubMed]
33. Angerosa, F. Influence of volatile compounds on virgin olive oil quality evaluated by analytical approaches and sensor panels. *Eur. J. Lipid Sci. Technol.* **2002**, *104*, 639–660. [CrossRef]
34. Morales, M.T.; Luna, G.; Aparicio, R. Comparative study of virgin olive oil sensory defects. *Food Chem.* **2005**, *91*, 293–301. [CrossRef]
35. Negro, C.; Aprile, A.; Luvisi, A.; Nicolì, F.; Nutricati, E.; Vergine, M.; Miceli, A.; Blando, F.; Sabella, E.; De Bellis, L. Phenolic Profile and Antioxidant Activity of Italian Monovarietal Extra Virgin Olive Oils. *Antioxidants (Basel)* **2019**, *8*, 161. [CrossRef] [PubMed]
36. Samaniego Sánchez, C.; Troncoso González, A.M.; García-Parrilla, M.C.; Quesada Granados, J.J.; López García de la Serrana, H.; López Martínez, M.C. Different radical scavenging tests in virgin olive oil and their relation to the total phenol content. *Anal. Chim. Acta* **2007**, *593*, 103–107. [CrossRef] [PubMed]
37. Zhou, D.Y.; Sun, Y.X.; Shahidi, F. Preparation and antioxidant activity of tyrosol and hydroxytyrosol esters. *J. Funct. Foods* **2017**, *37*, 66–73. [CrossRef]

38. Roche, M.; Dufour, C.; Mora, N.; Dangles, O. Antioxidant activity of olive phenols: Mechanistic investigation and characterization of oxidation products by mass spectrometry. *Org. Biomol. Chem.* **2005**, *3*, 423–430. [CrossRef] [PubMed]
39. Carrasco-Pancorbo, A.; Cerretani, L.; Bendini, A.; Segura-Carretero, A.; Del Carlo, M.; Gallina-Toschi, T.; Lercker, G.; Compagnone, D.; Fernández-Gutiérrez, A. Evaluation of the antioxidant capacity of individual phenolic compounds in virgin olive oil. *J. Agric. Food Chem.* **2005**, *53*, 8918–8925. [CrossRef]
40. Mateos, R.; Domínguez, M.M.; Espartero, J.L.; Cert, A. Antioxidant effect of phenolic compounds, alpha-tocopherol, and other minor components in virgin olive oil. *J. Agric. Food Chem.* **2003**, *51*, 7170–7175. [CrossRef]
41. Pliszka, B.; Huszcza-Ciołkowska, G.; Wierzbicka, E. Effects of solvents and extraction methods on the content and antiradical activity of polyphenols from fruits Actinidia arguta, Crataegus monogyna, Gaultheria procumbens and Schisandra chinensis. *Acta Sci. Pol. Technol. Aliment.* **2016**, *15*, 57–63. [CrossRef]
42. Schlesier, K.; Harwat, M.; Böhm, V.; Bitsch, R. Assessment of antioxidant activity by using different in vitro methods. *Free Radic. Res.* **2002**, *36*, 177–187. [CrossRef]
43. Kouka, P.; Priftis, A.; Stagos, D.; Angelis, A.; Stathopoulos, P.; Xinos, N.; Skaltsounis, A.L.; Mamoulakis, C.; Tsatsakis, A.M.; Spandidos, D.A.; et al. Assessment of the antioxidant activity of an olive oil total polyphenolic fraction and hydroxytyrosol from a Greek Oleaeuropea variety in endothelial cells and myoblasts. *Int. J. Mol. Med.* **2017**, *40*, 703–712. [CrossRef]
44. Li, S.; Tan, H.Y.; Wang, N.; Cheung, F.; Hong, M.; Feng, Y. The Potential and Action Mechanism of Polyphenols in the Treatment of Liver Diseases. *Oxid. Med. Cell Longev.* **2018**, *2018*, 8394818. [CrossRef]
45. Istvan, E. Statin inhibition of HMG-CoA reductase: A 3-dimensional view. *Atheroscler. Suppl.* **2003**, *4*, 3–8. [CrossRef]
46. Cichoz-Lach, H.; Michalak, A. Oxidative stress as a crucial factor in liver diseases. *World J. Gastroenterol.* **2014**, *20*, 8082–8091. [CrossRef] [PubMed]
47. Goya, L.; Mateos, R.; Bravo, L. Effect of the olive oil phenol hydroxytyrosol on human hepatoma HepG2 cells—Protection against oxidative stress induced by tert- butylhydroperoxide. *Eur. J. Nutr.* **2007**, *46*, 70–78. [CrossRef] [PubMed]
48. De Stefanis, D.; Scimè, S.; Accomazzo, S.; Catti, A.; Occhipinti, A.; Bertea, C.M.; Costelli, P. Anti-Proliferative Effects of an Extra-Virgin Olive Oil Extract Enriched in Ligstroside Aglycone and Oleocanthal on Human Liver Cancer Cell Lines. *Cancers (Basel)* **2019**, *11*, 1640. [CrossRef]
49. Suttnar, J.; Másová, L.; Dyr, J.E. Influence of citrate and EDTA anticoagulants on plasma malondialdehyde concentrations estimated by high-performance liquid chromatography. *J. Chromatogr. B Biomed. Sci. Appl.* **2001**, *751*, 193–197. [CrossRef]
50. Stiuso, P.; Bagarolo, M.L.; Ilisso, C.P.; Vanacore, D.; Martino, E.; Caraglia, M.; Porcelli, M.; Cacciapuoti, G. Protective Effect of Tyrosol and S-Adenosylmethionine against Ethanol-Induced Oxidative Stress of Hepg2 Cells Involves Sirtuin 1, P53 and Erk1/2 Signaling. *Int. J. Mol. Sci.* **2016**, *17*, 622. [CrossRef] [PubMed]
51. Katsoulieris, E.N. The olive leaf extract oleuropein exerts protective effects against oxidant-induced cell death, concurrently displaying pro-oxidant activity in human hepatocarcinoma cells. *Redox. Rep.* **2016**, *21*, 90–97. [CrossRef] [PubMed]
52. Fitó, M.; Cladellas, M.; de la Torre, R.; Martí, J.; Alcántara, M.; Pujadas-Bastardes, M.; Marrugat, J.; Bruguera, J.; López-Sabater, M.C.; Vila, J.; et al. Antioxidant effect of virgin olive oil in patients with stable coronary heart disease: A randomized, crossover, controlled, clinical trial. *Atherosclerosis* **2005**, *181*, 149–158. [CrossRef]
53. Covas, M.I.; Nyyssönen, K.; Poulsen, H.E.; Kaikkonen, J.; Zunft, H.J.; Kiesewetter, H.; Gaddi, A.; de la Torre, R.; Mursu, J.; Bäumler, H.; et al. The effect of polyphenols in olive oil on heart disease risk factors: A randomized trial. *Ann. Intern. Med.* **2006**, *145*, 333–341. [CrossRef]
54. Istvan, E.S.; Deisenhofer, J. Structural mechanism for statin inhibition of HMG-CoA reductase. *Science* **2001**, *292*, 1160–1164. [CrossRef]
55. Chaudhary, R.; Garg, J.; Shah, N.; Sumner, A. PCSK9 inhibitors: A new era of lipid lowering therapy. *World J. Cardiol.* **2017**, *9*, 76–91. [CrossRef] [PubMed]

antioxidants

MDPI

Article

Effects of Processing on Polyphenolic and Volatile Composition and Fruit Quality of Clery Strawberries

Stefania Garzoli [1], Francesco Cairone [1], Simone Carradori [2,*], Andrei Mocan [3], Luigi Menghini [2], Patrizia Paolicelli [1], Gunes Ak [4], Gokhan Zengin [4] and Stefania Cesa [1,*]

1 Department of Drug Chemistry and Technologies, University "Sapienza" of Rome, P.le Aldo Moro 5, 00185 Rome, Italy; stefania.garzoli@uniroma1.it (S.G.); francesco.cairone@uniroma1.it (F.C.); patrizia.paolicelli@uniroma1.it (P.P.)

2 Department of Pharmacy, "G. d'Annunzio" University of Chieti-Pescara, Via dei Vestini 31, 66100 Chieti, Italy; luigi.menghini@unich.it

3 Faculty of Pharmacy, "Iuliu Hațieganu" University of Medicine and Pharmacy, 8 Victor Babeș Street, 400012 Cluj-Napoca, Romania; Mocan.Andrei@umfcluj.ro

4 Department of Biology, Science Faculty, Selcuk University, Konya 42130, Turkey; akguneselcuk@gmail.com (G.A.); gokhanzengin@selcuk.edu.tr (G.Z.)

* Correspondence: simone.carradori@unich.it (S.C.); stefania.cesa@uniroma1.it (S.C.)

Received: 12 June 2020; Accepted: 13 July 2020; Published: 17 July 2020

Abstract: Strawberries belonging to cultivar Clery (*Fragaria x ananassa* (Duchesne ex Weston)), cultivated in central Italy were subjected to a multi-methodological experimental study. Fresh and defrosted strawberries were exposed to different processing methods, such as homogenization, thermal and microwave treatments. The homogenate samples were submitted to CIEL*a*b* color analysis and Head-Space GC/MS analysis to determine the impact of these procedures on phytochemical composition. Furthermore, the corresponding strawberry hydroalcoholic extracts were further analyzed by HPLC-DAD for secondary metabolites quantification and by means of spectrophotometric in vitro assays to evaluate their total phenolic and total flavonoid contents and antioxidant activity. These chemical investigations confirmed the richness in bioactive metabolites supporting the extraordinary healthy potential of this fruit as a food ingredient, as well as functional food, highlighting the strong influence of the processing steps which could negatively impact on the polyphenol composition. Despite a more brilliant red color and aroma preservation, non-pasteurized samples were characterized by a lower content of polyphenols and antioxidant activity with respect to pasteurized samples, as also suggested by the PCA analysis of the collected data.

Keywords: Clery strawberry; food processing; Polyphenols; multi-methodological evaluation; HS-GC/MS analysis; PCA

1. Introduction

The healthy potential associated with the daily consumption of vegetables and fruits has attracted increasing interest in the last two decades [1,2]. Particularly, berries are appreciated for their healthy potential in relation to their high content of polyphenols, such as cinnamic acids and flavonoids, between which contain, of particular interest, anthocyanins. Strawberries display a rich content in nutritive and non–nutritive bioactive components, correlated to antioxidant potential and disease prevention along with excellent characters of flavor, color and taste. All these aspects make them one of the preferred available fruits, with an annual production of about five million tons per year worldwide [3]. Many recent studies underlined the healthy potential of strawberries [4–6]. A dietetic regime in which strawberry derivatives were plenty and regularly introduced could prevent inflammation onset and promote the reduction of obesity–related disorders and even cardiovascular diseases and neurodegeneration [7].

Many different strawberry cultivars are available on the market in different countries, each with their own peculiarities, and there are many variables which could influence the final content of specific molecules and their health potential, such as humidity and pedoclimatic conditions, fertilizers and other growing parameters [5]. Moreover, these fruits are not always consumed fresh, especially in relation to their limited seasonal availability. In this view, a dominant role is played by the applied procedures to obtain processed products, such as juices and jams, whose impact on the labile components of phytocomplex must be strictly controlled in order to prevent degradation and fragmentation that can occur, for example, to phenolic compounds at high temperatures.

Among the 73 strawberry cultivars available in Italy (Gazzetta Ufficiale della Repubblica Italiana n. 217, 2017), and many other available worldwide, Clery strawberries, cultivated in central Italy, were chosen as the study material. Clery strawberry is a recent variety (1998) obtained by breeding "Sweet Charlie" and "Marmolada" cultivars and has been authorized for commercial use since 2002. Its main characters are its regular shape, sweetness and taste, largely appreciated by its consumers, coupled with its resistance to cold and adverse climate conditions [8]. This strawberry variety was previously the object of other research studies. Two studies showed the presence of polyphenolic compounds, among which catechins and procyanidins, quercetin and kaempferol glucuronides, ellagitannins, pelargonidin–3–glucoside and pelargonidin–3–rutinoside accounted for the most represented molecules [9,10].

Nowicka et al. [11], who recently analyzed fruits selected from 90 strawberry cultivars, concluded that ellagitannins and procyanidins deeply influence the antioxidant capacity of strawberries evaluated by the ABTS test. In the review by Afrin et al. [5], the highest reported components are represented by proanthocyanidins (up to 163 mg/100 g fresh weight [fw]), ellagitannins (up to 83 mg/100 g fw) and anthocyanins (up to 66 mg/100 g fw). The authors, who evaluated the health benefits, focusing on the existing clinical studies on cardiovascular disease, metabolic syndrome, obesity and diabetes, neuroprotection and antimicrobial activity, besides antioxidant and anti-inflammatory activity, also consider strawberries a rich source of phenolic acids (up to 13 mg/100 g fw). Finally, kaempferol and quercetin, whose bioactivity as neuroprotective agents and as enhancers of adiponectin secretion by PPARγ, was also largely documented [12,13], accounting for up to 3 and 5 mg/100 g fw, respectively. As reported by da Silva et al. [14], anthocyanin pigments were mainly represented by pelargonidin and cyanidin aglycones, mostly glycosylated by glucose and rutinose and less frequently by arabinose and rhamnose. Only rarely were some acylated anthocyanins also detected; therefore, despite a very high content of anthocyanins, the represented composition is quite simple. These latter components, besides the healthy properties, account for the brilliant and consumer-appealing red color, whereas other extraordinary flavoring compounds characterize the strawberry aroma, both representing a delicate issue in case of the applied processing.

The Clery fruit quality was evaluated in terms of ripening, pre–harvest and post–harvest conditions [15,16], and the impact of different genotypes, fertilizers and harvesting date on fruit yield, sugars, phenolic compounds and anthocyanins content was evaluated. As stated by Tomić et al. [15], if genotype–cultivar represents the determinant factor in fruit quality, ripening, pre– and post–harvest treatments, processing and storage deeply influence the phytocomplex modification and the change–related consequences; therefore, the aim of this work was to evaluate in which way processing modifies the strawberry quality. In this context, diverse processing steps were applied and differently combined. Some interesting quality parameters, significant both for the consumer choice and for the healthy strawberry potential, such as total polyphenol, flavonoid and phenolic acid content, anthocyanin content, flavor and color were chosen to be assayed by using different analytical techniques.

Starting with the increasing interest devoted in the last twenty years to the healthy properties generally recognized to polyphenols and conscious of their limiting factors due to bioavailability and human metabolism, our aim was to evaluate how and how deeply processing affects the starting food quality. The effects of similar processing methods were previously evaluated in blueberries and goji berries [17,18]. In the present work, the effects of different homogenization, thermal and microwave treatments applied on Clery strawberries of Italian origin were monitored in terms of color and aroma changes and as variations of the polyphenolic content and antioxidant activity.

2. Materials and Methods

2.1. Materials

Commercial samples of strawberry fruits, *Fragaria x ananassa* (Duchesne ex Weston) and Duchesne ex Rozier (syn *Fragaria x ananassa* Duchesne) of the patent cultivar "Clery", were purchased at the farm "Fragole di Carchitti" a local company that controls all the productive chain, from cultivation to sale (Palestrina, RM, 41,83274 °N, 12.88178 °E). Plant material was harvested as single crop on 28 May, 2019, macroscopically selected as ready for food consumption, which reflects the full ripening status, and immediately stored at 4–6 °C. The day after the harvest, the strawberries fruits started to be used as fresh material (series f, fresh) and treatments were completed in a week. After gentle cleaning to remove impurities and green parts, strawberries were carefully washed with tap water and dried on paper. All other material was immediately frozen at −80 °C and then stored at −18 °C and used for test just after the defreezing (series d) within two months from the harvest date. Bidistilled water, ethanol, 85% formic acid, acetonitrile RS and reference compounds for HPLC were purchased from Merck life Science s.r.l. (Milan, Italy).

2.2. Processing

The fresh or thawed out fruits were processed and extracted according to the synoptic workflow represented in Figure 1. For each homogenization process, an aliquot of approximately 25 g was used. Coming from the same farm and the same crop, picked as fully ripe and characterized by similar size (each about 7–10 g), the strawberries were considered a homogeneous sample and were weighed as whole fruits in order to prepare the aliquots.

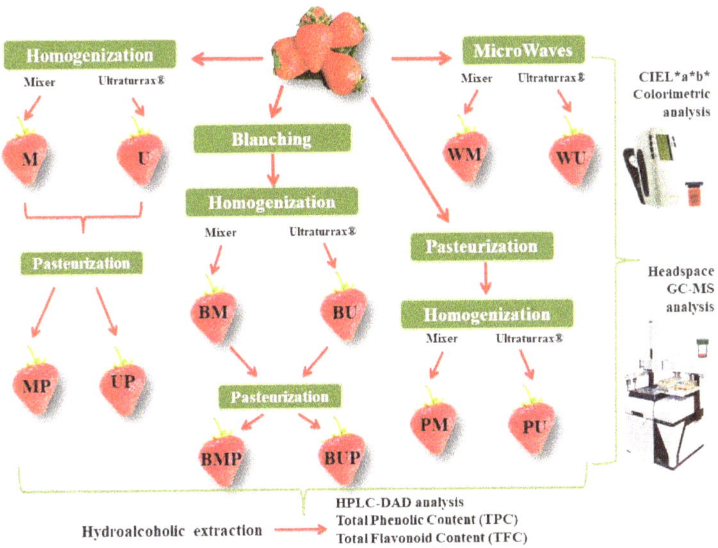

Figure 1. Flow chart of the applied treatments and the performed analyses on fully ripe fresh or defrosted Clery strawberries.

The blanching process consists in steamed plant material for 3 min at 85 °C, cooled at room temperature and then homogenized for 2 min either by a domestic mixer at 16,000 rpm (BM) or by a T18 Ultraturrax® homogenizer (IKA®, Staufen, Germany) at 10,000 rpm (BU) and subsequently submitted to a pasteurization treatment at 85 °C for 3 min (BMP, BUP). A parallel process starts with homogenization of plant material for 2 min (using domestic mixer or Ultraturrax® homogenizer, samples M and U, respectively) and the resulting purees heated at 85 °C for 3 min (samples MP and UP, respectively). The third approach starts with pasteurization followed by homogenization (as above described, samples PM and PU, respectively). The last experimental condition starts with pre-heating treatment in a domestic microwave oven (Samsung, T181) at 450 W for 1 min (final reached T by food was 70 ± 2 °C) and then samples were homogenized (as above described, samples WM and WU, respectively). All the treatments were performed in parallel on fresh and defrosted fruits. After each type of treatment, an aliquot of the purees was used for the colorimetric CIEL*a*b* and for the head space-gas chromatography/mass spectrometry (HS–GC/MS) analyses.

2.3. *pH Measurement*

Aliquots of about 1 g of the purees were transferred in conical tubes and centrifuged (5000 g for 10 min). The pH of the supernatant was determined using a Metrohm E632 pH–meter (Metrohm Italiana S.r.l., Rome, Italy).

2.4. *Colorimetric Analysis*

The homogenates were analyzed for their color character, with a colorimeter X–Rite (SP–62, GmbH, Regensdorf, Switzerland), equipped with D65 illuminant and an observer angle of 10°. Cylindrical coordinates C^*_{ab} and h_{ab} are calculated from a* and b* as previously indicated [18].

2.5. *HS–GC/MS Analyses*

Gas chromatographic/mass spectrometric (GC/MS) analysis was carried out using a Turbomass Clarus 500 GC–MS/GC-FID from Perkin Elmer instruments (Waltham, MA, USA). A Stabilwax fused–silica capillary column (Restek, Bellefonte, PA, USA) (60 m × 0.25 mm, 0.25 um film thickness) was used with helium as carrier gas (1.0 mL/min). GC oven temperature was kept at 40 °C and programmed to 220 °C at a rate of 5 °C/min, and kept constant at 220 °C for 5 min. All mass spectra were recorded in the electron impact ionization (EI) at 70 eV. Mass range was from 30 to 400 *m/z*. To examine the volatile fraction of strawberry samples, a Perkin–Elmer Headspace (HS) Turbomatrix 40 autosampler connected to a Clarus 500 GC-MS was used for the headspace analysis. This technique is ideal for sampling volatiles present in solid and liquid matrices [19,20].

The sampling procedure was thus carried out: sliced fruits and homogenates were collected into a 20 mL vial and tightly sealed immediately with crimp aluminum caps and 20 mm white rubber septa (Supelco, USA) using a vial crimper. With the intent to optimize the headspace procedure for the determination of a broader number of volatile organic compounds (VOCs) from strawberry samples, some parameters were adjusted as follows: thermostating temperature was 60 °C; thermostating time was 20 min; the pressurization time was 3.0 min; needle temperature was 90 °C; and the injection time was 0.3 min.

For the identification of the volatile fraction of strawberry samples, a Perkin-Elmer Headspace (HS) Turbomatrix 40 autosampler connected to a Clarus 500 GC–MS was used for the headspace analysis. This technique is ideal for sampling volatiles present in solid and liquid matrices [19,20]. With the intent to optimize the headspace sampling procedure for the determination of volatile organic compounds (VOCs), some parameters, such as equilibration time, temperature and head space analysis duration, were adjusted [21]. The sampling procedure was thus carried out: sliced fruits and homogenates were collected into a 20 mL vial and tightly sealed immediately with crimp aluminum caps and 20-mm white rubber septa (Supelco, PA, USA) using a vial crimper. The samples were incubated at 60 °C for 20 min prior the injection. The identification of the volatile main components was obtained by comparison of their linear retention indices (LRIs) and spectral mass with those reported in digital libraries data (Wiley 02 and Nist) of the GC/MS system. The LRI of each compound was calculated using a mixture of aliphatic hydrocarbons (C_8–C_{30}, Ultrasci) injected directly into GC injector under the same conditions described above. Relative percentages of the separated constituents were calculated from integration of the peak areas in the GC chromatograms without the use of an internal standard or correction factors using the same instrumentation with the FID detector configuration. GC/MS analysis of each strawberry sample was carried out twice.

2.6. Extraction of Polyphenols

According to Cesa et al. [17] with some modifications, an aliquot (5 g) of each sample was extracted for 1 h, at room temperature and in the dark, under stirring with a hydroalcoholic acidified mixture (15 mL of ethanol/5% formic acid in water, in the ratio 70:30, *v*/*v*). Then, the suspension was filtered, and the residue was washed. The separated solution was slightly concentrated under reduced pressure at 40 °C with a rotary evaporator. Filtered extract was adjusted to a final volume of 20 mL with the same solvent and directly analyzed by HPLC–DAD.

2.7. HPLC-DAD Analysis

HPLC analysis was carried out by a Perkin–Elmer (Waltham, MA, USA) apparatus equipped with a series LC 200 pump, a series 200 diode array detector and a series 200 autosampler. Data acquisition and processing were carried out with a Perkin–Elmer Totalchrom software. The chromatographic separation was performed as previously described [22]. In brief, a Luna RP18 column (250 × 4.6 mm, i.d. 5 µm) and a mobile phase, consisting of acetonitrile (A) and acidic water solution (B) in gradient, were used. The detection wavelengths were set at 520 nm for the detection of anthocyanins, at 360 nm for the detection of other flavonoids and at 280 nm for phenolic acids, catechins and other polyphenolic components. The injection volume for each extract was 10 µL. Pelargonidin-3–glucoside, identified at 520 nm in each sample, was quantified by an external-matrix matched calibration method on the basis of the area ratios respect to the pure chemical standard (R^2 = 0.9984). The other anthocyanins were calculated as the sum of all the chromatographic peaks identified at 520 nm. Calibration curves were built and used for quantitation of polyphenols, using, at 280 nm, epicatechin (R^2 = 0.9878), caffeic acid (R^2 = 0.9984), *p*–coumaric acid (R^2 = 0.9879) and ferulic acid (R^2 = 0.9974) and, at 360 nm, rutin (R^2 = 0.9986) and quercetin-3-D-galactoside (R^2 = 0.9999) as reference standards.

2.8. Total Phenolic Content (TPC), Total Flavonoid Content (TFC) and Antioxidant Assays

To obtain the total amount of these polyphenolic groups in the samples, colorimetric assays were used as described in our previous paper [23]. Gallic acid (GAE, Sigma–Aldrich, Germany) and rutin (RE, Sigma–Aldrich, Germany) were used as standards for phenols and flavonoids, respectively. To detect antioxidant and metal chelating properties, we used several chemical assays including different mechanisms, namely, radical scavenging, reducing power and metal chelating. Trolox (TE) and ethylenediaminetetraacetic acid (EDTA) were used as standard antioxidant compounds and experimental results are expressed as activity unit equivalents of these compounds [23].

2.9. Statistical Analysis

All analyses were carried out in quadruplicate. The data collected are presented as average values and completed by the standard deviations. One-way analysis of variance (ANOVA), followed by Tukey's post-hoc test, was employed to assess significant differences ($p < 0.05$) using GraphPad Prism version 5.01 for Windows (GraphPad Software, San Diego, CA, USA). The principal component analysis (PCA) was performed with XLSTAT Version 2020 software.

3. Results and Discussion

3.1. Processing and Aim of the Work

Two different grinding procedures followed or preceded by two different thermal treatments of pasteurization and steam blanching, or preceded by a microwave treatment, were applied to the selected fruits. A part of the selected fruits was processed immediately after the harvesting. Otherwise, these were frozen at −80 °C and stored at −20 °C, until the thawing and the immediate processing were performed. All the adopted treatments, freezing and thawing, grinding with a domestic mixer or with a highspeed industrial homogenizer, pasteurization and blanching performed at 85° C for 3 min and the microwave treatment performed at 450 W for 1 min represented some classical food processing procedures. Berries, such as strawberries, are often subjected to these treatments during the preparation of jellies or juices in order to reduce the bacterial load, as well as to prevent spontaneous and enzymatic oxidation that could affect organoleptic characters and other quality parameters. Even if the intent is to protect foodstuff, all the adopted steps could exert an influence on the polyphenolic and aromatic content. The aim of the present work was to compare the effects of different work processes and sequences on qualitative and quantitative composition and to define the optimal method to preserve the original color and aroma, which deeply influence the consumer's choice and the healthy peculiarities.

Previous works, performed on blueberries [17] and goji berries [18], indicated that significant differences could be found in terms of color between differently ground food samples and that more marked differences could be highlighted if the homogenization procedures are combined with different thermal procedures. Here, moreover, the microwave treatment was introduced and the thermal treatments were differently combined and the effects of freezing and thawing were also tested.

The two adopted analytic methods, CIEL*a*b* and HS-GC-MS, allowed us to analyze directly the fruit homogenates returning a sample photograph prior to any impact due to the extraction procedure. Moreover, the HPLC-DAD analysis of the hydroalcoholic extracts obtained by the homogenates, allowed us to complete and compare the data by the colorimetric analysis, giving information about any compositional change of the polyphenolic components. A correlation was finally attempted to be made among all the obtained results by the different used analytical procedures and the differently combined applied procedures.

3.2. pH Measurements

The pH of all the obtained homogenates was measured, to verify if differences among differently treated samples could justify consequences in terms of color change, being anthocyanins color expression deeply influenced by pH variation. All the measured pH ranged between 3.7 and 4.0, showing that the detected color changes cannot be attributed to pH variations.

3.3. Color Analysis

Color expressed by a foodstuff, besides giving information about its peculiar origin and the nature of its pigment content, could be deeply influenced by the processing. As is well known, fruits including strawberries, which are only available in a very limited seasonal period, are subjected, among many other berries, to significant work–up with the aim to obtain jellies, juices or other storable products. The result of these transformations impacts the organoleptic characters of the final foodstuff which, besides the great influence played on the consumer's choice, deeply modify the chemical composition.

The main pigments of strawberry phytocomplex are represented by anthocyanins and polyphenols, which also represent the most important molecules in terms of healthy properties. Therefore, information about the color change is still important as an index of the loss of quality. Analyses performed on all products obtained by different processes from fresh or defrosted samples gave the results reported in Table S1 (Supplementary material), through which it is possible to observe the wide range covered by the data. L* is the index of sample lightness, ranges between about 33.3

and 41.3 denoting browning or bleaching in dependence of the adopted processing; a*, whose positive values indicate the sample redness, ranges between about 17.1 and 21.9, and b* positive values, an index of yellowness, range between about 6.9 and 11.5. Color intensity (C^*_{ab}), ranging between 18.7 and 24.3, and moreover color tonality, ranging between 21.6 and 28.2, show the existing differences among samples, all deriving by the same crop. It could be exemplified comparing the reflectance curves recorded after the same process from fresh (f) or defrosted (d) samples or from different heating or homogenization procedure (Figure 2).

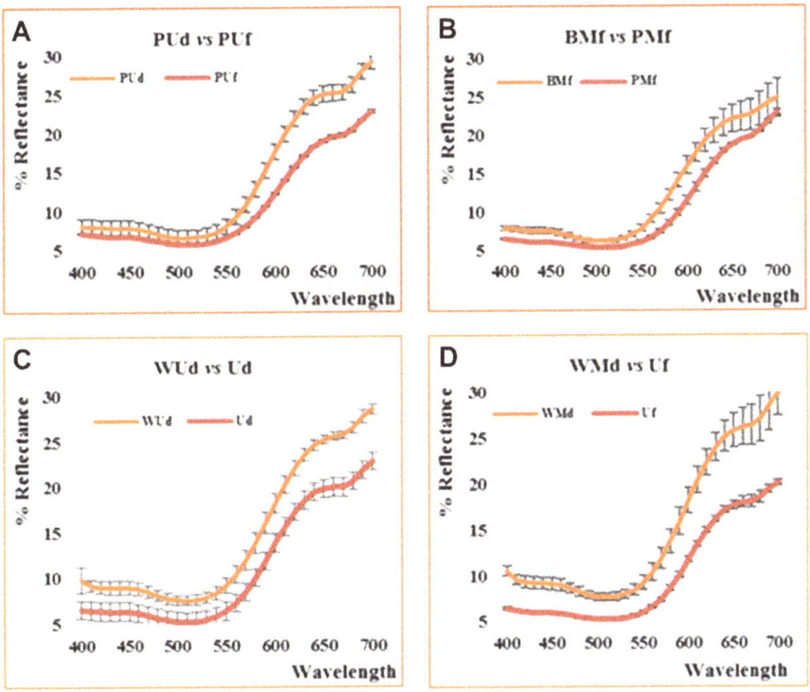

Figure 2. Reflectance curves of some selected homogenates: (**A**) comparison between PUd and PUf; (**B**) comparison between BMf and PMf; (**C**) comparison between WUd and Ud; (**D**) comparison between WMd and Uf. Each reported profile represents the mean of four measurements.

All these samples show significant differences, more marked in the region between 550 and 700 nm and in the panel "D" are reported the profiles of samples with the strongest difference at all wavelengths. However, if we consider the behavior of the different series in general (type of homogenizer, fresh or defrosted starting material, dry heated by pasteurization or blanched with steam or microwaves treated), the differences do not allow us to define the influence of specific factors on color parameters. It also shows results relevant to the high variability of the sample highlighted in the high range of standard deviations recorded. Each sample denotes a different trend (also further confirmed by the HS–GC–MS analyses) which makes it difficult to point out general conclusions. The only evidence is the rising of lightness in blanched and/or defrosted samples, as well as after the microwave treatment. All blanched and defrosted samples show a red orange color nuance, with respect to the red pink color of the fresh pasteurized or the not heated samples.

Previous papers focus the anthocyanins stability to pH change, thermal treatments, light exposure, polyphenol oxidase influence, browning and co-pigmentation [24,25]. On the other hand, no data are available on the effects of different treatments on the same batch of strawberries. Our previous experiments demonstrated that a correlation could be evidenced between selected color parameters and anthocyanins ratio in blueberries and zeaxanthin content in goji [17,18]. In this case, such a correlation could not be as well depicted by the obtained data. If the differences between BUf (130 μg/g fw anthocyanins) and PUf (260 μg/g) could be explained by the different anthocyanins content, it is not possible to explain the similar trends of BUf and PUd (130 μg/g), with respect to BUd (40 μg/g). As, to our knowledge, no results are reported in the literature about this context, we could only hypothesize that the steam blanching, although impacting on the anthocyanin degradation, could preserve the sample against the enzymatic browning. On the contrary, while protecting the anthocyanin content, the pasteurization process induces a slight browning process. Finally, the thawing, although impacting more drastically on anthocyanins content, preserves the samples color on the whole.

In any case, a direct correlation between color and anthocyanin content could not be shown and a more complex flavonoid composition (high content of flavanols with respect to flavonols, perhaps a small carotenoid content) and counteracting browning and/or bleaching processes [26] need to be taken into consideration. For this reason, the color appearance allows us to discriminate among differently treated samples, only if correlated with all the other monitored parameters, as further shown by the PCA analysis.

3.4. HS-GC/MS Analysis

Ten components, listed in Tables 1 and 2, were identified by HS–GC/MS of the different analyzed samples of Clery strawberries. In Table 1 the results of the analyses performed on fresh plant material are reported, comparing the two different homogenizations alone or after microwave or blanching treatments. Ethyl acetate and methyl butanoate were revealed in all samples, with the highest percentage of ethyl acetate (30.4%) in M sample, and the highest percentage of methyl butanoate (22.2%) in U. The most represented molecule in the only homogenized and microwave-treated samples (M, U, WM, WU) was methyl acetate, which accounted for about 35–40%. On the contrary, it was completely absent in the blanched samples (BM, BU), in which a high content of ethanol was found (60–70%). A low content of methyl hexanoate (6.9%) and of ethyl hexanoate (1.7%), in the sliced strawberries, has been found in the simply homogenized (U and M) samples.

Table 1. Fresh strawberry HS-GC peak area (%).

Components [1]	LRI [2]	LRI [3]	Sliced	U	M	WU	WM	BM	BU
Acetaldehyde	650	655	3.8	12.9	4.4	5.7	8.5	2.8	-
Methyl acetate	826	828	40.0	34.6	36.7	38.7	38.8	-	-
Acetone	838	842	1.9	9.9	4.4	6.9	4.9	-	11.5
Ethyl acetate	880	885	19.8	5.8	30.4	27.1	25.4	23.9	20.4
Methyl butanoate	986	985	17.7	22.2	9.6	7.9	9.2	2.4	9.1
Ethanol			nd	nd	nd	nd	nd	70.9	59.0
Ethyl butanoate	1040	1036	8.1	8.9	7.9	8.3	8.8	nd	nd
2-hydroxy propanamide	1114	*	-	-	2.6	-	4.4	-	-
Methyl hexanoate	1182	1187	6.9	2.1	3.5	-	-	-	-
Ethyl hexanoate	1234	1238	1.7	3.4	0.4	-	-	-	-
Sum			99.9	99.8	99.9	94.6	100.0	100.0	100.0

[1] Elution order on polar column; [2] linear retention indices (LRI) measured on polar column; [3] linear retention indices from literature; * LRI^{lit} not available; traces < 0.1%; sliced: untreated strawberry.

BM and BU showed a similar chemical composition, with the only difference in the acetaldehyde and acetone content. Sliced strawberry, M and U samples displayed the same chemical qualitative profile, but the M sample also contained a very small amount of 2-hydroxy propanamide.

In Table 2, the results obtained by the analysis of all the pasteurized samples are reported, comprising two defrosted samples (PMd and PUd). After the blanching treatment after the pasteurization, only few components were found and a high quantity of ethanol was expressed. The high percentage of ethanol in blanched or pasteurized samples before the application of the homogenization process (BM, BU, PM, PU, PMd and PUd) could be justified by the heating induction of an accelerated fermentation process during the next grinding process [27]. Methyl acetate, ethyl butanoate, 2–hydroxy propanamide, methyl and ethyl hexanoate had completely disappeared after the thermal treatment.

Table 2. HS-GC peak area (%) of fresh and defrosted (d) strawberries to which was applied a pasteurization process before or after homogenization or after blanching and homogenization.

Components [1]	LRI [2]	LRI [3]	MP	UP	PM	PU	BMP	BUP	PMd	PUd
Acetaldehyde	650	655	-	1.9	1.3	-	-	-	6.0	2.3
Acetone	838	842	57.4	8.6	0.4	26.9	53.1	76.4	-	0.8
Ethyl acetate	880	885	-	61.4	61.4	2.9	46.9	-	33.9	16.2
Ethanol	940	938	-	25.3	36.9	70.2	-	-	60.1	80.7
Methyl butanoate	986	985	42.6	2.8	-	-	-	23.6	-	-
Sum			100.0	100.0	100.0	100.0	100.0	100.0	100.0	100.0

[1] Elution order on polar column; [2] linear retention indices measured on polar column; [3] linear retention indices from literature.

All these samples are characterized for about the 90% by two only molecules which are represented by acetone and methyl butanoate (in MP and BUP), ethyl acetate and ethanol (in UP, PM, PMd, PUd), acetone and ethanol (in PU), acetone and ethyl acetate (in BMP), thus indicating a substantial modification and decrease of the aromatic character.

According to the literature, the more impacting molecules on strawberries' flavors are recognized in two furfural derivatives, ethyl hexanoate, hexanal, ethyl methyl butanoate and methyl butanoate [28]. The high impact on the volatile components, besides the specific cultivar, is also due to cultural techniques, shading and harvesting dates [29]. Regarding the ester derivatives, the impact on fruity and floral character is generally recognized. Although less important for the aroma character, methyl acetate, which we found as the most represented in the sliced strawberries, was also indicated by Watson et al. [28] as the highest peak in Elsanta cultivar. Ethyl hexanoate was well represented and found in seven of the eight strawberry cultivars analyzed by Oz et al. [29]; whereas methyl butanoate and ethyl butanoate were found in half of the samples, methyl butanoate and ethyl acetate (only in trace) were found in only one case. Ethyl hexanoate was also reported by Kafkas et al. in strawberry wine [30].

The decrease of these ester compounds with respect to the sliced strawberry used as comparator seems to represent a good index to evaluate the treatment impact on the aroma value. These modifications are more evident in the representation reported in Figures S1 and S2 (Supplementary Materials).

3.5. HPLC–Analysis

The hydroalcoholic extracts of all the homogenized samples were then subjected to HPLC–DAD analysis. The chromatograms registered at 280 nm, 360 nm and 520 nm are reported in Figure S3 (Supplementary Materials).

Many different hydroxycinnamic, flavonoid and anthocyanin molecules have been determined in the polyphenolic component of Clery strawberry cultivar, according to the different data shown by the literature [14,29,31]. By these reports, over twenty different anthocyanins were identified in different strawberry cultivars. The presence of anthocyanins in strawberry cultivars is confirmed in literature data. In particular, Kelebek et al. [31] observed in strawberries six different anthocyanins which were identified as cyanidin–3–glucoside, cyanidin–3–rutinoside, pelargonidin–3–glucoside, pelargonidin–3–rutinoside, pelargonidin–3–malonyl-glucoside and pelargonidin–3–acetyl–glucoside. In our analyzed samples, we could observe a very simple profile (Figure S3, Panel C), with regard to the anthocyanin content. A slightly more complex profile was shown by the chromatogram at 280 nm (Figure S3, Panel A) that allows the identification of three hydroxycinnamic acids (namely ferulic, caffeic and *p*–coumaric acid) and the two flavan–3–ols (namely catechin and epicatechin). A more complex chromatogram registered at 360 nm (Figure S3, Panel B), in which only quercetin–3–D–galactoside was identified, but some peaks remained unresolved after direct comparison with pure standard or data on literature [31,32]. In Table 3, the data of the quantitative HPLC analyses on the fresh and the defrosted series are reported. Anthocyanins were quantified by a calibration curve, built on the pure reference standard of pelargonidin–3–glucoside, and were expressed as the sum of all the peaks areas monitored at 520 nm.

As shown in Table 3, the total amount of anthocyanins, expressed as μg of pelargonidin–3–glucoside/g fw, is much higher in the fresh, not frozen, samples than in defrosted ones (d). In particular, in the fresh series, the amounts of anthocyanins range between about 120 μg/g fw in BM and 260 μg/g fw in PU, whereas, in the defrosted series (d), it varies from about 30 μg/g fw in UP to about 160 μg/g fw in U.

Table 3. HPLC-DAD quantitative analysis.

	M	U	BM	BU	BMP	BUP	PM	PU	MP	UP	WM	WU
Catechin	-	253.7	-	108.9	98.1	135.1	177.5	210.8	165.3	96.6	153.4	178.4
Epicatechin	236.4	251.4	181.2	195.6	237.6	288.8	184.4	316.4	168.5	229.5	195.9	220.8
Caffeic acid	11.8	10.8	11.7	12.2	10.0	18.2	8.4	18.5	7.2	14.4	17.8	48.9
p–Coumaric acid	5.6	5.9	-	7.6	13.1	17.1	4.8	15.1	11.2	16.5	8.2	13.4
Ferulic acid and derivatives *	75.8	134.4	73.7	141.6	140.6	180.7	56.2	197.3	103.9	173.2	133.1	200.6
Rutin	-	4.8	1.1	3.4	6.1	5.4	6.5	13.5	9.8	10.8	7.1	7.1
Flavonols **	19.3	18.2	1.6	12.2	10.8	15.4	8.4	8.5	28.4	14.1	12.8	13.2
Anthocyanins ***	216.6	219.1	123.5	131.4	131.7	144.6	237.9	258.5	227.2	222.6	223.5	151.7
Sum	565.5	898.3	398.8	612.9	648.0	805.3	684.1	1038.6	721.5	777.7	751.8	834.1
	Md	**Ud**	**BMd**	**BUd**	**BMPd**	**BUPd**	**PMd**	**PUd**	**MPd**	**UPd**	**WMd**	**WUd**
Epicatechin	202.3	194.8	200.7	73.9	133.5	141.4	157.8	126.5	222.6	192.7	148.6	290.2
Caffeic acid	38.6	77.2	20.0	5.8	31.7	5.7	31.0	88.7	47.9	54.5	nd	nd
p-Coumaric acid	20.4	31.7	22.8	26.2	25.3	29.9	23.2	34.3	19.7	62.2	21.3	31.8
Ferulic acid and derivatives *	216.2	242.5	167.2	83.2	92.7	90.4	134.4	207.8	227.9	120.9	122.9	178.8
Flavonols **	5.3	21.7	5.9	6.9	8.6	9.0	8.3	13.4	10.5	28.1	10.3	5.4
Anthocyanins ***	64.9	158.3	85.2	37.9	50.6	36.8	69.4	129.7	76.8	30.4	82.6	58.8
Sum	547.7	726.2	501.8	233.8	342.4	313.2	424.1	600.4	605.4	488.8	385.7	565.0

Results are expressed in μg/g of fresh weight. The RSD value, evaluated on triplicates, was < 5%. Catechin and Rutin, although detected, were not revealed in the defrosted samples. * Expressed as ferulic acid. ** Expressed as quercetin-3-D-galactoside. *** Expressed as pelargonidin–3–glucoside.

Buendia et al. [33] reported that the total anthocyanin content of different strawberry cultivars ranged from 202 to 466 μg/g fw, whereas in a study conducted on the fruits of cv. Clery during ripening [32], the total anthocyanin content in full ripe fruits is about 430 μg/g fw. Gasperotti et al. [9] reported an anthocyanin content of about 360 μg/g fw.

By our results, it is clearly shown, both in the fresh and in the thawed series, that significant differences of anthocyanins amounts are associated to the different applied treatments. It could be observed that, on average, the amounts of anthocyanins in the pasteurized samples (PM, PU, MP and UP) are much higher than in samples subjected to blanching (BM, BU, BMP and BUP). More specifically, in the fresh series, the average amount of anthocyanins in the pasteurized samples is almost double (about 236 μg/g fw) than in the blanched samples (about 134 μg/g fw), in which a decrease in anthocyanin content of about 44% is observed.

In a previous work, it was reported the high impact of a steam blanching treatment, with respect to a pasteurization process, both performed similarly to what we did. Although they hypothesized a browning due to anthocyanins polymerization, we did not see such effect on color appearance. On the other hand, we think the direct exposition of the fruits to the penetrating power of steam could have a different effect with respect to the dry heating of fruits. This could act on polyphenol oxidase inactivation with prevention of enzymatic browning, nevertheless impacting on anthocyanin stability. It must be also taken in consideration that different anthocyanins were here represented and many other variables could influence their modification, such as vitamin C content, other polyphenolic content and so on [34].

On the contrary, in the d series, the average amount of anthocyanins in pasteurized samples was only slightly higher (77 vs. 53 μg/g fw) than in the blanched samples, showing that the freezing and thawing process provokes an evident anthocyanin content loss and a partially equilibrating effect. In fact, there is a loss of about 70% of the anthocyanin content in thawed samples compared to fresh treatments, with the exception of Ud and BMd where there is a loss of only 30%.

With regards to the procyanidins in the fresh samples, catechin ranged between about 100 (in UP) and about 250 μg/g fw (in U), whereas, it completely lacks in M and BM, as well as in the defrosted samples. Epicatechin, more represented, varies between 170 and 315 in the fresh series, whereas it decreases to 75 and 290 in the d series, with a loss of catechin content of about 63% in thawed samples. Therefore, higher values of flavan-3-ols were confirmed in the not thawed series, as for the anthocyanin content, and also if a different variability is shown between different couples of samples. Much lower are the contents of *p*–coumaric acid (5–60 μg/g fw) and caffeic acid (about 10–90 μg/g fw), whereas the ferulic acid derivatives are a little more represented (60–240 μg/g fw). In the flavonoid scaffold, monitored at 360 nm, a very small content of rutin was quantified (between 1.1 and 13.5 μg/g fw) in the fresh samples, but it was not detected in the d series.

The sum of the areas related to the other shown peaks, among which kaempferol, was expressed as quercetin–3–D–galactoside. These quantified flavonoids ranged between 2 and 28 μg/g fw in the fresh series and was equally represented in the d series (5–28 μg/g fw). All these results are, as general consideration, in agreement with other data published in literature which report ranges of 15–60 μg/g fw for flavanols, 15–30 μg/g fw for flavonols, 20 35 μg/g fw for cinnamic acids, 10–50 μg/g fw for phenolic acids and less than 10 μg/g fw for quercetin and kaempferol [31–33,35].

By the overview of the all obtained results, it could be possible to conclude that, if the thawing has caused an important decrease in the amounts of anthocyanins, it did not impact on the hydroxycinnamic content. Rather, the average content of caffeic, ferulic and *p*–coumaric acids in the d series seems to be perfectly preserved, if related to the content found after other treatments.

On the contrary, the freezing and defrosting cause a severe impact on catechin and rutin that results in secondary effects from a quantitative point of view and results as a relevant modification of qualitative profile. The amount of other flavonoid metabolites (e.g., epicatechin) are not significantly affected by processes applied [36].

As far as concerns the homogenization process, a generally higher polyphenolic content can be observed in samples subjected to the Ultraturrax® treatment, in both series (Figure 3).

In particular, it seems that the different type of homogenization (domestic mixer or Ultraturrax®) does not affect the anthocyanin and flavonol content as much as the content of flavanols and phenolic acids, which account for an average loss of 35% in series f treated with domestic mixer, much less marked in thawed samples (about 4.5%). The hydroxycinnamic and flavonoid contents are influenced by the different carried out thermal treatments, although to a less drastic extent if compared to anthocyanins, indicating that these last pigments are much more sensitive, particularly to the thawing.

The effect of thawing on anthocyanins degradation was previously reported and discussed by Holzwarth et al. [37], which also reported the different impact of thawing at different temperature and time conditions. Anthocyanins degradation between about 7% and 21% was shown, dependently on the different applied method. These results only partially agree with ours, by which a decrease of anthocyanins of 27% was shown in the sample only homogenized in Ultraturrax® after thawing. Although frozen at -80 °C, avoiding the tissue destruction, and thawed at room temperature in less than an hour for their small size, much higher degradation was shown for all the other defrosted samples, confirming the importance of the adopted working process in the whole.

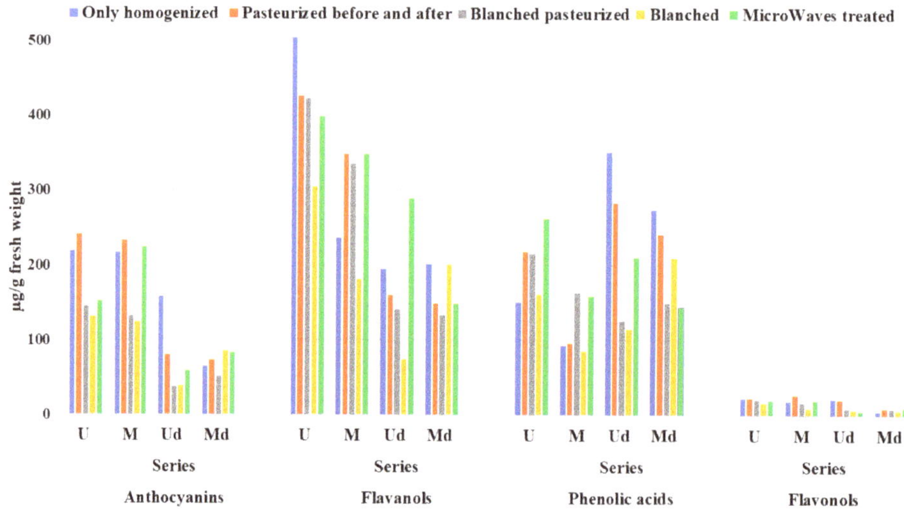

Figure 3. Effects of the different treatments applied to the strawberry cultivar Clery, U and M series, on the phytocomplex.

3.6. *Total Phenolic Content and Total Flavonoid Content Evaluation*

In recent years, the chemical and biological properties of phenolic compounds are one of the most attractive subjects in the scientific studies. In this sense, we investigated the total phenolic and flavonoid content of the strawberry samples. As can be seen in Table 4, the highest amounts of phenolics and flavonoids were determined in MP (19.28 mg GAE/g and 2.53 mg RE/g, respectively). The lowest levels of phenolics and flavonoids were noted in WU (11.34 mg GAE/g) and BUP (0.83 mg RE/g), respectively.

Table 4. Total phenolic and flavonoid contents in the strawberry samples *.

Samples	TPC (mg GAE/g)	TFC (mg RE/g)
BUP	14.63 ± 0.22 [d]	0.83 ± 0.05 [f]
BMP	12.66 ± 0.13 [f]	1.14 ± 0.04 [d]
U	12.99 ± 0.10 [f]	0.91 ± 0.04 [f]
M	13.69 ± 0.12 [e]	1.54 ± 0.06 [bc]
UP	14.39 ± 0.15 [d]	1.17 ± 0.03 [d]
MP	19.28 ± 0.06 [a]	2.53 ± 0.03 [a]
PU	16.38 ± 0.13 [c]	1.06 ± 0.07 [de]
PM	17.27 ± 0.21 [b]	1.59 ± 0.05 [b]
WU	11.34 ± 0.15 [h]	1.19 ± 0.03 [d]
WM	13.83 ± 0.11 [e]	1.19 ± 0.05 [d]
BU	12.84 ± 0.03 [f]	0.96 ± 0.05 [ef]
BM	11.97 ± 0.10 [g]	1.08 ± 0.05 [de]
Ud	12.16 ± 0.08 [g]	1.16 ± 0.06 [d]
Md	13.67 ± 0.15 [e]	1.40 ± 0.05 [c]

* Values are reported as mean ± S.D. TPC: total phenolic content; TFC: total flavonoid content; GAE: gallic acid equivalent; RE: rutin equivalent; different letters indicate significant differences in the samples ($p < 0.05$).

In Table 5, the results relative to the antioxidant and antiradical properties exerted by the treated samples are reported. They are more consistent with TPC and TFC values than HPLC data, thus highlighting that other non-detected components could participate to these properties. As a general trend, all the samples displayed a different profile on the basis of the treatment(s).

Homogenization with Ultraturrax® led to the lowest results in terms of metal chelating activity and in the phosphomolybdenum assay (Ud), especially if previously subjected to the microwave irradiation (WU). As regards their antioxidant ability, investigated by five in vitro methods, the microwave treatment and the Ultraturrax® homogenization had detrimental effects. The latter was partially tolerated only if associated to pasteurization or blanching approaches. The different impact of freezing can be seen comparing U–Ud and M-Md samples; the Ultraturrax® homogenization must be preferred with respect to the domestic mixer if the samples are defrozen. Homogenization-pasteurization should be privileged to pasteurization-homogenization. Collectively, the best results according to the different mechanisms of antioxidant potential were obtained with the MP and PM samples suggesting that processing conditions strongly impact on strawberry content and healthy potential.

Table 5. Antioxidant properties of the strawberry samples *.

Samples	DPPH (mg TE/g)	ABTS (mg TE/g)	CUPRAC (mg TE/g)	FRAP (mg/TE)	MCA (mg EDTAE/g)	PHD (mmol TE/g)
BUP	27.92±0.53 ^{b}	38.19 ± 0.19 ^{c}	56.45 ± 0.33 ^{c}	38.81 ± 0.57 ^{c}	7.73 ± 0.09 ^{a}	1.22 ± 0.13 ^{abc}
BMP	23.65 ± 0.32 ^{c}	33.50 ± 0.13 ^{e}	49.55 ± 0.24 ^{ef}	33.58 ± 0.19 ^{ef}	7.06 ± 0.06 ^{ab}	1.09 ± 0.03 ^{bc}
U	21.73 ± 0.24 ^{de}	31.47 ± 0.41 ^{f}	48.57 ± 0.40 ^{f}	32.52 ± 0.38 ^{f}	4.57 ± 0.78 ^{ef}	1.06 ± 0.07 ^{bc}
M	21.11 ± 0.35 ^{e}	31.73 ± 0.53 ^{f}	48.85 ± 0.18 ^{ef}	33.22 ± 0.29 ^{ef}	3.89 ± 0.32 ^{f}	1.13 ± 0.14 ^{abc}
UP	27.17 ± 0.50 ^{b}	36.35 ± 0.93 ^{d}	54.92 ± 0.31 ^{c}	37.47 ± 0.61 ^{d}	3.65 ± 0.20 ^{f}	1.10 ± 0.04 ^{abc}
MP	33.20 ± 0.96 ^{a}	45.41 ± 0.35 ^{a}	72.26 ± 0.42 ^{a}	48.11 ± 0.50 ^{a}	6.87 ± 0.21 ^{abc}	1.27 ± 0.09 ^{ab}
PU	26.79 ± 0.26 ^{b}	38.19 ± 0.28 ^{c}	61.02 ± 0.67 ^{b}	39.15 ± 0.16 ^{c}	6.09 ± 0.12 ^{bcd}	1.13 ± 0.03 ^{abc}
PM	27.29 ± 0.61 ^{b}	40.98 ± 0.85 ^{b}	61.76 ± 1.05 ^{b}	42.15 ± 0.34 ^{b}	6.98 ± 0.48 ^{abc}	1.17 ± 0.08 ^{abc}
WU	18.47 ± 0.51 ^{f}	25.52 ± 0.66 ^{i}	35.30 ± 0.18 ^{g}	26.90 ± 0.44 ^{h}	1.91 ± 0.14 ^{g}	0.95 ± 0.07 ^{c}
WM	21.70 ± 0.50 ^{de}	32.52 ± 0.09 ^{ef}	50.38 ± 1.31 ^{e}	36.31 ± 0.17 ^{d}	2.18 ± 0.44 ^{g}	1.03 ± 0.04 ^{bc}
BU	21.64 ± 0.09 ^{de}	28.49 ± 0.45 ^{gh}	52.79 ± 0.72 ^{d}	34.41 ± 0.50 ^{e}	5.98 ± 0.25 ^{cd}	1.23 ± 0.11 ^{ab}
BM	20.95 ± 0.50 ^{e}	29.87 ± 0.17 ^{g}	47.98 ± 0.61 ^{f}	30.59 ± 0.47 ^{g}	6.92 ± 0.30 ^{abc}	1.26 ± 0.16 ^{ab}
Ud	21.47 ± 0.24 ^{de}	28.16 ± 0.81 ^{h}	48.33 ± 0.25 ^{f}	32.40 ± 0.24 ^{f}	1.64 ± 0.24 ^{g}	1.38 ± 0.12 ^{a}
Md	22.73 ± 0.09 ^{cd}	32.25 ± 0.32 ^{ef}	52.42 ± 0.39 ^{d}	34.26 ± 0.55 ^{e}	5.57 ± 0.57 ^{de}	1.30 ± 0.07 ^{ab}

* Values are reported as mean ± S.D. TE: trolox equivalent; EDTAE: EDTA equivalent; MCA: metal chelating assay; PHD: phosphomolybdenum assay; different letters indicate significant differences in the samples ($p < 0.05$).

3.7. Principal Component Analysis (PCA)

The PCA was carried out on all the data collected on the different series of samples in order to better find a correlation among the large variability of the reported data and samples. The values have been scaled, using XLSTAT 2020 software, with the unit variance scale. After scaling, all the parameters contributed in the same way on the variance and all variables acquired equal weight for the PCA. As shown in Figure 4, the x–axis represents the first PCA dimension (F1) covering the 44% of the total variance, whereas y–axis is the second PCA dimension (F2, 18% of the total variance). The red vectors indicate the investigated variables. The vector lengths are an index of the representativeness of the investigated PCA dimensions.

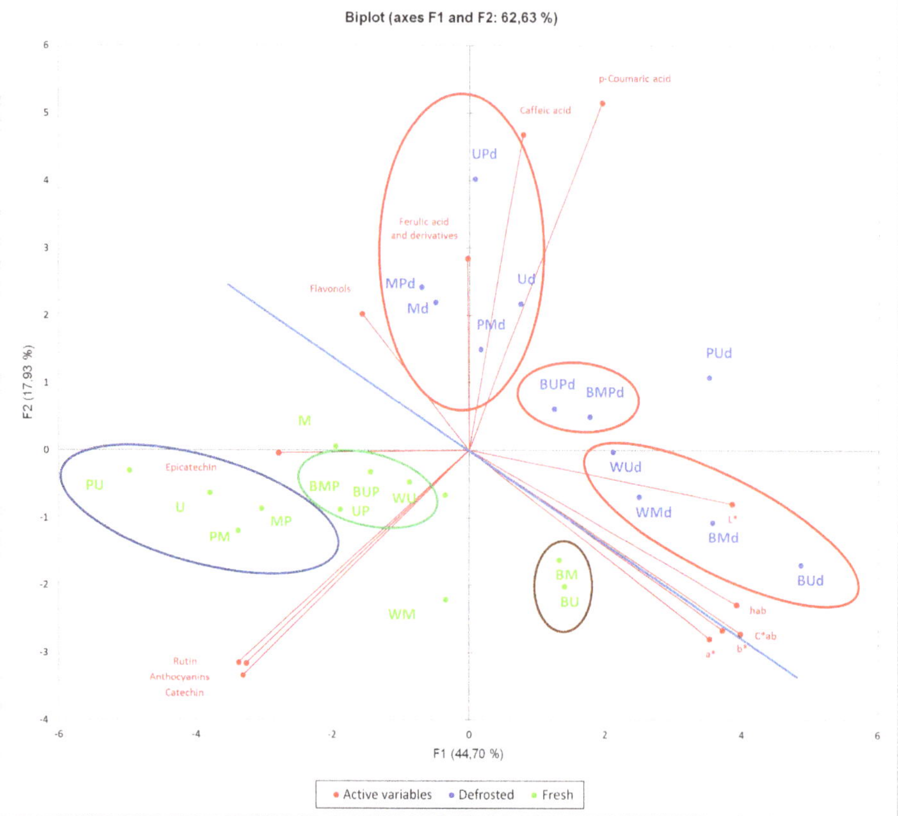

Figure 4. Principal component analysis (PCA) plots of the all strawberry Clery samples analyzed.

The most significant variables to discriminate between fresh and defrosted samples are represented by phenolic acids content, which is higher in the defrosted samples, whereas flavonoids are more represented in the fresh ones. The CIEL*a*b* parameters seem to contribute less. Considering the different analyzed variables (Figure 4), a clear separation between fresh (green) and defrosted (blue) samples could be revealed. Moreover, in both series, a separation into different groups could be underlined.

This could be better explained by Figure S4 (in Supplementary Materials) and in Figure 5, in which the effect of different treatments is compared within the fresh sample series. The pasteurized samples represent the most interesting ones in terms of TPC, TFC, antioxidant and antiradical effects with respect to the corresponding homogenates samples. Both the blanching and microwave treatment induce a relevant impact in quality parameters, while the pasteurization performed immediately after blanching (green samples) seems to preserve from modification induced by the blanching process.

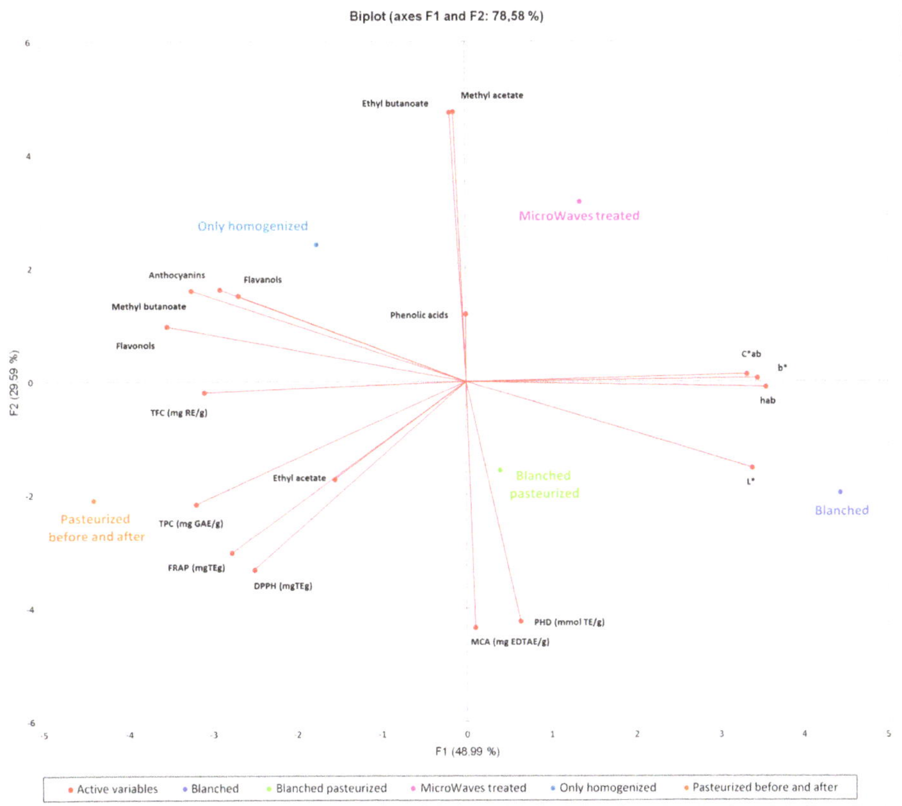

Figure 5. Principal component analysis (PCA) plots of the fresh analyzed samples evaluated as series.

4. Conclusions

The healthy potential of the consumption of fruits and vegetables is strongly related to the industrial and domestic processing treatments aiming at improving the shelf life of seasonal products and limiting the loss of bioactive components.

Overall, in treated strawberry samples, the best flow chart in terms of bioactives preservation is represented by the mild pasteurization procedure, performed before or after the homogenization process. In this regard, probably related to the swollen tissue that represents the edible part of the fruit, the use of domestic mixer results in an efficient homogenization system, supporting the possibility to activate productive chain with economic and friendly-use instruments. The pasteurization, both performed before or after homogenization and performed after blanching plus homogenization, determines a darker reddish color product, which could account for a very slight browning. We can speculate that this color change is related to enzyme mediated oxidative processes, for example, polyphenol oxidase, that was only partially inactivated by the blanching treatments or by the microwave treatment, in any case performed before the homogenization process. In conclusion, not pasteurized samples are characterized by a red brilliant, perhaps more pleasant color, which significantly affects the economic value of these products on the market and better preservation of some aroma compounds, whereas the pasteurized samples account for the best nutraceutical properties, maintaining a higher amount of polyphenolic compounds and their antioxidant potential. The PCA analysis of the collected data further corroborated the more suitable processing for this natural matrix.

Supplementary Materials: The following are available online at http://www.mdpi.com/2076-3921/9/7/632/s1, Figures S1 and S2: Quantified (%) single compounds of analyzed strawberry samples. Figure S3. Chromatograms of the strawberry hydroalcoholic extracts. Figure S4. Principal component analysis (PCA) plots of the fresh analyzed samples. Table S1: Colorimetric CIEL*a*b* parameters of strawberries homogenates.

Author Contributions: Conceptualization, S.C. (Simone Carradori), L.M., S.C. (Stefania Cesa); Data curation, G.Z., S.G.; Funding acquisition, S.C. (Simone Carradori); Investigation, A.M., G.Z; Methodology, S.G., P.P., F.C., G.A.; Project administration, S.C. (Simone Carradori); Resources, S.C. (Stefania Cesa), S.G.; Software, F.C. Supervision, A.M; Writing—original draft preparation, S.C. (Simone Carradori), S.C. (Stefania Cesa); Writing, review and editing, L.M. All authors have read and agreed to the published version of the manuscript.

Funding: This work was financially supported by funding from the "Sapienza" University of Rome.

Conflicts of Interest: The authors declare no conflict of interest.

References

1. Kearney, J. Food consumption trends and drivers. *Philos. Trans. R. Soc. B Biol. Sci.* **2010**, *365*, 2793. [CrossRef] [PubMed]
2. Kaur, C.; Kapoor, H.C. Antioxidants in fruits and vegetables—The millennium's health. *Int. J. Food Sci. Technol.* **2001**, *36*, 703. [CrossRef]
3. Pradas, I.; Medina, J.J.; Ortiz, V.; Moreno-Rojas, J.M. 'Fuentepina' and 'Amiga', two new strawberry cultivars: Evaluation of genotype, ripening and seasonal effects on quality characteristics and health-promoting compounds. *J. Berry Res.* **2015**, *5*, 157–171. [CrossRef]
4. Kan, Y.A.W.; Tzvetkov, N.T.; Gokhan, Z.; Dongdong, W.; Suowen, X.; Goranka, M. The berries on the top. *J. Berry Res.* **2019**, *9*, 125–139.
5. Afrin, S.; Gasparrini, M.; Forbes-Hernandez, T.Y.; Reboredo-Rodriguez, P.; Mezzetti, B.; Varela-López, A.; Giampieri, F.; Battino, M. Promising health benefits of the strawberry: A focus on clinical studies. *J. Agric. Food Chem.* **2016**, *64*, 4435–4449. [CrossRef]
6. Alvarez-Suarez, J.M.; Giampieri, F.; Tulipani, S.; Casoli, T.; Di Stefano, G.; González-Paramás, A.M.; Santos-Buelga, C.; Busco, F.; Quiles, J.L.; Cordero, M.D.; et al. One-month strawberry-rich anthocyanin supplementation ameliorates cardiovascular risk, oxidative stress markers and platelet activation in humans. *J. Nutr. Biochem.* **2014**, *25*, 289–294. [CrossRef]
7. Del Rio, D.; Rodriguez-Mateos, A.; Spencer, J.P.; Tognolini, M.; Borges, G.; Crozier, A. Dietary (poly)phenolics in human health: Structures, bioavailability, and evidence of protective effects against chronic diseases. *Antioxid. Redox Signal.* **2013**, *18*, 1818–1892. [CrossRef]
8. Bogunović, I.; Duralija, B.; Gadže, J.; Kisić, I. Biostimulant usage for preserving strawberries to climate damages. *Hortic. Sci.* **2015**, *42*, 132–140. [CrossRef]
9. Gasperotti, M.; Masuero, D.; Mattivi, F.; Vrhovsek, U. Overall dietary polyphenol intake in a bowl of strawberries: The influence of *Fragaria* spp. in nutritional studies. *J. Funct. Foods* **2015**, *18*, 1057–1069. [CrossRef]
10. Šaponjac, V.T.; Gironés-Vilaplana, A.; Djilas, S.; Mena, P.; Ćetković, G.; Moreno, D.A.; Čanadanović-Brunet, J.; Vulić, J.; Stajčić, S.; Vinčić, M. Chemical composition and potential bioactivity of strawberry pomace. *RSC Adv.* **2015**, *5*, 5397–5405. [CrossRef]
11. Nowicka, A.; Kucharska, A.Z.; Sokół-Łętowska, A.; Fecka, I. Comparison of polyphenol content and antioxidant capacity of strawberry fruit from 90 cultivars of *Fragaria × ananassa* Duch. *Food Chem.* **2019**, *270*, 32–46. [CrossRef] [PubMed]
12. Wein, S.; Behm, N.; Petersen, R.K.; Kristiansen, K.; Wolffram, S. Quercetin enhances adiponectin secretion by a PPAR-γ independent mechanism. *Eur. J. Pharm. Sci.* **2010**, *41*, 16–22. [CrossRef] [PubMed]
13. Silva, B.; Oliveira, P.J.; Dias, A.; Malva, J.O. Quercetin, kaempferol and biapigenin from *Hypericum perforatum* are neuroprotective against excytotoxic insults. *Neurotox. Res.* **2008**, *13*, 265–279. [CrossRef] [PubMed]
14. Da Silva, F.L.; Escribano-Bailón, M.T.; Alonso, J.J.P.; Rivas-Gonzalo, J.C.; Santos-Buelga, C. Anthocyanin pigments in strawberry. *LWT Food Sci. Technol.* **2007**, *40*, 374–382. [CrossRef]
15. Tomić, J.; Pesakovic, M.; Milivojevic, J.; Miletic, R.; Karaklajic-Stajic, Z.; Paunovic, S.M.; Milinkovic, M. Changes in anthocyanins and total phenols in fruit of three strawberry cultivars during five harvests. *Acta Hortic.* **2016**, *1139*, 633–638. [CrossRef]

16. Weber, N.; Schmitzer, V.; Jakopic, J.; Stampar, F. First fruit in season: Seaweed extract and silicon advance organic strawberry (*Fragaria* × *ananassa* Duch.) fruit formation and yield. *Sci. Hortic.* **2018**, *242*, 103–109. [CrossRef]
17. Cesa, S.; Carradori, S.; Bellagamba, G.; Locatelli, M.; Casadei, M.A.; Masci, A.; Paolicelli, P. Evaluation of processing effects on anthocyanin content and colour modifications of blueberry (*Vaccinium* spp.) extracts: Comparison between HPLC-DAD and CIELAB analyses. *Food Chem.* **2017**, *232*, 114–123. [CrossRef]
18. Patsilinakos, A.; Ragno, R.; Carradori, S.; Petralito, S.; Cesa, S. Carotenoid content of Goji berries: CIELAB, HPLC-DAD analyses and quantitative correlation. *Food Chem.* **2018**, *268*, 49–56. [CrossRef]
19. Oliva, A.; Costantini, S.; De Angelis, M.; Garzoli, S.; Bozovic, M.; Mascellino, M.T.; Vullo, V.; Ragno, R. High potency of *Maleleuca alternifolia* essential oil against multi-drug resistent Gram-negative bacteria and methicillin–resistant *Staphyloccocus aureus*. *Molecules* **2018**, *23*, 2584. [CrossRef]
20. Oliva, A.; Garzoli, S.; Sabatino, M.; Andreotti, E.; Tadic, V.; Costantini, S.; Ragno, R.; Bozovic, M. Chemical composition and antimicrobial activity of essential oil of *Helichrysum italicum* (Roth) G. Don fil. (Asteraceae). *Nat. Prod. Res.* **2020**, *34*, 445. [CrossRef]
21. Garzoli, S.; Turchetti, G.; Giacomello, P.; Tiezzi, A.; Laghezza Masci, V.; Ovidi, E. Liquid and Vapour Phase of Lavandin (*Lavandula* × *intermedia*) Essential Oil: Chemical Composition and Antimicrobial Activity. *Molecules* **2019**, *24*, 2701. [CrossRef] [PubMed]
22. Carradori, S.; Cairone, F.; Garzoli, S.; Fabrizi, G.; Iazzetti, A.; Giusti, A.M.; Menghini, L.; Uysal, S.; Ak, G.; Zengin, G.; et al. Phytocomplex Characterization and Biological Evaluation of Powdered Fruits and Leaves from *Elaeagnus angustifolia*. *Molecules* **2020**, *25*, 2021. [CrossRef] [PubMed]
23. Balli, D.; Cecchi, L.; Khatib, M.; Bellumori, M.; Cairone, F.; Carradori, S.; Zengin, G.; Cesa, S.; Innocenti, M.; Mulinacci, N. Characterization of arils juice and peel decoction of fifteen varieties of *Punica granatum* L.: A focus on anthocyanins, ellagitannins and polysaccharides. *Antioxidants* **2020**, *9*, 238. [CrossRef] [PubMed]
24. Malien-Aubert, C.; Dangles, O.; Amiot, M.J. Color stability of commercial anthocyanin-based extracts in relation to the phenolic composition. Protective effects by intra-and intermolecular copigmentation. *J. Agric. Food Chem.* **2001**, *49*, 170–176. [CrossRef] [PubMed]
25. Nunes, M.C.N.; Brecht, J.K.; Morais, A.M.; Sargent, S.A. Possible influences of water loss and polyphenol oxidase activity on anthocyanin content and discoloration in fresh ripe strawberry (cv. *Oso Grande*) during storage at 1 C. *J. Food Sci.* **2005**, *70*, S79–S84. [CrossRef]
26. Cesa, S.; Casadei, M.A.; Cerreto, F.; Paolicelli, P. Infant milk formulas: Effect of storage conditions on the stability of powdered products towards autoxidation. *Foods* **2015**, *4*, 487–500. [CrossRef] [PubMed]
27. Cutzu, R.; Bardi, L. Production of bioethanol from agricultural wastes using residual thermal energy of a cogeneration plant in the distillation phase. *Fermentation* **2017**, *3*, 24. [CrossRef]
28. Williams, A.; Ryan, D.; Guasca, A.O.; Marriott, P.; Pang, E. Analysis of strawberry volatiles using comprehensive two-dimensional gas chromatography with headspace solid-phase microextraction. *J. Chrom. B* **2005**, *817*, 97–107. [CrossRef]
29. Oz, A.T.; Baktemur, G.; Kargi, S.P.; Kafkas, E. Volatile compounds of strawberry varieties. *Chem. Nat. Compd.* **2016**, *52*, 507–509. [CrossRef]
30. Kafkas, E.; Cabaroglu, T.; Selli, S.; Bozdoğan, A.; Kürkçüoğlu, M.; Paydaş, S.; Başer, K.H.C. Identification of volatile aroma compounds of strawberry wine using solid-phase microextraction techniques coupled with gas chromatography–mass spectrometry. *Flavour Frag. J.* **2006**, *21*, 68–71. [CrossRef]
31. Kelebek, H.; Selli, S. Characterization of phenolic compounds in strawberry fruits by RP-HPLC-DAD and investigation of their antioxidant capacity. *J. Liq. Chromatogr. Relat. Technol.* **2011**, *34*, 2495–2504. [CrossRef]
32. Andrianjaka-Camps, Z.N.; Heritier, J.; Ancay, A.; Andlauer, W.; Carlen, C. Evolution of the taste-related and bioactive compound profiles of the external and internal tissues of strawberry fruits (*Fragaria* × *ananassa*) cv. 'Clery' during ripening. *J. Berry Res.* **2017**, *7*, 11–22. [CrossRef]
33. Buendia, B.; Gil, M.I.; Tudela, J.A.; Gady, A.L.; Medina, J.J.; Soria, C.; López, J.M.; Tomás-Barberán, F. HPLC-MS analysis of proanthocyanidin oligomers and other phenolics in 15 strawberry cultivars. *J. Agric. Food Chem.* **2010**, *58*, 3916–3926. [CrossRef] [PubMed]
34. Gancel, A.L.; Feneuil, A.; Acosta, O.; Pérez, A.M.; Vaillant, F. Impact of industrial processing and storage on major polyphenols and the antioxidant capacity of tropical highland blackberry (*Rubus adenotrichus*). *Food Res. Int.* **2011**, *44*, 2243. [CrossRef]

35. Chamorro, M.F.; Reiner, G.; Theoduloz, C.; Ladio, A.; Schmeda-Hirschmann, G.; Gómez-Alonso, S.; Jiménez-Aspee, F. Polyphenol Composition and (Bio) Activity of *Berberis* Species and Wild Strawberry from the Argentinean Patagonia. *Molecules* **2019**, *24*, 3331. [CrossRef]
36. Oszmiański, J.; Wojdyło, A.; Kolniak, J. Effect of L-ascorbic acid, sugar, pectin and freeze–thaw treatment on polyphenol content of frozen strawberries. *LWT-Food Sci. Technol.* **2009**, *42*, 581. [CrossRef]
37. Holzwarth, M.; Korhummel, S.; Carle, R.; Kammerer, D.R. Evaluation of the effects of different freezing and thawing methods on color, polyphenol and ascorbic acid retention in strawberries (*Fragaria* × *ananassa* Duch.). *Food Res. Int.* **2012**, *48*, 241. [CrossRef]

antioxidants

MDPI

Article

An Innovative Olive Pâté with Nutraceutical Properties

Pierpaolo Cavallo [1,2,*], Irene Dini [3,*], Immacolata Sepe [4], Gennaro Galasso [5], Francesca Luisa Fedele [6], Andrea Sicari [6], Sergio Bolletti Censi [6], Anna Gaspari [3], Alberto Ritieni [3,7], Matteo Lorito [8] and Francesco Vinale [9]

1 Dipartimento di Fisica, Università di Salerno, Via Giovanni Paolo II, 132, 84084 Fisciano, Salerno, Italy
2 ISC-CNR, Institute for Complex Systems, Via dei Taurini, 19, 00185 Roma, Italy
3 Dipartimento di Farmacia, Università degli Studi di Napoli "Federico II", 80131 Napoli, Italy; annagaspari@virgilio.it (A.G.); alberto.ritieni@unina.it (A.R.)
4 Diagnostica Cavallo—Centro Ricerca Albo MIUR, 84123 Salerno, Italy; ricerca@cavallo.net
5 Dipartimento di Medicina e Farmacia, Università di Salerno, 84081 Baronissi, Salerno, Italy; ggalasso@unisa.it
6 Linfa Scarl, University Spin Off, 80146 Napoli, Italy; francalisa@yahoo.com (F.L.F.); andrea@laboratoriolinfa.it (A.S.); sergio@laboratoriolinfa.it (S.B.C.)
7 UNESCO Chair of Health Education and Sustainable Development, University of Naples, 80131 Napoli, Italy
8 Dipartimento di Agraria, Università di Napoli "Federico II", 80055 Portici (NA), Italy; lorito@unina.it
9 Dipartimento di Medicina Veterinaria e Produzioni Animali, Università degli Studi di Napoli "Federico II", 80137 Napoli, Italy; frvinale@unina.it
* Correspondence: pcavallo@unisa.it (P.C.); irdini@unina.it (I.D.)

Received: 30 May 2020; Accepted: 30 June 2020; Published: 3 July 2020

Abstract: Food plays a central role in health, especially through consumption of plant-derived foods. Functional foods, supplements, and nutraceuticals are increasingly entering the market to respond to consumer demand for healthy products. They are foods, supplements, and ingredients which offer health benefits beyond the standard nutritional value. Some benefits are associated with phenolic compounds and phytochemicals with antioxidant properties. An olive pâté (OP) was added with antioxidants derived from olive mill wastewater (OMWW) to obtain a functional product rich in phenolic compounds. The olive pâté is produced from the ground olive pericarp, which shows an excellent natural antioxidant content. The OMWW is a waste product from oil processing, which is also rich in phenolic compounds. The result was a product rich in trans-resveratrol, OH tyrosol, and tyrosol in concentrations such as satisfying the European community's claims regarding the possible antioxidant action on plasma lipids with excellent shelf-life stability. The total phenolic content was assayed by a colorimetric method, the antioxidant activity by the ABTS [(2,2′-azino-bis (3-ethylbenzothiazoline-6-sulfonic acid)] test, the phenolic profile by Q Exactive Orbitrap LC-MS/MS. The shelf-life stability was confirmed by yeast, molds, and total microbial count, pH, and water activity determinations, and the best pasteurization parameters were determined. The palatability was judged as excellent.

Keywords: olive mill wastewater; olive oil; Olive Pâté; antioxidants; nutraceutical; Q Exactive Orbitrap LC-MS/MS

1. Introduction

Extra virgin olive oil (EVOO) is one of the highly valued products of the so-called 'Mediterranean Diet'. A diet indicated by Ancel Keys in the 1980s inspired by the eating habits of Italy and Greece in the 1960s. This diet suggests high consumption of olive oil, unrefined cereals, legumes, fish, vegetables and fruits, and moderate consumption of cheese, meat, and wine, to reduce the risk of some chronic diseases such as cardiovascular and cancer diseases [1]. The health-promoting action of the EVOO is linked to the phenolic fractions profile, which includes tyrosol (tyr), hydroxytyrosol (OH tyrosol), secoiridoids, and lignans [2].

The EVOO production includes washing of olives, followed by their crushing in a hammer mill to obtain 'pomace', that is a mixture of the crushed olive pericarp and stone; the pomace is then pressed to extract a liquid mix of oil and water, the olive mill wastewater (OMWW), which are finally separated to obtain the clear oil. The OMWW is rich in natural antioxidants. It contains 53% of the olives phenolic fraction [3,4]. These molecules can be recovered from OMWW with different methods [3,5–7]. Conventional separation techniques used for this purpose are the chromatography, extraction, centrifugation, and membrane separation.

New technologies used to decrease energy consumption and increase the extraction efficiency are electro-technologies (high voltages electrical and discharges pulsed electric fields), ultrasounds, microwaves, mechanical technologies (pressurized liquid extraction), extraction techniques with supercritical fluids, and filtration methods with reverse osmosis and tangential ultrafiltration systems. [8]. The phenolic fraction recovered from OMWW may be used to improve the content of antioxidants in foods. This "reinforcement" is extensively studied [9–14].

In this study, an OMWW concentrated was added to OP, to obtain a reinforced olive pâté (ROP) rich in phenols. A new olive-derived nutraceutical product, made of natural ingredients, capable of supplementing the antioxidants in the diet, without changing dietary habits. Olive pâté (OP), sometimes named 'olive fruit paste', is an olive-derived food of the traditional Mediterranean gastronomy with a coarse texture. The OP main ingredients are more or less finely ground [15] prepared using only the pericarp; pâté differs from pomace, since the latter is made from pericarp and stone of the olives, and it is not appropriate for gastronomic use. OP can include other ingredients—like garlic, capers, etc.—and usually contains olive oil as a covering agent. OP is used as a condiment for pasta, hard-boiled eggs, salads, snacks, etc., and is growingly being recognized [16] not only for its gastronomic role, but also by its beneficial effects on human health.

Several studies are mainly focused on the chemical/microbiological characteristics and the sensory properties of OP [15–21]. No data describe the functional properties and the stability of OP products added with OMWW. The nutraceutical potential of the ROP was evaluated in terms of the total phenol content, antioxidant activity, and phenolic compound profile. The shelf-life of ROP was also tested.

2. Materials and Methods

2.1. OP and ROP Preparation

The commercial OMWN extract of *Olea europea* fruit was from Hydrovas 10 (Bionap, Belpasso, Italy), The OP preparation, resumed in Figure 1, included washing, separation of pericarp from stone, smashing and mixing with olive oil, packaging, and pasteurization.

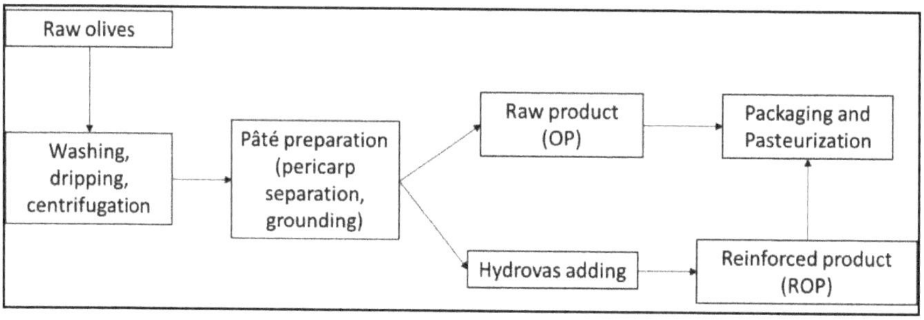

Figure 1. Reinforced olive pâté production process.

The ROP preparation was made by adding Hydrovas to the raw OP, to obtain two different final concentrations. The two concentration levels were 100 and 200 mg of phenols for 20 g of product, called 2X (10) and 4X (20) reinforced olive pâtés (2X ROP and 4X ROP), respectively.

2.2. Pasteurization Process

Pasteurization of OP and ROP was performed in a tunnel pasteurizer (TECNINOX, Parma, Italy) on the product previously packaged in glass containers. The thermal action was applied to the container, cap, or lid. The temperature of the box was raised with hot water showers at controlled temperatures and maintained for the time required for the pasteurization cycle. Subsequently, it was gradually lowered with water showers at decreasing temperature. Three different combinations of time and temperature: 86 °C for 20 min, 90 °C for 15 min, and 94 °C for 10 min were used.

2.3. Chemicals

Luteolin (≥98%; CAS number 491-70-3); apigenin (≥95%; CAS number 520-36-5); trans resveratrol (CAS number 501-36-0); oleuropein (≥98%; CAS number 36619-42-4); verbascoside (≥99%; CAS number 61276-17-3); isoverbascoside (≥95%; CAS number 61303-13-7); tyrosol (≥95%; CAS number 501-94-0); vanillic acid (≥97%; CAS number 121-34-6); cinnamic acid (≥99%; CAS number 140-10-3); ferulic acid (≥99%; CAS number 537-98-4); p-coumaric acid (≥98%; CAS number 501-98-4); 4-hydroxybenzoic acid (≥99%; CAS number 99-96-7); 3-hydroxybenzoic acid (≥99%; CAS number 99-06-9) were purchased from Sigma Aldrich (St. Louis, MO, USA); secologanoside (>95%) were bought from ChemFaces Biochemical Co., Ltd. (Hubei, China); hydroxytyrosol (>95%) were obtained by Indofine (Hillsborough, NJ, USA); oleuropein-aglycone monoaldehyde (>95%) were purchased from Extrasynthese (Genay, France). Ligstroside were purchased from Wuhan Golden Wing Industry & Trade Co., Ltd. (Wuhan, China), the ligstroside-aglycone monoaldehyde (>95%) was isolated by chromatographic method and identity by nuclear magnetic resonance (NMR). The other chemicals were the analytical (Sigma Aldrich, St. Louis, MO, USA).

2.4. Total Phenolic Compounds

The total phenol content was determined colorimetrically at 765 nm, using the Folin–Ciocalteau reagent Sigma Aldrich (St. Louis, MO, USA) as described by Singleton and coll [22]. 10 mL of a methanol/water solution (80:20 *v/v*) and Tween 20 (2% *v/v*) was mixed with 10 g of OMWW; the mixture was homogenized with the ULTRA-TURRAX® (Ika, Breisgau, Germania) at 25 °C to 15,000 RPM; then the homogenate was centrifuged for 10 min at 5000 RPM. The supernatant was collected and placed in the freezer for 24 h at −20 °C [23–28]. A Falcon (15 mL) was shaken by 2.5 mL H_2O, 625 µL methanolic extract, 625 µL of Folin-Ciocalteau's phenol reagent and, after 6 min, 6.25 mL of 7% Na_2CO_3 and 5 mL dd H_2O. After incubation for 90 min at room temperature, the absorbance of the reagent blank was determined at 760 nm by spectrophotometer (V-530 Jasco, Tokyo, Japan). The total concentration of polyphenols was expressed as mg/L of gallic acid. The determination of the polyphenol content was made both on unpasteurized and on heat-treated samples. All samples were analyzed in triplicate.

2.5. Antioxidant Activity Determination

The antioxidant activity was evaluated with ABTS colorimetric method [29]. A stock solution was obtained dissolving 9.6 mg of ABTS in 2.5 mL of water and adding 44 mL of a solution made by dissolving 37.5 mg of $K_2S_2O_8$, in 1 mL of water. The stock solution was kept in the dark at 4 °C for 8 h before use. The work solution was obtained from the stock solution by dilution using a 1:88 (*v/v*) ratio (it must measure between 0.7 and 0. 8 at 734 nm). Subsequently, 100 µL of sample and 1 mL of work solution were added, and A734 was measured exactly after 2 min and 30 s. (V-530 Jasco, Tokyo, Japan). The calibration curve was obtained using Trolox (Sigma Aldrich—St. Louis, MO, USA), and results were expressed as mmol Trolox/100 g. All biological samples were analyzed in triplicate.

2.6. UHPLC Operative Condition

An Agilent Technologies 1200 Series Ultra High Liquid Chromatograph (UHPLC) (Agilent, Santa Clara, CA, USA) equipped with pre-column Phenomenex (Torrance, CA, USA), and column Accucore aQ 2.6 µm 100 × 2.1 mm Thermo Scientific (Waltham, MA, USA) was used for experimental purposes. The injection volume was 5 µL. A gradient was employed as a mobile phase (Table 1).

Table 1. UHPLC mobile phase.

Minutes	Phase B% 100% Acetonitrile(ACN)	Phase A% H_2O Containing 0.1% (*v/v*) Acetic Acid (AcH)
0–5	5	95
6–25	40	60
25.1–27	100	0
27.1–35	5	95
35.1–45	0	100

2.7. Orbitrap UHPLC-MS/MS Operative Condition

All mass experiments were conducted at a Q Exactive Orbitrap LC-MS/MS (Thermo Fisher Scientific, Waltham, MA, USA) equipped with an ESI source (HESI II, Thermo Fisher Scientific, Waltham, MA, USA) operating in negative ion mode (ESI-). (Table 2) The accuracy and calibration of the Q Exactive Orbitrap LC-MS/MS were checked weekly. A reference standard mixture was purchased by Thermo Fisher Scientific (Waltham, MA, USA). The Xcalibur software v. 3.1.66.10 (Xcalibur, Thermo Fisher Scientific, Waltham, MA, USA) was used to analyze and process data.

Table 2. Mass operative setting.

Ion Source Parameters	
Spray Voltage	**−3.0 kV**
Sheath gas	(N_2 > 95%) 30
Auxiliary gas	(N_2 > 95%) 15
Capillary temperature	200 °C
S-lens	RF level 50
Auxiliary gas	305 °C
Analyzer Target SIM (Single Ion Monitoring) Parameter	
Automatic gain control (AGC)	target set at 1 and 6
Resolution	140,000 FWHM (full width at half maximum)
Scan rate	(100–500 *m/z*)

2.8. Shelf-Life Analysis

2.8.1. Microbiological Analysis

The total microbial count was enumerated on non-selective medium Plate Count Agar (PCA, Oxoid, Milan, Italy). Using a sterile procedure, 1 mg of sample was diluted with 9 mL of physiological saline (9 g/L NaCl) and were prepared serial dilution. The total microbial count was determined by pour plating of suitable dilution on with 15–18 mL of PCA medium and incubated at 30 °C for 48 h.

Yeast and Molds (YM) were enumerated on selective medium Sabouraud Dextrose Agar (SDA, Oxoid, Milan, Italy). The plates were incubated at 25 °C for 5 days. The number of bacteria and YM were expressed as colony forming units (CFU), at three different temperatures: 25, 37, and 55 °C.

2.8.2. pH and Water Activity

The pH of the olive pâté was measured using pH meter HI 9025 (Hanna Instruments, Woonsocket, RI, USA). 10 g samples were diluted in 10 mL of distilled water and were centrifuged for five minutes at 6000 RPM. The supernatant was filtered with a 45 µm Millex membrane (Merk, Darmstadt, Germany).

Water was measured using a water activity system TESTO 650 (Testo SpA, Milan, Italy).

2.8.3. Stress Test and Pasteurization Effects

The technological stability of the olive pâté was evaluated with the stress test. Two different types of samples (commercial and reinforced) were subjected to different temperatures (25, 37, 55 °C) for 19 days, and then tested at days 0, 2, 7, 14, 16, and 19 for pH and antioxidant activity, using the methods listed before.

Pasteurization effects were evaluated by assaying total phenols and antioxidant activity in samples of OP and in the two levels of ROP prepared (2X and 4X) treating them at three different combinations of time and temperature: 86 °C for 20 min, 90 °C for 15 min, and 94 °C for 10 min.

2.9. Palatability Test

The palatability of the two different olive pâté products, OP (commercial), and ROP (reinforced) was tested.

A spontaneous panel was formed by the researcher team members, who tasted 3.5 mL of the two products at room temperature, giving a judgement on a 1–10 scale (1 = worst, 10 = best) about texture, appearance, smell, taste, and after taste.

2.10. Statistical Analysis

Excel spreadsheet software, version 19.0 (Microsoft, Redmond, WA, USA) was used to perform statistical analyses.

3. Results

3.1. Phenolic Profile and Concentrations

A UHPLC-MS/MS method was employed to delineate the phenolic profile and dosage.

Eighteen phenolics—including three flavonoids, six phenolic acids, seven secoiridoids, and two phenolic alcohols—were characterized and dosed. Table S1, in the supplementary data, reported the chromatographic and spectroscopic parameters used to identify phenolics in samples. The quantification method was validated (AOAC, 2012) [30]. The matrix effect (ME, signal enhancement or suppression) was investigated by calculating the ratio percentage between the slopes of the matrix-matched calibration curve and the curve in solvent (Figure S1). The linearity was guaranteed by the coefficient of regression of the calibration curve close to 1. LODs and the LOQs range verified the sensitivity, triplicate injection of each phenolic standard, at seven different concentrations confirmed the intraday repeatability.

Flavonoids were the most representative phenolics in the OP and ROP samples, following by phenolic alcohols and then all the others (Table 3, Figure 2). The highest concentration in the raw OP was presented by *trans*-resveratrol, HT, and tyrosol, followed by vanillic acid, luteolin, 3-OH benzoic acid, secologanoside, and p-coumaric acid. The highest increment in concentration, 269.7% was presented by ligstroside-aglycone monoaldehyde, followed by apigenin with 96.4%, but both had a low initial level. The most significant increments were presented by molecules with high initial concentration, which were luteolin, more than doubled; tyrosol, 69.3% higher of baseline; *trans*-resveratrol, and HT, both over 40% of baseline. Only the phenolic acids were higher in OP than ROP.

The HPLC profiles of OP and ROP are reported in the Supplementary Data.

Table 3. Phenolic compounds in OP and ROP.

Group	Compound	OP (mg/kg)	ROP (mg/kg)	Delta	Delta% of Baseline
Flavinoids	*trans*-Resveratrol	278.1 ± 2.0	404.4 ± 10.2	126.2	45.4%
	Luteolin	9.9 ± 0.09	20.6 ± 0.1	10.7	107.8%
	Apigenin	0.5 ± 0.0	0.9 ± 0.0	0.45	96.4%
Secoiridoids	Ligstroside	0.2 ± 0.004	0.2 ± 0.004	0.038	20.5%
	Secologanoside	1.6 ± 0.016	1.9 ± 0.019	0.251	15.4%
	Verbascoside	0.6 ± 0.002	0.8 ± 0.009	0.186	31.1%
	Isoverbascoside	0.2 ± 0.004	0.4 ± 0.003	0.139	55.8%
	Oleuropein-aglycone monoaldehyde	0.5± 0.008	0.7 ± 0.011	0.156	30.6%
	Ligstroside-aglycone monoaldehyde	0.3 ± 0.026	1.2 ± 0.031	0.88	269.9%
	Oleuropein	0.4 ± 0.010	0.4 ± 0.002	0.042	10.2%
Phenolic Alcohol	Tyrosol	120.8 ± 6.6	204.5 ± 5.9	83.6	69.3%
	OH-Tyrosol	141.7 ± 3.1	201.1 ± 4.01	59.4	42.0%
Phenolic Acids	P-Coumaric	1.1 ± 0.014	0.5 ± 0.011	−0.619	−56.1%
	Cyinnamic	0.5 ± 0.017	0.01 ± 0.057	−0.522	−98.9%
	Ferulic	0.3 ± 0.002	0.06 ± 0.001	−0.25	−81.4%
	Vanillic	21.8 ± 0.055	9.0 ± 0.077	−12.7	−58.8%
	4-Hydroxybenzoic acid	1.0 ± 0.002	0.4 ± 0.007	−0.555	−57.8%
	3-Hydroxybenzoic acid	4.5 ± 0.061	2.8 ± 0.064	−1.66	−36.9%

OP = olive paté; ROP = reinforced olive pâté; Delta = absolute difference; Delta% of baseline = difference in per cent of the OP value.

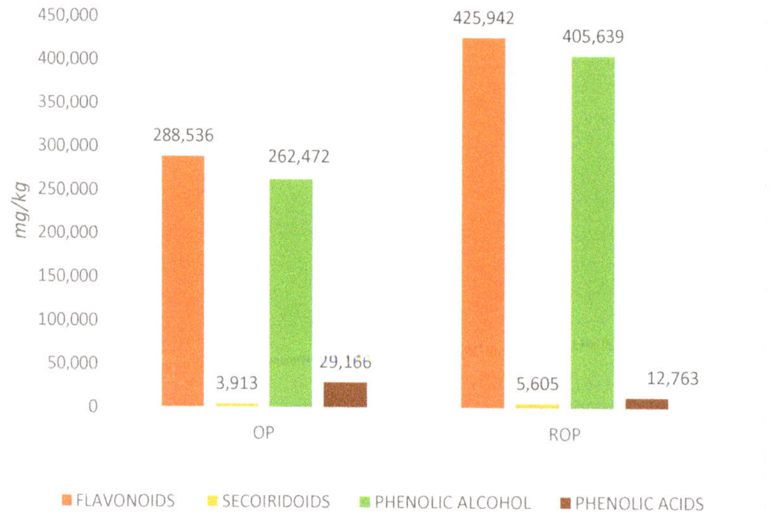

Figure 2. Total concentration of each class of phenolic compounds.

The total phenols concentrations were tested by the colorimetric method ≅765 nm, with the results expressed in mg/L of gallic acid; the results were OP = 2620 and ROP = 4750, with an increment of 81.2%. The antioxidant activity was tested by ABTS, with the results expressed in µmol/100 g of TROLOX; the results were OP = 715,131 and ROP = 975,951, with an increment of 36.4%.

To graphically compare the effects of reinforcement, in Figure 3 we present the data after normalization; the y axis shows this index number, given the OP base value at 100.

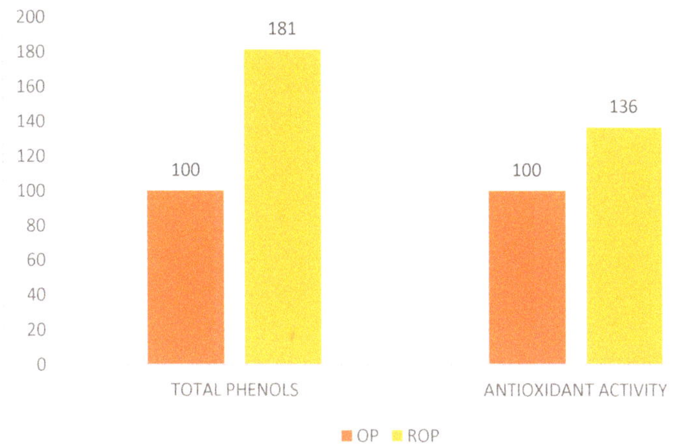

Figure 3. Comparison of the reinforcement effects (OP = 100).

It is clearly appreciable the higher effect of reinforcement on phenols rather than antioxidant activity.

3.2. Other Analytical Results

For pH and aW, OP and ROP had the same values, with pH = 4.39 tested by the potentiometric method, and aW = 0.91 measured in a sealed container.

The Total microbial and the Mold/yeast counts were negative for any kind of growth for both OP and ROP.

The stress tests results for pH are shown in Figure 4 Starting from the same initial value of 4.39, OP and ROP underwent a slight decrease, that was different for each product and condition of stress. The ROP obtained the best performance at 25 °C, which finished with a value of 4.37 at 19 days, while the worst result was obtained at 55 °C, even if with a pH reduction limited at 0.09 for the OP, with a final value at 4.28.

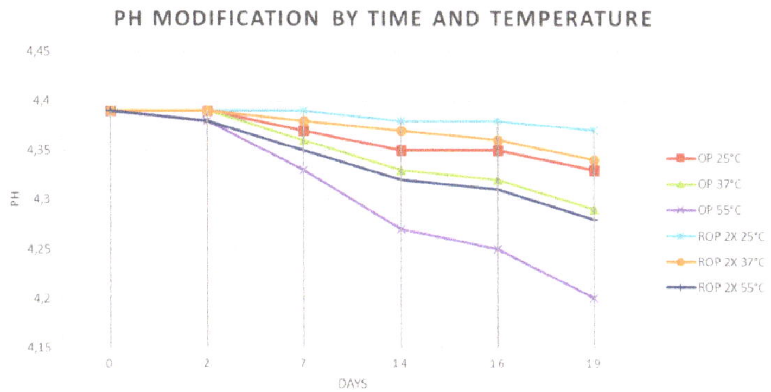

Figure 4. Results of the stress test for pH.

The stress tests results for antioxidant activity are shown in Figure 5 The ROP started from higher levels, and showed lower degradation than OP. Anyway, the maximum degradation obtained at 19 days was limited, as OP lost about 6% while ROP about 2% of the initial activity. For both, a significant degradation was shown only at the highest temperature of 55 °C.

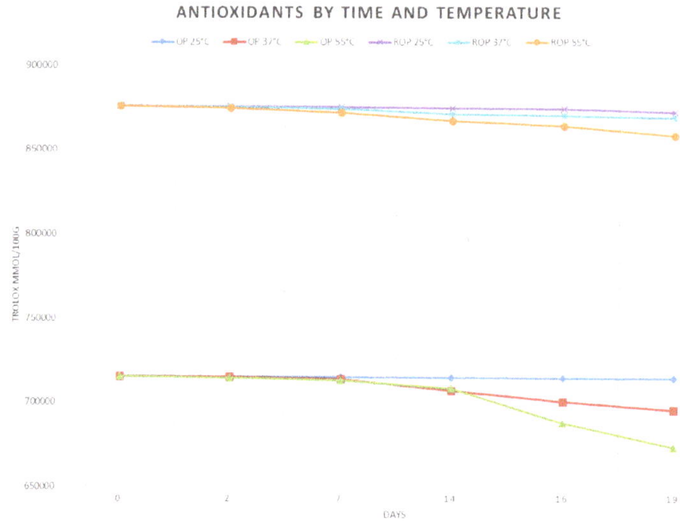

Figure 5. Results of the stress test for antioxidants.

The pasteurization effects were evaluated comparing three different combinations of time and temperature (t/t): 86 °C for 20 min, 90 °C for 15 min, and 94 °C for 10 min.

The contents of phenols expressed as Gallic acid, and antioxidants as TROLOX are reported in Table 4, comparing the pasteurized products with the raw, either in absolute values and in form of Index, considering the raw product = 100.

Table 4. Antioxidant activities after pasteurization under different t/t combinations.

Parameter	Pasteurization Combinations			
	Raw	**86 °C/20′**	**90 °C/15′**	**94 °C/10′**
OP-TROLOX μmol/100 g	715.31	455.308	515.308	435.308
ROP-TROLOX μmol/100 g	875.951	620.286	685.286	597.286
OP-Gallic Acid mg/L	2620	1820	1340	1520
ROP-Gallic Acid mg/L	4750	3550	2700	2850
OP-Index for TROLOX (Raw = 100)	100	63.7	72.1	60.9
ROP-Index for TROLOX (Raw = 100)	100	70.8	78.2	68.2
OP-Index for Gallic Acid (Raw = 100)	100	69.5	51.1	58.0
ROP-Index for Gallic Acid (Raw = 100)	100	74.7	56.8	60.0

The proportion of phenols and for antioxidants remaining after pasteurization was higher for ROP than OP, with a highly significant statistical significance ($p < 0.01$).

The palatability test gave excellent result: the average overall score was 9/10 for the OP, and 8.5/10 for the ROP. In detail, the scores for texture, appearance and smell were all excellent for both the products, while taste for ROP was equally considered excellent by those who appreciated its slightly soreness but a bit less by those who did not, and after taste perception followed the same judgment as the taste.

4. Discussion

The olive oil market is constantly growing, as it is considered a healthy product.

Unfortunately, oil production leads to the formation of environmentally harmful waste: solid and liquid residues are obtained with a high organic weight detrimental to the environment, and their natural residues vary according to the system used to extract the olive oil. For example, centrifugation systems (two-phase and three-phase) produce extra virgin olive oil, olive pomace (solid residue), and olive vegetation water (liquid residue).

Olives contain polyphenols [31,32], biomolecules with antioxidant activity able to decrease the risk of cancer, cardiovascular diseases, and inflammation [33–36]. Polyphenols are amphiphilic molecules, some of which are more soluble in water, others in oil; therefore, some classes are more concentrated in the extra virgin olive oil, others in the vegetation water fraction. Phenols are generally used to produce nutraceuticals, functional foods, and food additives [37–39].

In the present work, an olive pâté (OP) reinforced with concentrated OMWW was studied to obtain a nutraceutical product with a high phenol content and long shelf life. Olive pâté used in the Mediterranean diet is an olive fruit paste, made by pericarp only. For centuries, olive pâté has been produced only in small quantities, usually handcrafted, according to ethnic traditions of the Mediterranean area.

The process of reinforcement of foods with micronutrients is well established and widespread, with a set of WHO-FAO Guidelines [40] that have defined it. Some have concerned refined olive oil [41], and no specific experience appears to have been made with olive pâté even if are present experiences of fortification with phenols in many different foods.

Results showed that the addition of OMWW to OP strongly increased the total phenol content of ROP (81.2%), and little the antioxidant potential (36.4%). The slightest variation in the antioxidant activity may depend on the method used, as ABTS is known to underestimate the antioxidant potential of phenolic compounds with a complex structure and size, because the ring adducts—mainly secondary rings (Group 3)—interfere with phenol access to the ABTS•+, decreasing the reaction efficiency every OH added [42].

The inhomogeneous composition of the different classes of phenols is noteworthy; in particular, the very abundant concentration of trans-resveratrol, followed by OH tyrosol and tyrosol in ROP. In nature, resveratrol occurs in two stereoisomer forms: cis- and trans. The trans isomer is biologically more active than cis [43].

Resveratrol has antioxidant, coronary vasodilating, antihypertensive, neuroprotective, anti-inflammatory, and anticancer properties [44]. In vivo studies have demonstrated low bioavailability after oral intake. Therefore, the development of nutraceutical formulations with better pharmacologic properties is an exciting task. Resveratrol-enriched foods would allow a higher daily intake of resveratrol (therapeutically doses are ~1 g) than conventional foods [44]. Red wine is usually considered a good source of resveratrol, generally containing 0.361–1.972 mg/L, while ROP 404.388 mg/Kg fresh weight [45].

Moreover, ROP is an interesting source of the OH tyrosol (204.5102 mg/Kg) and tyrosol (201.129 mg/Kg), which have shown antidiabetic, anti-obesity, cardioprotective, antiatherogenic, neuroprotective, anticancer effects [46]. In ROP 2X, there is a concentration of OH-tyrosol (8.1 mg/20 g) higher than that required by EFSA (5 mg/20 g) to be able to attribute CLAIM "oil polyphenols contribute to the protection of blood lipids from oxidative stress" to olive oil (EU 432/2012).

OP and ROP were subjected to a pasteurization process to guarantee their safety, increase hygienic stability, interrupting the oxidative process, and extend their shelf life [47]. When the pasteurization process uses high temperature in short-time, bacterial spoilage is the most limiting factor in extending the shelf life of the processed products [48]. Aroma and taste attributes of the products change when microorganism's growth and consequential decrease consumer acceptability of the products. In this work, the microbial, mold, and yeast stability were checked, and no microorganisms were found to grow. This was partly due to the presence of high concentrations of phenols with antibacterial properties [25,26]. Moreover, the growth of bacteria surviving the pasteurization process was tested by monitoring the changes in pH [49].

The stability test for pH showed different performances between OP and ROP: the most stable preparation was ROP, with minimal reduction, from 4.37 to 4.36 for the worst storage case, 55 °C for 19 days, while in the same condition the OP showed a higher, even if still limited, variation, from 4.37 to 4.28.

Previous studies have shown that pasteurization can lead to loss of the antioxidant activity since phytochemicals with antioxidant properties are affected differently by heat treatments [50]. The stability test for antioxidant activity showed different performances between the two products, as ROP started from higher levels and showed a limited loss, about 2%, for the worst storage case, 55 °C for 19 days, while in the same condition the OP showed a higher, even if still limited, antioxidant activity loss of 6%. The reason may be referred to as the protective action of olive antioxidants on the stability of food [51].

We evaluated which pasteurization process guaranteed the best results to minimize the losses of phenolic compounds and antioxidant activity. The best performance was achieved pasteurizing at 90 °C for 15 min. In this condition, total phenols concentration was 95%, and antioxidant activity was 32.9%, higher in the ROP than OP product proving that the reinforcement process changes the biochemical characteristics of the olive pâté apart from enhancing its nutraceutical properties.

Finally, the excellent palatability of the ROP in comparison with the OP may suggest that this product, and probably other products of the same kind, are going to be well accepted by the healthy food market, and will have interesting commercial perspectives as they widen the possibilities of use of olive derived products in nearly any kind of diet.

5. Conclusions

In this study, we tested the possibility of producing a new nutraceutical product based on olive pâté and OMWW. We obtained a product rich in trans resveratrol, OH tyrosol, and tyrosol in concentrations such as to satisfy the claims of the European community regarding the possible antioxidant action on plasma lipids. The product shows good palatability, and good results in terms of stability, thus having an interesting market perspective either from both nutraceutical and commercial points of view.

Supplementary Materials: The following are available online at http://www.mdpi.com/2076-3921/9/7/581/s1, Figure S1: HPLC profiles of reinforced olive pâté (ROP-Upside chromatogram) and olive pâté (OP–Down side chromatogram); Table S1: Validation parameters of the UHPLC- MS/MS method of analysis.

Author Contributions: Conceptualization, F.V.; Formal analysis, I.D.; Funding acquisition, F.L.F., A.S. and S.B.C.; Investigation, I.D. and F.V.; Methodology, A.G. and M.L.; Resources, F.L.F., A.S. and A.R.; Supervision, M.L.; Validation, F.V.; Visualization, I.S., G.G., S.B.C. and A.R.; Writing—original draft, P.C., I.D. and F.V.; Writing—review and editing, P.C., I.D., A.G. and F.V. All authors have read and agreed to the published version of the manuscript.

Funding: The research was funded by MIURPON (grant number Linfa 03PE_00026_1; grant number Marea 03PE_00106); MIUR-GPS (grant number Sicura DM29156); POR FESR CAMPANIA 2014/2020- O.S. 1.1 (grant number Bioagro 559); UNISA, project 300391FFA18CAVAL.

Conflicts of Interest: The authors declare no conflict of interest.

References

1. Hu, F.B. The Mediterranean diet and mortality-olive oil and beyond. *N. Engl. J. Med.* **2003**, *348*, 2595–2596. [CrossRef]
2. Owen, R.W.; Giacosa, A.; Hull, W.E.; Haubner, R.; Würtele, G.; Spiegelhalder, B.; Bartsch, H. Olive-oil consumption and health: The possible role of antioxidants. *Lancet Oncol.* **2000**, *1*, 107–112. [CrossRef]
3. De Marco, E.; Savarese, M.; Paduano, A.; Sacchi, R. Characterization and fractionation of phenolic compounds extracted from olive oil mill wastewaters. *Food Chem.* **2007**, *104*, 858–867. [CrossRef]
4. Araújo, M.; Pimentel, F.B.; Alves, R.C.; Oliveira, M.B.P.P. Phenolic compounds from olive mill wastes: Health effects, analytical approach and application as food antioxidants. *Trends Food Sci. Technol.* **2015**, *45*, 200–211. [CrossRef]

5. Rahmanian, N.; Jafari, S.M.; Galanakis, C.M. Recovery and Removal of Phenolic Compounds from Olive Mill Wastewater. *J. Am. Oil Chem. Soc.* **2014**, *91*, 1–18. [CrossRef]
6. Takac, S.; Karakaya, A. Recovery of Phenolic Antioxidants from Olive Mill Wastewater. *Recent Pat. Chem. Eng.* **2009**, *2*, 230–237. [CrossRef]
7. Allouche, N.; Fki, I.; Sayadi, S. Toward a High Yield Recovery of Antioxidants and Purified Hydroxytyrosol from Olive Mill Wastewaters. *J. Agric. Food Chem.* **2004**, *52*, 267–273. [CrossRef]
8. Roselló-Soto, E.; Parniakov, O.; Deng, Q.; Patras, A.; Koubaa, M.; Grimi, N.; Boussetta, N.; Tiwari, B.K.; Vorobiev, E.; Lebovka, N.; et al. Application of Non-conventional Extraction Methods: Toward a Sustainable and Green Production of Valuable Compounds from Mushrooms. *Food Eng. Rev.* **2016**, *8*, 214–234. [CrossRef]
9. Pokorný, J. Natural antioxidants for food use. *Trends Food Sci. Technol.* **1991**, *2*, 223–227. [CrossRef]
10. Dewanto, V.; Wu, X.; Adom, K.K.; Liu, R.H. Thermal Processing Enhances the Nutritional Value of Tomatoes by Increasing Total Antioxidant Activity. *J. Agricand Food Chem.* **2002**, *50*, 3010–3014. [CrossRef]
11. Carlsen, M.H.; Halvorsen, B.L.; Holte, K.; Bøhn, S.K.; Dragland, S.; Sampson, L.; Willey, C.; Senoo, H.; Umezono, Y.; Sanada, C.; et al. The total antioxidant content of more than 3100 foods, beverages, spices, herbs and supplements used worldwide. *Nutr. J.* **2010**, *9*, 3. [CrossRef] [PubMed]
12. Savarese, M.; De Marco, E.; Falco, S.; D'Antuoni, I.; Sacchi, R. Biophenol extracts from olive oil mill wastewaters by membrane separation and adsorption resin. *Int. J. Food Sci. Technol.* **2016**, *51*, 2386–2395. [CrossRef]
13. Vissers, M.N.; Zock, P.L.; Roodenburg, A.J.C.; Leenen, R.; Katan, M.B. Olive Oil Phenols Are Absorbed in Humans. *Nutr. J.* **2002**, *132*, 409–417. [CrossRef]
14. Visioli, F.; Galli, C.; Bornet, F.; Mattei, A.; Patelli, R.; Galli, G.; Caruso, D. Olive oil phenolics are dose-dependently absorbed in humans. *FEBS Lett.* **2000**, *468*, 159–160. [CrossRef]
15. Cosmai, L.; Caponio, F.; Summo, C.; Paradiso, V.M.; Cassone, A.; Pasqualone, A. New formulations of olive-based patè: Development and quality. *Ital. J. Food Sci.* **2017**, *29*, 302–316. [CrossRef]
16. Alvarenga, N.B.; Lidon, F.J.C.; Silva, A.; Martins, G.; Cruz, T.; Palma, V.; Canada, J. Production and characterization of green and black olive paste using cream of animal and vegetable origins. *Emir. J. Food Agric.* **2012**, *24*, 12–16. [CrossRef]
17. Cosmai, L.; Campanella, D.; De Angelis, M.; Summo, C.; Paradiso, V.M.; Pasqualone, A.; Caponio, F. Use of starter cultures for table olives fermentation as possibility to improve the quality of thermally stabilized olive-based paste. *LWT* **2018**, *90*, 381–388. [CrossRef]
18. Nieto, A.; Grande Burgos, M.J.; GÁLvez, A.; Pérez Pulido, R. Preservation of Paste Obtained from Picual Green Olives by High Hydrostatic Pressure Treatment. *Czech J. Food Sci.* **2017**, *35*, 246–250. [CrossRef]
19. Aka-Kayguluoglu, A.; Akpinar-Bayizit, A.; Sahin-Cebeci, O.I. Evaluation of physicochemical and sensory properties of green olive pastes. *Indian J. Tradit. Knowl.* **2014**, *13*, 654–658.
20. Cosmai, L.; Campanella, D.; Summo, C.; Paradiso, V.M.; Pasqualone, A.; De Angelis, M.; Caponio, F. Shelf life of stored not pasteurized olive-based pâtés. *Ital. J. Food Sci.* **2015**, *28*, 28–32.
21. Escudero-Gilete, M.L.; Meléndez-Martínez, A.J.; Heredia, F.J.; Vicario, I.M. Optimization of olive-fruit paste production using a methodological proposal based on a sensory and objective color analysis. *Grasas Y Aceites* **2009**, *60*, 396–404.
22. Singleton, V.L.; Rossi, J.A. Colorimetry of total phenolics with phosphmolybdic phospotungstic acid reagents. *Am. J. Enol. Vitic.* **1965**, *16*, 144–158.
23. Montedoro, G.F.; Cantarelli, C. Investigation on the Phenolic Compounds of Virgin Olive Oils. *Riv. Ital. Sostanze Grasse* **1969**, *46*, 115–124.
24. Vazquez-Roncero, A. A Study of the Polar Compounds in Olive Oil by Gas Chromatography. *Grasas Y Aceites* **1980**, *31*, 309–316.
25. Solinas, M.; Cichelli, A. Determination of Phenolic Substances in Olive Oil by GLC and HPLC; Possible Role of Tyrosol in Determination of the Quantity of Virgin Oil in Blends with Refined Oil. *Riv. Soc. Ital. Sci. Aliment.* **1982**, *11*, 223–230.
26. Sudjana, A.N.; D'Orazio, C.; Ryan, V.; Rasool, N.; Ng, J.; Islam, T.; Riley, V.; Hammera, K.A. Antimicrobial activity of commercial *Olea europaea* (olive) leaf extract. *Int. J. Antimicrob. Agents* **2009**, *33*, 461–463. [CrossRef] [PubMed]
27. Obied, H.K.; Allen, M.S.; Bedgood, D.R.; Prenzler, P.D.; Robards, K.; Stockmann, R. Bioactivity and analysis of biophenols recovered from olive mill waste. *J. Agric. Food Chem.* **2005**, *53*, 823–837. [CrossRef] [PubMed]

28. Servili, M.; Baldioli, M.; Selvaggini, R.; Miniati, E.; Macchioni, A.; Montedoro, G. High-performance liquid chromatography evaluation of phenols in olive fruit, virgin olive oil, vegetation waters, and pmace and 1D-and 2D-nuclear magnetic resonance characterization. *J. Am. Oil Chem. Soc.* **1999**, *76*, 873–882. [CrossRef]
29. Re, R.; Pellegrini, N.; Proteggente, A.; Pannala, A.; Yang, M.; Rice-Evans, C. Antioxidant activity applying an improved ABTS radical cation decolorization assay. *Free Radic. Biol. Med.* **1999**, *26*, 231–1237. [CrossRef]
30. AOAC. Appendix F: Guidelines for Standard Method Performance Requirements (SMPR). In *AOAC Official Methods of Analysis*; AOAC INTERNATIONAL: Gaithersburg, MD, USA, 2012.
31. Dini, I.; Graziani, G.; Fedele, F.L.; Sicari, A.; Vinale, F.; Castaldo, L.; Ritieni, A. Effects of *Trichoderma* biostimulation on the phenolic profile of extra-virgin olive oil and olive oil by-products. *Antioxidants* **2020**, *9*, 284. [CrossRef]
32. Dini, I.; Graziani, G.; Gaspari, A.; Fedele, F.L.; Sicari, A.; Vinale, F.; Cavallo, P.; Lorito, M.; Ritieni, A. New Strategies in the Cultivation of Olive Trees and Repercussions on the Nutritional Value of the Extra Virgin Olive Oil. *Molecules* **2020**, *25*, 2345. [CrossRef] [PubMed]
33. Vita, J.A. Polyphenols and cardiovascular disease: Effects on endothelial and platelet function. *Am. J. Clin. Nutr.* **2005**, *81*, 292S–297S. [CrossRef] [PubMed]
34. Stoclet, J.C.; Chataigneau, T.; Ndiaye, M.; Oak, M.H.; El Bedoui, J.; Chataigneau, M.; Shini-Kerth, V.B. Vascular protection by dietary polyphenols. *Eur. J. Pharm.* **2004**, *500*, 299–313. [CrossRef] [PubMed]
35. Montesano, D.; Rocchetti, G.; Cossignani, L.; Senizza, B.; Pollini, L.; Lucini, L.; Blasi, F. Untargeted Metabolomics to Evaluate the Stability of Extra-Virgin Olive Oil with Added *Lycium barbarum* Carotenoids during Storage. *Foods* **2019**, *8*, 179. [CrossRef]
36. Lozano-Sánchez, J.; Bendini, A.; Di Lecce, G.; Valli, E.; Toschi, T.G.; Segura-Carretero, A. Macro and micro functional components of a spreadable olive by-product (pâté) generated by new concept of two-phase decanter. *Eur. J. Lipid Sci. Technol.* **2016**, *119*, 1600096. [CrossRef]
37. Gil-Chávez, J.G.; Villa, J.A.; Ayala-Zavala, F.J.; Heredia, B.J.; Sepulveda, D.; Yahia, E.M.; González-Aguilar, G.A. Technologies for extraction and production of bioactive compounds to be used as nutraceuticals and food ingredients: An overview. *Compr. Rev. Food Sci. Food Saf.* **2013**, *12*, 5–23. [CrossRef]
38. Dini, I.; Laneri, S. Nutricosmetics: A brief overview. *Phytother. Res.* **2019**, *33*, 3054–3063. [CrossRef]
39. Ruzzolini, J.; Peppicelli, S.; Andreucci, E.; Bianchini, F.; Scardigli, A.; Romani, A.; La Marca, G.; Nediani, C.; Calorini, L. Oleuropein, the Main Polyphenol of Olea europaea Leaf Extract, Has an Anti-Cancer Effect on Human BRAF Melanoma Cells and Potentiates the Cytotoxicity of Current Chemotherapies. *Nutrients* **2018**, *10*, 1950. [CrossRef]
40. Allen, L.H.; De Benoist, B.; Dary, O.; Hurrell, R. *World Health Organization. Guidelines on Food Fortification with Micronutrients*; World Health Organisation: Geneva, Switzerland, 2006; pp. 1–341.
41. Artajo, L.S.; Romero, M.P.; Morelló, J.R.; Motilva, M.J. Enrichment of Refined Olive Oil with Phenolic Compounds: Evaluation of Their Antioxidant Activity and Their Effect on the Bitter Index. *J. Agric. Food Chem.* **2006**, *54*, 6079–6088. [CrossRef]
42. Schaich, K.M.; Tian, X.; Xie, J. Hurdles and pitfalls in measuring antioxidant efficacy: A critical evaluation of ABTS, DPPH, and ORAC assays. *J. Funct. Foods* **2015**, *14*, 111–125. [CrossRef]
43. Morris, V.L.; Toseef, T.; Nazumudeen, F.B.; Rivoira, C.; Spatafora, C.; Tringali, C.; Rotenberg, S.A. Anti-tumor properties of *cis*-resveratrol methylated analogs in metastatic mouse melanoma cells. *Mol. Cell Biochem.* **2015**, *402*, 83–91. [CrossRef] [PubMed]
44. Weiskirchen, S.; Weiskirchen, R. Resveratrol: How much wine do you have to drink to stay healthy? *Adv. Nutr.* **2016**, *7*, 706–718. [CrossRef] [PubMed]
45. Jeandet, P.; Bessis, R.; Maume, B.F.; Sbaghi, M. Analysis of resveratrol in burgundy wines. *J. Wine Res.* **1993**, *4*, 79–85. [CrossRef]
46. Marković, A.K.; Torić, J.; Barbarić, M.; Brala, C.J.; Paiva-Martins, F. Hydroxytyrosol, Tyrosol and Derivatives and Their Potential Effects on Human Health. *Molecules* **2019**, *24*, 2001. [CrossRef]
47. Plazl, I.; Lakner, M.; Koloini, T. Modeling of temperature distributions in canned tomato based dip during industrial pasteurization. *J. Food Eng.* **2006**, *75*, 400–406. [CrossRef]
48. Boor, K.J. Fluid dairy product quality and safety: Looking to the future. *J. Dairy Sci.* **2001**, *84*, 1. [CrossRef]
49. Delaquis, P.J.; Stewart, S.; Toivonen, P.M.A.; Moyls, A.L. Effect of warm, chlorinated water on the microbial flora of shredded iceberg lettuce. *Food Res Int.* **1999**, *32*, 7–14. [CrossRef]

50. Oliveira, A.; Pintado, M.; Almeida, D.P.F. Phytochemical composition and antioxidant activity of peach as affected by pasteurization and storage duration. *LWT-Food Sci. Technol.* **2012**, *49*, 202–207. [CrossRef]
51. Salta, F.N.; Mylona, A.; Chiou, A.; Boskou, G.; Andrikopoulos, N.K. Oxidative Stability of Edible Vegetable Oils Enriched in Polyphenols with Olive Leaf Extract. *Food Sci. Technol. Int.* **2007**, *13*, 413–421. [CrossRef]

antioxidants

MDPI

Article

Innovative and Conventional Valorizations of Grape Seeds from Winery By-Products as Sustainable Source of Lipophilic Antioxidants

Ivana Dimić [1], Nemanja Teslić [2], Predrag Putnik [3], Danijela Bursać Kovačević [3], Zoran Zeković [1], Branislav Šojić [1], Živan Mrkonjić [1], Dušica Čolović [2], Domenico Montesano [4,*] and Branimir Pavlić [1,*]

1. Faculty of Technology, University of Novi Sad, Blvd. cara Lazara 1, 21000 Novi Sad, Serbia; ivana.dimic@live.com (I.D.); zzekovic@tf.uns.ac.rs (Z.Z.); sojic@tf.uns.ac.rs (B.Š.); zivan_mrkonjic@hotmail.com (Ž.M.)
2. Institute of Food Technology, University of Novi Sad, Blvd. cara Lazara 1, 21000 Novi Sad, Serbia; nemanja.teslic@fins.uns.ac.rs (N.T.); dusica.colovic@fins.uns.ac.rs (D.Č.)
3. Faculty of Food Technology and Biotechnology, University of Zagreb, Pierottijeva 6, 10000 Zagreb, Croatia; pputnik@alumni.uconn.edu (P.P.); dbursac@pbf.hr (D.B.K.)
4. Section of Food Science and Nutrition, Department of Pharmaceutical Sciences, University of Perugia, Via San Costanzo 1, 06126 Perugia, Italy

* Correspondence: domenico.montesano@unipg.it (D.M.); bpavlic@uns.ac.rs (B.P.)

Received: 25 May 2020; Accepted: 23 June 2020; Published: 1 July 2020

Abstract: The aim of this study was to valorize the oil recovery from red and white grape seeds (*Vitis vinifera* L.) that remains as by-product after the winemaking process. Oils were extracted by modern techniques, ultrasound assisted (UAE), microwave assisted (MAE) and supercritical fluid extraction (SFE), and compared to the Soxhlet extraction (SE). Firstly, SFE was optimized at different operating conditions: pressure (250–350 bar), temperature (40–60 °C), CO_2 flow rate (0.2, 0.3 and 0.4 kg h^{-1}), and particle size (315–800 μm and >800 μm). The highest extraction yields were achieved by SFE at the optimal conditions: 350 bar, 60 °C, 0.4 kg h^{-1}. Afterwards, SFE was compared to SE, UAE and MAE with respect to oil extraction yields, and analyzed for fatty acid composition and antioxidant capacity. Considering the general classification of fatty acids, it was found that samples had high content of polyunsaturated fatty acids, regardless of extraction technology. Tocopherol content was significantly influenced by all extraction methods, whereas UAE and MAE resulted in extracts richer with lipophilic antioxidants. In conclusion, modern extractions that are suited for industrial applications had better performance as compared to SE, as judging by the oil yield and quality.

Keywords: grape seed oil fatty acid; novel extraction; tocopherol; antioxidant activity; supercritical fluid; microwave assisted; ultrasound assisted; Soxhlet

1. Introduction

Grape processing generates large quantities of important agricultural and industrial wastes/by-products with potential to be reused for various purposes. It has been estimated that more than 0.3 kg of solid by-products is allocated per kg of mashed grape fruit during the processing [1]. The waste streams of wine production contain organic waste, greenhouse gases (CO_2, vaporous compounds etc.) and non-organic waste (diatomaceous earth, bentonite clay, perlite). In particular, organic waste, as grape pomace with seeds, pulp and skins, grape stems and leaves, represents about two thirds of entire solid waste [2,3]. The handling and disposal of this great amount of waste/by-products is a large environmental problem [3,4]. Currently, new processes for the controlled waste removal are being searched, targeting the conversion of the waste material and its incorporation into new bio-products with added value.

Seeds are the major material of the industrial processing of grape berry (e.g., found in pomace) and constitute about 7–20% of the weight of grapes processed [5]. Although grape seeds are mutually removed with the skins and vascular fruit tissues from the pomace, they can be easily separated through technological separation and sieving [6]. Although grape peels and stems do not have an economic background for industrial utilization, seeds are rich in bioactive antioxidants, and can be raw materials for the development of new foods, as natural extracts, pharmaceutical products [7,8] and cosmetics. Therefore, the production of grape seed oil contributes to the advance of waste management that could increase the financial income of the primary industrial process and sustainability [4].

The oil content of grape seeds from the literature was reported in the range of 13–15% depending on the variety and maturity of grapes [9]. The interest in grape seed oil as a functional food product has increased, particularly because of its high levels of lipophilic ingredients, such as vitamin E, unsaturated fatty acids (UFAs), and phytosterols [10] that possess greater antioxidant activity than hydrophilic ingredients [9]. Namely, grape seed oil was identified as a rich source of tocopherols, tocotrienols and unsaturated fatty acids, especially in polyunsaturated fatty acids (PUFAs), whereas linoleic acid (C18:2) was found as predominant (49.0–78.2% of total PUFAs) [4]. γ-tocotrienol was evidenced as the most abundant tocotrienol, followed by α-tocotrienol, while δ-tocotrienol was found in lower amounts [11]. Tocopherols from seed oils are α-, β-, γ-, and δ-tocopherol, with α-tocopherol as one of the most potent intracellular fat-soluble antioxidants due to its activity in inhibiting the peroxidation of polyunsaturated fatty acids in biological membranes [4]. A mixture of α-tocopherol and α- and γ-tocotrienol purified from grape seeds was more effective than other lipophilic grape seed fractions in neutralizing free and lipid peroxy radicals and in chelating prooxidant metals [12]. Therefore, grape seed oil extract can be considered a valuable source of natural liposoluble antioxidants, having potential health benefits [13,14].

On the industrial level, grape seed oils are mainly produced by traditional oil extraction methods, such as cold-press and solvent extraction [15,16]. Traditional processing often leads to a higher solvent consumption, longer extraction times, lower yields and poorer extraction quality [17]. As compared to Soxhlet extraction, cold pressing has potential for higher yields of fatty acids and tocopherols, without the assistance of heat and chemical treatment. Thus, cold-pressed oils are interesting raw materials for natural and safe food products favored by manufacturers and consumers [18].

Commonly, Soxhlet extraction employs *n*-hexane as solvent for grape oils, which is not selective and simultaneously removes non-volatile pigments and waxes. Consequently, the obtained extracts are dark, viscous and contaminated with the traces of toxic solvent [19]. Many contemporary extraction technologies avoid the negative impacts of thermal degradation and meet the criteria for the "green" extraction processes [20]. Aside from that, they offer energy savings, either minimize or avoid the use of organic solvents, shorten the processing time, reduce the temperature, enhance the mass transfer process, and increase the extraction yield with high quality extracts [21]. Extractions aided by ultrasound, microwave or high-pressure processing are being explored as alternative technologies for intensification of the extracted antioxidants from grape seeds.

Ultrasound-assisted extraction (UAE) is based on the acoustic cavitation phenomenon. The period of negative pressure during the ultrasound treatment causes bubbles, hence the origins of increased pressure and temperature with their subsequent collapse. When this happens, the resulting "shock waves" break the cellular walls and facilitate solvent penetration into plant materials which enhances extraction yields [22]. Microwave-assisted extraction (MAE) uses microwaves that are nonionizing, electromagnetic waves with the frequency between 300 MHz and 300 GHz. Here, electromagnetic waves are transformed to thermal energy, which induces heating of the matrix on inside and outside without thermal gradient. If a sufficient amount of thermal energy is generated, this local heating damages cell wall of plant matrix and causes leakage of target compounds into extraction medium.

In the literature, MAE was useful for extraction of biologically active substances with antioxidant properties from grape seeds [23,24]. Supercritical fluid extraction (SFE) represents another excellent alternative to conventional extraction with the potential to achieve comparable yields. Additionally, grape seed oils recovered by SFE are characterized by higher product quality that is similar to mechanical pressing [25]. Supercritical fluids, especially CO_2, have the gas-like properties, e.g., viscosity and diffusivity, and liquid-like properties, e.g., density and solvation power [26]. CO_2 is a green, low-cost, non-toxic and non-flammable solvent with critical pressure of 73 bar and temperature of 31 °C. It can be re-used in processing, hence its ability to reduce total energy costs in industry [27]. In addition, the residues of the solvent do not remain in the final product, because supercritical CO_2 can be completely eliminated by pressure reduction [9]. Co-solvents and modifiers (e.g., ethanol, methanol, acetone) may be added to improve the solubility of polar phytochemicals embedded in the cell wall [27]. Moreover, supercritical CO_2 ensures selectivity in the extraction of certain target compounds by varying operating conditions (e.g., temperature and pressure) [27], while an approximate economic estimate of industrial SFE scale-up from the laboratory is already available from the literature [28].

Since there are no abundant data in the literature for comparison of alternative vs. conventional extractions of grape seed oils with potential for industrial applications, the aim of this study was to compare Soxhlet against UAE, MAE and SFE concerning efficiency for obtaining high-quality extracts. In this work, we postulated that SFE is an important green alternative to organic solvent extraction for recovery of lipophilic antioxidants from winery waste streams by considering extraction parameters, in-depth chemical profiling, functional qualities and bioactivity of samples. Samples were compared in terms of total extraction yields, fatty acid profiling, tocopherol content, and antioxidant properties.

2. Materials and Methods

2.1. Plant Material

The industrial by-products of a various white and red grapes (*Vitis vinifera* L.) containing separated seeds were received from Kovačević Winery D.O.O. (Irig, Serbia). Red grape seed samples were a mixture of Cabernet Sauvignon, Merlot and Pinot noir with approximate ratio of 65:30:5 (*m*/*m*/*m*), while the white ration of grape varieties in the white seed sample was 60:30:10 (*m*/*m*/*m*) for Chardonnay, Sauvignon blanc and Riesling. The seeds were immediately milled in a domestic blender (Bosch, MMB21P0R/01, Germany) and subjected to extraction of bioactive antioxidants. Mean particle size of the sample was determined by sieving through the vibro-sieve set (CISA Cedaceria Industrial, Spain) with 0.1-, 0.315-, 0.8-, 2- and 4-mm pore size. The mean particle size for the red and white grape seeds was 0.578 and 0.566 mm, respectively. Red grape fractions had 0.315–0.8-mm particle size, and >0.8 mm were used to evaluate the granulation on SFE samples.

2.2. Chemicals and Reagents

Helium (>99.9997%) and carbon dioxide (99.9%) were purchased from Messer Technogas A.D., Novi Sad, Serbia. *n*-Hexane was purchased from Merck KgaA, Darmstadt, Germany. Ethanol was purchased from Sani-Hem D.O.O., Novi Bečej, Serbia. Methanol was purchased from Lach-ner Ltd., Neratovice, Chech Republic. Ethyl acetate was purchased from Zdravlje Leskovac, Leskovac, Serbia. 1,1-Diphenyl-2-picrylhydrazyl-hydrate (DPPH), 2,2′-azino-bis(3-ethylbenzothiazoline-6-sulfonic acid) diammonium salt (ABTS) and Trolox (6-hydroxy-2,5,7,8-tetramethylchroman-2-carboxylic acid) were purchased from Sigma Aldrich, St. Louis, MO, United States. Supelco 37 component fatty acid methyl esters (FAMEs) mix, DL-α tocopherol (99.9%), rac-β-tocopherol (99%), γ-tocopherol (97.3%) and δ-tocopherol (95.2%) were purchased from Supelco Inc., Bellefonte, PA, United States.

2.3. *Extraction Techniques*

2.3.1. Soxhlet Extraction (SE)

Grape seeds samples (30.0 g) were extracted using 120 mL of *n*-hexane in Soxhlet apparatus. Extraction was conducted for 6 h with 15 exchanges of the extracts and filtration of the solvent. Extraction solvent was then evaporated under vacuum at 40 °C. Obtained extracts were placed in a glass vials and kept at 4 °C until analysis. Extraction experiments were performed in triplicates and total extraction yield was expressed as mean ± standard deviation.

2.3.2. Ultrasound-Assisted Extraction

Ultrasound-assisted extraction was performed with sonication bath (EUP540A, Euinstruments, Paris, France) on constant frequency 40 kHz. Mass of 30.0 g of grape seeds sampled into a 500-mL glass flask and mixed with 300 mL of *n*-hexane. The condenser was put on glass flask to avoid solvent evaporation. The UAE was assessed using the modified method from the literature [29], with extraction conditions of $T = 50$ °C, $t = 40$ min, and sonication power at 60 W L^{-1}. Afterwards, the extracts were filtered and the extraction solvent was evaporated under vacuum at 40 °C. Obtained extracts were placed in glass vials and kept at 4 °C until analysis.

2.3.3. Microwave-Assisted Extraction

Microwave-assisted extraction was performed in experimental setup previously described [30]. Briefly, seeds samples of 10.0 g were mixed with 100 mL of *n*-hexane in glass flask, and placed in microwave extractor with connected condenser through a hole at the top of the casing. Matrix-to-solvent ratios were adjusted according to the cited references and limitations of the experimental setup. Extraction was performed at constant microwave irradiation power (600 W) for 15 min. The obtained extracts were filtered and solvent was removed under vacuum at 40 °C, then placed in glass vials and kept at 4 °C until analysis.

2.3.4. Supercritical Fluid Extraction

Supercritical fluid extraction was done at laboratory scale with high-pressure extraction equipment (HPEP, NOVA-Swiss, Effretikon, Switzerland). The main characteristics of supercritical fluid extractor were described in previous research [31]. Since, preliminary results shown that red varieties seeds contained higher amounts of oils, SFE operating conditions were initially optimized for red grape, and the best SFE conditions were applied to white grape. Red grape samples (100.0 g) were extracted for 4 h at different operating conditions: (i) pressures (250, 300 and 350 bar); (ii) temperatures (40, 50 and 60 °C); (iii) CO_2 flow rates (0.2, 0.3 and, i 0.4 kg h^{-1}); and (iv) particle size fractions (315–800 and >800 μm). A one-factor-at-a-time experimental design was used and sample labels and process conditions listed in Table 1. All extracts were collected in glass vials and stored at 4 °C until analysis.

Table 1. Design of an experiment for grape seeds oil isolation.

Sample	Extraction Technique	Process Conditions
	Red grape seeds (RGS)	
RGS-SFE1		250 bar, 40 °C, 0.3 kg h^{-1}
RGS-SFE2		300 bar, 40 °C, 0.3 kg h^{-1}
RGS-SFE3		350 bar, 40 °C, 0.3 kg h^{-1}
RGS-SFE4		350 bar, 50 °C, 0.3 kg h^{-1}
RGS-SFE5	Supercritical fluid extraction	350 bar, 60 °C, 0.3 kg h^{-1}
RGS-SFE6		350 bar, 60 °C, 0.2 kg h^{-1}
RGS-SFE7		350 bar, 60 °C, 0.4 kg h^{-1}
RGS-SFE315		350 bar, 60 °C, 0.4 kg h^{-1}, 315 < d < 800 µm
RGS-SFE800		350 bar, 60 °C, 0.4 kg h^{-1}, d > 800 µm
RGS-UAE	Ultrasound-assisted extraction	solvent: *n*-hexane, 40 kHz, 50 °C, 40 min, 60 W L^{-1}
RGS-MAE	Microwave-assisted extraction	solvent: *n*-hexane, 600 W, 15 min
RGS-SE	Soxhlet extraction	solvent: *n*-hexane, 6 h, 15 exchanges of extract
	White grape seeds (WGS)	
WGS-SFE	Supercritical fluid extraction	350 bar, 60 °C, 0.4 kg h^{-1}
WGS-UAE	Ultrasound-assisted extraction	solvent: *n*-hexane, 40 kHz, 50 °C, 40 min, 60 W L^{-1}
WGS-MAE	Microwave-assisted extraction	solvent: *n*-hexane, 600 W, 15 min
WGS-SE	Soxhlet extraction	solvent: *n*-hexane, 6 h, 15 exchanges of extract

2.3.5. Extraction Yield

The extraction yield for all applied extractions was calculated according to the following Equation (1):

$$Y[\%] = \frac{\text{mass of extracted oil}}{\text{mass of wheat germ}} \times 100 \quad (1)$$

2.4. *Chemical and Antioxidant Characterization of Grape Seeds Oils*

2.4.1. Fatty Acid Profiles

Fatty acid methyl esters were prepared from the extracted lipids using a method based on 14% boron trifluoride–methanol solution [32]. Nitrogen was used for drying and removing the solvent from fatty acid methyl esters. Obtained samples were analyzed on GC Agilent 7890A system with FID, automated liquid injection module, equipped with capillary column with silica gel (SP-2560, 100 m × 0.25 mm, I.D., 0.20 µm, Supelco Analytical, Bellefonte, PA, United States). Temperature regime during analysis was set as followed: initial temperature was 140 °C with hold of 5 min, heating up to 240 °C was with 2 °C/ min and hold on 240 °C was 5 min. Helium was used as carrier gas (flow rate = 1.26 mL min^{-1}). Fatty acid peaks in samples were identified by comparison with retention times of the standards from Supelco 37 component FAMEs mix and data from internal data library, based on earlier experiments and GC/MS analysis. The results were expressed as a mass of fatty acid or fatty acid group (g) per 100 g of oil.

2.4.2. Functional Quality

The functional quality of grape seed oils was determined by three indices obtained and calculated from fatty acid (FA) profiles. The ratio between hypocholesterolemic and hypercholesterolemic FAs (H/H) was calculated according to Equation (2) [33,34].

$$\frac{H}{H} = \frac{C18:1 + C18:2 + C18:3}{C14:0 + C16:0} \quad (2)$$

Furthermore, the atherogenicity index (AI) and thrombogenicity index (TI) were calculated according to the Equations (3) and (4) [33,35].

$$\mathrm{AI} = \frac{C14:0 + 4(C16:0)}{\sum MUFA + \sum \omega - 3 + \sum \omega - 6} \tag{3}$$

$$\mathrm{TI} = \frac{C14:0 + C16:0 + C18:0}{0.5(\sum MUFA) + 3\sum \omega - 3 + 0.5\sum \omega - 6 + \left(\frac{\sum \omega - 3}{\sum \omega - 6}\right)} \tag{4}$$

where C14:0 is myristic acid, C16:0 is palmitic acid, C18:0 is stearic acid, C18:1 is oleic acid, C18:2 is linoleic acid, C18:3 is α-linolenic acid. ∑MUFA is a sum of monounsaturated FAs, Σω-3 sum of the polyunsaturated ω-3 FAs and Σω-6 is sum of the polyunsaturated fatty ω-6 acids.

2.4.3. Tocopherols

Tocopherol content was determined by high-pressure liquid chromatography (HPLC), according to the modified method from the literature [36]. Samples were diluted in *n*-hexane and filtered through an RC 0.45-μm syringe filter (Agilent Technologies Inc., Böblingen, Germany). HPLC system (Agilent liquid chromatography series 1260) was equipped with quaternary pump, autosampler and fluorescence detector (Agilent Technologies Inc., Böblingen, Germany). Separation of tocopherols was carried out with normal-phase Luna® 5 μm Silica (2) 100ALC Column (250 × 4, 6 mm) analytical column (Phenomenex, Torrance, CA, United States) and 10 min isocratic analysis run with tetrahydrofurane/*n*-hexane mixture (4:96, *v*/*v*) as mobile phase with flow rate 1.3 mL min^{-1}. The column was thermostatted at 35 °C with an injection volume of 5 μL. Fluorescence detector was set at 290-nm excitation wavelength and 330-nm emission wavelength. For each tocopherol, standard stock solutions were prepared and treated as samples in the following fractions: α-tocopherol (0.5–50.0 ppm), β-tocopherol (0.5–50.0 ppm), γ-tocopherol (0.2–25.0 ppm), and δ-tocopherol (0.5–25 ppm). The external calibration curves were made and used for identification and quantification. The results were expressed as mg of tocopherol per g of grape seed oil (mg g^{-1} oil).

2.5. Determination of In Vitro Antioxidant Capacity

The in vitro antioxidant capacity was evaluated by two methods: DPPH (2,2-diphenyl-1-picrylhydrazyl) and ABTS (2,2′-azino-bis-(-3-ethylbenzothiazoline-6-sulfonic acid) diammonium salt).

2.5.1. DPPH Assay

The capacity of samples towards scavenging of 1,1-diphenyl-2-picrylhydrazyl-hydrate (DPPH) radicals was measured by published method [37] with slight modifications for lipid samples [38]. Briefly, methanol solution of DPPH reagent (65 μM) was freshly prepared and adjusted with methanol to reach 0.70 (± 0.02) absorbance. Volumes of 0.1 mL of samples were diluted in ethyl acetate, and mixed with 2.9 mL of DPPH reagent in a glass tubes and incubated in dark for 60 min. Blanks were prepared by mixing 0.1 mL of ethyl acetate and 2.9 mL of DPPH reagent. Free radical scavenging measurements were performed in triplicates at 517-nm wavelength by UV/Vis spectrophotometer (6300 Spectrophotometer, Jenway, Staffordshire, UK). Freshly prepared Trolox methanolic solutions (1.33–26.64 μM) were used for the calibration curves. The obtained results were expressed as μM of Trolox equivalents per g of grape seed oil (μM Trolox/g).

2.5.2. ABTS Assay

The ability of samples towards scavenging of ABTS radicals was measured by modified method from the literature [39]. Briefly, ABTS stock solution was freshly prepared from the mixture (1:1, *v*/*v*) of 2.45-mM potassium persulphate aqueous solution and 7-mM ABTS aqueous solution, then left to sit in a dark area at room temperature for the next 16 h. Stock solution was diluted using 300-mM acetate buffer (pH = 3.6) to reach 0.70 (± 0.02) absorbance. A volume of 0.1 mL of a sample (diluted in ethyl acetate) and ABTS reagent (2.9 mL) were mixed and incubated in a dark for 5 h. The blank was obtained by mixing 0.1 mL of ethyl acetate and 2.9 mL of ABTS reagent. Absorbance was measured at 734 nm in triplicates by UV/Vis spectrophotometer (6300 Spectrophotometer, Jenway, Staffordshire, UK). Freshly prepared Trolox ethanolic solutions (0.8–26.6 μM) were used for the calibration curve. The results were expressed as μM of Trolox equivalents per g of grape seed oil (μM Trolox/g).

2.6. Statistical Analysis

All experiments were performed in triplicates and results were presented as mean value ± standard deviation (SD), while significant levels were defined at $p \leq 0.05$ using Tukey's test. Statistical analysis was carried out using Statistica 10.0 (StatSoft Inc., Tulsa, OK, USA).

3. Results

3.1. Influence of SFE Operating Parameters on Total Extraction Yield

Preliminary research identified red grape seeds as richer in oil, so it was selected as a gauge for SFE optimization. SFE was conducted upon various extraction parameters of pressure with respect to the total extraction yield of red grape seed oil (Figure 1). Experimentally obtained SFE yields ranged from 7.46 to 12.23%. The highest yield was achieved at following operating conditions: pressure of 350 bar, temperature of 60 °C and CO_2 flow rate of 0.4 kg h^{-1}; therefore, these values were selected as the optimal parameters for further experimentations.

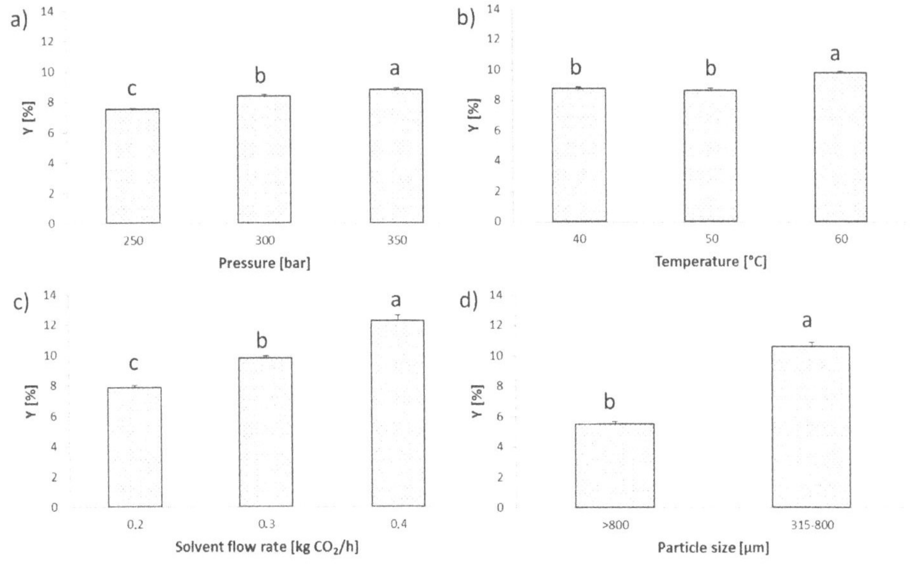

Figure 1. Influence of supercritical fluid extraction (SFE) extraction parameters on total extraction yield of grape seed oil: (**a**) Influence of pressure at fixed temperature (40 °C) and CO_2 flow rate (0.3 kg/h), (**b**) Influence of temperature at fixed pressure (350 bar) and CO_2 flow rate (0.3 kg/h), (**c**) Influence of CO_2 flow rate at fixed pressure (350 bar) and temperature (60 °C), and (**d**) Influence of particle size at 350 bar, 60 °C and 0.4 kg CO_2/h. * Different letters indicate significant difference ($p \leq 0.05$) between total extraction yields of grape seed oil with respect to sources of variation.

By increasing pressure from 250 to 350 bar, with other constant parameters (T = 40 °C and CO_2 flow rate of 0.3 kg h^{-1}), the extraction yield raised from 7.46 to 8.73% (Figure 1a). An increase in pressure at isothermal conditions caused the density of the supercritical CO_2 to increase as well, which improved its solvating power and dissolution rate, thus improving extraction efficiency [25].

The influence of temperature was observed at this constant pressure and CO_2 flow rate (0.3 kg h^{-1}) for different temperature values (40, 50 and 60 °C) (Figure 1b). The highest yield was found at 60 °C (9.80%) and the lowest at 50 °C (8.61%).

The impact of CO_2 flow rates on total SFE yield was studied at the constant pressure (350 bar) and temperature (60 °C). Extraction kinetics for CO_2 flow rates of 0.2, 0.3 and 0.4 kg h^{-1} is depicted in Figure 1c. The highest yield was achieved at the highest applied flow rate (12.23%) and the crucial differences were observed when flow rate of CO_2 increased.

A further aim in this study was to determine the influence of the particle size on total extraction yields that were previously detected as optimal (i.e., 350 bar, 60 °C, 0.4 kg h^{-1} CO_2). The influence of particle size on the yield was investigated in two sample fractions obtained by vibro-sifting. The first fraction consisted of red grape samples with particle size between 315 and 800 μm (R315-SFE) and the second one referred to the red grape with particles above 800 μm (R800-SFE). Figure 1d shows that reduced particle size lead to the significant increase in oil yield, which was 10.58% for R315-SFE fraction and 5.49% for R800-SFE fraction.

3.2. Influence of Different Extraction Techniques on Total Extraction Yield

After considering the influence of pressure, temperature, CO_2 flow rate and particle size, the optimal extraction parameters were selected, and supercritical CO_2 extraction was performed for red grape seeds (RGS) and white grape seeds (WGS) according to design of an experiment from Table 1. The comparative analysis between conventional (SE) and modern extraction techniques, including UAE, MAE and SFE, was also performed (Figure 2).

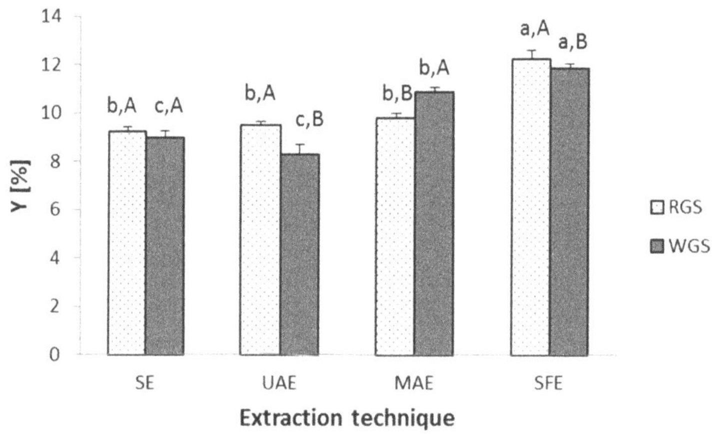

Figure 2. Total extraction yield for grape seed oils recovered by conventional and modern extraction techniques. * Different lowercase letters indicate significant difference ($p \leq 0.05$) between extraction techniques, while different uppercase letters indicate significant difference between red grape seeds (RGS) and white grape seeds (WGS).

The highest yield was achieved by supercritical CO_2 extraction at the optimal conditions. SFE yields were 12.23 and 11.86% for red and white grape seeds, respectively. Accordingly, it can be observed that SFE gave the highest total extraction yield as compared to the SE and to all other tested advanced techniques. MAE showed good extraction yield; however, no previous studies were found with confirming suitability of MAE to isolate oils from grape seeds. Therefore, the results from Figure 2 support the use of the MAE for this purpose as well. This is important, as MAE has the advantage over SE due to reducing the extraction time and allowing for lower temperatures that will decrease deterioration of thermolabile oily compounds in the process.

3.3. Fatty Acid Profile and Functional Quality

The fatty acid profile of grape seed oils from red and white grape varieties was investigated for all extraction techniques (Table 2). According to the results, it can be seen that samples contained saturated fatty acids (11.28–12.27%), monounsaturated fatty acids (13.53–18.62%), and polyunsaturated fatty acids (69.27–74.88%). Irrespective of extraction type (SE vs. SFE, UAE MAE) or grape variety (red vs. white), eight fatty acids were determined in all samples. Among saturated fatty acids, palmitic acid was predominant one (7.20–7.93%), followed by stearic acid (3.79–4.37%). Monounsaturated oleic acid was present between 13.39–18.47%. Linoleic acid was the most abundant polyunsaturated fatty acid in the investigated samples, accounting for the 68.61–74.15% of all present total fatty acids. Previously, Rodríguez and Ruiz [4] reported that polyunsaturated fatty acids (PUFAs) were predominant in the grape seed samples with 49.0–81.6%, followed by monounsaturated fatty acids (MUFAs) in a range from 13.9–29.1%, and saturated fatty acid (SFAs) that were in a range from 9.6–26.7%. Regarding the fatty acid profiles, palmitic acid (C16:0) dominated in the group of SFAs with 6.7–12.8%, followed by stearic acid (C18:0) that ranged from 2.5% to 15.0%. Concerning the MUFA contents, the oleic acid (C18:1) was found as the major contributor with 0.1–28.9% from the samples.

The chemical profile of fatty acids could be a useful parameter for the assessment of functional qualities of grape seed oil. The polyunsaturated/saturated fatty acids ratio (PUFA/SFA) is often used to measure indices for frequent cardiovascular disease syndromes (atherogenicity and thrombogenicity), since only three SFAs are hypercholesterolemic [35]. Hence, we calculated the hypo- and hypercholesterolemic fatty acids ratio (H/H), atherogenicity index (AI) and thrombogenicity index (TI) in Table 3.

H/H values ranged from 11.07 to 12.28 for red grape, and from 11.30 to 12.09 for white grape seed oils. The highest H/H in the samples was observed for SE, while for advanced extractions, white grape samples recovered by UAE had the highest H/H value. A higher level of this index is desirable for nutrition, since it expresses the effect of the fatty acids on cholesterol metabolism. For instance, healthy oils as linseed have higher H/H index than grape oils (13.24), sesame and olive oils have lower values [40].

AI values were 0.081–0.090 for red grape seed oil and 0.083–0.088 for white grape samples. TI values were in a range 0.242–0.268 for red grape seed oil and 0.256–0.268 for white grape. The literature reports lower values of AI and TI for linseed oil [40], while sweet cherry, pomegranate, pumpkin [41], sesame and olive oil have higher values [40].

Table 2. Relative content (%) of fatty acids in red and white grape seed oils obtained by different extraction techniques.

Fatty Acid	Palmitic (C16:0)	Palmitoleic (C16:1)	Stearic (18:0)	Oleic (C18:1n9C)	Linoleic (C18:2n6C)	γ-Linolenic (C18:3n6C)	α-Linolenic (C18:3n3C)	Heneicosanoic (C21:0)	Saturated Fatty Acids	Monounsaturated Fatty Acids	Polyunsaturated Fatty Acids	Unsaturated Fatty Acids	Ratio S/U
							Red grape seeds						
RGS-SFE	7.93	0.14	3.79	13.39	73.58	0.18	0.66	0.34	12.06	13.53	74.42	87.94	0.14
RGS-UAE	7.20	0.12	3.83	13.72	74.15	0.21	0.52	0.25	11.28	13.84	74.88	88.72	0.13
RGS-MAE	7.33	0.14	4.33	16.18	71.09	0.24	0.45	0.25	11.91	16.31	71.78	88.09	0.14
RGS-SE	7.42	0.13	3.93	14.03	73.45	0.23	0.55	0.26	11.61	14.16	74.23	88.39	0.13
							White grape seeds						
WGS-SFE	7.73	0.14	4.29	17.57	69.37	0.23	0.41	0.25	12.27	17.71	70.02	87.73	0.14
WGS-UAE	7.66	0.17	4.24	18.39	68.65	0.24	0.42	0.23	12.13	18.56	69.31	87.87	0.14
WGS-MAE	7.29	0.13	4.27	17.91	69.83	0.00	0.39	0.19	11.74	18.03	70.22	88.26	0.13
WGS-SE	7.51	0.16	4.37	18.47	68.61	0.25	0.40	0.24	12.11	18.62	69.27	87.89	0.14

extraction (12.28%). Jokić et al. [47] concluded that grape oil can be completely extracted by SFE at optimal operating conditions (P = 400 bar and T = 41 °C), resulting with the oil yield of 14.87%. However, here the yield of Soxhlet extraction with *n*-hexane was 14.96%. Bravi et al. [44] used a seed mixture of different red (Merlot, Cabernet Sauvignon, Cabernet Franc and Raboso) and white (Prosecco, Verduzzo, Pinot Grigio, Chardonnay, Pinot Bianco, Bianco) grape varieties in Soxhlet and SC-CO_2 extraction. The content of the oil that can be extracted with SC-CO_2 was 14.4% and was slightly lower than with hexane (15.4%). Prado et al. [28] achieved yield of 13.42% using SFE technique at 350 bar, 40 °C and 0.46 kg CO^2 h^{-1}. Da Porto et al. [48] reported Soxhlet extraction using *n*-hexane in a 1:12 ratio for 6 h and UAE in 1:8 ratio at 20 kHz and 50, 100 and 150 W for 30 min against the Soxhlet extraction of grape seeds of Raboso Piave variety. Here, an increase of the ultrasound power from 50 to 150 W caused the yield to jump from 11.42 to 14.08%. Anyhow, Soxhlet extraction had a higher yield of 14.64%, which can be explained by providing freshly condensed solvent for 6 h, while UAE was a batch system that lasted for 30 min. UAE at 150 W for 30 min increased the yield to approximately 14%, which is comparable to the yield of Soxhlet extraction at 70 °C for 6h [48]. In the study conducted by de Menezes et al. [17], oil from Burgundy variety grape seeds was obtained by the Soxhlet technique and UAE technique with hexane. The oil content was 16.28% for the Soxhlet technique and 11.60% for the UAE technique. Böger et al. [22] showed that UAE is useful technique for increasing of the oil yields while reducing the durations of the extractions, while using less solvent and obtaining the high quality oils.

It is important to note that fatty acid composition in grape seed oils may be highly influenced by the grape variety and growing conditions [49]. Since the SFE was selected as the technique with the highest potential for the oil recovery, it was further examined how its operating parameters affect the fatty acid profiling (Table S3). To that end, Prado et al. [28] investigated both lab- and pilot-scale SFE grape samples for Malbec and Cabernet Franc varieties. Although SFE parameters differed among pilot and lab experiments, it was still found that all extracted oils contained the linoleic (71.20%) and oleic acids (15.10%) as the main components. When considering saturated fatty acids, palmitic (8.13%) and stearic (4.05%) acids were determined as the most abundant. Coelho et al. [26] analyzed the fatty acid profiles of the SE/hexane samples and those extracted by SFE by varying different operating conditions. Highest percentages in samples accounted for linoleic (64.5–67.37%), oleic (19.18–20.64%), palmitic (7.38–8.22%) and stearic acids (4.33–5.61%). Judging by our data and the literature, it can be concluded that the fatty acid profile from all samples followed the expected content. However, notable differences in fatty acid profiling were found for seed samples recovered by different extraction techniques.

The obtained results of functional quality indices highlighted that grape seed oils recovered by SE and UAE expressed the best functional quality. On the other hand, it is important to mention that the aforementioned extracts were obtained using *n*-hexane, which can be evaporated from the oils, but they can still be contaminated by the traces of this organic solvent. In conclusion, SFE stands out as a successful technique for isolation of solvent-free grape seed oil with proper functional quality and without any traces of extraction solvents.

When observing the impact of pressure, it can be concluded that the higher pressure reduced tocopherol content, although higher pressures gave higher extraction yields. The increase in temperature resulted in elevated tocopherol content. While higher temperatures promote higher solubility of the solute and enhance mass transfer of solute from matrix to the SFE solvent, the lowest tocopherol content was found at 50 °C. This can be related to the crossover phenomenon at aforementioned temperature. Bravi et al. [44] have been observed for SFE extracts that an elevated temperature (40 vs. 80 °C) influenced increased α-tocopherol content, due to the higher solubility of α-tocopherol at 80 than at 40 °C. Samples with a reduced particle size (RGS315-SFE) exhibited higher tocopherol content than samples with larger particle size (7.57 vs. 4.60 mg 100 g^{-1}). This was expected, as particle size reduction leads to increased extraction efficiency, since the free surface area for mass transfer is increased and diffusional resistance in solid phase is decreased [44].

Since the optimal SFE parameters gave the highest overall yield, after adjusting for total tocopherols, results showed that SFE at given conditions was the most successful for exhaustion of grape seeds. In summary, red grape seed oils are richer in α- and γ-tocopherol and higher amount of total tocopherols can be recovered by SFE, but considerable attention must be paid to the proper adjustment of an extraction parameters.

Based on the data, it can be concluded that the antioxidant capacity is strongly dependent upon the applied extraction parameters. Moreover, oils obtained from red grape seeds had a higher antioxidant capacity as compared to those from white grape, probably because higher content of α-tocopherols was also in red grape samples. Tangolar et al. [47] have previously reported higher concentrations of α- and γ-tocopherol in the Cabernet Sauvignon variety than Chardonnay variety. Similar results were noted by Ben Mohamed et al. [9], who evaluated the bioactive compounds and antioxidant activities of six different grape seed oils.

Although several in vitro and in vivo studies have shown that tocotrienol-rich fractions from grape seeds are more potent antioxidants, another study documented that α-tocopherol had higher free radical scavenging activity than tocotrienol-rich fractions, and consisted of a mixture of γ-tocopherol and α- and γ-tocotrienol [12]. This was explained by the purity of tocotrienol-rich fractions, which contains approximately 6% tocols.

Among SFE and SE extracts, the SFE exhibited higher antioxidant capacity for both DPPH and ABTS assays. The opposite findings were reported by Wang et al. [50], who evaluated and compared in vitro antioxidant activities of unsaponifiable fractions of 11 kinds of edible vegetable oils (flaxseed, olive, grape seed, corn, soybean, sunflower seed, walnut, perilla, rapeseed, sesame, and camellia) by DPPH, ABTS and FRAP assays [50]. The authors identified grape seed oils with the lowest total antioxidant capability that might be attributed to the different processing techniques and different amounts of hydrophilic and lipophilic antioxidant content in the samples. Hence, confirming that oil extraction procedure is the crucial element that affect antioxidant activity and overall quality of extracts. A different study by Ben Mohamed et al. [9] found higher ABTS values than our study for red grape oils. Oils were recovered by both SFE and SE with 7.5–8.2 μM and 5.9–6.5 μM Trolox g^{-1}, respectively. White grape varieties recovered by SFE had the ABTS values of 4.9–6.0 μM Trolox g^{-1}, while SE samples had lower values for antioxidative activity of 4.4–4.9 μM Trolox g^{-1} [9]. In the work of Konuskan et al. [15], the highest DPPH radical scavenging activity was noted for Cabernet Sauvignon variety, which also makes up the highest part in red grape seed mixture used in this work. The authors also found that hydrophilic antioxidant values were unaffected by the extraction method, while lipophilic values were higher for the super critical CO_2-extracted oils. This suggested that the type of extraction, as well as the corresponding parameters, should be thoroughly considered for the isolation of oil from grape seeds in order to obtain desired antioxidant potential.

5. Conclusions

Grape seeds, as a by-product of wine industry, can be successfully valorized as a raw material for recovering oils with high-quality bioactive antioxidants. Modern extraction technologies, such as UAE, MAE and SFE, were compared to Soxhlet extraction for obtaining red and white grape seed oils. The SFE was the best method with respect to extraction yield at optimum processing parameters (350 bar, 60 °C and 0.4 kg h^{-1}), providing 12.23% and 11.86% yields for red and white grape seeds, respectively.

A fatty acid profiling of samples identified polyunsaturated fatty acids as dominating in this category of constituents (69.27–74.88%) with linoleic acid (68.61–74.15%) as major representative. Monounsaturated fatty acids were found in lower amounts (13.53–18.62%) where oleic acid was predominant compound (13.39–18.47%). Saturated fatty acids were detected in the lowest amounts ranging from 11.28–12.27%. Different extraction techniques did not alter fatty acid profiles in the samples; however, the application of SFE technology yielded appreciable quantities of tocopherols. The highest antioxidant potential (i.e., DPPH and ABTS) for red and white grape oils were observed for samples recovered by the UAE.

Based on the results, it can be concluded that the application of non-conventional extraction techniques was efficient for recovering of high-quality grape seed oils that were rich in lipid antioxidants. Thus, such extracts could be incorporated into different functional foods, pharmaceuticals or cosmetic products. Non-conventional techniques have environmental benefits, as they stand out as the "green" extractions, due to absence of organic solvents form the process. Hence, preventing numerous disadvantages of conventional alternatives, such as toxic residuals of organic solvents in the extracts, negative environmental impacts and flammability.

Supplementary Materials: The following are available online at http://www.mdpi.com/2076-3921/9/7/568/s1, Table S1: Influence of different extraction parameters on tocopherol yield in red grape seed oil samples (mg 100 g^{-1}); Table S2: Influence of different extraction techniques on tocopherol yield in grape seed oil (mg 100 g^{-1}); Table S3: Relative content of fatty acids (%) in all obtained samples.

Author Contributions: Conceptualization, B.P., Z.Z., D.B.K, and P.P.; methodology, I.D. and N.T.; software, N.T.; formal analysis, I.D. and Ž.M.; investigation, I.D., Ž.M. and D.Č.; resources, B.P., N.T. and D.B.K.; writing—original draft preparation, I.D. and B.P.; writing—review and editing, D.B.K., P.P., B.Š. and D.M.; supervision, Z.Z.; project administration, Z.Z. and D.B.K.; All authors have read and agreed to the published version of the manuscript.

Funding: This research was funded by Ministry of Education, Science and Technological Development, Republic of Serbia (451-03-68/2020-14/200134 and 451-03-68/2020-14/200222). P. Putnik and D. Bursać Kovačević wish to thank the Croatian Science Foundation for support through the funding of project number IP-2019-04-2105.

Conflicts of Interest: The authors declare no conflict of interest.

References

1. Duba, K.S.; Fiori, L. Supercritical CO_2 extraction of grape seed oil: Effect of process parameters on the extraction kinetics. *J. Supercrit. Fluids* **2015**, *98*, 33–43. [CrossRef]
2. Teixeira, A.; Baenas, N.; Dominguez-Perles, R.; Barros, A.; Rosa, E.; Moreno, D.A.; Garcia-Viguera, C. Natural bioactive compounds from winery by-products as health promoters: A review. *Int. J. Mol. Sci.* **2014**, *15*, 15638–15678. [CrossRef] [PubMed]
3. Bordiga, M.; Travaglia, F.; Locatelli, M.; Arlorio, M.; Coïsson, J.D. Spent grape pomace as a still potential by-product. *Int. J. Food Sci. Technol.* **2015**, *50*, 2022–2031. [CrossRef]
4. Rodríguez, J.M.L.; Ruiz, D.F. *Grape Seeds: Nutrient Content, Antioxidant Properties and Health Benefits*; Nova Science Publishers, Inc.: New York, NY, USA, 2016.
5. Dwyer, K.; Hosseinian, F.; Rod, M. The market potential of grape waste alternatives. *J. Food Res.* **2014**, *3*, 91. [CrossRef]
6. Domínguez, J.; Martínez-Cordeiro, H.; Lores, M. Earthworms and grape marc: Simultaneous production of a high-quality biofertilizer and bioactive-rich seeds. In *Grape and Wine Biotechnology*; IntechOpen: Rijeka, Croatia, 2016. [CrossRef]
7. Fidelis, M.; De Moura, C.; Junior, T.K.; Pap, N.; Mattila, P.H.; Mäkinen, S.; Putnik, P.; Kovačević, D.B.; Tian, Y.; Yang, B.; et al. Fruit seeds as sources of bioactive compounds: Sustainable production of high value-added ingredients from by-products within circular economy. *Molecules* **2019**, *24*, 3854. [CrossRef]
8. Guo, Y.; Huang, J.; Chen, Y.; Hou, Q.; Huang, M. Effect of grape seed extract combined with modified atmosphere packaging on the quality of roast chicken. *Poult. Sci.* **2020**, *99*, 1598–1605. [CrossRef]
9. Ben Mohamed, H.; Duba, K.S.; Fiori, L.; Abdelgawed, H.; Tlili, I.; Tounekti, T.; Zrig, A. Bioactive compounds and antioxidant activities of different grape (*Vitis vinifera* L.) seed oils extracted by supercritical CO_2 and organic solvent. *Lwt Food Sci. Technol.* **2016**, *74*, 557–562. [CrossRef]
10. Karaman, S.; Karasu, S.; Tornuk, F.; Toker, O.S.; Geçgel, Ü.; Sagdic, O.; Ozcan, N.; Gül, O. Recovery potential of cold press byproducts obtained from the edible oil industry: Physicochemical, bioactive, and antimicrobial properties. *J. Agric. Food Chem.* **2015**, *63*, 2305–2313. [CrossRef]
11. Lachman, J.; Hejtmánková, A.; Hejtmánková, K.; Horníčková, Š.; Pivec, V.; Skála, O.; Dědina, M.; Přibyl, J. Towards complex utilisation of winemaking residues: Characterisation of grape seeds by total phenols, tocols and essential elements content as a by-product of winemaking. *Ind. Crop. Prod.* **2013**, *49*, 445–453. [CrossRef]
12. Choi, Y.; Lee, J. Antioxidant and antiproliferative properties of a tocotrienol-rich fraction from grape seeds. *Food Chem.* **2009**, *114*, 1386–1390. [CrossRef]

13. Gupta, M.; Dey, S.; Marbaniang, D.; Pal, P.; Ray, S.; Mazumder, B. Grape seed extract: Having a potential health benefits. *J. Food Sci. Technol.* **2019**, *57*, 1205–1215. [CrossRef] [PubMed]
14. Patwardhan, M.; Morgan, M.T.; Dia, V.; D'Souza, D.H. Heat sensitization of hepatitis A virus and Tulane virus using grape seed extract, gingerol and curcumin. *Food Microbiol.* **2020**, *90*, 103461. [CrossRef] [PubMed]
15. Konuskan, D.B.; Kamiloglu, O.; Demirkeser, O. Fatty acid composition, total phenolic content and antioxidant activity of grape seed oils obtained by cold-pressed and solvent extraction. *Indian J. Pharm. Educ. Res.* **2019**, *53*, 144–150. [CrossRef]
16. Rombaut, N.; Savoire, R.; Thomasset, B.; Bélliard, T.; Castello, J.; Van Hecke, É.; Lanoisellé, J.-L. Grape seed oil extraction: Interest of supercritical fluid extraction and gas-assisted mechanical extraction for enhancing polyphenol co-extraction in oil. *Comptes Rendus Chim.* **2014**, *17*, 284–292. [CrossRef]
17. De Lopes Menezes, M.; Johann, G.; Diório, A.; Pereira, N.C.; da Silva, E.A. Phenomenological determination of mass transfer parameters of oil extraction from grape biomass waste. *J. Clean. Prod.* **2018**, *176*, 130–139. [CrossRef]
18. Al Juhaimi, F.; Özcan, M.M. Effect of cold press and soxhlet extraction systems on fatty acid, tocopherol contents, and phenolic compounds of various grape seed oils. *J. Food Process. Preserv.* **2018**, *42*. [CrossRef]
19. Passos, C.P.; Silva, R.M.; Da Silva, F.A.; Coimbra, M.A.; Silva, C.M. Supercritical fluid extraction of grape seed (*Vitis vinifera* L.) oil. Effect of the operating conditions upon oil composition and antioxidant capacity. *Chem. Eng. J.* **2010**, *160*, 634–640. [CrossRef]
20. Mwaurah, P.W.; Kumar, S.; Kumar, N.; Attkan, A.K.; Panghal, A.; Singh, V.K.; Garg, M.K. Novel oil extraction technologies: Process conditions, quality parameters, and optimization. *Compr. Rev. Food Sci. Food Saf.* **2019**, *19*, 3–20. [CrossRef]
21. Barba, F.J.; Zhu, Z.Z.; Koubaa, M.; Sant'Ana, A.S.; Orlien, V. Green alternative methods for the extraction of antioxidant bioactive compounds from winery wastes and by-products: A review. *Trends Food Sci. Technol.* **2016**, *49*, 96–109. [CrossRef]
22. Böger, B.R.; Salviato, A.; Valezi, D.F.; Di Mauro, E.; Georgetti, S.R.; Kurozawa, L.E. Optimization of ultrasound-assisted extraction of grape-seed oil to enhance process yield and minimize free radical formation. *J. Sci. Food Agric.* **2018**, *98*, 5019–5026. [CrossRef]
23. Krishnaswamy, K.; Orsat, V.; Gariépy, Y.; Thangavel, K. Optimization of microwave-assisted extraction of phenolic antioxidants from grape seeds (*Vitis vinifera*). *Food Bioprocess Technol.* **2012**, *6*, 441–455. [CrossRef]
24. Makarova, N.V.; Valiulina, D.F.; Eremeeva, N.B. Comparative studies of extraction methods of biologically-active substances with antioxidant properties from grape seed (*Vitis vinifera* L.). *Proc. Univ. Appl. Chem. Biotechnol.* **2020**, *10*, 140–148. [CrossRef]
25. Duba, K.S.; Fiori, L. Solubility of grape seed oil in supercritical CO_2: Experiments and modeling. *J. Chem. Thermodyn.* **2016**, *100*, 44–52. [CrossRef]
26. Coelho, J.P.; Filipe, R.M.; Robalo, M.P.; Stateva, R.P. Recovering value from organic waste materials: Supercritical fluid extraction of oil from industrial grape seeds. *J. Supercrit. Fluids* **2018**, *141*, 68–77. [CrossRef]
27. Khaw, K.-Y.; Parat, M.-O.; Shaw, P.N.; Falconer, J.R. Solvent supercritical fluid technologies to extract bioactive compounds from natural sources: A review. *Molecules* **2017**, *22*, 1186. [CrossRef]
28. Prado, J.M.; Dalmolin, I.; Carareto, N.D.D.; Basso, R.C.; Meirelles, A.J.A.; Vladimir Oliveira, J.; Batista, E.A.C.; Meireles, M.A.A. Supercritical fluid extraction of grape seed: Process scale-up, extract chemical composition and economic evaluation. *J. Food Eng.* **2012**, *109*, 249–257. [CrossRef]
29. Gayas, B.; Kaur, G.; Gul, K. Ultrasound-assisted extraction of apricot kernel oil: Effects on functional and rheological properties. *J. Food Process Eng.* **2017**, *40*, e12439. [CrossRef]
30. Zeković, Z.; Pintać, D.; Majkić, T.; Vidović, S.; Mimica-Dukić, N.; Teslić, N.; Versari, A.; Pavlić, B. Utilization of sage by-products as raw material for antioxidants recovery-Ultrasound versus microwave-assisted extraction. *Ind. Crop. Prod.* **2017**, *99*, 49–59. [CrossRef]
31. Pekić, B.; Zeković, Z.; Petrović, L.; Tolić, A. Behavior of (−)-α-bisabolol and (−)-α-bisabololoxides A and B in camomile flower extraction with supercritical carbon dioxide. *Sep. Sci. Technol.* **1995**, *30*, 3567–3576. [CrossRef]
32. Ivanov, D.; Čolović, R.; Bera, O.; Lević, J.; Sredanović, S. Supercritical fluid extraction as a method for fat content determination and preparative technique for fatty acid analysis in mesh feed for pigs. *Eur. Food Res. Technol.* **2011**, *233*, 343–350. [CrossRef]

33. Cunha, V.M.B.; Silva, M.P.d.; Sousa, S.H.B.d.; Bezerra, P.d.N.; Menezes, E.G.O.; Silva, N.J.N.d.; Banna, D.A.D.d.S.; Araújo, M.E.; Carvalho Junior, R.N.D. Bacaba-de-leque (*Oenocarpus distichus* Mart.) oil extraction using supercritical CO_2 and bioactive compounds determination in the residual pulp. *J. Supercrit. Fluids* **2019**, *144*, 81–90. [CrossRef]
34. Santos-Silva, J.; Bessa, R.J.B.; Santos-Silva, F. Effect of genotype, feeding system and slaughter weight on the quality of light lambs. *Livest. Prod. Sci.* **2002**, *77*, 187–194. [CrossRef]
35. Ulbricht, T.L.V.; Southgate, D.A.T. Coronary heart disease: Seven dietary factors. *Lancet* **1991**, *338*, 985–992. [CrossRef]
36. Eisenmenger, M.; Dunford, N.T.; Eller, F.; Taylor, S.; Martinez, J. Pilot-scale supercritical carbon dioxide extraction and fractionation of wheat germ oil. *J. Am. Oil Chem. Soc.* **2006**, *83*, 863–868. [CrossRef]
37. Brand-Williams, W.; Cuvelier, M.E.; Berset, C. Use of a free-radical method to evaluate antioxidant activity. *LWT Food Sci Technol.* **1995**, *28*, 25–30. [CrossRef]
38. Pavlić, B.; Bera, O.; Teslić, N.; Vidović, S.; Parpinello, G.; Zeković, Z. Chemical profile and antioxidant activity of sage herbal dust extracts obtained by supercritical fluid extraction. *Ind. Crop. Prod.* **2018**, *120*, 305–312. [CrossRef]
39. Re, R.; Pellegrini, N.; Proteggente, A.; Pannala, A.; Yang, M.; Rice-Evans, C. Antioxidant activity applying an improved ABTS radical cation decolorization assay. *Free Radic. Biol. Med.* **1999**, *26*, 1231–1237. [CrossRef]
40. Hashempour-Baltork, F.; Torbati, M.; Azadmard-Damirchi, S.; Peter Savage, G. Chemical, rheological and nutritional characteristics of sesame and olive oils blended with linseed oil. *Adv. Pharm. Bull.* **2018**, *8*, 107–113. [CrossRef] [PubMed]
41. Siano, F.; Straccia, M.C.; Paolucci, M.; Fasulo, G.; Boscaino, F.; Volpe, M.G. Physico-chemical properties and fatty acid composition of pomegranate, cherry and pumpkin seed oils. *J. Sci. Food Agric.* **2016**, *96*, 1730–1735. [CrossRef]
42. Tangolar, S.G.; Özogul, F.; Tangolar, S.; Yağmur, C. Tocopherol content in fifteen grape varieties obtained using a rapid HPLC method. *J. Food Compos. Anal.* **2011**, *24*, 481–486. [CrossRef]
43. Sabir, A.; Unver, A.; Kara, Z. The fatty acid and tocopherol constituents of the seed oil extracted from 21 grape varieties (Vitis spp.). *J. Sci. Food Agric.* **2012**, *92*, 1982–1987. [CrossRef] [PubMed]
44. Bravi, M.; Spinoglio, F.; Verdone, N.; Adami, M.; Aliboni, A.; D'Andrea, A.; De Santis, A.; Ferri, D. Improving the extraction of α-tocopherol-enriched oil from grape seeds by supercritical CO_2. Optimisation of the extraction conditions. *J. Food Eng.* **2007**, *78*, 488–493. [CrossRef]
45. Sovilj, M. Critical review of supercritical carbon dioxide extraction of selected oil seeds. *Acta Period. Technol.* **2010**, *10*, 105–120. [CrossRef]
46. Molero Gómez, A.; Pereyra López, C.; Martinez de la Ossa, E. Recovery of grape seed oil by liquid and supercritical carbon dioxide extraction: A comparison with conventional solvent extraction. *Chem. Eng. J. Biochem. Eng. J.* **1996**, *61*, 227–231. [CrossRef]
47. Jokić, S.; Bijuk, M.; Aladić, K.; Bilić, M.; Molnar, M. Optimisation of supercritical CO_2 extraction of grape seed oil using response surface methodology. *Int. J. Food Sci. Technol.* **2016**, *51*, 403–410. [CrossRef]
48. Da Porto, C.; Porretto, E.; Decorti, D. Comparison of ultrasound-assisted extraction with conventional extraction methods of oil and polyphenols from grape (*Vitis vinifera* L.) seeds. *Ultrason. Sonochemistry* **2013**, *20*, 1076–1080. [CrossRef] [PubMed]
49. Lachman, J.; Hejtmánková, A.; Táborský, J.; Kotíková, Z.; Pivec, V.; Střalková, R.; Vollmannová, A.; Bojňanská, T.; Dědina, M. Evaluation of oil content and fatty acid composition in the seed of grapevine varieties. *Lwt Food Sci. Technol.* **2015**, *63*, 620–625. [CrossRef]
50. Wang, S.; Yang, R.; Li, H.; Jiang, J.; Zhang, L.; Zhang, Q.; Li, P. Evaluation and comparison of in vitro antioxidant activities of unsaponifiable fraction of 11 kinds of edible vegetable oils. *Food Sci. Nutr.* **2018**, *6*, 2355–2362. [CrossRef]

antioxidants

MDPI

Article

An Environmentally Friendly Practice Used in Olive Cultivation Capable of Increasing Commercial Interest in Waste Products from Oil Processing

Irene Dini [1,*], Giulia Graziani [1,*], Francalisa Luisa Fedele [2], Andrea Sicari [2], Francesco Vinale [3,4], Luigi Castaldo [1,5] and Alberto Ritieni [1,6]

1 Department of Pharmacy, University of Naples Federico II, Via Domenico Montesano 49, 80141 Napoli, Italy; luigi.castaldo2@unina.it (L.C.); ritialb@unina.it (A.R.)
2 Linfa scarl, Via Zona Industriale Porto San Salvo, 89900 Vibo Valentia, Italy; ricerca@laboratoriolinfa.it (F.L.F.); andrea@laboratoriolinfa.it (A.S.)
3 Department of Veterinary Medicine and Animal Productions, University of Naples Federico II, Via Federico Delpino 1, 80137 Napoli, Italy; francesco.vinale@ipsp.cnr.it
4 Institute for Sustainable Plant Protection, National Research Council, Via Università 133, 80055 Portici (NA), Italy
5 Department of Clinical Medicine and Surgery, University of Naples Federico II, Via S. Pansini 5, 80141 Napoli, Italy
6 Unesco Chair for Health Education and Sustainable Development, 80131 Napoli, Italy
* Correspondence: irdini@unina.it (I.D.); giulia.graziani@unina.it (G.G.)

Received: 6 May 2020; Accepted: 29 May 2020; Published: 1 June 2020

Abstract: In the Rural Development Plan (2014–2020), the European Commission encouraged the conversion and supported the maintenance of organic farming. Organic olive oil (bioEVOO) production involves the use of environmentally sustainable fertilizers and the recycling of olive pomace (Pom) and olive vegetation waters (VW) to reduce the environmental impact of these wastes. An ecofriendly way to recycle olive wastes is to reuse them to extract bioactive compounds. In this study, the total phenolic compounds content, their profile and dosage, the antioxidant action in oil, pomace, and vegetation water was evaluated when the *Trichoderma harzianum* M10 was used as a biostimulant in agriculture. Two spectrophotometric tests (2,2-diphenyl-1-picrylhydrazyl (DPPH) and 2,2′-azinobis (3-ethylbenzothiazoline-6-sulfonic) acid (ABTS)) evaluated the antioxidant potential of samples, a spectrophotometric method estimated total phenolic content, and an Ultra-High-Performance Liquid Chromatography (UHPLC)–Orbitrap method evaluated the phenolics profile. Our results showed that the biostimulation improved the antioxidant potential and the total concentration of phenolics in the bioEVOO and bio-pomace (bioPom) samples and mainly enhanced, among all classes of phenolic compounds, the production of the flavonoids and the secoiridoids. Moreover, they demonstrated the *Trichoderma* action in the mevalonate pathway to produce phenols for the first time. The decisive action of the *Thricoderma* on the production of phenolic compounds increases the economic value of the waste materials as a source of bioactive compounds useful for the pharmaceutical, cosmetic, and food industries.

Keywords: *Trichoderma* spp.; EVOO; olive pomace; olive vegetation water; *Olea europea* var Leccino; HRMS-Orbitrap; phenolic identification; antioxidant activity

1. Introduction

The chemical composition of olives depends on the type of cultivar, pedoclimatic factors, and agricultural practices [1]. In general, olives contain oil (18–28%), the olive pulp (30–35%), and the vegetation water (40–50%) [2]. Olive oil extraction in olive mills by mechanical procedures determines some residues, solid and liquid, with a high organic weight detrimental to the environment. The nature of these wastes depends on the extraction system used to extract the olive oil. The most commonly employed are centrifugation systems (two-phase and three-phase) that produce extra virgin olive oil, a solid cake (olive pomace), and olive vegetation water [3]. The olive pomace is made in large amounts, leading to significant management problems. It contains fragments of skin, stone, pulp, olive kernel, a complex mixture of organic (lipids, carbohydrates, hemicellulose, cellulose, lignin, protein), inorganic compounds, (potassium, magnesium, calcium), and phenolic compounds [4]. Generally, it is further extracted (~2% of pomace by weight) by solid–liquid extraction (hexane), solvent recycling, and distillation. However, the extraction process produces potent pollutants to obtain the residual oil [5]. Ecofriendly ways to recycle pomace is to use it to produce biogas [4] as animal feed, insecticide, herbicide, and compost after thermal concentration [6]. Moreover, three-phase centrifugation and pressure systems produce significant liquid waste, which is called olive vegetation water (VW) or mill wastewater. VW properties vary significantly with the type of climatic conditions, the process, and region of origin. VW causes the disposal of environmental problems due to its phenolic composition and high organic load with limited biodegradability [7]. Olive oil, the olive pomace, and the olive vegetation water contain secoiridoids, phenolic alcohols, phenolic acids, and flavonoids [8–10]. They are phenolic compounds with high antioxidant properties. The antioxidants are compounds able to stop or prevent the oxidation of the substrate [11]. Antioxidants are needed to prevent the formation of the reactive oxygen and nitrogen species, which cause damage to DNA, proteins, lipids, and other biomolecules [12]. Phenolics are amphiphilic compounds. In the extracts, type and dosage vary according to the higher degree of lipophilia rather than affinity with water, the matrix, and technology used to extract them. Many techniques are used alone or in a combined form to extract the phenolic compounds from olive waste products. The extraction, membrane separation, centrifugation, and chromatographic methods are usually used for this purpose. The recent patent applications obtained lower energy consumption and higher extraction efficiency with non-conventional methods such as microwaves, ultrasounds, electrotechnologies (high voltages electrical and discharges pulsed electric fields), mechanical technologies (pressurized liquid extraction), and employing supercritical fluids as an alternative of organic solvents in extraction techniques, and using reverse osmosis and tangential ultrafiltration systems in place of conventional filtration methods in membrane methods [13]. Olive oil phenols have some functional, nutraceutical, and sensory properties closely related to their chemical structure [14,15]. A higher intake of phenolic compounds reduces the risk of cardiovascular diseases [16], determines hypoglycemia, hypocholesterolemia, and hypotension, prevents angiogenesis, inflammation [17,18], and cancer [19]. Unfortunately, the phenolic compounds in olive oil waste reduce the microbial growth, rendering the organic load resistant to degradation [20–22]. Therefore, an attractive way to valorize olive oil waste is the possibility of recovering phenol compounds from the pomace and the vegetation water in consideration of the growing interest of the pharmaceutical, nutraceutical, cosmetical, and food industry towards sources of phenolic compounds, and to make further processing residues more readily biodegradable. This study determines the possible impact on the antioxidant activity, phenolic content and profile in the olive waste products (pomace and vegetation water) when ecofriendly biostimulation of the olive trees with *Trichoderma* M10 is used as an agronomic strategy to use them as a resource of bioactive molecules for cosmetic, food, and pharmaceutical industries. *Trichoderma* is a saprophytic living fungus that stimulates the growth of the plants, adsorbs soil pollutants such as heavy metals, improves nutrient availability, interacts with processes involved in plant responses to stress, enhances the production of phenolics and induces systemic resistance [23–26]. Previous works studied *Trichoderma*'s ability to speed up the composting process of the olive pomace and the parameters that influence the composting process [27]. To date, the effects on the concentration and type of phenols that characterize the pomace and the vegetation water, obtained from the olive oil processing when *Trichoderma* fungi are used in olive tree agriculture, are not known.

2. Materials and Methods

2.1. Plant Material

Bioformulates was tested on *Olea europaea* var. Leccino. The trees (20-year-old) situated in the South-Western Calabria (Rombiolo, Vibo Valentia, Italy) were selected and marked.

Plant material was offered d by Dr. Andrea Sicari (Linfa Scarl, Vibo Valentia, Italy). Only plants in excellent phytosanitary and nutritional status were used for experimental purposes. Six treatments were applied every month, starting from February until July. One control sample (water treatment), and 10^6 ufc/mL of the living microbes and were applied as spray application on the leaves (10 L per row of which 5 L was sprayed and 5 L was drenching), and around the root system at 10 cm deep. Two times was replicated the field test.

2.2. Fungal Material

The strains *Trichoderma harzianum* (M10) (LGC Standards S.r.l. Sesto San Giovanni, Italy) were grown on potato dextrose agar medium (HiMedia, Laboratories Mumbai, Mumbai, India) and covered with sterilized mineral oil (Sigma Aldrich, St. Louis, MO, USA).

2.3. Oil Production

The oil samples were produced in a local three-phase mill. They were conserved in brown bottles without headspace and conserved at a constant temperature (10 ± 2 °C) until analysis.

2.4. Chemicals

Hydroxytyrosol was bought from Indofine (Hillsborough, NJ, USA), secologanoside was from ChemFaces Biochemical Co., Ltd. (Wuhan, China), all the other chemicals and standards were purchased from Sigma Aldrich (St. Louis, MO, USA) unless specified differently.

2.5. Analytical Methods

2.5.1. The Phenolics Extraction

The phenolic extraction method proposed by Vasquez Roncero [28] with some modification was carried out. An amount of 25 g of oil was extracted with hexane (25 mL). The organic fraction was treated with MeOH:H_2O/3:2 (*v/v*) (15 mL, three times). The extracts (three) were combined and extracted with 25 mL hexane. The hexane was dried at 40 °C in a rotary evaporator (Buchi, Switzerland); the residue was treated with 1 mL of MeOH, filtered through nylon filer (0.2 mm), frozen and stored (−18 °C) until analysis.

2.5.2. Q Exactive Orbitrap LC-MS/MS Method

Ultra-High-Performance Liquid Chromatography (UHPLC, Thermo Fisher Scientific, Waltham, MA, USA) was used to the dosage of the phenolics. The chromatographic instrument was provided with an autosampler device, a Dionex degassing system (Thermo Scientific™ Ultimate 3000, Waltham, MA, USA), a quaternary UHPLC pump (1250 bar), and a column (Accucore aQ 2.6 µm 100 × 2.1 mm Thermo Scientific, Waltham, MA USA, USA) in a thermostat column compartment (T = 30 °C). The mobile phase consisted of two phases: Phase A: acetic acid (0.1%), and phase B: 100% acetonitrile. The following gradient was used for experimental purposes: 5% phase B from 0 to 5 min, 40% phase B from 6 to 25 min, 100% phase B from 25.1 to 27 min, 5% phase B from 27.1 to 35 min, 0% phase B from 35.1 to 45 min. A flow rate of 0.4 mL/min operated.

A Thermo Fisher Scientific Orbitrap LC-MS/MS (Q Exactive, Waltham, MA, USA) was employed to characterize phenolic compounds. The spectrometer was provided with a HESI II (Thermo Fisher Scientific, Waltham, MA, USA). The ion source setting parameters were: spray voltage −3.0 kV, auxiliary gas (N_2 > 95%), sheath gas (N_2 > 95%), auxiliary gas heater temperature 305 °C, capillary temperature 200 °C, radiofrequency that captures and focuses the ions into a tight beam S-lens RF level 50. The MS detection was performed in full scan and targeted selected ion monitoring. Full scan acquisition parameters were scan rate 2 s^{-1}; scan range 100–1500 *m/z*; mass resolving power 35,000 full width at half maximum (at *m/z* 200); automatic gain control target 1×10^5 ions; maximum injection time of 200 ms. The SIM (selected ion monitoring acquisition) parameters were: 35,000 full widths and half maximum (at *m/z* 200) (resolution power); 15 s (time window); 1.2 *m/z* (quadrupole isolation window).

2.5.3. Method Validation of the Phenolics Dosage

The construction of a calibration curve was achieved using three different concentrations of each calibration standard.

The linearity of the method was obtained from the regression coefficient of the calibration curve.

Limits of detection (LODs) = $3 \times \frac{\text{standard deviation}}{angular\ coefficient}$

Limits of quantification (LOQs) = $10 \times \frac{\text{standard deviation}}{angular\ coefficient}$)

Intraday repeatability was performed by injecting each phenolic standard, three times, at seven different concentrations.

2.5.4. Total Polyphenol Content

Total phenol content was obtained by the Folin–Ciocalteu colorimetric method described previously by Gao et al. 2000 [29]. Extracts (0.1 mL) were added to H_2O (2 mL) and Folin–Ciocalteu reagent (0.2 mL) and were incubated at room temperature (3 min). Sequentially, 1 mL of the sodium carbonate (20%) was added, and the mixture was left (1 h) at room temperature. The total polyphenols were determined in a spectrophotometer (Lambda 25, PerkinElmer, Waltham, MA, USA) (λ = 765 nm). The results were expressed as mg gallic acid equivalents (GAE)/kg of sample. All determinations were performed in triplicate (n = 3).

2.5.5. The Antioxidant Activity Evaluation

DPPH Method

The radical-scavenging capacity was performed utilizing the 2.2-diphenyl-1-picrylhydrazyl (DPPH) method proposed by Brand-Williams et al. (1995) [30]. The phenolic extract (20 μL) was dissolved in 3 mL of DPPH solution (6×10^{-5} mol/L), and the spectrophotometric lecture was performed every 5 min at λ = 517 nm until the steady-state (spectrophotometer Lambda 25, PerkinElmer, Waltham, MA, USA).

ABTS Method

2,2′-azinobis (3-ethylbenzothiazoline-6-sulfonic) acid (ABTS) procedure proposed by Re et al. was used (1999) [31]. The stock solution of reagent was obtained mixed a solution A (9.6 mg ABTS in 2.5 mL water) and 44 mL of a solution B (37.5 mg $K_2S_2O_8$ in 1 mL H_2O). The stock solution was conserved for 8 h in the dark at 4 °C. The work solution was performed by diluting the stock solution [1:88 (*v*/*v*)]. The dilution of the work solution was adjusted depending on the measured absorbance at λ = 734 nm, until a value between 0.7 and 0.8. The sample (100 μL) and the work solution (1 mL) were mixed, and the absorbance (λ = 734) was measured (Lambda 25, PerkinElmer, Italy) after 2 min and 30 s. Three different concentrations of 6-hydroxy-2,5,7,8-tetramethylchroman-2-carboxylic acid (Trolox) solution were used to perform the calibration curve. Results were expressed as mmol Trolox equivalent (TE) kg^{-1} FW. Triplicate experiments were done for each sample.

2.6. Statistical Analysis

"Statistica" software version 7.0 (StatSoft, Inc. Tulsa, OK, USA) was used to perform statistical analyses.

3. Results

A UHPLC-MS/MS method was employed to delineate the phenolic profile in the Extra virgin olive oil (EVOO) and olive pomace. The dosage method was validated according to AOAC instructions (AOAC 2012) [32]. Table 1 showed the parameters used to validate it.

Table 1. Validation parameters of the Ultra-High-Performance Liquid Chromatography (UHPLC)–MS/MS method of analysis.

Phenolic Compounds	Linearity (mg/L)	R^2	LOD (mg/L)	LOQ (mg/L)	Intraday RSD % (n = 3), 50 mg/L
Phenolic Acids					
Vanillic acid	1–50	0.887	0.200	0.600	1.1
Cinnamic acid	1–50	0.991	0.200	0.600	0.9
Ferulic acid	1–50	0.912	0.100	0.300	1.7
p-Coumaric acid	1–50	1.000	0.100	0.300	1.8
4-Hydroxybenzoic acid	1–50	0.998	0.207	0.622	0.9
3-Hydroxybenzoic acid	1–50	0.995	0.205	0.622	1.1
Flavonoids and Lignans					
Luteolin	0.5–50	0.991	0.066	0.200	1.4
Apigenin	0.5–50	0.899	0.066	0.800	2.1
trans Resveratrol	0.5–5.0	0.898	0.090	0.200	1.8
(+)Pinoresinol	1–50	0.999	0.02	0.060	0.5
(+)1-Acetoxypinoresinol	1–50	0.899	0.233	0.700	1.5
Secoiridoids and Derivatives					
Oleuropein	1–50	0.991	0.166	0.500	5.0
Ligstroside	1–50	0.991	0.166	0.500	4.0
Secologanoside	1–50	0.967	0.333	1.000	2.1
Elenaic acid	1–50	0.991	0.333	1.000	0.7
Oleacein Oleuropein-aglycone monoaldehyde	1–50	0.998	1.000	3.000	2.1
Ligstroside-aglycone dialdehyde	1–50	0.899	0.416	1.250	3.0
Tyrosol	1–50	0.991	0.133	0.040	1.6
Hydroxytyrosol	1–50	0.992	0.666	2.000	3.0

3.1. The Phenolics Characterization

Seventeen phenolics, including two flavonoids, two phenolic alcohols, seven secoiridoids, and six phenolic acids, were identified and quantified. Table 2 reports the parameters used to identify phenolics in samples.

Table 2. Parameters used to characterize the phenolic compounds.

Phenolic Compounds	RT (min)	Formula	Theoretical m/z of Deprotonated Molecular Ions $[M-H]^-$	Experimental m/z of Deprotonated Molecular Ions $[M-H]^-$	Calculated Errors Δppm	Fragments	Collision Energy (eV)
Phenolic Acids							
Vanillic acid	4.30	$C_8H_8O_4$	167.03498	167.03522	1.44	152.01143	20
Cinnamic acid	11.54	$C_9H_8O_2$	147.04515	147.04536	1.43	103.04501	20
Ferulic acid	11.81	$C_{10}H_{10}O_4$	193.05063	193.05084	1.09	178.02685	20
p-Coumaric acid	9.71	$C_9H_{10}O_5$	163.04007	163.04028	1.29	119.05023	20
4-Hydroxybenzoic acid	2.57	$C_7H_6O_3$	137.02442	137.02456	1.02	93.03431	12
3-Hydroxybenzoic acid	2.88	$C_7H_6O_3$	137.02442	137.02458	1.17	93.03431	12
Flavonoids and Lignans							
Luteolin	19.07	$C_{15}H_{10}O_6$	285.04046	285.04106	2.10	133.02940	30
Apigenin	19.12	$C_{15}H_{10}O_5$	269.04555	269.04597	1.56	225.05592	35
trans Resveratrol	16.65	$C_{14}H_{12}O_3$	227.07137	227.07147	0.44	185.06082	30
(+) Pinoresinol	17.00	$C_{20}H_{22}O_6$	357.13436	357.13487	1.43	151.03961	40
(+) 1-Acetoxypinoresinol	19.10	$C_{22}H_{24}O_8$	415.13984	415.14007	0.55	415.13821	40

Table 2. *Cont.*

Phenolic Compounds	RT (min)	Formula	Theoretical *m/z* of Deprotonated Molecular Ions [M − H]⁻	Experimental *m/z* of Deprotonated Molecular Ions [M − H]⁻	Calculated Errors Δppm	Fragments	Collision Energy (eV)
			Secoiridoids and Derivatives				
Oleuropein	16.69	$C_{25}H_{32}O_{13}$	539.17701	539.17767	1.22	377.12393	20
Ligstroside	18.25	$C_{25}H_{32}O_{12}$	523.18210	523.18279	1.32	361.12914	12
Secologanoside	19.49	$C_{16}H_{21}O_{11}$	389.1092	389.109258	0.59	345.1195	12
Elenaic acid	13.14	$C_{11}H_{14}O_6$	241.07176	241.07212	1.49	209.04573	10
Oleacein	16.14	$C_{17}H_{20}O_6$	319.11871	319.11898	0.85	301.1082	15
Oleuropein-aglycone mono-aldehyde	21.25	$C_{19}H_{22}O_8$	377.12419	377.12442	0.61	345.09790	12
Ligstroside-aglycone dialdehyde	18.59	$C_{17}H_{20}O_5$	303.12380	303.12441	2.01	301.1082	12
Tyrosol	2.75	$C_8H_{10}O_2$	137.06080	137.06096	1.17	119.05022	12
Hydroxytyrosol	1.60	$C_8H_{10}O_3$	153.05572	153.05580	0.52	123.04561	12

3.2. The Phenolics Dosage

Tables 3–5 report the dosage of each phenolic compound found in the samples. The *Trichoderma* biostimulation improved the apigenin concentration in the EVOO, and the olive pomace (Pom), but decreased it together with luteolin in VW. The Pom and the VW samples did not contain lignans (Table 3).

Table 3. Flavonoids and lignans concentrations (mg/kg).

Compounds	Flavonoids			Lignans	
	Luteolin	Apigenin	*trans* Resveratrol	Pinoresinol	Acetoxipinoresinol
bioEVOO	7.317 ± 0.054	0.251 ± 0.005		0.203 ± 0.013	9.829 ± 0.035
EVOO	3.178 ± 0.046	0.228 ± 0.001		0.095 ± 0.007	4.344 ± 0.097
bioPom	110.371 ± 8.478	9.623 ± 1.011			
Pom	71.1713 ± 2.6	8.025 ± 0.27			
bioVWr	0.051 ± 0.002	0.008 ± 0.00	0.492 ± 0.0		
VW	1216.521 ± 57.985	154.388 ± 9.771	0.296 ± 0.001		

bioEVOO (Organic olive oil); EVOO (Extra virgin olive oil); Pom (pomace); bioPom (Organic pomace); VW (vegetation waters); bioVW (Organic vegetation waters).

The phenolic acid response to the biostimulation was like that seen for the other phenols; there was an increase in the concentration of each compound, but the increase was different from compound to compound (Table 4).

Table 4. Phenolic acids concentrations (mg/kg).

Compounds	4-Hydroxybenzoic Acid	3-Hydroxybenzoic Acid	Vanillic Acid	*p*-Coumaric Acid	Cinnamic Acid	Ferulic Acid
bioEVOO	0.883 ± 0.007	0.796 ± 0.004	7.05 ± 0.059	3.274 ± 0.024	0.482 ± 0.009	0.131 ± 0.001
EVOO	0.605 ± 0.007	0.27 ± 0.003	2.663 ± 0.012	1.422 ± 0.021	0.438 ± 0.002	0.064 ± 0.000
bioPom	0.657 ± 0.016	5.033 ± 0.516	22.104 ± 3.615	21.391 ± 1.769	0.206 ± 0.02	1.486 ± 0.153
Pom	0.331 ± 0.009	2.407 ± 0.100	10.121 ± 0.11	6.085 ± 0.447	0.301 ± 0.031	0.649 ± 0.044
bioVWr	3.587 ± 0.272	0.174 ± 0.021	0.331 ± 0.035	0.238 ± 0.001	0.469 ± 0.023	0.211 ± 0.023
VW	42.146 ± 1.14	27.259 ± 1.184	116.588 ± 19.641	163.859 ± 10.169	7.092 ± 0.659	14.132 ± 0.427

Our results confirmed the *Trichoderma*'s ability to increase the concentration of secoiridoids and their degradation products in EVOO, and established a similar activity in Pom, but not in VW (Tables 5 and 6).

Table 5. Secoiridoid compounds and their degradation product concentrations (mg/kg).

Compounds	Ligstroside	Oleuropein	Secologanoside	Elenaic Acid	Oleuropein-Aglycone di-Aldehyde	Ligstroside-Aglycone mono-Aldehyde	Tyrosol	Hydroxytyrosol
bioEVOO	0.009 ± 25.038	0.152 ± 2.6	0.307 ± 9.109	3.46 ± 6.552	344.531 ± 5.578	117.220 ± 2.866	105.91 ± 1.698	0.595 ± 17.946
EVOO	0.003 ± 2.205	0.099 ± 1.9	0.297 ± 1.635	7.58 ± 22.919	587.819 ± 5.041	157.254 ± 1.435	45.064 ± 6.736	0.152 ± 0.424
bioPom	0.763 ± 0.120	0.810 ± 0.09	27.724 ± 1.467	34.992 ± 0.802	17.492 ± 0.762	0.9144 ± 0.059	0.9144 ± 0.0059	8.481 ± 0.163
Pom	0.3093 ± 0.02	1.733 ± 0.005	0.519 ± 0.021	8.673 ± 0.275	2.247 ± 0.110	0.201 ± 0.0	0.201 ± 0.0	1.029 ± 0.001
bioVWr	0.0668 ± 0.003	0.484 ± 0.068	12.136 ± 0.473	0.815 ± 0.016	3.342 ± 0.111	0.0	0.014 ± 0.005	0.0
VW	3.007 ± 0.369	19.683 ± 1.245	892.645 ± 38.554	164.577 ± 8.116	9.721 ± 3.544	0.0	10.331 ± 0.989	22.678 ± 0.678

Table 6. Variation % of the concentration of each phenolic under biostimulation.

Compounds	Luteolin	Apigenin	Resveratrol	Pinoresinol	Acetoxypinoresinol	4-Hydroxybenzoic Acid	3-Hydroxybenzoic Acid	Vanillic Acid	*p*-Coumaric Acid	Cinnamic Acid	Ferulic Acid
bioEVOO	+130%	+10%	±	+114%	+126%	+46%	195%	165%	+130%	+10%	+105%
bioPom	−85%	+20%	+52%		±	+99%	+109%	+118%	+252%	−32%	+129%
bioVW	−100%	−100%	+66%			−92%	−99%	−100%	−100%	−93%	−99%

Compounds	Ligstroside	Oleuropein	Secologanoside	Elenaic Acid	Oleuropein-Aglycone di-Aldehyde	Ligstroside-Aglycone mono-Aldehyde	Tyrosol	Hydroxytyrosol
bioEVOO	+219%%	+68%	+3%	−51%	−41%	−26%	−77%	+290%
bioPom	+147%	−53%	+5242%	+304%	+679%	+355%	+395%	+724%
bioVW	−78%	−98%	−99%	−100%	−67%	NF	−100%	−100%

3.3. *Total Phenolic Concentration and Antioxidant Activity*

The biostimulation had a positive effect on the antioxidant activity measured with both methods (ABTS and DPPH). The ABTS method evaluated the antioxidant activity of the Pom and VW more than the DPPH method, and the opposite occurred in the EVOO samples.

4. Discussion

Liquid and solid olive processing waste contain high amounts of organic materials that are not easily degradable. When these wastes are put into the environment, they create odor nuisance, an oily shine, enhance the oxygen demand, and are toxic to plant life. Therefore, the direct release of olive processing waste is forbidden, and some actions must be required before discarding into the environment. Some studies showed that olive processing waste might also be considered as an economic resource. Some practices are proposed to recycle and reuse them; using them as starting material to extract beneficial products for human health such as antioxidants is interesting. Olive pomace and olive vegetation water are sources of phenols. The industry requires phenolic compounds to produce functional foods, supplements, food additives, and the formulation of cosmetics and drugs [5,33,34]. *Trichoderma* species promote the production of phytochemicals, including phenolic compounds [23,26], whose production varies according to the strain used [35]. Previous studies have discussed that the *Trichoderma* can enhance phenols in EVOO and olive leaves [1,35]. In this work, we tested the ability of the *Trichoderma* to increase the concentration of phenolic compounds in the olive pomace and the olive vegetation water. Moreover, we determined the concentration of each compound, considering that the interest in phenolic compounds from industry depends on their chemical structure to which biological action is linked. Phenolic profile and dosage were investigated by an HPLC–Orbitrap method validated in terms of linearity, precision, and sensitivity, as recommended by the AOAC (2012) guidelines [36]. The linearity of the method was confirmed by the coefficient of regression ($r \cong 1$) of the calibration curve. The sensitivity was verified by the inclusion of the concentration detected in the LODs and the LOQs range. The repeatability was confirmed by Relative Standard Deviation (RSD) values <6%. Phenolics identification was performed by comparing their mass spectra with those obtained by the standards analyses. The identification of the two hydroxybenzoic acid isomers was obtained through comparing the retention time and mass spectra with standards. The ligstroside identification, as it was not commercially available, was confirmed by comparing the chromatographic evidence and the spectroscopic data with those reported in the literature [37]. Biostimulation improved the total phenolic content in Pom and EVOO samples in accordance to our previous results [1]. On the other hand, biostimulation decreased the total phenolic concentration in VW. The main phenols found in the Pom and VW were the same as those in the EVOO, but their total concentrations, expressed as mmol Trolox/kg, were higher in the olive vegetation water control, followed by bioPom, Pom, bioVW, EVOO and bioEVOO. Among these, secoiridoids and their degradation products were the most concentrated compounds in the bioEVOO, the EVOO, and the VW samples. In contrast, the flavonoids were the most representative compounds in the bioPom, the Pom samples, and the bioVW. The lignans were found only in the bioEVOO and the EVOO. The resveratrol was in the bioVW and the VW samples. Therefore, the most variable phenolics were secologanoside, resveratrol, and lignans. Concerning secoiridoids fraction, secoiridoids biosynthesis in the plant occurs through two biosynthetic pathways: the shikimic pathway and the mevalonate pathway (Figure 1). Biostimulation with *Thricoderma* M10 enhanced the production of the oleuropein and ligstroside. It preserved them from the degradation during the malaxation process, as shown by their higher concentration and the lower concentrations of their degradation products (oleuropein-aglycone di-aldehyde, ligstroside-aglycone mono-aldehyde, tyrosol, and hydroxytyrosol) in the bioEVOO vs. the EVOO. Moreover, the higher concentration of ligstroside and secologanoside in the bioEVOO respect of the EVOO sample indicated the biostimulation's ability to enhance the secoiridoid biosynthesis mainly through the mevalonate pathway (Figure 2). Finally, the negative variation of the percentage content of oleuropein in the pomace and strongly positive increase of its precursor, the secologanoside, in the pomace were further confirmation (Table 6).

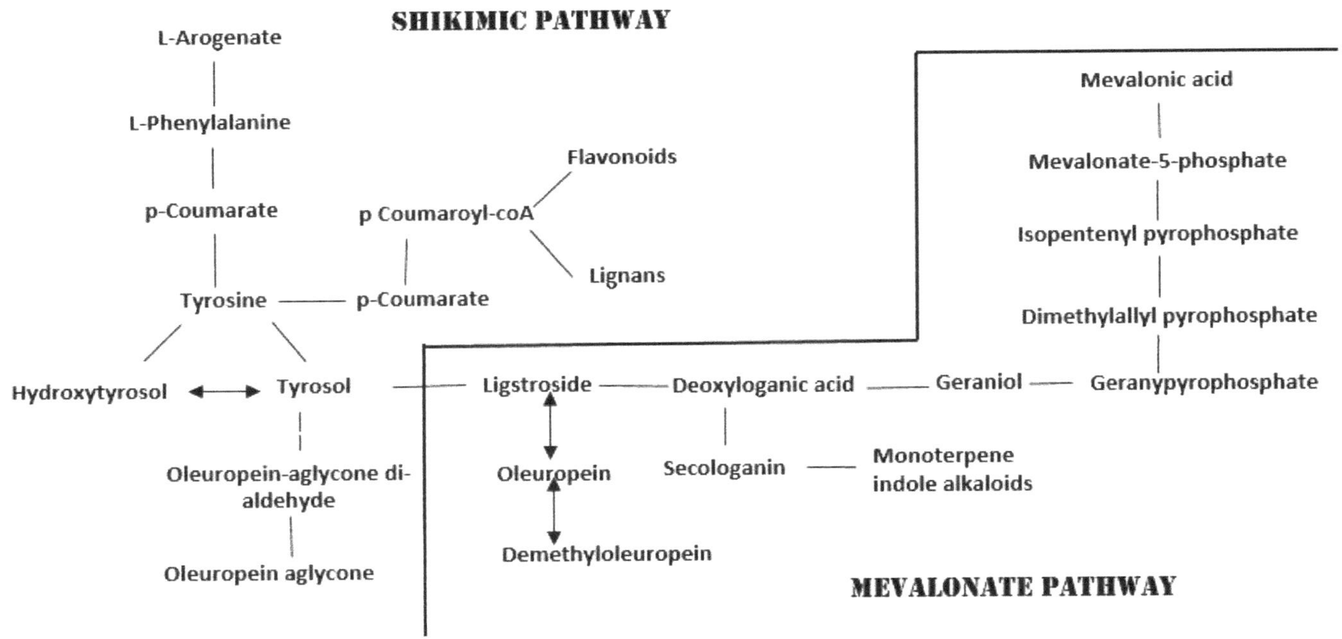

Figure 1. Secoiridoids biosynthesis [32].

Moreover, biocontrol agriculture enhanced the production of resveratrol, as shown by higher concentrations of the resveratrol and lower concentrations of its precursors (the cinnamic acid, and the *p*-coumaric acid) in the bioVW vs. VW sample (Figure 2). This datum is noteworthy since resveratrol has some nutraceutical properties, such as anti-inflammatory and anti-oxidative effects, and disturbs the start and progression of many illnesses such as some cancer types, and neurological and cardiovascular disorders through several mechanisms. In vitro and in vivo evidence confirmed the resveratrol's ability as a therapeutic agent [38].

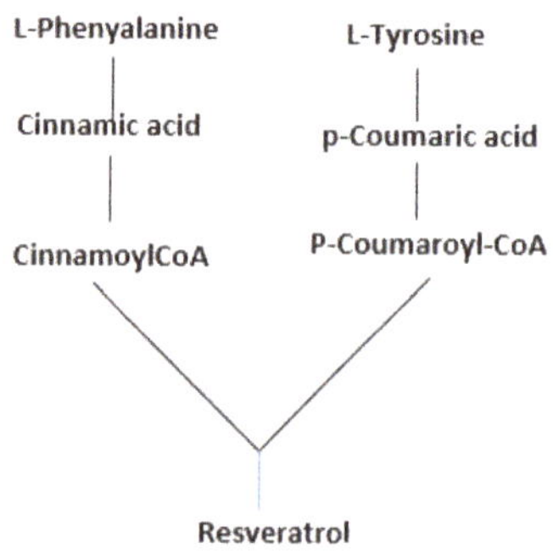

Figure 2. Secoiridoids biosynthesis [33].

The lignans were absent in the Pom and the VW samples, as they have high solubility in fats (logP 3.1) [39]. The improvement of total phenolics concentrations, determined using the *Trichoderma* in culture, is followed by an increase in the antioxidant activity in these samples. The values of the antioxidant activity, measured with the DPPH test, are overestimated in samples where the flavonoid concentration is high (Figure 3 and Table 3) [40]. It is clear that the use of biostimulants is useful to increase the concentration of phenolic compounds with antioxidant activity, not only in the EVOO, but also in the Pom, transforming the latter from an environmental problem into a source of bioactive molecules of nutraceutical, food, pharmaceutics, and cosmetic interest. Moreover, the *Thricoderma* improves the nutraceutical value of the bioEVOO, and decreases the losses of phenolic compounds in vegetation waters, favoring the transformation of phenolic alcohols into secoiridoids, lignans, and flavonoids which have higher properties for human health.

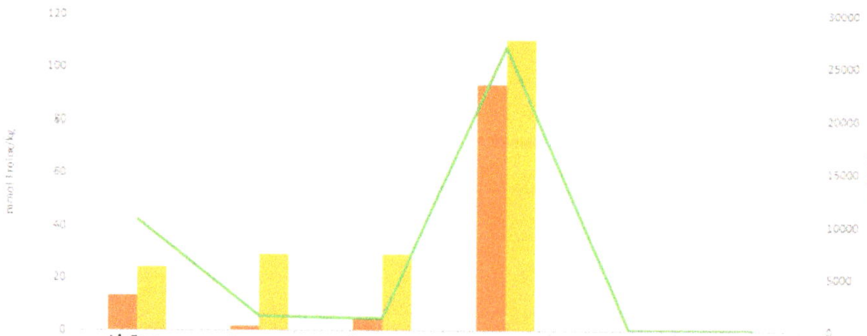

	bioPom	Pom	bioVW	VW	bioEVOO	EVOO
ABTS	13,499±1.090	24,569±1.022	10634,46±10.09	93,655±11.605	110,504±16.601	26984,28±0.001
DPPH	2,147±0.202	29,196±0.639	1501,767±118.172	0,522±0.084	0,46±0.045	142,413±5.410
FOLIN	5,02±0.157	29,196±4.520	1276,104±178.588	0,644±0.020	0,351±0.055	133,662±3.418

Figure 3. Antioxidant activities (mmol Trolox/kg) and total phenols content (mg GAE/kg).

5. Conclusions

For the first time, this study delineates the effects of the *Trichoderma* used in olive tree cultivation on the antioxidant activity and the phenol production in olive pomace and olive vegetation water. Our results confirmed the *Trichoderma*'s ability to improve the concentration of phenolics and antioxidant activity in EVOO, establishing the same ability in the olive pomace. Finally, they demonstrate that biostimulation principally determines, among the phenolic compounds, the biosynthesis of flavonoids and secoiridoids, two classes of phenolic compounds with well-known health properties, making olive pomace and olive vegetation water commercially appealing as a source of botanicals convenient for the food, cosmetic, and pharmaceutical industries. Finally, the *Trichoderma* action in the mevalonate pathway, producing phenols, was highlighted for the first time.

Author Contributions: Formal analysis, F.L.F.; formal analysis, F.V.; investigation, L.C.; investigation A.S.; data curation, G.G.; writing—original draft preparation, I.D.; project administration, A.R. All authors have read and agreed to the published version of the manuscript.

Funding: The following projects supported this work: MIUR (Ministry of Education, University and Research)–PON (National Operational Program) [grant number Linfa 03PE_00026_1], [grant number Marea 03PE_00106]; POR (Regional Operational Program) FESR (European Regional Development Fund) Campania 2014–2020 [grant number Bioagro NR. onc3-003189].

Conflicts of Interest: The authors declare no conflicts of interest.

References

1. Dini, I.; Graziani, G.; Fedele, F.L.; Sicari, A.; Vinale, F.; Castaldo, L.; Ritieni, A. Effects of *Trichoderma biostimulation* on the phenolic profile of extra-virgin olive oil and olive oil by-products. *Antioxidants* **2020**, *9*, 284. [CrossRef] [PubMed]
2. Niaounakis, M.; Halvadakis, C.P. *Olive-Mill Waste Management: Literature Review and Patent Survey*, 1st ed.; Typothito-George Dardanos: Athens, Greece, 2004; pp. 131–146.
3. Morillo, J.A.; Antizar-Ladislao, B.; Monteoliva-Sanchez, M.; Ramos-Cormenzana, A.; Russell, N.J. Bioremediation and biovalorisation of olive mill wastes. *Appl. Microbiol. Biotechnol.* **2009**, *82*, 25–39. [CrossRef] [PubMed]
4. Romero-García, J.M.; Niño, L.; Martínez-Patiño, C.; Álvarez, C.; Castro, E.; Negro, M.J. Biorefinery based on olive biomass. State of the art and future trends. *Bioresour. Technol.* **2014**, *159*, 421–432. [CrossRef] [PubMed]
5. Saviozzi, A.; Riffaldi, R.; Levi-Minzi, R.; Scagnozzi, A.; Vanni, G. Decomposition of vegetation-water sludge in soil. *Bioresour. Technol.* **1993**, *44*, 223–228. [CrossRef]
6. Qdais, H.A.; Alshraideh, H. Selection of management option for solid waste from olive oil industry using the analytical hierarchy process. *J. Mater. Cycles Waste Manag.* **2014**, *18*, 1–9. [CrossRef]
7. Elkacmi, R.; Kamil, N.; Bennajah, M. Separation and purification of high purity products from three different olive mill wastewater samples. *J. Environ. Chem. Eng.* **2017**, *5*, 829–837. [CrossRef]
8. Cioffi, G.; Pesca, M.S.; De Caprariis, P.; Braca, A. Phenolic compounds in olive oil and olive pomace from Cilento (Campania, Italy) and their antioxidant activity. *Food Chem.* **2010**, *121*, 105–111. [CrossRef]
9. Miraglia, D.; Esposto, S.; Branciari, R.; Urbani, S.; Servili, M.; Perucci, S.; Ranucci, D. Effect of a Phenolic extract from olive vegetation water on fresh salmon steak quality during storage. *Ital. J. Food Saf.* **2016**, *5*, 224–228. [CrossRef]
10. Papanikolaou, C.; Melliou, E.; Magiatis, P. Olive Oil Phenols. In *Functional Foods*; Intech Open: London, UK, 2019; pp. 9–13.
11. Halliwell, B. How to characterize a biological antioxidant. *Free Radic. Res. Commun.* **1990**, *9*, 1–32. [CrossRef] [PubMed]
12. Halliwell, B. Antioxidants in human health and disease. *Annu. Rev. Nutr.* **1996**, *16*, 33–50. [CrossRef] [PubMed]
13. Roselló-Soto, E.; Parniakov, O.; Deng, Q.; Patras, A.; Koubaa, M.; Grimi, N.; Boussetta, N.; Tiwari, B.K.; Vorobiev, E.; Lebovka, N.; et al. Application of Non-conventional Extraction Methods: Toward a Sustainable and Green Production of Valuable Compounds from Mushrooms. *Food Eng. Rev.* **2016**, *8*, 214–234. [CrossRef]
14. Saija, A.; Uccella, N. Olive biophenols: Functional effects onhuman wellbeing. *Trends Food Sci. Technol.* **2000**, *11*, 357–363. [CrossRef]

15. Dini, I.; Laneri, S. Nutricosmetics: A brief overview. *Phytother. Res.* **2019**, *33*, 3054–3063. [CrossRef] [PubMed]
16. Vita, J.A. Polyphenols and cardiovascular disease: Effects on endothelial and platelet function. *Am. J. Clin. Nutr.* **2005**, *81*, 292S–297S. [CrossRef]
17. Stoclet, J.C.; Chataigneau, T.; Ndiaye, M.; Oak, M.H.; El Bedoui, J.; Chataigneau, M.; Shini-Kerth, V.B. Vascular protection by dietary polyphenols. *Eur. J. Pharm.* **2004**, *500*, 299–313. [CrossRef]
18. Montesano, D.; Rocchetti, G.; Cossignani, L.; Senizza, B.; Pollini, L.; Lucini, L.; Blasi, F. Untargeted Metabolomics to Evaluate the Stability of Extra-Virgin Olive Oil with Added *Lycium barbarum* Carotenoids during Storage. *Foods* **2019**, *8*, 179. [CrossRef]
19. Ruzzolini, J.; Peppicelli, S.; Andreucci, E.; Bianchini, F.; Scardigli, A.; Romani, A.; La Marca, G.; Nediani, C.; Calorini, L. Oleuropein, the Main Polyphenol of Olea europaea Leaf Extract, Has an Anti-Cancer Effect on Human BRAF Melanoma Cells and Potentiates the Cytotoxicity of Current Chemotherapies. *Nutrients* **2018**, *10*, 1950. [CrossRef]
20. Bianco, A.; Muzzalupo, I.; Piperno, A.; Romeo, G.; Uccella, N. Bioactive derivatives of oleuropein from olive fruits. *J. Agric. Food Chem.* **1999**, *47*, 3531–3534. [CrossRef]
21. Capasso, R.; Evidente, A.; Schivo, L.; Orru, G.; Marcialis, M.A.; Cristinzio, G. Antibacterial polyphenols from olive oil mill wastewaters. *J. Appl. Bacteriol.* **1995**, *79*, 393–398. [CrossRef]
22. Ramos-Cormenzana, A.; Juárez-Jiménez, B.; Garcia-Pareja, M.P. Antimicrobial activity of olive mill waste-waters (alpechin) and biotransformed olive oil mill wastewater. *Int. Biodet. Biodeg.* **1996**, *38*, 283–290. [CrossRef]
23. Bulgari, R.; Franzoni, G.; Ferrante, A. Biostimulants application in horticultural crops under abiotic stress conditions. *Agronomy* **2019**, *9*, 306. [CrossRef]
24. Ertani, A.; Schiavon, M.; Muscolo, A.; Nardi, S. Alfalfa plant-derived biostimulant stimulate short-term growth of salt stressed Zea mays L. plants. *Plant Soil* **2013**, *364*, 145–158. [CrossRef]
25. Herrero, M.; Temirzoda, T.N.; Segura-Carretero, A.; Quirantes, R.; Plaza, M.; Ibañez, E. New possibilities for the valorization of olive oil by-products. *J. Chromatogr. A* **2011**, *1218*, 7511–7520. [CrossRef] [PubMed]
26. Pascale, A.; Vinale, F.; Manganiello, G.; Nigro, M.; Lanzuise, S.; Ruocco, M.; Marra, R.; Lombardi, N.; Woo, S.L.; Lorito, M. *Trichoderma* and its secondary metabolites improve yield and quality of grapes. *Crop Prot.* **2017**, *92*, 176–181. [CrossRef]
27. Haddadin, M.S.Y.; Haddadin, J.; Arabiyat, O.I.; Hattar, B. Biological conversion of olive pomace into compost by using *Trichoderma harzianum* and *Phanerochaete chrysosporium*. *Bioresour. Technol.* **2009**, *100*, 4773–4782. [CrossRef]
28. Vázquez Roncero, A. Les polyphenols de l'huile d'olive et leur influence sur les caracteristiques de l'huile. *Rev. Fr. Corps Gras.* **1978**, *25*, 21–26.
29. Gao, X.; Bjork, L.; Trajkovski, V.; Uggla, M. Evaluation of antioxidant activities of rosehip ethanol extracts in different test systems. *J. Agric. Food Chem.* **2000**, *80*, 2021–2027. [CrossRef]
30. Brand-Williams, W.; Cuvelier, M.E.; Berset, C. Use of a free radical method to evaluate antioxidant activity. *LWT Food Sci. Technol.* **1995**, *28*, 25–30. [CrossRef]
31. Re, R.; Pellegrini, N.; Proteggente, A.; Pannala, A.; Yang, M.; Rice-Evans, C. Antioxidant activity applying an improved ABTS radical cation decolorization assay. *Free Radic. Biol. Med.* **1999**, *26*, 1231–1237. [CrossRef]
32. AOAC. Appendix F: Guidelines for Standard Method Performance Requirements (SMPR). In *AOAC Official Methods of Analysis*; AOAC: Rockville, MD, USA, 2012.
33. Oksana, S.; Marian, B.; Mahendraand, R.; Bo, S.H. Plant phenolic compounds for food, pharmaceutical, and cosmeti s production. *J. Med. Plants Res.* **2012**, *6*, 2526–2539.
34. Tafuri, S.; Cocchia, N.; Carotenuto, D.; Vassetti, A.; Staropoli, A.; Mastellone, V.; Peretti, V.; Ciotola, F.; Albarella, S.; Del Prete, C.; et al. Chemical Analysis of Lepidium meyenii (Maca) and Its Effects on Redox Status and on Reproductive Biology in Stallions. *Molecules* **2019**, *24*, 1981. [CrossRef] [PubMed]
35. Dini, I.; Graziani, G.; Gaspari, A.; Fedele, F.L.; Sicari, A.; Vinale, F.; Cavallo, P.; Lorito, M.; Ritieni, A. New Strategies in the Cultivation of Olive Trees and Repercussions on the Nutritional Value of the Extra Virgin Olive Oil. *Molecules* **2020**, *25*, 2345. [CrossRef]
36. Klen, T.J.; Wondra, A.G.; Vrhovšek, U.; Mozetič, V.B. Phenolic profiling of olives and olive oil process-derived matrices using UPLC-DAD-ESI-QTOF-HRMS analysis. *J. Agric. Food Chem.* **2015**, *63*, 3859–3872. [CrossRef] [PubMed]

37. Alagna, F.; Mariotti, R.; Panara, F.; Caporali, S.; Urbani, S.; Veneziani, G.; Esposto, S.; Taticchi, A.; Rosati, A.; Rao, R.; et al. Olive phenolic compounds: Metabolic and transcriptional profiling during fruit development. *BMC Plant Biol.* **2012**, *12*, 162. [CrossRef] [PubMed]
38. Berman, A.Y.; Motechin, R.A.; Wiesenfeld, M.Y.; Holtz, M.K. The therapeutic potential of resveratrol: A review of clinical trials. *NPJ Precis. Oncol.* **2017**, *1*, 35. [CrossRef] [PubMed]
39. Thapa, S.B.; Pandey, R.P.; Park, Y.I.; Sohng, J.K. Biotechnological advances in resveratrol production and its chemical diversity. *Molecules* **2019**, *24*, 2571. [CrossRef] [PubMed]
40. Perecko, T.; Jancinova, V.; Drabikova, K.; Nosal, R.; Harmatha, J. Structure-efficiency relationship in derivatives of stilbene. Comparison of resveratrol, pinosylvin and pterostilbene. *Neuroendocrinol. Lett.* **2008**, *29*, 802–805. [PubMed]

MDPI
St. Alban-Anlage 66
4052 Basel
Switzerland
Tel. +41 61 683 77 34
Fax +41 61 302 89 18
www.mdpi.com

Antioxidants Editorial Office
E-mail: antioxidants@mdpi.com
www.mdpi.com/journal/antioxidants

www.ingramcontent.com/pod-product-compliance
Lightning Source LLC
Chambersburg PA
CBHW061457220926
51850CB00004B/112
9783036549521